Materials Science and Technology

Volume 3 A
Electronic and Magnetic Properties of Metals and Ceramics
Part I

VCH

Materials Science and Technology

Materials Science and Technology

A Comprehensive Treatment

Edited by
R.W. Cahn, P. Haasen, E. J. Kramer

Volume 3 A
Electronic and Magnetic Properties of Metals and Ceramics
Part I

Volume Editor: K. H. Jürgen Buschow

VCH Weinheim · New York · Basel · Cambridge

Editors-in-Chief:
Professor R.W. Cahn
University of Cambridge
Dept. of Materials Science
and Metallurgy
Pembroke Street
Cambridge CB2 3QZ, UK

Professor P. Haasen
Institut für Metallphysik
der Universität
Hospitalstraße 3/5
D-3400 Göttingen
Germany

Professor E. J. Kramer
Cornell University
Dept. of Materials Science
and Engineering
Bard Hall
Ithaca, NY 14853-1501, USA

Volume Editor:
Professor K. H. J. Buschow
Philips Research Laboratories
PO Box 80000
NL-5600 JA Eindhoven
The Netherlands

Published jointly by
VCH Verlagsgesellschaft mbH, Weinheim (Federal Republic of Germany)
VCH Publishers Inc., New York, NY (USA)

Editorial Directors: Dr. Christina Dyllick-Brenzinger, Karin Sora
Production Manager: Dipl.-Wirt.-Ing. (FH) H.-J. Schmitt
Indexing: Borkowski, Schauernheim
The cover illustration shows a semiconductor chip surface and is taken from the journal "Advanced Materials", published by VCH, Weinheim.

Library of Congress Card No.: 90-21936

British Library Cataloguing-in-Publication Data
Materials science and technology: Vol 3a.
Electronic and magnetic properties of metals and ceramics.
– (Materials science and technology: a comprehensive treatment)
I. Buschow, Kurt Heinz Jürgen II. Series
670
ISBN 3-527-26816-2

Die Deutsche Bibliothek – CIP-Einheitsaufnahme
Materials science and technology : a comprehensive treatment /
ed. by R. W. Cahn ... – Weinheim ; New York ; Basel ; Cambridge : VCH.
Vol. 3. Electronic and magnetic properties of metals and ceramics.
A (1991)
Electronic and magnetic properties of metals and ceramics /
vol. ed.: Kurt H. J. Buschow. – Weinheim ; New York ; Basel ; Cambridge : VCH.
(Materials science and technology ; Vol. 3)
ISBN 3-527-26813-8 (Weinheim ...)
ISBN 1-56081-190-0 (New York)
NE: Buschow, Kurt H. J. Hrsg.]
A (1991)
ISBN 3-527-26816-2 (Weinheim ...) Gb.
ISBN 0-89573-691-8 (New York) Gb.

Printed on acid-free and low chlorine paper

Composition, Printing and Bookbinding: Konrad Triltsch, Druck- und Verlagsanstalt GmbH, D-8700 Würzburg
Printed in the Federal Republic of Germany

Preface to the Series

Materials are highly diverse, yet many concepts, phenomena and transformations involved in making and using metals, ceramics, electronic materials, plastics and composites are strikingly similar. Matters such as transformation mechanisms, defect behavior, the thermodynamics of equilibria, diffusion, flow and fracture mechanisms, the fine structure and behavior of interfaces, the structures of crystals and glasses and the relationship between these, the motion or confinement of electrons in diverse types of materials, the statistical mechanics of assemblies of atoms or magnetic spins, have come to illuminate not only the behavior of the individual materials in which they were originally studied, but also the behavior of other materials which at first sight are quite unrelated.

This continual intellectual cross-linkage between materials is what has given birth to *Materials Science,* which has by now become a discipline in its own right as well as being a meeting place of constituent disciplines. The new Series is intended to mark the coming-of-age of that new discipline, define its nature and range and provide a comprehensive overview of its principal constituent themes.

Materials Technology (sometimes called Materials Engineering) is the more practical counterpart of Materials Science, and its central concern is the processing of materials, which has become an immensely complex skill, especially for the newer categories such as semiconductors, polymers and advanced ceramics but indeed also for the older materials: thus, the reader will find that the metallurgy and processing of modern steels has developed a long way beyond old-fashioned empiricism.

There exist, of course, other volumes and other series aimed at surveying these topics. They range from encyclopedias, via annual reviews and progress serials, to individual texts and monographs, quite apart from the flood of individual review articles in scientific periodicals. Many of these are essential reading for specialists (and those who intend to become specialists); our objective is not to belittle other sources in the cooperative enterprise which is modern materials science and technology, but rather to create a self-contained series of books which can be close at hand for frequent reference or systematic study, and to create these books rapidly enough so that the early volumes will not yet be badly out of date when the last ones are published. The individual chapters are more detailed and searching than encyclopedia or concise review articles, but less so than monographs wholly devoted to a single theme.

The Series is directed toward a broad readership, including not only those who define themselves as materials scientists or engineers but also those active in diverse disciplines such as solid-state physics, solid-state chemistry, metallurgy, construction engineering, electrical engineering and electronics, energy technology, polymer science and engineering.

While the Series is primarily classified on the basis of types of materials and their processing modes, some volumes will focus on particular groups of applications (Nuclear Materials, Biomedical Materials), and others on specific categories of properties (Phase Transformations, Characterization, Plastic Deformation and Fracture). Different aspects of the same topic are often treated in two or more volumes, and certain topics are treated in connection with a particular material (e.g., corrosion in one of the chapters on steel, and adhesion in one of the polymer volumes). Special care has been taken by the Editors to ensure extensive cross-references both within and between volumes, insofar as is feasible. A Cumulative Index volume will be published upon completion of the Series to enhance its usefulness as a whole.

We are very much indebted to the editorial and production staff at VCH for their substantial and highly efficient contribution to the heavy task of putting these volumes together and turning them into finished books. Our particular thanks go to Dr. Christina Dyllick on the editorial side and to Wirt. Ing. Hans-Jochen Schmitt on the production side. We are grateful to the management of VCH for their confidence in us and for their steadfast support.

Robert W. Cahn, Cambridge
Peter Haasen, Göttingen
Edward J. Kramer, Ithaca

April 1991

Preface to Volume 3 A

Volume 3 is intended to give a balanced survey of the extremely large field of *Electronic and Magnetic Properties of Metals and Ceramics*, viewed from the point of view of materials science and its interdisciplinary areas, comprising elements of solid state physics, solid state chemistry and modern technology.

Volume 3 cannot be regarded as a standard textbook on electronic and magnetic properties of metals and ceramics. Nor is it intended to be an exhaustive compilation or systematic arrangement of these properties for the many types of materials presently known. Rather it focuses on topics that can illustrate the main achievements made with regard to the understanding and description of fundamental phenomena and, equally important, with regard to technological applications. It proved not to be possible to accommodate all the topics that needed to be dealt with in a single volume. Thus, Volume 3 is composed of two parts, Volume 3 A and Volume 3 B. It is with great pleasure that I can introduce to you now the first of these two parts.

The first chapter deals with electronic structure calculations. The availability of high speed computers has made it possible to extend electronic structure calculations to systems of fairly large complexity, with the result that the corresponding computational models have penetrated more and more into the realm of materials science, where they are used not only to obtain a basic understanding of cohesion, magnetic properties and transport properties but also to deal with heats of formation and phase relationships.

Magneto-optical properties of materials, described in Chapter 2, have already held a prominent position in materials science for many decades. This chapter deals primarily with fundamental aspects, but several of the materials described here will be revisited in the chapter on high-density magneto-optical recording materials to be published in Volume 3 B.

Two chapters are concerned with transport properties. The first, Chapter 3, deals with the large variety of electronic transport properties of normal metals. Since superconductivity has become a field in its own right, superconducting properties of materials, placed in the proper framework of current theories and phenomenological models, are dealt with in a separate chapter (Chapter 4).

Magnetic materials have a long tradition in materials science as well as in fundamental physics. Two chapters are devoted to magnetic properties. Chapter 5 reviews the different types of magnetism encountered in metallic systems. Those of ceramic materials will be presented in a separate chapter in

Volume 3 B. The emphasis in both these two chapters is on fundamental aspects. The many applications of magnetic materials have led to several specializations, such as magnetic recording, permanent magnet materials, soft magnetic materials and invar alloys, all of which will be treated in Volume 3 B.

Comparatively young offspring in materials science are ultra-thin films and superlattices. Fundamental properties of these systems, mainly transport properties, magnetic properties and superconducting properties are described in Chapter 6. Finally, in Chapter 7, the reader is introduced into the exotic world of Fermi surfaces found in many metallic systems. This chapter, like Chapter 1, is of a strongly fundamental nature.

All the chapters described above have been written by leading authorities in the various fields of interest. For many of the authors it has not been easy to perform the task of writing their chapter on top of their daily work. I wish to thank all these authors for their many efforts to reduce delay of this Volume as far as possible. I also greatly acknowledge the help of Dr. Christina Dyllick of VCH who has guided and coordinated this valuable scientific project. Finally, I wish to express my gratitude to Professor Peter Haasen for the fruitful discussions we had about the contents of this Volume and for his continued interest and guidance.

K. H. Jürgen Buschow
Eindhoven, July 1991

Editorial Advisory Board

Electronic and Magnetic Properties of Metals and Ceramics

Edited by K. H. J. Buschow

Volume 3 A

Volume 3 B

Contributors to Volume 3 A

Professor Jack Bass
Department of Physics
and Astronomy
Michigan State University
East Lansing, MI 48824
USA
Chapter 3

Dr. Volker Eyert
Institut für Festkörperphysik
Technische Hochschule
Hochschulstraße 6
D-6100 Darmstadt
Germany
Chapter 1

Dr. Damien Gignoux
Centre National de Recherche
Scientifique
Laboratoire Louis Néel
25 Avenue des Martyrs
F-38042 Grenoble Cedex
France
Chapter 5

Dr. Takenari Goto
Research Institute for Scientific
Measurements
Tohoku University
Sendai 980
Japan
Chapter 7

Professor Tadao Kasuya
Tohoku University
Department of Physics
Aramaki
Aoba-ku
Sendai 980
Japan
Chapter 7

Dr. Peter H. Kes
Kamerlingh Onnes Laboratorium
University of Leiden
PO Box 9506
NL-2300 RA Leiden
The Netherlands
Chapter 4

Dr. John B. Ketterson
Department of Physics and
Astronomy
Materials Research Center
Northwestern University
2145 Sheridan Road
Evanston, IL 60208-3112
USA
Chapter 6

Professor Jürgen Kübler
Institut für Festkörperphysik
Technische Hochschule
Hochschulstraße 6
D-6100 Darmstadt
Germany
Chapter 1

Professor Yoshichika Onuki
Institute of Materials Science
University of Tsukuba
Sakura-mura
Ibaraki 305
Japan
Chapter 7

Dr. Joachim Schoenes
Laboratorium für Festkörperphysik
ETH Zürich
Hönggerberg/HPF
CH-8093 Zürich
Switzerland
Chapter 2

Professor Paul L. Rossiter
Department of Materials
Engineering
Monash University
Clayton
3168 Melbourne, Victoria
Australia
Chapter 3

Dr. Shengnian N. Song
Department of Physics and
Astronomy
Materials Research Center
Northwestern University
2145 Sheridan Road
Evanston, IL 60208-3112
USA
Chapter 6

Contents

1 Electronic Structure Calculations

Jürgen Kübler and Volker Eyert

Institut für Festkörperphysik, Technische Hochschule,
Darmstadt, Federal Republic of Germany

List of Symbols and Abbreviations

a	lattice constant
A, Γ, L, K, H, X, P, N, M, W	special points of high symmetry in the BZ
$\boldsymbol{a}, \boldsymbol{b}, \boldsymbol{c}$	real space basis vectors
$\boldsymbol{A}, \boldsymbol{B}, \boldsymbol{C}$	reciprocal space basis vectors
$\boldsymbol{A}$	vector potential
a_j, c_j, *or* $a_s(\boldsymbol{k})$	expansion coefficient in the wave function
a_0	$0.529177 \cdot 10^{-10}$ m, Bohr radius
B	bulk modulus
$\boldsymbol{B}$	magnetic induction
$B_{LL'}(\varepsilon, \boldsymbol{k})$	structure constant
c	velocity of light
$\lvert c \rangle$	core states
C_{ij}	elastic constants
$C_l = \omega_l$	center of band
$C_{LL'L''}$	Gaunt (or Clebsch-Gordan) coefficient
$D_l(\varepsilon)$	logarithmic derivative
D_L, $D_L^{(1)}$, $D_L^{(2)}$, $D^{(3)}$	parts constituting structure constant
E	energy
$E[n]$	functional for the total energy
e	electronic charge
E_{F}, $\boldsymbol{k}_{\mathrm{F}}$, $\boldsymbol{v}_{\mathrm{F}}$	Fermi energy, momentum and velocity, respectively
E_ν	characteristic LMTO energy
$f(\varepsilon)$	Fermi-Dirac distribution function
G, G_0	Green's function
g, f	radial functions
$\boldsymbol{G}_S$	reciprocal lattice vector, plus $\boldsymbol{k}$
H, $\mathscr{H}$	Hamiltonian (sometimes $\hat{H}$)
$\mathbf{H}$	Hamiltonian matrix
H_{D}	Dirac Hamiltonian
H_0, $\mathscr{H}_0$	Hamiltonian of non-interacting particles
H_{SO}	spin-orbit Hamiltonian
$\hbar$	Planck's constant divided by 2π
$\tilde{H}$, $\tilde{h}$, $\tilde{J}$, $\tilde{j}$, $\tilde{\chi}$	augmented functions
I	exchange constant
i	$\sqrt{-1}$ (but not when used as index)
J_L, H_L, N_L	wave functions related to j_l, h_l, n_l
J_l, n_l, $h_l^{\pm}$	spherical Bessel, Neumann, and Hankel functions, respectively
$\boldsymbol{k}$	reciprocal space vector in the first BZ
$\lvert \boldsymbol{K} \rangle$	plane wave state
k_{B}	Boltzmann constant
$\boldsymbol{K}_s$	reciprocal lattice vector
$\boldsymbol{L}, \boldsymbol{J}$	operator for angular momentum
l, m, $L = (l, m)$	angular momentum quantum numbers

m	electronic mass
m	magnetic moment
M_μ	ionic mass
M	mass parameter in Dirac theory
m^*	effective mass
N	number of cells in the crystal
$N(\varepsilon)$	density of states
$N_l(\varepsilon)$	partial density of states
n	electronic density
$\hat{\boldsymbol{n}}(\boldsymbol{r})$	electron density operator
$\hat{\boldsymbol{n}}_{\alpha\beta}$	electron-density operator in spin polarized case
$\tilde{n}_{\alpha\beta}$	density matrix
$n_\uparrow$, $n_\downarrow$	spin-up and spin-down electron densities
$\mathbf{O}$	overlap matrix
P	pressure
$\boldsymbol{p}$	momentum operator
P_l	potential function
$P_l(x)$	Legendre polynomial
$\boldsymbol{q}$	wave vector
q_l	partial charge
q_0	Ewald parameter
R	sphere radius
R	radial wave function
$\hat{\boldsymbol{r}}$	unit vector
$\boldsymbol{R}_i$	real space lattice vector
$\boldsymbol{r}_i$	electron position
r_s	density parameter
Ryd, Ry	13.6058 eV, Rydberg
S	sphere radius
S_D	entropy of disordered state
$\mathcal{S}_{LL'}$	LMTO structure constant
T	kinetic energy operator (sometimes $\hat{T}$)
$\mathcal{T}$	kinetic energy
$\mathbf{T}$, $\mathbf{t}$	transition matrix
T	temperature
T_i	translation operator
u	radial wave function
$U_{e\text{-}I}$	electron-ion interaction
$\boldsymbol{u}_{\boldsymbol{q}}$	displacement amplitude
$v_n(\Omega)$	many-body interaction potential
V, V_c	potential (sometimes $\hat{V}$)
V_I	interaction potential of the ions
$V_l = \omega_{l+}$	bottom of band
$\hat{\boldsymbol{x}}$, $\hat{\boldsymbol{y}}$, $\hat{\boldsymbol{z}}$	cartesian unit vectors
$\boldsymbol{X}_n$	ionic position

$\mathscr{Y}$	spin-angular function
Y_L	spherical harmonic
Z	atomic number
Z_V	valence
$\boldsymbol{\alpha}, \boldsymbol{\beta}$	operators used in Dirac theory
γ	Sommerfeld constant
γ_l	LMTO scale factor
δ_{ij}	Kronecker δ
$\delta(x)$	Dirac δ function
δ^*	restricted variation
ΔH	heat of formation
ε	band energy
ε_c	core state energy
ε_i	eigenenergies
$\varepsilon^{(H)}, \varepsilon^{(J)}$	Hankel and Bessel energy
ε_{XC}	exchange-correlation energy density
η_l	scattering phase shift
θ_D	Debye temperature
$\varkappa^2$	energy ε
$\varkappa$	Dirac quantum number
$\mathbf{\Lambda}$	KKR-ASA matrix
λ	electron-phonon enhancement factor
$\lambda_l^{(q)}$	energy moment of order q
μ_B	Bohr magneton
μ_l	scattering phase shift
ν	Poisson ratio
ξ_n	correlation function
$\varrho_{\alpha\beta}$	density operator
$\boldsymbol{\sigma}$	vector operator $(\sigma_x, \sigma_y, \sigma_z)$
Σ	self energy
σ_p	Ising spin-like variable
$\boldsymbol{\sigma}_x, \boldsymbol{\sigma}_y, \boldsymbol{\sigma}_z$	Pauli spin matrices
$\boldsymbol{\tau}_\nu$	atomic position in the unit cell
$\phi, \dot{\phi}$	LMTO basis function
ϕ_i	single particle states
χ, χ_0	susceptibility
χ_+, χ_-	bi-spinors
$\psi, \phi, \chi, \boldsymbol{\Phi}, \boldsymbol{\Psi}$	wave functions
ω_D, Π, Ω	wave-function expansion coefficients
$\omega_{\boldsymbol{q}}$	phonon frequency
Ω	volume of the crystal
Ω_{BZ}	volume of the BZ
Ω_{UC}	volume of the Wigner-Seitz cell
Ω_{MT}	volume of a muffin-tin sphere

AF	antiferromagnet
APW	augmented plane waves
ARUPS	angle-resolved photoemission
ASA	atomic sphere approximation
ASW	augmented spherical waves
a.u.	atomic units
b.c.c.	body-centered cubic
BZ	Brillouin zone
CPA	coherent potential approximation
CVM	cluster variation method
DFT	density functional theory
dHvA	de Haas-van Alphen
f.c.c.	face-centered cubic
GTO	Gaussian-type orbitals
h.c.p.	hexagonal closed packed
IS	interstitial
KKR	Korringa, Kohn, Rostoker
LAPW	linear augmented plane waves
LCAO	linear combination of atomic orbitals
LCGO	linear combination of Gaussian orbitals
LDA	local density approximation
LDF	local density functional
LMTO	linear combination of muffin-tin orbitals
LSDFA	local-spin-density-functional approximation
MST	multiple scattering theory
MT	muffin tin
NCPP	norm-conserving pseudopotential
NFE	nearly free electrons
NM	nonmagnetic
OPW	orthogonalized plane waves
PS	pseudopotential
PW	plane wave
RAPW	relativistic APW
RPA	random phase approximation
s.c.	simple cubic
SO	spin orbit
STO	Slater-type orbitals
XC	exchange correlation

1.1 Introduction

The determination of the electronic structure and related properties of condensed matter is an important field in modern solid state physics, and the progress made in the last two decades is truly impressive. This progress is due in part to the breathtaking development in computer hardware and software technologies. In addition, the progress in formulating the theory and methodologies for electronic structure calculations has led to vast improvements in our understanding of this quantum-mechanical many-body problem. Thus in the following chapter much space is devoted to theory and methodologies. Yet experimental tools, especially methods of spectroscopy, have also improved immensely and have thereby both accelerated the progress and expanded the size of this field. This growth presents a problem of scope. So before we really begin, we need to narrow down the field and define our perspective.

We will be chiefly concerned with periodic, three-dimensional metallic crystals. Toward the end, a certain class of insulators will be mentioned. We thus refrain from dealing with such interesting topics as imperfections, impurities, vacancies, surfaces, and thin films. This is unfortunate but necessary for the sake of brevity. Disordered or amorphous systems will only be dealt with in passing, which is again regrettable, but necessary. Furthermore, in the selection of methodologies we focus our attention nearly exclusively on ab initio methods which are based on density functional theory and which in all cases thus constitute a self-consistent field problem.

The style of the following is mostly tutorial except for the later Secs. 1.4.3 and 1.4.4 where we simply ran out of space. Indeed, a large portion of this chapter is the backbone of a course on solid state theory held repeatedly at the Technical University of Darmstadt. This is also the reason why this treatise should not be considered a review article giving a complete account of the literature although a great number of papers were included, starting roughly around 1980. The interested reader will find a reasonably complete review in an article by Koelling (1981) which includes references prior to 1980.

In large parts of the following, we will be concerned with the motion of electrons only, yet in Secs. 1.4.1.3 and 1.4.3.2 we will briefly discuss lattice degrees of freedom. We thus must ask why the motion of electrons and nuclei is considered separately, at least in a first approximation. The answer was given long ago by Born and Oppenheimer (1927) and can be stated quite simply: Because electrons are very light compared to nuclei, they move much more rapidly, and can follow the slower motions of the nuclei quite accurately. At the same time, the electron distribution determines the potential in which the nuclei move. This phenomenon is of considerable conceptual importance, and thus warrants a brief outline of its quantum-mechanical justification following Callaway (1974).

Thus we consider a system containing ions of mass M_μ with coordinates $\boldsymbol{X}_\mu$ and electrons of mass m with coordinates $\boldsymbol{r}_i$. The Hamiltonian for the system is

$$H = \sum_\mu \left[\left(-\frac{\hbar^2}{2M_\mu}\nabla_\mu^2\right) + \sum_{\nu>\mu} V_{\mathrm{I}}(\boldsymbol{X}_\mu - \boldsymbol{X}_\nu)\right] + \sum_i \left[\left(-\frac{\hbar^2}{2m}\nabla_i^2\right) + \sum_{j>i}\frac{e^2}{|\boldsymbol{r}_i - \boldsymbol{r}_j|} + \sum_\mu U_{\mathrm{e\text{-}I}}(\boldsymbol{r}_i - \boldsymbol{X}_\mu)\right] \tag{1-1}$$

The quantity $V_{\mathrm{I}}(\boldsymbol{X}_\mu - \boldsymbol{X}_\nu)$ is the interaction potential of the ions with each other, while $U_{\mathrm{e\text{-}I}}(\boldsymbol{r}_i - \boldsymbol{X}_\mu)$ represents the interaction between an electron at $\boldsymbol{r}_i$ and an ion at $\boldsymbol{X}_\mu$. It is natural to write this Hamiltonian as a sum of an ionic and an electronic part

$$H = H_{\mathrm{I}} + H_{\mathrm{e}} \tag{1-2}$$

in which H_{I} contains the first part of Eq. (1-1),

$$H_{\mathrm{I}} = \sum_\mu \left[-\frac{\hbar^2}{2M_\mu} \nabla_\mu^2 + \sum_{\nu>\mu} V_{\mathrm{I}}(\boldsymbol{X}_\mu - \boldsymbol{X}_\nu) \right] \tag{1-3}$$

and H_{e} contains the remainder, including the interaction of the electrons with the ions,

$$H_{\mathrm{e}} = \sum_i \left[-\frac{\hbar^2}{2m} \nabla_i^2 + \sum_{j>i} \frac{e^2}{|\boldsymbol{r}_i - \boldsymbol{r}_j|} + \right.$$
$$\left. + \sum_\mu U_{\mathrm{e_I}}(\boldsymbol{r}_i - \boldsymbol{X}_\mu) \right] \tag{1-4}$$

Now consider the Schrödinger equation for the electrons in the presence of *fixed* ions, first,

$$H_{\mathrm{e}}\, \Psi_K(\boldsymbol{X},\boldsymbol{r}) = E_K(\boldsymbol{X})\, \Psi_K(\boldsymbol{X},\boldsymbol{r}) \tag{1-5}$$

$\boldsymbol{X}$ and $\boldsymbol{r}$ denote the set of all ionic and electronic coordinates. The energy of the electronic system and the wave function of the electronic state depend on the ionic positions. In practice, we are unable to solve Eq. (1-5) exactly and must resort to approximation procedures. At this point, however, it is practical to proceed as if a complete set of solutions $\{\Psi_K(\boldsymbol{X},\boldsymbol{r})\}$ could be obtained. Then the wave functions for the entire system of electrons plus ions is to be expanded with respect to the Ψ as basis functions. Let Q denote the quantum numbers required to specify the total state of the system. The wave function is

$$\Phi_Q(\boldsymbol{X},\boldsymbol{r}) = \sum_K \chi_K(Q,\boldsymbol{X})\, \Psi_K(\boldsymbol{X},\boldsymbol{r}) \tag{1-6}$$

and Φ_Q must satisfy a Schrödinger equation with the full Hamiltonian (Eq. (1-1)), i.e.,

$$H\, \Phi_Q(\boldsymbol{X},\boldsymbol{r}) = \varepsilon_Q\, \Phi_Q(\boldsymbol{X},\boldsymbol{r}) \tag{1-7}$$

Let us assume the electronic functions Ψ_K can be normalized for all values of $\boldsymbol{X}$ and are orthogonal with respect to K for fixed $\boldsymbol{X}$. We can then substitute Eq. (1-6) into Eq. (1-7), use Eq. (1-5), and integrate over all $\boldsymbol{r}$'s to obtain a set of coupled equations for the functions χ_K of the form

$$\sum_K \{[H_{\mathrm{I}} + E_K(\boldsymbol{X})]\, \delta_{KK'} + C_{K'K}(\boldsymbol{X})\}\, \chi_K(Q,\boldsymbol{X})$$
$$= \varepsilon_Q\, \chi_{K'}(Q,\boldsymbol{X}) \tag{1-8}$$

The operator $C_{K'K}(\boldsymbol{X})$ has the form

$$C_{K'K}(\dot{\boldsymbol{X}}) = -\int \Psi_{K'}^*(\boldsymbol{X},\boldsymbol{r}) \sum_\mu \frac{\hbar^2}{2M_\mu} \cdot$$
$$\cdot [\nabla_\mu^2\, \Psi_K(\boldsymbol{X},\boldsymbol{r}) + 2\nabla_\mu\, \Psi_K(\boldsymbol{X},\boldsymbol{r}) \cdot \nabla_\mu]\, \mathrm{d}\boldsymbol{r} \tag{1-9}$$

In the lowest (or adiabatic) approximation, the coupling term $C_{K'K}(\boldsymbol{X})$ is ignored entirely. Then Eq. (1-8) is diagonal, indicating that the energy levels of the system of ions are determined by solving the Schrödinger equation

$$[H_{\mathrm{I}} + E_K(\boldsymbol{X})]\, \chi_K(Q,\boldsymbol{X}) = \varepsilon_Q\, \chi_K(Q,\boldsymbol{X}) \tag{1-10}$$

The Hamiltonian of Eq. (1-10) is obtained by adding to H_{I} (as given by Eq. (1-3)) the term $E_K(\boldsymbol{X})$, which thus is seen to represent a contribution to the potential energy of the ion system. This implies that the potential energy depends on the state of the electrons. However, this dependence should not be strong under usual circumstances, since normal conductivity processes in solids involve a rearrangement of only a few electrons near the Fermi surface. The term $C_{K'K}(\boldsymbol{X})$ couples different states of the ionic lattice. It is not diagonal in the electron wave vector and thus involves transitions between differing electronic states. The diagonal components of C contribute an additional term to the leading approximation, Eq. (1-10), which we will not consider here. The matrix element for a simultaneous transition of the lattice and the electronic system is

$$m_{K'K}(Q',Q) = \\ = \int \chi^*_{K'}(Q',X)\, C_{K'K}(X)\, \chi_K(Q,X)\, \mathrm{d}X \quad (1\text{-}11)$$

in which the χ are solutions of Eq. (1-10). We now turn to a discussion of the electronic part of the problem (Eq. (1-5)), but defer the treatment of the many-body aspects of this problem to Sec. 1.3.

1.2 Solution Methods for the Single-Particle Schrödinger Equation

1.2.1 Bloch's Theorem

Let us start by considering a crystal of some symmetry and a single electron moving in it. We do not need to specify here in detail whether or not this electron moves in the presence of a great number of other electrons; we simply assume we know an effective potential, $V(\boldsymbol{r})$, that acts on the electron. In Sec. 1.3 we will present a well-defined prescription to construct a meaningful potential.

According to the laws of quantum mechanics, the possible states of the electron are given by the solutions of the Schrödinger equation

$$\left(-\frac{\hbar^2}{2m}\nabla^2 + V(\boldsymbol{r})\right)\psi_i(\boldsymbol{r}) = \varepsilon_i\,\psi_i(\boldsymbol{r}) \quad (1\text{-}12)$$

where $\psi_i(\boldsymbol{r})$ is the wave function, ε_i the energy-eigenvalue, and i a label for quantum numbers to be specified below. The symmetry of the potential is the same as that of the crystal lattice. Thus its most prominent aspect is translational periodicity. It has been 60 years since Felix Bloch (1929) first discussed the consequences of this symmetry, and we now call the appropriate wave function the Bloch function. Its properties are of fundamental importance for solid state physics; we therefore take the time to establish them here.

First notice that if the potential acting on the electron at position $\boldsymbol{r}$ in the crystal is $V(\boldsymbol{r})$, then it is unchanged if the position is displaced by any translation vector $\boldsymbol{R}_j$, i.e.,

$$V(\boldsymbol{r}+\boldsymbol{R}_j) = V(\boldsymbol{r}) \quad (1\text{-}13)$$

The vectors $\boldsymbol{R}_j$ are customarily expressed by three independent translation vectors $\boldsymbol{a}$, $\boldsymbol{b}$, $\boldsymbol{c}$,

$$\boldsymbol{R}_j = l_j\,\boldsymbol{a} + m_j\,\boldsymbol{b} + n_j\,\boldsymbol{c} \quad (1\text{-}14)$$

where the quantities l_j, m_j, and n_j are integers. The vectors $\boldsymbol{a}$, $\boldsymbol{b}$, $\boldsymbol{c}$ are defined in a standard way by the crystal-lattice and are to be found among one of the possible 14 Bravais lattices in three dimensions, see, e.g., Kittel (1986). It is convenient to define a set of translation operators T_j having the property

$$T_j\,f(\boldsymbol{r}) = f(\boldsymbol{r}+\boldsymbol{R}_j) \quad (1\text{-}15)$$

where $f(\boldsymbol{r})$ is any function of position. The operators T_j obviously form a group and they commute with each other, and, because of Eq. (1-13), with the Hamiltonian

$$H = -\frac{\hbar^2}{2m}\nabla^2 + V(\boldsymbol{r}) \quad (1\text{-}16)$$

thus

$$[T_j, H] = 0 \quad (1\text{-}17)$$

As a consequence, the wave functions $\psi_i(\boldsymbol{r})$ can be chosen to be eigenfunctions of the energy *and* all the translations; i.e., besides Eq. (1-12) we have

$$T_j\,\psi_i(\boldsymbol{r}) = \psi_i(\boldsymbol{r}+\boldsymbol{R}_j) = \lambda_j\,\psi_i(\boldsymbol{r}) \quad (1\text{-}18)$$

where λ_j is the eigenvalue that describes the effect of the operator T_j on the function ψ_i. Next we require that the group representation be unitary, or, in other words, that the norm of ψ_i is unchanged by translations. Thus λ_j must be a complex number of modulus unity written

$$\lambda_j = \exp(\mathrm{i}\,\theta_j) \tag{1-19}$$

where θ_j is real.

Let us finally consider two translation operators, T_j and T_l, that act in succession. Obviously

$$T_j\, T_l\, \psi_i(\boldsymbol{r}) = \psi_i(\boldsymbol{r}+\boldsymbol{R}_j+\boldsymbol{R}_l) = \\ = \lambda_j\, \lambda_l\, \psi_i(\boldsymbol{r}) = \lambda_{j+l}\, \psi_i(\boldsymbol{r}) \tag{1-20}$$

and hence from Eq. (1-19)

$$\theta_j + \theta_l = \theta_{j+l} \tag{1-21}$$

This is satisfied if

$$\theta_j = \boldsymbol{k} \cdot \boldsymbol{R}_j \tag{1-22}$$

where $\boldsymbol{k}$ is an arbitrary vector that is the same for each operation. It characterizes the particular wave function $\psi_i(\boldsymbol{r})$ and is thus part of the subscript i. Allowing for other quantum numbers, n, besides $\boldsymbol{k}$, we may therefore replace $\psi_i(\boldsymbol{r})$ by $\psi_{n\boldsymbol{k}}(\boldsymbol{r})$ and the energy eigenvalue ε_i by $\varepsilon_{n\boldsymbol{k}}$. The Bloch function now satisfies

$$\psi_{n\boldsymbol{k}}(\boldsymbol{r}+\boldsymbol{R}_j) = \mathrm{e}^{\mathrm{i}\,\boldsymbol{k}\cdot\boldsymbol{R}_j}\, \psi_{n\boldsymbol{k}}(\boldsymbol{r}) \tag{1-23}$$

This is Bloch's theorem.

It is now apparent that the vector $\boldsymbol{k}$ is of great importance in describing the Bloch function. In fact, we see from Eq. (1-23) that it has the dimension of an inverse length. It is thus a vector in reciprocal space which is conveniently described using the *reciprocal lattice*. The latter is defined by vectors $\boldsymbol{K}_s$ obeying

$$\boldsymbol{K}_s \cdot \boldsymbol{R}_j = 2\pi\, n_{sj} \tag{1-24}$$

where the quantities n_{sj} are integers (positive, negative, or zero). Equation (1-24) holds for all translations $\boldsymbol{R}_j$; the end points of all the vectors $\boldsymbol{K}_s$ define a lattice of points, which is called the reciprocal lattice. Just as Eq. (1-24) expresses the translation $\boldsymbol{R}_j$ as a linear combination of the basic translations $\boldsymbol{a}$, $\boldsymbol{b}$, $\boldsymbol{c}$, one can express the reciprocal lattice vectors as linear combinations of fundamental translation vectors in reciprocal space $\boldsymbol{A}$, $\boldsymbol{B}$, $\boldsymbol{C}$, i.e.,

$$\boldsymbol{K}_s = g_{s1}\, \boldsymbol{A} + g_{s2}\, \boldsymbol{B} + g_{s3}\, \boldsymbol{C} \tag{1-25}$$

where the quantities g_{s1}, g_{s2}, g_{s3} are integers again. Equation (1-24) requires that $\boldsymbol{A}\cdot\boldsymbol{a} = 2\pi$, $\boldsymbol{B}\cdot\boldsymbol{b} = 2\pi$, $\boldsymbol{C}\cdot\boldsymbol{c} = 2\pi$, and all other scalar products vanish. These equations are satisfied if

$$\boldsymbol{A} = 2\pi\, \frac{\boldsymbol{b}\times\boldsymbol{c}}{\boldsymbol{a}\cdot(\boldsymbol{b}\times\boldsymbol{c})}, \quad \boldsymbol{B} = 2\pi\, \frac{\boldsymbol{c}\times\boldsymbol{a}}{\boldsymbol{a}\cdot(\boldsymbol{b}\times\boldsymbol{c})}, \\ \boldsymbol{C} = 2\pi\, \frac{\boldsymbol{a}\times\boldsymbol{b}}{\boldsymbol{a}\cdot(\boldsymbol{b}\times\boldsymbol{c})} \tag{1-26}$$

We have thus a formula for constructing the reciprocal lattice from the direct lattice vectors. Not doubt, the reader recognizes these last equations and is familiar with them from the theory of diffraction by crystals.

An immediate consequence of the above definitions follows for the wave vector $\boldsymbol{k}$ labeling the Bloch functions. To see this, consider two vectors $\boldsymbol{k}$ and $\boldsymbol{k}'$ which satisfy

$$\boldsymbol{k}' = \boldsymbol{k} + \boldsymbol{K}_s \tag{1-27}$$

where $\boldsymbol{K}_s$ is an arbitrary reciprocal lattice vector. Evidently, because of Eq. (1-24), one has

$$\mathrm{e}^{\mathrm{i}\,\boldsymbol{k}\cdot\boldsymbol{R}_j} = \mathrm{e}^{\mathrm{i}\,\boldsymbol{k}'\cdot\boldsymbol{R}_j} \tag{1-28}$$

for all vectors $\boldsymbol{R}_j$. Consequently, the wave functions $\psi_{n\boldsymbol{k}}(\boldsymbol{r})$ and $\psi_{n\boldsymbol{k}'}(\boldsymbol{r})$ possess the same translation eigenvalue λ. The wave vectors $\boldsymbol{k}$ and $\boldsymbol{k}'$ satisfying Eq. (1-27) are said to be equivalent and we adopt the convention that

$$\psi_{n\boldsymbol{k}}(\boldsymbol{r}) = \psi_{n\boldsymbol{k}+\boldsymbol{K}_s}(\boldsymbol{r}) \tag{1-29}$$

for any $\boldsymbol{K}_s$. This convention implies for the eigenvalue $\varepsilon_{n\boldsymbol{k}}$ of the Schrödinger equation,

$$H\, \psi_{n\boldsymbol{k}}(\boldsymbol{r}) = \varepsilon_{n\boldsymbol{k}}\, \psi_{n\boldsymbol{k}}(\boldsymbol{r}) \tag{1-30}$$

that

$$\varepsilon_{n\boldsymbol{k}} = \varepsilon_{n\boldsymbol{k}+\boldsymbol{K}_s} \tag{1-31}$$

and the domain of $\boldsymbol{k}$ is restricted to some unit cell in reciprocal space. It has become customary to choose a cell that has the full symmetry of the reciprocal lattice; it is called Brillouin zone and is constructed as follows: An arbitrary lattice point is chosen as the origin, and then the vectors connecting the origin with other lattice points are drawn. Next, the planes that are perpendicular bisectors of these vectors are constructed. The Brillouin zone (BZ) is the smallest volume containing the origin bounded by these planes. Useful for later examples are some of the most common BZ's, which are therefore illustrated here.

There are three Bravais lattices in the cubic system: the simple cubic (s.c.), the body centered cubic (b.c.c.), and the face centered cubic (f.c.c.) lattices. The s.c. Brillouin zone is just a cube. More interesting are the other cubic cases; the primitive translation vectors of a b.c.c. lattice are given by

$$\boldsymbol{a} = \frac{a}{2}(\hat{x}+\hat{y}-\hat{z}), \quad \boldsymbol{b} = \frac{a}{2}(-\hat{x}+\hat{y}+\hat{z}),$$

$$\boldsymbol{c} = \frac{a}{2}(\hat{x}-\hat{y}+\hat{z}) \tag{1-32}$$

and those of an f.c.c. lattice by

$$\boldsymbol{a} = \frac{a}{2}(\hat{x}+\hat{y}), \quad \boldsymbol{b} = \frac{a}{2}(\hat{y}+\hat{z}),$$

$$\boldsymbol{c} = \frac{a}{2}(\hat{z}+\hat{x}) \tag{1-33}$$

where a is the lattice constant and $\hat{x}$, $\hat{y}$, $\hat{z}$ are unit cartesian vectors. Using Eq. (1-26) we obtain the following for the primitive translation in reciprocal space:

$$\boldsymbol{A} = \frac{2\pi}{a}(\hat{x}+\hat{y}), \quad \boldsymbol{B} = \frac{2\pi}{a}(\hat{y}+\hat{z}),$$

$$\boldsymbol{C} = \frac{2\pi}{a}(\hat{x}+\hat{z}) \tag{1-34}$$

for the b.c.c. lattice and

$$\boldsymbol{A} = \frac{2\pi}{a}(\hat{x}+\hat{y}-\hat{z}), \quad \boldsymbol{B} = \frac{2\pi}{a}(-\hat{x}+\hat{y}+\hat{z}),$$

$$\boldsymbol{C} = \frac{2\pi}{a}(\hat{x}-\hat{y}+\hat{z}) \tag{1-35}$$

for the f.c.c. lattice. We see that the reciprocal lattice of b.c.c. is the f.c.c. lattice and vice versa. The BZ is now easily drawn and depicted in Fig. 1-1 for the f.c.c. lattice and in Fig. 1-2 for the b.c.c. lattice. For future reference we add the BZ of a hexagonal lattice in Fig. 1-3. Some special $\boldsymbol{k}$ points are labeled following the convention first introduced by Bouckaert et al. (1936) (see

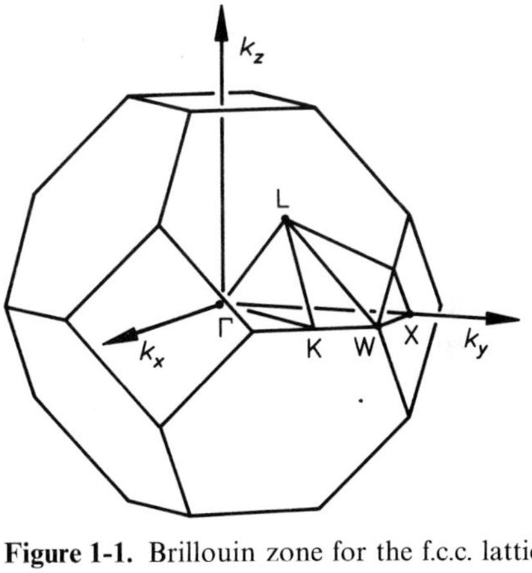

Figure 1-1. Brillouin zone for the f.c.c. lattice.

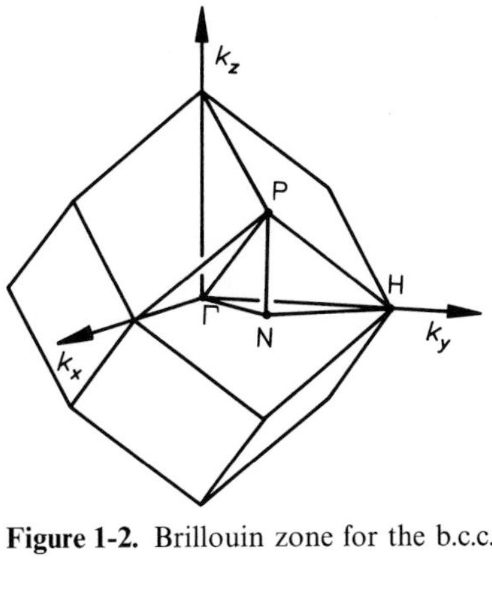

Figure 1-2. Brillouin zone for the b.c.c. lattice.

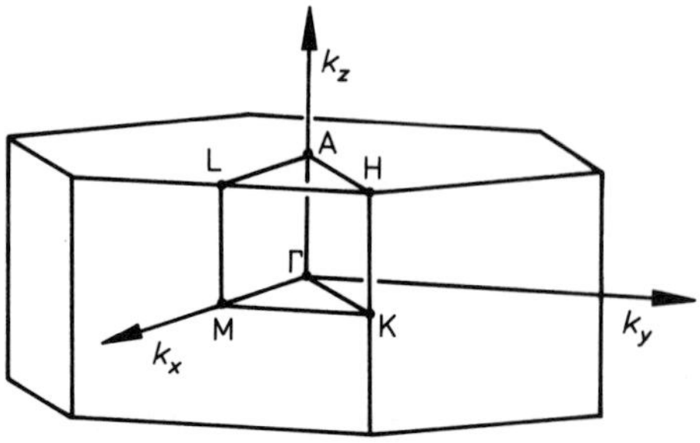

Figure 1-3. Brillouin zone for the hexagonal lattice.

also Bradley and Cracknell, 1972). These labels are usually used in plots of the band structure and in classifications of further symmetries of the electron states, because most interesting crystals possess more symmetry than the translational invariance which gives rise to Bloch functions. The additional symmetry operations, such as rotations and reflections, form groups again, the so-called point groups. These taken together with translations form the space group of a crystal. The corresponding operators commute with the Hamiltonian again and, therefore, lead to classifications of the electron states which, in general, are different for different $\boldsymbol{k}$ vectors. There is insufficient space to deal with this interesting but specialized subject here. But if forthcoming examples require it, the necessary explanations will be given later. The reader interested in more detail is referred to Koster (1957), Callaway (1964), Kovalev (1965), Bradley and Cracknell (1972), or Ludwig and Falter (1988), to name a few.

We close this section by determining the number of points in the BZ, and hence the density of states. This is done by imposing the so-called Born-von Kármán (or periodic) boundary condition on the Bloch functions by requiring

$$\psi_{n\boldsymbol{k}}(\boldsymbol{r}+N_1\boldsymbol{a}) = \psi_{n\boldsymbol{k}}(\boldsymbol{r}) = \psi_{n\boldsymbol{k}}(\boldsymbol{r}+N_2\boldsymbol{b}) = = \psi_{n\boldsymbol{k}}(\boldsymbol{r}+N_3\boldsymbol{c}) \tag{1-36}$$

where $N = N_1 \cdot N_2 \cdot N_3$ is the total number of primitive cells in the crystal. These rather artificial boundary conditions are adopted under the assumption that the bulk properties of the solid will not depend on this particular choice, which can therefore be made for reasons of analytical convenience. The Eq. (1-36) and Bloch's theorem together with Eq. (1-24) now imply

$$\exp(\mathrm{i}\,2\pi N_1 k_1) = \exp(\mathrm{i}\,2\pi N_2 k_2) = = \exp(\mathrm{i}\,2\pi N_3 k_3) = 1 \tag{1-37}$$

where k_1, k_2, k_3 are the components of $\boldsymbol{k}$ expressed as

$$\boldsymbol{k} = k_1\boldsymbol{A} + k_2\boldsymbol{B} + k_3\boldsymbol{C} \tag{1-38}$$

and hence the general form of the allowed $\boldsymbol{k}$ vectors is

$$\boldsymbol{k} = \frac{m_1}{N_1}\boldsymbol{A} + \frac{m_2}{N_2}\boldsymbol{B} + \frac{m_3}{N_3}\boldsymbol{C} \tag{1-39}$$

where m_1, m_2, m_3 are integers (positive, negative, or zero). It follows that the volume, d^3k, of $\boldsymbol{k}$ space per allowed value of $\boldsymbol{k}$ is the volume of the parallelepiped with edges $\boldsymbol{A}/N_1$, $\boldsymbol{B}/N_2$, and $\boldsymbol{C}/N_3$, i.e.,

$$\mathrm{d}^3k = \frac{\Omega_{\mathrm{BZ}}}{N} \tag{1-40}$$

where Ω_{BZ} is the volume of the BZ,

$$\Omega_{\mathrm{BZ}} = \boldsymbol{A}\cdot(\boldsymbol{B}\times\boldsymbol{C}) \tag{1-41}$$

It is easy to see with the basic relations (1-26) that

$$\Omega_{\mathrm{BZ}} = (2\pi)^3/\Omega_{\mathrm{UC}} \tag{1-42}$$

where Ω_{UC} is the volume of the unit cell,

$$\Omega_{\mathrm{UC}} = \boldsymbol{a}\cdot(\boldsymbol{b}\times\boldsymbol{c}) \tag{1-43}$$

Writing for the volume of the crystal,

$$\Omega = N\,\Omega_{\mathrm{UC}} \tag{1-44}$$

we see from Eqs. (1-40) and (1-42) that

$$\mathrm{d}^3k = \frac{(2\pi)^3}{\Omega} \tag{1-45}$$

This relation gives a simple prescription to convert a sum over $\boldsymbol{k}$ vectors to an integral:

$$\sum_k \rightarrow \frac{\Omega}{(2\pi)^3} \int d^3k = N \frac{\Omega_{UC}}{(2\pi)^3} \int d^3k \qquad (1\text{-}46)$$

Equation (1-46) states that the number of allowed wave vectors in the BZ is equal to the number of primitive cells in the crystal. A further application of Eq. (1-46) is an important formula for the *density of states*, $N(\varepsilon)$ (per spin), which is written as

$$N(\varepsilon) = \frac{1}{\Omega_{BZ}} \sum_n \int_{BZ} d^3k \, \delta(\varepsilon - \varepsilon_{n\boldsymbol{k}}) \qquad (1\text{-}47)$$

where $\delta(x)$ is the Dirac δ function, and the integral is taken over the BZ. For an actual computation of the density of states, certain numerical techniques must be used, for instance, the tetrahedron method. Since this is a rather specialized subject, we will not go into any details (see, e.g., Coleridge et al., 1982, or Skriver, 1984). The relation for $N(\varepsilon)$ itself, however, is quite plausible because the δ function does the counting of states, giving unity for the integral whenever $\varepsilon = \varepsilon_{n\boldsymbol{k}}$ and zero otherwise. The constant of proportionality is $1/\Omega_{BZ}$, ensuring that

$$\int_{\varepsilon_a}^{\varepsilon_b} N(\varepsilon) \, d\varepsilon = n_{ab} \qquad (1\text{-}48)$$

where n_{ab} is the number of bands in the interval $(\varepsilon_a, \varepsilon_b)$.

1.2.2 Plane Waves and Pseudopotentials

We now begin with the description of some important methods of solutions of the band-structure problem embodied in Eq. (1-12), focusing mainly on the underlying physics while leaving aside details which can be found in more specialized texts. To enable the reader to follow the literature in this field and for our own convenience we will in the following adopt atomic units (a.u.). These use the Bohr radius, $a_0 = \hbar^2/m e^2$ ($0.52918 \cdot 10^{-10}$ m) for the unit of length and the "Rydberg" $e^2/2a_0$ (13.6057 eV) for the unit of energy. For the units of length and energy to have the numerical value 1, one chooses $\hbar = 1$, $e^2 = 2$ and $m = \frac{1}{2}$. Another set of atomic units exists in which $\hbar$, m and e all have the numerical value 1; in this case the unit of energy is the "hartree" (27.2114 eV).

The solution of the band-structure problem will be attempted by an expansion of the Bloch function in terms of a suitable set of functions, $\{\chi_j(\boldsymbol{r})\}$:

$$\psi_{\boldsymbol{k}}(\boldsymbol{r}) = \sum_j a_j(\boldsymbol{k}) \, \chi_j(\boldsymbol{r}) \qquad (1\text{-}49)$$

The sum on j is *finite* in practice, and the set of functions, $\chi_j(\boldsymbol{r})$, is in general not orthonormal; completeness is another problem. The expansion coefficients, $a_j(\boldsymbol{k})$, are obtained by a variational procedure: one substitutes Eq. (1-49) into Eq. (1-12) multiplying from the left with $\psi^*_{\boldsymbol{k}}(\boldsymbol{r})$ and then integrating; the resulting expression is then varied; i.e., with H defined by Eq. (1-16) we can write

$$\delta \sum_{ij} a_i^*(\boldsymbol{k}) \, a_j(\boldsymbol{k}) \left\{ \int \chi_i^*(\boldsymbol{r}) \, H \, \chi_j(\boldsymbol{r}) \, d^3r - \varepsilon_{\boldsymbol{k}} \int \chi_i^*(\boldsymbol{r}) \, \chi_j(\boldsymbol{r}) \, d^3r \right\} = 0 \qquad (1\text{-}50)$$

which gives

$$\sum_j \left\{ \int \chi_i^*(\boldsymbol{r}) \, H \, \chi_j(\boldsymbol{r}) \, d^3r - \varepsilon_{\boldsymbol{k}} \int \chi_i^*(\boldsymbol{r}) \, \chi_j(\boldsymbol{r}) \, d^3r \right\} a_j(\boldsymbol{k}) = 0 \qquad (1\text{-}51)$$

This equation is called secular equation. Its detailed solution depends on the type of functions $\chi_j(\boldsymbol{r})$ chosen (they may or may not be energy dependent), but the eigenvalues, $\varepsilon_{\boldsymbol{k}}$, always follow by equating the determinant of $\{\ldots\}$ to zero and finding the roots.

1.2.2.1 Plane Waves

Returning to the basic expansion, we begin by making a seemingly simple choice and select a Fourier series. In the language of band theory this is called a plane-wave expansion; each plane wave is written as a Bloch function, $\exp[\mathrm{i}(\boldsymbol{k}+\boldsymbol{K}_s)\cdot\boldsymbol{r}]$, where $\boldsymbol{K}_s$ is a reciprocal lattice vector. It is easy to see that this is indeed a Bloch function by applying a translation and observing Eq. (1-24). Thus a valid expansion is a sum over reciprocal lattice vectors and has the form

$$\psi_{\boldsymbol{k}}(\boldsymbol{r}) = \frac{1}{\sqrt{\Omega}} \sum_s a_s(\boldsymbol{k})\, \mathrm{e}^{\mathrm{i}(\boldsymbol{k}+\boldsymbol{K}_s)\cdot\boldsymbol{r}} \tag{1-52}$$

where Ω is the volume of the crystal, and the band index was temporarily dropped in our notation. The factor $\Omega^{-1/2}$ ensures that $\psi_{\boldsymbol{k}}$ is normalized to unity in the volume Ω provided that

$$\sum_s |a_s(\boldsymbol{k})|^2 = 1 \tag{1-53}$$

To obtain the expansion coefficients $a_s(\boldsymbol{k})$ and the eigenvalues $\varepsilon_{\boldsymbol{k}}$ we set up the matrix of the Hamiltonian on the plane-wave basis as described by the Eqs. (1-50), (1-51). The resulting secular equation can be written as

$$\sum_s \{[(\boldsymbol{k}+\boldsymbol{K}_s)^2 - \varepsilon_{\boldsymbol{k}}]\,\delta_{st} + V(\boldsymbol{K}_t - \boldsymbol{K}_s)\} \cdot a_s(\boldsymbol{k}) = 0 \tag{1-54}$$

The solutions for the eigenvalues, $\varepsilon_{\boldsymbol{k}}$, and the eigenvectors, $a_s(\boldsymbol{k})$, thus constitute a standard linear algebra problem and are obtained by diagonalizing the matrix $\{\ldots\}$. The quantity $V(\boldsymbol{K})$ is the Fourier coefficient of the effective crystal potential. It is obtained by expressing the latter as a sum of identical terms, $V_c(\boldsymbol{r})$, centered on each unit cell of the crystal,

$$V(\boldsymbol{r}) = \sum_v V_c(\boldsymbol{r} - \boldsymbol{R}_v) \tag{1-55}$$

A straight forward calculation gives

$$V(\boldsymbol{K}) = \Omega_{\mathrm{UC}}^{-1} \int_{\Omega_{\mathrm{UC}}} \mathrm{e}^{\mathrm{i}\boldsymbol{K}\cdot\boldsymbol{r}}\, V_c(\boldsymbol{r})\, \mathrm{d}^3 r \tag{1-56}$$

where again use was made of Eq. (1-24).

Physically, we can view the potential $V_c(\boldsymbol{r})$ as something very closely related to an atomic potential or a sum of atomic potentials, if the unit cell contains more than one kind of atom. For a valence electron moving through the crystal it seems clear that the potential cannot be weak in the intra-atomic regions although it may well be so in the region between the atoms. In fact, for a solid made up of atoms with atomic number Z (for Al, e.g., $Z=13$) and valence Z_V ($Z_V=3$ for Al), the leading contribution to $V(\boldsymbol{r})$ very close to the nucleus is the Coulomb potential, $-Z e^2/r$, changing to $-Z_V e^2/r$ outside the core region because of electrostatic screening by the core electrons. Further out this potential becomes rather flat between two atoms because of periodicity. As a consequence, a great number of large Fourier coefficients are obtained from Eq. (1-56) that must be included in the secular Eq. (1-54). The wavelengths are very small (large $|\boldsymbol{K}_t - \boldsymbol{K}_s|$) near the nuclei and very large in the interstitial region. In a lucid description of the problem, Heine (1970), has indeed estimated the rank of the secular Eq. (1-54) to be about $10^6 \times 10^6$.

In view of this it is paradoxical that the electronic properties of at least the simple metals (Na, K, Rb, Ca, Al, etc.) seem to be experimentally well described by the free-electron model if one is willing to accept a small change in the apparent electron mass (see, e.g., Ashcroft and Mermin, 1976). To be more precise, this is remarkable for two reasons: the apparent size of the Fourier coefficients, which our physical reasoning somehow overestimates, and the largely invisible effect of Coulomb interactions between the electrons, which we also expect to be large. That these electron-electron interactions seem so weak (of course, there are prominent exceptions) will become clear in Sec. 1.3 when we describe the physics underlying the effective crystal potential and the single-particle approximation. The weak Fourier coefficients, on the other hand, are explained by the pseudopotential concept which we are leading up to now.

Following the experimental facts, let us therefore first accept free-electron behavior and set $V(\boldsymbol{K}) = 0$ for all reciprocal lattice vectors $\boldsymbol{K}$. However, we will assume the latter are still defined. This gives a rather artificial solid where there is no potential, but still a lattice; this is called the "empty lattice". The secular Eq. (1-54) is now trivial; it gives solutions, $\varepsilon_{\boldsymbol{k}}$, for a given $\boldsymbol{k}$ for each reciprocal lattice vector $\boldsymbol{K}_s$. The resulting ε versus $\boldsymbol{k}$ curves are plotted for an f.c.c. lattice in Fig. 1-4 along some typical directions in the BZ, the lettering in Figs. 1-1 and 1-2 corresponding with those of Figs. 1-4 and 1-5. These ε versus $\boldsymbol{k}$ curves are called "energy bands". Most states are highly degenerate and we indicate some of the degeneracies by the numbers appearing near the "bands".

To describe the band structure of the simple metals – at least in zeroth approximation – we first must ignore all core electrons. The valence electrons are then placed in the lowest energy states according to Fermi-Dirac statistics, which allows them to be filled up to the Fermi energy. For monovalent metals the lowest band in Figs. 1-4 and 1-5 becomes half-filled giving a Fermi energy of about 5 eV for the arbitrary, but convenient, choice of the lattice constant of $a = 2\pi\, a_0$. This band-filled ratio results from our counting of states following Eq. (1-46) when we include spin-degeneracy; i.e., each possible $\boldsymbol{k}$ state is occupied twice – once with a spin-up and once with a spin-down electron.

Proceeding one step further, we can empirically include a weak crystal potential in Eq. (1-54) and arrive at the nearly free electron model. Here, the crystal potential in Eq. (1-54) is treated conveniently by perturbation theory (Ashcroft and Mermin, 1976), and without doing any calculations we know that in lowest order it will lift the degeneracies of the bands. Since the lowest band is non-degenerate, the effect of the perturbation in the half-filled band will be strong above the Fermi energy, in Fig. 1-4, for instance, at the $\boldsymbol{k}$ points W, L, X, etc., where the degeneracies appear first. As a result the lowest-occupied band will be deformed only slightly giving rise to small changes in the apparent electron-mass, as is observed. But how can all this be justified?

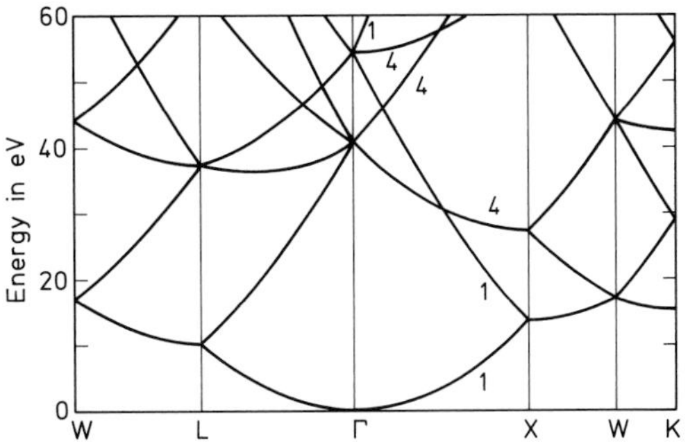

Figure 1-4. Free-electron dispersion curves for the f.c.c. lattice along certain symmetry lines of the Brillouin zone. The points indicated are the same as in Fig. 1-1.

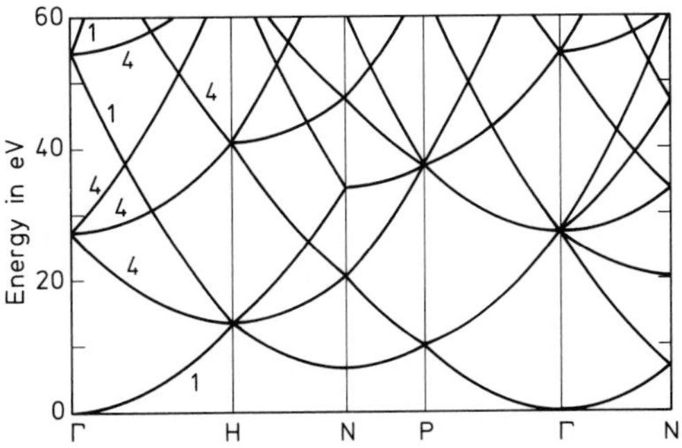

Figure 1-5. Same as Fig. 1-4 but for the b.c.c. lattice. The points now refer to Fig. 1-2.

The solutions of the crystal Schrödinger equation are difficult to obtain because one simultaneously copes with a very strong potential near the nuclei and a rather flat and weak one in between them (Heine, 1970). This is the reason why a plane-wave ansatz for a real crystal potential fails. Historically, two methods had been devised to deal with this problem by modifying the plane waves used in the expansion of Eq. (1-52) near the nuclei. One was Slater's (1937) augmented plane wave (APW) method, the other Herring's (1940) orthogonalized plane wave (OPW) method. While we will deal with Slater's APW method in some detail in Sec. 1.2.3 we use Herring's OPW's only to arrive at the pseudopotential concept (see, e.g., Callaway, 1964, 1974, for more details about the OPW method).

1.2.2.2 Pseudopotential

Using Dirac's notation to denote a plane wave by $|\boldsymbol{K}\rangle$ and a core state by $|c\rangle$, we write an OPW, $|\chi_{\boldsymbol{K}}\rangle$, in the form

$$|\chi_{\boldsymbol{K}}\rangle = |\boldsymbol{K}\rangle - \sum_c |c\rangle\langle c|\boldsymbol{K}\rangle \tag{1-57}$$

The sum extends over the core states, and we note as a reminder that the real-space notation is obtained from a scalar product with the bra vector $\langle \boldsymbol{r}|$, for instance,

$$\langle \boldsymbol{r}|\boldsymbol{K}\rangle = \frac{1}{\sqrt{\Omega}} \mathrm{e}^{\mathrm{i}\boldsymbol{K}\cdot\boldsymbol{r}} \tag{1-58}$$

Obviously, because of the orthogonality of different core states, we see that $\langle c|\chi_{\boldsymbol{K}}\rangle = 0$, thus $|\chi_{\boldsymbol{K}}\rangle$ is orthogonal to the core states. The OPW, Eq. (1-57), consists of a smooth part, $|\boldsymbol{K}\rangle$ (for not too large $\boldsymbol{K}$) and a part that possesses small-wavelength oscillations over the core region, $|c\rangle\langle c|\boldsymbol{K}\rangle$; one expects, therefore, that an expansion of the Bloch function of the form

$$|\psi_{\boldsymbol{k}}\rangle = \sum_s a_s(\boldsymbol{k})\,|\chi_{\boldsymbol{k}+\boldsymbol{K}_s}\rangle \tag{1-59}$$

will converge quite rapidly. This is also borne out by actual calculations. For Eq. (1-58) to be valid as it is stated, the OPW, $|\chi_{\boldsymbol{K}}\rangle$, must be a Bloch function, and, strictly speaking, this is not true unless the state $|c\rangle$ is constructed to have Bloch symmetry. This can formally be achieved by a linear combination of core states, i.e., by interpreting the ket $|c\rangle$ as

$$|c\rangle = \frac{1}{\sqrt{N}} \sum_\mu \mathrm{e}^{\mathrm{i}\boldsymbol{k}\cdot\boldsymbol{R}_\mu}\,|c_\mu\rangle \tag{1-60}$$

the sum extending over the lattice and $|c_\mu\rangle$ representing the core state c centered at $\boldsymbol{R}_\mu$. The $\boldsymbol{k}$ vector can be suppressed in the notation – as we have done – and no problems arise provided the core states centered on different atoms do not overlap. Proceeding now as in the plane-wave case, we calculate the secular equation analoguous to Eq. (1-54) and obtain

$$\sum_s \{[(\boldsymbol{k}+\boldsymbol{K}_s)^2 - \varepsilon_{\boldsymbol{k}}]\,\delta_{st} + V^{\mathrm{OPW}}(\boldsymbol{K}_t - \boldsymbol{K}_s)\}\cdot \cdot\, a_s(\boldsymbol{k}) = 0 \tag{1-61}$$

where the "OPW-Fourier coefficient" is obtained as

$$V^{\mathrm{OPW}}(\boldsymbol{K}_t - \boldsymbol{K}_s) = V(\boldsymbol{K}_t - \boldsymbol{K}_s) + + \sum_c (\varepsilon_{\boldsymbol{k}} - \varepsilon_c)\langle \boldsymbol{k}+\boldsymbol{K}_t|c\rangle\langle c|\boldsymbol{k}+\boldsymbol{K}_s\rangle \tag{1-62}$$

The quantity $V(\boldsymbol{K}_t - \boldsymbol{K}_s)$ is the Fourier coefficient given by Eq. (1-56), and in the derivation of Eq. (1-62) the core states were assumed to be eigenstates of H with energy ε_c. Up to this point we have been dealing with OPW theory; starting here, the derivation of the pseudopotential concept can be sketched in a few steps (roughly paralleling the historical development).

First, it is observed (Heine, 1970) that the OPW secular Eq. (1-61) has the same form as the plane wave (PW) secular Eq. (1-54). The latter is simply the Schrödinger equation

$$(T+V)\,|\psi_{\boldsymbol{k}}\rangle = \varepsilon_{\boldsymbol{k}}\,|\psi_{\boldsymbol{k}}\rangle \tag{1-63}$$

where T is the kinetic energy operator expressed in the PW representation. Similarly, we may reinterpret Eq. (1-61) as the plane-wave representation of some equation

$$(T+V_{ps})\,|\phi_{\boldsymbol{k}}\rangle = \varepsilon_{\boldsymbol{k}}\,|\phi_{\boldsymbol{k}}\rangle \tag{1-64}$$

where V_{ps} is a suitable operator such that its matrix elements between plane waves $|\boldsymbol{k}+\boldsymbol{K}_t\rangle$ and $|\boldsymbol{k}+\boldsymbol{K}_s\rangle$ are equal to V^{OPW} of Eq. (1-62). For the operator one obtains by inspection

$$V_{ps} = V + \sum_c (\varepsilon_{\boldsymbol{k}} - \varepsilon_c)\,|c\rangle\langle c| \tag{1-65}$$

The quantity V_{ps} is the pseudopotential and its essential property is that its eigenvalues in Eq. (1-64) are identical with those of the real potential for the valence bands, eliminating the core states from the states $|\phi_{\boldsymbol{k}}\rangle$ (Heine, 1970). It is weaker than the real potential V because V is attractive, and the added terms on the right-hand side of Eq. (1-65) are positive since the valence energies $\varepsilon_{\boldsymbol{k}}$ lie above the negative core energies ε_c. Phillips and Kleinman (1959) and Antoncik (1959) first expressed the OPW equation in the form Eq. (1-64) and thus started pseudopotential theory of the form presented here.

The determination of the pseudopotential from Eq. (1-65) is difficult in practice. Thus following the approach of Austin et al. (1962), we next show that there are many other forms and hence many possible implementations which may be more practical. Most importantly, however, we must obtain a pseudopotential which provides additional explanation of why the crystal potential appears as weak as it does. Thus we consider (Austin et al., 1962) a general (but non-hermitian) potential, V_A, defined by

$$V_A\,|\phi\rangle = \sum_c |c\rangle\langle F_c|\phi\rangle \tag{1-66}$$

Here the function $F_c = F_c(\boldsymbol{r})$ is quite arbitrary and $|c\rangle$, as before, denotes a core state. The ket $|\phi\rangle$ is supposed to be subject to the same boundary condition as $|\psi\rangle$ which we assume satisfies

$$H\,|\psi\rangle = \varepsilon\,|\psi\rangle \tag{1-67}$$

Now it is not difficult to show the eigenvalues, ε', of the problem

$$(H+V_A)\,|\phi\rangle = \varepsilon'\,|\phi\rangle \tag{1-68}$$

are the same as those of $|\psi\rangle$, i.e. $\varepsilon' = \varepsilon$, provided $|\psi\rangle$ and $|\phi\rangle$ are not orthogonal. One simply forms the scalar product of $\langle\psi|$ with Eq. (1-68) using $\langle\psi|c\rangle = 0$.

The freedom of choice supplied by the potential V_A is apparent now because F_c, chosen in some clever way, can be used in Eq. (1-68) and applied to a simple atomic case or the crystal-case as one wishes. An especially attractive form is obtained by demanding that, for the simpler atomic case, the potential V_A cancels the nuclear potential as best as possible then transferring this potential to the crystal, see e.g. Heine (1970). One can construct an appropriate F_c and write the pseudopotential with it as

$$V_{ps} = V - \sum_c |c\rangle\langle c|\,V \tag{1-69}$$

The cancellation is seen as follows: if the core functions were complete, we would have the well-known completeness relation

$$\sum_c |c\rangle\langle c| = 1 \tag{1-70}$$

and in this case V_{ps} given by Eq. (1-69) would be zero. But, of course, they are not, so the left hand side of Eq. (1-70) is smaller than unity and the cancellation although substantial, is not complete. One can obtain the cancellation more pictorially by approximating Eq. (1-69) to the form it has when it acts in Eq. (1-68) on a very smooth function, ϕ, which is assumed not to vary over the core region. Then taking ϕ out of the integral implied by $\langle c | V | \phi \rangle$ one obtains the local potential

$$V_{\mathrm{ps}}(\boldsymbol{r}) \cong V(\boldsymbol{r}) - \sum_c \left(\int \psi_c^* V \, \mathrm{d}^3 r \right) \psi_c(\boldsymbol{r}) \quad (1\text{-}71)$$

where $\psi_c(\boldsymbol{r}) = \langle \boldsymbol{r} | c \rangle$ and the sum on c only contains the angular-momentum component occurring in ϕ. The pseudopotential is shown in Fig. 1-6 for Na^+; it is written as $V_{\mathrm{ps}}(r) = Z_{\mathrm{ps}}(r)/r$ and $Z_{\mathrm{ps}}(r)$ is plotted for an s function. Outside the core $Z_{\mathrm{ps}} = -1$, the valence of Na^+, and inside the core the cancellation is quite apparent, but a divergence remains at the origin.

The general form of the pseudopotential – both mathematical and pictorial – now allows a justification of, or the motivation behind, many empirical or model potentials that were introduced with great success in the 1960's and 1970's to simplify electronic structure calculations (Cohen and Heine, 1970; Heine and Weaire, 1970; Harrison, 1966). They all eliminated the need to include atomic core states and the strong potentials that provide the core-state binding, thus justifying nearly-free electron behavior where it is observed. In detailed calculations the total effective potential was represented by just a few terms in a Fourier expansion, and the coefficients were adjusted to give agreement with some experimentally determined features of the energy bands. Alternatively, a simple function representing the ion-core potential was adjusted to yield the experimental ionization potential. One typically used a Coulomb tail at large distances from the nucleus changing to a state-dependent constant inside the core radius. Once the pseudopotential of the bare ion had been obtained, it could be planted into the electron gas, which then screens it to give the total pseudopotential of the solid. This screening was achieved by using a linear dielectric function method; it gave band structures in good agreement with experimental data if the higher Fourier components were set to zero.

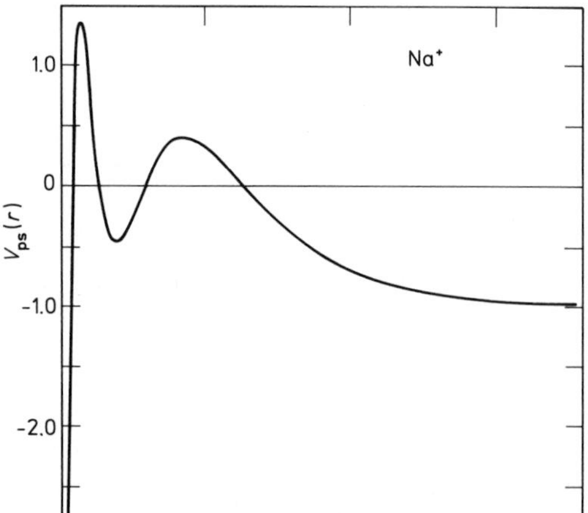

Figure 1-6. Pseudopotential for the s function as generated by a Na^+ ion. The pseudopotential is given by $V_{\mathrm{ps}}(r) = \dfrac{Z_{\mathrm{ps}}(r)}{r}$.

We do not want to go into more detail concerning the type of pseudopotentials discussed so far, rather we would now like to set the stage for the modern pseudopotentials used in self-consistent calculations that we are mainly interested in here.

1.2.2.3 Norm-Conserving Pseudopotentials

It was recognized quite early that the wave functions of the pseudopotentials present a certain problem. The normalized pseudo-wave function and the normalized "exact" function have the same shape in the region of space outside the cores, but have different amplitudes. This can be seen readily by writing out the norm of $|\psi_{\boldsymbol{k}}\rangle$ and assuming that the pseudo-wave function $|\phi_{\boldsymbol{k}}\rangle$ is normalized. One obtains

$$\langle \psi_{\boldsymbol{k}} | \psi_{\boldsymbol{k}} \rangle = 1 - \sum_c |\langle c | \phi_{\boldsymbol{k}} \rangle|^2 \quad (1\text{-}72)$$

Thus a normalized pseudo-wave function gives rise to a function $|\psi_{\mathbf{k}}\rangle$ that has a norm smaller than unity because a normalized $|\psi_{\mathbf{k}}\rangle$ would be $|\tilde{\psi}_{\mathbf{k}}\rangle$ with

$$|\tilde{\psi}_{\mathbf{k}}\rangle = |\psi_{\mathbf{k}}\rangle / [1 - \sum_c |\langle c|\phi_{\mathbf{k}}\rangle|^2]^{1/2} \qquad (1\text{-}73)$$

Hence, outside the core region, where $|\psi_{\mathbf{k}}\rangle$ and $|\phi_{\mathbf{k}}\rangle$ agree, the normalized $|\tilde{\psi}_{\mathbf{k}}\rangle$ is larger than the pseudo-wave function. This is also true for the pseudo-wave function of pseudopotentials of the general type of Eq. (1-66), because here the normalized $|\psi_{\mathbf{k}}\rangle$ can also be written as

$$|\psi_{\mathbf{k}}\rangle = C\,|\phi_{\mathbf{k}}\rangle + \sum_c \alpha_c\,|c\rangle \qquad (1\text{-}74)$$

with some finite coefficients α_c and $C \neq 1$; it thus has the same problem. This is quite serious if chemical aspects such as the bonding charge density between the atoms are desired. The same is true in the case of self-consistent calculations, where an incorrect distribution of charge between valence and core states will cause errors in the effective potential (see Sec. 3). An orthogonalization of the pseudo-wave function to the cores could remove these problems, at least in principle. But this is not only a difficult procedure; it also removes all the advantages of pseudopotentials which were invented to ignore the cores.

In building up the modern pseudopotentials, one first abandons any relation like Eq. (1-74); i.e., any notion of orthogonalizing to the core states is dropped. As before, however, it is the isolated atom (or ion) that is used to construct the pseudopotential. Following the work of Hamann (1979), Bachelet et al. (1982) and Hamann (1989), we require that the pseudo-wave function for the atom be nodeless, and when normalized, that it becomes identical to the true valence-wave function beyond some "core radius", R. The real atom and the "pseudo-atom" should, furthermore, have the same valence eigenvalues for some chosen electronic configuration (usually the ground state), and the pseudo-charge contained in the sphere with the radius R should be identical to the real charge in that sphere (Kerker, 1980). Hamann (1979) used the term "norm conserving" to describe pseudopotentials constructed in this fashion.

For the pseudopotential to be useful, its core portion must be *transferable* to other situations where the external potential has changed, such as in molecules, solids or surfaces. It appears that Topp and Hopfield (1973) were the first to use an identity derivable from the Wronskian of the radial Schrödinger equation for constructing a pseudopotential; this identity was used by Hamann (1979) to ensure the transferability of the norm-conserving pseudopotential (NCPP). It is obtained (Messiah, 1976; Callaway, 1964) by manipulating the Schrödinger equation for the radial function $u_l(\varepsilon, r)$ to some energy ε (not necessarily an eigenvalue),

$$\left[-\frac{\mathrm{d}^2}{\mathrm{d}r^2} + V_l(r)\right] u_l(\varepsilon, r) = \varepsilon\, u_l(\varepsilon, r) \qquad (1\text{-}75)$$

and, similarly, for $u_l(\varepsilon + \mathrm{d}\varepsilon, r)$. The potential V_l includes the centrifugal term $l(l+1)/r^2$, but otherwise is arbitrary. It must *not*, however, depend on the energy ε (as did most of the older pseudopotentials). One easily derives

$$\left[u_l^2(\varepsilon, r)\,\frac{\mathrm{d}}{\mathrm{d}\varepsilon}\frac{\mathrm{d}}{\mathrm{d}r}\ln u_l(\varepsilon, r)\right]_R = \\ = -\int_0^R u_l^2(\varepsilon, r)\,\mathrm{d}r \qquad (1\text{-}76)$$

The consequence of Eq. (1-76) is that for two potentials $V_l^{(1)}$ and $V_l^{(2)}$ the linear energy variation around ε of their scattering phase shifts at R is identical provided the solutions $u_l^{(1)}$ and $u_l^{(2)}$ with $u_l^{(1)}(R) = u_l^{(2)}(R)$

have the same charge inside the sphere of radius R. The connection of the scattering phase shifts with the logarithmic derivatives is elementary and their relation with the band-structure problem will become clear in Sec. 1.2.3. The requirement for the pseudo-wave function that it must agree with the full wave function for $r > R$ guarantees that the charge is identical for $r > R$, and thus that the scattering properties of the NCPP and the full potential have the same energy variation (to first order) when transferred to other systems (Bachelet et al., 1982).

To give the reader a specific example we close this section by outlining the construction of an NCPP in some detail essentially following Hamann (1989). The first step is to calculate the self-consistent local-density-functional (LDF) potential $V(r)$ for the *atom* of interest. (Sec. 1.3 will give details as to the LDF approximation and the definition of $V(r)$.) This completely defines $V_l(r)$ in Eq. (1-75) which in this step is meant to be an eigenvalue equation giving ε_l's and u_l's. The next step is the choice of a "core radius" r_{cl} which is empirically chosen to be ≈ 0.5 times the radius at which $u_l(r)$ has its outermost maximum. It is now possible to pick an outer radius at which the pseudo-wave function can be accurately converged to the full-potential wave function and $R_l = 2.5\,r_{cl}$ is found to be practical. Equation (1-75) is then integrated with $\varepsilon = \varepsilon_l$ from the origin and is stopped at R_l normalizing u_l such that

$$4\pi \int_0^{R_l} u_l^2(r)\,\mathrm{d}r = 1 \qquad (1\text{-}77)$$

The normalized values of $u_l(r)|_{r=R_l}$ and first derivatives $\frac{\mathrm{d}}{\mathrm{d}r} u_l(r)|_{r=R_l}$ are recorded.

Next, a so-called intermediate pseudopotential is defined by

$$V_{1l}(r) = [1 - f(r/r_{cl})]\,V(r) + c_l\, f(r/r_{cl}) + \frac{l(l+1)}{r^2} \qquad (1\text{-}78)$$

where as a cut-off function the use of

$$f(r/r_{cl}) = \exp[-(r/r_{cl})^{3.5}] \qquad (1\text{-}79)$$

has proven effective (Bachelet et al., 1982). One sees that V_{1l} converges to the potential $V(r)$ for $r > r_{cl}$. $V_{1l}(r)$ is now used in a radial Schrödinger equation to find the intermediate and nodeless wave function $w_{1l}(r)$ requiring the eigenvalues of the full potential and the intermediate pseudopotential to agree. The coefficients c_l are adjusted such that

$$\frac{\mathrm{d}}{\mathrm{d}r} \ln w_{1l}(r)|_{R_l} = \frac{\mathrm{d}}{\mathrm{d}r} \ln u_l(r)|_{R_l} \qquad (1\text{-}80)$$

Now a scale factor is found to make the two wave function types identical at R_l,

$$\gamma_l = u_l(R_l)/w_{1l}(R_l) \qquad (1\text{-}81)$$

and the final pseudo-wave function is constructed by adding a short-range norm-conserving term,

$$w_{2l}(r) = \gamma_l\,[w_{1l}(r) + \delta_l\, g_l(r)] \qquad (1\text{-}82)$$

δ_l is the smaller solution of the equation resulting from the condition that w_{1l} be normalized, i.e., from

$$\gamma_l^2 \int_0^{R_l} [w_{1l}(r) + \delta_l\, g_l(r)]^2\,\mathrm{d}r = 1 \qquad (1\text{-}83)$$

where $g_l(r)$ is chosen to be

$$g_l(r) = r^{l+1} f(r/r_{cl}) \qquad (1\text{-}84)$$

The final pseudopotential, V_{2l}, is found by inverting the Schrödinger equation

$$V_{2l}(r) = \varepsilon_l + \frac{1}{w_{2l}(r)} \frac{\mathrm{d}^2}{\mathrm{d}r^2} w_{2l}(r) \qquad (1\text{-}85)$$

which, with the parametrization given above, can be done analytically (Bachelet et al., 1982).

For either the functions $u_l(r)$ or $w_{2l}(r)$, the right-hand sides of the key identity, Eq. (1-76), are the same at $R = R_l$ by construction. This implies that the logarithmic derivatives at energy ε are the same to first order in $(\varepsilon - \varepsilon_l)$ – by a Taylor-series expansion –, so the scattering power of V_{2l} for l partial waves is the same as the full potential $V(r)$ to a good approximation at energies away from the selected ε_l's.

Finally, in an important last step, the pseudopotential $V_{2l}(r)$ is "unscreened" to give the ionic pseudopotential

$$V_l^{\text{ion}}(r) = V_{2l}(r) - V^{\text{H}}(n, r) - V_{\text{xc}}(n, r) \quad (1\text{-}86)$$

where V^{H} is the Hartree- and V_{xc} the exchange-correlation potential corresponding to the atomic density obtained from

$$n(r) = \sum_l n_l [w_{2l}(r)/r]^2 \quad (1\text{-}87)$$

(see Eq. (1-386)); n_l are the occupancies of the valence states.

As an example, the ionic pseudopotential for aluminium obtained by the above procedure and given by Bachelet et al. (1982) is depicted in Fig. 1-7. As in the older pseudopotentials, the cancellation in the core region is apparent, but no singularity remains at the origin. For a band calculation the ionic pseudopotential is now transferred to the solid and screened again by adding the Hartree- and exchange-correlation potentials corresponding to self-consistent, solid-state charge densities obtained from the Bloch functions by means of

$$n(\boldsymbol{r}) = \sum_{v\boldsymbol{k}} |\psi_{v\boldsymbol{k}}(\boldsymbol{r})|^2 \quad (1\text{-}88)$$

where the sum on the band index v and on the wave vector $\boldsymbol{k}$ extends over all the occupied states. We return to the self-consistency step in Sec. 1.3. A usual procedure is to construct the Bloch functions by a plane-wave expansion as outlined earlier

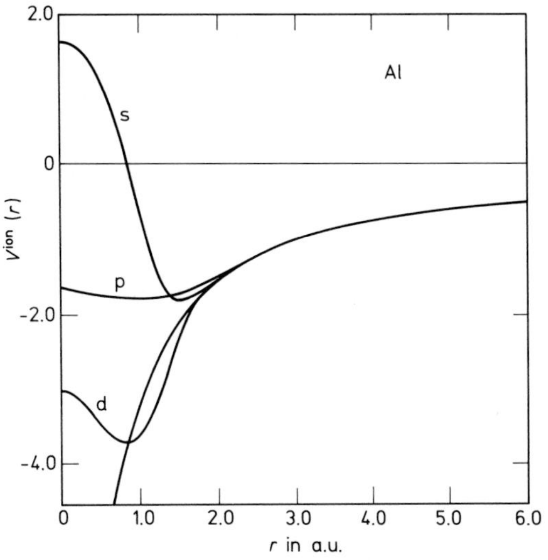

Figure 1-7. Norm-conserving pseudopotential for the s, p and d states of Al as given by Bachelet et al. (1982). Unlabeled curve corresponds to the bare ion potential $-\dfrac{Z_{\text{val}}(r)}{r}$.

by Eqs. (1-52) and (1-54), treating in the latter $V(\boldsymbol{K})$ as the plane-wave-matrix element of the screened pseudopotential. But this is by no means a necessary procedure; one may choose any of the other methods treated in the next sections. Note that no dielectric-function screening is needed any longer as in the case of the model or empirical pseudopotentials.

1.2.3 Augmented Plane Waves and Green's Functions

At the beginning of Slater's (1937) (Slater, 1965) augmented plane wave (APW) method there is an assumption about the form of the crystal potential. It is assumed to be spherically symmetric inside non-overlapping spheres centered at the atom sites and constant outside. Because of the particular shape of the potential, it is said to be a "muffin-tin" potential and the spheres (chosen as large as possible) are called "muffin-tin" spheres. The space outside the spheres we shall call interstitial space (IS) denoting the radius of the sphere centered at atom ν in the unit cell by S_ν. Although this is not a perfect representation of the potential actually found in crystals, it enables one to set up a rather rigorous solution of the Schrödinger equation. Having found these solutions, the departures of the potential from the assumed form can be treated as a perturbation.

1.2.3.1 Augmented Plane Waves (APW)

The basic expansion for the Bloch function is again written as

$$|\psi_{\boldsymbol{k}}\rangle = \sum_s a_s(\boldsymbol{k})\,|\chi_{\boldsymbol{k}+\boldsymbol{K}_s}\rangle \tag{1-89}$$

The (finite) sum on s extends over reciprocal lattice vectors as before, but the basis functions are constructed in a rather elaborate way. In the interstitial region of space, that is, for $\boldsymbol{r}$ in IS, they are plane waves

$$\langle \boldsymbol{r}|\chi_{\boldsymbol{k}+\boldsymbol{K}_s}\rangle = \mathrm{e}^{\mathrm{i}(\boldsymbol{k}+\boldsymbol{K}_s)\cdot\boldsymbol{r}} \tag{1-90}$$

because the potential is constant here. This, of course, satisfies the Bloch condition. Inside the sphere centered at $\boldsymbol{\tau}_\nu$ in the unit cell one writes

$$\langle \boldsymbol{r}|\chi_{\boldsymbol{k}+\boldsymbol{K}_s}\rangle = \sum_L C_{L\nu}(\boldsymbol{k}+\boldsymbol{K}_s)\,R_{l\nu}(\varepsilon,r)\,Y_L(\hat{\boldsymbol{r}}) \tag{1-91}$$

The quantity L denotes both the angular momentum and magnetic quantum numbers l and m; i.e., $\sum_L = \sum_l \sum_{m=-l}^{l}$, $Y_L(\hat{\boldsymbol{r}})$ are spherical harmonics, $Y_{lm}(\theta,\varphi)$, and the functions $R_{l\nu}(\varepsilon,r)$ are (numerical) solutions of the radial Schrödinger equation regular at the origin

$$-\frac{1}{r^2}\frac{\mathrm{d}}{\mathrm{d}r}\left(r^2\frac{\mathrm{d}R_{l\nu}}{\mathrm{d}r}\right) + \left[\frac{l(l+1)}{r^2} + V_\nu(r) - \varepsilon\right] R_{l\nu}(\varepsilon,r) = 0 \tag{1-92}$$

The energy, ε, is assumed to be a variable here and not an eigenvalue. The coefficients $C_{L\nu}(\boldsymbol{k}+\boldsymbol{K}_s)$ in Eq. (1-91) are obtained by requiring that the values of $\langle \boldsymbol{r}|\chi_{\boldsymbol{k}+\boldsymbol{K}_s}\rangle$ match continuously the plane waves given by Eq. (1-90) at the surface of the muffin-tin spheres. This is what is meant by augmentation; it is achieved mathematically by using the well-known identity

$$\mathrm{e}^{\mathrm{i}\boldsymbol{k}\cdot\boldsymbol{r}} = 4\pi \sum_L \mathrm{i}^l j_l(kr)\,Y_L^*(\hat{\boldsymbol{k}})\,Y_L(\hat{\boldsymbol{r}}) \tag{1-93}$$

where the quantities $j_l(kr)$ are spherical Bessel functions of order l which, we remind the reader, are admissible solutions of the radial Schrödinger equation (1-92) for a vanishing potential and $\varepsilon = k^2$. For the coefficients $C_{L\nu}(\boldsymbol{k}+\boldsymbol{K}_s)$ one easily obtains with Eq. (1-93)

$$\begin{aligned} C_{L\nu}(\boldsymbol{k}+\boldsymbol{K}_s) &= \\ &= 4\pi\,\mathrm{i}^l\,\mathrm{e}^{\mathrm{i}(\boldsymbol{k}+\boldsymbol{K}_s)\cdot\boldsymbol{\tau}_\nu} j_l(|\boldsymbol{k}+\boldsymbol{K}_s|\,S_\nu)\cdot \\ &\quad\cdot Y_L^*(\widehat{\boldsymbol{k}+\boldsymbol{K}_s})/R_{l\nu}(\varepsilon,S_\nu) \end{aligned} \tag{1-94}$$

This together with Eqs. (1-90) and (1-91) defines an augmented plane wave (APW), which, by construction, is discontinuous in *slope* on the muffin-tin sphere boundary. Therefore, the variational expression that yields the expansion coefficients $a_s(\boldsymbol{k})$ must be treated with some care.

Let us denote by H_0 the Hamiltonian underlying the APW method, i.e.,

$$H_0 = -\nabla^2 + V_{\mathrm{mt}} \tag{1-95}$$

where the muffin-tin potential V_{mt} is constant in interstitial space and has the value $V_\nu(r)$ inside the muffin-tin sphere centered at $\boldsymbol{\tau}_\nu$. The integral over the unit cell, $\int_{\mathrm{UC}} \psi_{\boldsymbol{k}}^*(\boldsymbol{r})\,H_0\,\psi_{\boldsymbol{k}}(\boldsymbol{r})\,\mathrm{d}^3r$, can now be written as the sum of an integral over the spherical volumes defined by the muffin-tin spheres (MT) and an integral extending over interstitial space; hence

$$\int_{UC} \psi_k^*(r)\, H_0\, \psi_k(r)\, d^3r = -\int_{MT} \psi_k^*(r)\, \nabla^2 \psi_k(r)\, d^3r - \int_{IS} \psi_k^*(r)\, \nabla^2 \psi_k(r)\, d^3r + \int_{UC} \psi_k^*(r)\, V_{mt}\, \psi_k(r)\, d^3r \tag{1-96}$$

Applying Green's identity to the first two terms on the right hand side, we derive

$$\int_{UC} \{[\nabla \psi_k^*(r)]\,[\nabla \psi_k(r)] + \psi_k^*(r)\, V_{mt}\, \psi_k(r) - \varepsilon\, \psi_k^*(r)\, \psi_k(r)\}\, d^3r = \int_{UC} \psi_k^*(r)\,(H_0-\varepsilon)\,\psi_k(r)\, d^3r + \int_S \psi_k^*(r^{(-)}) \left(\frac{\partial \psi_k(r^{(-)})}{\partial r} - \frac{\partial \psi_k(r^{(+)})}{\partial r}\right) dS \tag{1-97}$$

where the last term on the right-hand side is a surface integral over the muffin-tin spheres, $r^{(-)}$ and $r^{(+)}$ denoting positions just inside and outside, respectively, of the muffin-tin sphere boundaries (Loucks, 1967). It results from the above-mentioned discontinuity in slope.

It is important to note that the Euler-Lagrange equation obtained from varying the left-hand side of Eq. (1-97) is simply the Schrödinger equation (see, e.g., Merzbacher, 1970):

$$(H_0-\varepsilon)\,\psi_k(r) = 0 \tag{1-98}$$

It follows that the entire right-hand side of Eq. (1-97) must be varied to obtain the coefficients $a_s(k)$ of Eq. (1-89) and not only the integral $\int_{UC} \psi_k^*(r)\,(H_0-\varepsilon)\,\psi_k(r)\, d^3r$. Thus substituting Eq. (1-89) into the right-hand side of Eq. (1-97) and varying the coefficients $a_s^*(k)$ we obtain the algebraic equation

$$\sum_s \{\langle t|\, H_0-\varepsilon_k\, |s\rangle + \langle t|\, S\, |s\rangle\}\, a_s(k) = 0 \tag{1-99}$$

where we defined

$$\langle t|\, H_0-\varepsilon_k\, |s\rangle = \int_{UC} \chi_{k+K_t}^*(r)\,(H_0-\varepsilon_k)\,\chi_{k+K_s}(r)\, d^3r \tag{1-100}$$

and

$$\langle t|\, S\, |s\rangle = \int_S \chi_{k+K_t}^*(r) \left[\frac{\partial}{\partial r}\chi_{k+K_s}(r^{(-)}) - \frac{\partial}{\partial r}\chi_{k+K_s}(r^{(+)})\right] dS \tag{1-101}$$

Let us temporarily simplify the discussion by assuming there is only one atom per unit cell. Inside the muffin-tin sphere an APW is an eigenfunction of H_0 and hence in evaluating Eq. (1-100) only that part of the integral remains that extends over the interstitial region, i.e.,

$$\langle t|\, H_0-\varepsilon_k\, |s\rangle = [(k+K_s)^2-\varepsilon_k] \int_{IS} e^{i(K_s-K_t)\cdot r}\, d^3r \tag{1-102}$$

(here ε_k is counted from the constant interstitial potential). The remaining integral can be shown to be (Loucks, 1967)

$$\int_{IS} e^{i(K_s-K_t)\cdot r}\, d^3r = \Omega_{UC}\,\delta_{st} - 4\pi S^2 \frac{j_1(|K_s-K_t|\,S)}{|K_s-K_t|} \tag{1-103}$$

The calculation of the surface integral, Eq. (1-101) is straightforward. Using the addition theorem for spherical harmonics, dividing through by the volume of the unit cell, and collecting terms, we can write Eq. (1-94) in the form

$$\sum_s \{[(k+K_s)^2-\varepsilon_k]\,\delta_{ts} + V^{APW}(K_t,K_s)\} \cdot a_s(k) = 0 \tag{1-104}$$

where

$$V^{APW}(K_t,K_s) = \frac{4\pi S^2}{\Omega_{UC}} \sum_l (2l+1)\, P_l(\cos\theta_{ts}) \cdot j_l(|k+K_t|\,S)\, j_l(|k+K_s|\,S) \cdot \left[\frac{d}{dr}\ln R_l(\varepsilon,r) - \frac{d}{dr}\ln j_l(|k+K_s|\,r)\right]_S - [(k+K_s)^2-\varepsilon_k] \cdot \frac{4\pi S^2}{\Omega_{UC}} \frac{j_1(|K_s-K_t|\,S)}{|K_s-K_t|} \tag{1-105}$$

$P_l(x)$ is a Legendre polynomial of order l and θ_{ts} is the angle between the vectors $\boldsymbol{k}+\boldsymbol{K}_t$ and $\boldsymbol{k}+\boldsymbol{K}_s$. This form of the APW-equation is quite convenient for a discussion of the limit of nearly-free electron behavior. But there exists an alternative of the APW equation that is more practical for actual calculations; details can be found in the book by Loucks (1967).

Armed with the APW equations, we can repeat the discussion of nearly-free electron behavior. But we can broaden our understanding now by dealing semi-quantitatively with an important example where both nearly-free electron behavior and strong deviations from it occur, that is, the case of Copper.

We thus begin by following Slater (1965) and examine Eqs. (1-104) and (1-105), looking at the diagonal elements first. Since for small arguments, x, the asymptotic form of the spherical Bessel function is

$$j_l(x) \xrightarrow[x\to 0]{} x^l/(2l+1)!! \qquad (1\text{-}106)$$

and $P_l(1)=1$, we can write Eq. (1-105) for $\boldsymbol{K}_s=\boldsymbol{K}_t$ as

$$\begin{aligned} V^{\mathrm{APW}}(\boldsymbol{K}_s,\boldsymbol{K}_s) &= \\ &= \frac{4\pi S^2}{\Omega_{\mathrm{UC}}}\sum_l (2l+1)\, j_l^2(|\boldsymbol{k}+\boldsymbol{K}_s|\,S)\cdot \Delta D_l(\varepsilon) - \\ &- [(\boldsymbol{k}+\boldsymbol{K}_s)^2-\varepsilon_{\boldsymbol{k}}]\,\frac{4\pi S^3}{3\,\Omega_{\mathrm{UC}}} \qquad (1\text{-}107)\end{aligned}$$

where ΔD_l is the difference of the logarithmic derivatives

$$\Delta D_l(\varepsilon) = \frac{\mathrm{d}}{\mathrm{d}r}\ln R_l(\varepsilon,r)\bigg|_S - \frac{\mathrm{d}}{\mathrm{d}r}\ln j_l(|\boldsymbol{k}+\boldsymbol{K}_s|\,r)\bigg|_S \qquad (1\text{-}108)$$

We use perturbation theory (justifying this below) and replace the secular determinant by the product of its diagonal terms. If we set this equal to zero, we can set one such term equal to zero and read off from Eqs. (1-104) and (1-107)

$$\varepsilon_{\boldsymbol{k}} \simeq (\boldsymbol{k}+\boldsymbol{K}_s)^2 + \frac{4\pi S^2}{\Omega_{\mathrm{UC}}-\Omega_{\mathrm{MT}}}\cdot \sum_l (2l+1)\, j_l^2(|\boldsymbol{k}+\boldsymbol{K}_s|\,S)\,\Delta D_l(\varepsilon_{\boldsymbol{k}}) \qquad (1\text{-}109)$$

where Ω_{MT} is the volume of the muffin-tin sphere. The situation $\Delta D_l(\varepsilon)\simeq 0$ is that of nearly free electrons, because, as was stated after Eq. (1-93), the spherical Bessel functions are the solutions of the radial Schrödinger equation for the empty lattice, i.e., free electrons. Thus, in this case, Eq. (1-109) gives $\varepsilon_{\boldsymbol{k}}\simeq(\boldsymbol{k}+\boldsymbol{K}_s)^2$, which may be substituted into the non-diagonal matrix elements, $V^{\mathrm{APW}}(\boldsymbol{K}_t,\boldsymbol{K}_s)$; this allows the last term on the right hand side of Eq. (1-105) to be removed, giving $V^{\mathrm{APW}}(\boldsymbol{K}_t,\boldsymbol{K}_s)\simeq 0$ for $\boldsymbol{K}_t\neq\boldsymbol{K}_s$. This procedure justifies our use of perturbation theory.

The next step is to examine calculated values of the difference of the logarithmic derivatives, $\Delta D_l(\varepsilon)$, and to make their functional form plausible. The dashed curves in Fig. 1-8 are the logarithmic derivatives (times S) of free electrons, $S\,\frac{\mathrm{d}}{\mathrm{d}r}\ln j_l(kr)\Big|_{r=S}$, with $k^2=\varepsilon$ and $S=2.415\,a_0$ appropriate for Cu. The solid lines are estimated values compiled from the literature (Slater, 1965; Mattheis et al., 1968) of calculated logarithmic derivatives, $S\,\frac{\mathrm{d}}{\mathrm{d}r}\ln R_l(\varepsilon,r)\Big|_S$ which, except for $l=2$, are rather similar to those of free electrons. This remark applies only to the logarithmic derivatives at S and not to the radial functions themselves, which are completely different from the spherical Bessel functions further inside the atom. Here they show strong core oscillations because of orthogonality requirements. Thus the spherical Bessel functions can be considered as pseudo-functions, but only for $l\neq 2$. Hence we cannot really ignore the off-diagonal elements of $V^{\mathrm{APW}}(\boldsymbol{K}_t,\boldsymbol{K}_s)$ for

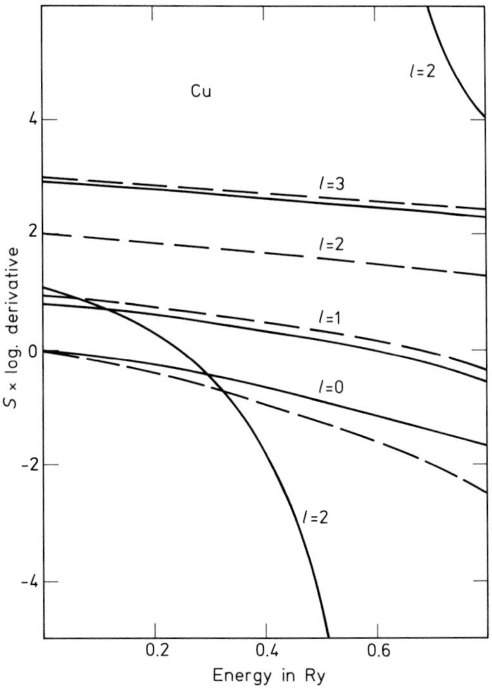

Figure 1-8. Logarithmic derivatives as a function of energy for copper (solid lines) and for free electrons (dashed lines).

the case of Cu. Proceeding thus with the numerical results, we examine the calculation by Burdick (1963), which is a rather famous case. It was obtained by the APW method using an empirical potential constructed by Chodorow (1939a, b) on the basis of Hartree and Hartree-Fock calculations for Cu^+.

Burdick's results are shown in Fig. 1-9 where we also compare them with measurements from angle-resolved photoemission experiments by Thiry et al. (1979). The agreement between the measured and calculated energies is quite astonishing – we shall further comment on this fact in Sec. 1.4.1.1. Here we want to look at the calculated band structure in some detail. It begins with $\varepsilon = 0$ Ry at the Γ point in the BZ (corresponding to $\boldsymbol{k} = (0,0,0)$); the label Γ_1 designates the same symmetry as an s wave function ($l = 0$). The parabolic $\varepsilon_{\boldsymbol{k}}$ curve is interrupted (we call it hybridized) by flatter states starting at Γ with labels $\Gamma_{25'}$ and Γ_{12}; these denote d-state symmetries. $\Gamma_{25'}$ is threefold degenerate and has transformation properties like xy, yz and zx, and Γ_{12} is twofold degenerate transforming like $x^2 - y^2$ and $2z^2 - x^2 - y^2$. (These products are a short notation of real $l = 2$ spherical harmonics.) The bands labeled Δ_5 and Λ_3 are twofold degenerate d states. We call further attention to the states $X_{4'}$ at $\varepsilon \approx 0.804$ Ry and $L_{2'}$ at $\varepsilon \approx 0.608$ Ry, where the band structure appears roughly parabolic again. These labels designate p symmetries ($l = 1$) whose logarithmic derivative is very close to the free-electron value (see Fig. 1-8). Indeed, with the lattice constant of Cu, $a = 6.8309\, a_0$, the free-electron value at X is $(2\pi/a)^2 = 0.846$ Ry, and at L it is $(\pi\sqrt{3}/a)^2 = 0.635$ Ry, which is reasonably close to the $X_{4'}$ and $L_{2'}$ values.

Concerning the non-d states, we emphasize that although some of the energies are close to free-electron values, the wave functions are far from the plane waves representing free electrons. Although they are

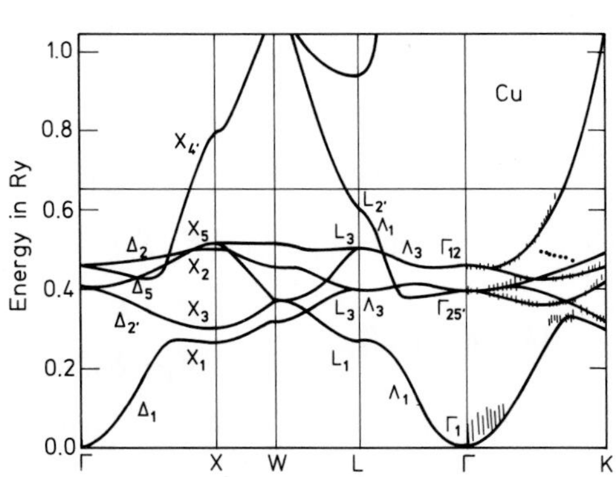

Figure 1-9. Band structure for Cu as calculated by Burdick (1963). Experimental results (vertical bars between ΓK) are angle-resolved photoemission data by Thiry et al. (1979). Black dots are presumably surface states.

described with good accuracy by a single plane wave between the atomic sphere, inside such a sphere the wave functions go through all the oscillations characteristic of atomic wave functions. The point is that a single plane wave joins smoothly onto the solution of the Schrödinger equation inside the sphere, the energy of this solution being the same as that of the plane wave (Slater, 1965).

We want to close this subsection on the APW method with some technical remarks. First, the number of basis functions that are required to obtain converged results is *not* small. Indeed, roughly 40 APW basis functions are needed for each atom in the unit cell. For compounds, the rank of the secular Eq. (1-104) can thus become rather large (Mattheiss et al., 1968). Its eigenvalues, $\varepsilon_{\boldsymbol{k}}$, are usually found by tracing the roots of the secular determinant. Since this is often done by triangularization, it is not hard to obtain the eigenvectors $a_s(\boldsymbol{k})$ belonging to each eigenvalue, $\varepsilon_{\boldsymbol{k}}$, by back-substitution and hence to obtain the corresponding wave function $\psi_{\boldsymbol{k}}(\boldsymbol{r})$ from Eq. (1-89). To normalize it one needs matrix elements of the form $\langle \chi_{\boldsymbol{k}+\boldsymbol{K}_t} | \chi_{\boldsymbol{k}+\boldsymbol{K}_s} \rangle$, which are readily obtained using the Eqs. (1-90), (1-91), and (1-94); one finds

$$\langle \chi_{\boldsymbol{k}+\boldsymbol{K}_t} | \chi_{\boldsymbol{k}+\boldsymbol{K}_s} \rangle = = \Omega_{\mathrm{UC}}\, \delta_{ts} + 4\pi \sum_{\nu=1}^{n} S_\nu^2\, \mathrm{e}^{i(\boldsymbol{K}_s - \boldsymbol{K}_t)\cdot \boldsymbol{\tau}_\nu} \cdot \left\{ \sum_l (2l+1)\, P_l(\cos\theta_{ts})\, j_l(|\boldsymbol{k}+\boldsymbol{K}_t|\, S_\nu) \cdot j_l(|\boldsymbol{k}+\boldsymbol{K}_s|\, S_\nu)\, I_{l\nu}(\varepsilon_{\boldsymbol{k}}) - \frac{j_1(|\boldsymbol{K}_t - \boldsymbol{K}_s|\, S_\nu)}{|\boldsymbol{K}_t - \boldsymbol{K}_s|} \right\} \tag{1-110}$$

where the symbols are defined as before (following Eq. (1-105)), and the quantity $I_{l\nu}(\varepsilon_{\boldsymbol{k}})$ denotes the integral

$$I_{l\nu}(\varepsilon_{\boldsymbol{k}}) = \frac{1}{S_\nu^2\, R_{l\nu}^2(\varepsilon_{\boldsymbol{k}}, S_\nu)} \int_0^{S_\nu} r^2\, R_{l\nu}^2(\varepsilon_{\boldsymbol{k}}, r)\, \mathrm{d}r \tag{1-111}$$

Because of Eq. (1-76) this quantity is also given by

$$I_{l\nu}(\varepsilon) = -\frac{\mathrm{d}}{\mathrm{d}\varepsilon} \frac{\mathrm{d}}{\mathrm{d}r} \ln R_{l\nu}(\varepsilon, r)|_{S_\nu} \tag{1-112}$$

One sees that the sum on l in Eq. (1-110) because of Eq. (1-111) is related to the probability of finding the electron inside the muffin-tin sphere, ν. The dominating term is recognizable by means of Eq. (1-112) as that with a large negative energy derivative of the logarithmic derivative at the sphere boundary, similar, for instance, to the d state in Cu (see Fig. 1-8).

Finally, if the self-consistent potential deviates from the assumed muffin-tin form in the interstitial region, one can define a potential $\Delta V(\boldsymbol{r})$, which describes this deviation, and add it to the Hamiltonian H_0 in Eq. (1-99). As a result one obtains an additional matrix element in V^{APW}, Eq. (1-105), that is, the plane-wave matrix element of ΔV. This is true because $\Delta V(\boldsymbol{r})$ is zero (or constant) inside the muffin-tin sphere. However, deviations from spherical symmetry inside the muffin-tin sphere are much harder to handle and are treated by perturbation theory (if at all) using the muffin-tin APW's as unperturbed basis functions. Not many calculations along these lines have been made so far (see, e.g., Kane, 1971). Although for most metals and intermetallic compounds this so-called shape approximation for the potential is seldom really serious, there are prominent exceptions, and problems occur, e.g., if shear distortions are considered. In these cases one must drop the muffin-tin approximation. Calculations that do are called "full potential" and in many modern studies are conducted using a linearized version of the APW method to be discussed in Sec. 1.2.4.

1.2.3.2 Green's Functions

We now turn to a description of multiple scattering theory (MST) which is a different and rather refined method to solve the single particle Schrödinger Eq. (1-12). It is also sometimes called Green's function or KKR method where KKR stands for the names Korringa (1947) and Kohn and Rostoker (1954), who first introduced this method into the electronic structure theory of crystalline solids. In contrast to the theories dealt with so far, MST strictly speaking does not find stationary values of the Hamiltonian but of another operator – the inverse transition matrix **T** for the entire crystal. Aside from its use with crystalline solids, it has been successfully applied to the study of the electronic structure of molecules and surfaces, and it is the basis of most theories of liquids, disordered systems, and impurities in various metallic and nonmetallic hosts. In all applications its content is the same and is easy to state (Williams, 1970; Williams and van Morgan, 1974): it is a prescription for obtaining the transition matrix **T** describing the scattering of the entire system in terms of the corresponding matrix **t** for individual scatterers. The central result of the theory can be expressed in a form independent of the details of the problem and is obtained as the result of summing all possible trajectories of the particle through the system. Thus if **G** describes free particle propagation from any individual scattering event to the next, we may describe multiple scattering by means of

$$\begin{aligned}\mathbf{T} &= \mathbf{t} + \mathbf{tGt} + \mathbf{tGtGt} + \ldots \\ &= \mathbf{t} + \mathbf{tG}(\mathbf{t} + \mathbf{tGt} + \mathbf{tGtGt} + \ldots) \\ &= \mathbf{t} + \mathbf{tGT}\end{aligned} \quad (1\text{-}113)$$

or

$$\mathbf{T} = (1 - \mathbf{tG})^{-1}\mathbf{t} \quad (1\text{-}114)$$

Stationary states (Messiah, 1978) of a system are given by the singularities of **T** and hence are obtained from the relation

$$\det(\mathbf{t}^{-1} - \mathbf{G}) = 0 \quad (1\text{-}115)$$

Application of this simple formula requires a suitable representation; this is where the real work starts.

It begins with the definition of the free-particle Green's function, $G_0(\varepsilon, \boldsymbol{r} - \boldsymbol{r}')$, which as usual is required to solve the equation

$$(\nabla^2 + \varepsilon)\, G_0(\varepsilon, \boldsymbol{r} - \boldsymbol{r}') = \delta(\boldsymbol{r} - \boldsymbol{r}') \quad (1\text{-}116)$$

where $\delta(\boldsymbol{r} - \boldsymbol{r}')$ is the Dirac δ function. A standing wave free-particle solution of this equation is easily constructed (Messiah, 1978) to be

$$G_0(\varepsilon, \boldsymbol{r} - \boldsymbol{r}') = -\frac{1}{4\pi} \frac{\cos(\varkappa |\boldsymbol{r} - \boldsymbol{r}'|)}{|\boldsymbol{r} - \boldsymbol{r}'|} \quad (1\text{-}117)$$

where $\varkappa^2 = \varepsilon$.

The use of the standing wave Green's function is appropriate to the determination of the stationary states for which the Schrödinger equation is written in integral form:

$$\psi_{\boldsymbol{k}}(\boldsymbol{r}) = \int \mathrm{d}^3 r'\, G_0(\varepsilon_{\boldsymbol{k}}, \boldsymbol{r} - \boldsymbol{r}')\, V(\boldsymbol{r}')\, \psi_{\boldsymbol{k}}(\boldsymbol{r}') \quad (1\text{-}118)$$

Applying the operator $(-\nabla^2 - \varepsilon_{\boldsymbol{k}})$ to this equation and using Eq. (1-116), one verifies directly that $\psi_{\boldsymbol{k}}(\boldsymbol{r})$ satisfies the Schrödinger equation.

There are many different ways to proceed from here. If one is willing to make the muffin-tin approximation at this early stage, one may consult the monographs by Callaway (1964) and (1974) or Segall and Ham (1968) and Ziman (1971) in addition to the original KKR literature. In fact, a very simple derivation will emerge later in Sec. 1.2.4 where we discuss linearized methods and their connection with multiple scattering theory. However, we will follow the derivation of Williams and van

Morgan (1974) because, although as presented here it requires the muffin-tin approximation as well, it can be applied with moderate effort to rather general potentials (Williams and van Morgan, 1974; Zeller, 1987). Still, the latter case looks quite formidable, so we will be content with a brief discussion only.

To simplify the notation, we assume there is one atom per unit cell; however, we will state the result for n atoms per unit cell at the end. Then, as in Eq. (1-55), we express the potential as a sum of identical, non-overlapping terms, $V_c(\boldsymbol{r})$, centered on each unit cell of the crystal

$$V(\boldsymbol{r}) = \sum_{\nu} V_c(r - \boldsymbol{R}_\nu) \quad (1\text{-}119)$$

This is still quite general. Substituting this form into Eq. (1-118) and bringing to the left the term corresponding to $R_\nu = 0$ in Eq. (1-119), we obtain

$$\psi_{\boldsymbol{k}}(\boldsymbol{r}) - \int_{\Omega_{\mathrm{UC}}} \mathrm{d}^3 r' \, G_0(\varepsilon_{\boldsymbol{k}}, \boldsymbol{r} - \boldsymbol{r}') \, V_c(\boldsymbol{r}') \, \psi_{\boldsymbol{k}}(\boldsymbol{r}') =$$
$$= \int_{\Omega_{\mathrm{UC}}} \mathrm{d}^3 r' \, G(\varepsilon_{\boldsymbol{k}}, \boldsymbol{k}, \boldsymbol{r} - \boldsymbol{r}') \, V_c(\boldsymbol{r}') \, \psi_{\boldsymbol{k}}(\boldsymbol{r}') \quad (1\text{-}120)$$

The Green's function on the right-hand side is obtained as the Fourier transformed Green's function G_0, i.e.,

$$G(\varepsilon_{\boldsymbol{k}}, \boldsymbol{k}, \boldsymbol{r} - \boldsymbol{r}') =$$
$$= \sum_{\nu \neq 0} G_0(\varepsilon_{\boldsymbol{k}}, \boldsymbol{r} - \boldsymbol{r}' - \boldsymbol{R}_\nu) \, \mathrm{e}^{\mathrm{i}\, \boldsymbol{k} \cdot \boldsymbol{R}_\nu} \quad (1\text{-}121)$$

In the next step, we require a solution, $\phi(\varepsilon, \boldsymbol{r})$, of the Schrödinger equation containing only the *single* cellular potential $V_c(\boldsymbol{r})$. Now, if the potential has the muffin-tin form, this solution is obtained from the radial Schrödinger equation (Eq. (1-92)) within the muffin-tin sphere just as in the APW case, i.e., $\phi(\varepsilon, \boldsymbol{r}) = \mathrm{i}^l R_l(\varepsilon, r) \, Y_L(\hat{\boldsymbol{r}})$. At the muffin-tin sphere radius, S, we match the inside solution, $R_l(\varepsilon, r)$, to proper scattering functions which are the spherical Bessel and Neumann functions $j_l(\varkappa r)$ and $n_l(\varkappa r)$:

$$R_l(\varepsilon, S) = C_l [j_l(\varkappa S) - \tan \eta_l \, n_l(\varkappa S)] \quad (1\text{-}122)$$

and

$$\frac{\mathrm{d}}{\mathrm{d}r} R_l(\varepsilon, r)|_S =$$
$$= C_l \left[\frac{\mathrm{d}}{\mathrm{d}r} j_l(\varkappa r) - \tan \eta_l \frac{\mathrm{d}}{\mathrm{d}r} n_l(\varkappa r) \right]_S \quad (1\text{-}123)$$

Here $\varkappa^2 = \varepsilon$ and $\eta_l = \eta_l(\varepsilon)$ is the scattering phase shift of angular momentum l (Messiah, 1978). It can be simply expressed with the logarithmic derivative

$$D_l(\varepsilon) = \frac{\mathrm{d}}{\mathrm{d}r} \ln R_l(\varepsilon, r)|_S \quad (1\text{-}124)$$

and Eqs. (1-122) and (1-123) as

$$\tan \eta_l = \left[\frac{\mathrm{d}}{\mathrm{d}r} j_l(\varkappa r)|_S - D_l(\varepsilon) j_l(\varkappa S) \right] \Big/ \left[\frac{\mathrm{d}}{\mathrm{d}r} n_l(\varkappa r)|_S - D_l(\varepsilon) \, n_l(\varkappa S) \right] \quad (1\text{-}125)$$

Thus the information about an atom contained in the logarithmic derivative is embodied as well in the scattering phase shift $\eta_l(\varepsilon)$, which we will see is related to the matrix **t** in Eq. (1-115).

Now the solution of the Schrödinger equation containing only the single cellular potential can formally be written for the general potential using the Green's function, G_0, again. That is, we know that

$$\phi(\varepsilon, \boldsymbol{r}) = \sum_L C_L(\varepsilon) \, J_{\varepsilon L}(\boldsymbol{r}) +$$
$$+ \int \mathrm{d}^3 r' \, G_0(\varepsilon, \boldsymbol{r} - \boldsymbol{r}') \, V_c(\boldsymbol{r}') \, \phi(\varepsilon, \boldsymbol{r}') \quad (1\text{-}126)$$

where the function $J_{\varepsilon L}(\boldsymbol{r})$ is defined through the spherical Bessel function and the spherical harmonics,

$$J_{\varepsilon L}(\boldsymbol{r}) = \mathrm{i}^l j_l(\varkappa r) \, Y_L(\hat{\boldsymbol{r}}) \quad (1\text{-}127)$$

The first term on the right-hand side of Eq. (1-126) is obviously the solution of the homogeneous equation. In the case of scattering solutions it must be added to the solution of the inhomogeneous equation, which is the second term on the right-hand side of Eq. (1-126). We will continue using the integral equation [Eq. (1-126)] but assume a spherically symmetric potential; then $\phi = \phi_L$ and we may expand the Bloch function $\psi_{\boldsymbol{k}}(\boldsymbol{r})$ in Eq. (1-120) in terms of the functions $\phi_L(\varepsilon, \boldsymbol{r})$; that is,

$$\psi_{\boldsymbol{k}}(\boldsymbol{r}) = \sum_L \gamma_L(\boldsymbol{k})\, \phi_L(\varepsilon_{\boldsymbol{k}}, \boldsymbol{r}) \tag{1-128}$$

Substituting this form into the right-hand side of Eq. (1-120), but using Eq. (1-126) on the left-hand side to simplify, we obtain

$$\begin{aligned} &\sum_L C_l(\varepsilon_{\boldsymbol{k}})\, J_{\varepsilon_{\boldsymbol{k}} L}(\boldsymbol{r})\, \gamma_L(\boldsymbol{k}) = \\ &= \sum_L \int d^3r'\, G(\varepsilon_{\boldsymbol{k}}, \boldsymbol{k}, \boldsymbol{r}-\boldsymbol{r}')\, V_c(\boldsymbol{r}') \cdot \\ &\quad \cdot \phi_L(\varepsilon_{\boldsymbol{k}}, \boldsymbol{r}')\, \gamma_L(\boldsymbol{k}) \end{aligned} \tag{1-129}$$

We will see subsequently that we can express the Green's function G in the form

$$\begin{aligned} &G(\varepsilon, \boldsymbol{k}, \boldsymbol{r}-\boldsymbol{r}') = \\ &= \sum_{LL'} J_{\varepsilon L}(\boldsymbol{r})\, B_{LL'}(\varepsilon, \boldsymbol{k})\, J^*_{\varepsilon L'}(\boldsymbol{r}') \end{aligned} \tag{1-130}$$

Using this equation to rewrite the right hand side of Eq. (1-129) and defining the quantity

$$S_l(\varepsilon) = \int d^3r\, J^*_{\varepsilon L}(\boldsymbol{r})\, V_c(\boldsymbol{r})\, \phi_L(\varepsilon, \boldsymbol{r}) \tag{1-131}$$

we obtain instead of Eq. (1-129):

$$\begin{aligned} &\sum_L C_l(\varepsilon_{\boldsymbol{k}})\, J_{\varepsilon_{\boldsymbol{k}} L}(\boldsymbol{r})\, \gamma_L(\boldsymbol{k}) = \\ &= \sum_{LL'} J_{\varepsilon_{\boldsymbol{k}} L}(\boldsymbol{r})\, B_{LL'}(\varepsilon_{\boldsymbol{k}}, \boldsymbol{k})\, S_{l'}(\varepsilon_{\boldsymbol{k}})\, \gamma_{L'}(\boldsymbol{k}) \end{aligned} \tag{1-132}$$

where we assumed the potential to be spherically symmetric again. We multiply by $Y^*_L(\hat{\boldsymbol{r}})$ and integrate over the solid angle to remove the sum on L. Furthermore, we define a more convenient unknown coefficient vector as

$$\alpha_L(\boldsymbol{k}) = S_l(\varepsilon_{\boldsymbol{k}})\, \gamma_L(\boldsymbol{k}) \tag{1-133}$$

to obtain the desired representation of Eq. (1-115) in the following form:

$$\sum_{L'} \{B_{LL'}(\varepsilon_{\boldsymbol{k}}, \boldsymbol{k}) - t_l^{-1}\, \delta_{LL'}\}\, \alpha_{L'}(\boldsymbol{k}) = 0 \tag{1-134}$$

where

$$t_l = t_l(\varepsilon) = S_l(\varepsilon)/C_l(\varepsilon) \tag{1-135}$$

This is the KKR equation. For spherically symmetric cell potentials, V_c, we can express the t matrix by the scattering phase shift, which we will do next at the same time starting the derivation of the important Eq. (1-130) for the Green's function.

In elementary scattering theories one uses a spherical-wave expansion of the free-particle Green's function, which we write as follows (Messiah, 1978):

$$\frac{\cos(\varkappa\,|\boldsymbol{r}-\boldsymbol{r}'|)}{|\boldsymbol{r}-\boldsymbol{r}'|} = -4\pi\,\varkappa \sum_L N_{\varepsilon L}(\boldsymbol{r}_>)\, J^*_{\varepsilon L}(\boldsymbol{r}_<) \tag{1-136}$$

where as before $\varkappa^2 = \varepsilon$ and, just like $J_{\varepsilon L}$ in Eq. (1-127), the quantity $N_{\varepsilon L}$ is defined with the spherical Neumann function, i.e.

$$N_{\varepsilon L}(\boldsymbol{r}) = \mathrm{i}^l\, n_l(\varkappa r)\, Y_L(\hat{\boldsymbol{r}}) \tag{1-137}$$

The vectors $\boldsymbol{r}_>$, $\boldsymbol{r}_<$ are the larger and the smaller of the vectors $\boldsymbol{r}$ and $\boldsymbol{r}'$, respectively. First we use relation (1-136) in Eq. (1-126) for $\boldsymbol{r}$ just outside the volume where V_c is finite. For the muffin-tin approximation this is $|\boldsymbol{r}| \geq S$ and Eq. (1-126) reads

$$\begin{aligned} \phi_L(\varepsilon, \boldsymbol{r}) = {} & C_l(\varepsilon)\, J_{\varepsilon L}(\boldsymbol{r}) - \\ & - \varkappa \sum_{L'} \left[\int d^3r'\, J^*_{\varepsilon L'}(\boldsymbol{r}')\, V_c(\boldsymbol{r}')\, \phi_L(\varepsilon, \boldsymbol{r}')\right] \cdot N_{\varepsilon L'}(\boldsymbol{r}) \end{aligned} \tag{1-138}$$

Identifying the integral with $S_l(\varepsilon)$ defined by Eq. (1-131) and comparing with Eq. (1-122), we read off

$$C_l(\varepsilon)/S_l(\varepsilon) = \varkappa \cot \eta_l(\varepsilon) \tag{1-139}$$

which allows us to write Eq. (1-134) in the standard KKR form:

$$\sum_{L'} [B_{LL'}(\varepsilon_{\boldsymbol{k}}, \boldsymbol{k}) + \varkappa \cot \eta_l(\varepsilon_{\boldsymbol{k}})\, \delta_{LL'}]\, \alpha_{L'}(\boldsymbol{k}) = 0 \tag{1-140}$$

What remains is a determination of the so-called structure constant $B_{LL'}(\varepsilon, \boldsymbol{k})$, which begins with Eq. (1-121) using Eq. (1-117) and Eq. (1-136). Because $|\boldsymbol{R}_\nu| > 0$ by definition is beyond the range of $\boldsymbol{r}$ and $\boldsymbol{r}'$ one sees at once that

$$G(\varepsilon, \boldsymbol{k}, \boldsymbol{r}-\boldsymbol{r}') = = -\varkappa \sum_L J^*_{\varepsilon L}(\boldsymbol{r}-\boldsymbol{r}') \sum_{\nu \neq 0} N_{\varepsilon L}(\boldsymbol{R}_\nu)\, \mathrm{e}^{\mathrm{i}\boldsymbol{k}\cdot\boldsymbol{R}_\nu} \tag{1-141}$$

The final relation we need is an expression for $J_{\varepsilon L}(\boldsymbol{r}-\boldsymbol{r}')$ in terms of $\boldsymbol{r}$ and $\boldsymbol{r}'$, separately. A manipulation of the expansion of plane waves in terms of spherical waves (Eq. 1-93)) indeed supplies the desired relation in the following form (Danos and Maximon, 1965; Williams et al., 1971):

$$J_{\varepsilon L}(\boldsymbol{r}-\boldsymbol{r}') = 4\pi \sum_{L'L''} C_{L''L'L}\, J^*_{\varepsilon L''}(\boldsymbol{r})\, J_{\varepsilon L'}(\boldsymbol{r}') \tag{1-142}$$

where $C_{L''L'L}$ is a Gaunt (or Clebsch-Gordon) coefficient given by

$$C_{L''L'L} = \int \mathrm{d}\hat{r}\, Y^*_{L''}(\hat{\boldsymbol{r}})\, Y_{L'}(\hat{\boldsymbol{r}})\, Y_L(\hat{\boldsymbol{r}}) \tag{1-143}$$

Substituting Eq. (1-142) into Eq. (1-141) and comparing with Eq. (1-130), we finally obtain

$$B_{LL'}(\varepsilon, \boldsymbol{k}) = = -4\pi\varkappa \sum_{L''} C_{LL'L''} \sum_{\nu\neq 0} \mathrm{e}^{\mathrm{i}\boldsymbol{k}\cdot\boldsymbol{R}_\nu} N_{\varepsilon L''}(\boldsymbol{R}_\nu) \tag{1-144}$$

This equation is seen to consist of a lattice sum over Neumann functions combined with spherical harmonics but contains no characteristics of the crystal potential beyond its symmetry and the lattice constant. Thus the KKR Eq. (1-140) separates structural properties from those of the atomic constituents. The latter are contained in the scattering phase shifts $\eta_l(\varepsilon)$ which, given a potential, follow by means of Eq. (1-125) from the logarithmic derivatives which, for example, may be tabulated analogously to those of Cu shown in Fig. 1-8.

The lattice sum occurring in Eq. (1-144) is not easy to evaluate because as it stands it is only slowly converging. However, it can be transformed to sums that are converging fast using the Ewald (1921) technique; for completeness and for later use we state the result (Kohn and Rostoker, 1954; Ham and Segall, 1961) extending it to the case of n atoms located at the positions $\boldsymbol{\tau}_\alpha$, $\alpha = 1, \ldots, n$, in the unit cell. One obtains

$$B_{LL'}(\boldsymbol{\tau}_\alpha - \boldsymbol{\tau}_\beta, \varepsilon, \boldsymbol{k}) = = 4\pi\varkappa \sum_{L''} C_{LL'L''}\, D_{L''}(\boldsymbol{\tau}_\alpha - \boldsymbol{\tau}_\beta, \varepsilon, \boldsymbol{k}) \cdot \mathrm{e}^{-\mathrm{i}\boldsymbol{k}\cdot(\boldsymbol{\tau}_\beta - \boldsymbol{\tau}_\alpha)} \tag{1-145}$$

where

$$D_L(\boldsymbol{\tau}_\alpha - \boldsymbol{\tau}_\beta, \varepsilon, \boldsymbol{k}) = = D_L^{(1)}(\boldsymbol{\tau}_\alpha - \boldsymbol{\tau}_\beta, \varepsilon, \boldsymbol{k}) + D_L^{(2)}(\boldsymbol{\tau}_\alpha - \boldsymbol{\tau}_\beta, \varepsilon, \boldsymbol{k}) + + \delta_{L,0}\, \delta_{\boldsymbol{\tau}_\alpha - \boldsymbol{\tau}_\beta, 0}\, D^{(3)} \tag{1-146}$$

Here

$$D_L^{(1)}(\boldsymbol{\tau}_\alpha - \boldsymbol{\tau}_\beta, \varepsilon, \boldsymbol{k}) = = (4\pi/\Omega_{\mathrm{UC}})\, \varkappa^{-l} \exp(\varepsilon/q_0) \cdot \sum_s \mathrm{e}^{\mathrm{i}\boldsymbol{K}_s\cdot(\boldsymbol{\tau}_\alpha - \boldsymbol{\tau}_\beta)}\, Y_L(\boldsymbol{K}_s + \boldsymbol{k})\, |\boldsymbol{K}_s + \boldsymbol{k}|^l \cdot \frac{\exp[-(\boldsymbol{K}_s + \boldsymbol{k})^2/q_0]}{\varepsilon - (\boldsymbol{K}_s + \boldsymbol{k})^2} \tag{1-147}$$

$$D_L^{(2)}(\boldsymbol{\tau}_\alpha - \boldsymbol{\tau}_\beta, \varepsilon, \boldsymbol{k}) = \pi^{-1/2}(-2)^{l+1}\, \mathrm{i}^l\, \varkappa^{-l} \cdot \sum_\nu{}' \exp[\mathrm{i}\boldsymbol{k}\cdot(\boldsymbol{R}_\nu - \boldsymbol{\tau}_\alpha + \boldsymbol{\tau}_\beta)]\, Y_L(\boldsymbol{R}_\nu + \boldsymbol{\tau}_\beta - \boldsymbol{\tau}_\alpha) \cdot |\boldsymbol{R}_\nu - \boldsymbol{\tau}_\alpha + \boldsymbol{\tau}_\beta|^l \int_{\sqrt{q_0}}^{\infty} z^{2l} \cdot \exp[-\tfrac{1}{4}(\boldsymbol{\tau}_\beta - \boldsymbol{\tau}_\alpha + \boldsymbol{R}_\nu)^2 z^2 + \varepsilon/z^2]\, \mathrm{d}z \tag{1-148}$$

and

$$D^{(3)} = -\frac{\sqrt{q_0}}{2\pi} \sum_{n=0}^{\infty} \frac{(\varepsilon/q_0)^n}{n!\,(2n-1)} \tag{1-149}$$

The symbol $\sum_\nu{}'$ is supposed to exclude from the summation the term $\boldsymbol{R}_\nu = \boldsymbol{\tau}_\alpha - \boldsymbol{\tau}_\beta$. The quantity q_0 is the Ewald parameter which may be chosen to achieve optimal convergence of the terms $D_L^{(1)}$ and $D_L^{(2)}$. The corresponding KKR equation appropriate to the case of n atoms per unit cell is

$$\sum_{L'} \sum_{\beta=1}^{n} \{B_{LL'}(\tau_\alpha - \tau_\beta, \varepsilon_k, k) + \\ + \varkappa \cot \eta_{l\beta}(\varepsilon_k)\, \delta_{LL'}\, \delta_{\alpha\beta}\}\, \alpha_{L'\beta}(k) = 0 \quad (1\text{-}150)$$

where $\eta_{l\beta}(\varepsilon)$ is the scattering phase shift of the atom numbered β, and $\alpha = 1, \ldots, n$.

Concerning the size of the secular equation, it is not a priori clear how many angular-momentum terms must be included in the Eqs. (1-140) or (1-150). But we expect that for higher angular momenta the logarithmic derivatives approach those of free electrons very quickly. This is, in fact, born out by experience and can be seen to be true for $l \geq 3$, for the example of Cu in Fig. 1-8. This, in turn, means that as

$$D_l(\varepsilon) \rightarrow \frac{\mathrm{d}}{\mathrm{d}r} \ln j_l(\varkappa r)|_S \quad (1\text{-}151)$$

the scattering phase shift becomes

$$\tan \eta_l \rightarrow 0 \quad (1\text{-}152)$$

(see Eq. (1-125)), and therefore, because of Eqs. (1-139) and (1-133)

$$\alpha_L(k) \rightarrow 0 \quad (1\text{-}153)$$

This obviously limits the size of the secular equation to rather small values. For instance, for elementary transition metals the maximal value of l can be chosen to be $l_{max} = 3$; i.e., the rank of the secular equation is 16.

The eigenvalues of the KKR equations (Eqs. (1-140) or (1-150)) must be obtained by root tracing, as in the case of the APW method (cf. the paragraph before Eq. (1-110)). With the eigenvalues obtained in this way, one determines the expansion coefficient α from the KKR equations and hence, using Eqs. (1-133) and (1-128), the wave function ψ. The wave function then yields the charge density by means of Eq. (1-88). Some technical questions concerning details of the construction of the wave function are addressed in the work of Segall and Ham (1968). We will return to the self-consistency step in Sec. 1.3.

We mentioned earlier that relaxing the muffin-tin approximation for the shape of the potential in the KKR method is possible but not easy. In its simplest incarnation it leads to a non-diagonal scattering matrix $t_{LL'}$, i.e., the relation, Eq. (1-135), connecting the quantities S, C, and t becomes a matrix equation (Williams and van Morgan, 1974) and, consequently, the KKR equation is non-diagonal not only in the structure-constant part, B, but also in the t-matrix part. The structure constants themselves, however, remain unchanged as one would expect from the remarks following Eq. (1-144). The convergence properties of these generalized KKR equations are not entirely clear, and, at least to our knowledge, no detailed applications to real physical problems have as yet appeared. Recent formal developments can be found, for instance, in papers by Gonis (1986), Zeller (1987), and Badralexe and Freeman (1987, 1989).

Except for the difficulties arising when the shape approximation of the potential is relaxed, the APW and KKR methods can be made arbitrarily accurate; but they are slow calculationally because the secular equations are nonlinear and thus require time-consuming root tracings. Linear methods, however, do exist and are important; therefore we will deal with them in the next section.

1.2.4 Linear Combination of Atomic Orbitals (LCAO) and Linear Methods

The introduction of linear methods into band theory was an important and far reaching step taken first some 15 years ago by Andersen (1975). It led to a large variety of calculations for realistic and complex systems and allowed considerable insight into their chemical and physical properties. At the same time these methods shifted the burden away from technical problems. Linear methods also supplied tools necessary to analyze and interpret the results of numerical calculations and are in this respect superior to the "classical" methods discussed in the previous two sections. A price, however, has to be paid: slight inaccuracies. Being proponents of linear methods ourselves, we are perhaps not unprejudiced and believe the price is not too high.

1.2.4.1 Linear Combination of Atomic Orbitals (LCAO)

Before we can deal with linear methods in detail, we must return to the general expansion introduced earlier in Eq. (1-49) and discuss what is called in rather general terms "linear combination of atomic orbitals" (LCAO). In this section we particularly mean energy-independent orbitals that are associated with a specific atom or ion at a well-defined position in the crystal but not necessarily related to eigenfunctions of the atom in question. Given such orbitals, $\chi_j(\boldsymbol{r}-\boldsymbol{R}_\nu-\boldsymbol{\tau})$, a Bloch function is written as

$$\chi_{j\boldsymbol{k}}(\boldsymbol{r}) = \sum_\nu \mathrm{e}^{\mathrm{i}\boldsymbol{k}\cdot\boldsymbol{R}_\nu}\,\chi_j(\boldsymbol{r}-\boldsymbol{R}_\nu-\boldsymbol{\tau}) \qquad (1\text{-}154)$$

where the summation extends over the Bravais lattice $\{\boldsymbol{R}_\nu\}$, $\boldsymbol{\tau}$ is a basis vector when the crystal is not primitive, and j is a combination of quantum numbers characterizing the orbital χ. The Bloch function defined by Eq. (1-154) is, of course, not yet an eigenfunction of the crystal-Schrödinger equation. To obtain it we use the expansion

$$\psi_{\boldsymbol{k}}(\boldsymbol{r}) = \sum_j a_j(\boldsymbol{k})\,\chi_{j\boldsymbol{k}}(\boldsymbol{r}) \qquad (1\text{-}155)$$

and determine the coefficients variationally as before, Eqs. (1-50) and (1-51). Let us here denote the matrix elements of the Hamiltonian by

$$\int \chi^*_{i\boldsymbol{k}}(\boldsymbol{r})\,H\,\chi_{j\boldsymbol{k}}(\boldsymbol{r})\,\mathrm{d}^3r \equiv \langle i\,|\,H\,|\,j\rangle \qquad (1\text{-}156)$$

and the overlap matrix by

$$\int \chi^*_{i\boldsymbol{k}}(\boldsymbol{r})\,\chi_{j\boldsymbol{k}}(\boldsymbol{r})\,\mathrm{d}^3r \equiv \langle i\,|\,j\rangle \qquad (1\text{-}157)$$

The secular equation is then

$$\sum_j \{\langle i\,|\,H\,|\,j\rangle - \varepsilon\langle i\,|\,j\rangle\}\,a_j = 0 \qquad (1\text{-}158)$$

for each i. The distinguishing feature of this secular equation is its linearity in energy, ε, because the orbitals χ, and hence the matrix elements, are independent of energy. Note that the dependence on $\boldsymbol{k}$ of the matrix elements does not mean energy dependence in this context. The convergence properties of Eq. (1-158), of course, depend strongly on the choice of the set of orbitals $\{\chi_j(\boldsymbol{r}-\boldsymbol{R})\}$. Here a number of possibilities exist. Roughly speaking, we may distinguish two: one is *augmented* in the vicinity of the atoms. This is similar to the APW or KKR methods, but the augmentation is energy independent. We will discuss this in detail later on in this section. The other type simply consists of a judicious choice of functions without augmentation. In both cases, however, the efficiency is largely determined by the ease with which multi-center integrals occurring in Eq. (1-158) can be evaluated. What we mean by this is the following: an integral of the type $\langle i\,|\,j\rangle$ contains terms such as

$$\int \chi^*_i(\boldsymbol{r}-\boldsymbol{R}_\nu-\boldsymbol{\tau})\,\chi_j(\boldsymbol{r}-\boldsymbol{R}_\mu-\boldsymbol{\tau}')\,\mathrm{d}^3r$$

Thus when the integration variable $\boldsymbol{r}$ ranges near the site $\boldsymbol{R}_\lambda+\boldsymbol{\tau}''$, but $\lambda \neq \nu$, $\lambda \neq \mu$, and $\mu \neq \nu$, then this integral is called a three-center term; the meaning of analogous two- and one-center integrals should be clear now. A determination of these multi-center terms can always be made numerically, but this is inefficient and time consuming. It becomes much more efficient when an expansion theorem exists which allows the orbital centered at some site $\boldsymbol{R}$ to be expressed in the crystal in terms of orbitals centered at some other site, $\boldsymbol{R}'$, i.e., for $\boldsymbol{R} \neq \boldsymbol{R}'$ and $\boldsymbol{r}$ in a sphere at $\boldsymbol{R}$ that does not overlap the same size sphere at $\boldsymbol{R}'$, we desire a relation

$$\chi_j(\boldsymbol{r}+\boldsymbol{R}') = \sum_i T_{ji}(\boldsymbol{R}', \boldsymbol{R})\, \chi_i(\boldsymbol{r}+\boldsymbol{R}) \qquad (1\text{-}159)$$

with mathematically well-defined coefficients T_{ji}. An example are the spherical Bessel, Neumann, and Hankel functions, and Eq. (1-142), e.g., can easily be written in the form of Eq. (1-159). Other sets of functions possessing an expansion theorem are Slater-type orbitals ($r^{\beta}\,\mathrm{e}^{-\alpha r}$) and Gaussian-type orbitals ($r^{\beta}\,\mathrm{e}^{-\lambda r^2}$) (GTO). The latter were (and still are) used extensively by Callaway (see, e.g., Wang and Callaway, 1978 and later examples in this chapter).

Eschrig (1989) has recently devised an LCAO scheme that uses Slater-type orbitals (STO) and yields a rather small secular equation. This is achieved by choosing STO's that are accurate, variational solutions of the atomic-like problem in the vicinity of the nuclei. In effect this is similar to augmentation, but the details are quite different. Since this method is rather new, more experience with it is needed before an impartial judgement can be made.

LCAO methods also constitute an important tool for *empirical* descriptions of band structures. In this case the detailed nature of the orbitals is left open, except for their angular dependence which is formulated with spherical harmonics, as before (see, e.g., Slater and Koster, 1954). The actual matrix elements occurring in Eq. (1-158) are simplified by neglecting, for instance, three-center terms and all interactions beyond next-nearest neighbors, and they are then treated as adjustable parameters to describe some measured or otherwise calculated features of the band structure (Harrison, 1980; Papaconstantopolous, 1986). These techniques are often called "empirical tight binding".

1.2.4.2 Energy Derivative of the Wave Function ϕ and $\dot{\phi}$

We now turn to a discussion of linear methods that use the concept of augmentation and begin with those developed by Anderson (1975), namely the linear-augmented-plane-wave (LAPW) and the linear-muffin-tin-orbital (LMTO) methods. Following a recent treatment by Anderson (1984), we assume the potential is spherically symmetric within some sphere of radius S but has a rather general form in the interstitial region; i.e., we write

$$V(\boldsymbol{r}) = \sum_{\nu,\tau} V_{c\tau}(|\boldsymbol{r}-\boldsymbol{R}_{\nu}-\boldsymbol{\tau}|) + V_i(\boldsymbol{r}) \qquad (1\text{-}160)$$

where $V_{c\tau}(\boldsymbol{r})$ is assumed to vanish *outside* the sphere of radius S_{τ}, and the potential in the interstitial, $V_i(\boldsymbol{r})$, is assumed to vanish *inside* these spheres; the vectors $\{\boldsymbol{R}_{\nu}\}$ define the Bravais lattice, and $\{\boldsymbol{\tau}\}$ the basis.

The energy dependence of the APW or KKR basis functions clearly stems from the solutions as a function of energy, ε, of the Schrödinger equation (1-92) for APW's, or Eq. (1-126) in multiple scattering theories. There are different possibilities to sidestep this energy dependence. Andersen achieves it by expanding the solution of the single-sphere Schrödinger equation in terms of $\phi(\boldsymbol{r})$ belonging to one, initially arbitrarily chosen energy, $\varepsilon = E_{\nu}$, i.e.,

$$(-\nabla^2 + V_c - E_{\nu})\,\phi(\boldsymbol{r}) = 0 \qquad (1\text{-}161)$$

and its energy derivative, $\frac{\partial}{\partial\varepsilon}\,\phi(\varepsilon,\boldsymbol{r})|_{\varepsilon=E_{\nu}}$, which is denoted by

$$\dot{\phi}(\boldsymbol{r}) = \frac{\partial}{\partial\varepsilon}\,\phi(\varepsilon,\boldsymbol{r})|_{\varepsilon=E_{\nu}} \qquad (1\text{-}162)$$

The motivation for this choice to expand the energy-dependent functions can superficially be argued to be the Taylor series, but a more physically compelling reason will be given later on in the ASW subsection.

Only a few facts and definitions are needed to make Andersen's scheme transparent. First, we see from the Schrödinger equation for $\phi(\varepsilon, \boldsymbol{r})$ that the equation for $\dot{\phi}$ is

$$(-\nabla^2 + V_c - E_v)\,\dot{\phi}(\boldsymbol{r}) = \phi(\boldsymbol{r}) \tag{1-163}$$

If the wave function ϕ is normalized to unity in the sphere, i.e.,

$$\langle \phi | \phi \rangle = 1 \tag{1-164}$$

defining the brackets appropriately, one sees at once that

$$\langle \phi | \dot{\phi} \rangle = 0 \tag{1-165}$$

hence ϕ and $\dot{\phi}$ are orthogonal. We next define the radial function belonging to $\phi(\varepsilon, \boldsymbol{r})$ by the symbol $\phi(\varepsilon, r)$. Note that we here distinguish these two functions only by their arguments. This differs from our notation in the previous section (see, e.g., Eq. (1-92)) and is done to keep the notation as close as possible to that of Andersen. Then the logarithmic derivative is defined by

$$D(\varepsilon) = \left[S \frac{\mathrm{d}}{\mathrm{d}r} \phi(\varepsilon, r)/\phi(\varepsilon, r) \right]_{r=S} \tag{1-166}$$

We can now expand any function having the logarithmic derivative D, which we denote by $\boldsymbol{\Phi}(D, r)$, in terms of ϕ and $\dot{\phi}$ defined by Eqs. (1-161) and (1-162) by

$$\boldsymbol{\Phi}(D, r) = \phi(r) + \omega\,\dot{\phi}(r) \tag{1-167}$$

Denoting the logarithmic derivatives of ϕ and $\dot{\phi}$ by $D\{\phi\}$ and $D\{\dot{\phi}\}$, respectively, we obtain by a simple calculation

$$\omega = \omega(D) = -\frac{\phi(S)}{\dot{\phi}(S)} \frac{D - D\{\phi\}}{D - D\{\dot{\phi}\}} \tag{1-168}$$

The four parameters occurring here, i.e., $\phi(S)$, $\dot{\phi}(S)$, $D\{\phi\}$, and $D\{\dot{\phi}\}$ are not independent. Substituting into the norm, Eq. (1-164), the Eq. (1-163) for ϕ and using Eq. (1-161) as well as Green's identity, one can derive

$$1 = S\,\phi(S)\,\dot{\phi}(S)\,[D\{\phi\} - D\{\dot{\phi}\}] \tag{1-169}$$

Using this relation and Eq. (1-168) for ω in Eq. (1-167), we obtain the following relation for the value of $\boldsymbol{\Phi}(D, r)$ at the sphere:

$$\boldsymbol{\Phi}(D, S) = [S\,\dot{\phi}(S)\,(D - D\{\dot{\phi}\})]^{-1} \tag{1-170}$$

With the help of these relations a useful parametrization of the logarithmic derivative function $D(\varepsilon)$ may finally be found (Andersen, 1984) by using $\boldsymbol{\Phi}(D, r)$ expanded above by ϕ and $\dot{\phi}$, Eq. (1-167) as a variational estimate to $\varepsilon(D)$, the inverse of $D(\varepsilon)$; a straightforward calculation gives

$$\begin{aligned} \varepsilon(D) &= \langle \boldsymbol{\Phi}(D) | -\nabla^2 + V_c | \boldsymbol{\Phi}(D) \rangle / \\ &\quad \langle \boldsymbol{\Phi}(D) | \boldsymbol{\Phi}(D) \rangle \\ &= E_v + \omega(D)/(1 + \omega^2(D) \langle \dot{\phi}^2 \rangle) \end{aligned} \tag{1-171}$$

or the less precise energy

$$\tilde{\varepsilon}(D) = E_v + \omega(D) \tag{1-172}$$

here

$$\langle \dot{\phi}^2 \rangle = \langle \dot{\phi} | \dot{\phi} \rangle \tag{1-173}$$

for which Andersen (1975) derives

$$\langle \dot{\phi}^2 \rangle = -\tfrac{1}{3}\,\ddot{\phi}(S)/\phi(S) \tag{1-174}$$

The value of these relations stems from the observation that each successive energy derivative of the function $\phi(r)$ decreases by an order of magnitude. This is illustrated in Fig. 1-10 by means of the d function of Y (taken from Anderson, 1975).

An explicit example for the parametrization given in Eq. (1-172) sheds light on the quality of these approximations. It is again obtained from free electrons and the simple

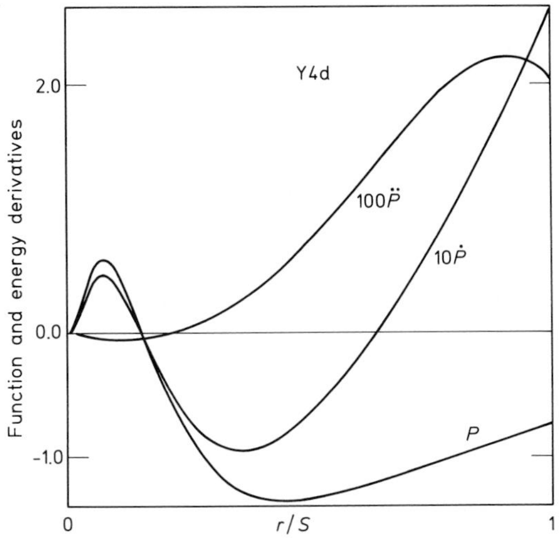

Figure 1-10. Radial d function, P, and energy derivatives, $\dot{P}$, $\ddot{P}$, of Y as given by Andersen (1975).

case of $l=0$, for which

$$\phi(\varepsilon, r) = \alpha \frac{\sin(\sqrt{\varepsilon}\, r)}{\sqrt{\varepsilon}\, r} \tag{1-175}$$

where α is obtained by normalizing in the sphere of radius S, Eq. (1-164) and

$$D(\varepsilon) = [x \cot x - 1]_{x = S\sqrt{\varepsilon}} \tag{1-176}$$

In Fig. 1-11 we compare the exact logarithmic derivative, Eq. (1-176) over a large energy range with the approximations (1-171) and (1-172) evaluated for an energy E_ν satisfying $D(E_\nu) = -1$.

After these preliminaries we are now in a position to define the energy-independent orbital $\chi_{jk}(r)$ (Eq. (1-154)) for the whole crystal given the functions ϕ and $\dot{\phi}$ in the spheres. Following Andersen (1984) in principle, we write

$$\chi_{jk}(\boldsymbol{r}) = \chi^{e}_{jk}(\boldsymbol{r}) + \sum_{L,\tau} \{\phi_{L\tau}(\boldsymbol{r}_\tau)\, \Pi_{L\tau j}(\boldsymbol{k}) + \dot{\phi}_{L\tau}(\boldsymbol{r}_\tau)\, \Omega_{L\tau j}(\boldsymbol{k})\} \tag{1-177}$$

where $\chi^{e}_{jk}(\boldsymbol{r})$ is a so-called envelope function that is defined to be nonzero only outside the spheres, and we have appended the angular momentum index, L, and the site index, τ, to ϕ and $\dot{\phi}$,

$$\phi_{L\tau}(\boldsymbol{r}_\tau) = \phi_{l\tau}(r_\tau)\, Y_L(\hat{\boldsymbol{r}}_\tau) \tag{1-178}$$

with $\boldsymbol{r}_\tau = \boldsymbol{r} - \boldsymbol{\tau}$. ϕ and $\dot{\phi}$ are assumed to be zero outside their respective spheres; the matrices $\boldsymbol{\Pi}$ and $\boldsymbol{\Omega}$ are constructed to have Bloch symmetry, i.e.,

$$\Pi_{L\tau + \boldsymbol{R}_\nu j}(\boldsymbol{k}) = \mathrm{e}^{\mathrm{i}\boldsymbol{k}\cdot\boldsymbol{R}_\nu}\, \Pi_{L\tau j}(\boldsymbol{k}) \tag{1-179}$$

The same is true for $\boldsymbol{\Omega}$, and the matrices are calculated from the requirement that

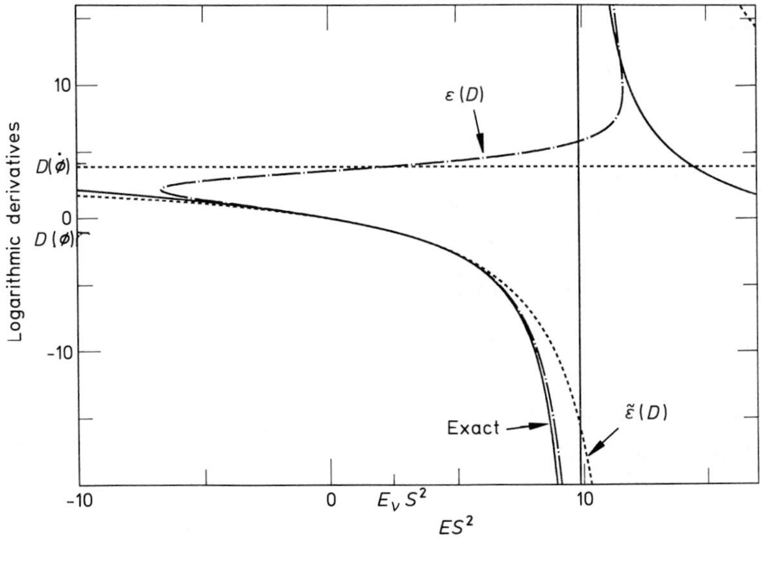

Figure 1-11. Exact logarithmic derivative $D(E)$ and the parametrizations $\varepsilon(D)$ and $\tilde{\varepsilon}(D)$ for free electrons with $l=0$ as given by Andersen (1975).

$\chi_{j\boldsymbol{k}}(\boldsymbol{r})$ is continuous and differentiable everywhere. Their determination becomes possible once the envelope function is specified. The procedure is thus quite general and we will explicitly complete the construction of $\chi_{j\boldsymbol{k}}(\boldsymbol{r})$ for the case of LAPW's and LMTO's. Obviously, other schemes are possible; for instance, the choice of a Slater orbital for the envelope function supplies a formalism to handle multi-center integrals and results in a linear method developed by Davenport (1984; Fernando et al., 1989 a, b). Unfortunately, there is not sufficient space to deal with this method here.

Before specifying the envelope function, we may determine the secular equation in terms of **Π** and **Ω**. Using the definitions for the integrals supplied by Eqs. (1-156) and (1-157), where we now imply integration over the unit cell, we easily derive the following result by means of the Schrödinger equation for ϕ and $\dot{\phi}$ (Eqs. (1-162) and (1-163)) and the orthogonality relation (Eq. (1-165)):

$$\langle i|H-E_\nu|j\rangle = \tag{1-180}$$
$$= \langle i|H-E_\nu|j\rangle^e + \sum_{L,\tau} \Pi^+_{L\tau i}(\boldsymbol{k})\,\Omega_{L\tau j}(\boldsymbol{k})$$

and

$$\langle i|j\rangle = \langle i|j\rangle^e + \sum_{L,\tau} \Pi^+_{L\tau i}(\boldsymbol{k})\,\Pi_{L\tau j}(\boldsymbol{k}) + $$
$$+ \sum_{L,\tau} \Omega^+_{L\tau i}(\boldsymbol{k})\,\langle \dot{\phi}^2_{L\tau}\rangle\,\Omega_{L\tau j}(\boldsymbol{k}) \tag{1-181}$$

where the notation $\langle\cdots\rangle^e$ denotes integration over the interstitial, the integrand containing the envelope function. The secular equation now becomes

$$\sum_j \{\langle i|H|j\rangle - \varepsilon\langle i|j\rangle\}\, a_j(\boldsymbol{k}) =$$
$$= \sum_j \{\langle i|H-E_\nu|j\rangle^e + \sum_{L,\tau} \Pi^+_{L\tau i}(\boldsymbol{k})\,\Omega_{L\tau j}(\boldsymbol{k}) -$$
$$- \varepsilon'\langle i|j\rangle\}\, a_j(\boldsymbol{k}) = 0 \tag{1-182}$$

where ε' is the energy counted from E_ν (which, in general, will depend on angular momentum).

1.2.4.3 Linear Augmented Plane Waves (LAPW)

With a choice of a plane wave for the envelope function we obtain the LAPW method. The quantum label j becomes a reciprocal lattice vector $\boldsymbol{K}_s$ and Bloch symmetrization (see Eq. (1-154)) is not necessary. We write for $\boldsymbol{r}$ outside the spheres

$$\chi^e_{\boldsymbol{K}_s\boldsymbol{k}}(\boldsymbol{r}) = e^{i(\boldsymbol{k}+\boldsymbol{K}_s)\cdot\boldsymbol{r}} \tag{1-183}$$

and augment this plane wave inside the spheres using the angular momentum expansion about the site $\boldsymbol{\tau}$ exactly as in Sec. 1.2.3:

$$e^{i\boldsymbol{K}\cdot\boldsymbol{r}} = e^{i\boldsymbol{K}\cdot\boldsymbol{\tau}}\, 4\pi \sum_L i^l j_l(Kr)\, Y^*_L(\hat{\boldsymbol{K}})\, Y_L(\hat{\boldsymbol{r}}) \tag{1-184}$$

Inside the sphere at $\boldsymbol{\tau}$ we augment using $\boldsymbol{\Phi}_{l\tau}(D_{l\tau K}, r)$ given by Eq. (1-167), where

$$D_{l\tau K} = [x\, j'_l(x)/j_l(x)]_{x=KS_\tau} \tag{1-185}$$

i.e.

$$\boldsymbol{\Phi}_{l\tau}(D_{l\tau K}, r) = \phi_{l\tau}(r) + \omega_{l\tau K}\,\dot{\phi}_{l\tau}(r) \tag{1-186}$$

and replace $j_l(Kr)$ in Eq. (1-184) by

$$j_l(Kr) \;\rightarrow\; j_l(KS_\tau)\,\boldsymbol{\Phi}_{l\tau}(D_{l\tau K}, r)/\boldsymbol{\Phi}_{l\tau}(D_{l\tau K}) \tag{1-187}$$

where we abbreviated

$$\boldsymbol{\Phi}_{l\tau}(D_{l\tau K}) = \boldsymbol{\Phi}_{l\tau}(D_{l\tau K}, S_\tau) \tag{1-188}$$

which is given by Eq. (1-170).

The quantities **Π** and **Ω** follow at once and are given by

$$\Pi_{L\boldsymbol{\tau}\boldsymbol{K}_s}(\boldsymbol{k}) = \tag{1-189}$$
$$= 4\pi\, i^l\, e^{i\boldsymbol{G}_s\cdot\boldsymbol{\tau}}\, j_l(G_s S_\tau)\, Y_L(\hat{\boldsymbol{G}}_s)/\boldsymbol{\Phi}_{l\tau}(D_{l\tau G_s})$$

and

$$\Omega_{L\boldsymbol{\tau}\boldsymbol{K}_s}(\boldsymbol{k}) = \Pi_{L\boldsymbol{\tau}\boldsymbol{K}_s}(\boldsymbol{k})\cdot\omega_{l\tau G_s} \tag{1-190}$$

where

$$\boldsymbol{G}_s = \boldsymbol{k} + \boldsymbol{K}_s \tag{1-191}$$

Now it is not hard to write out the secular equation in a fully general form, and the interested reader may do this as an exercise. We, however, specialize the treatment for simplicity in writing and assume an elementary solid, i.e., $\tau = 0$ (we drop this symbol from the notation). Furthermore, we use the muffin-tin approximation for the potential. The integral in the secular equation (Eq. (1-182)) containing the envelope function is then given by Eq. (1-102) and (1-103) in the APW section. For the remaining terms in Eq. (1-182) we use Eq. (1-190) and write

$$\sum_L \Pi^+_{LK_s} \Omega_{LK_t} - \varepsilon' \sum_L \Pi^+_{LK_s} \Pi_{LK_t} - \\ - \varepsilon' \sum_L \Omega^+_{LK_s} \langle \dot{\phi}^2_L \rangle \Omega_{LK_t} = \\ = \sum_L \Pi^+_{LK_s} (\omega_{lG_s} - \varepsilon' \beta_{lG_sG_t}) \Pi_{LK_t} \tag{1-192}$$

where

$$\beta_{lG_sG_t} = 1 + \langle \dot{\phi}^2_L \rangle \omega_{lG_s} \omega_{lG_t} \tag{1-193}$$

Next, we use the identity

$$\frac{1}{\Phi_l(D_{lK_s})} \frac{1}{\Phi_l(D_{lK_t})} = S \frac{D_{lK_s} - D_{lK_t}}{\omega_{lK_t} - \omega_{lK_s}} \tag{1-194}$$

which may be proved directly with the help of the Eqs. (1-168) to (1-170). If we now define the quantity

$$W^l_{st} = \frac{1}{\Omega_{UC}} \frac{\omega_{lK_t} - \omega_{lK_s}}{G^2_t - G^2_s} \sum_m \Pi_{LK_t} \Pi_{LK_t} \tag{1-195}$$

and insert the expression for Π, Eq. (1-189), we obtain

$$W^l_{st} = \frac{4\pi S}{\Omega_{UC}} (2l+1) P_l(\cos\theta_{st}) j_l(G_s S) j_l(G_t S) \\ \cdot \frac{D_{lK_s} - D_{lK_t}}{G^2_t - G^2_s} \tag{1-196}$$

(remember $L = (l, m)$). With the equation of the APW section one finds

$$\sum_l W^l_{st} = \frac{4\pi S^3}{\Omega_{UC}} \frac{j_1(|K_s - K_t| S)}{|K_s - K_t| S} \tag{1-197}$$

Defining two more abbreviations by

$$\Gamma^l_{st} = -(G^2_t - E_v) + \frac{G^2_t - G^2_s}{\omega_{lK_t} - \omega_{lK_s}} \cdot \omega_{lK_t} \tag{1-198}$$

and

$$\Delta^l_{st} = \beta_{lG_sG_t} \frac{G^2_t - G^2_s}{\omega_{lK_t} - \omega_{lK_s}} - 1 \tag{1-199}$$

we finally obtain

$$\langle s | H | t \rangle - \varepsilon \langle s | t \rangle = \\ = \Omega_{UC} \Big\{ [(k + K_t)^2 - \varepsilon] \, \delta_{st} + \\ + \sum_l W^l_{st} (\Gamma^l_{st} - \varepsilon' \Delta^l_{st}) \Big\} \tag{1-200}$$

This is the secular equation for the LAPW method, see also Andersen (1975) and (1984).

The major advantage of the LAPW method is its high accuracy and the relative ease with which it can treat a general potential. Andersen (1984) gives a simple description which, unfortunately, we have no space to repeat here. Another similar LAPW technique has been devised by Krakauer et al. (1979).

1.2.4.4 Linear Combination of Muffin-Tin Orbitals (LMTO)

We next turn to the "linear-muffin-tin-orbital method" (LMTO). It is obtained by requiring the envelope function to satisfy Laplace's equation (Andersen, 1975, 1984), i.e.,

$$\nabla^2 \chi^e_L(r) = 0 \tag{1-201}$$

In spherical geometry there are two well-known types of solutions, one type is regular at infinity and is given by

$$\chi^e_{L-}(r) = \left(\frac{r}{S}\right)^{-l-1} Y_L(\hat{r}) \tag{1-202}$$

the other type is regular at the origin and is given by

$$\chi^e_{L+}(\boldsymbol{r}) = \left(\frac{r}{S}\right)^l Y_L(\hat{\boldsymbol{r}}) \tag{1-203}$$

$S>0$ is arbitrary at this point. The expansion theorem (Eq. (1-159)), discussed in connection with multi-center integrals, is the well-known series of a static multipole potential

$$\left|\frac{|\boldsymbol{r}-\boldsymbol{R}|}{S}\right|^{-l-1} Y_L(\widehat{\boldsymbol{r}-\boldsymbol{R}}) = \tag{1-204}$$
$$= -\sum_{L'} \left(\frac{r}{S}\right)^{l'} Y_{L'}(\hat{\boldsymbol{r}}) \frac{1}{2(2l'+1)} \mathscr{S}_{L'L}(\boldsymbol{R})$$

connecting the two types of solutions, χ^e_{L-} and χ^e_{L+}. $\mathscr{S}$ is given by

$$\mathscr{S}_{L'L}(\boldsymbol{R}) = \tag{1-205}$$
$$= (4\pi)^{1/2}\, g_{L'L} \left(\frac{S}{R}\right)^{l'+l+1} Y^*_{l'+l,\, m'-m}(\hat{\boldsymbol{R}})$$

with $L=(l,m)$, $L'=(l',m')$ and

$$g_{L'L} = (-1)^{l+m+1} \cdot$$
$$\cdot 2\,[(2l'+1)(2l+1)(l'+l-m'+m)! \cdot$$
$$\cdot (l'+l+m'-m)!]^{1/2} \cdot$$
$$\cdot [(2l'+2l+1)(l'+m')!(l'-m')! \cdot$$
$$\cdot (l+m)!(l-m)!]^{-1/2} \tag{1-206}$$

(see Andersen, 1984). For simplicity our discussion is now once again restricted to the case of one atom per unit cell. The energy-independent orbital is thus written like Eq. (1-177):

$$\chi_{L\boldsymbol{k}}(\boldsymbol{r}) = \chi^e_{L-}(\boldsymbol{r}) + \tag{1-207}$$
$$+ \sum_{L'} [\phi_{L'}(\boldsymbol{r})\, \Pi_{L'L}(\boldsymbol{k}) + \dot{\phi}_{L'}(\boldsymbol{r})\, \Omega_{L'L}(\boldsymbol{k})]$$

and we determine the matrices $\boldsymbol{\Pi}$ and $\boldsymbol{\Omega}$ as follows: At $r=S$ the logarithmic derivative of χ^e_{L-} is $D=-l-1$. We therefore match the function χ^e_{L-} at S with the function $\boldsymbol{\Phi}_l(D=-l-1,r)/\boldsymbol{\Phi}_l(-)$, where we abbreviated

$$\boldsymbol{\Phi}_l(-) = \boldsymbol{\Phi}_l(D=-l-1,S) \tag{1-208}$$

In other words, for $r<S$ we replace

$$\left(\frac{r}{S}\right)^{-l-1} \rightarrow \boldsymbol{\Phi}_l(-l-1,r)/\boldsymbol{\Phi}_l(-) \tag{1-209}$$

and use suitable E_ν's to expand $\boldsymbol{\Phi}$ in terms of ϕ and $\dot{\phi}$ as given in Eq. (1-167). The sphere radius S will be specified a little later. If we place one orbital like Eq. (1-209) at the sphere centered at the origin, translational symmetry requires the same orbital to be placed at all other sites in such a way that they obey Bloch's theorem. Since the "tail" of one orbital centered at $R \neq 0$ is given in the central sphere by the expansion, Eq. (1-204), the tails of *all* orbitals superimpose with correct Bloch symmetry and are given by

$$-\sum_{L'} (r/S)^{l'}\, Y_{L'}(\hat{\boldsymbol{r}}) \frac{1}{2(2l'+1)} \mathscr{S}^{\boldsymbol{k}}_{L'L}$$

where we define

$$\mathscr{S}^{\boldsymbol{k}}_{L'L} = \sum_{\nu \neq 0} \mathrm{e}^{\mathrm{i}\boldsymbol{k}\cdot\boldsymbol{R}_\nu}\, \mathscr{S}_{L'L}(\boldsymbol{R}_\nu) \tag{1-210}$$

The quantities $\boldsymbol{\Phi}$ and $\boldsymbol{\Omega}$ are now obtained by matching the "heads", Eq. (1-209), and the tails above. The logarithmic derivatives of the functions χ^e_{L+} occurring in the tails are obviously $D=l$ and we abbreviate

$$\boldsymbol{\Phi}_l(D=l,S) = \boldsymbol{\Phi}_l(+) \tag{1-211}$$

Then for $r<S$ the desired orbital may be written as

$$\chi_{L\boldsymbol{k}}(\boldsymbol{r}) = \boldsymbol{\Phi}_l(-l-1,r)/\boldsymbol{\Phi}_l(-) - \tag{1-212}$$
$$-\sum_{L'} \boldsymbol{\Phi}_{L'}(l',r)/\boldsymbol{\Phi}_l(+) \frac{1}{2(2l'+1)} \mathscr{S}^{\boldsymbol{k}}_{L'L}$$

Using the relation of $\boldsymbol{\Phi}$ to ϕ and $\dot{\phi}$ in Eq. (1-167) and comparing Eq. (1-207) with Eq. (1-212), we can read off $\boldsymbol{\Pi}$ and $\boldsymbol{\Omega}$. The results are most compactly written by defining

$$\begin{aligned} \omega_{l-} &= \omega_l(-l-1) \\ \omega_{l+} &= \omega_l(l) \end{aligned} \tag{1-213}$$

$$\Delta_l = \omega_{l+} - \omega_{l-} \quad (1\text{-}214)$$

$$\tilde{\Phi}_l = \Phi_l(-)\sqrt{\frac{S}{2}} \quad (1\text{-}215)$$

and by using the identity

$$\omega_{l-} - \omega_{l+} = S(2l+1)\,\Phi_l(+)\,\Phi_l(-) \quad (1\text{-}216)$$

which is easily derived from Eqs. (1-168) and (1-170). One finds

$$\Pi_{L''L}(\boldsymbol{k}) = = \tilde{\Phi}_l^{-1}\,\delta_{LL''} + (\tilde{\Phi}_{l''}/\Delta_{l''})\,\mathcal{S}^{\boldsymbol{k}}_{L''L} \quad (1\text{-}217)$$

and

$$\Omega_{L''L}(\boldsymbol{k}) = \omega_{l''-}\,\tilde{\Phi}_{l''}^{-1}\,\delta_{LL''} + + (\tilde{\Phi}_{l''}\,\omega_{l''+}/\Delta_{l''})\,\mathcal{S}^{\boldsymbol{k}}_{L''L} \quad (1\text{-}218)$$

These quantities **Π** and **Ω** define a so-called linear-muffin-tin orbital (LMTO), and, apart from slight differences in notation, they agree with Skriver's (1984) expressions. Their specification is, however, still incomplete because the partitioning of space into interstitial and spherical regions has yet to be defined, in particular, the sphere radius S. We will do this next.

In closely packed solids each atom has 8 to 12 nearest neighbors, and the assumed muffin-tin (MT) form of the potential (see Sec. 1.2.3) is a good approximation. Far more convenient and sometimes even more accurate is an approximation in which the interstitial region is eliminated by enlarging the MT spheres and neglecting the slight overlap. The MT spheres in this approximation thus become the Wigner-Seitz atomic spheres (S the Wigner-Seitz radius) which are supposed to fill space completely (Wigner and Seitz, 1955). This is called the atomic sphere approximation (ASA), and in a b.c.c. crystal, for example, the radius S is related to the lattice constant, a, by

$$\frac{4\pi}{3}S^3 = \frac{1}{2}a^3 \quad (1\text{-}219)$$

(or $\frac{1}{4}a^3$ in the case of f.c.c. crystals).

In open structures the interstitial positions often have such high symmetry that both the atomic and the repulsive interstitial potential can be approximated by spherically symmetric ones, and the atomic and interstitial spheres together form a close packing. We will now use the ASA and continue by stating that the terms referring to the interstitial region and the non-muffin-tin part of the potential drop out, but, of course, the boundary conditions incorporated in the quantities Π and Ω remain. The secular matrix given in Eq. (1-182) (with Eq. (1-181)) then consists solely of products involving Π and Ω, and with the Eqs. (1-217) and (1-218) it is obtained as

$$\begin{aligned}
&\langle L|H|L'\rangle - \varepsilon\langle L|L'\rangle = \\
&= \{[\omega_- + E_\nu(1+\omega_-^2\langle\dot{\phi}^2\rangle)]/\tilde{\Phi}^2\}_l\,\delta_{LL'} + \\
&\quad + \{\{[\omega_+ + E_\nu(1+\omega_-\omega_+\langle\dot{\phi}^2\rangle)]/\Delta\}_l + \\
&\quad + \{\ldots\}_{l'}\}\,\mathcal{S}^{\boldsymbol{k}}_{LL'} + \\
&\quad + \sum_{L''}\mathcal{S}^{\boldsymbol{k}}_{LL''}\{\tilde{\Phi}^2\,[\omega_+ E_\nu(1+\omega_+^2\langle\dot{\phi}^2\rangle)]/ \\
&\qquad \Delta^2\}_{l''}\,\mathcal{S}^{\boldsymbol{k}}_{L''L'} - \\
&\quad - \varepsilon\{[1+\omega_-^2\langle\dot{\phi}^2\rangle]/\tilde{\Phi}^2\}_l\,\delta_{LL'} + \\
&\quad + \{\{[1+\omega_+\omega_-\langle\dot{\phi}^2\rangle]/\Delta\}_l + \{\ldots\}_{l'}\}\,\mathcal{S}^{\boldsymbol{k}}_{LL'} \\
&\quad + \sum_{L''}\mathcal{S}^{\boldsymbol{k}}_{LL''}\,[\tilde{\Phi}^2(1+\omega_+^2\langle\dot{\phi}^2\rangle)/\Delta^2]_{l''}\,\mathcal{S}^{\boldsymbol{k}}_{L''L'}
\end{aligned} \quad (1\text{-}220)$$

The terms without a factor of $\mathcal{S}$ are one-center terms, those with one factor of $\mathcal{S}$ are two-center terms, and those with two factors of $\mathcal{S}$ are three-center terms.

Although the errors introduced by the atomic-sphere approximation are unimportant for many applications, there are cases where energy bands of high accuracy are needed and where one should include the non-muffin-tin perturbations in some form. Skriver (1984) gives expressions which account to first order for the differences between the atomic spheres and the atomic polyhedron and correct for the neglecting of higher partial waves. These extra terms which are added to the LMTO matrices (Eq. (1-220)) are called the "combined correction terms". We will omit their treatment here but will discuss this problem in some detail in connection with the augmented spherical wave (ASW) method later on.

1.2.4.5 The Korringa-Kohn-Rostoker Atomic Sphere Approximation (KKR-ASA)

As it stands, the LMTO secular matrix is not particularly illuminating. An elegant treatment exists, however, that renders the LMTO method utterly transparent. It stems from Andersen (1973, 1975) and starts by constructing of an *energy-dependent* orbital, $\chi_L(\varepsilon, \boldsymbol{r})$, within the ASA. Again, the solutions of Laplace's equation are used to formulate the boundary conditions. In particular, one easily establishes that the orbital

$$\chi_L(\varepsilon, \boldsymbol{r}) = \phi_l(\varepsilon, S) \begin{cases} \dfrac{\phi_l(\varepsilon, r)}{\phi_l(\varepsilon, S)} Y_L(\hat{\boldsymbol{r}}) - \dfrac{D_l(\varepsilon) + l + 1}{2l+1} \left(\dfrac{r}{S}\right)^l Y_L(\hat{\boldsymbol{r}}), & r \leq S \\ \dfrac{l - D_l(\varepsilon)}{2l+1} \left(\dfrac{r}{S}\right)^{-l-1} Y_L(\hat{\boldsymbol{r}}), & r \geq S \end{cases} \quad (1\text{-}221)$$

is continuous and differentiable everywhere. Here $\phi_l(\varepsilon, r)$ is a solution of energy ε to the radial Schrödinger equation inside the atomic sphere and $D_l(\varepsilon)$ is its logarithmic derivative, Eq. (1-166), at $r = S$. As before, a Bloch function is constructed by placing such an orbital on each lattice site in the following, now familiar manner:

$$\chi_{L\boldsymbol{k}}(\varepsilon, \boldsymbol{r}) = \sum_{\nu} e^{i \boldsymbol{k} \cdot \boldsymbol{R}_\nu} \chi_L(\varepsilon, \boldsymbol{r} - \boldsymbol{R}_\nu) \quad (1\text{-}222)$$

Because of the expansion theorem (Eq. (1-204)) the tails of this Bloch function in the sphere at the origin can be expressed using $\mathscr{S}^{\boldsymbol{k}}_{LL'}$, Eq. (1-210), and one obtains from Eqs. (1-221) and (1-222) for $r \leq S$:

$$\chi_{L\boldsymbol{k}}(\varepsilon, \boldsymbol{r}) = \phi_l(\varepsilon, r) Y_L(\hat{\boldsymbol{r}}) - \phi_l(\varepsilon, S) \frac{D_l(\varepsilon) + l + 1}{2l+1} \left(\frac{r}{S}\right)^l Y_L(\hat{\boldsymbol{r}}) - \phi_l(\varepsilon, S) \frac{l - D_l(\varepsilon)}{2l+1} \sum_{\nu} \left(\frac{r}{S}\right)^{l'} \frac{1}{2(2l'+1)} \cdot Y_{L'}(\hat{\boldsymbol{r}}) \mathscr{S}^{\boldsymbol{k}}_{L'L} \quad (1\text{-}223)$$

This equation can be written a little more compactly by defining the quantities

$$P_l(\varepsilon) = 2(2l+1)[D_l(\varepsilon) + l + 1]/(D_l(\varepsilon) - l) \quad (1\text{-}224)$$

and

$$M_l(\varepsilon) = \phi_l(\varepsilon, S)(D_l(\varepsilon) - l)/(2l+1) \quad (1\text{-}225)$$

Then it becomes

$$\chi_{L\boldsymbol{k}}(\varepsilon, \boldsymbol{r}) = \phi_L(\varepsilon, \boldsymbol{r}) + \sum_{L'} \left(\frac{r}{S}\right)^{l'} \frac{1}{2(2l'+1)} Y_{L'}(\hat{\boldsymbol{r}}) \cdot \{\mathscr{S}^{\boldsymbol{k}}_{L'L} - P_l(\varepsilon)\delta_{L'L}\} M_l(\varepsilon) \quad (1\text{-}226)$$

A linear combination of these functions, as in Eq. (1-155), is now required to be a solution of the Schrödinger equation for the crystal. Since inside the atomic sphere the first term on the right-hand side of Eq. (1-226) is already a solution, the function

$$\psi_{\boldsymbol{k}}(\varepsilon, \boldsymbol{k}) = \sum_{L} a_L(\boldsymbol{k}) \phi_L(\varepsilon, \boldsymbol{r}) \quad (1\text{-}227)$$

is the desired result provided a corresponding linear combination of the second term on the right-hand side disappears:

$$\sum_{L'} \{\mathscr{S}^{\boldsymbol{k}}_{LL'} - P_l(\varepsilon)\delta_{LL'}\} M_{l'}(\varepsilon) a_{L'}(\boldsymbol{k}) = 0 \quad (1\text{-}228)$$

This set of linear, homogeneous equations has nontrivial solutions for the eigenvectors $a_L(\boldsymbol{k})$ at the energies $\varepsilon = \varepsilon_{\boldsymbol{k}}$ for which the determinant of the coefficient matrix vanishes:

$$\det [\mathcal{S}^{\boldsymbol{k}}_{LL'} - P_l(\varepsilon)\,\delta_{LL'}] = 0 \tag{1-229}$$

The condition embodied in Eq. (1-228) is for obvious reasons often called *tail cancellation,* and by comparing with Eq. (1-140) it is seen to be a KKR-type equation. Andersen calls Eq. (1-228) or Eq. (1-229) the KKR-ASA equations. However, the matrix $\mathcal{S}^{\boldsymbol{k}}_{LL'}$, does not depend on energy, in contrast to the corresponding KKR structure constants $B_{LL'}(\varepsilon, \boldsymbol{k})$. The entire energy dependence is contained in the term $P_l(\varepsilon)$ which is called the *potential function* since it involves the logarithmic derivatives and thereby the characteristics of the potential. The KKR-ASA equations therefore establish the link between the potential and the structure-dependent parts of the energy-band problem and provide the connection between ε and $\boldsymbol{k}$ which is the energy-band structure. In this approximation we deal with a boundary-value problem in which the lattice through the structure constants, $\mathcal{S}^{\boldsymbol{k}}_{LL'}$, imposes a $\boldsymbol{k}$-dependent and non-spherically symmetric boundary condition on the solutions $P_l(\varepsilon)$ inside the atomic Wigner-Seitz sphere (Skriver, 1984).

To obtain more insight into the band-structure problem we notice that the potential function, $P_l(\varepsilon)$, does not depend on the magnetic quantum number, m. Therefore, one may transform the structure matrix

$$\mathcal{S}^{\boldsymbol{k}}_{LL'} \equiv \mathcal{S}^{\boldsymbol{k}}_{lm,\,l'm'} \tag{1-230}$$

from the given representation to another one where the subblocks to each l are diagonalized and denote the $2l+1$ diagonal elements by $\mathcal{S}^{\boldsymbol{k}}_{li}$, $i = 1, \ldots, 2l+1$. This unitary transformation is independent of potential and atomic volume and leaves the form of the KKR-ASA equations invariant. If one therefore neglects hybridization, i.e., if one sets the elements of $\mathcal{S}^{\boldsymbol{k}}_{LL'}$ with $l \neq l'$ equal to zero, the "unhybridized bands" $[\varepsilon_{nli}(\boldsymbol{k})]$ are simply found as the nth solution of

$$P_l(\varepsilon) = \mathcal{S}^{\boldsymbol{k}}_{LL'} \tag{1-231}$$

For each value of l, the right-hand side of this equation is $2l+1$ functions of $\boldsymbol{k}$ which are called *canonical bands* and are illustrated for $l = 2$, i.e., d states, in Fig. 1-12 along a few lines of high symmetry in the Brillouin zone for f.c.c., b.c.c. and h.c.p. lattices. We see that at the center of the Brillouin zone Γ, in the case of f.c.c. and b.c.c. lattices, there are two energy levels, the lower of which is *triply* degenerate, and the upper *doubly* degenerate. The former level (as in the case of Cu, Fig. 1-9) comprises the xy, yz, and xz orbitals which are equivalent to one another in a cubic environment and are called T_{2g} (or $\Gamma_{25'}$) orbitals. The latter level comprises the $x^2 - y^2$ and $3z^2 - r^2$ orbitals called E_g (or Γ_{12}) which by pointing along the cubic axis are not equivalent to the T_{2g} orbitals. The degeneracy is partially lifted when going into the zone because eigenfunctions which are equivalent at $\boldsymbol{k} = 0$ may become non-equivalent for $\boldsymbol{k} \neq 0$ due to the translational phase factor $e^{i\boldsymbol{k}\cdot\boldsymbol{R}}$. The situation is somewhat more complicated for h.c.p. lattices (Fig. 1-12c) because there are two atoms in the unit cell and hence twice as many bands. We see six energy levels at Γ, the two lower ones being non-degenerate and the others doubly degenerate.

Equation (1-231) is a monotonic mapping of the canonical bands onto an energy scale specified by the nth branch of the potential function. Figure 1-13 illustrates such an unhybridized band structure for Osmium (Jepsen et al., 1975). The d bands

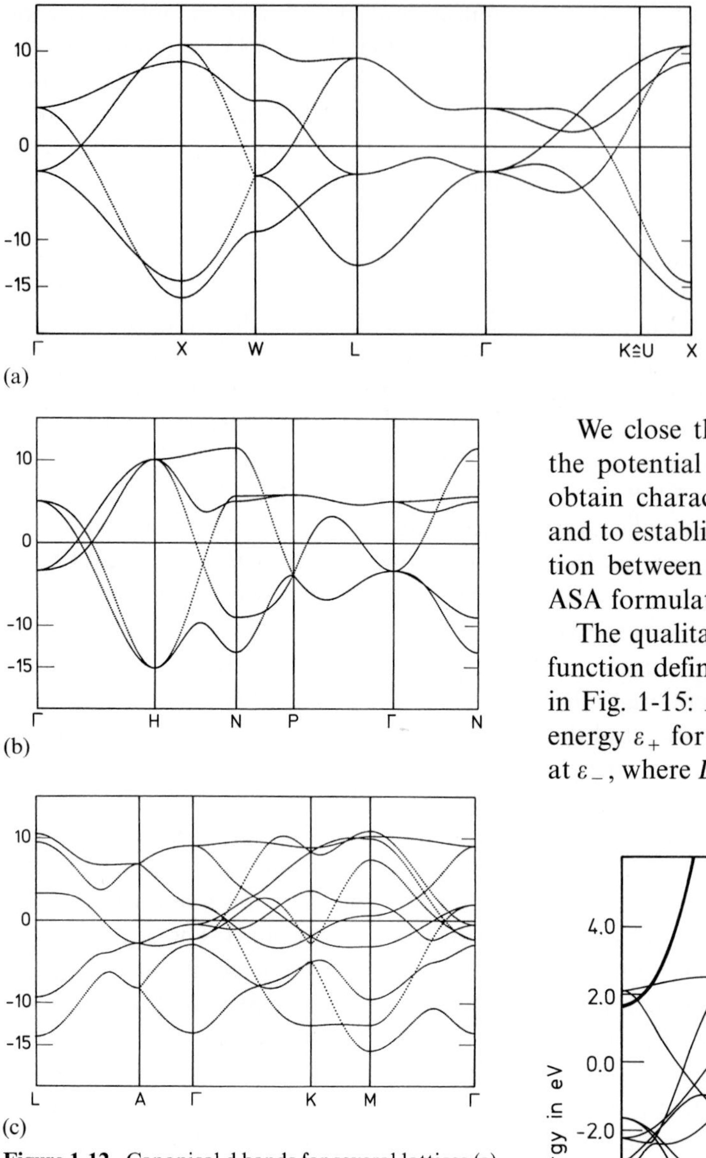

Figure 1-12. Canonical d bands for several lattices (a) f.c.c.; (b) b.c.c.; (c) h.c.p.

are clearly recognized and are seen to be those shown in Fig. 1-12c, with hardly any distortions. The role of hybridization is appreciated in Fig. 1-14, which shows the full band structure of Osmium (obtained by numerically solving Eq. (1-220) with a self-consistent potential that will be discussed in detail in Sec. 1.3).

We close this subsection by linearizing the potential function. This allows us to obtain characteristic *potential parameters* and to establish the mathematical connection between the LMTO and the KKR-ASA formulations.

The qualitative features of the potential function defined by Eq. (1-224) are shown in Fig. 1-15: $P_l(\varepsilon)$ has a singularity for an energy ε_+ for which $D_l(\varepsilon_+) = l$ and is zero at ε_-, where $D_l(\varepsilon_-) = -l-1$. With the ap-

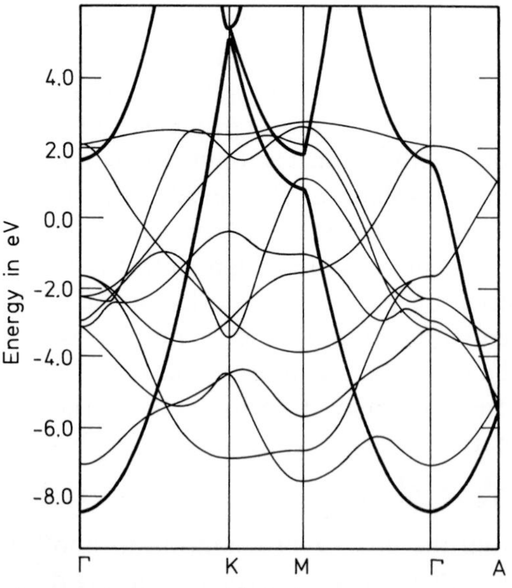

Figure 1-13. Non-relativistic band structure for Os without sp-d hybridization and combined correction terms as given by Jepsen et al. (1975). Thick lines indicate the sp bands.

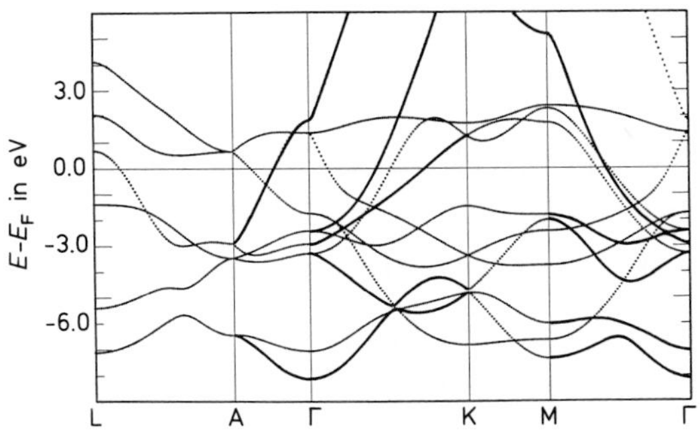

Figure 1-14. Non-relativistic band structure for Os including all hybridizations and the combined correction as obtained from the ASW. Thick lines indicate the bands with mainly sp character.

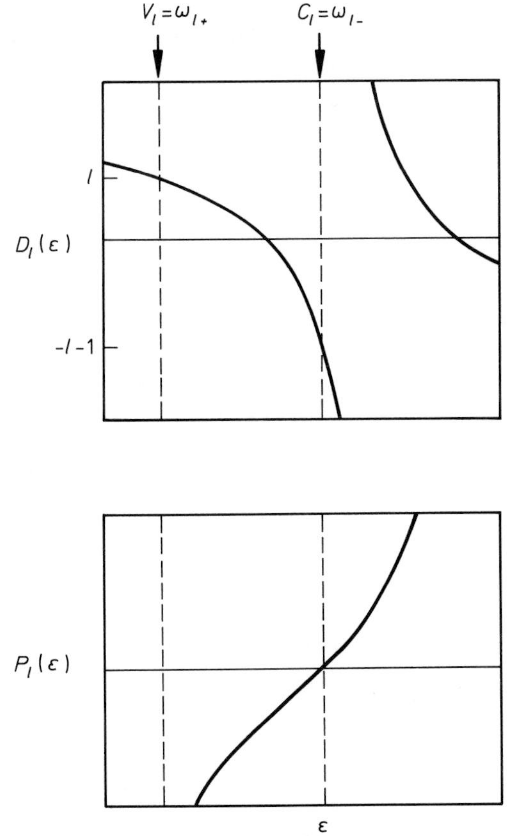

Figure 1-15. Qualitative features of the logarithmic derivative $D_l(\varepsilon)$ and the potential function $P_l(\varepsilon)$ and their connection to ω_{l+} and ω_{l-}.

proximate relationship between energy and $\omega(D)$, Eq. (1-172), the corresponding quantities ω are ω_{l+} and ω_{l-} defined by Eq. (1-213). Thus the simple expression

$$P_l(\varepsilon) \simeq \frac{1}{\gamma_l}(\varepsilon - \omega_{l-})/(\varepsilon - \omega_{l+}) \qquad (1\text{-}232)$$

has the correct analytical form, and γ_l is a scale factor which is related to typical LMTO parameters. To see this, first manipulate Eqs. (1-167) and (1-168) to eliminate $D\{\phi\}$ and $D\{\dot{\phi}\}$ in favor of $\boldsymbol{\Phi}_l(+)$ and $\boldsymbol{\Phi}_l(-)$ defined by Eqs. (1-211) and (1-208), and then derive the identity

$$\frac{\omega(D_l) - \omega_{l-}}{\omega(D_l) - \omega_{l+}} = \frac{\boldsymbol{\Phi}_l(-)}{\boldsymbol{\Phi}_l(+)} \cdot \frac{D_l + l + 1}{D_l - l} \qquad (1\text{-}233)$$

Using Eq. (1-172) and comparing with Eq. (1-232), we find that

$$\gamma_l^{-1} = 2(2l+1)\,\boldsymbol{\Phi}_l(+)/\boldsymbol{\Phi}_l(-) \qquad (1\text{-}234)$$

The potential parameter ω_{l-} is commonly denoted by C_l and it describes for $l>0$, the center of the band corresponding to angular momentum l. This is true because for $l>0$

$$\sum_{i=1}^{2l+1} \mathscr{S}_{li}^{\mathbf{k}=0} = 0 \qquad (1\text{-}235)$$

i.e., the center of gravity of the canonical bands is zero for $l>0$. This coincides with the value of $\varepsilon = C_l$, where $P_l = 0$ because of Eq. (1-231). The parameter ω_{l+} is often denoted by V_l, and the difference $C_l - V_l$ is related to the band width. Instead of listing this value to characterize a band, it is more common to define the so-called band-mass parameter through the relation

$$\mu_l^{-1} = \gamma_l (C_l - V_l)\, S^2 \qquad (1\text{-}236)$$

and to use this number as a potential parameter. It can be seen to be unity for free electrons, and it has the value $\mu_d = 4.4$ for the d bands of Osmium shown in Fig. 1-14, just to give an example.

With the help of these parameters, a simple picture emerges of the structure of unhybrized bands discussed in connection with Eq. (1-231). Inverting Eq. (1-232) to express $\varepsilon(P_l)$ we obtain easily

$$\varepsilon(P_l) = C_l + \frac{1}{\mu_l S^2} \frac{P_l}{1-\gamma_l P_l} \tag{1-237}$$

where μ_l is defined by Eq. (1-236). Inserting Eq. (1-231) leads to an explicit expression for unhybridized bands:

$$\varepsilon_{li}(\boldsymbol{k}) = C_l + \frac{1}{\mu_l S^2} \frac{\mathscr{S}_{li}^{\boldsymbol{k}}}{1-\gamma_l \mathscr{S}_{li}^{\boldsymbol{k}}} \tag{1-238}$$

The interpretation is that the pure l band is obtained from the canonical l-band structure by fixing the position through the parameter C_l, scaling it by $\mu_l S^2$, and distorting it by γ_l. This distortion is generally small in the case of d bands. Compare for instance, Fig. 1-12c with Fig. 1-13.

We complete this discussion by finally establishing the connection between the LMTO and KKR-ASA equation. It is, however, not an exact one, because we must drop the terms containing $\langle \dot{\phi}_L^2 \rangle$. Then, referring to Eqs. (1-180) and (1-181), we can write the Hamiltonian matrix $\mathbf{H}$ as

$$\mathbf{H} = \mathbf{\Pi}^+ \mathbf{\Omega} \tag{1-239}$$

and the overlap matrix, $\mathbf{O}$, as

$$\mathbf{O} = \mathbf{\Pi}^+ \mathbf{\Pi} \tag{1-240}$$

where we imply matrix multiplications, and the elements of $\mathbf{\Pi}$ and $\mathbf{\Omega}$ are given by Eqs. (1-217) and (1-218). The LMTO equations are then simply

$$\mathbf{H} - \varepsilon \mathbf{O} = \mathbf{\Pi}^+ (\mathbf{\Omega} - \varepsilon' \mathbf{\Pi}) \tag{1-241}$$

where ε' is counted from E_ν.

If we now define the KKR-ASA matrix as $\mathbf{\Lambda}$ with elements from Eqs. (1-228) and (1-232), (1-234):

$$\Lambda_{LL'} = \mathscr{S}_{LL'}^{\boldsymbol{k}} - 2(2l+1) \frac{\boldsymbol{\Phi}_l(+)}{\boldsymbol{\Phi}_l(-)} \frac{\varepsilon' - \omega_{l-}}{\varepsilon' - \omega_{l+}} \delta_{LL'} \tag{1-242}$$

and another matrix $\boldsymbol{\gamma}$ with elements

$$\gamma_{LL'} = \tilde{\boldsymbol{\Phi}}_l(\omega_{l+} - \varepsilon') \Delta_l^{-1} \delta_{LL'} \tag{1-243}$$

[cf. Eqs. (1-214) and (1-224)] we obtain by means of a simple multiplication

$$\boldsymbol{\gamma} \mathbf{\Lambda} = \mathbf{\Pi}^+ (\mathbf{\Omega} - \varepsilon' \mathbf{\Pi}) = \mathbf{H} - \varepsilon \mathbf{O} \tag{1-244}$$

Hence, the KKR-ASA matrix $\mathbf{\Lambda}$ is a factor of the LMTO matrix if the terms containing $\langle \dot{\phi}_L^2 \rangle$ are neglected as well as the combined correction terms. The potential parameters and the canonical bands are therefore useful tools for interpreting and documenting numerical results that are obtained by solving the more precise LMTO equations. The connection of the KKR-ASA equations with the exact KKR theory of Sec. 1.2.3 will be explained more thoroughly in the next section. For self-consistent calculations the electron density is needed; it is not hard to obtain in the ASA and will be discussed in the next section in connection with the augmented spherical wave method to which we now turn.

1.2.4.6 Augmented Spherical Waves (ASW)

The augmented spherical wave method (ASW) was developed by Williams et al. (1979), and without Andersen's LMTO technique it would not have been conceived. There are essentially two differences: a different choice of the envelope function and a different and simpler method for augmenting the basis functions. For the former solutions of the wave equation,

$$(\nabla^2 + \varepsilon_0)\, \chi_L^e(\boldsymbol{r}) = 0 \tag{1-245}$$

are used which, of course, coincide with the Laplace equation if $\varepsilon_0 = 0$ (see Eq. (1-201)). In spherical geometry there are again two types of solutions. For negative energies, the solutions that are regular at infinity are given by spherical Hankel functions $h_l^+(\varkappa r)$, $\varkappa = \sqrt{\varepsilon_0}$, and those regular at the origin are spherical Bessel functions $j_l(\varkappa r)$. For positive energies the spherical Hankel functions must be replaced by spherical Neumann functions, and the envelope functions in this case become identical with the basis used in multiple-scattering theories (see Sec. 1.2.3). The energy parameter, ε_0, obviously controls the degree of localization and it could, in principle, be used as a variational parameter. This was indeed done initially, but in all practical cases it was found that a choice of a fixed negative and small value ($\varepsilon_0 = -0.01$ Ry) gave best results for the electronic properties of solids at or near ambient pressure. We will therefore limit our discussion to the negative-energy case and – whenever possible – suppress the parameter $\varepsilon_0 = \varkappa^2$ in the notation.

For mathematical reasons it is convenient to define the envelope function in the following way

$$\chi_L^e(\boldsymbol{r}) = H_L(\boldsymbol{r}) \equiv \mathrm{i}^l \varkappa^{l+1} h_l^+(\varkappa r)\, Y_L(\hat{\boldsymbol{r}}) \quad (1\text{-}246)$$

The factor $\varkappa^{l+1}$ is introduced to cancel out the leading energy dependence for small values of $\varkappa r$, because here

$$h_l^+(x) = x^{-l-1} \frac{(2l+1)!!}{(2l+1)} \quad (1\text{-}247)$$

The expansion theorem, Eq. (1-159), which allows an easy treatment of multi-center integrals – and augmentation –, connects the two types of functions in the following way

$$H_L(\boldsymbol{r}+\boldsymbol{R}) = \sum_{L'} J_{L'}(\boldsymbol{r})\, B_{L'L}(\boldsymbol{R}) \quad (1\text{-}248)$$

Here we assume $\boldsymbol{R} \neq 0$ and

$$J_L(\boldsymbol{r}) = \mathrm{i}^l \varkappa^{-l} j_l(\varkappa r)\, Y_L(\hat{\boldsymbol{r}}) \quad (1\text{-}249)$$

the reason for the factor $\varkappa^{-l}$ is analogous to Eq. (1-246) (see Eq. (1-106)). The expansion coefficients $B_{LL'}(\boldsymbol{R})$ are the structure constants which arise in multiple-scattering theories (see Sec. 1.2.3) and are given by Williams et al. (1979)

$$B_{LL'}(\boldsymbol{R}) = 4\pi \sum_{L''} C_{LL'L''}\, \varkappa^{l+l'-l''} H_{L''}(\boldsymbol{R}) \quad (1\text{-}250)$$

where $C_{LL'L''}$ is a Gaunt coefficient

$$C_{LL'L''} = \int \mathrm{d}\hat{\boldsymbol{r}}\, Y_L^*(\hat{\boldsymbol{r}})\, Y_{L'}(\hat{\boldsymbol{r}})\, Y_{L''}(\hat{\boldsymbol{r}}) \quad (1\text{-}251)$$

and the LMTO structure constants $\mathscr{S}_{LL'}(\boldsymbol{r})$ can easily be seen to be a special case of $B_{LL'}(\boldsymbol{R})$ (see Skriver, 1984).

The mathematics of the augmentation process is straightforward. The spherical wave $H_L(\boldsymbol{r})$ is continued into the atomic sphere in which it is centered by a numerical solution of the Schrödinger equation which joins smoothly to $H_L(\boldsymbol{r})$ at the sphere radius. We may follow Andersen and call this "head augmentation". At atomic spheres where $H_L(\boldsymbol{r})$ is *not* centered, we use Eq. (1-248) and continue the spherical wave into the sphere by a *linear combination* of numerical solutions of the Schrödinger equation which joins smoothly to $H_L(\boldsymbol{r}-\boldsymbol{R})$ at the sphere radius. This we may call "tail augmentation". If we use the tilde to denote augmentation, we replace the head inside the sphere $|\boldsymbol{r}-\boldsymbol{R}_\nu| \leq S_\nu$ by the function

$$\tilde{H}_{L\nu}(\boldsymbol{r}-\boldsymbol{R}_\nu) = \\ = \mathrm{i}^l\, \tilde{h}_{l\nu}(|\boldsymbol{r}-\boldsymbol{R}_\nu|)\, Y_L(\widehat{\boldsymbol{r}-\boldsymbol{R}_\nu}) \quad (1\text{-}252)$$

where $\tilde{h}_{l\nu}(|\boldsymbol{r}-\boldsymbol{R}_\nu|)$ is a regular solution of the radial Schrödinger equation (1-92) inside the atomic sphere labelled ν satisfying boundary conditions

$$\frac{\mathrm{d}^n}{\mathrm{d}r^n} [\tilde{h}_{l\nu}(|\boldsymbol{r}-\boldsymbol{R}_\nu|) - \\ - \varkappa^{l+1} h_l^+(\varkappa|\boldsymbol{r}-\boldsymbol{R}_\nu|)]_{|\boldsymbol{r}-\boldsymbol{R}_\nu|=S_\nu} = 0\,, \\ n = 0, 1 \quad (1\text{-}253)$$

which makes the function continuous and differentiable across the spherical surface

$|\boldsymbol{r}-\boldsymbol{R}_\nu| = S_\nu$. The eigenvalues resulting from the head augmentation we denote by $\varepsilon_{l\nu}^{(H)}$. The tail of the spherical wave centered at $\boldsymbol{R}_\nu$ is replaced in the atomic sphere at $\boldsymbol{R}_\mu (\boldsymbol{R}_\mu \neq \boldsymbol{R}_\nu)$, $|\boldsymbol{r}-\boldsymbol{R}_\mu| \leq S_\mu$, by

$$\tilde{H}_{L\nu}(\boldsymbol{r}-\boldsymbol{R}_\nu) = = \sum_{L'} \tilde{J}_{L'\mu}(\boldsymbol{r}-\boldsymbol{R}_\mu)\, B_{L'L}(\boldsymbol{R}_\mu - \boldsymbol{R}_\nu) \qquad (1\text{-}254)$$

where

$$\tilde{J}_{L'\mu}(\boldsymbol{r}-\boldsymbol{R}_\mu) = = \mathrm{i}^l \tilde{j}_{l'\mu}(|\boldsymbol{r}-\boldsymbol{R}_\mu|)\, Y_{L'}(\widehat{\boldsymbol{r}-\boldsymbol{R}_\mu}) \qquad (1\text{-}255)$$

$\tilde{j}_{l\mu}(|\boldsymbol{r}-\boldsymbol{R}_\mu|)$ is the regular solution of the radial Schrödinger equation appropriate to the atomic sphere centered at $\boldsymbol{R}_\mu$ satisfying boundary conditions

$$\frac{\mathrm{d}^n}{\mathrm{d}r^n} [\tilde{j}_{l\mu}(|\boldsymbol{r}-\boldsymbol{R}_\mu|) - - \varkappa^{-l} j_l(\varkappa |\boldsymbol{r}-\boldsymbol{R}_\mu|)]_{|\boldsymbol{r}-\boldsymbol{R}_\mu| = S_\mu} = 0\,, \quad n = 0, 1 \qquad (1\text{-}256)$$

which makes the function continuous and differentiable across the spherical surface $|\boldsymbol{r}-\boldsymbol{R}_\mu| = S_\mu$. The eigenvalues resulting from the tail augmentation are denoted by $\varepsilon_{l\mu}^{(J)}$. Note that $\tilde{h}_{l\nu}(r)$ and $\tilde{j}_{l\nu}(r)$ are solutions of the same radial Schrödinger equation; they differ only in normalization and energy. The four functions $h_l^+(\varkappa r)$, $\tilde{h}_l(r)$, $j_l(\varkappa r)$, and $\tilde{j}_l(r)$ are compared in Fig. 1-16.

We may pause here and consider augmentation in some more detail by furnishing a connection of the ASW's with ϕ and $\dot{\phi}$. Fig. 1-17 illustrates qualitatively two augmented orbitals centered on two different sites. The tail augmentation achieved by the functions $\tilde{J}_{L\mu}$ is a feature not possessed by an otherwise perfectly chosen basis orbital that is a solution of Schrödinger's equation for one given atom or atomic sphere. Consider now the state of a homonuclear, diatomic molecule with the two functions depicted in Fig. 1-17. To describe this we form two different linear combinations: bonding and antibonding. Ignoring normalization and suppressing angular momentum indices in the notation, we express the bonding orbital as

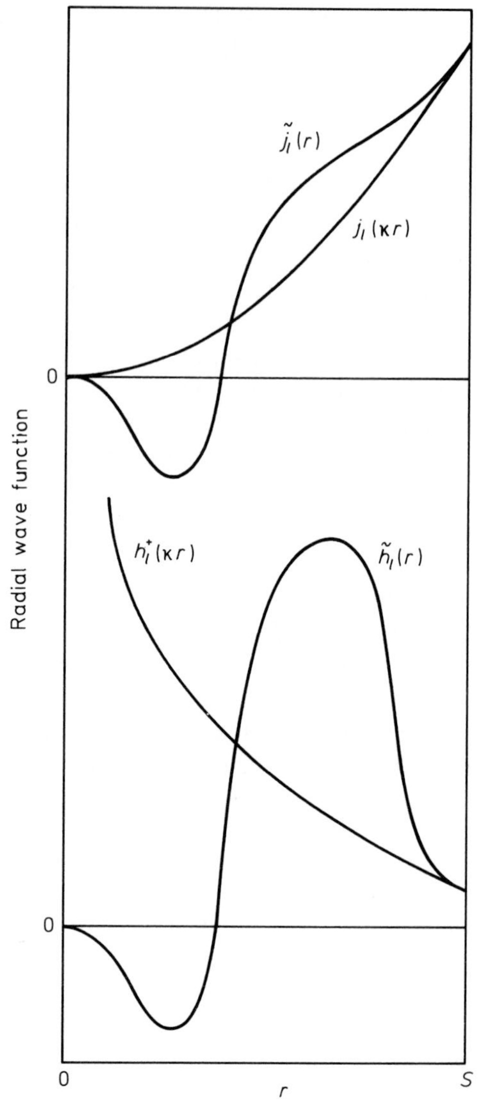

Figure 1-16. Schematic representation of the spherical Hankel and Bessel functions, $h_l^+(\varkappa r)$ and $j_l(\varkappa r)$, and the corresponding augmented functions, $\tilde{h}_l(r)$ and $\tilde{j}_l(r)$, all for $l = 2$ as given by Williams et al. (1979).

$$\boldsymbol{\Phi}_\mathrm{B}(\boldsymbol{r}) = \tilde{\chi}(\boldsymbol{r}) + \tilde{\chi}(\boldsymbol{r}-\boldsymbol{R}) \qquad (1\text{-}257)$$

where the parts on the right-hand side are meant to be those shown in Fig. 1-17. $\boldsymbol{\Phi}_\mathrm{B}$

has zero slope between the two nuclei and hence a finite charge density there which supplies the bonding. The antibonding orbital is

$$\boldsymbol{\Phi}_{\mathrm{A}}(\boldsymbol{r}) = \tilde{\chi}(\boldsymbol{r}) - \tilde{\chi}(\boldsymbol{r}-\boldsymbol{R}) \tag{1-258}$$

having a node between the two nuclei and hence zero charge density there. If the energy of the bonding orbital is ε_{B} and that of the antibonding is ε_{A}, then $\varepsilon_{\mathrm{B}} < \varepsilon_{\mathrm{A}}$. Let us next express the head and the tail, at the left nucleus in Fig. 1-17, say, by $\boldsymbol{\Phi}_{\mathrm{B}}$ and $\boldsymbol{\Phi}_{\mathrm{A}}$; we see trivially that

$$\tilde{\chi}(\boldsymbol{r}) = \tfrac{1}{2}\,[\boldsymbol{\Phi}_{\mathrm{B}}(\boldsymbol{r}) + \boldsymbol{\Phi}_{\mathrm{A}}(\boldsymbol{r})] \tag{1-259}$$

and

$$\tilde{\chi}(\boldsymbol{r}-\boldsymbol{R}) = \tfrac{1}{2}\,[\boldsymbol{\Phi}_{\mathrm{B}}(\boldsymbol{r}) - \boldsymbol{\Phi}_{\mathrm{A}}(\boldsymbol{r})] \tag{1-260}$$

We may use a single-site expansion for $\boldsymbol{\Phi}_{\mathrm{B}}$ and $\boldsymbol{\Phi}_{\mathrm{A}}$ in terms of ϕ and $\dot{\phi}$ (Eq. (1-167)) and simplify by means of the approximate relation between energy and ω (Eq. (1-172)). The result is

$$\tilde{\chi}(\boldsymbol{r}) = \phi(\boldsymbol{r}) + \tfrac{1}{2}\,(\varepsilon_{\mathrm{B}}+\varepsilon_{\mathrm{A}})\,\dot{\phi}(\boldsymbol{r}) \tag{1-261}$$

and

$$\tilde{\chi}(\boldsymbol{r}-\boldsymbol{R}) = \tfrac{1}{2}\,(\varepsilon_{\mathrm{B}}-\varepsilon_{\mathrm{A}})\,\dot{\phi}(\boldsymbol{r}) \tag{1-262}$$

where we count ε_{B} and ε_{A} from E_ν. This is the result of the digression: although the augmentation procedures underlying the ASW and LMTO methods differ in an essential way, the augmented tail is proportional to $\dot{\phi}(\boldsymbol{r})$ whereas the head is proportional to a linear combination of $\phi(\boldsymbol{r})$ and $\dot{\phi}(\boldsymbol{r})$. We now complete the description of the ASW method.

The augmented spherical waves are now defined in all regions of the polyatomic system; they are energy independent, continuous, and continuously differentiable. As before, we form a Bloch function, $\chi_{L\nu\boldsymbol{k}}(\boldsymbol{r})$, which we use in the variational procedure outlined by Eqs. (1-155) and (1-158); writing an arbitrary translation as a Bravais vector, $\boldsymbol{R}$, plus a basis vector, $\boldsymbol{\tau}_\nu$,

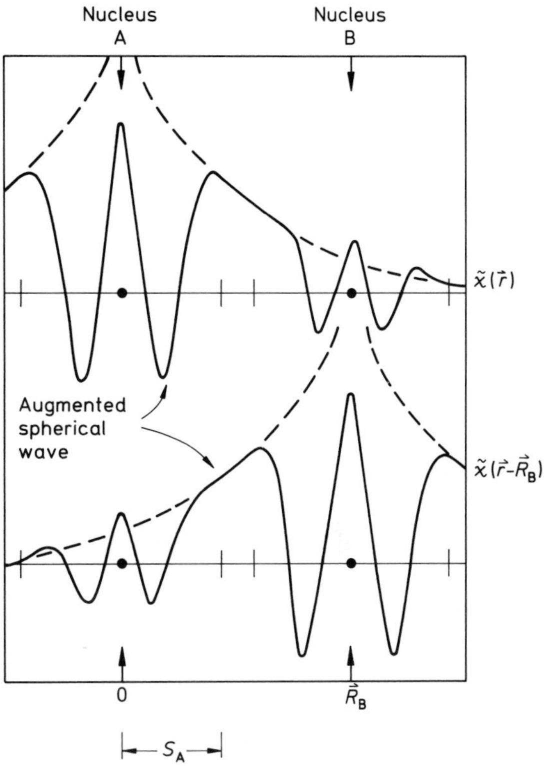

Figure 1-17. Schematic diagram of two ASW's centered at different sites.

$$\boldsymbol{R}_\nu = \boldsymbol{R} + \boldsymbol{\tau}_\nu \tag{1-263}$$

we define

$$\chi_{L\nu\boldsymbol{k}}(\boldsymbol{r}) = \sum_{\boldsymbol{R}} \mathrm{e}^{\mathrm{i}\,\boldsymbol{k}\cdot\boldsymbol{R}}\,\tilde{H}_{L\nu}(\boldsymbol{r}-\boldsymbol{\tau}_\nu-\boldsymbol{R}) \tag{1-264}$$

We are now in the position to determine the elements of the secular matrix which we denote as

$$\sum_{L'\nu'} \{\langle \nu\,\tilde{L}|\mathscr{H}|\tilde{L}'\,\nu'\rangle - \varepsilon_{\boldsymbol{k}}\langle \nu\,\tilde{L}|\tilde{L}'\,\nu'\rangle\}\cdot \cdot\, a_{L'\nu'}(\boldsymbol{k}) = 0 \tag{1-265}$$

where $a_{L\nu}(\boldsymbol{k})$ are the expansion coefficients that determine

$$\psi_{\boldsymbol{k}}(\boldsymbol{r}) = \sum_{L\nu} a_{L\nu}(\boldsymbol{k})\,\chi_{L\nu\boldsymbol{k}}(\boldsymbol{r}) \tag{1-266}$$

Furthermore, $\mathscr{H} = -\nabla^2 + V(\boldsymbol{r})$. The brackets $\langle \ldots \rangle$ indicate integration over all space, e.g.,

$$\langle \nu \tilde{L} | \tilde{L}' \nu' \rangle = = \int d^3 r \, \tilde{H}_L^* (\boldsymbol{r} - \boldsymbol{R}_\nu) \, \tilde{H}_{L'} (\boldsymbol{r} - \boldsymbol{R}_{\nu'}) \tag{1-267}$$

and the tilde denotes augmentation.

We eliminate the interstitial region by the ASA described before and can thus justify to approximate the integrals by a sum of contributions from each sphere,

$$\langle \nu \tilde{L} | \mathscr{H} | \tilde{L}' \nu' \rangle \simeq \sum_{\nu''} \langle \nu \tilde{L} | \mathscr{H} | \tilde{L}' \nu' \rangle_{\nu''} \tag{1-268}$$

where $\langle \ldots \rangle_{\nu''}$ indicates integration over the sphere centered at $\boldsymbol{R}_{\nu''}$. Although for actual calculations this approximation is never used, it nonetheless is quite instructive and leads to interpretable quantities, just as the KKR-ASA clarifies the LMTO-secular equation, as we hope we were able to demonstrate. Let us postpone this discussion, however, and proceed differently because with little additional effort we can include in the ASW secular equation terms known as "combined correction terms" in the LMTO method. We will then come back to this point and simplify the theory conceptually by exploring the ASW version of the KKR-ASA method.

Taking the effective potential to be zero in the interstitial region (which we are free to do if we later eliminate it) allows us to write the matrix element as follows

$$\langle \nu \tilde{L} | \mathscr{H} | \tilde{L}' \nu' \rangle = \langle \nu L | \mathscr{H}_0 | L' \nu' \rangle + + \sum_{\nu''} \{ \langle \nu \tilde{L} | \mathscr{H} | \tilde{L}' \nu' \rangle_{\nu''} - \langle \nu L | \mathscr{H}_0 | L' \nu' \rangle_{\nu''} \} \tag{1-269}$$

where $\mathscr{H}_0 = -\nabla^2$ denotes the free-particle Hamiltonian. In other words, we first construct matrix elements of $\mathscr{H}_0$ using unaugmented spherical waves everywhere; then we replace the contributions from the intraatomic regions by integrations over the full Hamiltonian and ASW's. The reason for this manipulation is that the integrations over the atomic spheres are performed using spherical-harmonic expansions, and the expansion of the differences in $\{\ldots\}$ in Eq. (1-269) converges much more rapidly than does the corresponding expansion of $\langle \nu \tilde{L} | \mathscr{H} | \tilde{L}' \nu' \rangle_{\nu''}$ alone. This improved l convergence represents a gain both in efficiency and in accuracy.

In addition, we should exploit the fact that in all integrals in Eq. (1-269) the states occurring are eigenfunctions of the relevant Hamiltonian, e.g.,

$$\langle \nu L | \mathscr{H}_0 | L' \nu' \rangle = \varkappa^2 \langle \nu L | L' \nu' \rangle \tag{1-270}$$

Furthermore, the integrals over the atomic spheres are of the three basic types previously discussed: one-center, two-center, or three-center terms. One-center contributions are those in which in Eq. (1-269) $\nu = \nu' = \nu''$; in this case the integral involving augmented functions is

$$\langle \nu'' \tilde{L} | \mathscr{H} | \tilde{L}' \nu' \rangle_{\nu''} = = \varepsilon_{l\nu}^{(H)} \langle \tilde{H}_{L\nu} | \tilde{H}_{L\nu} \rangle_\nu \, \delta_{LL'} \tag{1-271}$$

Two-center contributions are those in which $\nu = \nu'' \neq \nu'$ or $\nu \neq \nu'' = \nu'$; in this case

$$\langle \nu'' \tilde{L} | \mathscr{H} | \tilde{L}' \nu' \rangle_{\nu''} = = \varepsilon_{l\nu}^{(J)} \langle \tilde{H}_{L\nu''} | J_{L\nu''} \rangle_{\nu''} \, B_{LL'} (\boldsymbol{R}_{\nu''} - \boldsymbol{R}_{\nu'}) \tag{1-272}$$

and

$$\langle \nu \tilde{L} | \mathscr{H} | \tilde{L}' \nu' \rangle_{\nu''} = = B_{LL'}^{+} (\boldsymbol{R}_\nu - \boldsymbol{R}_{\nu''}) \langle \tilde{J}_{L'\nu''} | \tilde{H}_{L'\nu''} \rangle_{\nu''} \, \varepsilon_{l'\nu''}^{(H)} \tag{1-273}$$

where

$$B_{LL'}^{+} (\boldsymbol{R}_\nu - \boldsymbol{R}_{\nu''}) \equiv B_{L'L}^{*} (\boldsymbol{R}_{\nu''} - \boldsymbol{R}_\nu) \tag{1-274}$$

Three-center contributions are those in which $\nu \neq \nu'' \neq \nu'$; now one obtains

$$\langle \nu \tilde{L} | \mathscr{H} | \tilde{L}' \nu' \rangle_{\nu''} = = \sum_{L''} B_{LL''}^{+} (\boldsymbol{R}_\nu - \boldsymbol{R}_{\nu''}) \, \varepsilon_{l''\nu''}^{(J)} \langle \tilde{J}_{L''\nu''} | \tilde{J}_{L''\nu''} \rangle_{\nu''} \cdot \cdot B_{L''L'} (\boldsymbol{R}_{\nu''} - \boldsymbol{R}_{\nu'}) \tag{1-275}$$

The evaluation of the integrals in Eq. (1-269) involving unaugmented functions proceeds in the same way, except for the one-center term which combines with the integral over all space to give

$$\langle \nu L | L' \nu \rangle - \langle \nu L | L' \nu \rangle_\nu = \delta_{LL'} \langle H_L | H_L \rangle'_\nu \tag{1-276}$$

where

$$\langle H_L | H_L \rangle'_\nu = = \varkappa^{2l+2} \int_{S_\nu}^{\infty} r^2 \, \mathrm{d}r \, |h_l^+(\varkappa r)|^2 \tag{1-277}$$

i.e., the integral excludes the (diverging) part in the interior of the sphere. Finally, the integral over all space in Eq. (1-269) involving spherical waves centered at different nuclei is given by an energy derivative of the corresponding structure constant; i.e., for $\nu \neq \nu'$

$$\langle \nu L | L' \nu' \rangle = = \dot{B}_{LL'}(\boldsymbol{R}_\nu - \boldsymbol{R}_{\nu'}) \equiv \frac{\mathrm{d}}{\mathrm{d}\varkappa^2} B_{LL'}(\boldsymbol{R}_\nu - \boldsymbol{R}_{\nu'}) \tag{1-278}$$

To derive this equation one first differentiates the free-particle Schrödinger equation with respect to energy (like Eq. (1-163))

$$(\nabla^2 + \varkappa^2) \dot{H}_L(\boldsymbol{r} - \boldsymbol{R}_{\nu'}) + H_L(\boldsymbol{r} - \boldsymbol{R}_{\nu'}) = 0 \tag{1-279}$$

Then multiplying Eq. (1-279) by $H_L^*(\boldsymbol{r} - \boldsymbol{R}_\nu)$, subtracting the result from $\dot{H}_{L'}(\boldsymbol{r} - \boldsymbol{R}_{\nu'}) \cdot (\nabla^2 + \varkappa^2) H_L^*(\boldsymbol{r} - \boldsymbol{R}_\nu) = 0$, and integrating over all space provides a relationship betwen the desired integral and

$$\int \mathrm{d}^3 r \, \{\dot{H}_{L'}(\boldsymbol{r} - \boldsymbol{R}_{\nu'}) \nabla^2 H_L^*(\boldsymbol{r} - \boldsymbol{R}_\nu) - - H_L^*(\boldsymbol{r} - \boldsymbol{R}_\nu) \nabla^2 H_{L'}(\boldsymbol{r} - \boldsymbol{R}_{\nu'})\}$$

After integrating by parts, the remaining quantity becomes an integral over small spherical surfaces containing the points of singularity ($\boldsymbol{r} = \boldsymbol{R}_\nu$ and $\boldsymbol{r} = \boldsymbol{R}_{\nu'}$). The introduction of the structure-constant expansion (Eq. (1-248)) permits the surface integrals to be easily evaluated, giving us Eq. (1-278). The expression for the individual integrals which enter the representation for the Hamiltonian matrix elements can now be combined to complete the specification of the secular matrix:

$$\begin{aligned}
&\langle \nu \tilde{L} | \mathscr{H} | \tilde{L}' \nu' \rangle = \\
&= [\varepsilon_{l\nu}^{(H)} \langle \tilde{H}_{L\nu} | \tilde{H}_{L\nu} \rangle_\nu + \varkappa^2 \langle H_L | H_L \rangle'_\nu] \, \delta_{\nu\nu'} \delta_{LL'} \\
&+ \varkappa^2 \dot{B}_{LL'}(\boldsymbol{\tau}_\nu - \boldsymbol{\tau}_{\nu'}, \boldsymbol{k}) + \\
&+ B^+_{LL'}(\boldsymbol{\tau}_\nu - \boldsymbol{\tau}_{\nu'}, \boldsymbol{k}) \cdot \\
&\quad \cdot [\varepsilon_{l'\nu'}^{(H)} \langle \tilde{J}_{L'\nu'} | \tilde{H}_{L'\nu'} \rangle_{\nu'} - \varkappa^2 \langle J_{L'} | H_{L'} \rangle_{\nu'}] \\
&+ [\varepsilon_{l\nu}^{(J)} \langle \tilde{H}_{L\nu} | \tilde{J}_{L\nu} \rangle_\nu - \varkappa^2 \langle H_L | J_L \rangle_\nu] \cdot \\
&\quad \cdot B_{LL'}(\boldsymbol{\tau}_\nu - \boldsymbol{\tau}_{\nu'}, \boldsymbol{k}) + \\
&+ \sum_{\nu''} \sum_{L''} B^+_{LL''}(\boldsymbol{\tau}_\nu - \boldsymbol{\tau}_{\nu''}, \boldsymbol{k}) \cdot \\
&\quad \cdot [\varepsilon_{l''\nu''}^{(J)} \langle \tilde{J}_{L''\nu''} | \tilde{J}_{L''\nu''} \rangle_{\nu''} - \varkappa^2 \langle J_{L''} | J_{L''} \rangle_{\nu''}] \\
&\quad \cdot B_{L''L'}(\boldsymbol{\tau}_{\nu''} - \boldsymbol{\tau}_{\nu'}, \boldsymbol{k})
\end{aligned} \tag{1-280}$$

where translational symmetry has been exploited; i.e., the sum indicated in Eq. (1-264) has been taken to yield the structure constants appropriate to energy-band theory,

$$B_{LL'}(\boldsymbol{\tau}_\nu - \boldsymbol{\tau}_{\nu'}, \boldsymbol{k}) = = \sum_{\boldsymbol{R}} \mathrm{e}^{\mathrm{i}\boldsymbol{k} \cdot \boldsymbol{R}} B_{LL'}(\boldsymbol{\tau}_\nu - \boldsymbol{\tau}_{\nu'} - \boldsymbol{R}) \tag{1-281}$$

Here $B_{LL'}(\boldsymbol{x})$ is defined to be zero when $\boldsymbol{x}$ vanishes and the basis vectors $\boldsymbol{\tau}$ are obtained from Eq. (1-263). The required structure constants are those given by Eqs. (1-145) to (1-148) of the KKR theory, except for a factor of $\varkappa^{l+l'}$, with $\varepsilon_0 = \varkappa^2$, and with one more small exception: When the energy parameter $\varkappa^2$ is negative, the cosine in Eq. (1-136) together with the Neumann function must be replaced by an exponential and a Hankel function, respectively. The net result of this difference is that $D^{(3)}$ (Eq. (1-149)) is given by

$$D^{(3)} = -2\sqrt{q_0} \left[\mathrm{e}^{-\varkappa^2/q_0} + \left(\frac{-\pi \varkappa^2}{q_0} \right)^{1/2} + + \frac{2\varkappa^2}{q_0} \sum_{n=0}^{\infty} \frac{\varkappa^2/q_0}{n!(2n+1)} \right] \tag{1-282}$$

The normalization matrix $\langle \nu \tilde{L} | \tilde{L}' \nu' \rangle$, is obtained from Eq. (1-280) by setting the energies $\varkappa^2$, $\varepsilon_{l\nu}^{(H)}$, and $\varepsilon_{l\nu}^{(J)}$ equal to unity. Furthermore, a simple manipulation of the Schrödinger equation for the augmented Bessel and Hankel functions, $\tilde{j}_l$ and $\tilde{h}_l^+$, shows that

$$\langle \tilde{H}_{L\nu} | \tilde{J}_{L\nu} \rangle_\nu = \langle \tilde{J}_{L\nu} | \tilde{H}_{L\nu} \rangle_\nu = = (\varepsilon_{l\nu}^{(H)} - \varepsilon_{l\nu}^{(J)})^{-1} \tag{1-283}$$

so that like LMTO's the ASW secular matrix involves only four potential-dependent quantities (for each l and ν): $\varepsilon_{l\nu}^{(H)}$, $\varepsilon_{l\nu}^{(J)}$, $\langle \tilde{H}_{L\nu} | \tilde{H}_{L\nu} \rangle_\nu$, and $\langle \tilde{J}_{L\nu} | \tilde{J}_{L\nu} \rangle_\nu$.

The integrals involving augmented functions require one-dimensional (radial) numerical integrations, and those involving unaugmented spherical Bessel and Hankel functions can be found in standard mathematical texts. The form of the secular equation is similar to the LMTO equation (1-220), but the more precise integration embodied in Eq. (1-269) supplies extra terms which are comparable to the "combined correction terms" mentioned earlier. They lead to fast convergence because the difference converges faster than the separate terms in the brackets of the sum in Eq. (1-280). Conceptually simpler, however, is an approximate form of the ASW equations. Although it is never used in numerical calculations, this form can be linked with KKR theory and in this way converted to a combination of physically transparent terms. This parallels closely Andersen's KKR-ASA treatment.

KKR from ASW

We begin by simplifying the ASW equations. Referring to Eq. (1-263) and (1-264), we first ignore the basis, $\boldsymbol{\tau}_\nu$. Then, using the ASA and omitting the envelope function (except for the boundary condition which it supplies through Eqs. (1-253) and (1-256)), we may write the function corresponding to Eq. (1-264) as

$$\chi_{L\boldsymbol{k}}(\boldsymbol{r}) = \tilde{H}_L(\boldsymbol{r}) + \sum_{L'} \tilde{J}_{L'}(\boldsymbol{r})\, B_{L'L}(\boldsymbol{k}) \tag{1-284}$$

where the quantities $\tilde{H}$ and $\tilde{J}$ are defined to be nonzero in the atomic sphere only. $B_{L'L}(\boldsymbol{k})$ is the structure constant corresponding to Eq. (1-281). The secular equation is then written down by inspection:

$$\begin{aligned}\langle \tilde{L} | \mathscr{H} | \tilde{L}' \rangle - \varepsilon \langle \tilde{L} | \tilde{L}' \rangle &= \\ = \delta_{LL'} (\varepsilon_l^{(H)} - \varepsilon) \langle \tilde{H}_L | \tilde{H}_L \rangle &+ \\ + B_{LL'}^+(\boldsymbol{k}) (\varepsilon_{l'}^{(H)} - \varepsilon) \langle \tilde{J}_{L'} | \tilde{H}_{L'} \rangle &+ \\ + (\varepsilon_l^{(J)} - \varepsilon) \langle \tilde{H}_L | \tilde{J}_L \rangle B_{LL'}(\boldsymbol{k}) &+ \\ + \sum_{L''} B_{LL''}^+(\boldsymbol{k}) (\varepsilon_{l''}^{(J)} - \varepsilon) \langle \tilde{J}_{L''} | \tilde{J}_{L''} \rangle B_{L''L'}(\boldsymbol{k})\end{aligned} \tag{1-285}$$

It should be compared with Eq. (1-280) and would have been obtained by the preceeding steps had we used the simple expression for the matrix element given by Eq. (1-268). We will come back to Eq. (1-285) but first repeat what was called KKR-ASA. This time we will construct *energy-dependent* orbitals, $\chi_l(\varepsilon, r)$, but instead of using solutions of Laplace's equation to formulate the boundary conditions, we will use solutions of the wave equation (1-245), initially employing a fixed value for the energy parameter $\varkappa = \sqrt{\varepsilon_0}$. Exactly as before (Eq. (1-221)) one easily establishes that the orbital

$$\chi_L(\varepsilon, \boldsymbol{r}) = \phi_l(\varepsilon, S) \begin{cases} \dfrac{\phi_l(\varepsilon, r)}{\phi_l(\varepsilon, S)} Y_L(\hat{\boldsymbol{r}}) - \dfrac{D_l(\varepsilon) - D\{h_l^+\}}{D\{j_l\} - D\{h_l^+\}} \dfrac{j_l(\varkappa r)}{j_l(\varkappa S)} Y_L(\hat{\boldsymbol{r}}), & r \le S \\ -\dfrac{D_l(\varepsilon) - D\{j_l\}}{D\{j_l\} - D\{h_l^+\}} \dfrac{h_l^+(\varkappa r)}{h_l^+(\varkappa S)} Y_L(\hat{\boldsymbol{r}}), & r \ge S \end{cases} \tag{1-286}$$

is continuous and differentiable everywhere, $\phi_l(\varepsilon, r)$ is a solution of energy ε to the radial Schrödinger equation inside the atomic sphere, and $D_l(\varepsilon)$ is its logarithmic derivative, whereas

$$D\{h_l^+\} = x \left. \frac{h_l^{+\prime}(x)}{h_l^+(x)} \right|_{x=\varkappa S} \tag{1-287}$$

and

$$D\{j_l\} = x \left. \frac{j_l'(x)}{j_l(x)} \right|_{x=\varkappa S} \tag{1-288}$$

h_l^+ and j_l are the spherical Hankel and Bessel functions, respectively, and $\varkappa = \sqrt{\varepsilon_0}$. Now we need not write down any more details but can repeat the KKR-ASA derivation that follows Eq. (1-221): Forming a Bloch sum with the orbital given by Eq. (1-286) and requiring that the tails cancel exactly, we obtain

$$\sum_{L'} \{B_{LL'}(\boldsymbol{k}) + P_l(\varepsilon)\, \delta_{LL'}\}\, M_{l'}(\varepsilon)\, a_{L'} = 0 \tag{1-289}$$

which replaces Eq. (1-228); but P_l and M_l are now given by

$$P_l(\varepsilon) = \frac{\varkappa^{2l+1}\, h_l^+(\varkappa S)}{j_l(\varkappa S)} \cdot \frac{D_l(\varepsilon) - D\{h_l^+\}}{D_l(\varepsilon) - D\{j_l\}} \tag{1-290}$$

and

$$M_l(\varepsilon) = -\,\mathrm{i}^{-l}\, \varkappa^{-l}\, j_l(\varkappa S)\, \phi_l(\varepsilon, S) \cdot [D_l(\varepsilon) - D\{j_l\}] \tag{1-291}$$

which, except for a common factor, can easily be seen to approach the analogous terms given by Eqs. (1-224) and (1-225), when $\varkappa \to 0$. The KKR-ASA equation thus becomes

$$\det \{B_{LL'}(\boldsymbol{k}) + P_l(\varepsilon)\, \delta_{LL'}\} = 0 \tag{1-292}$$

Obviously, just like the case $\varkappa = 0$, the B's give rise to *canonical bands*. In fact, Fig. 1-12 was obtained using B's rather than S's with a very small value of $\varkappa$. The *exact KKR* equation (Eq. (1-140)) is obtained if we add the further requirement that the energy ε is equal to the energy parameter ε_0; for positive values of ε the Hankel functions must then be replaced by spherical Neumann functions and with the help of Eqs. (1-125) and (1-290), one derives easily

$$P_l(\varepsilon) = \cot \eta_l(\varepsilon) \tag{1-293}$$

where η_l is the scattering phase shift. The structure constants $B_{LL'}$, of course, depend on energy in this case and are equal to the KKR structure constants (except for a factor of $\varkappa^{l+l'}$ which is a matter of definition). Returning to the approximation with a constant energy parameter, the potential function given by Eq. (1-290) must still be linearized, the potential parameters appropriate for the ASW method extracted, and the mathematical connection between the ASW and this version of the KKR-ASA formulations established.

The qualitative features of the potential function defined by Eq. (1-290) are again like those shown in Fig. 1-15, except that the singularity occurs for an energy where $D_l(\varepsilon) = D\{j_l\}$ which by definition is the Bessel energy $\varepsilon_l^{(J)}$, and the zero occurs at the Hankel energy $\varepsilon_l^{(H)}$. A useful linearization (Sticht, 1986) is obtained explicitly if the logarithmic derivative, $D_l(\varepsilon)$, is expanded in a Taylor series. In the numerator of $P_l(\varepsilon)$ we use the exact relation

$$\frac{\mathrm{d}}{\mathrm{d}\varepsilon} D_l(\varepsilon)\big|_{\varepsilon=\varepsilon_l^{(H)}} = -\frac{\langle \tilde{h}_l | \tilde{h}_l \rangle}{\varkappa^{2l+2}\, S\, h_l^+(\varkappa S)^2} \tag{1-294}$$

which is obtained from Eq. (1-76) which relates the norm $\langle \tilde{h}_l | \tilde{h}_l \rangle$ of a radial wave function, $\tilde{h}_l(r)$, to its logarithmic derivative. Using a finite-difference approximation for $\frac{\mathrm{d}}{\mathrm{d}\varepsilon} D_l(\varepsilon)$ in the denominator and the Wronskian relation for Bessel and Hankel functions, we obtain

$$P_l(\varepsilon) \approx \Delta_l \langle \tilde{h}_l | \tilde{h}_l \rangle \frac{\varepsilon - \varepsilon_l^{(H)}}{\varepsilon - \varepsilon_l^{(J)}} \tag{1-295}$$

where

$$\Delta_l = \varepsilon_l^{(H)} - \varepsilon_l^{(J)} = \langle \tilde{j}_l | \tilde{h}_l \rangle^{-1} \tag{1-296}$$

(compare with Eq. (1-283)). Thus, the potential parameter C_l of the LMTO method becomes the Hankel energy $\varepsilon_l^{(H)}$ which thus gives the center of a band corresponding to angular momentum l; the parameter V_l becomes the Bessel energy $\varepsilon_l^{(J)}$ and the distortion factor γ_l is $\Delta_l^{-1} \langle \tilde{h}_l | \tilde{h}_l \rangle^{-1}$, the scaling factor being given by Eq. (1-236). The interpretation is exactly the same as that following Eq. (1-238), which we therefore need not repeat here.

To establish the connection between the simplified ASW secular Eq. (1-285), for which we use the notation $\mathbf{H} - \varepsilon \mathbf{O}$ again, we define the matrix $\mathbf{\Lambda}$ with elements

$$\Lambda_{LL'} = B_{LL'}(\boldsymbol{k}) + \Delta_l \langle \tilde{h}_l | \tilde{h}_l \rangle \frac{\varepsilon - \varepsilon_l^{(H)}}{\varepsilon - \varepsilon_l^{(J)}} \delta_{LL'} \tag{1-297}$$

which is the KKR matrix with the approximate form of the potential function $P_l(\varepsilon)$ given by Eq. (1-295). Then with the further definition of the matrix Y having elements

$$y_{LL'} = \Delta_l^{-1} (\varepsilon_l^{(J)} - \varepsilon) \, \delta_{LL'} + B_{LL'}^{+} \cdot \cdot \langle \tilde{j}_{l'} | \tilde{j}_{l'} \rangle (\varepsilon_l^{(J)} - \varepsilon) \tag{1-298}$$

we see by carrying out a simple matrix multiplication that

$$\mathbf{H} - \varepsilon \mathbf{O} = Y \mathbf{\Lambda} \tag{1-299}$$

provided the relation

$$\langle \tilde{h}_l | \tilde{h}_l \rangle \Delta_l^2 \langle \tilde{j}_l | \tilde{j}_l \rangle = 1 \tag{1-300}$$

is approximately satisfied. We indeed find this relation numerically fulfilled when l is not too large so that this version of KKR-ASA is also an approximate factor of the simplified ASW secular equation (1-285).

We close this section with a brief discussion of the construction of the charge density within the ASW and LMTO methods (Williams et al., 1979; Andersen, 1984). It is well known that the Rayleigh-Ritz variational procedure gives rather good estimates of the energies but does poorly on the wave functions. We therefore do not use the coefficients $a_{L\nu}$ that are eigenvectors of the secular equation (1-265) together with the basis functions $\chi_{L\nu}$ (Eq. (1-264)) directly in constructing the electron density, but rather emphasize the role of the most reliable aspect of the Rayleigh-Ritz procedure: the variationally determined eigenenergies. To do this we take the electron density to have the form it would have in an accurate (KKR or APW) calculation, i.e.,

$$n_\nu(r_\nu) = n_\nu^{(c)}(r_\nu) + \sum_l \int^{\varepsilon_F} \mathrm{d}\varepsilon \, N_{l\nu}(\varepsilon) \, R_{l\nu}^2(\varepsilon, r_\nu) \tag{1-301}$$

where $n_\nu(r_\nu)$ is the spherical average of the electron density in the νth atomic sphere; $r_\nu = |\boldsymbol{r} - \boldsymbol{R}_\nu|$; $n_\nu^{(c)}(r_\nu)$ is the contribution of the core levels; $R_{l\nu}(\varepsilon, r_\nu)$ is a normalized solution of the Schrödinger equation appropriate to the νth sphere; and $N_{l\nu}(\varepsilon)$ is the valence-electron state density decomposed according to angular momentum and atomic site. The energy integration extends over the occupied part of the valence band. Although it is clear that the *total* valence state density, $N(\varepsilon)$ (Eq. (1-47)), requires only a knowledge of the eigenenergies, the *partial* state densities, $N_{l\nu}(\varepsilon)$, appearing in our representation of the electron density require the decomposition of the norm of each eigenstate, i.e.,

$$N_{l\nu}(\varepsilon) = \frac{1}{\Omega_{\mathrm{BZ}}} \sum_n \int_{\mathrm{BZ}} \mathrm{d}^3k \, \delta(\varepsilon - \varepsilon_{n\boldsymbol{k}}) \, q_{l\nu n}(\boldsymbol{k}) \tag{1-302}$$

where the quantities $q_{l\nu n}(\boldsymbol{k})$ (to be determined in due course) are the angular momentum l and site ν decompositions of the single electron norm associated with each eigenstate,

$$\sum_{l\nu} q_{l\nu n}(\boldsymbol{k}) = 1 \tag{1-303}$$

so that

$$N(\varepsilon) = \sum_{l\nu} N_{l\nu}(\varepsilon) \tag{1-304}$$

(The notation is that introduced in connection with Eq. (1-47).) It might seem therefore that the representation of the electron density in terms of the partial state densities merely exchanges one problem for another, i.e., the l and ν decomposition of each eigenstate requires wave-function information which we wanted to avoid because it is given somewhat unreliably by the Rayleigh-Ritz procedure. Still, the present procedure is warranted because it exploits the fact that the normalization of a radial wave function varies more slowly with energy than the wave function itself. This can be seen if the central relation, which connects the norm with the energy derivative of the logarithmic derivative (Eq. (1-76)) is differentiated with respect to energy; the result is found to involve second energy derivatives of the wave function whose size, which was previously discussed, was found to drop off rather fast (see Fig. 1-10).

The required decomposition of the normalization of each state, $q_{l\nu n}(\boldsymbol{k})$, is obtained in a straightforward manner and involves only a small correction that stems from the overlap of the atomic spheres and the associated problem of the convergence of the spherical-harmonic expansions. With the coefficients $a_{L\nu n}(\boldsymbol{k})$ that diagonalize the secular problem and overlap matrix we know that

$$\sum_{L\nu} \sum_{L'\nu'} a^*_{L\nu n}(\boldsymbol{k}) \langle \nu \tilde{L} | \tilde{L}' \nu' \rangle a_{L'\nu' n}(\boldsymbol{k}) = 1 \tag{1-305}$$

Setting all the energies appearing in Eq. (1-280) ($\varepsilon^{(H)}_{l\nu}$, $\varepsilon^{(J)}_{l\nu}$, $\varkappa^2$) to unity provides an explicit expression for the normalization matrix $\langle \nu \tilde{L} | \tilde{L}' \nu' \rangle$, which when substituted into Eq. (1-305) allows the normalization to be written as a single summation over atomic sites and angular momenta, plus a small quantity $\delta_n(\boldsymbol{k})$:

$$\begin{aligned} 1 = \sum_{L\nu} \{ & a^*_{L\nu n}(\boldsymbol{k}) \langle \tilde{H}_{L\nu} | \tilde{H}_{L\nu} \rangle_\nu \, a_{L\nu n}(\boldsymbol{k}) \\ & + a^*_{L\nu n}(\boldsymbol{k}) \langle \tilde{H}_{L\nu} | \tilde{J}_{L\nu} \rangle_\nu \, A_{L\nu n}(\boldsymbol{k}) \\ & + A^*_{L\nu n}(\boldsymbol{k}) \langle \tilde{J}_{L\nu} | \tilde{H}_{L\nu} \rangle_\nu \, a_{L\nu n}(\boldsymbol{k}) \\ & + A^*_{L\nu n}(\boldsymbol{k}) \langle \tilde{J}_{L\nu} | \tilde{J}_{L\nu} \rangle_\nu \, A_{L\nu n}(\boldsymbol{k}) \} \\ & + \delta_n(\boldsymbol{k}) \end{aligned} \tag{1-306}$$

where the coefficients $A_{L\nu n}(\boldsymbol{k})$ are given by

$$A_{L\nu n}(\boldsymbol{k}) = \\ = \sum_{L'\nu'} B_{LL'}(\boldsymbol{\tau}_\nu - \boldsymbol{\tau}_{\nu'}, \boldsymbol{k}) \, a_{L'\nu' n}(\boldsymbol{k}) \tag{1-307}$$

Thus, if it were not for the quantity $\delta_n(\boldsymbol{k})$, Eq. (1-306) would constitute the required ν and l decomposition of the normalization (see Eqs. (1-303) and (1-304)) and the sum of terms in the curly brackets would be $q_{l\nu n}(\boldsymbol{k})$; $\delta_n(\boldsymbol{k})$ is the exactly computed contribution to the normalization due to unaugmented spherical waves minus the integral over the atomic spheres of the spherical-harmonic expansion of the same quantity. In the ASA and for closely packed systems it is thus rather small – generally of the order of a few percent. There is no unique way to distribute this quantity over the various atomic sites and the angular momenta; an acceptable approximation, however, is to omit it all together and to renormalize the sum over the curly brackets in Eq. (1-306) to unity.

The final step in the construction of the charge density is self-evident if we remember that the energy dependence of the solutions to the radial Schrödinger equation is approximately linear. In the *LMTO method*, $R_{l\nu}(\varepsilon, r_\nu)$ in Eq. (1-301) is therefore expanded in a Taylor series in $\varepsilon - E_\nu$. Defining energy moments as

$$\bar{\lambda}_{l\nu}^{(q)} = \int^{\varepsilon_F} (\varepsilon - E_{\nu l})^q \, N_{l\nu}(\varepsilon) \, d\varepsilon \tag{1-308}$$

one easily sees that the valence charge density (the second term on the right-hand side of Eq. (1-301)) can be written as

$$\begin{aligned} n_\nu^{\text{val}}(r_\nu) = \sum_l \{ & n_{l\nu} R_{l\nu}^2(E_{\nu l}, r_\nu) \\ & + 2\bar{\lambda}_{l\nu}^{(1)} R_{l\nu}(E_{\nu l}, r_\nu) \dot{R}_{l\nu}(E_{\nu l}, r_\nu) \\ & + \bar{\lambda}_{l\nu}^{(2)} [\dot{R}_{l\nu}^2(E_{l\nu}, r_\nu) + R_{l\nu}(E_{\nu l}, r_\nu) \ddot{R}_{l\nu}(E_{\nu l}, r_\nu)] \} \end{aligned} \tag{1-309}$$

where

$$n_{l\nu} = \bar{\lambda}_{l\nu}^{(0)} = \int^{\varepsilon_F} N_{l\nu}(\varepsilon) \, d\varepsilon \tag{1-310}$$

is the partial charge in state l of the νth atomic sphere. Only the first term on the right-hand side of Eq. (1-309) contributes net charge to the sphere. The remaining terms merely redistribute the spectral weight of the charge within that sphere. This is true because integration of Eq. (1-309) over the sphere gives

$$\begin{aligned} & \int_0^{S_\nu} n_\nu^{\text{val}}(r_\nu) \, r_\nu^2 \, dr_\nu \equiv \langle n_\nu^{\text{val}} \rangle = \\ & = \sum_l \{ n_{l\nu} + 2\bar{\lambda}_{l\nu}^{(1)} \langle R_{l\nu} \dot{R}_{l\nu} \rangle + \\ & \qquad + \bar{\lambda}_{l\nu}^{(2)} [\langle \dot{R}_{l\nu}^2 \rangle + \langle R_{l\nu} \ddot{R}_{l\nu} \rangle] \} \\ & = \sum_l n_{l\nu} \end{aligned} \tag{1-311}$$

on account of the orthogonality of $R_{l\nu}$ and $\dot{R}_{l\nu}$ (see Eq. (1-165)) and because

$$\begin{aligned} 0 & = \frac{\partial}{\partial \varepsilon} \langle R_{l\nu}(\varepsilon, r_\nu) \dot{R}_{l\nu}(\varepsilon, r_\nu) \rangle = \\ & = \langle \dot{R}_{l\nu}^2 \rangle + \langle R_{l\nu} \ddot{R}_{l\nu} \rangle \end{aligned} \tag{1-312}$$

In the *ASW method* (where energy derivatives are not used directly) each partial state density is sampled with *two* characteristic energies and corresponding weights. In particular, energy moments are defined in a similar manner to Eq. (1-308),

$$\lambda_{l\nu}^{(q)} = \int^{\varepsilon_F} \varepsilon^q \, N_{l\nu}(\varepsilon) \, d\varepsilon \tag{1-313}$$

Then, for each site ν and angular momentum l, two energies $E_{l\nu}^{(1)}$, $E_{l\nu}^{(2)}$ and two weights $W_{l\nu}^{(1)}$, $W_{l\nu}^{(2)}$ are required to satisfy

$$\sum_{i=1}^{2} W_{l\nu}^{(i)} [E_{l\nu}^{(i)}]^q = \lambda_{l\nu}^{(q)} \tag{1-314}$$

where $q = 0, 1, 2, 3$. For the two pairs of unknown weights and energies there are four equations that can be solved analytically. In Fig. 1-18 we illustrate the relation of the weights and energies to the s and d state densities corresponding to the band structure of Os shown in Fig. 1-14. The valence charge density is now given by the compact representation

$$n_\nu^{\text{val}}(r_\nu) = \sum_l \sum_{i=1}^{2} W_{l\nu}^{(i)} R_{l\nu}^2(E_{l\nu}^{(i)}, r_\nu) \tag{1-315}$$

where the radial functions $R_{l\nu}$ are numerical solutions to the Schrödinger equation (Eq. (1-92)), with energies

$$\varepsilon = E_{l\nu}^{(i)} \tag{1-316}$$

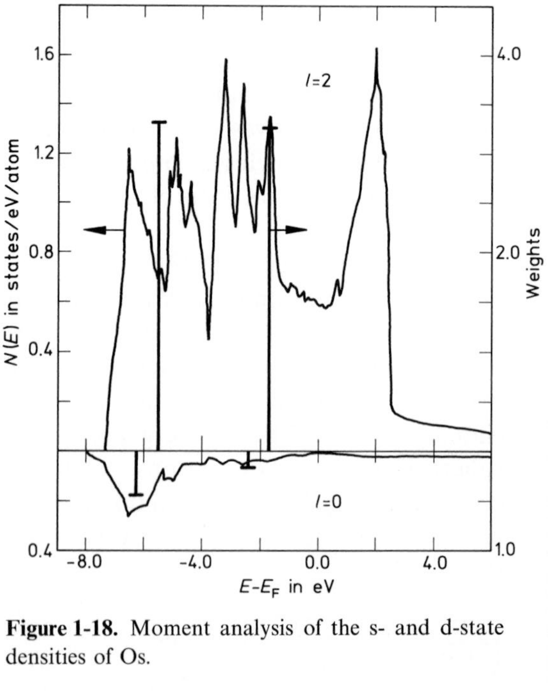

Figure 1-18. Moment analysis of the s- and d-state densities of Os.

obtained from Eqs. (1-314). It is conceptually preferable to think of the environment of a given atom as characterized by *boundary conditions* which the intra-atomic wave functions are obliged to satisfy, rather than by sampling-energies. The effective one-electron potential and the radial Schrödinger equation containing it indeed imply an unambiguous relationship between the energies $E_{l\nu}^{(i)}$ and the logarithmic derivative $D_{l\nu}^{(i)}$ of the corresponding solution at the surface of the atomic sphere, i.e.,

$$D_{l\nu}^{(i)} = \left[r \frac{\mathrm{d}}{\mathrm{d}r} R_{l\nu}(\varepsilon, r) \Big/ R_{l\nu}(\varepsilon, r) \right]_{r = S_\nu, \varepsilon = E_{l\nu}^{(i)}} \tag{1-317}$$

This characterization of the environment in terms of electronic configurations, $\{W_{l\nu}^{(i)}\}$, and boundary conditions, $\{D_{l\nu}^{(i)}\}$, imposed at finite radii, $\{S_\nu\}$, provides a link with intuitive theories of metallic bonding and crystal structure. For instance, the basis of the theory of metallic cohesion of Wigner and Seitz (1955) is the difference between atomic boundary conditions and those characteristic of partially filled energy bands. Brewer's (1967) analysis of crystal-structure preference is based exclusively on electronic configurations, i.e.,

$$n_{l\nu} = \sum_{i=1}^{2} W_{l\nu}^{(i)} \tag{1-318}$$

Other examples will be dealt with in Sec. 1.4. We should stress the rigorous link established in this analysis between these intuitively appealing concepts and ab initio self-consistent field calculations. Furthermore, this analysis also has a practical virtue: it decouples intra- and interatomic self-consistency. In other words, once the $W_{l\nu}^{(i)}$ and $D_{l\nu}^{(i)}$ have been produced by a band calculation, the atomic calculations they specify can be iterated to self-consistency before another band calculation is performed. In this way every band calculation performed in the course of obtaining interatomic self-consistency is based on atomic potentials which are internally self-consistent. This fact substantially reduces the number of band calculations required for total self-consistency. We add that these virtues unfortunately disappear to a large extend if the atomic sphere approximation and the shape approximation for the potential are abandoned.

1.2.5 Relativistic Corrections

The preceding discussion was exclusively based on the Schrödinger equation and non-relativistic quantum mechanics. The spin of the electron was therefore largely ignored except when states were counted invoking the Pauli principle. Spin as a dynamical variable (or observable) does, however, play an important role in determining a wide variety of properties of solids; thus without proper account of the electron spin our theory is incomplete.

One could start by simply adding spin to the Schrödinger theory, i.e., by attaching an additional quantum number s to the state function. s can be "up" or "down", or + or −, but nothing else changes until a magnetic field is turned on, which is known to supply a Zeeman term. The latter can indeed be conveniently obtained in the Schrödinger theory provided one *postulates* (following Feynman) that the kinetic energy operator for an electron is

$$\mathcal{T} = (\boldsymbol{\sigma} \cdot \boldsymbol{p})(\boldsymbol{\sigma} \cdot \boldsymbol{p})/2m \tag{1-319}$$

where $\boldsymbol{\sigma}$ is a vector operator whose components are the Pauli-spin matrices, $\boldsymbol{\sigma}_x$, $\boldsymbol{\sigma}_y$, and $\boldsymbol{\sigma}_z$, and $\boldsymbol{p} = -\mathrm{i}\hbar\nabla$ is the momentum operator. The Pauli matrices are characterized by the commutation relations

$$\begin{aligned} [\boldsymbol{\sigma}_x, \boldsymbol{\sigma}_y] &= 2\mathrm{i}\,\boldsymbol{\sigma}_z, \\ [\boldsymbol{\sigma}_y, \boldsymbol{\sigma}_z] &= 2\mathrm{i}\,\boldsymbol{\sigma}_x, \\ [\boldsymbol{\sigma}_z, \boldsymbol{\sigma}_x] &= 2\mathrm{i}\,\boldsymbol{\sigma}_y \end{aligned} \tag{1-320}$$

and are explicitly given in the standard representation by

$$\boldsymbol{\sigma}_x = \begin{pmatrix} 0 & 1 \\ 1 & 0 \end{pmatrix}, \quad \boldsymbol{\sigma}_y = \begin{pmatrix} 0 & -\mathrm{i} \\ \mathrm{i} & 0 \end{pmatrix},$$
$$\boldsymbol{\sigma}_z = \begin{pmatrix} 1 & 0 \\ 0 & -1 \end{pmatrix} \tag{1-321}$$

One can easily prove a useful formula,

$$(\boldsymbol{\sigma} \cdot \boldsymbol{A})\,(\boldsymbol{\sigma} \cdot \boldsymbol{B}) = \boldsymbol{A} \cdot \boldsymbol{B} + \mathrm{i}\,\boldsymbol{\sigma} \cdot (\boldsymbol{A} \times \boldsymbol{B}) \tag{1-322}$$

where $\boldsymbol{A}$ and $\boldsymbol{B}$ are any vector operators. With the help of this formula we see at once that Eq. (1-319) is also

$$\mathscr{T} = (\boldsymbol{p})^2/2\,m \tag{1-323}$$

which is the usual expression for the kinetic energy. However, when a magnetic field is turned on, one must make the substitution

$$\boldsymbol{p} \rightarrow \boldsymbol{p} - \frac{e}{c}\boldsymbol{A}$$

Now the two forms of the kinetic energy are no longer the same and one obtains

$$\mathscr{T} = \frac{1}{2m}\boldsymbol{\sigma} \cdot \left(\boldsymbol{p} - \frac{e}{c}\boldsymbol{A}\right) \boldsymbol{\sigma} \cdot \left(\boldsymbol{p} - \frac{e}{c}\boldsymbol{A}\right)$$
$$= \frac{1}{2m}\left(\boldsymbol{p} - \frac{e}{c}\boldsymbol{A}\right)^2 - \mu_{\mathrm{B}}\,\boldsymbol{\sigma} \cdot \boldsymbol{B} \tag{1-324}$$

where

$$\mu_{\mathrm{B}} = \frac{e\,\hbar}{2\,m\,c} \tag{1-325}$$

is the Bohr-magneton and the magnetic field is

$$\boldsymbol{B} = \nabla \times \boldsymbol{A} \tag{1-326}$$

The Zeeman term for the spin is, of course, the last term of Eq. (1-324); furthermore, if the term $\left(\boldsymbol{p} - \frac{e}{c}\boldsymbol{A}\right)^2$ is evaluated, one obtains a diamagnetic contribution ($\sim \boldsymbol{A}^2$) and the Zeeman term for the orbital moment (if the magnetic field is assumed to be uniform). We will ignore the diamagnetic term throughout – this is admissible, at least in the present context. The orbital contribution, however, is more of a problem. If one is willing to accept the usual argument that the orbital moment is *quenched* in the solid state, then we may ignore this term, too, and we are left with the spin-Zeeman term $-\mu_{\mathrm{B}}\,\boldsymbol{\sigma} \cdot \boldsymbol{B}$ only. We may leave it at that for the moment but will come back to this difficult question later (Sec. 1.4.4.2).

Now, with only the spin-Zeeman term left, we may assume without loss of generality that the field is oriented in the z direction, then

$$\boldsymbol{\sigma} \cdot \boldsymbol{B} = \sigma_z\,B_z \tag{1-327}$$

and the Bloch function acquires a factor χ_{S} which is a Pauli bi-spinor and is eigenstate to σ_z, with

$$\chi_+ = \begin{pmatrix} 1 \\ 0 \end{pmatrix} \quad \text{and} \quad \chi_- = \begin{pmatrix} 0 \\ 1 \end{pmatrix}$$

This rather simple generalization of Schrödinger theory is extremely useful as it allows the treatment of metallic magnetism. Indeed, in Sec. 1.3 we will demonstrate how an "effective magnetic field" is generated which shows up just like the Zeeman term above, but it should not be confused with a real magnetic field (in fact, this "effective magnetic field" is not derivable from the curl of an effective vector potential). The net effect, though, is that the spin-up electrons move in a different potential than the spin-down electrons; this leads to spin-polarization and, under well-defined conditions, to the formation of stable magnetic moments, as we will show in some detail in Sec. 1.4.

We could thus be content with this theory, were it not for two more important phenomena: spin-orbit coupling and a genuine relativistic effect that becomes important for heavy elements – roughly those with atomic numbers larger than that of Ag ($Z > 47$). Here, the s states are most strongly affected, as they allow the electrons to be close to the nucleus where, broadly speaking, they have high velocities. We therefore must turn to the Dirac equation (Rose, 1961; Messiah, 1976; Sakurai, 1967) which we will discuss briefly, emphasizing an approximate treatment that has become quite popular in band-structure theory.

We begin with the Dirac Hamiltonian, H_D, for an electron moving in an effective potential V and no vector potential, initially:

$$H_D = c\,\boldsymbol{\alpha} \cdot p + \beta m c^2 + V \tag{1-328}$$

where $\boldsymbol{\alpha}$ is a vector operator whose components are written using the Pauli-spin matrices, Eq. (1-321), as

$$\boldsymbol{\alpha}_k = \begin{pmatrix} 0 & \sigma_k \\ \sigma_k & 0 \end{pmatrix}, \quad k = 1, 2, 3 \tag{1-329}$$

where $k = 1$ corresponds to the x component, $k = 2$ to the y-component, etc. The matrix $\boldsymbol{\beta}$ is given by

$$\boldsymbol{\beta} = \begin{pmatrix} \mathbf{I} & 0 \\ 0 & -\mathbf{I} \end{pmatrix} \tag{1-330}$$

where

$$\mathbf{I} = \begin{pmatrix} 1 & 0 \\ 0 & 1 \end{pmatrix} \tag{1-331}$$

is the unit (2×2) matrix. The Dirac Hamiltonian acts on a *four-component* wave function ψ which may be written in terms of two *two-component* functions ϕ and χ as

$$\psi = \begin{pmatrix} \phi \\ \chi \end{pmatrix} \tag{1-332}$$

where in the case of electrons (positive-energy solutions) ϕ is the "large" and χ is the "small" component. The eigenvalue problem,

$$H_D \psi = \varepsilon \psi \tag{1-333}$$

is, with Eqs. (1-329) and (1-330), at once seen to lead to a set of coupled equations for ϕ and χ, namely

$$\begin{aligned} c(\boldsymbol{\sigma} \cdot \boldsymbol{p})\chi &= (\varepsilon - V - m c^2)\phi \\ c(\boldsymbol{\sigma} \cdot \boldsymbol{p})\phi &= (\varepsilon - V + m c^2)\chi \end{aligned} \tag{1-334}$$

Since it was the radial Schrödinger equation that played such an important role in the previous subsections, we next sketch the derivation of the radial Dirac equation. So let the potential be spherically symmetric. Then, in order to obtain the correct classifications (or quantum numbers) of the solutions of the Dirac equation, we must list the constants of motion, i.e., those operators that commute with H_D (and with themselves). They are derived in appropriate text books, see e.g. Messiah (1978) or Sakurai (1967), where one finds that the total angular momentum is one such operator:

$$\boldsymbol{J} = \boldsymbol{L} + \hbar\,\boldsymbol{\Sigma}/2 \tag{1-335}$$

It is a sum of the orbital angular momentum, $\boldsymbol{L}$, and the spin

$$\boldsymbol{\Sigma} = \begin{pmatrix} \boldsymbol{\sigma} & 0 \\ 0 & \boldsymbol{\sigma} \end{pmatrix} \tag{1-336}$$

But $\boldsymbol{L}$ and $\boldsymbol{\Sigma}$ separately are *not* constants of motion. Another operator is $\mathbf{K}$ which is explicitly given by

$$\mathbf{K} = \begin{pmatrix} \boldsymbol{\sigma} \cdot \boldsymbol{L} + \hbar & 0 \\ 0 & -\boldsymbol{\sigma} \cdot \boldsymbol{L} - \hbar \end{pmatrix} \tag{1-337}$$

Since $\mathbf{K}$ also commutes with $\boldsymbol{J}$, the eigenfunctions of H_D in a central potential are eigenfunctions of $\mathbf{K}$, $\boldsymbol{J}^2$, and $\boldsymbol{J}_z$. The corresponding eigenvalues are denoted by $-\varkappa\hbar$, $j(j+1)\hbar^2$, and $j_z\hbar$, but $\varkappa$ and j are related by

$$\varkappa = \pm\left(j + \frac{1}{2}\right) \tag{1-338}$$

as one can easily show. Thus $\varkappa$ is a nonzero integer which can be positive or negative.

Pictorially speaking, the sign of $\varkappa$ determines whether the spin is antiparallel ($\varkappa > 0$) or parallel ($\varkappa < 0$) to the total angular momentum in the non-relativistic limit. To simplify the notation, instead of attaching all the quantum numbers ($\varkappa, j, j_z$) to the wave function, we merely write

$$\psi = \begin{pmatrix} \phi \\ \chi \end{pmatrix} = \begin{pmatrix} g(r) & \mathcal{Y}_{jl}^{j_z} \\ \mathrm{i} f(r) & \mathcal{Y}_{jl'}^{j_z} \end{pmatrix} \tag{1-339}$$

where g and f are radial functions, and where $\mathcal{Y}_{jl}^{j_z}$ stands for a normalized spin-angular function (or r-independent eigenfunction of $\boldsymbol{J}^2$, J_z, $\boldsymbol{L}^2$, and $\boldsymbol{S}^2$) formed by the combination of the Pauli spinor with the spherical harmonics of order l. Explicitly,

$$\mathcal{Y}_{jl}^{j_z} = \pm \sqrt{\frac{l + j_z + \frac{1}{2}}{2l+1}}\, Y_l^{j_z - 1/2} \begin{pmatrix} 1 \\ 0 \end{pmatrix} + \\ + \sqrt{\frac{l \mp j_z + \frac{1}{2}}{2l+1}}\, Y_l^{j_z + 1/2} \begin{pmatrix} 0 \\ 1 \end{pmatrix} \tag{1-340}$$

where $j = \pm \frac{1}{2}$. The relations among $\varkappa$, j, l and l' are:

$$\varkappa = j + \frac{1}{2} \text{ then } l = j + \frac{1}{2},\; l' = j - \frac{1}{2} \tag{1-341}$$

$$\varkappa = -\left(j + \frac{1}{2}\right) \text{ then } l = j - \frac{1}{2},\; l' = j + \frac{1}{2}$$

Now an easy calculation yields the radial Dirac equations: one starts with Eq. (1-333) using (1-339) and the relation

$$\boldsymbol{\sigma} \cdot \boldsymbol{p} \equiv \frac{(\boldsymbol{\sigma} \cdot \boldsymbol{x})}{r^2} (\boldsymbol{\sigma} \cdot \boldsymbol{x})(\boldsymbol{\sigma} \cdot \boldsymbol{p}) \\ = \frac{(\boldsymbol{\sigma} \cdot \boldsymbol{x})}{r} \left(-i\hbar \frac{\mathrm{d}}{\mathrm{d}r} + \mathrm{i}\, \boldsymbol{\sigma} \cdot \boldsymbol{L} \right) \tag{1-342}$$

and obtains

$$\hbar c \left(\frac{\mathrm{d}f}{\mathrm{d}r} + \frac{(1-\varkappa)}{r} f \right) = -(\varepsilon - V - mc^2)\, g \\ \hbar c \left(\frac{\mathrm{d}g}{\mathrm{d}r} + \frac{(1+\varkappa)}{r} g \right) = (\varepsilon - V + mc^2)\, f \tag{1-343}$$

The band-structure problem with this full four-component formalism can now be set up and solved numerically following the various procedures outlined in the previous subsections. Thus, the relativistic version of the APW method was obtained early by Loucks (1967), and the relativistic KKR theory is described in great detail by Weinberger (1990). Ebert (1988) developed a relativistic LMTO method, and Takeda (1979) a relativistic ASW method. A treatment of any of these fully relativistic procedures is beyond the scope of this chapter, but an approximate and physically transparent version is not. It is generally known as the scalar-relativistic approximation (Koelling and Harmon, 1977; Takeda, 1978; MacDonald et al., 1980) and it supplies an equation that formally looks like the Schrödinger equation slightly modified. The spin-orbit interaction is separated out explicitly and is normally treated variationally or by perturbation theory, while the new wave equation (often called scalar-relativistic wave equation, which looks quite like the Pauli equation but is not identical with it) simply replaces the Schrödinger equation whenever it appears in the procedures outlined before.

To summarize the salient features of this approximation, we start by defining another energy origin, i.e.,

$$\varepsilon' = \varepsilon - mc^2 \tag{1-344}$$

then the factor multiplying f on the right-hand side of Eq. (1-343) can formally be written as $2Mc^2$, if

$$M = m + \frac{\varepsilon' - V}{2c^2} \tag{1-345}$$

and the second of the Eq. (1-343) gives

$$f = \frac{\hbar}{2Mc} \left(\frac{\mathrm{d}g}{\mathrm{d}r} + \frac{(1+\varkappa)}{r} g \right) \tag{1-346}$$

Substituting this into the first of the Eq. (1-343) and carrying out the differentiations, we obtain

$$-\frac{\hbar^2}{2M}\frac{1}{r^2}\frac{\mathrm{d}}{\mathrm{d}r}\left(r^2\frac{\mathrm{d}g}{\mathrm{d}r}\right)+\left[V+\frac{\hbar^2}{2M}\frac{\varkappa(\varkappa+1)}{r^2}\right]g$$
$$-\frac{\hbar^2}{4M^2c^2}\frac{\mathrm{d}V}{\mathrm{d}r}\frac{\mathrm{d}g}{\mathrm{d}r}-$$
$$-\frac{\hbar^2}{4M^2c^2}\frac{\mathrm{d}V}{\mathrm{d}r}\frac{(1+\varkappa)}{r}g=\varepsilon' g \qquad (1\text{-}347)$$

For each of the cases specified in Eq. (1-341), the factor $\varkappa(\varkappa+1)$ is seen to be

$$\varkappa(\varkappa+1)=l(l+1) \qquad (1\text{-}348)$$

Thus apart from the last two terms on the left hand side, Eq. (1-347) has an appearance like the radial Schrödinger equation. Of course, care must be taken with M, which, as Eq. (1-345) shows, is not a constant. No approximation has been made yet. The formal mass term M is sometimes called "mass-velocity" term, the term $\frac{\mathrm{d}V}{\mathrm{d}r}\frac{\mathrm{d}g}{\mathrm{d}r}$ is known as the Darwin term, and the last term on the left hand side of Eq. (1-347) is the spin-orbit coupling term; this is true because

$$(\boldsymbol{\sigma}\cdot\boldsymbol{L}+\hbar)\,\phi=-\varkappa\,\hbar\,\phi \qquad (1\text{-}349)$$

We now make the essential approximation and drop the spin-orbit term from the radial Eq. (1-347) (and from Eq. (1-346)). We may call these approximate functions $\tilde{g}$ and $\tilde{f}$ (not to be confused with augmentation), i.e. we require

$$-\frac{\hbar^2}{2M}\frac{1}{r^2}\frac{\mathrm{d}}{\mathrm{d}r}\left(r^2\frac{\mathrm{d}\tilde{g}}{\mathrm{d}r}\right)+\left[V+\frac{\hbar^2}{2M}\frac{l(l+1)}{r^2}\right]\tilde{g}$$
$$-\frac{\hbar^2}{4M^2c^2}\frac{\mathrm{d}V}{\mathrm{d}r}\frac{\mathrm{d}\tilde{g}}{\mathrm{d}r}=\varepsilon'\,\tilde{g} \qquad (1\text{-}350)$$

and

$$\tilde{f}=\frac{\hbar}{2Mc}\frac{\mathrm{d}\tilde{g}}{\mathrm{d}r} \qquad (1\text{-}351)$$

The latter is needed for the proper normalization

$$\int(\tilde{g}^2+\tilde{f}^2)\,r^2\,\mathrm{d}r=1 \qquad (1\text{-}352)$$

It may also be used to simplify the numerical procedure to solve Eq. (1-350). But this is more of a technical detail. In passing, we note that by expanding M we easily obtain the Pauli equation from Eq. (1-350).

The Eq. (1-350) is the scalar-relativistic radial equation, and it may be used in place of the Schrödinger equation in any of the methods described in the previous subsections; no further changes are needed. When this is done, and the band structure is calculated for a metal containing light elements, very few changes occur, in contrast to heavy metals. This is illustrated in Fig. 1-19 where we show the band structure of Os obtained by the ASW method using the scalar relativistic equation (it is essentially identical with the LMTO results by Jepsen et al., 1975). The relativistic band structure of Fig. 1-19 should be compared with the corresponding non-relativistic one which can be seen in Fig. 1-14. The d bands are easily recognized with the help of the canonical bands shown in Fig. 1-12c and are seen to be nearly identical. Quite conspicuous, however, is the change of the s band, the bottom of the conduction band, which is lowered by approximately 0.3 Ry. This is due to the mass-velocity and Darwin terms.

It should be obvious now that one needs to formulate the corrections that are brought about when the spin-orbit coupling term is included in a final step of the calculations. The most satisfactory way seems to be the following (MacDonald et al., 1980): Using the radial functions $\tilde{g}$ and $\tilde{f}$, the four-component wave function is first written out as

$$\tilde{\psi}=\begin{pmatrix}\tilde{\Phi}\\ \tilde{\chi}\end{pmatrix} \qquad (1\text{-}353)$$

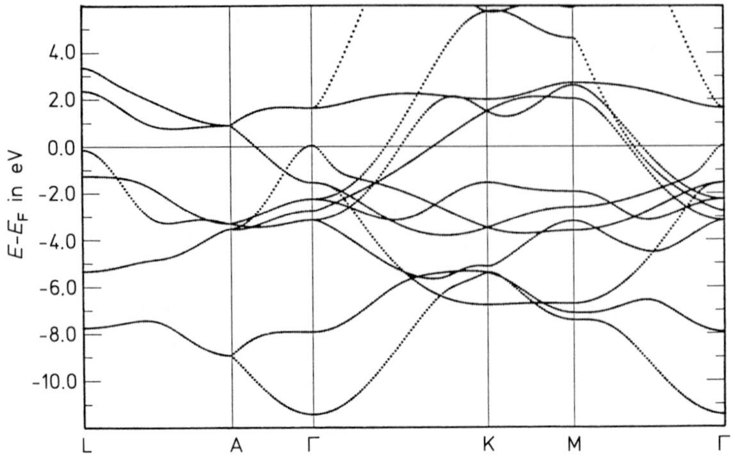

Figure 1-19. Band structure for Os as calculated with the ASW within the scalar-relativistic approximation.

where $\tilde{\phi}$ is assumed to be a pure spin state, i.e.,

$$\tilde{\phi} = \tilde{g}\, Y_L\, \chi_S \tag{1-354}$$

where $\chi_+ = \begin{pmatrix}1\\0\end{pmatrix}$ and $\chi_- = \begin{pmatrix}0\\1\end{pmatrix}$. Then $\tilde{\chi}$ is obtained, using Eq. (1-334),

$$\tilde{\chi} = \mathrm{i}\left(\frac{\boldsymbol{\sigma}\cdot\boldsymbol{x}}{r}\right)\left(-\tilde{f} + \frac{1}{2\,M\,c\,r}\,\tilde{g}\,\boldsymbol{\sigma}\cdot\boldsymbol{L}\right) Y_L\,\chi_S \tag{1-355}$$

The spin-orbit correction is now obtained by applying the Dirac Hamiltonian H_D, Eq. (1-328), to the wave function $\tilde{\psi}$, Eq. (1-353). The details of this calculation are somewhat tedious but straightforward and give

$$H_D\,\tilde{\psi} = \varepsilon'\,\tilde{\psi} + H_{SO}\,\tilde{\psi} \tag{1-356}$$

where

$$H_{SO} = \frac{\hbar}{(2\,M\,c)^2}\,\frac{1}{r}\,\frac{\mathrm{d}V}{\mathrm{d}r}\begin{pmatrix}(\boldsymbol{\sigma}\cdot\boldsymbol{L})\,I\\ 0\end{pmatrix} \tag{1-357}$$

This is the spin-orbit coupling operator that represents a measure of the extent to which the function $\tilde{\psi}$ fails to be a true solution of the spherical-potential Dirac equation. In most applications the matrix element of H_{SO} is computed on whatever basis is used in the particular method and is then added to the Hamiltonian matrix in the variational procedure. It is clear that this breaks the degeneracy of the formerly spin-degenerate bands. Details can be found in the paper by Andersen (1975) and a very clear account of spin-orbit coupling in the LAPW method in the paper by MacDonald et al. (1980). The latter authors also compare their results with results of a full-four-component RAPW calculation and find excellent agreement. Thus it seems that this approximate treatment is fully satisfactory.

In Fig. 1-20 the role of spin-orbit coupling (SO) on the band structure of Os is illustrated, and the results of ASW calculations shown (compare with Fig. 1-19). In the ASW method, the SO Hamiltonian, Eq. (1-357), is included in the matrix element $\langle s\,\nu\,\tilde{L}\,|\,\mathscr{H}\,|\,\nu'\,L'\,s'\rangle$ to be used with Eq. (1-268) (s, s' are spin-indices). The one center term is dominant, but small corrections result from the two- and three-center terms. Quite conspicuous is the SO splitting near the Fermi energy at the points A, Γ, K, and M. Closer inspection shows that the Fermi surface is different in the two calculations corresponding to Figs. 1-19 and 1-20. For instance, the intersections of the bands with the Fermi energy near Γ in

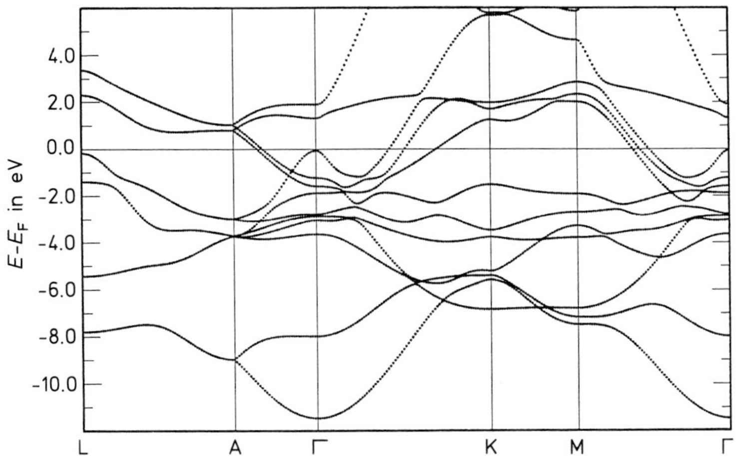

Figure 1-20. The same as in Fig. 1-19 but now including spin-orbit coupling.

Fig. 1-19 is absent in Fig. 1-20. SO coupling thus removes an orbit which is predicted in a calculation without SO coupling. We may compare this with experimental knowledge of the Fermi surface and do this in Table 1-1 and Fig. 1-21 for Os (Schneider, 1991). The latter shows cuts of the Fermi surface with principle planes in the Brillouin zone and is derived from the calculation corresponding to Fig. 1-20 (including SO coupling). Just as in the first calculations for this system by Jepsen et al. (1975), we conclude again that our treatment of relativistic effects is fully satisfactory.

Table 1-1. Fermi surface areas for Os. The theoretical results are based on ASW calculations including spin-orbit coupling. All values in 10^2 T.

Direction		S[a]	K[b] (exp)
U7h	$[10\bar{1}0]\,\alpha$	1.7	1.4
	$[11\bar{2}0]\,\alpha$	2.2	2.2
	$[0001]\,\alpha$	3.0	2.9
KM8h	$[10\bar{1}0]\,\delta$	111.1	110.0
	$[11\bar{2}0]\,\delta$	110.5	109.0
	$[11\bar{2}0]\,\varepsilon$	69.6	68.0
	$[0001]$	–	–
Γ 9e	$[10\bar{1}0]\,\gamma$	157.9	160.0
	$[11\bar{2}0]\,\gamma$	165.7	168.0
	$[0001]\,\gamma$	206.2	205.0
Γ 10e	$[10\bar{1}0]\,\beta$	119.1	124.0
	$[11\bar{2}0]\,\beta$	129.7	133.0
	$[0001]\,\beta$	145.9	148.0
	$[0001]\,\beta$	150.1	153.0

[a] Schneider (1991), [b] Kamm and Anderson (1970).

We close this subsection with a short discussion of the *spin-polarized* band structure, mentioned before in connection with the Schrödinger equation and encountered in the field of metallic magnetism. This problem appears to be more complex in a relativistic formalism than in the non-relativistic one.

A desirable approach would start with the Dirac equation for an electron moving in a scalar effective potential, V, *and* an effective vector potential $\boldsymbol{A}_{\text{eff}}$:

$$H_{\text{D}} = c\,\boldsymbol{\alpha} \cdot \left(\boldsymbol{p} - \frac{e}{c}\boldsymbol{A}_{\text{eff}}\right) + \beta m c^2 + V \tag{1-358}$$

Recent attempts with this form of starting point were made by Eschrig et al. (1985) and Diener and Gräfenstein (1989). But only in the framework of *non-relativistic* density functional theory has an effective vector potential been derived in a completely satisfactory way (Vignale and Rasolt, 1987, 1988). On the other hand, rela-

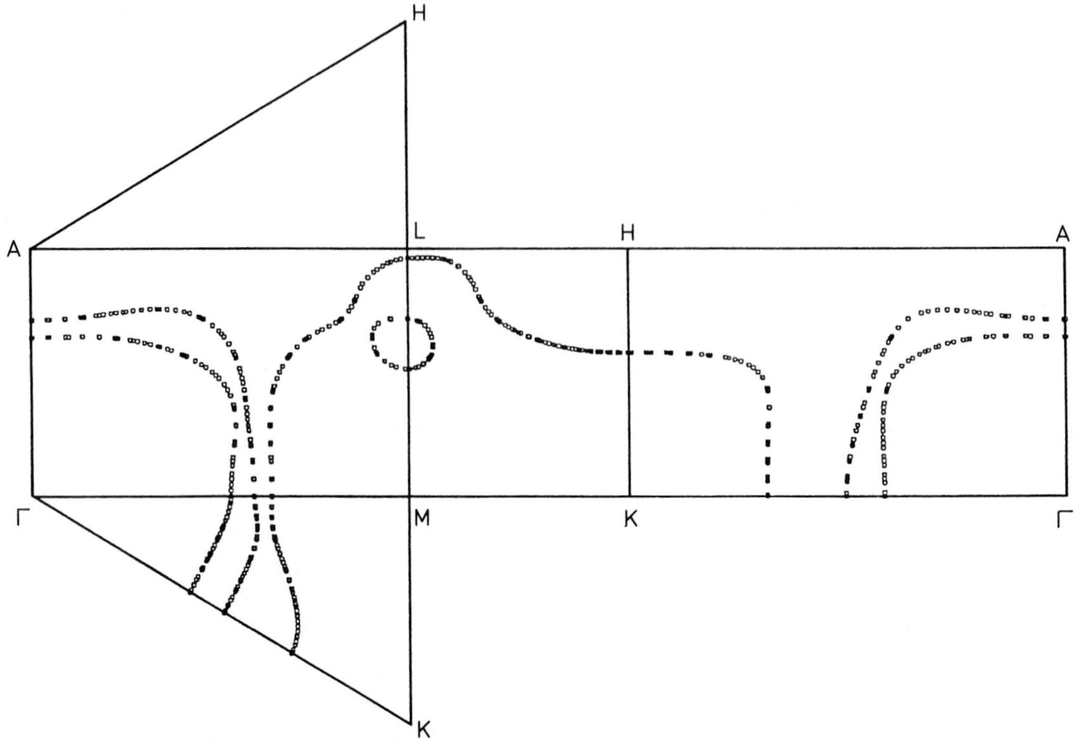

Figure 1-21. Sections of the Fermi surface for Os in the irreducible wedge of the Brillouin zone as obtained by an ASW calculation (Schneider, 1991).

tivistic theories seem non-contradictory only when they work with a spin-polarized scalar effective potential and no effective vector-potential (see MacDonald and Vosko, 1979; Doniach and Sommers, 1982; Cortona et al., 1985). Physically, one retains a coupling of the exchange-correlation potential to the electron spin only discarding a possible coupling to currents and hence to orbital magnetic moments:

$$H_{\mathrm{D}} = c\,\boldsymbol{\alpha} \cdot \boldsymbol{p} + \beta\, m\, c^2 + V + \beta\, \Sigma_z\, V_1 \quad (1\text{-}359)$$

where V is the effective potential that is independent of the spin direction, Σ_z is the z component of $\boldsymbol{\Sigma}$ (Eq. (1-336)), and V_1 could be called an effective (exchange correlation) "magnetic field" but with all the precautions we called attention to in the beginning of this subsection (following Eq. (1-327)).

The consequences of these approximations are not known at present (1990). Strange et al. (1984) first used Eq. (1-359) in a full self-consistent KKR band-structure calculation, but even on this level further approximations must be made which have their origin in the fact that spin is not a good quantum number in the Dirac theory. A thorough discussion of these facts and consequences can be found in a paper by Feder et al. (1983). Unfortunately, they also affect the scalar relativistic approximation. It can be achieved, however, if the spin-dependent potential, V_1, is set to zero in the equation for the small component χ, the first of the Eq. (1-334). In this case there are the same steps as before that must be taken to derive the equation for $\tilde{g}$, which now reads

$$-\frac{\hbar^2}{2M}\frac{1}{r^2}\frac{\mathrm{d}}{\mathrm{d}r}\left(r^2\frac{\mathrm{d}\tilde{g}_s}{\mathrm{d}r}\right)+$$
$$+\left[V+sV_1+\frac{\hbar^2}{2M}\frac{l(l+1)}{r^2}\right]\tilde{g}_s$$
$$-\frac{\hbar^2}{4M^2c^2}\frac{\mathrm{d}V}{\mathrm{d}r}\cdot\frac{\mathrm{d}\tilde{g}_s}{\mathrm{d}r}=\varepsilon_s'\,\tilde{g}_s \qquad (1\text{-}360)$$

where $s=\pm 1$: $+1$ for spin-up and -1 for spin-down electrons. M, Eq. (1-345), $\frac{\mathrm{d}V}{\mathrm{d}r}$, and H_{SO}, Eq. (1-357), do not contain the effect of spin polarization, i.e., the term V_1. This is, of course, a consequence of setting V_1 equal to zero in the small component. Again, the consequences of this approximation are not presently known, but we believe they are unimportant. We see that the scalar relativistic radial equation must be solved for each spin-direction separately. The spin-orbit matrix element is then treated as before and results in a coupling of the two spin-spaces so that, when everything is done, the bands are no longer labeled by "spin-up" and "spin-down". Later examples will shed more light on the type of solutions.

1.3 Derivation of the Single-Particle Schrödinger Equation: Density Functional Theory

1.3.1 General

In this section we briefly discuss the foundations of density functional theory (DFT) (Hohenberg and Kohn, 1964; Kohn and Sham, 1965). It constitutes a reduction of the complicated many-body problem to an effective single-particle problem and is thus central to band theory; the reader is referred to the numerous reviews available. Examples include von Barth (1983, 1984), Kohn and Vashishta (1983), Williams and von Barth (1983), and Callaway and March (1984).

We begin by writing down the general Hamiltonian for the interacting crystal electrons within the Born-Oppenheimer approximation as

$$H=T+V+U \qquad (1\text{-}361)$$

where

$$T=-\sum_{i=1}^{N}\nabla_i^2 \qquad (1\text{-}362)$$

$$V=\sum_{i=1}^{N}v_{\mathrm{ext}}(\boldsymbol{r}_i) \qquad (1\text{-}363)$$

$$U=\sum_{\substack{i,j=1\\ i\neq j}}^{N}v_{\mathrm{el-el}}(\boldsymbol{r}_i-\boldsymbol{r}_j) \qquad (1\text{-}364)$$

with

$$v_{\mathrm{el-el}}(\boldsymbol{r}_i-\boldsymbol{r}_j)=1/|\boldsymbol{r}_i-\boldsymbol{r}_j|$$

(given in atomic units). The quantity T denotes the kinetic energy, V the external potential, and U the Coulomb interaction energy of the N-particle system. We define the electron-density operator by

$$\hat{n}(\boldsymbol{r})=\sum_{i=1}^{N}\delta(\boldsymbol{r}-\boldsymbol{r}_i) \qquad (1\text{-}365)$$

from which we obtain the electron density as

$$n(\boldsymbol{r})=\langle\Phi|\,n(\boldsymbol{r})\,|\Phi\rangle \qquad (1\text{-}366)$$

where $|\Phi\rangle$ is a many-body state.

The external potential consists of the potential due to the ions located at (static) positions $\boldsymbol{R}_\mu$ and possibly other external fields. Thus

$$v_{\mathrm{ext}}(\boldsymbol{r})=\sum_{\mu}v_{\mathrm{ion}}(\boldsymbol{r}-\boldsymbol{R}_\mu)+v_{\mathrm{field}}(\boldsymbol{r}) \qquad (1\text{-}367)$$

Although numerous techniques exist to deal with this many-body problem, there is no exact solution. Hohenberg and Kohn (1964) established that the electron density is the central variable and proved the following two theorems:

1. The total *ground state* energy, E, of any many-electron system is a functional of the particle density $n(\boldsymbol{r})$:

$$E[n] = F[n] + \int n(\boldsymbol{r})\, v_{\text{ext}}(\boldsymbol{r})\, \mathrm{d}^3\boldsymbol{r} \quad (1\text{-}368)$$

where $F[n]$ itself is a functional of the density $n(\boldsymbol{r})$, but is otherwise independent of the external potential.
2. For any many-electron system the functional $E[n]$ for the total energy has a minimum equal to the ground-state energy at the ground-state density.

This second theorem is of practical importance as it leads to a variational principle. To prove these theorems we will turn to the work of Levy (1979, 1982), who overcame some of the restrictions of the Hohenberg-Kohn theory concerning, for instance, the non-degeneracy of the ground state, by defining a functional

$$O[n] = \inf_{|\boldsymbol{\Phi}\rangle \in M(n)} \langle \boldsymbol{\Phi} | O | \boldsymbol{\Phi} \rangle \quad (1\text{-}369)$$

for a physical observable O (we will assume the infimum to exist, which is quite reasonable for the functionals of interest). The set $M(n)$ contains all those many-body wave functions which yield the density n. It might be further limited by certain conditions, e.g., that this set picks out only wave functions giving the ground state to an external potential or corresponding to an N-particle state. These situations normally are referred to as v representability or N representability, respectively. It should be noted, that the Levy construction not only is a natural generalization of the Hohenberg-Kohn ansatz of a v representable density but also allows for spin polarized systems or relativistic calculations (von Barth, 1983, 1984).

Using the Levy functional to set up the functionals of the operators H, V, and $F = T + U$ for an N-representable density, the validity of the first theorem can be seen directly (one uses Eqs. (1-365) and (1-366) to obtain the second term on the right-hand side of Eq. (1-368) as the functional of V). The proof of the second theorem is carried out in the following way: We denote by $|\Phi_0\rangle$, n_0, and E_0 the ground state, its density, and its energy, respectively (or any ground state in the case of degeneracy). $|\Phi_n\rangle$ may be any state yielding the density $n(\boldsymbol{r})$ and minimizing the functional $F[n]$ and thus $E[n]$. Now we have to show that in this case E_n and n are the same as E_0 and n_0.

Since E_0 was defined as the ground-state energy, we have

$$E[n] \geq E_0 \quad (1\text{-}370)$$

for any density $n(\boldsymbol{r})$. On the other hand, if we calculate $F[n]$ with the ground state according to Eq. (1-369), we obtain

$$F[n_0] = \inf_{|\Phi\rangle} \langle \Phi | T + U | \Phi \rangle \leq$$
$$\leq \langle \Phi_0 | T + U | \Phi_0 \rangle \quad (1\text{-}371)$$

and thus by adding the functional of the external potential

$$E[n_0] \leq E_0 \quad (1\text{-}372)$$

The expressions (1-370) and (1-372) can be fulfilled simultaneously only if $n = n_0$ and thus $E[n] = E[n_0]$. From this it follows immediately that

$$E[n_0] = E_0 \quad (1\text{-}373)$$

Although the basic facts of density functional theory are stated now, we still need a key to its application. Such a key was given by Kohn and Sham (1965), who used the variational principle implied by the minimal properties of the energy functional to derive single particle Schrödinger-like equations.

Following Kohn and Sham, we first split the functional $F[n]$ into three parts

$$F[n] = $$
$$= T[n] + \iint n(\boldsymbol{r})\, v_{\text{el-el}}(\boldsymbol{r}-\boldsymbol{r}')\, n(\boldsymbol{r}')\, \mathrm{d}^3r'\, \mathrm{d}^3\boldsymbol{r} +$$
$$+ E_{xc}[n] \qquad (1\text{-}374)$$

which describe the kinetic, the Hartree, and the exchange correlation energy. In contrast to the Hartree integral, an explicit form of the other functionals, T and E_{xc}, is not known in general. We will ignore this problem for now and proceed in writing down the Euler-Lagrange equation following from the variational principle:

$$\frac{\delta E[n]}{\delta n(\boldsymbol{r})} + \mu \frac{\delta(N - \int n(\boldsymbol{r})\, \mathrm{d}^3\boldsymbol{r})}{\delta n(\boldsymbol{r})} = 0 \qquad (1\text{-}375)$$

where μ is a Lagrange multiplier taking care of particle conservation. Inserting Eq. (1-374) into the Euler-Lagrange equation we get

$$\frac{\delta T[n]}{\delta n(\boldsymbol{r})} + v_{\text{ext}}(\boldsymbol{r}) + 2\int v_{\text{el-el}}(\boldsymbol{r}-\boldsymbol{r}')\, n(\boldsymbol{r}')\, \mathrm{d}^3\boldsymbol{r}' +$$
$$+ \frac{\delta E_{xc}[n]}{\delta n(\boldsymbol{r})} - \mu = 0 \qquad (1\text{-}376)$$

The functional $E_{xc}[n]$ is still not known, but concerning the kinetic energy, Kohn and Sham realized that the above equation can be interpreted as the Euler-Lagrange equation of noninteracting particles moving in an external, effective potential, given by

$$v_{\text{eff}}(\boldsymbol{r}) = v_{\text{ext}}(\boldsymbol{r}) +$$
$$+ 2\int v_{\text{el-el}}(\boldsymbol{r}-\boldsymbol{r}')\, n(\boldsymbol{r}')\, \mathrm{d}^3\boldsymbol{r}' + v_{xc}(\boldsymbol{r}) \qquad (1\text{-}377)$$

where the exchange correlation potential,

$$v_{xc}(\boldsymbol{r}) = \frac{\delta E_{xc}[n]}{\delta n(\boldsymbol{r})} \qquad (1\text{-}378)$$

contains all the many-body effects. So they took advantage of the fact that the case of noninteracting electrons allows an exact computation of the particle density and the kinetic energy as

$$n(\boldsymbol{r}) = \sum_{\substack{i=1 \\ \varepsilon_i \le E_F}}^{N} |\phi_i(\boldsymbol{r})|^2 \qquad (1\text{-}379)$$

$$T_0[n] = -\sum_{\substack{i=1 \\ \varepsilon_i \le E_F}}^{N} \int \phi_i^*(\boldsymbol{r})\, \nabla^2 \phi_i(\boldsymbol{r})\, \mathrm{d}^3\boldsymbol{r} \qquad (1\text{-}380)$$

Note that we implicitly introduced Fermi statistics when filling the single-particle states ϕ_i in Eq. (1-379) up to the Fermi energy E_F, and that we assumed these states to be normalized. Furthermore, we used the symbol T_0 to designate the kinetic energy of noninteracting electrons, thus parts of the kinetic energy are contained in the exchange correlation energy E_{xc}, for which we use the same symbol as in Eq. (1-374), for simplicity in writing.

We next derive the single-particle equations which we used in the foregoing section. We start with Eq. (1-375) and write down the Euler-Lagrange equation again, this time in terms of the single-particle wave functions:

$$\sum_{i=1}^{N} \left[\frac{\delta E[n]}{\delta \phi_i^*(\boldsymbol{r})} + \varepsilon_i \frac{\delta(N - \int n(\boldsymbol{r})\, \mathrm{d}^3\boldsymbol{r})}{\delta \phi_i^*(\boldsymbol{r})} \right] \delta\phi_i^* = 0 \qquad (1\text{-}381)$$

where we introduced Lagrange multipliers for every ϕ_i. Since the wave functions are independent of each other, the expression in brackets must vanish for each ϕ_i. Now using Eqs. (1-377) to (1-380), we obtain

$$\frac{\delta T_0[n]}{\delta \phi_i^*(\boldsymbol{r})} + v_{\text{eff}}(\boldsymbol{r}) \frac{\delta n(\boldsymbol{r})}{\delta \phi_i^*(\boldsymbol{r})} +$$
$$+ \varepsilon_i \frac{\delta(N - \int n(\boldsymbol{r})\, \mathrm{d}^3\boldsymbol{r})}{\delta \phi_i^*(\boldsymbol{r})} =$$
$$= [-\nabla^2 + v_{\text{eff}}(\boldsymbol{r}) - \varepsilon_i]\, \phi_i(\boldsymbol{r}) = 0 \qquad (1\text{-}382)$$

This effective single-particle equation is often called the Kohn-Sham equation. It is a single-particle Schrödinger equation with the external potential replaced by the effective potential (Eq. (1-377)), which depends on the density. The density itself, according to Eq. (1-379), depends on the single-particle states ϕ_i. The Kohn-Sham equation thus constitutes a self-consistent field problem.

The Kohn-Sham equation furthermore allows us to derive an alternative expression for the total energy. It follows from Eq. (1-382) by multiplying with ϕ_i^*, summing over all i, and integrating over all space:

$$T_0[n] + \int v_{\text{eff}}(\boldsymbol{r})\, n(\boldsymbol{r})\, \mathrm{d}^3\boldsymbol{r} - \sum_{\substack{i=1 \\ \varepsilon_i \le E_F}}^{N} \varepsilon_i = 0 \tag{1-383}$$

Now combining this with Eqs. (1-368), (1-374), and (1-377) we obtain the result

$$\begin{aligned} E[n] = & \sum_{\substack{i=1 \\ \varepsilon_i \le E_F}}^{N} \varepsilon_i - \\ & - \iint n(\boldsymbol{r})\, v_{\text{el-el}}(\boldsymbol{r}-\boldsymbol{r}')\, n(\boldsymbol{r}')\, \mathrm{d}^3\boldsymbol{r}'\, \mathrm{d}^3\boldsymbol{r} - \\ & - \int v_{\text{xc}}(\boldsymbol{r})\, n(\boldsymbol{r})\, \mathrm{d}^3\boldsymbol{r} + E_{\text{xc}}[n] \end{aligned} \tag{1-384}$$

The total energy thus consists of the sum of the eigenvalues minus the so-called "double-counting" terms.

Although density-functional theory as outlined above provides a scheme to reduce the entire many-body problem to a Schrödinger-like effective single-particle equation, the physical meaning of the eigenvalues ε_i is not clear. These eigenvalues have, in fact, been used very often and with success to interpret excitation spectra. But there are also cases which are problematic, as we will see. Although we cannot decide whether this failure is inherent to density-functional theory or simply a consequence of the local-density approximation (which will be discussed next), we still may present some results here, which we hope will shed some light on the problem.

Let us start with a relation due to Slater (1971, 1972) and Janak (1978) which connects the eigenvalues ε_i to occupation number changes of the orbitals ϕ_i:

$$\varepsilon_i = \frac{\partial \tilde{E}}{\partial n_i} \tag{1-385}$$

Here, $\tilde{E}$ denotes a proper generalization of the total energy as given by Eq. (1-368), allowing for fractional occupations. Its construction may be readily achieved by modifying the expressions (1-379) and (1-380) for the electron density and the kinetic energy

$$\tilde{n}(\boldsymbol{r}) = \sum_{\substack{i=1 \\ \varepsilon_i \le E_F}}^{N} n_i |\phi_i(\boldsymbol{r})|^2 \tag{1-386}$$

$$\tilde{T}_0(\tilde{n}) = - \sum_{\substack{i=1 \\ \varepsilon_i \le E_F}}^{N} n_i \int \phi_i^*(\boldsymbol{r})\, \nabla^2 \phi_i(\boldsymbol{r})\, \mathrm{d}^3\boldsymbol{r} \tag{1-387}$$

where n_i can vary between 0 and 1. We should note, however, that because of this somewhat loose procedure, the density – besides not being N representable – also lacks the v representability if orbitals other than the highest-occupied orbitals are depleted. In such cases the total energy may not be obtained by solving an effective single-particle equation. The problem can be circumvented by fixing the occupation numbers according to the laws of Fermi-Dirac statistics or by turning completely to the finite temperature extension of density-functional theory as introduced by Mermin (1965). But this is beyond the scope of our chapter.

Next, writing down $\tilde{E}$ in complete analogy with Eq. (1-384), inserting n_i into the first term on the right-hand side and using for the density Eq. (1-386), we differentiate the expression and easily prove Eq. (1-385). Equation (1-385) states that if the occupation of the ith orbital is changed by an infinitesimal amount δn_i, then the total energy of the system changes by $\varepsilon_i\, \delta n_i$. This still does not fully clarify the connection of the quantities ε_i with the excitation energies of the system for two reasons. First, real excitations involve whole electrons, thus the excitation energy does not correspond simply to the first derivative of the total energy, but to an entire Taylor series (assuming it converges),

$$\tilde{E}(n_i + \Delta n_i) = \tilde{E}(n_i) + (\delta \tilde{E}/\delta n_i)\,\Delta n_i + + \tfrac{1}{2}(\delta^2 \tilde{E}/\delta n_i^2)\,(\Delta n_i)^2 + \ldots \qquad (1\text{-}388)$$

The importance of second- and higher-order terms, which correspond to relaxation, depends primarily on the localization of the electron density of the states from which the electron was excited and, hence, should not be too important for delocalized continuum states. Second, there are the complications discussed after Eq. (1-387) whose importance is not easily assessed.

Another somewhat different discussion of the same problem stems from von Barth (1983, 1984) who considered the homogeneous, interacting electron gas where the quasi-particle excitation spectrum can formally be obtained from the Dyson equation and can be written in terms of a momentum- and energy-dependent (and in general complex) self-energy as

$$\tilde{\varepsilon}(\boldsymbol{k}) = \boldsymbol{k}^2 + \Sigma(\boldsymbol{k}, \tilde{\varepsilon}(\boldsymbol{k})) \qquad (1\text{-}389)$$

On the other hand when using density-functional theory to calculate the band energies for the electron gas a $\boldsymbol{k}$-independent effective potential is involved which supplies a free electron dispersion curve shifted by a constant. Since density functional theory correctly describes the Fermi level E_F, as has been shown by Kohn and Sham (1965) (see also Kohn and Vashishta, 1983), we may conclude that

$$\varepsilon(\boldsymbol{k}) = \boldsymbol{k}^2 + \Sigma(k_F, E(\boldsymbol{k}_F)) \qquad (1\text{-}390)$$

Comparison of these two dispersion relations yields the result that the density-functional band structure is quite accurate near the Fermi energy provided the variations of the self-energy with $\boldsymbol{k}$ and energy are small. This condition is presumably fulfilled in most of the conventional metals, as will be discussed in Sec. 1.4, but is not, for instance, in the broad class of heavy-fermion systems. The extent of the deviations of the self-energy from its value at k_F and E_F may be characterized by the effective mass m^* which is defined by the relation

$$\frac{m^*}{m} = \frac{1 - \dfrac{\partial \Sigma(\boldsymbol{k}, \omega)}{\partial \omega}}{1 + \dfrac{m}{k_F}\dfrac{\partial \Sigma(\boldsymbol{k}, \omega)}{\partial \boldsymbol{k}}} \qquad (1\text{-}391)$$

and which indeed may be several hundreds of the electron mass in heavy-fermion systems (for a review see for instance Fulde et al., 1988; or Grewe and Steglich, 1990).

Summarizing the previous discussion, we can state that we may not be too far off when interpreting the calculated eigenvalues as excitation energies. Nevertheless, caution is necessary. Strictly speaking, this statement applies not only to density-functional theory but also to a greater extent to the local-density approximation, an approximation required in order to obtain a practical computational scheme. However, before turning to its discussion, we will extend density functional theory to magnetic systems.

1.3.2 Spin Polarization

Apparently, von Barth and Hedin (1972) and Rajagopal and Callaway (1973) were the first to generalize the density-functional theory to include effects of spin polarization. They used a matrix formalism to represent the density and the external potential instead of single variables. We omit the basic proofs here, since they are very similar to those given in the previous subsection, and concentrate on facts that are typical for spin-density-functional theory.

Recall in Sec. 1.2.5, in the simplest version of a theory for magnetic materials we attached bispinors χ_+ and χ_- to the wave function. Such a two component spinor function consequently requires a 2×2 matrix for the Hamiltonian which may again be written in three parts representing the kinetic energy, the external potential, and the Coulomb interaction of the electrons. The elements of these matrices are given by

$$T_{\alpha\beta} = -\,\delta_{\alpha\beta} \sum_{i=1}^{N} \nabla_i^2 \tag{1-392}$$

$$V_{\alpha\beta} = \sum_{i=1}^{N} v_{\alpha\beta}^{\text{ext}}(\boldsymbol{r}_i) \tag{1-393}$$

$$U_{\alpha\beta} = \sum_{\substack{i,j=1\\ i\neq j}}^{N} v_{\text{el-el}}(\boldsymbol{r}_i - \boldsymbol{r}_j) \tag{1-394}$$

where α and β are spin indices. Comparing this with the corresponding equations of the preceding subsection, we see that it is only the external potential, Eq. (1-393), that needs special attention. Furthermore, if we generalize the definition of the electron-density operator (Eq. (1-365)) by

$$\hat{\boldsymbol{n}}_{\alpha\beta}(\boldsymbol{r}) = \delta_{\alpha\beta} \sum_{i=1}^{N} \delta(\boldsymbol{r} - \boldsymbol{r}_i) \tag{1-395}$$

then the electron density is obtained as in Eq. (1-366), where the many-body state is now assumed to be given by a spinor function. But we would like to go a little further and include in our discussion also those cases where the spin state cannot be described by a single, global quantization axis. According to the laws of quantum mechanics such a mixed state is describable by the density operator $\varrho_{\alpha\beta}$ taking the place of the spinor function. The electron density is then given by

$$n(\boldsymbol{r}) = \text{Tr}\,\langle \varrho\, \hat{\boldsymbol{n}}(\boldsymbol{r})\rangle \tag{1-396}$$

where the symbol Tr denotes the trace. For a complete description of the state of an itinerant magnetic system we need the spin density in addition to the electron density, Eq. (1-396). For this we define the two-by-two matrix

$$\tilde{n}(\boldsymbol{r}) = \langle \varrho\, \hat{\boldsymbol{n}}(\boldsymbol{r})\rangle \tag{1-397}$$

which is frequently called the density matrix, but should not be confused with the term density operator appropriate for ϱ. In general, the quantity $\tilde{n}(\boldsymbol{r})$ will not be diagonal, but it can be locally diagonalized, and in this case the difference of the eigenvalues supplies the spin density (see, e.g., Sticht et al., 1989; or Vignale and Rasolt, 1987, 1988 for a somewhat different formulation).

It should now be clear that the generalized Levy functional depends on the full density matrix. Thus the definition

$$O[\tilde{n}] = \inf_{\varrho \in M(\tilde{n})} \text{Tr}\,\langle \varrho\, O\rangle \tag{1-398}$$

is a straightforward extension of the functional given in Eq. (1-369), the only difference being that the constrained search for the infimum now includes all density operators ϱ which yield the required density matrix $\tilde{n}$.

We can now set up the total energy and subsequently derive the single-particle equation. Thus we proceed by replacing Eq. (1-368) by

$$E[\tilde{n}] = F[\tilde{n}] + V[\tilde{n}] \tag{1-399}$$

where

$$\begin{aligned} V[\tilde{n}] &= \inf_{\varrho \in M(\tilde{n})} \text{Tr}\,\langle \varrho\, V\rangle = \\ &= \inf_{\varrho \in M(\tilde{n})} \text{Tr} \sum_{\beta} \int \sum_{i=1}^{N} v_{\alpha\beta}^{\text{ext}}(\boldsymbol{r}_i)\, \varrho_{\beta\alpha}(\boldsymbol{r}_1 \ldots \boldsymbol{r}_N) \cdot \\ &\quad \cdot \mathrm{d}^3\boldsymbol{r}_1 \ldots \mathrm{d}^3\boldsymbol{r}_N = \\ &= \inf_{\varrho \in M(\tilde{n})} \text{Tr} \sum_{\beta} \int \sum_{i=1}^{N} v_{\alpha\beta}^{\text{ext}}(\boldsymbol{r})\, \delta(\boldsymbol{r} - \boldsymbol{r}_i) \cdot \\ &\quad \cdot \varrho_{\beta\alpha}(\boldsymbol{r}_1 \ldots \boldsymbol{r}_N)\, \mathrm{d}^3\boldsymbol{r}_1 \ldots \mathrm{d}^3\boldsymbol{r}_N\, \mathrm{d}^3\boldsymbol{r} = \\ &= \sum_{\alpha\beta} \int v_{\alpha\beta}^{\text{ext}}(\boldsymbol{r})\, \tilde{n}_{\beta\alpha}(\boldsymbol{r})\, \mathrm{d}^3\boldsymbol{r} \end{aligned} \tag{1-400}$$

As we have already mentioned, we may now benefit from the universality of the proofs of the Hohenberg-Kohn theorems in the preceding subsection and directly turn to the derivation of the single-particle equations. It should be pointed out, however, that the variational properties of the total energy this time apply to all components of the density matrix. Therefore when defining the functional $F[\tilde{n}]$ by

$$F[\tilde{n}] = T_0[\tilde{n}] + \iint n(\boldsymbol{r})\, v_{\text{el}-\text{el}}(\boldsymbol{r}-\boldsymbol{r}')\, n(\boldsymbol{r}') \cdot \mathrm{d}^3\boldsymbol{r}'\, \mathrm{d}^3\boldsymbol{r} + E_{\text{xc}}[\tilde{n}] \tag{1-401}$$

we obtain a matrix form for the effective potential according to

$$v_{\alpha\beta}^{\text{eff}}(\boldsymbol{r}) = \frac{\delta(E[\tilde{n}] - T_0[\tilde{n}])}{\delta \tilde{n}_{\beta\alpha}(\boldsymbol{r})} = \tag{1-402}$$

$$= v_{\alpha\beta}^{\text{ext}}(\boldsymbol{r}) + 2\,\delta_{\alpha\beta} \int v_{\text{el}-\text{el}}(\boldsymbol{r}-\boldsymbol{r}')\, n(\boldsymbol{r}')\, \mathrm{d}^3\boldsymbol{r}' + v_{\alpha\beta}^{\text{xc}}(\boldsymbol{r})$$

where

$$v_{\alpha\beta}^{\text{xc}}(\boldsymbol{r}) = \frac{\delta E_{\text{xc}}[\tilde{n}]}{\delta \tilde{n}_{\beta\alpha}(\boldsymbol{r})} \tag{1-403}$$

is the exchange-correlation potential.

The Kohn-Sham concept of assuming independent particles moving in an effective potential leads to the following extensions for the density matrix and the kinetic energy:

$$\tilde{n}_{\beta\alpha}(\boldsymbol{r}) = \sum_{\substack{i=1 \\ \varepsilon_{i\alpha},\, \varepsilon_{i\beta} \le E_{\text{F}}}}^{N} \phi_{i\beta}(\boldsymbol{r})\, \phi_{i\alpha}^{*}(\boldsymbol{r}) \tag{1-404}$$

$$T_0[\tilde{n}] = -\sum_{\alpha} \sum_{\substack{i=1 \\ \varepsilon_{i\alpha} \le E_{\text{F}}}}^{N} \int \phi_{i\alpha}^{*}(\boldsymbol{r})\, \nabla^2 \phi_{i\alpha}(\boldsymbol{r})\, \mathrm{d}^3\boldsymbol{r} \tag{1-405}$$

where the Fermi energy is determined according to

$$n(\boldsymbol{r}) = \sum_{\alpha} \sum_{\substack{i=1 \\ \varepsilon_{i\alpha} \le E_{\text{F}}}}^{N} |\phi_{i\alpha}(\boldsymbol{r})|^2 \tag{1-406}$$

Finally, the generalized Euler-Lagrange equation is

$$\sum_{i=1}^{N} \sum_{\beta} \left(\frac{\delta E[\tilde{n}]}{\delta \phi_{i\beta}^{*}(\boldsymbol{r})} + \varepsilon_{i\beta} \frac{\delta(N - \int n(\boldsymbol{r})\, \mathrm{d}^3\boldsymbol{r})}{\delta \phi_{i\beta}^{*}(\boldsymbol{r})} \right) \cdot \delta\phi_{i\beta}^{*} = 0 \tag{1-407}$$

where the $\varepsilon_{i\beta}$ once again secure particle conservation. Since the single-particle orbitals $\phi_{i\beta}$ are independent of each other, we obtain

$$\sum_{\beta} [-\delta_{\alpha\beta} \nabla^2 + v_{\alpha\beta}^{\text{eff}}(\boldsymbol{r}) - \varepsilon_{i\alpha}\, \delta_{\alpha\beta}]\, \phi_{i\beta}(\boldsymbol{r}) = 0\,, \quad \alpha = 1, 2 \tag{1-408}$$

As a consequence of the mixed state, we assumed that we generally no longer have separate equations for the spin-up and spin-down components of the spinor function but rather a set of two coupled equations. However, for most cases we deal with in this chapter, the exchange-correlation potential, $v_{\alpha\beta}^{\text{eff}}(\boldsymbol{r})$, is diagonal.

We close this subsection by writing down an expression for the total energy similar to Eq. (1-384),

$$E[\tilde{n}] = \sum_{\alpha} \sum_{\substack{i=1 \\ \varepsilon_{i\alpha} \le E_{\text{F}}}}^{N} \varepsilon_{i\alpha} - \iint n(\boldsymbol{r})\, v_{\text{el}-\text{el}}(\boldsymbol{r}-\boldsymbol{r}')\, n(\boldsymbol{r}')\, \mathrm{d}^3\boldsymbol{r}'\, \mathrm{d}^3\boldsymbol{r} - \sum_{\alpha\beta} \int v_{\alpha\beta}^{\text{xc}}(\boldsymbol{r})\, \tilde{n}_{\beta\alpha}(\boldsymbol{r})\, \mathrm{d}^3\boldsymbol{r} + E_{\text{xc}}[\tilde{n}] \tag{1-409}$$

1.3.3 The Local-Density Approximation (LDA)

Our discussion in the preceding sections was of completely theoretical nature without regard to practical applications. We now turn to an approximation scheme to deal with the exchange correlation functional which is both accurate and useful. This is the local-density approximation (LDA) where the homogeneous, interacting electron gas serves to model the exchange-correlation energy in the form

$$E_{xc}[n] = \int n(\boldsymbol{r})\, \varepsilon_{xc}(n(\boldsymbol{r}))\, d^3\boldsymbol{r} \tag{1-410}$$

The symbol ε here should not be confused with an eigenenergy, $\varepsilon_{xc}(n)$ is a function of the density instead of a functional and Eq. (1-410) may be viewed as dividing the inhomogeneous electron system into small "boxes", each containing a homogeneous interacting electron gas with a density $n(\boldsymbol{r})$ appropriate for the "box" at $\boldsymbol{r}$.

From Eqs. (1-378) and (1-410) we derive for the exchange-correlation potential

$$v_{xc}(\boldsymbol{r}) = \left[\frac{d}{dn}\{n\, \varepsilon_{xc}(n)\}\right]_{n=n(\boldsymbol{r})} \tag{1-411}$$

The local-density approximation may be readily generalized to the spin polarized case. Assuming the eigenvalues of the density matrix to be $n_\uparrow(\boldsymbol{r})$ and $n_\downarrow(\boldsymbol{r})$ the exchange-correlation energy is written as (von Barth and Hedin, 1972)

$$E_{xc}[\tilde{n}] = \int n(\boldsymbol{r})\, \varepsilon_{xc}(n_\uparrow(\boldsymbol{r}), n_\downarrow(\boldsymbol{r}))\, d^3\boldsymbol{r} \tag{1-412}$$

This results in the following form of the exchange-correlation potential

$$v_\alpha^{xc}(\boldsymbol{r}) = \left[\frac{\partial}{\partial n_\alpha}\{n\, \varepsilon_{xc}(n_\uparrow, n_\downarrow)\}\right]_{\substack{n_\uparrow = n_\uparrow(\boldsymbol{r}) \\ n_\downarrow = n_\downarrow(\boldsymbol{r})}} \tag{1-413}$$

where $\alpha = \uparrow$ or $\downarrow$.

For simplicity we will deal with ferromagnetic or simple antiferromagnetic systems in the following where the diagonal form of the density matrix is self-evident (but see Sticht et al., 1989 for more general cases).

To obtain an explicit expression for the exchange-correlation-energy density, $\varepsilon_{xc}(n_\uparrow, n_\downarrow)$, we initially neglect correlation and begin with the Hartree-Fock approximation for a spin-polarized gas of electrons (von Barth and Hedin, 1972) in atomic units (see Sec. 1.2.2):

$$\varepsilon_x(n_\uparrow, n_\downarrow) = -6\left(\frac{3}{4\pi}\right)^{1/3} \frac{1}{n}\left(n_\uparrow^{4/3} + n_\downarrow^{4/3}\right), \tag{1-414}$$

where the density is $n = n_\uparrow + n_\downarrow$.

This is merely the exchange-energy density, thus correlation gives rise to another term $\varepsilon_c(n_\uparrow, n_\downarrow)$ that must be added to the exchange term to give

$$\varepsilon_{xc}(n_\uparrow, n_\downarrow) = \varepsilon_x(n_\uparrow, n_\downarrow) + \varepsilon_c(n_\uparrow, n_\downarrow). \tag{1-415}$$

Now, the total energy, and hence ε_c, of a homogeneous but interacting electron gas is known only approximately. But it is known numerically to an accuracy which makes Eq. (1-412) an exceedingly useful approximation. Different parametrizations of the numerical results of different authors exist. In the following we will give the parametrization of Moruzzi et al. (1978) which embodies work by Singwi et al. (1970), Hedin and Lundqvist (1971), and von Barth and Hedin (1972). However, other parametrizations exist, and are used, for instance, that by Vosko et al. (1980), based on Monte-Carlo calculations by Ceperley and Alder (1980) and, in addition, the extensions to the relativistic case, which go back to Rajagopal (1980), MacDonald and Vosko (1979), and Ramana and Rajagopal (1983).

Our implementation of the LDA proceeds briefly as follows: First we write

$$\varepsilon_x(n_\uparrow, n_\downarrow) = \varepsilon_x^P(r_S) + (\varepsilon_x^F(r_S) - \varepsilon_x^P(r_S))\, f(n_\uparrow, n_\downarrow) \tag{1-416}$$

$$\varepsilon_c(n_\uparrow, n_\downarrow) = \varepsilon_c^P(r_S) + (\varepsilon_c^F(r_S) - \varepsilon_c^P(r_S))\, f(n_\uparrow, n_\downarrow) \tag{1-417}$$

where the function f is given by

$$f(n_\uparrow, n_\downarrow) = \frac{1}{2^{4/3}-2}\left[\left(\frac{2n_\uparrow}{n}\right)^{4/3} + \left(\frac{2n_\downarrow}{n}\right)^{4/3} - 2\right] \tag{1-418}$$

where again $n = n_\uparrow + n_\downarrow$. It vanishes when the spin polarization is zero, i.e., $n_\uparrow = n_\downarrow = n/2$. The functions describing exchange in Eq. (1-416) are

$$\varepsilon_x^P(r_S) = -\frac{\varepsilon_x^0}{r_S}$$
$$\varepsilon_x^0 = \frac{3}{2\pi}\left(\frac{9\pi}{4}\right)^{1/3} = 0.91633 \text{ Ry} \tag{1-419}$$

$$\varepsilon_x^F(r_S) = 2^{1/3}\,\varepsilon_x^P(r_S) \tag{1-420}$$

which reproduce the Hartree-Fock results. Here we defined a quantity r_S that is connected with the electron density as

$$\frac{4\pi}{3} r_S^3\, a_0^3 = \frac{1}{n} \tag{1-421}$$

where a_0 is the Bohr radius. Correlation is described by the functions

$$\varepsilon_c^P(r_S) = -\,c_P\, F\left(\frac{r_S}{r_P}\right) \tag{1-422}$$

$$\varepsilon_c^F(r_S) = -\,c_F\, F\left(\frac{r_S}{r_F}\right) \tag{1-423}$$

where we have defined a function

$$F(z) = (1+z^3)\ln\left(1+\frac{1}{z}\right) + \frac{z}{2} - z^2 - \frac{1}{3} \tag{1-424}$$

The correlation energy is further modeled by the set of parameters

$$c_P = 0.0504\,, \quad r_P = 30$$
$$c_F = 0.0254\,, \quad r_F = 75 \tag{1-425}$$

as given by von Barth and Hedin (1972) or

$$c_P = 0.045\,, \quad r_P = 21 \tag{1-426}$$
$$c_F = \frac{c_P}{2} = 0.025\,, \quad r_F = 2^{4/3}\, r_P = 52.91668$$

as given by Moruzzi et al. (1978). The connection between c_F and c_P as well as r_F and r_P given in Eq. (1-426) refers to the RPA scaling.

From the above equations the exchange-correlation potential can be calculated by means of Eq. (1-413):

$$\begin{aligned} v_\alpha^{xc}(n_\uparrow, n_\downarrow) &= \\ &= \left[\frac{4}{3}\varepsilon_x^P(r_S) + \gamma\,(\varepsilon_c^F(r_S) - \varepsilon_c^P(r_S))\right]\left(\frac{2n_\alpha}{n}\right)^{1/3} + \\ &+ \mu_c^P(r_S) - \gamma\,(\varepsilon_c^F(r_S) - \varepsilon_c^P(r_S)) + \\ &+ \left[\mu_c^F(r_S) - \mu_c^P(r_S) - \frac{4}{3}(\varepsilon_c^F(r_S) - \varepsilon_c^P(r_S))\right]\cdot \\ &\cdot f(n_\uparrow, n_\downarrow) \end{aligned} \tag{1-427}$$

with

$$\mu_c^P(r_S) = -\,c_P \ln\left(1 + \frac{r_P}{r_S}\right) \tag{1-428}$$

$$\mu_c^F(r_S) = c_F \ln\left(1 + \frac{r_F}{r_S}\right) \tag{1-429}$$

and

$$\gamma = \frac{4}{3}\,\frac{1}{2^{1/3}-1} \tag{1-430}$$

This completes the specification of the effective potential. The local density approximation is exact in the limit of a constant density or density matrix, respectively. In addition, we would expect it to be quite good for slowly varying densities but not when dealing with realistic systems as atoms or solids where the density depends strongly on position. Surprisingly, this is not the case. Instead, experience shows that this approximation does work well in a great number of realistic cases.

Despite its success, there have been numerous attempts to improve the LDA. The earliest attempt came from Kohn and Sham themselves. Like the von Weizsäcker correction to the Thomas-Fermi theory (March, 1983), they proposed to add a gradient correction term to the exchange correlation energy. However, attempts along these lines were not successful, and in many cases the good agreement that is obtained by using the simple approximation above was spoiled. The most probable reason for this is that the gradient expansions proposed so far were not converged.

The question remains as to why the LDA does as well as it does. Examples will be found in the next section. This question has received a great deal of attention, especially in connection with investigations of the spatial form of the exchange-correlation potential and the exchange-correlation hole. Examples include the work of Gunnarsson and Lundqvist (1976), as well as that of Kohn and Vashishta (1983) or Callaway and March (1984). We do not have enough space to go into any mathematical detail here and thus refer the reader to a qualitative discussion comparing experimental and theoretical data at the end of Sec. 1.4.2.2. The theoretical investigation into density functional theory and refined approximations is still ongoing.

1.4 Selected Case Studies

When the two broad fields covered in the last two sections are combined to give both a computational scheme and a set of interpretable parameters, we can begin to grasp the interrelationship of many physical properties of solids. We will make an attempt to uncover these interrelationships, but we will be neither complete nor are all aspects as yet fully understood. Still, we hope a physical picture emerges that will help our understanding of a part of the properties of metals and metallic compounds.

Instead of separately considering the hundred basic building blocks of nature that are arranged in the Periodic Table of the Elements, which give matter such a wide range and variety of physical properties, we will assume that the reader is familiar with the structure of the table and its underlying reason, as well as the reason for the difference of the chemical behavior of the various elements. In essence, we will limit our attention to elementary metals (Secs. 1.4.1 and 1.4.2) and only very briefly discuss metallic compounds (Secs. 1.4.3 and 1.4.4); in the former case we will distinguish simple metals from transition metals. An earlier review of this subject was carried out by Williams and von Barth (1983).

1.4.1 Elementary Nonmagnetic Materials

1.4.1.1 Band Structure

Simple Metals

In Sec. 1.2.2.1 we referred to the paradoxical fact that some metallic properties such as electrical or thermal conductivity as well as the heat capacity of the simple metals can be experimentally understood by regarding the valence electrons as a gas of non-interacting particles that obey Fermi statistics and are free to travel throughout the metal without being affected much by the parent ions. In fact, we showed that these ions give rise to weak Fourier coefficients in a plane wave basis that are explained by the *Pseudopotential* concept (Sec. 1.2.2.2). The band structure is roughly similar to that of the "empty lattice" which was given for an f.c.c. structure in Fig. 1-4 and for a b.c.c. structure in Fig. 1-5. The energy, however, must be rescaled with the correct lattice constant in each case. For instance, examining aluminum in some detail, we know that it has an f.c.c. structure and a lattice constant of $a=7.6$ a.u. Thus the lowest (doubly degenerate) state at X where $\boldsymbol{k}=\dfrac{2\pi}{a}(1,0,0)$ takes the value $(4\pi/a)^2=9.3$ eV and the Fermi energy will take the value of approximately 11.6 eV (which on the scale of Fig. 1-4 ap-

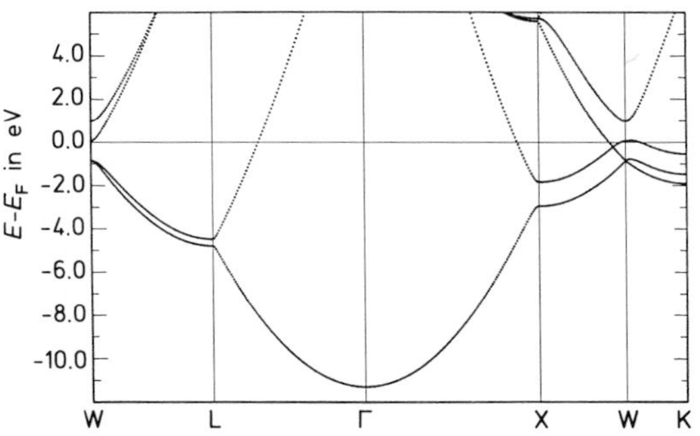

Figure 1-22. Band structure of Al as calculated with the ASW method.

pears at roughly 20 eV – three valence electrons).

The self-consistent band structure of Al is shown in Fig. 1-22; it was obtained by Moruzzi et al. (1978) (or equivalently by the ASW method which we have used frequently in this section for a quick orientation) and is based on the local-density-functional approximation for exchange and correlation. We see that the degeneracy is lifted at the zone boundaries, e.g., at the points L and X. In an attempt to clarify this band structure somewhat further, the pseudopotential concept in its simplest form is extremely useful. However, we do not follow the historical process (Heine and Abarenkov, 1964) by constructing the energy bands of Al using (empirical) pseudopotentials, but rather use the information that Fig. 1-22 supplies for an estimate of the size of the relevant Fourier components, $V(\boldsymbol{K})$, defined by Eq. (1-56) with $V_c(\boldsymbol{r})$ replaced by the pseudopotential. Concentrating first at the X point and remembering that the degenerate states in the empty lattice are

$$\varepsilon_X^{(1)} = \boldsymbol{k}_X^2 \quad \text{and} \quad \varepsilon_X^{(2)} = (\boldsymbol{k}_X + \boldsymbol{K}_1)^2 \tag{1-431}$$

where

$$\boldsymbol{k}_X = \frac{2\pi}{a}(1, 0, 0) \tag{1-432}$$

and

$$\boldsymbol{K}_1 = \frac{2\pi}{a}(-2, 0, 0) \tag{1-433}$$

we make the simplest expansion possible and use only two plane waves in Eq. (1-52), i.e.,

$$\psi_{\boldsymbol{k}_X}(\boldsymbol{r}) = \frac{1}{\sqrt{\Omega}}\left(a_0\, e^{i\,\boldsymbol{k}_X \cdot \boldsymbol{r}} + a_1\, e^{i(\boldsymbol{k}_X + \boldsymbol{K}_1)\cdot \boldsymbol{r}}\right) \tag{1-434}$$

The secular determinant from Eq. (1-54) is then simply

$$\begin{vmatrix} \boldsymbol{k}_X^2 - \varepsilon & V(\boldsymbol{K}_1) \\ V(\boldsymbol{K}_1) & (\boldsymbol{k}_X + \boldsymbol{K}_1)^2 - \varepsilon \end{vmatrix} = 0 \tag{1-435}$$

which gives

$$\varepsilon_X = \boldsymbol{k}_X^2 \pm V(\boldsymbol{K}_1) \tag{1-436}$$

Thus the pseudopotential has opened up a gap in the free-electron band structure of the size of

$$\varepsilon_X^{\text{gap}} = 2\,|V(\boldsymbol{K}_1)| \tag{1-437}$$

Because the gap at X is seen to be about 1 eV, the magnitude of the Fourier component is ≈ 0.5 eV (which is in agreement with Heine and Abarenkov, 1964). With the same kind of arguments a somewhat smaller value is obtained for another Fourier component $|V(\boldsymbol{K}_2)|$ with $\boldsymbol{K}_2 = \frac{2\pi}{a}(1, 1, 1)$ from the gap at the L point. These values are small compared with the Fermi energy of about 11 eV in Al and, therefore, the band structure $\varepsilon_{\boldsymbol{k}}$ and the density of states it defines by means of Eq. (1-47) are nearly-free-electron-like (NFE) to a very good approximation. Had we used an ionic potential in Eq. (1-56) and calculated $V(\boldsymbol{K}_1)$ from first principles, a value of about -5 eV would have resulted, which is obviously not appropriate for the valence electrons.

The NFE behavior has been observed experimentally in studies of the *Fermi sur-*

face. We do not want to discuss this intriguing subject here any further. The interested reader will find a concise treatment in an article by Pettifor (1984) and much more detailed information in Harrison (1966), Heine and Weaire (1970), and Kittel (1986). The latter is also referred to for a discussion of transport properties and concepts such as holes and effective masses. But we do want to compare the calculated band structure with experimental data and have chosen to do so using angle-resolved photoemission (ARUPS) data of the type we have already shown in connection with Cu, Fig. 1-9. For Al the experimental band structure was obtained from ARUPS by Levinson et al. (1983) who compared their measurements with calculations by Singhal and Callaway (1977). Parts of their results are shown in Fig. 1-23, where we see generally good agreement between measured and calculated data except when looking at them with higher energy resolution along the lines X to W (Fig. 1-23b). A word of caution is, however, in order.

Since in a photoemission experiment light is used to excite electrons from states below the Fermi energy to states high above it, strictly speaking one is not measuring ground-state properties but rather properties of the excited states or, in other words, of quasiparticles. These quasiparticles may be different from the electrons in the ground state, and we remind the reader of the discussion appropriate to this situation found at the end of Sec. 1.3.1.

Returning to the simple metals, we follow Pettifor (1984) and show in Fig. 1-24 their densities of states, $N(\varepsilon)$, which have been computed from first principles by Moruzzi et al. (1978) using the KKR method. We see that Na, Mg, and Al across a period and Al, Ga, and In down a group are good NFE metals, because their densities of states show only very small perturbations of the free-electron density of states that is easily obtained from Eq. (1-47) to be

$$N(\varepsilon) = \frac{3}{2} \frac{Z_v}{E_F^{3/2}} \sqrt{\varepsilon} \qquad (1\text{-}438)$$

where Z_v is the valence and E_F the Fermi energy. However, we see that Li and Be show very strong deviations from the square-root behavior. This is so because Li and Be have no p-core electrons, and consequently there is no repulsive pseudopotential for the valence 2p electrons. There is of course one for the 2s electrons – this is an example of a non-local pseudopotential (Sec. 1.2.2.3) (if one chooses to use this

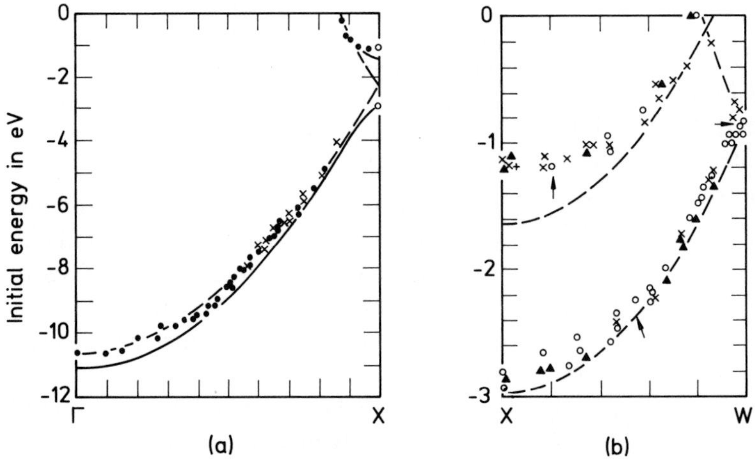

Figure 1-23. Experimental band structure for Al as obtained by Levinson et al. (1983) from ARUPS, compared to calculations by Singhal and Callaway (1977). (a) Bands from Γ to X: solid lines correspond to the calculation by Singhal and Callaway; dashed lines indicate a free-electron calculation with an effective mass $m^* = 1.1\,m$. (b) Bands from X to W: dashed curves from Singhal and Callaway.

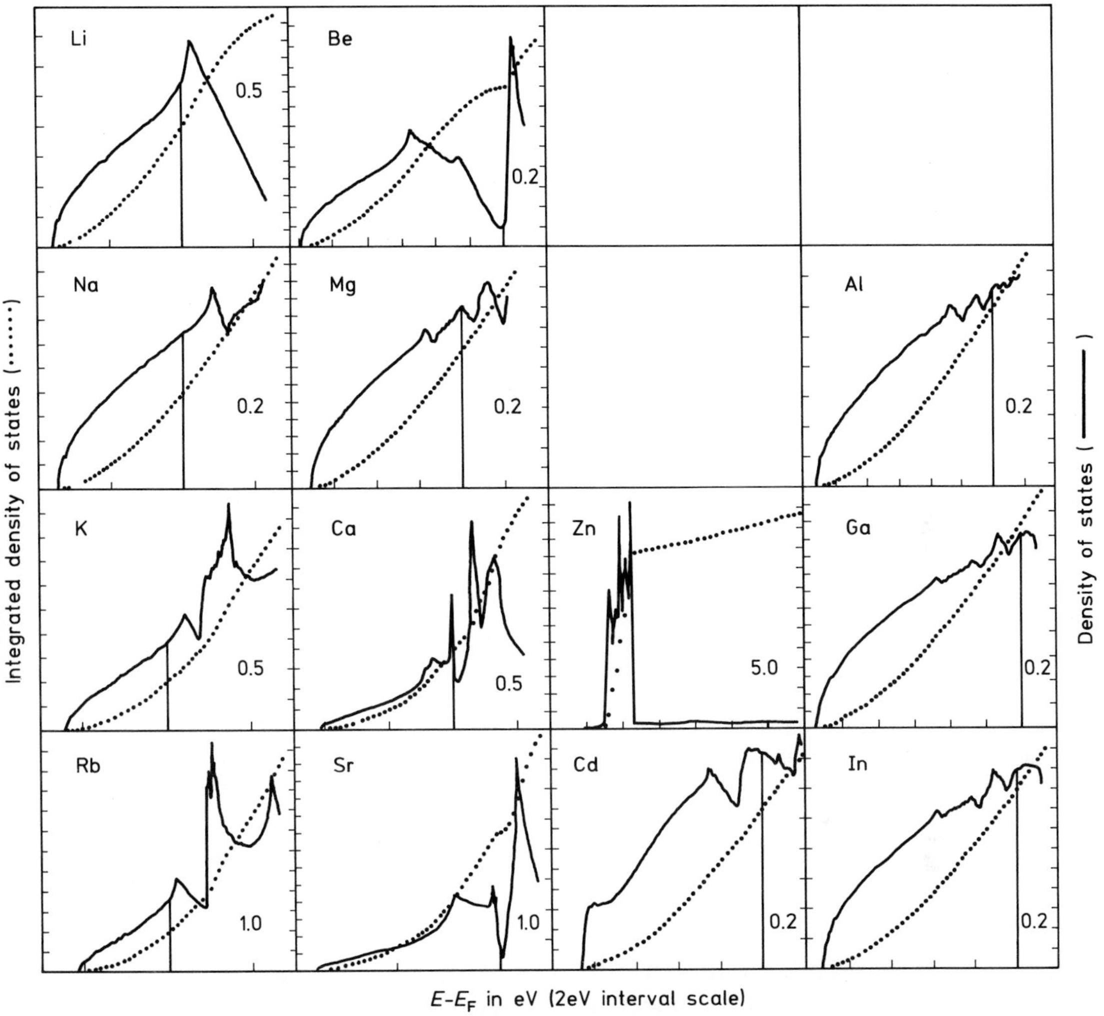

Figure 1-24. Densities of states of selected simple metals after Moruzzi et al. (1978). Vertical lines indicate the Fermi energy.

language). The Fourier components of the potential for the p electrons thus are large and this leads to very large band gaps; we therefore look at Be in some more detail. Its band structure, shown in Fig. 1-25, was computed by Chou et al. (1983) using *norm-conserving pseudopotentials* and the correct h.c.p. structure – in the calculations of Moruzzi et al. (1978), h.c.p. systems were treated as if they had the f.c.c. structure. We see that the bottom of the band is free-electron like but, indeed, large gaps occur at the Fermi energy, E_F, where p states set in. The density of states also shown in Fig. 1-25 is very similar to that depicted in Fig. 1-24 for f.c.c. Be. In fact, the band gaps in different directions of the Brillouin zone are seen to be nearly large enough to open a gap in the density of states at E_F, thereby leading to semiconducting behavior. Chou et al. (1983) also show interesting charge density plots which are reproduced in Fig.

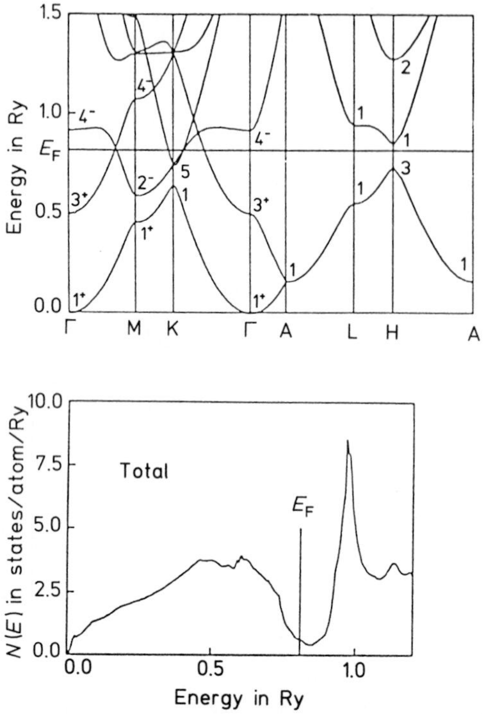

Figure 1-25. Band structure and density of states for Be calculated in the h.c.p. structure. Reprinted with permission from *Phys. Rev. B28,* Chou, M. Y., Lam, P. K., Cohen, M. L. (1983), *Ab initio study of structural and electronic properties of beryllium,* p. 4179.

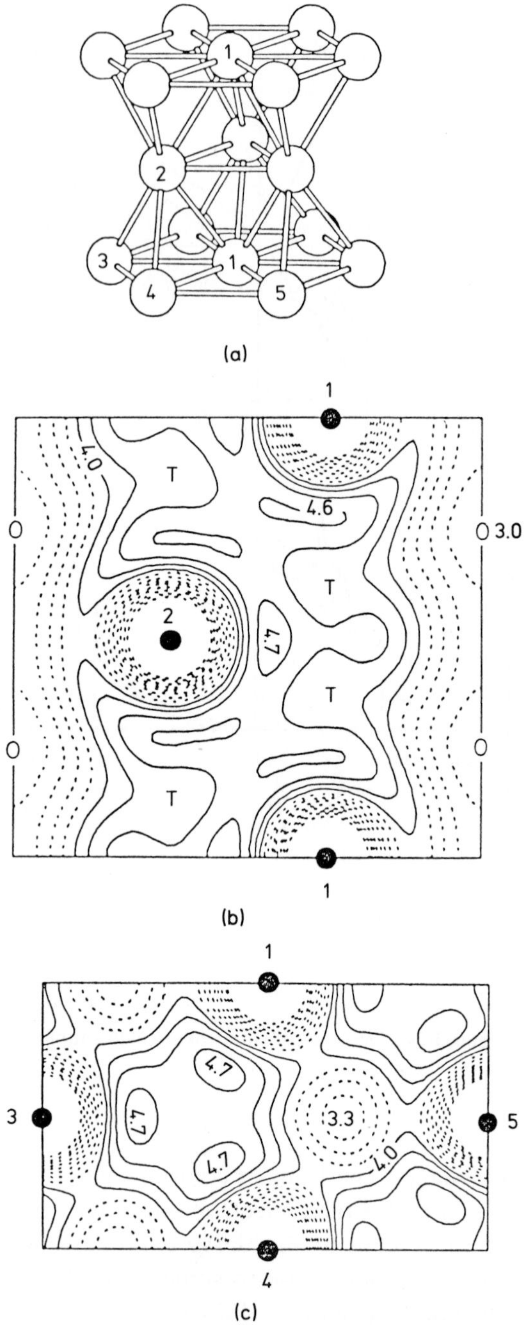

Figure 1-26. (a) Ball-and-stick model for the h.c.p. structure. Valence charge density for Be: (b) contour plot in the [11$\bar{2}$0]-plane; (c) contour plot in the [0001]-plane (see text as well). Reprinted with permission from *Phys. Rev. B28,* Chou, M. Y., Lam, P. K., Cohen, M. L. (1983), *Ab initio study of structural and electronic properties of beryllium,* p. 4179.

1-26. The two planes shown are defined by the numbers at the atomic positions which correspond to those given in the model of the structure. Contours with a charge density smaller than 4.0 are shown with dashed lines, those larger than 4.0 with solid lines. This figure clearly shows the deviation from (uniform) free-electron behavior which gives a charge density of 4.0 electrons per cell volume (two valence electrons for each of the two atoms in h.c.p.). Modern X-ray diffraction studies of Be could be used for comparison with the calculated charge densities; they exist but do not seem to be published (Chou et al., 1983).

In the heavier alkalis K and Rb and alkaline earths Ca and Sr, the occupied bands are affected by the presence of the 3d or 4d states which lie just above the Fermi level, E_F. Therefore d states are admixed into the occupied states leading again to departure from nearly free electron behavior; this effect is so strong for Sr that the hybridized bottom of the d band has moved below E_F and has nearly opened up a gap as in Be. In fact, theoretically, it requires only 0.3 GPa of pressure to turn Sr into a semiconductor, which is in reasonable agreement with experimental data (Jan and Skriver, 1981). Recent calculations for Ba (Chen et al., 1988b) also show the proximity of the 5d band to the Fermi level and the ensuing strong departure from NFE behavior which leads to an interesting pressure-induced b.c.c.-h.c.p. phase transition, see Sec. 1.4.1.2. The group IIB elements Zn and Cd, on the other hand, have strongly distorted valence states because of *filled* d bands in contrast to the empty ones in the case of K, Rb, Ca, and Sr. Figure 1-24 thus demonstrates that not all simple metals display NFE behavior and, in particular, care needs to be taken with Li and Be and the group II elements on either side of the transition metal series (Pettifor, 1984).

Transition Metals

Transition metals are characterized by a partially filled d band which becomes more and more occupied as we move in between the group II elements from the left to the right in the Periodic Table. In zeroth order the bands can be described within the tight binding approximation (Sec. 1.2.4.1) (e.g. Pettifor, 1984), but we choose to discuss them using first principles calculations like those of Moruzzi et al. (1978) or ASW calculations which were produced for this chapter. These we want to supplement with canonical bands shown in Fig. 1-12 in order to recognize the "fingerprint" of the d states.

The gross feature of the band structure of the transition metals is easily stated: as we move across the series, the d states are filled, and the bands first widen slightly moving up in energy, then they move down and become narrower. Figure 1-27 shows the trend of the most important band parameters for the 4d transition metals. It is inspired by a figure of Pettifor (1977a, b) to which it bears great similarity, but the details are different since we used parameters from fully hybridized, self-consistent scalar-relativistic ASW calculations and

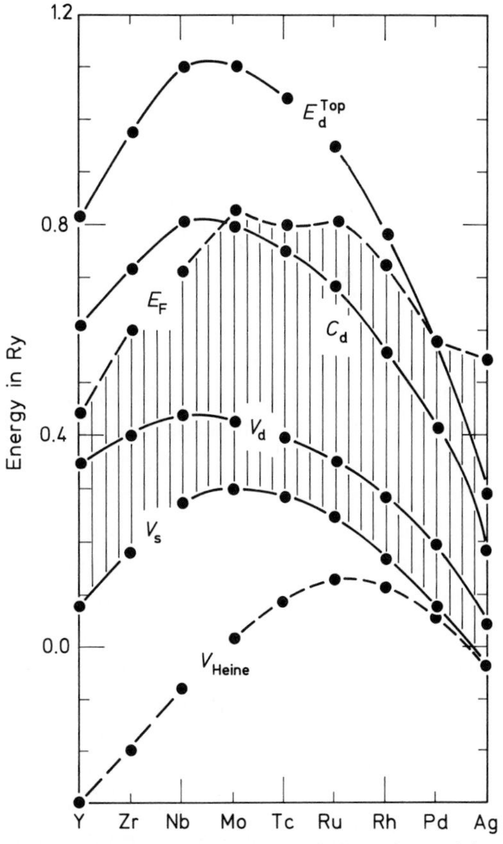

Figure 1-27. Band parameters of the 4d transition metals: Fermi energy E_F; top, center and bottom E_d^{TOP}, C_d, and V_d of the d band; bottom of the s band V_s and the Heine power law V_{Heine}.

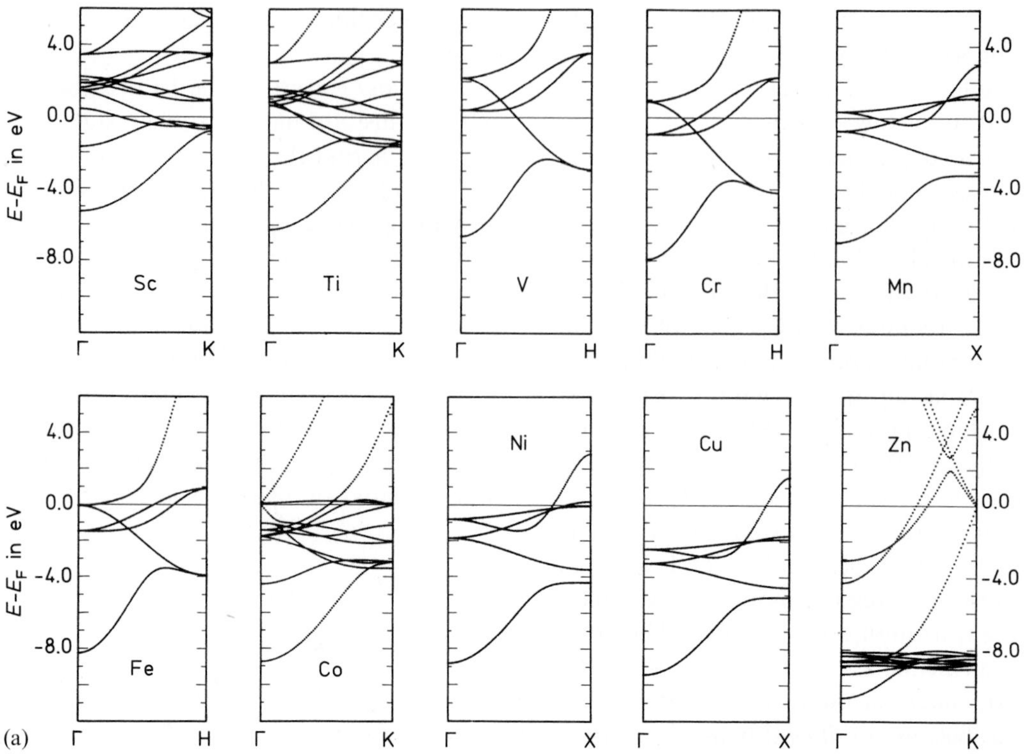

Figure 1-28. Parts of the band structures for the transition metals calculated with the ASW method at the experimental lattice constants taken from Landolt-Börnstein (1973). (a) 3d metals, (b) 4d metals, (c) 5d metals (La is assumed to have the h.c.p. structure and Hg is calculated within an s.c. lattice).

slightly different definitions of the band parameters. The bottom of the s band is denoted by V_s; it is the Bessel function energy as explained in Sec. 1.2.4.6 and agrees closely with the lowest Γ_1 state in the band structure. The quantity V_d is the bottom of the d band which is obtained as the energy of the bonding state (zero logarithmic derivative) in the self-consistent potential. C_d is the center of the d band which is very close to the equivalent quantity in LMTO theory (Sec. 1.2.4.5), but it is here the Hankel function energy (see Sec. 1.2.4.6). E_F denotes the Fermi energy and E_d^{TOP} is the top of the d band which is obtained as the energy of the antibonding state (infinite logarithmic derivative) in the self-consistent potential and is slightly higher in energy than the corresponding band parameter in Pettifor's (1977a, b) analysis. The quantities V_d and E_d^{TOP} are thus the energies obtained from so-called Wigner-Seitz (1955) boundary conditions. The trends in the 3d and 5d series are approximately the same as in the 4d series. The bands are, however, narrower for 3d electrons, as their radial wave function is more contracted and has one fewer node. On the other hand the bands are slightly wider for the 5d electrons, since their radial wave functions overlap more and have one more node than the 4d electrons.

The parabolic trend that is so clearly visible in Fig. 1-27 is explained as follows: on the left-hand side, bonding states are occupied that accumulate charge mostly

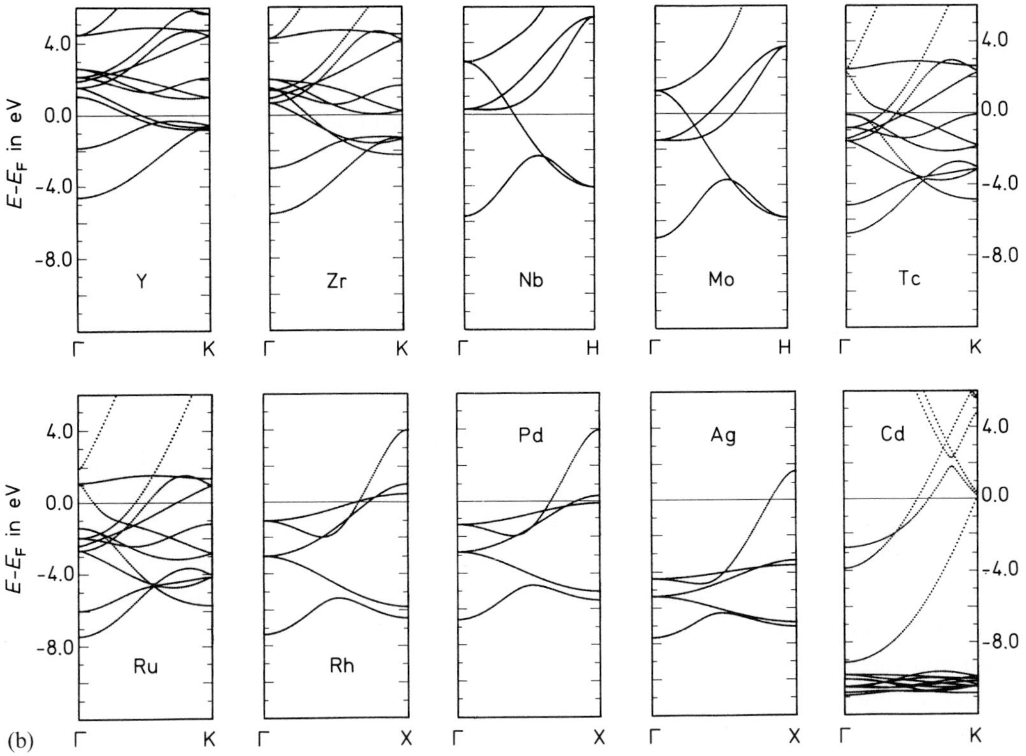
E-E_F in eV
4.0
0.0
-4.0
-8.0
Y
Zr
Nb
Mo
Tc
Γ K
Γ K
Γ H
Γ H
Γ K
Ru
Rh
Pd
Ag
Cd
Γ K
Γ X
Γ X
Γ X
Γ K
(b)

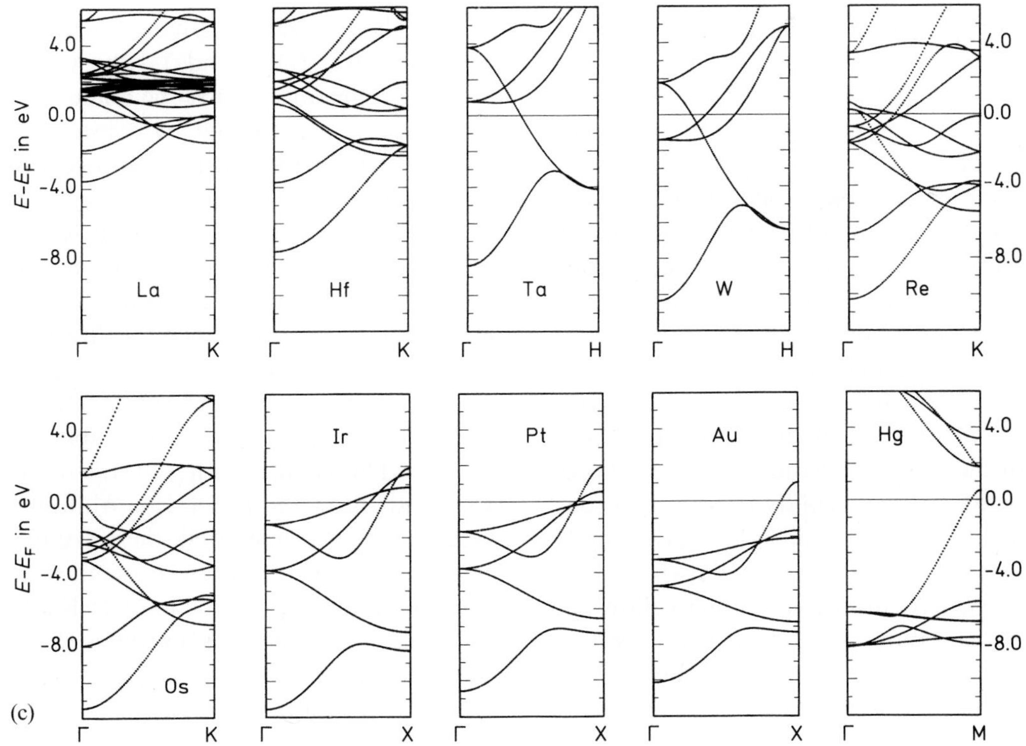
E-E_F in eV
4.0
0.0
-4.0
-8.0
La
Hf
Ta
W
Re
Γ K
Γ K
Γ H
Γ H
Γ K
Os
Ir
Pt
Au
Hg
Γ K
Γ X
Γ X
Γ X
Γ M
(c)

in between two atoms, thus keeping the charge *away* from the nuclei. As a consequence the lattice shrinks and the bands move up. Obviously, the equilibrium Wigner-Seitz (or atomic) radii, S, must reflect this behavior – and they do, as the curve V_{Heine} shows, which because of Heine's (1967) power law is an energy given by

$$V_{\text{Heine}} = \text{const} + C/S^5 \qquad (1\text{-}439)$$

where the constant and C are chosen arbitrarily so that the curve neatly fits into Fig. 1-27. In the middle of the series non-bonding and further to the right anti-bonding states are occupied. Since the nuclear charge is only poorly screened it pulls electrons *toward* the nuclei, thus lowering the energy of the states and decreasing the overlap and thereby narrowing the bands. At the same time the lattice expands. This intimate connection of the band structure with cohesive properties is further analyzed in the next subsection. Finally, the Fermi energy, E_F, displays a parabolic trend because of the rising *and* the filling of the d band and then decreases on account of the lowering in the center of gravity of the d band.

Figure 1-28 depicts small sections of the band structure of all transition metals. It was obtained as in Fig. 1-27 using self-consistent scalar-relativistic ASW calculations with experimental lattice constants and structure information (Landolt-Börnstein, 1971). The d bands can be clearly recognized in all cases if one compares them with the canonical bands in Fig. 1-12 and remembers the role of hybridization as depicted in Figs. 1-13 and 1-14. Corresponding densities of states are shown in Fig. 1-29. We have made a selection to allow the same structure type to occur down a group, but different ones to occur across the period. It is seen that each structure type has a characteristic "fingerprint" which is similar to unhybridized canonical d-state densities: this can be seen if Fig. 1-29 is compared with a canonical d-state density, for instance that given by Andersen and Jepsen (1977). But the features are generally sharper in Fig. 1-29 than in canonical densities of states because hybridization with sp bands repels band of the same symmetry, thereby creating local gaps and hence more pronounced structure. An early and interesting explanation of the different structure types in the density of states of transition metal d electrons was attempted by Mott (1964). What we are concerned about is the pronounced minimum in the case of the b.c.c. density of states which is in contrast to f.c.c. and h.c.p. This is an important feature which shows up in phase stabilities (Sec. 1.4.1.3) and, as Mott (1964) pointed out, in b.c.c. metallic alloys, but not in f.c.c. or h.c.p. alloys. Let us concentrate on the difference between the b.c.c. and f.c.c. density of states and consult the canonical d-band structure shown in Figs. 1-12a and 1-12b. Mott's original argument can now be made with confidence: he started out by considering unhybridized d bands, just as they occur in Fig. 1-12b along Γ to H. Along other directions, however, the bands hybridize, resulting in two narrow bands at the bottom and two narrow bands at the top being crossed by *one* band as in for instance, the direction Γ to N, or even Γ to P, where for reasons of symmetry (space-group representations) the lower and upper states are doubly degenerate. This is different in the case of the f.c.c. bands shown in Fig. 1-12a. Although the bands hybridize as well, there are *two* bands that connect the bottom and the top, for example, along the direction Γ to K and X, in such a way that the bottom and top states are more mixed up and not separated in two groups as in b.c.c.

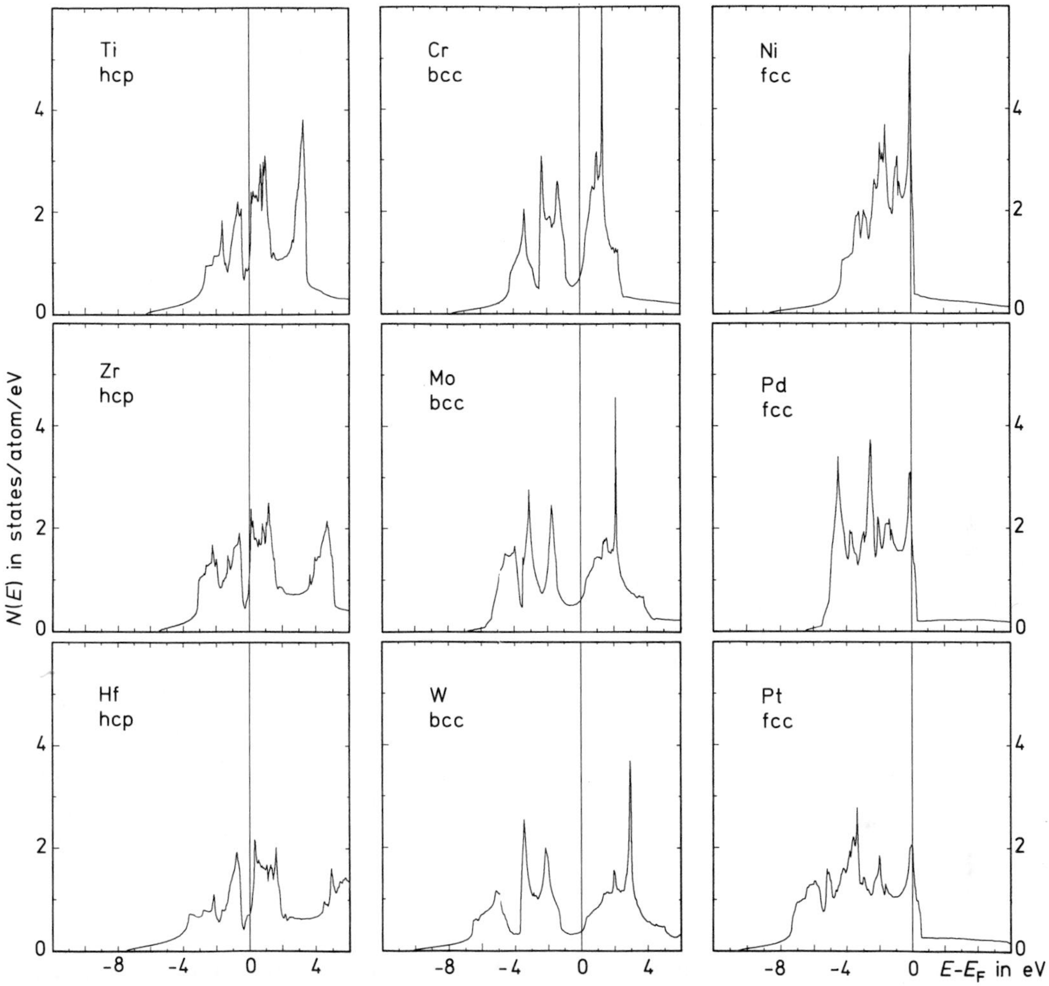

Figure 1-29. Densities of states of selected transition metals (3d to 5d from top to bottom and in the sequence h.c.p-b.c.c.-f.c.c. from left to right). ASW method used.

In Fig. 1-30 we finally show calculated values of the density of states at the Fermi energy for the transition metals using data from Papaconstantopoulos (1986). The obvious, huge variations reflect, of course, the peak structure of the density of states shown in Fig. 1-29, and they come about because these states are successively filled moving the Fermi energy across the bands. The values marked with a cross (×) are experimental Sommerfeld constants, γ (Kittel, 1986) (converted to state-density values) which are obtained from the specific heat at low temperatures and are

$$\gamma = \tfrac{1}{3}\,\pi^2\,k_B^2\,\tilde{N}(E_F) \qquad (1\text{-}440)$$

where k_B is Boltzmann's constant and $\tilde{N}(E_F)$ is the quasiparticle density of states at the Fermi energy. It is usually written in terms of the band density of states by the simple relation

$$\tilde{N}(E_F) = (1+\lambda)\,N(E_F) \qquad (1\text{-}441)$$

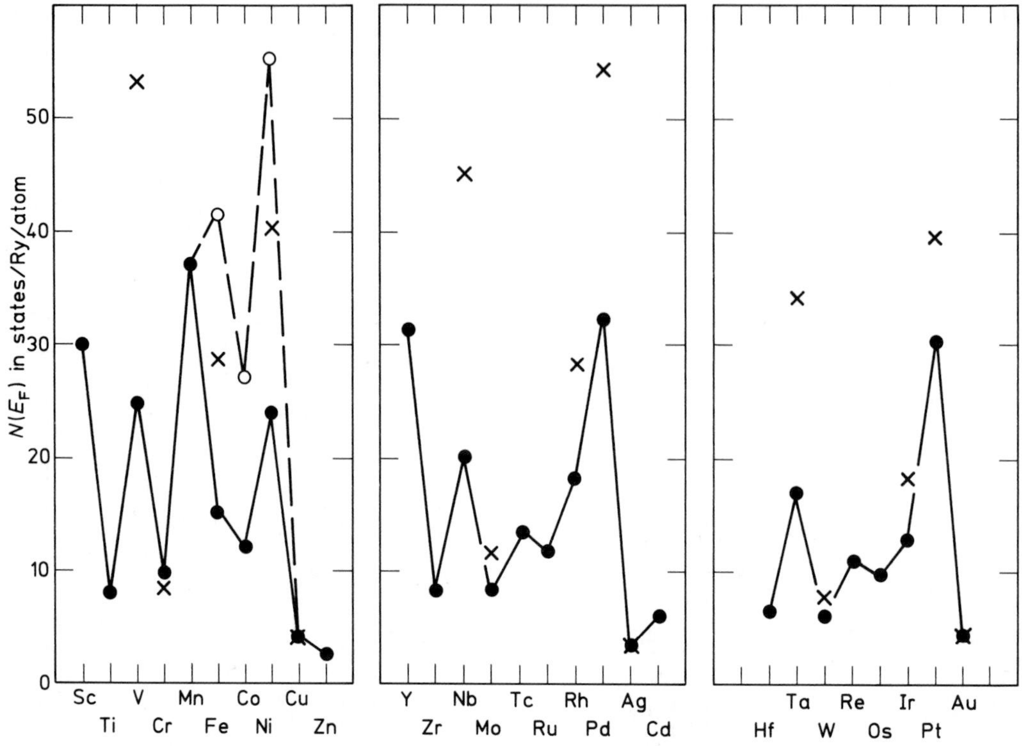

Figure 1-30. Densities of states at the Fermi energy of the transition metals based on data from Papaconstantopoulos (1986). Experimental values marked by crosses are calculated from the Sommerfeld constants as given by Kittel (1986). The dashed curve corresponds to the respective data from the non-spin-polarized calculations by Moruzzi et al. (1978).

and λ is known as the "electron-phonon" enhancement factor which is so prominent in the McMillan formula for the superconducting transition temperature. The connection is certainly not that simple, however, and a recent short survey on this matter appeared by Allen (1987). Thus, for instance, the sizeable enhancement for Pd ($Z=46$) and Pt ($Z=78$) does not correlate with the superconducting transition temperature at all because it is zero experimentally. Recent first principles calculations of $N(E_F)$ for Nb (Jani et al., 1988), Mo (Jani et al., 1989) and Pd (Chen et al., 1989), for which the linear combination of Gaussian orbitals method was used, are in very good agreement with the values given in Fig. 1-30. We also show values of $N(E_F)$ for assumed non-magnetic Fe, Co, and Ni as open circles connected with dashed lines calculated by Moruzzi et al. (1978). In contrast to this are the other values which are a sum of spin-up and spin-down contributions – we will come back to the case of spin-polarized energy bands in detail in Sec. 1.4.2. But now we turn to a more thorough and quantitative treatment of bonding trends that were seen to be so intimately connected with the band-structure trends discussed in connection with Fig. 1-27. Of course, we realize that the preceding description of the band structures of elementary metals is highly incomplete; thus, for instance, we have not shown any

Fermi surfaces and related them to experimental information except for Os in Sec. 1.2.5, Fig. 1-21, and Table 1-1. We also have not given enough information on photoemission data except for Al in Fig. 1-23 and Cu in Fig. 1-9. Also omitted are calculations of the optical conductivity using band-structure data that have been quite successful. The interested reader should consult the literature here: For Fermi surfaces, Mackintosh and Andersen (1980) and, e.g., Jani et al. (1988, 1989) and Chen et al. (1989), the latter also giving theoretical and experimental data of optical conductivities; and for photoemission, Campagna et al. (1979), Cardona and Ley (1979), Hüfner (1979), Schaich (1979), Shirley (1979), Smith (1979), Steiner et al. (1979), Wertheim et al. (1979). The subsection that follows rests heavily on a paper by Williams et al. (1980 a); a slightly different point of view is taken by Pettifor (1984) and Andersen (1984), but the basic facts are undoubtedly the same.

1.4.1.2 Cohesive and Elastic Properties

Metallic cohesion has presented a challenge to theorists since the development of quantum mechanics. The early work of Wigner and Seitz (1933, 1934, 1955) provided an understanding of cohesion in simple metals, whereas in the case of transition metals, Friedel (1968) elucidated much later the important role played by d band "covalency". In the 1970s, quantitative work employing self-consistent field calculations, based on Slater's X_α method (Averill, 1972; Conklin et al., 1972, 1973; Trickey et al., 1973) and on the local-density approximation (Janak et al., 1975), showed that a complete theory was in reach, but by themselves these calculations shed little light on the microscopic mechanisms responsible for cohesion. They did, however, demonstrate that a single formulation could describe cohesion in a broad variety of systems – from simple and transition metals to rare-gas solids.

For the simple metals some progress can be achieved in understanding bulk properties such as the equilibrium atomic volume and structural stability by using the total energy of a uniform gas of interacting electrons, according to the LDA (see Sec. 1.3.3), treating the ionic lattice in first-order perturbation theory (Heine and Weaire, 1970; Girifalco, 1976; Harrison, 1980; Pettifor, 1984). But the uniform electron gas cannot provide reliable cohesive energies, which require an accurate comparison with the *free* atom whose wave functions are not describable by weakly perturbed plane waves. It is necessary, therefore, to perform *similar* calculations in both the free atom and the bulk. In this way, total energies are obtained whose difference is the cohesive energy, apart from a small contribution due to zero-point vibrations. This was done for the simple metals in the early calculations of Wigner and Seitz, and for the prototypical case of Na, their results are easily explained; when the Na atoms are brought together, the energy of the 3s atomic level is lowered because the boundary condition of its wave function changes from zero value at infinity to zero slope at the Wigner-Seitz radius (the Wigner-Seitz bonding boundary condition). The energy gain was computed to be −3.05 eV. Filling up this band, one pays a kinetic energy price of $\frac{3}{5}E_F$, where E_F is the Fermi energy, or 1.9 eV. The difference is −1.15 eV which is nearly identical with the experimental value for the cohesive energy. Interestingly, the agreement is fortuitous because the exchange-correlation contributions, the so-called double-counting terms in Eq. (1-384), were not considered, but in this case they cancel numerically. Still, these early calculations isolated an important bonding mechanism: the lowering of the atomic valence level when the atoms condense.

In 1977 Gelatt et al. (1977) analyzed binding in the entire 3d and 4d transition series using the "renormalized-atom" concept (Watson et al., 1970). This was an approximate scheme that we did not deal with in the preceding section because we can now get more precise but equivalent information from an LMTO or ASW calculations. Gelatt's analysis identified the role, not only of d-band covalency, but of *configuration* changes, i.e., the change of the distribution of valence charge among s, p and d states, and s-d hybridization as well. In an extensive set of self-consistent field calculations using the KKR method and the local-density approximation to electronic exchange and correlation effects, Moruzzi et al. (1978) demonstrated that such calculations can accurately describe the lattice constant, by minimizing the total energy, as well as the cohesive energy and compressibility of the full range of metals, given only the atomic number. The results of these calculations for the 3d and 4d transition series are reproduced in Fig. 1-31. With the credibility of the fundamental theoretical framework established, subsequent efforts have been made toward simplifying the calculations and decomposing them into intuitively understandable components (Anderson, 1975, 1984; Williams et al., 1979; Pettifor, 1976, 1977a, b, 1978a, b).

Beginning with a more detailied analysis now, we first notice in Fig. 1-31 a striking departure from the otherwise close agreement with experiment seen in the case of the magnetic 3d transition metals Cr, Mn, Fe, Co, and Ni. (The reader should concentrate on the Wigner-Seitz radius or the bulk modulus; the departures in the cohesive energies occur for a different reason.) We will show later (Sec. 1.4.2) that most of the deviation of the paramagnetic calculations from experiment can be ascribed to the dilation of the metal lattice which accompanies the development of magnetic order. Thus postponing the treatment of the magnetic metals to Sec. 1.4.2, we next introduce the hydrostatic pressure as an intuitively satisfying concept to analyze cohesion. The pressure is, of course, the negative volume derivative of the total energy, E,

$$P = -\,\partial E/\partial \Omega \tag{1-442}$$

and, as such, does not seem to supply much more insight than the total energy itself. But an alternative expression exists for the pressure which allows for an unambiguous identification of the role played in the binding process by electrons differing in angular momentum. We pause to sketch the derivation of this pressure formula that has been given by a number of authors using a variety of different methods (Liberman, 1971; Janak, 1974; Pettifor, 1976; Nieminen and Hodges, 1976; Mackintosh and Andersen, 1980; Heine, 1980; Methfessel and Kübler, 1982; Christensen and Heine, 1985).

We write the total energy as

$$E = \int^{E_F} \varepsilon\, N(\varepsilon)\, \mathrm{d}\varepsilon - E_1 \tag{1-443}$$

where $N(\varepsilon)$ is the density of states, and E_1 is the double-counting term which, from Eq. (1-384) in the local density approximation, is given by

$$E_1 = \iint_{\Omega} \frac{n(\boldsymbol{r})\, n(\boldsymbol{r}')}{|\boldsymbol{r}-\boldsymbol{r}'|}\, \mathrm{d}^3r\, \mathrm{d}^3r' + \int_{\Omega} \left[n^2 \frac{\mathrm{d}\varepsilon_{\mathrm{xc}}}{\mathrm{d}n} \right]_{n=n(\boldsymbol{r})} \mathrm{d}^3 r \tag{1-444}$$

where $\varepsilon_{\mathrm{xc}}$ is the exchange-correlation-energy density. One is tempted to decompose the total energy into contributions coming

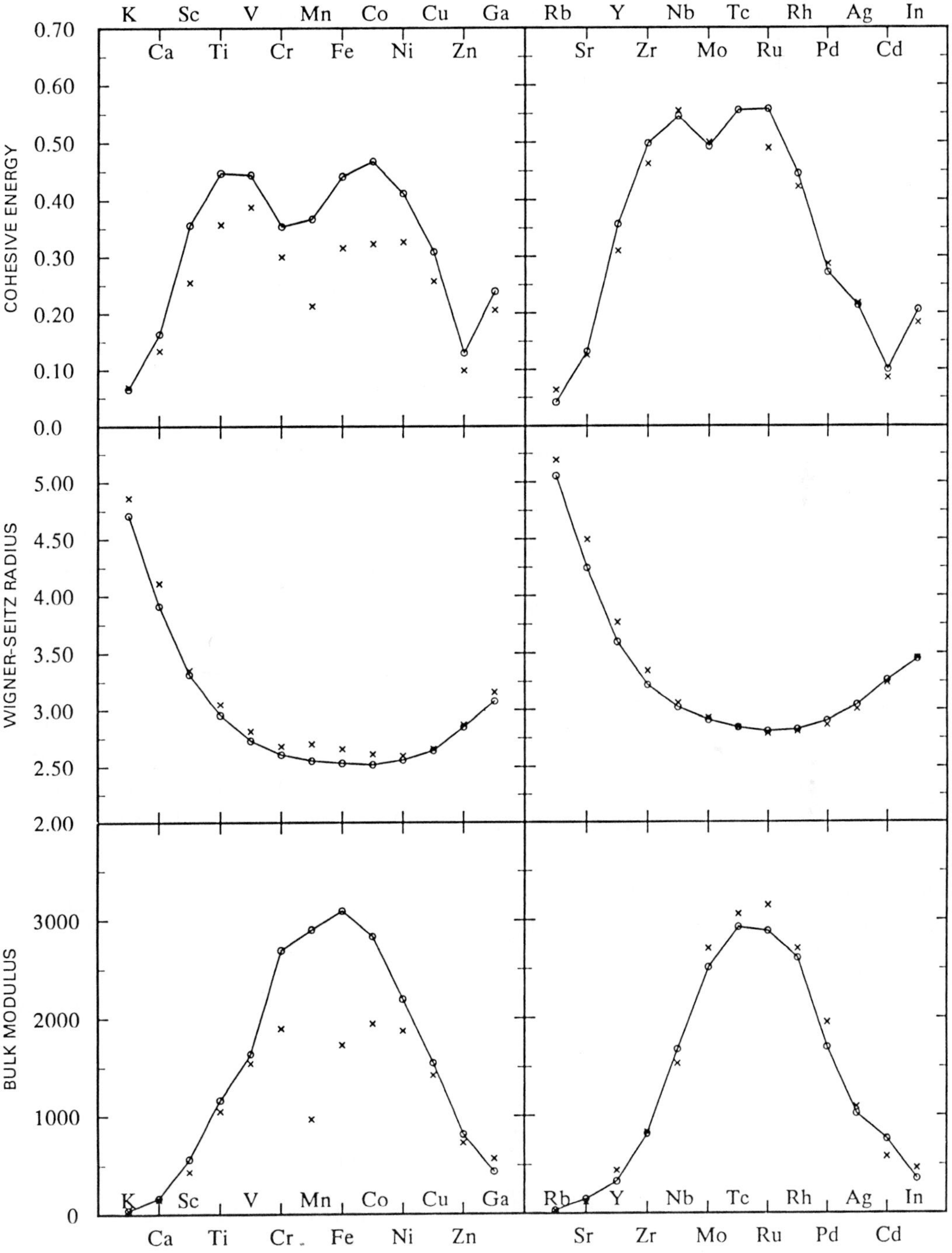

Figure 1-31. Bulk moduli in kbar, Wigner-Seitz radii in atomic units and cohesive energies in Ry per atom of selected metals. Crosses indicate experimental values. Reprinted with permission from Moruzzi, V. L., Janak, J. F., Williams, A. R. (1978), *Calculated electronic properties of metals.* Pergamon Press PLC.

from different angular momenta because the density of states can be so decomposed in this way (see Eq. (1-304)). But it is the double counting term above that makes this decomposition ambiguous. Thus at first, we calculate generally the first-order change in E when the system is subjected to some perturbation; for example, a perturbation of the boundary conditions, of the occupation counts, of the external potential or of the shape and size of the volume Ω. We assume these changes to be parametrized by some set of quantities X_1, $X_2, \ldots$ called $\{X_i\}$; the self-consistent total energy is then a function of these parameters and the aim is to calculate

$$\delta E = \sum_j \frac{\partial E}{\partial X_j} \delta X_j \tag{1-445}$$

For a given disturbance, specified by some set of $\{\Delta X_i\}$, we consider the system as moving to new self-consistency in two steps. In the first step, the effective potential v_{eff}, Eq. (1-377), is held fixed while the Schrödinger equation is solved to the new parameter set $\{X_i + \delta X_i\}$ giving rise to new occupation numbers and eigenvalues. Denoting this restricted variation by the symbol δ^*, we obtain the entire variation by adding δ^{SC} that is obtained in a second step in which the potential is allowed to relax to self-consistency keeping the parameter set fixed at $\{X_i + \delta X_i\}$, i.e.,

$$\delta = \delta^* + \delta^{\text{SC}} \tag{1-446}$$

Thus, for instance, the change of the single-particle term in Eq. (1-443),

$$E_0 = \int^{E_{\text{F}}} \varepsilon N(\varepsilon)\, \mathrm{d}\varepsilon \tag{1-447}$$

is

$$\delta E_0 = \delta^* E_0 + \delta^{\text{SC}} E_0 \tag{1-448}$$

and the change of the entire total energy is

$$\delta E = \delta^* E_0 + \delta^{\text{SC}} E_0 - \delta E_1 \tag{1-449}$$

The essential observation now is that the last two terms on the right-hand side cancel in first order leaving only a surface term when the volume changes, i.e.,

$$\delta E = \delta^* E_0 - \int_{\partial\Omega} n^2 \frac{\mathrm{d}\varepsilon_{\text{xc}}}{\mathrm{d}n} \delta \boldsymbol{S} \cdot \mathrm{d}\boldsymbol{S} \tag{1-450}$$

where $\mathrm{d}\boldsymbol{S}$ is the surface element and $\delta\boldsymbol{S}$ describes the change in Ω in the sense that a point $\boldsymbol{S}$ is taken to $\boldsymbol{S} + \delta\boldsymbol{S}$ by the perturbation. The proof of the cancellation is not difficult; one writes out the terms δE_1 and δV and uses first-order perturbation theory on the term $\delta^{\text{SC}} E_0$ (see, e.g., Methfessel and Kübler, 1982).

Before we finally determine $\delta^* E_0$, we should note that the basic idea behind Eq. (1-450) is the same as in Andersen's local force theorem (Heine, 1980; Mackintosh and Andersen, 1980); namely, that in linear approximation the interaction drops out of the total energy change. In particular, the self-consistency step can be ignored, except for the small surface term in the case of a volume change. We therefore have a mathematical justification for thinking in terms of independent electrons and can generally calculate forces with Eq. (1-450).

The hydrostatic pressure, Eq. (1-442), is now obtained from Eq. (1-450) by

$$-P = \frac{\partial E}{\partial \Omega} = \frac{1}{3\Omega} \frac{\partial E}{\partial \ln S} \tag{1-451}$$

i.e.

$$3P\Omega = \frac{\partial^*}{\partial \ln S} \int^{E_{\text{F}}} \varepsilon N(\varepsilon)\, \mathrm{d}\varepsilon + + \int_{\partial\Omega} n^2 \frac{\mathrm{d}\varepsilon_{\text{xc}}}{\mathrm{d}n} \mathrm{d}\boldsymbol{S} \cdot \frac{\partial \boldsymbol{S}}{\partial \ln S} \tag{1-452}$$

Assuming we know the exact *energy-dependent* wave function in the atomic sphere (like in the derivation of the density, Eq. (1-301)), we determine the first term on the right-hand side in Eq. (1-452) using

$$\frac{\mathrm{d}n_l(\varepsilon)}{\mathrm{d}\varepsilon} = N_l(\varepsilon) \tag{1-453}$$

and an integration by parts

$$-\frac{\partial^*}{\partial \ln S} \int^{E} \varepsilon\, N(\varepsilon)\, \mathrm{d}\varepsilon = = \sum_l \int^{E_\mathrm{F}} \frac{\delta^* n_l(\varepsilon)}{\delta \ln S}\, \mathrm{d}\varepsilon \tag{1-454}$$

Furthermore, using the Wronskian for the radial wave function $R_l(\varepsilon, r)$, we obtain with Eq. (1-453)

$$3P\Omega = -\sum_l \int^{E_\mathrm{F}} N_l(\varepsilon)\, S\, R_l^2(\varepsilon, S)\, \frac{\delta^* D_l}{\delta \ln S}\, \mathrm{d}\varepsilon + + \int_{\partial\Omega} n^2 \frac{\mathrm{d}\varepsilon_\mathrm{xc}}{\mathrm{d}n}\, \mathrm{d}\boldsymbol{S} \cdot \frac{\partial \boldsymbol{S}}{\partial \ln S} \tag{1-455}$$

and finally, upon calculating the change of the logarithmic derivative, D_l, directly with the radial Schrödinger equation, we obtain the desired angular momentum decomposition of the pressure in the form

$$3P_l\Omega = \int^{E_\mathrm{F}} N_l(\varepsilon)\, S\, R_l^2(\varepsilon, S) \cdot \cdot \left[D_l(D_l+1) - l(l+1) + S^2(\varepsilon - V(S)) + + S^2 n \frac{\mathrm{d}\varepsilon_\mathrm{xc}}{\mathrm{d}n}\bigg|_{r=S} \right] \mathrm{d}\varepsilon \tag{1-456}$$

The LMTO or ASW forms of this equation are easily written out, in the former case by expanding R_l using $\dot{R}_l$, and in the latter by using weights and sampling energies as in the determination of the density (Mackintosh and Andersen, 1980; Williams et al., 1979). The total pressure is, of course,

$$3P\Omega = \sum_l 3P_l\Omega \tag{1-457}$$

Returning now to the main topic of this subsection, we take an overall look at the trends that are so conspicuous in Fig. 1-31 and show in Fig. 1-32 partial pressures, $3P_l\Omega$, computed by an LMTO approximation to Eq. (1-456) and taken from Andersen et al. (1985). It is quite apparent that for the 4d and 5d transition elements (and for the 3d transition elements), the dominating attractive force is due to the d electrons which are therefore responsible for the parabolic trends. Going into more detail, we reproduce in Fig. 1-33 results obtained by Williams et al. (1980a) for $3P\Omega$ of Mo using self-consistent ASW calculations. The pressure is decomposed into contributions of the 4d and the delocalized 5s and 5p electrons (the latter are added and indicated by 5sp) using Eq. (1-456). Fig. 1-33 quite distinctly illustrates again the attractive power of the d electrons and their dominant contribution to the cohesive energy relative to that of the mobile 5s and 5p electrons. Note that because of Eqs. (1-451) and (1-457) the cohesive energy is the area under both curves. It is therefore important that the calculation is carried out up to the atomic limit if it is to supply the cohesive energy. Not further that equilibrium (zero total pressure) is achieved by cancelling the attractive d pressure and the repulsive non-d pressure. The analysis of cohesion due to Friedel (1968) suggests that once the d shell is filled, as in a noble metal, it no longer contributes to cohesion. This is not exactly true, as can be seen by Fig. 1-34 in which results obtained by Williams et al. (1980a) are reproduced for Cu. While the filled d shell does not dominate cohesion as an open d shell does, it nevertheless makes a substantial contribution. Note that even in Cu it is the compression of the mobile electrons which provides the repulsive force that balances the attractive force of the d electrons at equilibrium. The intuitively attractive notion that the *core* electrons are responsible for the repulsion at equilibrium is thus incorrect.

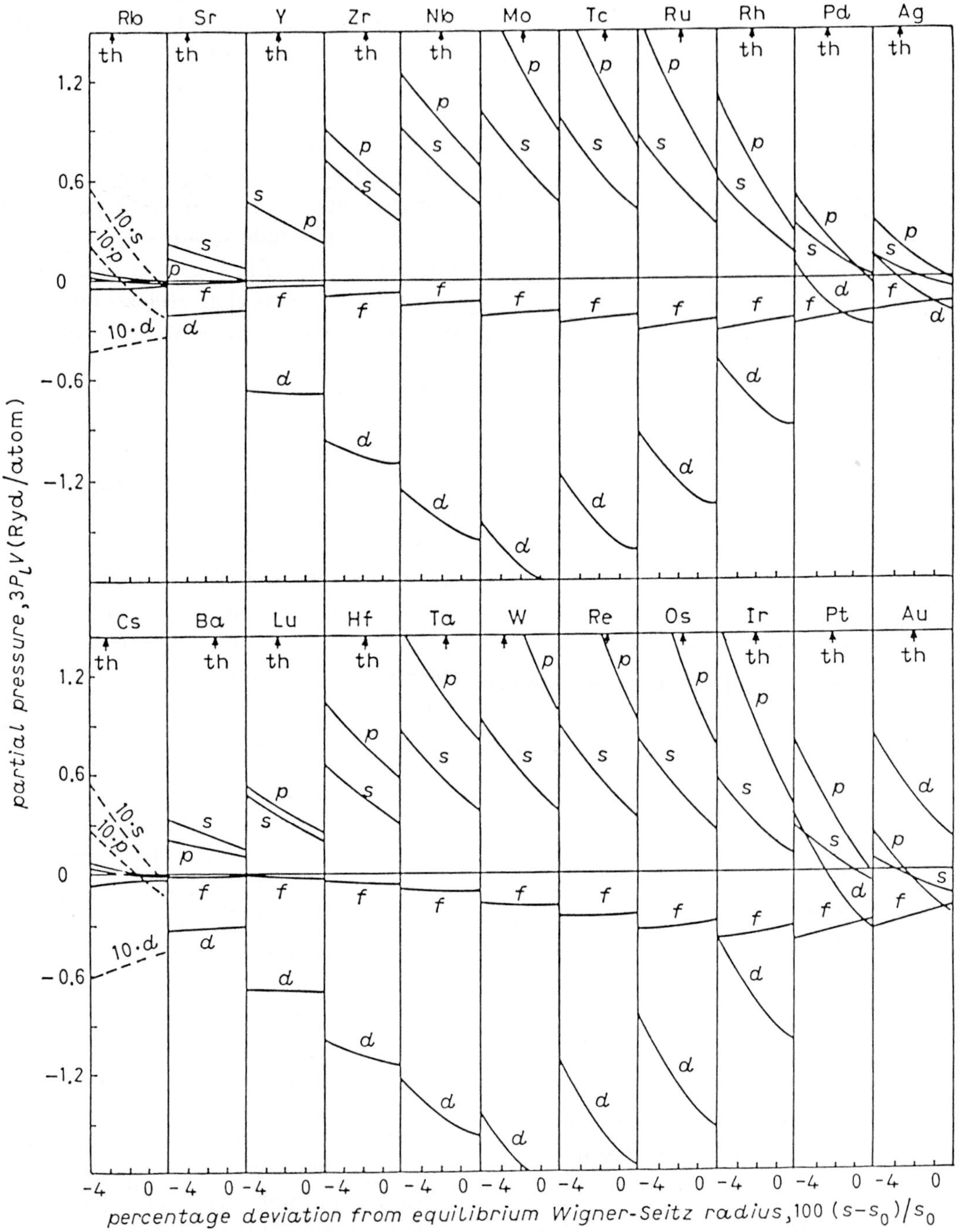

Figure 1-32. Partial pressures calculated for the Rb and Cs series as functions of $\frac{(s-s_0)}{s_0}$, where s_0 is the experimentally observed equilibrium radius. th indicates the calculated equilibrium Wigner-Seitz radius. Reprinted with permission from Andersen, O. K., Jepsen, O., Glötzel, D. (1985), in: *Highlights of Condensed Matter Theory, Proceedings of the International School of Physics "Enrico Fermi", Course LXXXIX:* Bassani, F., Fumi, F., Tosi, M. P. (Eds.). Amsterdam: North Holland, pp. 59–176.

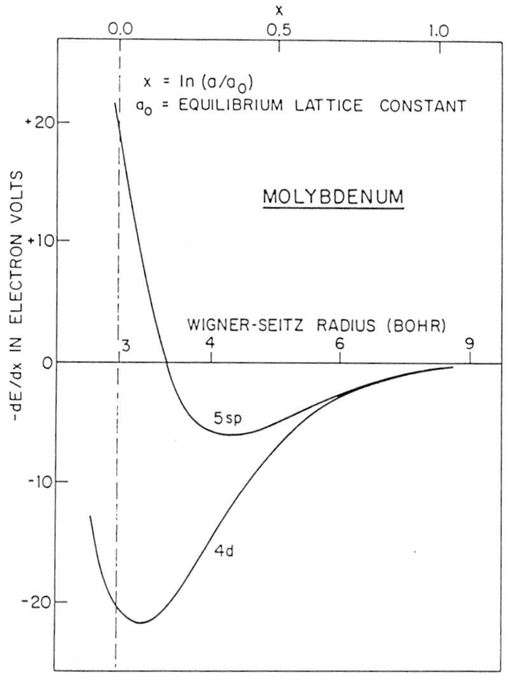

Figure 1-33. Pressure versus lattice constant curve for Mo separated into the sp and the d contributions. Reprinted by permission, *Theory of Alloy Phase Formation:* Bennett, L. H. (Ed.) (1980). Warrendale (PA): The Metallurgical Society of AIME.

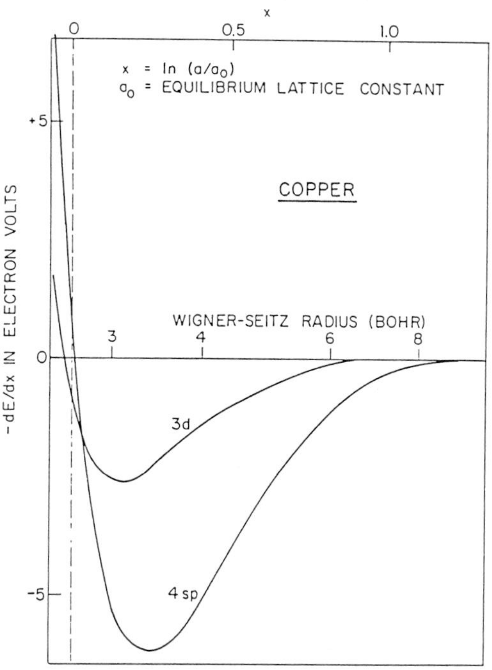

Figure 1-34. Pressure versus lattice constant curve for Cu separated into the sp and the d contributions. Reprinted by permission, *Theory of Alloy Phase Formation:* Bennett, L. H. (Ed.) (1980). Warrendale (PA): The Metallurgical Society of AIME.

Elastic Constants

We now turn to a mechanical property of metals, i.e, the compressibility, or rather to its inverse which is the bulk modulus whose behavior was shown in Fig. 1-31. It is simply proportional to the volume derivative of the pressure

$$B = -\,\Omega \frac{\partial P}{\partial \Omega} \tag{1-458}$$

taken at equilibrium, and Fig. 1-33 for Mo (to a lesser extent Fig. 1-34 for Cu) indicates that the compression of the mobile electrons is primarily responsible for the bulk modulus. This is even true in general. This remarkable fact is revealed by Fig. 1-35, which is taken from Williams et al. (1980a) and shows the calculated bulk modulus (Moruzzi et al., 1978) against the density parameter r_s (which is the radius of a sphere that contains one electron, see Eq. (1-421)), where r_s corresponds to the (assumed constant) electron density between the metal atoms. Important is the extent to which the points plotted in Fig. 1-35 fall near the solid curve, because the curve describes the bulk modulus as a function of the same density parameter but for the uniform *gas of interacting electrons,* obtained by derivatives of Eq. (1-384) in the LDA. Fig. 1-35 therefore indicates that the bulk modulus reflects the energy required to compress the interstitial electrons. The dashed curve further simplifies the interpretation by removing from the solid curve

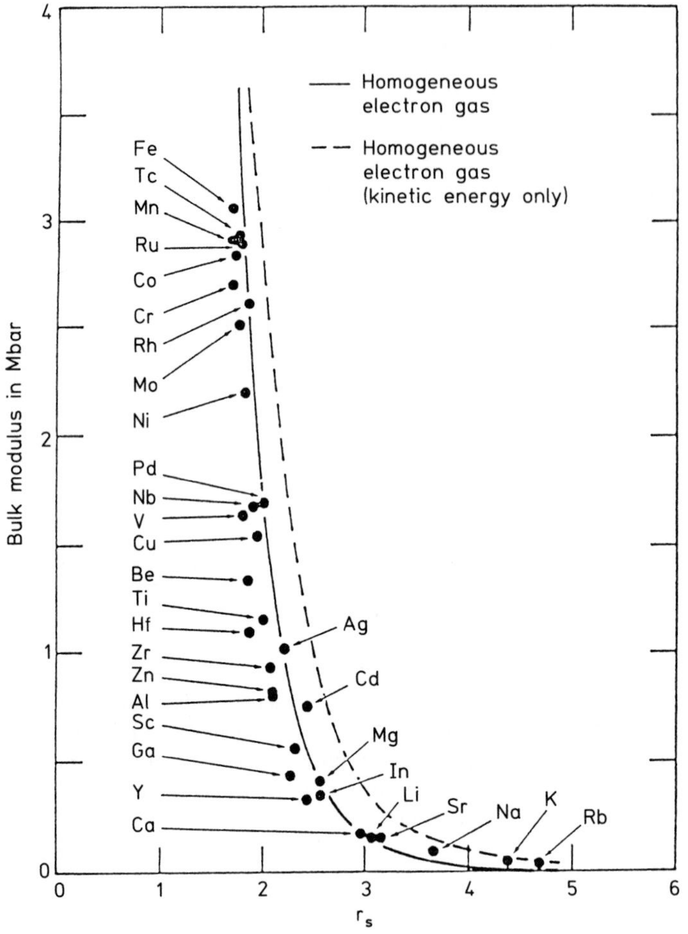

Figure 1-35. Bulk modulus versus density parameter r_s for selected metals. Reprinted by permission, *Theory of Alloy Phase Formation:* Bennett, L. H. (Ed.) (1980). Warrendale (PA): The Metallurgical Society of AIME.

the contributions of exchange and correlation. The obvious fact that exchange and correlation make only a small contribution to the bulk modulus indicates further that it is the *kinetic energy* of the interstitial electrons that is primarily responsible. Collecting all facts, we thus understand the parabolic trend of the bulk modulus in Fig. 1-31 as an effect of optimal binding in the half-filled d shell which leads to maximal compression of the non-d electrons in the interstitial region and hence to the maximum of the bulk modulus in roughly the middle of the series. The lower portion of Fig. 1-32 shows that the situation is not different in the 5d transition metal series, and, indeed, the calculated and measured cohesive energy, bulk moduli, and Wigner-Seitz radii are derived qualitatively, as in Fig. 1-31 (see Brooks and Johansson, 1983; Davenport et al., 1985 and Andersen et al., 1985).

At this point, it seems appropriate to discuss the other elastic constants, i.e., C_{11}, C_{12}, and C_{44} for cubic systems or, for instance, the corresponding five for hexagonal systems, i.e., C_{13} and C_{33} in addition to those above. The discussion is short because band-structure theory has not been applied widely here in spite of its ability to produce total energies satisfactorily. Early attempts, like those of Carlsson et al.

(1980), Ashkenazi (1982), Dacorogna et al. (1982), and Christensen (1984) showed that all methods which used spherizing approximations for the potential (e.g., the atomic sphere or Wigner-Seitz approximation) needed substantial corrections if shear or longitudinal constants were desired. This is intuitively clear because distortions which break the symmetry are expected to give directed forces, and hence any total-energy changes depend sensitively on non-spherical potentials. Thus even when the distortions are chosen such that they conserve the volume, the precision required to evaluate $\delta^* \int^{E_F} \varepsilon N(\varepsilon)\,d\varepsilon$ (see the force theorem, Eq. (1-450)) is quite demanding and can only be achieved when no shape approximations on the potential are made. Thus the apparent success of the older calculations did depend to a large extent on rather ad hoc correction terms which we do not want to judge here; hence it will not be discussed further.

In some quite recent calculations no shape approximations were made. Thus the elastic constants were successfully computed for some carbides and nitrides (Chen et al., 1988a; Price and Cooper, 1989) which must be discussed elsewhere. In applications to elementary metals, Chou et al. (1983) and Chelikowsky et al. (1986) used norm-conserving self-consistent pseudopotentials, and Lu et al. (1987) a shape unrestricted LAPW method. Their results, which all concern hexagonal metals, are collected in Table 1-2 where the calculated numbers are also compared with experimental data.

Note that in all cases the calculated cohesive energy is too low, i.e., the binding energy is overestimated. As a consequence, the lattice constants are somewhat too small and the bulk moduli too large. Gunnarsson and Jones (1985) have analyzed this fact and found that it is due to the local-density-functional (LDF) approximation. The description of the exchange energy here, which depends on the angular characteristics and nodal structure of the orbitals, seems to be unsatisfactory, and it is this feature that is not equally well treated in the atomic- and solid-state calculations in the LDF approximation. Very recent attempts to remedy these defects are emerging (Bagno et al., 1989). In Table 1-2, unfortunately, beyond the bulk modulus, there is only Poisson's ratio, ν, which is connected with an *average* of the elastic constants; thus none of the hexagonal constants have been calculated. The situation is the same for the cubic metals (except for the older calculations). Here a calculation of Poisson's ratio would be more interesting because it together with the bulk mod-

Table 1-2. Calculated and measured ground-state properties of some hexagonal metals. Experimental values in parentheses: E_{coh} cohesive energy, B bulk modulus, ν Poisson ratio.

	Be [a]	Ti [b]	Zr [b]	Ru [c]
a (Å)	2.25 (2.285)	2.866 (2.951)	3.145 (3.232)	2.68 (2.71)
c (Å)	3.57 (3.585)	4.547 (4.685)	5.116 (5.147)	4.16 (4.28)
c/a	1.586 (1.569)	1.586 (1.588)	1.627 (1.593)	1.552 (1.579)
E_{coh} (eV/atom)	3.60 (3.32)	6.42 (4.85)	7.40 (6.25)	7.70 (6.62)
B (Mbar)	1.31 (1.27)	1.27 (1.05)	0.986 (0.833)	3.51 (3.21)
ν	0.05 (0.05)	0.32 (0.26)	0.34 (0.29)	0.31 (0.29)

[a] Chou et al. (1983); [b] Lu et al. (1987); [c] Chelikowsky et al. (1986).

ulus determines the two constants C_{11} and C_{12}, since for cubic metals

$$\nu = C_{12}/(C_{11} + C_{12}) \qquad (1\text{-}459)$$

and

$$B = \tfrac{1}{3}(C_{11} + 2\,C_{12}) \qquad (1\text{-}460)$$

Empirically, Poisson's ratio does not vary much for the cubic metals and is $\nu \approx 0.4$ on the average (see Kittel, 1986). Thus the trend of C_{11} and C_{12} roughly follows that of the bulk modulus B. It would be interesting to examine the physical reason for this connection, and high-precision calculations of the type discussed in this chapter could be enlightening.

To prepare for the discussion of the next subsection we remind the reader that the elastic constants are related to the velocity of sound in matter. In other words, the elastic constants are important in understanding the long wavelength behavior of *phonons*. For instance, in cubic metals C_{11} is proportional to the square of the velocity of a longitudinal phonon propagating in the [100] direction, and C_{44} to that of a transverse phonon, the velocities in other directions being connected with certain linear combinations of the elastic constants (Kittel, 1986). To put it differently, the slope of the dispersion relations of the acoustic phonons near the center of the Brillouin zone are given by the elastic constants. As one moves out further into the zone, these dispersion relations deviate more and more from linear behavior, and the energy of the phonons begins to reflect the restoring forces on a microscopic scale. The dispersion relations can, for instance, develop anomalies which may signal phase instabilities and hence structural phase transitions. These short wavelength phonons are thus intimately connected with questions of structural phase stability, and they have been determined recently in some prominent examples using first-principle, total-energy calculations and have been connected with band-structure features. We will next to turn to this interesting application of band theory.

1.4.1.3 Structural Phase Stability and Phonons

To begin we must discuss questions of crystal phase stability in quite general terms and ask if the theoretical methods described in this chapter suffice to explain the crystal structure trends of elementary metals that are observed to occur in the Periodic Table of the Elements. The most prominent example of this phenomenon are the three transition-metal series which exhibit the same h.c.p.→b.c.c.→h.c.p.→f.c.c. sequence as the d bands become progressively filled. We must, however, exclude and treat separately the magnetic 3d metals (see Sec. 1.4.2.1). This subject received a great deal of attention in the past and we refer the interested reader to a collection of references by Skriver (1985) on these more historical aspects which are interesting but not of concern to us in this context.

Crystal Phase Stability

At low temperature, the crystal structure of a metal is determined by the total energy E; in addition, there is a small contribution E_0, from zero-point motion which may be neglected for the following reason: the zero-point energy in the Debye model is given by

$$E_0 = \tfrac{9}{8}\,k_B\,\theta_D \qquad (1\text{-}461)$$

where k_B is Boltzmann's constant and θ_D is the Debye temperature. The latter is observed to vary by at most 10 K (Gschneidner, 1964) between different structures of the same metal, and thus the corresponding change, ΔE_0, of the zero-point energy is about 0.1 mRy which is more than an order of magnitude smaller than the energy differences we will be concerned with here. Hence, if the stability of some structure is to be determined with respect to some reference structure, which we here take to be f.c.c., the total energy per atom is calculated for both phases and the energy difference is formed,

$$\Delta E = E_{\text{b.c.c.}} - E_{\text{f.c.c.}} \quad \text{or}$$
$$\Delta E = E_{\text{h.c.p.}} - E_{\text{f.c.c.}} \qquad (1\text{-}462)$$

In Fig. 1-36 we show the b.c.c.-f.c.c. (upper part) and h.c.p.-f.c.c. (lower part) structural energy differences for the 4d transition metal series. Here the open circles are enthalpy differences derived from phase-diagram studies (Miedema and Niessen, 1983); they are usually accepted as the experimental counterpart to the calculated values ΔE, Eq. (1-462), shown as solid dots. The latter were obtained by ASW calculations minimizing the total energies, $E_{\text{b.c.c.}}$ and $E_{\text{f.c.c.}}$, separately, with respect to the volume (Esposito et al., 1980; Kübler, 1980d). The other border of the shaded area is defined by LMTO calculations of Skriver (1985) who used the force theorem, Eq. (1-450), at equal volumes (the experimental one), i.e.,

$$\delta E = \delta^* \int^{E_F} \varepsilon N(\varepsilon)\, d\varepsilon \qquad (1\text{-}463)$$

and the self-consistent f.c.c. potential in a b.c.c. geometry to obtain the energy differences. The agreement between these two types of calculations is apparent and satisfactory. Therefore we have chosen to show the h.c.p.-f.c.c. total-energy differences from Skriver's calculations (obtained in complete analogy to the b.c.c.-f.c.c. differences) in Fig. 1-36, and in Fig. 1-37 those of 3d and 5d transition metal series also from Skriver's (1985) calculations.

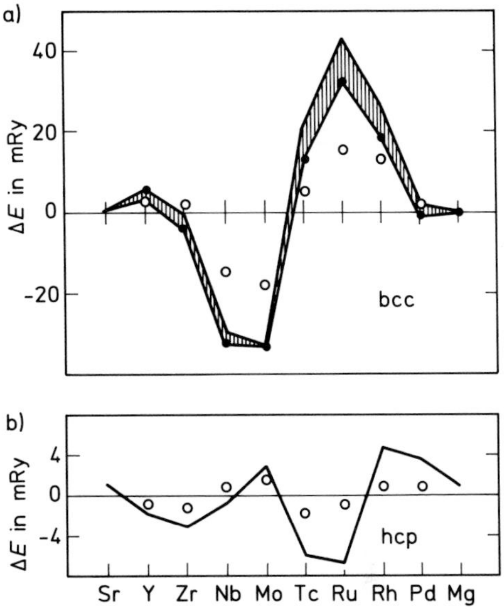

Figure 1-36. Total energy of the b.c.c. (a) and h.c.p. structure (b), relative to the total energy of the f.c.c. structure for the 4d transition metals as given by Skriver (1985).

The most conspicuous feature of Figs. 1-36 and 1-37 is the pronounced b.c.c.-stability in the middle of the series. Since the force-theorem calculations of Skriver give the trend correctly (i.e., ΔE is well represented by first order changes, Eq. (1-463)), we may interpret the b.c.c. stability as the signature of the density-of-states trend and relate it to the pronounced minimum in the b.c.c. density of states that separates the bonding from the antibonding states in the middle of the series (see Fig. 1-29 and its interpretation).

The results depicted in Figs. 1-36 and 1-37 are qualitatively similar to those shown by Pettifor (1984) calculated by means of density of states from canonical bands. This supports the interpretation given above. – The structural energy differences for the 5d transition metals are also confirmed by augmented Slater-type orbital calculations by Davenport et al.

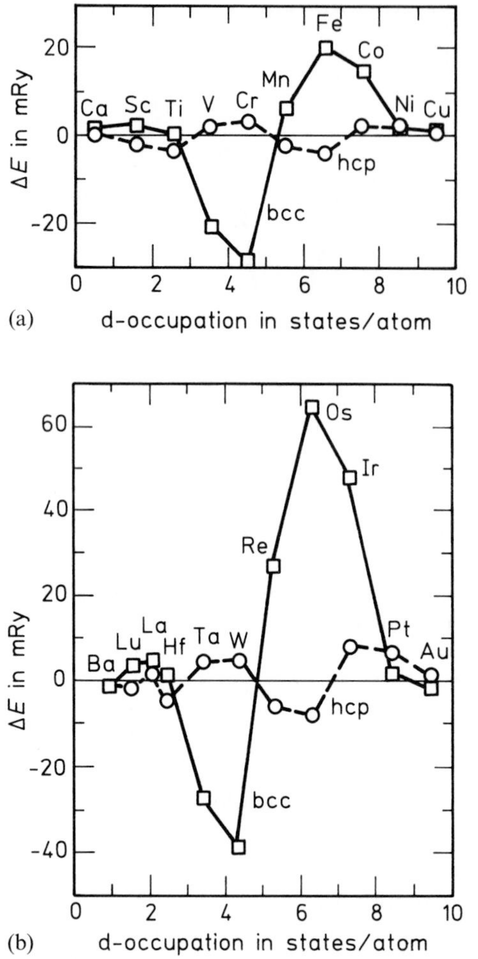

Figure 1-37. Total energy differences of different lattices for the 3d (a) and 5d transition metals (b) as given by Skriver (1985).

(1985) who used self-consistent total-energy differences (and not the force-theorem), see also Fernando et al. (1990). However, despite the fact that the theoretical values agree within ≈25%, and that the crystal structures of 27 metals are correctly obtained, the calculated structural energy differences are found to be as much as a factor of 3 to 5 larger than the enthalpy differences given by Miedema and Niessen (1983). The reason for this discrepancy is not known at present. A single calculation exists, however, for *W* by Jansen and Freeman (1984) who use a *full-potential* LAPW method and give a value of 42 mRy in favor of the b.c.c. structure. This is in good agreement with Skriver's value (Fig. 1-37) and thus indicates that the discrepancy is not likely to be due to the muffin-tin or Wigner-Seitz approximations for the potential. One cannot exclude a genuine failure of the local-density approximation. But, neither can one exclude the possibility that the "experimental" values of Miedema and Niessen (1983) are uncertain since they are model dependent and therefore could possess rather large error bars.

Skriver (1985), besides the systems above, treated other classes of metals like the alkali metals, the alkaline-earth metals, the lanthanides and the light acitinides. We do not want to go into more depth here, but rather take one brief look at the third-period simple metals (Na, Mg, Al) which were studied by McMahan and Moriarty (1983) using a combination of generalized pseudopotentials and LMTO methods. We only want to call attention to an interpretative part and show in Fig. 1-38 the density of states for Al at the experimental volume for the f.c.c. (solid line), h.c.p. (dashed line), and b.c.c. structures (dotted line), all obtained from a self-consistent f.c.c.-Al potential, but the latter two in h.c.p. and b.c.c. geometries. Assuming again the total-energy changes are linear and hence well described by the force theorem, Eq. (1-463), we can see by visual inspection that at two-electron occupation, i.e., Mg, the h.c.p. phase is favored in agreement with experiment because of the low value of the h.c.p. density of states. At one-electron occupation (Na) and slightly above no structural-energy differences are discernable and the free-electron behavior is obvious. Indeed, experimentally Na has the b.c.c. structure at room temperature

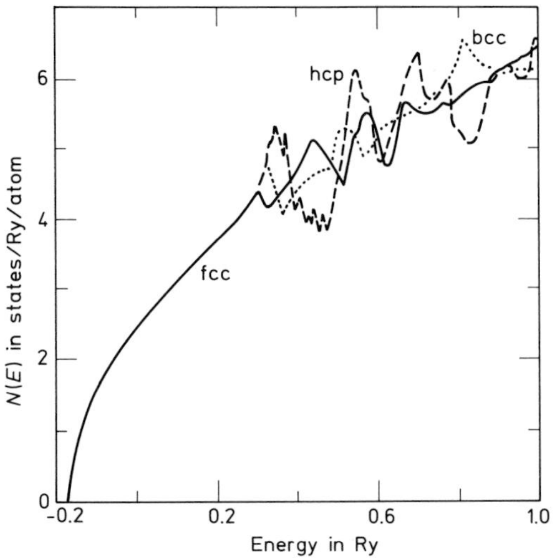

Figure 1-38. Density of states for Al, calculated for different structures as given by McMahan and Moriarty (1983).

but undergoes a Martensitic transformation (this term is to be described in detail a little later) below 35 K to a close packed 9R structure, the latter being closely related to h.c.p. At low temperature, both the 9R phase and the related h.c.p. structure are frequently observed and the crystal is said to possess mixed phases. Very precise pseudopotential calculations by Ye et al. (1990) indeed find the h.c.p. structure very close to f.c.c. but lower in energy than the b.c.c. structure by 0.073 mRy (i.e., 1 meV) per atom. Going from Mg toward Al, i.e., to the three-electron occupation, Fig. 1-38 clearly shows that a peak in the h.c.p. density of states must be traversed which works against the h.c.p. phase in favor of the f.c.c. phase in the case of Al (Mott and Jones, 1936).

Having established credibility in band-structure theory to properly describe structural phase stability, we now turn to a treatment of microscopic electronic changes as atoms are moved, first discussing the concept of frozen phonons, then returning again to the question of phase stability. Our exposition is based to an overwhelming extent on work by Harmon and coworkers (Ho et al., 1984a, b; Chen et al., 1985, 1988b; Ye et al., 1990).

Phonons

Although a good knowledge of the theory of lattice dynamics would be helpful (see for instance, the large collection of references given by Ho et al., 1984a, and the various perspectives), not more than a rudimentary familiarity of the basics is really needed as the concepts used here are rather elementary. The approach to be outlined involves the precise determination of the crystalline total energy as a function of the lattice displacement associated with a particular phonon, i.e., a quantized lattice vibration. The method which is commonly referred to as the frozen-phonon method utilizes first-principles band-structure techniques to obtain the total energy for a *frozen-in position* of the lattice. The phonon frequency can then be obtained from the resultant total-energy curve. The only approximations made are the local-density approximation, as is the case throughout this chapter, and the Born-Oppenheimer approximation, which was outlined in the introduction (Sec. 1.1). We emphasize that information concerning phonon anharmonicity and possible lattice instabilities can and will be obtained because of the non-perturbative nature of the calculations.

In the Born-Oppenheimer (1927) approximation, the wave function of the electrons is assumed to adjust instantaneously to any lattice distortion; i.e., it is assumed that the electrons are always in the ground state defined by the instantaneous ionic configuration. The total energy of the electrons in the ground state (Eq. (1-5)) is a function of the ionic position as we have seen at many places in this section. It yields a potential for the ionic motion if the potential of the ion-ion Coulomb interactions is added, and it is this potential that is calculated in the frozen-phonon method. When the crystal is distorted with atomic displacements corresponding to a particular phonon mode that has a wave vector commensurate with the reciprocal lattice (i.e., equal to $\boldsymbol{K}_s/n$, where $\boldsymbol{K}_s$ is a reciprocal lattice vector and n an integer), the resultant deformed lattice can be viewed as a different crystal structure with (normally) reduced symmetry. This usually implies a larger unit cell, and accurate calculations are limited by the total number of atoms per cell (approximately 15–20). The band-structure calculations needed to obtain the ground-state energy must, of course, be self-consistent, and the calculational technique must be able to treat general potentials for the same reasons that were explained in connection with the elastic constants (end of Sec. 1.4.1.2). At first we restrict out attention to small lattice displacements; in this case we may use an expansion of second order in the displacement and express the total-energy change per atom as

$$\Delta E_{\boldsymbol{q}} = \frac{1}{2} M \, \omega_{\boldsymbol{q}}^2 \, \boldsymbol{u}_{\boldsymbol{q}}^2 \, \frac{1}{N} \sum_i \cos^2 (\boldsymbol{q} \cdot \boldsymbol{R}_i^{(0)} + \delta_{\boldsymbol{q}}) \tag{1-464}$$

Here $\boldsymbol{q}$ denotes the wave vector of the frozen phonon, M is the atomic mass, $\omega_{\boldsymbol{q}}$ is the phonon frequency, $\boldsymbol{u}_{\boldsymbol{q}}$ is the amplitude of the wave (the vector denoting the direction of polarization), N is the number of atoms in the crystal, $\boldsymbol{R}_i^{(0)}$ are the equilibrium positions, and $\delta_{\boldsymbol{q}}$ is a phase factor. When $\boldsymbol{q}$ is at a zone boundary, Eq. (1-464) gives

$$\Delta E_{\boldsymbol{q}} = \tfrac{1}{2} M \, \omega_{\boldsymbol{q}}^2 \, \boldsymbol{u}_{\boldsymbol{q}}^2 \tag{1-465}$$

and for arbitrary $\boldsymbol{q}$ (not at the zone boundary)

$$\Delta E_{\boldsymbol{q}} = \tfrac{1}{4} M \, \omega_{\boldsymbol{q}}^2 \, \boldsymbol{u}_{\boldsymbol{q}}^2 \tag{1-466}$$

These formulas are valid for monatomic crystals and serve to obtain the phonon frequency, $\omega_{\boldsymbol{q}}$, from energy changes quadratic in $\boldsymbol{u}_{\boldsymbol{q}}$. However, the calculations are not restricted to small displacements so that effects of anharmonicity, become accessable.

An early calculation of zone-boundary phonon frequencies for Al was carried out this way by Lam and Cohen (1982) who used norm-conserving pseudopotentials. They demonstrated that these frequencies can be obtained very accurately, in some cases within 1% to 2% of the experimental values. Only in one case (out of six) did the calculated value deviate by 10% from the experimentally determined frequency.

More interesting is a set of calculations by Harmon and co-workers, who also used norm-conserving pseudopotentials but a mixed basis of plane waves and Gaussian orbitals. Their calculations (Ho et al., 1982, 1984a, b; Chen et al., 1985) were for the metals Zr, Nb, and Mo for which important results are summarized in Table 1-3. Paying attention first to the N point phonon frequencies, we notice the generally good agreement between the calculated and measured frequencies. The result "unstable" in the case of Zr refers to the facts illustrated in Fig. 1-39, where the total energy change is plotted versus the displacement for the two transverse modes. Thus "unstable" refers to the negative curvature at the origin, the minimum occurring for a finite displacement. The measured value quoted in Table 1-3 was obtained at high temperatures where Zr is indeed b.c.c., but it undergoes a phase transformation to the h.c.p. structure as the temperature is lowered below ≈1100 K. It should be noted

Table 1-3. Comparison of the calculated and measured values of the frequencies of the N-point, (1, 0, 0) π/a, transverse vibrational modes in Mo, Nb, and the high-temperature b.c.c. phase of Zr (Chen et al., 1985), and in addition, the longitudinal H-point, (1, 0, 0) π/a, (denoted by $L(\frac{2}{3}, \frac{2}{3}, \frac{2}{3})$) phonon modes in Mo and Nb (Ho et al., 1984a, b). The theoretical values are stated to be converged to ± 0.1 THz.

Phonon frequency (THz)	Mo		Nb		Zr (b.c.c.)	
	Exp.	Calc. [a]	Exp.	Calc. [a]	Exp.	Calc. [a]
T_1 at N ($\boldsymbol{u} \parallel \langle 1\bar{1}0 \rangle$)	5.73 ± 0.06	5.8	3.93 ± 0.06	4.3	1.00 ± 0.05	unstable
T_2 at N ($\boldsymbol{u} \parallel \langle 001 \rangle$)	4.56 ± 0.06	4.0	5.07 ± 0.10	5.1	3.94 ± 0.07	3.6
L at H ($\boldsymbol{u} \parallel \langle 100 \rangle$)	5.51 ± 0.05	5.0	6.49 ± 0.10	6.4		
$L(\frac{2}{3}, \frac{2}{3}, \frac{2}{3})$	6.31 ± 0.04	6.1	3.57 ± 0.06	3.6	(1.2) [b]	

[a] as quoted by Chen et al. (1985) or Ho et al. (1984); [b] value in parentheses taken from Fig. 1 of Ho et al. (1984a).

that the measured frequency is very small; it is the so-called "soft mode" that indicates the phase transition, and the calculated instability signals the same fact. This b.c.c.-h.c.p. transition is called *martensitic:* Rather than diffusing, the atoms involved in this transformation move collectively, which means that only a few transformation coordinates or "order parameters" need to be considered. For the b.c.c.-h.c.p. transition this was established a long time ago by Burgers (1934); the order parameters and the special role that the N-point T_1 phonon mode plays are easily pictured (Chen et al., 1988b). In Fig. 1-40a the arrows show the direction of the displacements for the atoms in the b.c.c. cell for this phonon. This pattern is repeated throughout the crystal as indicated, so that there is a simple periodic structure which differs in symmetry from the original b.c.c. cell. When two (110) planes (one is shaded in part (a)) are placed into the plane of the paper, Fig. 1-40b, one sees that stationary T_1 displacements of suitable size yield a nearly hexagonal structure; it is only the angle θ which is 109.5° that has to be changed to $\theta' = 120°$, Fig. 1-40c, and this is accomplished by shear strain displacements. These two coordinates (the shear strain and the phonon displacement) are all that is necessary to obtain the h.c.p. structure. – The complete energy surface as

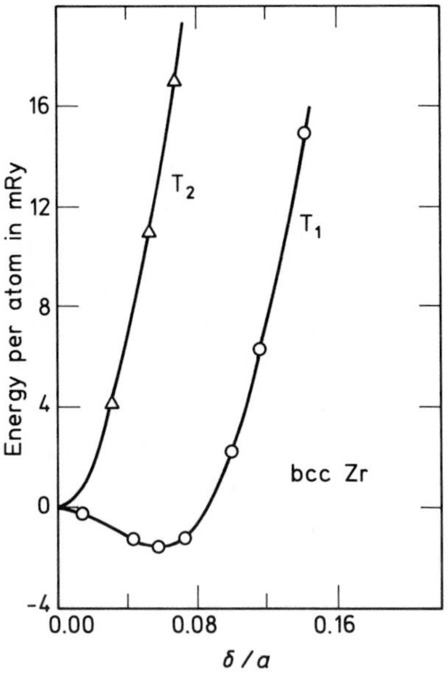

Figure 1-39. Total-energy change versus displacement for two transverse modes for b.c.c. Zr as given by Chen et al. (1985).

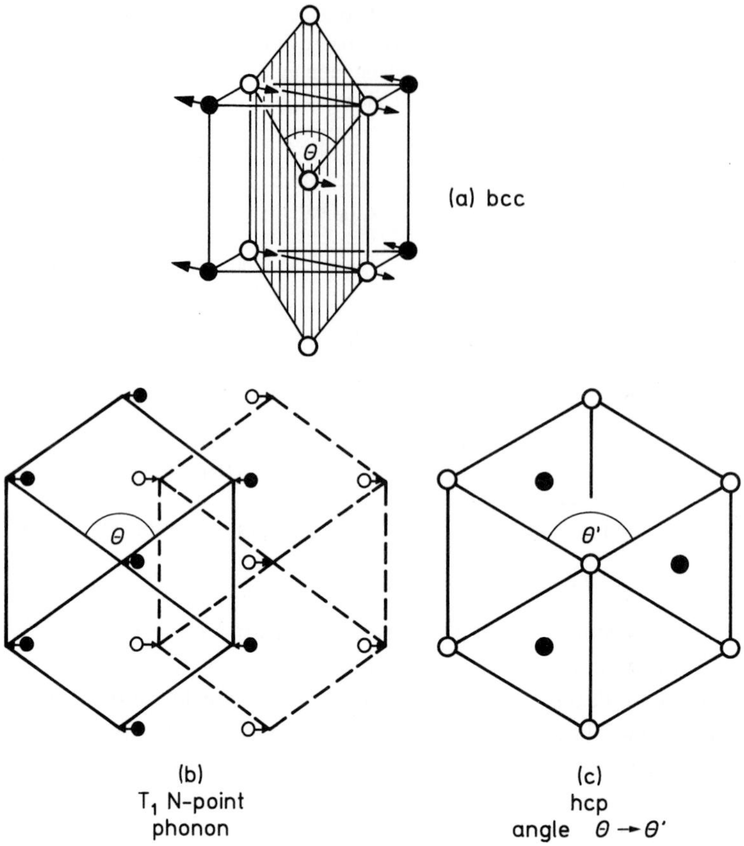

Figure 1-40. b.c.c.-h.c.p. phase transition coordinates for Ba as given by Chen et al. (1988 b). For detailed explanation see text.

a function of the two coordinates (order parameters) has not yet been determined for the case of Zr, but very recent calculations for Ba under pressure (Chen et al., 1988 b; Ho and Harmon, 1990) and for Na (Ye et al., 1990) have revealed these surfaces and firmly established the energetics of the path to equilibrium as sketched in Fig. 1-40.

Returning to the data collected in Table 1-3, we observe that the calculated and measured H-point phonon frequencies agree nicely in the case of Nb, but the agreement is less satisfactory in the case of Mo. A plot of the energy bands of Mo in the distorted structure for displacements corresponding to this phonon reveals a large band splitting at the Fermi level between bands with large and oppositely directed Fermi velocities (Ho et al., 1984). The reason for this is a substantial *nesting* feature of the b.c.c. energy bands near the Fermi level. In isoelectronic Cr this same nesting feature will be discussed in some more detail in Sec. 1.4.2.3; there it gives rise to a spin-density wave with a wave vector near H. The strong coupling in Mo of the electronic states with the phonon might well mean that the Born-Oppenheimer approximation breaks down. Indeed, this led Fu et al. (1983) to investigate the approximation, but they found it well justified. Rather, effects caused by the many-body renormalization of electronic states near the Fermi energy were observed to be of the same order of magnitude as the discrepancy we noticed in the Table. Details are given in the paper by Fu et al. (1983).

The $L(\frac{2}{3},\frac{2}{3},\frac{2}{3})$ phonon, finally, is interesting because it is again intimately connected with an incipient phase transformation in Zr that is of some importance. For Ho et al. (1984a) the motivation for studying the longitudinal mode they state was the marked difference in the frequencies of the metals apparent in Table 1-3. For Nb the longitudinal (111) phonon branch exhibits a dip near the $(\frac{2}{3},\frac{2}{3},\frac{2}{3})$ position, whereas the same branch in Mo is flat. In Zr the dip is even more pronounced than in Nb. As mentioned before, Zr undergoes a phase transition to the h.c.p. structure below ≈1100 K. However, there is a competing transformation to the so-called ω phase by alloying or by the application of high pressure. The atomic displacements for the phase transition from b.c.c. to ω are in the same directions as the polarization vectors for the $L(\frac{2}{3},\frac{2}{3},\frac{2}{3})$ phonon mode. The distortion of the b.c.c. unit cell that is caused by it is best seen by considering three neighboring atomic planes normal to the [111] direction, as shown in Fig. 1-41 taken from Ho et al. (1984a); the distortion corresponds to leaving every third plane (labeled plane 1) stationary and moving the remaining pair of adjacent planes (labeled 2 and 3) toward each other or apart. When planes 2 and 3 fall together as depicted in Fig. 1-41b, the structure is known as the ω phase. It, incidentally, can be obtained from the hexagonal ZrB_2 structure by replacing the boron atoms by Zr.

The calculated total energy changes for Mo, Nb, and Zr as a function of lattice displacement corresponding to the $L(\frac{2}{3},\frac{2}{3},\frac{2}{3})$ mode are shown in Fig. 1-42, also taken from Ho et al. (1984a). The frequencies obtained from the analysis of the curvature near zero displacement are the values given in Table 1-3. Ho et al. (1984a)

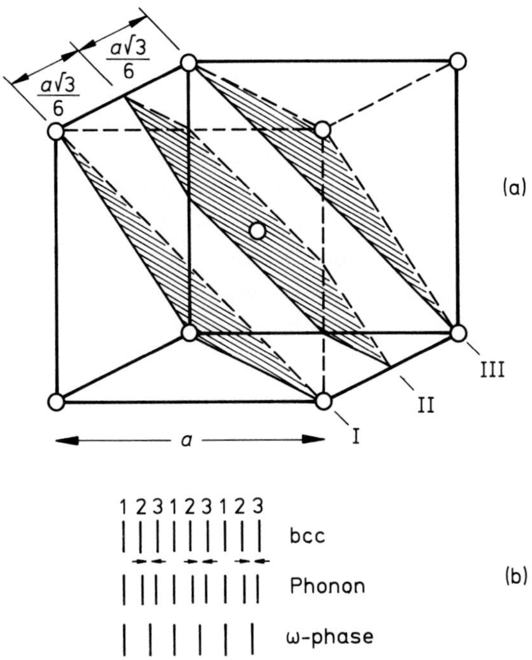

Figure 1-41. Distortion of the b.c.c. unit cell of Zr corresponding to the $L(\frac{2}{3},\frac{2}{3},\frac{2}{3})$-phonon as given by Ho et al. (1984a). For detailed explanation see text.

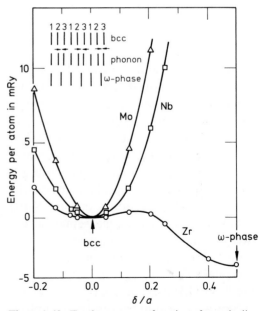

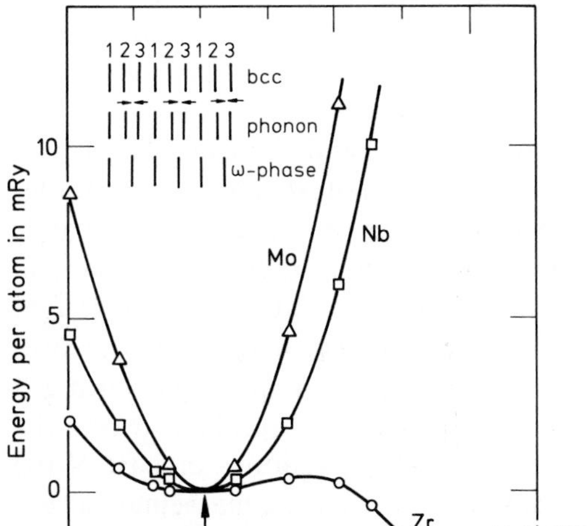

Figure 1-42. Total energy as a function of atomic displacements corresponding to the $L(\frac{2}{3},\frac{2}{3},\frac{2}{3})$-phonon for Mo, Nb, and Zr as given by Ho et al. (1984a).

also obtained the effects of anharmonicity. But of particular interest is the curve calculated for b.c.c. Zr; the calculated potential well is strongly anharmonic and at zero displacement the system is metastable. The total-energy minimum occurs when two of the b.c.c. (111) planes coincide to form the ω phase. The energy difference calculated is 0.045 eV per atom. We emphasize that temperature effects have not been included in the calculation, but the stability of the b.c.c. phase at high temperature is *thought* to be due to either its higher lattice vibrational entropy or else the temperature dependence of the strong third-order anharmonic term in the potential.

We may finally ask if calculations of the type discussed here can shed light on the physical reason for the vastly different phonon frequencies in Zr, Nb, and Mo and thus can provide a model to understand the ω phase in the alloys of the group-IV B metals for which there have been numerous studies (Hickman, 1969; Sass, 1972). One approach is to abandon the excessive wealth of numbers that the computer produces and revert to a simplified model that can be treated analytically. This was in fact done by Heine and Samson (1983) in a much broader context. They used a rather general theorem on tight-binding bands and showed among other things (we will come back to them later) that a threefold superlattice, for which the ω phase provides an example, can be stable when a d band is about one-third full, as it is in the case of Zr. The question – or part of the question – thus finds a mathematical answer in a model of the electronic structure of the type that also applies to the sequence of crystal structures shown, for instance, in Fig. 1-36. One may, on the other hand, try to cope with the numbers and extract physical explanations that are not blurred by simplifications contained in the model. This is not easy, and a truly heuristic model is still missing, but the meticulous use of the force theorem by Ho et al. (1983) and the analysis of the charge density changes by Ho et al. (1984a) have led to an explanation that can be summarized as follows: The systematic increase of the longitudinal phonon frequency as the number of valence electrons is increased going from Zr to Mo suggests that this anomaly is *not* associated with a sharp feature in the Fermi surface but is caused by an electronic structure effect which shows up as an increase in the bonding d-like charge density directed along the nearest-neighbor (111) direction of the b.c.c. lattice. The development of directional bonding from additional occupied d states causes the frequencies to increase because it gives rise to bond-bending forces which help to restore the equilibrium position much like in covalently bonded semiconductors. The transition to the ω phase involves large displacements, and the driving mechanism is traced to shifts of the band structure near the Fermi level. These shifts cause band occupation changes resulting in charge density changes that lead to delocalized sp electrons in the high-density plane (the coinciding planes 2 and 3) (Gooding et al., 1991).

The application of band theory for the precise determination of total energies is not old and is certainly of great interest in the physics of materials. To our knowledge, however, there are not many such studies concerning metals other than those considered in this subsection. This, we believe, is likely to change as our understanding improves, and more supercomputers as well as good algorithms will become available.

1.4.2 Elementary Magnetic Materials

The aim of this subsection is to summarize our knowledge of magnetism in the transition metals, focusing mainly on ground-state properties as they are obtained from applications of band theory within the local spin-density-functional approximation. We will initially be concerned with the ferromagnetic metals but will later on broaden the horizon and include antiferromagnetic metals as well. Since the problem of the Curie temperature still poses difficulties, we will only briefly comment on it, but we will in addition include some newer developments.

When we use the term "band theory" we obviously do not mean independent, itinerant electrons, as exchange and correlation is contained in the effective potential (Sec. 1.3). In some framework of definition (Mott, 1964), however, the electrons are itinerant since they participate in the states at the Fermi surface. In spite of this, they do not necessarily form delocalized magnetic moments, but they can conspire to such an extent that the moments are fairly well localized. The term "Stoner theory" that we will use in the following is not meant in the way originally proposed in its entirety (Stoner, 1938, 1939), but is used for the ground state only, for which it supplies a criterion for the ferromagnetic instability based on *intra-atomic* exchange, much in the way as Stoner originally postulated.

1.4.2.1 Stoner Theory

In principle, the approach to be taken to obtain a ferromagnetic moment from a band-structure calculation is quite simple; one numerically solves the Schrödinger equation (or its scalar relativistic counterpart) for each spin direction, $s = +1$ and $s = -1$, separately, i.e.,

$$(-\nabla^2 + V_0 + s\,V_1)\,\psi_{sk} = \varepsilon_{sk}\,\psi_{sk} \qquad (1\text{-}467)$$

where V_0 is the spin-independent effective potential, and V_1 the part that depends on the spin (one easily derives V_0 and V_1 using Eq. (1-427)). These calculations must be started with a different charge density for the two spin directions; i.e., a non-zero magnetic moment, defined as

$$m = n_\uparrow - n_\downarrow \qquad (1\text{-}468)$$

must be assumed initially. Here

$$n_s = \int_\Omega \sum_k |\psi_{sk}(\boldsymbol{r})|^2\,\mathrm{d}^3r \qquad (1\text{-}469)$$

is the integrated charge density of spin s and the sum, as usual, extends over the lowest states filling them up to a common Fermi energy so that the cell is neutral, i.e.,

$$Z_\mathrm{V} = n_\uparrow + n_\downarrow \qquad (1\text{-}470)$$

if Z_V denotes the valence. The calculation must be made self-consistent and converges either to a value of $m = 0$ or to a non-zero value in the cases of Fe, Co, and Ni (and also Gd). If the calculations are conducted properly, then the value computed for m will agree closely with the experimental magnetic moments – how closely, we will see later on –, and the band structure will be different for the two spin directions, enabling us to define the band splitting, called exchange splitting, $\Delta\varepsilon_k$, and an exchange constant I by means of

$$\langle \Delta\varepsilon_k \rangle = I \cdot m \qquad (1\text{-}471)$$

Although actual calculations are indeed conducted in this way, it is more instructive to start from a nonmagnetic situation and ask for the conditions that must be fulfilled for a magnetic moment to develop. So, just as a phonon can indicate a phase transformation if the displacement conforms with the order parameter, a small imbalance in the spin density can be made to determine if the nonmagnetic case is stable or not. Let us thus suppose we change the spin density slightly so that a small nonzero magnetic moment, m, results. This requires that the spins be flipped, increasing the number of up-spin electrons, say, at the expense of the down-spin electrons, thus raising and lowering the up- and down-spin Fermi energies, respectively. This costs kinetic energy which increases by $\frac{1}{2} m^2/N(E_F)$. On the other hand, exchange energy is gained, which acts only between like-spin electrons, and this number grows; the gain in exchange energy amounts to $-\frac{1}{2} I m^2$. Thus the total energy counted from the nonmagnetic value is

$$E = \tfrac{1}{2}\,[m^2/N(E_F)]\,[1 - I\,N(E_F)] \qquad (1\text{-}472)$$

Clearly the system is unstable to ferromagnetic ordering if

$$I\,N(E_F) \geq 1 \qquad (1\text{-}473)$$

which is the famous Stoner criterion. Since the inverse of the second derivative of the total energy with respect to the magnetization is the uniform spin susceptibility, χ, we obtain

$$\chi = N(E_F)/[1 - I\,N(E_F)] \qquad (1\text{-}474)$$

which is the well-known exchange-enhanced susceptibility.

The foregoing arguments should not be taken as proof of the relations of Eqs. (1-472) to (1-474); they are only meant to make them plausible. A valid derivation is somewhat more intricate, and on the basis of spin-density-functional theory, this was made by Vosko and Perdew (1975) and independently by Gunnarsson (1976). Important is that their derivations give an expression for the "exchange integral", I, in terms of the exchange-correlation functional E_{xc}; this is

$$I = \int \mathrm{d}^3 r\, \gamma^2(\boldsymbol{r})\, K(\boldsymbol{r}) \qquad (1\text{-}475)$$

Here $K(\boldsymbol{r})$ is a functional deriviative of E_{xc} with respect to magnetization:

$$K(\boldsymbol{r}) = -\frac{1}{2}\left(\frac{\partial^2 E_{xc}}{\partial m^2}\right)_{m(\boldsymbol{r})=0} \qquad (1\text{-}476)$$

and the quantity γ is defined as

$$\gamma(\boldsymbol{r}) = \sum_k |\psi_k(\boldsymbol{r})|^2\, \delta(E_F - \varepsilon_k)/N(E_F) = \\ = \frac{1}{N(E_F)}\left(\frac{\partial n(\boldsymbol{r},\varepsilon)}{\partial \varepsilon}\right)_{\varepsilon = E_F} \qquad (1\text{-}477)$$

where the function $n(\boldsymbol{r},\varepsilon)$ is the charge density, Eq. (1-379), as though the Fermi energy were at energy ε. Janak (1977) has evaluated the exchange integral I (sometimes also called Stoner parameter) for 32 metals using self-consistent energy-band calculations, and following Eastman et al. (1979) we graph the Stoner product $I \cdot N(E_F)$ against the atomic number in Fig. 1-43 using his values of I and the values of $N(E_F)$ given by Moruzzi et al. (1978). Clearly, the condition for spontaneous ferromagnetic order is fulfilled for Fe, Co, and Ni which experimentally are the only ferromagnetic metals in this set. This is a significant achievement because the only inputs to Janak's calculations were the atomic numbers and the crystal structures. By considering the Stoner parameter I and the state density separately, Janak (1977) showed that the absence of ferromagnetism in the 4d transition series results from the reduction of *both* $N(E_F)$ and I; if either were as large as their 3d counterparts, ferromagnetism would occur in the 4d series.

We may now return to Fig. 1-31 in Sec. 1.4.1.2 and discuss groundstate properties of Fe, Co, and Ni. Since the calculations represented in this figure ignored magnetic effects, it is important to notice the much closer agreement with experiment in the 4d series compared with the 3d series, which provides an indication of the importance of magnetic effects for such properties. These remarks can be made quantitative by calculating the pressure from Eq. (1-472). Since the Stoner parameter, I, is found to be roughly independent of the volume, the

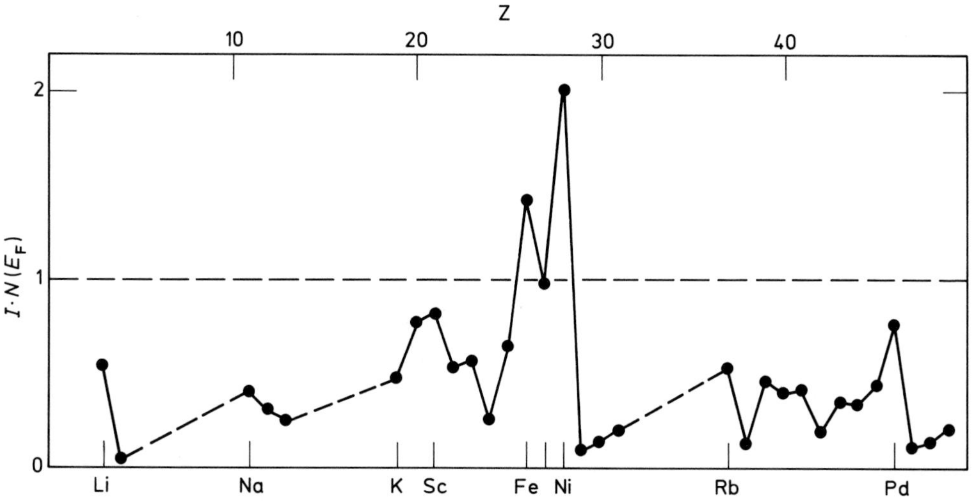

Figure 1-43. Stoner parameter for selected metals (data taken from Moruzzi et al., 1978).

pressure due to magnetic order known as the magnetic pressure is

$$P_M = -\left(\frac{\partial \mathscr{T}}{\partial \Omega}\right)_M \tag{1-478}$$

where $\mathscr{T}$ is the kinetic energy

$$\mathscr{T} = \tfrac{1}{2}\, m^2/N(E_F) \tag{1-479}$$

Assuming for simplicity Heine's scaling law for d bands, $N(E_F) \propto S^5$, where S is the Wigner-Seitz radius, we obtain from Eq. (1-479)

$$P_M = \tfrac{5}{3}\,\mathscr{T}/\Omega \tag{1-480}$$

Thus, when the magnetic moment develops, the magnetic pressure expands the lattice noticeably, thereby reducing the kinetic-energy price. The results of self-consistent calculations verify this quantitatively as Fig. 1-44, taken from Moruzzi et al. (1978) shows. (The values shown for Cr and Mn are estimates using Eq. (1-480) and $m = 0.45$, $m = 2.4$ for Cr and Mn, respectively. We emphasize that these metals are *not* ferromagnetic.) Concerning the bulk modulus we remind the reader of the trend study in Sec. 1.4.1.2, especially Fig. 1-35; it is then obvious that the magnetically induced increase in the atomic volume reduces the compression to which the s electrons are subjected, thereby sharply reducing the bulk modulus, as demonstrated in Fig. 1-45, which is also taken from Moruzzi et al. (1978). Finally, the de-

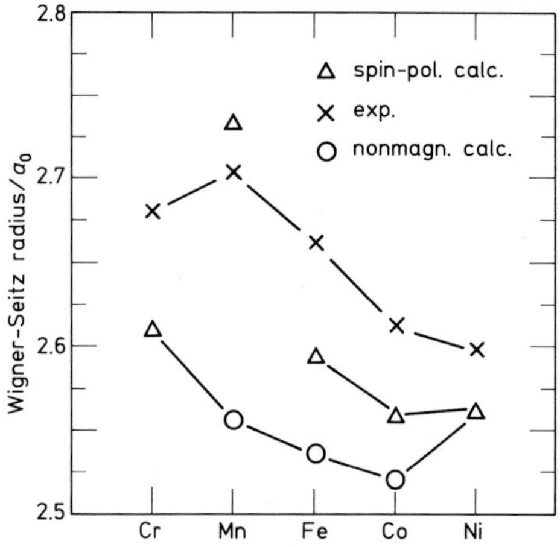

Figure 1-44. Wigner-Seitz radii for magnetic 3d metals as given by Moruzzi et al. (1978); crosses (×), circles (○), and triangles (△) correspond to experimental values, nonmagnetic calculations, and spin-polarized calculations, respectively.

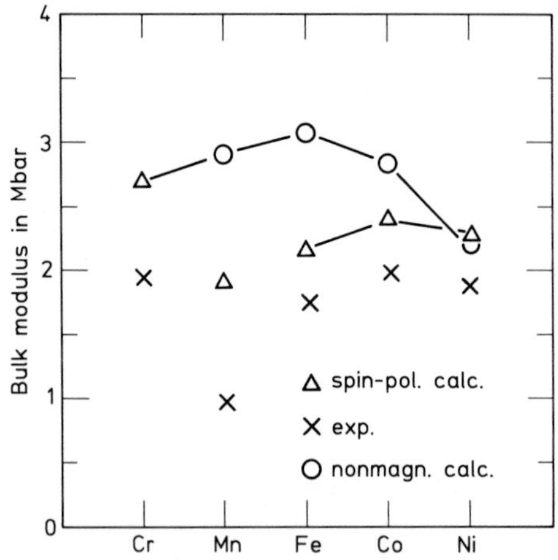

Figure 1-45. Bulk moduli for certain 3d metals as given by Moruzzi et al. (1978); symbols as in Fig. 1-44.

scription of other ground-state properties is summarized in Fig. 1-46, taken from Eastman et al. (1979). The agreement with experiment is good. The magnetic moments at the calculated equilibrium volumes (used throughout this discussion) are given almost exactly. The magnetic moment and the hyperfine field at the observed lattice constant were calculated by Wang and Callaway (1977), and Callaway and Wang (1977). The chemical trend is well described in each case and is put into a broader perspective in Sec. 1.4.4.1; the calculated magnitudes of the other quantities shown are quite reasonable. Eastman et al. (1979) comment in their review that the ability of such calculations to describe the hyperfine field and its variation with pressure is particularly noteworthy because the hyperfine field (in T: 52.4 times the spin density at the nucleus in atomic units) would seem to be an intrinsically non-local effect. That is, the spin density at the nucleus is due to s electrons which have been polarized by the magnetization in the valence 3d shell. One might think that the non-locality of the exchange interaction is of importance in this context; it is in this sense that the success of the *local* spin density treatment was felt to be somewhat surprising.

The mechanical implications of magnetic order seen in this subsection are also fundamental to our understanding of the equilibrium volumes of the actinides and transition-metal monoxides. We demonstrate the relevance of our remark with a famous figure taken from the work of Andersen (1984) which we include as Fig. 1-47; it needs no further comment.

We close this subsection by briefly taking up one more aspect of Stoner theory, which concerns the Curie temperature. Stoner interpreted the temperature at

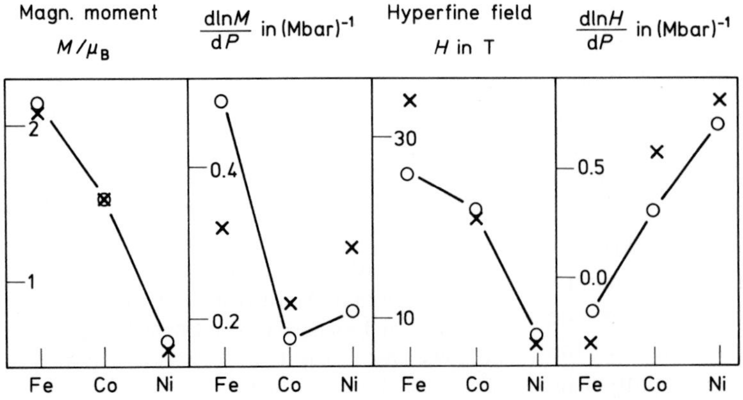

Figure 1-46. Ground-state magnetic properties of Fe, Co, and Ni; crosses and circles indicate experimental and theoretical values, respectively as given by Eastman et al. (1979).

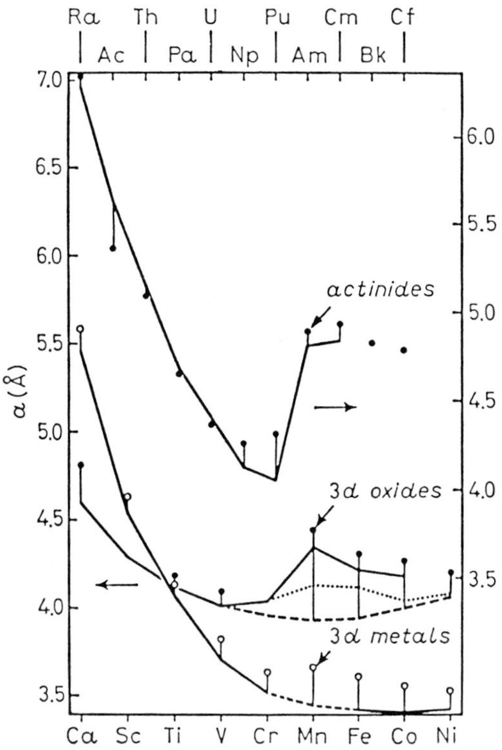

Figure 1-47. Experimental and theoretical lattice constants for the series of 3d metals (Mackintosh et al., 1980), the 3d monoxides (Andersen et al., 1979), and the 5f actinide metals (Skriver et al., 1978; Johansson et al., 1980a; Johansson et al., 1981). For the monoxides having the NaCl structure, a is the lattice constant, while for the elemental metals, having various closely packed crystal structures, a is the lattice constant of the f.c.c. structure with the proper atomic volume. Dots correspond to experiment, and solid, dotted, and dashed curves to nonferro- or antiferromagnetic, ferromagnetic, and nonmagnetic calculations, respectively. Reprinted with permission from Andersen, O. K., Jepsen, O., Glötzel, D. (1985), in: *Highlights of Condensed Matter Theory, Proceedings of the International School of Physics "Enrico Fermi", Course LXXXIX:* Bassani, F., Fumi, F., Tosi, M. P. (Eds.). Amsterdam: North Holland, pp. 59–176.

which the equation

$$I \int_{-\infty}^{\infty} \frac{\partial f}{\partial \varepsilon} N(\varepsilon)\, \mathrm{d}\varepsilon + 1 = 0 \qquad (1\text{-}481)$$

is satisfied as the Curie temperature. The function $f = f(\varepsilon)$ is the Fermi-Dirac distribution and at $T = 0$ Eq. (1-481) reduces to the Stoner criterion, Eq. (1-473). There is now overwhelming evidence, mainly from experimental work but also strongly from theory, that the temperature defined by Eq. (1-481) is *not* the Curie temperature. We will briefly return to this difficult problem in Sec. 1.4.2.4.

1.4.2.2 Band Structure of Ferromagnetic Metals

A great number of band-structure calculations for magnetic metals has appeared in the past, like the early work of Asano and Yamashita (1973) and that of Callaway and co-workers (see Callaway, 1981; Moruzzi et al., 1978; Hathaway et al., 1985; Johnson et al., 1984, and many others). They all show features like the band structure of *ferromagnetic Fe* of Callaway and Wang (1977) shown in Fig. 1-48, where the spin-up bands (majority) are drawn as solid lines, and the spin-down bands (minority) as dashed lines. The exchange splitting is obvious, but is not the same for all bands and all $\boldsymbol{k}$ points, so that the average $\langle \Delta\varepsilon_k \rangle$ (Eq. (1-471)) must be suitably defined and is ≈ 1.5 eV for states near the Fermi energy in the case of Fe (Fig. 1-48). Hence with a calculated magnetic moment of $m = 2.2\ (\mu_B)$, the Stoner parameter becomes $I \approx 0.68$ eV using Eq. (1-471). This is somewhat larger than Janak's (1977) value (0.46 eV), and, indeed, full agreement should not be expected since the latter originated from a formula (Eq. (1-475)) which was derived assuming an infinitesimally small magnetization.

After seeing in the last subsection that the ground-state properties are well described by self-consistent band theory, we will now enquire into the validity of the band structure itself, i.e., compare it directly with experimental information. This

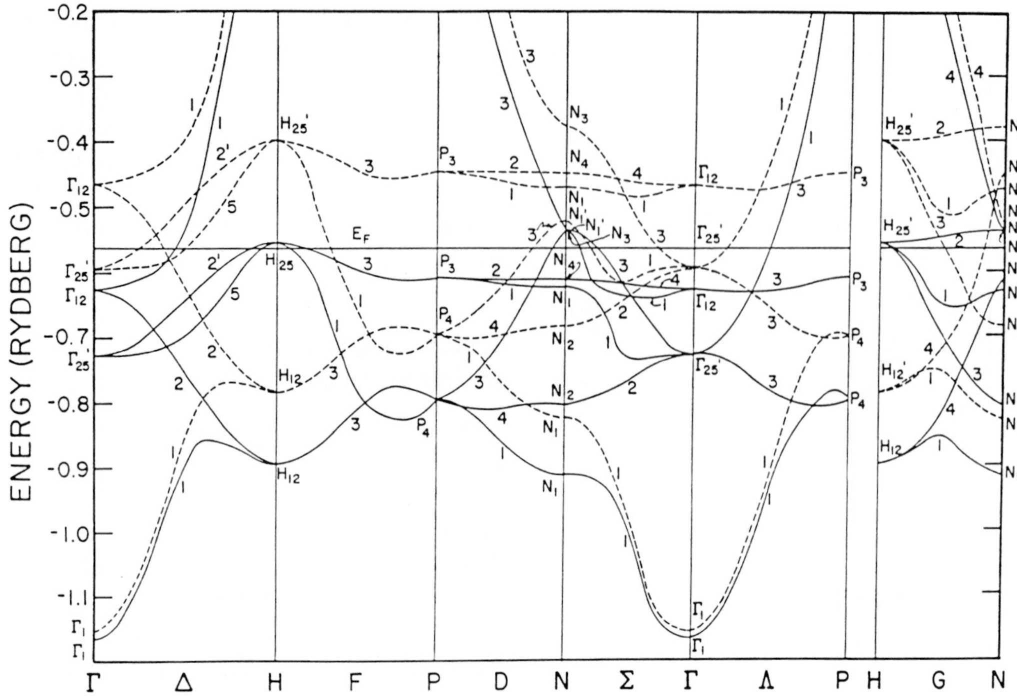

Figure 1-48. Energy bands for ferromagnetic Fe. Reprinted with permission from *Phys. Rev. B16,* Callaway, J., Wang, C. S., *Energy bands in ferromagnetic iron,* 1977, p. 2095.

comparison could include measurements of Fermi surfaces (via the de Haas-van Alphen effect) and energy-band dispersions (via photoemission methods). We remind the reader of our words of warning in Sec. 1.3.1, where excited states were discussed. Nevertheless, we will make this comparison and will come back to additional remarks along similar lines shortly.

The Fermi surfaces of Fe, Co, and Ni have been studied in great detail, but showing their interesting geometrical structures would take up too much space here. Thus we refer the reader to the articles by Eastman et al. (1979), Lonzarich et al. (1980), and Coleman et al. (1981). In general terms it can be said, however, that the experimental surfaces are in agreement with the band-structure calculations except for minor, but interesting, details; a recent discussion of the Fermi surface of Fe can be found in a paper by Hathaway et al. (1985).

Angle-resolved photoemission data, on the other hand, reveal that the band structure obtained on the basis of the local-spin-density-functional approximation (LSDFA) supplies a description of the single-particle excitations that is reasonable for Fe, less so for Co, and much less so for Ni (Eastman et al., 1979). We first look at the case of Fe in some detail and compare experimental results with calculated bands in Fig. 1-49, taken from Eastman et al. (1979). The experimental-band dispersions are from photoemission measurements of Eastman et al. (1980a, b) (circles and crosses) and crossings at the Fermi energy (solid and hollow triangles for majority and minority bands, respectively) are from de Haas-van Alphen experiments. The calculated bands (solid and dashed lines) are the self-consistent LSDFA-bands of Callaway and Wang (1977) shown before in Fig. 1-48. It is interesting that the exchange splitting at the P-point has been observed to decrease from $\Delta = 1.5$ eV at 293 K to about 1.2 eV at 973 K which is just a little short of the Curie temperature of $T_c = 1043$ K. It is thus most likely that the exchange splitting does not disappear (at least not locally) at T_c. Quite extensive photoemission measurements by Turner et al. (1984) allow for comparison with the data of Eastman et al. (1980a, b) and in addition furnish new data at the N point in the BZ. We thus can undertake a comprehensive comparison of a number of different band-structure results both with themselves and with the experimental data; this is done in Table 1-4. The columns denoted by M and H_1 contain band structure obtained at the same, theoretical equilibrium lattice constant ($a = 5.23$ a.u.), but in M (Moruzzi et al., 1978) the KKR method was used, and in H_1 (Hathaway et al., 1985) the full potential (i.e., *not* shape-approximated, see Sec. 1.2.4.4) LAPW method was employed. The columns denoted by C and H_2, on the other hand, contain band structure ob-

Table 1-4. Negative binding energies of Fe and exchange splittings determined by experiment and compared with four different LSDFA-calculations. All values are in eV measured from the Fermi energy.

k point	M [a]	C [b]	H_1 [c]	H_2 [c]	T [d] (exp)
$\langle\Gamma_{1\uparrow\downarrow}\rangle$	8.42	8.12	8.83	8.23	8.15 ± 0.20
$\Gamma_{25'\uparrow}$	2.48	2.25	2.51	2.26	2.35 ± 0.10
$\Gamma_{12\uparrow}$	0.97	0.86	0.99	0.94	0.78 ± 0.10
$\Gamma_{25'\downarrow}$	0.45	0.43	0.69	0.34	0.27 ± 0.05
$H_{12\uparrow}$	5.17	4.50	5.28	4.64	3.80 ± 0.30
$H_{12\downarrow}$	3.71	2.99	3.71	2.99	2.50 ± 0.30
$P_{4\uparrow}$	3.50	3.17	3.57	3.18	3.20 ± 0.10
$P_{3\uparrow}$	0.68	0.53	0.72	0.72	0.60 ± 0.08
$P_{4\downarrow}$	1.95	1.83	2.17	1.71	1.85 ± 0.10
$N_{1\uparrow}$	5.24	4.75	5.41	4.82	4.50 ± 0.23
$N_{2\uparrow}$	3.65	3.27	3.74	3.32	3.00 ± 0.15
$N_{1\uparrow}$	0.94	0.86	0.96	0.93	0.70 ± 0.08
$N_{4\uparrow}$	0.72	0.69	0.76	0.76	0.70 ± 0.08
$N_{1\downarrow}$	3.92	3.60	4.16	3.52	3.60 ± 0.20
$N_{2\downarrow}$	1.82	1.62	2.09	1.58	1.40 ± 0.10
Exchange splittings					
$\Gamma_{25'}$	2.03	1.82	1.82	1.92	2.08 ± 0.10
H_{12}	1.46	1.51	1.57	1.65	1.30 ± 0.30
P_4	1.55	1.34	1.39	1.47	1.35 ± 0.10
N_2	1.83	1.65	1.64	1.74	1.60 ± 0.15
Simple averages of exchange splittings					
	1.72	1.58	1.60	1.69	1.58 ± 0.3

[a] Moruzzi et al. (1978); [b] Callaway and Wang (1977); [c] Hathaway et al. (1985); [d] Turner et al. (1984).

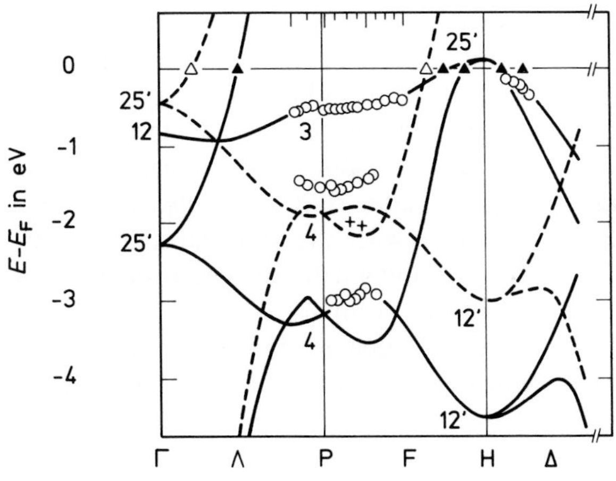

Figure 1-49. Experimental band structure for Fe as given by Eastman et al. (1979). Solid and open triangles denote band crossings at the Fermi surface determined from dHvA data, and solid and dashed lines indicate the bands calculated by Callaway and Wang (1977) for majority and minority spin, respectively.

tained at the observed experimental lattice constant ($a = 5.406$ a.u.). The method of H_2 (Hathaway et al., 1985) is that of H_1, and C (Callaway and Wang, 1977) the LCGO (linear combination of Gaussian orbitals) method; the latter makes no shape approximation in the potential either. It is not clear why the columns C and H_2 are as different as they are, and the agreement of C with the experimental data T (Turner et al., 1984) is to be considered excellent. The error, however, that the local spin-density-functional approximation makes in the determination of the equilibrium volume (we discussed this before in Sec. 1.4.1.2 in connection with Table 1-2) is amplified in the band structure which is seen to react very sensitively to changes in the lattice constant. It must be taken as an empirical fact that the calculations at the experimental lattice constant describe the photoemission data best. This success is still quite impressive, but Kisker (1983) remarks that the good agreement is fortuitous deriving his statement from an analysis of *spin-polarized* photoemission measurements. It is a little hard to see at present why this should be true on the basis of the limited data supplied by Kisker, especially in view of the many data points involved in Table 1-4. Perhaps this issue is not fully resolved. For a very recent account see the paper by Sakisaka et al. (1990).

The situation for ferromagnetic Co (which is h.c.p.) is somewhat less satisfactory. On the experimental side, there are the photoemission data by Himpsel and Eastman (1980), but they are limited since they give the band structure only along the hexagonal z-axis. Other measurements exist, but unfortunately, we know of none (except for those of Himpsel and Eastman) that have been analyzed to allow a comparison with calculated band structures. Himpsel and Eastman compared their results with the calculations of Moruzzi et al. (1978) for f.c.c. Co. Although, as they comment, this did lead to a successful analysis, the procedure is not quite convincing in all respects. It is true that as early as 1980 a self-consistent calculation for h.c.p. Co did not exist. Thus Himpsel and Eastman had no other choice but to compare measurements on an h.c.p.-crystal with calculations for an assumed f.c.c.-system, but their statement that the h.c.p. band structure along the direction Γ to A (see Fig. 1-3 for the BZ) is the same as the f.c.c. band structure along the direction Γ to L provided the latter is folded about the midpoint between Γ and L is not quite true. To shed light on this problem, we carried out a self-consistent ASW calculation for h.c.p. Co at the experimental lattice constant and obtained both the h.c.p. and f.c.c. band structure using for the latter the potentials converged in the h.c.p. geometry. Our results are depicted in Fig. 1-50 and are shown separately for spin-up and spin-down – for simplicity only showing the section of the band structure that is relevant for a comparison with the experimental results which are also provided in Fig. 1-50. Visual inspection shows to what extent the assumption of Himpsel and Eastman is justified: it holds for all states except those within a range of 2 eV directly below the Fermi energy, E_F. In particular, the state labeled Γ_{12} in cubic symmetry is shifted considerably in hexagonal symmetry, and for the down-spin electrons just below E_F, we find a state that we labeled Γ_{6+} which fits nicely with the state labeled $\Gamma_{12\downarrow}$ by Himpsel and Eastman. The results of our numerical comparison are collected in Table 1-5 which in addition to the results of Moruzzi et al. (1978) also contains the experimental values. The labels given are those of Himpsel and Eastman, and in the

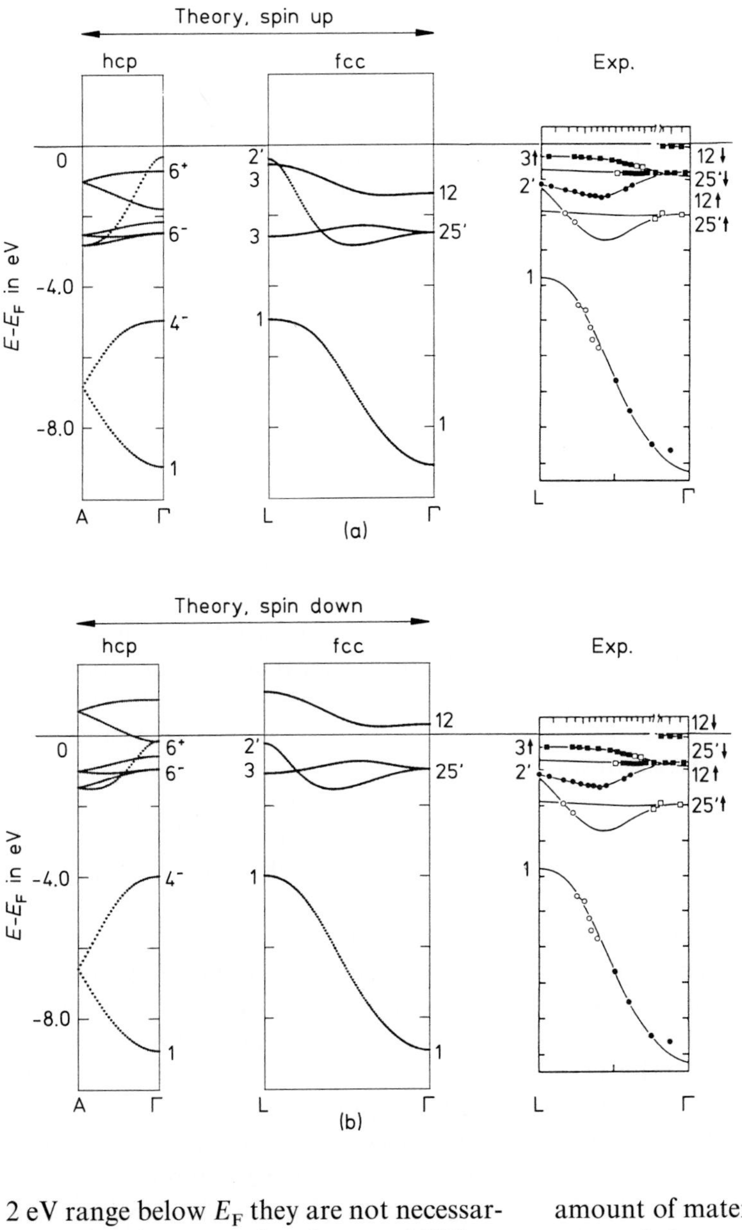

Figure 1-50. Energy bands for Co, calculated with the ASW method for the h.c.p. and f.c.c. structure for majority (a) and minority spin (b); the experimental band structure as given by Himpsel et al. (1980).

2 eV range below E_F they are not necessarily the labels appropriate for our ASW results. The differences of our calculations compared to those of Moruzzi et al., in addition to the problems in the 2 eV range below E_F, arise because they used a different theoretical equilibrium lattice constant. Our results are in very good agreement with LMTO results of Jarlborg and Peter (1984) and, were it not for the limited amount of material to compare, the agreement of theory with experiment is to be considered satisfactory.

The amount of literature on the electronic structure of Ni is overwhelming. Without trying to be complete, we mention on the theoretical side the work of Callaway (1981 and ref. therein), Moruzzi et al. (1978), Anderson et al. (1979), and on the experimental side Eastman et al. (1979),

Table 1-5. Binding energies of Co and exchange splittings determined by experiments and compared with two different LSDFA calculations. All values are in eV measured from the Fermi energy.

k point f.c.c.	***k*** point h.c.p.	M[a]	ASW	H[b] (exp.)
$\langle\Gamma_1\rangle$	Γ_1^+	−8.90	−9.0	−8.7 ±1.0
$\Gamma'_{25\uparrow}$	Γ_1^+, Γ_6^-	−2.65	−2.46	−2.0 ±0.3
$\Gamma_{12\uparrow}$	Γ_6^+	−1.43	−0.72	−0.9 ±0.2
$\Gamma'_{25\downarrow}$	Γ_1^+, Γ_6^-	−1.14	−0.94	−0.8 ±0.1
$\Gamma_{12\downarrow}$	Γ_6^+	+0.22	−0.17	−0.05±0.05
$\langle L_1\rangle$	Γ_4^-	−4.75	−4.47	−3.8 ±0.5
$L_{3\uparrow}$	Γ_5^+	−2.76	−2.15	−1.9 ±0.4
$\langle L'_2\rangle$	Γ_3^+	−0.23	−1.0	−1.5 ±0.2
$L_{3\downarrow}$	Γ_5^+	−1.28	−0.58	−0.75±0.2
$L_{3\uparrow}$	Γ_5^-	−0.53	−0.31	−0.35±0.05
Exchange splittings				
$\Gamma_{25'}$		1.51	1.52	1.2 ±0.3
Γ_{12}		1.65	0.55	0.85±0.2
L_3		1.48	1.57	1.15±0.4
Simple averages				
		1.55	1.21	1.07±0.4

[a] Moruzzi et al. (1978); [b] Himpsel and Eastman (1980).

Dietz et al. (1978), Maetz et al. (1982), Kisker (1983), and Mårtensson and Nilsson (1984). The situation is quickly summarized by Fig. 1-51, where parts (a) and (b) are taken from Mårtensson and Nilsson, and (c) is from the LSDFA-KKR calculation of Moruzzi et al. (1978). This time it is rather immaterial whether we take a band structure calculated at the theoretical equilibrium lattice constant (as that of Moruzzi et al.) or one calculated at the experimental volume and compare with the experimental data; the situation remains the same and is as follows: In Fig. 1-51 a we see in the upper left hand corner two bands marked with arrows that designate the spin directions. The splitting seen is thus the exchange splitting, and it is ≈0.3 eV, a value verified by all other experimental work cited above; this should be compared with the calculated value of ≈0.5 eV that, likewise, is verified by all other calculations cited above.

Mårtensson and Nilsson (1984) fitted their extensive photoemission data with bands that they call semiempirical. They are shown as solid lines in Fig. 1-51 a and b. In the former they are compared with photoemission data by Himpsel et al. (1979) (circles), in the latter with minority bands of Moruzzi et al. (1978) (Fig. 1-51 c) that are slightly shifted; one clearly sees that the measured d-band width is smaller by about 30% compared with the calculated width. This is also in good agreement with other work. There is no universally accepted explanation of this discrepancy, but certain observations made as early as 1979 still seem plausible (Eastman et al., 1979): In the local-density-functional approximation the dynamical behavior of an electron can be viewed as being that of a neutral quasiparticle; i.e., an analysis of the treatment of exchange and correlation in this theory (Kohn and Vashishta, 1983) shows that as an electron moves through the system, it carries with it a perfectly neutralizing "hole" in the distribution of other electrons. This hole is allowed by the theory to breath as it moves; i.e., it is spatially small where the local electron density is large and expands when the electron to which it is pinned moves to regions of low average-electron density. But, the description of the hole is very approximate; the hole is rigidly pinned to the electron, and its form is approximated by homogeneous-electron-gas results. This implies in particular that when the correlated motion of the electron is atomic in character – as is, for example, the correlated motion responsible for multiplet structure in free atoms – the local-density theory becomes inappropriate. The boundary of the adequacy of this simple description of correlation ap-

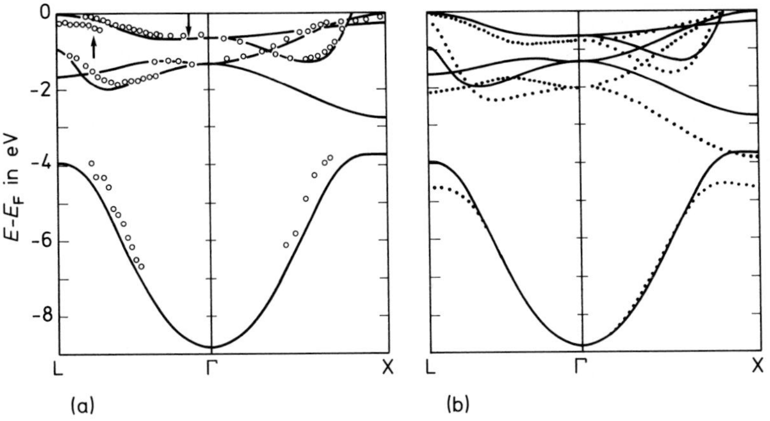

Figure 1-51. Energy bands for Ni: comparison of the semiempirical band structure given by Mårtensson et al. (1984) (a) to the experimental values (circles) given by Himpsel et al. (1979) and (b) to the calculated bands by Moruzzi et al. (1978). (c) Spin-polarized band structure as given by Moruzzi et al. (1978) (thick and thin lines for majority and minority spin).

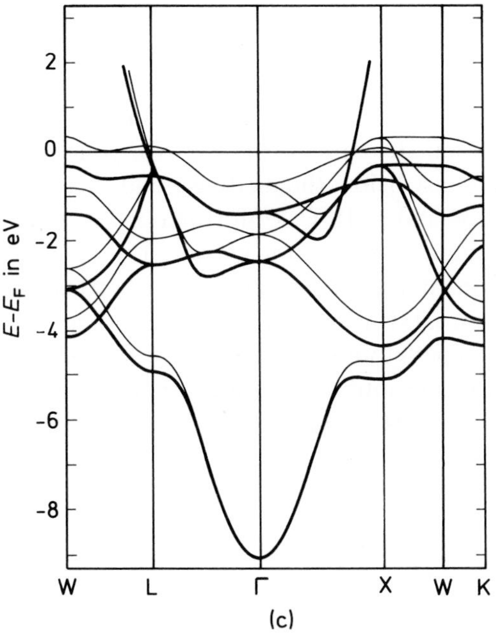

pears to be in the 3d transition metals. Nickel, from several points of view, appears to be a case in which two-particle effects cannot be ignored. So the self-energy of the hole, left as the final state in the photoemission process, seems to be responsible for the small exchange splitting (Liebsch, 1979), and it is this effect, in addition to the narrower d bands, that self-consistent band-structure calculations based on the local-spin-density approximation fail to reproduce.

1.4.2.3 Antiferromagnetic Metals and Competing Phases

We turn to the subject of antiferromagnetism in the pure metals after first posing the following question: why is antiferromagnetism confined to the elements near the middle of the 3d transition series (Cr, Mn, and f.c.c. Fe). This question has been answered in a variety of ways in recent years (Hamada, 1981; Heine and Samson, 1983) and earlier by Moriya (1964). We referred previously to the work of Heine and Samson in connection with the stability of the ω phase (Sec. 1.4.1.3) and commented that they used a general theorem on tight-binding bands to answer questions on the stability of various phases. They found generally that the metals in the middle of the transition metal series have a tendency toward antiferromagnetism, ordered AB alloy structures and lattice distortions based on a twofold superlattice. The physics of the part of this statement pertaining to magnetism can be explained in its simplest form as follows: An electron travelling through an antiferromagnetic crystal experiences exchange-correlation forces that point in opposite directions on two sublattices, and that polarize each sublattice in such a way that on one sublattice the magnetization (Eq. (1-468)) is positive, and on the other it is negative. The situation can therefore be compared with the states of a diatomic molecule (Williams et al., 1982). Leaving aside all the wiggles of wave functions that were pictured in Fig. 1-17, we show in Fig. 1-52 the asymmetry introduced into the states of a diatomic molecule by the energy difference between the atomic states which interact to form bonding and anti-bonding hybrids. An important aspect of this physical effect is the ionic component of covalent bonding. It reflects the energy gained by concentrating the charge associated with the bonding molecular orbital on the atom possessing the more attractive atomic level. The connection between ionicity and antiferromagnetism is that the exchange interaction can lower the energy of the states of a given spin on one of the sublattices. The bonding molecular orbitals (or band states) exploit these exchange-energy differences by placing more charge there. In the same way, the bonding molecular orbital (or band states) for the *other* spin concentrates charge on the *other* atom or sublattice. The concentration of the different spins on different atoms or sublattices is what we mean by antiferromagnetism. It should be noted, however, that the antibonding molecular orbitals (or band states) concentrate charge in precisely the reverse manner. Therefore, to exploit the exchange interaction in this way, the system must preferentially occupy the bonding levels, and this is the connection between antiferromagnetism and the half-filled d band. We substantiate our explanation by means of the density of states from a self-consistent ASW calculation for b.c.c. Mn.

Manganese exists in four allotropic forms, but only one has the simple b.c.c. crystal structures. This is δ-Mn which exists just below the melting point, between 1134 and 1245 °C. Unfortunately, its magnetic moment is not known. The calculation at the experimental lattice constant ($a = 3.081$ Å) gives a magnetic moment of $3\,\mu_B$ per atom (Kübler, 1980b). The sublattices are interpenetrating simple cubic, as in the CsCl structure, and the density of

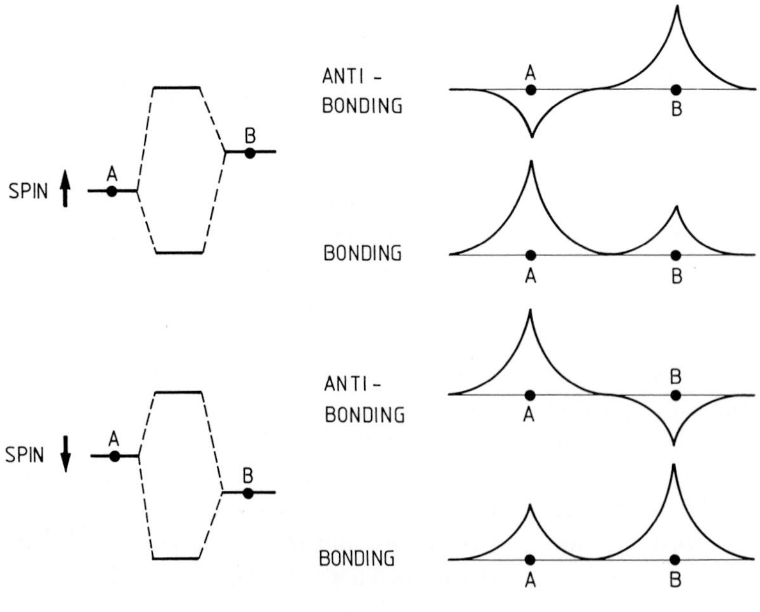

Figure 1-52. Schematic drawing of bonding and antibonding orbitals.

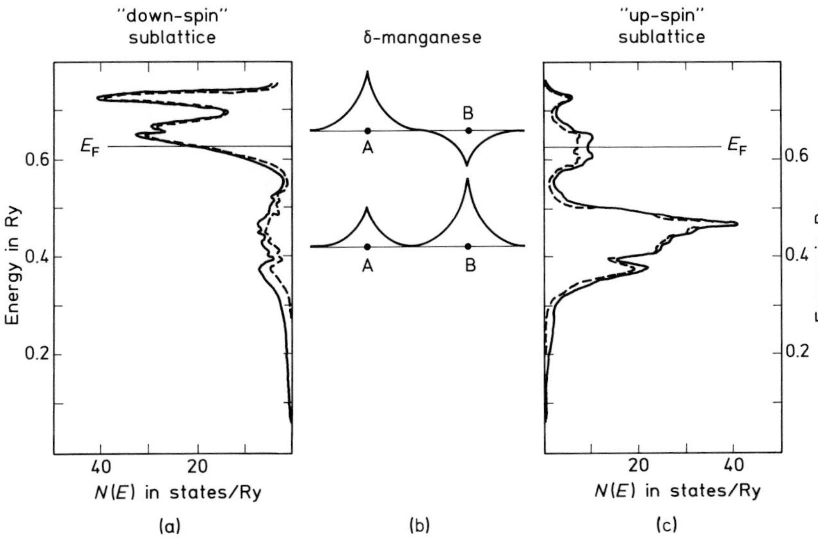

Figure 1-53. Density of states for δ-manganese separated for the *spin-down* (a) and *spin-up* (c) sublattices, (b) orbitals taken from Fig. 1-52.

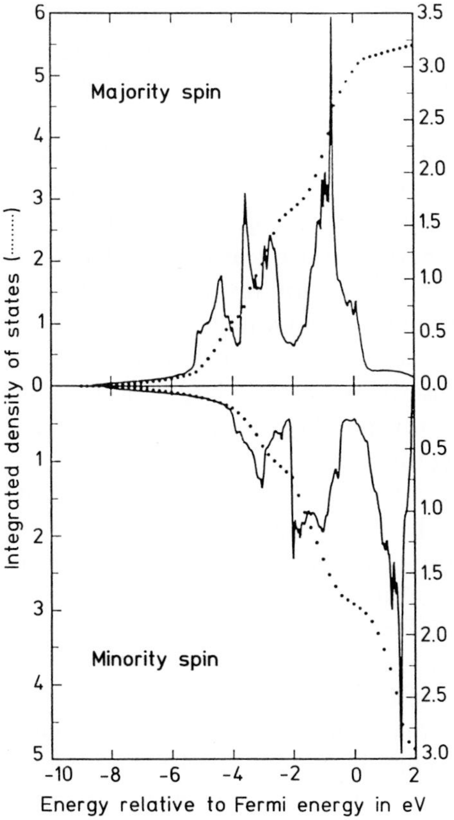

states is shown in Fig. 1-53, where in part (b), we repeat the schematic wave functions from Fig. 1-52. This helps to identify part (a) as the spin-down density of states which are the minority spin electrons on sublattice A and the majority spin electrons on sublattice B, or, equivalently (see Fig. 1-52) to identify part (c) as the spin-up density of states which are the majority spin electrons on sublattice A and the minority spin electrons on sublattice B. This symmetry leads to a spin-degenerate band structure in contrast to ferromagnetic bands. The difference between the antiferromagnetic and the ferromagnetic cases are further stressed by comparing Fig. 1-53 with Fig. 1-54 which shows the density of states of ferromagnetic Fe. It is clear that Fig. 1-54 con-

Figure 1-54. Density of states and integrated density of states for Fe. Reprinted with permission from Moruzzi, V. L., Janak, J. F., Williams, A. R., *Calculated electronic properties of metals* (1978). Pergamon Press PLC.

sists of essentially one curve which is shifted by the exchange splitting in opposite directions for the two kinds of spins, whereas the curves of Fig. 1-53 are distinct because of shifted spectral weight. We will come back to the case of Mn again after discussing Cr.

The case of Chromium is of special interest. As early as 1962 Lomer (1962) proposed Cr to be antiferromagnetic on account of a particular band-structure feature, namely, "nesting", which was referred to already in the case of phonon frequencies in Mo (Sec. 1.4.1.3 in connection with Table 1-3). But before going into any detail and exposing this mechanism, we want to stress an argument by Heine and Samson (1983) that brings us back to the beginning of this subsection: If one has Fermi surfaces with good nesting, then the states near the Fermi energy are particularly significant and will affect the fine tuning, as we will see, or even provide extra help that makes the difference in Cr either being magnetic or nonmagnetic; however, this effect does not alter the overriding general point that Cr, having a half-filled d band, will have an inherent tendency toward changes with a twofold superlattice, i.e., antiferromagnetism.

Beginning now with the concept of Fermi surface nesting, we use moderately new band-structure information and show in Fig. 1-55 the results by Laurent et al. (1981), which were obtained using self-consistent LCGO calculations for nonmagnetic Cr. The Fermi-surface cross sections in the (001) plane shown here consist of an electron surface centered at the point Γ and a hole-surface centered at H. We may ignore the other details at present. Since the band structure obtained by Laurent et al. is in rather good agreement with an ASW-calculation (Kübler, 1980c) that we will use for further considerations, we do not show their band-structure results here but the reader may glance at Fig. 1-56 a to verify our Fermi-surface assignments. It is seen from Fig. 1-55 that there are parallel portions of electron and hole surfaces that can be brought to coincidence by a rigid shift of one part onto the other; the necessary translation may, for instance, be defined by the vector $\boldsymbol{q}$ drawn in Fig. 1-55. This Fermi-surface feature is not uncommon; it exists, for instance, in Mo and other, more complicated systems, and it is called *nesting*. It is important physically because it can have a large effect on the spin susceptibility χ, or any generalized susceptibility.

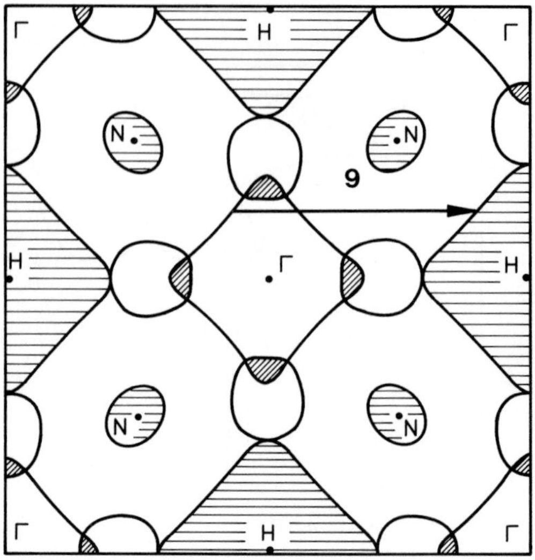

Figure 1-55. Fermi-surface cross section in the [001]-plane for Cr as given by Fry et al. (1981).

We cannot go into great detail here and refer the interested reader instead to the treatise of Jones and March (1973), for instance, but in an attempt to make the physics plausible, we first remind the reader of the uniform spin-susceptibility given in Eq. (1-474). Its divergence signals a ferromagnetic, i.e., infinite-wavelength instability. If we next ask for the response of the system to a small perturbation of

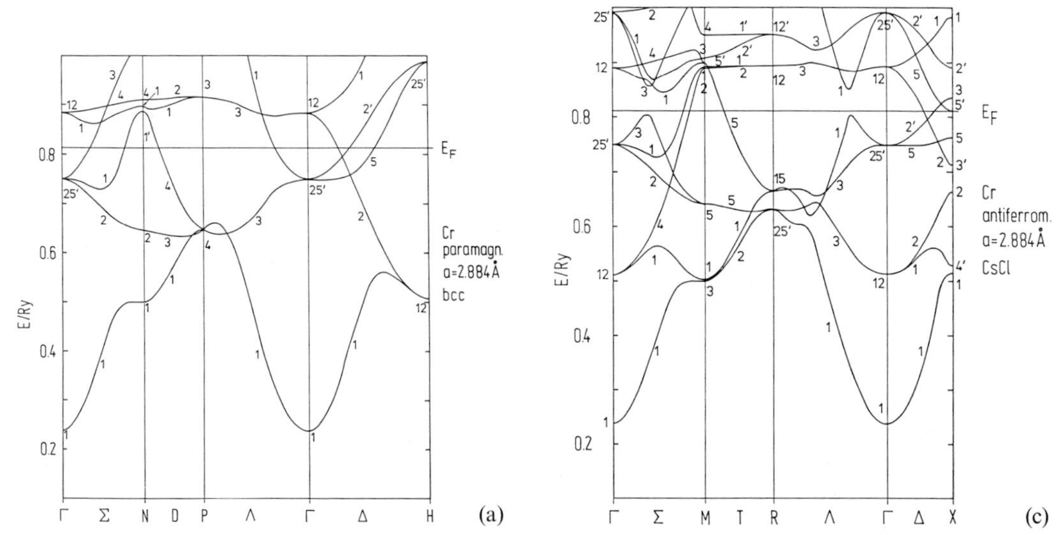

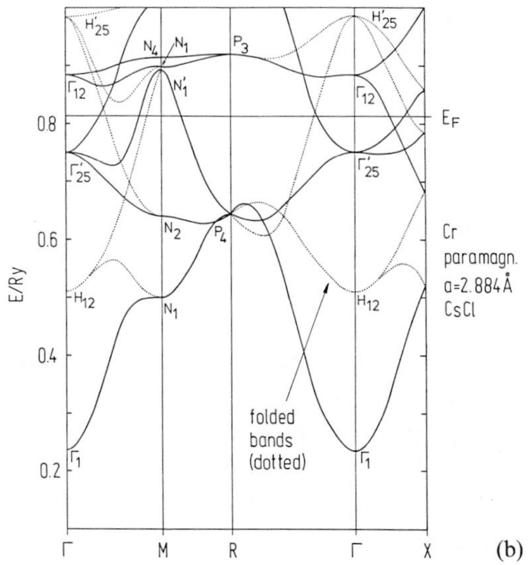

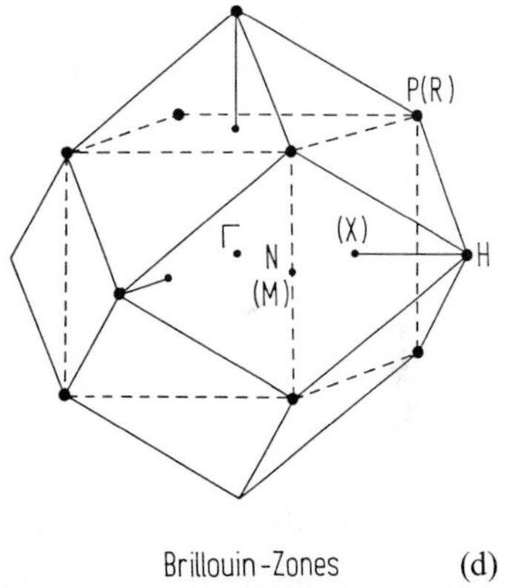

Figure 1-56. Energy bands for Cr: (a) paramagnetic calculation, b.c.c. structure, (b) paramagnetic calculation, CsCl structure, (c) antiferromagnetic calculation, CsCl structure, (d) the Brillouin zones corresponding to the b.c.c. and the CsCl structure.

finite wavelength that may be designated by a wave vector $\boldsymbol{q}$, we are led to the spin susceptibility of the form

$$\chi(\boldsymbol{q}) = \frac{\chi_0(\boldsymbol{q})}{1 - I\,\chi_0(\boldsymbol{q})} \tag{1-482}$$

where the function $\chi_0(\boldsymbol{q})$ tends to the density of states at the Fermi energy as $\boldsymbol{q}$ goes to zero. A well-known formula for $\chi_0(\boldsymbol{q})$ is

$$\chi_0(\boldsymbol{q}) = \sum_{\boldsymbol{k}\,\nu\,\mu} \frac{[f(\varepsilon_{\nu\boldsymbol{k}}) - f(\varepsilon_{\mu\boldsymbol{k}-\boldsymbol{q}})]}{\varepsilon_{\mu\boldsymbol{k}-\boldsymbol{q}} - \varepsilon_{\nu\boldsymbol{k}} + \mathrm{i}\,\delta} \cdot |\langle \boldsymbol{k}\,\nu\,|\,\mathrm{e}^{\mathrm{i}\,\boldsymbol{q}\cdot\boldsymbol{r}}\,|\,\boldsymbol{k}-\boldsymbol{q}\,\mu\rangle|^2 \tag{1-483}$$

(see, e.g., Jones and March, 1973 and many other volumes on theoretical solid state physics). In Eq. (1-483) ν and μ are band indices, δ is an infinitesimal, and $f(\varepsilon)$ is the Fermi-Dirac distribution function; the matrix-element is between Bloch functions in an obvious notation. It should now be clear that for a very low temperature and for the case of nesting Fermi surfaces the susceptibility $\chi_0(\boldsymbol{q})$ can be very large for $\boldsymbol{q}$ vectors that define a "nesting translation" of the type indicated in Fig. 1-55. This is true because the denominator of Eq. (1-483) is near zero for many values of $\boldsymbol{k}$ in the sum, and the numerator is near unity. Obviously, because of Eq. (1-482), the spin susceptibility $\chi_0(\boldsymbol{q})$ can diverge for these values of $\boldsymbol{q}$ and we obtain a generalized Stoner condition

$$I\,\chi_0(\boldsymbol{q}) \geq 1 \qquad (1\text{-}484)$$

that signals a magnet instability. Indeed, a glance at Fig. 1-55 reveals that the nesting wave vector is nearly of the same length as the distance Γ to H, i.e., half a reciprocal lattice vector. Hence the magnetic instability is likely to occur when the magnetization of ferromagnetic planes alternates up and down as we go along one of the cubic axes; this is an antiferromagnetic arrangement with which we began our discussion in this subsection. Of course, these arguments must be augmented with real calculations to prove that the instability in fact occurs. If it does, then general experience is that the symmetry of the ground state formed conforms with the instability (but this must not necessarily be so). It is interesting that the value of $\boldsymbol{q}$ is not exactly one half of a reciprocal lattice vector but is somewhat less. Therefore, in the observed structure, although it is as stated, the *magnitude* of the magnetization of adjacent planes varies sinusoidally with a long period that is *incommensurate* with the lattice: A spin-density wave exists either in the form of a transverse or a longitudinal standing wave, the periodicity being given by a $\boldsymbol{q}$ vector that is $\boldsymbol{q} = (2\pi/a)\,(0, 0, 0.96)$. A band-structure calculation for this type of ground state is as yet beyond our capability, so the calculations that exist ignore this modulation and assume a simple antiferromagnetic spin arrangement. Experimentally, the nesting wave vector can apparently be changed by alloying Cr with small amounts of Mn (e.g., $Cr_{0.99}Mn_{0.01}$). The vector $\boldsymbol{q}$ becomes one half of a reciprocal lattice vector, and the resulting magnetic order indeed assumes the simple (001) antiferromagnetic spin arrangement, which is thus very close to the real ground state of Cr.

A classical paper on the band structure of Cr is that by Asano and Yamashita (1967). The first self-consistent calculation on the basis of the local approximation to density-functional theory, however, was made by Kübler (1980 c). The results are shown in Fig. 1-56; for nonmagnetic Cr, the energy bands are shown again in Fig. 1-56 b but are now folded into the simple cubic BZ appropriate for the CsCl structure, the twofold superlattice necessary to describe the antiferromagnetic spin arrangement. The evolution of the band structure of the latter, shown in Fig. 1-56 c, is now readily apparent; a gap opens along the directions Σ and Λ and degeneracies are lifted at the zone boundaries, most noticeable at R and X. The latter is of some interest since a comparison of Fig. 1-56 b and c shows that a Fermi surface centered at X is present in the nonmagnetic case but has disappeared (or become very small) in the antiferromagnetic band structure (see the state labeled $X_{5'}$ at E_F). Singh et al. (1988) measured two-dimensional angular correlations of the positron annihilitation radiation from Cr in its paramagnetic and antiferromagnetic phases, as well as from paramagnetic $Cr_{0.95}V_{0.05}$ alloys, and found a clear difference between Fermi-surface topologies in the two phases. They attributed this difference convincingly to the above-noted band-structure changes. Furthermore, angle resolved photoemission measurements by Johansson et al. (1980) are in good agreement with the occupied band structure along the line Σ, except near Σ_3 where they seem to be much less dispersive.

The calculated sublattice magnetization in 0.59 μ_B per atom, which is in agreement with the maximum value obtained from neutron diffraction measurements (Bacon, 1962), but in view of the spin-density wave, it may be somewhat too large. Indeed, Skriver (1981) finds a smaller value in similar LMTO calculations and, consequently, smaller splittings at the zone boundaries. This, for instance, leads to a Fermi surface that is centered at X in the magnetic state but is smaller in diameter than it is in the nonmagnetic case. Kulikov and Kulatov (1982) also calculated the band structure and ground-state properties of Cr under pressure and under normal conditions using a linearized computational scheme that is an outgrowth of the KKR method. Their results are in good agreement with those of Kübler (1980c). The calculated cohesive energy is larger in absolute value by $\approx 13\%$ compared with a measured value. Thus we see the same effect of overbinding here as in the case of the ferromagnetic metals (Sec. 1.4.2.1) that seems to be a defect of the local-density-functional approximation (see also the discussion in Sec. 1.4.1.2 in connection with Table 1-2). As a consequence of this, the calculated equilibrium volume is underestimated by 3% in Kübler's calculation, and the bulk modulus is obtained too large – in Kübler's calculation by $\approx 12\%$ and in Skriver's by $\approx 24\%$. Unpublished calculations made later by one of us showed that there are some subtle convergence problems in the sublattice magnetization near the theoretical equilibrium volume that could be related with the occurrence of the spin-density wave. It would therefore be of great interest to study such a spin-density wave with self-consistent band-structure techniques employing perhaps spin-spirals of a type proposed and used for different purposes by Sandratskii (1986), Sandratskii and Guletskii (1986), and Heine et al. (1990). The entire subject of the electronic and magnetic structure of Cr has recently been thoroughly reviewed by Fawcett (1988).

Antiferromagnetism of manganese in the observed ground-state crystal structure has not received much theoretical attention, since it is exceedingly complex. But the high-temperature γ (f.c.c.) phase can be retained at low temperatures by rapid quenching, provided it is doped with a sufficient concentration of impurities such as carbon, nickel or iron (Honda et al., 1976), or its properties can be inferred from extrapolations to $x \to 0$ of f.c.c. alloys $Mn_{1-x}Fe_x$ (Yamaoka et al., 1974). It was therefore studied by Cade in a number of publications (Cade, 1980, 1981a, 1981b), using the LMTO method. He first assumed a collinear antiferromagnetic spin arrangement that consists of ferromagnetic planes alternating up and down as one moves along the z-axis. The corresponding two-fold superlattice is the CuAu structure which is tetragonal with a lattice constant of $a_{\text{f.c.c.}}/\sqrt{2}$ and $c = a_{\text{f.c.c.}}$, where $a_{\text{f.c.c.}}$ is the lattice constant of γ (f.c.c.) Mn. This is called antiferromagnetic order AF I. Another possible ordering consists of ferromagnetic planes alternating up and down as one moves along one of the body-diagonals of the (chemical) cubic cell. This is called order AF II. Finally, Cade (1981b) suggested a non-collinear arrangement where the spins point along the four different cubic body diagonals forming a tetrahedral order that must be described with *four* interpenetrating simple cubic lattices. We have repeated Cade's calculations (Kübler, 1983) but have also obtained the total energy as a function of the volume in both b.c.c. and f.c.c. crystal structures. Our results are collected in Fig. 1-57, which shows that b.c.c. Mn (Fig. 1-57 a) is antiferromagnetic at large volumes but becomes unstable at $S \approx 2.67$ u.a. losing its moment there. However, a ferromagnetic solution is seen to extend down to the theoretical equilibrium at about $S \approx 2.6$ a.u. One should notice the shallow (metastable) minimum of the antiferromagnetic solution that has a very low bulk modulus. It could be that this type of antiferromagnetic manganese can be stabilized epitaxially on a suitable substrate; i.e., the situation could be comparable to ferromagnetic f.c.c. Fe, which will be briefly described below. The case of γ (f.c.c.) Mn, on the other hand, is shown in Fig. 1-57 b. At theoretical equilibrium, $S \approx 2.6$ a.u. again, it is antiferromagnetic with a rather small sublattice magnetization. This agrees with more recent calculations by Duschanek et al. (1989). The γ phase is seen to be the stable form of the two shown in Fig. 1-57 since its total energy is lower than in b.c.c. Mn. The antiferromagnetic arrangement AF II is obviously less favorable than AF I. Recently, we obtained the total energy for the noncollinear spin arrangement discussed by Cade (1981 b) by a method that is described in a paper by Sticht et al. (1989). At two different volumes, we find the total energy to be between AF I and AF II; we could thus conclude that Mn has, in agreement with Heine's argument, a two-sublattice or collinear antiferromagnetic order, but to be certain one should investigate a larger range of volumes. Unfortunately, in the case of manganese the overbinding error that has already been mentioned repeat-

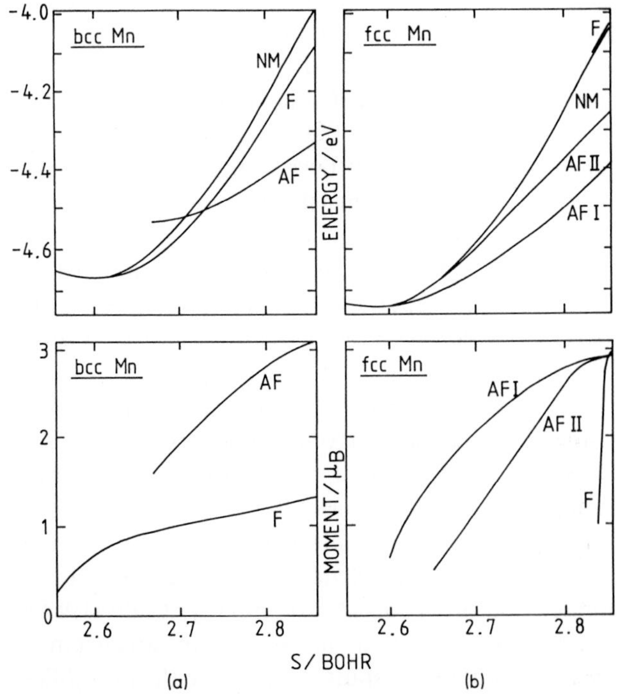

Figure 1-57. Total energy and magnetic moment versus Wigner-Seitz radius for Mn in the b.c.c. (a) and the f.c.c. structure (b).

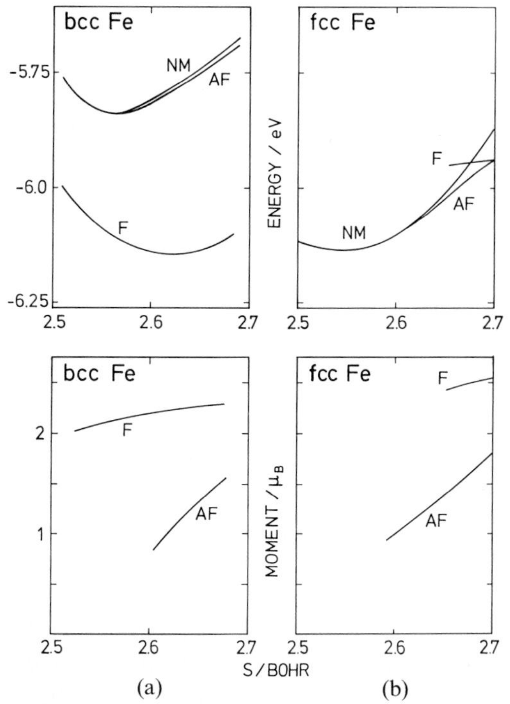

Figure 1-58. Total energy and magnetic moment versus Wigner-Seitz radius for Fe in the b.c.c. (a), f.c.c. (b), and the h.c.p structure (c), for nonmagnetic (NM), ferromagnetic (F), and antiferromagnetic (AF) spin alignments.

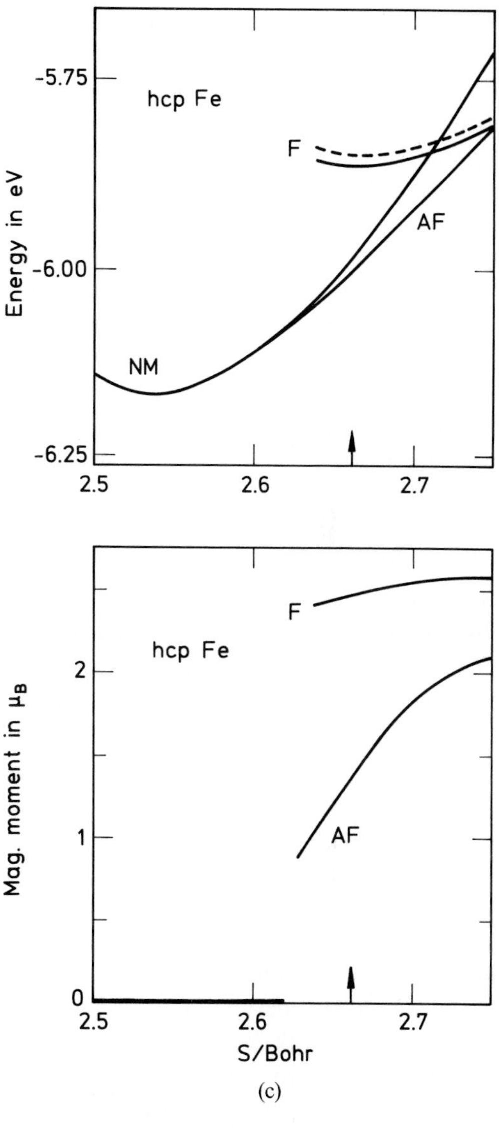

edly seems to be particularly bad. This can be seen in the work of Moruzzi et al. (1978) (Fig. 1-31), but for the case at hand, it leads to a calculated equilibrium Wigner-Seitz radius of $S \approx 2.6$ a.u. that is drastically smaller than the value of $S \approx 2.7$ a.u. suggested by experiment – admittedly, somewhat indirectly. The consequence for the calculated sublattice magnetization can easily be read off from Fig. 1-57b: $\approx 0.9\ \mu_B$ compared with the experimental estimate of $\approx 2.2\ \mu_B$, which is surprisingly close to the value calculated for the experimental Wigner-Seitz radius, $S \approx 2.7$ a.u. We believe these observations imply that the quantitative contents of Fig. 1-57a and b can only be accepted with serious reservations. The qualitative features, however, could be valuable as a guide to reality.

We close this subsection with a brief discussion of the phase stability of Fe at absolute zero temperature and the different forms of magnetism that can occur (Kübler, 1983, 1989). But the last sentence of the paragraph above could easily apply to this case as well, perhaps in somewhat milder form. We compare in Fig. 1-58 the total energies as a function of the volume of nonmagnetic (NM), antiferromagnetic (AF), and ferromagnetic (F) iron in (a) b.c.c., (b) f.c.c., and (c) h.c.p. crystal structures; Fig. 1-58 a to c also show the corresponding calculated magnetic moments. We recall (see Fig. 1-37) that Fe, when assumed nonmagnetic, is calculated to be stable in the h.c.p. structure just like Ru and Os down the group. The stabilization in the observed b.c.c. form must therefore be achieved by its being magnetic. Fig. 1-58 a shows, indeed, that the energy gained from magnetism is substantial and the reasons for this were given in Sec. 1.4.2.1. Unfortunately, the energy gain is not quite enough: nonmagnetic or antiferromagnetic (AF I in this case) γ (f.c.c.) Fe has about the same total energy but ε (h.c.p.) Fe is lower by about 0.05 eV per atom, and it is nonmagnetic at the theoretical equilibrium volume. At a slightly expanded volume there is a low total energy, low-moment antiferromagnetic state *and* a high total energy, high-moment ferromagnetic state. Qualitatively, this is exactly as in γ (f.c.c.) Fe (see Fig. 1-57 b and e), and similar to b.c.c. Mn (Fig. 1-58 a), where, however, the roles of antiferromagnetic and ferromagnetic states are interchanged.

The low total energy of h.c.p. Fe must be considered a (weak) failure of the local density function approximation since Fe at low temperatures *is* b.c.c. *and* magnetic in reality. On the basis of other calculations, this statement was made before by Wang et al. (1985). Nevertheless, the results of these types of calculations have received a good deal of attention by a number of workers in this field. This seems to be due to two facts. *First,* Weiss (1963) postulated the existence of two magnetic states, γ_1 and γ_2, for f.c.c. Fe in an attempt to treat the thermodynamics of the α (b.c.c.) $\rightarrow \gamma$ (f.c.c.) phase transition. The γ_1 state was argued to be antiferromagnetically ordered with a moment of about 0.5 μ_B and a Néel temperature around 80 K, while the γ_2 state was ferromagnetic with a moment of about 2.8 μ_B and a Curie temperature around 1800 K, the two states being energetically separated by some 0.05 eV. Roy and Pettifor (1977) were the first to show on the basis of Stoner theory and band calculations that two states do, indeed, exist, but they found both to be ferromagnetic. Figure 1-58, in contrast, reveals two states for both f.c.c. (and h.c.p.) Fe, as postulated by Weiss, but the energetic separation is much larger than assumed. This two-state hypothesis thus seems to hold true, and it plays an important role also in models that are conceived to explain the anomalous thermodynamic behavior of INVAR (Williams et al., 1982; Wagner, 1989; Mohn et al., 1989; Moroni and Jarlborg, 1989). *Second,* with the pioneering work of Prinz (1985) it was demonstrated that magnetic layers can be produced using molecular-beam epitaxy by appropriate matching of the substrate to the desired structure and lattice constant of the epitaxial layer. A recent example of this is h.c.p. Fe grown on Ru substrates (Maurer et al., 1989). It would seem plausible that magnetic states having a low bulk modulus, like the high-moment states in b.c.c. Mn (Fig. 1-57), f.c.c. Fe, and h.c.p. Fe (Fig. 1-58), make particularly suitable epitaxial films. – All this apparently stimulated recent work by, for instance, Bagayoko and Callaway (1983), Moruzzi et al. (1986), Krasko (1987, 1989), Fuster et al. (1988), Marcus and Moruzzi (1988), and Podgorny (1989), who investigated various features of stability and band structures. Among other things, we find it noteworthy (Bagayoko and Callaway, 1983; Kübler, 1989) that the sudden occurrence of the high-moment state of f.c.c. Fe and h.c.p. Fe as the volume is increased is connected with the magnetism being strong, i.e., with all majority-spin d-electron states being occupied. In other words, when we let the volume decrease, the d bands get wider and the high-moment state disappears entirely as the top of the majority d band becomes unoccupied.

1.4.2.4 Open Problems

We will now briefly discuss two prominent open problems which are believed to be – at least in part – resolvable by the methods of this chapter but which still lack convincing answers. One problem involves a ground-state property, namely, the magnetocrystalline anisotropy. The other one does not and concerns the determination of the Curie temperature.

The problem of calculating the magnetocrystalline anisotropy has a long history (Brooks, 1940) that we do not want to go into here because it is sufficiently well documented in some recent, fully detailed papers on this subject, in particular, those by Fritsche et al. (1987), Strange et al. (1989 a, b) and Daalderop et al. (1990). They all agree that the magnetocrystalline anisotropy is mediated by the spin-orbit (S.O.) interaction and thus needs a relativistic treatment of the effective single-particle equation (Sec. 1.2.5). Since the orbital degrees of freedom are coupled to the crystal lattice, the S.O. interaction will give a dependence of the total energy on the choice of the direction of the magnetization and since, furthermore, the total-energy differences are small (experimentally $\approx 10^{-6}$ to 10^{-5} eV/atom) one expects that using the "force theorem", Eq. (1-447) with Eq. (1-450), will be sufficiently precise. But, although all workers claim a high numerical precision and indeed do obtain the correct order of magnitude for Fe, Co, and Ni, they do not predict – or they only partially predict – the correct easy axes of magnetization for these ferromagnets. We believe that a possible reason for this might be the missing relativistic corrections in the density functional expression (Sec. 1.3.3) and/or, as Jansen (1988 a, b) suggested, the Breit interaction term that up to now has not yet been included in the density functional theory.

Turning now to the second topic of this subsection, we begin by pointing out that the broad subject of metallic magnetism at finite temperatures is conveniently described by a collection of papers edited by Capellmann (1987). However, the detailed calculation of the Curie temperature, T_c, for the ferromagnetic 3d metals and their compounds, using suitable information from band-structure and total-energy calculations, is a serious problem that has been tackled using roughly four different approaches: (1) an interesting semiempirical theory was developed by Mohn and Wohlfarth (1987), who determined T_c of Fe, Co, and Ni as well as of their compounds with Y (including $Y_2Fe_{14}B$) by including the effects of spin fluctuations in the long wavelength limit via a renormalization of Landau coefficients. They found that their numerical values depend on the use of recent calculations of correlation effects (Olés and Stollhoff, 1986) and on the band structure, and obtained reasonable agreement with experimental values of T_c. (2) "The fluctuating local band theory" is discussed, for example, by Wang et al. (1982), but for the sake of brevity, we will refrain from giving any details. (3) There are theories that in one way or another utilize disordering techniques from alloy theories; an example is the paper by Oguchi et al. (1983 a) or the monumental treatise by Gyorffy et al. (1985). (4) Closest to our approach in this chapter (see also Sec. 1.4.3.2) is the use of total-energy differences from various differently ordered magnetic structures including superlattices. An early, but crude, attempt for compounds is given in a paper by Kübler et al. (1983) and later for Fe by Peng and Jansen (1988), who obtained rather reasonable values of T_c from calculated Heisenberg exchange constants. However, it might be seriously questioned whether or not Heisenberg exchange constants together with the Heisenberg Hamiltonian are a good model to describe finite-temperature properties of Fe, Co, and Ni. After all, it is well known that the exchange constants depend on the type of order assumed (see Sec. 1.4.2.3 for extreme cases, and for other recent accounts see, e.g., Luchini and Heine 1989; and Heine et al., 1990). Still, if the angle chosen between two adjacent spins is not too large and if methods using noncollinear spin arrangements are used to calculate total-energy differences, the Curie temperatures obtained in this manner are all reasonable estimates, in contrast to the temperature for which Eq. (1-481) is satisfied. Examples are calculations by Liechtenstein et al. (1987), Sandratskii and Guletskii (1986, 1989a, b), and Köhler et al. (1991), who also obtain quite reasonable estimates for the spin-wave stiffness constants.

1.4.3 Nonmagnetic Compounds and Alloys

The number of band-structure calculations for *compounds,* both metallic and nonmetallic, is truly large; hence an in-depth treatment of this interesting field is beyond the scope of this chapter, and in contrast to our treatment of elementary metals in the previous sections, we must change the style of our presentation and continue by introducing the reader to a selected portion of the literature going deeper only in some cases of particular interest. For this selection our guideline is the following question: what is new and different compared to elementary metals? Clearly, a prominent topic is the application of band theory to elucidate the chemical bond in metallic compounds and, in particular, trends in the heats of formation and their prediction. Connected with this is the important topic of the theoretical determination of phase diagrams; this field is just beginning to be conquered by the methods described in this chapter. Other topics concern trends in the properties of compounds whose superconducting, magnetic, and other properties are of special interest, both in basis science and in technological applications. We begin with a brief discussion of the chemical bond in metallic compounds. An enlightening review has been written by Pettifor (1987).

1.4.3.1 Heats of Formation

Some time ago Miedema and co-workers (Boom et al., 1976; Miedema, 1976) developed a highly successful empirical theory that reproduces the observed signs of the heat of formation of about 500 alloys of the simple and transition metals. The theory assumes that this wealth of experimental data is described by the interplay of only two constituent properties: a quantity φ that is thought to be the electronegativity, and n, the electron density at the boundary separating atomic cells in the constituent. The heat of formation $\Delta H(\mathrm{A,B})$ (the total energy of the compound, A B, minus those of the constituents, A and B) is, according to Miedema, simply

$$\Delta H(\mathrm{A,B}) = Q\,[n(\mathrm{A}) - n(\mathrm{B})]^{2/3} - P\,[\varphi(\mathrm{A}) - \varphi(\mathrm{B})]^2 \qquad (1\text{-}485)$$

with positive constants P and Q. Pettifor (1978b, 1979) derived an expression for ΔH of transition-metal alloys within the tight-binding approximation that has the form

$$\Delta H = f(\bar{Z})\,(\Delta Z)^2 \qquad (1\text{-}486)$$

where $\bar{Z}$ and ΔZ are the average and difference in number of valence d electrons, respectively. The prefactor $f(\bar{Z})$ is negative for $\bar{Z}$ lying near the middle of the series, and it becomes positive toward the edges. The trends obtained are in good agreement with Miedema's formula, Eq. (1-485), but the microscopic interpretation is different. In fact, according to Pettifor the origin of the negative values of f is the *mismatch* of the *d-band centers* of the constituents, which replaces the second term on the right•hand side of Eq. (1-485). This is so because the common d band of the compound is wider than the separate d bands of the constituents if their centers are not degenerate, and the parabolic bonding trend had the same origin as the trends in the cohesive energies which were discussed in detail in Sec. 1.4.1.2. But Pettifor also identified the reason for the tendency toward phase separation, i.e., for the positive values of f in Eq. (1-486) or for the first term on the right-hand side of Eq. (1-485); when the d bands of the constituents merge to form the common d band, the latter's average value represents, in general, a loss of d-band width for one of the constituents, and the bond-energy reduction caused by this loss of width is not balanced by the corresponding gain is width on the other constituent. Hence, the physical quantity responsible for the tendency toward phase separation is the *mismatch* between the constituent *d-band widths.*

Pettifor's (1979) simple formula (1-486) and its physical ingredients rest on the tight-binding approximation. The extent to which this model is true can be established by self-consistent calculations of the band structures, electron densities, state densities, and heats of formation of the type discussed in this chapter. Early attempts (Kübler, 1978) for ScX, with X = Mo, Ru, Rh, Pd, and Ag, do show the above-mentioned trends in the heats of formations and the band structures, but were not systematic enough to expose the mechanisms of Pettifor's theory. Williams et al. (1980b), however, undertook an extensive calculation for the 28 binary combinations of the eight 4d-transition metals Y through Pd in assumed CsCl (b.c.c.-coordinated) and AuCuI (f.c.c.-coordinated) structures. Their results are shown in Fig. 1-59, where $\Delta H\,(\mathrm{A,B})/(\Delta Z)^2$ is plotted versus $\bar{Z}$ and compared with Miedema's empirical results. Since the heats of formation can be expanded in powers of ΔZ,

$$\Delta H\,(\mathrm{A,B}) = \\ = f(\bar{Z})\,(\Delta Z)^2 + g\,(\bar{Z})\,(\Delta Z)^4 + \ldots \qquad (1\text{-}487)$$

Fig. 1-59 shows to what extent the entire set of 28 heats of formation is described by the quadratic term and the single, ΔZ-independent function $f(\bar{Z})$. Important is the overall similarity of the curves in Fig. 1-59, in particular, the fact that compounds with $\bar{Z}$ in the middle of the transition series form ($\Delta H < 0$), while those toward the edges do not. Concerning the details, Williams et al. (1980b) found that both the crystal structure and the ΔZ dependence of $\Delta H/(\Delta Z)^2$ can be traced back to the dependence of the total energy on geometry-related variations in the d-band state density – through the sum of single-particle energies. This is the source of rapid variations in ΔH with atomic number (similar to the facts controlling the phase stability of elementary metals discussed in Sec. 1.4.1.2). It therefore controls the convergence of the series in Eq. (1-487) and, hence, the ΔZ dependence of the curves in Fig. 1-59. Compounds corresponding to small ΔZ probe the rapid variation of the sum, whereas

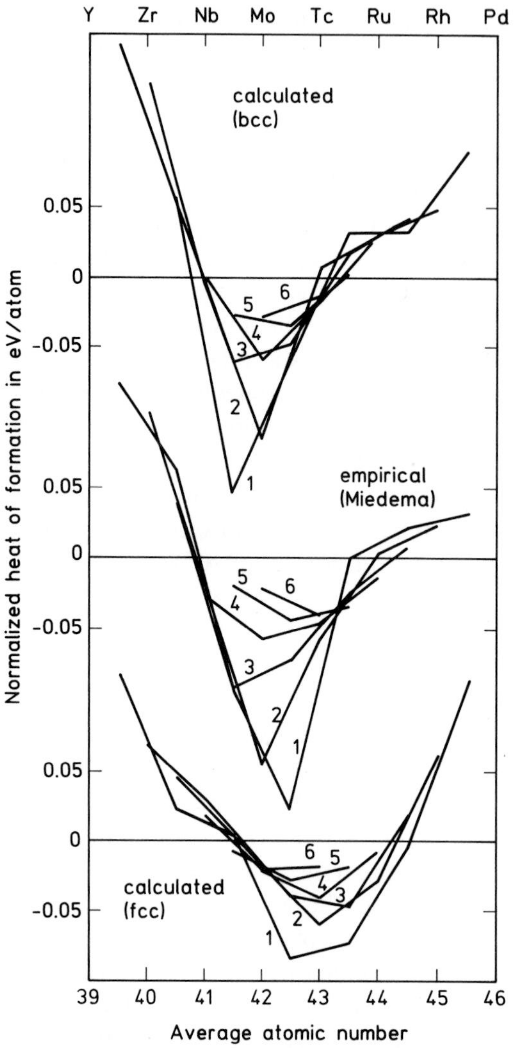

Figure 1-59. Normalized heats of formation versus the average atomic number as given by Williams et al. (1980b). Integers indicate the common value of ΔZ of points connected by lines.

those corresponding to large ΔZ sample its overall parabolic dependence on the number of the d electrons.

We next turn to the factors responsible for the bonding of transition metals to non-transition metals. This was first investigated by Gelatt et al. (1983) whose work is distinguished from previous studies on carbides, nitrides and oxides by Ern and Switendick (1965) and Neckel et al. (1976) by their using of self-consistent total energies. They analyzed the calculated chemical trends in terms of the s-, p-, and d-like state densities of the compounds and the constituents, and they found rather different types of electronic structure and bonding when the atomic s and p levels of the non-transition metal lie above, near, or below the energy of the transition metal d level. This crossover of levels is illustrated in Fig. 1-60, which is taken from the work of Gelatt et al. (1983) and shows the electronic structure of the typical cases of PdLi, PdB, and PdF. Referring the reader to the paper by Gelatt et al. for a detailed discussion and to Williams et al. (1983a) for a short review, we state the salient facts by pointing out that there are two important ingredients to the understanding of both the electronic structure and the heat of compound formation: The first is the expansion of the transition-metal lattice due to the insertion of the p-element, and the second is the interaction between the valence d shell on the transition metal with the valence sp shell of the light element and the extent to which covalent hybridization of states can develop. Clearly, the latter is seen to be most pronounced in the case of PdB where the levels cross. It thus constitutes a typical example of metallic covalent bonding. On the other hand, the compound PdF is a typical case of ionic bonding, because the low-lying p states of F are seen to be completely filled. The reduced palladium d-band width, which is a consequence of the lattice expansion, is most obvious in the cases of PdLi and PdF, where the ensuing loss of d-band bonding is thus considerable. Both the energy lost due to the lattice expansion and the energy gain from bond formation were explicitly calculated by Gelatt et al. (1983) for a series of 4d transition elements in compounds with

Li, B, and F. Their combined results are shown in Fig. 1-61, which is again in qualitative agreement with the theory of Miedema (described by a formula like Eq. (1-485) but with an added term on the right-hand side); however, as before, the physical mechanisms are different.

Of interest in this context are calculations by Koenig et al. (1984) of alkali metal-gold compounds which naturally extend the calculations by Gelatt et al. (1983) down the alkali group and give the trends of the electronic structure as well as the heats of formation. Furthermore, calculations by Mohn and Pettifor (1988) of the electronic structure of the transition-metal monoborides across the 3d period, TMB (TM = Sc, Ti, ..., Cu), for five different structure types give the preferred crystallographic modification which confirms the experimental trend. The binding mechanism in the FeB and CrB structure types exhibit not only transition-metal-boron interactions but also strong boron-boron bonds along the chains which are responsible for their stability. The binding energies and electronic structure of Ni with Al were determined by Hackenbracht and Kübler (1980) and the substitution of Al by V in Ni_3Al was investigated by Xu et al. (1987). Compounds of Rh with Al and of Ga with In were considered by Verbeek et al. (1983) and of TiAl and VAl alloys by Chubb et al. (1988). The latter calculations constitute a step, although small, toward understanding the reduced embrittlement of Ll_0 Ti-Al alloys which accompanies the introduction of small concentrations of V.

Transition metal hydrides, especially the connection of the electronic structure with the heats of formation and their dependence on different types of interstitial positions, were investigated by means of self-consistent calculations by Williams et al.

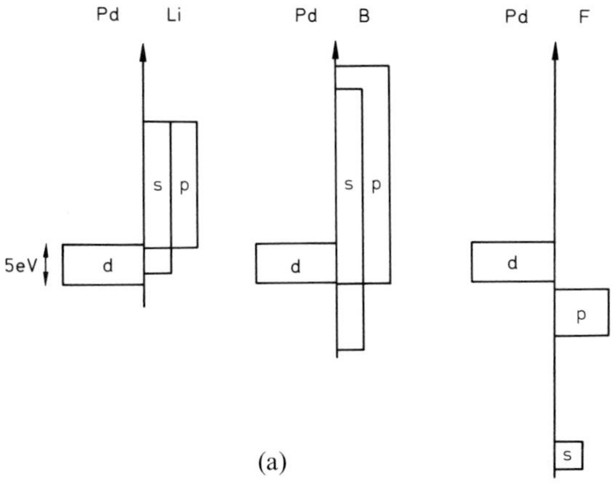

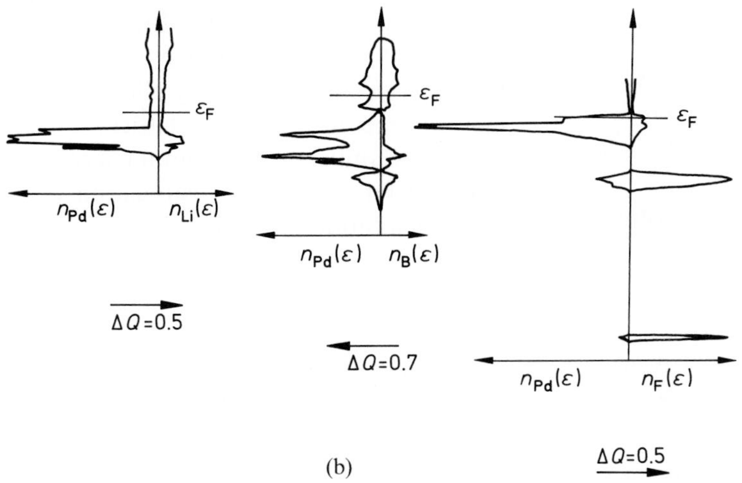

Figure 1-60. Electronic structure of compounds: (a) Constituent band limits and (b) calculated compound state densities, after Gelatt et al. (1983).

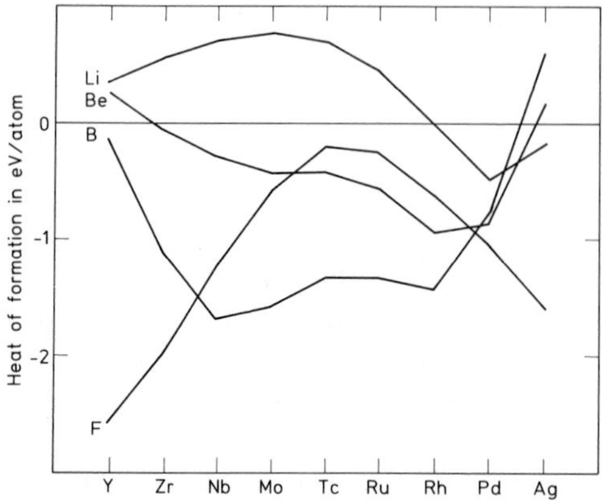

Figure 1-61. Variation of heat of formation with the number of valence electrons. Corresponding to elements indicated, after Gelatt et al. (1983).

(1979), Methfessel and Kübler (1982), Chan and Louie (1983), Gupta (1984), Ho et al. (1984b), Tao et al. (1986), and, most recently, by Min and Ho (1989).

Bonding- and total-energy studies for carbides and nitrides of Ti, V, Nb, and Mo have been done recently by Blaha et al. (1985), Blaha and Schwarz (1987), Chen et al. (1988a), and Price and Cooper (1989), and of Y by Zhukov et al. (1988). Furthermore, interesting calculations and enlightening discussions of the band structure and the chemical bond in intermetallic compounds *not* containing transition metals can be found in treatises by Schmidt (1985), Hafner and Weber (1986), and Guo et al. (1989). Finally, the band structure and cohesion of the refractory compounds has been reviewed by Zhukov et al. (1989).

1.4.3.2 Toward Alloy Phase Diagrams

Bonding in extended systems has so far only been described for periodic crystals with a small number of atoms per unit cell. If a description of more general systems is desired, one is faced with the problem that the lack of order in nonperiodic solids, as in solid solutions (chemical disorder), or in amorphous solids (structural disorder), makes it difficult to solve the density-functional equations in a systematic way. In this case one may adopt the coherent-potential approximation (CPA) and related methods which deal with random alloys (Ehrenreich and Schwartz, 1976; Gautier et al., 1975; Sigli et al., 1986; Hawkins et al., 1987; Turchi et al., 1987, 1988; Bieber and Gautier, 1986, 1987; Johnson et al., 1986; Weinberger, 1990; Johnson et al., 1990) or supercell band calculations for ordered alloys. The CPA and related methods are exceedingly interesting topics but, unfortunately, are outside the scope of this chapter. We will, however, briefly discuss supercell methods, in particular, a cluster expansion that is due to Sanchez and de Fontaine (1981). Connolly and Williams (1983) showed that this expansion not only allows the possibility of describing chemically disordered solids by density-functional theory and the techniques covered in this chapter, but also allows the introduction of temperature (through the methods of statistical mechanics) and the subsequent description of phase diagrams. To be precise, these should be called coherent phase diagrams because of limitations to transformations and phase changes which conserve the basic lattice framework of the crystalline solid solution (de Fontaine, 1979).

Proceeding now with the basis of the theory, we consider a binary alloy consisting of atoms A and B on a lattice of fixed symmetry and express the total energy per unit cell referred to the segregation limit (heat of solution at $T=0$) as

$$\Delta E(\Omega) = \sum_n v_n(\Omega)\, \xi_n \qquad (1\text{-}488)$$

where $v_n(\Omega)$ are many-body interaction potentials, ξ_n are multisite correlation functions defined on an nth-order cluster, Ω is the volume, and the sum is over all cluster types on a fixed lattice. The correlation function can be written as

$$\xi_n = \frac{1}{N_n} \sum_{\{p_i\}} \sigma_{p_1} \sigma_{p_2} \dots \sigma_{p_n} \tag{1-489}$$

where σ_{p_i} is an Ising spin-like variable, which takes the values $+1$ or -1, depending on whether the lattice point p_i is occupied by an A or a B atom. The sum is over all nth-order clusters of a given type in the lattice. N_n is the total number of such clusters. The expansion of Eq. (1-488) is useful only if it converges fast, i.e., if only low order clusters need to be considered. Note that Eq. (1-488) formally corresponds to the Ising model in a magnetic field if all terms with $n \geq 3$ are omitted, but experience shows that typically clusters up to order $n=4$ are needed. Now, Eq. (1-488) is first used to derive the interaction potentials $v_n(\Omega)$ by means of electronic-structure calculations for a selected set of ordered structures – an example is given below. In a second step, it is assumed that the potentials $v_n(\Omega)$ are transferable to the disordered state at $T=0$ and are unchanged (or nearly so) even when $T>0$. For the finite temperature behavior, one then uses methods of statistical mechanics to evaluate the entropy and thermal averages of ξ_n, i.e., $\langle \xi_n \rangle$; examples are the cluster variation method (CVM) (for recent treatises see Sanchez et al., 1984 and Mohri et al., 1985) or Monte Carlo simulations (Binder et al., 1981; Styer et al., 1986).

Following Connolly and Williams (1983, 1984), we now consider as an example the case of chemical disorder on a face-centered cubic lattice. If we assume that only clusters consisting entirely of nearest neighbors are important, then the cluster sum of Eq. (1-488) truncates at the fourth order. Hence, there are five interaction potentials $v_n(\Omega)$, $n=0, 1, 2, 3, 4$, to be determined from density-functional calculations for which one may choose the ordered structures f.c.c. for the metal A, f.c.c. for the metal B, the $L1_2$ structure (Cu_3Au type) for the compounds A_3B and AB_3 and the $L1_0$ structure (CuAuI type) for the compound AB. The cluster correlation functions, ξ_n, $n=0, 1, 2, 3, 4$, are then easily worked out with Eq. (1-489). In particular, $\xi_0=1$ for all structures (it is the structure independent term); ξ_1 is given by

$$\xi_1 = \tfrac{1}{4}(\sigma_1+\sigma_2+\sigma_3+\sigma_4) \tag{1-490}$$

and is the "point" correlation function for σ_i, $i=1, \dots, 4$, on the vertices of the nearest-neighbor tetrahedron. ξ_2 (equal to 1/6 times the sum of all products $\sigma_i \sigma_j$, $i \neq j$) corresponds to nearest-neighbor pairs, ξ_3 to nearest-neighbor triangles, and ξ_4 to the nearest-neighbor tetrahedron, i.e.,

$$\xi_4 = \sigma_1 \sigma_2 \sigma_3 \sigma_4 \tag{1-491}$$

The numerical values of ξ_n, given by Connolly and Williams (1983), allow the total-energy curves for the ordered compounds to be inverted giving

$$v_n(\Omega) = \sum_{m=0}^{4} (\xi^{-1})_{nm} \, \Delta E_m(\Omega) \tag{1-492}$$

where the index m denumerates the structure types (A, A_3B, AB, AB_3, and B), and $\Delta E_m(\Omega)$ are the heats of formation determined from density-functional theory as a function of the volume, Ω.

The heats of formation, ΔE_D, of disordered AB alloys at $T=0$ are now obtained from Eq. (1-488) if the correlation function appropriate for a random alloy $A_{1-x}B_x$ is used; this is

$$\xi_{n,D} = (1-2x)^n \tag{1-493}$$

and, using the values of $v_n(\Omega)$ obtained from Eq. (1-492), we obtain

$$\Delta E_D(\Omega) = \sum_{n=0}^{4} (1-2x)^n \, v_n(\Omega) \tag{1-494}$$

where x is the relative concentration of the B atom in $A_{1-x}B_x$.

Connolly and Williams (1983, 1984) were also the first to apply this formalism to real materials by calculating the heats of formation for all 28 binary compounds that can be formed from the 4d transition metals Y through Pd, using the LDA and the ASW method. They obtained the interaction potentials and found that they decrease rapidly with cluster complexity, indicating a convergence of the cluster expansion of the kind assumed initially. Terakura et al. (1987), who applied this formalism to investigate the alloy-phase stability of Cu-Ag, Cu-Au, and Ag-Au systems, give the potentials, $v_0(\Omega)$ to $v_n(\Omega)$, providing a clear graphic demonstration of the convergence. These calculations, furthermore, show that the *qualitative* features are correctly predicted, for instance, ordered compounds for Cu-Au, a broad miscibility gap for Cu-Ag and complete solubility below melting for Ag-Au. Prior to the paper by Terakura et al. (1987), these distinct properties remained largely unexplained.

Returning to the calculations by Connolly and Williams (1983, 1984), we first mention their finding that the triangle and tetrahedron potentials, v_3 and v_4, are very small for systems such as PdAg and CuAg, in which directional bonding is relatively unimportant and are larger in systems such as YPd for which the average number of d electrons indicates an approximately half-filled d shell and directional bonding. Second, and more important, is the fact that Connolly and Williams succeeded in calculating the *dominant* features of the phase diagrams of the 28 4d transition metal alloy systems. These calculations were approximations, in that Connolly and Williams, in addition to the formalism described above, used the entropy of the completely disordered state,

$$S_D = -k_B[(1-x)\ln(1-x) + x\ln x] \quad (1\text{-}495)$$

where k_B is Boltzmann's constant, yet the agreement with measured phase diagrams is generally very good.

An in-depth study of the phase diagrams of the Cu-Au, Cu-Ag, and Ag-Au systems is described by Wei et al. (1987), who discuss the assumptions necessary to reproduce the experimental phase diagrams in detail and, furthermore, make an attempt to understand them in terms of the electronic properties and atomic-scale structure of the constituent ordered phases; this was also attempted by Terakura et al. (1989). Takizawa et al. (1989) studied with encouraging success the nine alloy systems that can be formed using Cu, Ag, and Au, on the one hand, and Ni, Pd, and Pt on the other. Epitaxial effects on coherent phase diagrams of alloys, in this case $Cu_{1-x}Au_x$ and $GaAs_xSb_{1-x}$, were investigated by Wood and Zunger (1989). Finally, generalizing the cluster-expansion formalism described above to include b.c.c. systems and using the CVM, Sluiter et al. (1990) calculated and discussed the complete phase diagram of the Al-Li system. It seems certain that calculations of the kind discussed in this subsection will become increasingly prominent in future applications of band theory to questions concerning alloy-phase stability and complicated phase diagrams.

1.4.3.3 Challenging Models

The discovery of high-temperature superconductivity in the compound $La_{2-x}Ba_xCuO_{4-y}$ by Bednorz and Müller (1986) and later in $YBa_2Cu_3O_{7-y}$ by Wu et al. (1987) has led to a great number of band-structure calculations for these and related systems. Obviously, the challenge was – and still is – to uncover the reason for the striking and exceptional properties of these compounds.

Fortunately, Pickett (1989) has reviewed all but the most recent developments in our understanding of their electronic structure, so we will not discuss this interesting topic here any further. But one remark seems in order: the local approximation to density functional theory discussed in Sec. 1.3 coupled with any of the calculational tools treated in Sec. 1.2 apparently cannot yet describe satisfactorily the electronic structure of these oxides, especially those parts of their phase diagrams where they are insulating and magnetic. Thus, progress is still to be made here.

The field of ternary superconductors was most actively investigated in the early 1980s, and research was mainly carried out on two classes of materials: the ternary molybdenum chalcogenides of the type $PbMo_6S_6$, often referred to as "Chevrel phases", and the ternary rhodium borides of the type $ErRh_4B_4$ (Fischer and Maple, 1982). Of particular interest was the interplay of magnetic order with superconductivity. Electron band-structure calculations for the ternary molybdenum chalcogenides are reviewed by Nohl et al. (1982) who, in particular, examine the relationship of the band structure to important physical properties of these compounds such as their high superconducting transition temperatures. Freeman and Jarlborg (1982), on the other hand, review the connection of the band structure with magnetism for both the molybdenum chalcogenides and the rhodium borides, whereas Jarlborg and Freeman (1982) deal with the electronic structure, spin polarization, and high critical fields in Chevrel compounds.

A brief discussion of the vast field of binary superconductors, especially the A15 compounds (β tungstens) and C15 compounds (Laves phases), will take up a little more space.

The A15 compounds, *after* the high-temperature oxide superconductors mentioned in the beginning of this subsection, include materials having the second-highest superconducting transition temperatures, T_c, known. The most prominent example is Nb_3Sn which is also of great technological importance. Many of these compounds show unusual or anomalous low-temperature behavior of electronically derived properties such as elastic constants, Knight shifts, electrical resistivity, and magnetic susceptibility. Furthermore, two of the "high-T_c" members of the A15 family, V_3Si with $T_c = 17$ K and Nb_3Sn with $T_c = 18$ K, undergo martensitic cubic-to-tetragonal structural phase transitions as the temperature is lowered to T_M, which is slightly higher than T_c. Band-structure calculations were done as early as 1965 and for a thorough discussion of the literature, we refer the reader to the comprehensive review by Klein et al. (1978) that also contains the first results of self-consistent calculations. These were carried out using the local-density-functional approximation (LDA) and the APW method (refined by removal of muffin-tin restrictions of the interstitial potential), and include the compounds V_3X and Nb_3X, with X = Al, Ga, Si, Ge, and Sn. They are found to possess very flat bands that evolve from the two-fold degenerate d state Γ_{12} near or at E_F and give rise to sharp peaks in the density of states as a function of the energy.

It is not surprising that the rich low-temperature properties of the A15s have led to various theoretical models to explain their physical mechanisms; Weger and Goldberg (1973) give an extensive review. It is significant that the calculations of Klein et al. (1978) – and all later ones to be cited subsequently – allow a critical assessment of these models and lead to the conclusion that the basic assumptions of any of the models cannot be justified by these ab-initio calculations. Yet these calculations – without any adjustable parameters – are in agreement with experimental data. Similarly, for instance, the LMTO calculations and the measurements of angular correlation of positron annihilation radiation both give consistent pictures of the Fermi surface of V_3Si (Jarlborg et al., 1983 b). Furthermore the de Haas-van Alphen frequencies in the tetragonal, low-temperature phase of Nb_3Sn are nicely explained by ASW calculations (Wolfrat et al., 1985). Thus we believe that calculations based on parameter-free electronic band-structure calculations do have predictive powers, and that determinations of the electron-phonon coupling parameters such as those of Klein et al. (1979), Radousky et al. (1982), and Jarlborg et al. (1983 a) are of considerable value.

The martensitic, cubic-to-tetragonal phase transitions mentioned above are, of course, not only found in some of the A15s, but also occur, for instance, in simpler compounds having the CsCl (B2) structure. Thus at some time it was believed that the phase transition in LaAg alloyed with some In (Ihrig et al., 1973) was due to a "band-Jahn-Teller effect". In this case, the lifting of subband degeneracy at a single point in k space (here the Γ_{12}-level, which, like the A15-compounds, is near the Fermi energy) was assumed to be the driving mechanism. That this is most likely untrue was revealed by measurements and band-structure calculations by Niksch et al. (1987), who showed that the deformation potential coupling lifts the degeneracy of equivalent Fermi surface pockets and thus leads to softening of the Γ_3 elastic constant. The situation may be similar in VRu (Asada et al., 1985), and the premartensitic phenomena in NiTi are most likely Fermi surface effects as well (Zhao et al., 1989).

The Laves (C15) intermetallic compounds also host interesting superconductors and unusual magnetic materials. The electronic band structure of the ferromagnet $ZrFe_2$ and the superconductor ZrV_2 was obtained by Klein et al. (1983) using self-consistent APW calculations, while de Groot et al. (1980) and Jarlborg et al. (1981) presented calculations for the unusual ferromagnet $ZrZn_2$, and van der Heide (1984) compared de Groot's calculation successfully with measured optical properties of $ZrZn_2$. Finally, the differences in the electronic structure between the ferromagnet $ZrZn_2$ and the superconductor ZrV_2 received particular attention in highly precise calculations by Huang et al. (1988).

We close this subsection by pointing out that it is of considerable interest to understand the possible role of spin fluctuations in suppressing the superconducting transition temperature in a number of materials. An example that was pointed out earlier in Sec. 1.4.1.1 is Pd; another one is conceivably the enhanced paramagnetic $TiBe_2$ which also has the C15 structure. This complex of questions received serious attention in theoretical papers by Rietschel and Winter (1979) and Rietschel et al. (1980), and by calculations carried out by Pictet et al. (1987).

1.4.4 Magnetic Intermetallics and Compounds

As in Sec. 1.4.3, our review of the band-structure properties of this important class of materials must unfortunately be brief and is therefore incomplete. Although we will essentially exclude materials containing rare earths and actinides, the number of transition-metal compounds and alloys that remains is still large. Their magnetic properties, in all cases, are due to itinerant electrons. The meaning of the term "itinerant" is explained at the beginning of Sec. 1.4.2. This statement is, of course, often challenged, and controversial cases of great conceptual interest exist and must be discussed.

We begin with a rather detailed discussion of magnetization data referring the reader later to the relevant literature for other topics of interest. Saturation magnetization is a ground-state property and is thus amenable to spin-polarized energy-band calculations of the type described in Secs. 1.2 and 1.3. The data are most conveniently displayed by the Slater-Pauling curve (Bozorth, 1951) where the saturation magnetization of ferromagnets is plotted versus the electron-to-atom ratio. However, Williams et al. (1983c, 1984) and Malozemoff et al. (1983, 1984a, b) have shown another way of displaying these data which results in a more "universal" curve and is physically more revealing. Therefore, we will essentially use their concept and call it the Slater-Pauling-Friedel curve.

1.4.4.1 The Slater-Pauling-Friedel Curve

The approach of Williams et al. (1983c, 1984) and Malozemoff et al. (1983, 1984a, b) (let us temporarily call this group of authors WM) is based on band-structure calculations coupled with Friedel's (1958) interpretation of the Slater-Pauling curve and an extension of the Friedel picture by Terakura and Kanamori (1971) and Terakura (1977a, b). The analysis is remarkably simple and is as follows: As stated earlier (Eq. (1-468)), the magnetization m (in units of μ_B) is the net excess of majority-spin electrons, $n_\uparrow$,

$$m = n_\uparrow - n_\downarrow$$

while neutrality requires (Eq. (1-470)) that the corresponding sum equal the number of valence electrons Z_V

$$Z_V = n_\uparrow + n_\downarrow$$

The origin of the simplicity of the analysis is that very often *either* $n_\uparrow$ *or* $n_\downarrow$ can be accurately estimated on the basis of general considerations. In such cases, the neutrality condition (Eq. (1-470)) can be used to eliminate from the expression for the magnetization $n_\uparrow$ or $n_\downarrow$, whichever is more difficult to estimate, giving either

$$m = 2n_\uparrow - Z_V \tag{1-496}$$

when $n_\uparrow$ is easily estimated or

$$m = Z_V - 2n_\downarrow \tag{1-497}$$

when $n_\downarrow$ is easily estimated. It is important to add that in the case of compounds or alloys, all quantities m, $n_\uparrow$, $n_\downarrow$, and Z_V are atom averages. Thus, for instance, if the alloy is $A_{1-x}B_x$, then

$$m = m_A(1-x) + m_B\, x \tag{1-498}$$

where m_A is the magnetic moment of atom A, and m_B that of atom B. The same applies for Z_V. Of course, the neutrality condition concerns the entire unit cell, not necessarily the separate constituents.

As WM demonstrate, for overwhelmingly many alloys, it is the number of the majority-spin electrons, $n_\uparrow$, that is easily estimated and *nearly constant*. Thus Eq. (1-496) describes the right-hand side of the traditional Slater-Pauling curve having a downward 45° slope, whereas the left-hand side with an upward 45° slope is described by Eq. (1-497) for a smaller number of cases, provided that $n_\downarrow$ is nearly constant. The near constancy of $n_\uparrow$ in the former case or $n_\downarrow$ in the latter is a general effect of the band structure as was argued by WM. In particular, from the results given in Sec. 1.4.2.2 it is clear that for Ni and Co the spin-up d band lies entirely below the Fermi level and therefore contributes precisely five electrons to $n_\uparrow$. Ni and Co, therefore, are called "strong" magnets. WM now define the concept of magnetic valence, Z_m, by this integer contribution, $n_{d\uparrow}$, as

$$Z_m = 2n_{d\uparrow} - Z_V \quad (1\text{-}499)$$

Since elements to the right of Fe in the periodic table possess filled up-spin d bands, the magnetic valence of Co, Ni, Cu, Zn, ..., Br are, for example, simply 1, 0, −1, −2, ..., −7. In the late transition metals, the sp band contributes ≈0.3 electrons to $n_\uparrow$, giving via Eq. (1-496) Co and Ni magnetic moments of $\approx 1.6\,\mu_B$ and $\approx 0.6\,\mu_B$, respectively, i.e.,

$$m = Z_m + 0.6 \quad (1\text{-}500)$$

The same prescription (Williams et al., 1983c) gives Fe a magnetic valence of 2, and the fact that the magnetization of Fe is $2.2\,\mu_B$, i.e., less $2.6\,\mu_B$, reveals the magnetic weakness of Fe: the penetration of the up-spin d band by the Fermi level, see Fig. 1-54. On the right-hand side of Fig. 1-62, we plot as a straight line the magnetic valence, Z_m, plus 0.6, but Z_m decreases going to the right. This is opposite to the convention introduced by WM but gives the curve the appearance of the traditional Slater-Pauling curve. The band structure of any combination of Fe, Co, Ni, and Cu is very complicated, due to the hybridization of the various d bands with one another and with the sp bands. However, the magnetization of these alloys follows closely the straight line in Fig. 1-62 because the number of sp electrons remains near 0.3 being locked in an sp-hybridization valley (or Fano-type resonance, Terakura (1977a, b)). More interesting, still, are alloys of Fe, Co, and Ni with *early* transition elements. They certainly possess d electrons, however, when atoms of magnetic Fe, Co, and Ni are replaced by early transition-metal atoms, the effect is to reduce $n_{d\uparrow}$, the number of up-spin d electrons, not continuously, but by precisely five. This crucial observation, made by Friedel (1958), usually gives a magnetic valence of $-Z_V$ for the early transition metals. For the reason given above, the number of sp electrons still remains near 0.3. Consider, for instance, the CoCr alloy system. In the conventional Slater-Pauling curve the magnetization of these alloys appears as a dramatic departure from regular behavior, whereas in Fig. 1-62 the magnetic valence of $Z_m = -6$ for Cr and $Z_m = 1$ for Co places all values close to the right-hand straight line. With VAu_4 and $MnAu_4$, Fig. 1-62 contains some exotic cases which appear here when the magnetic valence of Mn is chosen to be 3 and that of V to be 5 (Kübler, 1984a, b). Further examples of the utility of this type of analysis are given by Williams et al. (1983c) and Malozemoff et al. (1984a, b), who include many cases involving non-transition-metal solutes and amorphous alloys.

Before we continue to highlight the literature of some particular band structures of magnetic compounds, we turn to the left-hand side of Fig. 1-62 where the straight line is given by Eq. (1-497) with $n_\downarrow = 3$. The listing of cases in Fig. 1-62 is not complete, but those given follow closely Eq. (1-497) and are all b.c.c. coordinated compounds and alloys. Band-structure calculations for $CrFe_3$ (Kübler, 1984a) show that the value of $n_\downarrow = 3$ comes about because the Fermi level is pinned in the deep down-spin density-of-states valley typical of b.c.c.-coordinated systems (see, e.g. Fig. 1-29 and its interpretation and also Fig. 1-54). This is also found in Heusler alloys having the $L2_1$ structure (Ishida et al., 1982; Kübler et al., 1983) but is particularly pronounced in Heusler alloys having the $C1_b$ structure (like NiMnSb) which have metallic majority-spin electrons and insulating minority-

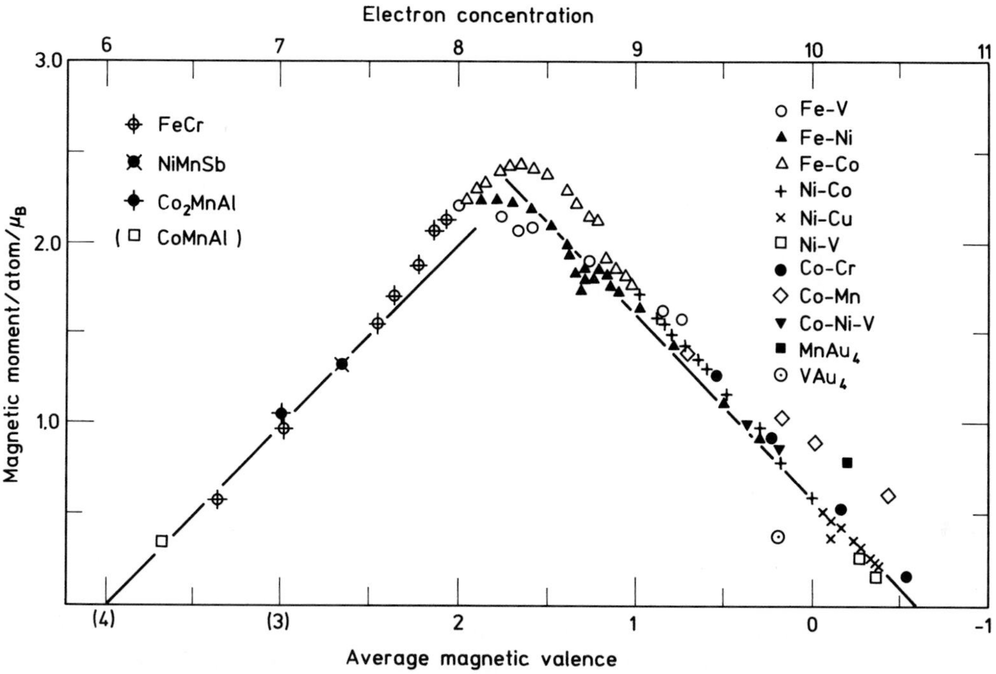

Figure 1-62. Slater-Pauling-Friedel curve. The data on the right are plotted versus magnetic valence, Eq. (1-500). The data on the left are plotted versus electron concentration according to Eq. (1-497).

spin electrons, i.e., 100% spin-polarized electrons at the Fermi level. This unusual band structure was first discovered by de Groot et al. (1983a) for NiMnSb and PtMnSb which were called "half-metallic ferromagnets". The compound PtMnSb is of great interest because of its exceptionally high magneto-optical Kerr effect (de Groot et al., 1983b; Wijngaard et al., 1989). A number of experiments were conducted to verify this unusual band structure, for instance, by Bona et al. (1985), Hanssen and Mijnarends (1986), and Kisker et al. (1987). Little doubt remains about its correctness. Other $C1_b$ Heusler alloys having the same band structure are described by Kübler (1984a) and de Groot and Buschow (1986).

Returning now to the right-hand side of the Slater-Pauling-Friedel curve and starting with alloys possessing large magnetizations, we point out the band-structure calculations by Schwarz et al. (1984), Joynt and Heine (1984), and Victora et al. (1984), who succeeded in explaining the distribution of the magnetic moments in FeCo-alloys, in particular in Fe_3Co, FeCo, and $FeCo_3$ for which Ebert (1989) calculated the conductivity tensor that presumably determines the size of the magneto-optical Kerr effect. As mentioned at the end of Sec. 1.4.2.3, the band structure and its connection with models for the unusual properties of INVAR alloys ($Fe_{1-x}Ni_x$) were the topic of papers by Williams et al. (1982), Mohn et al. (1989), and Moroni and Jarlborg (1989). Wagner (1989), on the basis of band-structure calculations, attempted a formulation of a general theory for the thermal and magneto-mechanical properties of INVAR. Of further interest are calculations for $TiFe_{1-x}Co_x$ by Kübler

(1980a) and by Schadler and Weinberger (1986); for FeV-systems by Hamada et al. (1984); for $MnPt_3$, $FePd_3$, and Fe_3Pt by Hasegawa (1985); and for $FePd_3$, again by Kuhnen and da Silva (1987). The very good agreement between the APW calculations of Hasegawa and the LMTO calculations of Kuhnen and da Silva is remarkable. In connection with the high-performance permanent magnet $Nd_2Fe_{14}B$ calculations have been made by Gu and Ching (1987) and Jaswal (1990) and very recently by Nordström et al. (1991). Very systematic calculations for YFe systems were carried out by Coehoorn (1989) and later by the same author (Coehoorn, 1990) for $YFe_{12-x}M_x$ (M = Ti, V, Cr, Mn, Mo, and W). The role which orbital contributions play in the formation of magnetic moments of YFe_2, FeAl, Fe_3Al, $ThFe_5$, and Fe_2P was investigated recently by Eriksson et al. (1990b).

In Sec. 1.4.3.3 we pointed out that the Laves (C15 and C14) phases have received considerable attention. Here we want to draw attention to band-structure work on magnetic Laves phases, in particular those involving Fe and Co. Examples include calculations by Klein et al. (1983) for $ZrFe_2$ and $ZrCo_2$; for YCo_2 by Schwarz and Mohn (1984); for YM_2 (M = Mn, Fe, Co and Ni) by Yamada et al. (1984); for $ZrMn_2$, $TiFe_2$, and $NbFe_2$ by Ishida et al. (1985); for $ScFe_2$ by Ishida and Asano (1985); and for $(Sc_{1-x}Ti_x)Fe_2$ and $(Zr_{1-x}Nb_x)Fe_2$ by Asano and Ishida (1989).

We close this subsection with a brief description of the literature concerning compounds of the magnetic transition metals with non-transition elements. To begin, there is the experimentally well-studied case of Ni_3Al; its band structure and magnetization was first obtained by Hackenbracht and Kübler (1980) using self-consistent LDA-calculations, and later by Buiting et al. (1983) emphasizing Fermi-surface properties. Sigfussen et al. (1984) determined the Fermi surface by the de Haas-van Alphen effect and found good agreement with the calculations for non-magnetic Ni_3Al provided they are suitably modified to include spin-orbit and exchange splittings; Van der Heide et al. (1985) compared these calculations with spectroscopic ellipsometry. The electronic and magnetic properties of MnSb were determined in great detail by Coehoorn et al. (1985). The system Fe_3Si received attention by Williams et al. (1982) and later again by Garba and Jacobs (1986). The compounds Fe_4N and Mn_4N were studied by Matar et al. (1988); Dijkstra et al. (1989a) discovered that $KCrSe_2$ is another half-metallic ferromagnet, and the same seems to be true of CrO_2 (Schwarz, 1986; but see also Kämper et al., 1987). Mohn (1988) treated the transition-metal semi-borides, and Dijkstra et al. (1989b) ferromagnetic chromium tellurides. Stenholm et al. (1989), finally, calculated the magnetic behavior of $(Fe_{1-x}Ni_x)_2P$.

1.4.4.2 Antiferromagnets

All evidence indicates that itinerant electron theory of the type discussed in this chapter also describes metallic antiferromagnetic compounds exceedingly well. We will give some examples but then briefly venture into the controversial field of transition-metal oxides that traditionally are considered to be Mott-Hubbard insulators for which band theory is believed to break down completely (Mott, 1974; Brandow, 1977).

For a series of Heusler alloys it was demonstrated some time ago (Kübler et al., 1983) that the total energy obtained within the LDA from self-consistent band-structure calculations correctly predicts which of the alloys order ferromagnetically or antiferromagnetically and why. In this context it is important to note that these Heusler alloys (e.g., Cu_2MnSn, Pd_2MnIn) are considered to be "ideal local-moment" systems. The calculations revealed why an itinerant-electron theory can describe local-moment systems of this type. The reason is that the Mn spin-up d states are found to be completely occupied, and the band widths indicate that they are just as delocalized as the d electrons of Cu or Pd. The Mn spin-down states, however, are nearly empty. Put differently, the spin-up d electrons of the Mn atoms join those of the Cu or Pd atoms in forming a common d band, whereas the spin-down d electrons are almost completely excluded from the Mn sites. The result of this localized exclusion is an equally localized region of magnetization. We therefore find localized magnetic moments composed of itinerant electrons (Williams et al., 1982; Kübler et al., 1983).

More recently, Sébilleau et al. (1989) considered antiferromagnetic $NaCrS_2$, but in a series of calculations Dijkstra et al. (1989 b, c) made an exhaustive trend study of chromium compounds such as CrTe, Cr_3Te_4, and Cr_2Te_3 as well as CrS, CrSe, Cr_3Se_4, and CrSb. They also correctly predicted the type of magnetic order, except for one case which may in reality possess an antiferromagnetic structure more complicated than assumed in the calculations. Such complications may come about through canted, helical, or, quite generally, non-collinear spin arrangements of a type apparently first proposed by Kouvel and Kasper (1963). In fact, self-consistent band-structure calculations and the total energies were obtained for a number of metallic, *non-collinear* antiferromagnets by Kübler et al. (1988 a, b) and Sticht et al. (1989). The metals theoretically investigated include γ-FeMn, $RhMn_3$, $PtMn_3$, Mn_3GaN, and Mn_3Sn. It is again manganese which dictates the local properties by means of locally excluded states comparable to the Heusler alloys. The total energy is minimized in a non-collinear tetrahedral structure in the case of γ-FeMn; in triangular spin arrangements in the cases of $RhMn_3$, $PtMn_3$, and Mn_3GaN (Kübler et al., 1988 a, b); and in a variety of other triangular structures in Mn_3Sn (Sticht et al., 1989).

Beginning now with the controversial cases, we first point out calculations of the electronic and magnetic structure of magnetite, Fe_3O_4. This is a spinel type ferrimagnet with a Curie temperature of 860 K. In a face centered cubic lattice of O^{2-} ions, Fe ions are located at two crystallographically different sites, tetrahedrally (A) and octahedrally (B) coordinated. The average valence of Fe at the A sites is 3+ and at the B sites 2.5+. The conductivity is high at room temperature, and a phase transition occurs at about 125 K, below which the compound becomes insulating. From a localized electron point of view, the high conductivity of Fe_3O_4 is due to electron hopping between the Fe^{2+} and Fe^{3+} ions occupying the B sites. The low-temperature transition is believed to be an order-disorder transition on these Fe ions. However, from an itinerant electron point of view, the high conductivity is a consequence of a partially filled d band of the B site Fe ions. The transition to the insulating state is presumably brought about by a band splitting on account of electron correlations or the electron-phonon interaction (Mott, 1967).

Yanase and Siratori (1984) were the first to calculate the band structure of Fe_3O_4 self-consistently using the APW method. Somewhat later de Groot and Buschow (1986) used self-consistent ASW calculations to obtain nearly identical results that support the picture of itinerant electrons in the high-temperature phase of Fe_3O_4. The band structure is half-metallic and thus gives an integer value for the magnetization of Fe_3O_4 which is in agreement with experiment. A photoemission study by Siratori et al. (1986) by and large seems to lend credibility to the calculated band structure, but the low-temperature phase has not found an unambiguous explanation yet and thus remains an open problem and a serious challenge for band theory.

The transition-metal monoxides MnO, FeO, CoO, and NiO possess a special conceptual significance because the microscopic origin of their insulating nature and antiferromagnetic ordering has been and remains a fundamental question of solid-state physics. A series of articles (Anderson, 1963; Mott, 1974; Brandow, 1977) proclaimed the issue settled some time ago, but new measurements (Sawatzky and Allen, 1984; McKay and Henrich, 1984; Hüfner et al., 1984; Hüfner, 1984, 1985; Riesterer et al., 1986) and calculations (Oguchi et al., 1983b; Terkura et al., 1984a, b) have reopened the issue. The fundamental question is easily stated as follows: The transition metal elements Mn, Fe, Co, and Ni possess between seven and ten valence electrons. Therefore, in an ionic compound, such as a monoxide, the Fermi level necessarily falls within the high density of d states which one might think would produce a metal. Why then are the monoxides insulators? Answers to this question fall into two distinct classes. One is the localized, or Mott-Hubbard, insulator picture (Mott, 1974; Brandow, 1977) with which the famous superexchange (Anderson, 1963) mechanism can be associated. The alternative conceptual picture is the itinerant or band picture (Slater, 1974; Wilson, 1970) which accounts well for such ground-state properties as their lattice constant, see Fig. 1-47, their cohesive energy (Anderson et al., 1979; Yamashita and Asano, 1983), their antiferromagnetism (even the crystalline direction in which the magnetization varies), and their insulating behavior for MnO and NiO, at least at 0 K (Terakura et al., 1984a, b). However, excitation properties like photoemission and optical absorption are serious problems that are not easily explained by ground-state band structures. The nature of the difficulty is perhaps similar to the problems encountered for the ferromagnetic metals, especially as described at the end of Sec. 1.4.2.2, but more enhanced in the case of the monoxides. The effects of electron correlations are not small and they can, in fact, be estimated by means of "Hubbard's U", which is approximately given by the third term on the right-hand side of Eq. (1-388) (for $\Delta n = 1$). An estimate by means of supercell calculation by Kübler and Williams (1986) gives $U \approx 4$ eV for NiO, but Norman and Freeman (1986) obtain a larger value (≈ 7.9 eV), treating differently the screening of the d states.

In contrast to MnO and NiO is CoO, where treating the antiferromagnetic order by the standard local spin-density approximation does not lead to an energy gap at the Fermi energy. Still, from angle-resolved photoemission spectra, there is evidence of itinerant behavior of the Co 3d electrons (Brooks et al., 1989). It is rewarding, therefore, that Norman (1990) succeeded in verifying a suggestion by Terakura et al. (1984b) that, by means of an orbital-polarization correction, a population imbalance in the t_{2g} band can lead to a gap. This calculation became possible because Eriksson et al. (1989, 1990a, b) developed a formal version of this correction that, however, still lacks a solid basis in the framework of density-functional theory.

1.5 Conclusion

The size of this chapter and the number of references given show clearly that in the last fifteen years *ab initio* methods based on the density functional formalism and the local density approximation for exchange and correlation have accumulated an impressive record of calculations. We have mainly focused our attention on bulk metals but would like to remind the reader of an equal – or even larger – amount of material on the electronic and structural properties of molecules, clusters, surfaces, thin films, impurities, etc., that concern both metals *and* semiconductors and also include thermal and kinetic properties. For many properties of such systems, numerical accuracies of 10% or better are systematically obtained, and with the continuous improvements in computer codes and hardware, these techniques, as Soler and Williams (1990) point out, may soon become an important tool in materials design, and, perhaps, even in the design of drugs. The term "design" has not yet played an important role, and, consequently, was not specifically mentioned in the context of this chapter. Instead, we stressed the importance of *ab initio* methods in developing conceptual understanding, in supporting or dismissing physical models, and in supplying tools that considerably aid the experimentalists in interpreting their data. In order to extend the scope of these methods to design problems, the calculations must be practical, and for this reason it will frequently be necessary to obtain not only the total energy but also the *forces* on the atoms. This then should result in a method in which the electronic *and* atomic degrees of freedom are simultaneously computed and relaxed, letting the atoms move to their equilibrium positions instantaneously. The first pioneering work along these lines in the field of semiconductor physics is the quantum-mechanical molecular dynamics calculations by Car and Parinello (1985) and the optimizations of complex structures by Ballone et al. (1988). Initial attempts at first-principles molecular dynamics for *metals* are the papers by Fernando et al. (1989a, b) and Woodward et al. (1989). The elegant treatment of forces within the APW method by Soler and Williams (1990), in addition to other results, has provided impressive data for phonon frequencies of Cu. Certainly, the 1990s will see a great deal of activity in this field.

1.6 Acknowledgements

One of the authors (J. K.) would like to express his gratitude to the Institut Romand de Recherche Numérique en Physique des Matériaux (IRRMA) in Lausanne (Switzerland) for enabling a two-months stay during which a substantial part of this work was completed. In particular he would like to thank Prof. M. Peter, Dr. T. Jarlborg, Mr. E. Moroni, and Prof. A. Baldereschi. The excellent technical assistance by Mrs. A. Hanna-Daoud and Mrs. B. Knell is gratefully acknowledged. We have received support by the Deutsche Forschungsgemeinschaft, under the auspices of Sonderforschungsbereich 252, Darmstadt/Frankfurt/Mainz.

1.7 References

Allen, P. B. (1987), *Phys. Rev. B36,* 2920.
Andersen, O. K. (1973), *Solid State Commun. 13,* 133–136.
Andersen, O. K. (1975), *Phys. Rev. B12,* 3060.
Andersen, O. K. (1984), in: *The Electronic Structure of Complex Systems:* Phariseau, P., Temmerman, W. M. (Eds.). New York: Plenum Press, pp. 11–66.
Andersen, O. K., Jepsen, O. (1977), *Physica 91B,* 317.

Andersen, O. K., Skriver, H. L., Nohl, H., Johansson, B. (1979), *Pure Appl. Chem. 52,* 93.

Andersen, O. K., Jepsen, O., Glötzel, D. (1985), in: *Highlights of Condensed Matter Theory, Proceedings of the International School of Physics "Enrico Fermi", Course LXXXIX:* Bassani, F., Fumi, F., Tosi, M. P. (Eds.). Amsterdam: North Holland, pp. 59–176.

Anderson, P. W. (1963), in: *Solid State Physics, Vol. 14:* Seitz, F., Turnbull, D. (Eds.). New York: Academic Press, pp. 99–214.

Anderson, J. R., Papaconstantopoulos, D. A., Boyer, L. L., Schirber, J. E. (1979), *Phys. Rev. B20,* 3172.

Antoncik, E. (1959), *J. Phys. Chem. Solids 10,* 314.

Asada, T., Hoshino, T., Kataoka, M. (1985), *J. Phys. F15,* 1497.

Asano, S., Ishida, S. (1989), *J. Phys.: Cond. Matt. 1,* 8501.

Asano, S., Yamashita, J. (1967), *J. Phys. Soc. Japan 23,* 714.

Asano, S., Yamashita, J. (1973), *Prog. Theor. Phys. 49,* 373.

Ashcroft, N. W., Mermin, N. D. (1976), *Solid State Physics.* Philadelphia: Holt-Saunders.

Ashkenazi, J. (1982), *Phys. Rev. B26,* 1512.

Austin, B. J., Heine, V., Sham, L. J. (1962), *Phys. Rev. 127,* 276.

Averill, F. W. (1972), *Phys. Rev. B6,* 3637.

Bachelet, G. B., Hamann, D. R., Schlüter, M. (1982), *Phys. Rev. B26,* 4199–4228.

Bacon, G. E. (1962), *Neutron Diffraction.* Oxford: Clarendon Press; p. 274.

Badralexe, E., Freeman, A. J. (1987), *Phys. Rev. B36,* 1378.

Badralexe, E., Freeman, A. J. (1989), *Phys. Rev. B40,* 1981–1983.

Bagayoko, D., Callaway, J. (1983), *Phys. Rev. B28,* 5419.

Bagno, P., Jepsen, O., Gunnarsson, O. (1989), *Phys. Rev. B40,* 1997.

Ballone, P., Andreoni, W., Car, R., Parinello, M. (1988), *Phys. Rev. Lett. 60,* 271.

Barth, von U. (1984a), in: *The Electronic Structure of Complex Systems:* Phariseau, P., Temmerman, W. M. (Eds.). New York: Plenum Press; pp. 67–140.

Barth, von U. (1984b), in: *Many-Body Phenomena at Surfaces:* Langreth, D., Suhl, H. (Eds.). Orlando: Academic Press, pp. 3–50.

Barth, von U., Hedin, L. (1972), *J. Phys. C5,* 1629–1642.

Bednorz, J. G., Müller, K. A. (1986), *Z. Phys. B64,* 189.

Bieber, A., Gautier, F. (1986), *Acta Metall. 34,* 2291.

Bieber, A., Gautier, F. (1987), *Acta Metall, 35,* 1839.

Binder, K., Lebowitz, J. L., Phani, M. K., Kalos, M. H. (1981), *Acta Metall. 29,* 1655.

Blaha, P., Redinger, J., Schwarz, K. (1985), *Phys. Rev. B31,* 2316.

Blaha, P., Schwarz, K. (1987), *Phys. Rev. B36,* 1420.

Bloch, F. (1929), *Z. Phys. 52,* 555.

Bona, G. L., Meier, F., Taborelli, M., Bucher, E., Schmidt, P. H. (1985), *Solid State Commun. 56,* 391.

Boom, R., de Boer, F. R., Miedema, A. R. (1976), *J. Less. Com. Met. 46,* 271.

Born, M., Oppenheimer, R. (1927), *Ann. Phys. (Leipzig) 84,* 457.

Bouckaert, L. P., Smoluchowski, R., Wigner, E. P. (1936), *Phys. Rev. 50,* 58.

Bozorth, R. M. (1951), *Ferromagnetism.* New York: van Nostrand, p. 441.

Bradley, C. J., Cracknell, A. P. (1972), *The Mathematical Theory of Symmetry in Solids.* Oxford: Clarendon Press.

Brandow, B. H. (1977), *Adv. Phys. 26,* 651.

Brewer, L. (1967), in: *Phase Stability in Metals and Alloys:* Rudman, P. S., Stringer, J., Jaffee, R. I. (Eds.). New York: McGraw-Hill, p. 39.

Brookes, N. B., Law, D. S.-L., Warburton, D. R., Wincott, P. L., Thornton, G. (1989), *J. Phys.: Cond. Matt. 1,* 4267.

Brooks, H. (1940), *Phys. Rev. 58,* 909.

Brooks, M. S., Johansson, B. (1983), *J. Phys. F13,* L197.

Buiting, J. J. M., Kübler, J., Mueller, F. M. (1983), *J. Phys. F13,* L179.

Burdick, G. A. (1963), *Phys. Rev. 129,* 138–150.

Burgers, W. G. (1934), *Physica 1,* 561.

Cade, N. A. (1980), *J. Phys. F10,* L187.

Cade, N. A. (1981a), *J. Phys. F11,* 2399.

Cade, N. A. (1981b), *Institute of Physics Conference Series No. 55, Ch. 5.* Rhodes, P. (Ed.). Bristol, London: The Institute of Physics.

Callaway, J. (1964), *Energy Band Theory.* New York: Academic Press.

Callaway, J. (1974), *Quantum Theory of the Solid State. Parts A and B.* New York: Academic Press.

Callaway, J. (1981), *Institute of Physics Conference Series, No. 55.* Rhodes, P. (Ed.). Bristol, London: The Institute of Physics.

Callaway, J., March, N. H. (1984), in: *Solid State Physics, Vol. 38:* Seitz, F., Turnbull, D., Ehrenreich, H. (Eds.). Orlando: Academic Press; pp. 136–223.

Callaway, J., Wang, C. S. (1977), *Phys. Rev. B16,* 2095.

Campagna, M., Wertheim, G. K., Baer, Y. (1979), in: *Photoemission in Solids II:* Ley, L., Cardona, M. (Eds.). Berlin: Springer, pp. 217–260.

Capellmann, H. (Ed.) (1987), *Metallic Magnetism.* Berlin: Springer.

Car, R., Parinello, M. (1985), *Phys. Rev. Lett. 55,* 2471.

Cardona, M., Ley, L. (1979), in: *Photoemission in Solids I:* Cardona, M., Ley, L. (Eds.). Berlin: Springer, pp. 1–104.

Carlsson, A. E., Gelatt, C. D. Jr., Ehrenreich, H. (1980), *Phil. Mag. A41,* 241.

Ceperley, D. M., Alder, B. J. (1980), *Phys. Rev. Lett. 45,* 566.

Chelikowsky, J. R., Chan, C. T., Louie, S. G. (1986), *Phys. Rev. B34,* 6656.
Chan, C. T., Louie, S. G. (1983), *Phys. Rev. B27,* 3325.
Chen, Y., Fu, C.-L., Ho, K.-M., Harmon, B. N. (1985), *Phys. Rev. B31,* 6775.
Chen, J., Boyer, L. L., Krakauer, H., Mehl, M. J. (1988a), *Phys. Rev. B37,* 3295.
Chen, Y., Ho, K.-M., Harmon, B. N. (1988b), *Phys. Rev. B37,* 283.
Chen, H., Brener, N. E., Callaway, J. (1989), *Phys. Rev. B40,* 1443.
Chodorow, M. I. (1939a), *Ph.D. Thesis* (MIT), unpublished.
Chodorow, M. I. (1939b), *Phys. Rev. 55,* 675.
Chou, M. Y., Lam, P. K., Cohen, M. L. (1983), *Phys. Rev. B28,* 4179.
Christensen, N. E. (1984), *Solid State Commun. 49,* 701.
Christensen, N. E., Heine, V. (1985), *Phys. Rev. B32,* 6145.
Chubb, S. R., Papaconstantopoulos, D. A., Klein, B. M. (1988), *Phys. Rev. B38,* 12120.
Coehoorn, R. (1989), *Phys. Rev. B39,* 13072.
Coehoorn, R. (1990), *Phys. Rev. B41,* 11790.
Coehoorn, R., Haas, C., de Groot, R. A. (1985), *Phys. Rev. B31,* 1980.
Cohen, M. L., Heine, V. (1970), in: *Solid State Physics, Vol. 24:* Ehrenreich, H., Seitz, F., Turnbull, D. (Eds.). Orlando: Academic Press, pp. 37–248.
Coleman, R. V., Lowrey, W. H., Polo, J. A. jr. (1981), *Phys. Rev. B23,* 2491.
Coleridge, P. T., Molenaar, J., Lodder, A. (1982), *J. Phys. C15,* 6943.
Conklin, J. B., Averill, F. W., Hattox, T. M. (1972), *J. Phys.* (Paris) *Suppl. C3, 33,* 213.
Conklin, J. B., Averill, F. W., Hattox, T. M. (1973), *J. Phys. Chem. Sol. 34,* 1627.
Connolly, J. W. D., Williams, A. R. (1983), *Phys. Rev. B27,* 5169.
Connolly, J. W. D., Williams, A. R. (1984), in: *The Electronic Structure of Complex Systems:* Phariseau, P., Temmerman, W. M. (Eds.). New York: Plenum Press, pp. 581–592.
Cortona, P., Doniach, S., Sommers, C. (1985), *Phys. Rev. A31,* 2842.
Daalderop, G. H. O., Kelly, P. J., Schuurmans, M. F. H. (1990), *Phys. Rev. B41,* 11919.
Daalderop, G. H. O., Kelly, P. J., Schuurmans, M. F. H., Jansen, H. J. F. (1989), *J. Phys.* (Paris) *Suppl. C8, 12,* 93.
Dacarogna, M., Ashkenazi, J., Peter, M. (1982), *Phys. Rev. B26,* 1527.
Danos, M., Maximon, L. C. (1965), *J. Math. Phys. 6,* 766–778.
Davenport, J. W. (1984), *Phys. Rev. B29,* 2896.
Davenport, J. W., Watson, R. E., Weinert, M. (1985), *Phys. Rev. B32,* 4883.
de Fontaine, D. (1979), in: *Solid State Physics, Vol. 34:* Seitz, F., Turnbull, D., Ehrenreich, H. (Eds.). Orlando: Academic Press, pp. 73–274.
de Groot, R. A., Koelling, D. D., Mueller, F. M. (1980), *J. Phys. F10,* L235.
de Groot, R. A., Mueller, F. M., van Engen, P. G., Buschow, K. H. J. (1983a), *Phys. Rev. Lett. 50,* 2024.
de Groot, R. A., Mueller, F. M., van Engen, P. G., Buschow, K. H. J. (1983b), *J. Appl. Phys. 55,* 2151.
de Groot, R. A., Buschow, K. H. J. (1986), *J. Magn. Magn. Mat. 54–57,* 1377.
Diener, G., Gräfenstein, J. (1989), *J. Phys.: Cond. Matt. 1,* 8445.
Dietz, E., Gerhardt, U., Maetz, C. J. (1978), *Phys. Rev. Lett. 40,* 892.
Dijkstra, J., van Bruggen, C. F., Haas, C., de Groot, R. A. (1989a), *Phys. Rev. B40,* 7973.
Dijkstra, J., van Bruggen, C. F., Haas, C., de Groot, R. A. (1989b), *J. Phys.: Cond. Matt. 1,* 9163.
Dijkstra, J., Weitering, H. H., van Bruggen, C. F., Haas, C., de Groot, R. A. (1989), *J. Phys.: Cond. Matt. 1,* 9141.
Doniach, S., Sommers, C. (1982), in: *Proceedings of the International Conference on Valence Fluctuations of Solids:* Falicov, L. M., Hauke, W., Maple, M. B. (Eds.). Amsterdam: North Holland.
Duschanek, H., Mohn, P., Schwarz, K. (1989), *Physica B161,* 139.
Eastman, D. E., Himpsel, F. J., Knapp, J. A. (1980a), *Phys. Rev. Lett. 44,* 95.
Eastman, D. E., Himpsel, F. J., Knapp, J. A. (1980b), *Phys. Rev. Lett. 45,* 498.
Eastman, D. E., Janak, J. F., Williams, A. R., Coleman, R. V., Wendin, G. (1979), *J. Appl. Phys. 50,* 7423.
Ebert, H. (1988), *Phys. Rev. B38,* 9390.
Ebert, H. (1989), *Physica B161,* 175.
Ehrenreich, H., Schwartz, L. M. (1976), in: *Solid State Physics, Vol. 31:* Ehrenreich, H., Seitz, F., Turnbull, D. (Eds.). New York: Academic Press, pp. 149–286.
Eriksson, O., Johansson, B., Brooks, M. S. S. (1989), *J. Phys.: Cond. Matt. 1,* 4005.
Eriksson, O., Johansson, B., Brooks, M. S. S. (1990a), *J. Phys.: Cond. Matt. 2,* 1529.
Eriksson, O., Nordström, L., Pohl, A., Severin, L., Boring, A. M., Johansson, B. (1990b), *Phys. Rev. B41,* 11807.
Ern, V., Switendick, A. C. (1965), *Phys. Rev. 137,* A1927.
Eschrig, H. (1989), *Optimized LCAO Method and the Electronic Structure of Extended Systems.* Berlin: Springer.
Eschrig, H., Seifert, G., Ziesche, P. (1985), *Solid State Commun. 56,* 777.
Esposito, E., Carlsson, A. E., Ling, D. D., Ehrenreich, H., Gelatt, C. D. jr. (1980), *Phil. Mag. A41,* 251.
Ewald, P. P. (1921), *Ann. Phys. 64,* 253.
Fawcett, E. (1988), *Rev. Mod. Phys. 60,* 209.
Feder, R., Rosicky, F., Ackermann, B. (1983), *Z. Phys. B52,* 31.
Fermi, E. (1928), *Z. Phys. 48,* 73.

Fernando, G. W., Davenport, J. W., Watson, R. E., Weinert, M. (1989a), *Phys. Rev. B40,* 2757–2766.

Fernando, G. W., Qian, G.-X., Weinert, M., Davenport, J. W. (1989b), *Phys. Rev. B40,* 7985.

Fernando, G. W., Watson, R. E., Weinert, M., Wang, Y. J., Davenport, J. W. (1990), *Phys. Rev. B41,* 11813.

Fischer, Ø., Maple, M. B. (1982), in: *Superconductivity in Ternary Compounds I:* Fischer, Ø., Maple, M. B. (Eds.). Berlin: Springer, pp. 1–24.

Freeman, A. J., Jarlborg, T. (1982), in: *Superconductivity in Ternary Compounds II:* Fischer, Ø., Maple, M. B. (Eds.). Berlin: Springer, pp. 167–200.

Friedel, J. (1958), *Nuovo Cimento 10,* Suppl. 2, 287.

Friedel, J. (1968), in: *The Physics of Metals:* Ziman, J. M. (Ed.). Cambridge: University Press, pp. 340–408.

Fritsche, L., Noffke, J., Eckardt, H. (1987), *J. Phys. F17,* 943.

Fry, J. L., Brener, N. E., Laurent, D. G., Callaway, J. (1981), *J. Appl. Phys. 52,* 2101.

Fu, C.-L., Ho, K.-M., Harmon, B. N., Liu, S. H. (1983), *Phys. Rev. B28,* 2957.

Fulde, P., Keller, J., Zwicknagl, G. (1988), in: *Solid State Physics, Vol. 41:* Ehrenreich, H., Turnbull, D. (Eds.). San Diego: Academic Press, pp. 1–150.

Fuster, G., Brener, N. E., Callaway, J., Fry, J. L., Zhao, Y. Z., Papaconstantopoulos, D. A. (1988), *Phys. Rev. B38,* 423.

Garba, E. J. D., Jacobs, R. L. (1986), *J. Phys. F16,* 1485.

Gautier, F., Ducastelle, F., Giner, J. (1975), *Phil. Mag. 31,* 1373.

Gelatt, C. D. jr., Ehrenreich, H., Watson, R. E. (1977), *Phys. Rev. B15,* 1613.

Gelatt, C. D. jr., Williams, A. R., Moruzzi, V. L. (1983), *Phys. Rev. B27,* 2005.

Girifalco, L. A. (1976), *Acta Metall. 24,* 759.

Gonis, A. (1986), *Phys. Rev. B33,* 5914–5916.

Gooding, R. J., Ye, Y. Y., Chan, C. T., Ho, K.-M., Harmon, B. N. (1991), *Phys. Rev. B43,* 13626.

Grewe, N., Steglich, F. (1990), in: *Handbook of the Physics and Chemistry of Rare Earths, Vol. 14:* Gschneidner, K. A. jr. (Ed). Amsterdam: North Holland.

Gschneidner, K. A. (1964), in: *Solid State Physics, Vol. 16:* Ehrenreich, H., Seitz, F., Turnbull, D. (Eds.). Orlando: Academic Press, pp. 275–426.

Gu, Z.-Q., Ching, W. Y. (1987), *Phys. Rev. B36,* 8530.

Gunnarsson, P. (1976), *J. Phys. F6,* 587.

Gunnarsson, O., Jones, R. O. (1985), *Phys. Rev. B31,* 7588.

Gunnarsson, O., Lundquist, B. I. (1976), *Phys. Rev. B13,* 4274.

Guo, X.-Q., Podloucky, R., Freeman, A. J. (1989), *Phys. Rev. B40,* 2793.

Gupta, M. (1984), *J. Less Common Met. 101,* 35.

Gyorffy, B. L., Pindor, A. J., Staunton, J. B., Stocks, G. M., Winter, H. (1985), *J. Phys. F15,* 1337.

Hackenbracht, D., Kübler, J. (1980), *J. Phys. F10,* 427.

Hafner, J., Weber, W. (1986), *Phys. Rev. B33,* 747.

Ham, F. S., Segall, B. (1961), *Phys. Rev. 124,* 1786–1796.

Hamada, N. (1981), *J. Phys. Soc. Japan 50,* 77.

Hamada, N., Terakura, K., Yanase, A. (1984), *J. Phys. F14,* 2371.

Hamann, D. R. (1979), *Phys. Rev. Lett. 42,* 662.

Hamann, D. R. (1989), *Phys. Rev. B40,* 2980–2987.

Hanssen, K. E. H. M., Mijnarends, P. E. (1986), *Phys. Rev. B34,* 5009.

Harrison, W. A. (1966), *Pseudopotentials in the Theory of Metals.* New York: Benjamin.

Harrison, W. A. (1980), *Electronic Structure and the Properties of Solids.* San Francisco: Freeman.

Hasegawa, A. (1985), *J. Phys. Soc. Japan 54,* 1477.

Hathaway, K. B., Jansen, H. J. F., Freeman, A. J. (1985), *Phys. Rev. B31,* 7603.

Hawkins, R. J., Robbins, M. O., Sanchez, J. M. (1987), *Phys. Rev. B33,* 4782.

Hedin, L., Lundquist, B. I. (1971), *J. Phys. C4,* 2064.

Heide van der, P. A. M., Baelde, W., de Groot, R. A., de Vroomen, A. R., Mattocks, P. G. (1984), *J. Phys. F14,* 1745.

Heide van der, P. A. M., Buiting, J. J. M., ten Dam, L. M., Schreurs, L. W. M., de Groot, R. A., de Vroomen, A. R. (1985), *J. Phys. F15,* 1195.

Heine, V. (1967), *Phys. Rev. 153,* 673.

Heine, V. (1970), in: *Solid State Physics, Vol. 24:* Ehrenreich, H., Seitz, F., Turnbull, D. (Eds.). Orlando: Academic Press, pp. 1–36.

Heine, V. (1980), in: *Solid State Physics, Vol. 35:* Ehrenreich, H., Seitz, F., Turnbull, D. (Eds.). Orlando: Academic Press, pp. 1–123.

Heine, V., Abarenko, I. (1964), *Phil. Mag. 9,* 451.

Heine, V., Liechtenstein, A. I., Mryasov, O. N. (1990), *Europhys. Lett. 12,* 545.

Heine, V., Samson, J. H. (1983), *J. Phys. F13,* 2155.

Heine, V., Weaire, D. (1970), in: *Solid State Physics, Vol. 24:* Ehrenreich, H., Seitz, F., Turnbull, D. (Eds.). Orlando: Academic Press, pp. 249–463.

Herring, C. (1940), *Phys. Rev. 57,* 1169.

Hickman, B. S. (1969), *J. Mater. Sci. 4,* 554.

Himpsel, F. J., Eastman, D. E. (1980), *Phys. Rev. B21,* 3207.

Himpsel, F. F., Knapp, J. A., Eastman, D. E. (1979), *Phys. Rev. B19,* 2919.

Ho, K.-M., Harmon, B. N. (1990), *Mat. Science and Eng. A127,* 155.

Ho, K.-M., Fu, C.-L., Harmon, B. N. (1983), *Phys. Rev. B28,* 6687.

Ho, K.-M., Fu, C.-L., Harmon, B. N., Weber, W., Hamann, D. R. (1982), *Phys. Rev. Lett. 49,* 673.

Ho, K.-M., Fu, C.-L., Harmon, B. N. (1984a), *Phys. Rev. B29,* 1575.

Ho, K.-M., Tao, H.-J., Zhu, X.-Y. (1984b), *Phys. Rev. Lett. 53,* 1586.

Hohenberg, P., Kohn, W. (1964), *Phys. Rev. 136,* B864.

Honda, N., Tanji, Y., Nakagawa, Y. (1976), *J. Phys. Soc. Japan 41,* 1931.

Huang, M., Jansen, H. J. F., Freeman, A. J. (1988), *Phys. Rev. B37,* 3489.
Hüfner, S. (1979), in: *Photoemission in Solids II:* Ley, L., Cardona, M. (Eds.). Berlin: Springer, pp. 173–216.
Hüfner, S. (1984), *Z. Phys. B58,* 1.
Hüfner, S. (1985), *Z. Phys. B61,* 135.
Hüfner, S., Osterwalder, J., Riesterer, T., Hulliger, F. (1984), *Solid State Commun. 52,* 793.
Ihrig, H., Vigren, D. T., Kübler, J., Methfessel, S. (1973), *Phys. Rev. B8,* 4525.
Ishida, S., Akazawa, S., Kubo, Y., Ishida, J. (1982), *J. Phys. F12,* 1111.
Ishida, S., Asano, S. (1985), *J. Phys. Soc. Japan 54,* 4688.
Ishida, S., Asano, S., Ishida, J. (1985), *J. Phys. Soc. Japan 54,* 3925.
Jan, J.-P., Skriver, H. L. (1981), *J. Phys. F11,* 805.
Janak, J. F. (1974), *Phys. Rev. B9,* 3985.
Janak, J. F. (1977), *Phys. Rev. B16,* 255.
Janak, J. F. (1978), *Phys. Rev. B18,* 7165–7168.
Janak, J. F., Moruzzi, V. L., Williams, A. R. (1975), *Phys. Rev. B12,* 1257.
Jani, A. R., Brener, N. E., Callaway, J. (1988), *Phys. Rev. B38,* 9425.
Jani, A. R., Tripathi, G. S., Brener, N. E., Callaway, J. (1989), *Phys. Rev. B40,* 1593.
Jansen, H. L. F. (1988 a), *J. Appl. Phys. 64,* 5604.
Jansen, H. L. F. (1988 b), *Phys. Rev. B38,* 8022.
Jansen, H. L. F., Freeman, A. J. (1984), *Phys. Rev. B30,* 561.
Jarlborg, T., Freeman, A. J. (1982), *J. Magn. Magn. Mat. 27,* 135.
Jarlborg, T., Peter, M. (1984), *J. Magn. Magn. Mat.* 42, 89.
Jarlborg, T., Freeman, A. J., Koelling, D. D. (1981), *J. Magn. Magn. Mat. 23,* 291.
Jarlborg, T., Junod, A., Peter, M. (1983 a), *Phys. Rev. B27,* 1558.
Jarlborg, T., Manuel, A. A., Peter, M. (1983 b), *Phys. Rev. B27,* 4210.
Jaswal, S. S. (1990), *Phys. Rev. B41,* 9697.
Jepsen, O., Andersen, O. K., Mackintosh, A. R. (1975), *Phys. Rev. B12,* 3084.
Johansson, B., Skriver, H. L., Martensson, H. L., Andersen, O. K., Glötzel, D. (1980 a), *Physica B102,* 12.
Johansson, L. I., Petersson, L.-G., Bergren, K.-F., Allen, J. W. (1980 b), *Phys. Rev. B22,* 3294.
Johansson, B., Skriver, H. L., Andersen, O. K. (1982), in: *Physics of Solids under High Pressure:* Schilling, J. S., Shelton, R. N. (Eds.). Amsterdam: Elsevier, p. 245.
Johnson, D. D., Nicholson, D. M., Pinski, F. J., Gyorffy, B. L., Stocks, G. M. (1986), *Phys. Rev. Lett. 56,* 2088.
Johnson, D. D., Nicholson, D. M., Pinski, F. J., Gyorffy, B. L., Stocks, G. M. (1990), *Phys. Rev. B41,* 9701.
Johnson, W. B., Anderson, J. R., Papaonstantopoulos, D. A. (1984), *Phys. Rev. B29,* 5337.
Jones, W., March, N. H. (1973), *Theoretical Solid State Physics.* London: Wiley, pp. 417 ff.
Joynt, R., Heine, V. (1984), *J. Magn. Magn. Mat. 45,* 74.
Kamm, G. N., Anderson, J. R. (1970), *Phys. Rev. B2,* 2944.
Kämper, K. P., Schmidt, W., Güntherodt, G., Gambino, R. J., Ruf, R. (1987), *Phys. Rev. Lett. 59,* 2788.
Kane, E. O. (1971), *Phys. Rev. B4,* 1917–1925.
Kerker, G. P. (1980), *J. Phys. C13,* L189–L194.
Kisker, E. (1983), *J. Phys. Chem. 87,* 3597.
Kisker, E., Carbone, C., Flipse, C. F., Wassermann, E. F. (1987), *J. Magn. Magn. Mat. 70,* 21.
Kittel, C. (1986), *Introduction to Solid State Physics.* New York: Wiley.
Klein, B. M., Boyer, L. L., Papaconstantopoulos, D. A., Mattheiss, L. F. (1978), *Phys. Rev. B18,* 6411.
Klein, B. M., Boyer, L. L., Papaconstantopoulos, D. A. (1979), *Phys. Rev. Lett. 42,* 530.
Klein, B. M., Pickett, W. E., Papaconstantopoulos, D. A., Boyer, L. L. (1983), *Phys. Rev. B27,* 6721.
Koelling, D. D., Harmon, B. N. (1977), *J. Phys. C10,* 3107.
Koelling, D. D. (1981), *Rep. Prog. Phys. 44,* 139–212.
Köhler, H., Sticht, J., Kübler, J. (1991), *Physica B172,* 79.
König, C., Christensen, N. E., Kollar, J. (1984), *Phys. Rev. B29,* 6481.
Kohn, W., Rostoker, N. (1954), *Phys. Rev. 94,* A1111.
Kohn, W., Sham, L. J. (1965), *Phys. Rev. 140,* A1133.
Kohn, W., Vashishta, P. (1983), in: *Theory of the Inhomogeneous Electron Gas:* Lundquist, S., March, N. H. (Eds.). New York: Plenum Press, pp. 79–147.
Korringa, J. (1947), *Physica 13,* 392.
Koster, G. F. (1957), in: *Solid State Physics, Vol. 5:* Seitz, F., Turnbull, D. (Eds.). Orlando: Academic Press, pp. 173–256.
Kouvel, J. S., Kasper, J. S. (1963), *J. Phys. Chem. Sol. 24,* 529.
Kovalev, O. V. (1965), *Irreducible Representations of Space Groups.* New York: Gordon and Breach.
Krakauer, H., Posternak, M., Freeman, A. J. (1979), *Phys. Rev. B19,* 1706–1719.
Krasko, G. L. (1987), *Phys. Rev. B36,* 8565.
Krasko, G. L. (1989), *Solid State Commun. 70,* 1099.
Kübler, J. (1978), *J. Phys. F8,* 2301.
Kübler, J. (1980 a), *J. Magn. Magn. Mat. 15–18,* 859.
Kübler, J. (1980 b), *J. Magn. Magn. Mat. 20,* 107.
Kübler, J. (1980 c), *J. Magn. Magn. Mat. 20,* 277.
Kübler, J. (1980 d), unpublished.
Kübler, J. (1981), *Phys. Lett. 81A,* 81.
Kübler, J. (1983), in: *Proceedings of the Institute von Laue-Langevin, Workshop on 3d Metallic Magnetism, Grenoble:* Givord, D., Ziebeck, K. (Eds.). Unpublished.
Kübler, J. (1984 a), *J. Magn. Magn. Mat. 45,* 415.
Kübler, J. (1984 b), *Physica 127B,* 257.
Kübler, J. (1989), *Solid State Commun. 72,* 631.

Kübler, J., Williams, A. R. (1986), *J. Magn. Magn. Mat. 54–57,* 603.
Kübler, J., Höck, K.-H., Sticht, J., Williams, A. R. (1988a), *J. Appl. Phys. 63,* 3482.
Kübler, J., Höck, K.-H., Sticht, J., Williams, A. R. (1988b), *J. Phys. F18,* 469.
Kübler, J., Williams, A. R., Sommers, C. B. (1983), *Phys. Rev. B28,* 1745.
Kuhnen, C. A., da Silva, E. Z. (1987), *Phys. Rev. B35,* 370.
Kulikov, N. I., Kulatov, E. T. (1982), *J. Phys. F12,* 2291.
Lam, P. K., Cohen, M. L. (1982), *Phys. Rev. B25,* 6139.
Landolt-Börnstein, New Series (1971), *Group III: Crystal and Solid State Physics, Vol. 6: Structure Data of Elements and Intermetallic Phases.* Berlin: Springer.
Laurent, D. G., Callaway, J., Fry, J. L., Brener, N. E. (1981), *Phys. Rev. B23,* 4977.
Levinson, H. J., Greuter, F., Plummer, E. W. (1983), *Phys. Rev. B27,* 727.
Levy, M. (1979), *Proc. Natl. Acad. Sci. (USA) 76,* 6062–6065.
Levy, M. (1982), *Phys. Rev. A26,* 1200–1208.
Liberman, D. A. (1971), *Phys. Rev. B3,* 2081.
Liebsch, A. (1979), *Phys. Rev. Lett. 43,* 1431.
Liechtenstein, A. I., Katsnelson, M. I., Antropov, V. P., Gubanov, V. A. (1987), *J. Magn. Magn. Mat. 67,* 65.
Lomer, W. M. (1962), *Proc. Phys. Soc. London 86,* 489.
Lonzarich, G. G. (1980), in: *Electrons at the Fermi-surface:* Springford, M. (Ed.). Cambridge: University Press, pp. 225–318.
Loucks, T. L. (1967), *Augmented Plane Wave Method.* New York: Benjamin.
Lu, Z.-W., Singh, D., Krakauer, H. (1987), *Phys. Rev. B36,* 7335.
Luchini, M. U., Heine, V. (1989), *J. Phys.: Cond. Matt. 1,* 8961.
Ludwig, W., Falter, C. (1988), *Symmetries in Physics.* Berlin: Springer.
MacDonald, A. H., Pickett, W. E., Koelling, D. D. (1980), *J. Phys. C13,* 2675.
MacDonald, A. H., Vosko, J. (1979), *J. Phys. C12,* 2977.
Mackintosh, A. R., Andersen, O. K. (1980), in: *Electrons at the Fermi Surface:* Springford, M. (Ed.). Cambridge: University Press, pp. 149–224.
Maetz, C. J., Gerhardt, U., Dietz, E., Ziegler, A., Jelitto, R. J. (1982), *Phys. Rev. Lett. 48,* 1686.
Malozemoff, A. P., Williams, A. R., Terakura, K., Moruzzi, V. L., Fukamichi, K. (1983), *J. Magn. Magn. Mat. 35,* 192.
Malozemoff, A. P., Williams, A. R., Moruzzi, V. L. (1984a), *Phys. Rev. B29,* 1620.
Malozemoff, A. P., Williams, A. R., Moruzzi, V. L., Terakura, K. (1984b), *Phys. Rev. B30,* 6565.
March, N. H. (1983), in: *Theory of the Inhomogeneous Electron Gas:* Lundquist, S., March, N. H. (Eds.). New York: Plenum Press, pp. 1–78.
Marcus, P. M., Moruzzi, V. L. (1988), *J. Appl. Phys. 63,* 4045.
Mårtensson, H., Nilsson, P. O. (1984), *Phys. Rev. 30,* 3047.
Matar, S., Mohn, P., Demazeau, G., Siberchicot, B. (1988), *J. Phys. France 49,* 1761.
Mattheiss, L. F., Wood, J. H., Switendick, A. C. (1968), in: *Methods in Computational Physics, Vol. 8:* Alder, B., Fernbach, S., Rotenberg, M. (Eds.). New York: Academic Press, pp. 64–148.
Maurer, M., Ousset, J. C., Ravet, M. F., Piecuch, M. (1989), *Europhys. Lett. 9,* 803.
McKay, J. M., Henrich, V. E. (1984), *Phys. Rev. Lett. 53,* 2343.
McMahan, A. K., Moriarty, J. A. (1983), *Phys. Rev. B27,* 3235.
Mermin, N. D. (1965), *Phys. Rev. 137,* A1441.
Merzbacher, E. (1970), *Quantum Mechanics.* New York: Wiley.
Messiah, A. (1976), *Quantum Mechanics, Vol. 1.* Amsterdam: North Holland
Messiah, A. (1978), *Quantum Mechanics, Vol. 2.* Amsterdam: North Holland.
Methfessel, M., Kübler, J. (1982), *J. Phys. F12,* 141.
Miedema, A. R. (1976), *Philips Tech. Rev. 36,* 217.
Miedema, A. R., Niessen, A. K. (1983), *Comput. Coupling Phase Diagrams and Thermochem. (CALPHAD) 7,* 27.
Min, B. J., Ho, K.-M. (1989), *Phys. Rev. B40,* 7532.
Mohn, P. (1988), *J. Phys. C21,* 2841.
Mohn, P., Pettifor, D. G. (1988), *J. Phys. C21,* 2829.
Mohn, P., Wohlfarth, E. P. (1987), *J. Phys. F17,* 2421.
Mohn, P., Schwarz, K., Wagner, D. (1989), *Physica B161,* 153.
Mohri, T., Sanchez, J. M., de Fontaine, D. (1985), *Acta Metall. 33,* 1171.
Moriya, T. (1964), *Solid State Commun. 2,* 239.
Moroni, E. G., Jarlborg, T. (1989), *Physica B161,* 115.
Moruzzi, V. L., Janak, F. F., Williams, A. R. (1978), *Calculated Electronic Properties of Metals.* New York: Pergamon Press.
Moruzzi, V. L., Marcus, P. M., Schwarz, K., Mohn, P. (1986), *Phys. Rev. B34,* 1784.
Mott, N. F. (1964), *Adv. Phys. 13,* 325.
Mott, N. F. (1967), *Adv. Phys. 16,* 49.
Mott, N. F. (1974), *Metal Insulator Transitions.* London: Taylor and Francis.
Mott, N. F., Jones, H. (1936), *The Theory of the Properties of Metals and Alloys.* Oxford: University Press, ch. 7, pp. 240–315.
Neckel, A., Rastl, P., Eibler, R., Weinberger, P., Schwarz, K. (1976), *J. Phys. C9,* 579.
Nieminen, R. M., Hodges, C. H. (1976), *J. Phys. F6,* 573.
Niksch, M., Lüthi, B., Kübler, J. (1987), *Z. Phys. B68,* 291.

Nohl, H., Klose, W., Andersen, O. K. (1982), in: *Superconductivity in Ternary Compounds I:* Fischer, Ø., Maple, M. B. (Eds.). Berlin: Springer, pp. 165–221.

Nordström, L., Johansson, B., Brooks, M. S. S. (1991), *J. Appl. Phys.*, in press.

Norman, M. R. (1990), *Phys. Rev. Lett. 64,* 1162.

Norman, M. R., Freeman, A. J. (1987), *Phys. Rev. B33,* 8896.

Oguchi, T., Terakura, K., Hamada, N. (1983a), *J. Phys. F13,* 145.

Oguchi, T., Terakura, K., Williams, A. R. (1983b), *Phys. Rev. B28,* 6443.

Oleś, A. M., Stollhoff, G. (1986), *J. Magn. Magn. Mat. 54–57,* 1045.

Papaconstantopoulos, D. A. (1986), *Handbook of the Band Structure of Elemental Solids.* New York: Plenum Press.

Peng, S. S., Jansen, H. L. F. (1988), *J. Appl. Phys. 64,* 5607.

Pettifor, D. G. (1976), *Commun. Phys. 1,* 141.

Pettifor, D. G. (1977a), *J. Phys. F7,* 613.

Pettifor, D. G. (1977b), *J. Phys. F7,* 1009.

Pettifor, D. G. (1978a), *J. Phys. F8,* 219.

Pettifor, D. G. (1978b), *Solid State Commun. 28,* 621.

Pettifor, D. G. (1979), *Phys. Rev. Lett. 42,* 846.

Pettifor, D. G. (1984), in: *Physical Metallurgy, 3³ d revised and enlarged ed.:* Cohen, R. W., Haasen, P. (Eds.). Amsterdam: North Holland, ch. 3.

Pettifor, D. G. (1987), in: *Solid State Physics, Vol. 40:* Ehrenreich, H., Turnbull, D. (Eds.). Orlando: Academic Press, pp. 43–92.

Phillips, J. C., Kleinman, L. (1959), *Phys. Rev. 116,* 287.

Pickett, W. E. (1989), *Rev. Mod. Phys. 61,* 433–512.

Pictet, O., Jarlborg, T., Peter, M. (1987), *J. Phys. F17,* 221.

Podgorny, M. (1989), *J. Magn. Magn. Mat. 78,* 352.

Price, D. L., Cooper, B. R. (1989), *Phys. Rev. B39,* 4945.

Prinz, G. A. (1985), *Phys. Rev. Lett. 54,* 1051.

Radousky, H. B., Jarlborg, T., Knapp, G. S., Freeman, A. J. (1982), *Phys. Rev. B26,* 1208.

Rajagopal, A. K. (1980), *Adv. Chem. Phys. 41,* 59.

Rajagopal, A. K., Callaway, J. (1973), *Phys. Rev. B7,* 1912.

Ramana, M. V., Rajagopal, A. K. (1983), *Adv. Chem. Phys. 54,* 231.

Riesterer, T., Schlapbach, L., Hüfner, S. (1986), *Solid State Commun. 57,* 109.

Rietschel, H., Winter, H. (1979), *Phys. Rev. Lett. 43,* 1256.

Rietschel, H., Winter, H., Reichardt, W. (1980), *Phys. Rev. B22,* 4284.

Rose, M. E. (1961), *Relativistic Electron Theory.* New York: Wiley.

Roy, D. M., Pettifor, D. G. (1977), *J. Phys. F7,* L183.

Sakisaka, Y., Maruyama, T., Kato, H., Aiura, Y., Yanashima, H. (1990), *Phys. Rev. B41,* 11865.

Sakurai, J. J. (1967), *Advanced Quantum Mechanics.* New York: Addison Wesley.

Sanchez, J. M., de Fontaine, D. (1981), in: *Structure and Bonding in Crystals, Vol. 2:* O'Keefe, M., Navrotski, A. (Eds.). New York: Academic Press, pp. 117–132.

Sanchez, J. M., Ducastelle, F., Gratias, D. (1984), *Physica 128A,* 334.

Sandratskii, L. M. (1986), *phys. stat. sol. (b) 135,* 167.

Sandratskii, L. M., Guletskii, P. G. (1986), *J. Phys. F16,* L43.

Sandratskii, L. M., Guletskii, P. G. (1989a), *J. Magn. Magn. Mat. 79,* 306.

Sandratskii, L. M., Guletskii, P. G. (1989b), *phys. stat. sol. (b) 154,* 623.

Sass, S. L. (1972), *J. Less Common. Met. 28,* 157.

Sawatzky, G. A., Allen, J. W. (1984), *Phys. Rev. Lett. 53,* 2239.

Schadler, G., Weinberger, P. (1986), *J. Phys. F16,* 27.

Schaich, W. L. (1979), in: *Photoemission in Solids I:* Cardona, M., Ley, L. (Eds.). Berlin: Springer, pp. 105–134.

Schmidt, P. C. (1985), *Z. Naturforsch. 40a,* 335.

Schneider, B. (1991), Diplomarbeit TH Darmstadt, unpublished.

Schwarz, K. (1986), *J. Phys. F16,* L21.

Schwarz, K., Mohn, P. (1984), *J. Phys. F14,* L129.

Schwarz, K., Mohn, P., Blaha, P., Kübler, J. (1984), *J. Phys. F14,* 2659.

Sébilleau, D., Guo, G. Y., Temmerman, W. M. (1989), *J. Phys.: Cond. Matt. 1,* 5653.

Segall, B., Ham, F. S. (1968), in: *Methods in Computational Physics:* Alder, B., Fernbach, S., Rotenberg, M. (Eds.). New York: Academic Press, pp. 251–293.

Shirley, D. A. (1979), in: *Photoemission in Solids I:* Cardona, M., Ley, L. (Eds.). Berlin: Springer, pp. 165–196.

Sigfuson, T. I., Bernhoeft, N. R., Lonzarich, G. G. (1984), *J. Phys. F14,* 2141.

Sigli, C., Kosugi, M., Sanchez, J. M. (1986), *Phys. Rev. Lett. 57,* 253.

Singh, A. K., Manuel, A. A., Walker, E. (1988), *Europhys. Lett. 6,* 67.

Singhal, S. P., Callaway, J. (1977), *Phys. Rev. B16,* 1744.

Singwi, K. S., Sjölander, A., Tosi, M. P., Land, R. H. (1970), *Phys. Rev. B1,* 1044.

Siratori, K., Suga, S., Taniguchi, M., Soda, K., Kimura, S., Yanase, A. (1986), *J. Phys. Soc. Japan 55,* 690.

Skriver, H. L. (1981), *J. Phys. 11,* 97.

Skriver, H. L. (1984), *The LMTO Method.* Berlin: Springer.

Skriver, H. L. (1985), *Phys. Rev. B31,* 1909.

Skriver, H. L., Andersen, O. K., Johansson, B. (1978), *Phys. Rev. Lett. 41,* 42.

Slater, J. C. (1937), *Phys. Rev. 51,* 846.

Slater, J. C. (1965), *Quantum Theory of Molecules and Solids, Vol. 2*. New York: McGraw-Hill.
Slater, J. C. (1972), *Adv. Quantum Chem. 6,* 1.
Slater, J. C. (1974), *The self-consistent Field for Molecules and Solids, Vol. 4*. New York: McGraw-Hill.
Slater, J. C., Koster, G. F. (1954), *Phys. Rev. 94,* 1498.
Slater, J. C., Wood, J. H. (1971), *Int. J. Quantum Chem. Suppl. 4,* 3.
Sluiter, M., de Fontaine, D., Guo, X. Q., Podloucky, R., Freeman, A. J. (1990), *Phys. Rev. B42,* 10460.
Smith, N. V. (1979), in: *Photoemission in Solids I:* Cardona, M., Ley, L. (Eds.). Berlin: Springer, pp. 237–264.
Soler, J. M., Williams, A. R. (1990), *Phys. Rev. B42,* 9728.
Steiner, P., Höchst, H., Hüfner, S. (1979), in: *Photoemission in Solids II:* Ley, L., Cardona, M. (Eds.). Berlin: Springer, pp. 349–372.
Stenholm, J., Eriksson, O., Johansson, B., Noläng, B. (1989), *J. Phys.: Cond. Matt. 1,* 7329.
Sticht, J. (1986), unpublished.
Sticht, J., Höck, K.-H., Kübler, J. (1989), *J. Phys.: Cond. Matt. 1,* 8155.
Stoner, E. C. (1938), *Proc. Roy. Soc. 165,* 372.
Stoner, E. C. (1939), *Proc. Roy. Soc. 169,* 339.
Strange, P., Staunton, J. B., Gyorffy, B. L. (1984), *J. Phys. C17,* 3355.
Strange, P., Ebert, H., Staunton, J. B., Gyorffy, B. L. (1989a), *J. Phys.: Cond. Matt. 1,* 3947.
Strange, P., Staunton, J. B., Ebert, H. (1989b), *Europhys. Lett. 9,* 169.
Styer, D. F., Phani, M. K., Lebowitz, J. L. (1986), *Phys. Rev. B34,* 3361.
Takeda, T. (1978), *Z. Physik B32,* 43.
Takeda, T. (1979), *J. Phys. F9,* 815.
Takizawa, S., Terakura, K., Mohri, T. (1989), *Phys. Rev. B39,* 5792.
Tao, H.-J., Ho, K.-M., Zhu, X.-Y. (1986), *Phys. Rev. B34,* 8394.
Terakura, K. (1977a), *J. Phys. F7,* 1773.
Terakura, K. (1977b), *Physica 91B,* 162.
Terakura, K., Kanamori, J. (1971), *Progr. Theor. Phys. 46,* 1007.
Terakura, K., Williams, A. R., Oguchi, T., Kübler, J. (1984a), *Phys. Rev. B30,* 4734.
Terakura, K., Williams, A. R., Oguchi, T., Kübler, J. (1984b), *Phys. Rev. Lett. 52,* 1830.
Terakura, K., Mohri, T., Oguchi, T. (1989), *Materials Science Forum 37,* 39.
Terakura, K., Oguchi, T., Mohri, T., Watanabe, K. (1987), *Phys. Rev. B35,* 2169.
Thiry, P., Chandesris, D., Lecante, J., Guillot, C., Pinchaux, R., Pétroff, Y. (1979), *Phys. Rev. Lett. 43,* 82.
Thomas, L. H. (1926), *Proc. Cambridge Phil. Soc. 23,* 542.
Topiol, S., Zunger, A., Ratner, M. A. (1977), *Chem. Phys. Lett. 49,* 367–373.
Topp, W. C., Hopfield, J. J. (1973), *Phys. Rev. B7,* 1295–1303.
Trickey, S. B., Green, F. R. jr., Averill, F. W. (1973), *Phys. Rev. B8,* 4822.
Turchi, P. E. A., Sluiter, M., de Fontaine, D. (1987), *Phys. Rev. B36,* 3161.
Turchi, P. E. A., Stocks, C. M., Butler, W. H., Nicholson, D. M., Gonis, A. (1988), *Phys. Rev. B37,* 5982.
Turner, A. M., Donoho, A. W., Erskine, J. L. (1984), *Phys. Rev. B29,* 2986.
Verbeek, B. H., Rompa, H. W. A. M., Larsen, P. K., Methfessel, M. S., Mueller, F. M. (1983), *Phys. Rev. B28,* 6774.
Victora, R. H., Falicov, L. M., Ishida, S. (1984), *Phys. Rev. B30,* 3896.
Vignale, G., Rasolt, M. (1987), *Phys. Rev. Lett. 59,* 2360.
Vignale, G., Rasolt, M. (1988), *Phys. Rev. B37,* 10685.
Vosko, S. H., Perdew, J. P. (1975), *Can. J. Phys. 53,* 385.
Vosko, S. H., Wilk, L., Nusair, M. (1980), *Can. J. Phys. 58,* 1200.
Wagner, D. (1989), *J. Phys.: Cond. Matt. 1,* 4635.
Wang, C. S., Callaway, J. (1977), *Phys. Rev. B15,* 298.
Wang, C. S., Callaway, J. (1978), *Comput. Phys. Commun. 14,* 327–365.
Wang, C. S., Klein, B. M., Krakauer, H. (1985), *Phys. Rev. Lett. 54,* 1852.
Wang, C. S., Prange, R. E., Korenman, V. (1982), *Phys. Rev. B25,* 5766.
Watson, R. E., Ehrenreich, H., Hodges, L. (1970), *Phys. Rev. Lett. 24,* 829.
Weger, M., Goldberg, I. B. (1973), in: *Solid State Physics, Vol. 28:* Ehrenreich, H., Seitz, F., Turnbull, D. (Eds.). New York: Academic Press, pp. 2–177.
Wei, S.-H., Mbaye, A. A., Ferreira, L. G., Zunger, A. (1987), *Phys. Rev. B36,* 4163.
Weinberger, P. (1990), *Electron Scattering Theory for Ordered and Disordered Matter*. Oxford: Clarendon Press.
Weiss, R. J. (1963), *Proc. Phys. Soc. 82,* 281.
Wertheim, G. K., Citrin, P. H. (1979), in: *Photoemission in Solids I:* Cardona, M., Ley, L. (Eds.). Berlin: Springer, pp. 197–236.
Wigner, E. P., Seitz, F. (1933), *Phys. Rev. 43,* 804.
Wigner, E. P., Seitz, F. (1934), *Phys. Rev. 46,* 509.
Wigner, E. P., Seitz, F. (1955), in: *Solid State Physics, Vol. 1:* Seitz, F., Turnbull, D. (Eds.). Orlando: Academic Press, pp. 97–126.
Wijngaard, J. H., Haas, C., de Groot, R. A. (1989), *Phys. Rev. B40,* 9318.
Williams, A. R. (1970), *Phys. Rev. B1,* 3417.
Williams, A. R., van Morgan, W. (1974), *J. Phys. C7,* 37–60.
Williams, A. R., von Barth, U. (1983), in: *Theory of the Inhomogeneous Electron Gas:* Lundquist, S., March, N. H. (Eds.). New York: Plenum Press, pp. 189–307.

Williams, A. R., Hu, S. M., Jepsen, D. W. (1971), in: *Computational Methods in Band Theory:* Marcus, P. M., Janak, J. F., Williams, A. R. (Eds.). New York: Plenum Press, pp. 157–177.

Williams, A. R., Kübler, J., Gelatt, C. D. jr. (1979), *Phys. Rev. B19,* 6094.

Williams, A. R., Gelatt, C. D. jr., Janak, J. F. (1980a), in: *Theory of Alloy Phase Formation:* Bennett, L. H. (Ed.). New York: American Institute of Mining, Metallurgical and Petroleum Engineers, pp. 40–64.

Williams, A. R., Gelatt, C. D. jr., Moruzzi, V. L. (1980b), *Phys. Rev. Lett. 44,* 429.

Williams, A. R., Moruzzi, V. L., Gelatt, C. D. jr., Kübler, J., Schwarz, K. (1982), *J. Appl. Phys. 53,* 2019.

Williams, A. R., Gelatt, C. D. jr., Connolly, J. W. D., Moruzzi, V. L. (1983a), in: *Alloy Phase Diagrams:* Bennett, L., Massalski, T. B., Giessen, B. C. (Eds.). New York: North Holland, pp. 17–28.

Williams, A. R., Moruzzi, V. L., Gelatt, C. D. jr., Kübler, J. (1983b), *J. Magn. Magn. Mat. 31–34,* 88.

Williams, A. R., Moruzzi, V. L., Malozemoff, A. P., Terakura, K. (1983c), *IEEE Trans. Magn. 19,* 1983.

Williams, A. R., Malozemoff, A. P., Moruzzi, V. L., Matsui, M. (1984), *J. Appl. Phys. 55,* 2353.

Wilson, T. M. (1970), *Int. J. Quant. Chem. Symp. 3,* 757.

Wolfrat, J. C., Menovsky, A. A., Roeland, L. W., ten Cate, H., Koster, C. H. A., Mueller, F. M. (1985), *J. Phys. F15,* 297.

Wood, D. M., Zunger, A. (1989), *Phys. Rev. B40,* 4062.

Woodward, C., Min, B. I., Benedek, R., Garner, J. (1989), *Phys. Rev. B39,* 4853.

Wu, M. K., Ashburn, J. R., Torng, C. J., Hor, P. H., Meng, R. L., Gao, L., Huang, Z. J., Wang, Y. Q., Chu, C. W. (1987), *Phys. Rev. Lett. 58,* 908.

Xu, J.-H., Oguchi, T., Freeman, A. J. (1987), *Phys. Rev. B36,* 4186.

Yamada, H., Inoue, J., Terao, K., Kanda, S., Shimizu, M. (1984), *J. Phys. F14,* 1943.

Yamaoka, T., Mekata, M., Takaki, H. (1974), *J. Phys. Soc. Japan 36,* 438.

Yamashita, J., Asano, S. (1983), *J. Phys. Soc. Japan 52,* 3514.

Yanase, A., Siratori, K. (1984), *J. Phys. Soc. Japan 53,* 312.

Ye, Y. Y., Chan, C. T., Ho, K.-M., Harmon, B. N. (1990), *Int. J. Supercomputer Appl. 4,* 111.

Zeller, R. (1987), *J. Phys. C20,* 2347–2360.

Zhao, G.-L., Leung, T. C., Harmon, B. N., Keil, M., Müllner, M., Weber, W. (1989), *Phys. Rev. B40,* 7999.

Zhukov, V. P., Medvedeva, N. I., Novikov, D. L., Gubanov, V. A. (1988), *Phys. Stat. Sol. (b) 149,* 175.

Zhukov, V. P., Medvedeva, N. I., Gubanov, V. A. (1989), *Phys. Stat. Sol. (b) 151,* 407.

Ziman, J. M. (1971), in: *Solid State Physics, Vol. 26:* Ehrenreich, H., Seitz, F., Turnbull, D. (Eds.). Olando: Academic Press, pp. 1–101.

General Reading

Ashcroft, N. W., Mermin, N. D. (1976), *Solid State Physics.* Philadelphia: Holt-Saunders.

Callaway, J. (1974), *Quantum Theory of the Solid State, Parts A and B.* New York: Academic Press.

Dreizler, R. M., de Providência, J. (1983), *Density Functional Methods in Physics.* New York: Plenum Press.

Dreizler, R. M., Gross, E. K. U. (1990), *Density Functional Theory.* Berlin: Springer.

Harrison, W. A. (1980), *Electronic Structure and the Properties of Solids.* San Francisco: Freeman.

Loucks, T. L. (1967), *Augmented Plane Wave Method.* New York: Benjamin.

Lundquist, S., March, N. H. (1983), *Theory of the Inhomogeneous Electron Gas.* New York: Plenum Press.

Phariseau, P., Temmerman, W. M. (1984), *The Electronic Structure of Complex Systems.* New York: Plenum Press.

Skriver, H. L. (1984), *The LMTO Method.* Berlin: Springer.

Springford, M. (1980), *Electrons at the Fermi Surface.* Cambridge: University Press.

2 Magneto-Optical Properties of Metals, Alloys and Compounds

J. Schoenes

Laboratorium für Festkörperphysik, ETH Zürich, Zürich, Switzerland

List of Symbols and Abbreviations

$\boldsymbol{B}$	magnetic flux density
c	speed of light
$\boldsymbol{D}$	electric displacement
$\boldsymbol{E}$	electric field strength
E_{exch}	exchange splitting
E_{F}	Fermi energy
$\bar{e}$	electron charge
e	elementary charge
$\hat{f}$	restoring force constant
$\boldsymbol{g}$	wave vector of light
$\boldsymbol{H}$	magnetic field strength
H_{c}	coercive field
H_{s}	switching field
h	Planck constant
$I(t)$	intensity
J	total quantum momentum
$\boldsymbol{J}$	current density
J_{n}	Bessel function of order n
K	absorption coefficient
K_{n}	anisotropy constant
k	absorption index
k_+, k_-	absorption index for right-circularly and left-circularly polarized light
k	Boltzmann constant
L	orbital quantum momentum
$\boldsymbol{M}$	magnetization per unit volume
M_{L}	quantum number
m	magnetic quantum number
m	mass
m^*	effective mass
n	real refractive index
n_+, n_-	real refractive index for right-circularly and left-circularly polarized light
$\tilde{n}_{\pm}$	complex refractive index for right-circularly (upper-sign) and left-circularly polarized light (lower sign)
n_{D}	refractive index of a dielectric interference layer
$\boldsymbol{P}$	electric polarization per unit volume
Q, Q_0	magneto-optical constant and its amplitude
R	reflectance
r_+, r_-	amplitude ratio of the reflected to the incident wave for right- and left-circularly polarized light
S_{T}	total spin
T	temperature
T_{C}	Curie temperature

T_c	superconducting transition temperature
T_{comp}	compensation temperature
T_N	Néel temperature
T_s	switching temperature
t	time
v_0	Fermi velocity
$\boldsymbol{\alpha}_1$	real polarizability tensor
$\lvert\alpha\uparrow\rangle$, $\lvert\beta\uparrow\rangle$	quantum states
γ	damping term
Δ	phase or retardation
$\tilde{\boldsymbol{\varepsilon}}$, $\tilde{\varepsilon}$	complex dielectric tensor and function
ε_{ij}	matrix element of $\tilde{\boldsymbol{\varepsilon}}$
ε_0	permittivity of vacuum
ε_1, ε_2	real and imaginary parts of $\tilde{\varepsilon}$
ε_F	magnetic circular dichroism
ε_K	Kerr ellipticity
ε_{opt}	interband contribution to ε_1 for $\omega \to 0$
ζ	arctan of the ellipticity η
η	ellipticity
θ_F, θ_K	Faraday rotation, Kerr rotation
λ	wavelength
λ_t	wavelength of interband transition
μ_0	permeability of vacuum
μ_B	Bohr magnetron
$\boldsymbol{\pi}_x$, $\boldsymbol{\pi}_y$	kinetic momentum operators
$\boldsymbol{\pi}^{\pm}$	$\boldsymbol{\pi}_x \pm i\boldsymbol{\pi}_y$
$\tilde{\boldsymbol{\sigma}}$, $\tilde{\sigma}$	complex conductivity tensor and function
$\boldsymbol{\sigma}_1$	real conductivity tensor
σ_1, σ_2	real and imaginary parts of $\tilde{\sigma}$
σ_{ij}	matrix element of $\tilde{\boldsymbol{\sigma}}$
τ, τ_s	normal and skew-scattering lifetime
$\boldsymbol{\chi}_1$	real susceptibility tensor
ψ	azimuth angle
Ω	skew-scattering frequency
ω	angular frequency of light
ω_0	resonance frequency in absence of a magnetic field
ω_c	cyclotron resonance frequency
ω_L	Larmor frequency
ω_{min}	frequency of minimal reflectivity
ω_P	plasma frequency
ω_{SO}	spin-orbit frequency
ASW	augmented spherical wave
DRAM	dynamical random access memories
f.c.c.	face-centered cubic

f.u.	formula unit
KKR-CPA	Korringa-Kohn-Rostoker–Coherent Potential Approximation
LMTO	linear muffin-tin-orbital
RIG	rare earth iron garnet
TE, TEM	transverse electric and magnetic
XPS	X-ray photoemission spectroscopy
YIG	yttrium iron garnet

2.1 Introduction

Magneto-optics is the study of the interaction of electromagnetic radiation with magnetized matter. The term magneto-optical properties is generally used in a more restricted sense to mean those properties of matter which manifest themselves as Faraday or magneto-optical Kerr effect. In this chapter, we adopt this latter definition; i.e., we will not deal with inelastic magneto-optical effects, like Raman scattering, nor with nonlinear magneto-optical effects or the inverse Faraday effect. The *magnetic circular birefringence*, which has been known as the *Faraday rotation* since the discovery of this first genuine magneto-optical effect by Michael Faraday (1846), describes the rotation of the polarization plane of linearly polarized light on transmission through matter magnetized in the direction of light propagation (Fig. 2-1 a). In the same experimental arrangement, one can observe a second magneto-optical effect, namely, the *magnetic circular dichroism*. This effect, which is also called *Faraday ellipticity*, is proportional to the difference of the indices of absorption for right- and left-hand circularly polarized light induced by the magnetization.

If the magnetization is perpendicular to the propagation direction of light one speaks of the *Voigt* or *Cotton-Mouton configuration*. The difference in the refractive indices for light polarized parallel and perpendicular to the magnetic field gives rise to the *transverse* or *linear magnetic birefringence*. The difference in the absorption indices for the two polarization directions leads to the *transverse* or *linear dichroism*. The linear magnetic birefringence and dichroism are to lowest-order quadratic functions of the magnetization, while the circular effects are to lowest-order linear functions. Therefore, the former effects can be used neither to read the magnetization direction of domains nor for magneto-optical storage of data. However, they are suitable for studying magnetic phase transitions.

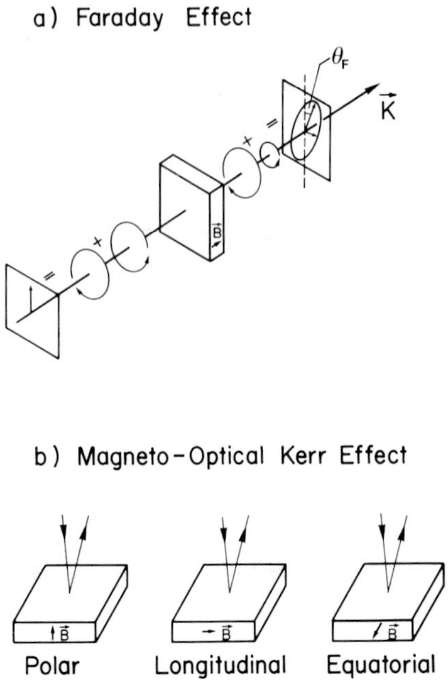

Figure 2-1. Scheme of the Faraday effect (a) and the magneto-optical Kerr effect (b). In the latter effect one distinguishes polar, longitudinal, and equatorial configurations, depending on the relative orientation of the field, the reflecting surface, and the plane of incidence.

Most metals and ceramics are in the solid state under normal conditions. Thus, electromagnetic radiation can also be reflected by these materials. The changes of the polarization on reflection which is induced by the magnetization are known by the term *magneto-optical Kerr effect* (Kerr, 1877). The adjective "magneto-optical" is added to distinguish these effects from those due to an electric field, that were also discovered by Kerr. If the magnetization is parallel or antiparallel to the light propagation direction one speaks of the *polar* magneto-optical Kerr effect. This effect is closest to the Faraday effect in transmission since it uses the same configuration. As for the Faraday effect, the polar Kerr signal is proportional to the magnetization. If the magnetization is perpendicular to the direction of light propagation, one has to distinguish for non-normal incidence the cases where the magnetization is parallel and orthogonal to the plane of incidence (Fig. 2-1 b). Freiser (1968) has discussed the magnetization dependence of the magneto-optical Kerr effect for the two latter configurations. In particular, he has pointed out that in the equatorial configuration and for oblique incidence there is an effect proportional to the magnetization despite the fact that the magnetization is perpendicular to the light propagation direction. This effect has been used a few times to derive magneto-optical properties, but the large majority of the available data have been collected using the polar Kerr effect.

We have already mentioned that Kerr-effect measurements are generally limited to solids, although measurements on liquids are conceivable. Faraday-rotation measurements have been reported for gases, but these do not fall in the scope of a book dedicated to the electronic and magnetic properties of metals and ceramics. Magneto-optical measurements have also been performed on normal, i.e., non-magnetically-ordered metals, semiconductors, and insulators. The signals for diamagnetic or paramagnetic metals are in most cases very small, and in fact silver and aluminum mirrors are used commonly as reference materials for their nearly zero Kerr signal. The photon energy dependence of these spectra, however, is very instructive for magnetic materials and some examples will be presented. Since the discovery of microwave absorption in InSb by Dresselhaus et al. (1955), magneto-optical spectroscopy has become an important technique for the determination of band parameters in classical (non-magnetically-ordered) semiconductors. Except for a few cross-links, we will not deal with these materials, and the interested reader is referred to the existing reviews and handbook articles, e.g., Pidgeon (1980). However, the topic of this Volume being metals *and* ceramics, the magneto-optical properties of glasses will be discussed. The main focus of the present chapter is to give an overview on the magneto-optical properties of magnetic metals, compounds, and alloys.

This topic has received much attention over the last decade, primarily due to the realization of erasable magneto-optical memories. Insulating magnetic materials, like iron garnets, play an important role as nonreciprocal elements in communication techniques. In addition, magneto-optical spectroscopy has contributed in the last years to an improved understanding of the electronic structure of a wide class of materials and of many basic problems ranging from surface magnetism to exchange coupling in magnetic superconductors. The actual interest in magneto-optical properties manifests itself also in several recent handbook articles. Thus, Buschow (1988) has reviewed the magneto-optical properties of alloys and intermetallic compounds and Reim and Schoenes (1990) have treated the aspects of magneto-optical spectroscopy of f-electron systems. Repetition of these articles will be avoided as far as possible by treating in more detail those materials which have not been incluced in the above-mentioned monographs and by condensing and referencing those which have been given ample consideration.

This chapter is organized as follows: In the following section the theory of magneto-optics, including the macroscopic and microscopic models, will be outlined. Section 2.3 will cover experimental aspects. The discussion of materials begins in Sec. 2.4 with metals and glasses which do not order magnetically. In Secs. 2.5–2.8, the different classes of magnetic materials will be reviewed under the headings transition-element materials, lanthanide materials, actinide materials, and mixed 3d–4f and 3d–5f materials, respectively. Finally, Sec. 2.9 will deal with the two most important applications of magneto-optics.

2.2 Theory

The analysis of magneto-optical spectra is generally performed in two steps. In a first step, the measured Faraday- or Kerr-effect spectra are transformed into the off-diagonal components of the complex dielectric tensor $\tilde{\boldsymbol{\varepsilon}}(\omega)$ or the complex conductivity tensor $\tilde{\boldsymbol{\sigma}}(\omega)$. This transformation is to be dealt with in the first subsection. The second step consists in the development of microscopic models either on the basis of atomic models or on the basis of band-structure calculations to reproduce the experimental $\tilde{\varepsilon}_{xy}(\omega)$ or $\tilde{\sigma}_{xy}(\omega)$ spectra and will be discussed in Sec. 2.2.2. As atomic theories generally use Gaussian (cgs) units, it has become a custom to use these units all along magneto-optical literature (Argyres, 1955; Bennett and Stern, 1965; Pershan, 1967; Suits, 1972; Erskine and Stern, 1973a; Buschow, 1988; Reim and Schoenes, 1990). Thus, the unit for all the elements of the optical conductivity tensor is generally s^{-1}. The conversion of this unit in the SI unit $(\Omega\,m)^{-1}$ is performed by dividing the value in s^{-1} by 9×10^9*. The dielectric tensor elements are dimensionless in both unit systems. To be consistent with the other chapters in this series, the theory will be developed in the following using the SI system. All the theoretical models treat the low-field case, where Landau quantization can be neglected.

* This is true for the magneto-optical quantities σ_{1xy} and σ_{2xy} and the optical quantity σ_{1xx}. However, $\sigma_{2xx}\,(\mathrm{SI}) = \sigma_{2xx}\,(\mathrm{c.g.s.})/9 \times 10^9 - \omega\,\varepsilon_0$.

2.2.1 Macroscopic Theory

2.2.1.1 The Optical Functions

The interaction of electromagnetic radiation with an uncharged, polarizable, magnetizable and conducting medium is described by Maxwell's equations:

$$\mathrm{rot}\,\boldsymbol{H} = \dot{\boldsymbol{D}} + \boldsymbol{J} \tag{2-1}$$

$$\mathrm{rot}\,\boldsymbol{E} = -\dot{\boldsymbol{B}} \tag{2-2}$$

$$\mathrm{div}\,\boldsymbol{B} = 0 \tag{2-3}$$

$$\mathrm{div}\,\boldsymbol{D} = 0 \tag{2-4}$$

The dielectric displacement $\boldsymbol{D}$ and the magnetic induction $\boldsymbol{B}$ are related to the electric field and the magnetic field, respectively, by

$$\boldsymbol{D} = \varepsilon_0\,\boldsymbol{E} + \boldsymbol{P} \tag{2-5}$$

$$\boldsymbol{B} = \mu_0\,(\boldsymbol{H} + \boldsymbol{M}) \tag{2-6}$$

where $\boldsymbol{P}$ is the electric polarization, $\boldsymbol{M}$ is the magnetization per unit volume, $\mu_0 = 4\pi \times 10^{-7}\,\mathrm{Vs\,A^{-1}\,m^{-1}}$, and $\varepsilon_0 = 1/\mu_0 c^2 = 8.854 \times 10^{-12}\,\mathrm{As\,V^{-1}\,m^{-1}}$. Neglecting nonlinear effects in the material response, i.e. for radiation fields that are not too large, one sets

$$\boldsymbol{J} = \boldsymbol{\sigma}_1\,\boldsymbol{E} \tag{2-7}$$

$$\boldsymbol{P} = \boldsymbol{\alpha}_1\,\boldsymbol{E} \tag{2-8}$$

$$\boldsymbol{M} = \boldsymbol{\chi}_1\,\boldsymbol{H} \tag{2-9}$$

Equation (2-7) is *Ohm's law*, which relates the current density $\boldsymbol{J}$ to the electric field $\boldsymbol{E}$ via the real conductivity tensor $\boldsymbol{\sigma}_1$. This 3×3 tensor $\boldsymbol{\sigma}_1$, like the real polarizability tensor $\boldsymbol{\alpha}_1$ and the real magnetic susceptibility tensor χ_1, reduces for isotropic materials in the absence of external fields to a scalar. After substitution of Eqs. (2-5) to (2-9) into Eqs. (2-1) and (2-2), a first simplification results from the introduction of the real dielectric function $\varepsilon_1 = 1 + \alpha_1/\varepsilon_0$ and the real permeability $\mu_1 = 1 + \chi_1$, giving:

$$\mathrm{rot}\, \boldsymbol{H} = \varepsilon_0\, \varepsilon_1\, \dot{\boldsymbol{E}} + \sigma_1\, \boldsymbol{E} \tag{2-10}$$

$$\mathrm{rot}\, \boldsymbol{E} = -\mu_0\, \mu_1\, \dot{\boldsymbol{H}} \tag{2-11}$$

A second formal simplification results for Eq. (2-10) by the introduction of either a complex dielectric function $\tilde{\varepsilon} = \varepsilon_1 - \mathrm{i}\,\varepsilon_2$ or a complex conductivity function $\tilde{\sigma} = \sigma_1 + \mathrm{i}\,\sigma_2$. With $\boldsymbol{E} \sim \boldsymbol{E}_0\, \mathrm{e}^{\mathrm{i}\omega t}$ and

$$\tilde{\varepsilon} = \varepsilon_1 - \mathrm{i}\,\sigma_1/\omega\, \varepsilon_0 \tag{2-12}$$

we obtain:

$$\mathrm{rot}\, \boldsymbol{H} = \varepsilon_0\, \tilde{\varepsilon}\, \dot{\boldsymbol{E}} \tag{2-13}$$

If one introduces a complex conductivity, it is customary to define it as:

$$\tilde{\sigma} = \sigma_1 + \mathrm{i}\,\omega\, \varepsilon_0\, \varepsilon_1 \tag{2-14}$$

which leads to the interrelation:

$$\tilde{\sigma} = \sigma_1 + \mathrm{i}\,\sigma_2 = \mathrm{i}\,\omega\, \varepsilon_0 (\varepsilon_1 - \mathrm{i}\,\varepsilon_2) = \\ = \mathrm{i}\,\omega\, \varepsilon_0\, \tilde{\varepsilon} \tag{2-15}$$

The combination of Eqs. (2-11) and (2-13) allows the derivation of the wave equations:

$$\Delta \boldsymbol{E} = \mu_0\, \mu_1\, \varepsilon_0\, \tilde{\varepsilon}\, \ddot{\boldsymbol{E}} \tag{2-16}$$

$$\Delta \boldsymbol{H} = \mu_0\, \mu_1\, \varepsilon_0\, \tilde{\varepsilon}\, \ddot{\boldsymbol{H}} \tag{2-17}$$

Plane waves satisfying Eqs. (2-16) and (2-17) are, respectively,

$$\boldsymbol{E} = \boldsymbol{E}_0\, \mathrm{e}^{\mathrm{i}(\omega t - \boldsymbol{g} \cdot \boldsymbol{r})} \tag{2-18}$$

$$\boldsymbol{H} = \boldsymbol{H}_0\, \mathrm{e}^{\mathrm{i}(\omega t - \boldsymbol{g} \cdot \boldsymbol{r})} \tag{2-19}$$

where $\boldsymbol{r}$ is the propagation direction and $\boldsymbol{g}$ is the wave vector which points along the propagation direction and whose length is complex and equal to $(n - \mathrm{i}\,k)\,\omega/c$. Here, n and k are the real refractive index and the absorption index, respectively. Substituting Eq. (2-18) into (2-16) leads to the following connection of the complex dielectric function and the optical functions:

$$\tilde{\varepsilon}\, \mu_1 = (n - \mathrm{i}\,k)^2 = \tilde{n}^2 \tag{2-20}$$

Equation (2-20) deserves special comment. There exists in the literature some disagreement about whether in magnetic materials μ_1 can be set equal to 1 at optical frequencies or whether the permeability is a tensor which, in the presence of an external magnetic field, should have complex diagonal as well as off-diagonal elements different from 1. The former point of view has been taken by Landau and Lifshitz (1960) and Pershan (1967), both arguing on theoretical grounds that at optical frequencies $\mu_1 = 1$. The latter point of view has been defended by the Cologne school (Clemens and Jaumann, 1963; Burkhard and Jaumann, 1970; Bonenberger, 1978). Their arguments are based on measurements of at least four different magneto-optical quantities, for example, the Faraday rotation and ellipticity and the polar Kerr rotation and ellipticity (Bonenberger, 1978) or the complex polar as well as the complex equatorial Kerr effect (Burkhard and Jaumann, 1970). The idea behind these experiments is to overdetermine $\tilde{\varepsilon}$ if $\mu_1 = 1$ and thus, to produce an incompatibility which, indeed, they claim to have observed. However, some caution appears to be necessary regarding these experimental results. Thus, the data reported for the Kerr effect of iron, cobalt, and nickel films (Burkhard and Jaumann, 1970) do not agree with those reported by other authors like Krinchik and Artem'ev (1968) or van Engen (1983). The polar Kerr data for EuS films by Bonenberger (1978) also deviate considerably from those obtained for analogous EuO and EuSe single crystals

(Schoenes and Kaldis, 1987). Potential difficulties arise from oxidation and deterioration of the films as well as in the reduction of the data. Thus, measurements of the Kerr rotation on a transparent thin film always contain contributions from the Faraday effect of those parts of the light which are reflected on the interfaces film support and support vacuum, plus multiple reflected beams. Conversely, in Faraday-effect measurements there are contributions from the Kerr effect, and this Kerr signal is different for the interface vacuum film and film support. Wittekoek et al. (1975) have performed Faraday-rotation and Faraday-ellipticity measurements on films of $Y_3Fe_5O_{12}$ and polar Kerr-rotation and Kerr-ellipticity measurements on bulk samples of $Y_3Fe_5O_{12}$. They computed $\tilde{\boldsymbol{\varepsilon}}$ for both sets of data setting $\mu_1=1$ and obtained very satisfying agreement when some of the above-mentioned corrections were made (see Sec. 2.5.3.3).

Adopting here the most widely accepted assumption that $\mu_1=1$, Eq. (2-20) reduces for the real and imaginary parts of $\tilde{\varepsilon}$ to:

$$\varepsilon_1 = n^2 - k^2 \tag{2-21}$$

$$\varepsilon_2 = 2nk \tag{2-22}$$

2.2.1.2 The Faraday Effect and the Kerr Effect

On application of an external magnetic field or at the presence of a spontaneous magnetization, an anisotropy is induced in the material which will first be assumed isotropic in the absence of a field. The case of naturally anisotropic materials will be considered later. If the field or the magnetization points in the z-direction, the complex dielectric tensor acquires the form:

$$\tilde{\boldsymbol{\varepsilon}} = \begin{pmatrix} \tilde{\varepsilon}_{xx} & \tilde{\varepsilon}_{xy} & 0 \\ -\tilde{\varepsilon}_{xy} & \tilde{\varepsilon}_{xx} & 0 \\ 0 & 0 & \tilde{\varepsilon}_{zz} \end{pmatrix} \tag{2-23}$$

Alternatively, if one uses a description in terms of the optical conductivity, one obtains

$$\tilde{\boldsymbol{\sigma}} = \begin{pmatrix} \tilde{\sigma}_{xx} & \tilde{\sigma}_{xy} & 0 \\ -\tilde{\sigma}_{xy} & \tilde{\sigma}_{xx} & 0 \\ 0 & 0 & \tilde{\sigma}_{zz} \end{pmatrix} \tag{2-24}$$

with $(i, j = x, y, z)$:

$$\tilde{\varepsilon}_{ij} = -\mathrm{i}\,\tilde{\sigma}_{ij}/\omega\,\varepsilon_0 \tag{2-25}$$

$$\tilde{\sigma}_{ij} = \sigma_{1ij} + \mathrm{i}\,\sigma_{2ij} \tag{2-26}$$

$$\tilde{\varepsilon}_{ij} = \varepsilon_{1ij} - \mathrm{i}\,\varepsilon_{2ij} \tag{2-27}$$

The incoming linearly polarized wave in the Faraday configuration can be written as the sum of a right-circularly and a left-circularly polarized wave with complex refractive indices $\tilde{n}_\pm = n_\pm - \mathrm{i}\,k_\pm$. The Faraday rotation θ_F after propagation of a distance l_0 in the medium is given by:

$$\theta_F = \frac{\omega\, l_0}{2c}(n_+ - n_-) \tag{2-28}$$

The magnetic circular dichroism ε_F, similarly, is proportional to the difference of the absorption indices for right- and left-circularly polarized light:

$$\varepsilon_F = \frac{\omega\, l_0}{2c}(k_+ - k_-) \tag{2-29}$$

Traditionally, the Faraday rotation is said to be positive if the rotation sense of the main axis of polarization for $\boldsymbol{g} \parallel \boldsymbol{B}$ is the same as the positive current which induces the magnetic field $\boldsymbol{B}$ in the coil. This is compatible with the definition of a right-circularly polarized wave as:

$$\boldsymbol{E}_+ = \mathrm{Re}\,\{\boldsymbol{E}_0(\hat{\mathbf{e}}_x + \mathrm{i}\,\hat{\mathbf{e}}_y)\,\mathrm{e}^{\mathrm{i}(\omega t - \boldsymbol{g}_+ \cdot z)}\} \tag{2-30}$$

and a left-circularly polarized wave as:

$$\boldsymbol{E}_- = \mathrm{Re}\,\{\boldsymbol{E}_0(\hat{\mathbf{e}}_x - \mathrm{i}\,\hat{\mathbf{e}}_y)\,\mathrm{e}^{\mathrm{i}(\omega t - \boldsymbol{g}_- \cdot z)}\} \tag{2-31}$$

where $\hat{\mathbf{e}}_x$ and $\hat{\mathbf{e}}_y$ are unit vectors in the x and y direction, respectively. If we define right and left current densities $\boldsymbol{J}_+$ and $\boldsymbol{J}_-$

similarly to Eqs. (2-30) and (2-31), respectively, Ohm's law and Eq. (2-24) require that

$$\tilde{\sigma}_{\pm} = \tilde{\sigma}_{xx} \mp \mathrm{i}\, \tilde{\sigma}_{xy} \tag{2-32}$$

Combining Eqs. (2-32) and (2-25) and using $\tilde{n}_{\pm}^2 = \tilde{\varepsilon}_{\pm}$, we obtain

$$\tilde{n}_{+}^2 - \tilde{n}_{-}^2 = -2\tilde{\sigma}_{xy}/\omega\, \varepsilon_0 \tag{2-33}$$

which, after expansion of Eqs. (2-28) and (2-29) with $2\bar{n} = n_{+} + n_{-}$ and $2\bar{k} = k_{+} - k_{-}$, respectively, leads to

$$\theta_{\mathrm{F}} = -\frac{l_0}{2c\,\varepsilon_0}\left(\frac{\bar{n}\,\sigma_{1xy} - \bar{k}\,\sigma_{2xy}}{\bar{n}^2 + \bar{k}^2}\right) \tag{2-34}$$

$$\varepsilon_{\mathrm{F}} = -\frac{l_0}{2c\,\varepsilon_0}\left(\frac{\bar{k}\,\sigma_{1xy} - \bar{n}\,\sigma_{2xy}}{\bar{n}^2 + \bar{k}^2}\right) \tag{2-35}$$

Conversely, the off-diagonal-conductivity elements are obtained from the measured functions $\bar{n}$, $\bar{k}$, θ_{F}, and ε_{F} through

$$\sigma_{1xy} = -2c\,\varepsilon_0\,[\bar{n}\,\theta_{\mathrm{F}} - \bar{k}\,\varepsilon_{\mathrm{F}}]/l_0 \tag{2-36}$$

$$\sigma_{2xy} = 2c\,\varepsilon_0\,[\bar{k}\,\theta_{\mathrm{F}} + \bar{n}\,\varepsilon_{\mathrm{F}}]/l_0 \tag{2-37}$$

To derive the analogous relations for the polar Kerr effect, we start from Fresnel's equations, which relate the amplitudes $r_{\pm}$ and the phases $\Delta_{\pm}$ of the reflected waves with the optical functions $n_{\pm}$ and $k_{\pm}$ for right- and left-circularly polarized light. For normal incidence these are as follows (Born, 1964):

$$r_{\pm} = |r_{\pm}|\,\mathrm{e}^{\mathrm{i}\Delta\pm} = \frac{n_{\pm} - \mathrm{i}\,k_{\pm} - 1}{n_{\pm} - \mathrm{i}\,k_{\pm} + 1} \tag{2-38}$$

The Kerr rotation θ_{K} and the Kerr ellipticity ε_{K} are given by

$$\theta_{\mathrm{K}} = -\tfrac{1}{2}(\Delta_{+} - \Delta_{-}) \tag{2-39}$$

$$\varepsilon_{\mathrm{K}} = -\frac{|r_{+}| - |r_{-}|}{|r_{+}| + |r_{-}|} \tag{2-40}$$

For Kerr signals below a few degrees, one can make the approximations $\sin(\Delta_{+} - \Delta_{-}) \approx \Delta_{+} - \Delta_{-}$, $\cos(\Delta_{+} - \Delta_{-}) \approx 1$ and $(r_{+} - r_{-})^2 \ll 2r_{+}r_{-}$, giving

$$\theta_{\mathrm{K}} - \mathrm{i}\,\varepsilon_{\mathrm{K}} = \mathrm{i}\frac{\tilde{r}_{+} - \tilde{r}_{-}}{\tilde{r}_{+} + \tilde{r}_{-}} = \mathrm{i}\frac{\tilde{n}_{+} - \tilde{n}_{-}}{\tilde{n}_{+}\tilde{n}_{-} - 1} \tag{2-41}$$

Inserting (2-33) after an expansion of Eq. (2-41) with $2\bar{n} = n_{+} + n_{-}$ leads to:

$$\theta_{\mathrm{K}} = \frac{1}{\omega\,\varepsilon_0}\left(\frac{B\sigma_{1xy} + A\,\sigma_{2xy}}{A^2 + B^2}\right) \tag{2-42}$$

$$\varepsilon_{\mathrm{K}} = \frac{1}{\omega\,\varepsilon_0}\left(\frac{A\sigma_{1xy} - B\,\sigma_{2xy}}{A^2 + B^2}\right) \tag{2-43}$$

with the coefficients

$$A = \bar{n}^3 - 3\bar{n}\,\bar{k}^2 - \bar{n} \tag{2-44}$$

$$B = -\bar{k}^3 + 3\bar{n}^2\,\bar{k} - \bar{k} \tag{2-45}$$

Again, we express the off-diagonal conductivity as a function of the measured quantities n, k, ε_{K} and θ_{K} and obtain:

$$\sigma_{1xy} = \omega\,\varepsilon_0\,[B\,\theta_{\mathrm{K}} + A\,\varepsilon_{\mathrm{K}}] \tag{2-46}$$

$$\sigma_{2xy} = \omega\,\varepsilon_0\,[A\,\theta_{\mathrm{K}} - B\,\varepsilon_{\mathrm{K}}] \tag{2-47}$$

All the magneto-optical equations derived up to this stage apply to materials that are isotropic in the xy plane in the absence of an external field or a spontaneous magnetization. In other words, they are valid for amorphous, random-polycrystalline and cubic materials. In addition, they can be used for tetragonal and hexagonal systems if the propagation direction of light is parallel to the crystallographic c-axis. For other configurations and for lower symmetries, the relations may become very complicated. Fumagalli (1990) has derived expressions for the case of anisotropic crystals which in the absence of a magnetization have a diagonal conductivity tensor, i.e., $\tilde{\sigma}_{xx} \neq \tilde{\sigma}_{yy} \neq \tilde{\sigma}_{zz}$ and $\tilde{\sigma}_{ij} = 0\,(i \neq j)$. In the presence of a magnetization along the z direction and for an incoming wave travelling along the z direction and being polarized in the x direction, the conductivity tensor has the form

$$\tilde{\boldsymbol{\sigma}} = \begin{pmatrix} \tilde{\sigma}_{xx} & \tilde{\sigma}_{xy} & 0 \\ -\tilde{\sigma}_{xy} & \tilde{\sigma}_{yy} & 0 \\ 0 & 0 & \tilde{\sigma}_{zz} \end{pmatrix} \quad (2\text{-}48)$$

The complex Kerr rotation then becomes

$$\theta_K = \frac{1}{\omega\,\varepsilon_0}\left[\frac{(B-\Delta B)\,\sigma_{1xy} + (A-\Delta A)\,\sigma_{2xy}}{(A-\Delta A)^2 + (B-\Delta B)^2}\right] \quad (2\text{-}49)$$

$$\varepsilon_K = \frac{1}{\omega\,\varepsilon_0}\left[\frac{(A-\Delta A)\,\sigma_{1xy} + (B-\Delta B)\,\sigma_{2xy}}{(A-\Delta A)^2 + (B-\Delta B)^2}\right] \quad (2\text{-}50)$$

where A and B are formally given by Eqs. (2-44) and (2-45), but with optical constants averaged over the x and y direction and

$$\Delta A = \bar{n}\,(\bar{n}_y - \bar{n}_x) - \bar{k}\,(\bar{k}_y - \bar{k}_x) \quad (2\text{-}51)$$

$$\Delta B = \bar{n}\,(\bar{k}_y - \bar{k}_x) + \bar{k}\,(\bar{n}_y - \bar{n}_x) \quad (2\text{-}52)$$

If the conductivity tensor $\tilde{\boldsymbol{\sigma}}$ is not diagonal in the absence of a magnetization, the occurrence of a magnetization will lead to off-diagonal elements which are sums of intrinsic and magnetic-field-induced parts. Then, relations of the form of Eqs. (2-49) and (2-50) do not hold any longer and, in addition, the complete conductivity tensor for $\boldsymbol{M}=0$ has to be determined to derive the magnetization-induced part to $\tilde{\sigma}_{xy}$. To my knowledge, this task has never been undertaken for magnetic materials, but the very large majority of the investigations has been performed on isotropic samples.

2.2.2 Microscopic Models

2.2.2.1 The Becquerel Equation

The first quantitative description of the Faraday rotation started from the picture of Drude and Lorentz, who assumed that the electron is elastically bound to the nucleus. In the presence of an external magnetic field $\boldsymbol{B}$ the equation of motion for an elementary charge e takes the form:

$$m\,\frac{\mathrm{d}^2\boldsymbol{r}}{\mathrm{d}t^2} + m\,\gamma\,\frac{\mathrm{d}\boldsymbol{r}}{\mathrm{d}t} + \hat{f}\,\boldsymbol{r} = e\,\boldsymbol{E} + e\left[\frac{\mathrm{d}\boldsymbol{r}}{\mathrm{d}t}\times\boldsymbol{B}\right] \quad (2\text{-}53)$$

where m is the mass of the electron, $\hat{f}$ is the restoring force constant, and γ is a phenomenological damping constant. With an Ansatz of the form of Eqs. (2-30) and (2-31) for a right- and left-circularly polarized wave, respectively, one obtains for N oscillators per unit volume

$$\tilde{\varepsilon}_{\pm} = 1 + \frac{N\,e^2}{m\,\varepsilon_0\,(-\omega^2 + \mathrm{i}\,\omega\,\gamma + \omega_0^2 \pm \omega\,\omega_c)} \quad (2\text{-}54)$$

with the resonance frequency of the system in the absence of a magnetic field $\omega_0 = \hat{f}/m$ and the cyclotron resonance frequency $\omega_c = e\,B/m$. The application of the static magnetic field has displaced the resonance frequencies by:

$$\omega_{0_{\pm}} - \omega_0 = \pm\,\omega_c/2 = \pm\,\omega_L, \quad (2\text{-}55)$$

where ω_L is the Larmor frequency. For electrons with charge $-e$, ω_c and ω_L are negative and $n_{+}(\omega) = n(\omega - \omega_L)$. Thus Eq. (2-28) transforms to

$$\theta_F = \frac{\omega\,l_0}{2c}\,[n(\omega - \omega_L) - n(\omega + \omega_L)] \approx$$

$$\approx -\,\frac{\omega\,l_0}{c}\,\omega_L\,\frac{\mathrm{d}n}{\mathrm{d}\omega} \quad (2\text{-}56)$$

which is called the Becquerel equation (Becquerel, 1897). This expression predicts a frequency dependence of the Faraday rotation which is proportional to the derivative of the refractive index. Analogously, the magnetic circular dichroism will have a line shape corresponding to the derivative of the absorption index. One speaks then of a *diamagnetic line shape* or a *diamagnetic rotation*. As we shall see later, this type of line shape occurs often in glasses but is not exclusive to diamagnetic materials.

2.2.2.2 Diamagnetic and Paramagnetic Rotation

An important improvement in the theory arose after the Zeeman effect had been discovered. In fact, it showed that the splittings due to an applied magnetic field are generally not equal to the Larmor frequency, but depend on the quantum numbers. In addition, the oscillator strength for right- and left-circularly polarized light does not need to be identical and may also vary as a function of temperature. Therefore, the Faraday rotation should contain, besides a term proportional to $\partial n/\partial\omega$, two more terms proportional to the derivative of the refractive index with respect to the oscillator strength f and the occupation of the ground state, respectively. Schütz (1936) has derived the expression:

$$\theta_{\mathrm{F}} = -\frac{N e^2 f l_0 \omega_{\mathrm{L}}}{m n \varepsilon_0 c}\left\{a \frac{\omega^2}{(\omega_0^2-\omega^2)^2} + \frac{b}{2}\frac{\omega^2}{\omega_0(\omega_0^2-\omega^2)\,\omega_{\mathrm{so}}} + \frac{c}{2}\frac{\omega^2 \hbar}{\omega_0(\omega_0^2-\omega^2)\,kT}\right\} \qquad (2\text{-}57)$$

where a, b, and c are correction factors of order unity to take into account the ratio of Zeeman to Larmor splitting, and $\hbar\omega_{\mathrm{so}}$ is the spin-orbit splitting. The first term in Eq. (2-57) is the corrected Becquerel expression, the second term is called Darwin term, and the third one is known as paramagnetic rotation. The frequency dependence of these 2 latter terms is identical to that of the refractive index. Consequently, the frequency dependence of the corresponding magnetic circular dichroism will be directly proportional to the absorption index. The above theory has been applied quite successfully to the analysis of Faraday and magnetic circular dichroism spectra of diluted magnetic ions in solids, for example, Pr^{3+} (Ferré et al., 1970), Nd^{3+} (Boccara et al., 1969 a), or Eu^{2+} (Boccara et al., 1969 b). It has also been used sometimes for concentrated systems if the absorption bands were narrow as in alkali nickel fluorides (Ferré et al., 1973, Pisarev et al., 1974).

For solids with broader bands, models have been developed (Boswarva et al., 1962; Kolodziejczak et al., 1962; Mavroides, 1972; Piller, 1972) which start from an equation of motion of the type in Eq. (2-53) for every interband transition $\hbar\omega_{\mathrm{k}}$, where k is the wave vector in the solid. The evaluation of the phenomenological oscillator strength requires, of course, the application of the perturbation theory. For simple bands and in the limit of long wavelength this approach is useful, but for more general cases the model is too simple.

2.2.2.3 The Classical Model for Intraband Transitions

The classical theory is more adequate to describe the magneto-optical properties of free carriers. One starts again from an equation of the type in Eq. (2-53), but deletes the term for the restoring force $\hat{f}\,\boldsymbol{r}$ and replaces the electron mass m by an effective mass m^*. For the complex dielectric function $\tilde{\varepsilon}_\pm$ one obtains

$$\tilde{\varepsilon}_\pm = 1 + \frac{N e^2}{m^* \omega \varepsilon_0(-\omega \pm \omega_{\mathrm{c}} + \mathrm{i}\gamma)} \qquad (2\text{-}58)$$

whose real part is

$$\varepsilon_{1\pm} = 1 + \frac{N e^2(-\omega \pm \omega_{\mathrm{c}})}{m^* \omega \varepsilon_0[(-\omega \pm \omega_{\mathrm{c}})^2 + \gamma^2]} \qquad (2\text{-}59)$$

Using Eq. (2-28) the Faraday rotation is calculated to be

$$\theta_{\mathrm{F}} = \frac{l_0 N e^2 \omega_{\mathrm{c}}}{2 c \bar{n} m^* \varepsilon_0} \cdot \frac{\omega_{\mathrm{c}}^2 - \omega^2 + \gamma^2}{[(\omega+\omega_{\mathrm{c}})^2+\gamma^2][(-\omega+\omega_{\mathrm{c}})^2+\gamma^2]} \qquad (2\text{-}60)$$

For frequencies larger than the cyclotron frequency (note that for 10 T, $\hbar\,\omega_c \approx 1$ meV), the Faraday rotation becomes:

$$\theta_F = \frac{l_0 N e^2 \omega_c}{2 c \bar{n} m^* \varepsilon_0} \frac{(-\omega^2+\gamma^2)}{(\omega^2+\gamma^2)^2} \quad (2\text{-}61)$$

For $\omega \gg \gamma$ this equation expresses the well-known proportionality of θ_F to ω^{-2} (or λ^2) for free carriers. It contrasts with the frequency dependence of the Faraday rotation for bound electrons in the low-frequency limit, which according to Eq. (2-57) is ω^2. This different frequency dependence has been used quite widely to separate the interband- and intraband contributions in doped semiconductors like Si, Ge, indium and gallium pnictides, zinc and cadmium chalcogenides (for reviews see for example Amzallag, 1970; Madelung, 1962) and also in magnetic semiconductors like doped EuO (Schoenes and Wachter, 1974). In this latter investigation, the authors also determined the absorptive and dispersive parts of the dielectric function of free carriers in the absence of the magnetic field. Setting $\omega_c = 0$ in Eq. (2-59) and calling the contribution of the interband transitions to ε_1, ε_{opt} one derives

$$\varepsilon_2 = N e^2 \gamma / m^* \omega\, \varepsilon_0 (\omega^2+\gamma^2) \quad (2\text{-}62)$$

$$\varepsilon_1 = \varepsilon_{opt} - N e^2 / m^* \varepsilon_0 (\omega^2+\gamma^2) \quad (2\text{-}63)$$

Equations (2-61) to (2-63) constitute a system of three equations with the three unknowns N, m^* and γ. As they enter in different relations in these equations ($\omega_c = e B / m^*$), the system can be solved, thus allowing the complete determination of the transport properties of the free carriers from exclusively optical and magneto-optical measurements.

Another interesting relation can be derived for the reflectivity spectra due to free carriers. If we neglect the damping in Eq. (2-63) and introduce the plasma frequency $\omega_p = (N e^2 / m^* \varepsilon_0)^{1/2}$, we can write

$$\varepsilon_1 = \varepsilon_{opt} - \omega_p^2/\omega^2 \quad (2\text{-}64)$$

The optical reflectivity will be minimal for $\varepsilon_1 = 1$, which is called *the plasma minimum.* It occurs at a frequency ω_{min} which we use to express the plasma frequency

$$\omega_p^2 = (\varepsilon_{opt} - 1)\,\omega_{min}^2 \quad (2\text{-}65)$$

In the presence of an external magnetic field the plasma minimum is shifted, and its frequency depends on the rotation sense of the circular wave. From Eqs. (2-59) and (2-63) one derives

$$\varepsilon_1 = \varepsilon_{opt} + \omega_p^2 / (\pm\omega_c - \omega_{min\pm})\,\omega_{min\pm} \quad (2\text{-}66)$$

After substitution of Eq. (2-65) into Eq. (2-66), a quadratic equation for the plasma minimum in a field is obtained, whose solutions are

$$\omega_{min\pm 1,2} = [\pm\omega_c \pm (\omega_c^2 + 4\omega_{min}^2)^{1/2}]/2 \quad (2\text{-}67)$$

The positive solutions in the limit $4\omega_{min}^2 \gg \omega_c^2$ are

$$\omega_{min\pm} \approx \omega_{min} \pm \omega_c/2 \quad (2\text{-}68)$$

Equation (2-68) shows that the splitting of the plasma minimum for right- and left-circularly polarized light gives directly the cyclotron resonance frequency, i.e., the effective mass of the free carriers. If this is known, the plasma frequency unambiguously determines the free-carrier density. Fitting the reflectivity curves rather than considering only the frequencies of the minima, the third transport parameter, i.e., the damping, can also be derived. An excellent example for the application of this method on doped semiconductors has been presented by Palik et al. (1962) for n-type InSb.

2.2.2.4 Interband Transitions in Magnetically Ordered Materials

The magneto-optical signals in ferromagnets and often also in ferrimagnets can be orders of magnitude larger than in diamagnetic or paramagnetric materials. The rather limited Zeeman splitting that a common external field may produce is much too small to account for the observed effects. It was Hulme (1932) who first proposed that the spin-orbit interaction should be considered to explain *ferromagnetic* rotations. The spin-orbit interaction couples the momentum of the electron with its spin and thus provides the link for the circularly-polarized electromagnetic waves to couple to the magnetization of the material. The interesting feature of magneto-optics is that the relevant magnetization is not the net magnetization of the sample but the spin polarization of the initial and final states of the optical transition. The spin-orbit interaction has two effects: it splits some degenerate states and it modifies the wave functions. The former case has been considered, among others, by Hulme (1932), Argyres (1955), and Bennett and Stern (1965), while the latter case has been treated by Kittel (1951). Following here the treatment by Bennett and Stern (1965) and Erskine and Stern (1973a), one writes the absorptive part of σ_{xy} as the sum of separate contributions from spin-up and spin-down states, i.e., one neglects spin-flip transitions. For spin-up states Erskine and Stern (1973a) derived

$$\sigma_{2xy_\uparrow}(\omega)=\frac{\pi e^2}{4\hbar\omega m^2 V}\cdot \sum_{\alpha,\beta}\{|\langle\beta\uparrow|\boldsymbol{\pi}^-|\alpha\uparrow\rangle|^2-|\langle\beta\uparrow|\boldsymbol{\pi}^+|\alpha\uparrow\rangle|^2\}\cdot \delta(\omega_{\alpha\beta\uparrow}-\omega) \tag{2-69}$$

where the operators $\boldsymbol{\pi}^{\pm}=\boldsymbol{\pi}_x\pm \mathrm{i}\,\boldsymbol{\pi}_y$ are linear combinations of the kinetic momentum operator and the sum is a double sum to be taken for every occupied spin-up state $\alpha\uparrow$ over all unoccupied spin-up states $\beta\uparrow$. Because spin-flip transitions are neglected, Erskine and Stern (1973a) also drop the spin-orbit term in the kinetic-momentum operator and use $\boldsymbol{\pi}^{\pm}=\mathrm{i}\,\omega\, m(x\pm \mathrm{i}\,y)$. The transition from the atom to the solid is performed by a sum-rule argument and by restricting the evaluation to the total weight of the transitions from states a into states b, of which α and β are the respective sublevels. With the joint spin polarization

$$\sigma_j=\frac{n_{\alpha\uparrow}n_{\beta\uparrow}-n_{\alpha\downarrow}n_{\beta\downarrow}}{n_{\alpha\uparrow}n_{\beta\uparrow}+n_{\alpha\downarrow}n_{\beta\downarrow}} \tag{2-70}$$

the total weight becomes

$$\langle\sigma_{2xy}\rangle=\frac{\pi N e^2}{4\hbar}\cdot \sum_{\alpha,\beta}[\omega_{\alpha\beta}^-|(x-\mathrm{i}\,y)_{\alpha\beta}|^2-\omega_{\alpha\beta}^+|(x+\mathrm{i}\,y)_{\alpha\beta}|^2]\cdot \sigma_j n_\alpha n_\beta \tag{2-71}$$

Here $N=1/V$ is the density of oscillators, and n_α and n_β are the occupied and empty states per oscillator, respectively. For larger bands it is appropriate to multiply Eq. (2-71) with the ratio of the spin-orbit energy to the band width and the ratio of the exchange splitting to the band width (Erskine and Stern, 1973a; Schoenes, 1975; Reim and Schoenes, 1990).

While the above model has led to a semi-quantitative interpretation of the magneto-optical properties of several lanthanide and actinide materials that are dominated by rather narrow and spin-conserving f→d transitions (Reim and Schoenes, 1990), it is much less suitable for transition metals. Here, bands are wide and the concept of a total weight is not very helpful to interpret structures in the magneto-optical spectra. In addition, electric-dipole and spin-forbidden transitions appear to play a non-negligible role (Wang and Callaway, 1974; Singh et al., 1975). The latter two groups of authors used a modified form of Eq. (2-69) in combination with band-structure calculations to compute the magneto-optical spectra of Ni (Wang and Callaway) and Fe (Singh et al.). The agreement with experiment is moderate for Fe and less satisfactory for Ni, presumably due to a larger self-energy term in Ni. Smith et al. (1982), also using a bandstructure calculation but a different technique in computing σ_{2xy}, came to results for Ni which agree even less with experiment. Recently, an improvement of the theory has been realized that may prove to be very effective. Instead of treating the spin-orbit splitting as a perturbation, fully relativistic calculations have been performed. Thus, Daalderdop et al. (1988) used a Hamiltonian in a fully relativistic localized basis which was obtained by a Fourier fitting procedure. Ebert et al. (1988) put spin-polarization and the relativistic effects on equal footing by describing the electronic structure of the ferromagnetic transition metals in a spin-polarized version of the relativistic LMTO (linear muffin-tin-orbital) method. It remains to be seen whether these sophisticated calculations will lead to a better understanding of the magneto-optical properties of complex systems and whether they will allow one to predict, for example, potential candidates for applications like thermo-magnetic writing.

2.2.2.5 Intraband Transitions in Magnetically Ordered Materials

A characteristic feature of metals is the presence of *free carriers*. These provide generally large contributions to the optical conductivity in the energy range from zero to several eV. In ferromagnetic metals these free carriers can equally contribute to the off-diagonal conductivity. In principle, a spin-polarized and relativistic band-structure calculation would be the ideal starting point to compute also this intraband part to the magneto-optical spectra. In practice, however, the interpretation of experimental data still relies on phenomenological theories. The most widely used theory is probably that of Erskine and Stern (1973a) with the correction by Reim et al. (1984a). In this model the conductivity is calculated considering an electric field of the form

$$\boldsymbol{E}(t) = E_0\, \delta(t)\, \hat{\mathbf{x}} \tag{2-72}$$

where $\hat{\mathbf{x}}$ is a unit vector in the x direction. This field shifts each state from its original value $\boldsymbol{k}_0$ to a new value from which it relaxes back to its original value. In the relaxation-time approximation, Erskine and Stern (1973a) write

$$-\frac{\mathrm{d}\boldsymbol{k}}{\mathrm{d}t} = \frac{\boldsymbol{k}}{\tau} + \frac{\boldsymbol{s}\times\boldsymbol{k}}{\tau_\mathrm{s}} \tag{2-73}$$

where τ and τ_s characterize the normal and the skew-scattering lifetime, respectively. It may be worthwhile to mention here that τ and τ_s play different roles. In fact, the integration of Eq. (2-73) requires for the first term on the right side of the equality sign an expression of the form $\mathrm{e}^{-t/\tau}$, i.e., τ is a relaxation time, and for the second term (due to the vector product) periodic functions with the argument t/τ_s, i.e., τ_s is a period. For the integration of Eq. (2-73) $\boldsymbol{s}$ is placed along the z direction and $\boldsymbol{k}$ and the current $\boldsymbol{J}$ are calculated along the y direction. The conductivity as a function of frequency is obtained by Fourier transformation to be

$$\tilde{\sigma}_{xy}(\omega) = \omega_\mathrm{p}^2\, \varepsilon_0 \langle \sigma_z \rangle \left[\frac{-\Omega}{\Omega^2 + (\gamma + \mathrm{i}\,\omega)^2} + \right. \\ \left. + \frac{|P_0|}{e\, V_0} \left(1 - \frac{\mathrm{i}\,\omega(\gamma + \mathrm{i}\,\omega)}{\Omega^2 + (\gamma + \mathrm{i}\,\omega)^2} \right) \right] \tag{2-74}$$

where $\Omega = s/\tau_s$ and $s = \pm 1$ for spin-up and spin-down states, $\gamma = \tau^{-1}$ is the damping term, v_0 is the Fermi velocity, $|P_0|$ is the maximum value of the dipole moment $\boldsymbol{P}(\boldsymbol{k})$ per unit cell due to spin-orbit coupling, $\omega_p^2 = N e^2/m^* \varepsilon_0$ is the square of the plasma frequency of the free carriers, and $\langle \sigma_z \rangle = (n_\uparrow - n_\downarrow)/(n_\uparrow + n_\downarrow)$ is the spin polarization. For the absorptive part of σ_{xy} one obtains

$$\sigma_{2xy}(\omega) = \omega_p^2 \varepsilon_0 \langle \sigma_z \rangle \cdot \left[\frac{2\gamma \Omega \omega}{(\Omega^2 + \gamma^2 - \omega^2)^2 + 4\gamma^2 \omega^2} - \frac{|P_0|}{e V_0} \frac{\omega \gamma (\Omega^2 + \gamma^2 + \omega^2)}{(\Omega^2 + \gamma^2 - \omega^2)^2 + 4\gamma^2 \omega^2} \right] \tag{2-75}$$

Two features of this equation merit attention. The first is the different sign of the two terms and the second is the different frequency dependence. The first term is formally identical to the classical Drude term

$$\sigma_{2xy}(\omega) = \omega_p^2 \varepsilon_0 \frac{2\gamma \omega_c \omega}{(\omega_c^2 + \gamma^2 - \omega^2)^2 + 4\gamma^2 \omega^2)} \tag{2-76}$$

which is derived from Eq. (2-58) with Eqs. (2-25) and (2-32). However, the cyclotron resonance frequency of the classical formula is replaced by the skew-scattering frequency in the magnetic material and the signal becomes proportional to the spin polarization. This is analogous to the situation for the Hall effect, which for magnetic materials is generally dominated by the anomalous term proportional to the magnetization. For $\omega \gg \Omega, \gamma$, the first term in Eq. (2-75), displays a ω^{-3} dependence, while the second term becomes proportional to ω^{-1}. Therefore, for many magnetic metals the second term is dominant in the energy range of a few eV, and one often plots $\omega \sigma_{2xy}(\omega)$ to suppress the frequency dependence of the free carriers in the discussion of interband effects. However, we will also meet materials for which the first term in Eq. (2-75) dominates in the eV range.

2.3 Experimental Aspects

The goals for performing magneto-optical measurements are manifold and the available samples may have different forms, shapes, and properties. Consequently, the techniques used to obtain the desired information are numerous. Those who are interested, for example, in applications such as erasable magneto-optical disks often limit their magneto-optical measurements to the determination of rotation and/or ellipticity angles for one or a few laser wavelengths. If in addition the measurement is only performed at room temperature, the required experimental setup is very simple. Others who use magneto-optics as a spectroscopic method need to perform the magneto-optical measurements over a wide spectral range, which generally implies the use of a monochromator; i.e., they face intensity and focusing problems. Then, in order to interpret these data in terms of the electronic structure, the normal optical constants must also be determined, at least over the same spectral range. Often these measurements are performed at cryogenic temperatures, which means that the light beam has to pass various windows. These windows are subject to the magnetic field which changes the state of polarization of the light beam. The windows cause multiple reflections and sometimes very serious problems due to thermal stress-induced birefringence. If the material is exotic or chemically unstable, additional problems will occur, for example, because of badly reflecting surfaces of pressed samples or necessary coatings.

2.3.1 Measurement Techniques

The complete determination of the Faraday effect or the polar Kerr effect requires the measurement of the rotation and the ellipticity. In most cases the measurement of the ellipticity is converted into a second rotation measurement by using the *Senarmont principle* (Theocaris and Gdoutos, 1979). The simplest way to illustrate the Senarmont principle is to make use of the Poincaré sphere (Fig. 2-2). On this sphere the azimuth ψ (angle between the main axis and the x-axis) of the elliptically polarized light is plotted as angle 2ψ from the x-axis on the "equator" of the sphere. The ellipticity η is the ratio of the short axis b to the long axis a of the ellipse. It defines an angle $\zeta = \operatorname{arctg} \eta$ which is plotted as an angle 2ζ from the equator on the "meridian" with given 2ψ. The northern and southern "hemisphere" correspond to right- and left-elliptically polarized light, respectively. The effect of a linear phase shifter, such as a Soleil-Babinet compensator with a retardation Δ is represented on the Poincaré sphere as a rotation Δ about the axis formed by the fast and the slow axis of the phase shifter. Hence, if the slow and fast axes of the phase shifter are parallel to the two axes of the ellipse, and the retardation is set equal to $\lambda/4 = 90°$, the elliptically polarized light is transformed into linearly polarized light at an azimuth ζ to the slow axis of the phase shifter equal to the original ellipticity. Thus, the problem of measuring an ellipticity is transformed into a second measurement of an angle after the $\lambda/4$ phase shifter is introduced in the light path and has been properly adjusted.

The methods to determine an angle of rotation can be divided in roughly three groups (Schoenes, 1981): In the first group the angle is measured more or less directly by rotating a linear polarizer (often the analyzer) to compensate for the rotation of the sample. As indicator for the correct rotation of the polarizer, one uses nowadays a detector and an amplifier and zeros the signal. A phase-sensitive detector can be used to increase the sensitivity if the polar-

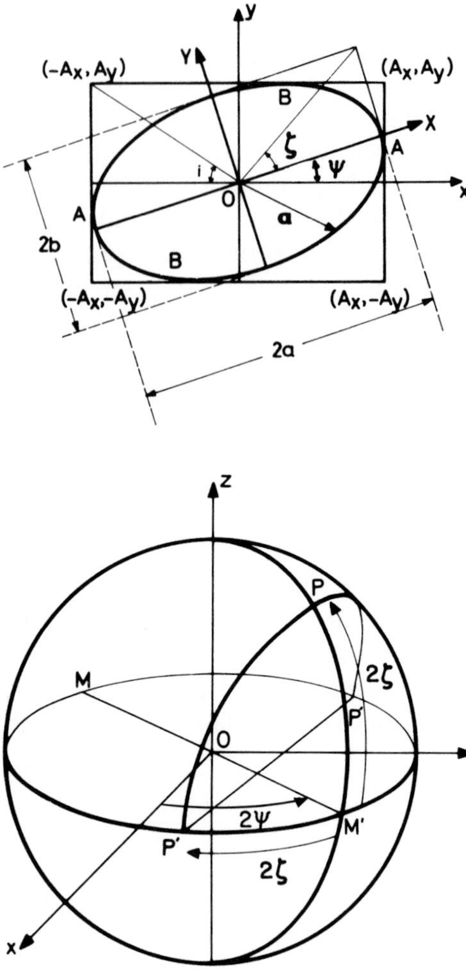

Figure 2-2. Representation of elliptically polarized light (top) on the Poincaré sphere (bottom) and illustration of the principle of Senarmont. The quarter-wave plate, whose axes are set parallel to the axes of the elliptically polarized light, represented by the point P on the Poincaré sphere, transforms the elliptically polarized light into linearly polarized light, represented by the point P′ (after Theocaris and Gdoutos, 1979).

ization state of the light is modulated, for example, with a Faraday modulator or a photo-elastic (piezo) modulator (Schoenes, 1984; Reim, 1986). A further improvement of the resolution can be achieved if the crossed position of the analyzer is determined from a fit of several points on both sides of the intensity minimum. Such a system with computer-controlled adjustments of polarizers and compensators via stepping motors has been designed to study the magneto-optical Kerr spectra of magnetic superconductors down to He^3 temperatures in fields up to 13 T (Fumagalli et al., 1990). Figure 2-3 shows a scheme of this setup (Fumagalli, 1990). The resolution depends somewhat on the photon energy due to varying intensities on the detector and the different sensitivity of the detectors. In the 1–5 eV range it is better than 0.01° even after the light beam has passed through six windows.

The second class of measurements can be called phase methods. If the rotation angles are rather large and a resolution of 0.1° is sufficient, these methods allow a quick determination of the rotation (Suits, 1971; Schoenes, 1979). Figure 2-4 shows schematically equipment used to measure the energy, temperature, and field dependence of the Faraday rotation of europium chalcogenide films (Schoenes; 1979). A

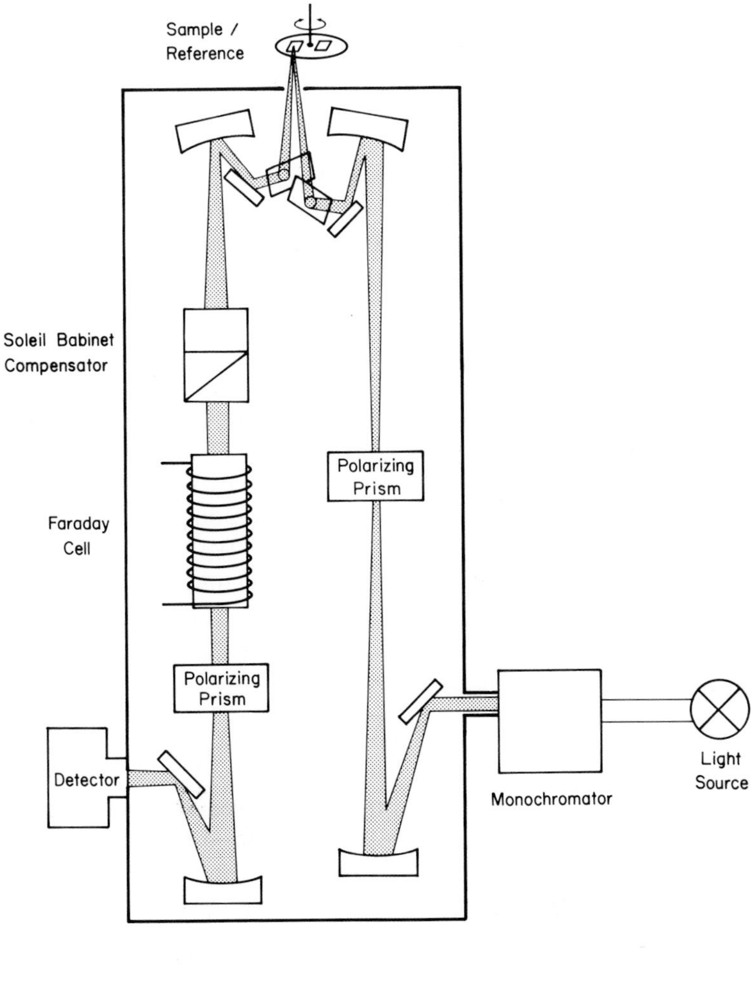

Figure 2-3. Scheme of a setup to measure the rotation and ellipticity with a computer-controlled compensation method (after Fumagalli, 1990).

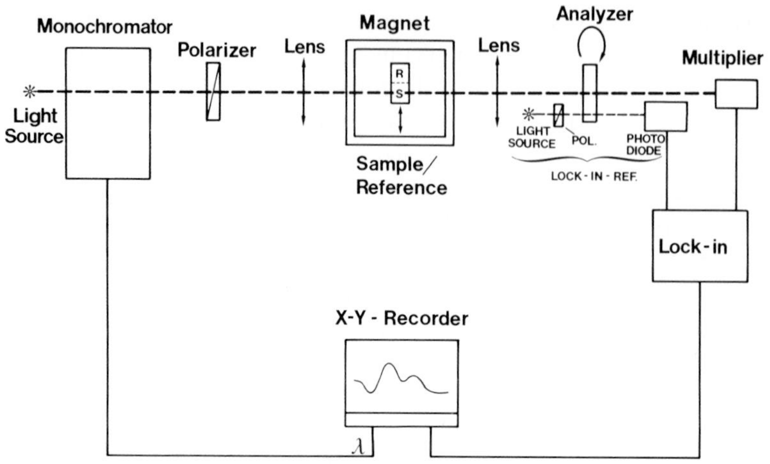

Figure 2-4. Scheme of a phase method to register the photon energy dependence of the rotation on an $x-y$ recorder (after Schoenes, 1979).

lock-in amplifier with a phase output measures at the doubled frequency of the spinning analyzer a phase difference between the sample and the reference beam which is equal to twice the rotation of the film. At the same time the amplitude output of the lock-in amplifier allows the determination of the absorption of the film.

Spinning analyzers, Faraday modulators, and piezomodulators are also used in the third group of measurements, which I call *amplitude*, or *intensity*, methods. These methods are generally highly sensitive and less laborious than the comparably sensitive angle methods. Their disadvantages reside in the large change in intensity that generally occurs when an extended photon energy range is to be covered, and in spurious effects on the measured intensity. The common principle of these methods is the following: The intensity on the detector for linearly polarized light falling under an azimuth ψ relative to the transmitting direction onto the analyzer is proportional to $\sin^2\psi$. If the azimuth is modulated in the form $\phi=\phi_0 \sin \Omega t$ with a Faraday modulator, the intensity after the analyzer becomes

$$I(t)=I_0 \sin^2[\psi+\phi_0 \sin \Omega t] \tag{2-77}$$

which after some algebra leads to

$$\begin{aligned} I(t)=\frac{I_0}{2}\{&1-J_0(2\phi_0)\cos 2\psi+\\ &+2J_1(2\phi_0)\sin\Omega t\sin 2\psi-\\ &-2J_2(2\phi_0)\cos(2\Omega t)\cos 2\psi+\ldots\}\end{aligned} \tag{2-78}$$

where $J_n(2\phi_0)$ are Bessel functions of order n. The expression shows that the signal with frequency Ω is proportional to $\sin 2\psi$, which for small rotations can be approximated by 2ψ.

Another intensity method makes use of a Wollaston prism. This prism splits an incoming light beam in two rays with orthogonal polarizations. If the incoming light is linearly polarized with an azimuthal polarization along the bisectrix of the optical axes of the Wollaston prism, the two split beams have equal intensity. If the azimuthal polarization is rotated by an angle ψ, the intensity of the two split beams becomes

$$I_{1,2}=I_0\cos^2(45°\pm\psi)\approx\frac{I_0}{2}(1\pm 2\psi) \tag{2-79}$$

One alternative, used for example for reading of magneto-optical disks, is to operate the system with one detector for every split beam and to form the sum

$I_1 + I_2 = I_0$ and the difference $I_1 - I_2 = \pm 2 I_0 \psi$ of the two signals. The ratio of these quantities gives directly 2ψ. A second alternative involves recombining the two beams on the same detector after the beams have passed through a spinning analyzer. For an angular frequency Ω of the spinning analyzer, the intensity of the 2Ω signal is

$$I(2\Omega) = I_0 \psi \cos(2\Omega t) \qquad (2\text{-}80)$$

The last intensity method to be mentioned is that used to measure the dynamic Faraday (or Kerr) rotation (Schoenes, 1975). In this method (Fig. 2-5) one applies an ac magnetic field $\boldsymbol{B} = \boldsymbol{B}_0 \sin(\Omega t)$ directly to the sample. The rotation from the sample is $\theta = \theta_0 \sin(\Omega t)$ and after passing through an analyzer oriented at 45° to the polarizer the intensity is

$$I = I_0 \cos^2(45° + \theta) \qquad (2\text{-}81)$$

For small rotations – note that B_0 is generally only of the order 0.01 T – we have to lowest order in θ

$$I = I_0/2 - I_0 \theta_0 \sin(\Omega t) \qquad (2\text{-}82)$$

If the incoming monochromatic light is in addition modulated by a chopper or an oscillating diaphragm as shown in Fig. 2-5, I_0 can be held constant over a certain photon energy range, and $\theta(\omega)$ is directly proportional to the lock-in amplifier output at frequency Ω. Besides the extreme sensitivity of the method – a resolution of 10^{-4} degree has been achieved (Schoenes, 1975) –, other advantages are the very low fields, which help to avoid unwarranted magnetic phase transitions, and the possibility to study the field derivative of the spin polarization as a function of the applied magnetic field.

2.3.2 Samples

The best-suited experimental method for studying the magneto-optical properties of a material largely depends also on the shape of the sample. If thin films can be prepared, one can choose a thickness which allows sufficient light to be transmitted and which enables Faraday-effect measurements to be performed. Thin films generally have the advantage of smoothness of the surface, i.e., little light scattering, uniform thickness, and large surface areas.

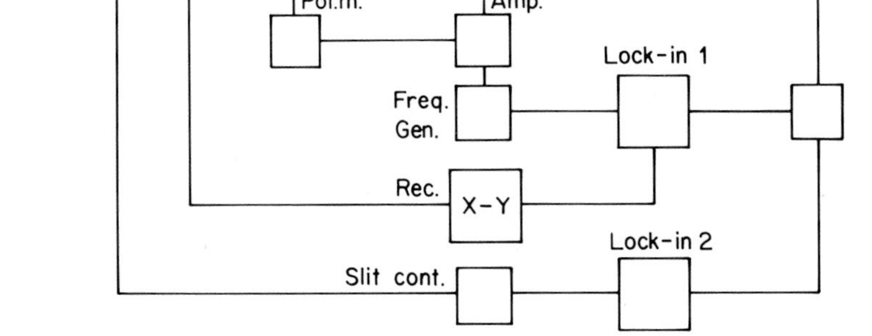

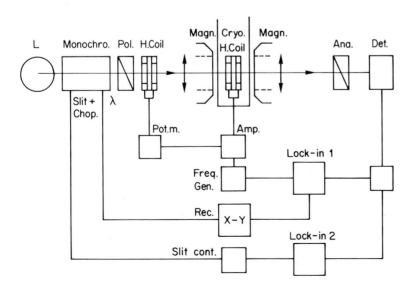

Figure 2-5. Scheme of an experimental setup to measure the dynamic Faraday rotation. The sample is placed in the gap of a Helmholtz coil (after Schoenes, 1975).

For alloys such as the transition metal rare-earth alloys, thin films offer the practical advantage of rather easy and controlled preparation by sputtering or multi-electron-gun evaporation. Disadvantages of thin films are the frequently occurring birefringence due to different thermal contraction of the film and the substrate and a large susceptibility to corrosion. Attention should also be paid to the possibility that some part of the detected light may have been the subject of multiple reflection and rotation. This is particularly the case if the absorption is low and a laser with a high correlation length is used. The problem of corrosion is often encountered by the deposition of a protective layer on top of the magneto-optical film. New problems then arise because of interference, absorption, and the reduction of the data for a multilayer system. The cleanest way to avoid these difficulties in such a situation is an in-situ measurements (Erskine and Stern, 1973a), and it is astonishing that this is not carried out more frequently for thin films.

Of course, thin films can also be used to perform Kerr-effect measurements. A prerequisite is that the film is thick enough not to transmit any substantial part of the incoming light. The problems with corrosion are of similar origin but are more serious than for Faraday-effect measurements because one does not average over the full thickness of the sample.

Single crystals are often too thick to allow Faraday-effect measurements in their absorbing spectral region. Therefore, single crystals are generally studied by use of the Kerr effect. If the crystals cleave easily, corrosion problems can be avoided by cleaving the crystals either in vacuum or in an inert gas atmosphere and performing the measurements in situ. If the crystals can not be cleaved, one may deposit a protective layer on a smooth and clean surface, and one will arrive at the problems mentioned before for the coated films. Advantages of single crystals are often their purity, perfection, and good characterization, which are manifested in a higher reproducibility of the data. A drawback is often the small size of single crystals, which necessitates a reduction of the size of the light spot leading to a reduction of the light intensity and the resolution.

Many magneto-optical measurements have been performed on bulk polycrystalline or glassy samples. In general, these samples can not be cleaved, and the required smooth surface is prepared by mechanical or electro-chemical polishing. If the materials are chemically unstable, the corrosion problems tend to be more serious than for the two cases discussed before. But even without corrosion, problems may arise from the surface preparation. The electro-chemical polishing may dissolve more of one component than another. Mechanical polishing introduces scratches that cause light scattering at short wavelengths, and it may induce structural, electronic, or magnetic phase transitions. An expedient to these changes may be a subsequent annealing, which, however, bears new risks of sample deterioration.

2.4 Non-Magnetic Metals and Glasses

2.4.1 Non-Magnetic Metals

The magneto-optical effects in non-magnetic metals, often also called normal metals, are quite small and there exist only few data. Majorana (1944) was apparently the first to measure the polar Kerr effect of Al, Ag, Au, Pt and Bi. Later on, measurements with much higher precision have been reported for the same materials by

Stern et al. (1964), McGroddy et al. (1965) and Schnatterly (1969). I should mention that Stern et al. (1964) reserve the term "polar magneto-optical Kerr effect" for magnetic materials and use for normal metals the expression "polar reflection Faraday effect", while Schnatterly (1969) speaks of "magnetoreflection". Figure 2-6 shows from top to bottom the spectra for Ag, Au, and Cu obtained by Schnatterly (1969) on evaporated thin films and normalized to a field of 1 T. The ellipticity can be calculated from the measured $\Delta R/R$ values for every energy with the relation $\varepsilon_K \approx -\Delta R/4R$ which follows from Eq. (2-40), while the phase-difference spectrum $\Delta\theta(\omega) = -2\theta_K$ is the result of a Kramers-Kronig transformation of the $\Delta R/R$ spectrum:

$$\Delta\theta(\omega) = \frac{\omega^2}{\pi} P \int_0^\infty \frac{(\Delta R/R)(\omega')}{\omega'(\omega'^2 - \omega^2)} \mathrm{d}\omega' \quad (2\text{-}83)$$

The spectra for gold and particularly those for silver have model character. Ehrenreich and Philipp (1962) have analyzed the diagonal optical poperties of Ag and Cu and have shown that below the coupled plasma frequency $\hbar\omega_p^* = 3.8$ eV the optical properties of Ag can perfectly be described with the Drude model for free carriers. The corresponding model for the magneto-optical properties has been presented in Sec. 2.2.2.3 and, indeed, the $\Delta R/R$ spectrum can be reproduced by this theory. At about 4 eV more substantial deviations occur indicating contributions from interband transitions (Schnatterly, 1969). For gold the magneto-optical spectra still show all the characteristics of free electron behavior, but the plasma edge is less steep due to a stronger spectral overlap of inter- and intraband transitions. Therefore the maximum in the ellipticity curve of Au is broader and about ten times smaller than in Ag, and a fit with a free-electron model gives much less satisfying agreement. In copper this trend continues and the magneto-optical spectra are now dominated by interband transitions. The data for aluminum of Stern et al. (1964) cover the range from 1.5 to 3 eV. At 1 T the rotation is of the order 4×10^{-4} deg ($\equiv 7 \times 10^{-6}$ rad) with a minimum near 1.9 eV. The small rotation of these metals, their high reflectivity and the chemical stability makes gold and aluminum very suitable as zero-rotation reference materials for Kerr measurements of magnetic materials.

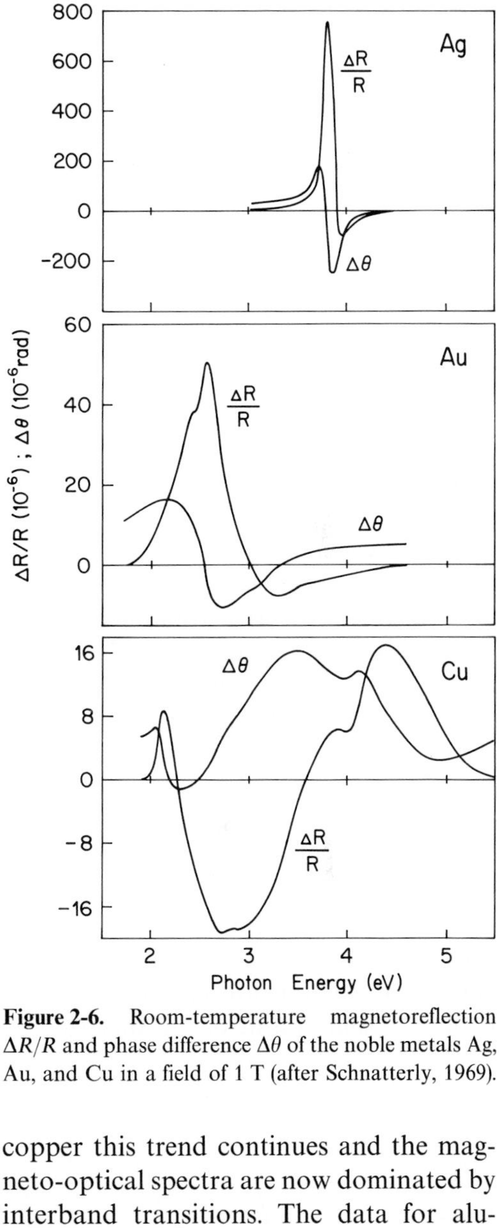

Figure 2-6. Room-temperature magnetoreflection $\Delta R/R$ and phase difference $\Delta\theta$ of the noble metals Ag, Au, and Cu in a field of 1 T (after Schnatterly, 1969).

2.4.2 Glasses

Glasses are playing an increasingly important role in telecommunication in the demand for Faraday rotators as switches and nonreciprocal elements. Glasses are also generally used in Faraday modulators in experimental setups (see Sec. 2.3). In addition to the size of the rotation, the determinant property for the usefulness of a material is its transparency. This latter aspect is at the root of the wide use of fused quartz for Faraday cells and piezo-modulators. Figure 2-7 displays the photon-energy dependence of the specific Faraday rotation (the Verdet constant) for Herasil I fused quartz (Schoenes, 1971). For energies smaller than 3 eV, glasses containing heavier elements have larger rotations and are still transparent. As an example, Fig. 2-7 also includes data for a heavy flint glass (Schott SF 59). Both materials have a positive Faraday rotation of diamagnetic line shape (see Sec. 2.2.2.2), but at 3 eV the specific rotation of the flint glass is about one order of magnitude larger than that of quartz.

Weber (1982) has examined a large number of lanthanide glasses as potential candidates for isolators for high power lasers. Figure 2-8 shows a fit of the spectrum at room temperature for a terbium borosilicate glass (Hoya FR-5) with the expression $K/(\lambda_t^2-\lambda^2)$ which corresponds to the paramagnetic rotation term in Eq. (2-57). The dominance of the paramagnetic rotation is due to the presence of the paramagnetic ion Tb^{3+} in this glass which has also been corroborated by the observation of a

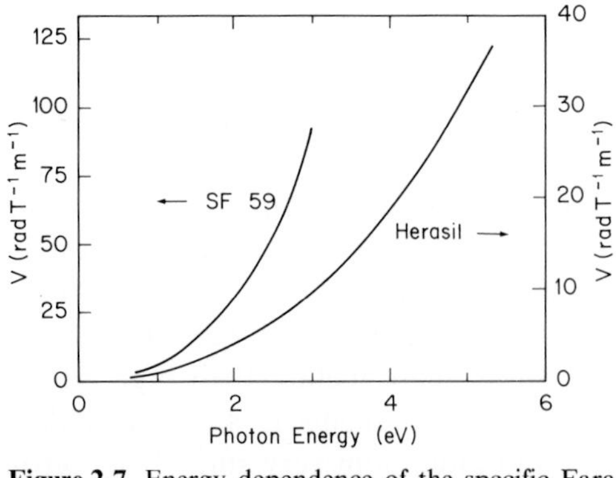

Figure 2-7. Energy dependence of the specific Faraday rotation of fused quartz (Herasil I) and of a heavy flint glass (Schott SF 59).

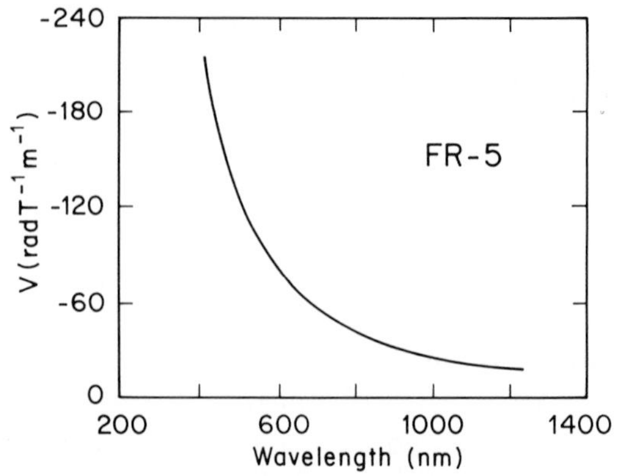

Figure 2-8. Fit of the wavelength dependence of the specific Faraday rotation of a terbium borosilicate glass (Hoya FR-5) at 300 K with the expression $K/(\lambda_t^2-\lambda^2)$ (after Weber, 1982).

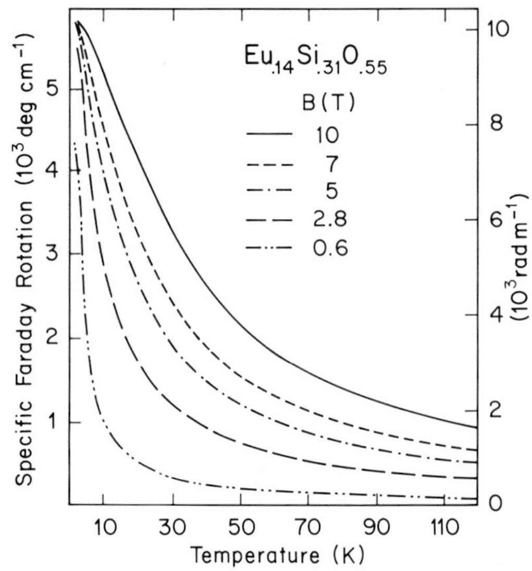

Figure 2-9. Temperature dependence of the Faraday rotation per unit length of $Eu_{0.14}Si_{0.31}O_{0.55}$ at 800 nm in various fields (after Schoenes et al., 1979).

strong increase of the rotation with decreasing temperature (Davis and Bunch, 1984). Another very effective paramagnetic lanthanide ion is Eu^{2+}. The temperature dependence of the Faraday rotation at a fixed photon energy for a glass of composition $Eu_{0.14}Si_{0.31}O_{0.55}$ (Fig. 2-9) provides evidence of the Curie behavior of the paramagnetic rotation term (Schoenes et al., 1979). Despite the large rotation of these paramagnetic glasses, their figure of merit θ_F/K at room temperature is smaller than that of the heavy flint glass due to an absorption coefficient K of the order of $100\ m^{-1}$ (Weber, 1982).

2.5 Transition-Element Materials

2.5.1 Elemental 3d Metals

Owing to their ferromagnetic ordering temperature above room temperature, the magneto-optical properties of iron, cobalt and nickel have been studied for more than 100 years. Jaggi et al. (1962) have collected the Faraday-rotation and polar Kerr-effect data as well as meridional and equatorial Kerr-effect data from 1886 to 1960. Difficulties in preparation of thin films and consecutive deterioration have led to large discrepancies in the reported Faraday-rotation data. For iron and cobalt the average specific Faraday rotation at saturation in the visible spectral range is approximately $2 \times 10^5\ deg/cm = 3.5 \times 10^5\ rad\ m^{-1}$. For nickel it is about half as large.

Buschow (1988) has compared the Kerr spectra for Fe, Co, and Ni reported by Krinchik and Artem'ev (1968), Clemens and Jaumann (1963), Burkhard and Jaumann (1970), and van Engen (1983). He noted a satisfactory agreement between the polar Kerr spectra determined by Krinchik and Artem'ev (1968) and van Engen (1983), while those of Jaumann and collaborators showed marked differences. Recent data obtained on sputtered films by Choe et al. (1987) and Weller et al. (1988) confirm the spectral dependence found by Krinchik and Artem'ev and van Engen, although deviations of the absolute value up to 0.1° may exist. Figure 2-10a displays the polar Kerr effect at room temperature for Fe, Co and Ni as determined by van Engen (1983). The rotation is dominantly negative with two extrema in the energy range from 0.2 to 4.5 eV. The dominant extrema, located between 1 and 1.5 eV reach values of approximately −0.53°, −0.37° and −0.15° in Fe, Co, and Ni, respectively, which scales roughly with the relative magnetization of these three materials. The spectrum of nickel is somewhat different from the others, showing changes of sign near 0.9 and again near 4.1 eV, while the Kerr rota-

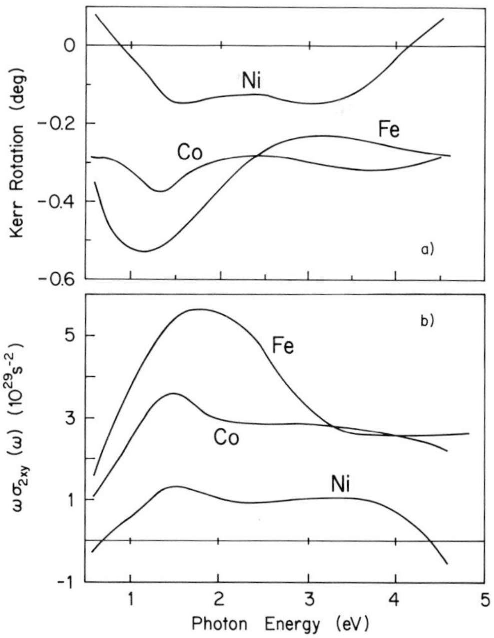

Figure 2-10. Polar Kerr rotation (a) and the product of the angular frequency and the absorptive part of the off-diagonal conductivity (b) of Fe, Co, and Ni for magnetic saturation (after van Engen, 1983; cited by Buschow, 1988).

tion of Fe and Co becomes positive only below about 0.3 eV (Krinchik and Artem'ev, 1968).

The photon energy dependence of the equatorial Kerr effect of Fe, Co, and Ni has also been measured by Burkhard and Jaumann (1970), Voloshinskaya and Bolotin (1974), Krinchik and Artem'ev (1968), Krinchik and Nurmukhamedov (1965), and Krinchik and Nuralieva (1959). Since these data depend on the angle of incidence, the comparison between different measurements is less straightforward and generally requires the computation of the off-diagonal components of the dielectric or conductivity tensor. Krinchik and coworkers also calculated the polar Kerr rotation from the equatorial Kerr spectra and optical constants taken from the literature. The agreement is quite good if one considers the uncertainty in this choice of optical constants.

The longitudinal Kerr effect of Ni has been measured by Martin et al. (1965) and Yoshino and Tanaka (1969). Again, a direct comparison with the polar Kerr effect is difficult, but the latter authors claim that the transformation of their data show the sign change close to 1 eV.

In discussing magneto-optical spectra in terms of electronic structure, the optical conductivity should be computed. To suppress the energy dependence of the intraband contribution, many authors plot $\omega\,\sigma_{2xy}(\omega)$. Figure 2-10b follows this scheme and displays the corresponding spectra for Fe, Co, and Ni as calculated from van Engen's data. Erskine and Stern (1973b) have concluded from similar data that the negative sign of $\omega\,\sigma_{2xy}(\omega)$ below 0.5 eV is characteristic of transitions involving minority spin electrons. Using in addition the theoretical result of Cooper (1965), who assigned the low-energy structure to transitions between minority spin bands at the symmetry point L_2, Erskine and Stern argued that the magneto-optical spectra predict a negative spin polarization of the d electrons in the vicinity of E_F. The change of sign of $\omega\,\sigma_{2xy}(\omega)$ at 0.5 eV – the intraband contribution was estimated not to alter this energy significantly – is attributed to a change from dominantly negative to positive spin polarization 0.5 eV below the Fermi energy. The absence of negative parts in the $\omega\,\sigma_{2xy}(\omega)$ spectra of Fe and Co is explained by contributions of p → d transitions which become more important as more empty d states are available. Wang and Callaway (1974) and Singh et al. (1975) have computed the diagonal and off-diagonal conductivity of Ni and Fe, respectively, using a self-consistent LCAO (linear combination of atomic orbitals) calculation including spin-orbit coupling. Better agreement with the experimental magneto-optical results is obtained for Fe, and Fig. 2-11 shows the computed (Singh et al., 1975) and the experimental data of Krinchik and Artem'ev (1968) which are similar to those of van Engen (1983). Note that the calculation of $\omega\,\sigma_{2xy}(\omega)$ of Fe, Ni and Co by Erskine and Stern (1973b) from the data of Krinchik and Artem'ev (1968) appears to be in error by a factor π and that Singh et al. (1975) have overlooked in the plot by Erskine and Stern the change of scale by a factor of 5 between Ni and Fe. The dominant peak in the off-diagonal conductivity spectrum of Fe around 1.7 eV appears to come from spin-up states in the vicinity of the symmetry point N of the Brillouin zone. The minimum occurring in the theoretical spectrum near 5.7 eV is tentatively assigned by Singh et al. to transitions from the minority-spin lower s-p band into the minority-spin d bands above E_F. It is interesting to note that the diagonal conductivity in the experimental spectra as well as in the computation displays its dominant

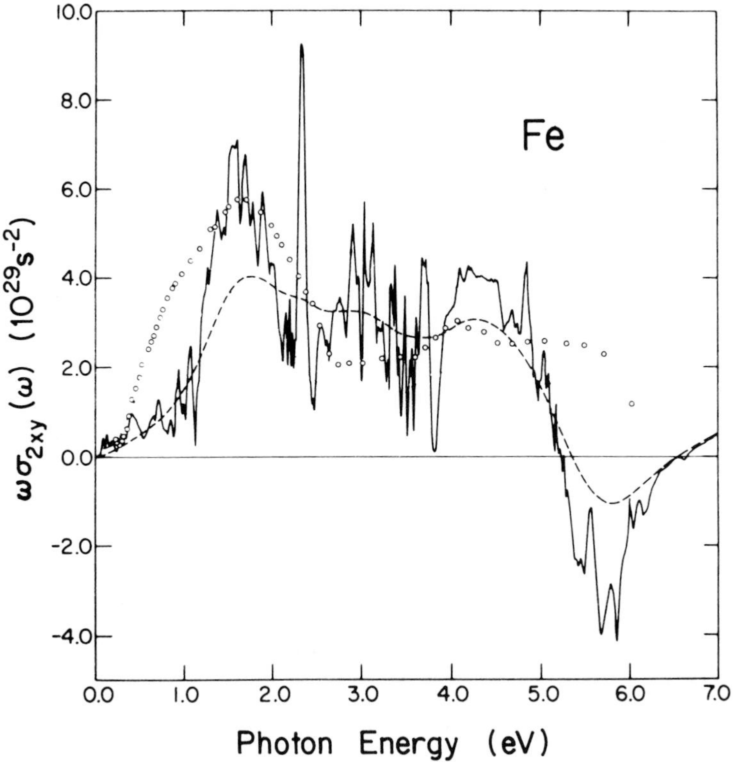

Figure 2-11. Comparison of theoretical and experimental spectra of the product of the angular frequency and the absorptive part of the off-diagonal conductivity of Fe for magnetic saturation. Full curve calculation of the interband part in the sharp limit; dashed curve: same calculation including a phenomenological damping of 0.3 eV. Circles: Experimental points of Krinchik and Artem'ev (1969) (after Singh et al., 1975).

maximum near 2.5 eV. The computation assigns this peak to a transition between exchange-split d states in the vicinity of the zone face. In the absence of spin-orbit coupling this transition would be forbidden, and also in the off-diagonal conductivity this transition is strongly suppressed. The agreement between theory and experiment is less satisfying for Ni. The comparison of the diagonal conductivity shows differences in peak position of up to 1 eV. Yet, the computation corroborates the net dominance of minority spin states in the range from 0 to 0.5–1 eV below the Fermi energy. Initial results for Fe using a spin-polarized relativistic version of the linear-muffin-tin orbital method (Ebert, 1989) are not in as good agreement with the experiment as the computation of Singh et al. In particular, it predicts a negative σ_{2xy} below 1 eV and a maximum near 3 eV, two features which are not found experimentally.

2.5.2 Alloys with 3d Elements

Numerous alloys with 3d elements have been studied. In this section, results for intra-3d alloys will be discussed followed by alloys of 3d elements with other non-magnetic elements. Alloys with lanthanides and actinides will be presented only after pure lanthanides and actinides have been discussed. Also multilayers with 3d elements will be dealt with in Sec. 2.9.

2.5.2.1 Intra-3d Alloys

Jaggi et al. (1962) have compiled the older data, including single-wavelength polar Kerr values for alloys of Fe with Mn, Co, and Ni and alloys of Co with Cr and

Ni. More recently the system Fe–Co has been investigated by Prinz et al. (1981), van Engen (1983), and Weller et al. (1988). Prinz et al. (1981) studied the longitudinal Kerr effect for the two photon energies 1.1 and 2.0 eV on evaporated polycrystalline thin films. They found an increase of the signal above the rotation of the single constituents in the middle concentration range. Van Engen (1983) investigated the polar Kerr effect at 2 eV on polycrystalline arc-cast samples. He found a maximum rotation near 50 at.% Co, whereas the saturation magnetization peaks were near 25 at.% Co. The spectral dependence of the polar Kerr rotation of Fe–Co alloys was studied from 0.5 to 5 eV by Weller et al. (1988). Figure 2-12 shows their Kerr spectra for various polycrystalline ($\approx$100 nm thick) thin films deposited in ultrahigh vacuum on fused-quartz substrates by ion-beam sputtering. It can be seen that

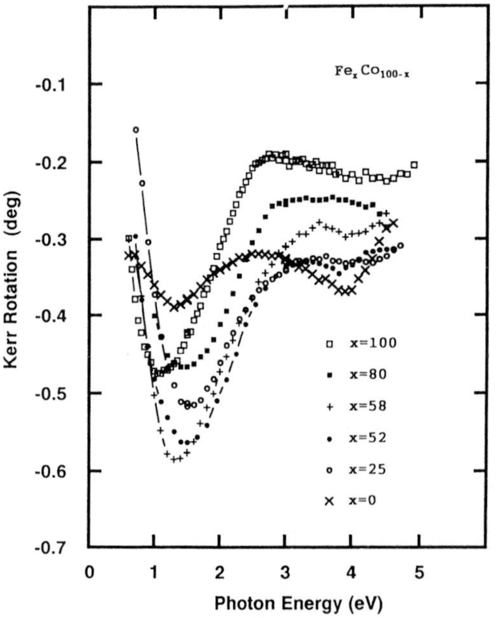

Figure 2-12. Room-temperature polar Kerr rotation of various polycrystalline Fe_xCo_{100-x} films (after Weller et al., 1988).

indeed at 2 eV the sample closest to the middle concentration has the largest rotation. However, there is also an energy shift of the rotation peak, and the highest absolute value is found for the sample $Fe_{0.58}Co_{0.42}$. Weller et al. have compared the Kerr spectra with ultraviolet photoemission spectroscopy and results of a spin-polarized version of the KKR-CPA (Korringa-Kohn-Rostocker–Coherent Potential Approximation) band structure calculation method. From the observation of similar shifts in the three kinds of spectra, they conclude that the dominant peak in the magneto-optical Kerr spectra involves excitations of majority d states.

Longitudinal Kerr rotation spectra have been reported for vacuum (2×10^{-5} Torr) evaporated Fe–Ni films with Ni concentrations ranging from 48 to 100 at.% Ni by Yoshino and Tanaka (1969). Unfortunately, neither the off-diagonal conductivity nor the polar Kerr rotation have been reported to allow an unambiguous discussion of the trends.

Voloshinskaya and Fedorov (1973) investigated the polar and the equatorial Kerr effect of Fe–V alloys from 0.07 to 3 eV. They observed a strong decrease of the polar Kerr rotation with increasing V concentration as expected in the simplest picture for the substitution of magnetic iron by nonmagnetic vanadium. For the lowest vanadium concentration of 0.035 at.%, Voloshinskaya and Fedorov report a polar Kerr rotation of 0.8° near 1 eV which is well above the rotation observed by other authors for pure Fe (see Sec. 2.5.1) and therefore may come into question.

The polar Kerr effect and the optical reflectivity of a Co–Cr alloy has been studied by Abe et al. (1982). They performed rotation and ellipticity measurements on a film with 21.5 at.% Cr sputtered on an anodized aluminum substrate for photon energies from 1.5 to 3 eV. The absolute value of the ellipticity increased continuously with increasing energy to reach $\approx 0.05°$ at 3 eV, while the absolute value of the rotation showed a soft minimum near 2.2 eV and grew to $\approx 0.08°$ at the limits of the measurements. Thus, except for alloys of Co and Fe, alloying of two transition elements appears to reduce the magneto-optical Kerr effect.

2.5.2.2 Fe Alloys with Non-3d Elements

Buschow et al. (1983) have studied at room temperature the polar Kerr effect, the crystal structure, the lattice parameters and the magnetization of more than 200 alloys and intermetallic compounds. For selected materials the rotation, and in fewer cases also the ellipticity, were measured from 0.6 to 4.4 eV, while for other materials the rotation was determined for the two laser wavelengths 633 nm (1.96 eV) and 830 nm (1.49 eV). From these 200 materials more than 50 are Fe based. The iron-based alloys comprise combinations with s and p electron elements like Al, Ga, Si, and Ge and noble metals like Pd and Pt. For the latter, the Kerr rotation spectra were measured. These showed rather small changes of shape in the vicinity of the low-energy peak near 1.2 eV on substitution of Fe with Pd or Pt, but interestingly enough an increase of the Kerr rotation near 4 eV. Thus, Buschow et al. found above 3.5 eV a higher rotation in $Fe_{0.5}Pt_{0.5}$ than in pure Fe. This increase at high energies is ascribed to contributions from 4d (Pd) and 5d (Pt) electrons, the larger spin-orbit interaction in Pt having a greater effect on alloying with Pt than with Pd. Brändle et al. (1991 b) have tried to unravel the Pd contribution by investigating the Pd-rich concentration range. Figure 2-13 shows the Kerr rotation and ellipticity spectra for an

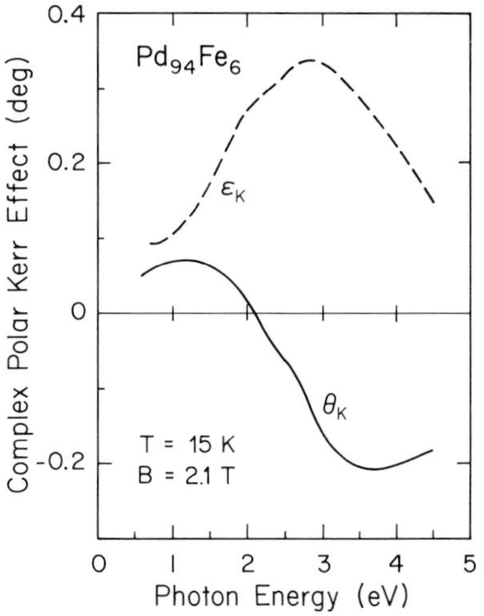

Figure 2-13. Kerr rotation and ellipticity of a $Pd_{0.94}Fe_{0.06}$ film at $T = 15$ K for an applied field of 2.1 T (after Brändle et al., 1991 b).

ion-beam-sputtered film of $Pd_{0.94}Fe_{0.06}$. The Pd atoms in this film carry a moment of 0.3 μ_B, which is the largest moment that, according to the investigation of Cable et al. (1965), can be induced by magnetic exchange with Fe or Co. Because of the low Fe concentration, the Curie temperature is below room temperature, and the measurements have been performed at 15 K in a field of 2.1 T. The difference relative to the spectra for iron (Fig. 2-10 a) is evident. At low energies the rotation is positive and the dominant peak of θ_K occurs near 3.8 eV. Figure 2-14 shows the off-diagonal conductivity for the same film, an extrapolation of σ_{1xy} to the Hall value at zero energy and a Kramers-Kronig transformation of this σ_{1xy} spectrum to obtain the low-energy part of $\sigma_{2xy}(\omega)$. One observes a large negative contribution at lower energies, a zero crossing at 1.8 eV, and a positive maximum near 3 eV. If this spectrum is multiplied with ω to suppress the energy dependence of the free carriers, the latter maximum moves to higher energies. To interpret these data, one should bear in mind

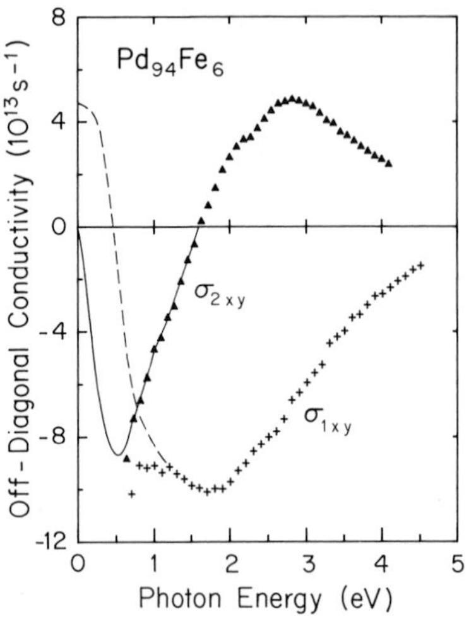

Figure 2-14. Off-diagonal conductivity of $Pd_{0.94}Fe_{0.06}$ computed fom the data in Fig. 2-13 and the optical constants (after Brändle et al., 1991 b).

that pure Pd metal is known to have ≈ 0.36 holes per Pd atom in its d band. An exchange splitting E_{exch} of a few tenths of an electronvolt of a band with a simple parabolic density of states and width $\Delta E \gg E_{exch}$ will lead in a rigid-band model to a dominance of minority spin states at E_F, to an equal density of spin-up and spin-down states near $E_F - \Delta E/2$, and to a maximum spin polarization near $E_F - 3\Delta E/2$. If one neglects final state effects, $\sigma_{2xy}(\omega)$ can be interpreted directly in terms of this model with $\Delta E \approx 4-5$ eV.

2.5.2.3 Co Alloys with Non-3d Elements

The comprehensive study of Buschow et al. (1983) contains also more than 50 cobalt-based materials. The rotation was found to decrease compared to pure Co in all cases, except for the alloys with Pt. Yet, the rotation normalized with the magnetization increases in several cases, on account of the even faster decreases of the magnetization with the substitution of Co. This beautifully demonstrates that while the Kerr rotation for a considered material is in most cases proportional to the magnetization or sublattice magnetization, the coefficients relating the absolute values of the Kerr effect and the magnetization depend on the electronic structure and are therefore material specific.

Because compound formation already takes place in most binary Co systems at a rather high Co concentration, solid solutions often only form in a limited concentration range. Exceptions are the Co–Pt and the Co–Pd systems. Figure 2-15 displays the data by Buschow et al. (1983) for the latter system. A reduction of the low-energy peak near 1.3 eV roughly proportional to the cobalt dilution can be observed. However, near 4 eV the structure remains nearly constant and thus becomes the dominant structure in the spectrum of $Pd_{0.5}Co_{0.5}$. Similar behavior is found on substitution with Pt, while the substitution with Ga leads to a strong suppression and even a sign reversal for larger Ga concen-

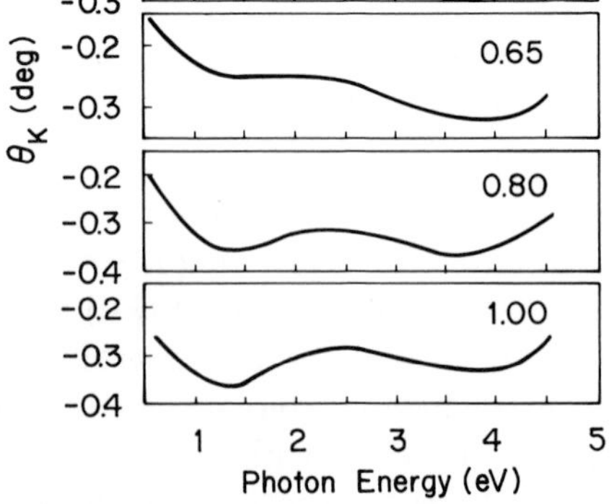

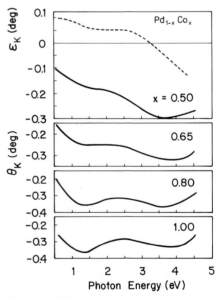

Figure 2-15. Room-temperature Kerr-rotation spectra of some $Pd_{1-x}Co_x$ alloys and compounds. For PdCo the Kerr ellipticity is also shown (dashed line) (after Buschow et al., 1983).

trations above 4 eV. As for the Fe alloys with Pd and Pt, the 4 eV structure is assigned to transitions involving majority-spin d electrons of the noble metals. This view was corroborated by a recent study on ion-beam sputtered films of Pt–Co in which Weller et al. (1991) also computed $\omega\,\sigma_{2xy}(\omega)$ and found that in $Pt_{0.73}Co_{0.27}$ the peak near 4 eV is as large as the one near 1.3 eV. Combinations of Co with the noble metals Pd and Pt and also Cu, Ag, and Au have received much attention in the last years in the form of multilayers. These data will be discussed in Sec. 2.9, which is especially devoted to applications.

2.5.2.4 Ni Alloys with Non-3d Elements

The number of Ni alloys with non-magnetic elements which are ferromagnetic at room temperature is considerably smaller than those formed by Fe and Co. This reflects the fact that the 3d band of pure nickel is relatively more filled and that an additional charge transfer from the alloy partner further reduces the magnetic moment. Buschow et al. (1983) have investigated alloys of the form $Ni_{0.92}M_{0.08}$, where M is (in increasing order of the f.c.c. (A_1) lattice constant) Cu, Al, Si, Ga, Ge, V, Pd, Pt, Nb, Ta, Au, Sn, In, Ti, W, and Cr. Figure 2-16 shows the effect of alloying Ni with s, p metals or with Cu. The shape of the low-energy peak near 1.5 eV is marginally affected, but its intensity decreases with alloying. On the other hand, the peak at 3.2 eV is depressed more strongly on alloying with Sn or Al than with Cu. Alloying with 8 at.% Pt, on the contrary, leads to a small increase of the Kerr rotation, especially at higher energies. Hence, the trend is similar to the alloys of Fe and Co with noble metals. Alloys of Ni with Al and Pd have also been investigated by Voloshinskaya and Fedorov (1973) and

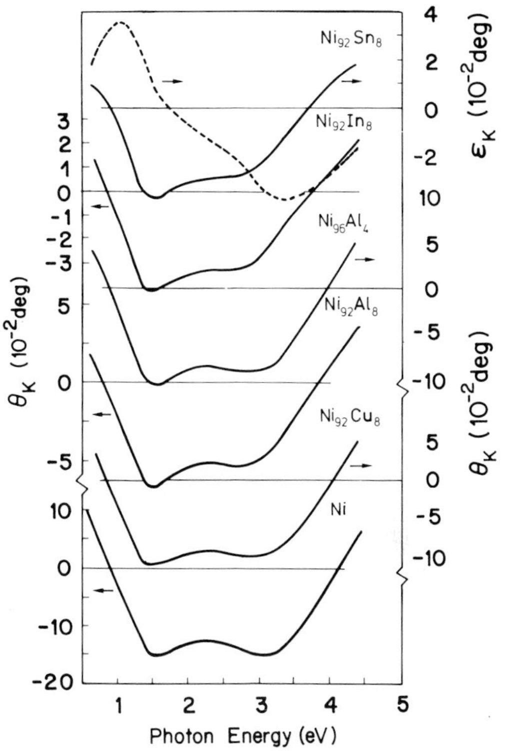

Figure 2-16. Room-temperature Kerr-rotation spectra of some Ni-base alloys. For the alloy $Ni_{92}Sn_8$ the ellipticity is also shown (dashed line) (after Buschow et al., 1983).

Voloshinskaya et al. (1974), who made an analysis of the low-energy data in terms of intraband transitions. Assuming two groups of d electrons and using for each an expression of the form of the first term in Eq. (2-75), they satisfactorily fitted their $\varepsilon_{1xy}(\omega)$ and $\varepsilon_{2xy}(\omega)$ spectra between 0.1 and 3 eV. Both models, the qualitative interband and the quantitative intraband model, have sufficient numbers of parameters, so that it is not possible at present to discriminate one in favor of the other.

2.5.2.5 3d Metal-Base Amorphous Films and Alloys

The disordered state of matter has attracted much attention over the last two decades. Basic questions relating to the electronic structure of liquid metals (see Chap. 9 in Volume 3 B of this Series) or symmetry problems in quasicrystals (see Volume 1 of this Series) and to materials engineering – the development of materials with the combination of desired mechanical, electrical, magnetic, and thermal properties – are the driving forces of this expanding field. In magneto-optics, this twofold interest in disordered materials also exists, but judging from the number of publications, there is a net concentration on amorphous films for applications such as thermomagnetic writing. To the knowledge of the author, no magneto-optical data have been published so far for liquid metals. Disordered solids can be prepared by fast cooling (quenching) from the liquid or the vapor phase. In the former case, one generally obtains self-supporting disks or ribbons, and in the latter case thin films deposited on a substrate. Both methods allow the formation of many more compositions than in the crystalline state. A big advantage of amorphous films in thermomagnetic writing is the absence of grain boundaries reducing the noise level. A prerequisite for using the polar Kerr effect for reading the magnetically stored information in an optical disk is the occurrence of a perpendicular magnetic anisotropy. This anisotropy can be generated in some amorphous magnetic films, presumably by atomic short-range ordering during growth (see Sec. 2.8.2). Rare-earth transition metal alloys show particularly appropriate anisotropy properties and will be dealt after a discussion of the pure rare-earth metals and compounds.

The transverse magneto-optical Kerr effect in amorphous Fe_xSi_{1-x} for x between 0.64 and 0.80 has been reported by Afonso et al. (1980) in the energy range of 1.2–5 eV. They also annealed the films at 500 °C in vacuum and remeasured the Kerr effect. Afonso et al. report the surprising result that the amorphous films have larger rotations than the annealed crystalline films and propose that the release of some orbital quenching in the amorphous state leads to an enhancement of the spin-orbit coupling and therefore to a larger Kerr effect. In apparent contrast to these data are the results of Ray and Tauc (1980) on the liquid-quenched metallic glass $Fe_{0.8}B_{0.2}$. These authors also measured the transverse (= equatorial in Fig. 2-1 b) Kerr effect and the optical reflectivity in order to compute $\omega\,\sigma_{2xy}(\omega)$ and to compare it with the corresponding spectrum of pure Fe reported by Erskine and Stern (1973 b). Ray and Tauc noted similarities in the shape of the spectrum, but also a reduction by a factor five of their values compared to the data of Erskine and Stern for Fe, which greatly exceeds the reduction in magnetization by about 25%. The solution to this discrepancy is the erroneous factor π in the calculation of Erskine and Stern (see Sec. 2.5.1). Figure 2-17 shows the comparison of the $Fe_{0.8}B_{0.2}$ results with the corrected data for Fe. The system Fe–B has also been studied by Buschow and van Engen

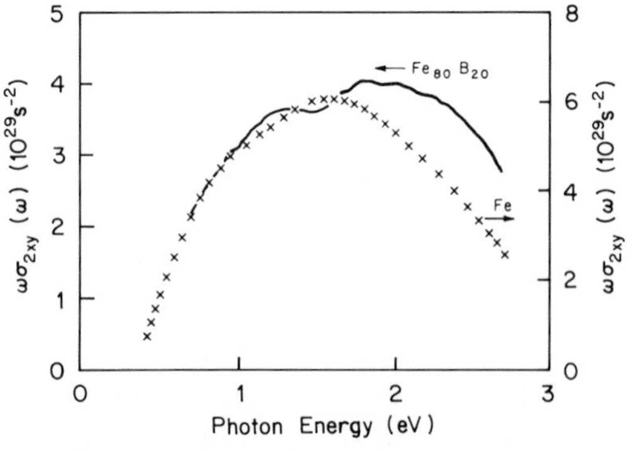

Figure 2-17. The product of the angular frequency and the absorptive part of the off-diagonal conductivity of the metallic glass $Fe_{0.8}B_{0.2}$ (after Ray and Tauc, 1980) and Fe.

(1981 a) and Katayama and Hasegawa (1982). The former group measured the polar Kerr effect on vapor-deposited films for Fe concentrations ranging from 40 to 100 at.%. At 1.95 eV (the He–Ne laser wavelength 633 nm) Buschow and van Engen observed a maximum for an Fe concentration of 70 and 80 at.% for air-incident and substrate-incident light, respectively. These rotation values are slightly higher than those of crystalline Fe at the same energy. Since the full spectrum has not been measured by Buschow and van Engen, it can not be determined whether the larger value at one wavelength is due to a shift of a peak or to an increase of the magneto-optical activity. It appears, however, in agreement with the findings of Katayama and Hasegawa (1982), that the changes in the magneto-optical properties between the amorphous and the crystalline state of Fe_xB_{1-x} are small.

Afonso et al. (1981) also investigated amorphous Co_xSi_{1-x} films. In contrast to Fe_xSi_{1-x}, Afonso et al. observed a smaller transverse Kerr effect than reported for crystalline Co by Krinchik and Artem'ev (1968). The difference increases with increasing substitution of Co by Si, and the structures wash out. Buschow and van Engen (1981 b, c) investigated amorphous alloys of Co with B and with Mg, respectively. They found that the Kerr rotation decreases faster with the substitution by boron than by magnesium, which is at least in qualitative agreement with the change in magnetization.

2.5.3 Transition-Element Compounds

A large variety of compounds with transition elements exist, extending from intermetallic to ionic compounds. The magneto-optical properties of intermetallic compounds containing transition elements have been reviewed in great detail by Buschow (1988). Their magneto-optical behavior often resembles that of alloys formed with the same elements. In these cases I will not repeat the discussion but refer the reader to the foregoing subsection and to the aforementioned review by Buschow. In a few cases, e.g., the Heusler alloys, the magneto-optical properties are substantially different and these compounds will be presented in detail. The more ionic compounds, even when they are metallic, present new features which are characteristic for the symmetry of the material, and which require interpretations including crystal field effects. Regarding the order of presentation, preference is given here to the *magnetic* element at the expense of an ordering according to the number of constituents. The first three transition elements are Sc, Ti, and V. Although a few Ti and V compounds do order magnetically and some V compounds, for example, VAu_4 (Adachi et al., 1980), VPt_3 (Jesser et al., 1981), $BaVS_{3-x}$ (Massenet et al., 1978), and $K_5V_3F_{14}$ (Cros et al., 1977) even show ferromagnetic or ferrimagnetic ordering, magneto-optical data are scare for these materials (see for example magnetic circular dichroism data for VI_2 by Boudewijn and Haas, 1980), and the next subsection starts with Cr.

2.5.3.1 Cr Compounds

The simplest ferromagnetic Cr compound is CrTe. It crystallizes in the hexagonal NiAs structure, has a T_C of 61 °C and its easy axis of magnetization is along the hexagonal *c*-axis (Mayer, 1958). Thin films of CrTe have been prepared by sequential electron-beam evaporation of Cr–Te–Cr on cleaved mica substrates by Comstock and Lissberger (1970). These authors measured the absorption coefficient between

0.5 and 2.5 eV and the Faraday rotation from 1.2 to 2.3 eV. In the investigated energy ranges, the absorption coefficient decreases monotonically with decreasing energy from 2×10^5 to $0.7 \times 10^5\ cm^{-1}$ and the Faraday rotation decreases from $0.5 \times 10^5\ deg\ cm^{-1}$ ($0.87 \times 10^5\ rad\ m^{-1}$) to about $0.4 \times 10^5\ deg\ cm^{-1}$ (1.27×10^5 rad m^{-1}). Hence the Faraday rotation is smaller than in Fe and Co (Sec. 2.5.1) by a factor of approximately 4, but Comstock and Lissberger note that the saturation magnetization of their film was also smaller than the $206 \times 10^3\ A\ m^{-1}$ reported for the bulk single crystal by a factor of 2.5.

Atkinson (1977) prepared films of Cr_3Te_4, which has a monoclinic structure. He determined the diagonal and off-diagonal dielectric tensor elements for photon energies from 2 to 3.1 eV using ellipsometric and longitudinal Kerr-effect measurements. The longitudinal Kerr effect is reported to lead to a signal-to-noise ratio which is smaller than that of the Faraday rotation in MnBi by a factor of 50 and is also worse than that of CrTe.

In contrast to the aforementioned metallic Cr compounds, $CrCl_3$, $CrBr_3$, and CrI_3 are insulators. $CrCl_3$ and $CrBr_3$ have a hexagonal structure of the BiI_3 type below 238 and 400 K, respectively (Morosin and Narath, 1964; Hulliger, 1976). At room temperature, CrI_3 appears to crystallize in the $AlCl_3$-type structure, which is also the structure type of $CrCl_3$ and $CrBr_3$ above their structural phase transitions (Hulliger, 1976). $CrBr_3$ was the first ferromagnetic insulator discovered (Tsubokawa, 1960). Its T_C amounts to 32.5 K, while for CrI_3 $T_C = 68$ K. $CrCl_3$ orders antiferromagnetically in zero field with $T_N = 16.8$ K. However, a small field is sufficient to align the spins in a ferromagnetic-like manner. The optical absorption and the Faraday rotation has been measured on thin crystal

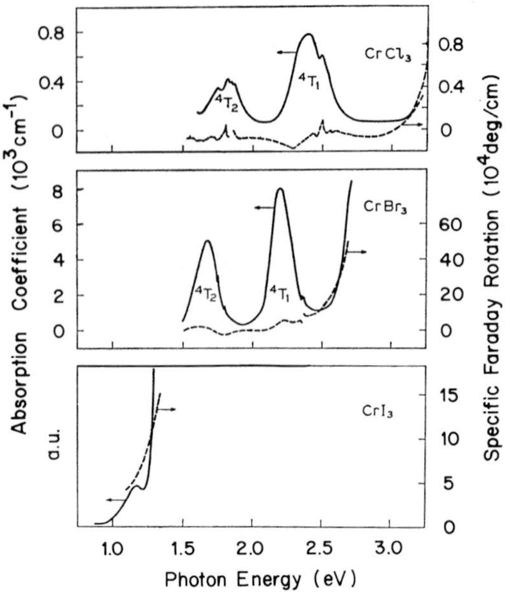

Figure 2-18. Absorption coefficient and Faraday rotation per unit length in the weakly absorbing region of $CrCl_3$, $CrBr_3$, and CrI_3 at 1.5 K and in a saturating field (after Dillon et al., 1966).

plates of the three halides by Dillon et al. (1966). Figure 2-18 reproduces these data. The energy range on the high-energy side is limited by large absorption marking the onset of interband absorption. This edge shifts from above 3 eV in $CrCl_3$ to 1.3 eV in CrI_3. At the edge, the Faraday rotation for magnetic saturation reaches values of the order $5 \times 10^5\ deg\ cm^{-1}$ ($8.7 \times 10^5\ rad\ m^{-1}$) in $CrBr_3$, and it is expected to increase further at higher energies. Below the interband absorption edge, Dillon et al. observed two weak transitions in the chloride and bromide, which they assigned to intra $3d^3$ transitions of Cr^{3+}. The chromium ions are surrounded by a nearly regular octahedron of halogen ions. The leading term in the expansion of the crystal field is therefore cubic and the following term is trigonal. The fit of the experimental spectra with the crystal-field model leads to the assignment of the two peaks to transitions

$^4A_2 \rightarrow {}^4T_2$ and $^4A_2 \rightarrow {}^4T_1$. More recent data by Pedroli et al. (1975) confirm these assignments. Dillon et al. (1966) also developed a molecular-orbital model to interpret the observed circular dichroism at the absorption edge of $CrBr_3$ that they attribute to a charge-transfer transition from a valence band orbital to an e_g^* orbital. Figure 2-19 shows the polar Kerr rotation at 1.5 K of $CrBr_3$ as measured in the fundamental absorption region by Jung (1965). The maximum rotation amounts to 3.5° at 2.9 eV with a second negative extremum of −2.5° at 3.3 eV. The former peak is assigned to the transfer of an electron from one of the 4pπ-type orbitals of Br^- into 3d Cr^{3+} orbitals, while the latter peak is thought to correspond to the transfer from 4pσ-type orbitals into the same Cr final states. Borghesi et al. (1981) have performed thermoreflectance measurements on $CrBr_3$ in the energy range 2.5 to 4 eV. They found that the dominant modulation mechanism is electron-phonon interaction in the paramagnetic temperature range and exchange interaction in the ferromagnetic phase.

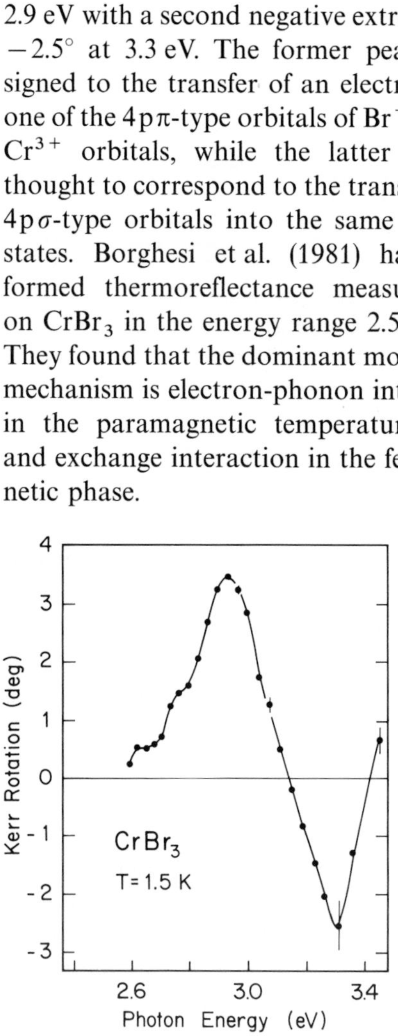

Figure 2-19. Polar Kerr rotation of magnetically saturated $CrBr_3$ in the fundamental absorption region (after Jung, 1965).

CrO_2 is a technologically very important material, serving in the form of needle shaped powders for magnetic recording. It crystallizes in a tetragonal structure of the rutile (TiO_2) type and orders ferromagnetically at 390 K. Chamberland (1977) has written a comprehensive review on the preparation and the physical and chemical properties of CrO_2. As he notes single crystals exceeding a length of 10 to 70 μm can only be grown at extreme pressure and high temperature. This has limited the number of investigations and the agreement between the reported results, leaving many open questions regarding the electronic structure of this compound. Chase (1974) has measured the optical reflectivity from 0.1 to 6 eV on oriented films of CrO_2 grown on rutile substrates. He found a metallic behavior with $\hbar\omega_p \approx 2$ eV. Assuming one free carrier per Cr, this decoupled plasma frequency gives $m^* = 10\,m$. Resistivity measurements on single crystals showed values as low as $1.4 \times 10^{-6}\,\Omega$m at room temperature (Chamberland), corroborating the presence of a substantial carrier concentration at E_F. On the other hand, magnetization measurements (Siratori and Iida, 1960; Flippen, 1963) indicate an ordered moment of $2\,\mu_B$ which points to a localized character of the 3d electrons.

One approach to this problem uses the model of Goodenough (1965, 1971), which is displayed in Fig. 2-20. It shows on the left side, from bottom to top, the 3d, 4s, and 4p states of Cr^{4+}, and on the right side the 2s 2p states of O^{2-}. The octahedral field splits the 3d states into lower-lying t_{2g} and higher-lying e_g states. Because the structure is tetragonal, the t_{2g} states are further split by the orthorhombic component of the crystalline field into a $t_{2g\parallel}$ and two quasi-degenerate $t_{2g\perp}$ orbitals. The middle column of Fig. 2-20 shows the states which result from combining these

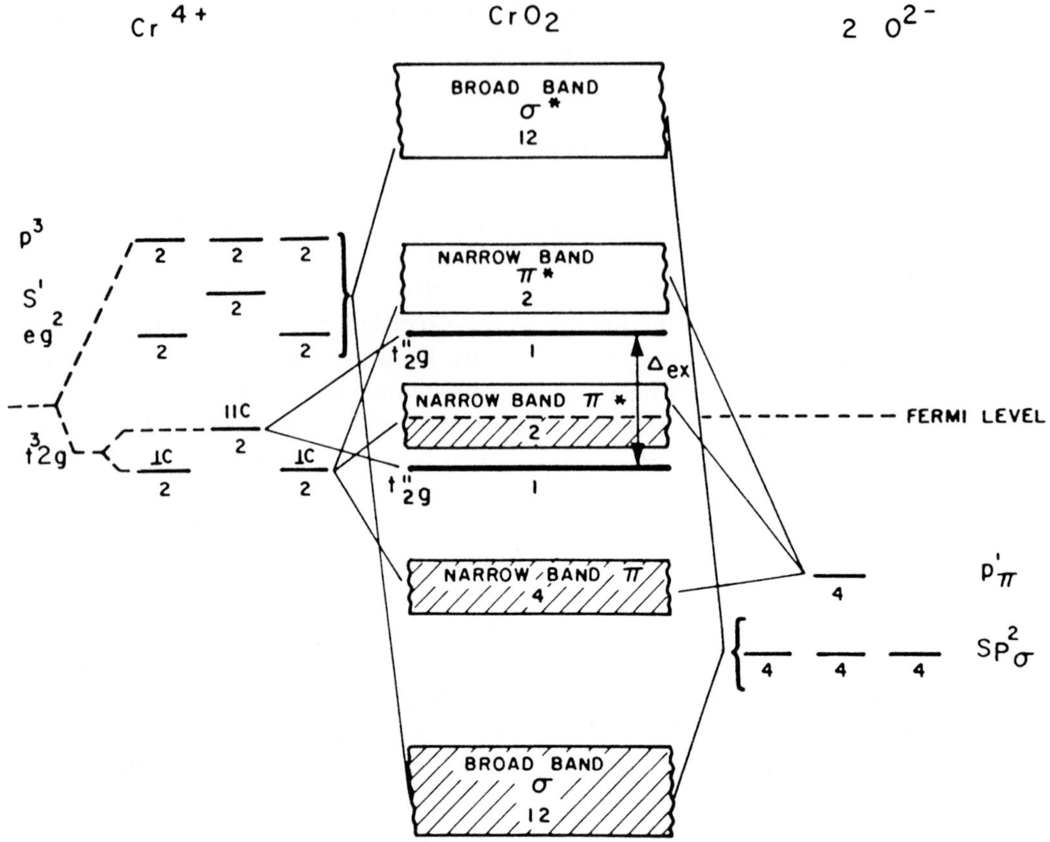

Figure 2-20. Energy level scheme for CrO_2 (after Goodenough, 1965; and Chamberland, 1977).

atomic orbitals to molecular orbitals after an exchange splitting of the d states Δ_{ex} has been added. The 18 molecular electrons (6 from Cr and 12 from O_2) are now filled into the bands from the bottom. Near the Fermi energy the diagram shows a $dt_{2g\perp}$ derived π^* band which is spin polarized and half filled. The second d electron occupies the localized $t_{2g\parallel}$ state just below the narrow π^* band. Thus the result of this scheme is a metal with a narrow band at E_F contributing 1 μ_B and an other localized state contributing the second Bohr magneton to the ordered moment.

More recently, Schwarz (1986) has performed a self-consistent spin-polarized band structure calculation predicting CrO_2 as a half-magnetic ferromagnet. The term half-magnetic ferromagnet was introduced first by de Groot et al. (1983) for NiMnSb. For both materials the band structure calculations predict densities of states which are metallic for majority-spin electrons and semiconducting for minority-spin electrons. In an attempt to verify this prediction for CrO_2, Kämper et al. (1987) have performed a spin-resolved photoemission study on polycrystalline CrO_2 films. They found a spin polarization of nearly 100% near 2 eV below the Fermi energy, but because of a very low intensity closer to E_F, they could not determine the spin polarization from E_F to $E_F - 2$ eV. Despite the metallic resistivity, this low emission intensity is taken as evidence that the magnetism is not of the Stoner-Wohlfarth band type but of a localized nature. Clearly, more experimental data are needed to solve this puzzling problem.

Magneto-optical data have apparently only been published by Stoffel as early as 1969. Stoffel (1969) determined with a vacuum ellipsometer the optical and magneto-optical constants $N = i n k$ and $Q = Q_0 \, e^{-iq}$, respectively, in the wavelength range from 400 to 1000 nm. Figure 2-21 shows a transformation of these spectra in polar Kerr-rotation and -ellipticity spectra (Brändle, 1990a). A negative extremum can be seen in the ellipticity spectrum near 1.4 eV reaching 0.27°. At this energy the rotation is strongly energy dependent and it is expected that it peaks near 1 eV. A thorough study (Brändle, 1990b) of the prefactors A and B defined in Eqs. (2-44) and (2-45) shows that the extrema in Fig. 2-21 are enhanced by the optical constants, but that

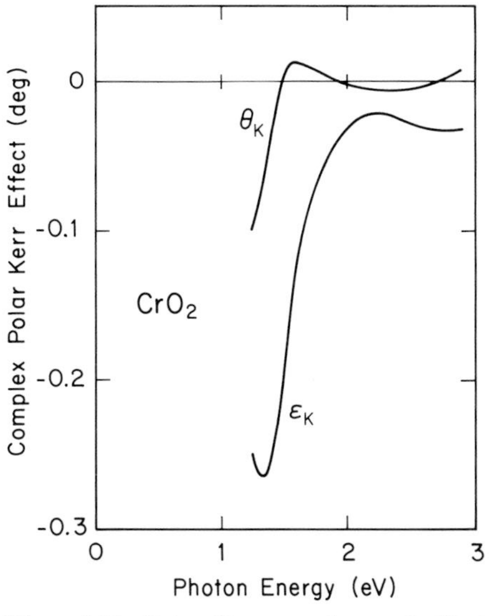

Figure 2-21. Polar Kerr rotation and ellipticity of CrO_2 as computed from the magneto-optical constants determined by Stoffel (1969). The temperature and the field were not given, but presumably the sample is saturated (after Brändle, 1990a).

the energy position corresponds approximately to those in σ_{1xy}, σ_{2xy}, and σ_{1xx} and therefore are related to interband transitions.

Chromium also forms the oxide Cr_2O_3, which is a spin 3/2 antiferromagnet below 307 K. Krichevtsov et al. (1988) have studied *nonreciprocal effects* induced by the application of electric and magnetic fields to oriented crystals. For magnetic fields, the nonreciprocal effect studied corresponds to the Faraday effect. The signal is quite small, reaching $\approx 0.05°$/kA at the lowest temperatures. From a comparison with susceptibility measurements, the authors conclude that the observed rotation reflects the competition of a positive diamagnetic and a negative paramagnetic contribution.

Among the ternary compounds containing magnetic transition elements, four large classes have attracted special attention for their magneto-optical properties. These comprise spinels, garnets, orthoferrites, and Heusler alloys. *Spinels* have the general formula $M_A^{2+}(M_B^{3+})_2X_4$, where M_A^{2+} and M_B^{3+} are di- and trivalent cations, respectively, and X stands for a chalcogen O, S, Se or Te. In normal spinels the divalent cations M_A occupy the tetrahedral holes (A-sites) and the trivalent cations M_B occupy octahedral holes (B-sites) of the cubic close packing of the X anions. Besides these normal spinels, *inverse spinels* also exist, where the cation M_A has exchanged its position with one of the two M_B cations. The most famous representative for an inverse spinel is magnetite $Fe_3O_4 \equiv Fe^{2+}(Fe^{3+})_2O_4$. Von Philipsborn and Treitinger (1980) have published a comprehensive structure table showing which combinations of elements M_A, M_B, and X crystallize in which structure type. Chromium occurs only in the role of the cation M_B, while the cation M_A

can be Mg, Mn, Fe, Co, Ni, Cu, Zn, Cd, and Hg. Not all chalcogens combine with these cations to form spinels. Thus for the combination magnesium–chromium only the oxide $MgCr_2O_4$ exists, and the tellurides are also rare. The magneto-optical investigations of the chromium spinels have concentrated on those compounds which are ferromagnetic and semiconducting, i.e., cadmium and mercury sulfides and selenides, and on those metallic compounds which have ferromagnetic ordering temperatures above room temperature, i.e., $CuCr_2S_4$ ($T_C = 377$ K), $CuCr_2Se_4$ ($T_C = 430$ K), and $CuCr_2Te_4$ ($T_C \approx 350$ K). A third group, which we will discuss later, are spinels containing Co in addition to Cr, because their magneto-optical signals depend strongly on the former element.

Ferromagnetic semiconductors attracted large interest in the 1960s and 1970s, and their physical properties have been reviewed by Methfessel and Mattis (1968), Haas (1970), Wachter (1972), Kasuya (1972), Nagaev (1975), Nolting (1979), to cite only a few. The interest in these materials is due to the strong influence of magnetic ordering on electrical and optical properties. As an example, Fig. 2-22 displays the optical density versus photon energy for various temperatures above and below $T_C = 84.5$ K of a single crystal sample of $CdCr_2S_4$ (Harbeke and Pinch, 1966). With decreasing temperatures a shift of the absorption edge to higher photon energy is observed. This shift is largest in the neighborhood of T_C and is accompanied also with a change of the edge shape in this temperature range. $CdCr_2Se_4$ shows an exchange-induced shift of the absorption edge to lower photon energy (Harbeke and Pinch, 1966; Busch et al., 1966) similar to the effect in the europium chalcogenides to be discussed in Sec. 6 of this chapter. The Faraday rotation and the polar magneto-optical Kerr effect of $CdCr_2S_4$ have been investigated by Wittekoek and Rinzema (1971). Figure 2-23 displays Kerr rotation and ellipticity spectra taken in a field of 0.5 T which saturates the magnetization. For the discussion of the spectra, Wittekoek and Rinzema (1971) divide the spectra in three energy ranges: a region from 680 to 520 nm, where at least three sharp transitions labeled 1, 2, and 3 occur, then a region of little magneto-optical activity extending from 520 to 420 nm, and finally a strong and broad transition centered at 360 nm and labeled 4 in Fig. 2-23. Transition 1 shows a red-shift with decreasing temperature, contrary to the absorption edge observed in the absorption measurements on $CdCr_2S_4$ but in agreement with the general behavior of $CdCr_2Se_4$. This led to the conclusion that the absorption edge observed in Fig. 2-22 is due to a different

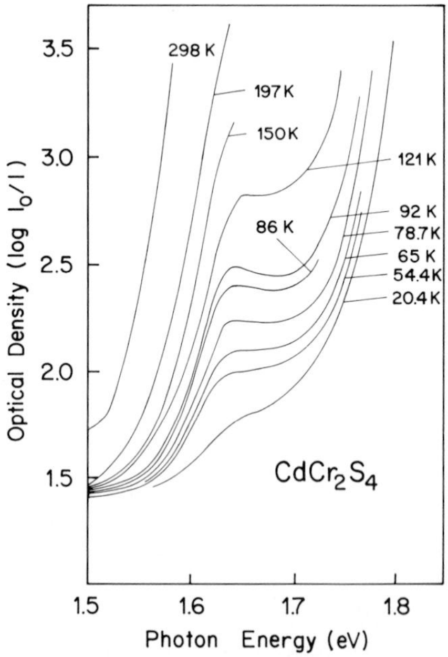

Figure 2-22. Energy dependence of the optical density of a $CdCr_2S_4$ crystal for various temperatures above and below $T_C = 84.5$ K (after Harbeke and Pinch, 1966).

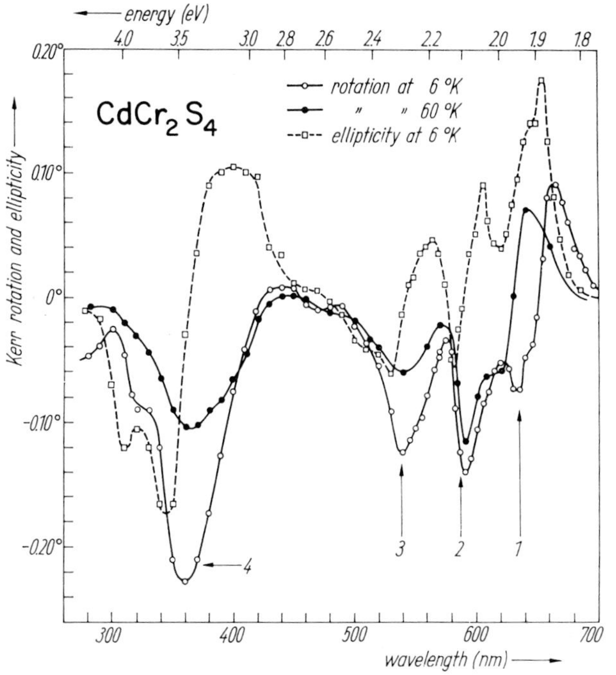

Figure 2-23. Kerr rotation and ellipticity of $CdCr_2S_4$ in a field of 0.5 T (after Wittekoek and Rinzema, 1971).

transition than the absorption edge in the other chalcogenides, probably a weak crystal field transition of the Cr^{3+} ion. The interpretation of peak 1 in Fig. 2-23 is still being debated. Lehmann et al. (1971) proposed a transition to an F-center-like state where a sulfur vacancy is filled with a chromium 4s electron. Goodenough (1969) proposed a transition from the exchange-split spin-down valence band to the low-spin $^3T_{1g}$ state of Cr^{2+}. Haas (1970) favored a model in which the conduction band formed predominantly by 4s orbitals of Cr^{3+} is lower in energy than the d-states of Cr^{2+} and the red-shifting transition takes place between the top of the valence band and this conduction band. Peaks 2 and 3 in Fig. 2-23 are assigned to crystal field transitions $^4A_2 \rightarrow {}^2T_2$ and $^4A_2 \rightarrow {}^4T_1$, respectively (Wittekoek and Rinzema, 1971). The broad and intense peak 4, finally, is attributed to a charge-transfer transition from a sulfur nonbonding orbital to an antibonding e_g^* state from chromium.

The magneto-optical polar Kerr effect of $CdCr_2Se_4$ has been reported by Bongers et al. (1969). Figure 2-24 shows the energy dependence of the rotation at 4, 120, and 140 K and of the ellipticity at 80 K. The Curie temperature of $CdCr_2Se_4$ is 130 K. The maximum rotation is more than a factor of 4 larger than in $CdCr_2S_4$, which has been attributed to the larger spin-orbit splitting of the Se-4p states compared to the S-3p states, in support of the assignment $p \rightarrow de_g$ (Wittekoek and Rinzema, 1971). On the other hand, Bongers et al. (1969) assign to this dominant peak a transition from the Se-4p valence band into Cr-4s states forming the conduction band. The structures marked 2 and 3 in Fig. 2-24 have been tentatively assigned by the same authors to transitions 3d (Cr^{3+}) → 4s and 4p (Se) → 3d (Cr^{2+}), respectively. The quantity $\Delta R/R = (R^+ - R^-)/[(R^+ + R^-)/2]$,

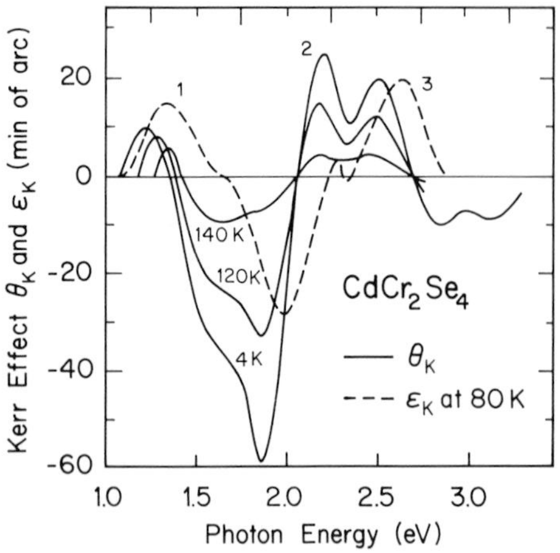

Figure 2-24. Kerr rotation and ellipticity of $CdCr_2Se_4$. The field was not indicated, but should be of the order of 1 T (after Bongers et al., 1969).

called *reflectance magneto-circular dichroism*, was studied in detail by Sato and Teranishi (1983) for $CdCr_2Se_4$ single crystals and by Ahrenkiel et al. (1971) for both $CdCr_2S_4$ and $CdCr_2Se_4$. Zvara et al. (1979) measured the optical reflectivity from 0.5 to 12 eV and discussed the results in terms of a semiempirical ionic model. The absorption of polarized light in thin films was investigated by Edelman et al. (1983), and an interpretation in terms of transitions $Cr^{3+} + e^- \rightarrow Cr^{2+}$ and intra Cr^{3+} crystal-field transitions was offered. The band-structure calculation of Kambara et al. (1980) was used by Sanford and Nettel (1981) to relate the photoferromagnetism of $CdCr_2Se_4$ and the mechanism of its ground-state ferromagnetic coupling.

$HgCr_2S_4$ and $HgCr_2Se_4$ are two additional semiconducting chromium spinels showing ferromagnetic ordering below 36 and 106 K, respectively. Harbeke et al. (1968) have studied the shift of the absorption edge as a function of temperature on powdered as well as on monocrystalline samples of $HgCr_2S_4$. They observed (Fig. 2-25) with 0.36 eV the largest red-shift of any magnetic semiconductor and noted that this shift extends from the paramagnetic to the ferromagnetic and to the metamagnetic phase occurring below 25 K. Faraday rotation measurements have mostly concentrated on the weakly absorbing energy range below the absorption edge of the insulating chromium spinels, but measurements at higher energies have also been performed on thin films of $CdCr_2Se_4$ (Edelman and Dustmuradov, 1980). In the infrared, below the absorption edge, generally two contributions exist: a wavelength-independent term caused by ferromagnetic resonance and a component varying as $(\lambda - \lambda_0)^{-2}$ due to the electronic transitions above the absorption edge. While these two components compensate largely in $CdCr_2Se_4$ (Wittekoek and Rinzema, 1971), a specific Faraday rotation of 1000° cm^{-1}

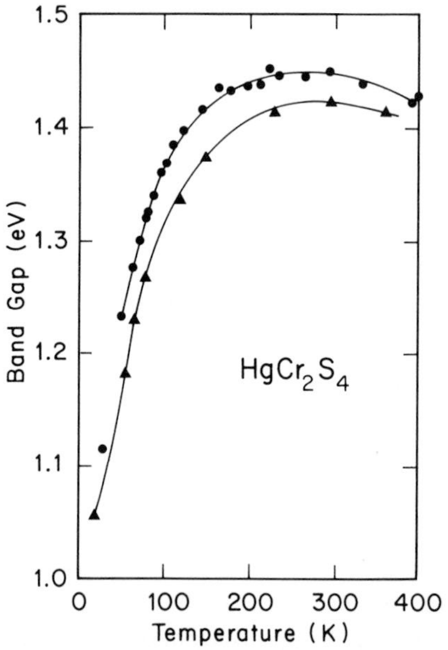

Figure 2-25. Shift of the absorption edge of $HgCr_2S_4$ as a function of temperature. The triangles are from absorption measurements on thin monocrystalline plates and the dots are from diffuse reflection measurements on powdered samples (after Harbeke et al., 1968).

is observed in $HgCr_2Se_4$ at the CO_2 laser line of 10.6 μm (Lee et al., 1971).

The zinc-chromium spinels order antiferromagnetically. Nevertheless, a substantial red-shift of the absorption edge has been found in $ZnCr_2Se_4$ (Busch et al., 1966; Lehmann et al., 1971). The spin structure of this compound is helical for $T < T_C = 20$ K and fields below a critical field amounting to 6.4 T at 4.2 K (Plumier, 1966).

Among the chromium spinels the copper-chromium spinels with the chalcogens S, Se, and Te play a specific role. While $CuCr_2O_4$ – like many other chromium spinels and particularly most oxides – is an insulator which orders antiferromagnetically at $T_N = 135$ K, $CuCr_2S_4$, $CuCr_2Se_4$, and $CuCr_2Te_4$ are metallic and order ferromagnetically with Curie temperatures well above room temperature. This metallic character and the magnetic moment of approximately 5 μ_B per formula unit, i.e., for two Cr atoms, has prompted various models for the electronic structure of these materials. A first model was introduced by Lotgering (1965), who made the assumption that Cu occurs as a monovalent, diamagnetic ion, and that one Cr ion is trivalent and the other is tretravalent. The ferromagnetism and conduction were attributed to double exchange between the Cr^{3+} and the Cr^{4+} ions and the magnetic moment of 5 μ_B/f.u. was explained with a parallel alignment of the Cr^{3+} and Cr^{4+} spin moment. The model of Goodenough (1967a) assumes that chromium occurs only as Cr^{3+}, but that the covalent mixing between the copper d states and the selenium p and s states is sufficiently strong to form bands. The Fermi level is assumed to fall in a t_{2g}^* subband which accommodates on hole per formula unit. Thus, Cu is formally divalent. Since neutron scattering experiments (Robbins et al., 1967; Colominas, 1967) neither showed two chromium sublattices with different magnetic moments, nor a moment on the Cu site, these early models were modified in the years that followed. Lotgering and Stapele (1968) restricted the validity of Lotgering's model to temperatures above T_C and introduced for $T < T_C$ a temperature-dependent polarization of the valence band which partially compensates the moment from the Cr ions. This allows a reduction of the assumed amount of Cr^{4+} to conform with a moment of 5.5 ± 0.3 μ_B/f.u. at the Cr sites (Colominas, 1967) and a net moment of ≈ 5 μ_B/f.u. Goodenough (1967b) introduced an overlap of the Cu t_{2g}^*, with the top of the valence band allowing more itineracy to account for the absence of a magnetic moment at the Cu site. Band-structure calculations by Ogata et al. (1982) corroborate Goodenough's (1967b) model, but open questions remain in regard to the small number of hole-like carriers derived from Hall effect measurements (Lotgering, 1965; Valiev et al., 1972; Koroleva and Shalimova, 1979; Tsurkan et al., 1985; Groń et al., 1990).

The optical and magneto-optical properties of $CuCr_2Se_4$ and $CuCr_2Se_{3.7}Br_{0.3}$ have been measured by Brändle et al. (1990a, 1991a). Figure 2-26 shows the reflectivity as a function of photon energy for these two compounds (Brändle et al., 1991a). Metallic behavior with rather steep plasma edges is observable, shifting to lower energies on substitution of Se by Br. This indicates a reduction of the free carriers in agreement with hole conduction. The inset of Fig. 2-26 displays the prefactors which relate the Kerr rotation and ellipticity to the off-diagonal conductivity terms σ_{1xy} and σ_{2xy} as expressed in Eqs. (2-42) and (2-43). It can be seen that these prefactors peak near 1 eV, allowing some enhancement of the Kerr signal to be ex-

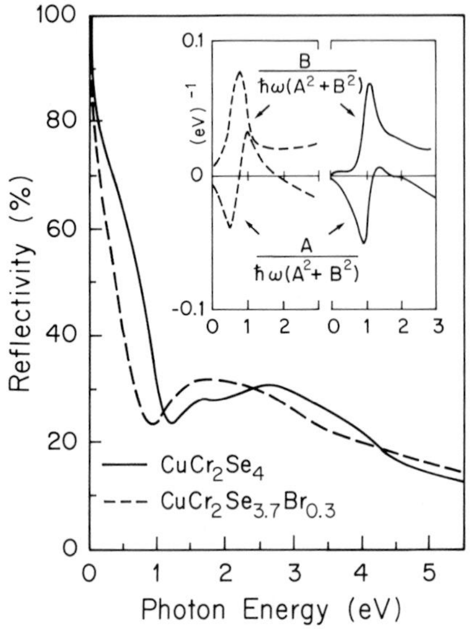

Figure 2-26. Near-normal incidence reflectivity of single crystals of $CuCr_2Se_4$ and $CuCr_2Se_{3.7}Br_{0.3}$ at room temperature. The inset shows for the two materials the energy dependence of the prefactors relating the Kerr effect with the off-diagonal conductivity (after Brändle et al., 1991 a).

pected in this energy range. Figure 2-27 displays the polar Kerr rotation and ellipticity as measured at room temperature in a field of 2 T (Brändle et al., 1991 a). The anticipated peaks do indeed occur, reaching values of nearly 1.2°. These are among the highest room-temperature values in this spectral range for any material. The computation of the off-diagonal conductivity (Fig. 2-28) evidences that the Kerr signal is not an effect of the prefactors alone (i.e., the optical constants), but that a magneto-optically active transition exists in this spectral range. From a comparison of the peak energies in the diagonal and off-diagonal conductivity, it can be concluded that in $CuCr_2Se_4$ the transition at 0.8 eV is of paramagnetic line shape, and the transitions at 1.5 and 2.6 eV of diamagnetic line shape. In $CuCr_2Se_4$ the transition at 0.8 eV acquires a diamagnetic contribution. The strength of this transition indicates the participation of highly spin-polarized states like the Cr^{3+} 3d states.

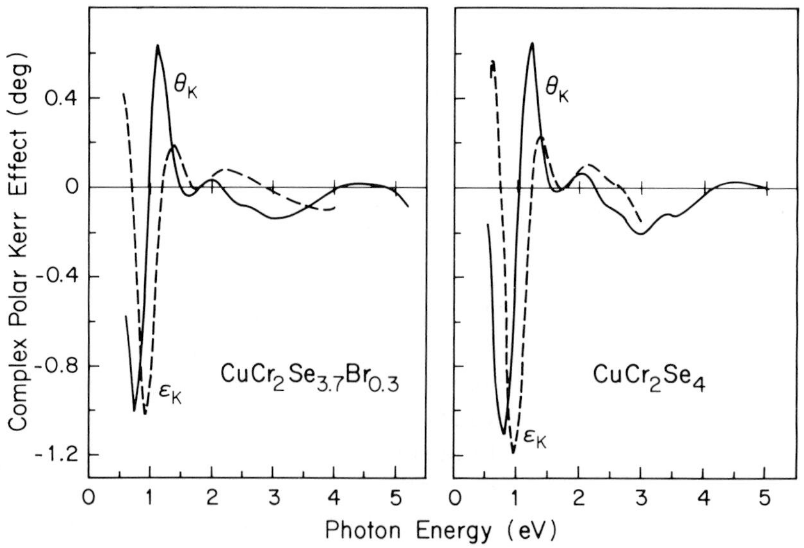

Figure 2-27. Kerr rotation θ_K and ellipticity ε_K of single crystals of $CuCr_2Se_{3.7}Br_{0.3}$ (left side) and $CuCr_2Se_4$ (right side) at room temperature and a field of 2 T (after Brändle et al., 1991 a).

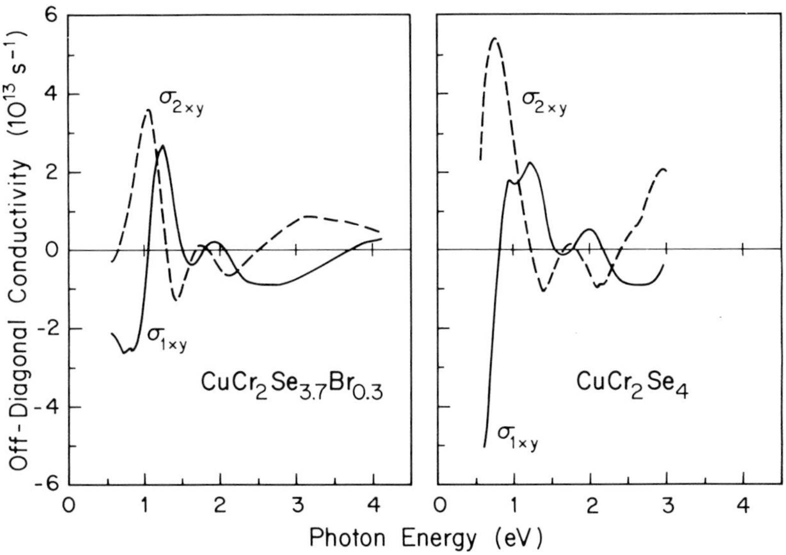

Figure 2-28. Off-diagonal conductivity of single crystals of $CuCr_2Se_{3.7}Br_{0.3}$ (left) and $CuCr_2Se_4$ (right), as computed from the spectra in Fig. 2-27 and the optical constants derived from the reflectivity spectra in Fig. 2-26 (after Brändle et al., 1991 a).

The sensitivity to the valence band speaks for the participation of the states near E_F. Thus the peak at 0.8 eV most probably corresponds to a charge transfer transition from the Cr^{3+} 3d($^4A_{2g}$) ground state into empty valence states just above E_F. For the diamagnetic transitions at 1.5 and 2.6 eV, Brändle (1990b) considers orbital promotion processes as have been used before by Zhang et al. (1983) to interpret magneto-optical data on Fe_3O_4 and Mg- and Li-substituted Fe_3O_4. Figure 2-29 displays the coupling scheme for these $3d^3 \rightarrow 3d^2 4s4p$ transitions. The ground states of Cr^{3+} in a cubic crystal field is the singlet $^4A_{2g}$. This term has $M_L = 0$ and therefore is magneto-optically inactive. The $3d^2(^3F)$ states, on the other hand, are split by the same crystal field into two triplets $^3T_{1g}$ and $^3T_{2g}$ and one singlet $^3A_{2g}$. Again $M_L = 0$ for the singlet, while the orbital moment of the triplets has been reduced by the crystal field to $M_L = 1$. Thus, two transitions with diamagnetic line shape, corresponding to

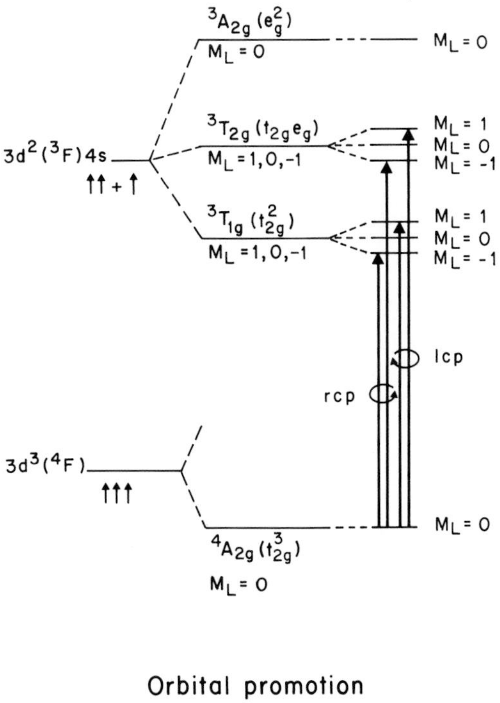

Figure 2-29. Coupling scheme for $3d^3 \rightarrow 3d^2(4s\,4p)$ transitions in copper-chromium spinels (after Brändle, 1990 b).

$\Delta M_L = \pm 1$, are expected. The coupling with one 4s or 4p electron in the final states allow for electron and spin conservation and enhances the oscillator strength by several orders of magnitude compared to intra $3d^3$ transitions.

The Faraday rotation and the magnetic circular dichroism of $CuCr_2Se_4$ have been measured by Edelman and Dustmuradov (1980) on the films. Figure 2-30 shows their Faraday rotation spectrum together with a calculation of the Faraday rotation and ellipticity from the Kerr spectra and the optical constants (Brändle, 1990 b).

A last class of Cr compounds, to be mentioned briefly, are ferromagnetic insulators containing formally divalent chromium. They have the general formula A_2CrX_4, where A is either an organic or an inorganic unipositive cation and X is a halide ion. The ordering temperature is in the range 50–60 K. Day (1979) has thoroughly investigated the compounds where A is K or Rb. He found strong changes in transmission color on cooling the materials from room temperature to liquid He temperature, due to the suppression of several crystal-field transitions. Thus, a dark olive green crystal of K_2CrCl_4 becomes pale yellow at low temperatures.

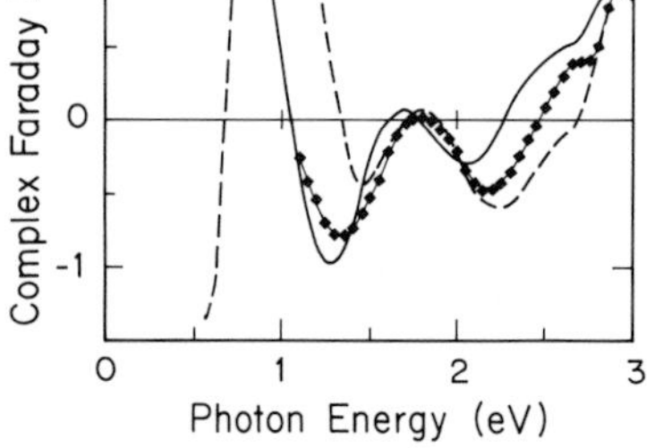

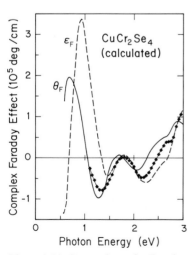

Figure 2-30. Comparison of a Faraday rotation spectrum measured on thin films of $CuCr_2Se_4$ (squares) (Edelman and Dustmuradov, 1980), with a spectrum computed from the Kerr data obtained on single crystals (full line). Also shown is the computed Faraday ellipticity (dashed line) (after Brändle, 1990 b).

2.5.3.2 Mn Compounds

Manganese forms with the pnictogens As, Sb, and Bi hexagonal 1:1 compounds with the NiAs structure. They all order ferromagnetically with ordering temperatures increasing from about room temperature in MnAs ($T_C = 305$ to 318 K) to ≈ 555 K in MnSb, to 633 K in MnBi. The optical and magneto-optical properties of MnAs films were investigated by Stoffel and Schneider (1970) in view of applications in beam-addressable memories. They measured the change in the state of polarization of light after reflection from the magnetized film surface at an angle of incidence of 60° and calculated from this change the optical constants, the longitudinal Kerr effect, the Faraday effect, and the off-diagonal element of the dielectric tensor. In the measured spectral range from 0.48 to 0.92 μm, the reflectivity increases from 0.33 to 0.59, indicating metallic character. The absolute value of the *complex Faraday effect*, defined as $|\phi_F| = (\theta_F + \varepsilon_F)^{1/2}$ (probably at room temperature and not magnetically saturated), is reproduced in Fig. 2-31. It peaks at 0.8 μm with a maximum value of 7.5×10^4 deg cm^{-1} ($\equiv 1.3 \times 10^5$ rad m^{-1}). Figure 2-32 displays the polar Kerr rotation and the ellipticity of MnAs computed from the data of Stoffel and Schneider. Both θ_K and ε_K show extrema in the neighborhood of the peak in the Faraday rotation, i.e., near 1.5 eV. The polar Kerr rotation at the peak amounts to $-0.14°$ compared with 0.06° for the longitudinal Kerr effect. For this longitudinal Kerr effect

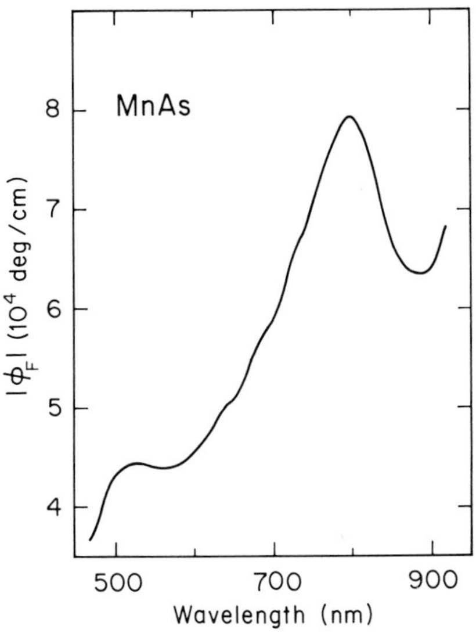

Figure 2-31. The absolute value of the complex Faraday effect as measured on MnAs films (probably at room temperature and not saturated; $T_C = 305$ to 318 K) (after Stoffel and Schneider, 1970).

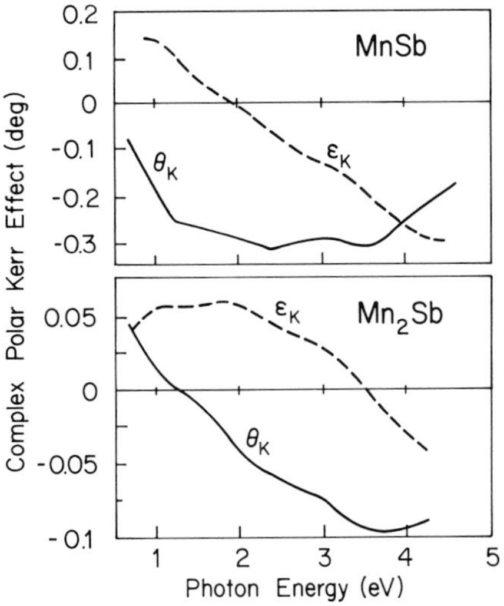

Figure 2-33. Room-temperature Kerr rotation (full line) and ellipticity (dashed line) of polycrystalline samples of MnSb (top) and Mn_2Sb (bottom). The field is 1.6 T (after Buschow et al., 1983).

the extremum occurs for the ellipticity at 1.72 eV and reaches $-0.15°$.

The absorption and Faraday rotation of MnSb has been studied by Sawatzky and Street (1971). They report a weakly de-

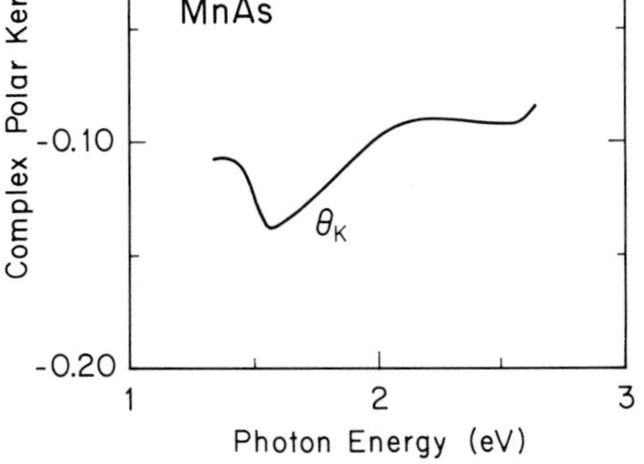

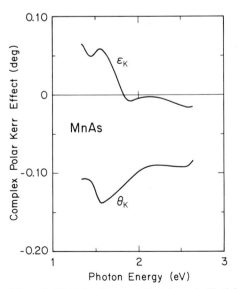

Figure 2-32. Polar Kerr rotation θ_K and ellipticity ε_K of MnAs as computed from the spectra in Fig. 2-31.

creasing absorption coefficient with energy decreasing from 6 to 1 eV. The Faraday rotation at room temperature varies in the same spectral range between 2.1×10^5 and 2.8×10^5 deg cm^{-1} ($\equiv 3.7 \times 10^5$ to 4.9×10^5 rad m^{-1}) with flat maxima near 1.3 and 2.5 eV. The observation of a Faraday signal which is roughly 4 times as large as in MnAs most probably reflects the fact that at room temperature the magnetic field of an electromagnet is sufficiently high to saturate a material with a T_C of 550 K, but not with a T_C of 310 K. Sawatzky and Street noted an instability of MnSb against thermal cycling although no first-order phase transition occurs in the temperature range of interest for thermomagnetic writing. The room-temperature polar Kerr effect of MnSb has been studied by Buschow et al. (1983). Figure 2-33 shows the results of this investigation together with data for tetragonal Mn_2Sb. The largest absolute value of the rotation in MnSb amounts to 0.3°, which is larger than for MnAs by only a factor of 2. Note, however, that the energy range of large rotation is much wider in MnSb, so that the ratio of the integrated spectra should come closer to the ratio of the magnetization at room temperature. Although no attempt has been made in any of the cited investigations to interpret the spectra, it appears that the broadening of the spectrum on going from MnAs to MnSb corresponds to the broadening of the valence bands in a series of pnictides of chalcogenides (Schoenes, 1984). Various authors have investigated the Kerr rotation of Mn–Sb compounds with a variable Mn/Sb ratio. Yoshioka et al. (1988) found for the composition $Mn_{54}Sb_{46}$ a rotation spectrum which was shifted to higher values by about 0.05° as compared with stoichiometric MnSb. Carey et al. (1990) reported maximum Kerr rotations at $\lambda = 633$ nm for Mn contents ranging from 45 to 50 at.% depending on whether the film was deposited on glass or an Mn or Sb sublayer. A crucial point for applications is the existence of perpendicular magnetic anisotropy. While MnSb films have generally an in-plane anisotropy, the study of Mn/Sb multilayers (Yoshioka et al., 1988) and of sublayers (Carey et al., 1990) has shown that perpendicular anisotropy can be achieved. The small rotation of Mn_2Sb (Fig. 2-33) can be accounted for by the magnetization which amounts to less than 1/3 of that of MnSb.

MnBi is the Mn pnictide which has received most attention since its favorable magneto-optical properties were recognized (Williams et al., 1957; Mayer, 1958, 1960; Adachi, 1961; Chen and Gondo, 1964; Chen et al., 1968, 1970; Unger and Räth, 1971; Feldtkeller, 1972; Atkinson and Lissberger, 1974; Egashira and Yamada, 1974; Wang, 1990). Figure 2-34 shows the Kerr rotation, the Faraday rotation, and the reflectivity for MnBi films prepared by successive vapor-deposition of Bi and Mn layers and in situ annealing at 300 °C (Egashira and Yamada, 1974). From 0.45 to 0.65 μm the Kerr rotation varies approximately linearly from −0.55 to −0.73°. The Faraday rotation and the reflectivity also increase with increasing wavelength. Since the thickness of the film

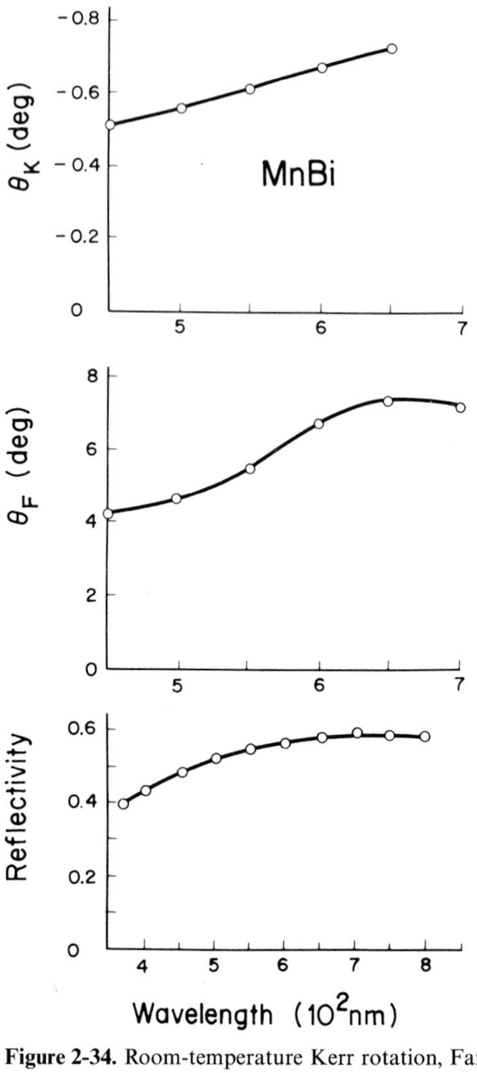

Figure 2-34. Room-temperature Kerr rotation, Faraday rotation, and reflectivity for MnBi films prepared by successive vapor-deposition of Bi and Mn layers and subsequent annealing at 300 °C (after Egashira and Yamada, 1974).

is not given, the specific Faraday rotation can not be derived. However, this quantity has been determined, for instance, by Chen et al. (1968) for MnBi films deposited on freshly cleaved mica (Fig. 2-35). Mica as a film substrate was thought to be necessary to allow for oriented growth with the hexagonal *c*-axis perpendicular to the substrate. But later Feldtkeller (1972) and Egashira and Yamada (1974) succeeded in growing *c*-axis-oriented films also on glass substrates. Since the *c*-axis is the axis of easy magnetization, these films have the desired perpendicular anisotropy. A severe problem for the use of MnBi for thermomagnetic writing is the existence of a crystallographic phase transition above 628 K (Andresen et al., 1967; Chen and Stutius, 1974). The low-temperature unit cell of MnBi possesses two vacant interstitial sites. At high temperatures a phase decomposition of MnBi into $Mn_{1.08}Bi$ and free Bi occurs, and 10–15% of the Mn atoms occupy these interstitial sites. Chen et al. (1970) have prepared films of MnBi in the high-temperature phase by quenching the normal films from above 360 °C. Compared to the low-temperature (normal) phase films, they observed for the quenched films a reduction of the figure of merit – defined for a Faraday rotator as twice the specific Faraday rotation divided by the absorption coefficient – by a factor of up to 3, depending on the wavelength. Although Chen et al. found only a reduction in the magnetization of 26% for their films, measurements by Heikes (1955) revealed a reduction of the magnetic moment per Mn atom from 3.95 to 1.75 μ_B when MnBi is quenched. Chen et al. (1970) also studied the possibility of using the high-temperature phase for thermomagnetic writing. While the lower T_C of this phase is more favorable, they noted that the quenched film gradually annealed back to the low-temperature phase.

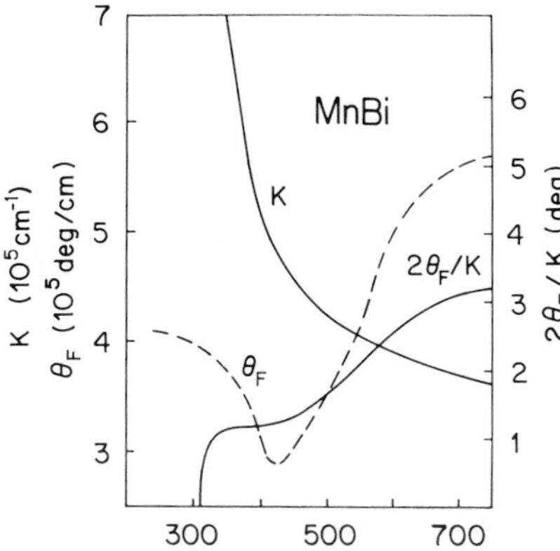

Figure 2-35. Absorption, Faraday rotation, and figure of merit for MnBi films deposited on mica (after Chen et al., 1968).

Many attempts have been made to improve the properties of MnBi by varying the stoichiometry or by substitution. Egashira and Yamada (1974) found that stoichiometric MnBi has the largest Kerr rotation. Unger et al. (1972) and Egashira et al. (1977) reported that the substitution of Mn with Ti reduces the transformation rate from the metastable high-temperature phase to the stable low-temperature phase. Katsui (1976a) and Shibukawa (1977) studied Mn–Cu–Bi films to take advantage of the reduced Curie temperature of 180 °C. They note a rather good chemical stability and a Kerr rotation of 0.4°, which can be enhanced by a coating with a dielectric. Katsui et al. (1976) investigated $Mn_x(Cu_{1-y}Ni_y)_zBi_x$, where $x+y+z=1$. The Curie temperature varied between 100 and 200 °C and a maximum figure of merit $2F/\alpha = 1.1°$ compared to ≈ 3 for the low-temperature phase of MnBi was achieved. The Mn–Cu–Bi-type films generally have an f.c.c. structure, but internal stress appears to induce uniaxial anisotropy. Katsui (1976b) improved the rectangularity of the hysteresis loop by substituting up to 30 at.% of Cu with Dy, which is known to have a large single-ion anisotropy. Crystalline films with f.c.c. structure and compositions $Mn_6Cu_4Bi_4$ (probably $Mn_3Cu_4Bi_4$ + Mn-oxide) and $Mn_5Ni_2Bi_4$ have been prepared by Gomi et al. (1984) by sputtering and subsequent annealing. These authors found large Faraday rotations at $\lambda = 633$ nm of 2.2×10^5 deg cm^{-1} ($\equiv 3.85 \times 10^5$ rad m^{-1}) in $Mn_6Cu_4Bi_4$ and 1.4×10^5 deg cm^{-1} ($\equiv 2.44 \times 10^5$ rad m^{-1}) in $Mn_5Ni_2Bi_4$ with a perpendicular anisotropy in the former material.

A huge increase of the Kerr rotation up to 2.04° was achieved by substitution of Al and Si into MnBi (Wang, 1990). The films were prepared by consecutive vapor-deposition of Al, Mn + Bi and SiO_2 from three different tantalum boats and subsequent annealing at about 350 °C for 3–5 h. A maximum Kerr rotation was reported for the composition $Mn_{1.0}Bi_{0.8}Al_{0.5}Si_{2.5}$. Figure 2-36 shows the spectral dependence of the polar Kerr rotation and the reflectivity (Wang, 1990). Since the rotation maximum does not occur near a reflectivity minimum, Wang excludes that the large rotation is due to an enhancement effect from

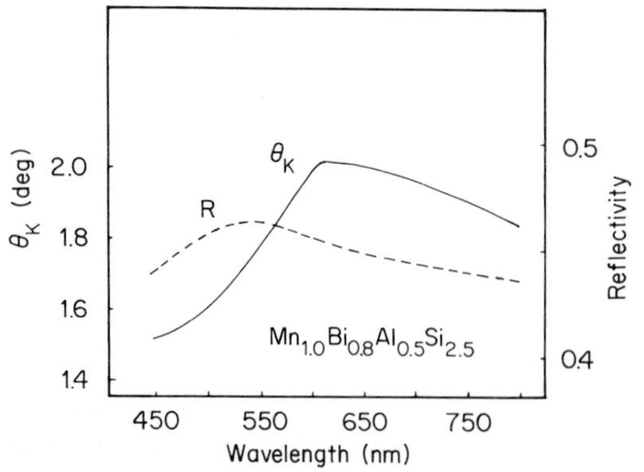

Figure 2-36. Room-temperature polar Kerr rotation and absorption of an $Mn_{1.0}Bi_{0.8}Al_{0.5}Si_{2.5}$ film (after Wang, 1990).

a dielectric oxide layer or a plasma edge effect, but speculates that it reflects changes of the electronic structure near E_F toward half-metallic ferromagnetism. However, open questions remain regarding the structure, the homogeneity, and the fact that the magnetization is reduced by a factor of 2.4 compared to normal MnBi. Clearly, more experiments are needed to corroborate this record room-temperature Kerr rotation.

Despite the great potential for applications of MnBi and homologues, very few attempts have been made to relate the magneto-optical properties to the electronic structure. Coehoorn et a. (1985) and Coehoorn and de Groot (1985) performed self-consistent spin-polarized band-structure calculations for MnSb and MnBi, respectively. For MnBi Coehoorn and de Groot compute the site, spin- and angular-momentum-decomposed densities of states in the ferromagnetic state. They find a peak for Mn 3d spin-up states 2.5 eV below the Fermi energy, and for the Mn 3d spin-down states, 0.5 eV above E_F. Magneto-optically active transitions are primarily expected from these 3d spin-up states into empty Bi 6p spin-up states and from occupied Bi 6p spin-down states into empty Bi 6d spin-down states. Misemer (1988) has studied the effect of spin-orbit interaction and exchange-splitting on the off-diagonal conductivity of MnBi. Yet, the result is rather schematic. Another result, which according to the author's knowledge has not been satisfactorily explained, is the extremely strong temperature dependence of the Kerr rotation of a MnBi film reported by Chen and Gondo (1964) for $T < 300$ K.

Other binary Mn compounds studied include Mn_5Ge_3 (Sawatzky, 1971) Mn_2Ga, Mn_3In, $MnNi_3$, MnNi, MnV, Mn_2Sn, $MnPt_3$, Mn_5Ge_2, and Mn_2Sb (Buschow et al., 1983). All these materials show small Kerr signals at room temperature, often well below 0.1°. Spectra have been published for only a few exceptions such as Mn_5Ge_3 and Mn_2Sb (see Fig. 2-33). Recently, a more complete study has been performed on $MnPt_3$ (Brändle et al., 1991 c). Figure 2-37 displays the rotation and ellipticity spectra from 0.6 to 5.2 eV for an arc-melted sample with composition $Mn_{28}Pt_{72}$ and a magnetic moment of 3 μ_B/Mn at room temperature in a field of 2 T. From these data and the optical constants the off-diagonal conductivity has been computed and Fig. 2-38 shows the product of the absorptive part of σ_{xy} and the photon frequency. As argued in Sec. 2.2.2.5, this type of plot suppresses the frequency dependence of the free-carrier contribution. Thus Fig. 2-38 evidences the contribution of at least two interband transitions in the considered energy range. From the sign, it can be concluded that the low-energy peak involves majority spin states, i.e., transitions of Mn 3d↑ electrons, while the structure at higher energies is probably due to transitions into empty 3d↓ states.

The ternary compound MnAlGe crystallizes in the same tetragonal structure as Mn_2Sb. It is a uniaxial ferromagnet with a Curie temperature $T_C = 185\,°C$. Sawatzky and Street (1973) reported room-temperature measurements of the optical absorption and the Faraday rotation of sputtered films. The rotation shows a weak energy dependence with a maximum value of 8.3×10^4 deg cm^{-1} ($\equiv 1.45 \times 10^5$ rad m^{-1}) at 2.3 eV. The films, however, were not oriented with the c-axis perpendicular to the substrate, nor was the applied field sufficiently large to overcome the magnetocrystalline anisotropy. Therefore, an even larger rotation can be expected in a c-axis-oriented film.

Among the ternary Mn compounds the Heusler-type alloys and the closely related compounds with the cubic $C1_b$ structure

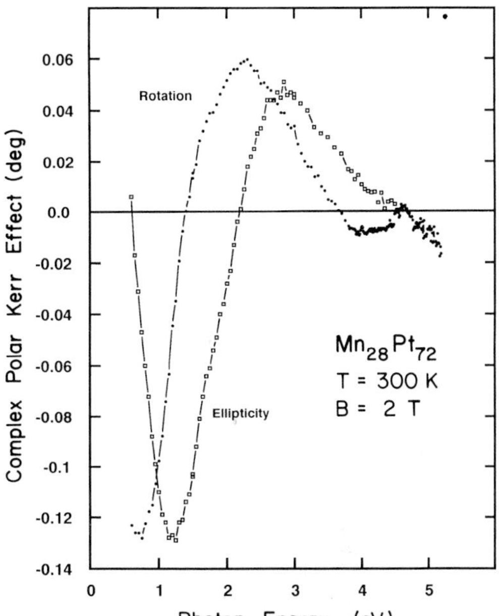

Figure 2-37. Room-temperature complex polar Kerr effect of polycrystalline $Mn_{0.28}Pt_{0.72}$ for a field of 2 T (after Brändle et al., 1991 c).

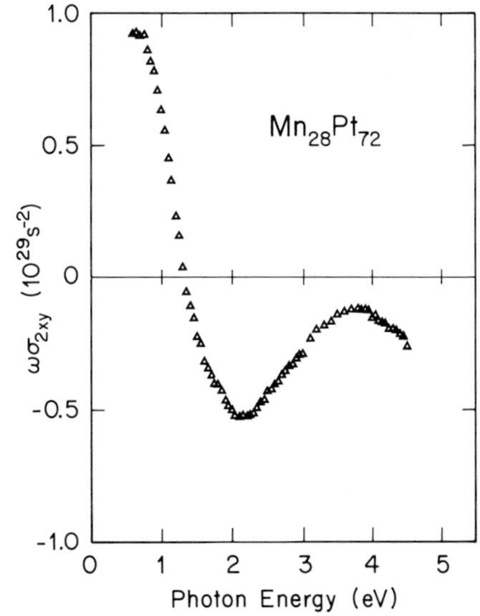

Figure 2-38. The product of the angular frequency and the absorptive part of the off-diagonal conductivity of $Mn_{0.28}Pt_{0.72}$ computed from the spectra in Fig. 2-37 and the optical constants (after Brändle et al., 1991 c).

have received the attention. The *Heusler alloys*, with the general formula X_2YZ, crystallize in the cubic $L2_1$ structure type (Heusler, 1903; Buschow and van Engen, 1981 d). The latter authors have studied the magnetic and magneto-optical properties of several X_2YZ alloys with X being Mn, Fe, Co, Ni, or Cu; Y being a second transition element or Nb, Zr, Ta, or Hf; and Z being either Al or Ga. The manganese compounds Mn_2VAl and Mn_2VGa were found to have Kerr-rotation signals at the two wavelengths 633 and 830 nm below 0.01°. For those Heusler alloys where Mn occupies the Y site, the size of the Kerr signal depends on the X atom. Thus, for Cu_2MnAl no rotation was found, while for Fe_2MnAl, Co_2MnAl, and Co_2MnGa Kerr rotations up to 0.11° were observed. In order to determine the relative importance of the Ni and Mn atoms in the magneto-optical spectra of the Heusler alloy Ni_2MnSn, Buschow et al. (1984) investigated the series $Ni_{3-x}Mn_xSn$. Only minor changes were observed in the shape of the spectra, but the intensity peaked near $x = 1.2$.

The largest room-temperature Kerr-rotation values have been observed in PtMnSb. The $C1_b$ structure of this compound evolves from that of the X_2YZ Heusler compounds by removing one of the X atoms. The first magneto-optical Kerr-effect measurements on PtMnSb, PtMnSn, PdMnSb, and NiMnSb were performed by van Engen et al. (1983). Figure 2-39 displays the data obtained by these authors on a polycrystalline PtMnSb sample prepared by arc melting and polished with diamond paste. A rotation peak is observable at 720 nm with a signal of −1.27°. Many research groups have investigated the magneto-optical properties of PtMnSb films prepared either by sputtering (Ohyama et al., 1985, 1987; Shouji et al., 1986; Takanashi et al., 1987) or by electron-beam evaporation (Inukai et al., 1986, 1987; Shiomi et al., 1987). In these latter cases the three components were vapor-deposited one after the other and an annealing procedure was necessary to form the compound. Generally, these films show larger Kerr rotation than reported by van

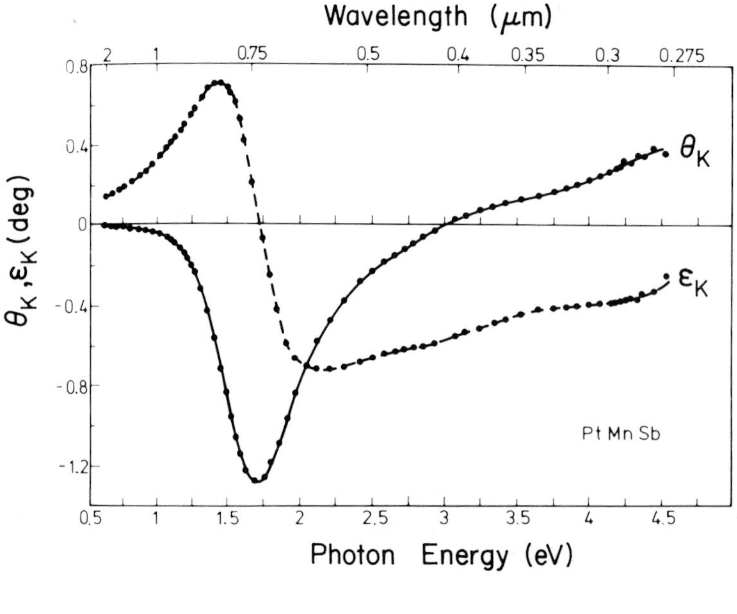

Figure 2-39. Room-temperature polar Kerr rotation θ_K and ellipticity ε_K measured on polycrystalline arc-melted samples of PtMnSb (after van Engen et al., 1983).

Engen et al. (1983). Inukai et al. (1987) have investigated the influence of the composition and found maximum $|\theta_K|$ of 2° for a composition $Pt_{31}Mn_{38}Sb_{31}$. The effect of annealing was subsequently also studied on bulk samples (Takanashi et al., 1988, 1990). Figure 2-40 shows the results of Takanashi et al. (1988) for an as-polished sample and one annealed at 500 °C for two hours. Based on the fact that the reflectivity changes little by annealing and on Rutherford back scattering and Auger electron analysis, one can exclude the possibility that the increase in the Kerr signal is due to an enhancement by an oxide surface layer. Instead, a study in fields of up to 15 T corroborates that the annealing decreases the magnetic anisotropy near the surface that is caused by the polishing process (Takanashi et al., 1990).

De Groot et al. (1984) and de Groot and Buschow (1986) have related the Kerr signal of PtMnSb to the special electronic structure of this class of materials. A self-consistent scalar-relativistic spin-polarized

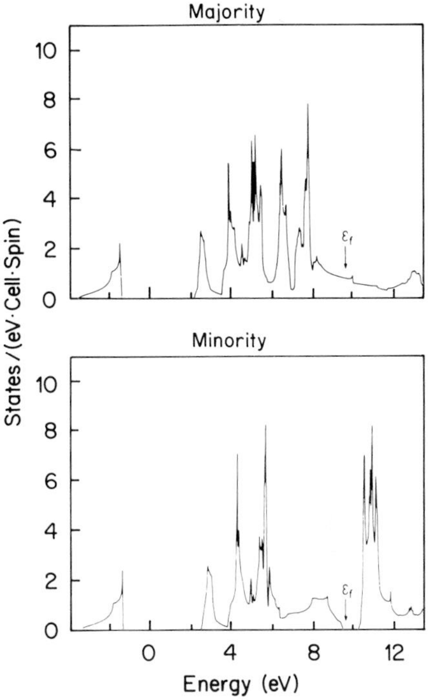

Figure 2-41. Spin-resolved density of states obtained for PtMnSb in a self-consistent, scalar-relativistic, spin-polarized augmented spherical wave calculation (after de Groot et al., 1984).

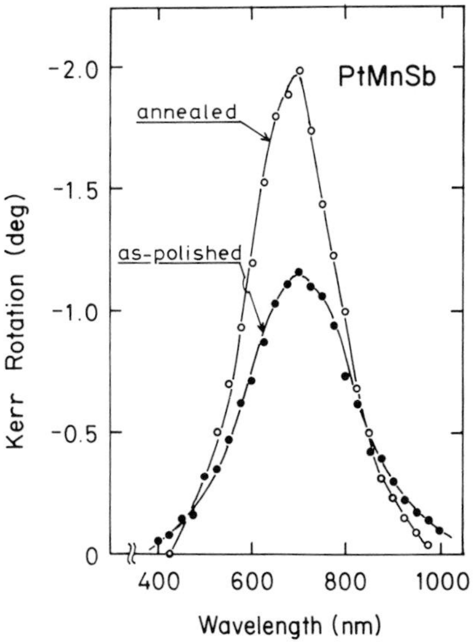

Figure 2-40. The effect of annealing on the Kerr rotation of PtMnSb (after Takanashi et al., 1988, 1990).

augmented spherical wave (ASW) calculation leads to the result that for the majority spin direction PtMnSb is a metal, while for the minority spin direction a gap opens, and the Fermi energy falls into this gap. Figure 2-41 shows the density of states for the majority and minority spin direction, as computed by de Groot et al. (1984). The states at the top of the valence band are primarily composed of Sb p states. Since E_F falls near the top of this valence band for the minority spin direction, the spin-orbit interaction will push the $m = +1$ state above E_F, while the $m = 0$ and -1 states will stay below E_F. This excludes excitations with left-circularly polarized light from $m = +1$ into the conduction band with symmetry Γ_1 and $m = 0$ (formed from Pt and Mn s states), while excitations from

$m = -1$ into the same final state are possible for right-circularly polarized light.

The values obtained for the polar Kerr effect in PtMnSn, PdMnSb, and NiMnSb (van Engen et al., 1983) do not exceed 0.2°. An initial argument presented for the reduced Kerr signal of PdMnSb as compared to PtMnSb has been that the spin-orbit coupling constant of Pd is smaller than that of Pt by a factor of 3. Later discussions (de Groot and Buschow, 1986) include arguments on the exact position of the Fermi energy, the interaction with third bands, and the joint density of states, but a quantitative description is still lacking.

Since the discovery of high-T_c superconductivity by Bednorz and Müller (1986), lanthanum-strontium perovskites have advanced to one of the most investigated class of materials. Yet, some of them had already received attention for their magneto-optical properties ten years earlier. Popma and Kamminga (1975) studied the polar magneto-optical Kerr rotation of $Bi_xLa_{0.7-x}Sr_{0.3}MnO_3$ samples prepared by usual ceramic techniques. Figure 2-42 shows the results for $La_{0.7}Sr_{0.3}MnO_3$ and $Bi_{0.3}La_{0.4}Sr_{0.3}MnO_3$. It can be seen that the substitution with Bi enhances considerably the Kerr rotation at low temperatures. Yet, the Curie temperature decreases almost linearly from 380 K for $La_{0.7}Sr_{0.3}MnO_3$ to 280 K for $Bi_{0.5}La_{0.2}Sr_{0.3}MnO_3$ and the magnetic moment per formula unit a 4 K decreases from 3.4 to 2.2 μ_B, thus explaining the roughly constant value of the rotation at room temperature.

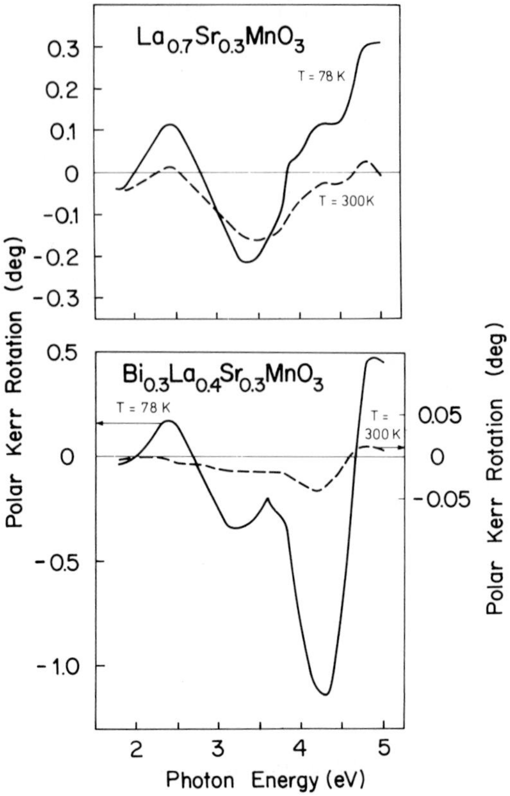

Figure 2-42. Polar Kerr rotation of ceramic samples of $La_{0.7}Sr_{0.3}MnO_3$ (top) and $Bi_{0.3}La_{0.4}Sr_{0.3}MnO_3$ (bottom) at 300 and 78 K. The applied field (not given) does not saturate the former sample at $T =$ 300 K and induces an even lower magnetization at 300 K in the latter sample on account of its lower T_C (after Popma and Kamminga, 1975).

2.5.3.3 Fe Compounds

The magnetic iron compounds are so numerous that a comprehensive discussion of their magneto-optical properties would require a separate chapter. Keeping this subsection to a size comparable to the others unavoidably results in the omission of valuable material. No discussion will be given of low-dimensional materials such as $FeCl_2$ (see, e.g., Gebhardt, 1976), rhombohedral FeF_3 (see, e.g., Wolfe et al., 1970), as well as hematite or other variations of Fe_2O_3, for which the reader is referred to a collection of data in Landolt-Börnstein (Jaggi et al., 1962) or to the more recent paper by Blazey (1974). Even with the latter exclusions, the ferrites will fill the largest part of this subsection. First, some non-oxide compounds will be discussed.

Sato et al. (1987) studied the Kerr rotation and the reflectance magneto-circular dichroism of Fe_7Se_8. This material crystallizes in the hexagonal NiAs-type structure with a superstructure associated with the ordering of vacant Fe sites. The Kerr effect was found to be below 2.7×10^{-2} deg at room temperature.

Van Engelen and Buschow (1990a) investigated the Kerr rotation and ellipticity of Fe_3X compounds, where X is one of the metalloids boron, phosphorus, germanium, and silicon. These materials have Curie temperatures ranging from 638 K (Fe_3Ge) to 770 K (Fe_3B) and a uniaxial crystallographic structure which makes them potential candidates for thermomagnetic recording (see also Sec. 2.5.2.5 on amorphous films and alloys). Fe_3B does not exist in the Fe–B phase diagram. Therefore, van Engelen and Buschow added some Nd to enhance the glass-forming tendency and prepared an amorphous ribbon using a melt spinner. The results of their investigation at room temperature are reproduced in Fig. 2-43. Rotations of up to 0.49° can be observed. If the rotation is normalized to the respective magnetization, a substantial enhancement is noticeable for Fe_3Ge and Fe_3Sn, i.e., for the heavy metalloids. This is related to the increase of the spin-orbit splitting with atomic number.

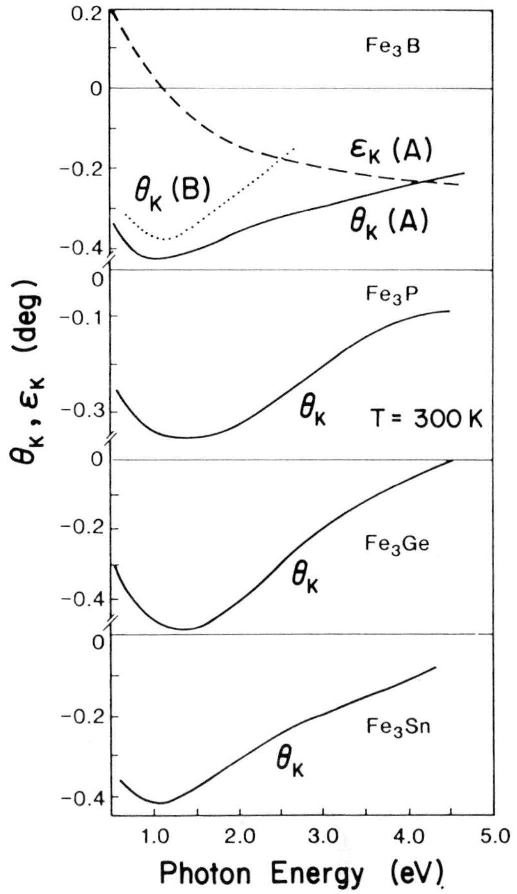

Figure 2-43. Room-temperature Kerr rotation of (from top) Fe_3B, Fe_3P, Fe_3Ge, and Fe_3Sn in a field of ≈ 1.2 T. For Fe_3B the letters A and B refer to neodymium-doped and pure samples, respectively. For the former sample the ellipticity is also shown (after van Engelen and Buschow, 1990a).

Magnetite (Fe_3O_4) is probably the most studied binary iron compound. It undergoes a metal-nonmetal phase transition at 119 K known as *Verwey transition* (Verwey and Haayman, 1941). Fe_3O_4 is a cubic ferrite with an inverse spinel structure. One of the two Fe^{3+} ions occupies the tetrahedral A-site, and the other Fe^{3+} ion occupies one of the two octahedral B-sites. The Fe^{2+} ion is located on the second octahedral site. To indicate this occupation, one often writes Fe_3O_4 as $(Fe^{3+})\{Fe^{2+}Fe^{3+}\}O_4$. The magnetic moments of Fe^{2+} and Fe^{3+} on the octahedral sites are parallel to each other and antiparallel to the moment of the Fe^{3+} ion on the tetrahedral site. In addition to the older investigations (see Jaggi et al., 1962), the magneto-optical properties of Fe_3O_4 have been investigated by Muret (1974), Müller and Buchenau (1975), Matsumoto et al. (1978), Šimša et al. (1980a), and Zhang et al. (1981a, 1983). Figure 2-44 displays the room-temperature Kerr rotation and ellipticity spectra ob-

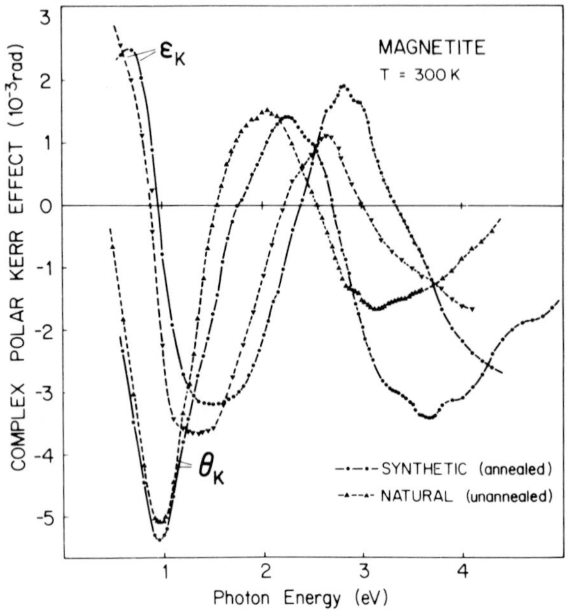

Figure 2-44. Room-temperature polar Kerr rotation and ellipticity of synthetic and natural magnetite. The magnetic monocrystal is annealed, and the natural monocrystal is unannealed (after Zhang et al., 1981 a).

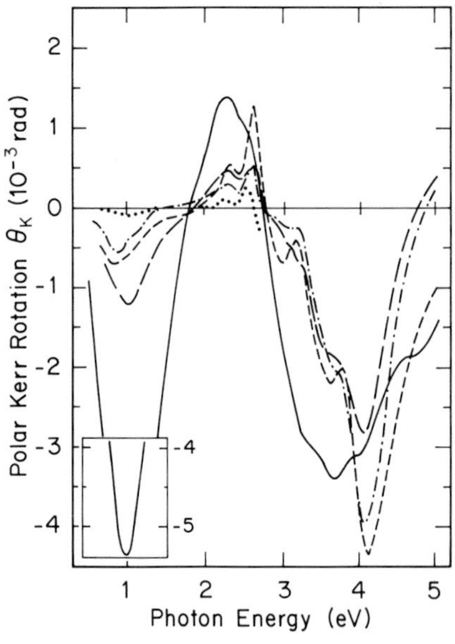

Figure 2-45. Room-temperature polar Kerr rotation of:
$Fe_3O_4 = (Fe^{3+})\{Fe^{2+}Fe^{3+})O_4$ (full line),
$(Mg^{2+}_{0.24}Fe^{3+}_{0.76})\{Mg^{2+}_{0.76}Fe^{2+}_{0.22}Fe^{3+}_{1.02}\}O_{3.89}$ (long-dashed line),
$(Mg^{2+}_{0.10}Fe^{3+}_{0.90})\{Mg^{2+}_{0.90}Fe^{2+}_{0.10}Fe^{3+}_{1.00}\}O_{3.95}$ (dot-dashed line),
$(Fe^{3+}_{1.0})\{Li^{1+}_{0.5}Fe^{2+}_{0.13}Fe^{3+}_{1.37}\}O_{3.94}$ (short-dashed line),
$(Fe^{3+}_{1.0})\{Li^{1+}_{0.5}Fe^{3+}_{1.5}\}O_4$ (dotted line).
The first four samples are annealed, and the fifth is unannealed (after Zhang et al., 1983).

tained by Zhang et al. (1981 a) on natural as well as on synthetic magnetite crystals. The largest rotation (absolute value) is found near 1 eV and amounts to 0.3° ($\equiv 5.3 \times 10^{-3}$ rad). Although the spectra are quite structured, measurements on magnesium- and lithium-substituted magnetite have revealed that large compensations of signals from the three Fe ions per formula unit do exist (Zhang et al., 1981 b, 1983). Figure 2-45 demonstrates this compensation by comparing the rotation spectra of pure magnetite and various types of substituted magnetite. One recognizes also that the dominant peak near 1 eV decreases strongly with the substitution of Fe^{2+} with non-magnetic Li^{1+} or Mg^{2+}. From the seven compounds studied, Zhang et al. (1981 b, 1983) could empirically relate the magneto-optical structures to the Fe ions on the three sublattice sites. Using a crystal-field-splitting model for the initial and final 3d states of orbital promotion excitations $3d^n \rightarrow 3d^{n-1}$ 4(sp), they were then able to explain the line shape, the

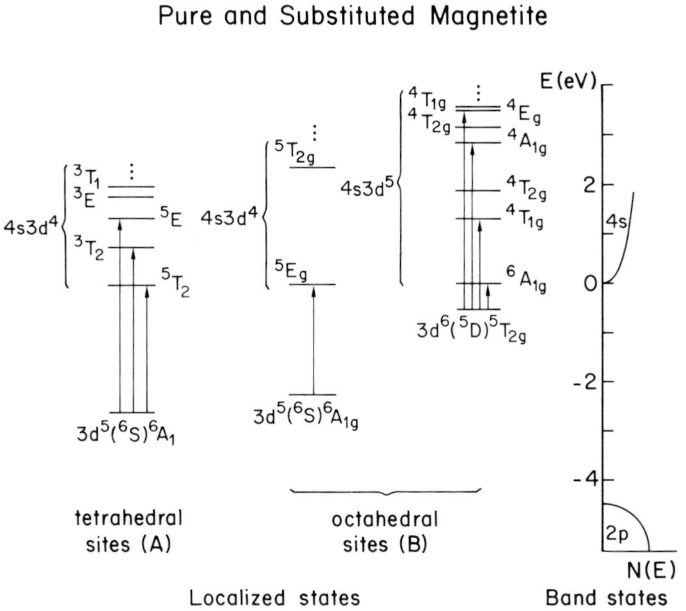

Figure 2-46. Coupling scheme for $3d^n \rightarrow 3d^{n-1}4s4p$ transitions in pure and substituted magnetite (after Zhang et al., 1983).

sign, and the relative intensity of eight transitions. Figure 2-46 displays this energy level scheme.

The Verwey transition of Fe_3O_4 has been studied with the equatorial Kerr effect by Buchenau and Müller (1972) and with the polar Kerr rotation by Matsumoto et al. (1978). The latter authors report an abrupt change of the Kerr signal, either increasing or decreasing, depending on the wavelength (Fig. 2-47). Unfortunately, a precise interpretation for this interesting behavior is not given.

The infrared Faraday rotation of Li-substituted magnetite $Li_{0.5}Fe_{2.5}O_4$, Mg-substituted magnetite $MgFe_2O_4$, as well as $NiFe_2O_4$ was invstigated by Zanmarchi and Bongers (1969). The rotation is found to be a superposition of a strongly dispersive part caused by electronic transitions and a nearly constant part from magnetic resonance. From an analysis of the Ni ferrite data it is concluded that the contribution from Ni peaks at 1.17 μm, and that it corresponds to a $^3A_2 \rightarrow {}^3T_2$ transition of Ni^{2+} in an oxygen octahedron. The polar Kerr effect for a (111) face of Ni ferrite has been measured from 1.8 to 5.6 eV by Kahn et al. (1969) (see Fig. 2-59 in Sec. 2.5.3.5 on Ni compounds). Compared with pure

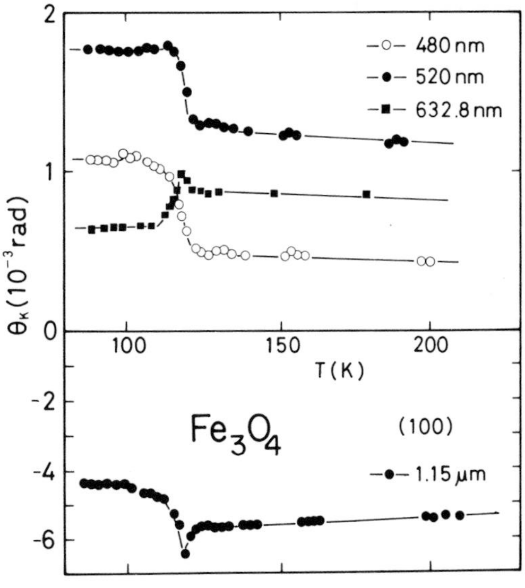

Figure 2-47. Temperature dependence of the polar Kerr effect of magnetite near the Verwey transition for different wavelengths (after Matsumoto et al., 1978).

magnetite, the spectrum shows weak structures below 2.2 eV, which are attributed to the Ni^{2+} sublattice. The substitution with manganese has been studied by Šimša et al. (1980 b). Figure 2-48 displays their Kerr rotation spectra for $Mn_xFe_{3-x}O_4$ with x varying from 0 to 1.6. A strong decrease of the rotation peak is observed at 1 eV, and a larger negative rotation near 3 eV. The Kerr effect of $Mn_{0.4}Zn_{0.3}Fe^{2+}_{0.3}Fe_2O_4$ has been studied by Martens et al. (1983), who found a rotation below 0.2°.

The largest amount of data exists for Co-substituted magnetite. Coburn et al. (1974) and Ahrenkiel et al. (1975 a) were the first to study the magneto-optical properties of cobalt ferrites. Coburn et al. observed a maximum Kerr rotation of 0.2° at 0.8 eV in hot-pressed polycrystalline samples of $CoFe_2O_4$ with $T_C = 790$ K. However, they noted that $CoFe_2O_4$, as Fe_3O_4, is an inverse spinel and that a substantial increase of the Kerr signal should be expected if divalent cobalt enters on the tetrahedral lattice site (Ahrenkiel et al., 1975 a). This situation can be realized by the addition of certain trivalent ions such as Rh^{3+}, Cr^{3+} and Al^{3+}. If M^{3+} designates such ions, a normal spinel with the configuration $Co^{2+}\{Fe^{3+}M^{3+}\}O_4$ is obtained. Indeed, Ahrenkiel et al. (1975 a) observed at 290 K an increase of the Kerr rotation by a factor of 3 for $CoRhFeO_4$ compared to $CoFe_2O_4$, although T_C had decreased to 350 K. Later, Martens et al. (1982) and Peeters and Martens (1982) measured the Kerr rotation of a $CoFe_2O_4$ single crystal and of Al-substituted ceramic samples of cobalt ferrites (Fig. 2-49). Peeters and Martens found for the (111) surface a maximum Kerr rotation of 0.5° at 0.75 eV. The substitution with Al enhances this peak to 0.9° and produces changes in the structure near 2 and 4 eV. While the changes near 4 eV can be simply related to the reduced concentration of Fe^{3+} on octahedral sites, the changes at 0.8 and 2 eV are more subtle. Peeters and Martens ar-

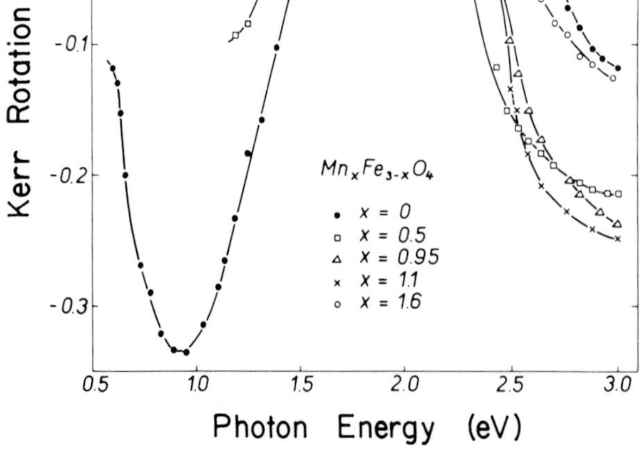

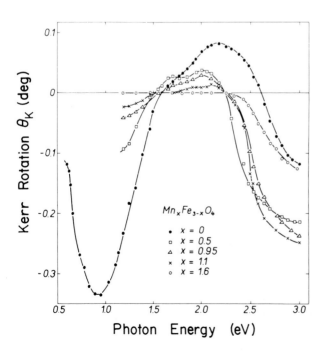

Figure 2-48. Room-temperature polar Kerr rotation of various manganese ferrites in a saturating field of 1 T (after Šimša et al., 1980 b).

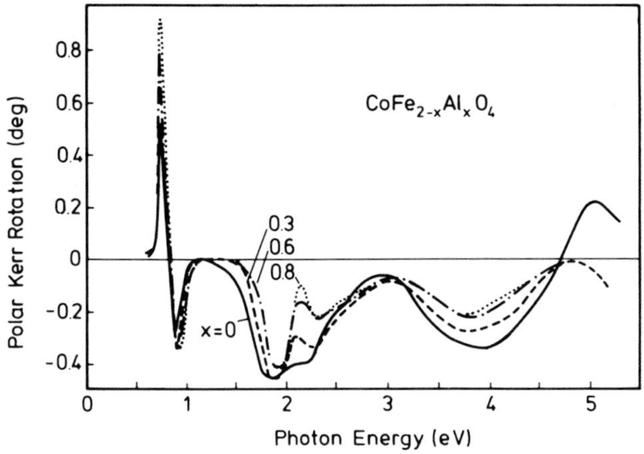

Figure 2-49. Room-temperature polar Kerr rotation of polycrystalline aluminum-substituted cobalt ferrites in a field of 1 T (after Peeters and Martens, 1982).

gue that in $CoFe_2O_4$ some Co^{2+} ions already occupy tetrahedral sites, and that the substitution with Al increases this percentage. The dispersive Kerr signals at 0.8 and 2 eV are then assigned to $^4A_2 \rightarrow {}^4T_1(F)$ and $^4A_2 \rightarrow {}^4T_1(P)$ crystal-field transition of Co^{2+} ions on tetrahedral sites, respectively. Yet, from a study of $Co_xFe_{3-x}O_4$, Peeters and Martens conclude that the structure near 2 eV should also include contributions from a charge transfer $Co^{2+} \rightarrow Fe^{3+}$ both on octahedral sites. The Kerr ellipticity, the Faraday rotation, and the possibility to enhance the rotation by dielectric interference films have also been investigated (Martens and Peeters, 1983; Martens and Voermans, 1984). Finally, Martens (1986) studied the effect of substituting Rh^{3+}, Mn^{3+}, and $Ti^{4+} + Co^{2+}$ for iron in cobalt ferrites, but the improvements of the magneto-optical properties are minor. Partly the same substitutions with Al^{3+} and Mn^{3+}, but also with Cr^{3+}, have been investigated on ceramic samples by Abe and Gomi (1982). Figure 2-50 reproduces their Kerr-rotation spectra at room temperature. While the spectrum for $CoFe_2O_4$ is quite similar to that in the previous figure and consequently a factor of 2.5 larger than that reported by Ahren-

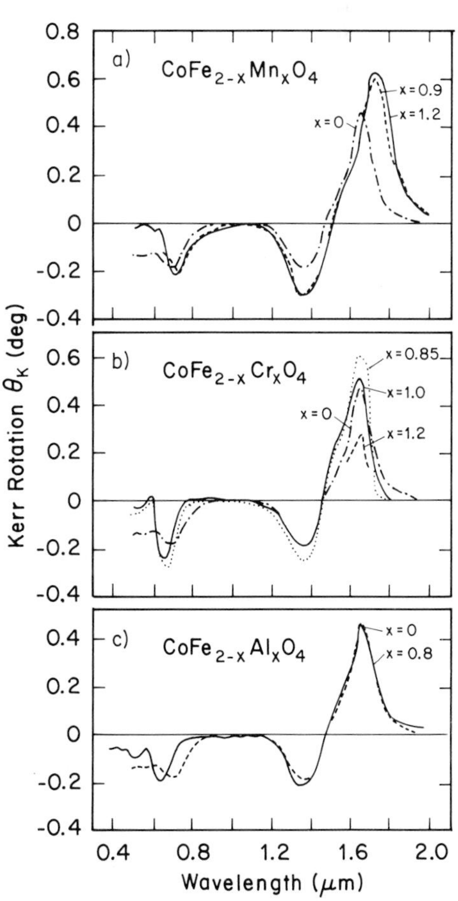

Figure 2-50. Room-temperature polar Kerr effect of ceramic cobalt ferrites substituted with manganese (a), chromium (b), and aluminum (c) (after Abe and Gomi, 1982).

kiel, Abe and Gomi do not observe an increase of the low-energy peak (1.7 μm) upon Al substitution. However, they find an increase of this rotation peak on substitution with Mn and for certain concentrations of Cr. These differences in data between three different groups for pure and substituted $CoFe_2O_4$ may indicate that the percentage of Co^{2+} on tetrahedral sites depends much on the preparation, annealing, and quenching conditions.

The second class of ferric oxide compounds to be discussed are the *garnets.* These are cubic insulators with the general formula $(M_A^{3+})_3(M_B^{3+})_5O_{12}$. M_A is the larger element, in most cases a lanthanide or Y. M_B is considerably smaller, e.g., a 3d transition metal or Ga and Al. In this subsection we are of course concerned primarily with Fe, i.e., with $Y_3Fe_5O_{12}$ and non-rare-earth substitutions. The discussion of mixed 4f–3d compounds is reserved for Sec. 2.8 of this chapter. Two of the five Fe atoms per formula unit occupy octahedral sites; the other three occupy tetrahedral sites. The magnetic coupling between these sublattices is antiferromagnetic, thus reducing the resulting magnetic moment per formula unit to that of a single Fe^{3+} ion, i.e., to 5 μ_B for magnetic saturation. The Néel temperature of ferrimagnetic $Y_3Fe_5O_{12}$, abbreviated as YIG, is 559 K. Dillon (1959) was probably the first to observe the large Faraday rotations in this material. The first Kerr spectrum for a single crystal of YIG and many other ferric oxide compounds were reported by Kahn et al. (1969). Figure 2-51 shows data obtained ten years later by Višňovský et al. (1979a), which, except for more structures in the photon energy range 3–4 eV, are very much like those of Kahn et al. To separate the contributions from the two sublattices, Višňovský et al. (1979a) also determined the Kerr rotation for lithium ferrite

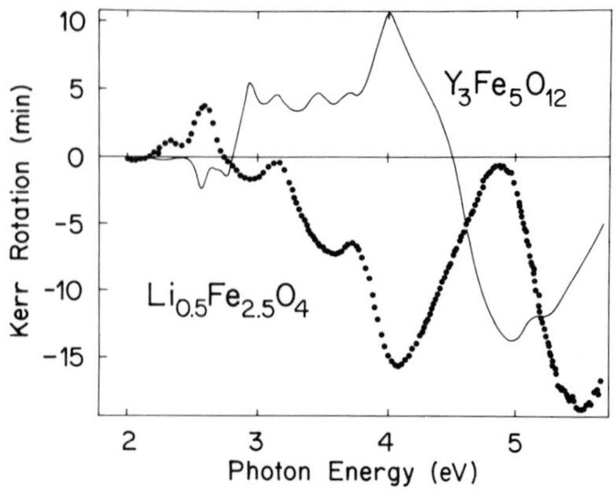

Figure 2-51. Room-temperature polar Kerr rotation of single crystals of yttrium iron garnet, $Y_3Fe_5O_{12}$, and lithium ferrite, $Li_{0.5}Fe_{2.5}O_4$ (after Višňovský et al., 1979).

$Li_{0.5}Fe_{2.5}O_4$. As discussed above for the inverse spinel, the Fe^{3+} sublattice magnetization from the octahedral site is oriented parallel to the net magnetization in $Li_{0.5}Fe_{2.5}O_4$, while it is antiparallel to the net magnetization in the garnet. Conversely, the tetrahedral sublattice magnetization is antiparallel to the net magnetization in $Li_{0.5}Fe_{2.5}O_4$ and parallel to it in YIG. Thus, one expects opposite signs for the magneto-optical contributions of both sublattices when comparing Li ferrite and YIG. Indeed, such a trend appears in Fig. 2-51. The deviations from exactly mirror-like spectra reflect mostly the different ratio of tetrahedral and octahedral sites and, to a minor extent, differences in dimensions and symmetries of the oxygen polyhedra in the two compounds.

On account of the transparency and the availability also of epitaxially grown YIG films, much of the magneto-optical work has been performed in transmission. To cite only a few, one may quote the works of Crossley et al. (1969), Wettling et al. (1973), Canit et al. (1974), Scott et al. (1975), Wittekoek et al. (1975), and Višňovský et al. (1979b). Figure 2-52 shows a logarithmic

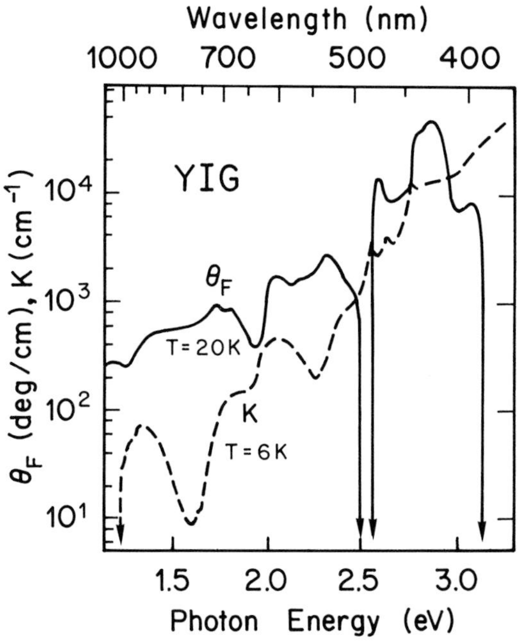

Figure 2-52. Logarithmic plot of the Faraday rotation and absorption of $Y_3Fe_5O_{12}$ in the region of weak absorption. The temperature is 20 and 6 K, respectively (after Wettling et al., 1973).

plot of the Faraday rotation and the absorption between 1.1 and 3.3 eV at 20 and 6 K, respectively, as reported by Wettling et al. (1973). Immediately above the energy range of complete transparency between 0.25 and 1.25 eV, it shows rather large rotations in regions of weak absorption, giving a $|\theta_F|/K$ of $\approx 80°$ at 1.6 eV. The gross line of the interpretation is that the narrow lines are due to crystal-field transitions of the Fe^{3+} ion which are superimposed on the wing of much stronger charge-transfer transitions situated above 2.5 eV (Wittekoek et al., 1975). An extensive study of the magnetic circular dichroism and the Faraday rotation has been performed by Scott et al. (1975) who made a detailed analysis of line shapes. They also showed that the circular dichroism, being a difference of absorptive quantities, allow for better resolution of single transitions. The thorough investigation of Wittekoek et al. (1975) deserves a special note. In addition to the study of bismuth-substituted YIG, to be discussed below, these authors compared for pure YIG the complex off-diagonal tensor element derived from Faraday-effect and Kerr-effect measurements. The Faraday rotation and ellipticity were measured on thin films prepared by liquid-phase epitaxy and the data were corrected for the influence of the reflection at the two interfaces. The Kerr rotation and ellipticity were measured on polycrystalline bulk samples. In Fig. 2-53 the two sets of data

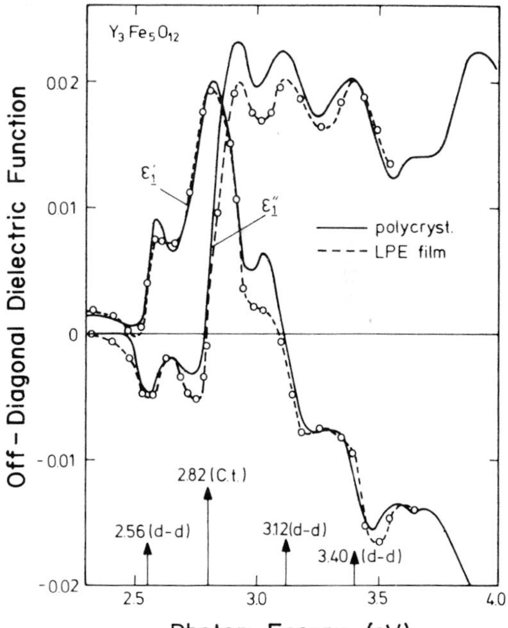

Figure 2-53. Comparison of the complex off-diagonal dielectric-tensor element computed from polar-Kerr-effect measurements on polycrystalline samples (full line) and from Faraday-effect measurements on thin films (dashed line). ε_1' is the dispersive part and ε_1'' is the absorptive part of the off-diagonal dielectric-tensor element. The arrows at the bottom of the figure indicate assignments to charge transfer (C.t.) and crystal-field transitions (d-d) (after Wittekoek et al., 1975).

are compared with an agreement which is better than anticipated, empirically justifying setting the permeability tensor $\boldsymbol{\mu}$ to 1 in Eq. (2-20).

An intriguing observation in YIG is that above magnetic saturation the absolute value of the Faraday rotation decreases with increasing field (Kharchenko et al., 1969; Krinchik and Gushchina, 1969; Pisarev et al., 1976, 1977; Guillot and Le Gall, 1976; Kido et al., 1977). Figure 2-54 reproduces the result by Kharchenko

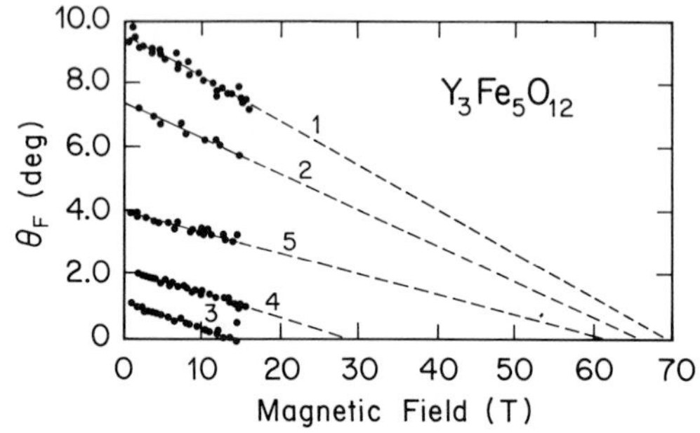

Figure 2-54. Field dependence of the Faraday rotation in $Y_3Fe_5O_{12}$ at room temperature. The wavelengths are as follows: (1) 546 nm; (2) 560 nm; (3) 632.8 nm; (4) 650 nm; and (5) 700 nm (after Kharchenko et al., 1969).

et al. (1969) obtained at room temperature for various wavelengths between 550 and 760 nm (2.25 to 1.63 eV). Pisarev et al. (1976, 1977) explained this surprising effect by assuming magnetic-field-dependent magneto-optical coefficients.

Another effect observed on YIG which is worth mentioning is the angular variation of induced anisotropy (Tucciarone, 1977; Antonini et al., 1978). These authors showed that by a measurement of the linear dichroism as a function of the direction of a saturating magnetic field, the symmetry, and therefore the site of Fe^{2+} centers in doped YIG, can be determined.

Buhrer (1969a, 1970), Wittekoek et al. (1973), Lacklison et al. (1973), and Takeuchi et al. (1973) were among the first to notice that the substitution of Bi and some lanthanides considerably enhances the magneto-optical signals of YIG. A very detailed study of the Faraday and Kerr effect as a function of the Bi content has been performed by Wittekoek et al. (1975). Figure 2-55 reproduces their Kerr rotation results on bulk polycrystalline samples and Fig. 2-56 their Faraday rotation results on liquid-phase epitaxy films. As x increases from 0 to 1 in $Y_{3-x}Bi_xFe_5O_{12}$, the negative Kerr rotation peak at 2.7 eV increases from $-0.02°$ to $-0.6°$, and the positive peak near 3.7 eV increases from 0.18° to 0.45°. Correspondingly, the Faraday rotation develops two huge peaks at 2.4 and 3.3 eV. For the latter, Wittekoek et al. find for $x = 0.5$ a specific Faraday rotation of nearly 2×10^5 deg cm^{-1} ($\equiv 3.5$ rad m^{-1}). This discovery has prompted hundreds of papers on the substitution of YIG, not only with one element, but with several elements at the same time, and the interested reader is referred to the proceedings of the various magnetism conferences or to the review by Le Gall et al. (1987).

For thermodynamic reasons and due to lattice constants the complete substitution of Y with Bi in $Y_3Fe_5O_{12}$ is not possible in thermodynamic equilibrium. The experimental solid-solution limit is $Y_{1.12}Bi_{1.88}Fe_5O_{12}$ (Geller and Colville, 1975). Okuda et al. (1990) succeeded to circumvent the above limit by growing epitaxial films of $Bi_3Fe_5O_{12}$ on a single-crystal garnet substrate using a reactive ion-beam sputtering technique. In agreement with the extrapolation of the rate of -2.1×10^4 deg cm^{-1} per Bi atom per formula unit derived by Hansen et al. (1984) for the Faraday rotation at 633 nm, Okuda et al. measured a rotation of -7.2×10^4 deg cm^{-1} ($\equiv -1.25 \times 10^5$ rad m^{-1}) in $Bi_3Fe_5O_{12}$ at room temperature. At

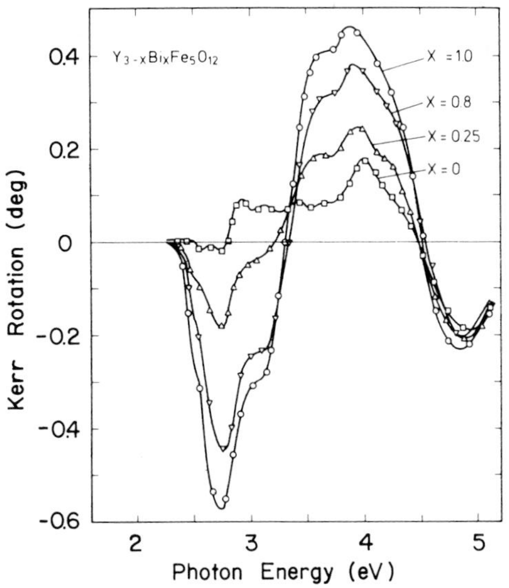

Figure 2-55. The effect of bismuth substitution on the room-temperature Kerr effect of bulk polycrystalline yttrium iron garnet. The saturating field is 0.2 T (after Wittekoek et al., 1975).

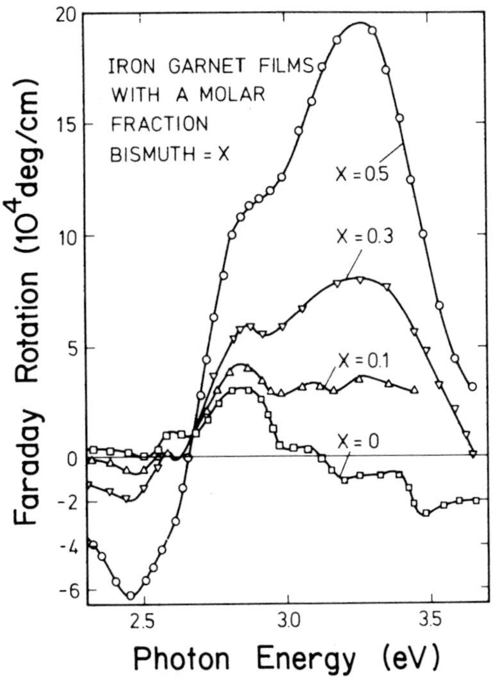

Figure 2-56. The effect of bismuth substitution on the room-temperature Faraday rotation of yttrium iron garnet films grown by liquid-phase epitaxy (after Wittekoek et al., 1975).

500 nm the Faraday rotation amounts to -3.7×10^5 deg cm^{-1} ($\equiv -6.45 \times 10^5$ rad m^{-1}) at 77 K and -2.35×10^5 deg cm^{-1} ($\equiv -4.1 \times 10^5$ rad m^{-1}) at 300 K, but the absorption has also increased. The Kerr spectrum measured by Okuda et al. at room temperature is shown in Fig. 2-57. It resembles the spectra for lower Bi concentrations (Fig. 2-55), but the absolute value has increased up to a maximum of nearly 1.2°; i.e., it has doubled compared to $BiY_2Fe_5O_{12}$. This puts $Bi_3Fe_5O_{12}$ in the small class of materials with room temperature Kerr effects exceeding 1°, with the advantage as potential candidate for a third generation of magneto-optical memories that the maximum occurs at a shorter wavelength than in its competitors. Despite the potential for applications of Bi-substituted YIG, the electronic structure of YIG and the mechanism leading to the strong enhancement with Bi substitution is not fully understood. Many different types

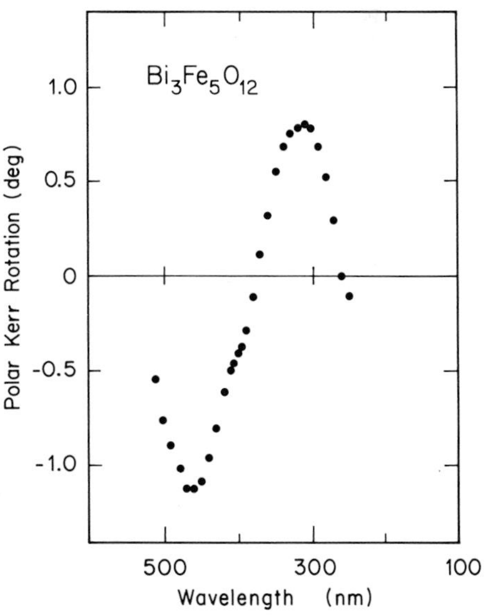

Figure 2-57. Room-temperature polar Kerr rotation of an epitaxial $Bi_3Fe_5O_{12}$ film (after Okuda et al., 1990).

of transitions have been put forward, but most investigators agree that the large spin-orbit splitting of the Bi 6p states (≈ 2.1 eV) should play a decisive role in the enhancement of the magneto-optical activity.

A third class of ferric oxide compounds is formed by the *orthoferrites.* These have the chemical formula $M^{3+}FeO_3$, where M^{3+} stands for a trivalent group III transition element or a rare-earth ion. The crystal structure is orthorhombic and corresponds to a distorted perovskite type. All Fe^{3+} ions are crystallographically equivalent but their coordination octahedron $[FeO_6]$ has a lower symmetry than in the garnets because of distortion. The magnetic exchange between nearest-neighbor Fe^{3+} ions is antiferromagnetic. Yet, a slight canting of the spins, due to antisymmetric exchange, produces a weak magnetization parallel to the *c*-axis (except for $SmFeO_3$, where $\boldsymbol{M} \parallel \boldsymbol{a}$). Initial magneto-optical Kerr effect measurements on 11 different orthoferrites were reported in the pioneering work of Kahn et al. (1969). At the same time Tabor and Chen (1969) used $YFeO_3$ to study the effect of light propagation through a material displaying both Faraday rotation and birefringence. Later, the magneto-optical and birefringent optical properties were investigated by Tabor et al. (1970), Clover et al. (1971), Chetkin et al. (1975), Gomi et al. (1979), and Višňovský et al. (1984). Figure 2-58 reproduces the polar Kerr effect spectra obtained by the latter authors on a natural (001) surface of a flux-grown $YFeO_3$ single crystal at room temperature. The signal is small, but the spectra are more structured than those for the garnet $Y_3Fe_5O_{12}$. This reflects, on the one hand, the fact that the lower symmetry lifts the degeneracy of certain states, and on the other, that similar to the situation in Fe_3O_4 (Zhang et al., 1983), a partial compensation of the contributions from octahedral and tetrahedral sites exists in $Y_3Fe_5O_{12}$.

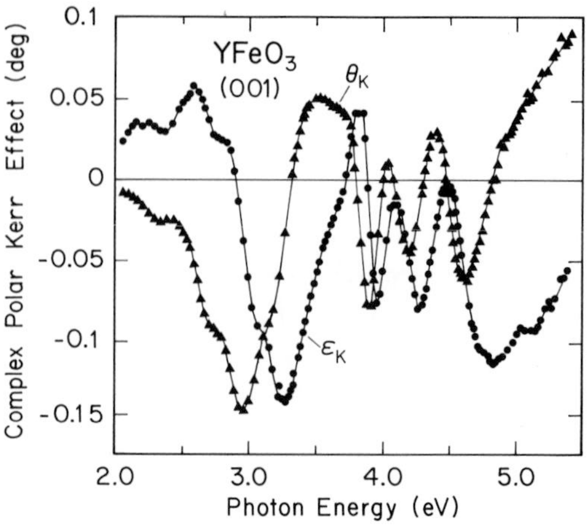

Figure 2-58. Room-temperature Kerr rotation θ_K and ellipticity ε_K of yttrium orthoferrite, $YFeO_3$ (after Višňovský et al., 1984).

Abe et al. (1980) have investigated the Faraday rotation and the birefringence of Fe_3BO_6. This ferric borate resembles orthoferrites by its orthorhombic structure and its weak ferromagnetism. Abe et al. reported a Faraday rotation at 0.62 μm of 1400 deg cm^{-1} along the ***a***-axis, which is about three times as large as that along the *c*-axis.

Another anisotropic Fe compound whose magneto-optical properties have been investigated is $FeBO_3$ (Jantz et al., 1976; Wolfe et al., 1970). This iron borate is an easy-plane weak ferromagnet ($T_C =$ 348 K), crystallizing in the rhombohedral calcite structure. Its absorption, Faraday rotation, and magnetic linear birefringence have been reported by Jantz et al. (1976). Its absorption is weak in the visible, and the Faraday rotation peaks at 2.6 eV with a specific rotation of 4270 deg cm^{-1} ($\equiv 7.5 \times 10^3$ rad m^{-1}).

Magnetoplumbites are ternary ferric oxides with the chemical formula

$M^{2+}Fe_{12}O_{19}$, where M^{2+} is Pb or Ca, Sr, and Ba. They crystallize in the hexagonal $PbFe_{12}O_{19}$ structure and order ferrimagnetically with Néel temperatures over 700 K. Their strong magnetic anisotropies along the *c*-axis make them suitable for permanent-magnet applications. Early magneto-optical measurements have been performed by Kahn et al. (1969), Zanmarchi and Bongers (1969), and Drews and Jaumann (1969). More recent studies were carried out by Shono et al. (1982a) on single crystals of the Ba, Sr, and Pb magnetoplumbites and by Watada (1987) on Ba-magnetoplumbite substituted with Co–Ti. Figure 2-59 displays the polar Kerr rotation spectrum reported by Kahn et al. (1969) for a (001) face of $PbFe_{12}O_{19}$ together with the results for the inverse spinel $NiFe_2O_4$ at about 87% of magnetic saturation. Kahn et al. associate the peaks at 4 and 5 eV with transitions from Fe^{3+} in octahedral and tetrahedral sites, respectively.

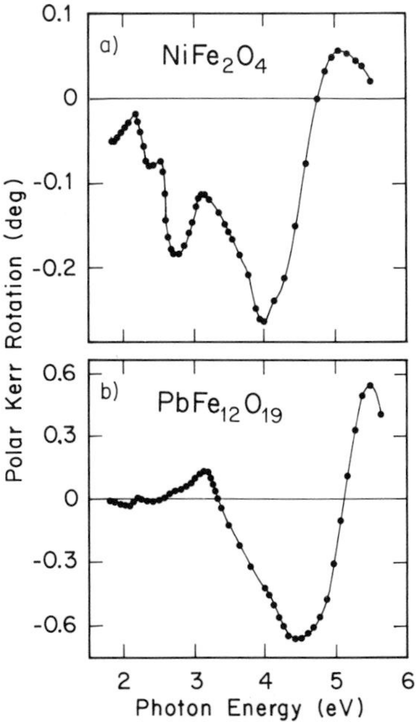

Figure 2-59. Room-temperature Kerr rotation of single crystals of nickel ferrite, $NiFe_2O_4$ (a) and magnetoplumbite, $PbFe_{12}O_{19}$ (b). The applied field provided about 87% of saturation magnetization in $NiFe_2O_4$ and 93% in $PbFe_{12}O_{19}$ (after Kahn et al., 1969).

The last class of materials to be mentioned in this subsection deals with *ordered perovskites* $M_2^{2+}Fe^{3+}Mo^{5+}O_6$, where M^{2+} is Ca, Sr, or Ba. They are ferrimagnets with ordering temperatures on the order of 450 K. Shono et al. (1981, 1982b) have reported reflectivity and polar Kerr effect spectra on freshly-polished hot-pressed ceramics. In the Kerr rotation spectra they observed a dispersion-like line centered near 0.8 μm which coincides approximately with the plasma minimum observed in the reflectivity spectrum. Shono et al. repeated the measurements one month after the polishing and found an increase of the Kerr rotation by about 50%, so that the peak rotation for Sr_2FeMoO_6 reached 0.9°. The authors suggest that this increase is due to some oxidation of the surface. Indeed, the reflectivity minimum appears to decrease, allowing a stronger enhancement of the Kerr rotation by smaller optical constants. A quantitative analysis, however, is lacking.

2.5.3.4 Co Compounds

In Sec. 2.5.2.3 we pointed to the large number of binary Co alloys and intermetallic compounds that have been investigated at room temperature by Buschow et al. (1983). Co_3B and Co_2Mg are particularly worth mentioning; their Kerr rotations amount to 0.25 and 0.28°, respectively, which are among the largest Kerr rotations of the materials listed by Buschow et al. Other compounds studied in more detail include Ga–Co and Hf–Co intermetallics. In general, Buschow et al. (1983) found a decrease of the Kerr signal with decreasing Co content and only minor changes in the shape of the spectra.

Magneto-optical spectra with more pronounced structures have been observed in the binary compound CoS_2 (Sato and Teranishi, 1982). CoS_2 crystallizes in the cubic pyrite structure and orders ferromagnetically at ≈ 120 K. From a measurement of the reflectance magneto-circular dichroism at 4 K on a single crystal, Sato and Teranishi have computed the Kerr rotation and find a sharp peak at 0.82 eV with a rotation of 1.1°. An inspection of the reflectivity (Sato and Teranishi, 1981) shows that this peak energy is close to the plasma minimum, indicating that some enhancement effect may occur. Figure 2-60 displays the absorptive (ε_1'') and the dispersive (ε_1') part of the off-diagonal dielectric tensor element, where an interband transition can clearly be seen at 0.8 eV. Sato and Teranishi assign the structure to a transition $^2E \rightarrow {}^2T_1$ in the $3d^7$ low-spin configuration of Co^{2+}.

Among the ternary Co compounds, the largest amount of magneto-optical data exists for the spinel-type structure. Those cobalt spinels which also contain Fe have already been discussed in the previous subsection, while those containing Cr, for example, have been reserved for this subsection because their magneto-optical signal originates exclusively from cobalt. $CoCr_2O_4$ and $CoCr_2S_4$ are (normal) spinels which order ferrimagnetically below about 96 and 235 K, respectively. They are both insulators. Figure 2-61 displays the reflectance magneto-circular dichroism $\Delta R/R$ and the reflectivity R as reported by Ahrenkiel et al. (1973) on hot-pressed powders of $CoCr_2S_4$ at 80 K. The leveling-off of the reflectivity at long wavelengths indicates the nonmetallic character. The reflectance magneto-circular dichroism shows a huge peak of -30% at 1.6 μm. With the relation for the ellipticity $\tan \varepsilon_K = -\Delta R/4R$, a maximum Kerr ellipticity of 4.2° is calculated. Ahrenkiel et al. (1974) have measured and also computed the Kerr rotation using the Kramers-Kronig relation and find a peak rotation of $\approx -4°$. The Faraday rotation was calculated to amount to 1.3×10^6 deg cm^{-1} ($\equiv 2.27 \times 10^6$ rad m^{-1}). The large magneto-optical signal in this energy range is assigned to

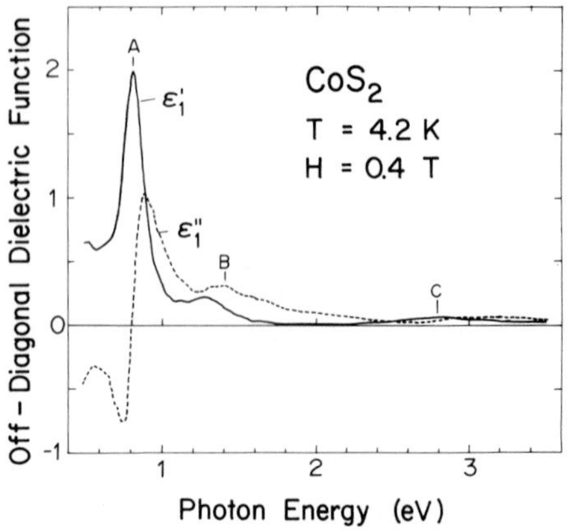

Figure 2-60. Dispersive (ε_1') and absorptive (ε_1'') parts of the off-diagonal dielectric tensor element of CoS_2 single crystals at 4.2 K (after Sato and Teranishi, 1982).

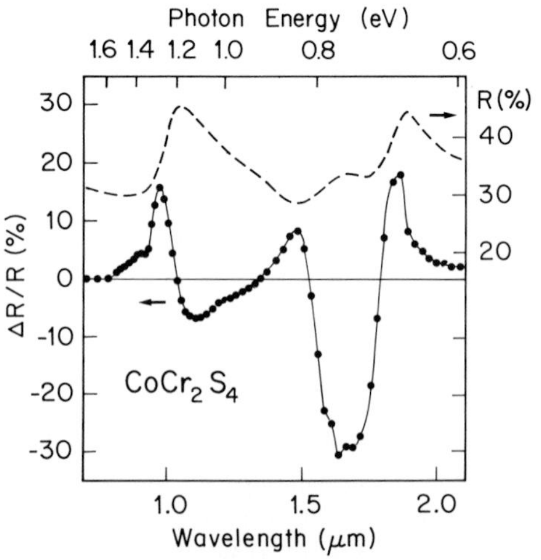

Figure 2-61. Reflectance circular dichroism (full line) and specular reflectance (dashed line) of $CoCr_2S_4$ ceramic at 80 K (after Ahrenkiel et al., 1973).

transitions between crystal-field-split level of Co^{2+} in tetrahedral symmetry. The free-ion ground state for the $3d^7$ configuration is $^4F_{9/2}$ with a ^4P term lying about 1.8 eV higher in energy. The tetrahedral crystal field (T_d) splits the ^4F term into $^4A_2(F)$, $^4A_2(F)$, and $^4T_1(F)$. Spin-orbit interaction splits the excited $^4T_1(F)$ state into Γ_6, Γ_8, and $\Gamma_8 + \Gamma_7$, while the ground state term $^4A_2(F)$ transforms like Γ_8. The exchange splits the ground state further into Zeeman multiplets with $J = -3/2$, $-1/2$, $1/2$, $3/2$, as well as the excited states. At low temperatures ($kT \ll E_{exc}$) only the ground state with $J = -3/2$ will be occupied, and the selection rules for right- and left-circularly polarized light give rise to the circular dichroism (Fig. 2-62). Ahrenkiel et al. (1975 b) have also studied in detail the transition near 1 μm in the substituted spinels $Co_xCd_{1-x}Cr_2S_4$. With increasing x they observed a red shift of the transition energy and a nonlinear increase of the intensity. For $x = 0.25$, 0.75, and 1 the Kerr ellipticity of the peak varies from -3.9 to -4.3 and to $-1.9°$. Ahrenkiel et al. (1975 b) assign this peak to a transition from the $^4A_2(F)$ ground state into the $^4T_1(P)$ term and argue that the larger spatial extent of the P terms compared to the F terms makes transitions into the $^4T_1(P)$ state more sensitive to overlap effects.

In $CoCr_2S_4$ as well as in the substituted Co ferrites shown in Fig. 2-50 the magnetization of the Co sublattice is antiparallel to the applied field (for this reason the ground state in Fig. 2-62 is $J = -3/2$). However, if x in $CoRh_xFe_{2-x}O_4$ exceeds a value between 1.2 to 1.5, the sublattice magnetization on the iron sublattice is smaller than that of Co, and the latter will be parallel to the external field. The same is true for $CoCr_2O_4$ because the Cr sublattice is ordered by a spiral spin arrangement allowing a predominance of the magnetization from the tetrahedrally coordinated Co sublattice. Figure 2-63 displays the reflectance magneto-circular dichroism of hot-pressed powders of $CoRh_{1.5}Fe_{0.5}O_4$ and $CoCr_2O_4$ at 80 K (Ahrenkiel and Coburn, 1975). The inversion about the abscissa of the spectra can be recognized

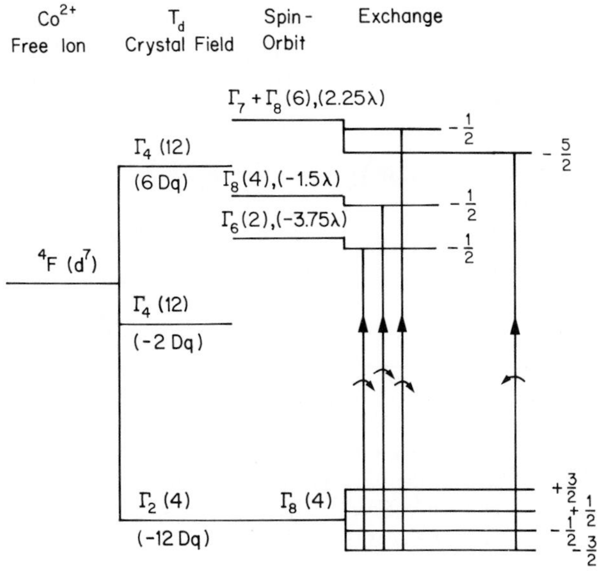

Figure 2-62. Energy levels for Co^{2+} subject to a tetrahedral crystal field, spin-orbit and exchange interaction, and dipole-allowed optical transitions for right- and left-circularly polarized light (after Ahrenkiel et al., 1973).

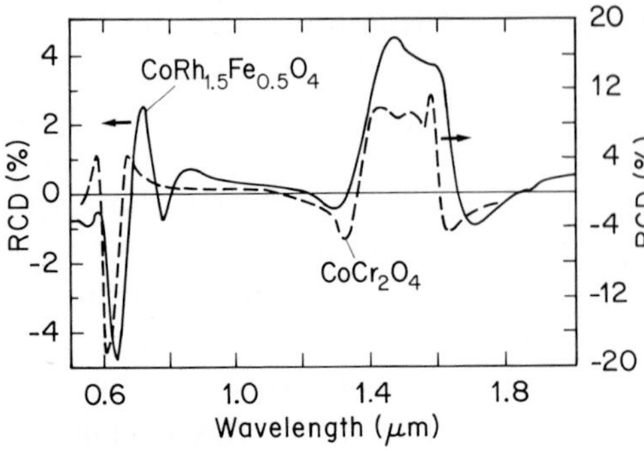

Figure 2-63. Reflectance circular dichroism of ceramic samples of $CoCr_2O_4$ (dashed line) and $CoRh_{1.5}Fe_{0.5}O_4$ (full line) at 80 K (after Ahrenkiel and Coburn, 1975).

when compared to that for $CoCr_2S_4$. A second important difference is the reduction of the absolute values by a factor of about three going from the sulfide to the oxide. As evidenced by absorption measurements, this difference stems from the oscillator strength and is also found if one compares $Co^{2+}(T_d)$ transitions in Co-doped ZnS and ZnO. It indicates that much of the oscillator strength originates from the overlap of the cobalt d states with the ligand orbitals.

Among the Heusler alloys with the general chemical formula X_2YZ, Buschow and van Engen (1981 d) have studied the magneto-optical properties for X = Co, Z = Al or Ga and Y = V, Cr, Mn, Fe, Zr, Nb, Mo, Ta, and Hf. The polar Kerr rotation is generally below 0.1°, except for Y = Fe, where the Kerr rotation and the ellipticity reach maximum values of ≈ 0.4°. The same behavior is also found if Co and Fe interchange their respective sites, i.e., in Fe_2CoZ alloys. In his review, Buschow (1988) discusses these Heusler alloys in some detail as well as Co_2FeGe and Co_2FeSi, which show similar Kerr rotations.

The transverse Kerr rotation for so-called ferroxplana-type ferrites has been studied by Voekov and Zheltukhin (1980). *Ferroxplana-type* compounds are hexagonal barium ferrites like $BaM_2Fe_{16}O_{27}$, $Ba_2M_2Fe_{12}O_{22}$, $Ba_3M_2Fe_{24}O_{41}$, and $Ba_4M_2Fe_{36}O_{60}$, with M being a divalent cation such as Zn, Fe or Co. Figure 2-64 displays the transverse Kerr effect for angles of incidence of 65° and 70° for $BaZn_2Fe_{16}O_{27}$ and $BaCo_2Fe_{16}O_{27}$ (Voekov and Zheltukhin, 1980) The substitution with cobalt clearly produces a peak near 2 eV with an angle of rotation of $\approx 8 \cdot 10^{-3}$ rad ($\equiv 0.46°$).

2.5.3.5 Ni and Cu Compounds

The magneto-optical properties of Ni and Cu compounds have received much less attention than the Mn, Fe, or Co compounds. This reflects, on the one hand, the fact that the increasing filling of the 3d shell leads in general to a decrease of the magnetic moment and the ordering temperature, and on the other, that many of the simpler Ni compounds order antiferro-

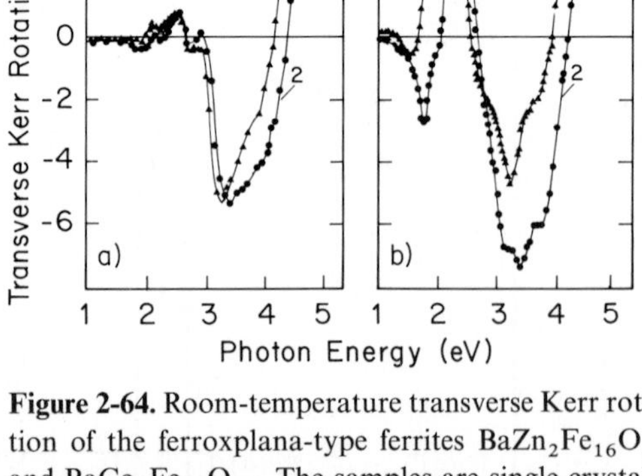

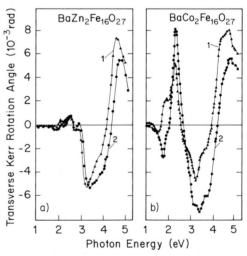

Figure 2-64. Room-temperature transverse Kerr rotation of the ferroxplana-type ferrites $BaZn_2Fe_{16}O_{27}$ and $BaCo_2Fe_{16}O_{27}$. The samples are single crystals, and the angles of incidence are 65° (1) and 70° (2) (after Voekov et al., 1980).

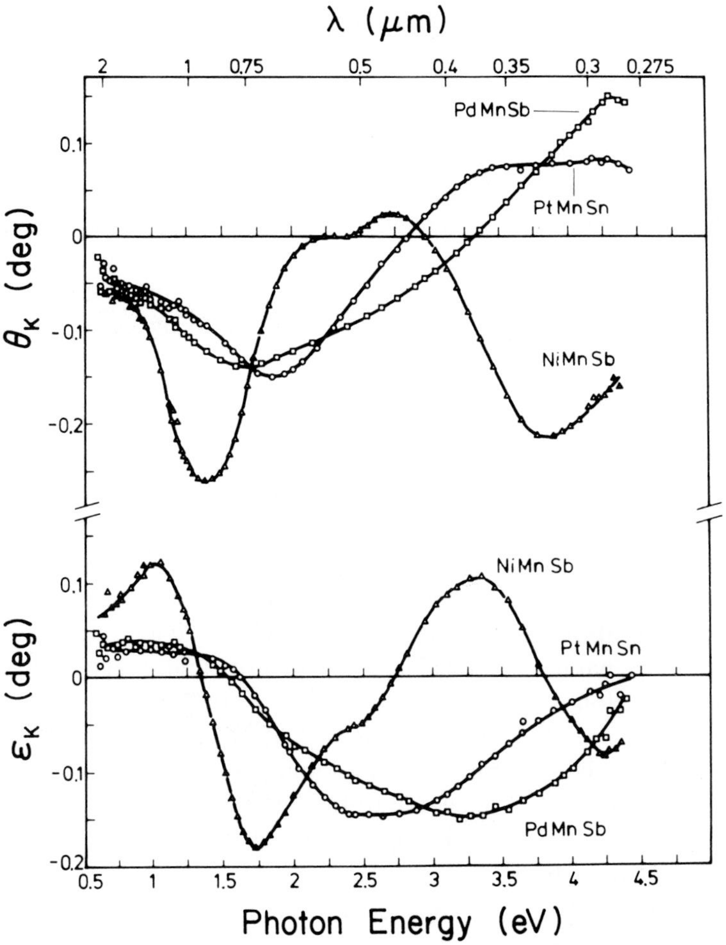

Figure 2-65. Room-temperature polar Kerr rotation (top) and ellipticity (bottom) of polycrystalline bulk samples of PdMnSb, NiMnSb and PtMnSn (after van Engen et al., 1983).

magnetically, while those with monovalent Cu have no permanent magnetic moment at all.

One of the simple Ni compounds, for example, is NiO. It crystallizes in the NaCl-type structure and orders antiferromagnetically at ≈ 524 K. While Kerr spectra have not been published for this compound to my knowledge, linear magnetic birefringence measurements with lasers have been performed to study the long-range order parameter (Germann et al., 1974).

Among the ternary Ni compounds, we discussed the Ni ferrites in Sec. 2.5.3.3. The results of Fig. 2-59 indicate that Ni contributes somewhat to the Kerr rotation spectrum of $NiFe_2O_4$ below 2 eV. For the Heusler alloys $Ni_{3-x}Mn_xSn$ the contribution from Ni is only minor. Of more interest is the Kerr rotation of NiMnSb, which crystallizes in the Heusler $C1_b$ structure. Figure 2-65 displays the polar Kerr rotation and ellipticity of NiMnSb together with the corresponding spectra of PdMnSb and PtMnSn (van Engen et al., 1983). Although the size of the signal is similar for the three materials, the spectra of NiMnSb are more structured and for $\hbar\omega < 2.5$ eV more closely resemble those found in PtMnSb (see Fig. 2-39). Band structure calculations (de Groot and Buschow, 1986) indicate that, as for PtMnSb, the Fermi energy falls into a gap of the minority spin density of states in NiMnSb, while for PtMnSn both spin directions should show normal metallic behavior. The difference between NiMnSb and PtMnSb is quantitative. In the former E_F falls in the middle of the gap, while in the latter it falls just above the top of the valence band, allowing the spin-orbit splitting to push the $m=+1$ state above E_F. In a more recent calculation the different size of the magneto-optical effect was attributed to final states (Wijngaard et al., 1989). These authors found that scalar relativistic effects like mass-velocity and Darwin terms influence the energy of the Γ_1 state more strongly with Pt 6s than with Ni 4s character and move it for PtMnSb to about 1 eV above the top of the minority-spin valence band, while in NiMnSb the energy separation is ≈ 4.2 eV.

Quite some work has been done on alkali-nickel fluorides (see, for example, Pisarev et al., 1974). $KNiF_3$ and Rb_2NiF_4 are antiferromagnets with Néel temperatures of 246 and 90 K, respectively. $RbNiF_3$ orders ferrimagnetically with $T_N = 139$ K. Pisarev et al. have determined and analyzed the optical absorption and the magnetic circular and linear dichroism for these Ni compounds in the temperature range of 6–300 K. From the fine structure which they observed for the line near 2 eV they concluded that the ${}^3A_2 \rightarrow {}^1E^a$ transition has components originating from exciton, exciton-phonon, and exciton-phonon-magnon transitions.

The absorption and Faraday rotation of $(CH_3NH_3)_2CuCl_4$ has been investigated by Arend et al. (1975). This material belongs to a class of *quasi-two-dimensional* materials with the general chemical formula $(C_nH_{2n+1}NH_3)_2M^{2+}X_4$ with $n = 0$, 1, 2, ..., 10, M^{2+} being a divalent 3d ion (mostly Cu or Mn), and X being either Cl or Br. For $(CH_3NH_3)_2CuCl_4$ the intra- and interlayer exchanges are both positive leading to ferromagnetic ordering below ≈ 9 K, while for $n \geqq 2$ the interlayer exchange is negative and the ordering is antiferromagnetic. Figure 2-66 reproduces the Faraday rotation spectra reported by Arend et al. (1975) above and below the Curie temperature. The peaks at 0.9 and 0.75 μm have been assigned to transitions $\Gamma_{t3} \rightarrow \Gamma_{t4}$ and Γ_{t5}, respectively, where the initial and final states are the crystal-field-split states of the free-ion $(3d^9)\,{}^2D$ state for a tetragonal distortion of the Cl octahedron surrounding the Cu^{2+} ion.

The *high-T_c superconductors* can be viewed as doped antiferromagnetic insulators containing Cu^{2+} (e.g., La_2CuO_4, $YBa_2Cu_3O_{6.5}$). In principle, magneto-optics should allow the study of the interactions of Cu with its neighbors and their changes on doping. However, difficulties arise due to the large natural anisotropies of the optical properties (Schoenes et al.,

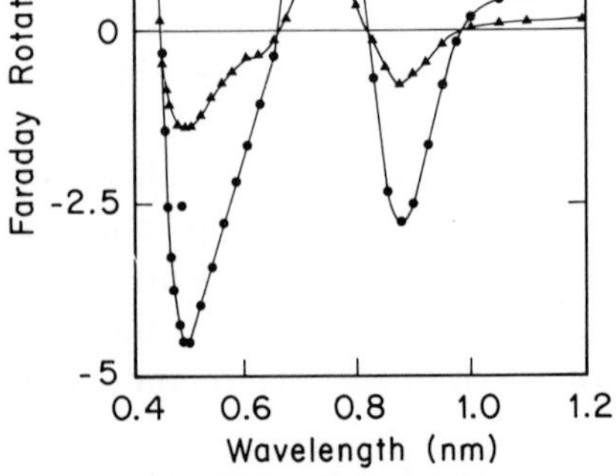

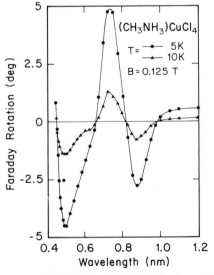

Figure 2-66. Faraday rotation of a $(CH_3NH_3)CuCl_4$ single crystal above and below the Curie temperature of 9 K (after Arend et al., 1975).

1990a) and the twinning in several of these materials. Nevertheless, subtle magneto-optic-like measurements have been performed to test the anyon-superconductivity theory for these oxide superconductors. Lyons et al. (1990) reported the observation of a reflectance circular dichroism (i.e., a polar Kerr ellipticity) without applied field and spontaneous magnetization. On the other hand, Spielman et al. (1990) in an even more sensitive circular birefringence experiment on $YBa_2Cu_3O_7$ thin films, did not observe a nonreciprocal rotation of the polarization direction. Hence, it is not clear at the moment whether the concept of anyon superconductivity applies for high-T_c superconductors, but it is to be expected that many more of these crucial magneto-optic experiments will be performed in the future.

2.6 Lanthanide Materials

The magnetic properties of lanthanide materials are closely related to the partially filled 4f shell. The main difference between 4f and 3d electrons is the stronger localization of the former. As a result the ordering temperatures of lanthanide materials are generally one to two orders of magnitude smaller, thereby excluding pure lanthanide systems (with no 3d elements) from many practical applications. On the other hand, the strong localization makes the description of the f states in crystal-field models more appropriate. Thus, with the exception of certain "exotic" Ce, Pr, Sm, Tm, and Yb compounds, the magnetic properties are well understood theoretically. Several lanthanide systems have acquired model character for their magnetic and magneto-optical properties. Although many striking magneto-optical results have been reported for some lanthanide materials, other lanthanides have not been studied at all so far, because the 4f binding energy is too large to allow an optical excitation in the energy range up to 6 eV which is commonly used in magneto-optics. In the first subsection we will see the only exception, namely Gd, for which magneto-optical measurements extending up to 12 eV have been reported. As mentioned in the introduction, the magneto-optical properties of f systems have been reviewed comprehensively by Reim and Schoenes (1990), so that the present and the next section on actinides will give more of a summary than a full account of the available data.

2.6.1 Elemental 4f Metals

Among the 14 lanthanides, the magneto-optical properties have only been investigated for Gd, Tb and Dy. The most complete set of data exists for Gd (Lambeck et al., 1963; Erskine and Stern, 1973a; Erskine, 1976, 1977). While Lambeck et al. measured the Faraday rotation, Erskine et al. determined the longitudinal Kerr effect under an angle of incidence of 30°. From the part which is linear with the magnetization, and from the optical constants, Erskine and Stern (1973a) computed the off-diagonal conductivity. Figure 2-67 shows the results for $\sigma_{1xx}(\omega)$ and $\omega\sigma_{2xy}(\omega)$, including the data obtained using synchrotron radiation (Erskine, 1976, 1977). The dashed horizontal line is an estimate of the free-electron contribution to $\omega\sigma_{2xy}$ (see Sec. 2.2.2.5). The peaks at 2 and 4 eV are assigned to $p \rightarrow d$ transitions, the double peak structure being a consequence of a covalency splitting of the 5d band. The structures near 6 and 8 eV have been assigned to $f \rightarrow d$ transitions; this time the doubling being attributed to various screening mechanisms for the hole created in the emission process (Erskine, 1976, 1977).

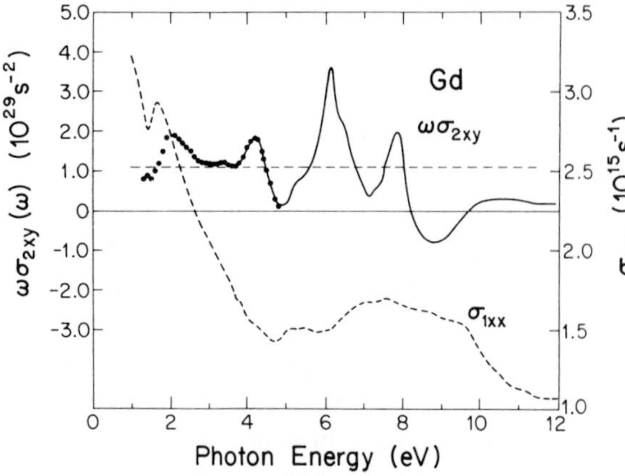

Figure 2-67. Absorptive part of the diagonal conductivity σ_{1xx} and product of the angular frequency and the absorptive part of the off-diagonal conductivity $\omega\sigma_{2xy}$ for polycrystalline Gd films (after Erskine, 1976, 1977).

Reim and Schoenes (1990) have argued that this doubling may also reflect the covalency splitting of the 5d band.

The product of the off-diagonal conductivity and the photon frequency has been reported for Tb and Dy in the energy range from 1 to 4 eV (Erskine, 1975). Few attempts have been made to interpret these spectra, but it appears that a prominent peak above 4 eV in Dy could be due to an f→d transition.

2.6.2 Lanthanide Compounds

The large majority of magneto-optical work on lanthanide systems has been performed on compounds. Few data exist for lanthanide alloys containing only elements other than transition elements. Because these mixed 3d−4f (and also 3d−5f) systems are reserved to Sec. 2.8 of this chapter, the few data on pure lanthanide alloys will be mentioned in the present subsection under compounds.

2.6.2.1 Light Lanthanide Compounds

Like most lanthanides, cerium forms NaCl-type compounds with the chalcogens S, Se, and Te and with the pnictogens, N, P, As, Sb, and Bi. Except for CeN, which appears to have an unstable valence (Baer and Zürcher, 1977; Patthey et al., 1986), they all order antiferromagnetically with Néel temperatures ranging from 2.2 to 25 K. Figure 2-68 displays the complex Kerr effect for CeSb, $SeSb_{0.75}Te_{0.25}$, and CeTe (Reim et al., 1986). At 2 K a field of 5 T is sufficient to reach the saturation moment of 2.06 μ_B/Ce in CeSb, while $CeSb_{0.75}Te_{0.25}$ and CeTe are in an intermediate spin state with a reduced moment. The Kerr signals are larger than in any 3d systems, and in fact the rotation of CeSb is with 14° the largest rotation of any material in this spectral range. Comparing the three compounds, one observes a shift of the dominant structure from 2 eV in CeTe to 1 eV in $CeSb_{0.75}Te_{0.25}$ and to below the limit of the measurements of 0.5 eV in CeSb. The computation of $\sigma_{1xy}(\omega)$ and $\sigma_{2xy}(\omega)$ shows dominant structures at the same energies excluding a pure intraband or enhancement effect (Schoenes, 1987). On the other hand, the diagonal conductivity shows peaks also at the same energies with oscillator strengths as computed for f→d transitions using atomic wave functions (Schoenes and Reim, 1985) and thus evidences that the optical and magneto-optical structures are due to $4f^1 \rightarrow 4f^0\,5d$ transitions. The determination of the initial-state energies in Ce compounds has been one of the most debated issues in the physics of rare earths and intermediate valence, and the magneto-optical results have contributed substantially to resolving apparent contradictions.

Magneto-optics has also been applied to study the 4f state in $CeRh_3B_2$ (Schoenes

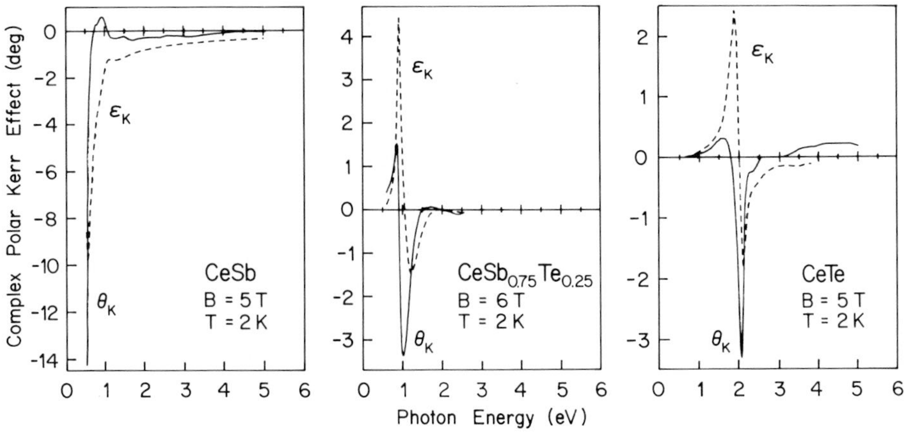

Figure 2-68. Polar Kerr rotation θ_K and ellipticity ε_K of CeSb, $CeSb_{0.75}Te_{0.25}$, and CeTe at 2 K. Note that in this figure and many of the following on the Kerr effect of lanthanide and actinide compounds the signs have been reversed compared to the originals to keep a uniform definition of positive rotation throughout this chapter (after Reim et al., 1986).

et al., 1991). With $T_C = 115$ K, this material has the highest Curie temperature of any known compound of Ce with a non-magnetic element, which raised the question of the possibility of itinerant ferromagnetism from rhodium 4d electrons (Dhar et al., 1981). The fact that the substitution of La for Ce leads to the disappearance of magnetic order and to superconductivity in $LaRh_3B_2$, however, argues against such an interpretation (Shaheen et al., 1985). If one assumes that magnetism is connected with Ce, the small ordered moment of $\approx 0.38\ \mu_B$/f.u. remains to be explained. The magneto-optical Kerr measurements at 10 K and in a field of 4 T show signals of only $\approx 0.1°$, but the $\sigma_{2xy}(\omega)$ spectrum clearly reveals peaks at 1.0 and 1.3 eV which are assigned to $4f^1(^2F_{7/2}) \rightarrow 4f^0\,5d$ and $4f^1(^2F_{5/2}) \rightarrow 4f^0\,5d$ transitions, respectively (Schoenes et al., 1991). It can then be concluded that the 4f states in $CeRh_3B_2$ should be described in a localized picture with a Kondo effect reducing the magnetic moment.

The $f^2(^3H_4)$ free-ion configuration of Pr^{3+} tends to give rise to a non-magnetic singlet ground state in an octahedral crystal field. Yet, the application of large magnetic fields allows one to induce sufficient magnetization for performing magneto-optical measurements. Polar Kerr-effect measurements have been reported for PrSb single crystals (Schoenes et al., 1990b). At 15 K a field of 10 T induces a magnetic moment of 0.9 μ_B/f.u., and a Kerr ellipticity of up to $-1.4°$ has been measured near the plasma minimum of 0.4 eV. However, the main goal of these measurements was the determination of the $f \rightarrow d$ transition energy to check whether Pr^{3+} ions in lower site symmetry may be suited for an increase of the magneto-optical signal in other materials. Figure 2-69 shows the real part of the diagonal as well as the real and the imaginary part of the off-diagonal conductivity. From a comparison of these spectra and the identification of all but one structure below 3 eV as intra $4f^2$ transitions, Schoenes et al. (1990b) concluded that the $4f^2 \rightarrow 4f^1\,5d$ transition occurs at 1.4 eV. This is 1 eV higher than in CeSb, in very nice agreement with the difference in binding energy derived in a su-

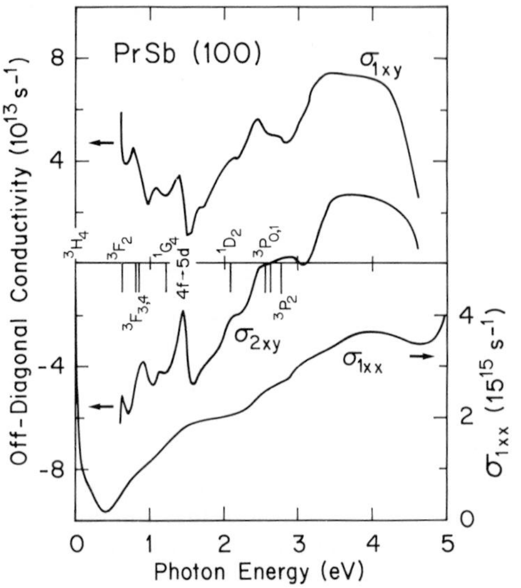

Figure 2-69. Absorptive (σ_{2xy}) and dispersive (σ_{1xy}) parts of the off-diagonal conductivity and absorptive part of the diagonal conductivity σ_{1xx} of PrSb single crystals. The off-diagonal conductivity has been computed from polar Kerr-effect measurements performed at 15 K in a field of 10 T. The vertical bars indicate the energy separation of the excited states from the $^3H_4$ ground state of the f^2 free-ion configuration (after Schoenes et al., 1990 b).

percell calculation (Norman et al., 1985). Thus, Pr^{3+} might indeed be a suitable ion for incorporation in a material for applications.

The next lanthanide is neodymium. All its monochalcogenides and monopnictides order antiferromagnetically, except NdN which appears to order ferromagnetically (Hulliger, 1979). The optical and magneto-optical properties of NdS have been reported by Brändle et al. (1989). Figure 2-70 displays the reflectivity as measured on a (100) oriented surface of a cleaved crystal. The spectrum is very characteristic for the monochalcogenides of trivalent lanthanides. It shows a steep plasma edge which generally shifts to lower energies on going from the sulfide to the selenide and to the telluride due to the decrease of the free-carrier concentration per volume unit with increasing lattice constant. For energies higher than the deep reflectivity minimum, the spectrum is dominated by transitions from the valence band formed by anion p states and cation d and s states into the conduction band formed primarily by cation 5d states. Figure 2-71 shows the complex polar Kerr rotation of the same material at 15 K in a field of 10 T. The

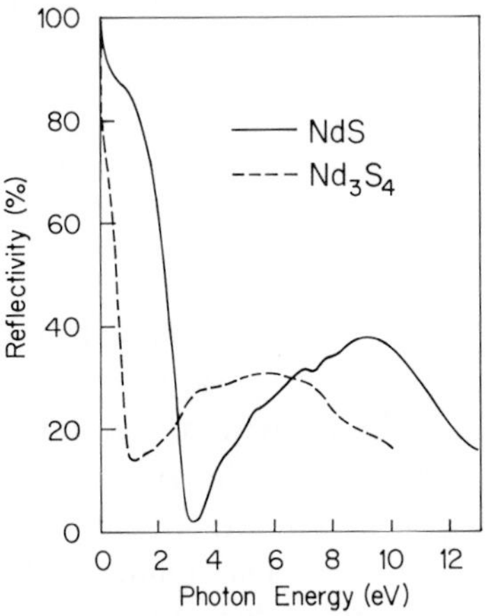

Figure 2-70. Room-temperature normal incidence of cleaved NdS single crystals (after Brändle et al., 1989) and polished Nd_3S_4 single crystals (after Schoenes et al., 1988 a).

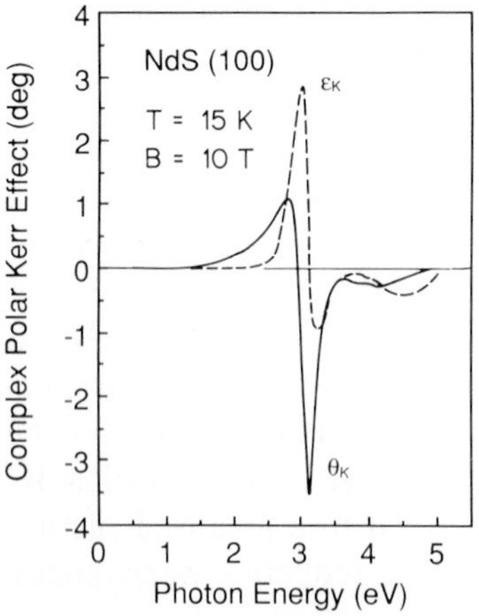

Figure 2-71. Polar Kerr rotation θ_K and ellipticity ε_K of NdS single crystals at 15 K and in a field of 10 T (after Brändle et al., 1989).

magnetization under these external parameters amounts to 0.81 μ_B/f.u., which is approximately 1/4 of the free-ion value of 3.27 μ_B/f.u. The Kerr spectra are dominated by structures at 3 eV, i.e., near the plasma minimum. Additional smaller peaks occur near 4.5 eV. The computation of the off-diagonal conductivity (Fig. 2-72) allows the assignment of these various structures. One recognizes that the smaller structures near 4.5 eV in θ_K and ε_K have transformed to the dominant structures in σ_{1xy} and σ_{2xy}, and these are assigned to the $4f^3 \rightarrow 4f^2 5d$ transitions. However, the large Kerr signal near the plasma minimum transforms into minor structures and, therefore, is primarily due to resonances of the prefactors A and B in Eqs. (2-42) and (2-43) (Brändle et al., 1989).

Neodymium and sulfur also form the ferromagnetic compound Nd_3S_4, with $T_C = 47$ K. Its optical and magneto-optical properties have been investigated on polished single crystals by Schoenes et al. (1988a). The reflectivity spectrum (Fig. 2-70) displays a plasma minimum near 1 eV which reflects the smaller free-carrier concentration of formally 1/3 e^-/Nd compared to NdS with 1 e^-/Nd. Again the complex Kerr effect (Fig. 2-73) shows the largest signal near this plasma minimum and weaker but well-resolved structures in the neighborhood of 4.5 eV. As in NdS, the computation of $\tilde{\sigma}_{xy}$ reveals that the magneto-optically active transition, i.e., the $4f^3 \rightarrow 4f^2 5d$ transition occurs near 4 eV, and that the peaks near 1 eV in θ_K and ε_K are primarily effects of the optical constants (Schoenes et al., 1988a). Figure 2-74 shows an atomic coupling scheme for the $4f^3 \rightarrow 4f^2 5d^1$ transition which explains the S-shaped (i.e., diamagnetic) type of magneto-optical signal in the absorptive spectra. The various interaction terms in the final state are considered in order of decreasing energy. The largest splitting is due to the cubic part of the crystal field on the 5d

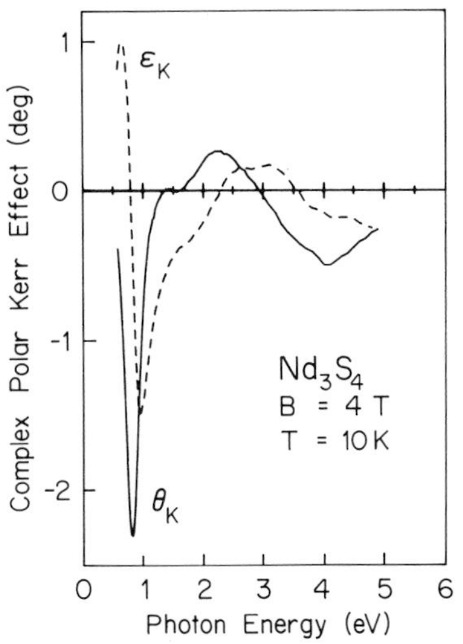

Figure 2-73. Polar Kerr rotation θ_K and ellipticity ε_K of Nd_3S_4 single crystals at 10 K in a field of 4 T (after Schoenes et al., 1988 a).

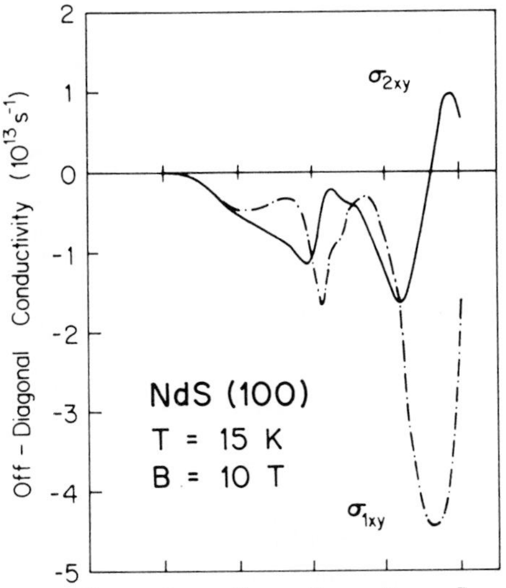

Figure 2-72. Dispersive (σ_{1xy}) and absorptive (σ_{2xy}) parts of the off-diagonal conductivity of NdS, as computed from the data in Figs. 2-70 and 2-71 (after Brändle et al., 1989).

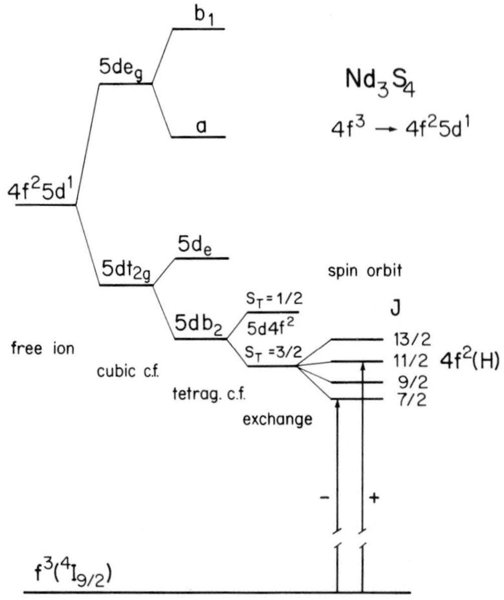

Figure 2-74. Coupling scheme for the $4f^3 \rightarrow 4f^25d^1$ transition in Nd_3S_4 and dipole-allowed transitions for right- and left-circularly polarized light.

state, followed by the tetragonal part. The third term is the energy difference between a parallel and antiparallel orientation of the excited electron to the remaining $4f^2$ state. Spin conservation requires transitions into the state with $S_T = 3/2$. The fourth term is the spin-orbit interaction between the total spin and the orbital momentum of the $4f^2$ state. Dipole-allowed transitions require $\Delta J = 0, \pm 1$. While $\Delta J = 0$ leads to transitions which have similar weight for right- and left-circularly polarized light, the transitions with $\Delta J = -1$ and $+1$ give a negative and positive signal for σ_{2xy}, respectively.

McGuire and Gambino (1987) have studied the Faraday rotation in amorphous films of binary alloys of the lanthanides Pr, Nd, Gd, Tb, and Dy with Au, Cu, and Al. Many of these films are found to show *asperomagnetic order*, which means that the atomic moments are dispersed over a hemisphere because of random crystal fields (Coey et al., 1981). The magnetization observed at 3 T is about half that expected for a parallel alignment of all atomic moments. Figure 2-75 shows the Faraday rotation spectra for the Nd alloys. Nd is the lanthanide which gives the largest signals in this series. McGuire and Gambino (1987) find a broad peak at a wavelength corresponding approximately to the energy separation of the empty $4f^4$ state from E_F, as derived for Nd metal by Lang et al. (1981), and assign the Faraday rotation peak to transitions of 5d electrons into this $4f^4$ state.

The next and last light lanthanide is samarium. The Sm monochalcogenides at-

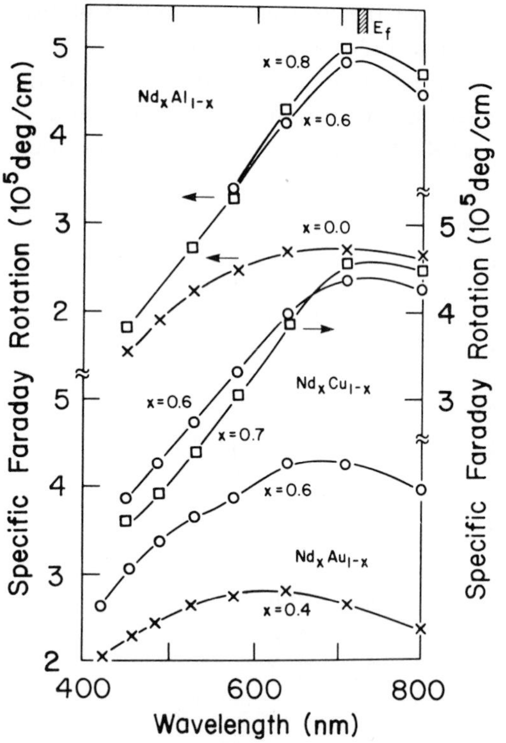

Figure 2-75. Wavelength dependence of the Faraday rotation per unit length of amorphous films of Nd_xAl_{1-x}, Nd_xCu_{1-x}, and Nd_xAu_{1-x} at 1.6 K in a field of 3.4 T. The wavelength corresponding to the energy difference between the energy of the empty f state and E_F is marked on the top of the figure (after McGuire and Gambino, 1987).

tracted large interest in the 1970s after the discovery of a semiconductor → metal phase transition under pressure in SmS (Jayaraman et al., 1970). It was shown that in the metallic phase the valence of Sm is intermediate between two and three, i.e. the ground state is a linear combination of $4f^5$ and $4f^6$ wave functions. The intermediate-valence phases in the Sm monochalcogenides are non-magnetic, and no magneto-optical spectrum has been reported. However, magnetic-circular-dichroism spectra have been measured on semiconducting SmTe (Suryanarayanan et al., 1972). The free-ion ground state of Sm^{2+} is $4f^6(^7F_0)$ and, consequently, the magneto-optical signal is very small. Nevertheless, Suryanarayanan et al. detected two S-shaped structures near 3 and 3.7 eV which they assigned to transitions $4f^6(^7F_0) \rightarrow 4f^5(^6H)5de_g$ and $4f^6(^7F_0) \rightarrow 4f^5(^6F)5de_g$, respectively.

2.6.2.2 Lanthanide Compounds with a Half-Filled f-Shell

This subsection is mostly devoted to Eu^{2+} compounds, because to my knowledge Gd^{3+} compounds of non-magnetic elements have not been studied by magneto-optical means. Amorphous alloys of Gd with Al, Cu, and Au will be discussed shortly at the end of this subsection. The largest amount of magneto-optical work on lanthanides has been performed on the europium monochalcogenides. Since the discovery of ferromagnetism in semiconducting EuO by Matthias et al. (1961), these materials have acquired model character for their magnetic as well as for their magneto-optical properties. The first striking magneto-optical effect discovered was the exchange-induced red shift of the absorption edge (Busch et al., 1964). Subsequently, very large rotations of the plane of polarization were observed in the ferromagnetic phases (Greiner and Fan, 1966; Suits et al., 1966). Figure 2-76 shows more recent data of the Faraday rotation for (100)-textured EuS films under external parameters which warrant magnetic saturation (Schoenes, 1979). Also shown is the Faraday ellipticity as computed from the Faraday rotation spectrum with the Kramers-Kronig relation and in agreement with direct measurements by Ferré

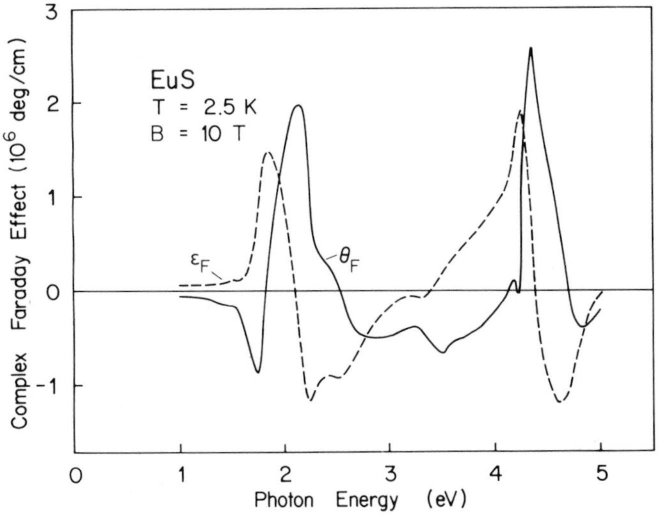

Figure 2-76. Faraday rotation θ_F and ellipticity ε_F per unit length of magnetically saturated EuS films (after Schoenes, 1979).

(1974). The specific Faraday rotation reaches a maximum value of 2.5×10^6 deg cm^{-1} ($\equiv 4.4 \times 10^6$ rad m^{-1}), which, together with similar values for EuSe and EuTe (Schoenes, 1979), is the largest value for any material in this spectral range. The shape of the magneto-optical spectra has been explained by an atomic coupling scheme (Schoenes, 1975) similar to that for Nd_3S_4 displayed in Fig. 2-74. Dipole-allowed, strongly polarization-dependent transitions occur from the $4f^7(^8S_{7/2})$ ground state into $4f^6 5d^1$ states with $J = 5/2$ and $9/2$. The $5d^1$ state is split by the octahedral crystal field into t_{2g} and e_g states leading to the two groups of structures centered near 2 and 4 eV in EuS. The shoulder on the high-energy side of the S-shaped curve has been assigned to a spin-flip transition with $\Delta J = 0$. This last assignment has also gained support from the observation that in the antiferromagnetic phases of EuTe (Schoenes, 1975; Schoenes and Wachter, 1977a) and EuSe (Schoenes and Wachter, 1977b) this peak grows while the net magnetization of the sample decreases. Figure 2-77 displays the corresponding data for EuTe. In the bottom part, one can see that in a rather small field a Faraday rotation peak grows at 2.8 eV when the temperature is lowered below the Néel temperature of 9.6 K. In the upper part of the figure, this Faraday rotation peak is suppressed if on application of a large magnetic field the antiferromagnetic order is destroyed and the spins are aligned. This behavior can either be interpreted in the *Slater model*, which postulates an antiferromagnetic superlattice-band splitting, or in terms of a model in which the on-site spin-flip transition becomes a neighbor-site spin-conserving transition in the antiferromagnetic phase. Schoenes (1975) has also decomposed the σ_{2xy} spectra computed from the Faraday rotation measurements of EuS, EuSe, and EuTe into single transition components and compared the oscillator strength of $f \rightarrow d$ transitions with theoretical values using atomic wave functions. The agreement was

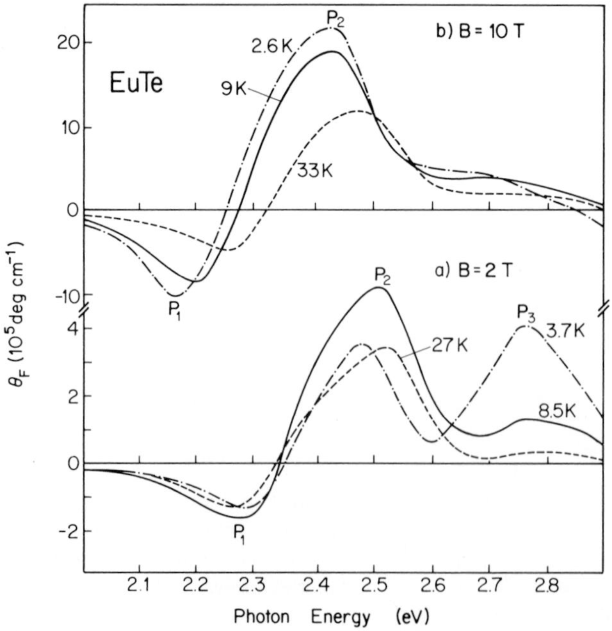

Figure 2-77. Faraday rotation per unit length at temperatures above and below the Néel temperature of 9.6 K of EuTe. The lower curves are for a field of 2 T which is far below the spin-alignment field, while for $B = 10$ T the spins are aligned (after Schoenes and Wachter, 1977a).

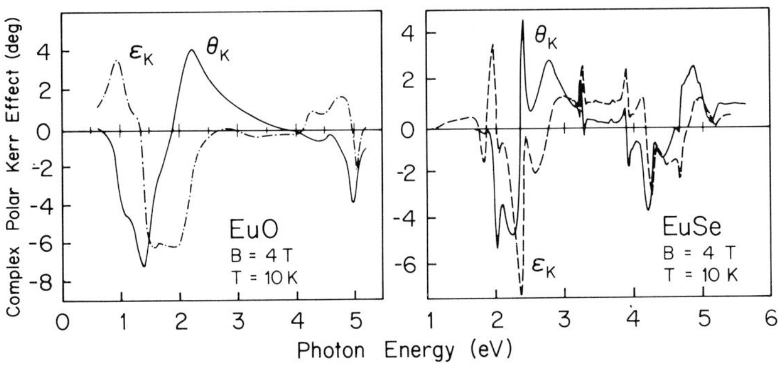

Figure 2-78. Polar Kerr rotation θ_K and ellipticity ε_K of single crystals of EuO (left side) and EuSe (right side) at 10 K and in a field of 4 T. These external parameters saturate the magnetization in EuO but not in EuSe (after Schoenes and Kaldis, 1987).

found to be good, corroborating the assignments performed on energy and lineshape arguments.

The polar Kerr effect has been measured on (100)-oriented single crystals of EuO (Wang et al., 1986) and EuSe (Schoenes and Kaldis, 1987). As can be seen in Fig. 2-78, the EuSe spectra obtained on single crystals display narrower structures, but the interpretation within the coupling scheme derived earlier is still valid. The largest peak in EuO reaches a rotation value of about $-7°$. The external parameters used to obtain the data for EuSe displayed in Fig. 2-78 do not saturate the sample. At saturation, i.e., at 2 K and in a field of 4 T, the largest rotation is 9.7° (Schoenes and Kaldis, 1987). Greiner and Fan (1966) reported for the longitudinal Kerr effect extrema of $-2.5°$ in EuO and $-1.5°$ in EuS. In addition to these classical magneto-optical effects, there exist studies of magneto-reflectance (Feinleib et al., 1969; Pidgeon et al., 1969; Scouler et al., 1969; Güntherodt, 1974), thermo-reflectance (Mitani and Koda, 1973), electro-reflectance (Löfgren et al., 1974), and magneto-absorption (Busch et al., 1974; Freiser et al., 1968; Wachter, 1972; Schoenes, 1975, 1979; Schoenes and Nolting, 1978; Mitani et al., 1975; Llinares et al., 1973; Ferré, 1974). Finally, I would like to mention the use of Faraday-rotation measurements for the study of critical exponents near the Curie temperature of EuO by Huang et al. (1974) and Huang and Ho (1975) and of EuS by Berkner (1975).

Magnetic semiconductors can also be doped. One goal was to try to increase the ferromagnetic ordering temperature to above room temperature. Although nearly a doubling of the Curie temperature from 69 K in pure EuO to 135 K in EuO doped with 3.4 at.% Gd was achieved (Shafer and McGuire, 1968), doping failed to raise T_C to temperatures desired for many applications. The other goal was to study the exchange between conduction electrons and local moments, which led to some interesting results for our basic understanding of magnetism (see, e.g., Methfessel and Mattis, 1968; Kasuya, 1972; Schoenes and Wachter, 1974; Nolting and Oles, 1981).

The high rotatory power of Eu^{2+} has manifested itself also in the europium doping of CaF_2 (Shen and Bloembergen, 1964), SrS and KBr (Methfessel and Mattis, 1968), silicate glasses (Schoenes et al., 1979), or the matrix of EuF_2 (Suits et al., 1966) and Eu_2SiO_4 (Shafer et al., 1963). This rotatory power has also been used for the first time to investigate magnetic superconductors by magneto-optical means (Fumagalli et al., 1990). The class of materials studied are the Chevrel phases $Eu_{1-x}Sn_xMo_6S_{8-y}Se_y$, which, depending on the exact values of x and y, can be superconducting below critical temperatures ranging up to about 7 K or may show magnetic-field-induced superconductivity. Figure 2-79 shows the phase diagram for $Eu_{0.75}Sn_{0.25}Mo_6S_{8-y}Se_y$ as reported by Rossel et al. (1985). The magnetic-field-induced superconducting phase is seen to occur for samples with $y = 0.8$ at temperatures below 1 K and fields exceeding several tesla. Figure 2-80 displays results of some of the Kerr-rotation and -ellipticity measurements performed on these compounds at temperatures as low as 0.5 K and fields as high as 12 T (Fumagalli et al., 1990; Fumagalli and Schoenes, 1991). The polar Kerr effect is rather weak, but the spectra clearly show the characteristic line shape of $4f^7 \rightarrow 4f^6 5d$ transitions. As in the europium chalcogenides the presence of two S-type structures reflects the crystal-field splitting of the 5d final state into t_{2g} and e_g states, which is determined to be 0.9 eV. A comparison of the intensity shows that the magneto-optical signal is smaller than expected after normalization with the Eu^{2+} concentration indicating a smaller f–d overlap than in the europium chalcogenides. The temperature and field dependence of the magneto-optical signal allows the study of the variation of the spin polarization of the Eu^{2+} state. The different behavior at different photon energies led to the discovery of a hidden transition which was confirmed by Kerr-effect measurements on $PbMo_6S_8$ and which could

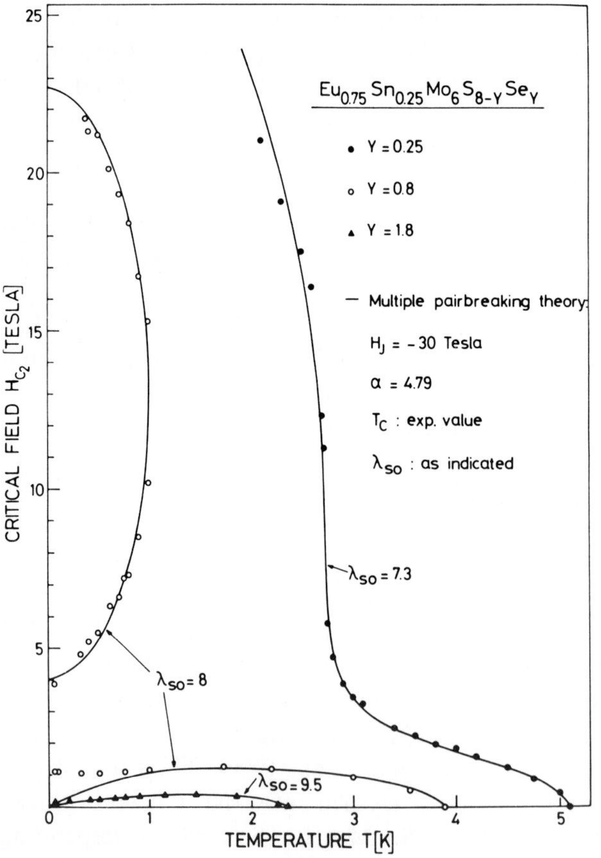

Figure 2-79. Critical field H_{c2} as a function of temperature for three samples of the series $Eu_{0.75}Sn_{0.25}Mo_6S_{8-y}Se_y$. The solid lines are fits to the data using a multiple pair-breaking theory (after Rossel et al., 1985).

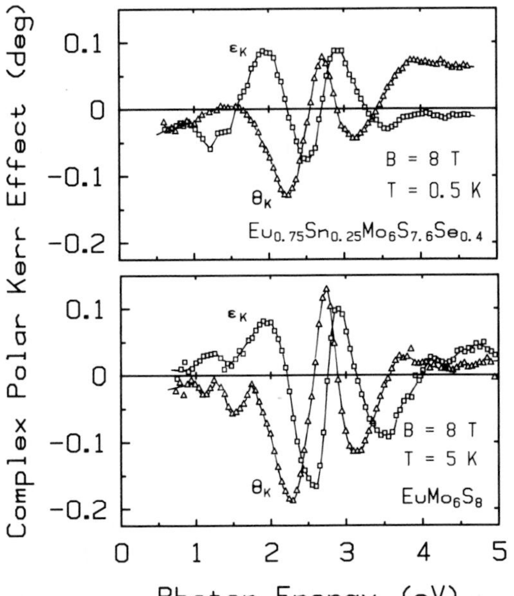

Figure 2-80. Polar Kerr rotation θ_K and ellipticity ε_K of the Chevrel phases $Eu_{0.75}Sn_{0.25}Mo_6S_{7.6}Se_{0.4}$ (top) and $EuMo_6S_8$ (bottom). The field is 8 T and the temperature is 0.5 and 5 K, respectively (after Fumagalli et al., 1990).

be identified as Mo 4d → S 3p transition (Fumagalli, 1990; Fumagalli and Schoenes, 1991).

As mentioned already in the previous subsection, McGuire and Gambino (1987) have also studied the Faraday rotation of amorphous films of Gd_xAl_{1-x}, Gd_xCu_{1-x}, and Gd_xAu_{1-x}, where $x = 0.5, 0.6$, and 0.7. The Faraday rotation shows very broad maxima at energies varying between 1.8 and 2.5 eV with the alloying element and with x. The absolute value of the Faraday rotation increases with increasing x in most cases. Since the empty $4f^8$ state is about 4 eV above E_F in Gd metal (Lang et al., 1981), a similar interpretation as for the Nd alloys, i.e., in terms of d→f transitions, is excluded and McGuire and Gambino speculate that the Faraday rotation peak may correspond to d→p transitions.

2.6.2.3 Heavy Lanthanide Compounds

In Sec. 2.4.2 some data for glasses containing Tb^{3+} have already been presented. More data exist for alloys of heavy lanthanides with transition elements to be discussed in Sec. 2.8. From the large number of heavy lanthanide chalcogenides and pnictides only a few Tm and Yb compounds have been studied by magneto-optical methods. In particular, the thulium chalcogenides have received much attention due to the occurrence of intermediate valence in TmSe. Figure 2-81 displays the polar Kerr effect of TmS and TmSe single

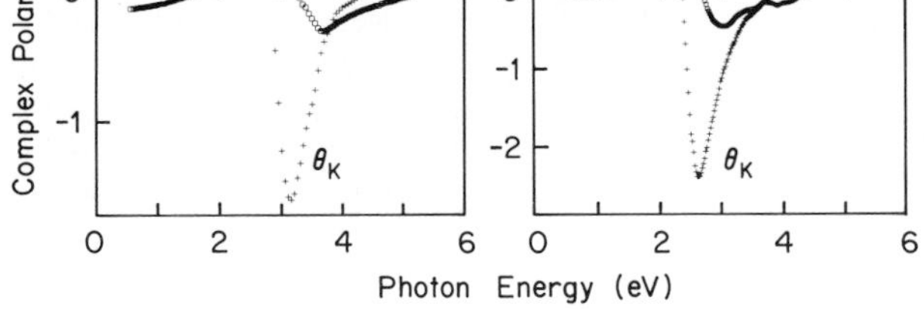

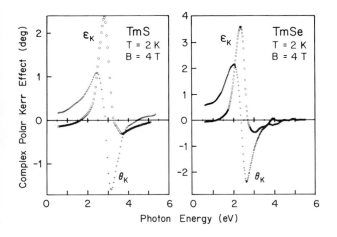

Figure 2-81. Polar Kerr rotation θ_K and ellipticity ε_K of single crystals of TmS and TmSe at 2 K and in a field of 4 T (after Reim et al., 1984 a).

crystals at 2 K and a field of 4 T (Reim et al., 1984a). In integer trivalent TmS as well as in intermediate valent TmS the Kerr spectra are dominated by one structure occurring near the respective deep plasma minima in the reflectivity spectra. Reim et al. have fitted the θ_K and ε_K spectra with a Drude-model-like expression. However, in TmS, for example, a plasma-edge splitting as large as 30 meV had to be assumed to reproduce Kerr signals of the order of 2°, compared to only 10^{-2} deg found in Ag. The off-diagonal conductivity $\tilde{\sigma}_{xy}(\omega)$ was also computed and is shown in Fig. 2-82 together with the calculated curves using Eq. (2-76) and the corresponding real part of $\tilde{\sigma}_{xy}$ (Reim et al., 1984a; Schoenes et al., 1985). The absence of substantial structure in $\tilde{\sigma}_{xy}$ at the energies of the dominant structures in θ_K and ε_K evidences that the latter are not determined by interband transitions. The excellent fits with Eq. (2-76), on the contrary, indicate that the magneto-optical signal is due to a different plasma energy for right- and left-circularly polarized light. The calculations give splittings of 30 and 60 meV for TmS and TmSe, respectively, compared to 0.4 meV expected from the classical cyclotron resonance frequency in the same field of 4 T. This led Reim et al. (1984a) to the conclusion that the driving mechanism is the exchange coupling between conduction electrons and local moments, and not just the direct action of the applied field on the free carriers.

TmTe is a semiconductor with a gap of ≈ 0.3 eV and divalent Tm. On substitution of Te with Se, the pseudo-binary compounds $TmSe_{1-x}Te_x$ remain semiconducting for $x > 0.4$, while for $x < 0.2$ they are metallic (Boppart and Wachter, 1984). For $0.2 < x < 0.4$ a miscibility gap exists (Kaldis et al., 1982). Figure 2-83 displays the complex polar Kerr effect for semiconducting $TmSe_{0.32}Te_{0.68}$ (Schoenes et al., 1985). The spectra are competely different from those of the metallic Tm compounds discussed above. A variety of structures are present, which, after transformation into the off-diagonal conductivity and a comparison with the diagonal conductivity, have been assigned to $4f^{13} \rightarrow 4f^{12}5d$ transitions.

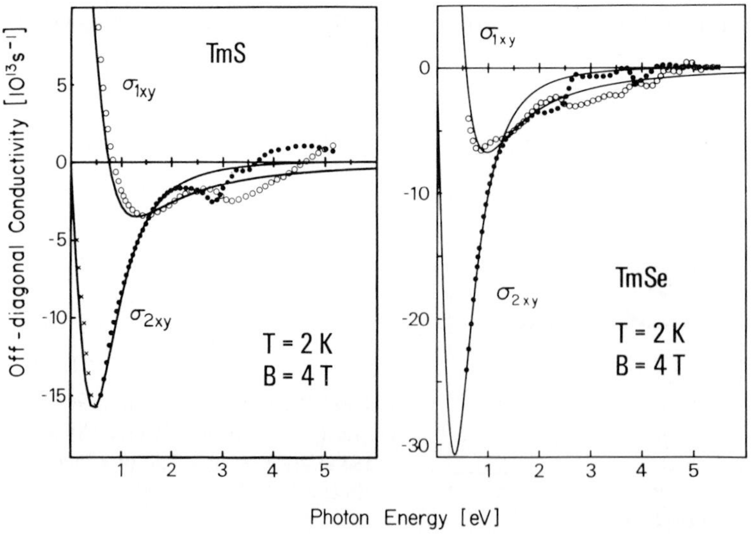

Figure 2-82. Dispersive (σ_{1xy}) and absorptive (σ_{2xy}) part of the off-diagonal conductivity of TmS and TmSe as derived from the Kerr spectra in Fig. 2-81 and the optical constants. The full lines are fits with the first term on the right side of Eq. (2-74) (after Reim et al., 1984 a, and Schoenes et al., 1985).

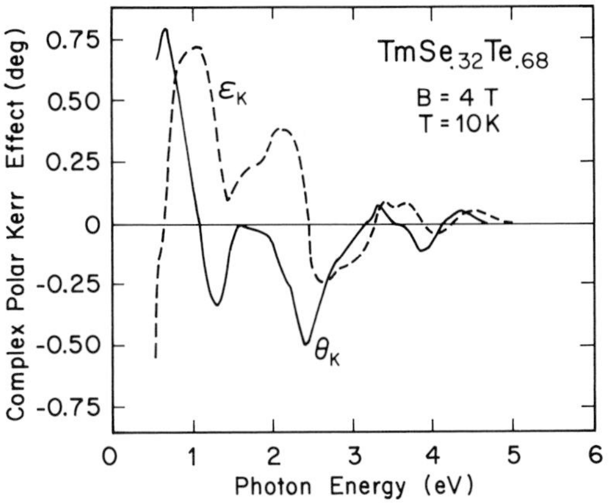

Figure 2-83. Polar Kerr rotation θ_K and ellipticity ε_K of semiconducting $TmSe_{0.32}Te_{0.68}$ at 10 K and in a field of 4 T (after Schoenes et al., 1985).

Suryanarayanan et al. (1970, 1974) have studied the optical absorption and the magnetic circular dichroism of YbSe and YbTe, respectively. They found these materials to be insulators, in agreement with the simple ionic model that in these diamagnetic compounds Yb is divalent and has a completely filled 4f shell. The circular dichroism spectrum of YbTe displays three structures which Suryanarayanan et al. (1974) assign to $4f^{14} \rightarrow 4f^{13}\,5d\,(t_{2g}, e_g)$ transitions. The spin-orbit interaction splits the $4f^{13}$ state into $F_{7/2}$ and $F_{5/2}$ producing a total of four transitions of which two are nearly degenerate in energy. Thus, the Yb chalcogenides provide another excellent example demonstrating that magneto-optics depend as much on the final as on the initial state of the optical transition. While in the europium chalcogenides the ground state is magnetic but has no spin-orbit splitting, so that the spin-orbit splitting in the final state is responsible for the large magneto-optical effects, the ytterbium chalcogenides are non-magnetic and the final state provides spin-polarization as well as spin-orbit splitting.

2.7 Actinide Materials

In the actinide series the 5f shell is gradually filled. If one normalizes the spatial extent of the 3d, 4f, and 5f atomic orbitals by the corresponding lattice spacings in transition, lanthanide, and actinide metals, actinides fall between transition metals and lanthanides (Freeman and Koelling, 1974). This means that as a first approximation the 5f electrons should be less localized than the 4f electrons but more localized than the 3d electrons. The ordering temperatures are expected to be higher than in lanthanide systems and lower than in 3d materials. This crude approach finds corroboration in several actinide materials, but can not be general as it neglects bonding and hybridization effects.

A severe drawback for the study of solid-state properties in general and magneto-optical properties in particular is the scarcity and the radioactivity of most actinides. Because thorium has no 5f electron and thorium systems therefore are non-magnetic, only uranium compounds have so far been investigated by magneto-optical methods. Yet, the detection of large magneto-optical signals found in several uranium compounds has led to investigations of many different types of compounds, and the following discussion will be grouped according to these types. Uranium metal is non-magnetic, and has not been studied.

2.7.1 Uranium Dioxide

UO_2 crystallizes in the cubic CaF_2-type structure and orders antiferromagnetically at 30.8 K. It is an insulator , but the determination of the gap energy and the nature of the states near the gap were the object of keen debates until the beginning of the 1980s (Schoenes, 1980a, 1984). Magneto-optics have contributed substantially to unravel the electronic structure of this compound. Reim and Schoenes (1981) measured the Faraday rotation and the magnetic circular dichroism on thin UO_2 films between 1.5 and 5.5 eV. From these data and the optical constants derived from a Kramers-Kronig transformation of reflectivity data on single crystals (Schoenes, 1978) they computed the off-diagonal conductivity. Figure 2-84 displays the absorptive part of the off-diagonal conductivity and the absorption coefficient of UO_2. To interpret these data, Reim and Schoenes (1981) considered the $5f^2 \rightarrow 5f^1 6d^1$ transition in the two approximations of $L-S$ and $j-j$ coupling. While the former coupling can not account for the two main structures observed near 3.3 and 4.2 eV, the latter coupling leads to the relative intensities shown as vertical bars in Fig. 2-84, which indeed fit the experimental data very well. Thus, UO_2 constitutes an example for localized 5f electrons, but the strong spin-orbit coupling necessitates that the final $5f^1 6d^1$ state is described in $j-j$ coupling, contrary to the situation in lanthanide compounds. Figure 2-85 shows the density-of-states scheme as derived from optical and magneto-optical measurements (Schoenes, 1984). The gap is formed by transitions from the localized $5f^2$ state into the $6de_g$ sub-band with a coupling between the excited electron and the remaining $5f^1$ state. The transitions from the p valence band into the 6d conduction band set in near 5 eV.

2.7.2 NaCl-Type Compounds

Uranium forms with the chalcogens S, Se, and Te as well as the pnictogens N, P, As, Sb and Bi face-centered-cubic compounds of the NaCl type. The monochalcogenides order ferromagnetically with ordering temperatures decreasing from

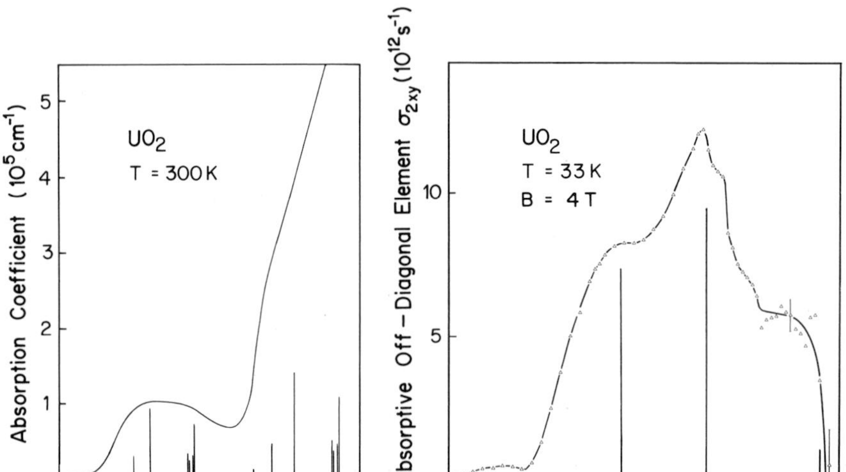

Figure 2-84. Absorption coefficient (left side) and absorptive part of the off-diagonal conductivity σ_{2xy} (right side) of UO_2. The vertical bars indicate the relative oscillator strengths of transitions from the $^3H_4$ ground state of the $5f^2$ state into the final $5f^1 6d^1$ states as computed in $j-j$ coupling. For the absorption the transitions are shown for the $5f^1 6de_g$ and $5f^1 6dt_{2g}$ final states, while for σ_{2xy} only the transitions into the lower crystal-field split states fall in the energy range of the measurements (after Reim and Schoenes, 1981).

177 K in US, to 160 K in USe, to 102 K in UTe. The monopnictides order antiferromagnetically with ordering temperatures increasing with increasing lattice constant from 53 K in UN to 285 K in UBi (Lam and Aldred, 1974; Schoenes, 1980 b; Rossat-Mignod et al., 1984). The optical properties (Schoenes, 1980 b, 1984) and the electrical transport properties (Schoenes et al., 1984) indicate that the monochalcogenides are metals and the monopnictides are preferably described as semimetals. In a simplified model the different magnetic behavior can then be viewed as follows: In the pnictides the antiferromagnetic superexchange via the anions dominates. The addition of formally one electron when we move from the Va to the VIa group elements leads to a ferromagnetic exchange of the Ruderman-Kittel-Kasuya-Yosida type which overcompensates the antiferromagnetic superexchange. In the uranium monochalcogenides, the 5f electrons hybridize with the 6d conduction electrons and form a narrow band at E_F. In the uranium monopnictides, hybridization of the 5f electrons with the anion p-states, so-called $p-f$ mixing, plays an important role (Suzuki et al., 1982). With increasing lattice constant, the 5f band narrows and may eventually be better described in a localized picture for the heaviest anions (Schoenes, 1984).

The polar magneto-optical Kerr rotation and ellipticity has been measured on cleaved single crystals of uranium monochalcogenides by Reim et al. (1984 b). Figure 2-86 displays these Kerr spectra and the derived off-diagonal conductivity for $T = 15$ K and a saturating field of 4 T. A remarkable feature is the size of the Kerr signal, reaching up to 4° in these metallic compounds. On going from US to UTe, the spectra become simpler and narrower. A comparison with the diagonal conductivity

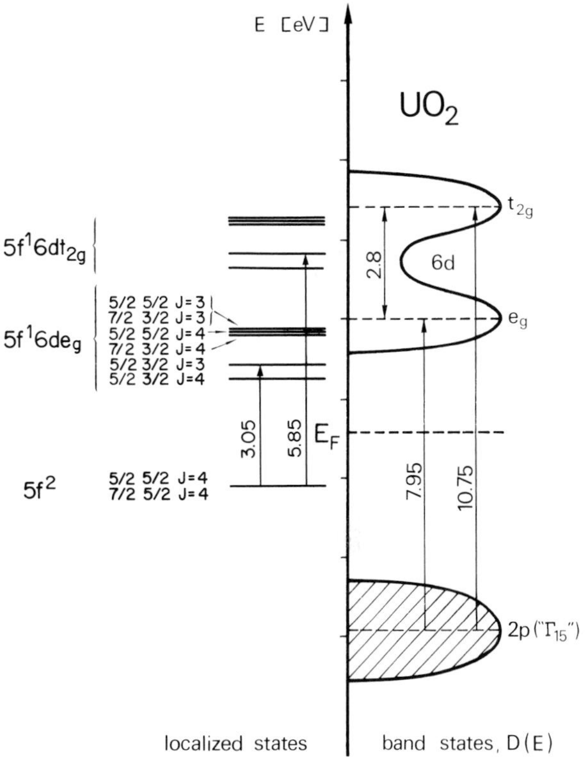

Figure 2-85. Energy level scheme of UO_2 derived from optical and magneto-optical measurements (after Schoenes, 1984).

shows that the spectra consist of an intraband contribution and two interband transitions. An estimate of the former is indicated in Fig. 2-86 as solid and dashed lines for σ_{2xy} and σ_{1xy}, respectively. The two interband contributions are assigned to an f→d transition with a diamagnetic line shape in σ_{2xy} at lower photon energy and a d→f transition with a paramagnetic line shape in σ_{2xy} at higher photon energy (Reim et al., 1984 b). On account of the increasing localization of the f state with the atomic number of the anion, the d→f transition weakens from US to USe, to UTe, explaining the simpler spectra for UTe. The sign of the free-carrier contribution indicates a negative spin polarization of the 6d conduction electrons, which, together with

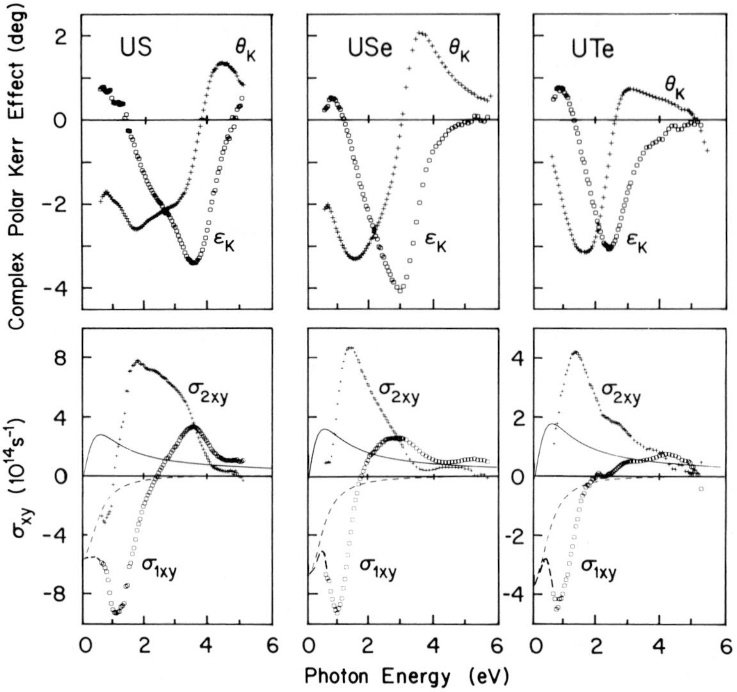

Figure 2-86. Complex polar Kerr effect (top row) and complex off-diagonal conductivity (bottom row) of the uranium monochalcogenides. The samples are cleaved single crystals and are magnetically saturated ($T=15$ K, $B=4$ T). The full and dashed lines are estimates of the free-carrier contributions to σ_{2xy} and σ_{1xy}, respectively (after Reim et al., 1984 b).

the observation of a magnetization-induced red-shift of the f→d transition energy, leads to the conclusion that the orbital moment dominates the spin moment of the 5f state.

The magnetic phase diagram of the uranium nonopnictides is rather complex (Rossat-Mignod et al., 1984). In UAs a field of 10 T is sufficient to induce a ferrimagnetic phase, and a Kerr rotation peak with a rotation value of about $-1°$ has been observed at 20 K (Reim et al., 1984c). Under similar external parameters UP is still in an antiferromagnetic phase and shows a rotation of $-0.5°$ (Reim, 1986). USb is such a hard antiferromagnet that Kerr effect measurements were not performed. However, the substitution of some 10 at.% of either Sb by Te or U by Th or Y leads to ferromagnetic phases with nearly unchanged transition temperatures. Figure 2-87 displays the Kerr rotation of $USb_{0.8}Te_{0.2}$ at 15 K and a field of 4 T (Reim et al., 1984d). In a cleaved crystal, Reim et al. found a maximum rotation of $-6°$ which is twice as large as in UTe. This increase goes along with an increase of the magnetization from 1.91 to 2.58 μ_B/f.u.

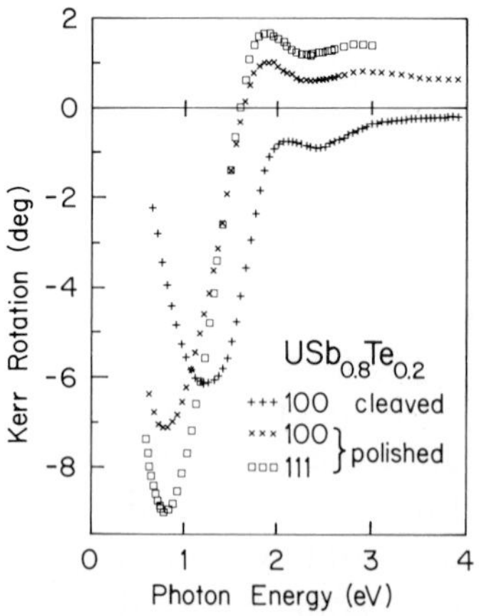

Figure 2-87. Kerr rotation of single crystals of $USb_{0.8}Te_{0.2}$ for two crystal orientations and different preparation methods of the reflecting surface. The samples are saturated at 15 K in a field of 4 T (after Reim et al., 1984 d).

and a narrowing of the Kerr rotation peak. Because the magnetization is a factor $\sqrt{3}$ larger in the [111] direction than along [100], Reim et al. (1984d) measured also on a (111)-oriented crystal and found a rotation exceeding $-9°$. Next to CeSb, this is the largest Kerr rotation in this spectral range. When making this comparison, it should be noted that the peak energy of $USb_{0.8}Te_{0.2}$ is twice that of CeSb, and that the former material is ferromagnetic below 204 K, while CeSb is antiferromagnetic below 16 K, and the record rotation occurs only if a field of 5 T is applied at a temperature of 2 K to saturate the sample.

The effect of the grinding and polishing, necessary to prepare a (111) surface, was studied by measurements on (100)-oriented crystals, whose surfaces were either prepared by cleaving or polishing. Indeed, polishing of a (100) surface of $USb_{0.8}Te_{0.2}$ produces a shift of the rotation peak to lower energy and an increase of the rotation by about 1°. This is not a general trend, but reflects very specific electronic structure properties of this system which can be interpreted either in the context of a complex band-structure or as a transition from localized f-electron behavior for an antimony content above 0.8 to an itinerant behavior for an antimony content below 0.8 (Reim, 1986).

2.7.3 Th_3P_4- and PbFCl-Type Compounds

As most lanthanides, uranium also forms compounds in the b.c.c. Th_3P_4-type structure. While Th_3P_4 and Th_3As_4 are diamagnetic semiconductors (Schoenes et al., 1983), U_3P_4 and U_3As_4 are ferromagnetic metals with ordering temperatures of 138 and 198 K and saturation moments of 1.39 and 1.83 μ_B/U, respectively (Burlet et al., 1981). A remarkable property of these materials is the large magnetic anisotropy manifesting itself in magnetic (Buhrer, 1969b; Trzebiatowski et al., 1971), magnetorestrictive (Bielov et al., 1973) and electrical measurements (Henkie, 1980). Takegara et al. (1981) and Suzuki et al. (1982), performing band-structure calculations, related this anisotropy to strong p–f mixing which, by a comparison of optical data for the corresponding Th and U compounds, was determined to be 0.85 eV (Schoenes et al., 1983). Reim and Schoenes (1983) and Schoenes and Reim (1986) have reported magneto-optical Kerr measurements for (112)-oriented and polished single crystals of U_3P_4 and U_3As_4. Figure 2-88 displays the spectra obtained at 15 K in a field of 4 T. A strong increase of the absolute value of the Kerr rotation is observable with decreasing energy, reaching nearly 6° in U_3As_4 at the limit of the mea-

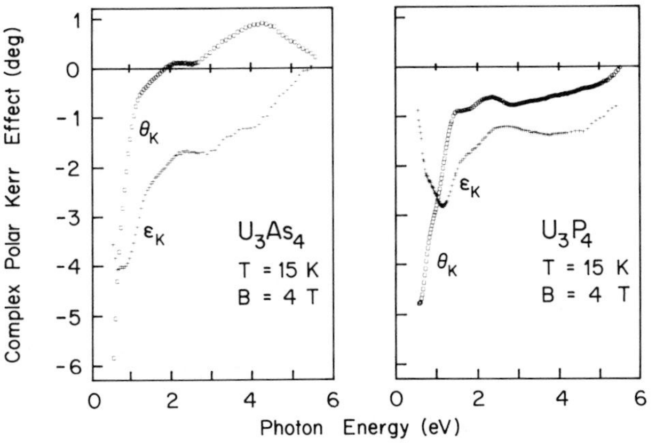

Figure 2-88. Polar Kerr rotation θ_K and ellipticity ε_K of (112)-polished U_3As_4 and U_3P_4 at magnetic saturation (after Reim and Schoenes, 1983, and Schoenes and Reim, 1986).

surements at 0.5 eV. The computation of the off-diagonal conductivity shows a ω^{-1} dependence of σ_{2xy} superimposed by several weak transitions (Schoenes and Reim, 1986).

This behavior is nicely demonstrated by plotting $\omega\sigma_{2xy}(\omega)$ in Fig. 2-89 together with the absorptive part of the diagonal conductivity element of U_3P_4. One recognizes both the correspondence of structures in σ_{1xx} and σ_{2xy} and the fact that the interband structure in $\omega\sigma_{2xy}$ are superimposed on a dominant free-carrier contribution (solid line in Fig. 2-89). Because σ_{1xx} of U_3P_4 shows basically the same structures as Th_3P_4, only shifted by 0.85 eV to lower energies, these peaks can not correspond to pure f→d transitions, but have been assigned to excitations of strongly mixed p and f states into conduction d bands, corroborating once more the p–f mixing model.

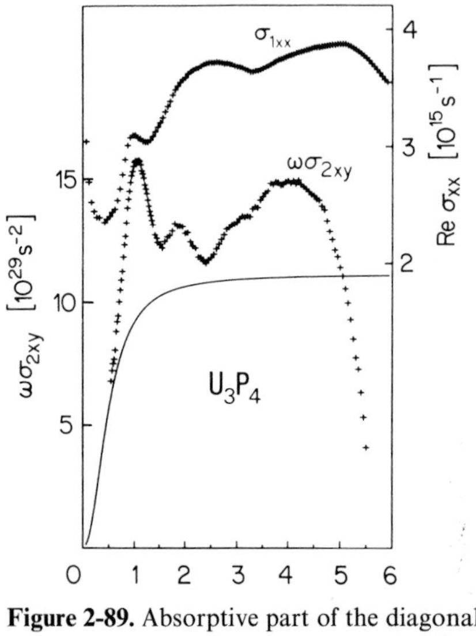

Figure 2-89. Absorptive part of the diagonal conductivity σ_{1xx} and product of the angular frequency ω and the absorptive part of the off-diagonal conductivity σ_{2xy} of U_3P_4 at magnetic saturation. The full line is an estimate of the free-carrier contribution to $\omega\sigma_{2xy}$ (after Schoenes and Reim, 1986).

The large Kerr signals observed in uranium pnictide single crystals has motivated McElfresh et al. (1990) to study amorphous U_xSb_{1-x} films. The highest Curie temperature found is 135 K for $x=0.45$ and the magnetic moments are $\approx 1.5\ \mu_B/U$. At saturation, the film with $x=0.48$ has a Faraday rotation of 2×10^6 deg cm^{-1} ($\equiv 3.5\times10^6$ rad m^{-1}). The largest Kerr rotation of nearly $(-)3.5°$ is observed for a film with $x=0.51$. The photon-energy dependence between 1.55 and 2.76 eV is weak, leading McElfresh to conclude that the magneto-optical effects are dominated by intraband effects, and that the coordination of the U atoms is similar to that in U_3Sb_4 rather than to that in USb.

The last material to be described in this section on actinide compounds containing no 3d element is UAsSe. This ternary compound crystallizes in the tetragonal PbFCl structure, which consists of layers stacked in the *c*-direction with the sequence As–U–Se–Se–U–As. UAsSe orders ferromagnetically near 110 K with a saturation moment at low temperatures of 1.36 μ_B/U and a neutron moment of 1.5 μ_B/U (Leciejewicz and Zygmunt, 1972). The magnetic moments are aligned parallel to the *c*-axis. The optical reflectivity and the magneto-optical polar Kerr effect have been measured by Reim et al. (1985). The optical and also the electrical transport properties (Schoenes et al., 1988 b) indicate only weak metallic conductivity at room temperature and an anomalous decrease of the conductivity with decreasing temperature. The results of the polar Kerr effect measurements at 10 K and in a field of 5 T are depicted in Fig. 2-90 (Reim and Schoenes, 1990). The rotation shows a negative peak at 3 eV with a rotation of about −1.4° and a shoulder near 1 eV. The computation of the off-diagonal conductivity (Fig. 2-90) (Reim et al., 1985) reveals that

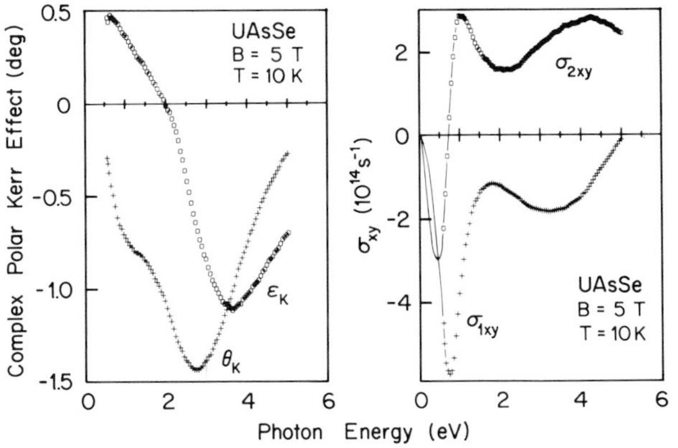

Figure 2-90. Complex polar Kerr rotation (left side) and complex off-diagonal conductivity (right side) of UAsSe single crystals at magnetic saturation (after Reim and Schoenes, 1990).

these structures correspond to a transition with diamagnetic line shape at 0.73 eV and a transition with paramagnetic line shape at 4.3 eV. The low-energy structure has been assigned to an f→d transition and the high-energy structure to excitations of bonding p–d electrons from the top of the valence band into spin-polarized f states.

2.8 Mixed 3d–4f and 3d–5f Materials

After the discussion of pure 3d, 4f, and 5f systems in the previous sections, we are now prepared to disentangle the contributions of these elements in complex magnetic systems containing d and f elements at the same time. These mixed systems play the dominant role in today's applications of magneto-optics. Roughly speaking, the 3d transition element is needed to provide the desired ordering temperature above room temperature, and the 4f and eventually 5f element provides the necessary perpendicular anisotropy (see Sec. 2.9). As guideline for the order of presentation of these systems, the ionicity will be used. In other words, insulating compounds which have well-structured spectra will be discussed first, followed by *ionic* metals, then by intermetallics and finally by a selection of the numerous work on amorphous films.

2.8.1 Mixed 3d–4f and 3d–5f Compounds

A new aspect in mixed d–f materials is the existence of two or more sublattices which are occupied by magnetic elements with very different exchange interactions. The rare-earth iron garnets (RIG), to be discussed first, provide an excellent example. As can be seen for GdIG in Fig. 2-91, the magnetic moment of the Gd^{3+} sublattice is antiparallel to the resulting moment of the Fe^{3+} ions. While the strong exchange between the Fe^{3+} ions determines the ordering temperature, the weak exchange between the Gd^{3+} ions alone would only lead to a much lower ordering temperature. However, for thermodynamic reasons, only a single ordering temperature can exist, and the Gd^{3+} sublattice starts to order together with the Fe^{3+} sublattice but with a weaker increase of the magnetization with decreasing temperature than the Fe^{3+} sublattice. Because the saturation magnetization of the Gd^{3+} sublattice ($3 \times 7\ \mu_B$/f.u.) is larger than the one resulting from the two Fe^{3+} sublattices ($5\ \mu_B$/f.u.), a temperature exists where the net magnetization of the ordered phase vanishes. This temperature is called com-

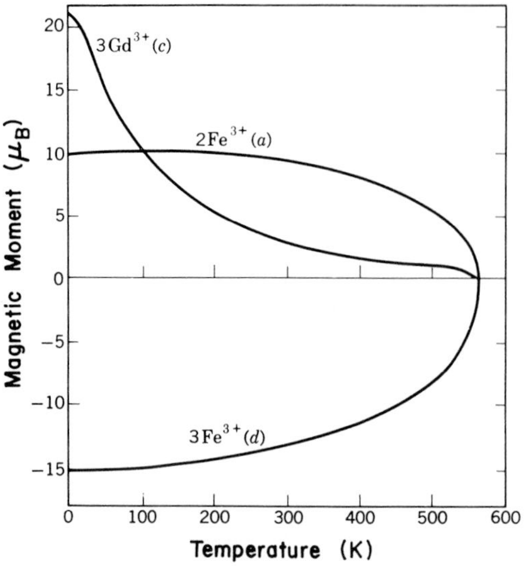

Figure 2-91. Temperature dependence of the magnetization of the sublattices in $Gd_3Fe_5O_{12}$ as calculated by Pauthenet (1958). At $T_{comp} = 280$ K, the net magnetization is 0.

pensation temperature (T_{comp}) and has a value of 280 K in $Gd_3Fe_5O_{12}$.

As Kahn et al. (1969) and Crossley et al. (1969) noted, the Kerr and Faraday rotations should change their sign on going through T_{comp}, because in an applied field all the moments are reversed at T_{comp}. Crossley et al. (1969) have investigated the Faraday rotation of GdIG and TbIG at 1.15 μm as a function of temperature. The substitution of Y with Gd leads to a decrease in the Faraday rotation at room temperature. With decreasing temperature the Faraday rotation first decreases, then reverses its sign at T_{comp} and increases again from negative to positive values with a zero crossing near 150 K. In TbIG the rotation at room temperature is larger than in YIG and increases with decreasing temperature down to T_{comp}. At T_{comp} the Faraday rotation becomes negative with increasing negative values for decreasing temperatures. Because the lanthanide sublattice magnetization is rather small at room temperature, stronger changes of the magneto-optical spectra necessitate magneto-optically very active transitions of the f element. As discussed in the section on lanthanides, these are known to occur below about 4 eV only in Ce^{3+}, Pr^{3+}, Nd^{3+}, Eu^{2+}, and Tm^{2+}.

Wemple et al. (1973) have investigated a large number of lanthanide-doped YIG crystals for their usefulness as magneto-optical light modulators at 1.064 μm and find, indeed, the largest negative rotation values for substitutions with Pr^{3+} and Nd^{3+}. Figure 2-92 displays a more recent plot of the Faraday rotation of Ce-, Pr-, Nd-, and Bi-substituted YIG at the two wavelengths 1150 and 633 nm (LeGall et al., 1987). The curves show that the substitution with Ce is even more effective than that with Bi, but the largest Ce concentration which could be substituted was 0.06 out of 3 Y atoms.

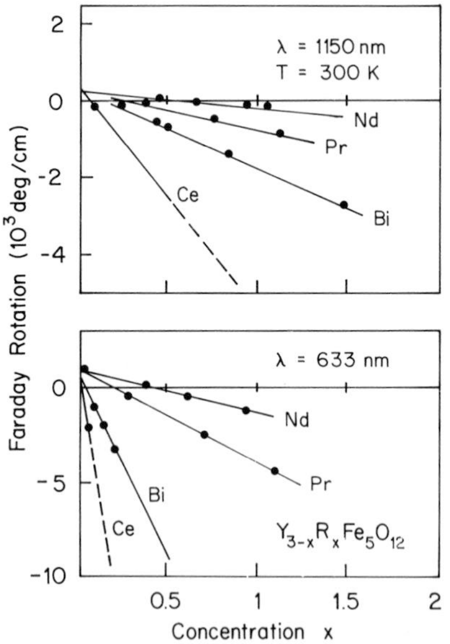

Figure 2-92. Variation of the Faraday rotation of $Y_{3-x}R_xFe_5O_{12}$ with the concentration x for R = Ce, Pr, Nd, and Bi at the wavelengths 1150 nm (top) and 633 nm (bottom). Note that the data for Ce at 633 nm are in contradiction to those shown in Fig. 2-94 (after le Gall et al., 1987).

For Pr and Nd, 1.12 and 1.05 atoms per formula unit were the respective limits for substitution. However, after the growth of fully substituted PrIG and NdIG films by liquid-phase epitaxy had become possible, Dillon et al. (1987) reported the Faraday rotation measurements reproduced in Fig. 2-93. The signals are somewhat larger than anticipated from a linear extrapolation of the results for low substitution rates. For PrIG the spectra show many narrow lines which resemble those found in PrSb (Sec. 2.6.2.1). Also the growth of films with much higher Ce substitution for Y has been achieved in the meantime.

Figure 2-94 displays the Faraday rotation spectra published by Gomi et al. (1988 a) and Itoh (1989) for rf-sputtered films of $Y_{3-x}Ce_xFe_5O_{12}$ with x varying from 0 to 1. One recognizes that the common sign change of the Faraday rotation occurs at low energies with substitution, but also that the Faraday rotation becomes positive again above 1.4 eV. This is in contrast to the effect of substitution with Bi, Pr, and Nd and appears to result from a strong transition occurring near 1.5 eV. The similarity in energy, line shape, and intensity with CeTe is very suggestive for an assignment as a $4f^1 \rightarrow 4f^0\,5d^1$ transition. For applications, the huge increase of the Faraday rotation is of primary importance. At 1 eV the rotation of $Y_2CeFe_5O_{12}$ exceeds 10^4 deg cm^{-1} ($\equiv 1.75 \times 10^4$ rad m^{-1}) which is half an order of magnitude larger than that of $Y_2BiFe_5O_{12}$ and is twice the value extrapolated from the low Ce substitutions. At 2 eV, the substitution of one Ce and one Bi atom per formula unit results in a Faraday rotation of $\approx 2 \times 10^4$ deg cm^{-1}, but with positive sign for Ce and a negative sign for Bi substitution. The former sign is in contradiction to the extrapolation in Fig. 2-92.

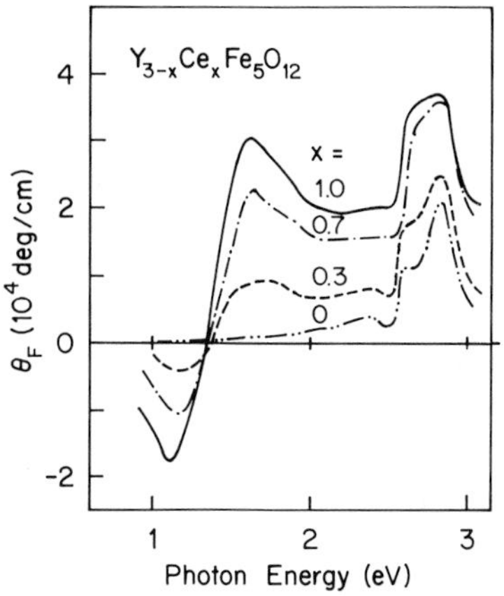

Figure 2-94. Faraday rotation per unit length and for magnetic saturation of films of $Y_{3-x}Ce_xFe_5O_{12}$ grown by r.f. sputtering (after Gomi et al., 1988 a, and Itoh, 1989).

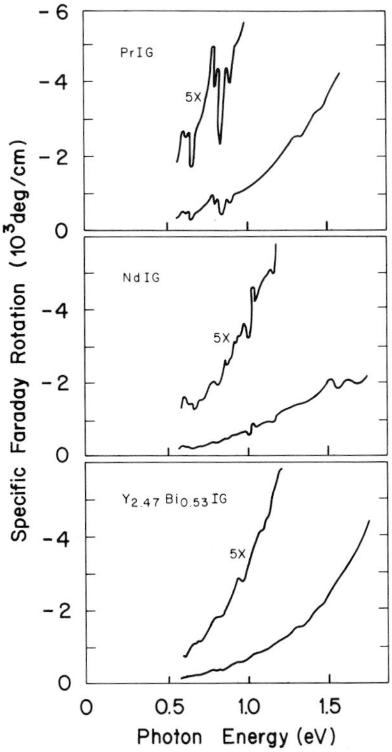

Figure 2-93. Faraday rotation per unit length and for magnetic saturation of films of $Pr_3Fe_5O_{12}$ (top), $Nd_3Fe_5O_{12}$ (middle), and $Y_{2.47}Bi_{0.53}Fe_5O_{12}$ (bottom) grown by liquid-phase epitaxy (after Dillon et al., 1987).

Many garnets with substitutions of more than one element have been studied to improve either on the optical transparency, the magneto-optical effects, the coercivity, the squareness of the hysteresis loop, the grain size of polycrystalline films, or the corrosion resistance. Aside from containing iron, oxygen, and bismuth, several of these garnets also contain Nd, Dy, Co, Al, Ge, Ga, Ba, or W (Itoh et al., 1985, 1987; Shono et al., 1988; Gomi et al., 1988 b; Itoh and Kryder, 1988; Itoh, 1989).

A second class of ferric oxides which offers the possibility to incorporate lanthanides are the rare-earth orthoferrites. Kahn et al. (1969) reported the room-temperature polar Kerr effect from 2 to 5.5 eV for $RFeO_3$ compounds with R = Y, Sm, Eu, Gd, Tb, Dy, Ho, Er, Tm, Yb, and Lu. Except for $SmFeO_3$ and a few small structures occurring in some compounds, all spectra look like that of $YFeO_3$ displayed in Fig. 2-58. As argued for the garnets this insensibility reflects the small magnetization of the lanthanide sublattice and the lack of dipole-allowed magneto-optical active transitions of the heavy lanthanides in the investigated spectral range. $PrFeO_3$, on the other hand, shows substantial changes compared to $YFeO_3$, but no assignments have been made (Višňovský et al., 1984).

The next class of compounds to be discussed are ternary uranium compounds of the general chemical formula UT_2X_2, where T is a 3d transition element and X is either an element of the Va or IVa group of the periodic table. For T = Fe and X = Si or Ge, these compounds are Pauli paramagnets down to at least 1.8 K (Szytuła et al., 1988). With Mn as the transition element, they are ferromagnetic at room temperature, with T_C as high as 377 K in UMn_2Si_2 and 390 K in UMn_2Ge_2. However, the uranium sublattice acquires a substantial magnetization only below 90 and 150 K, respectively (Szytuła et al., 1988). Magneto-optical Kerr measurements performed at room temperature (van Engelen et al., 1988) show moderate (negative) Kerr rotations of less than 0.1° above 1 eV and an increase up to 0.3° for the lowest energy of 0.6 eV. Although this signal is larger than that for the reference materials $LaMn_2Si_2$ and $LaMn_2Ge_2$, it is much smaller than in materials with magneto-optical active f → d transitions. Such a material is UCu_2P_2. The optical and magneto-optical properties of this hexagonal compound have been investigated by Schoenes et al. (1989). It is a so-called *bad* metal with less than 0.2 car./f.u. Because the Cu atom carries no magnetic moment, the ferromagnetic ordering relies on the U–U exchange, and T_C is below room temperature. Yet, the observed 216 K make UCu_2P_2 the compound with the highest Curie temperature originating solely from uranium. Figure 2-95 displays the Kerr rotation and ellipticity at 10 K in a field of 4 T as well as the off-diagonal conductivity computed from these data and the optical constants. The rotation shows a peak rotation of −3.5° near 1 eV, which is 35 times greater than the rotation of UMn_2Si_2 at the same energy but at room temperature. The off-diagonal conductivity can be decomposed into two interband transitions, marked A and B in Fig. 2-95. The low-energy transition near 0.8 eV is of diamagnetic line shape in σ_{2xy} and paramagnetic line shape in σ_{1xy}, which is typical for f → d transitions, while transition B is of paramagnetic line shape in σ_{2xy}, which is typical for d → f transitions. Fumagalli et al. (1988) and Schoenes et al. (1989) have also investigated the magneto-optical Kerr effect of $UCuP_2$ and $UCuAs_2$. These tetragonal compounds order ferromagnetically at 76 and 131 K, respectively. The optical and

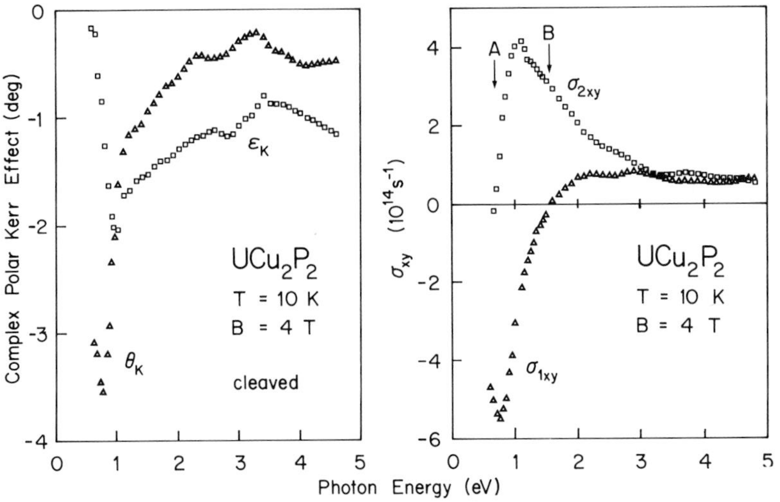

Figure 2-95. Complex polar Kerr effect (left side) and complex off-diagonal conductivity (right side) of UCu_2P_2 at magnetic saturation (after Schoenes et al., 1989).

magneto-optical spectra are qualitatively similar to those of UCu_2P_2, but the maximum rotation is a factor two smaller.

A somewhat similar compound, namely the $C1_b$ phase material NiUSn (same structure as PtMnSb discussed in Sec. 2.5.3.2), has received attention from the theoretical side. Daalderop et al. (1988) have computed Kerr rotation spectra for this material assuming ferromagnetic ordering, and they predict rotation peaks of 5°. Unfortunately, NiUSn orders antiferromagnetically and a rough estimate of the field necessary to align the moments gives values of the order 200 T, preventing a check of the theoretical prediction.

Another ternary uranium compound, namely $UFe_{10}Si_2$, has recently been studied independently by van Engelen and Buschow (1990b) and Brändle et al. (1990b). Figure 2-96 compares the Kerr spectra of the two groups. The agreement is quite satisfying considering the different magnetic fields used in the two measurements and the lack of phase purity. Van Engelen and Buschow (1990b) compare their rotation data with $RFe_{10}Si_2$, where R stands for Gd, Tb, Dy, and Lu. They note that the uranium compound has the largest rotation and conclude that the uranium sublattice contributes to the Kerr effect. Brändle et al. (1990b), on the other hand, compute the off-diagonal conductivity spectra and show that the similar rotation peak value of $UFe_{10}Si_2$ and pure Fe is due

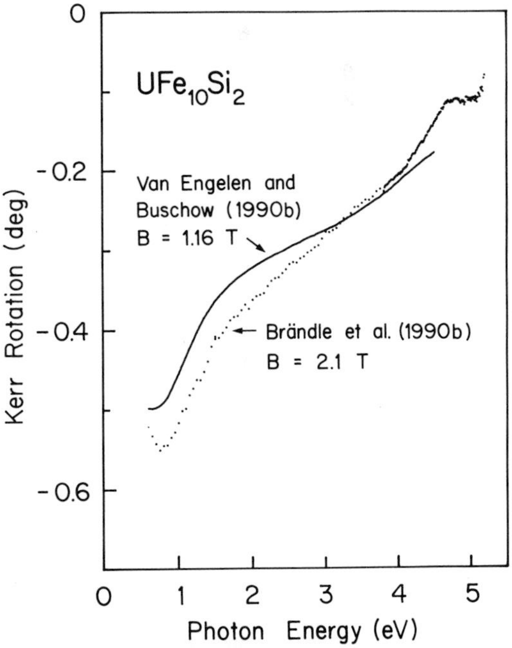

Figure 2-96. Comparison of the Kerr-rotation spectra of $UFe_{10}Si_2$ obtained by van Engelen and Buschow (1990 b) and Brändle et al. (1990 b).

to a compensation of the lower magnetization in $UFe_{10}Si_2$ by a narrower magneto-optically active transition. The investigation of van Engelen and Buschow further includes compounds $RFe_{10}V_2$, where R = Nd, Gd, Dy, and Er, and compounds $R_2Fe_{14}C$, where R = Nd, Gd and Lu. The Kerr rotation spectra all look quite similar with the exception that for Nd some extra contribution appears near 4 eV. From the discussion of Nd compounds in Sec. 2.6.2.1 it becomes evident that this contribution can be assigned to $f^3 \rightarrow f^2d$ transitions. $Nd_2Fe_{14}B$ attracted much attention when its record maximum energy product was discovered by Croat et al. (1984) and Sagawa et al. (1984). Van Engelen and Buschow (1987) have determined the polar Kerr effect of this permanent magnet material and the parent compounds $R_2Fe_{14}B$, where R = La, Ce, Gd, and Lu, and $R_2Co_{14}B$, where R = La and Gd. Figure 2-97 reproduces the results for the former group of materials. Again, the neodymium compound displays the largest values of $|\theta_K|$ and an extra contribution near 4 eV.

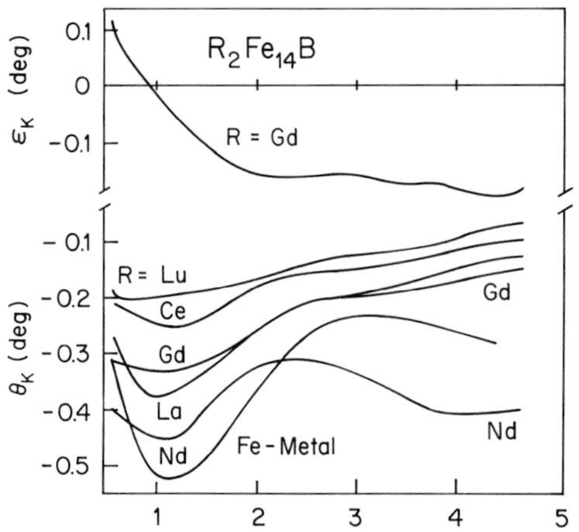

Figure 2-97. Room-temperature polar Kerr rotation θ_K and ellipticity ε_K of some $R_2Fe_{14}B$ compounds and comparison with the rotation of Fe. The field is ≈ 1.2 T (after van Engelen and Buschow, 1987).

After normalization of the rotation with the magnetization at room temperature, the differences between the various $R_2Fe_{14}B$ compounds are somewhat reduced, and van Engelen and Buschow conclude that the contribution from the lanthanides to the Kerr rotation is weak in the low-energy range.

The binary compounds RCo_5 were among the first mixed 3d–4f materials investigated by magneto-optical methods after Nesbitt et al. (1962) reported a Kerr rotation of 2° in $SmCo_5$. Stoffel and Strnat (1965) did not corroborate this large value, and Buschow and van Engen (1984) also found $|\theta_K|$ values below 0.17° in YCo_5, $LaCo_5$, and $GdCo_5$.The magneto-optical properties of UCo_5 were determined in a similar way. Deryagin and Andreev (1976) claimed that $UCo_{5.3}$ has a Kerr rotation of 2–3° in the visible region of the spectrum. However, Brändle et al. (1990c) showed that the rotation of UCo_5 is only of the order of 0.25°, in agreement with the absence of a uranium moment in this intermetallic compound. Besides the quoted RCo_5 compounds, Buschow and van Engen (1984) reported the Kerr rotation of YCo_3, Y_2Co_7, Y_2Co_{17}, La_2Co_7, $LaCo_{13}$, $GdCo_2$, $GdCo_3$, and Gd_2Co_{17} at room temperature. With the exception of $GdCo_2$, the general trend reflects a decrease in Kerr rotation with decreasing Co concentration.

The polar and the equatorial magneto-optical Kerr effect of several RFe_2 Laves phases have been reported by Katayama and Hasegawa (1982) and Sharipov et al. (1986), respectively. Figure 2-98 displays the polar Kerr spectra of Katayama and Hasegawa (1982). Except for $GdFe_2$, the spectra have similar shapes with a pronounced minimum near 4.3 eV. This shape is very different from that of pure Fe, and an interpretation is pending.

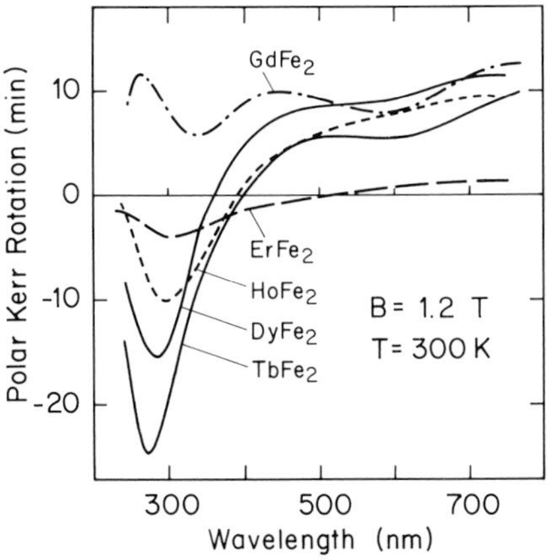

Figure 2-98. Room-temperature polar Kerr rotation of some RFe_2 Laves phases. The field is 1.2 T (after Katayama and Hasegawa, 1982).

2.8.2 Amorphous Films

Most materials actually used for magneto-optical recording consist of amorphous lanthanide–Fe–Co films. The number of investigations, correspondingly, is enormous, so that only a few can be cited here. The breakthrough of this class of materials is more associated with practical reasons rather than its unique magneto-optical performance. One of the practical and economical advantages of lanthanide–Fe–Co films is the capability of preparing these amorphous films by sputtering onto glass or plastic substrates. And since, in contrast to oxides, no subsequent heat treatment is necessary, these substrates can be pregrooved to provide a tracking signal for the tracking servo in the optical head. The amorphous nature of the alloys allows an optimization of the writing temperature by an adjustment of the constituents and also helps to avoid noise from grain boundaries. Of course, a major concern in amorphous alloys is the generation of a sufficiently large perpendicular anisotropy. Chaudhari et al. (1973) demonstrated that amorphous films of Gd–Co and Gd–Fe with large enough perpendicular anisotropy can be prepared by rf or dc sputtering. Gambino et al. (1974) discussed the origin of the anisotropy, and by a process of elimination, concluded that atomic short-range ordering is the main source of anisotropy in sputtered amorphous Gd–Co films. An estimate of the magnitude of the anisotropy indicates that 10^{26} to 10^{27} Co–Co pairs m^{-3}, i.e., only a relatively small fraction of pairs, should be preferentially oriented to account for the observed anisotropy. Later studies with different lanthanides have shown that the single-ion anisotropy also plays a significant role. Figure 2-99 demonstrates this effect by showing the uniaxial anisotropy constant K_u as a function of the lanthanide element substituted in amorphous films of $(Gd_{0.75}R_{0.25})_{0.19}Co_{0.81}$ (Y. Suzuki et al., 1987). From the rather small difference between the data before and after removal of the film from the substrate, one recognizes also that stress-induced anisotropy is of minor importance. The magnetic properties of amorphous rare-earth transition-metal alloys are manifold. Generally, light rare-earth elements couple ferromagnetically whereas heavy rare-earth elements

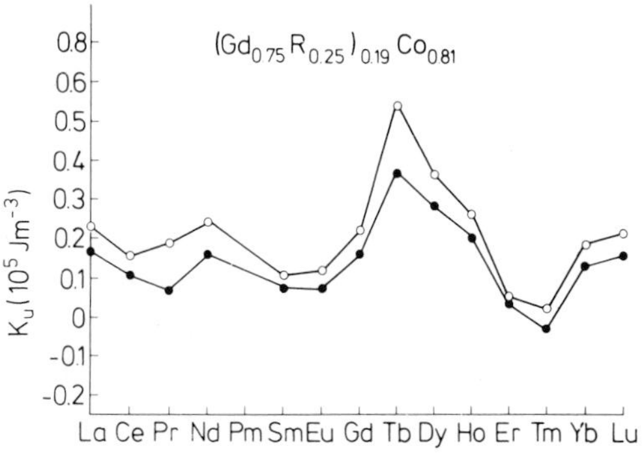

Figure 2-99. Uniaxial anisotropy constant of amorphous $(Gd_{0.75}R_{0.25})_{0.19}Co_{0.81}$ films before (open circles) and after (full dots) removal of the substrate (after Y. Suzuki et al., 1987).

couple antiferromagnetically to the 3d moments. In the latter case, a compensation of the moments can occur and the compensation temperature can be adjusted with the composition. Because of the strong sensitivity of the Fe–Fe exchange to the interatomic separation, amorphous rare-earth iron alloys show a compensation temperature less frequently than alloys with cobalt.

The Kerr signals of the amorphous rare-earth transition-element alloys are rather moderate and, in general, decrease with increasing rare-earth content. Figure 2-100 reproduces, as an example, the data of Choe et al. (1987) at a wavelength of 633 nm. To counterbalance the decreasing signal from the transition element, the lanthanide element should have a magnetic moment at room temperature and a magneto-optically active transition in the desired spectral range. As we have seen in the section on lanthanides, the second requirement can possibly be fulfilled with Ce^{3+}, Pr^{3+}, Nd^{3+}, Eu^{2+}, and Tm^{2+}. The data of Choe et al. (1987) displayed in Fig. 2-101 show that neodymium can indeed lead to an improvement of the Kerr signal at short wavelength, while cerium only slows down the decrease of the rotation with increasing lanthanide content. Similar behavior has been observed by Weller and Reim (1989) for $Nd_x(Fe,Co)_{1-x}$. Gambino and McGuire (1986) studied the substitution with neodymium and praseodymium and reported a Kerr rotation of 0.59° at 633 nm for a material with 20 at.% Pr and 40 at.% Fe and Co. Suzuki and Katayama (1986) studied the magnetic and magneto-optical properties of Nd–Fe–Co and Pr–Co films. At 500 nm the largest Kerr rotation is 0.55° for a composition $Nd_{40}(Fe_{0.75}Co_{0.25})_{60}$. Pr–Fe films were studied by the same group in the following year (T. Suzuki et al., 1987). Weller and Reim (1989) also investigated $Pr_x(Fe,Co)_{1-x}$ and $Eu_x(Fe,Co)_{1-x}$. For the former amorphous films, they found an increase of the Kerr rotation between 4 and 5 eV, while in the latter material no sign of europium contribution was observed, although XPS data indicate that europium is in a divalent state in the amorphous films. Amorphous films of Nd–Co and Nd–Fe have also received interest for their Faraday rotation. McGuire et al.

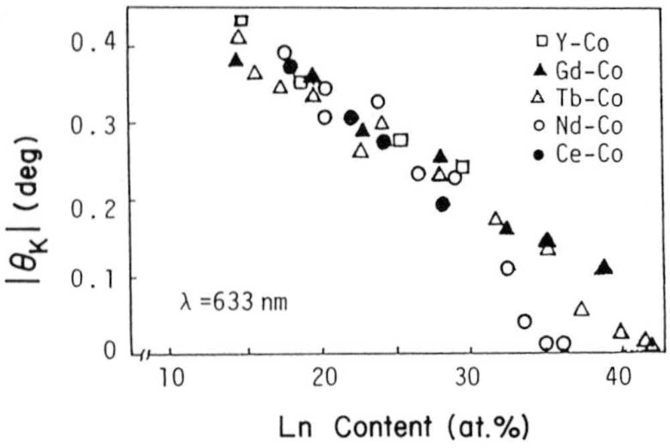

Figure 2-100. Variation of the room-temperature Kerr rotation at 633 nm with the lanthanide content in various amorphous rare-earth transition-element films (after Choe et al., 1987).

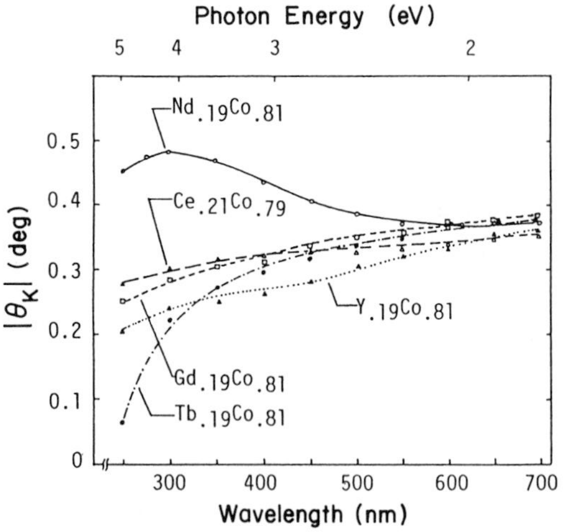

Figure 2-101. Room-temperature Kerr-rotation spectra for some lanthanide-cobalt films with cobalt concentrations of 81 and 79 at.%. The field is 1.3 T (after Choe et al., 1987).

(1987) reported a maximum rotation of 5×10^5 deg cm^{-1} ($\equiv 8.7 \times 10^5$ rad m^{-1}) for $Nd_{75}Co_{25}$ at 1.8 eV.

Both the Faraday and the Kerr rotation were determined on $(Gd,Tb)_{1-x}(Fe,Co)_x$ amorphous alloys by Hansen et al. (1987), who made an analysis of the signals assuming that the rotation can be expressed as a sum of the rotation due to the transition element and the lanthanide element. Prior investigations on films containing two or three of the above four elements include those by Imamura and Mimura (1976), Višňovský et al. (1976), Urner-Wille et al. (1978), Katayama and Hasegawa (1982), Togami (1982), Tsunashima et al. (1982), Conell and Allen (1983), Sato and Togami (1983), Buschow and van Engen (1984), Imamura et al. (1985), Gambino et al. (1986), McGuire et al. (1986), Wolniansky et al. (1986), and Choe et al. (1987). More recent investigations are those by Suzuki et al. (1988) and Weller and Reim (1989). Figure 2-102, taken from the work of Suzuki et al. (1988), gives an impression of the variation possibilities for amorphous rare-earth transition-element alloys and their Kerr rotation spectra. Additional substitutions were concentrated on the heavy metals Bi, Sn, and Pb (Urner-Wille et al., 1980; Masui et al., 1984). As for the garnets, these heavy metals tend to increase the magneto-optical signal in the rare-earth transition-element alloys. The mechanism is probably a transfer of spin-orbit splitting from heavy metal to transition-metal states.

2.9 Applications

The treatment of the various materials in Secs. 2.4–2.8 demonstrate that a large part of magneto-optical research has been directed toward the development of magneto-optical disks. The second most important application of magneto-optics is nonreciprocal elements. This section reviews the principles of these two main applications of magneto-optics and discusses some of the problems and recent developments which have not been addressed in the previous sections.

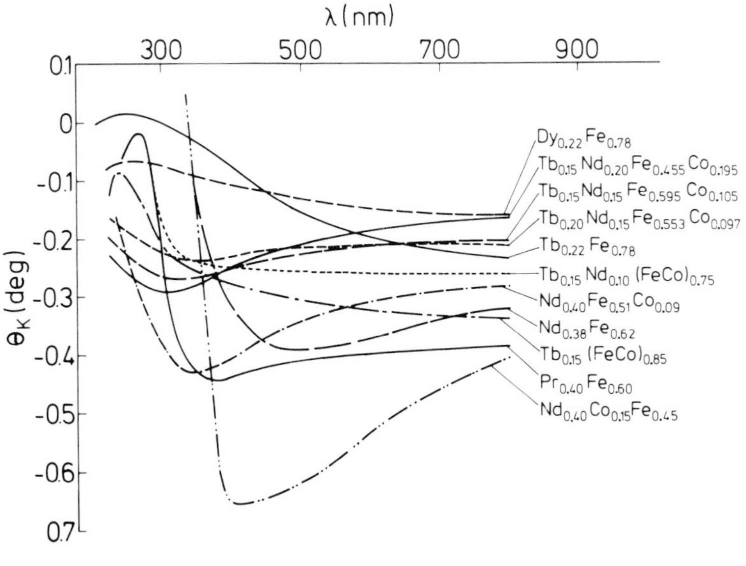

Figure 2-102. Room-temperature Kerr-rotation spectra of a variety of amorphous rare-earth transition-metal films with up to four different elements (after Suzuki et al., 1988).

2.9.1 Principles of Magneto-Optical Recording

Magneto-optical recording combines the advantages of magnetic and optical storage techniques. The information is stored in magnetic domains, while writing and reading is performed with a laser beam. Compared to semiconductor memories with dynamic random access (DRAM), magneto-optical memories have the advantage of lower cost per bit but the disadvantage of a much longer access time. Compared to the two other main types of laser-addressable high-density recording devices, i.e., hole burning and vitrification, magneto-optical recording provides unlimited cyclability. The strongest competitor for magneto-optical recording in the field of overwritable permanent storage is the *Winchester technology,* with a capacity of $\geqq 10^{11}$ bit m^{-2}, a medium access time of the order of 1 to 10 ms, and a low cost per bit. In this technique, the information is also stored magnetically, but writing and reading is done with an induction coil. Originally, magneto-optical recording had the advantage that smaller magnetic bits could be used, since the minimum useful size of these bits is determined by the wavelength of the laser. Yet, the introduction of thin-film technology for the production of induction heads has considerably increased the bit density per unit area in Winchester systems. In addition, by stacking several disks on the same spindle one can obtain a very high storage density per unit volume. A problem which remains is the very short distance between the disk and the head of ≈ 100 nm, which in the case of a head crash leads to a loss of stored information. Magneto-optical storage does not suffer from this problem and allows the exchange of disks by the consumer.

The writing of a bit is performed by heating with a pulse (≈ 10 mW, 60 ns) of a focused solid-state laser an area of the disk above a critical temperature, known as the switching temperature T_s, and reversing the magnetization direction of this area in a magnetic field H_s. Figure 2-103 (Hansen, 1990) shows that for good switching performance the coercive field H_c should decrease from a value much larger than the switching field H_s at room temperature to a value smaller than H_s at the switching temperature. The temperature dependence of H_c sketched in Fig. 2-103 corresponds to that needed for Curie-point writing. A second possibility is compensation-point writing in ferrimagnetic materials possessing a compensation temperature. Because H_c is proportional to the ratio of the uniaxial anisotropy constant K_u and the magnetization M, the coercive field tends to infinity at the compensation temperature T_{comp}. If T_{comp} is near room temperature, H_c (300 K) is much larger than H_s. With increasing temperature, H_c decreases and falls below H_s at a temperature which may

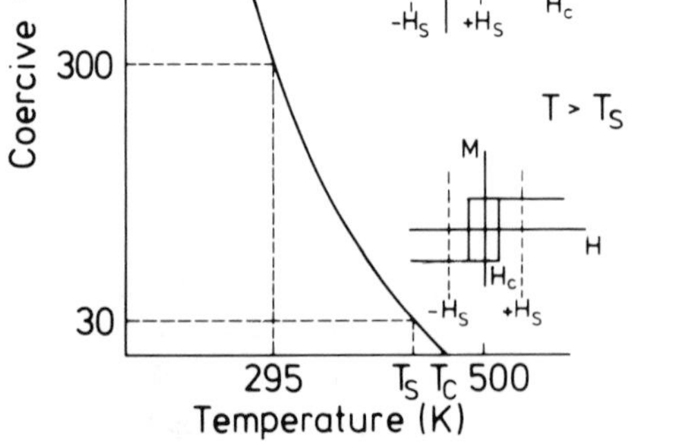

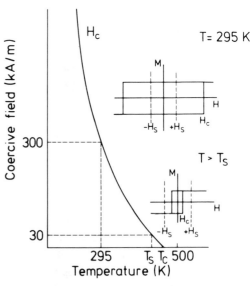

Figure 2-103. Schematic representation of the temperature variation of the coercive field for an amorphous rare-earth transition metal alloy (after Hansen et al., 1990).

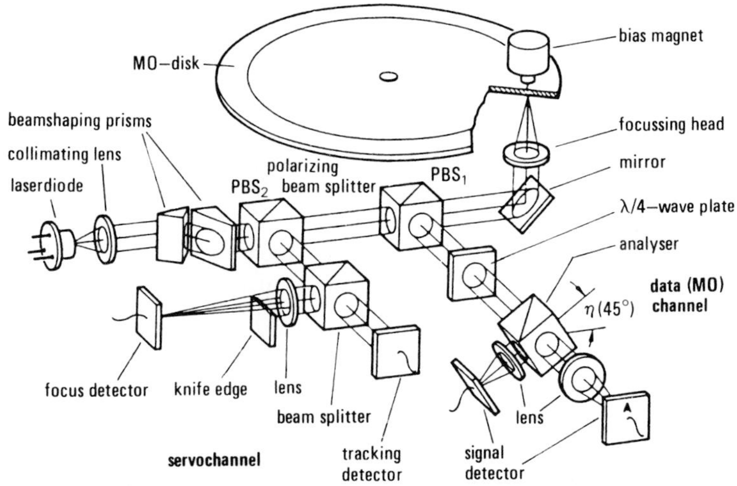

Figure 2-104. Scheme of a setup for reading and writing a magneto-optical disk (after Vieth et al., 1989).

be well below the ordering temperature T_N. A serious drawback for compensation-point writing is the necessity of a high homogeneity of T_{comp} and of the composition over the whole disk.

A major concern in magneto-optical recording is the write speed. In principle, direct overwriting is possible by field modulation at constant laser power. However, reversing the magnetic field over the whole disk is a rather slow process and limits the switching frequency to below 10 MHz (Tanaka et al., 1987; Hansen, 1990). Laser modulation at constant magnetic field is considerably faster and currently allows transfer rates of several Mbits s^{-1}. This rate could be more than doubled if direct overwrite was introduced. At present, the laser beam passes twice over the disk. In the first pass the old data are erased on the whole sector, and in the second pass the new data are written. Kryder (1990) has reviewed various techniques proposed by his group and others to allow direct overwriting. One method (Shieh and Kryder, 1986) uses ferrimagnetic rare-earth transition-element films with a compensation point a few tens of degrees higher than room temperature. In these films micrometer size domains can be nucleated by locally heating the film with no external field but making use of the demagnetizing field. Erasing is performed with a lower-intensity laser pulse, using again the demagnetizing field. In another approach (Ruigrok et al., 1988; Miyamoto et al., 1989) the bits are written with a magnetic head flying a few micrometers above the magnetic film. A third class of approaches necessitates multilayers with different Curie temperatures and coercive forces (Saito et al., 1987).

A standard setup for the magneto-optical write-read system is sketched in Fig. 2-104 (Vieth et al., 1989). It uses a set of beamsplitters and two detectors in the data channel, the signals of which are fed into a differential amplifier for optimum read-out performance (see in Sec. 2.3.1 the Wollaston-prism method). Two other detectors are used for focusing and tracking. Reading can either be done by use of the polar magneto-optical Kerr effect or the Faraday effect. To allow for double-sided storage one must deposit in the latter case a reflecting film under the magneto-optic film and pass the light twice through the film. In practice, the Kerr effect is used more frequently, and metallic films have better heat dissipation than insulating films.

2.9.2 Multilayer Systems for Magneto-Optical Recording

In addition to the above-mentioned requirements for the material (i.e., large magneto-optical signal, T_C, T_N and/or T_{comp} in the desired temperature range, uniaxial anisotropy and strong temperature dependence of the coercivity), other important factors include the size of the domains, the size of the microcrystalline grains, and the chemical stability. Some of the latter problems, if they exist, can be reduced by the deposition of a protective layer. As shown in the schemes of Fig. 2-105 (Bloomberg and Conell, 1987), a protective layer made of a dielectric can also improve the magneto-optical performance if its thickness and refractive index are correctly chosen. In a magneto-optical disk the light generally penetrates through the substrate. Since the light beam is not focused at the outer surface of the substrate, the disk becomes insensitive to dust or finger-prints. The left-hand side of Fig. 2-105 displays a thin-film multilayer consisting of a dielectric, the magneto-optical medium and a protective layer. Sufficient thickness of the magneto-optical film is chosen, so that no light reaches the protective layer; the system is therefore called a bilayer. Reim and Weller (1988a) have prepared two bilayers consisting of both a 100-nm-thick $Tb_{22}(Fe_{0.85}Co_{0.15})_{78}$ film and a dielectric AlN film either 50 or 20 nm in thickness. Figure 2-106 shows the measured apparent Kerr rotation for the two bilayer structures and a calculation using the optical constants and the thicknesses of the different films. For an AlN film thickness of 50 nm, Reim and Weller found an enhancement of the rotation angle by a factor of about 2 at a wavelength slightly lower than expected from the first-order approach, which states that the retardation $\Delta = 2 n_D d$ should be equal to $\lambda/2$. Unfortunately, this factor of 2 can not be fully transferred to the performance of the disk because the dielectric layer leads to a reduction of the reflectivity which also enters into the *figure of merit*. A short comment is necessary regarding the definition of figure of merit in this context. For technical applications, the signal-to-noise ratio is a crucial quantity. If the dominant noise is shot noise and a read system resembling that in Fig. 2-104 is used, the signal-to-noise ratio is proportional to $R^{1/2}(\theta_K^2 + \varepsilon_K^2)^{1/2}$. This quantity is often called figure of merit but differs from the definition $R(\theta_K^2 + \varepsilon_K^2)$, which is derived from a consideration of the intensity trans-

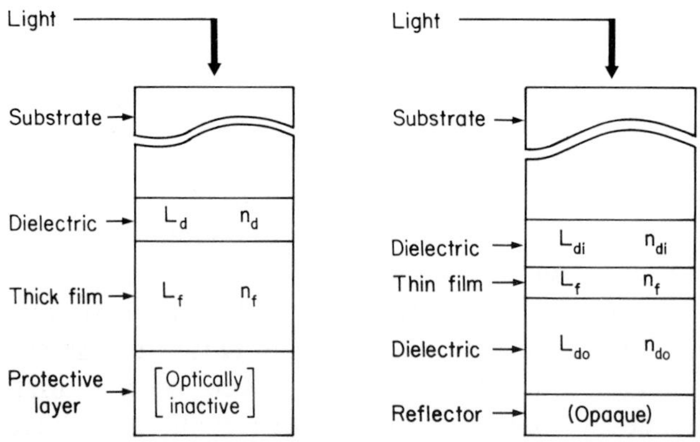

Figure 2-105. Scheme of a magneto-optical bilayer (left side) and a quadrilayer (right side) structure (after Bloomberg and Conell, 1987).

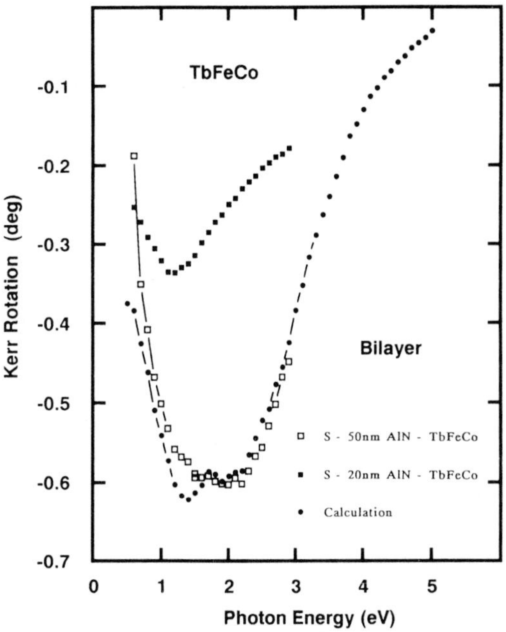

Figure 2-106. Experimental substrate (quartz) incident Kerr rotation of a bilayer consisting of a 50-nm-thick AlN film and a 100-nm-thick $Tb_{0.22}(Fe_{0.85}Co_{0.15})_{0.78}$ film (empty squares). The data are compared with experimental results for a bilayer in which the dielectric film thickness is below the interference limit (full squares) and a calculation taking into account multiple reflections (dots) (after Reim and Weller, 1988).

mitted by an analyzer with an azimuth $-\theta_K$ relative to the reflected elliptically polarized beam. In another study of the enhancement of the Kerr signal with dielectric layers, Weller and Reim (1989) find that for a 200-nm-thick film of $Tb_{20}(Fe_{0.85}Co_{0.15})_{80}$, which is covered with a 89-nm-thick AlN layer and no substrate on top, $|\theta_K|$ increases from ≈ 0.3 to 1.5°, but R decreases from $\approx 55\%$ to 13%. The result is an enhancement of the figure of merit $R^{1/2}(\theta_K^2 + \varepsilon_K^2)^{1/2}$ from 0.21 to 0.5°. Morishita et al. (1988) reported for a system consisting of a FeTb layer sandwiched between two silicon layers a rotation of $-25°$, but at almost zero reflectivity. To further increase the figure of merit, quadrilayers like those displayed on the right side of Fig. 2-105 can be produced. In this system interference enhancement occurs on both sides of the magneto-optical film, and Faraday as well as Kerr rotation contribute. Reim and Weller (1988 a) reported a maximum $|\theta_K|$ of 1.8°, which should be compared to the 0.6° for the bilayer with the incident light through the substrate, because the substrate reduces the apparent rotation by roughly a factor of 2. For an optimized system consisting of a glass substrate/AlN (70 nm)/$Tb_{24}(Fe_{0.88}Co_{0.12})_{76}$ (20 nm) / AlN (30 nm) / Cu (40 nm), Reim (1990) found $|\theta_K|$ to be 2.2° and a figure of merit to be 0.78°.

We have seen in the previous sections that several metals with steep reflectivity edges at the plasma frequency (CeSb, NdS, Nd_3S_4, $CuCr_2S_4$) display rather narrow Kerr rotation peaks at these energies because of a resonance-like behavior of the coefficients A and B relating the Kerr rotation with the off-diagonal conductivity in Eq. (2-42). This effect is sometimes called enhancement, although experiments or an analysis of the off-diagonal conductivity and the optical constants shows (Schoenes, 1987) that a nonmetal with the same magneto-optical transitions generally has an even larger Kerr rotation. Thus, in reality, the free carriers in a magnetic metal reduce the Kerr signal due to their contribution to the optical constants, but near the plasma edge the latter may become small enough to restore similar Kerr values as in the nonmetallic analogue. Exceptions are TmS and TmSe, which show a genuine intraband magneto-optical effect (Schoenes and Reim, 1988). Ohta et al. (1983) appear to have been the first to use a metallic bilayer structure to combine the magneto-optical activity of one metal with the small optical constants near the plasma edge of another metal. They found that at $\lambda = 633$ nm Gd–

Tb–Fe films of ≈ 20 nm thickness show an increased Kerr rotation if they are deposited on Cu, Ag, or Au but no increase if they are deposited on Al. In a series of papers, Katayama and co-workers (see, e.g., Katayama et al., 1986, 1988) have reported magneto-optical Kerr studies for multilayers consisting of magneto-optically active Fe or Co films and layers of the noble metals Ag, Au, and Cu. Figure 2-107 displays results for bilayer structures consisting of Fe films of thicknesses ranging from 3.7 to 30 nm on top of a thick Cu film (Katayama et al., 1988). A maximum rotation can be seen near the plasma edge of Cu for a Fe film thickness of 13.1 nm. As has been pointed out by Sato and Kida (1988) and emphasized by Nies and Kessler (1990), the observed thickness dependence of the Kerr spectra can be explained well by multiple reflections and wave superposition within the Fe film, and model calculations using the matrix method to handle Fresnel's equations are in good agreement with the experimental results. Bilayers of Tb–Fe–Co on Cu and on Ag have been studied by Reim and Weller (1988 b). These authors report a Kerr rotation of -1° at 3.7 eV for an 11-nm-thick $Gd_{11}Tb_{11}(Fe_{0.80}Co_{0.20})_{78}$ film rf-sputtered on top of a thick Ag film. The same group also performed measurements both for Fe–Co deposited on Ag, Au, Cu, TiN, and ZrN, and for $Tb_{16}(Fe_{0.80}Co_{0.20})_{84}$ deposited on LaB_6 (Weller and Reim, 1989). For the latter bilayer, they found a rotation peak of -1.6° at 1.8 eV. Katayama et al. (1986), Weller and Reim (1989) and Hashimoto et al. (1989) have also investigated multilayers with several periods. The enhancement of the figure of merit is nearly the same as it is for the bilayers. However, interesting new possibilities arise with the reduction of the single-layered thickness. The thinner magnetic films tend to have a larger perpendicular anisotropy, and the increase of the interface volume allows the use of exchange coupling to optimize the performance of the magneto-optical media (Wakabayashi et al., 1989; Nakamura et al., 1989; Zeper et al., 1989; Ochiai et al., 1989; Hashimoto et al., 1989; Ferré et al., 1990). It is possible that metallic multilayers will be the basis for the second generation of magneto-optical disks.

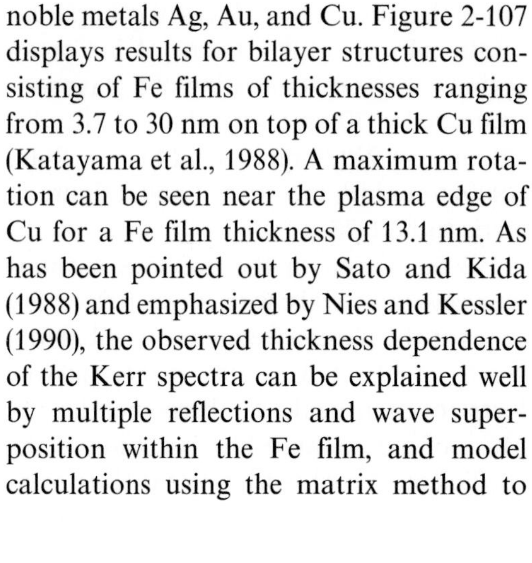

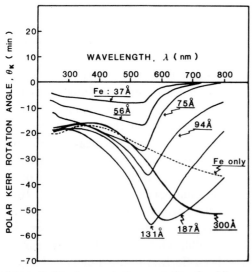

Figure 2-107. Apparent Kerr rotation for bilayers consisting of Fe films deposited on top of a 240-nm-thick Cu film. The varying thickness of the Fe film is given in Å. For comparison, the spectrum for pure Fe is also displayed (after Katayama et al., 1988).

2.9.3 Nonreciprocal Elements

The *Rayleigh light trap*, sketched in Fig. 2-108, is the basis for optical isolators or so-called nonreciprocal elements. It allows light to transmit the device in one direction and reject light traveling in the opposite direction. With the increasing importance of optical data transmission in communication, the need for nonreciprocal elements has grown very rapidly. The nonreciprocal

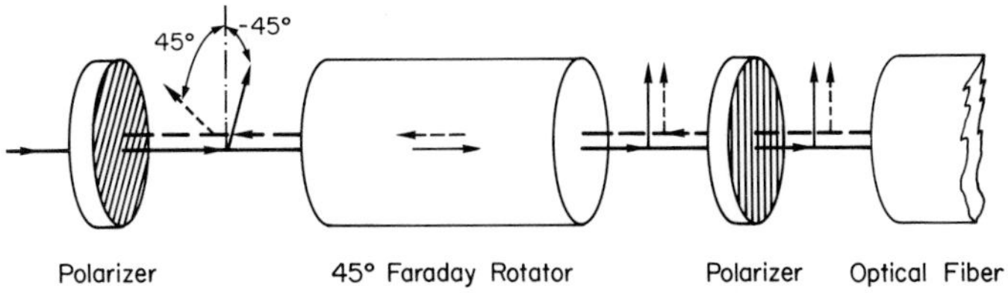

Figure 2-108. Scheme of the Rayleigh light trap. Light which is reflected from optical components in the light path on the right side of the second polarizer is blocked on its way back to the light source by the first polarizer.

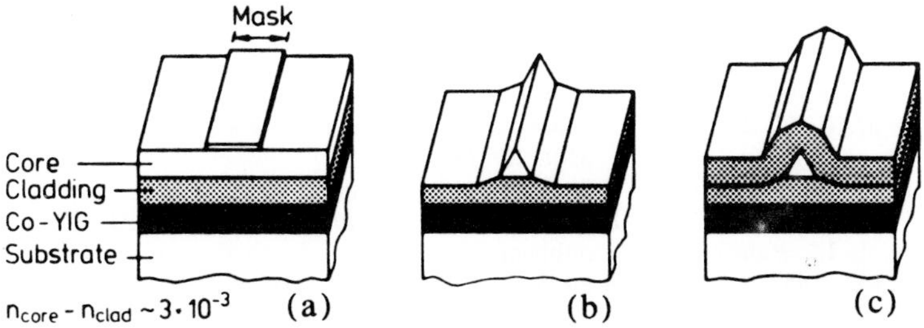

Figure 2-109. Scheme of the fabrication of buried channel waveguides in YIG: (a) planar single-mode waveguide; (b) etched rib waveguide; (c) buried channel waveguide (after Pross et al., 1988).

element is needed to prevent feedback from the optical system into the laser. Such interactions can lead to fluctuations of the frequency, the phase, and/or the amplitude of the laser and, therefore, to an increase of the bit-error rate. In coherent optical systems with frequency coding an optical isolation of 60 dB is required (Vodhanel et al., 1986).

Micro-optic isolators containing a bulk YIG crystal between two polarizers have been in use since the mid 1980s (Sairasaki et al., 1986). At a wavelength of 1.3 μm the crystal length needed to rotate the polarization direction by 45° in YIG is about 2.3 mm. The magnetic field necessary to saturate YIG amounts to a few 10^{-3} T. To allow optical insulation independently of the incoming polarization direction, the light beam can be split into two components with orthogonal polarization which are both rotated and combined again afterward. Often, a reciprocal −45° rotator is added to the nonreciprocal 45° rotator to restore the initial polarization direction of the transmitted beam, while the reflected one is rotated by 90° (Schmitt and Dammann, 1987). Microoptic isolators are not well suited for coupling to an optical fiber system and therefore waveguide isolators have been developed. Figure 2-109 shows the three steps for fabricating buried channel waveguides in YIG (Pross et al., 1988). The typical thickness of each individual layer is about 5 μm and the magneto-optically active material is either $Y_{2.8}Bi_{0.2}Fe_{4.3}Ga_{0.7}O_{12}$ or $Y_{2.7}Bi_{0.3}Fe_{4.7}Al_{0.3}O_{12}$, grown by liquid-phase epitaxy on a gadolinium-gallium-garnet wafer. The propagating modes in thin films and these channel waveguides are TE (transverse electric) and TM (transverse magnetic) waves. The effective conversion of one wave in another requires phase matching, the realization of which is one of the current research efforts in this field. Another goal is the integration of the nonreciprocal and the reciprocal element in one waveguide (see, e.g., Koshizuka, 1986; or Schmitt and Dammann, 1987).

2.10 References

Abe, M., Gomi, M. (1982), *J. Appl. Phys. 53,* 8172–8174.

Abe, M., Gomi, M., Nomura, S. (1980), *Jpn. J. Appl. Phys. 19,* 1329–1332.

Abe, M., Shono, K., Kobayashi, K., Gomi, M., Nomura, S. (1982), *Jpn. J. Appl. Phys. 21,* L22–L24.

Adachi, K. (1961), *J. Phys. Soc. Jpn. 16,* 2187–2206.

Adachi, K., Matsui, M., Fukuda, Y. (1980), *J. Phys. Soc. Jap. 48,* 62–70.

Afonso, C. N., Lagunas, A. R., Briones, F., Giron, S. (1980), *J. Magn. Magn. Mat. 15–18,* 833–834.

Afonso, C. N., Giron, S., Lagunas, A. R., Vincent, J. L. (1981), *IEEE Trans. Magn. MAG-17,* 2849–2851.

Ahrenkiel, R. K., Coburn, T. J. (1975), *IEEE Trans. Magn. MAG-11,* 1103–1108.

Ahrenkiel, R. K., Moser, F., Lyu, S. L., Pidgeon, C. R. (1971), *J. Appl. Phys. 42,* 1452–1453.

Ahrenkiel, R. K., Lee, T. H., Lyu, S. L. Moser, F. (1973), *Solid State Commun. 12,* 1113–1115.

Ahrenkiel, R. K., Coburn, T. J., Carnall, E. Jr. (1974), *IEEE Trans. Magn. MAG-10,* 2–7.

Ahrenkiel, R. K., Coburn, T. J., Pearlman, D., Carnall, E. Jr., Martin, T. W., Lyu, S. L. (1975a), *AIP Conf. Proc. 24,* 186–187.

Ahrenkiel, R. K., Lyu, S. L., Coburn, T. J. (1975b), *J. Appl. Phys. 46,* 894–899.

Amzallag, E. (1970), *Ann Phys. 5,* 27–35.

Andresen, A. F., Hälg, W., Fischer, P., Stoll, E. (1967), *Acta Chem. Scand. 21,* 1543.

Antonini, B., Geller, S., Paoletti, A., Paroli, P., Tucciarone, A. (1978), *Phys. Rev. Lett. 41,* 1556–1558.

Arend, H., Schoenes, J., Wachter, P. (1975), *phys. stat. sol. (b) 69,* 105–112.

Argyres, P. N. (1955), *Phys. Rev. 97,* 334–345.

Atkinson, R. (1977), *Thin Solid Films 47,* 177–186.

Atkinson, R. Lissberger, P. H. (1974), *Int. J. Magn. 6,* 227–241.

Baer, Y., Zürcher, Ch. (1977), *Phys. Rev. Lett. 39,* 956–959.

Becquerel, H. (1897), *Comptes Rend. 125,* 679.

Bednorz, J. G., Müller, K. A. (1986), *Z. Phys. B64,* 189–193.

Bennett, H. S., Stern, E. A. (1965), *Phys. Rev. 137,* A448–A461.

Berkner, D. D. (1975), *Phys. Lett. A54,* 396–398.

Bielov, K. P., Henkie, Z., Dimitrievsky, D. S., Levitin, R. Z., Trzebiatowski, W. (1973), *Zh. Eksp. & Teor. Fiz. 64,* 1351.

Blazey, K. W. (1974), *J. Appl. Phys. 45,* 2273–2280.

Bloomberg, D. S., Conell, G. A. N. (1987), in: *Magnetic Recording*. New York: McGraw-Hill, Vol. 2, pp. 305–399.

Boccara, A. C., Ferré, J., Badoz, J. (1969a), *phys. stat. sol. 36,* 601–608.

Boccara, A. C., Ferré, J., Briat, B., Billardon, M., Badoz, J. P. (1969b), *J. Chem. Phys. 50,* 2716–2718.

Bonenberger, R. (1978), *Dissertation. Technische Hochschule Aachen,* unpublished.

Bongers, P. F., Haas, C., van Run, A. M. J. G., Zanmarchi, G. (1969), *J. Appl. Phys. 40,* 958–963.

Boppart, H., Wachter, W. (1984), in: *Moment Formation in Solids*. New York: Plenum Publ., pp. 229–235.

Borghesi, A., Guizetti, G., Samoggia, G., Reguzzoni, E. (1981), *Phys. Rev. Lett. 47,* 538–541.

Born, M. (1964), *Principles of Optics*. London: Springer.

Boswarva, I. M., Howard, R. E., Lidiard, A. D. (1962), *Proc. Roy. Soc. A 269,* 125–141.

Boudewijn, P. R., Haas, C. (1980), *J. Magn. Magn. Mat. 15–18,* 787–788.

Brändle, H. (1990a), unpublished.

Brändle, H. (1990b), *PhD thesis, ETH Zürich,* unpublished.

Brändle, H., Schoenes, J., Hulliger, F. (1989), *Helv. Phys. Acta 62,* 199–202.

Brändle, H., Schoenes, J., Wachter, P., Hulliger, F., Reim, W. (1990a), *Appl. Phys. Lett. 56,* 2602–2603.

Brändle, H., Schoenes, J., Hulliger, F., Reim, W. (1990b), *IEEE Trans. Magn. MAG-26,* 2795–2797.

Brändle, H., Schoenes, J., Hulliger, F., Reim, W. (1990c), *J. Magn. Magn. Mat. 83,* 29–31.

Brändle, H., Schoenes, J., Wachter, P., Hulliger, F., Reim, W. (1991a), *J. Magn. Magn. Mat. 93,* 207–210.

Brändle, H., Reim, W., Weller, D., Schoenes, J. (1991b), *J. Magn. Magn. Mat. 93,* 220–224.

Brändle, H., Schoenes, J., Hulliger, F., Reim, W. (1991c), to be published.

Buchenau, U., Müller, I. (1972), *Solid State Commun. 11,* 1291–1293.

Buhrer, C. F. (1969a), *J. Appl. Phys. 40,* 4500–4502.

Buhrer, C. F. (1969b), *J. Phys. & Chem. Solids 30,* 1273–1276.

Buhrer, C. F. (1970), *J. Appl. Phys. 41,* 1393–1394.

Burkhard, H., Jaumann, J. (1970), *Z. Phys. 235,* 1–19.

Burlet, P., Rossat-Mignod, J., Troć, R., Henkie, Z. (1981), *Solid State Commun. 39,* 745–749.

Busch, G., Junod, P., Wachter, P. (1964), *Phys. Lett. 12,* 11–12.

Busch, G., Magyar, B., Wachter, P. (1966), *Phys. Lett. 23,* 438–440.

Buschow, K. H. J. (1988), *Handbook on Ferromagnetic Materials,* Vol. 4. Amsterdam: North-Holland, p. 493.

Buschow, K. H. J., van Engen, P. G. (1981a), *J. Appl. Phys. 52,* 3557–3561.

Buschow, K. H. J., van Engen, P. G. (1981b), *Mat. Res. Bull. 16,* 1177–1184.

Buschow, K. H. J., van Engen, P. G. (1981 c), *Solid State Commun. 39,* 1–3.
Buschow, K. H. J., van Engen, P. G. (1981 d), *J. Magn. Magn. Mat. 25,* 90–96.
Buschow, K. H. J., van Engen, P. G. (1984), *Philips J. Res. 39,* 82–93.
Buschow, K. H. J., van Engen, P. G., Jongebreur, R. (1983), *J. Magn. Magn. Mat. 38,* 1–22.
Buschow, K. H. J., van Engen, P. G., de Mooij, D. B. (1984), *J. Magn. Magn. Mat. 40,* 339–347.
Cable, J. W., Wollan, E. O., Koehler, W. C. (1965), *Phys. Rev. 138,* A755–A759.
Canit, J. C., Badoz, J., Briat, B., Krishnan, R. (1974), *Solid State Commun. 15,* 767–770.
Carey, R., Thomas, B. W. J., Bains, G. S. (1990), *J. Magn. Magn. Mat. 83,* 21–22.
Chamberland, B. L. (1977), *CRC Critical Reviews in Solid State & Materials Sciences 7,* 1–31.
Chase, L. L. (1974), *Phys. Rev. B10,* 2226–2231.
Chaudhari, P., Cuomo, J. J., Gambino, R. J. (1973), *Appl. Phys. Lett. 22,* 337–339.
Chen, D., Gondo, Y. (1964), *J. Appl. Phys. 35,* 1024–1025.
Chen, D., Stutius, W. E. (1974), *IEEE Trans. Magn. MAG-10,* 581–586.
Chen, D., Ready, J. F., Bernal, G. E. (1968), *J. Appl. Phys. 39,* 3916–3927.
Chen, D., Aagard, R. L., Liu. T. S. (1970), *J. Appl. Phys. 41,* 1395–1396.
Chetkin, M. V., Shcherbakov, Y. I., Volenko, A. P., Shevchuk, L. D. (1975), *Sov. Phys. JETP 40,* 509–511.
Choe, Y. J., Tsunashima, S., Katayama, T., Uchiyama, S. (1987), *J. Magn. Soc. Jpn. 11, Suppl. No. S1,* 273–276.
Clemens, K. H., Jaumann, J. (1963), *Z. Phys. 173,* 135–148.
Clover, R. B., Rayl, M., Gutman, D. (1971), *AIP Conf. Proc. 5,* 264–268.
Coburn, T. J., Ahrenkiel, R. K., Carnall, E. Jr., Pearlman, D. (1974), *AIP Conf. Proc. 18,* 1118–1121.
Coehoorn, R., de Groot, R. A. (1985), *J. Phys. F: Met. Phys. 15,* 2135–2144.
Coehoorn, R., Haas, C., de Groot, R. A. (1985), *Phys. Rev. B31,* 1980–1986.
Coey, J. M. D., McGuire, T. R., Tissier, B. (1981), *Phys. Rev. B24,* 1261–1273.
Colominas, C. (1967), *Phys. Rev. 153,* 558–560.
Comstock, R. L., Lissberger, P. H. (1970), *J. Appl. Phys. 41,* 1397–1398.
Connell, G. A. N., Allen, R. (1983), *J. Magn. Magn. Mat. 31–34,* 1516–1518.
Cooper, B. R. (1965), *Phys. Rev. 139,* A1505–A1514.
Croat, J. J., Herbst, J. F., Lee, R. W., Pinkerton, F. E. (1984), *J. Appl. Phys. 55,* 2078–2082.
Cros, C., Dance, J. M., Grenier, J. C., Wanklyn, B. M., Garrad, B. J. (1977), *Mater. Res. Bull. (USA) 12,* 415–419.
Crossley, W. A., Cooper, R. W., Page, J. L., van Stapele, R. P. (1969), *Phys. Rev. 181,* 896–904.
Daalderop, G. H. O., Mueller, F. M., Albers, R. C., Boring, A. M. (1988), *J. Magn. Magn. Mat. 74,* 211–218.
Davis, J. A., Bunch, R. M. (1984), *Appl. Optics, 23,* 633–636.
Day, P. (1979), *Accounts of Chemical Research 12,* 236–243.
de Groot, R. A., Buschow, K. H. J. (1986), *J. Magn. Magn. Mat. 54–57,* 1377–1380.
de Groot, R. A., Mueller, F. M., van Engen, P. G., Buschow, K. H. J. (1983), *Phys. Rev. Lett. 50,* 2024–2027.
de Groot, R. A., Mueller, F. M., van Engen, P. G., Buschow, K. H. J. (1984), *J. Appl. Phys. 55,* 2151–2154.
Deryagin, A. V., Andreev, A. V. (1976), *Sov. Phys. JETP 44,* 610–613.
Dhar, S. K., Malik, S. K., Vijayaraghavan, R. (1981), *J. Phys. C14,* L321–L324.
Dillon, J. F., Jr. (1959), *J. Phys. Radium 20,* 374–377.
Dillon, J. F., Jr., Kamimura, H., Remeika, J. P. (1966), *J. Phys. Chem. Solids 27,* 1531–1549.
Dillon, J. F., Jr., Albiston, S. D., Fratello, V. J. (1987), *J. Magn. Soc. Jpn. 11, Suppl. S1,* 241–244.
Dresselhaus, G., Kip, A. F., Kittel, C. (1955), *Phys. Rev. 128,* 368–384.
Drews, U., Jaumann, J. (1969), *Z. Angew. Phys. 26,* 48–50.
Ebert, H. (1989), *Proc. ISOMES '89,* Duisburg, March 1989.
Ebert, H., Strange, P., Gyorffy, B. L. (1988), *J. Physique 39,* C8-31–C8-36.
Edelman, I. S., Dustmuradov, G. (1980), *Magnitnye Poluprovodniki i ikh Svoista, Physical Institute, Krasnoyarsk,* 161–170.
Edelman, I. S., Dustmuradov, G., Kononov, V. P. (1983), *phys. stat. sol. (b) 117,* 351–354.
Egashira, K., Yamada, T. (1974), *J. Appl. Phys. 45,* 3643–3648.
Egashira, K., Katsui, A., Shibukawa, A. (1977), *Rev. Electr. Commun. Lab. 25,* 163–171.
Ehrenreich, H., Philipp, H. R. (1962), *Phys. Rev. 128,* 1622–1629.
Erskine, J. L. (1975), *AIP Conf. Proc. 24,* 190–194.
Erskine, J. L. (1976), *Phys. Rev. Lett. 37,* 157–160.
Erskine, J. L. (1977), *Physica B89,* 83–90.
Erskine, J. L., Stern, E. A. (1973 a), *Phys. Rev. B8,* 1239–1255.
Erskine, J. L., Stern, E. A. (1973 b), *Phys. Rev. Lett. 30,* 1329–1332.
Faraday, M. (1846), *Phil. Trans. Roy. Soc. 136,* 1.
Feinleib, J., Scouler, W. J., Dimmock, J. O., Hanus, J., Reed, T. B., Pidgeon, C. R. (1969), *Phys. Rev. Lett. 22,* 1385–1387.
Feldtkeller, E. (1972), *IEEE Trans. Magn. MAG-8,* 481–486.
Ferré, J. (1974), *J. Physique 35,* 781–801.

Ferré, J., Boccara, A. C., Briat, B. (1970), *J. Physique 31,* 631–636.

Ferré, J., Pisarev, R. V., Harding, M. J., Badoz, J., Kizhaev, S. A. (1973), *J. Phys. C: Solid State Phys. 6,* 1623–1638.

Ferré, J., Pénissard, G., Marlière, C., Renard, D., Beauvillain, P., Renard, J. P. (1990), *Appl. Phys. Lett. 56,* 1588–1590.

Flippen, R. B. (1963), *J. Appl. Phys. 34,* 2026–2032.

Freeman, A. J., Koelling, D. D. (1974), in: *The Actinides: Electronic Structure and Related Properties.* New York: Academic Press, Vol. 1, pp. 51–108.

Freiser, M. J. (1968), *IEEE Trans. Magn. MAG-4,* 152–161.

Freiser, M. J., Holtzberg, F., Methfessel, S., Pettit, G. D., Shafer, M. W., Suits, J. G. (1968), *Helv. Phys. Acta 41,* 832–837.

Fumagalli, P. (1990), *Dissertataion ETH-Zürich,* unpublished.

Fumagalli, P., Schoenes, J., Kaczorowski, D. (1988), *Solid State Commun. 65,* 173–176.

Fumagalli, P., Schoenes, J., Decroux, M., Fischer, O. (1990), *J. Appl. Phys. 67,* 5035–5037.

Fumagalli, P., Schoenes, J. (1991), to be published in Phys. Rev. B.

Gambino, R. J., McGuire, T. R. (1986), *J. Magn. Magn. Mat. 54–57,* 1365–1370.

Gambino, R. J., Chaudhari, P., Cuomo, J. J. (1974), *AIP Conf. Proc. 18,* 578–592.

Gambino, R. J., McGuire, T. R., Plaskett, T. S., Reim, W. (1986), *IEEE Trans. Magn. MAG-22,* 1227–1229.

Gebhardt, W. (1976), *J. Magn. Magn. Mat. 3,* 129–142.

Geller, S., Colville, A. A. (1975), *AIP Conf. Proc. 24,* 372–373.

German, K. H., Maier, K., Strauss, E. (1974), *Solid State Commun. 14,* 1309–1311.

Gomi, M., Abe, M., Nomura, S. (1979), *Jpn. J. Appl. Phys. 18,* 739–742.

Gomi, M., Kajima, M., Abe, M. (1984), *J. Appl. Phys. 55,* 2170–2172.

Gomi, M., Satoh, K., Abe, M. (1988a), *Jpn. J. Appl. Phys. 27,* L1536–L1538.

Gomi, M., Satoh, K., Abe, M. (1988b), *J. Appl. Phys. 63,* 3642–3644.

Goodenough, J. B. (1965), *Bull. Soc. Chim. Fr.,* 1200–1207.

Goodenough, J. B. (1967a), *Colloque Intern. du C.N.R.S. No. 157,* Orsay 1965, Editions du C.N.R.S., p. 1200–1207.

Goodenough, J. B. (1967b), *Solid State Commun. 5,* 577–580.

Goodenough, J. B. (1969), *J. Phys. Chem. Solids 30,* 261–280.

Goodenough, J. B. (1971), *Progress in Solid State Chemistry.* Oxford: Pergamon, Vol. 5, p. 145.

Greiner, J. H., Fan, G. J. (1966), *J. Appl. Phys. Lett. 9,* 27–29.

Groń, T., Duda, H., Warcczewski, J. (1990), *Phys. Rev. B41,* 12424–12431.

Guillot, M., Le Gall, H. (1976), *phys. stat. sol. (b) 77,* 121–130.

Güntherodt, G. (1974), *Phys. Condens. Matter 18,* 37–78.

Haas, C. (1970), *Crit. Rev. Solid State Sci. 1,* 47–98.

Hansen, P. (1990), *J. Magn. Magn. Mat. 83,* 6–12.

Hansen, P., Tolksdorf, W., Witter, K., Robertson, J. M. (1984), *IEEE Trans. Magn. MAG-20,* 1099–1104.

Hansen, P., Hartmann, M., Witter, K. (1987), *J. Magn. Soc. Jpn. 11, Suppl. S1,* 257–260.

Harbeke, G., Pinch, H. (1966), *Phys. Rev. Lett. 17,* 1090–1092.

Harbeke, G., Berger, S. B., Emmenegger, F. P. (1968), *Solid State Commun. 6,* 553–555.

Hashimoto, S., Ochiai, Y., Aso, K. (1989), *Jpn. J. Appl. Phys. 28,* L1824–L1826.

Heikes, H. R. (1955), Phys. Rev. 99, 446–447.

Henkie, Z. (1980), *Physica B102,* 329ff.

Heusler, F. (1903), *Verh. Dtsch. Phys. Ges. 5,* 219.

Huang, C. C., Ho, J. T. (1975), *Phys. Rev. B12,* 5255–5260.

Huang, C. C., Pindak, R. S., Ho, J. T. (1974), *Solid State Commun. 14,* 559–562.

Hulliger, F. (1976), *Structural Chemistry of Layer-Type Phases,* Dordrecht-Holland/Boston: D. Reidel Publ. Comp.

Hulliger, F. (1979), in: *Handbook of the Physics and Chemistry of Rare Earths.* Amsterdam: North-Holland, Chap. 33, p. 153.

Hulme, H. R. (1932), *Proc. Roy. Soc. A 135,* 237–257.

Imamura, N., Mimura, Y. (1976), *J. Phys. Soc. Jpn. 41,* 1067–1068.

Imamura, N., Tanaka, S., Tanaka, F., Nagao, Y. (1985), *IEEE Trans. Magn. MAG-21,* 1607–1612.

Inukai, T., Matsuoka, M., Ono, K. (1986), *Appl. Phys. Lett. 49,* 52–53.

Inukai, T., Sugimoto, N., Matsuoka, M., Ono, K. (1987), *J. Magn. Soc. Jpn. 11, Suppl. S1,* 217–220.

Itoh, A. (1989), *Jpn. J. Appl. Phys. 28, Suppl. 28-3,* 15–20.

Itoh, A., Kryder, M. H. (1988), *Appl. Phys. Lett. 53,* 1125–1126.

Itoh, A., Unozawa, K., Shinohara, T., Nakada, M., Inoue, F., Kawanishi, J. (1985), *IEEE Trans. Magn. MAG-21,* 1672–1674.

Itoh, A., Toriumi, Y., Ishii, T., Nakada, M., Inoue, F., Kawanishi, K. (1987), *IEEE Trans. Magn. MAG-23,* 2964–2966.

Jaggi, R., Methfessel, S., Sommerhalder, R. (1962), in: *Landolt-Börnstein.* Vol. II, part 9, pp. 1-180 to 1-199.

Jantz, W., Sandercock, J. R., Wettling, W. (1976), *J. Phys. C: Solid State Phys. 9,* 2229–2240.

Jayaraman, A., Narayanamurti, V., Bucher, E., Maines, R. G. (1970), *Phys. Rev. Lett. 25,* 1430–1433.

Jesser, R., Bieber, A., Kuentzler, R. (1981), *J. Physique 42,* 1157–1166.
Jung, W. (1965), *J. Appl. Phys. 36,* 2422–2426.
Kahn, F. J., Pershan, P. S., Remeika, J. P. (1969), *Phys. Rev. 186,* 891–918.
Kaldis, E., Fritzler, B., Spychiger, H., Jilek, E. (1982), *Proc. Int. Conf. on Valence Instabilities,* Zürich. Amsterdam: North-Holland, p. 131.
Kambara, T., Oguchi, T., Gondaira, I. K. (1980), *J. Phys. C: Solid State Phys. 13,* 1493–1511.
Kämper, K. P., Schmitt, W., Güntherodt, G., Gambino, R. J., Ruf, R. (1987), *Phys. Rev. Lett. 59,* 2788–2791.
Kasuya, T. (1972), *Crit. Rev. Solid State Sci. 3,* 131–164.
Katayama, T., Hasegawa, K. (1982), in: *Rapidly Quenched Metals IV:* Japan Inst. of Metals, Sendai, pp. 915–918.
Katayama, T., Awano, H., Nishihara, Y. (1986), *J. Phys. Soc. Jpn. 55,* 2539–2542.
Katayama, T., Suzuki, Y., Awano, H., Nishihara, Y., Koshizuka, N. (1988), *Phys. Rev. Lett. 60,* 1426–1429.
Katsui, A. (1976a), *J. Appl. Phys. 47,* 3609–3611.
Katsui, A. (1976b), *J. Appl. Phys. 47,* 4663–4665.
Katsui, A., Shibukawa, A., Terui, H., Egashira, K. (1976), *J. Appl. Phys. 47,* 5069–5071.
Kerr, J. (1877), *Phil. Mag. 3,* 321.
Kharchenko, N. F., Belyi, L. I., Tutakina, O. P. (1969), *Sov. Phys.-Sol. State 10,* 2221–2223.
Kido, G., Miura, N., Kawauchi, K., Oguro, I., Dillon, J. F., Jr., Chikazumi, S. (1977), *Physica 89 B,* 147–149.
Kittel, C. (1951), *Phys. Rev. A 83,* 208.
Kolodziejczak, J., Lax, B., Nishina, Y. (1962), *Phys. Rev. 128,* 2655–2660.
Koroleva, L. I., Shalimova, M. A. (1979), *Sov. Phys. Solid State 21,* 266–269.
Koshizuka, N. (1986), *Jpn. Ann. Rev. Electr. Comp. & Telecommun. 21,* 75–87.
Krichevtsov, B. B., Pavlov, V. V., Pisarev, R. V. (1988), *Sov. Phys. JETP 67,* 378–384.
Krinchik, G. S., Artem'ev, V. A. (1968), *Sov. Phys. JETP 26,* 1080–1085.
Krinchik, G. S., Gushchina, S. A. (1969), *Sov. Phys. JETP 28,* 257–259.
Krinchik, G. S., Nuralieva, R. D. (1959), *Sov. Phys. JETP 9,* 724–726.
Krinchik, G. S., Nurmukhamedov, G. M. (1965), *Sov. Phys. JETP 21,* 22–25.
Kryder, M. H. (1990), *J. Magn. Magn. Mat. 83,* 1–5.
Lacklison, D. E., Scott, G. B., Ralph, H. I., Page, J. L. (1973), *IEEE Trans. Magn. MAG-9,* 457–460.
Lam, D. J., Aldred, A. T. (1974), in: *The Actinides: Electronic Structure and Related Properties.* New York: Academic Press, Vol. 1, pp. 109–179.
Lambeck, M., Michel, L., Waldschmidt, M. (1963), *Z. Angew. Phys. 15,* 369–371.
Landau, L. D., Lifshitz, E. M. (1960), *Electrodynamics of Condensed Media.* New York: Pergamon.
Lang, J. K., Baer, Y., Cox, P. A. (1981), *J. Phys. F. Metal. Phys. 11,* 121–138.
Leciejewicz, J., Zygmunt, A. (1972), *phys. status sol. (a) 13,* 657–670.
Lee, T. H., Coburn, T., Gluck, R. (1971), *Solid State Commun. 9,* 1821–1824.
LeGall, H., Guillot, M., Marchand, A., Nomi, Y., Artinian, M., Desvignes, J. M. (1987), *J. Magn. Soc. Jpn. 11, Suppl. S 1,* 235–240.
Lehman, H. W., Harbeke, G., Pinch, H. (1971), *J. Physique Colloque 32,* C 1-932 to C 1-933.
Llinares, C., Monteil, E., Bordure, G., Paparoditis, C. (1973), *Solid State Commun. 13,* 205–208.
Löfgren, K.-E., Tuomi, T., Stubb, T. (1974), *Solid State Commun. 14,* 1285–1286.
Lotgering, F. K. (1965), *Proc. Intern. Conf. Magnetism,* Nottingham. The Institute of Physics and the Royal Society, London, 533–537.
Lotgering, F. K., van Stapele, R. P. (1968), *J. Appl. Phys. 39,* 417–423.
Lyons, K. B., Kwo, J., Dillon, J. F., Jr., Espinosa, G. P., McGlashan-Powell, M., Ramirez, A. P., Schneemeyer, L. F. (1990), *Phys. Rev. Lett. 64,* 2949–2952.
Majorana, Q. (1944), *Nuovo Cimento 2,* 1.
Madelung, O. (1962), in: *Landolt-Börnstein.* Berlin: Springer, Vol. II, part. 9.
Martens, J. W. D. (1986), *J. Appl. Phys. 59,* 3820–3823.
Martens, J. W. D., Peeters, W. L. (1983), *SPIE Proc. 420,* 231–235.
Martens, J. W. D., Voermans, A. B. (1984), *IEEE Trans. Magn. MAG-20,* 1007–1012.
Martens, J. W. D., Peeters, W. L., Erman, M. (1982), *Solid State Commun. 41,* 667–669.
Martens, J. W. D., Peeters, W. L., Nederpel, P. Q. J. (1983), *Surf. Sciences 135,* 334–340.
Martin, D. H., Neal, K. F., Dean, T. J. (1965), *Proc. Phys. Soc. 86,* 605–614.
Massenet, O., Buder, R., Since, J. J., Schlenker, C., Mercier, J., Kelber, J., Stucky, D. G. (1978), *Mat. Res. Bull. 13,* 187–195.
Masui, S., Kobayashi, T., Tsunashima, S., Uchiyama, S., Sumiyama, K., Nakamura, Y. (1984), *IEEE Trans. Magn. MAG-20,* 1036–1038.
Matthias, B. T., Bozorth, R. M., van Vleck, J. H. (1961), *Phys. Rev. Lett. 7,* 160–161.
Matsumoto, S., Goto, T., Syono, Y., Nakagawa, Y. (1978), *J. Phys. Soc. Jpn. 44,* 162–164.
Mavroides, J. G. (1972), in: *Optical Properties of Solids.* Amsterdam-London: North-Holland.
Mayer, L. (1958), *J. Appl. Phys. 29,* 1454–1456.
Mayer, L. (1960), *J. Appl. Phys. 31,* 384S–385S.
McElfresh, M. W., Plaskett, T. S., Gambino, R. J., McGuire, T. R. (1990), *Appl. Phys. Lett. 57,* 730–732.
McGroddy, J. C., McAlister, A. J., Stern, E. A. (1965), *Phys. Rev. 139,* A1844–A1848.
McGuire, T. R., Gambino, R. J. (1987), *J. Magn. Soc. Jpn. 11, Suppl. S 1,* 261–264.

McGuire, T. R., Gambino, R. J., Bell, A. E., Sprokel, G. J. (1986), *J. Magn. Magn. Mat. 54–57,* 1387–1388.
McGuire, T. R., Gambino, R. J., Plaskett, T. S., Reim, W. (1987), *J. Appl. Phys. 61,* 3352–3354.
Methfessel, S., Mattis, D. C. (1968), *Magnetic Semiconductors,* in: *Hdb. der Physik XVIII/1.* Heidelberg-Berlin-New York: Springer.
Misemer, D. K. (1988), *J. Magn. Magn. Mat. 72,* 267–274.
Mitani, T., Koda, T. (1973), *Phys. Lett. A 43,* 137–138.
Mitani, T., Ishibashi, M., Koda, T. (1975), *J. Phys. Soc. Jpn. 38,* 731–738.
Miyamoto, H., Niihara, T., Sukeda, H., Takahashi, M., Nakao, T., Ojima, M., Ohta, N. (1989), *J. Appl. Phys. 66,* 6138–6143.
Morishita, T., Sato, R., Sato, K., Kida, H. (1988), *J. Physique Colloque 49,* C8-1741–C8-1742.
Morisin, B., Narath, A. (1964), *J. Chem. Phys. 40,* 1958–1967.
Müller, I., Buchenau, U. (1975), *Physica B80,* 69–74.
Muret, P. (1974), *Solid State Commun. 14,* 1119–1122.
Nagaev, E. L. (1975), *Sov. Phys. Usp. 18,* 863–892.
Nakamura, K., Tsunashima, S., Iwata, S., Uchiyama, S. (1989), *IEEE Trans. Magn. MAG-25,* 3758–3760.
Nesbitt, E. A., Williams, H. J., Wernick, J. H., Sherwood, R. C. (1962), *J. Appl. Phys. 33,* 1674–1678.
Nies, R., Kessler, F. R. (1990), *Phys. Rev. Lett. 64,* 105.
Nolting, W. (1979), *phys. stat. sol. (b) 96,* 11–54.
Nolting, W., Oles, A. M. (1981), *Phys. Rev. 23,* 4122–4128.
Norman, M. R., Koelling, D. D., Freeman, A. J. (1985), *Phys. Rev. B 32,* 7748–7752.
Ochiai, Y., Hashimoto, S., Aso, K. (1989), *Jpn. J. Appl. Phys. 28,* L659–L660.
Ogata, F., Hamajiama, T., Kambara, T., Gondaira, K. I. (1982), *J. Phys. C: Solid State Phys. 15,* 3483–3492.
Ohta, K., Takahashi, A., Deguchi, T., Hyuga, T., Kobayashi, S., Yamaoka, H. (1983), *SPIE 382,* 252.
Ohyama, R., Abe, J., Matsubara, K. (1985), *IEEE Transl. J. Magn. Soc. Jpn. TJMJ-1,* 122–123.
Ohyama, R., Koyanagi, T., Matsubara, K. (1987), *J. Appl. Phys. 61,* 2347–2352.
Okuda, T. Katayama, T., Satoh, K., Oikawa, T., Yamamoto, H., Koshizuka, N. (1990), *Proc. 5th. Symp. Magnetism & Mag. Mat.,* Taipei, April 1989: Huang, H. L., Kuo, P. C. (Eds.). Singapore: World Scientific, p. 61.
Palik, E. D., Teitler, S., Henvis, B. W., Wallis, R. F. (1962), in: *Proc. Int. Conf. Physics of Semiconductors, Exeter,* pp. 288–294.
Patthey, F., Cattarinussi, S., Schneider, W.-D., Baer, Y., Delley, B. (1986), *Europhys. Lett. 2,* 883–889.
Pauthenet, R. (1958), *Ann de Phys. 3,* 424–462.
Pedroli, G., Pollini, I., Spinolo, G. (1975), *J. Phys. C: Solid State Phys. 8,* 2317–2322.
Peeters, W. L., Martens, J. W. D. (1962), *J. Appl. Phys. 53,* 8178–8180.
Pershan, P. S. (1967), *J. Appl. Phys. 38,* 1482–1490.
Pidgeon, C. R. (1980), *Handbook on Semiconductors.* Amsterdam: North-Holland, Vol. 2, p. 223.
Pidgeon, C. R., Feinleib, J., Scouler, W. J., Hanus, J., Dimmock, J. O., Reed, T. B. (1969), *Solid State Commun. 7,* 1323–1326.
Piller, H. (1972), in: *Semiconductors and Semimetals, Vol. 8.* New York-London: Academic Press, p. 103.
Pisarev, R. V., Fèrré, J., Petit, R. H., Krichevtsov, B. B., Syrnikov, P. P. (1974), *J. Phys. C: Solid State Phys. 7,* 4143–4163.
Pisarev, R. V., Schoenes, J., Wachter, P. (1976), *Helv. Phys. Acta 49,* 722.
Pisarev, R. V., Schoenes, J., Wachter, P. (1977), *Solid State Commun. 23,* 657–659.
Plumier, R. (1966), *J. Physique 27,* 213–219.
Popma, T. J. A., Kamminga, M. G. J. (1975), *Solid State Commun. 17,* 1073–1075.
Prinz, G. A., Maisch, W. G., Lubitz, P., Forrester, D. W., Krebs, J. J. (1981), *IEEE Trans. Magn. MAG-17,* 3232–3234.
Pross, E., Tolksdorf, W., Dammann, H. (1988), *Appl. Phys. Lett. 52,* 682–684.
Ray, S., Tauc, J. (1980), *Solid State Commun. 34,* 769–772.
Reim, W. (1986), *J. Magn. Magn. Mat. 58,* 1–47.
Reim, W. (1990), priv. Commun.
Reim, W., Schoenes, J. (1981), *Solid State Commun. 39,* 1101–1104.
Reim, W., Schoenes, J. (1983), *Helv. Phys. Acta 56,* 916–917.
Reim, W., Schoenes, J. (1990), in: *Handbook on Ferromagnetic Materials.* Amsterdam: North-Holland, Vol. 5, pp. 133–236.
Reim, W., Weller, D. (1988 a), *J. Physique Colloque 49,* C8-1959–C8-1960.
Reim, W., Weller, D. (1988 b), *Appl. Phys. Lett. 53,* 2453–2454.
Reim, W., Hüsser, O. E., Schoenes, J., Kaldis, E., Wachter, P., Seiler, K. (1984 a), *J. Appl. Phys. 55,* 2155–2157.
Reim, W., Schoenes, J., Vogt, O. (1984 b), *J. Appl. Phys. 55,* 1853–1855.
Reim, W., Schoenes, J., Vogt, O. (1984 c), *Phys. Rev. B 29,* 3252–3258.
Reim, W., Schoenes, J., Wachter, P. (1984 d), *IEEE Trans. Magn. MAG-20,* 1045–1047.
Reim, W., Schoenes, J., Hulliger, F. (1985), *Physica 130 B,* 64–65.
Reim, W., Schoenes, J., Hulliger, F., Vogt, O. (1986), *J. Magn. Magn. Mat. 54–57,* 1401–1402.
Rossel, C., Meul, H. W., Decroux, M., Fischer, Ø., Remenyi, G., Briggs, A. (1985), *J. Appl. Phys. 57,* 3099–3103.
Robbins, M., Lehmann, H. W., White, J. G. (1967), *J. Phys. Chem. Solids 28,* 897–902.

Rossat-Mignod, J., Lander, G. H., Burlet, P. (1984), in: *Handbook on the Physics and Chemistry of the Actinides*. Amsterdam: North-Holland, Vol. 1, pp. 415–512.
Ruigrok, J. J. M., Greidanus, F. J. A. M., Godlieb, W. F., Spruit, J. H. M. (1988), *J. Appl. Phys. 63,* 3847–3849.
Sagawa, M., Fujimura, S., Togawa, N., Yamamoto, H., Matsuura, Y. (1984), *J. Appl. Phys. 55,* 2083–2087.
Sairasaki, M., Fukushima, N., Nakajima, N., Asama, K. (1986), *Proc. 12th Europ. Conf. Optical Commun.* Telefonica, Subdirrecion General de Technologia, Barcelona, Vol. 2, p. 3.
Saito, J., Sato, M., Matsumoto, H., Akasaka, H. (1987), *Proc. Intern. Symp. on Optical Memory, Kyoto,* p. 149.
Sanford, N., Nettel, S. J. (1981), *J. Appl. Phys. 52,* 3542–3545.
Sato, K., Kida, H. (1988), *J. Physique Colloque 49,* C8-1779–C8-1780.
Sato, K., Teranishi, T. (1981), *J. Phys. Soc. Jpn. 50,* 2069–2072.
Sato, K., Teranishi, T. (1982), *J. Phys. Soc. Jpn. 51,* 2955–2961.
Sato, K., Teranishi, T. (1983), *Il Nuovo Cim. 2D (6),* 1803–1808.
Sato, K., Togami, Y. (1983), *J. Magn. Magn. Mat. 35,* 181–182.
Sato, K., Kida, H., Kamimura, T. (1987), *J. Magn. Soc. Jpn. 11 Suppl. S1,* 113–116.
Sawatzky, E. (1971), *J. Appl. Phys. 42,* 1706–1707.
Sawatzky, E., Street, G. B. (1971), *IEEE Trans. Magn. MAG-7,* 377–380.
Sawatzky, E., Street, G. B. (1973), *J. Appl. Phys. 44,* 1789–1792.
Schmitt, H.-J., Dammann, H. (1987), *Physik in unserer Zeit 18,* 130–136 (Weinheim: VCH Verlagsges.).
Schnatterly, S. E. (1969), *Phys. Rev. 183,* 664–667.
Schoenes, J. (1971), unpublished.
Schoenes, J. (1975), *Z. Physik B 20,* 345–368.
Schoenes, J. (1978), *J. Appl. Phys. 49,* 1463–1465.
Schoenes, J. (1979), *J. Magn. Magn. Mat. 11,* 102–108, and unpublished.
Schoenes, J. (1980a), *Phys. Reports 63,* 301–336.
Schoenes, J. (1980b), *Phys. Reports 66,* 187–212.
Schoenes, J. (1981), *lecture monograph: "Magnetooptik" ETH-Zürich,* unpublished.
Schoenes, J. (1984), in: *Handbook on the Physics and Chemistry of the Actinides,* Vol. 1. Amsterdam: North-Holland, pp. 341–413.
Schoenes, J. (1987), *J. Magn. Soc. Jpn. 11, Suppl. S1,* 99–105.
Schoenes, J., Kaldis, E. (1987), *Proc. Int. Symp. of Magnetic Materials,* Sendai, Japan, April 8–11, 1987. Singapore: World Scientific, pp. 542–545.
Schoenes, J., Nolting, W. (1978), *J. Appl. Phys. 49,* 1466–1468.
Schoenes, J., Reim, W. (1985), *J. Less Common. Met. 112,* 19–25.
Schoenes, J., Reim, W. (1986), *J. Magn. Magn. Mat. 54–57,* 1371–1376.
Schoenes, J., Reim, W. (1988), *Phys. Rev. Lett. 60,* 1988.
Schoenes, J., Wachter, P. (1974), *Phys. Rev. B 9,* 3097–3105.
Schoenes, J., Wachter, P. (1977a), *Physica 86–88 B,* 125–126.
Schoenes, J., Wachter, P. (1977b), *Phys. Lett 61 A,* 68–70.
Schoenes, J., Kaldis, R., Thöni, W., Wachter, P. (1979), *phys. stat. sol. (a) 51,* 173–181.
Schoenes, J., Küng, M., Hauert, R., Henkie, Z. (1983), *Solid State Commun. 47,* 23–27.
Schoenes, J., Frick, B., Vogt, O. (1984), *Phys. Rev. B 30,* 6578–6585.
Schoenes, J., Hüsser, O. E., Reim, W., Kaldis, E., Wachter, P. (1985), *J. Magn. Magn. Mat. 47&48,* 481–484.
Schoenes, J., Reim, W., Studer, W., Kaldis, E. (1988a), *J. Physique Colloque 49,* C8-333–C8-334.
Schoenes, J., Bacsa, W., Hulliger, F. (1988b), *Solid State Commun. 68,* 287–289.
Schoenes, J., Fumagalli, P., Rüegsegger, H., Kaczorowski, D. (1989), *J. Magn. Magn. Mat. 81,* 112–120.
Schoenes, J., Karpinski, J., Kaldis, E., Keller, J., de la Mora, P. (1990a), *Physica C 166,* 145–150.
Schoenes, J., Brändle, H., Weber, A., Hulliger, F. (1990b), *Physica B 163,* 496–498.
Schoenes, J., Fumagalli, P., Krieger, M., Hulliger, F. (1991), to be published.
Schütz, W. (1936), in: *Handbuch der Experimentalphysik XVI. 1.* Leipzig: Akad. Verlagsges.
Schwarz, K. (1986), *J. Phys. F: Met. Phys. 16,* L211–L215.
Scott, G. B., Lacklison, D. E., Ralph, H. I., Page, J. L. (1975), *Phys. Rev. B 12,* 2562–2571.
Scouler, W. J., Feinleib, J., Dimmock, J. O., Pidgeon, C. R. (1969), *Solid State Commun. 7,* 1685–1690.
Shafer, M. W., McGuire, T. R. (1968), *J. Appl. Phys. 39,* 588–590.
Shafer, M. W., McGuire, T. R., Suits, J. C. (1963), *Phys. Rev. Lett. 11,* 251–252.
Shaheen, S. A., Schilling, J. S., Shelton, R. N. (1985), *Phys. Rev. B 31,* 656–659.
Sharipov, Sh. M., Mukimov, K. M., Ernazarova, L. A. (1986), *phys. status solidi (b) 134,* K59–K62.
Shen, Y. R., Bloembergen, N. (1964), *Phys. Rev. 133,* A515–A520.
Shibukawa, A. (1977), *Jpn. J. Appl. Phys. 16,* 1601–1604.
Shieh, H.-P. D., Kryder, M. H. (1986), *Appl. Phys. Lett. 49,* 473–475.
Shiomi, S., Ito, A., Masuda, M. (1987), *J. Magn. Soc. Jpn. 11, Suppl. S1,* 221–224.
Shono, K., Abe, M., Gomi, M., Nomura, S. (1981), *Jpn. J. Appl. Phys. 20,* L426–L428.
Shono, K., Gomi, M., Abe, M. (1982a), *Jpn. J. Appl. Phys. 21,* 1451–1454.

Shono, K., Abe, M., Gomi, M., Nomura, S. (1982b), *Jpn. J. Appl. Phys. 21,* 1720–1722.

Shono, K., Kano, H., Koshino, N., Ogawa, S. (1988), *J. Appl. Phys. 63,* 3639–3641.

Shouji, M., Nagai, A., Murayama, N., Obi, Y., Fujimori, H. (1986), *IEEE Transl. J. Magn. Soc. Jpn. TJMJ2,* 381–382.

Šimša, Z., Legall, H., Široky, P. (1980a), *physica status sol. (b) 100,* 665–670.

Šimša, Z., Široky, P., Kolaček, J., Brabers, V. A. M. (1980b), *J. Magn. Magn. Mat. 15–18,* 775–776.

Singh, M., Wang, C. S., Callaway, J. (1975), *Phys. Rev. B 11,* 287–294.

Siratori, K., Iida, S. (1960), *J. Phys. Soc. Japan 15,* 210–211.

Smith, N. V., Lässer, R., Chiang, S. (1982), *Phys. Rev. B 25,* 793–805.

Spielman, S., Fesler, K., Eom, C. B., Geballe, T. H., Fejer, M. M., Kapitulnik, A. (1990), *Phys. Rev. Lett. 65,* 123–126.

Stern, E. A. McGroddy, J. C., Harte, W. E. (1964), *Phys. Rev. 135,* A1306–A1314.

Stoffel, A. M. (1969), *J. Appl. Phys. 40,* 1238–1239.

Stoffel, A. M., Schneider, J. (1970), *J. Appl. Phys. 41,* 1405–1407.

Stoffel, A. M., Strnat, K. J. (1965), *Proc. Intermag. Conf. Paper 2.5.*

Suits, J. C. (1971), *Rev. Sc. Instr. 42,* 19–22.

Suits, J. C. (1972), *IEEE Trans. Magn. MAG-8,* 95–105.

Suits, J. C., Argyle, B. E., Freiser, M. J. (1966), *J. Appl. Phys. 37,* 1391–1397.

Suryanarayanan, R., Paparoditis, C., Ferré, J., Briat, B. (1970), *Solid State Commun. 8,* 1853–1855.

Suryanarayanan, R., Paparoditis, C., Ferré, J., Briat, B. (1972), *J. Appl. Phys. 43,* 4105–4108.

Suryanarayanan, R., Ferré, J., Briat, B. (1974), *Phys. Rev. B 9,* 554–557.

Suzuki, T., Katayama, T. (1986), *IEEE Trans. Magn. MAG-22,* 1230–1232.

Suzuki, T., Takagi, S., Niitsuma, N., Takegahara, T., Kasuya, T., Yanase, A., Sakakibara, T., Date, M., Markowski, P. J., Henkie, Z. (1982), *Proc. Int. Symp. High Field Magnetism:* Date, M. (Ed.). Amsterdam: North-Holland.

Suzuki, T., Murakami, A., Katayama, T. (1987), *IEEE Trans. Magn. MAG-23,* 2958–2960.

Suzuki, T., Lin, C.-J., Bell, A. E. (1988), *IEEE Trans. Magn. MAG-24,* 2452–2454.

Suzuki, Y., Takayama, S., Kirino, F., Ohta, N. (1987), *IEEE Trans. Magn. MAG-23,* 2275–2277.

Szytuła, A., Siek, S., Leciejewicz, J., Zygmunt, A., Ban, Z. (1988), *J. Phys. Chem. Solids 49,* 1113–1118.

Tabor, W. J., Chen, F. S. (1969), *J. Appl. Phys. 40,* 2760–2765.

Tabor, W. J., Anderson, A. W., van Uitert, L. G. (1970), *J. Appl. Phys. 41,* 3018–3021.

Takanashi, K., Fujimori, H., Shoji, M., Nagai, A. (1987), *Jpn. J. Appl. Phys. 26,* L1317–L1319.

Takanashi, K., Fujimori, H., Watanabe, J., Shoji, M., Nagai, A. (1988), *Jpn. J. Appl. Phys. 27,* L2351–L2353.

Takanashi, K., Watanabe, J., Kido, G., Fujimori, H. (1990), *Jpn. J. Appl. Phys. 29,* L306–L307.

Takegara, K., Yanase, A., Kasuya, T. (1981), *Proc. 4th Int. Conf. Crystal Field and Structure Effects in f-Electron Systems.* Warsaw.

Takeuchi, H., Shinagana, K., Taniguchi, S. (1973), *Jpn. J. Appl. Phys. 12,* 465.

Tanaka, F., Tanaka, S., Imamura, N. (1987), *Jpn. J. Appl. Phys. 26,* 231–235.

Theocaris, P. S., Gdoutos, E. E. (1979), *Matrix Theory of Photoelasticity.* Berlin: Springer.

Togami, Y. (1982), *IEEE Trans. Magn. MAG-18,* 1233–1237.

Trzebiatowski, W., Henkie, Z., Bielow, K. P., Dmitrievskii, A. S., Levitin, R. Z., Popov, Y. F. (1971), *Zh. Eksp. & Teor. Fiz. 61,* 1522.

Tsurkan, V. V., Ratseev, S. A., Tezlevan, V. E., Radautsan, S. I. (1985), *Progr. Crystal Growth & Charact. 10,* 385–389.

Tsubokawa, I. (1960), *J. Phys. Soc. Japan 15,* 1664–1668.

Tsunashima, S., Masui, S., Kobayashi, T., Uchiyama, S. (1983), *J. Appl. Phys. 53,* 8175–8177.

Tucciarone, A. (1977), *Lett. Nuovo Cimento 20,* 275–281.

Unger, W. K., Räth, R. (1971), *IEEE Trans. Magn. MAG-7,* 885–890.

Unger, W. K., Wolfgang, E., Harms, H., Handek, H. (1972), *J. Appl. Phys. 43,* 2875–2880.

Urner-Wille, M., te Velde, T. S., van Engen, P. G. (1978), *phys. status solidi (a) 50,* K29–K31.

Urner-Wille, M., Hansen, P., Witter, K. (1980), *IEEE Trans. Magn. MAG-16,* 1188–1193.

Valiev, L. M., Kerimov, I. G., Babaev, S. K., Namazov, Z. M. (1972), *Phys. Stat. Sol. (a) 13,* 231–234.

van Engelen, P. P. J., Buschow, K. H. J. (1987), *J. Magn. Magn. Mat. 66,* 291–293.

van Engelen, P. P. J., Buschow, K. H. J. (1990a), *J. Less-Common Met. 159,* L1–L4.

van Engelen, P. P. J., Buschow, K. H. J. (1990b), *J. Magn. Magn. Mat. 84,* 47–51.

van Engelen, P. P. J., de Mooij, D. B., Buschow, K. H. J. (1988), *IEEE Trans. Magn. MAG-24,* 1728–1730.

van Engen, P. G. (1983), *Thesis Technical University Delft,* unpublished, cited by Buschow (1988).

van Engen, P. G., Buschow, K. H. J., Jongebreur, R., Erman, M. (1983), *Appl. Phys. Lett. 42,* 202–204.

Verwey, E. J. W., Haayman, P. W. (1941), *Physica 8,* 979.

Vieth, M., Weissenberger, V., Reim, W., Winkler, S. (1989), *SPIE 1139,* 130–135.

Višňovský, S., Knappe, B., Prosser, V., Müller, H. R. (1976), *phys. status solidi (a) 38,* K53–K56.

Višňovský, S., Krishnan, R., Prosser, V., Nguyen Phu Thuy, Streda, I. (1979a), *Appl. Phys. 18,* 243–247.

Višňovský, S., Canit, J. C., Briat, B., Krishnan, R. (1979b), *J. Physique 40*, 73–77.
Višňovský, S., Wanklyn, B. M., Prosser, V. (1984), *IEEE Trans. Magn. MAG-20*, 1054–1056.
Vodhanel, R. S., Gimblett, J. L., Curtis, L., Cheung, N. K. (1986), *Proc. 12th Europ. Conf. Optical Commun.*, p. 339.
Voekov, D. V., Zheltukhin, A. A. (1980), *Izv. Akad. Nauk. SSSR 44*, 1480.
Voloshinskaya, N. M., Bolotin, G. A. (1974), *Fiz. Met. & Metalloved. 38*, 975–984.
Voloshinskaya, N. M., Fedorov, G. V. (1973), *Fiz. Met. & Metalloved. 36*, 946–956.
Voloshinskaya, N. M., Sasovskaya, I. I., Noskov, M. M. (1974), *Fiz. Met. & Metalloved. 38*, 1134–1138.
von Philipsborn, H., Treitinger, L. (1980), in: *Landolt-Börnstein, Neue Serie, Gruppe III*. Heidelberg-New York: Springer, Band 12, p. 54.
Wachter, P. (1972), *Crit. Rev. Solid State Sci. 3*, 189–240.
Wakabayashi, H., Notarys, H., Suits, J. C., Suzuki, T. (1989), IBM Research Rep. RJ 6851.
Wang, C. S., Callaway, J. (1974), *Phys. Rev. B 9*, 4897–4907.
Wang, Y. J. (1990), *J. Magn. Magn. Mat. 84*, 39–46.
Wang, H.-Y., Schoenes, J., Kaldis, E. (1986), *Helv. Phys. Acta 59*, 102–105.
Watada, A. (1987), *Proc. Int. Symp. on the Physics of Magnetic Materials*, Sendai, Japan. Singapore: World Scientific, pp. 546–549.
Weber, M. J. (1982), *Faraday Rotator Materials*. Lawrence Livermore Laboratory (California, USA).
Weller, D., Reim, W. (1989), *Appl. Phys. A 49*, 599–618.
Weller, D., Reim, W., Ebert, H., Johnson, D. D., Pinski, F. J. (1988), *J. Physique 49, Colloque*, C 8-41 to C 8-42.
Weller, D., Reim, W., Spörl, K., Brändle, H. (1991), *J. Magn. Magn. Mat. 93*, 183–193.
Wemple, S. H., Dillon, J. F., Jr., van Uitert, L. G., Grodkiewicz, W. H. (1973), *Appl. Phys. Lett. 22*, 331–333.
Wettling, W., Andlauer, B., Koidl, P., Schneider, J., Tolksdorf, W. (1973), *phys. stat. sol. (b) 59*, 63–70.
Wijngaard, J. H., Haas, C., de Groot, R. A. (1989), *Phys. Rev. B 40*, 9318–9320.
Williams, H. J., Sherwood, R. C., Foster, F. G., Kelley, E. M. (1957), *J. Appl. Phys. 28*, 1181–184.
Wittekoek, S., Rinzema, G. (1971), *phys. stat. sol. (b) 44*, 849–860.
Wittekoek, S., Popma, T. J. A., Robertson, J. M., Bongers, P. F. (1973), *AIP Conf. Proc. 10*, 1418–1427.
Wittekoek, S., Popma, T. J. A., Robertson, J. M., Bongers, P. F. (1975), *Phys. Rev. B 12*, 2777–2788.
Wolfe, R., Kurtzig, A. J., LeGraw, R. C. (1970), *J. Appl. Phys. 41*, 1218–1224.
Wolniansky, P., Chase, S., Rosenvold, R., Ruane, M., Mansuripur, M. (1986), *J. Appl. Phys. 60*, 346–351.
Yoshino, T., Tanaka, S.-I. (1969), *Optics Commun. 1*, 149–152.
Yoshioka, N., Koshimura, M., Ono, M., Takahashi, M., Miyazaki, T. (1988), *J. Magn. Magn. Mat. 74*, 51–58.
Zanmarchi, G., Bongers, P. F. (1969), *J. Appl. Phys. 40*, 1230–1231.
Zeper, W. B., Greidanus, F. J. A. M., Garcia, P. F., Fincher, C. R. (1989), *J. Appl. Phys. 65*, 4971–4975.
Zhang, X. X., Schoenes, J., Wachter, P. (1981a), *Solid State Commun. 39*, 189–192.
Zhang, X. X., Schoenes, J., Wachter, P. (1981b), *J. Magn. Magn. Mat. 24*, 202–205.
Zhang, X. X., Schoenes, J., Reim, W., Wachter, P. (1983), *J. Phys. C: Solid State Phys. 16*, 6055–6072.
Zvara, M., Prosser, V., Schlegel, A., Wachter, P. (1979), *J. Magn. Magn. Mat. 12*, 219–226.

General Reading

Advances in Magneto-Optics (1987), *Proc. of Intern. Symp. on Magneto-Optics, April 1987, Kyoto, Japan*, in: *J. Magn. Soc. Jpn., Vol. 11, Suppl. S 1*.
Buschow, K. H. J. (1988), *"Magneto-Optical Properties of Alloys and Intermetallic Compounds"*, in: *Handbook on Ferromagnetic Materials*, Vol. 4, Ch. 5. Amsterdam: North-Holland.
Proc. Intern. Conf. on Magneto-Optics, Sept. 1976, Zürich, Switzerland (1977), *Physica 89 B + C*.
Proc. Magneto-Optical Recording Intern. Symp., April 1991, Tokyo, Japan (1991), to be published.
Reim, W., Schoenes, J. (1990), *"Magneto-Optical Spectroscopy of f-Electron Systems"*, in: *Handbook on Ferromagnetic Materials*, Vol. 5, Ch. 2. Amsterdam: North-Holland.
Schoenes, J. (1984), *"Optical and Magneto-Optical Properties"*, in: *Handbook on the Physics and Chemistry of the Actinides*, Vol. 1, Ch. 5. Amsterdam: North-Holland.

3 Electronic Transport Properties of Normal Metals

Paul L. Rossiter

Department of Materials Engineering, Monash University, Melbourne, Australia

Jack Bass

Department of Physics and Astronomy, Michigan State University,
East Lansing, MI, U.S.A. and
Max-Planck-Institut für Festkörperforschung, Hochfeld Magnetlabor, Grenoble, France

List of Symbols and Abbreviations

a	lattice constant
$\boldsymbol{A}$	vector in a trial function, analogous $\boldsymbol{E}$
A	cross-sectional area
A, B	atomic species
A, A'	parameters
$\boldsymbol{B}, B, B_z$	magnetic field (magnetic induction), z-component of $\boldsymbol{B}$
C	heat capacity per unit volume
C_A, C_B	fraction of atomic species A, B
C_e	electron flow contribution to the heat capacity per unit volume
C_g	phonon flow contribution to the heat capacity per unit volume
C_i	concentration of impurity atoms
c_i	number of atoms in the ith coordination shell of radius r_i
d	film, foil or wire thickness; mean grain diameter
$D(\varepsilon)$	density of states
e	electron charge
$\boldsymbol{E}, E_x, E_y$	electric field, x- and y-component
E_H	Hall field
$\boldsymbol{E}_L$	longitudinal electric field
$\exp(-2M)$	Debye-Waller factor
$f(\boldsymbol{k}, \boldsymbol{r}, t)$	distribution function
$\boldsymbol{F}$	force
$f_0(\boldsymbol{k}, \boldsymbol{r})$	Fermi-Dirac distribution function
F_E	force per unit volume due to electric field $\boldsymbol{E}$
f_k	contribution of state $\boldsymbol{k}$ to the distribution function
F_{flow}	force per unit volume associated with electron and phonon flow
$g(\boldsymbol{k}, \boldsymbol{r}, t)$	change in distribution function $f_k - f_0$
$\boldsymbol{G}$	reciprocal lattice vector
g_k	contribution of state $\boldsymbol{k}$ to the change in distribution function
$h, \hbar$	Planck constant, devided by 2π
$\boldsymbol{I}_q$	total heat current
$\boldsymbol{I}, I_x$	current, x-component
$\boldsymbol{j}, j_x$	current density of charge, x-component
$\boldsymbol{j}_q$	current density of heat
$\boldsymbol{k}, \boldsymbol{k}', k_x, k_y, k_z$	wave vectors, spatial components of the wave vector $\boldsymbol{k}$
k_B	Boltzmann constant
k_F	magnitude of the Fermi wave vector
$d\boldsymbol{k}$	volume element of $\boldsymbol{k}$-space
l	mean free path
L	distance between two measuring points (wire, foil), over which the voltage V_x is generated
$l_{(loc)}$	local mean free path
l_∞	bulk mean free path
L-MR	longitudinal magnetoresistance

m	electron mass
M	atomic mass
M	magnetic long range order parameter (spin-spin correlation parameter)
m_i	magnetic short range order parameter (spin-spin correlation parameter)
m^*	effective mass (of an electron)
m_1^*, m_2^*	effective mass in a two-band model
M_S	saturation magnetization
MR	magnetoresistance
n	number of atoms per unit volume
n_e	density of electrons
n_h	density of holes
N	number of unit cells
N	dislocation unit density
N_e	number of electrons/atom
N_h	number of holes/atom
p	parameter describing the fraction of electrons scattered in a specular (i.e., non-diffuse) manner
$P_{kk'}$	probability of transition from state $\boldsymbol{k}$ to state $\boldsymbol{k}'$
$-P^a/B$	adiabatic Nernst-Ettinghausen coefficient
$\boldsymbol{q}$	scattering wave vector
$\boldsymbol{Q}$	phonon wave vector
$\boldsymbol{q}_p$	scattering wave vector related to a peak in the structure factor
q_{min}	minimal magnitude of wave vector $\boldsymbol{q}$ in an Umklapp process
$\boldsymbol{r}$	distance vector
r	wire radius
R	electrical resistance
r_C	cyclotron radius
r_i	site i
R_H	Hall coefficient
R_S	spontaneous Hall coefficient
R_0	ordinary Hall coefficient
s	Bragg-Williams long range order parameter
S, $\mathbf{S}$	thermopower, tensor coefficient
$\mathfrak{s}(\boldsymbol{q})$	structure factor
$\boldsymbol{S}$, S, $\boldsymbol{S}_i$	electron spin, spin at site i
S_c	out-of-plane thermopower
S_{ab}	in-plane thermopower
S_d	electron diffusion component
S_g	phonon drag component
S_p	temperature dependent thermopower of an ideally pure metal
S_0	impurity thermopower
S_{TOT}	total thermopower (Nordheim-Gorter rule)
$d\boldsymbol{S}$	area element of Fermi surface
t	time
t	foil thickness
t	empirical specular coefficient

T	absolute temperature
T_c	critical temperature
T_M	melting point
T_N	Néel temperature
T-MR	transverse magnetoresistance
$\nabla_r T$	temperature gradient in the direction of $\boldsymbol{r}$
U	energy density per unit volume
$u_{Q,j}$	displacement associated with lattice vibration mode Q, j
V	number of vacancies per cm^3
$\boldsymbol{v}(\boldsymbol{k})$	velocity of an electron in state $\boldsymbol{k}$
v_F	velocity of an electron at the Fermi surface
v_i	volume fraction of phase i in a composite structure
V_{21}	thermoelectric voltage
V_x, V_y	voltage generated in x, y-direction
$\langle v_x \rangle$	average drift velocity in x-direction
W	foil width
$w(\boldsymbol{r})$	atomic potential
$w_i(\boldsymbol{q})$	form factor of an impurity atom
$\bar{w}(\boldsymbol{q})$	form factor of average lattice
$\bar{w}(\boldsymbol{r}')$	average lattice potential
$W(\boldsymbol{r})$	total scattering potential
$w_A(\boldsymbol{q}), w_B(\boldsymbol{q})$	form factor of atom A, B
$\bar{x}^2$	mean square thermal displacement (Einstein model)
y^α	fraction of α sites
z	valency
α	sublattice
α	ratio of the probability of phonon-electron scattering to the probability of all types of scattering
α_i	average value of α_{ij} for a shell of radius r_i of atoms
α_{ij}	Warren-Cowley short range parameter between sites i, j
$\boldsymbol{\gamma}$	$= -S\boldsymbol{\sigma}$
γ^c	electronic specific heat
$\boldsymbol{\delta}_i$	total atomic displacement at site r_i
ε	mean shear strain
ε	energy of an electron
ε_F	Fermi energy
ε_k	energy of an electron with wave vector $\boldsymbol{k}$
Θ_D	Debye temperature
Θ_E	Einstein temperature
Θ_R	resistivity characteristic temperature
$\varkappa, \boldsymbol{\varkappa}$	thermal conductivity, tensor coefficient
λ_k	de Broglie wavelength of state $\boldsymbol{k}$
μ	coefficient of the Thomson effect
μ	chemical potential
ν	grain size dependent parameter

Π, $\mathbf{\Pi}$	coefficient of the Peltier effect, tensor coefficient
ϱ, $\boldsymbol{\varrho}$	resistivity, tensor coefficient
ϱ_{H}	Hall resistivity
ϱ_{ij}	element of the resistivity tensor $\boldsymbol{\varrho}$
$\varrho_{\mathrm{m}}(T)$	magnetic contribution to the residual resistivity
ϱ_{p}	resistivity of the ideal metal
$\varrho_{\mathrm{p}}(T)$	temperature dependent phonon scattering resistivity
ϱ_0	impurity contribution to resistivity ("residual resistivity")
ϱ_{TOT}	total resistivity
ϱ_{∞}	bulk resistivity
$\varrho_{\Theta_{\mathrm{D}}}$	resistivity at Debye temperature
$\varrho_{\uparrow}, \varrho_{\downarrow}$	contribution of spin-up or spin-down electrons to resistivity
$\varrho_{\uparrow\downarrow}$	contribution of spin-flip processes to resistivity
$\varrho_{0\uparrow}, \varrho_{0\downarrow}$	residual magnetic contribution to resistivity
$\varrho_{\mathrm{p}\uparrow}, \varrho_{\mathrm{p}\downarrow}, \varrho_{\mathrm{p}\uparrow\downarrow}$	phonon magnetic scattering contribution to resistivity
$\boldsymbol{\sigma}$, σ_{ij}	conductivity tensor, element i, j
σ_1, σ_2	conductivities in a two-band model
σ_{∞}	bulk conductivity
σ_i^{A}, σ_j^{B}	site occupation parameter (regarding a species A, species B, atom at site r_i, r_j)
$\{\sigma_i^{\mathrm{A}}\}$	complete set of parameters σ_i^{A}
$\langle\sigma_i^{\mathrm{A}}\rangle$	site occupation average
$\langle\sigma_i^{\mathrm{A}}\rangle^{\alpha}$	site occupation average over the α sublattice
$\langle\sigma_i^{\mathrm{A}}\sigma_j^{\mathrm{B}}\rangle$	probability of finding an A-B atom pair on a pair of sites i and j
τ	general conduction electron relaxation time
τ_1, τ_2	relaxation times in a two-band model
τ_k	relaxation time dependent upon $\boldsymbol{k}$
$\tau_{\mathrm{p}e}$	relaxation time inversely proportional to the probability for phonon-electron scattering process
$\tau_{\mathrm{p}0}$	relaxation time inversely proportional to the probability for all types of scattering other than phonon-electron
ω	lattice vibration frequency
ω_{c}	cyclotron frequency
$\omega_{Q,j}$	frequency of lattice vibration mode Q, j
Ω	specimen volume
Ω_0	volume of unit cell

b.c.c.	body-centered cubic
BZ	Brillouin zone
emf	electromotive force
f.c.c.	face-centered cubic
FS	Fermi surface
GP	Guinier-Preston
MR	magnetoresistance
NE	Nernst-Ettingshausen-Effect

3.1 Introduction

High electrical and thermal conductivities are among the characteristic features of metals. In this chapter we examine how the electrons in metals transport charge and heat under the influence of an applied electric field $\boldsymbol{E}$ and/or a temperature gradient, $\nabla_r T = \Delta T/\Delta x$. Here ΔT is a temperature difference applied over a small distance Δx, and boldface indicates a vector.

In zero magnetic field, the quantities of main interest in this chapter will be the electrical resistivity, ϱ, and the thermopower, S. When a magnetic field, $\boldsymbol{B}$, is applied, we shall focus upon the magnetoresistance ($MR = [\varrho(\boldsymbol{B}) - \varrho(\boldsymbol{B}=0)]/\varrho(\boldsymbol{B}=0)$, and the Hall coefficient R_H. The third of the (generally) independent transport coefficients, the thermal conductivity, $\varkappa$, will be discussed only briefly, in terms of the Wiedemann-Franz law, which relates it to ϱ at low and high temperatures. The reader interested in more detailed information about $\varkappa$ is referred to Ziman (1965, 1972).

In this introduction, we briefly describe how each of these four quantities is measured, and then discuss the general linear transport equations relating the current density of charge, $\boldsymbol{j}$, and the current density of heat, $\boldsymbol{j}_q$, to $\boldsymbol{E}$ and $\nabla_r T$. In the following sections of the article we outline the microscopic theory of electronic transport in metals and examine a variety of experimental data.

3.1.1 Electrical Resistance, R, and Resistivity, ϱ

The electrical resistance of a metallic sample R is normally measured by injecting a known current I_x through the metal of interest in the form of a wire (Fig. 3-1 a) or foil (Fig. 3-1 b) having cross-sectional area $A = \pi r^2$ or Wt, respectively, where r is

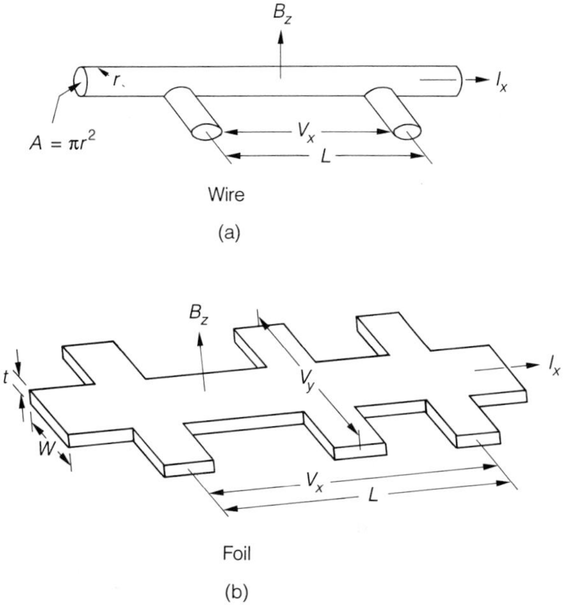

Figure 3-1. Specimen configuration for measuring the electrical resistance of a wire (a) and foil (b). Arrangement shown in (b) also allows determination of the Hall voltage V_y.

the wire radius and W and t the foil width and thickness. From Ohm's law:

$$R = V_x/I_x \tag{3-1}$$

where V_x is the emf (or voltage) generated over length l. R is an extensive quantity which doubles if l is doubled or A is halved. The intrinsic property of the metal is ϱ, related to R by:

$$\varrho = R(A/l) \tag{3-2}$$

provided that R consists of homogeneous material and that $j_x = I_x/A$ is uniform across the sample area. The resistivity ϱ is discussed in Sec. 3.5.

3.1.2 Thermopower, S

There are three experimental thermoelectric phenomena with three corresponding coefficients: the Seebeck effect (thermopower coefficient S); the Thomson effect (coefficient μ), and the Peltier effect (coefficient Π). Only the Thomson effect can be measured on a single metal. It involves the reversible generation of heat when current flows in a conductor while a temperature gradient is present (i.e., the production of heat changes to absorption if the direction of either the current or the temperature – but not both at once – is reversed). In the Peltier effect, heat is reversibly generated at the junction between two different metals when a current flows through the junction. In the Seebeck effect, a thermoelectric voltage is generated when a temperature difference is applied between the two ends of thermocouple (Fig. 3-2) under conditions where no current flows in the circuit. These three coefficients are connected by the Kelvin relations:

$$S_1 = \int_0^T (\mu_1/T)\,\mathrm{d}T \qquad (3\text{-}3\,\mathrm{a})$$

$$\mu_1 = T\,(\partial S_1/\partial T) \qquad (3\text{-}3\,\mathrm{b})$$

$$\Pi_{12} = T\,(S_2 - S_1) \qquad (3\text{-}3\,\mathrm{c})$$

so that complete knowledge of any one yields complete knowledge of all three. S is normally the easiest to measure.

When a temperature difference, $\Delta T = T_\mathrm{H} - T_\mathrm{L}$, is applied across a thermocouple,

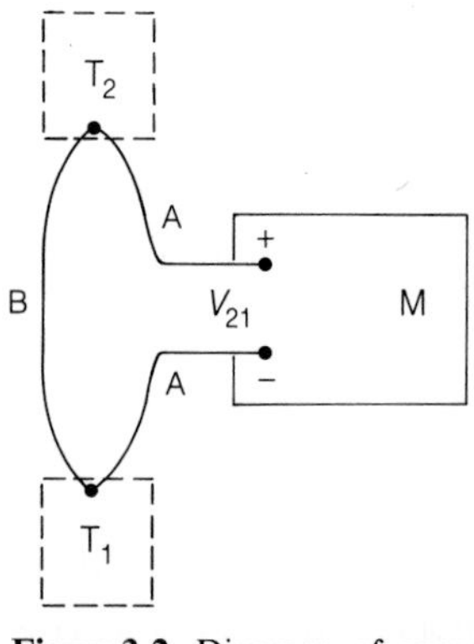

Figure 3-2. Diagram of apparatus usually used for measuring thermoelectric (Seebeck) voltage V_{21}. M is an instrument for measuring potential.

the resulting thermoelectric voltage is:

$$V_{21} = \int_{T_\mathrm{L}}^{T_\mathrm{H}} [S_2(T) - S_1(T)]\,\mathrm{d}T = \int_{T_\mathrm{L}}^{T_\mathrm{H}} S_2(T)\,\mathrm{d}T - \int_{T_\mathrm{L}}^{T_\mathrm{H}} S_1(T)\,\mathrm{d}T \qquad (3\text{-}4)$$

where S_1 and S_2 are the thermopowers of metals 1 and 2, respectively. If both metals are completely homogeneous, the integrals depend only upon the materials and the end temperatures, T_H and T_L, and not upon the details of how the temperature varies along each of the two arms of the couple. If ΔT is small, then

$$S_2 - S_1 = V_{21}/\Delta T \qquad (3\text{-}5)$$

Equation (3-5) shows that S_1 for a sample with unknown properties can only be determined if S_2, the thermopower of a reference material, is known. At very low temperatures, the reference material can be chosen to be a superconductor, the thermopower of which is zero in its superconducting state. The new high temperature superconductors can be used to above 90 K (Uher, 1987). Pb has been calibrated as a reference standard up to 350 K (Roberts, 1977). Above 350 K there is no universal standard, but Au, Cu, and Pt have been used (Blatt et al., 1976).

The thermopower S consists of two very different parts: S_d, the electron diffusion component, results from a differential flow of electrons from one end of the metal to the other under the influence of a temperature gradient, with the phonons in the metal taken to be in thermal equilibrium; S_g, the phonon-drag component, results from the “dragging along” of electrons by the non-equilibrium phonon flow from the hot end of the metal to the cool end that is caused by the temperature gradient. In pure metals, S_g often dominates S from a few K up to above 100 K. In contrast to S, phonon-drag produces only negligible contributions to ϱ or $\varkappa$, except for the temperature dependent portion of ϱ in high purity alkali metals in the vicinity of 1 K. The quantities S_d and S_g are discussed in Sec. 3.8.

3.1.3 Magnetoresistance, *MR*, and Hall Coefficient, R_H

MR and R_H are traditionally obtained from measurements on a foil sample having the shape shown in Fig. 3.1 b. A current I_x is passed through the sample from left to right, a magnetic field B_z is applied normal to the sample, and the voltages V_x and V_y are measured. The direction of $\boldsymbol{B}$ is then reversed, and the two voltages measured again. The *MR* is defined to be an even function of $\boldsymbol{B}$, and is determined from the average of the values of V_x for the two field directions. R_H is defined to be an odd function of $\boldsymbol{B}$, and is determined from the difference between the two values of V_y. This reversing procedure is one of the steps necessary to minimize systematic errors in *MR* and R_H. Erroneous results can be obtained if adequate care is not taken to ensure: spatial uniformity of $\boldsymbol{B}$; good alignment of the sample in the field; a uniform current flow through the sample; etc.

Since $V_x = E_x l$ and $I_x = j_x W t$, we have $\varrho(B) = V_x W t / I_x l$ for each of the two terms to be averaged in determining the *MR*. Since $V_y = E_y W$, each term for R_H must have the form $R_H = V_y t / B I_x$. R_H is defined with $\boldsymbol{B}$ divided out, because in many cases V_y is proportional to $\boldsymbol{B}$.

MR and R_H are discussed in Sec. 3.6 and Sec. 3.7, respectively.

3.1.4 General Transport Equations and Coefficients

The formal transport equations for metals provide linear relations among four vector quantities: $\boldsymbol{E}$, $\nabla_r T$, the electrical current density $\boldsymbol{j}$, and the heat current density $\boldsymbol{j}_q = \boldsymbol{I}_q / A$, where $\boldsymbol{I}_q$ is the total heat current. These equations can be written in two different forms.

The first form defines the quantities that are normally directly measured. The vectors $\boldsymbol{E}$ and $\boldsymbol{j}_q$ are related to the vectors $\boldsymbol{j}$ and $\nabla_r T$ by tensor coefficients (also designated by bold symbols):

$$\boldsymbol{E} = \boldsymbol{\varrho}(\boldsymbol{B}) \cdot \boldsymbol{j} + \mathbf{S}(\boldsymbol{B}) \cdot \nabla_r T \qquad (3\text{-}6\,a)$$

$$\boldsymbol{j}_q = \boldsymbol{\Pi}(\boldsymbol{B}) \cdot \boldsymbol{j} - \boldsymbol{\varkappa}(\boldsymbol{B}) \cdot \nabla_r T \qquad (3\text{-}6\,b)$$

where we have indicated that the coefficients are generally functions of $\boldsymbol{B}$. The thermal conductivity $\boldsymbol{\varkappa}$ is determined by applying $\boldsymbol{j}_q$ and measuring $\nabla_r T$ while holding $\boldsymbol{j} = 0$. Similarly, ϱ (i.e. R) is determined by applying $\boldsymbol{j}$ and measuring $\boldsymbol{E}$ while holding $\nabla_r T = 0$ and $\mathbf{S}$ is determined by applying $\nabla_r T$ and measuring $\boldsymbol{E}$ while holding $\boldsymbol{j} = 0$. Due to the Onsager relation, $\boldsymbol{\Pi}(\boldsymbol{B}) = T\,\mathbf{S}(-\boldsymbol{B})$, there are only three independent tensors in Eqs. (3-6): $\boldsymbol{\varrho}$, $\boldsymbol{\varkappa}$, and $\mathbf{S}$.

The second pair of equations defines the quantities that are calculated. $\boldsymbol{E}$ and $\nabla_r T$ are treated as "forces" and $\boldsymbol{j}$ and $\boldsymbol{j}_q$ as derived "flows". These equations have the advantage of being directly comparable with the Boltzmann transport equation which we discuss in Sec. 3.2.

$$\boldsymbol{j} = \boldsymbol{\sigma}(\boldsymbol{B}) \cdot \boldsymbol{E} + \boldsymbol{\gamma}(\boldsymbol{B}) \cdot \nabla_r T \qquad (3\text{-}7\,a)$$

$$\boldsymbol{j}_q = \boldsymbol{\gamma}(-\boldsymbol{B})\,T \cdot \boldsymbol{E} - \boldsymbol{\varkappa}'(\boldsymbol{B}) \cdot \nabla_r T \qquad (3\text{-}7\,b)$$

Both Eqs. (3-6) and (3-7) reduce to scalars in cubic metals in the absence of a magnetic field. Comparison of Eqs. (3-7) with Eqs. (3-6) in scalar form shows that $\varkappa' = \varkappa - S^2/\sigma$ and $\gamma = -S\sigma$. In metals, S^2/σ is almost always negligibly small compared to $\varkappa$. The differences between Eqs. (3-6) and (3-7) are of little importance when the coefficients are simply scalars, since converting from, e.g., ϱ to σ, simply involves scalar inversion. When the full tensors must be retained, however, the conversion from one form to another involves matrix inversion, which is usually much more complex. Writing $\boldsymbol{\varrho}$ as a tensor with elements ϱ_{ij} where $i,j = x, y, z$, the definitions of the MR and R_H given above yield:

$$MR = [\{\varrho_{xx}(\boldsymbol{B}) + \varrho_{xx}(-\boldsymbol{B})\}/2 - \varrho(\boldsymbol{B}=0)]/\varrho(\boldsymbol{B}=0) \quad \text{(3-8 a)}$$

$$R_\mathrm{H} = [\varrho_{yx}(\boldsymbol{B}) - \varrho_{xy}(-\boldsymbol{B})]/2B \quad \text{(3-8 b)}$$

3.2 The Boltzmann Transport Equation

3.2.1 Background

If an electric field is applied to a conductor, all electrons not in completely filled bands are displaced at a uniform rate in $\boldsymbol{k}$-space. Scattering by defects tends to restore the electrons to their equilibrium distribution. The nature of the actual distribution is thus determined by a dynamical balance between the acceleration of the electrons in the applied fields and their scattering by the lattice. A distribution function $f(\boldsymbol{k}, \boldsymbol{r}, t)$ may be used to describe the location of the state of an electron in six-dimensional $(\boldsymbol{k}, \boldsymbol{r})$ phase space. It is defined so that $(4\pi^3)^{-1} f(\boldsymbol{k}, \boldsymbol{r}, t)\, \mathrm{d}\boldsymbol{k}\, \mathrm{d}\boldsymbol{r}$ is the number of electrons which lie in an element $\mathrm{d}\boldsymbol{r}$ of real space and $\mathrm{d}\boldsymbol{k}$ of wave vector space at time t, such that $f=1$ if all states are filled or $f=0$ if all are empty. If the electron gas is at equilibrium at some temperature, f is then simply the Fermi-Dirac distribution function, designated f_0.

In the presence of applied electric and magnetic fields $\boldsymbol{E}$ and $\boldsymbol{B}$, but without scattering, any particular state will move through phase space according to semi-classical equations but its occupancy will not change (i.e., if it was initially filled it will remain so). However, if scattering occurs, an electron can discontinuously change its momentum from some initial state $\boldsymbol{k}$ to a final state $\boldsymbol{k}'$ and the time derivative of f is given by

$$\frac{\mathrm{d}f}{\mathrm{d}t} = \left.\frac{\partial f}{\partial t}\right|_{\text{scatt}} \quad \text{(3-9)}$$

where the last term is the change in f due to that scattering process.

Expanding the left-hand-side in terms of its partial derivatives gives

$$\frac{\partial f}{\partial t} + \frac{\partial f}{\partial \boldsymbol{r}}\frac{\mathrm{d}\boldsymbol{r}}{\mathrm{d}t} + \frac{\partial f}{\partial \boldsymbol{k}}\frac{\mathrm{d}\boldsymbol{k}}{\mathrm{d}t} = \left.\frac{\partial f}{\partial t}\right|_{\text{scatt}} \quad \text{(3-10)}$$

Since

$$\hbar\frac{\partial \boldsymbol{k}}{\partial t} = \boldsymbol{F} \quad \text{(3-11)}$$

where $\boldsymbol{F}$ is any applied force, Eq. (3-10) may be written as

$$\frac{\partial f}{\partial t} + \boldsymbol{v}(\boldsymbol{k})\frac{\partial f}{\partial \boldsymbol{r}} + \frac{e}{\hbar}[\boldsymbol{E} + \boldsymbol{v}(\boldsymbol{k}) \times \boldsymbol{B}]\frac{\partial f}{\partial \boldsymbol{k}} = \left.= \frac{\partial f}{\partial t}\right|_{\text{scatt}} \quad \text{(3-12)}$$

where $\boldsymbol{v}(\boldsymbol{k})$ is the velocity of electrons in state $\boldsymbol{k}$ and e the electronic charge, which is one form of the celebrated Boltzmann equation. Solution of this equation for some specific system will give the distribution function in the presence of external fields or temperature gradients. This function is sufficient to determine the electrical and thermal conductivities and all thermoelectric effects, including their dependence upon magnetic fields. However finding such a solution is not a trivial matter. The difficulty lies in the complexity of the scattering term, which involves the rates of transition from all states to some particular state $\boldsymbol{k}'$, for example, and these will in turn depend upon the occupation numbers of these states. This term will thus generally involve an integral over all values of $\boldsymbol{k}'$ with the distribution function itself appearing in the integrand, leading to computational difficulties.

3.2.2 Linearised Boltzmann Equation

Under steady state conditions, the deviation in f from the equilibrium electron distribution is usually only small. By expressing f explicitly in terms of this deviation $g(\boldsymbol{k}, \boldsymbol{r}, t)$

$$f(\boldsymbol{k}, \boldsymbol{r}, t) = f_0(\boldsymbol{k}, \boldsymbol{r}) + g(\boldsymbol{k}, \boldsymbol{r}, t) \tag{3-13}$$

one may substitute f_0 for f on the left hand side of Eq. (3-12) and then retain only the lowest power of $f - f_0$ that does not vanish in the scattering term. This procedure leads to a linearised form of the Boltzmann equation:

$$\boldsymbol{v}(\boldsymbol{k}) \frac{\partial f_0}{\partial t} \nabla_r T + e \frac{\partial f_0}{\partial \varepsilon_k} \boldsymbol{v}(\boldsymbol{k}) \cdot \boldsymbol{E} = \tag{3-14}$$

$$= \left.\frac{\partial f_k}{\partial t}\right|_{\text{scatt}} - \boldsymbol{v}(\boldsymbol{k}) \cdot \frac{\partial g_k}{\partial \boldsymbol{r}} - \frac{e}{\hbar} (\boldsymbol{v}(\boldsymbol{k}) \times \boldsymbol{B}) \cdot \frac{\partial g_k}{\partial \boldsymbol{k}}$$

where ε_k is the energy of an electron of wave vector $\boldsymbol{k}$ (see, e.g., Ziman, 1972; p. 213), and the subscript k has been added to f and g to remind of their explicit dependence upon $\boldsymbol{k}$.

Such a linearised form is in accordance with the observed linear response of the electron and thermal fluxes to electric and magnetic fields and thermal gradients. Equation (3-14) is the starting point for many derivations of the electrical and thermal conductivities. However, its solution still requires evaluation of the scattering term. If $P_{kk'}$ is defined as the probability of transition from some state $\boldsymbol{k}$ (assumed to be filled) to some other state $\boldsymbol{k}'$ (assumed to be empty), the scattering term may be written as

$$\left.\frac{\partial f_k}{\partial t}\right|_{\text{scatt}} = \frac{\Omega}{(2\pi)^3} \int (g_{k'} - g_k)\, P_{kk'}\, \mathrm{d}\boldsymbol{k}' \tag{3-15}$$

where Ω is the specimen volume.

Evaluation of Eq. (3-15) then requires determination of the transition probability, often with the assistance of first order perturbation theory.

3.2.3 Relaxation Time Approximation

Instead of evaluating the transition probability $P_{kk'}$ directly, one may make the phenomenological assumption that the distribution will relax exponentially in time if the field is switched off. In this case, Eq. (3-15) simply becomes

$$\left.\frac{\partial f_k}{\partial t}\right|_{\text{scatt}} = -\frac{f_k - f_0}{\tau_k} = -\frac{g_k}{\tau_k} \tag{3-16}$$

τ_k is called the relaxation time, and may vary with the magnitude of $\boldsymbol{k}$ but not its direction. This approximation is rigorous if the scattering is purely elastic (i.e., $|\boldsymbol{k}| = |\boldsymbol{k}'|$) and the Fermi surface is spherical (Ashcroft and Mermin, 1976). In many situations, the asssumption of isotropic scattering over a spherical Fermi surface is not valid and more general solutions must be found. Nevertheless, as the approximation leads to considerable simplification, it is used here to illustrate the evaluation of the transport coefficients from the linearised Boltzmann Eq. (3-14).

Electrical Conductivity. In the presence of a uniform electric field, but no magnetic field or thermal gradient, Eq. (3-14) reduces to

$$e \frac{\partial f_0}{\partial \varepsilon_k} \boldsymbol{v}(\boldsymbol{k}) \cdot \boldsymbol{E} = \frac{-g_k}{\tau_k} \tag{3-17}$$

The current density $\boldsymbol{j}(\boldsymbol{r}, t)$ is obtained by summing the contributions $e\,\boldsymbol{v}(\boldsymbol{k})$ for all electrons:

$$j(r,t) = \frac{e}{4\pi^3} \int v(k)\, f(k,r,t)\, dk = $$
$$= \frac{e}{4\pi^3} \int v(k)\, g_k\, dk \tag{3-18}$$

[from Eq. (3-13)]. Substitution of g_k from Eq. (3-17) ultimately leads to

$$j(r,t) = \frac{e^2}{4\pi^3 \hbar} \int \tau_k\, v(k)\, [v(k) \cdot E]\, \frac{dS}{|v(k)|} \tag{3-19}$$

where the integration over a volume dk of k-space has been transformed into an integration over an element of area of the Fermi surface dS.

Comparison of Eq. (3-19) with Eq. (3-7a) ($\nabla_r T = 0$) leads to an expression for components of the conductivity tensor

$$\sigma_{ij} = \frac{e^2 \tau_k}{4\pi^3 \hbar} \int \frac{v_i(k)\, v_j(k)\, dS}{|v(k)|} \tag{3-20}$$

The tensor is symmetric and can be reduced to diagonal form giving the three principal conducting coefficients parallel to the principal axis of the crystal. In cubic crystals the tensor becomes a scalar and, for free electrons, Eq. (3-20) becomes the simple Drude expression

$$\sigma = \frac{1}{\varrho} = \frac{n_e e^2 \tau}{m} \tag{3-21}$$

where n_e is the electron density and m the electron mass.

Thermoelectric Effects. If the temperature gradient is non-zero, Eqs. (3-14) and (3-16) give

$$v(k) \frac{\partial f_0}{\partial T} \nabla_r T + e \frac{\partial f_0}{\partial \varepsilon_k} v(k) \cdot E = \frac{-g_k}{\tau_k} \tag{3-22}$$

This may be substituted into Eq. (3-18) to give the electric current, which is now Eq. (3-19) plus an additional term

$$\frac{e^2}{4\pi^3 \hbar} \int \tau_k\, v(k) \left(v(k) \frac{\partial f_0}{\partial T} \nabla_r T \right) dk \tag{3-23}$$

i.e. a temperature gradient $\nabla_r T$ alone produces an electric current through the thermoelectric effect. Comparison with Eq. (3-7a) (with $E = 0$) allows determination of γ.

Electronic Thermal Conductivity. An additional effect of a temperature gradient is to cause a flow of heat. The heat current density j_q corresponding to the electrical current density j [Eq. (3-18)] is then

$$j_q = \frac{1}{4\pi^3} \int (\varepsilon_k - \mu)\, v(k)\, g_k\, dk \tag{3-24}$$

since the heat current is just the electron flux times the difference between the kinetic energy per electron ε_k and the chemical potential μ of the electron system.

By analogy with the previous derivations, one obtains the electronic thermal conductivity (see, e.g., Ziman, 1972). In cubic materials

$$\varkappa = \frac{\pi^2}{3} \frac{n_e \tau}{m} k_B^2 T \tag{3-25}$$

where k_B is the Boltzmann constant.

Comparison with Eq. (3-21) allows definition of the Wiedemann-Franz ratio

$$\frac{\varkappa}{\sigma} = \frac{\pi^2}{3} \frac{k_B^2 T}{e^2} \tag{3-26}$$

Because it was derived from the relaxation time approximation, Eq. (3-26) is valid only at very low or very high temperatures (see, e.g., Ziman, 1972).

Hall Effect. In the presence of both electric and magnetic fields, Eq. (3-14) becomes

$$e \frac{\partial f_0}{\partial \varepsilon_k} v(k) \cdot E = $$
$$= \frac{-g_k}{\tau_k} - \frac{e}{\hbar} (v(k) \times B) \cdot \frac{\partial g_k}{\partial k} \tag{3-27}$$

This equation cannot be solved directly for g_k. However, by using a trial function of the same form as Eq. (3-17):

$$g_k = -e\,\tau \frac{\partial f_0}{\partial \varepsilon_k} \boldsymbol{v}(\boldsymbol{k}) \cdot \boldsymbol{A} \tag{3-28}$$

Equation (3-27) becomes

$$\boldsymbol{v}(\boldsymbol{k}) \times \boldsymbol{E} = = \boldsymbol{v}(\boldsymbol{k}) \times \boldsymbol{A} + \frac{e\,\tau}{m} (\boldsymbol{v}(\boldsymbol{k}) \times \boldsymbol{B}) \cdot \boldsymbol{A} \tag{3-29}$$

which has a solution

$$\boldsymbol{E} = \boldsymbol{A} + \frac{e\,\tau}{m} (\boldsymbol{B} \times \boldsymbol{A}) \tag{3-30}$$

By analogy with Eq. (3-7a)

$$\boldsymbol{j} = \sigma \boldsymbol{A} \tag{3-31}$$

and Eq. (3-30) becomes

$$\boldsymbol{E} = \frac{1}{\sigma} \boldsymbol{j} + \frac{e\,\tau}{m\,\sigma} \boldsymbol{B} \times \boldsymbol{j} \tag{3-32}$$

This equation shows that there are two components to the field: the normal longitudinal field parallel to $\boldsymbol{j}$ and, if $\boldsymbol{B}$ is perpendicular to $\boldsymbol{j}$ a transverse field know as the Hall field, which may be written as

$$E_{\mathrm{H}} = \frac{1}{n_e\,e} |\boldsymbol{B}|\,|\boldsymbol{j}| = R_{\mathrm{H}} |\boldsymbol{B}|\,|\boldsymbol{j}| \tag{3-33}$$

In the relaxation time approximation, it has the same sign as the charge carrier:

$$R_{\mathrm{H}} = \frac{1}{n_e\,e} \tag{3-34}$$

A more general analysis of R_{H} is given in Sec. 3.7.

Magnetoresistance. The relaxation time approximation does not predict any dependence of conductivity upon the magnetic field since the longitudinal electric field (i.e., that component of $\boldsymbol{E}$ parallel to $\boldsymbol{j}$) is given simply by $\boldsymbol{E}_{\mathrm{L}} = \frac{1}{\sigma} \boldsymbol{j}$ [see Eq. (3-32)]. However, if there is more than one band of carriers, or if the scattering is not isotropic,

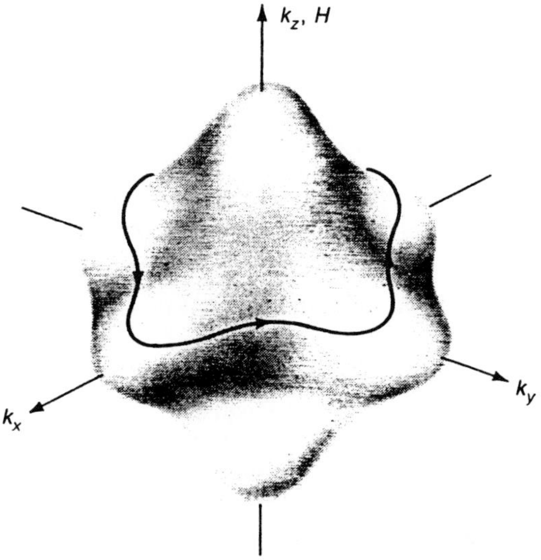

Figure 3-3. Intersection of a surface of constant energy with a plane perpendicular to the magnetic field. The arrow indicates the direction of motion along the orbit if the levels enclosed by the surface have lower energy than those outside (from Ashcroft and Mermin, 1976).

there does appear a component of the longitudinal electric field that depends upon $\boldsymbol{B}$. This is just the magnetoresistance introduced in Sec. 3.1 and discussed further in Sec. 3.6.

The detailed behavior of the conduction electrons in a magnetic field is further complicated by the fact that the Lorentz force $\boldsymbol{v}(\boldsymbol{k}) \times \boldsymbol{B}$ causes them to orbit around the Fermi surface as shown in Fig. 3-3. Whether or not they traverse complete orbits depends upon the relative magnitudes of the scattering rate (τ_k) and the time to complete an orbit (the inverse of the cyclotron frequency ω_c). We return to this aspect of the problem in Sec. 3.4.

3.3 Scattering Mechanisms

A perfect, infinite lattice would produce no obstacle to the passage of an electric current. Only deviations from a perfect periodicity of the lattice potential cause the scattering of electrons from initial to final states leading to a reduced conductivity. These deviations may be separated into those which do not have any intrinsic dependence upon thermally induced atomic displacements ("static" atomic or magnetic disorder) and those directly resulting from the thermal motion of atoms (thermally induced disorder).

3.3.1 Static Atomic Disorder

One form of static disorder results from atomic displacements such as occur around point and extended defects, from interstitials, or from atomic size effects in alloys containing atoms with different atomic radii. Such disturbances are characterised by the total atomic displacements produced.

In alloys there is an additional disruption to the periodicity of the lattice potential due to imperfect atomic ordering over the lattice sites. (A structure with perfect long range atomic order would again have no resistivity.)

The state of atomic order is determined by the distribution of atoms over the lattice sites and may be described by some suitable set of site occupation parameters, such as

$$\sigma_i^{\mathrm{A}} = \begin{cases} 1 & \text{if an A atom is at site } \boldsymbol{r}_i \\ 0 & \text{if not} \end{cases}$$

A complete set of these parameters $\{\sigma_i^{\mathrm{A}}\}$, one for each lattice site, would give the exact configuration of the system. However, it is not possible or even useful to specify such a set and some suitable averages must be used. The average over single sites, written as $\langle\sigma_i{}^{\mathrm{A}}\rangle$, is the probability of finding an A atom at any site and is given simply by the atomic fraction of A atoms, C_{A}. By itself, such a parameter contains no information about the distribution of atoms, although if a long range ordered lattice is decomposed (somewhat artificially) into sublattices, it can give some description of the state of order.

The Bragg-Williams long range order parameter, s, is defined as

$$s = \frac{\langle\sigma_i{}^{\mathrm{A}}\rangle^\alpha - C_{\mathrm{A}}}{1 - y^\alpha} \tag{3-35}$$

where $\langle\sigma_i{}^{\mathrm{A}}\rangle^\alpha$ is the site occupation average over the α sublattice and y^α is the fraction of α sites. For a completely random alloy, $\langle\sigma_i{}^{\mathrm{A}}\rangle^\alpha = C_{\mathrm{A}}$ and $s = 0$. For complete long range order (only possible at stoichiometric composition) $\langle\sigma_i{}^{\mathrm{A}}\rangle^\alpha = 1$, $y^\alpha = C_{\mathrm{A}}$ and $s = 1$. However, the rate of conduction electron scattering is determined by deviations in periodicity only over a distance of the order of the conduction electron mean free path $l = v_{\mathrm{F}}\,\tau$, where v_{F} is the velocity of the electron at the Fermi surface. Thus, if antiphase domain boundaries occur on a scale much larger than l, they do not produce a large resistivity although they may reduce s to near zero. It is thus necessary to consider the atomic distribution on a more local scale.

A suitable average of the site occupation parameters over pairs of sites, $\langle\sigma_i^{\mathrm{A}}\,\sigma_j^{\mathrm{B}}\rangle$, defines the probability of finding an A-B pair. Such pairwise correlations determine the diffraction effects of an imperfectly ordered lattice and thus determine the electrical transport properties (although much larger clusters may be necessary to determine the equilibrium atomic distribution itself).

The pairwise average serves to define the useful Warren-Cowley short range order parameter

$$\alpha_{ij} = 1 - \frac{\langle\sigma_i{}^{\mathrm{A}}\,\sigma_j^{\mathrm{B}}\rangle}{C_{\mathrm{A}}\,C_{\mathrm{B}}} \tag{3-36}$$

If like near neighbors are preferred (clustering) $\alpha_{ij} > 0$, whereas if unlike near neighbors are preferred (short range ordering) $\alpha_{ij} < 0$. From the definition of σ_i^A and Eq. (3-36), $\alpha_{ij} = 0$ if the atoms are randomly distributed. Long range order is defined by non-zero values of α_{ij} for large separations of sites i and j (Cowley, 1975).

Alternative ways of describing local atomic arrangements, such as composition waves (Cahn, 1961, 1962, 1968) and ordering waves in reciprocal space (Kachaturyan, 1979), may also be used directly in calculations of the electric resistivity (Rossiter, 1987).

If the scale of a decomposition (e.g., precipitation, spinodal decomposition, antiphase domains), exceeds l, it is necessary to regard each region and the boundary as separate elements and develop models consisting of arrays of these elements to adequately represent the microstructure. We return to this discussion in Sec. 3.5.

As noted above, the periodicity of the lattice potential is disrupted at internal interfaces (i.e., grain or interphase boundaries). If the scale of the microstructure is larger than l, these may be regarded as a different "phase" as far as electron scattering is concerned. Similarly, external surfaces constitute a severe disruption to the potential and so also lead to scattering of the conduction electrons. This is particularly important in thin fibres or fine wires that have one or more dimensions comparable to or smaller than l.

The sensitivity of the scattering process to the scale of the structure or microstructure presents a significant complication. In principle, the Boltzmann equation should be solved within each characteristic region to obtain a "local" mean free path $l_{(\mathrm{loc})}$. However, the scattering term in the Boltzmann equation depends upon the scattering matrix, which in turn depends upon the size of region $\sim l_{(\mathrm{loc})}$ over which the potential is averaged. Thus a complete solution to the problem strictly requires a self-consistent solution to the equation.

3.3.2 Thermally Induced Disorder

At any realistic temperature, thermal excitation of the lattice will cause dynamic atomic displacements. As far as the conduction electrons are concerned, the adiabatic approximation holds and the lattice may be considered to be fixed or "frozen" into this distorted configuration. The electron scattering is thus determined by the size of the displacements. At temperatures well above the Debye temperature Θ_D, the vibrations of each atom may be regarded as independent. The Einstein model then leads to mean square thermal displacements given by

$$\bar{x}^2 = \frac{\hbar^2 T}{M k_B \Theta_E^2} \tag{3-37}$$

where Θ_E is the characteristic Einstein temperature (related to the natural frequency of vibration ω by $\hbar\omega = k_B \Theta_E$) and M the atomic mass. However, at lower temperatures, excitations become collective and a lattice wave (or phonon) description is more appropriate (Debye model). In this case, the mean square displacement of some particular mode described by a wave vector $\boldsymbol{Q}$ is

$$\overline{|u_{Q,j}|^2} = \frac{2}{M\omega_{Q,j}^2}\left[\frac{\hbar\omega_{Q,j}}{\exp(\hbar\omega_{Q,j}/k_B T) - 1}\right] \tag{3-38}$$

where $\omega_{Q,j}$ is the frequency of that mode (linked to $\boldsymbol{Q}$ via the dispersion relationship). The total atomic displacement is then found by summing over all modes. However, it is again only necessary to consider the 'thermal' disorder on a scale $\le l$. This means that only phonons with wavelength $2\pi/|\boldsymbol{Q}| \le l$ determine the electrical transport properties.

An extreme case of spatial atomic disorder occurs in amorphous or glassy alloys. Whilst no long range structural order exists, the well-defined atomic sizes and closest distances of approach result in definite local correlations.

The structure of amorphous alloys is considered in more detail in Chapter 9 of this Volume.

3.3.3 Magnetic Disorder

Conduction electrons are also scattered by magnetic moments, so that any deviation from perfect periodicity in an array of these moments will also contribute to the resistivity. The magnetic structure may be described in terms of spin-spin correlation parameters, in direct analogy to the atomic order parameters described above, or alternatively by the use of spin waves. For example, one may define a magnetic long range order parameter by

$$M = \frac{\langle \boldsymbol{S}_0 \rangle \cdot \langle \boldsymbol{S}_i \rangle}{S(S+1)} \tag{3-39}$$

where $\boldsymbol{S}_0$ is now used to designate the spin at site 0 and $S = |\boldsymbol{S}|$. The corresponding short range order parameter is

$$m_i = \frac{\langle \boldsymbol{S}_0 \boldsymbol{S}_i \rangle - \langle \boldsymbol{S}_0 \rangle \cdot \langle \boldsymbol{S}_i \rangle}{S(S+1)} \tag{3-40}$$

However, unlike the atomic distribution, which may not be in thermodynamic equilibrium because of limited atomic diffusion, the spin system is in thermal equilibrium; thus the separation into static and dynamic excitations is not particularly useful.

Finally, competing magnetic interactions (e.g., a mixture of ferromagnetic and antiferromagnetic interactions) may give rise to complex magnetic structures, commonly termed spin glasses or cluster glasses.

3.4 Electrons in a Magnetic Field

3.4.1 Cyclotron Orbits and $\omega_c \tau$

The transport properties of a metal in a magnetic field are profoundly affected by the detailed structure of the Fermi surface (FS) of the metal in two different ways. First, different FSs yield different relative amounts of Umklapp and Normal scattering (i.e., scattering with and without participation of a reciprocal lattice vector, respectively). Examples of electron-phonon Normal and Umklapp scattering are shown schematically in Fig. 3-4. Such scattering differences are generally most important at low and intermediate fields, where they can profoundly affect both the magnitudes of R_H and MR and the sign of R_H. Second, different FSs have very different topologies. These topologies are most important at high fields, where they lead to different field dependences of both R_H and MR.

A fundamental quantity for characterizing electronic transport in the presence of $\boldsymbol{B}$

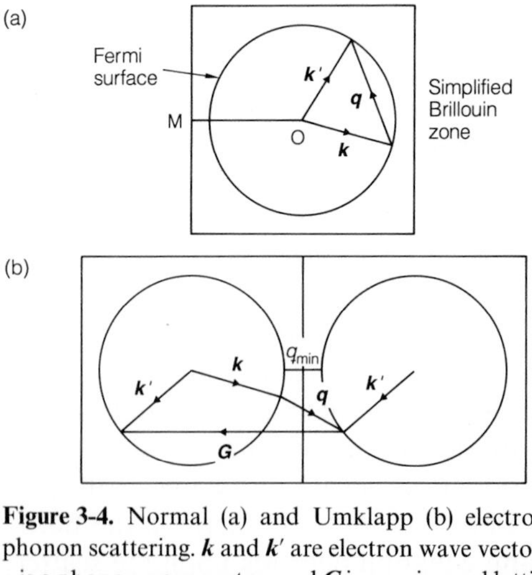

Figure 3-4. Normal (a) and Umklapp (b) electron-phonon scattering. $\boldsymbol{k}$ and $\boldsymbol{k}'$ are electron wave vectors, $\boldsymbol{q}$ is a phonon wave vector, and $\boldsymbol{G}$ is a reciprocal lattice vector. For the normal electron-phonon scattering process it is $\boldsymbol{k}+\boldsymbol{q}=\boldsymbol{k}'$ and for the Umklapp process it is $\boldsymbol{k}+\boldsymbol{q}+\boldsymbol{G}=\boldsymbol{k}'$.

is the dimensionless parameter $l/r_c = \omega_c \tau$, which specifies the average number of times an electron traverses a given cyclotron orbit between scattering events. Here l is the electron mean free path, τ is the average scattering time, r_c is the cyclotron radius, and ω_c is the cyclotron frequency given by (Abrikosov, 1972; Ashcroft and Mermin, 1976)

$$\omega_c = (h/2\pi)(\partial A/\partial \varepsilon) \qquad (3\text{-}41)$$

where h is Planck's constant and $\partial A/\partial \varepsilon$ is the energy derivative of the cross-sectional area of the cyclotron orbit in k-space.

(1) Low magnetic fields: $\omega_c \tau \ll 1$. In this limit, electrons traverse only a small fraction of a cyclotron orbit before being scattered. Their behavior is determined primarily by the scattering processes and by the properties of the FS in the local vicinity of their wave vectors before and after scattering.

(2) Intermediate magnetic fields: $\omega_c \tau \approx 1$. Here, electrons traverse about one full orbit between scattering events. Both scattering processes and surface topology are important, making this usually the most difficult region of $\boldsymbol{B}$ in which to analyze data.

(3) High magnetic fields: $\omega_c \tau \gg 1$. In this limit, electrons traverse their cyclotron orbits many times between scattering events. The field dependences of both the MR and R_H (i.e., whether MR is proportional to B^0, B^1, or B^2, and whether R_H is proportional to B^0 or B^{-2}) are determined solely by the topological structure of the FS in the plane in $\boldsymbol{k}$-space perpendicular to $\boldsymbol{B}$, with details of scattering affecting only the magnitudes of the transport properties.

At low temperatures and high magnetic fields, quantum oscillations appear in transport properties due to Landau quantization of the electronic energy levels. These occur in a fourth limit:

(4) The landau level quantization limit: $\omega_c \tau > k_B T$, where k_B is Boltzmann's constant and T is the absolute temperature.

3.4.2 Fermi Surface (FS) Topologies

We briefly describe the most important different FS topologies that appear in metals and the different types of cyclotron orbits that can occur. Chapter 2 of this Volume contains more information about the FSs and Brillouin Zones (BZ) of metals.

3.4.2.1 The One-Electron Free-Electron Fermi Surface

The FS of a one-electron free-electron metal is the simplest possible FS, namely a perfect sphere. The alkali metals in their body-centered cubic (b.c.c.) crystal structures have nearly spherical FS's, with only small dimples and bulges. A two-dimensional schematic drawing of the repeated zone scheme for a free electron metal with a simple-cubic structure was shown in Fig. 3-4. In a magnetic field $\boldsymbol{B}$, the only orbits on such a FS are circles. These orbits are closed, in that, in the absence of scattering, $\boldsymbol{k}$ traverses exactly the same orbit over and over again.

3.4.2.2 The Fermi Surface of Copper

The next simplest FSs are those of the one electron, f.c.c. noble metals, Cu, Ag, and Au. Figure 3-5a shows the FS and the 1st BZ boundaries of f.c.c. Cu, and Fig. 3-5b shows this same FS in the repeated zone scheme. Unlike the b.c.c. alkali metals, the FSs of these metals contact the BZ boundaries in the region around the [111] direction in $\boldsymbol{k}$-space. Although the FS still consists of only a single "sheet", this contact leads both to different closed orbits for different directions of $\boldsymbol{B}$, and to open

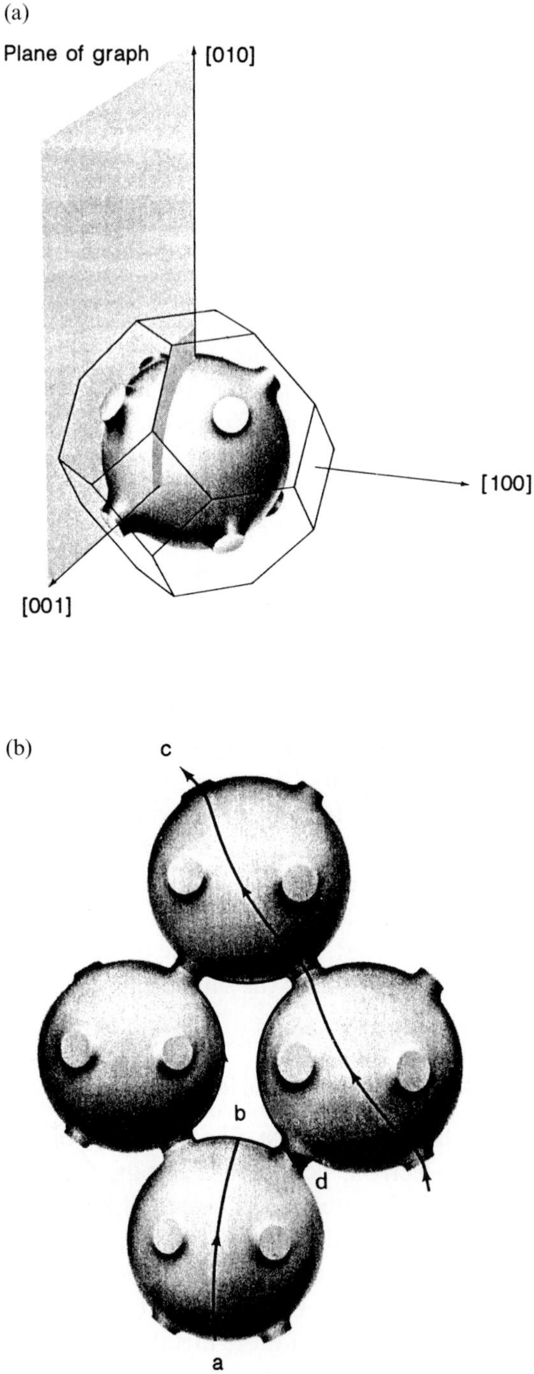

Figure 3-5. (a) The Fermi surface and first Brillouin zone boundaries of Cu. (b) Open orbits c and closed electron-like orbits a and d and hole-like orbits b for electrons in a uniform magnetic field (from Ashcroft and Mermin, 1976).

orbits (i.e., orbits in the repeated zone scheme that never close upon themselves). Closed orbits a and d enclose only filled electron states (these are called "electron-like" orbits), whereas closed orbit b encloses only empty states (a "hole-like" orbit). Certain directions of $\boldsymbol{B}$ yield open orbits, such as orbit c.

3.4.2.3 The Fermi Surface for 3 Free-Electrons per Atom and for Aluminium

Metals with more than one conduction electron/atom generally have more complex FS's, containing two or more different sheets, each of which can support closed and/or open orbits. As an example of such complexity, Fig. 3-6 shows the FS and BZ boundaries in the reduced zone scheme for 3 free-electrons/atom in a f.c.c. structure. The 1st BZ is full, the 2nd and 3rd are partially filled, and the 4th contains tiny electron pockets. In the real 3-electron metal Al, these tiny 4th zone pockets disappear, leaving a FS consisting of two separate parts: a 2nd zone "hole" sheet surrounding ≈ 2 holes/atom ($N_h \approx 2$), and a 3rd zone "electron" sheet surrounding ≈ 1 electron/atom ($N_e \approx 1$). Figure 3-7 shows a slice through the extended zone scheme in Al. At points W, the 2nd and 3rd zones are so close together that a large enough B pointing nearly along $\langle 001 \rangle$ can cause electrons to quantum mechanically tunnel between the second and third zone portions of the FS, thereby leading to open orbits. This phenomenon is called "magnetic breakdown".

3.4.3 High Field Forms of MR and R_{H}

The FS examples given above are all for metals that in the free electron approximation have odd numbers of conduction electrons per atom (i.e., 1 or 3), and thus odd numbers per primitive unit cell (which for

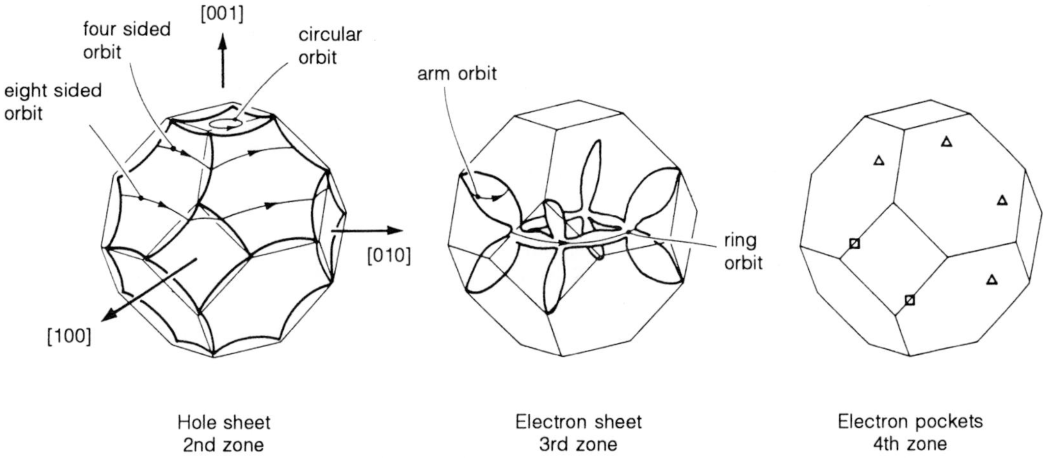

Figure 3-6. The Fermi surface of a trivalent, face-centered-cubic metal (such as Al) as given by the Harrison construction. Some of the possible types of electron orbits are shown (from Hurd, 1972).

these metals contains only one atom). Since each BZ can accommodate 2 electrons/atom, whatever changes in the Fermi surface occur when the BZ's are added, the numbers of electrons and holes per atom in such metals must remain unequal; e.g., for Al, $N_h - N_e = 1$. Metals for which $N_e \neq N_h$ are called "uncompensated". In contrast, metals or alloys that in the free electron approximation have an even number of conduction electrons per primitive unit cell, must have $N_e = N_h$. Such materials are called "compensated". The alternative high field forms of MR and R_{H} for different FS topologies are listed in Table 3-1 and illustrated in Fig. 3-8. Note the generally

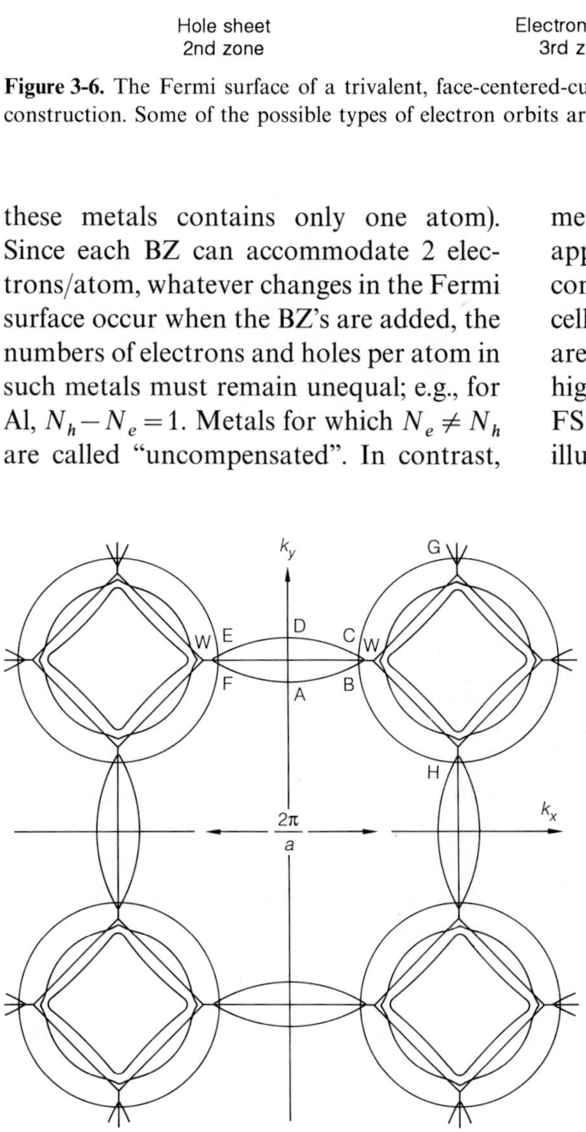

Figure 3-7. Central [001] section through the aluminium Fermi surface (after Ashcroft, 1963).

Table 3-1. Types of behavior of the Hall and transverse magnetoresistance effects in metals (Hurd, 1972).

Type	Magnetoresistance [b]	Hall coefficient	Nature of orbits in planes normal to the applied field direction; state of compensation [e]
Behavior in the high-field condition			
Single Crystal Sample			
1	$\Delta\varrho/\varrho(0) \rightarrow$ saturation	$R \propto (n_e - n_h)^{-1}$	all closed; $N_e \neq N_h$
2	$\Delta\varrho/\varrho(0) \rightarrow B^2$ (transverse) $\Delta\varrho/\varrho(0) \rightarrow$ saturation (longitudinal)	$R \rightarrow const$ [a]	all closed; $N_e = N_h$
3	$\Delta\varrho/\varrho(0) \rightarrow$ saturation	$R \propto (n_e - n_h)^{-1}$	negligible number of open orbits; $N_e \neq N_h$
4	$\Delta\varrho/\varrho(0) \rightarrow B^2 \sin^2\theta$ [c]	$R \rightarrow const$ [a]	open orbits in one direction only
5	$\Delta\varrho/\varrho(0) \rightarrow$ saturation	$R \propto B^{-2}$	open orbits in more than one direction
Polycrystalline sample			
6	$\Delta\varrho/\varrho(0) \propto B$	$R \rightarrow const$ [a]	all types if the crystallites have random orientations
Behavior in the low-field condition			
Single Crystal or Polycrystalline Sample			
7	$\Delta\varrho/\varrho(0) \propto B^2$	$R \rightarrow const$ [d]	it is irrelevant what type of orbit predominates

[a] The Hall coefficient is not related to the effective number of carriers in any simple manner; [b] Transverse magnetoresistance except as indicated for type 2; [c] Note that $\Delta\varrho/\varrho(0) \rightarrow$ saturation when θ, the angle between the applied electric field and the axis of the open orbit in real space, is zero; [d] The Hall coefficient depends upon the anisotropy of the dominant electron scattering process and the electron's velocity and effective mass at each point on the Fermi surface; [e] N_e and N_h are, respectively, the number of electrons and holes per unit cell of the Bravais lattice; n_e and n_h are, respectively, the density of electrons and holes in real space.

different behaviors for compensated and uncompensated metals. For further details, see Abrikosov (1972, 1988).

3.5 Electrical Resistivity, ϱ

All of the scattering mechanisms discussed in Sec. 3.3 may be regarded as producing one or both of two effects: (i) a change in the nature of the total scattering potential $W(\boldsymbol{r})$ at the 'defect' (whether a foreign atom or vacancy), and (ii) a disruption to the spatial periodicity of the lattice potential. If the disturbances are not too large, first order perturbation theory allows a fairly straightforward evaluation of the scattering associated with the defect. Equation (3-20) then leads to an expression of the form (see e.g., Rossiter, 1987):

$$\varrho = \frac{3\pi m^2 \Omega_0 N}{4\hbar^3 e^2 k_F^6} \int_0^{2k_F} |\langle \boldsymbol{k}+\boldsymbol{q}|\, W(\boldsymbol{r})\,|\boldsymbol{k}\rangle|^2 q^3\, \mathrm{d}q \tag{3-42}$$

where N and Ω_0 are the number and volume of unit cells, k_F the magnitude of the Fermi wave vector (see Chapter 2), the integrand $\langle \boldsymbol{k}+\boldsymbol{q}|\, W(\boldsymbol{r})\,|\boldsymbol{k}\rangle$ represents the matrix elements of the total scattering potential $W(\boldsymbol{r})$ and the integration is over the magnitude of the scattering wave vector $\boldsymbol{q}$ defined by

$$\boldsymbol{q} = \boldsymbol{k} - \boldsymbol{k}' \tag{3-43}$$

The scattering potential may be factorised into two terms

$$\langle \boldsymbol{k}+\boldsymbol{q}|\, W(\boldsymbol{r})\,|\boldsymbol{k}\rangle = \mathfrak{s}(\boldsymbol{q})\,\langle \boldsymbol{k}+\boldsymbol{q}|\, w(\boldsymbol{r})\,|\boldsymbol{k}\rangle \tag{3-44}$$

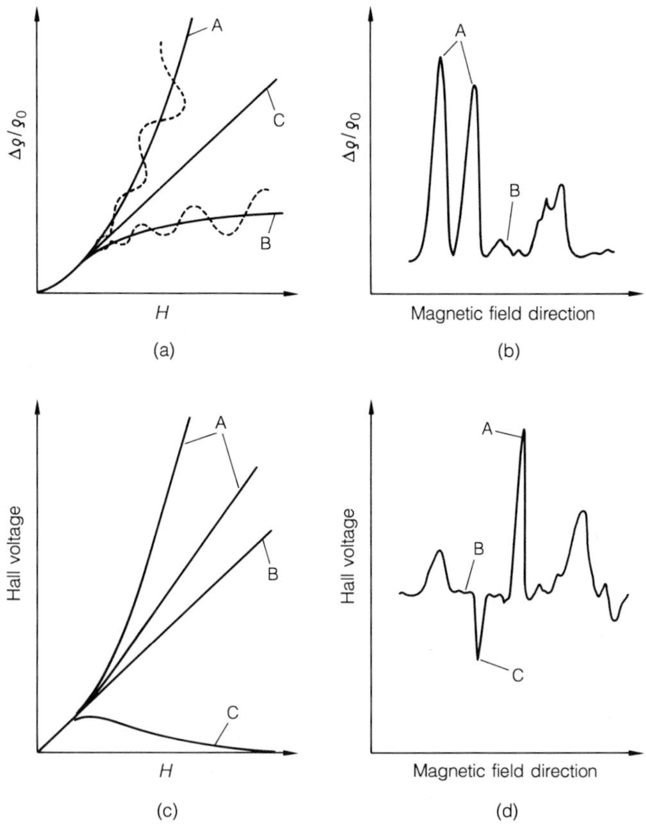

Figure 3-8. Types of field dependence observed experimentally for the transverse magnetoresistance and the Hall voltage (from Hurd, 1972). In Fig. 3-8 a and 3-8 b: A illustrates a quadratic (i.e., B^2) high field variation in $\Delta\varrho$; B illustrates saturation in $\Delta\varrho$; and C illustrates the linear (i.e., B) variation in $\Delta\varrho$; that is generally observed in polycrystalline samples. Quantum oscillations (dashed curves in 3-8 a) may be superimposed upon behaviors A and C for single crystal samples. In Fig. 3-8 c, and 3-8 d: A indicates high field linear variations of different magnitudes; B illustrates a linear variation extending all the way down to $B=0$; and C illustrates a B^{-1} high field variation (from Hurd, 1972).

where $\mathfrak{s}(\boldsymbol{q})$ is the structure factor

$$\mathfrak{s}(\boldsymbol{q}) = \frac{1}{N} \sum_{\boldsymbol{r}} \exp(-i\,\boldsymbol{q} \cdot \boldsymbol{r}) \qquad (3\text{-}45)$$

and depends only upon atom positions, and $\langle \boldsymbol{k}+\boldsymbol{q} |\, w(\boldsymbol{r}) \,| \boldsymbol{k} \rangle$ [often written simply as $w(\boldsymbol{q})$] is the form factor that is determined entirely by the atomic potential $w(\boldsymbol{r})$ associated with site $\boldsymbol{r}$.

Thus the two different scattering effects described above (nature of the potential and location of the potential) occur explicitly in the resistivity expression. However, it must be emphasised that Eq. (3-42) and the simple separation of Eq. (3-44) are only strictly valid for metals with a simple electronic structure (i.e., free electron-like bands) and isotropic scattering over a spherical Fermi surface. These requirements are not satisfied in many metals and alloys of interest, particularly those involving transition metals. It is then necessary to use more complex methods of analysis and it must be admitted that there is still much theoretical work to be done. Nevertheless, the basic concept of a resistivity produced by deviations in periodicity of the lattice potential remains valid.

If the scattering mechanisms are independent and the scattering isotropic, the contributions to the total resistivity are simply additive. In an alloy this leads to Matthiessen's rule

$$\varrho_{\mathrm{TOT}} = \varrho_0 + \varrho_{\mathrm{p}}(T) \qquad (3\text{-}46)$$

where all of the static mechanisms are collected into the 'residual' resistivity ϱ_0 and $\varrho_p(T)$ is the temperature dependent phonon scattering term. The assumption of independent scattering is only partly true and there is a large body of work devoted to studying deviations from this rule (see e.g., Bass, 1972).

3.5.1 Pure Metals

3.5.1.1 Static Displacements

The presence of a vacancy at some site $\boldsymbol{r}_i$ has the effect of removing the average lattice potential $\bar{w}(\boldsymbol{r}_i)$ from that site (neglecting here the effect of charge redistribution) (Omini, 1980) and causing lattice distortions around that site leading to a change in the structure factor $\mathfrak{s}(\boldsymbol{q})$. The former produces a positive contribution to the resistivity but this is reduced somewhat by the effects of lattice relaxation. Values for the resistivity due to 1 at.% of vacancies in some pure metals are given in Table 3-2.

Table 3-2. Electrical resistivity due to one atomic percent of vacancies in the metals indicated.

Metal	ϱ_0 (μΩ cm)/at.%	References
K	1.9–2.1	Benedek and Baratoff (1971)
Sn	4.4	Sun and Ohring (1976)
Al	1.0–3.3	Benedek and Baratoff (1971)
Cu	1.0–3.0	Simmons and Balluffi (1960)
Ag	1.3	Dugdale (1977)
Au	1.5	Dugdale (1977)

From these results it appears that the vacancy-specific resistivity is approximately 3×10^{-21} V μΩ cm for a density of V vacancies per cm^3. For a typical metal, the equilibrium vacancy density is around 1×10^{13}/cm^3 at 300 °C and 2.7×10^{16}/cm^3 at 600 °C. Since these represent vacancy concentrations of $\sim 10^{-8}$ at.% at 300 °C and $\sim 10^{-5}$ at.% at 600 °C, the actual contribution of the equilibrium concentration of vacancies to ϱ_0 is very small indeed.

The resistivity contribution due to dislocations may similarly be considered as due to the potential discontinuity at the core and the effects of atomic displacements in the surrounding strain field. Calculations based upon the latter alone produce resistivities which are generally some orders of magnitude too low, leading Brown (1977) to propose core scattering as the dominant effect. Some experimental data are compared with the prediction of Brown's s-wave resonance model in Table 3-3.

It is interesting to note that the dislocation density in a heavily cold worked metal is of order 10^{12}/cm^2, yielding a maximum dislocation resistivity contribution of ~ 0.1 μΩ cm.

On a more microscopic level, the effects of cold work upon the resistivity have been considered by Broom (1954) and van

Table 3-3. Dislocation specific resistivities per unit dislocation density N (measured in lines/cm^2). These figures represent averages over all dislocation orientations (Brown, 1977, 1982).

Metal	ϱ_0/N in 10^{-13} μΩ cm^3	T in K	Theory (Brown, 1977)
K	4.0	4.2	8.0
Cu	1.6 ± 0.2	4.2	1.3
Ag	1.9	4.2	1.9
Au	2.6	4.2	1.9
Be	34.0	80.0	28.0
Cd	24.0	80.0	25.0
Al	1.7 ± 0.3	4.2	1.8
Zr	100.0	80.0	40.0
Ti	100.0	4.2	29.0
Pb	1.1	80.0	4.2
Bi	2×10^5	1.3	1.7×10^5
Mo	5.8	4.2	3.7
W	7.5	4.2	7.4
Pt	9.0	80.0	4.0
Fe	10.0 ± 4.0	80.0	1.9
Ni	10.0	80.0	1.1
Rh	32.0	80.0	1.0

Bueren (1957), generally leading to expressions of the form

$$\Delta\varrho_0 = a\,\varepsilon^b \qquad (3\text{-}47)$$

where ε is the mean shear strain ($2.24\,\Delta l/l$), with $a \sim 0.04-0.16\,\mu\Omega$ cm and $b \sim 1.3-1.5$. For example, a 25% tensile strain in a Cu wire produces an increase in resistivity of $\sim 0.015\,\mu\Omega$ cm.

The resistivity due to grain boundaries has been considered using a number of models, including an array of scattering potentials (Mayadas and Schatzkes, 1970) or an array of dislocations (Brown, 1977, 1982). The importance of the change in orientation of the lattice has been discussed by Ziman (1960) and Guyot (1970). Calculations which take into account the detailed positions of the atoms at the boundary via $\mathfrak{s}(\boldsymbol{q})$ also appear to give good agreement with experimental data (Lormand, 1982). Some results for a number of metals are given in Table 3-4. Also shown in that table are calculated values based simply upon the assumption that the grain boundary is represented by an array of dislocations and that these act as independent scatterers.

Table 3-4. Grain boundary specific resistivities in terms of area of boundary per unit volume S (measured in cm^2/cm^3) (Brown, 1982).

Metal	ϱ_0/S in $10^{-12}\,\Omega\,cm^2$	T in K	Theory (Brown, 1977) $10^{-12}\,\Omega\,cm^2$
Cu	2.4	4.2	2.7
Au	3.5	300.0	3.9
Cd	17.0	4.2	32.0
Al	1.1 – 2.4	4.2	2.6
Pd	0.3, 1.3	5.0 – 20.0	1.4
Bi	6.9×10^4	77.0	17.0×10^4
W	20.0	77.0	22.0
Fe	80.0 – 160.0	4.2	33.0
Ni	140.0	77.0	17.0

Various mechanisms have been proposed for the scattering of conduction electrons at free surfaces, which will become important in thin foils or fine wires. These include microscopic surface roughness and localised surface charges (see e.g., Ziman, 1960; Greene, 1964). Such approaches often lead to the notion of a specularity parameter p that describes the fraction of electrons which are scattered in a specular (i.e., non-diffuse) manner. The resistivity or conductivity of a foil or wire can then be written in one of two forms

$$\frac{\varrho_0(d,T)}{\varrho_\infty} = 1 + A(p, l_\infty, d)\,\frac{l_\infty}{d} \qquad (3\text{-}48)$$

or

$$\frac{\sigma_0(d,T)}{\sigma_\infty} = 1 - A'(p, l_\infty, d)\,\frac{l_\infty}{d} \qquad (3\text{-}49)$$

where the parameters $A(p, l_\infty, d)$ and $A'(p, l_\infty, d)$ depend upon the model used, d is the foil or wire thickness, l_∞ is the bulk mean free path and ϱ_∞ and σ_∞ the corresponding bulk values of the resistivity and conductivity. The dependence of A' upon d/l_∞ for some different models is shown in Fig. 3-9. Note that this size effect scattering [Eqs. (3-48) or (3-49)], represents an explicit deviation from Matthiessen's rule since it depends explicitly upon T.

This form of analysis may also be applied to grain boundaries, in which case it is of no consequence whether the electron is specularly reflected back into the grain or transmitted through the boundary, provided that its velocity in the direction of $\boldsymbol{E}$ is the same. The grain boundary resistivity may then be expressed in the form of Eq. (3-48) where d is now the mean grain diameter. The corresponding variation of $A(p, l_\infty, d)$ with d/l_∞ is shown in Fig. 3-10. This size effect is demonstrated in Fig. 3-11 which shows the normalised grain boundary resistivity as a function

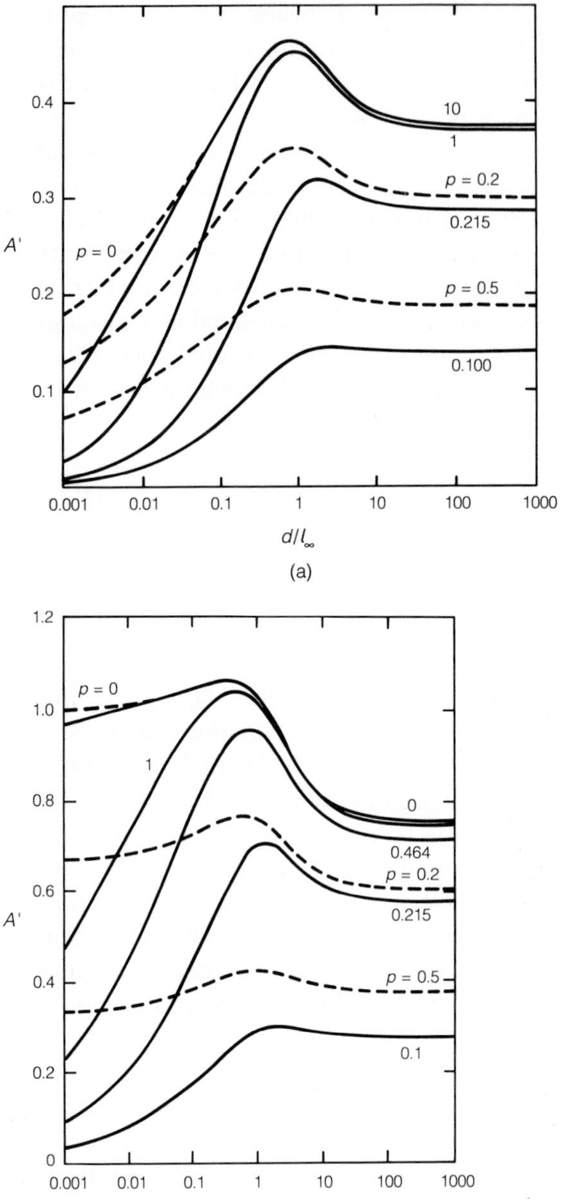

Figure 3-9. The parameter $A'(l_\infty, d)$ as a function of d/l_∞ for foils (a), and wires (b), derived from the Soffer model for the values of h/λ_k indicated by the numbers on the curves. (h is the local deviation in surface height and λ_k the de Broglie wavelength of state $\boldsymbol{k}$.) Also shown (dashed lines) are the Fuchs-Sondheimer and Dingle results for the values of p indicated (from Rossiter, 1987; data of Sambles and Elsom, 1980; Sambles and Priest, 1982).

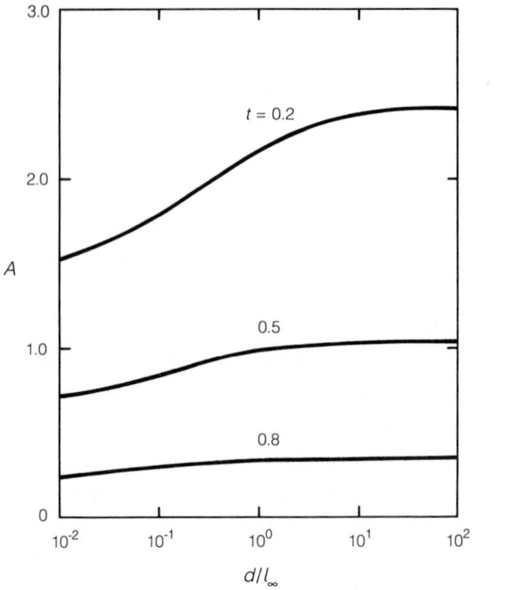

Figure 3-10. The parameter A as a function of d/l_∞ from the grain boundary scattering model for various values of the transmission coefficient t (from Rossiter, 1987).

of the grain size dependent parameter $\nu = \dfrac{d}{l_\infty \ln(1/t)}$, where t is an empirical specular transmission coefficient.

3.5.1.2 Thermal Effects

The effects of thermal scattering may in principle be incorporated directly into the structure factor by allowing the atom to be displaced away from the lattice site $\boldsymbol{r}_\mathrm{i}$ by some distance $\boldsymbol{\delta}_\mathrm{i}$. The thermal displacement $\boldsymbol{\delta}_\mathrm{i}$ is obtained from a sum over all phonon wavevectors $\boldsymbol{Q}$ and modes. Such information is available either from a solution of the dynamical matrix based upon known force constants (see Pal, 1973; Kumar, 1975) or from experimentally determined phonon dispersion curves (see, e.g., Darby and March, 1964).

An alternative approach is to write the structure factor directly in terms of the dy-

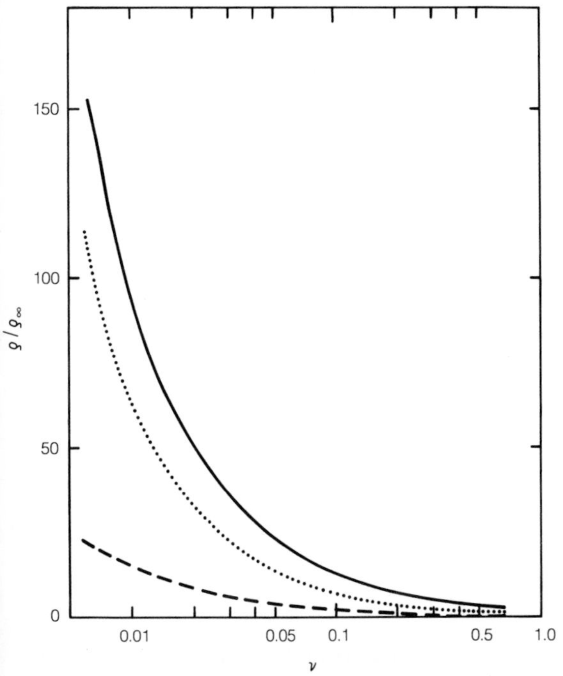

Figure 3-11. The normalised grain boundary resistivities as a function of the grain size dependent parameter ν, solid line $\varrho_g(\nu)/\varrho_\infty$; dotted line $\varrho_\perp(\nu)/\varrho_\infty$; dashed line $\varrho_\parallel(\nu)/\varrho_\infty$ where $\varrho_\perp$ and $\varrho_\parallel$ are the resistivities perpendicular and parallel to the current flow, ϱ_g, the average resistivity for a three-dimensional array of boundaries and ϱ_∞ the resistivity for infinite boundary spacing (from Rossiter, 1987).

namical matrix (Robinson and Dow, 1968) allowing $\varrho_p(T)$ to be determined directly from known force constants. Once again, for all but the simple metals, it is necessary to go beyond the first order perturbation theory approach.

Some experimental results for the alkali and noble metals are shown in Fig. 3-12a and b and indicate that reasonable agreement between theory and experiment is now possible. A compilation of results for a number of pure metals is given in Bass (1982b).

A simpler though less satisfactory approach is to use the Debye (or Einstein) model to evaluate the thermal displacements as discussed in Sec. 3.3.4, leading to the general prediction of $\varrho_p(T) \propto T^5$ at low temperatures ($T < \Theta_D$) and $\varrho_p(T) \propto T$ at higher temperatures, in general agreement with experiment for free electron type metals, although the predicted magnitudes of the resistivity are usually not correct and U-processes are ignored. Nevertheless, this approach leads to a useful phenomenological expression due to Bloch (1930), Grüneisen (1933), and Wilson (1937).

$$\frac{\varrho_p(T)}{\varrho_{\Theta_D}} = 4.225 \left(\frac{T}{\Theta_D}\right)^5 \int_0^{\Theta_D/T} \frac{z^5 \, dz}{(e^z - 1)(1 - e^{-z})} \tag{3-50}$$

where ϱ_{Θ_D} is the resistivity at the Debye temperature. In practice, a better fit between experiment and theory is obtained by allowing Θ_D in Eq. (3-50) to vary slightly from the actual Debye temperature (as obtained from specific heat measurements, for example). The corrected characteristic "resistivity" temperatures Θ_R for some metals are shown in Table 3-5 and the results plotted in Fig. 3-13. Some other results are given in Meaden (1965).

Finally, it is interesting to note that at room temperature $\varrho_p(T) \sim 1\ \mu\Omega$ cm which is at least one order of magnitude larger than the contribution from lattice defects, even in severely cold worked materials.

3.5.2 Alloys

Substitutional and interstitial impurities lead to changes in the form factor as well as the structure factor through the local distortions of the lattice. In dilute alloys, one finds that the residual resistivity contribution varies linearly with impurity concentration

$$\varrho_0 \propto C_i \tag{3-51}$$

where C_i is the concentration of the impurity atoms. This expression may be ob-

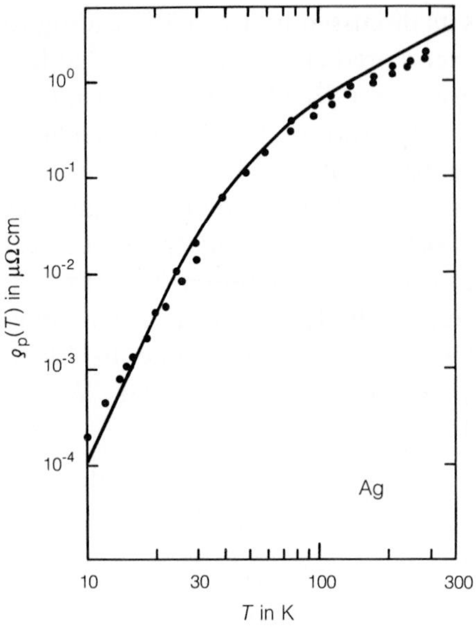

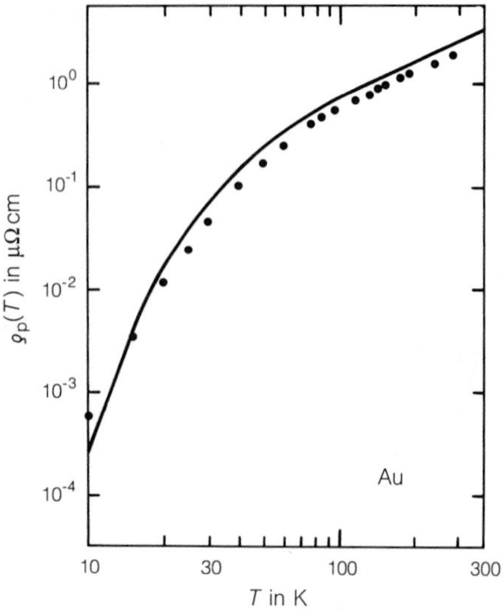

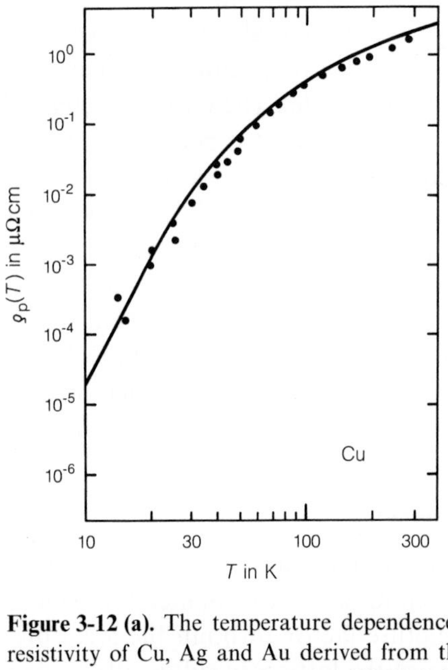

Figure 3-12 (a). The temperature dependence of the resistivity of Cu, Ag and Au derived from the form factors of Borchi and de Gennaro (1970) (solid line). The solid circles represent a collection of experimental data (from Rossiter, 1987; data of Kharoo et al., 1978).

tained directly from Eq. (3-42) since, if the effects of lattice strain are neglected,

$$\varrho_0 = \frac{3\pi m^2 \Omega_0}{4\hbar^3 e^2 k_F^6} C_i \int_0^{2k_F} |w_i(\boldsymbol{q}) - \bar{w}(\boldsymbol{q})|^2 q^3 \, dq \tag{3-52}$$

where $w_i(\boldsymbol{q})$ and $\bar{w}(\boldsymbol{q})$ are the form factors for the impurity and host respectively. Some experimental data for various impurities are shown in Fig. 3-14 and a collection is given in Bass (1982 b, 1985). Considerable effort has been spent attempting more detailed calculations of such impurity resistivities taking into account lattice distortions (see, e.g., Singh et al., 1977; Yamamoto et al., 1973; Popovic et al., 1973).

From Eq. (3-52) one expects a variation of ϱ_0 with the square of the magnitude of the scattering potential. Substitution of form factors for some simple model potentials (such as screened Coulomb or empty core potentials) leads to the result that the impurity resistivity should depend upon the square of the difference in valency Δz between the solute and host

$$\varrho_0 \propto \Delta z^2 \tag{3-53}$$

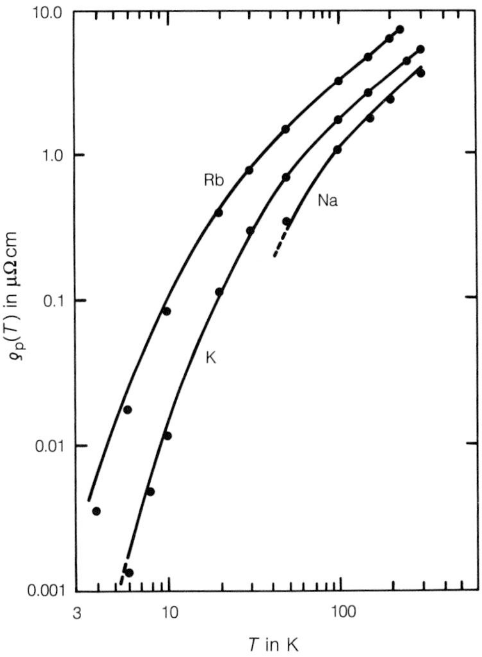

Figure 3-12 (b). Resistivity (at constant volume) as a function of temperature for Na, K and Rb. The lines are theory and solid circles are experimental data (from Rossiter, 1987; using data of Hayman and Carbotte, 1971).

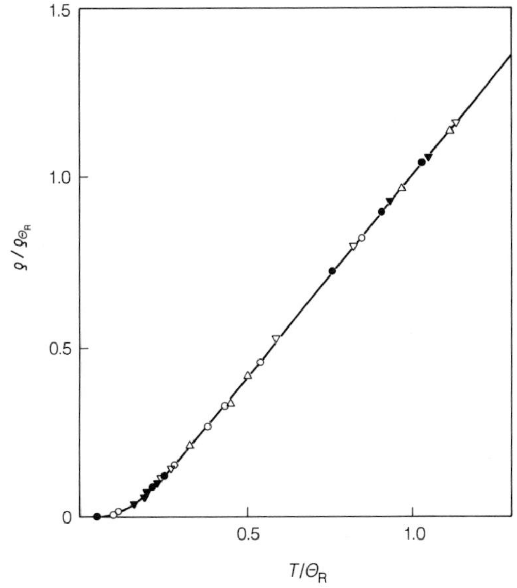

Figure 3-13. The temperature dependence of the resistivity ϱ/ϱ_Θ plotted against T/Θ_R. ●, Li, $\Theta_R = 363$ K; ○, Na, $\Theta_R = 202$ K; ▽, Cu, $\Theta_R = 333$ K; △, Au, $\Theta_R = 175$ K; ▼, Pb, $\Theta_R = 86$ K (from Olsen, 1962). These values of Θ_R differ from those shown in Table 3-5 as the data has been fitted to lower temperatures.

This is the Linde-Norbury rule and some experimental data are shown in Fig. 3-15.

However this form of behavior is found only in free-electron like metals. If transition metal impurities are added to Al or Cu hosts, for example, a bound state associated with the unfilled 3 d shell and localised in the vicinity of the impurity may form. If these lie at energies within the conduction band, an electron may temporarily become bound in the "virtual" bound state, effectively giving rise to a large scattering effect. As we go along the transition series Sc, Ti, V, Cr, Mn, Fe, Co, Ni this virtual bound state passes through the Fermi level as shown in Fig. 3-16. This leads to a 'resonance' in the scattering as shown in Fig. 3-17 with ϱ_0 peaking at Cr in which the 3 d shell is just half filled.

Table 3-5. Some representative high temperature values for Θ_R (valid for $\Theta_R/3$ to Θ_R) compared with corresponding values for Θ_D (Meaden, 1965).

Metal	Θ_D in K	Θ_R in K
Be	1000	1240
Na	160	195
Mg	325	340
Al	385	395
K	100	110
Ca	225	~145
Ti	355	342
Cr	450	485
Cu	320	320
Zn	245	175
Pd	300	270
Ag	220	200
Cd	165	130
W	315	333
Pt	225	240
Au	185	200
Pb	88	~100

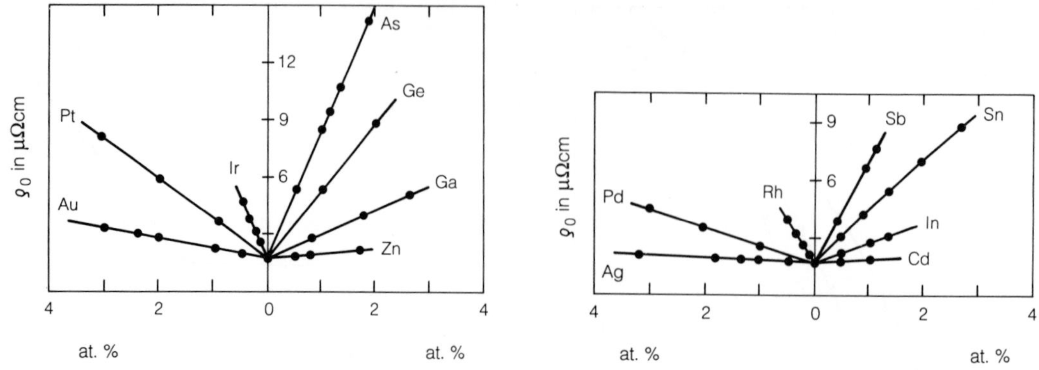

Figure 3-14. Resistivity as a function of concentration for dilute alloys of copper with the impurities indicated (from Rossiter, 1987; using the data of Linde, 1932).

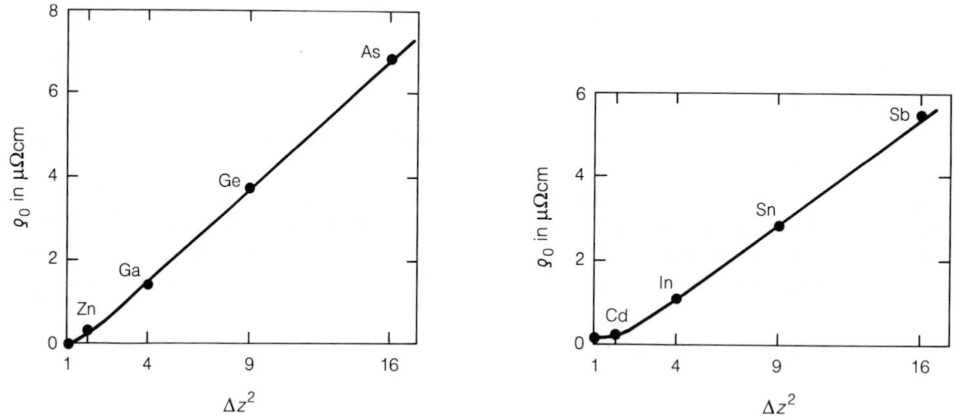

Figure 3-15. The change in resistivity for one at.% impurity in Cu as a function of $|\Delta z|^2$ (from Rossiter, 1987; using the data of Linde, 1932).

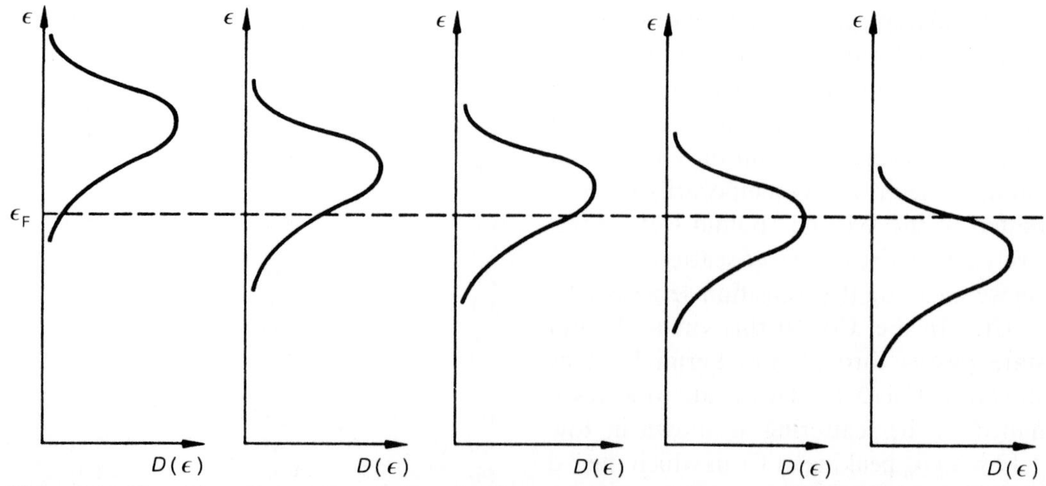

Figure 3-16. Virtual bound state shown schematically as we go from Sc towards Ni (from Dugdale, 1977).

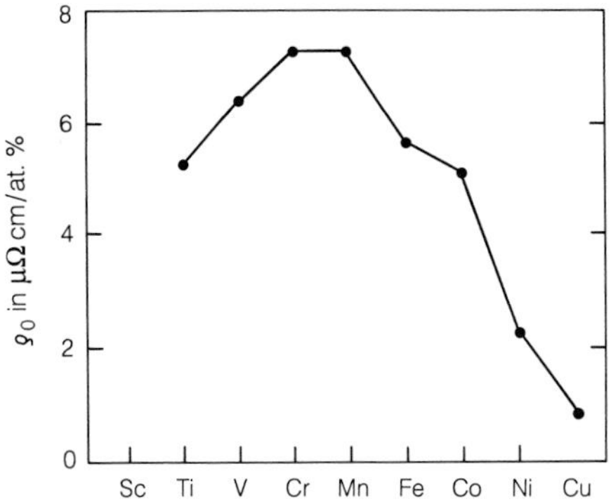

Figure 3-17. Resistivity per at.% impurity for 3d transition metals in Al host (from Babic et al., 1972, 1973).

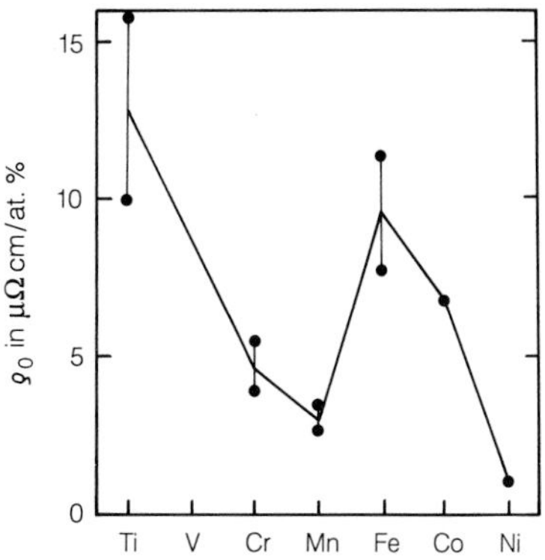

Figure 3-18. Residual resistivity of 1 at.% 3d impurity in Cu (from Podloucky et al., 1980).

In the case of a Cu host, two peaks are usually found to occur, as shown in Fig. 3-18. This is because the d-shell is divided into two spin-split subshells of different energy (see, e.g., Heeger, 1969). As the different impurities from Ti to Ni are added, the electrons first fill one d subshell which can hold five electrons and then fill the other subshell with the opposite spin direction. However, at very low temperatures the magnetic moments of the impurity are effectively quenched and the curve exhibits only a single peak, as shown in Fig. 3-19. A more detailed summary of some of the theoretical models developed to explain the behavior is given in Rossiter (1987).

In more concentrated binary alloys the form factor of interest turns out to be the difference in the form factors of the alloy components in which case

$$\varrho_0 = \frac{3\pi m^2 \Omega_0}{4\hbar^3 e^2 k_F^6} N \cdot \int_0^{2k_F} |\mathfrak{s}^d(\boldsymbol{q})|^2 \, |w_A(\boldsymbol{q}) - w_B(\boldsymbol{q})|^2 \, q^3 \, dq \tag{3-54}$$

where $\mathfrak{s}^d(\boldsymbol{q})$ is the structure factor which distributes the A and B atoms over the lattice.

Since the form factor is squared in Eq. (3-54), a random alloy containing one atomic percent of A in B should lead to the same resistivity of one percent B in A. This is known as Mott's rule and some results are shown in Table 3-6.

The short and long range atomic correlations discussed in Sec. 3.4.4 directly affect the structure factor leading to expressions of the form

$$\varrho_0 = \frac{3\pi m^2 \Omega_0}{4\hbar^3 e^2 k_F^6} C_A C_B \cdot \int_0^{2k_F} \sum_i c_i \alpha_i \frac{\sin(q r_i)}{q r_i} |w_A(\boldsymbol{q}) - w_B(\boldsymbol{q})|^2 q^3 \, dq \tag{3-55}$$

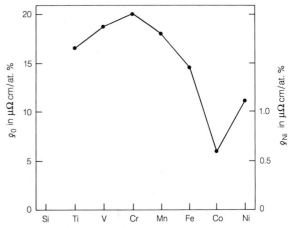

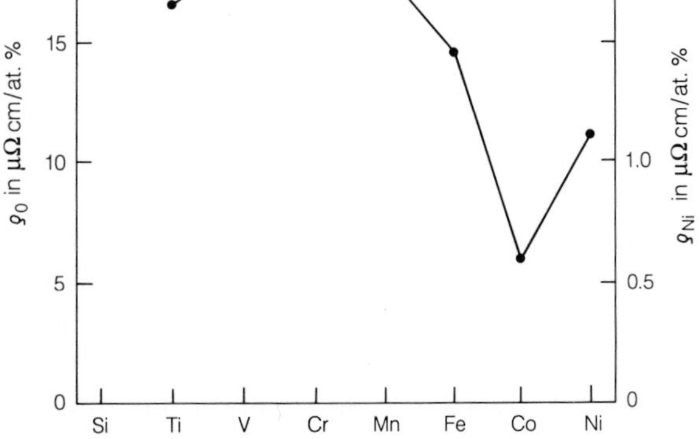

Figure 3-19. The residual resistivity due to 1 at.% 3d transition metal impurity in copper at low temperatures (from Lodder et al., 1982).

Table 3-6. Change in resistivity due to one atomic percent solute in the alloy systems indicated. Also shown are the ratios of atomic volumes. The data are from results cited in Mott and Jones (1936), Blatt (1968) and Rossiter (1987).

Alloy components A	B	$\Delta\varrho$ in $\mu\Omega$ cm/at.% 1% A in B	1% B in A	Ratio of atomic volumes
Cu	Au	0.45	0.55	0.70
Cu	Ag	0.07	0.14	0.70
Ag	Au	0.36	0.36	1.01
Mg	Cd	0.3–0.4	0.68	1.17
Pd	Pt	0.6	0.7	0.97

where c_i is the number of atoms in the ith co-ordination shell of radius r_i. In a random alloy there is no order, $\alpha_0 = 1$ and $\alpha_i = 0$ for all $i > 0$. Equation (3-54) then gives

$$\varrho_0 \propto C_A C_B \tag{3-56}$$

which is known as Nordheim's rule.

In concentrated alloys containing transition metals, it is necessary to take into account the composition dependence of the bandstructure and also allow for the fact that the charge carriers may occupy (and scatter between) a number of energy bands. The early approach to this problem due to Mott (see Mott and Jones, 1936) was based upon the idea of rigid s and d bands which are filled up as the number of available electrons is increased by alloying. Scattering of electrons between s states and from s to d states was allowed leading to an explanation of the observed variations of ϱ_0 and $\varrho_p(T)$ with composition in Pd-Au and Pd-Ag alloys, for example (Figs. 3-20, 3-21). However, more recent first principles calculations by Butler and Stocks (1984) have thrown some doubt upon this simple analysis, suggesting that the predominant charge carriers in Pd-rich alloys had essentially d character, for example.

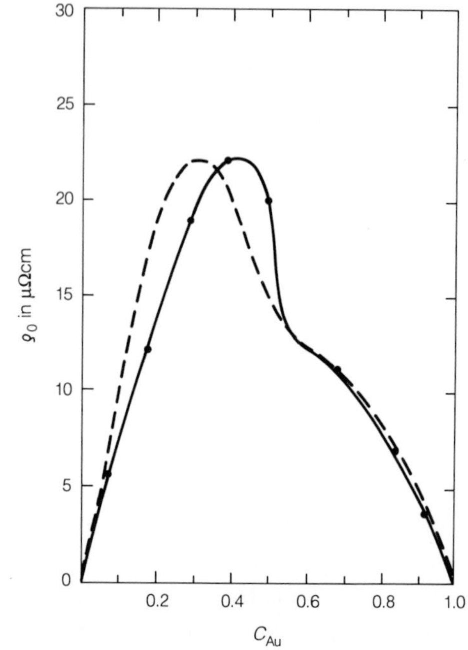

Figure 3-20. Experimental (solid line) and theoretical (dashed line) composition dependence of the residual resistivity of Pd-Au (from Rossiter, 1987; data of Coles and Taylor, 1962).

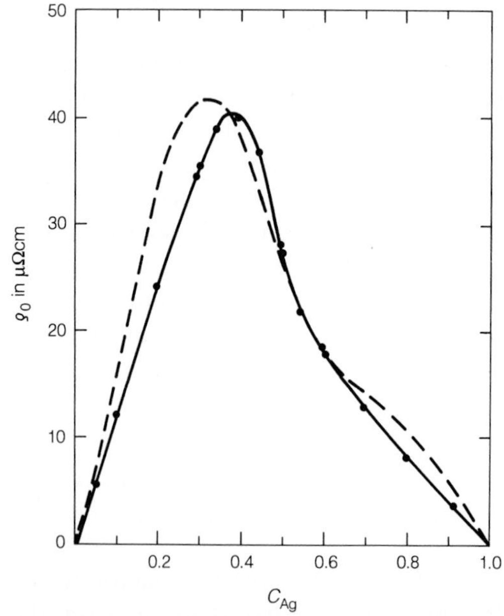

Figure 3-21. Experimental (solid line) and theoretical (dashed line) dependence of the residual resistivity of Pd-Ag (from Rossiter, 1987; data of Kim and Flanagan, 1967).

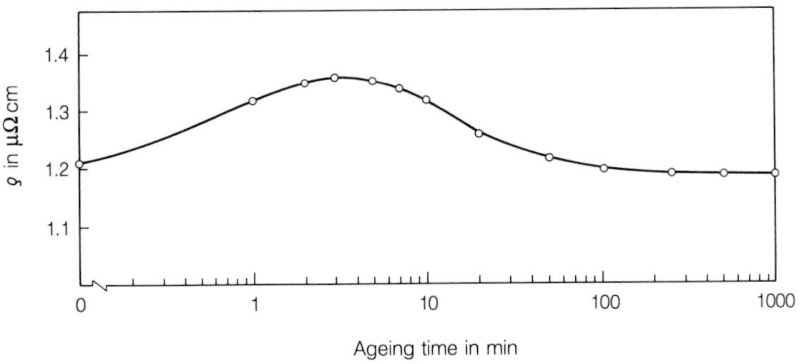

Figure 3-22. Electrical resistivity measured at 4.2 K for Al–4.9 at.% Zn as a function of ageing time (ageing temperature 20 °C) (from Osamura et al., 1973).

The presence of short range atomic ordering or clustering may lead to either an increase or a decrease in the resistivity from the random alloy value (Rossiter and Wells, 1971). However, when the size of the ordered or clustered region becomes comparable with l, any further increase in the degree of atomic correlation (whether clustering or ordering) will cause the resistivity to decrease, since both represent a decrease in the degree of disorder on a scale of l.

This form of behavior is typified in the changes in resistivity that occur during GP zone formation and growth, as shown in Fig. 3-22.

In the case of long range order, Eqs. (3-35) and (3-55) lead to

$$\varrho_0(s) = \varrho_0(0)\,(1 - s^2) \qquad (3\text{-}57)$$

where $\varrho_0(s)$ is the residual resistivity of an alloy with long range order s. However, such ordering may also lead to changes in the electronic band structure which will also appear as an order-dependent change to the temperature coefficient of resistivity. Such an effect can lead to a marked difference between results determined at temperature, and those extrapolated to 0 K, or obtained by quenching, as shown in Fig. 3-23 for example. Furthermore, complete long range ordering ($s = 1$) is only possible at stoichiometric compositions and so one obtains a composition dependence of the residual resistivity of fully ordered specimens as shown in Fig. 3-24.

In the limit of a complete decomposition into two perfectly ordered or pure metal phases, the only remaining disorder will occur at the interfaces. Normally, however, the different phases do not have vanishing resistivities and it is necessary to assume some model of the microstructure. The resistivity (or conductivity) of the composite structure must then lie between upper and

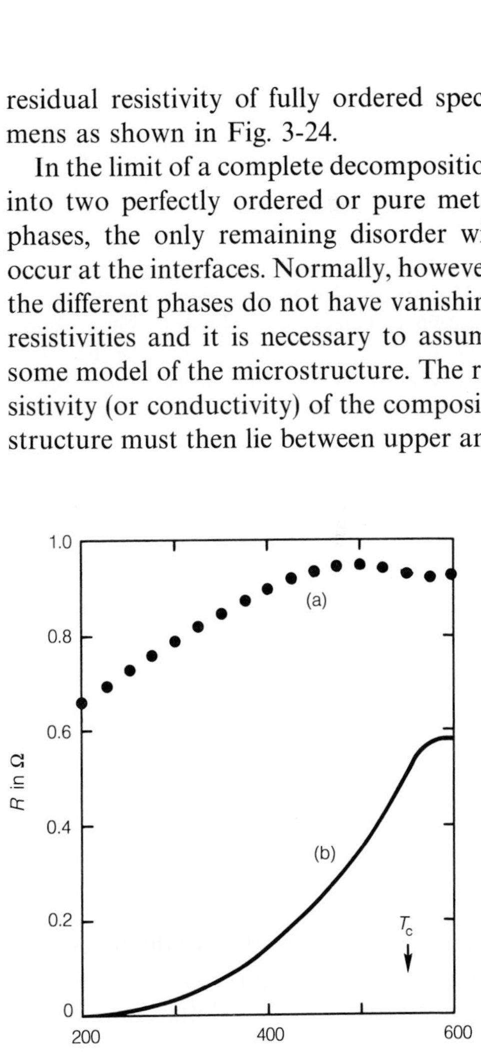

Figure 3-23. The variation of resistance of Fe_3Al as a function of temperature: (a) at temperature; (b) quenched (from Rossiter, 1987; using the data of Cahn and Feder, 1960).

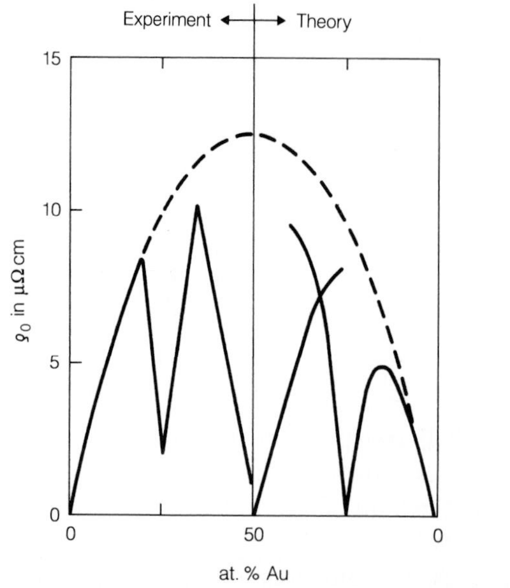

Figure 3-24. Variation of resistivity with composition for long range ordered Cu-Au alloys. The experimental results of Johansson and Linde (1936) are shown in the full curve on the left, whereas the theoretical curve is shown on the right-hand side. The resistivity of a disordered alloy is shown as the dashed curve. The theoretical curve is scaled so that the disordered resistivities are the same at Cu-50% Au. The experimental data has been reduced by 1.7 μΩ cm to give $\varrho = 0$ for pure Cu (from Rossiter, 1987).

lower bounds:

$$\sigma_u = \sigma_2 + \frac{v_1}{\dfrac{1}{\sigma_1 - \sigma_2} + \dfrac{v_2}{3\sigma_2}} \tag{3-58}$$

$$\sigma_l = \sigma_1 + \frac{v_2}{\dfrac{1}{\sigma_2 - \sigma_1} + \dfrac{v_1}{3\sigma_1}} \tag{3-59}$$

where σ_i and v_i are the conductivity and volume fraction of phase i and we have assumed that $\sigma_2 > \sigma_1$. Better predictions are possible if some knowledge about the shape of the embedded phase is available, as in the effective medium theories (see, e.g., Reynolds and Hough, 1957; Landauer, 1978) or if some geometrical model of the microstructure is analysed in terms of series and parallel resistances (Watson et al., 1975). Some results for a two phase mixture of Mg_2Pb–Pb are shown in Fig. 3-25.

Magnetic and Nearly-Magnetic Metals and Alloys

Scattering from disordered magnetic moments in the transition metals is complicated by their decidedly non-free-electron bandstructure and the associated interband scattering effects. The conduction electrons may be classified into two groups having spin directions either parallel (spin "up", ↑) or antiparallel (spin "down", ↓) to the local magnetisation. Elastic scattering of the electrons will not change the popula-

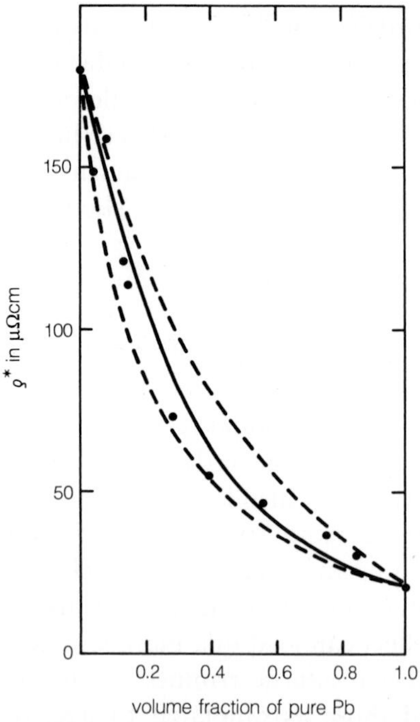

Figure 3-25. Resistivity of two-phase Mg_2Pb-Pb mixture as a function of volume fraction of Pb: dashed lines, upper and lower bounds; solid line, effective medium theory for random spheres; points, experimental data (from Rossiter, 1987; using the data of Hashin and Shtrikman, 1962).

tion of the sub-bands, but inelastic (spin-flip) scattering will result in a transfer of momentum between the two groups of electrons.

In the simplest approximation it is assumed that the conduction band contains nearly free electrons. The current is then carried by the spin-up and spin-down bands in parallel (Fert and Campbell, 1976; Dorleijn, 1976). Solution of the Boltzmann equation for each band then leads to

$$\varrho_{\mathrm{TOT}} = \frac{\varrho_\uparrow \varrho_\downarrow + \varrho_{\uparrow\downarrow}(\varrho_\uparrow + \varrho_\downarrow)}{\varrho_\uparrow + \varrho_\downarrow + 4\varrho_{\uparrow\downarrow}} \tag{3-60}$$

where $\varrho_{\uparrow\downarrow}$ describes the contributions of the spin flip processes and each resistivity $\varrho_\uparrow$, $\varrho_\downarrow$ is given by the sum of residual and phonon scattering contributions:

$$\varrho_{\mathrm{TOT}\uparrow} = \varrho_{0\uparrow} + \varrho_{\mathrm{p}\uparrow}(T) \tag{3-61}$$

and similarly for $\varrho_\downarrow$ (i.e., Matthiessen's rule is assumed valid within each sub-band). Note that the spin-disorder resistivity (see Sec. 3.3.3) is contained in $\varrho_{0\uparrow}$ and $\varrho_{0\downarrow}$ and that different sub-band resistivities will lead to a dependence of ϱ_{TOT} upon the direction of the current in relation to the direction of magnetisation.

The phonon contribution to the resistivities is complicated by bandstructure effects. At low temperatures, the excitations within the spin system are spin waves (magnons), which lead to inelastic scattering. This mechanism introduces a temperature dependent resistivity $\varrho_{\mathrm{p}\uparrow\downarrow}(T)$ that must freeze out as $T \to 0$ K. Other possible contributions to $\varrho_{\uparrow\downarrow}$ include $\uparrow$electron-$\downarrow$electron interactions and spin-orbit interactions, although these are expected to be negligible in all but very dilute alloys. Thus, as $T \to 0$, all temperature dependent parts also go to zero and

$$\varrho_0 = \frac{\varrho_{0\uparrow}\varrho_{0\downarrow}}{\varrho_{0\uparrow} + \varrho_{0\downarrow}} \tag{3-62}$$

Note that even though each of the sub-bands is assumed to obey Matthiessen's rule, the total resistivity does not.

At high temperatures there will be complete spin mixing and

$$\varrho_{\mathrm{TOT}} = \tfrac{1}{4}\left[\varrho_{0\uparrow} + \varrho_{0\downarrow} + \varrho_{\mathrm{p}\uparrow}(T) + \varrho_{\mathrm{p}\downarrow}(T)\right] \tag{3-63}$$

By varying temperature and composition it is possible to separate out the different contributions (Durand and Gautier, 1970; Loegel and Gautier, 1971; Fert and Campbell, 1976). Some results for Ni are shown in Figs. 3-26, 3-27 and Table 3-7.

The variation of the spin-disorder resistivity in each sub band $\varrho_{0\uparrow}$ and $\varrho_{0\downarrow}$ may be written in terms of the magnetic long range order parameter [Eq. (3-39)] as

$$\varrho_{0\uparrow} = \varrho_{\infty\uparrow}(1 - M) \tag{3-64}$$

where $\varrho_{\infty\uparrow} = Const_\uparrow\, S(S+1)$.

By analogy with the Bragg-Williams theory, a rough idea of the variation of M

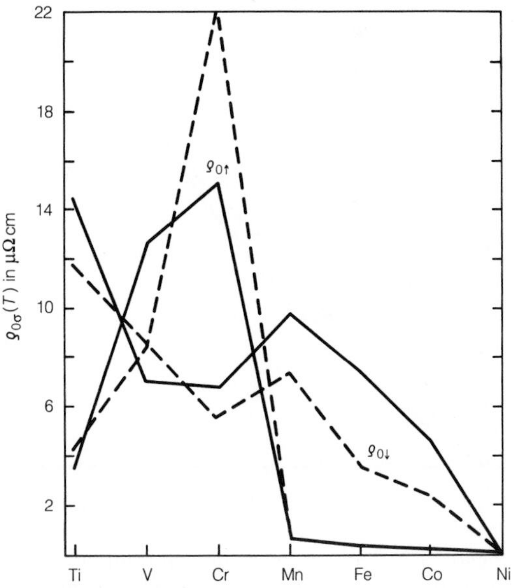

Figure 3-26. The residual resistivity of 3d impurities in Ni for each spin direction: solid lines, Fert and Campbell (1976); dashed lines, Dorleijn and Miedema (1975) (from Rossiter, 1987).

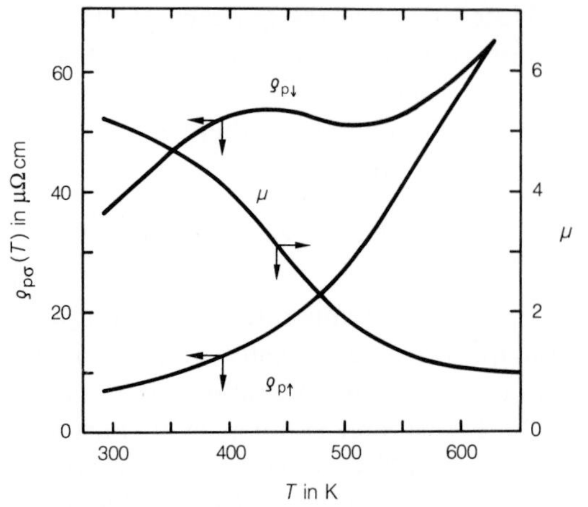

Figure 3-27. Temperature dependence of the spin-up and spin-down resistivity and the ratio $\mu = \varrho_{p\downarrow}(T)/\varrho_{p\uparrow}(T)$ for pure Ni (from Rossiter, 1987; using the data of Kaul, 1977).

Table 3-7. Values of the parameters used to interpret the sub-band resistivities of Ni-based alloys (Fert and Campbell, 1976).

Temperature K	$\varrho_{\uparrow\downarrow}(T)$ μΩ cm	$\varrho_0(T)$ μΩ cm	$\varrho_{p\uparrow}(T)$ μΩ cm	$\varrho_{p\downarrow}(T)$ μΩ cm
77	0.9±0.3	0.32	0.38	1.9
200	5.0±2.0	2.5	3.0	15.0
300	11.0±4.0	5.4	6.7	27.0

with temperature may be derived from the equivalent (single-site) molecular field approximation resulting in the variation of ϱ_0 with T for a ferromagnet of the form shown in Fig. 3-28.

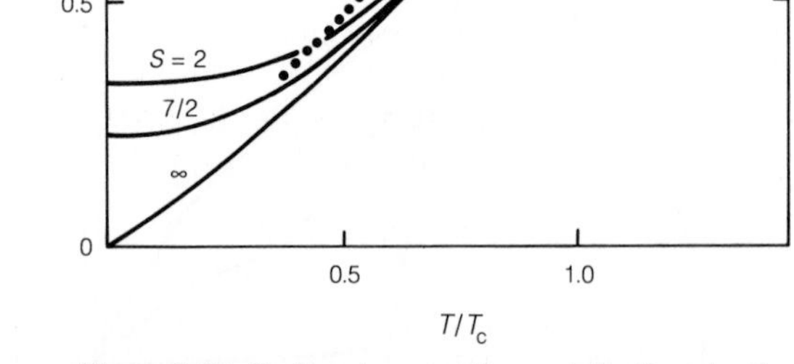

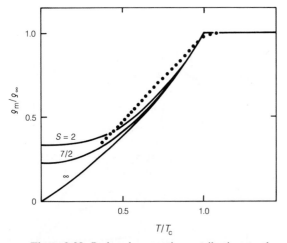

Figure 3-28. Reduced magnetic contribution to the residual resistivity as a function of reduced temperature T/T_c given by the molecular field approximation for $S = 2$, $S = 7/2$ and $S = \infty$. The points are experimental data for Gd corrected for the phonon scattering contribution (from de Gennes and Friedel, 1958).

As in the case of long range atomic ordering (see Fig. 3-23), long range antiferromagnetic ordering may introduce new energy gaps at superlattice Brillouin zone boundaries which will modify this behavior and possibly result in an increase in ϱ_{TOT} just below T_N (see, e.g., Miwa, 1962, 1963; Elliott and Wedgwood, 1963; Ausloos, 1976). Experimental data given in Meaden (1965) and Schröder (1983) indicate a rich variation in the types of behaviors observed. Probably the safest generalisation that can be made is that there will usually be a change in gradient of ϱ_{TOT} at the magnetic transition temperature. Again by analogy with the atomic case, short range magnetic correlations are expected to play an important role in the vicinity of T_c or T_N (Rossiter and Wells, 1971). The corresponding expression for resistivity is then Eq. (3-55) with α_i replaced with m_i [see Eq. (3-40)] and $w_A(\boldsymbol{q}) - w_B(\boldsymbol{q}) = const$ (assuming a delta function scattering potential).

Adding impurities to a magnetic host will lead to virtual bound state behavior for each sub-band, as shown in Fig. 3-26 for example. In more concentrated alloys the situation becomes quite complex since it is likely that all contributions to ϱ_{TOT} are interdependent. Theories which attempt to explain this behavior are quite complex, as they must necessarily take into account the bandstructures of the alloys under consideration. Some results have been obtained with the coherent potential approximation for Ni-Fe (Akai, 1977), Ni-Cu (Brouers et al., 1973), Ni-Co (Muth and Christoph, 1981), Ni-Mn (Muth, 1983a), Ni-Cr, Ni-Mo, Ni-W, Ni-Ru, Ni-V (Muth, 1983b).

In a ferromagnetic material just above T_c, or a nearly magnetic metal at low temperatures, there will be fluctuations in which the spins within some small region are aligned for a time longer than the characteristic conduction electron relaxation time τ ($\sim 10^{-14}$ s). Such localised regions of magnetic alignment may also be associated with strongly magnetic impurities and composition fluctuations (or precipitates) in nearly magnetic alloys. The calculation of the resistivity for a magnetic cluster generally follows that for atomic clusters. However, here the degree of magnetic order within the cluster and the cluster size may both change with temperature, leading to a wide range of possible behaviors.

For example, Pd is nearly magnetic and scattering of conduction electrons by fluctuations into the ferromagnetic state at low temperatures are thought to contribute to the observed strong T^2-dependence in the resistivity. If a magnetic impurity such as Ni is added, it enhances these fluctuations leading to a large increase in the coefficient of the T^2-term. A different behavior is observed with Cr impurities in Pd or Fe in Cu, for example (one of the celebrated Kondo systems). Here the resistivity is observed to decrease with increasing temperature, pass through a minimum, and then increase as the phonon scattering becomes dominant (see, e.g., Fig. 3-29). Whether the resistivity initially increases or decreases with increasing temperature depends upon the electronic structures of the host and impurity (Rivier and Zitkova, 1971). A summary of the type of behavior is given in Table 3-8.

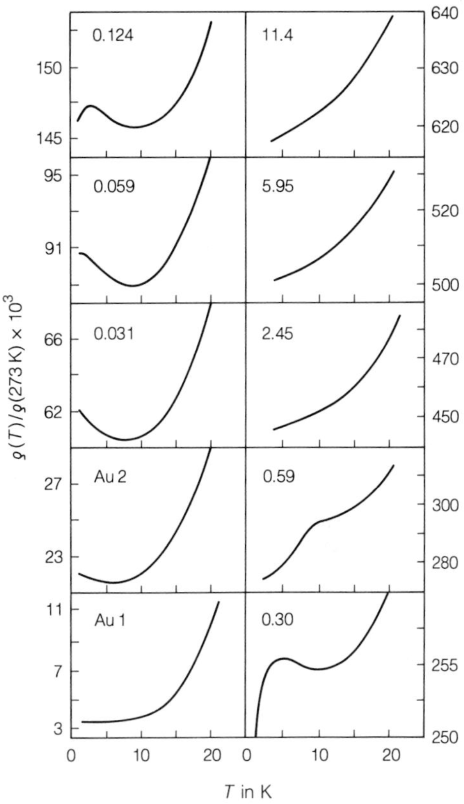

Figure 3-29. Reduced resistivities of Au and Au-Mn alloys as a function of temperature below 20 K. The numbers on each graph give the atomic percentage of Mn solute (from Meaden, 1965).

Competing ferromagnetic and antiferromagnetic interactions can give rise to an amorphous spin structure – a spin glass. At low temperatures a $T^{3/2}$-behavior has been attributed to scattering from long wavelength magnon modes (Rivier and Adkins, 1975), damped ferromagnetic modes (Fischer, 1979), and discrete excitations (Campbell et al., 1982). Such behavior is shown in Fig. 3-30. However a T^2-behavior is also sometimes observed, supposedly due to random spin reversals at higher temperatures. In many such systems spin fluctuation effects are also important, leading to low temperature resistivity minima and maxima (Schilling et al., 1976).

Table 3-8. Nature of impurity states in a number of alloy systems. "Anderson" solutes produce resistivity minima while "Wolff" solutes exhibit T^2-behavior (Rossiter, 1987).

Solvent	Solute	
	Anderson	Wolff
Al	Cr	
	Mn	
Zn	Mn	
	Fe	
Cu	Ti	
	V?	
	Cr	
	Mn	
	Fe	
	Co	
	Ni (concentrated)	
Rh	Cr	Mn
		Fe
		Co
		Ni
Pd	V	Mn?
	Cr	Fe
	Mo	Co
	Ru	Ni
	Rh	
	Ag (concentrated)	
	Pt (concentrated)	
	Au (concentrated)	
	U	
	Np	
Au	Ti	
	V	
	Cr	
	Mn	
	Fe	
	Co	
	Ni	

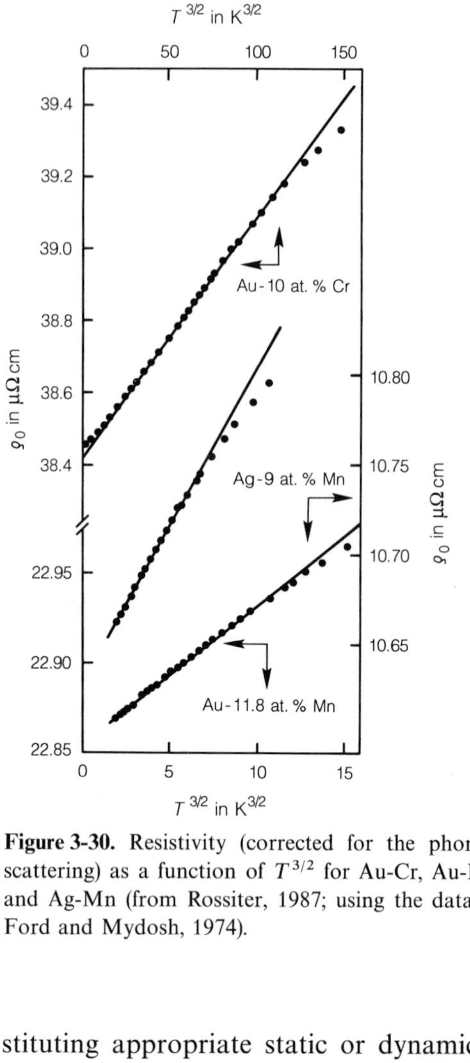

Figure 3-30. Resistivity (corrected for the phonon scattering) as a function of $T^{3/2}$ for Au-Cr, Au-Mn and Ag-Mn (from Rossiter, 1987; using the data of Ford and Mydosh, 1974).

3.5.3 Additional Examples: Amorphous Alloys

Amorphous alloys exhibit a wide range of behaviors (see, e.g., Mizutani, 1983; Naugle, 1984). Some experimental results are shown in Figs. 3-31 and 3-32.

In many nearly free electron materials, the resistivity may be evaluated by substituting appropriate static or dynamical structure factors into Eq. (3-44) and then using Eq. (3-42). However, this diffraction approach is not valid if $l \sim$ lattice spacing since the electron wavepackets and associated wavevectors $\boldsymbol{k}$ are no longer clearly defined and one would expect a localised (rather than extended) state model to be more realistic. The transition should occur where $\varrho_{\text{TOT}} \approx 100-160\,\mu\Omega$ cm. (The same statement applies to high resistance crystalline materials, see, e.g., Rossiter, 1987.) Nevertheless, where the model may be applied it can explain the observed positive and negative temperature coefficients of re-

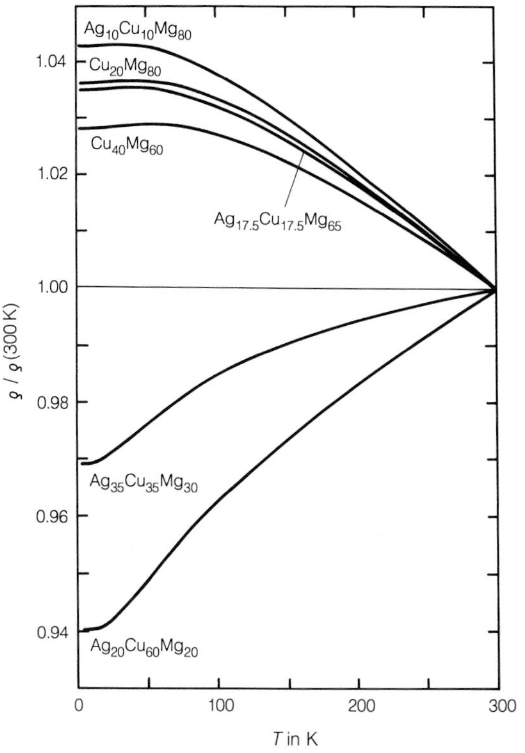

Figure 3-31. Temperature dependence of the electrical resistivity of amorphous $(Ag_{0.5}Cu_{0.5})_{100-x}$ Mg_x alloys (from Rossiter, 1987; using the data of Mizutani and Yoshino, 1984).

sistivity. In particular, Fig. 3-33 gives a schematic representation of the structure factor at two different temperatures.

The resistivity given by Eq. (3-42), is heavily weighted by the q^3 term in favour of the $q = 2k_F$ region. Thus, for systems with $2k_F$ (1) indicated on the diagram, the resistivity will increase with increasing temperature, whereas with $2k_F$ (2) a negative $d\varrho/dT$ will result. From this simple argument it is expected that the behavior is determined by the position of $2k_F$ in relation to the peak in the structure fracture at q_p. Such a correlation is in fact observed, as illustrated in Fig. 3-34.

At temperatures well below Θ_D two competing effects appear. The phonon scattering goes as $+T^2$ but the static structure factor is modified by a Debye-Waller factor: $\exp(-2M)$ (see, e.g., Meisel and Cote, 1983), the importance of which depends upon the size of ϱ_0. At low temperatures $M \propto T^2$ and so there is a direct competition between the $+T^2$ phonon term and $-T^2$-Debye-Waller term. If ϱ_0 is

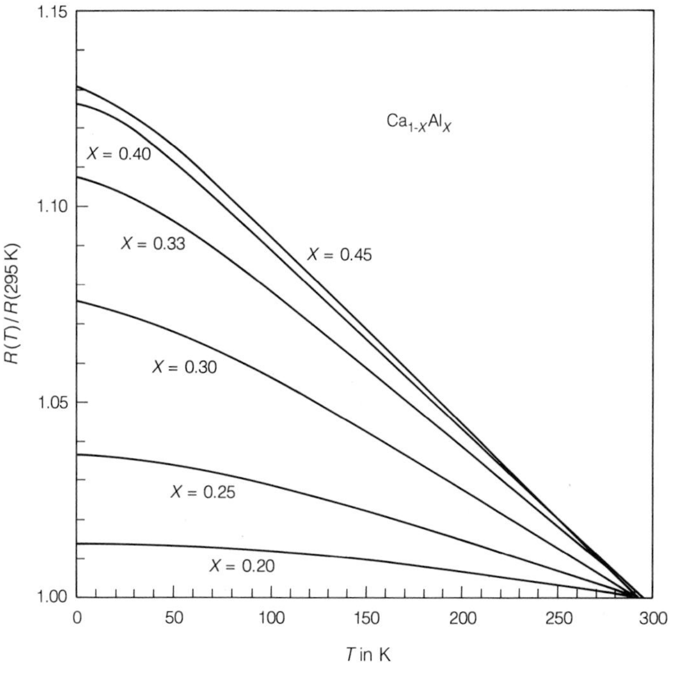

Figure 3-32. Temperature dependence of the resistance for $Ca_{1-x}Al_x$ metallic glass (from Naugle, 1984).

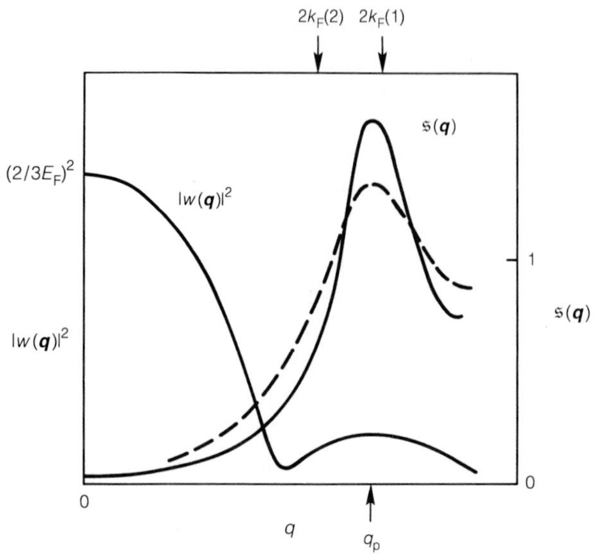

Figure 3-33. Schematic representation of the factors $|w(\boldsymbol{q})|^2$ and $\mathfrak{s}(\boldsymbol{q})$ that appear in the integrand of the resistivity integral. The structure factor $\mathfrak{s}(\boldsymbol{q})$ is represented by the full line at the temperature T_1, and the dashed line at $T_2 > T_1$ (from Rossiter, 1987).

small, the phonon term is dominant and gives an overall $+T^2$ dependence, but if ϱ_0 is large a $-T^2$ dependence will prevail.

The resistivity of Iron and Nickel-based ferromagnetic amorphous alloys tends to vary as $T^{3/2}$ at temperatures less than $\sim T_c/2$ and then as T^2 up to T_c. The former results from spin wave scattering, which breaks down into random spin reversals at higher temperatures leading to the T^2 variation. Some typical results are shown in Fig. 3-35.

Finally, many amorphous metals exhibit resistivity minima at low temperatures, as shown in Fig. 3-36. This behavior does not appear to depend upon whether the material is paramagnetic or ferromagnetic and behavior is insensitive to composition suggesting that it is a fundamental property of the disordered lattice rather than some specific fluctuation effect. The reader is referred to Cochrane and Strom-Olsen (1977) and Thomas (1984) for further discussion of the effect.

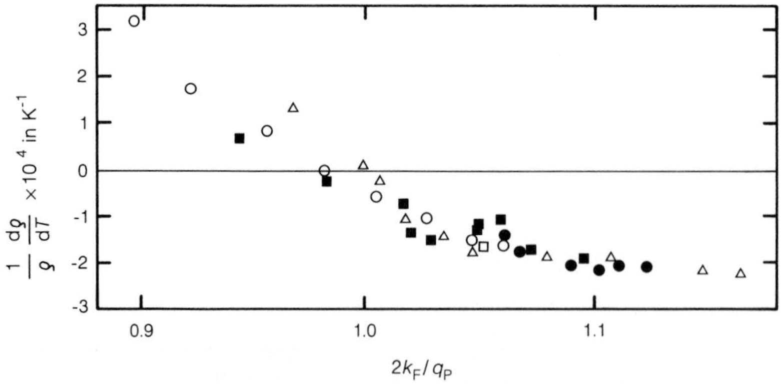

Figure 3-34. $\mathrm{d}\varrho/\mathrm{d}T$ as a function of $2k_F/q_p$ for a number of simple metallic glasses: ○, Ag-Cu-Mg; △, Ag-Cu-Al; ■, Ag-Cu-Ge; ●, Mg-Zn; □, Mg-Cu (from Rossiter, 1987; using the data of Mizutani, 1983).

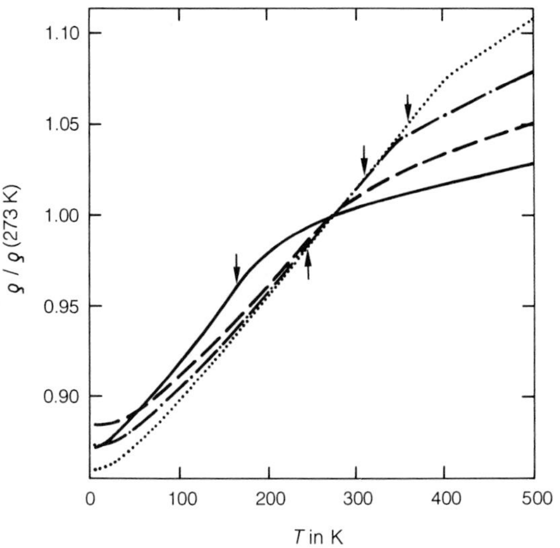

Figure 3-35. The temperature dependence of the resistivity of some $Fe_xNi_{80-x}B_{19}Si$ metallic glasses: solid line, $x=10$; dashed line, $x=13$; dot-dash line, $x=16$; dotted line, $x=20$. The Curie points are marked by an arrow (from Rossiter, 1987; using the data of Bohnke and Rosenberg, 1980).

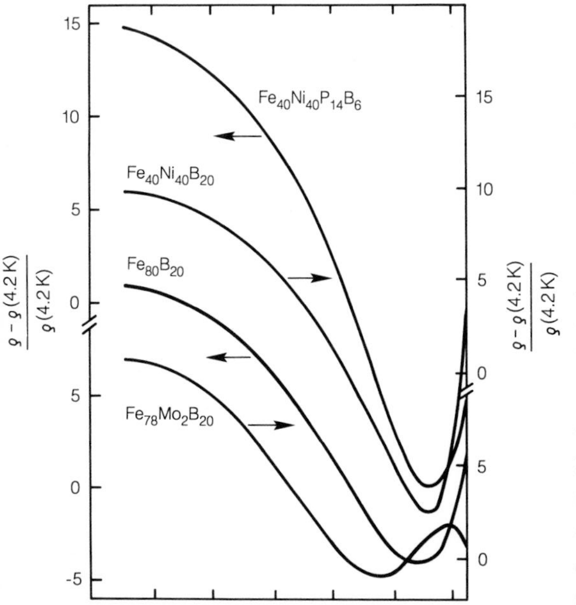

Figure 3-36. Normalised resistivity $[\varrho(T)-\varrho(4.2\,K)]/\varrho(4.2\,K)$ as a function of temperature. The lower three curves have been offset vertically for clarity (from Rossiter, 1987; using the data of Cochrane, 1978).

3.6 Magnetoresistance (*MR*): Transverse (*T*-) and Longitudinal (*L*-)

There are two separate magnetoresistances (*MR*): Transverse (*T-MR*), where the magnetic field is applied perpendicular to the current direction; and Longitudinal (*L-MR*), where the field is applied parallel to the current.

3.6.1 Kohler's Rule

If the electrons on the spherical Fermi surface of a free electron metal are all scattered with a single relaxation time τ, then both the *T-MR* and *L-MR* are zero as already noted in Sec. 3.2.3. This occurs because the average force $e\,E_y$ produced on any given electron by the Hall Voltage V_y exactly cancels the Lorenz Force $e\langle v_x\rangle B_z$ acting upon the same electron. Here B_z is the applied magnetic field pointing in the z-direction and $\langle v_x\rangle$ is the electron's average drift velocity in the x-direction. If the electrons occupy a more complex FS, with different values of the effective mass m^* and different relaxation times for electrons on different parts of the FS, then the single Hall field cannot simultaneously cancel the different Lorenz forces felt by different electrons, and non-zero *T-MR* and *L-MR* both appear.

If the electron scattering on such a more complex Fermi surface can still be approximated by a single relaxation time the *MR* should be a function only of the unitless quantity $\omega_c\tau$ defined in Sec. 3.4, and the data for a given metal should fall on a single curve when the *MR* is plotted as a function of $\omega_c\tau$. Such behavior is called Kohler's rule and is found to be experimentally valid under somewhat more general conditions than just described. Figure 3-37 shows how Kohler's rule applies to the

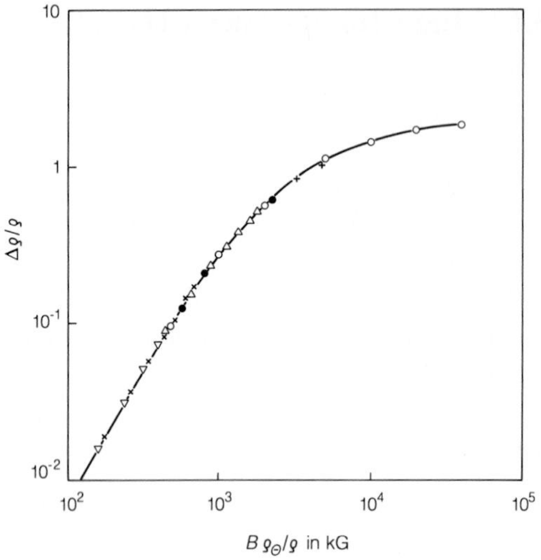

Figure 3-37. Magnetoresistance in polycrystalline indium in a transverse magnetic field showing agreement with Kohler's rule. Some points affected by boundary scattering have been suppressed. ▽, In 1 : 14 K, ϱ/ϱ (273 K) = 0.024; ×, In 6 : 20 K, ϱ/ϱ (273 K) = 0.023; △, In 6 : 14 K, ϱ/ϱ (273 K) = 0.0086; +, In 6 : 4.2 K, ϱ/ϱ (273 K) = 0.0012; ○, In 2 : 4.2 K, ϱ/ϱ (273 K) = 0.00007; ●, In 2 : 2 K, ϱ/ϱ (273 K) = 0.00003 (from Olsen, 1962).

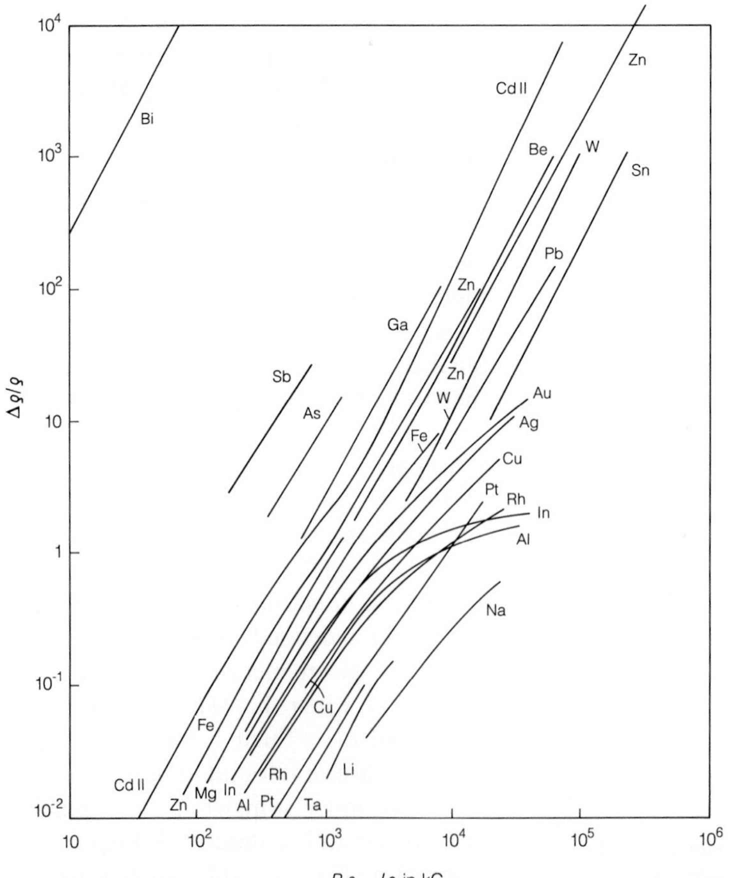

Figure 3-38. The transverse magnetoresistance of polycrystalline metals in the reduced Kohler diagram. $\Delta\varrho/\varrho$ is plotted against $\boldsymbol{B}\varrho_{\Theta_D}/\varrho$ where ϱ is the resistivity in zero field, and ϱ_{Θ_D} that at $T = \Theta_D$ in zero field (from Olsen, 1962).

T-MR of polycrystalline samples of In at temperatures ranging from 2 K to 20 K. Similar plots of the *T-MR* and the *L-MR* of a variety of polycrystalline metals are shown in Figs. 3-38 and 3-39 respectively. The abscissa scale in these figures is proportional to $\omega_c \tau$.

3.6.2 Low Magnetic Fields ($\omega_c \tau \ll 1$)

At very low fields, the general theory of *MR* predicts a B^2-increase of the *MR* independent of the Fermi surface topology and whether the sample is a single crystal or polycrystalline. This form automatically satisfies the Onsager relation requirement that the *MR* must be an even function of ***B***. Examples of approximately B^2-variations are shown in the low field regions of Figs. 3-37 and 3-38.

3.6.3 High Magnetic Fields ($\omega_c \tau \gg 1$)

In single crystal samples, the high field *L-MR* becomes independent of *B* ("saturates") for all FS topologies and for both compensated ($N_e = N_h$) and uncompensated ($N_e \neq N_h$) metals. The *T-MR*, in contrast, can either become independent of *B* or vary as B^2, depending upon the topology of the FS and the orientation of ***B*** with respect to the crystallographic axes; the permitted alternatives are listed in Table 3-1.

For an uncompensated metal with no open orbits (see Sec. 3.4.3), the *T-MR* "saturates" for all directions of ***B*** when $\omega_c \tau \gg 1$. If the Fermi surface contains open orbits, then the *T-MR* saturates when there are no open orbits in the plane perpendicular to ***B***, increases as B^2 when there is one band of open orbits in the plane

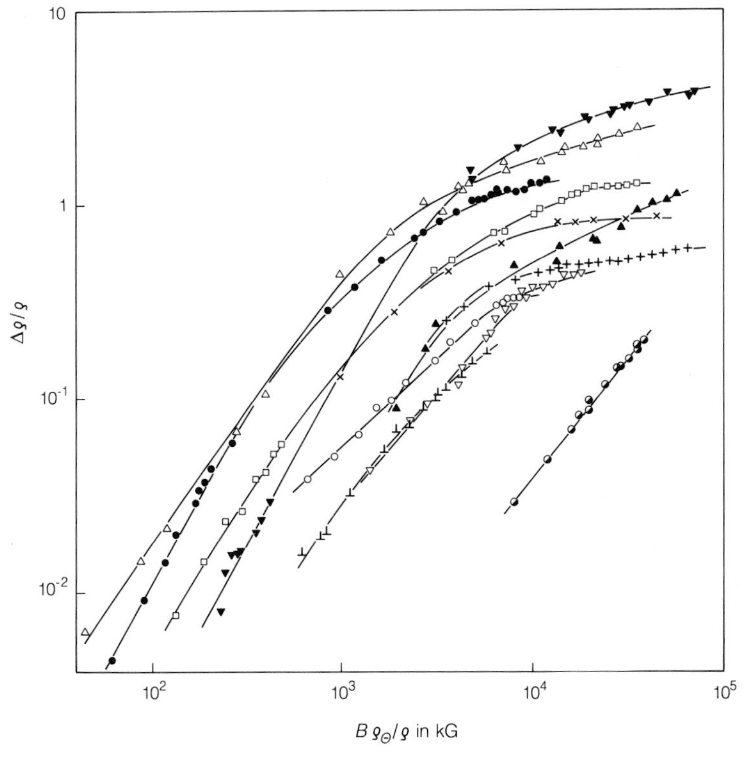

Figure 3-39. The magnetoresistance of polycrystalline metals in longitudinal fields. ϱ is the zero field resistivity at the temperature of observation, and ϱ_Θ that at $T = \Theta_D$. The measurements were made in fields up to 200,000 gauss. ◑ Li, $\Theta_D = 430$ K; + Al, $\Theta_D = 410$ K; ▼ Sn, $\Theta_D = 160$ K; ○ Ni, $\Theta_D = 410$ K; □ Ag, $\Theta_D = 220$ K; × In, $\Theta_D = 100$ K; ▲ Pb, $\Theta_D = 90$ K; ▽ Pt, $\Theta_D = 240$ K; ● Au, $\Theta_D = 320$ K; △ Zn, $\Theta_D = 240$ K; ⊥ Fe, $\Theta_D = 355$ K (from Olsen, 1962).

perpendicular to $\boldsymbol{B}$, and saturates again when there are two perpendicular open orbits in the plane perpendicular to $\boldsymbol{B}$. Saturation and B^2-variations for different field directions are illustrated in Fig. 3-40a for the uncompensated intermetallic compound $AuGa_2$, which can be produced in a highly ordered state that permits the high field condition to be reached. The specific dependence upon field direction is shown in Fig. 3-40b.

For a compensated metal with no open orbits or one band of open orbits, the *T-MR* is proportional to B^2 independent of field orientation, except that the *T-MR* saturates when the open orbit direction in $\boldsymbol{k}$-space is exactly the direction of current flow in the sample. The B^2 behavior for compensated metals is illustrated in Fig. 3-41 for Pb, Cd and W.

Finally, in polycrystalline samples that contain open orbits, an approximately linear *T-MR* may result from averaging of B^0 and B^2 behaviors on different orbits. Figure 3-42 shows that the *T-MR* for polycrystalline Au varies approximately linearly with B at high fields.

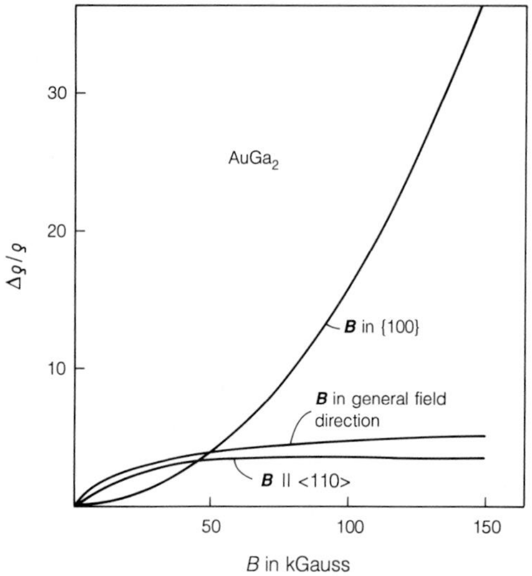

Figure 3-40 (a). $\Delta\varrho/\varrho$ of $AuGa_2$ versus B for various directions of $\boldsymbol{B}$ (from Longo et al., 1969).

3.6.4 Additional Examples

Deviations from the Ideal: The analyses just given are rigorous for a completely

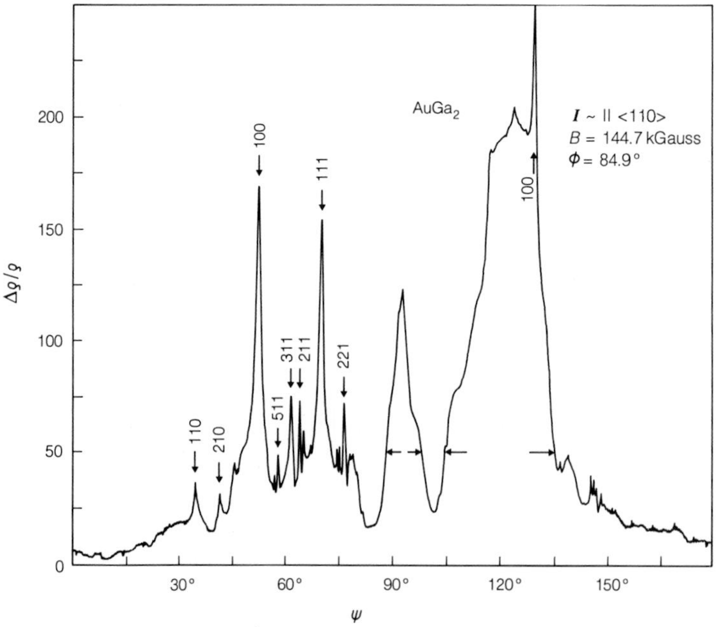

Figure 3-40 (b). $\Delta\varrho/\varrho$ versus magnetic field direction ψ for a $\langle 110 \rangle$ oriented crystal with $\varrho(300\,K)/\varrho(4\,K) = 904$ (from Longo et al., 1969).

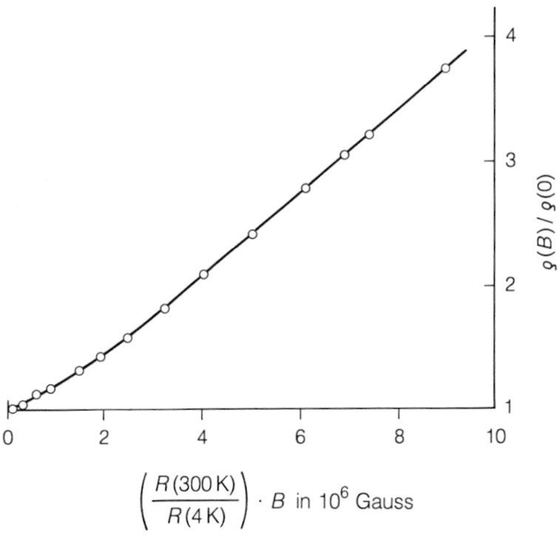

Figure 3-42. Magnetoresistance of Au (from Chambers, 1956).

homogeneous sample in which the current density is uniform across the sample area and the magnetic field is uniform along the sample length. If the current density is perturbed, for example by defects in the sample or on its surface, or the magnetic field is not uniform, the *T-MR* can become linear in *B*. Such behavior is illustrated in Fig. 3-43 for bulk and surface defects.

Quantum Oscillations: At low temperatures and high magnetic fields, quantum oscillations (called Shubnikov-de Haas oscillations) are seen in single crystals of high purity metals, as illustrated in Fig. 3-44. The periods of these oscillations are determined by the cross-sectional areas of the Fermi surface perpendicular to the direction of ***B***.

Size-Effects: If a transverse ***B*** is directed in the plane of a thin, high purity foil,

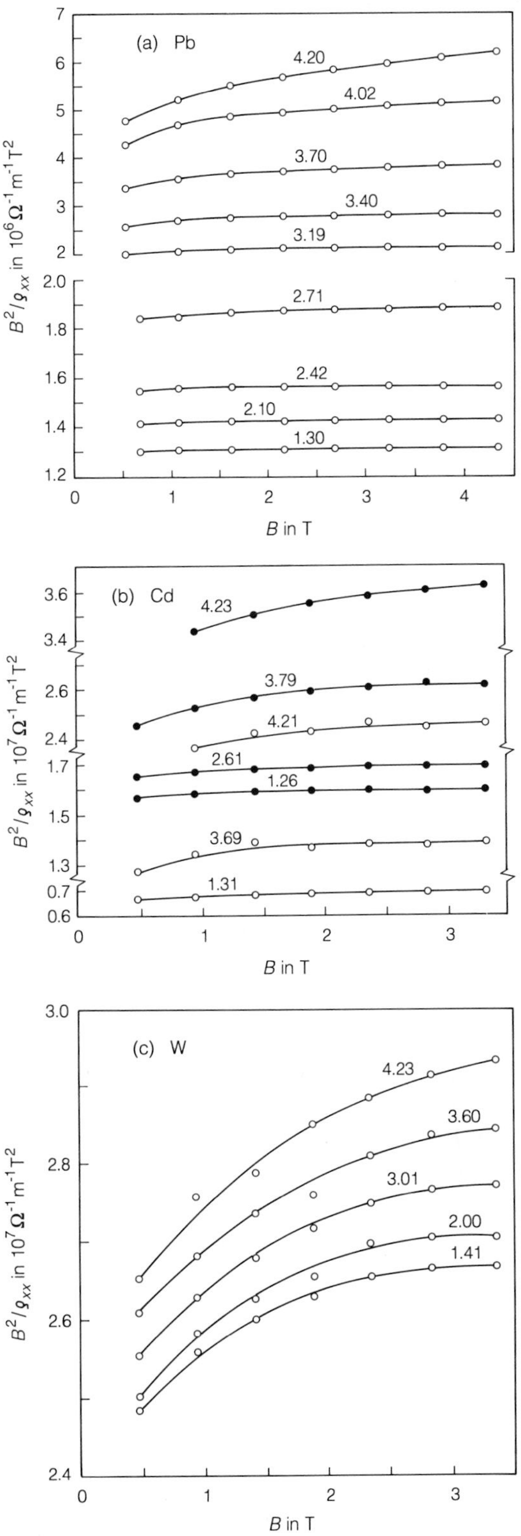

Figure 3-41. The quantity B^2/ϱ_{xx} plotted as a function of magnetic field ***B*** for Pb (a), Cd (b) and W (c) at various temperatures. There are two sets of data for Cd corresponding to CdI (open circles) and CdII (closed circles). Notice the scale change in (a) and the scale breaks in (b) (from Fletcher, 1977).

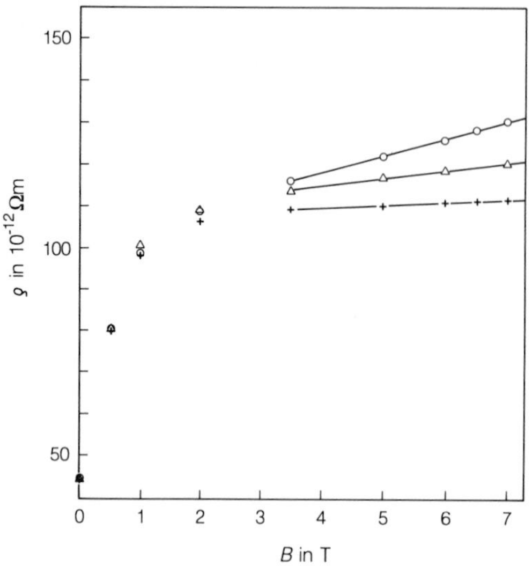

Figure 3-43 (a). Resistivity ϱ as a function of the magnetic field $\boldsymbol{B}$ for a sample without macroscopic voids (crosses), the same sample with one cylindrical void every millimetre (circles), and a sample with one cylindrical void every two millimetres (triangles) (from Beers et al., 1978).

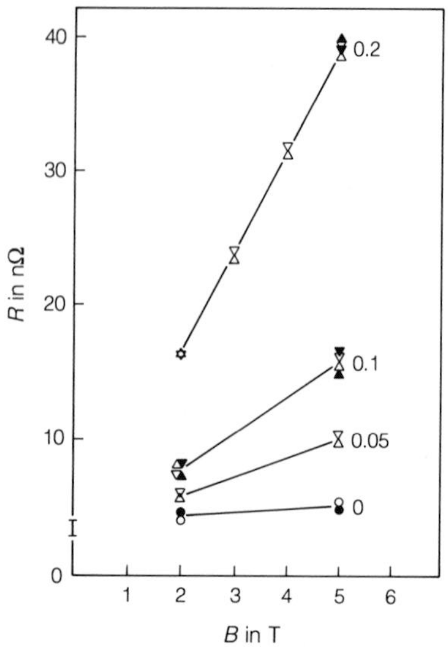

Figure 3-43 (b). Measurement of resistance as a function of field for three grooves of different depth (Δ) and two flat regions on an annealed 4 mm wide sample. The numbers shown are values of Δ in mm. Upright and inverted triangles indicate results for opposite-field directions. Solid and open symbols indicate measurements on opposite sides of the sample. The circles, which describe data for the flat regions, represent averages after field reversal. The two flat pieces gave results too close together to separate on this graph (from Bruls et al., 1985).

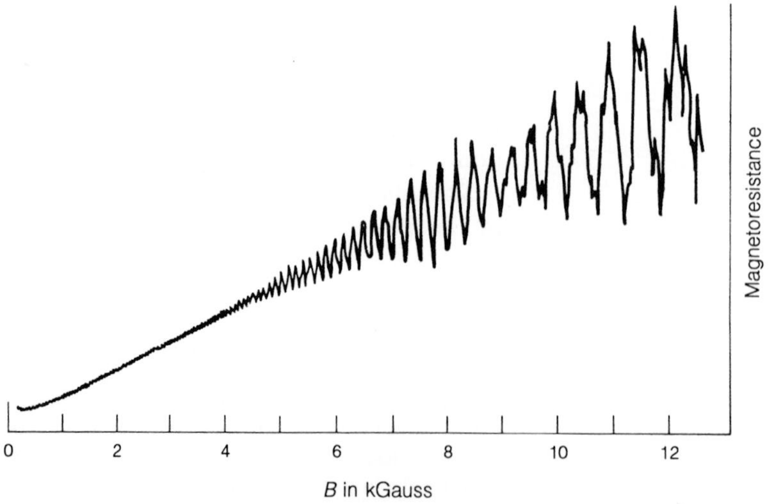

Figure 3-44. Shubnikov-de Haas oscillations in the magnetoresistance of gallium vs. field at 1.3 K (from Ashcroft and Mermin, 1976).

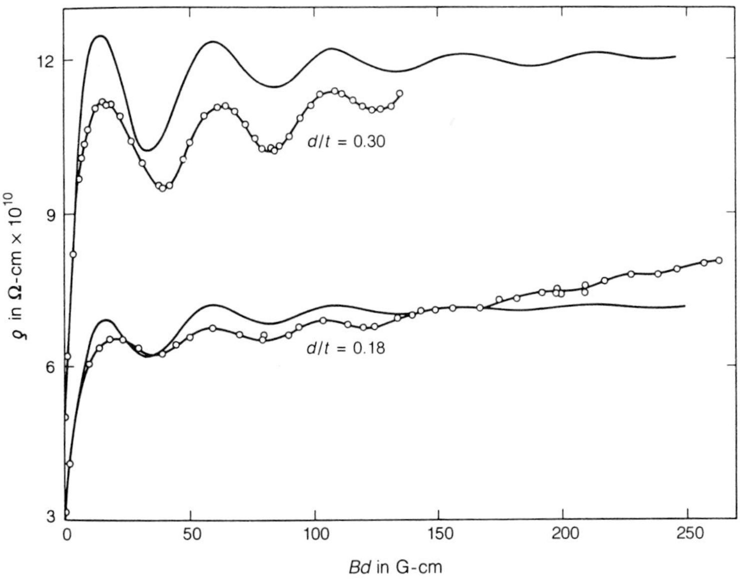

Figure 3-45. Sondheimer oscillations in the resistivity of thin aluminium films. The abscissa is normalized – the values are ***B****d* in G-cm, where *d* is the film thickness. The ratio *d*/*t* for the two film samples is shown in the figure. The solid curves are calculated using Sondheimer's theory and assuming an average electron Fermi momentum of 1.33×10^{-19} g-cm/sec (from Blatt, 1968; using the data of Forsvoll and Holwech, 1964).

Sondheimer oscillations appear in the *T-MR* as illustrated in Fig. 3-45. The period of these oscillations is determined by the average Fermi momentum of the electrons. If ***B*** points along the axis of a thin wire, the resulting *L-MR* first increases slightly with increasing ***B*** and then decreases as illustrated in Fig. 3-46. The decrease begins when the electron's cyclotron orbit becomes comparable to the wire radius; thereafter, increasing the field reduces the number of electrons that reach the wire surface, and thus reduces the contribution of surface scattering to ϱ.

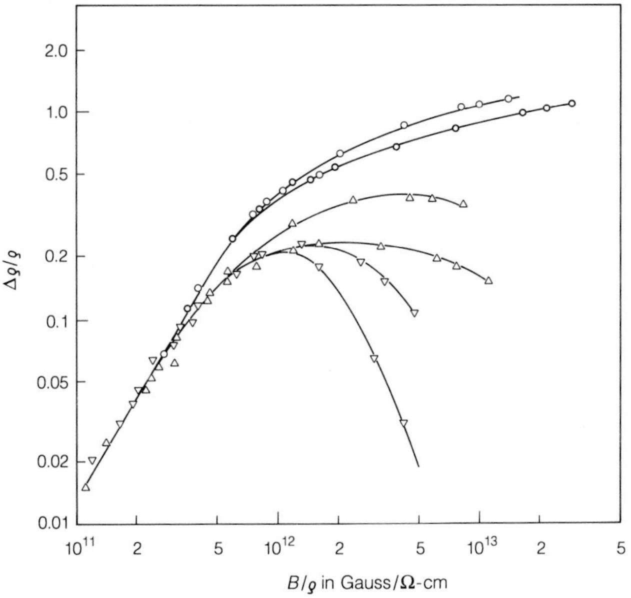

Figure 3-46. Longitudinal magnetoresistance of thin indium wires. ○ In 1, $d = 1.48$ mm; △ In 3, $d = 0.282$ mm; ▽ In 5, $d = 0.094$ mm (from Blatt, 1968; using data of Wyder, 1965).

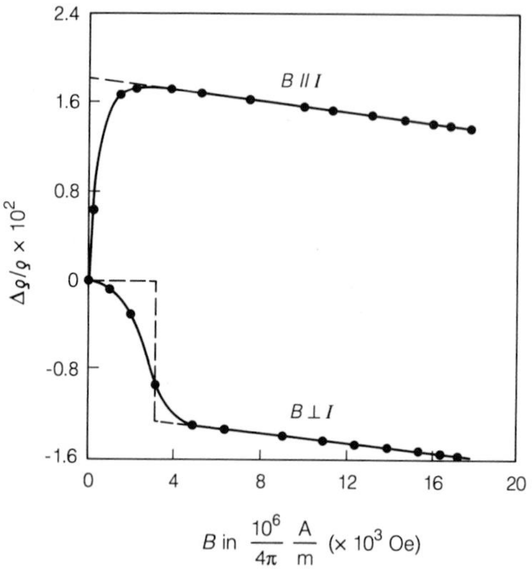

Figure 3-47. Variation of Ni resistivity as a function of a magnetic field (from Chikazumi, 1964; using the data of Englert, 1932).

Ferromagnets: The *MR* of ferromagnets can be complex and anomalous. For example, Fig. 3-47 shows that the initial values of the *T-MR* for Ni are negative, and those of the *L-MR* are positive. At fields high enough to align all of the magnetic domains parallel to the field, both the *T-MR* and *L-MR* decrease with increasing field, probably due primarily to magnetostrictive effects (see, e.g., Chikazumi, 1964).

3.7 Hall Effect, R_H

3.7.1 Free Electrons

The free-electron model of a metal provides the simplest picture of the physical phenomenon underlying the Hall effect. The current density j_x in a foil of width W is given by $n_e e \langle v_x \rangle$. When a magnetic field B_z is applied in the negative z-direction, it produces on the electron stream a force $e \langle v_x \rangle B_z$, which is directed in the y-direction and causes the electrons to try to flow also in this direction. Since the sample has finite width, the electrons can only pile up on the edge of the foil, thereby generating a charge separation in the foil that leads to an electric field E_y. The electronic system comes into dynamic equilibrium when this field produces a force just large enough to balance the magnetic force, i.e.

$$e E_y = e \langle v_x \rangle B_z \tag{3-65}$$

Substituting $j_x = n_e e \langle v_x \rangle$ into the right hand side of Eq. (3-65) and defining $R_H = E_y/(J_x B_z)$ yields

$$R_H = 1/n_e e \tag{3-66}$$

as already derived under similar assumptions with the Boltzmann transport equation in Sec. 3.2.3. Equation (3-66) predicts R_H to be directly proportional to the electron charge (i.e., strictly negative) and inversely proportional to the electron density.

3.7.2 Beyond Free Electrons

Table 3-9a lists values of R_H at room temperature for several metals and semimetals, and Table 3-9b lists approximate normalized values of R_H at low temperatures and high magnetic fields for several uncompensated metals (i.e. $N_e \neq N_h$). Equation (3-66) gives the correct order of magnitude of R_H for all of these metals, but rarely gives precisely the correct value, and sometimes gives even the wrong sign. These deviations from Eq. (3-66) arise because different sets of electrons in a metal have different effective masses and are scattered in very different ways, neither of which was taken into account in the free electron model.

The Boltzmann equation analysis given in Sec. 3.2 can be generalized to allow for different effective masses on different parts of the Fermi surface. It is useful to consider this expression in two extreme limits: the

Table 3-9 a. Comparison of observed Hall constants with those calculated on free electron theory (Hurd, 1972).

Metal	Method	Experimental R_H in 10^{-24} CGS units	Assumed carriers per atom	Calculated $-1/nec$ in 10^{-24} CGS units
Li	conv.	−1.89	1 electron	−1.48
Na	helicon	−2.619	1 electron	−2.603
	conv.	−2.3		
K	helicon	−4.946	1 electron	−4.944
	conv.	−4.7		
Rb	conv.	−5.6	1 electron	−6.04
Cu	conv.	−0.6	1 electron	−0.82
Ag	conv.	−1.0	1 electron	−1.19
Au	conv.	−0.8	1 electron	−1.18
Be	conv.	+2.7	–	–
Mg	conv.	−0.92	–	–
Al	helicon	+1.136	1 hole	+1.135
	conv.	−0.43		
In	helicon	+1.774	1 hole	+1.780
As	conv.	+50.0	–	–
Sb	conv.	−22.0	–	–
Bi	conv.	−6000.0	–	–

Table 3-9 b. Hall coefficients of selected elements in moderate high fields (Ashcroft and Mermin, 1976).

Metal	Valence	$-1/R_H nec$
Li	1	0.8
Na	1	1.2
K	1	1.1
Rb	1	1.0
Cs	1	0.9
Cu	1	1.5
Ag	1	1.3
Au	1	1.5
Be	2	−0.2
Mg	2	−0.4
In	3	−0.3
Al	3	−0.3

low field limit ($\omega_c \tau \ll 1$) and the high field limit ($\omega_c \tau \gg 1$).

3.7.3 Low Magnetic Fields

The simplest generalization is a two-band model, with relaxation times τ_1 and τ_2, conductivities σ_1 and σ_2, and effective masses m_1^* and m_2^* where τ and σ are positive quantities but m^* can be positive (electron-like) or negative (hole-like). This model yields (Olsen, 1962)

$$R_H = e[(\tau_1 \sigma_1/m_1^*) + (\tau_2 \sigma_2/m_2^*)]/(\sigma_1^2 + \sigma_2^2) \tag{3-67}$$

For a free-electron Fermi surface with single values of τ, σ, and m^*, Eq. (3-67) reduces to Eq. (3-66), since $\sigma = (n_e e^2 \tau)/m^*$. In general, however, the sign of R_H depends on a balance between the two bands.

3.7.4 High Magnetic Fields

Three different cases occur in this limit, as listed in Table 3-1.

(a) For an uncompensated metal ($N_e \neq N_h$) with no open orbits, R_H takes the value

$$R_H = 1/(n_e - n_h)\, e \tag{3-68}$$

which is simply a generalization of Eq. (3-34) or Eq. (3-66) to take into account that the Fermi surfaces of real metals can contain both electron-like and hole-like portions. For one-electron metals such as the alkali and noble metals, where $N_e = 1$ and $N_h = 0$, Eq. (3-68) reduces to Eq. (3-66), and the numbers in Table 3-9 b are nearly unity. For Al, in contrast, the 2nd zone surface contains ≈ 2 holes/atom, and the 3rd zone surface contains ≈ 1 electron/atom. In high magnetic fields, Eq. (3-68) thus predicts R_H to be $-1/3$ of the free electron value. In low fields, in contrast, electrons execute only a small fraction of a cyclotron orbit before being scattered, and thus most of them never collide with a BZ boundary. The BZ boundaries can then be neglected, and Eq. (3-68) would reduce to Eq. (3-66) if the details of scattering were unimportant. The high field value of R_H in Al should be well defined, but the low field value changes as the dominant scatterer changes. This behavior is illustrated in Fig. 3-48.

For uncompensated metals, and compensated metals with open orbits, the situation is more complex.

(b) In most cases, R_H still becomes constant at high fields, but the value of this constant depends upon details of both scattering and the Fermi surface.

(c) When open orbits are present in more than one direction, R_H decreases as $(1/B)$ as $\boldsymbol{B}$ increases.

3.7.5 Additional Examples, Including Magnetic Metals

The temperature dependences of R_H for metals and alloys can be quite complex. As illustrations, Figs. 3-49 and 3-50 show data for high purity Cu and dilute Cu-based alloys, respectively.

Both ϱ and R_H change upon melting. For simple metals, the changes are usually not large, as illustrated in Fig. 3-51 for Li.

R_H can exhibit different behaviors in crystalline (c), amorphous (a) and liquid (l) forms of the same metal or alloy, as illustrated in Fig. 3-52 for a $Mg_{0.7}Zn_{0.3}$ alloy.

Although quantum oscillations of all kinds occur in R_H under the same conditions that they occur in MR, they are usually very small. An exception is quantum

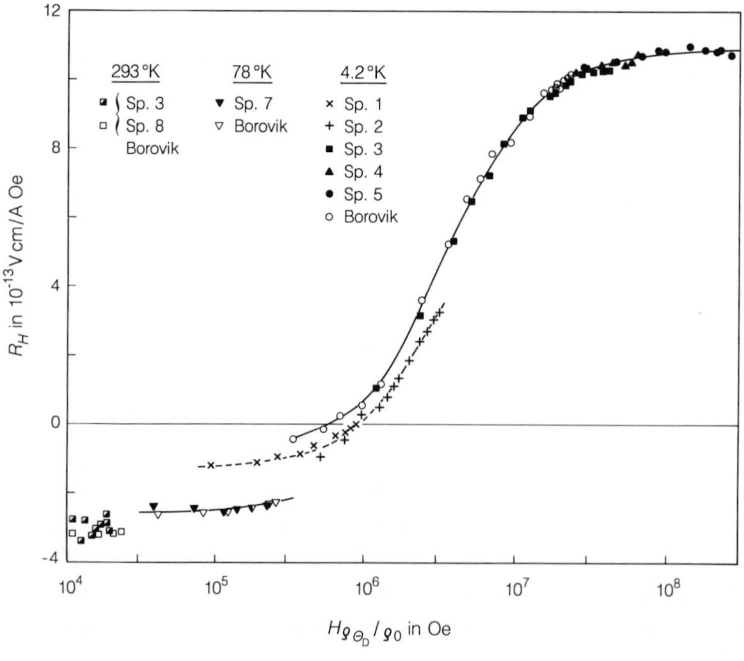

Figure 3-48. Hall coefficient of Al as a function of the reduced magnetic field measured through the high-field/low-field transition. Sp1 and Sp2 refer, respectively, to Al + 0.4 at.% Zn and Al + 0.1 at.% Zn. The other specimen numbers refer to pure Al specimens of various thicknesses (from Hurd, 1972; using the data of Forsvoll and Holwech, 1964).

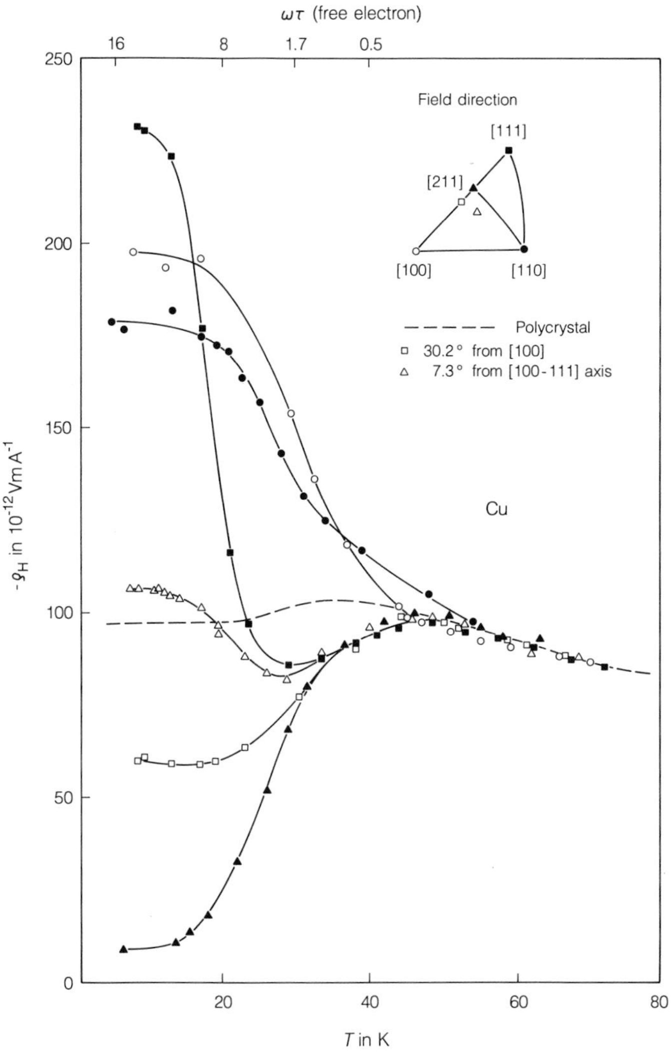

Figure 3-49. Temperature dependence of the Hall resistivity at 15.17 kG for a crystal of Cu with $\varrho(300\,\mathrm{K})/\varrho(4\,\mathrm{K}) = 2312$ and $\boldsymbol{B}$ along the directions indicated (from Hurd, 1972; using data of Hurd and Alderson, 1970).

size effects in the semi-metal Bi, illustrated in Fig. 3-53.

An additional phenomenon, called the anomalous Hall effect, has made Hall effect measurements important for studies of metals containing large, localized magnetic moments (e.g., ferromagnets), due to an additional contribution, called the anomalous Hall effect. Figure 3-54 a shows schematically what is expected for the Hall resistivity (i.e., $R_H B$) as a function of $\boldsymbol{B}$ in such a metal, and Fig. 3-54 b illustrates the differences in behavior seen in the ferromagnetic and paramagnetic states of Tb. Formally, a single magnetic domain in a ferromagnet has a spontaneous Hall resistivity $4\pi M_s R_s$ whenever a current is passed through it, even in zero applied $\boldsymbol{B}$. In an initially unmagnetized bulk sample, however, many small domains are present, with magnetizations pointing in different directions. The application of $\boldsymbol{B}$ causes these domains to increasingly align as $\boldsymbol{B}$ increases, which makes the spontaneous

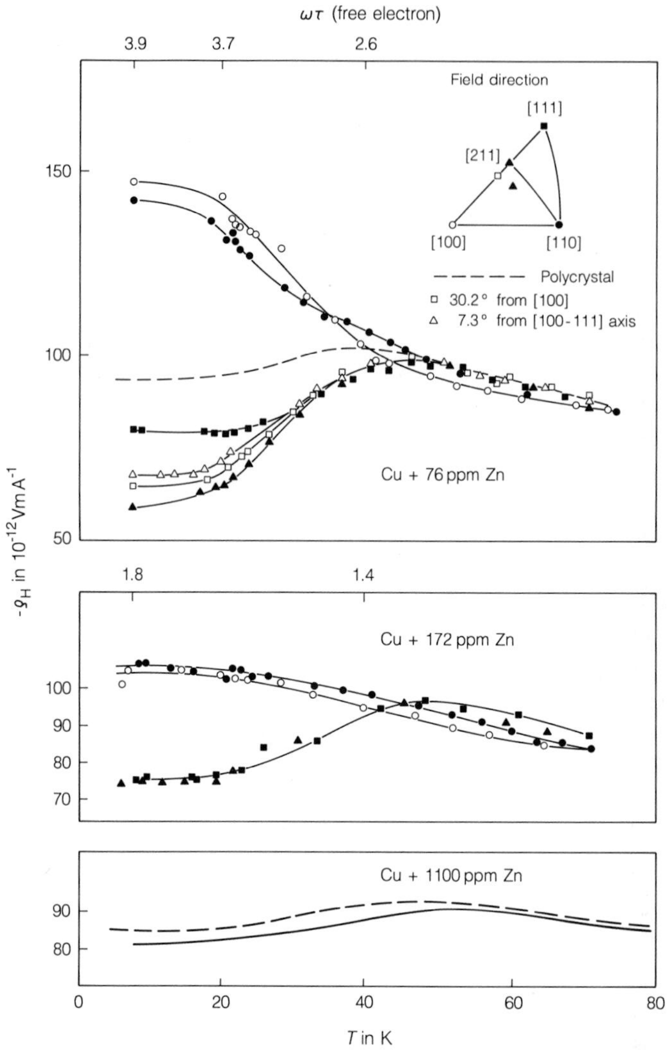

Figure 3-50. Temperature dependence of the Hall resistivity at 15.17 kG for crystals of Cu containing Zn and having ***B*** along the directions indicated (from Hurd, 1972; using data of Hurd and Alderson, 1970).

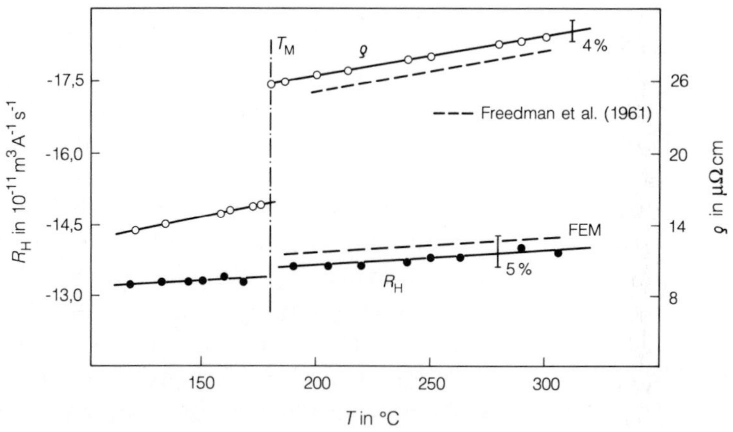

Figure 3-51. Hall coefficient and electrical resistivity of solid and liquid Li. T_M = melting point and FEM = free electron value (from Künzi and Güntherodt, 1980).

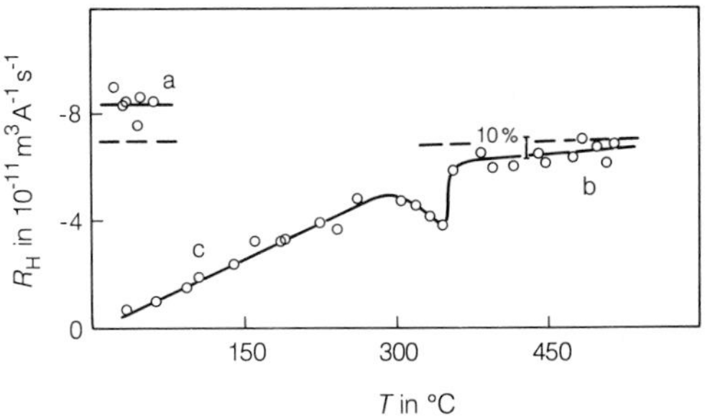

Figure 3-52. Hall coefficient of an amorphous (a), a liquid (b) and a crystalline (c) $Mg_{70}Zn_{30}$ alloy; dashed line indicates the free electron values for the Hall-coefficient (from Künzi and Güntherodt, 1980).

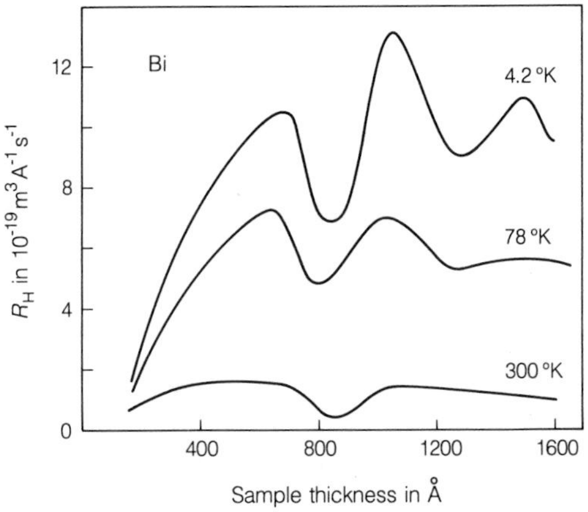

Figure 3-53. Quantum size effects in the Hall coefficient of Bi films (from Hurd, 1972; using data from Ogrin et al., 1966).

Hall resistivity visible. Figure 3-55 shows that the ordinary and spontaneous Hall resistivities have different temperature dependences.

3.8 Thermopower, *S*

3.8.1 The Temperature Dependence of *S* in Pure Metals

Figure 3-56 shows the thermopower S for several pure metals (for more complete data see Foiles, 1985). For each metal, the data can be divided into two components (see Sec. 3.1.2). The first, S_d, is approximately linear in T, as indicated by the dashed lines. The second, S_g, passes through a maximum value well below room temperature.

3.8.2 Rough Derivations of S_d and S_g

The presence of a temperature gradient, $\nabla_r T$, causes both electrons and phonons to flow from the hot end of the metal to the cold end. In each case, we define an effective "force per unit volume" associated with this flow that acts on electrons:

$$F_{\text{flow}} = \mathrm{d}U/\mathrm{d}x = (\mathrm{d}U/\mathrm{d}T)(\mathrm{d}T/\mathrm{d}x) = = C(\mathrm{d}T/\mathrm{d}x) \quad (3\text{-}69)$$

where U and C are the energy density and heat capacity per unit volume, respectively. In Sec. 3.1 we noted that S is measured under conditions of no net current flow. This means that the electrons being "pushed" by the force of Eq. (3-69) must pile up at one end of the sample to produce an electric field E that opposes further electron flow. The force per unit volume due to this field acting on the electron system must be:

$$F_E = n_e e E \quad (3\text{-}70)$$

For electron flow alone, writing $C = C_e$ and equating the forces of Eqs. (3-69) and (3-70) gives:

$$S_d = E/(\mathrm{d}T/\mathrm{d}x) \sim C_e/n_e e = = \pi^2 (k_B/e)(k_B T/\varepsilon_F) \quad (3\text{-}71)$$

where k_B is Boltzmann's constant and ε_F is the Fermi energy. Equation (3-71) predicts that S_d should increase linearly with T. For a typical $\varepsilon_F \approx 1$ eV, the coefficient in Eq. (3-71) is $\sim 10^{-8}$ V/K^2, leading to room temperature values of $S_d \sim 1$ μV/K^2.

For phonon flow, we initially assume that the phonons collide only with elec-

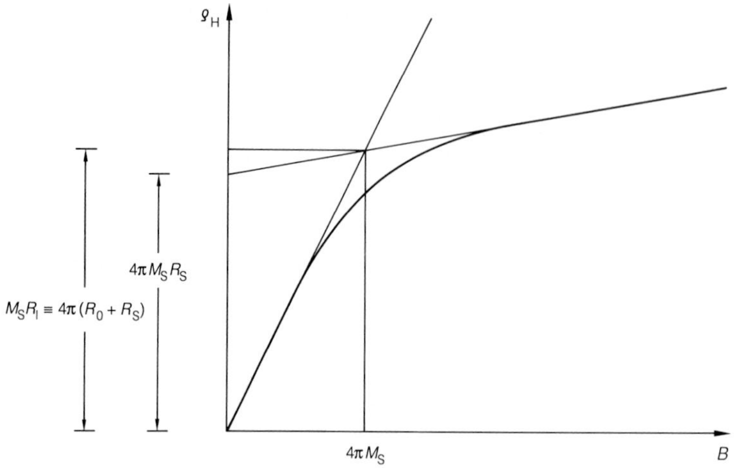

Figure 3-54 (a). Schematic behavior of the Hall resistivity ϱ_H as a function of magnetic induction $\boldsymbol{B}$ in a metal showing appreciable magnetization (from Hurd, 1972). M_s is the spontaneous magnetization, R_0 and R_S are constants called, respectively, the ordinary and spontaneous Hall coefficients.

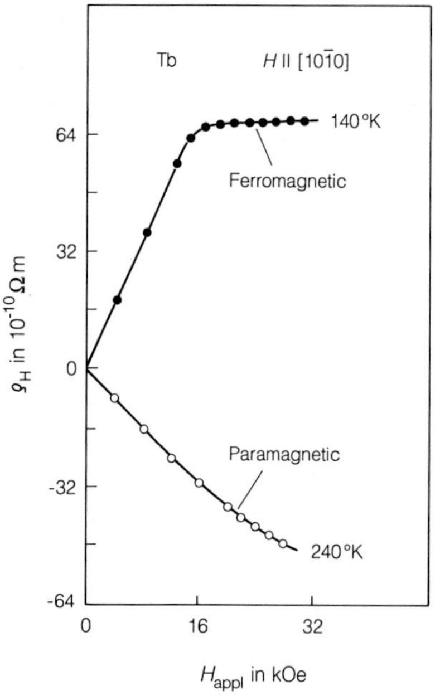

Figure 3-54 (b). Illustrating the typical dependence of the Hall resistivity on the applied field strength in the paramagnetic and ferromagnetic states (from Hurd, 1972; using data from Rhyne, 1969).

trons, thereby driving them along with their own motion. As above, we then find:

$$S_g = E/(\mathrm{d}T/\mathrm{d}x) \sim C_g/n_e e \quad (3\text{-}72)$$

Aside from a factor of 1/3, this is the correct term for free electrons interacting with Debye phonons via normal scattering processes alone.

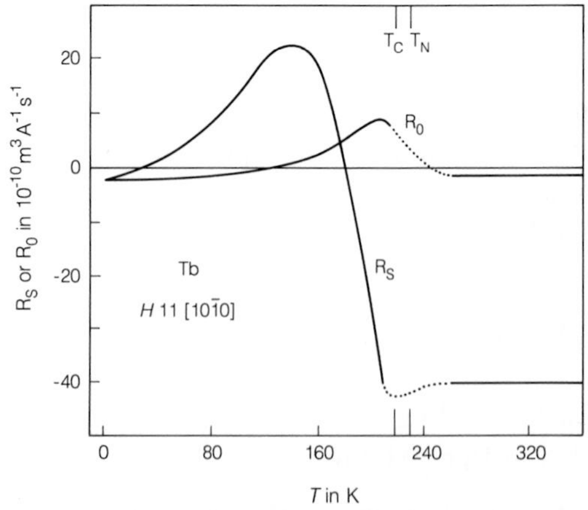

Figure 3-55. Temperature dependence of the ordinary and spontaneous Hall coefficients in a single crystal of Tb (from Hurd, 1972; using data from Rhyne, 1969).

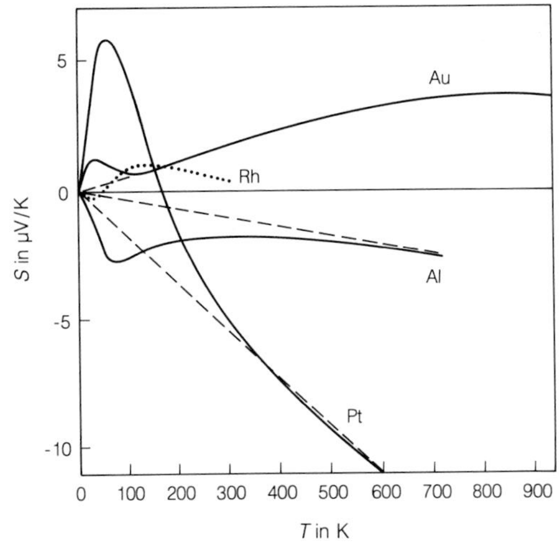

Figure 3-56. The thermopowers S of the metals gold (Au), aluminium (Al), platinum (Pt) (solid curves) and rhodium (Rh) (dotted curve) as a function of temperature. For Pt, Al and Au, the differences between the solid curves and the dashed lines indicate the magnitude of the phonon-drag components S_g (from Bass, 1982a).

Since phonons also scatter off of other phonons, impurities, etc., Eq. (3-72) must be multiplied by a factor α, the ratio of the probability for phonon-electron scattering (the only scattering that drives the electrons along with the phonons) to the probability for all scattering (including electron-phonon scattering). Writing the probability for phonon-electron scattering as inversely proportional to one relaxation time, τ_{pe}, and the probability for all other types of scattering as inversely proportional to another, τ_{p0}, gives:

$$S_g \sim (C_g/n_e e)\,\alpha \sim (C_g/n_e e)\,[\tau_{p0}/(\tau_{p0}+\tau_{pe})] \qquad (3\text{-}73)$$

At temperatures low compared to the Debye temperature Θ_D (i.e., $T \ll \Theta_D$): phonon-electron scattering dominates, τ_{pe} is short compared to τ_{p0}, and the ratio of the τ's cancels out. This leaves $S_g = C_g/n_e e$, so that $S_g = (12\pi^4/5)\,(k_B/e)\,(T/\Theta_D)^3$. For $T > \Theta_D$, C_g becomes constant at the classical value $3 n k_B$, where n is the number of atoms per unit volume in the metal. τ_{p0} also becomes temperature independent and small compared to τ_{pe}, and τ_{pe} varies as T^{-1}. S_g then decays as T^{-1}. Inserting appropriate values for τ_{pe} and τ_{p0}, reveals that as T increases from $T = 0$ K, the S_g of Eq. (3-73) first increases as T^3, then passes through a maximum value of order $\sim \mu V/K^2$ at $T \approx \Theta_D/5$, and finally decreases as T^{-1}.

Combining S_d from Eq. (3-71) with S_g from Eq. (3-73), yields an S containing one term linear in T with room temperature values of order $\mu V/K$, and another which peaks at $\sim \Theta_D/5$ with maximum values also of order $\mu V/K$. This combination correctly describes both the forms and magnitudes of the data of Fig. 3-56. The model predicts, however, that both terms will be strictly negative, whereas Fig. 3-56 shows that each can be either negative or positive.

This simple model has neglected: (1) details of the scattering of electrons on real FSs by realistic phonons and by other scatterers; (2) the presence of "Umklapp" scattering events, which can cause the scattered electron to end up moving oppositely to the direction of the incoming phonon; and (3) contributions to S_d from higher order scattering events and from many-body effects. The first two items provide mechanisms for changes in the sign of S_g, and all three provide mechanisms for changes in the sign and magnitude of S_d.

3.8.3 More Detailed Analysis of S_d

Direct application of the Boltzmann transport equation described in Sec. 3.2, yields for S_d (e.g., Barnard, 1972):

$$\begin{aligned} S_d &= (\pi^2 k_B^2 T/3 e \varepsilon_F)\,[\mathrm{d}\ln\sigma(\varepsilon)/\mathrm{d}\ln\varepsilon]_{\varepsilon_F} = \\ &= (\pi^2 k_B^2 T/3 e \varepsilon_F)\cdot \\ &\cdot [\mathrm{d}\ln n_e/\mathrm{d}\ln\varepsilon + \mathrm{d}\ln\tau/\mathrm{d}\ln\varepsilon]_{\varepsilon_F} \qquad (3\text{-}74) \end{aligned}$$

where ε is the electron's energy, $\sigma(\varepsilon)$ is the energy dependent conductivity and the derivatives are to be evaluated at ε_F. The second expression involves the single relaxation time approximation. Since for a spherical FS, $n_e \propto \varepsilon^{3/2}$, the first term in Eq. (3-74) is simply the generalization of Eq. (3-71). The second term in Eq. (3-74) contains the effects of scattering alluded to at the end of Sec. 3.8.2. The simplest way to understand the physical sources of the two terms in Eq. (3-74) is to examine the Fermi distributions at the two ends of the metallic sample of interest. If T_H and T_L are the temperatures at the high and low temperature ends of a bar, the resultant Fermi distributions are as shown in exaggerated fashion in Fig. 3-57. This diagram shows that there are more "high energy" electrons (i.e., energy greater than ε_F) at T_H than at T_L, and more "low energy" electrons (energy less than ε_F) at T_L than at T_H. Thus, if both n_e and τ are larger at higher electron energies, the energy derivative of each will

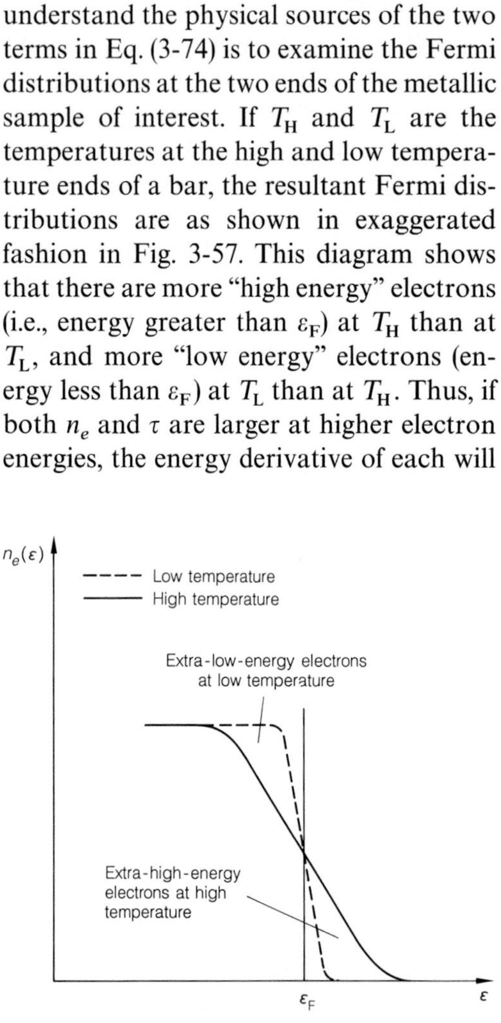

Figure 3-57. The variation with energy ε of the number of conduction electrons n_e in a metal in the vicinity of the Fermi energy ε_F for two different temperatures. In the interests of simplicity, a small variation of ε_F with temperature has been neglected in drawing this figure (from Bass, 1982a).

cause electrons to flow more easily from the hot end of the metal to the cold end, thereby building up an excess of electrons at the cold end. These electrons will produce an electric field which will grow until it just counteracts the flow due to the two "forces" just described. This is exactly what happens on a spherical FS with simple scattering, and yields the normal negative S_d. If, however, the energy dependences of n_e or τ are opposite to those assumed above, then electrons will tend to pile up at the hot end of the metal, yielding a positive contribution to S_d. This can occur, for example, for n_e, in the case where the FS contacts the BZ boundary.

Even in the free-electron model, the magnitude of the linear coefficient in S_d changes with temperature. The simplest models yield $\tau \propto \varepsilon^{3/2}$ for electron-phonon scattering (which is dominant at high temperatures), but $\tau \propto \varepsilon^{-1/2}$ for electron-impurity scattering) which is dominant at low temperatures). Combining each of these terms with $n_e \propto e^{3/2}$, we find that the coefficient of S_d decreases by 1/3 from high to low temperatures.

More generally, $\sigma(\varepsilon)$ is a complex integral over the FS which depends upon how $\tau(\boldsymbol{k})$ varies from place to place, and it is not possible to separate out terms related simply to n_e and τ. A complete calculation of S_d must also include contributions from higher order scattering events (i.e., events involving virtual scattering contributions) and corrections due to many-body-effects.

Adding impurities to a metal generally reduces S_g and changes the slope of S_d, as illustrated in Fig. 3-58 (see also Foiles, 1985). The change in slope of S_d can be understood in terms of the Nordheim-Gorter rule:

$$S_{TOT} = [\varrho_p S_p + \varrho_0 S_0]/\varrho_{TOT} = \\ = (\varrho_p/\varrho_{TOT})(S_p - S_0) + S_0 \qquad (3\text{-}75)$$

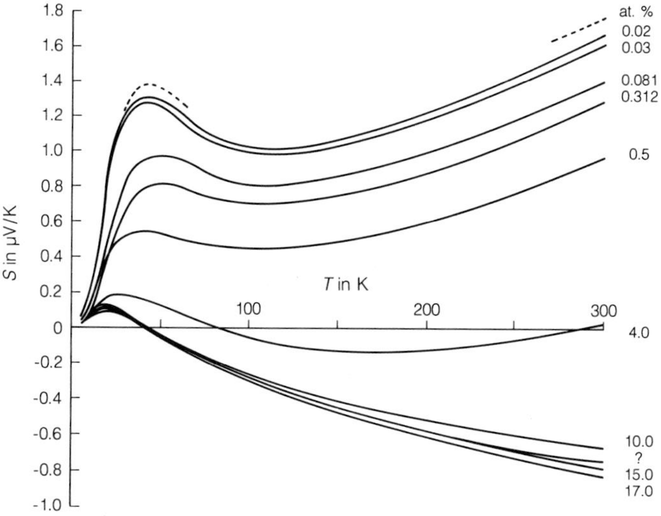

Figure 3-58. Measured thermopowers as a function of temperature for the Ag-Hg system (from Craig and Crisp, 1978). The concentrations of Hg in at.% are shown adjacent to each curve.

which is derived from Eq. (3-74) by assuming that the impurities represent an additional scatterer for which ϱ can be treated by Matthiessen's rule [Eq. (3-46)].

In this equation, S_{TOT} is the total thermopower, S_{p} is the temperature dependent thermopower of the ideally pure metal, and S_0 is the impurity thermopower. The far right hand side of Eq. (3-75) predicts that a plot of S_{TOT} versus $1/\varrho_{\mathrm{TOT}}$ should yield a straight line with slope $\varrho_{\mathrm{p}}(S_{\mathrm{p}} - S_0)$ and ordinate axis intercept S_0. Figure 3-59 illustrates such behavior. The reductions in S_{g} shown in Fig. 3-58 are usually attributed to quenching of phonon-drag due to impurity scattering, which drives the phonons back toward thermal equilibrium before they can "drag" the electrons along with them.

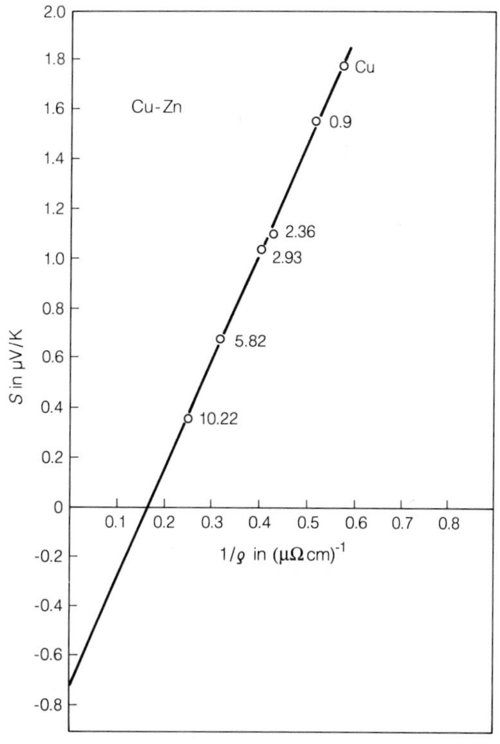

Figure 3-59. A Gorter-Nordheim plot for copper-zinc alloys at 300 K (from Blatt, 1968; using data of Henry and Schroeder, 1963).

3.8.4 Additional Examples

The thermopower of a single crystal of a non-cubic metal can be very different along different crystallographic axes, as illustrated in Fig. 3-60.

The thermopower of a superconductor drops to zero in the superconducting state. This fact is illustrated in Fig. 3-61, along with evidence that S can be very different

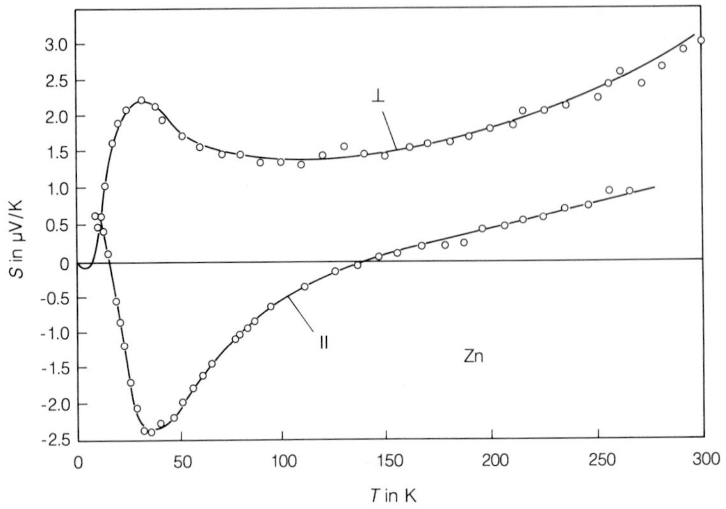

Figure 3-60. The thermopower of Zn parallel and perpendicular to the hexagonal axis (from Rowe and Schroeder, 1970).

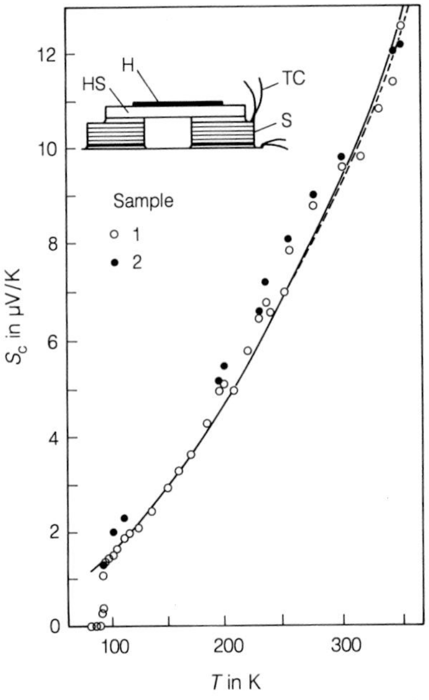

Figure 3-61 (a). Variation of the out-of-plane thermopower S_c with temperature in two crystals of $YBa_2Cu_3O_7$ (of dimensions $0.5 \times 0.4 \times 0.25$ and $0.4 \times 0.4 \times 0.20$ mm). The broken line is an empirical fit to $S_c = C(T/T_0)^\alpha$ ($C = 1.556\ \mu V/K$, $\alpha = 1.65$, $T_0 = 100$ K); the solid line is a fit to $S_c = AT + BT^2$ ($A = 7.6 \times 10^{-3}\ \mu V/K^2$, $B = 8 \times 10^{-5}\ \mu V/K^3$). Below 250 K the two curves cannot be distinguished. The inset shows the sample mounting configuration. A 1 kΩ heater comprised of a film (H) evaporated onto a sapphire substrate (HS) is glued to one face of both crystals (S). Thermocouples (TC) are copper-Constantan (from Wang and Ong, 1988).

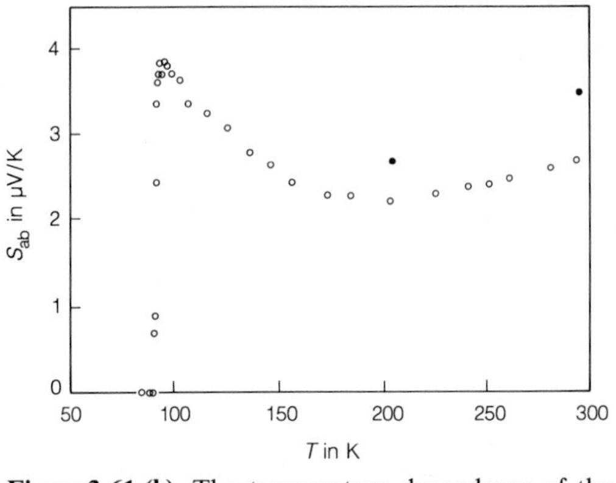

Figure 3-61 (b). The temperature dependence of the in-plane thermopower S_{ab} of $YBa_2Cu_3O_7$ crystal ($1.2 \times 0.4 \times 0.08\ mm^3$ in dimensions). The peak between T_c and 150 K is similar to peaks seen in some ceramic samples, but is absent in Fig. 3-61 (a). Chromel-Constantan thermocouples were used for this geometry. A limited amount of data (solid circles) from another crystal ($1.1 \times 0.5 \times 0.07\ mm^3$ in size) is also shown (from Wang and Ong, 1988).

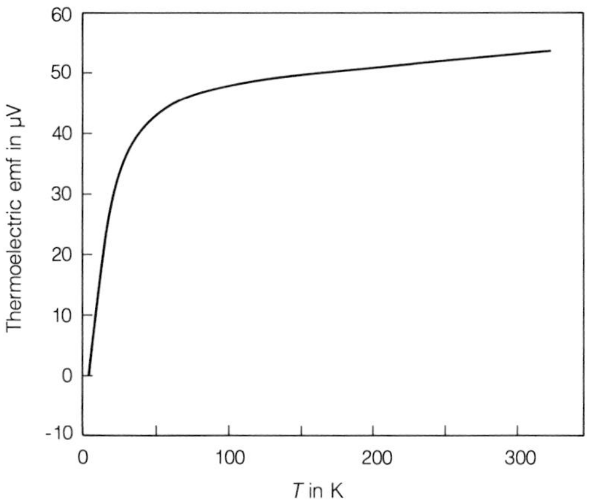

Figure 3-62. The thermoelectric emf of a thermocouple formed from pure annealed and pure cold-worked copper. The cold junction reference temperature is 4.2 K (from Kropschot and Blatt, 1959).

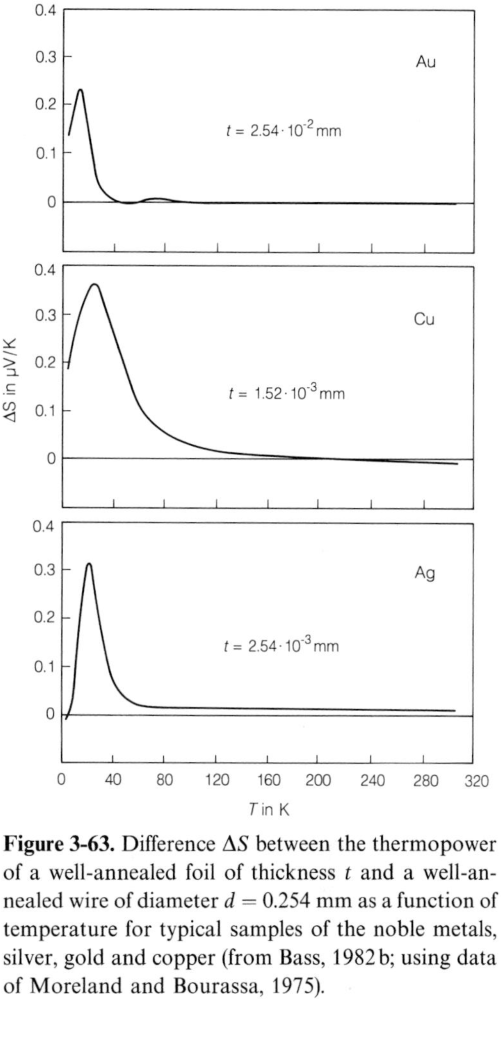

Figure 3-63. Difference ΔS between the thermopower of a well-annealed foil of thickness t and a well-annealed wire of diameter $d = 0.254$ mm as a function of temperature for typical samples of the noble metals, silver, gold and copper (from Bass, 1982b; using data of Moreland and Bourassa, 1975).

along the two different axes of single crystals of the very anisotropic high temperature superconductors.

Straining or thinning a metal can produce significant changes in S. Such changes are illustrated in Figs. 3-62 and 3-63, respectively.

For a liquid metal, S generally has a magnitude similar to that for the solid; the temperature dependent component of S normally varies approximately linearly with T and the coefficient of this linear term can have the opposite sign from that in the solid. These behaviors are illustrated in Fig. 3-64.

Like the resistivity (see Sec. 3.5.2), the thermopower undergoes characteristic changes when a ferromagnet becomes paramagnetic. The behavior for Ni is shown in Fig. 3-65.

The scattering of electrons by defects and disorder in non-magnetic amorphous metals is so large that phonon-drag is completely quenched. Examples of S for several non-magnetic amorphous metals are given in Fig. 3-66. The change in slope of S with increasing temperature is attributed to the disappearance with increasing temperature of many-body electron-phonon mass enhancement (Howson and Gallagher, 1988).

When a magnetic impurity, such as Fe, is dissolved in a non-magnetic noble metal, such as Au, a "giant" low temperature anomaly (also called a Kondo anomaly, cf., Sec. 3.5.2) appears in S. Compare, for example, the curve of S for 0.054% Fe in Cu in Fig. 3-67 with curve 1 of S for nominally pure Cu.

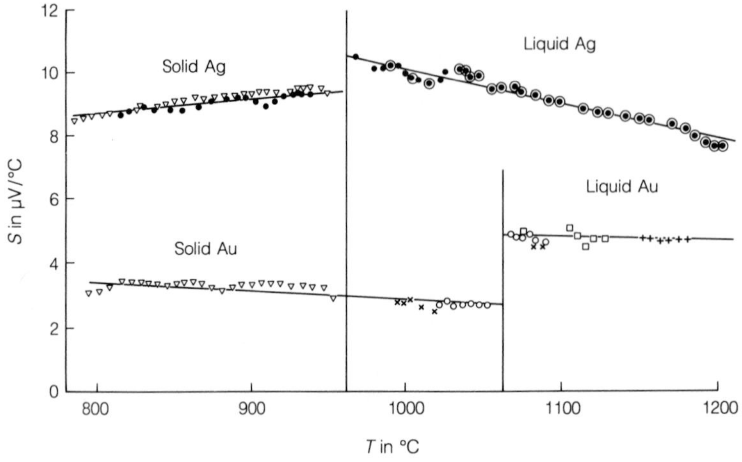

Figure 3-64. The changes in the thermopowers of gold (Au) and silver (Ag) upon melting (from Howe and Enderby, 1967).

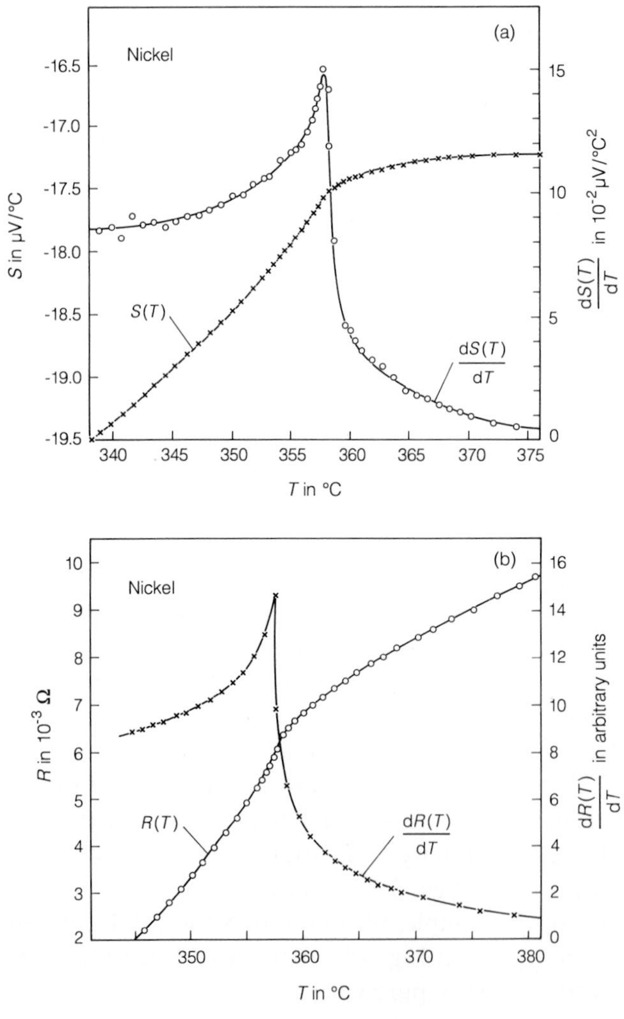

Figure 3-65. Thermoelectric power (a) and resistance (b) of nickel near the Curie temperature (from Tang et al., 1974).

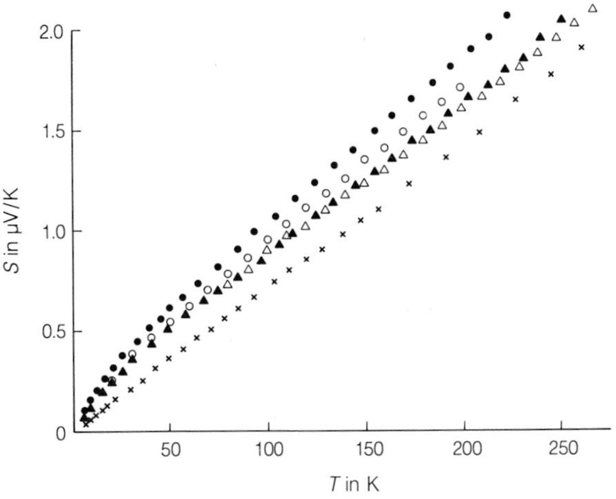

Figure 3-66. Measured thermopowers for three Cu-Zr systems: ●, $Cu_{45}Zr_{55}$; ○, $Cu_{27.5}Zr_{72.5}$; △, ▲, $Cu_{60}Zr_{40}$; and for (×) $Cu_{50}Zr_{40}Fe_{10}$ (from Gallagher and Grieg, 1982).

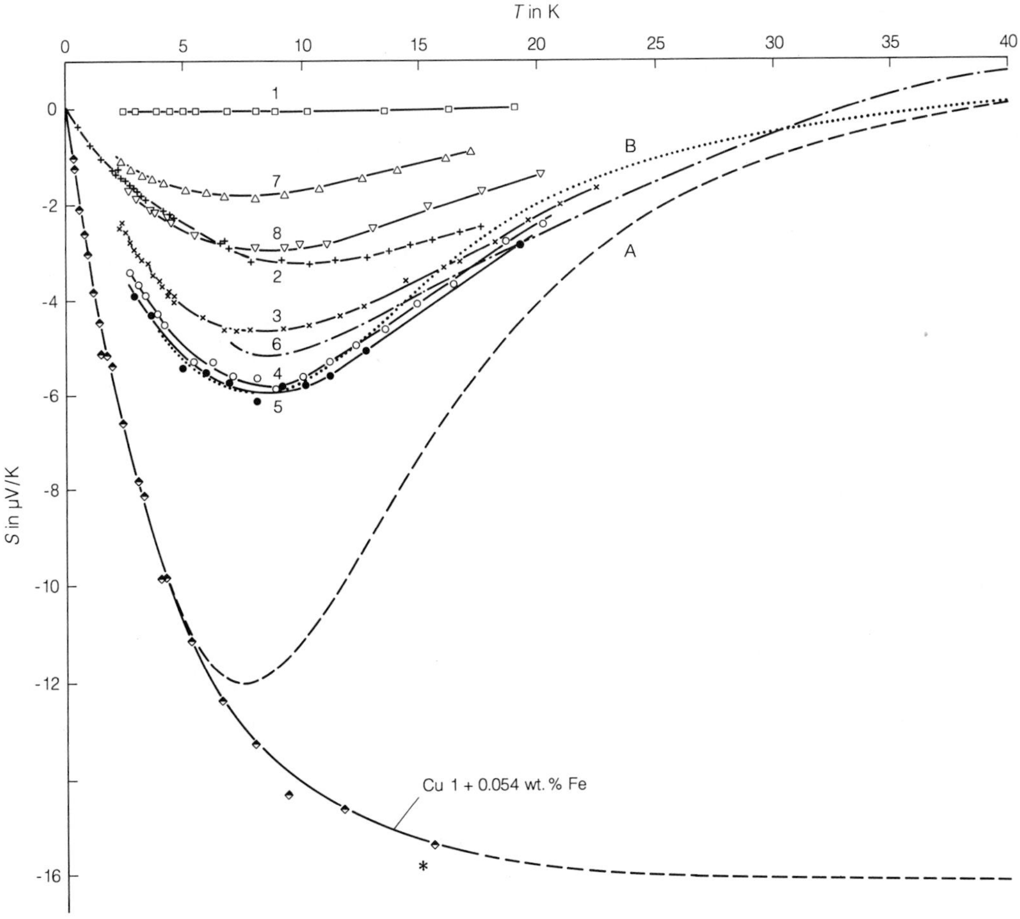

Figure 3-67. The low temperature thermopowers of various samples of copper containing very small concentrations of iron. Sample 1 is most representative of pure copper, because the iron is present as oxide (from Gold et al., 1960).

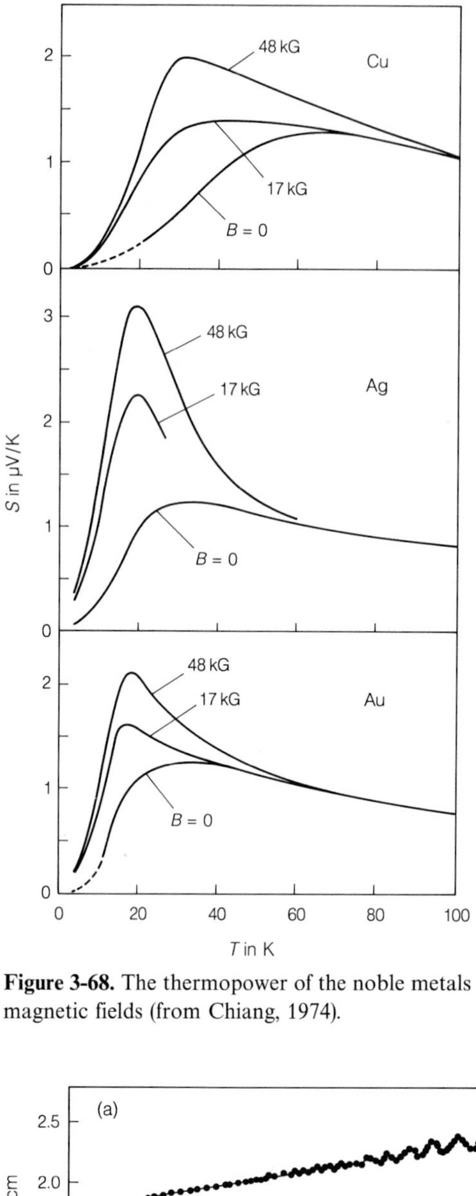

Figure 3-68. The thermopower of the noble metals in magnetic fields (from Chiang, 1974).

3.8.5 Magnetic Field Effects

Application of a magnetic field to a pure metal can produce significant changes in S, both in the quasiclassical and quantum regimes defined in Sec. 3.4. Figure 3-68 illustrates changes in S at higher temperatures. Figure 3-69 a contains an example of the unusually large oscillations in S that appear in single crystals of Al in the quantum regime due to magnetic breakdown. Figure 3-69 b shows that the breakdown oscillations are much smaller in the MR than in S.

In Sec. 3.7 it was shown that the Hall coefficient of Al changes sign when a magnetic field is applied at low temperatures, a phenomenon attributed to a change from electron-like to hole-like behavior with increasing $\boldsymbol{B}$. Figure 3-70 shows that S in dilute Al-based alloys displays the same behavior.

Application of a magnetic field also leads to more complex thermomagnetic effects. For example, when a field $\boldsymbol{B}$ is present in the z-direction, application of a temperature gradient in the x-direction gives rise to a voltage in the y-direction. If the sample is thermally isolated, this phenomenon is called the adiabatic Nernst-Ettingshausen

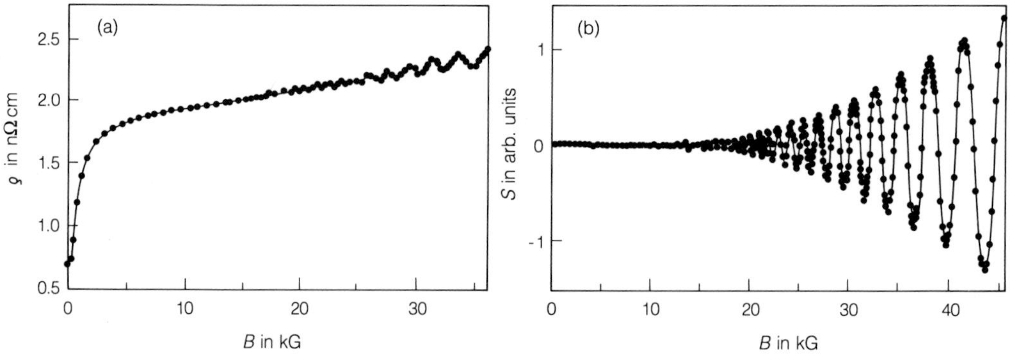

Figure 3-69. Magnetic field dependence of magnetoresistance and thermoelectric power for $\boldsymbol{B}$ 2° off a ⟨001⟩ direction: (a) magnetoresistance; (b) thermoelectric power. One unit in the vertical scale represents about 5 μV per degree (from Kesternich and Papastaikouidis, 1974).

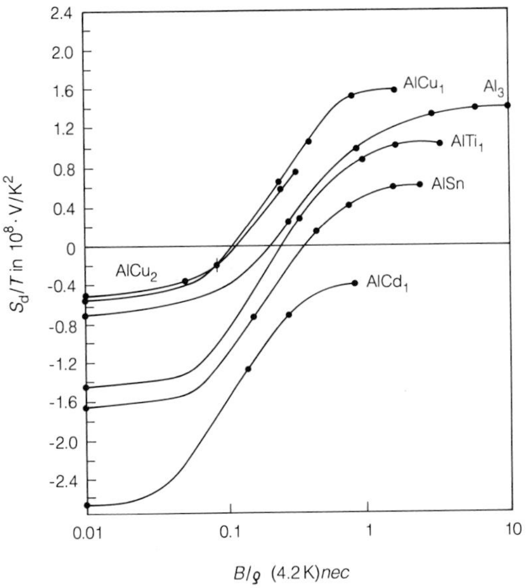

◀ **Figure 3-70.** The variation with magnetic field $\boldsymbol{B}$ of the low temperature electron-diffusion thermopower S_d of aluminium (Al) and various dilute aluminium-based alloys. To obtain a temperature independent quantity, S_d has been divided by the absolute temperature T. $\boldsymbol{B}$ has been normalized to remove the effects of varying impurity concentrations (from Averback et al., 1973).

(NE) effect. If the sample is immersed in a constant temperature bath, the voltage is more nearly representative of the isothermal NE effect. Figure 3-71 shows adiabatic NE measurements on Al below 4 K. The intercept of the data with the ordinate-axis yields the electron-diffusion component of the NE effect, which can be calculated for Al with high accuracy with no adjustable parameters. The data of Fig. 3-71 were used to show that the NE effect contains an electron-phonon mass enhancement.

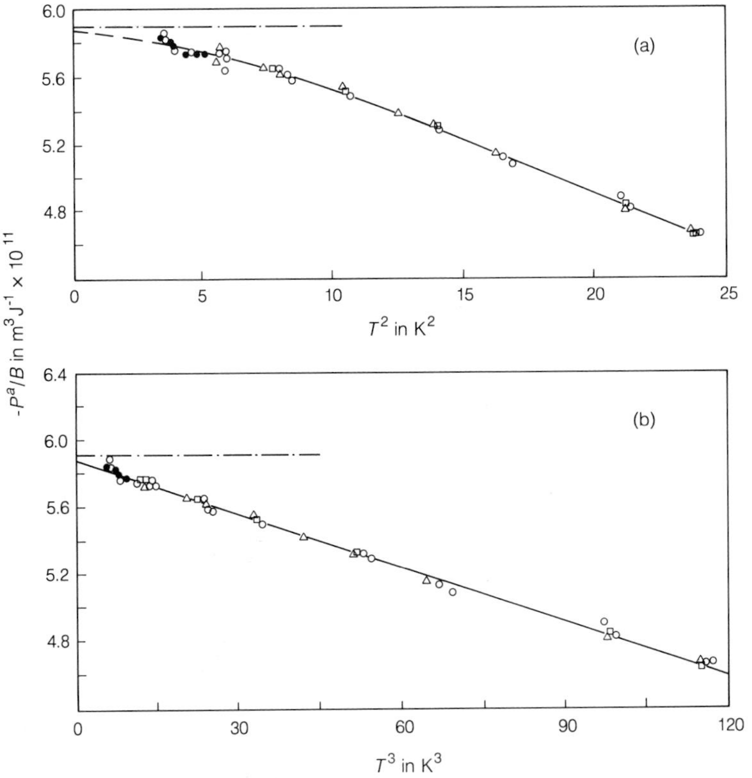

Figure 3-71. Adiabatic Nernst-Ettinghausen coefficient ($-P^a/B$) as a function of T^2 (a) and T^3 (b) for magnetic fields of 1.5 T, △; 1.8 T, □; 2.0 T, ○; and 2.1 T, ●. The broken lines indicate the value predicted from the electronic specific heat γ^c which contains the enhancement factor $1+\lambda$ (from Thaler et al., 1978).

3.9 References

Abrikosov, A. A. (1972), *Introduction to the Theory of Normal Metals, SSP Supplement 12*. New York: Academic Press.

Abrikosov, A. A. (1988), *Fundamentals of the Theory of Metals*. Amsterdam: North Holland Physics Publishing.

Akai, H. (1977), *Physics 86-8*, 539.

Ashcroft, N. W. (1963), *Phil. Mag. 8*, 2055.

Ashcroft, N., Mermin, N. D. (1976), *Solid State Physics*. New York: Holt, Reinhart and Winston.

Ausloos, M. (1976), *J. Phys. F. (Met. Phys.) 6*, 1723.

Averbrack, R. S., Stephan, C. H., Bass, J. (1973), *J. Low Temp. Phys. 12*, 319.

Babic, E., Krsnik, R., Leontic, B., Ocko, M., Vucic, Z., Zoric, I., Girst, E. (1972), *Solid St. Comm. 10*, 691.

Babic, E., Krsnik, R., Rizzuto, C. (1973), *Solid St. Comm. 13*, 1027.

Barnard, R. D. (1972), *Thermoelectricity in Metals and Alloys*. London: Taylor and Francis.

Bass, J. (1972), *Adv. in Phys. 21*, 431.

Bass, J. (1982a), in: *McGraw-Hill Encyclopaedia of Science and Technology, Vol. 13, Thermoelectricity*. New York: McGraw-Hill, p. 295.

Bass, J. (1982b), *Electrical Resistivity of Pure Metals and Dilute Alloys, in: Landolt-Börnstein Tables, New Series III, 15a:* Hellwege, K. H., Olson, J. L. (Eds.). Berlin: Springer Verlag.

Bass, J. (1985), *Electrical Resistivity of Pure Metals and Dilute Alloys, in: Landolt-Börnstein Tables, New Series III, 15b:* Hellwege, K. H., Olson, J. L. (Eds.). Berlin: Springer Verlag.

Beers, C. J., Van Dongen, J. C. M., Van Kempen, H., Wyder, P. (1978), *Phys. Rev. Lett. 40*, 1194.

Benedek, R., Baratoff, A. (1971), *J. Phys. Chem. Sol. 32*, 1015.

Blatt, F. J. (1968), *Physics of Electronic Conduction in Solids*. New York: McGraw-Hill.

Blatt, F. J., Schroeder, P. A., Foiles, C. L., Greig, D. (1976), *Thermoelectric Power of Metals*. New York: Plenum Press.

Blatt, F. J., Schroeder, P. A. (Eds.) (1978), *Thermoelectricity in Metallic Conductors*. New York: Plenum Press.

Bloch, F. (1930), *Phys. 59*, 208.

Bohnke, G., Rosenberg, M. (1980), *J. de Phys. Colloque 41*, C8-481.

Borchi, G., de Gennaro, S. (1970), *Phys. Lett. 32A*, 301.

Broom, T. (1954), *Adv. Phys. 3*, 26.

Brouers, F., Vedyayev, A. V., Giorgino, M. (1973), *Phys. Rev. B7*, 380.

Brown, R. A. (1977), *J. Phys. F (Metal Phys.) 7*, 1477

Brown, R. A. (1982), *Can. J. Phys. 60*, 766.

Bruls, G. J. C., Bass, J., Van Gelder, A. P., Van Kempen, H., Wyder, P. (1985), *Phys. Rev. B32*, 1927.

Butler, W. H., Stocks, G. M. (1984), *Phys. Rev. B29*, 4217.

Cahn, J. W. (1961), *Acta Met. 9*, 795.

Cahn, J. W. (1962), *Acta Met. 10*, 179.

Cahn, J. W. (1968), *Trans. Met. Soc. AIME 242*, 166.

Cahn, R. W., Feder, R. (1960), *Phil. Mag. 5*, 451.

Campbell, I. A., Ford, P. J., Hamzic, A. (1982), *Phys. Rev. B26*, 5195.

Chambers, R. G. (1956), *Proc. Roy. Soc. A238*, 344.

Chiang, C. K. (1974), Ph.D. Thesis, Michigan State University (unpublished).

Chikazumi, S. (1964), *Physics of Magnetism*. New York: John Wiley & Sons.

Cochrane, R. W. (1978), *J. de Phys. Colloque 39*, 6-1540.

Cochrane, R. W., Strom-Olsen, J. O. (1977), *J. Phys. F (Met. Phys.) 7*, 1799.

Coles, B. R., Taylor, J. C. (1962), *Proc. Roy. Soc. A267*, 139.

Cowley, J. M. (1975), *Diffraction Physics*. Amsterdam: North Holland Physics Publishing.

Craig, R., Crisp, S. (1978), *Thermoelectricity in Metallic Conductors:* Blatt, F. J., Schroeder, P. A. (Eds.). New York: Plenum Press, p. 51.

Darby, J. K., March, N. H. (1964), *Proc. Phys. Soc. 84*, 591.

de Gennes, P. G., Friedel, J. (1958), *J. Phys. Chem. Sol. 4*, 71.

Dorleijn, J. W. F. (1976), *Philips Res. Rep. 31*, 287.

Dorleijn, J. W. F., Miedema, A. R. (1975), *J. Phys. F (Met. Phys.) 5*, 487.

Dugdale, J. S. (1977), *The Electrical Properties of Metals and Alloys*. London: Arnold.

Durand, J., Gautier, F. (1970), *J. Phys. Chem. Solids 31*, 2773.

Elliott, R. J., Wedgwood, F. A. (1963), *Proc. Phys. Soc. 81*, 846.

Englert, E. (1932), *Ann. Physik 14*, 569.

Fert, A., Campbell, I. A. (1976), *J. Phys. F (Met. Phys.) 6*, 849.

Fischer, K. H. (1979), *Z. Physik B34*, 45.

Fletcher, R. (1977), *Sol. St. Comm. 21*, 1139.

Foiles, C. L. (1985), *Thermoelectricity in Metals and Dilute Alloys, in: Landolt-Börnstein Tables, New Series III, 15b:* Hellwege, K. H., Olsen, J. L. (Eds.). Berlin: Springer Verlag.

Ford, P. J., Mydosh, J. A. (1974), *J. de Phys. Colloque 35*, C4-241.

Forsvoll, K., Holwech, I. (1964), *Phil. Mag. 9*, 435.

Gallagher, B. L., Greig, D. (1982), *J. Phys. F. 12*, 1721.

Gold, A. V., MacDonald, D. K. C., Pearson, W. B., Templeton, I. M. (1960), *Phil. Mag. 5*, 765.

Greene, R. F. (1964), *Surface Sci. 2*, 101.

Grüneisen, E. (1933), *Ann. Phys. 16*, 530.

Guyot, P. (1970), *Phys. Stat. Sol. 38*, 409.

Hashin, Z., Shtrikman, S. (1962), *J. Appl. Phys. 33*, 3125.

Hayman, B., Carbotte, J. B. (1971), *Can. J. Phys. 49*, 1952.

Heeger, A. J. (1969), *Solid St. Phys. 23,* 283.
Henry, W. G., Schroeder, P. A. (1963), *Can. J. Phys. 41,* 1076.
Howe, R. A., Enderby, J. E. (1967), *Phil. Mag. 16,* 467.
Howson, M. A., Gallagher, B. L. (1988), *Phys. Rep. 170,* 267.
Hurd, C. M. (1972), *The Hall Effect in Metals and Alloys.* New York: Plenum Press.
Hurd, C. M., Alderson, J. E. A. (1970), *J. Phys. Chem. Sol. 32,* 175.
Johansson, C. H., Linde, J. O. (1936), *Ann. Phys. 25,* 1.
Kaul, S. N. (1977), *J. Phys. F (Met. Phys.) 7,* 2091.
Kesternich, W., Papastaikouidis, C. (1974), *Phys. Stat. Sol. 64,* K41.
Khachaturyan, A. G. (1979), *Prog. Mats. Sci. 22,* 1.
Kharoo, H. L., Gupta, O. P., Hemkar, M. P. (1978), *Phys. Rev. B18,* 5419.
Kim, M. J., Flanagan, W. F. (1967), *Acta Met. 15,* 747.
Kropschot, R. H., Blatt, F. J. (1959), *Phys. Rev. 116,* 617.
Kumar, S. (1975), *Indian J. Phys. 49,* 615.
Künzi, H. U., Güntherodt, H.-J. (1980), *The Hall Effect and Its Applications:* Chien, C. J., Westgate, C. R. (Eds.). New York: Plenum Press.
Landauer, R. (1978), *AIP Conf. Proc. No. 40:* Garland, J. C., Tanner, D. B. (Eds.). New York: American Institute of Physics, p. 2.
Linde, J. O. (1932), *Ann. Phys. 15,* 219.
Lodder, A., Boerrigter, P. M., Braspenning, P. J. (1982), *Phys. Stat. Sol. (b) 114,* 405.
Loegel, B., Gautier, F. (1971), *J. Phys. Chem. Sol. 32,* 2723.
Longo, J. T., Schroeder, P. A., Sellmyer, D. J. (1969), *Phys. Rev. 182,* 658.
Lormond, G. (1982), *J. de Phys. Colloque 33,* C6-283.
Mayadas, A. F., Shatzkes, M. (1970), *Phys. Rev. B1,* 1382.
Meaden, G. T. (1965), *Electrical Resistance of Metals.* London: Heywood Books.
Meisel, L. V., Cote, P. J. (1983), *Phys. Rev. B27,* 4617.
Miwa, H. (1962), *Prog. Theor. Phys. 28,* 209.
Miwa, H. (1963), *Prog. Theor. Phys. 29,* 477.
Mizutani, U. (1983), *Prog. Mats. Sci. 28,* 97.
Mizutani, U., Yoshino, K. (1984), *J. Phys. F (Met. Phys.) 14,* 1179.
Moreland, R. F., Bourassa, R. R. (1975), *Phys. Rev. B12,* 3991.
Mott, N. F., Jones, H. (1936), *The Theory of the Properties of Metals and Alloys.* Oxford: Clarendon Press.
Muth, P. (1983a), *Phys. Stat. Sol. (b) 118,* K117.
Muth, P. (1983b), *Phys. Stat. Sol. (b) 118,* K137.
Muth, P., Christoph, V. (1981), *J. Phys. F (Met. Phys.) 11,* 2119.
Naugle, D. G. (1984), *J. Phys. Chem. Sol. 45,* 367.
Ogrin Yu, F., Luytskii, V. N., Elinson, M. I. (1966), *Sov. Phys. JETP Letters 3, 71,* 114.
Olsen, J. L. (1962), *Electronic Transport in Metals.* New York: John Wiley & Sons.
Omini, M. (1980), *Phil. Mag. B42,* 31.
Osamura, K. O., Hiraoka, Y., Murakami, V. (1973), *Phil. Mag. 28,* 809.
Pal, S. (1973), *J. Phys. F (Met. Phys.) 3,* 1296.
Podloucky, R., Zeller, R., Dederichs, P. H. (1980), *Phys. Rev. B22,* 5777.
Popovic, Z., Carbotte, J. P., Piercy, G. R. (1973), *J. Phys. F (Met. Phys.) 3,* 1008.
Reynolds, J. A., Hough, J. M. (1957), *Proc. Phys. Soc. 70,* 769.
Rhyne, J. J. (1969), *J. Appl. Phys. 40,* 1001.
Rivier, N., Adkins, K. (1975), *J. Phys. F (Met. Phys.) 5,* 1745.
Rivier, N., Zitkova, J. (1971), *Adv. Phys. 20,* 143.
Roberts, R. B. (1977), *Phil. Mag. 36,* 91.
Robinson, J. E., Dow, J. D. (1968), *Phys. Rev. 171,* 815.
Rossiter, P. L., Wells, P. (1971), *J. Phys. C (Solid St. Physics) 4,* 354.
Rossiter, P. L. (1987), *The Electrical Resistivity of Metals and Alloys.* Cambridge: University Press.
Rowe, V. A., Schroeder, P. A. (1970), *J. Phys. Chem. Sol. 31,* 1.
Sambles, J. R., Elsom, K. C. (1980), *J. Phys. F (Met. Phys.) 10,* 1487.
Sambles, J. R., Priest, T. W. (1982), *J. Phys. F (Met. Phys.) 12,* 1971.
Schilling, J. S., Ford, P. J., Larsen, U., Mydosh, J. A. (1976), *Phys. Rev. B14,* 4368.
Schröder, K. (1983), in: *CRC Handbook of Electrical Resistivities of Binary Metallic Alloys.* Boca Raton: CRC Press, p. 1.
Simmons, R. O., Baluffi, R. W. (1960), *Phys. Rev. 117,* 62.
Singh, N., Singh, J., Prakash, S. (1977), *Phys. Stat. Sol. (b) 79,* 787.
Sun, P. H., Ohring, M. (1976), *J. Appl. Phys. 47,* 478.
Tang, S. H., Kitchens, T. H., Cadieu, F. J., Craig, P. P. (1974), *Proc. LT-13, 4.* New York: Plenum Press, p. 385.
Thaler, B. J., Fletcher, R., Bass, J. (1978), *J. Phys. F 8,* 131.
Thomas, N. (1984), *J. Phys. C (Solid St. Phys.) 17,* L59.
Uher, C. (1987), *J. Appl. Phys. 62,* 4636.
van Bueren, H. G. (1957), *Philips Res. Reports 12,* 190.
Wang, Z. Z., Ong, P. P. (1988), *Phys. Rev. B38,* 7160.
Watson, W. G., Hahn, W. C., Kraft, R. W. (1975), *Met. Trans. A 6A,* 151.
Wilson, A. H. (1937), *Proc. Cambridge Phil. Soc. 33,* 371.
Wyder, P. (1965), *Phys. Kondens. Materie (1965) 3,* 263.
Yamamoto, R., Doyama, M., Takai, O., Fukusako, T. (1973), *J. Phys. F (Met. Phys.) 3,* 1134.

Ziman, J. M. (1960), *Electrons and Phonons*. Cambridge: University Press.
Ziman, J. M. (1972), *Electrons and Phonons, 2nd Ed.* Oxford: Clarendon Press.

General Reading

Abrikosov, A. A. (1988), *Fundamentals of the Theory of Metals*. Amsterdam: North Holland Physics Publishing.
Ashcroft, N., Mermin, N. D. (1976), *Solid State Physics*. Philadelphia: Holt Reinhart and Winston.
Barnard, R. D. (1976), *Thermoelectricity in Metals and Alloys*. London: Taylor and Francis.
Blatt, F. J. (1968), *Physics of Electronic Conduction in Solids*. New York: McGraw-Hill.
Blatt, F. J., Schroeder, P. A., Foiles, C. L., Greig, D. (1976), *Thermoelectric Power of Metals*. New York: Plenum Press.
Chien, C. J., Westgate, C. R. (Eds.), *The Hall Effect and Its Applications*. New York: Plenum Press.
Hurd, C. M. (1972), *The Hall Effect in Metals and Alloys*. New York: Plenum Press.
Kittel, C. (1971), *Introduction to Solid State Physics, 4th Ed.* New York: John Wiley & Sons.
Rossiter, P. L. (1987), *The Electrical Resistivity of Metals and Alloys*. Cambridge: Cambridge University Press.
Ziman, J. M. (1965), *Principles of the Theory of Solids*. Cambridge: Cambridge University.
Ziman, J. M. (1972), *Electrons and Phonons, 2nd Ed.* Cambridge: Cambridge University.

4 Superconductivity

Peter H. Kes

Kamerlingh Onnes Laboratorium, Leiden University, Leiden, The Netherlands

List of Symbols and Abbreviations

$\boldsymbol{A}(r)$	vector potential
A	wire cross section
a_0	$\equiv a_\triangle$
$a_\square$, $a_\triangle$	lattice parameter for square, hexagonal lattice structure
$\boldsymbol{b}$, b	local magnetic field, magnitude
b	$\equiv B/B_{c2}$
B	magnetic induction
B	flux density
$\boldsymbol{b}_B$, b_B	Burgers vector, magnitude
B_0	$=(3\,\Phi_0/4\,d_0^2)^{1/2}$
b_{co}	normalized crossover field from 2DCP to 3DCP
B_{c2}	"upper" magnetic induction
B_{2D}	crossover field
$C(T)$	specific heat
$C_n(T)$	normal state specific heat
c_{11}	element of elastic tensor describing compressional deformations
c_{44}	element of elastic tensor describing tilt
c_{66}	element of elastic tensor describing shear distortions
d	film thickness
D	demagnetizing factor
D	(electron) diffusion constant
D	diameter of dislocation loops
D	grain size
d_0	maximal distance of attractive vortex interaction
$\boldsymbol{e}$	local electric field
e	electron charge
$\boldsymbol{E}$	electric field
E_{ks}	excitation energy of an electron in state k in a superconductor
E_{kn}	excitation energy of an electron in state k in a normal conductor
$f(r)$	reduced order parameter
f	magnitude of force exerted by a pin
F	free energy
$\boldsymbol{f}_d$, f_d	driving force, magnitude
F_n, F_M, F_s	F of the normal, Meissner, and superconducting state
$f_n(T)$, $f_s(T)$	free energy density of the normal, superconducting state
f_p	maximum pinning force of single defect
F_p	bulk pinning force
F_{sn}	$=F_s-F_n$
f_{th}	threshold value of f_p
$f_{n0}(T)$	normal state free energy density in zero field
$g(T, H)$	Gibbs free energy density
$g(\boldsymbol{r})$	displacement correlation function

G	Gibbs free energy
G_M, G_{ms}	Gibbs free energy for Meissner (M), mixed state (ms)
g_s, g_n	Gibbs free energy density for superconducting and normal state
G_0	$= e^2/\hbar$
g_{sn}	$g_s - g_n$
$\hbar$	Planck constant/2π
$\boldsymbol{H}$, H	magnetic field, magnitude
H_c	thermodynamic critical field
H_i	internal magnetic field
H_p	pair breaking field
H_0	thermodynamic critical field at zero temperature
H_{c1}	lower critical field
H_{c2}	upper critical field
H_{c3}	"nucleation" field
$H_{c\perp}$	nucleation field perpendicular to a thin film
$H_{c\parallel}$	nucleation field parallel to a thin film
$\boldsymbol{j}$	current density
j_c	critical current density
$\boldsymbol{j}_s$	supercurrent density
$\boldsymbol{k}$, k	wavevector, magnitude
k_B	Boltzmann constant
k_F	Fermi wavevector
k_0	$= 2\pi/a_0$
$K_0(z)$	zero-order Hankel function of imaginary argument
k_{Br}	radius of the circular approximation for the first Brillouin zone of the VL
$k_\perp$	$= (k_x^2 + k_y^2)^{1/2}$
ℓ	mean free path
L_c	longitudinal correlation length
m	electron mass
M	isotopic mass
$\boldsymbol{M}(T, \boldsymbol{H})$, M	magnetization density, magnitude
ΔM	irreversibility of the magnetization curves
m^*	effective electron mass
M_{rev}	reversible magnetization
$N(E)$	density of states
$N(0)$	density of states at ε_F
n_p	concentration of pinning centers
n_s	number density of electrons in the ground state
$\boldsymbol{p}$	momentum
Q	latent heat
$\boldsymbol{r}$, $\boldsymbol{r}'$	distance vector
$\boldsymbol{R}$	$= \boldsymbol{r} - \boldsymbol{r}'$
R	radius of a solid cylinder
R_c	perpendicular correlation length
$\boldsymbol{r}_d$	position of a defect

r_f	range of pinning force
R_n	normal state resistance per unit area of junction
$s(T, H)$	entropy density
s_s, s_n	entropy density of a superconductor, normal conductor
t	time
t	$\equiv T/T_c$
T	absolute temperature
T_c	critical temperature
T_1^{-1}	nuclear spin relaxation rate
$\boldsymbol{u}(\boldsymbol{r})$, u	vortex displacement in plane perpendicular to the field, magnitude
U_p	pin potential
U_v	energy of a single vortex per unit length
$\boldsymbol{v}$	velocity
V	potential
V_c	$\approx R_c^2 \cdot L_c$, volume of correlated region
v_F	Fermi velocity
$\boldsymbol{v}_L$, v_L	flux-flow velocity
v_s	superfluid velocity
w	width of a thin film or wire
W	pin correlation function for small pins (pin "strength")
W	width of the weak pinning path
α	ultrasonic absorption
$\alpha(T)$, $\beta(T)$	material parameter
α_n, α_s	normal and superconductive state ultrasonic absorption
α_L	Labusch parameter
β_A	Abrikosov parameter
$\beta_{A\square}$, $\beta_{A\triangle}$	β_A according to square, hexagonal lattice
γ	Sommerfeld constant
Γ	anisotropy parameter
Δ	energy gap parameter
$\Delta(\boldsymbol{r})$	pair potential
Δ_0	energy gap parameter at $T=0$
$\delta_2(\boldsymbol{r})$	two-dimensional delta function
ε_F	Fermi energy
ε_k	energy of an electron of state k
ε_v	volume dilatation
η	flux-flow viscosity
$\varkappa$	GL parameter
$\varkappa_0$	GL parameter of a pure superconductor
$\varkappa_1$, $\varkappa_2$, $\varkappa_3$	Maki parameters
λ	wavelength
λ_d	magnetic penetration depth in dirty superconductor
λ_C	Campbell penetration depth
λ_L	London penetration depth

λ_{eff}	effective magnetic penetration depth
μ_0	vacuum permeability
$\xi(T)$	GL coherence length
ξ_n	penetration depth of Cooper pair tunneling into a normal conductor
ξ_s	penetration depth of normal state electron tunneling into a superconductor
ξ_0	Pippard (or BCS) coherence length
ϱ	resistivity
ϱ_d	dirt parameter
ϱ_f	flux-flow resistivity
ϱ_0	resistivity at $T=0$
σ_f	flux-flow conductivity
σ_n	normal state conductivity
$\sigma_{1D}, \sigma_{2D}, \sigma_{3D}$	1-, 2-, 3-dimensional dc paraconductivity
τ	lifetime
τ	shear strength of the VL
$\Delta\phi$	$=\phi_1-\phi_2$, phase difference of BCS wavefunctions in superconductors S and S′
$\phi(r)$	phase of BCS wavefunction ψ
Φ	magnetic flux
Φ_0	flux quantum
χ	susceptibility
χ_{2D}, χ_{3D}	2-, 3-dimensional fluctuation-contribution to susceptibility
ψ	wavefunction
ψ_0	amplitude of ψ
ψ_∞	value of ψ far from perturbation
ω	frequency
ω_D	Debye frequency
$\Omega_p(\boldsymbol{r}_d)$	pin potential of pinning center

BCS	Bardeen, Cooper and Schrieffer
CP	collective pinning theory
2DCP, 3DCP	2-, 3-dimensional CP
CVD	chemical vapor deposition
GL	Ginzburg-Landau
HTS	high-temperature superconductivity
LO	Larkin and Ovchinnikov
MPMQ	melt process-, melt quench procedure
ms	mixed state
n, s	normal-, superconducting
S, S′	superconductors
SIN	superconductor-insulator-normal conductor
SIS′	superconductor-insulator-superconductor
SQUID	superconducting quantum interference device
VL	vortex lattice

4.1 Introduction

When superconductivity was discovered in 1911, Kamerlingh Onnes immediately sensed the great practical use this phenomenon could have for society. Unfortunately, the superconducting state in a coil made out of Pb wire disappeared at a disappointingly low field value. Society had to wait more than five decades until the first commercial applications appeared. In these sixty years many elements were discovered to be superconducting, sometimes only under high pressures, with critical temperatures T_c ranging from 0.35 mK for Rh to 9.27 K for Nb. A large number of alloys and compounds were also found to be superconducting (Roberts, 1978; and Phillips, 1989) with Nb_3Ge holding the record of 23 K for a long time, until in 1986 a break-through was triggered by the discovery of Bednorz and Müller of high-temperature superconductivity in ceramic materials containing CuO_2 layers. In about one year T_c jumped to a value of 125 K in $Tl_2Ba_2Ca_2Cu_3O_{10}$.

Most high-temperature superconductors, which we like to define as materials with T_c's above 10 K, have large upper critical fields H_{c2} as well. Hence, the two intrinsic materials requirements for large-scale applications, mainly big magnets, are fulfilled for $T_c > 10$ K. Naturally, a $T_c >$ 77 K is preferable, but the economic gain is sometimes exaggerated. However, as for these strong-current applications an equally important, but non-intrinsic property is a large critical current density j_c at high flux densities B. A considerable materials-research effort is usually needed to exceed the practical lower limit of 2×10^8 A/m^2 in fields up to 15 T. Large current densities are also required in another potential field of superconducting technology, namely, the small-current applications in a variety of electronic devices. This time the fields encountered are much smaller and a quite different (thin-film) fabrication technique is exploited, but the materials involved are practically the same.

Interesting questions are, why is it that some materials have favorable properties, how can we manipulate them to improve the characteristics, and what are the theoretical limits? To answer these questions we have to understand the physics behind the relationship between electronics, superconductivity and materials properties. Within the scope of the series it is not possible to discuss these issues in great detail, apart from the fact that our understanding of some aspects is still limited, e.g. a microscopic explanation for high-temperature superconductivity (HTS) is still missing. In this chapter we concentrate on the fundamental issues related to the applicability of superconductivity, namely, a high T_c, a large H_{c2}, and a large $j_c(B)$. For this purpose we shall discuss some general elements of the microscopic BCS theory, and in more detail the phenomenological Ginzburg-Landau (GL) theory for non-uniform superconductors. In both cases we shall illustrate the theoretical predictions of electronic and magnetic properties with some experimental results. Some emphasis is given to the physics of Josephson junctions and the characteristic properties of the flux line lattice in the mixed state. Subsequently, the fundamentals of flux pinning are discussed, and finally, some materials requirements for technology. Both conventional and HTS are considered assuming that the phenomenology of the HTS can be described by the GL theory.

4.2 Bardeen, Cooper, Schrieffer (BCS) Theory and Electronic Properties

4.2.1 Cooper Pairs and Attractive Interaction

For a detailed theory of superconductivity one needs a thorough comprehension of a metal in the normal state. The question is how a delicate effect with a condensation energy of $\approx 10^{-7}$ eV per electron can arise amidst a strongly interacting (≈ 1 eV) many-body system of conduction electrons and vibrating ions (phonons). The first indication that the electron-phonon interaction is responsible came from the discovery of the isotope effect, i.e. when the isotopic mass M of the material is varied, the critical temperature T_c and the critical field at zero temperature H_0 change according to

$$T_c \propto M^{-\alpha} \propto H_0 \, (\alpha \simeq 0.5) \tag{4-1}$$

Notable deviations from the value $\alpha \simeq 0.5$ have been observed though. A clear description of how the relevant interactions can be separated, is given by Schrieffer (1964). The long-range Coulomb repulsion between two electrons is largely suppressed by screening due to strong correlation effects of the other electrons. Screening also effects the phonon frequencies and the electron-phonon interaction. It yields a small, attractive effective potential corresponding to *overscreening* of the Coulomb repulsion between electrons by the vibrating ions. In a simplified picture one can imagine how a fast moving electron leaves behind a slightly positive trace of displaced ions which attracts other electrons. This effect is most profitable for an electron which moves with the same velocity, but in the opposite direction as the first electron. Such a combination of oppositely moving electrons exchanging phonons is known as a Cooper pair. The total spin of the pair is usually zero giving rise to singlet superconductivity. A Cooper pair forms a bound state when it is added to the electron states (wavevectors $\boldsymbol{k}$) of a normal metal at $T=0$, so that all states are occupied up to k_F, the Fermi wavevector. The energy gain is $\Delta_{cp} \approx 2\hbar\omega_c \exp[-1/(N(0)\,|\lambda|)]$ for an attractive interaction which is characterized by $\lambda < 0$ in an energy band of width $2\hbar\omega_c$ around the Fermi energy ε_F. The density of states at ε_F is denoted by $N(0) = m^* k_F / 2\pi^2 \hbar$. The effective electron mass m^* is supposed to absorp all interactions and band structure effects common to both the normal and superconducting states. Cooper pairs form the essential elements for the superconducting groundstate.

4.2.2 Bardeen, Cooper, Schrieffer Groundstate

Bardeen, Cooper and Schrieffer (1957) showed that interacting electrons under influence of a weak attractive potential V condense into a groundstate consisting of pair states $(-\boldsymbol{k}, \uparrow), (\boldsymbol{k}, \downarrow)$ (the arrows denote the electron spins). The only requirement for V is that it is negative around the Fermi energy. BCS choose $V = -V_0$ with V_0 a positive constant, in a shell of width $2\hbar\omega_c \ll \varepsilon_F$ and zero elsewhere. For electron-phonon interaction the natural choice for ω_c is the Debye frequency ω_D. The energy required to excite an electron in $\boldsymbol{k}$ is

$$E_{ks} = +[(\varepsilon_k - \varepsilon_F)^2 + \Delta_0]^{1/2} \tag{4-2}$$

where $\varepsilon_k = \hbar^2 k^2 / 2m^*$ and Δ_0 is the energy gap parameter at $T=0$. For this simple form of V the relation between Δ_0 and the effective electron-phonon coupling constant $N(0)\,V_0$ is given by

$$\Delta_0 = 2\hbar\omega_D \exp[-1/N(0)\,V_0] \tag{4-3}$$

In Fig. 4-1 the energy spectrum for the normal excitations in the superconducting

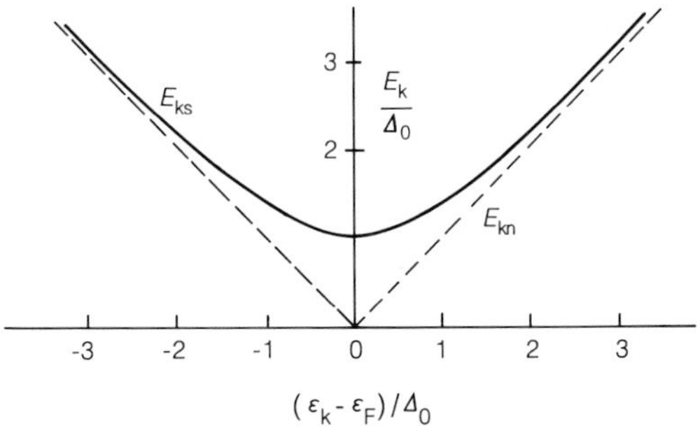

Figure 4-1. Energy spectrum of elementary excitations in the normal (E_{kn}) and superconducting (E_{ks}) states about the Fermi energy (full and dashed lines, respectively).

state represented by Eq. (4-2) is compared with the spectrum in the normal state given by $E_{kn} = |\varepsilon_k - \varepsilon_F|$. Only in the vicinity of the Fermi surface there is a difference. In experiments where no electrons are injected into the superconductor, an energy gap $2\Delta_0$ is observed, e.g. in the case of infrared absorption. The density of states related to $E_{ks,n}$ is shown in Fig. 4-2. The enhancement at the gap edge is clearly observable in the tunnel conductance of an oxide barrier between a normal metal and a superconductor.

With increasing k the occupation number of the pair states gradually decreases from one to zero in a band of width Δ_0 about ε_F. Interpreting this as the energy uncertainty, it follows that the lifetime τ of a Cooper pair is of the order $\hbar/\Delta_0$. The distance of coherent interaction between the two electrons is therefore

$$\xi_0 = \frac{\hbar v_F}{\pi \Delta_0} \tag{4-4}$$

This is the Pippard coherence length; the coefficient π^{-1} follows from the BCS theory.

4.2.3 Finite Temperatures

At increasing temperatures normal excitations are created to an increasing extend, thereby deminishing the available phase space for the pair states which in turn leads to a reduction of the energy gap as shown in Fig. 4-3. Below $T = 0.4\,T_c$, Δ is almost constant, while near T_c it decreases sharply according to

$$\Delta(T)/\Delta_0 = 1.74(1-t)^{1/2}, \quad T \approx T_c \tag{4-5}$$

with $t \equiv T/T_c$. For the simple form of V it follows that $2\Delta_0/k_B T_c = 3.52$ where k_B is the Boltzmann constant. For most superconducting elements the BCS prediction agrees well with the experimental results as

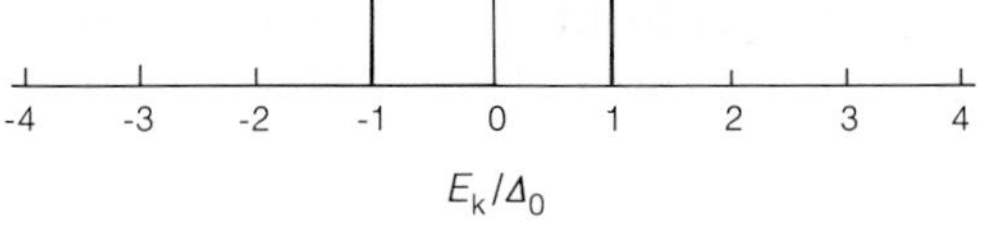

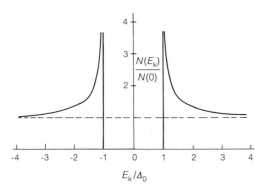

Figure 4-2. Density of states in the superconducting and normal states about the Fermi energy (full and dashed lines, respectively).

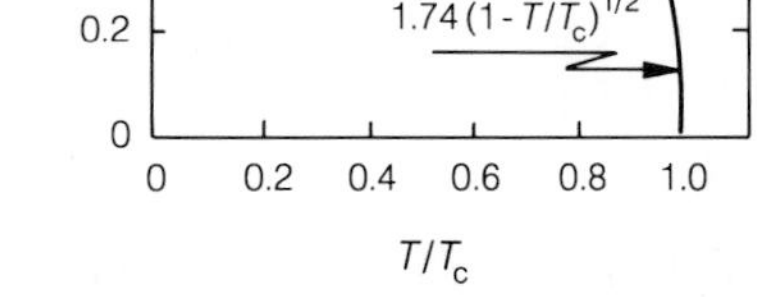

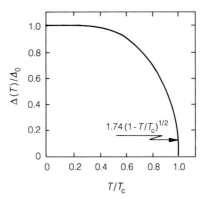

Figure 4-3. Reduced energy gap as a function of reduced temperature.

Table 4-1. Characteristic data of some elemental superconductors.

Superconductor	$2\Delta_0/k_B T_c$	T_c (K)	$\mu_0 H_0$ (mT)	$\Delta C/\gamma T_c$
BCS	3.53	–	–	1.43
Al	3.3–3.53	1.20	9.9	1.44
Cd	3.40	0.56	3.0	1.36
Ga	3.50	1.09	5.1	1.42
Hg	4.30	4.15	41.1	2.37
In	3.60	3.40	29.3	1.73
La	3.2–3.7	4.90	79.8	1.50
Nb	3.65	9.23	198.0	1.87
Pb	4.21	7.19	80.3	2.71
Sn	3.54	3.72	30.5	1.60
Ta	3.63	4.48	83.0	1.59
Tl	3.60	2.39	17.1	1.50
V	3.45	5.30	102.0	1.49
Zn	3.40	0.875	5.3	1.27

can be seen in Table 4-1. However, in the case of stronger electron-phonon interaction, i.e. $N(0)\,V_0 \geq 0.5$, this ratio increases, e.g. for Pb and Hg (see Scalapino, 1969). From Eq. (4-3) it is clear that T_c also should increase with growing $N(0)\,V_0$. This holds indeed, but up to a point where the crystal structure becomes unstable and the material transfers to a different phase with much lower T_c. A clear discussion of this subject and the upper boundary of T_c is given by Phillips (1989).

The normal excitations obey Fermi-Dirac statistics. Knowing $\Delta(T)$, the energy spectrum at finite T follows from replacing Δ_0 by $\Delta(T)$ in Eq. (4-2), giving $E_k(T)$. Substituting $E_k(T)$ into the Fermi-Dirac distribution function one can compute the free energy F_s and related thermodynamic quantities of the superconducting state. The energy difference, e.g., with the normal state free energy F_n defines the thermodynamic critical field H_c through (Sec. 4.2.7)

$$\tfrac{1}{2}\,\mu_0 H_c^2\, V = F_n(T) - F_s(T) \qquad (4\text{-}6)$$

where $\mu_0 = 4\pi \times 10^{-7}$ Vs/Am is the vacuum permeability and V the volume of the superconductor. The deviations from the empirical relation

$$H_c \approx H_0(1-t^2) \qquad (4\text{-}7)$$

are less than four percent. Exact BCS results are

$$H_c/H_0 = 1 - 1.06\,t^2, \quad t \approx 0 \qquad (4\text{-}8\,\text{a})$$

$$H_c/H_0 = 1.74\,(1-t), \quad t \approx 1 \qquad (4\text{-}8\,\text{b})$$

Values of H_0 are listed in Table 4-1 (note: $\mu_0 H_0$ of 1 T corresponds to H_0 of 10^4 Oe). One can obtain a rough estimate for H_0 by noting that $\mu_0 H_0/T_c \approx 10$–20 mT/K. This approximate proportionally follows from the relations $\Delta_0 = 1.76\, k_B T_c$ and

$$N(0)\,\Delta_0^2 = \mu_0 H_0^2 \qquad (4\text{-}9)$$

The left-hand side expresses the fact that the condensation energy of one Cooper pair is of order Δ_0 and that a fraction Δ_0/ε_F of the electrons is involved in the transition to the superconducting state, i.e. the electrons in a shell Δ_0 centered around ε_F.

The electron contribution to the specific heat shows two features related to the opening of the energy gap when the temperature is lowered through T_c. With respect to the normal state specific heat $C_n = \gamma T$, with $\gamma = 2\pi^2 k_B^2 N(0)/3$[1], C_s at T_c shows a jump given by $(C_s - C_n)/\gamma T_c = 1.43$ followed by an exponential decrease roughly proportional to $\exp(-\Delta_0/k_B T)$. The phonon specific heat does not change in the superconducting state. One has to be aware of low-temperature anomalies in the phonon contribution which have frequently been misinterpreted as deviations of C_s from the BCS behavior. See for comparison experimental equilibrium properties with the BCS theory: Meservey and Schwartz (1969).

1 In the superconductivity literature it is common practice to use the density of states per spin.

4.2.4 Electronic Properties

The absorption α of a longitudinal ultrasonic sound wave of frequency $\ll 2\Delta_0/\hbar$, so that pair breaking can be neglected for $T<T_c$, will decrease with the reduced amount of normal excitations below T_c. The theoretical result $\alpha_s/\alpha_n = 2/[1+\exp(\Delta/k_B T)]$ is experimentally well obeyed. A similar exponential decay is predicted and observed for the electronic contribution in the thermal conductivity. In addition, the phonon contribution shows an increase below T_c, because the scattering by electrons disappears. This is not always observed when other phonon-scattering mechanisms (e.g., lattice defects and crystal boundaries) dominate.

In computing effects like the ultrasonic absorption or the thermal conductivity it is essential to know the scattering amplitudes of the normal excitations from a state $(\boldsymbol{k}_i, \uparrow)$ to a state $(\boldsymbol{k}_f, \uparrow)$ or $(\boldsymbol{k}_f, \downarrow)$. A subtlety expressing the essence of the pairing theory, is the occurrence of four different coherence factors in the scattering amplitudes which are specific for the scattering process considered, e.g. electron-phonon scattering for the ultrasonic absorption and thermal conductivity, and spin-flip scattering for nuclear-spin relaxation, and the process related to electromagnetic absorption, see Schrieffer (1964) and Tinkham (1975). The reason of this important effect is that a normal excitation is created by either putting it in a state $(\boldsymbol{k}, \uparrow)$, or by annihilating the occupied pair state $(-\boldsymbol{k}, \downarrow)$ leaving its mate in $(\boldsymbol{k}, \uparrow)$. In the former case the state should be empty, and not occupied by a pair state. Both processes add coherently and should be summed before being squared to obtain the scattering rate. The final result may differ considerably from the behavior of the ultrasonic absorption. A striking example confirming the BCS theory, is the initial *increase* below T_c of the nuclear spin relaxation rate $1/T_1$, the "Hebel-Schlichter peak". It arises from a different combination of the coherence factors which now reveals the increase of the density of states at the gap edge (Fig. 4-2). At lower temperatures the opening of the gap overrules this effect suppressing T_1^{-1} to zero. Similar behavior is expected for the absorption of electromagnetic radiation. New effects arise in this case when the frequencies are large enough to break up Cooper pairs into two normal excitations, i.e. for $\hbar\omega \geq 2\Delta$. An extensive discussion of electromagnetic absorption and the response to a static magnetic field is given by Tinkham (1975). Many experimental results of nonequilibrium properties are compared with the BCS theory by Ginsberg and Hebel (1969). For further reading on the microscopic foundations of the BCS theory the reader is referred to Schrieffer (1964), Rickayzen (1964), and Kuper (1968).

4.2.5 London-Pippard Theory

The fact that superconductivity is a quantum phenomenon on a macroscopic scale is best illustrated by the behavior in a static or low frequency magnetic field as discovered by Meissner and Ochsenfeld in 1933. They showed that in addition to a *perfect conductor*, a superconductor also is a *perfect diamagnet*, i.e. a superconductor in an applied field expels the field when cooled from the normal state through T_c. This behavior can be well described by postulating two equations in addition to the Maxwell equations as has been shown in 1935 by F. and H. London. F. London (1950) gave the quantum motivation for the postulated equations. If the superconducting electrons condense into the groundstate described by a wavefunction ψ

and if for some reason this wavefunction is rigid (it retains its groundstate properties) when a field is applied, then one can use a theorem of Bloch which states that the net momentum $\boldsymbol{p}=m\boldsymbol{v}+e\boldsymbol{A}$ in the groundstate is zero, to arrive at $\langle \boldsymbol{v}_s \rangle = -e\boldsymbol{A}/m$. The vector potential $\boldsymbol{A}=\nabla\times\boldsymbol{b}$ with $\boldsymbol{b}$ the *local* magnetic field, $-e$ is the electron charge and $\langle v_s \rangle$ is the local average velocity of the superconducting electrons. Introducing the number density of electrons in the ground state $n_s \propto |\psi|^2$ we have for the supercurrent density

$$\boldsymbol{j}_s = n_s e \langle \boldsymbol{v}_s \rangle = -n_s e^2 \boldsymbol{A}/m \equiv -\boldsymbol{A}/\Lambda \qquad (4\text{-}10)$$

By taking the time derivative and the curl of both sides the two London equations are obtained

$$\boldsymbol{e} = \partial/\partial t\,(\Lambda \boldsymbol{j}_s) \qquad (4\text{-}11)$$

$$\boldsymbol{b} = -\nabla \times (\Lambda \boldsymbol{j}_s) \qquad (4\text{-}12)$$

This form being not gauge-invariant, requires a particular gauge choice for $\boldsymbol{A}$ known as the London gauge. It requires that $\nabla \cdot \boldsymbol{A} = 0$ (charge conservation in isolated superconductors), and that $\boldsymbol{A} \to 0$ in the interior of bulk samples. In the case of a supercurrent through the surface this should be related to the normal component of $\boldsymbol{A}$ by Eq. (4-10). In combination with

$$\nabla \times \boldsymbol{b} = \mu_0 \boldsymbol{j} \qquad (4\text{-}13)$$

Eq. (4-12) leads to the equation

$$\nabla^2 \boldsymbol{b} = \frac{\boldsymbol{b}}{\lambda_L^2} \qquad (4\text{-}14)$$

where λ_L denotes the London penetration depth:

$$\lambda_L = \left(\frac{m}{\mu_0 n_s e^2} \right)^{1/2} \qquad (4\text{-}15)$$

For a thick sample with a field $\mu_0 H$ applied parallel to the surface, it follows from Eq. (4-14) that the local field decreases exponentially, like $b = \mu_0 H \exp(-x/\lambda_L)$, where x denotes the distance from the surface. Because $\lambda_L \approx 100$ nm, Eq. (4-14) explains the Meissner-Ochsenfeld effect. Note that in a thin sample of dimension $< \lambda_L$, the field will not be entirely expelled. The penetration depth depends on temperature via n_s. An often used empirical relation originating from the two-fluid description of superconductivity by Gorter and Casimir (1934 a, b), is

$$\lambda_L(T) \approx \lambda_L(0)\,(1-t^4)^{-1/2} \qquad (4\text{-}16)$$

Eq. (4-10) relates the *local* vector potential to the *local* supercurrent density. Since the superconducting coherence is a *non-local* phenomenon with length scale ξ_0, the value of the vector potential should be averaged over a sphere of radius ξ_0 about the position $\boldsymbol{r}$ where one wants to know $\boldsymbol{j}_s$. This has been first realised by Pippard in 1953 (see Tinkham, 1975) who suggested the following nonlocal relation

$$\boldsymbol{j}_s(\boldsymbol{r}) = -\frac{3}{4\pi \xi_0 \Lambda} \int \frac{\boldsymbol{R}[\boldsymbol{R} \cdot \boldsymbol{A}(\boldsymbol{r}')]}{|\boldsymbol{R}|^4} \cdot e^{-|\boldsymbol{R}|/\xi_0}\, d^3 r', \quad \boldsymbol{R} = \boldsymbol{r}' - \boldsymbol{r} \qquad (4\text{-}17)$$

This expression tells us that an electron travelling from $\boldsymbol{r}'$ to $\boldsymbol{r}$ remembers the value of $\boldsymbol{A}$ it experienced at $\boldsymbol{r}'$. When $\lambda_L(0) \gg \xi_0$, the vector potential does not change over a distance ξ_0 and can be taken out of the integral yielding Eq. (4-10). From Table 4-2, where for some elements the values of $\lambda_L(0)$ and ξ_0 are collected, it follows that for most elements Pippard's equation should be used. Assuming $\xi_0 \gg \lambda_L(0)$ and that both $\boldsymbol{b}$ and $\boldsymbol{A}$ still drop exponentially at the surface, but now over a typical distance $\lambda_{eff} \ll \xi_0$, the integral yields $(4\pi/3)\,\lambda_{eff} \boldsymbol{A}$. Combining this with $\boldsymbol{j}_s \approx -(\mu_0 \lambda_{eff}^2)^{-1} \boldsymbol{A}$, an estimate for λ_{eff} is obtained, namely $\lambda_{eff}^3 \approx \xi_0 \lambda_L^2(0)$.

Table 4-2 shows that $\xi_0 \approx 0.1$ to $1\ \mu m$. Therefore, ξ_0 can be larger than the electron mean free path ℓ for elastic scattering, especially in alloys. Elastic scattering does not erase the phase coherence of the Cooper pairs, but it will reduce the distance ξ over which $\boldsymbol{A}(\boldsymbol{r}')$ contributes to $\boldsymbol{j}_s(\boldsymbol{r})$ (see for a discussion of this point De Gennes, 1966; p. 219). Pippard conjectured $\xi^{-1} = \xi_0^{-1} + (\alpha \ell)^{-1}$, with α an adaptable parameter which turned out to be $\alpha = 1.33$. This ξ should replace ξ_0 in Eq. (4-17). In the case of extremely small ℓ, known as the dirty limit ($\ell \ll \xi_0$), integration of Eq. (4-17) yields $(4\pi/3)\,\alpha \ell$ which gives for the penetration depth in this limit

$$\lambda_d = \lambda_L(0)\,[\xi_0/(\alpha \ell)]^{1/2} \qquad (4\text{-}18)$$

This is an important result for superconducting technology, as we will see later. It shows that λ_d increases with alloying. On the other hand, one can trim the electron density and density of states by alloying in such a way that T_c goes up and v_F goes down resulting in a small value of ξ_0, see Gladstone et al. (1969). The condition $\lambda_d \gg \xi_0$ is fulfilled for all technical superconductors. For a more extensive discussion and comparison with experiment we refer to Meservey and Schwartz (1969) and Tinkham (1975).

Table 4-2. Penetration depth and coherence length for some elemental superconductors in nm.

Superconductor	$\lambda_L(0)$ theoretical	$\lambda_L(0)$ experimental	ξ_0
Al	16	51	1360
Cd	110	130	760
In	40	40–64	250–350
Nb	39	31	39
Pb	37	39–63	51–96
Sn	34	52	94
V	39	38	46

4.2.6 Tunneling and the Josephson Effects

Interesting features occur when a superconductor is separated from a normal conductor by a very thin (≈ 2 nm) insulating layer, usually a native oxyde. Evidently, at zero temperature and a bias voltage $|V| < \Delta_0$ this device (SIN tunnel junction) does not conduct. As soon as $|V| > \Delta_0$, however, a steep increase of the conductance is detected, Fig. 4-4a, because there are single-electron states available at the gap-edge in the superconductor. The conductance measures the density of states in the superconductor and its behavior is thus reminiscent to that of $N(E)$ in Fig. 4-2 if one replaces E/Δ by eV/Δ, see Fig. 4-4b. In fact, the real measurements show more features at elevated voltages which are related to the fact that the gap equation depends on the details of the electron-phonon interaction and the phonon modes of the superconductor. These details were ignored in the BCS theory. By using an inversion scheme (see McMillan and Rowell, 1969) tunnel experiments provide valuable information on the phonon spectrum. At higher temperatures the contribution of the thermally excited normal excitations will smear out the sharp features of the $T = 0$ effects, as depicted by the dashed lines in Fig. 4-4.

Superconductor-insulator superconductor (SIS′) tunnel junctions behave similarly as regards the tunneling of single electrons. As can be easily understood, conductance anomalies now occur at both $eV = |\Delta_1 - \Delta_2|$ and $\Delta_1 + \Delta_2$ for superconductors with different gaps, Fig. 4-4c. In addition to single-electron tunneling, pair tunneling should as well be possible in an SIS′ junction, although it seemed as if the pair-tunneling amplitude would be much smaller. Josephson in 1962 showed that in fact the amplitudes for single-particle and pair tun-

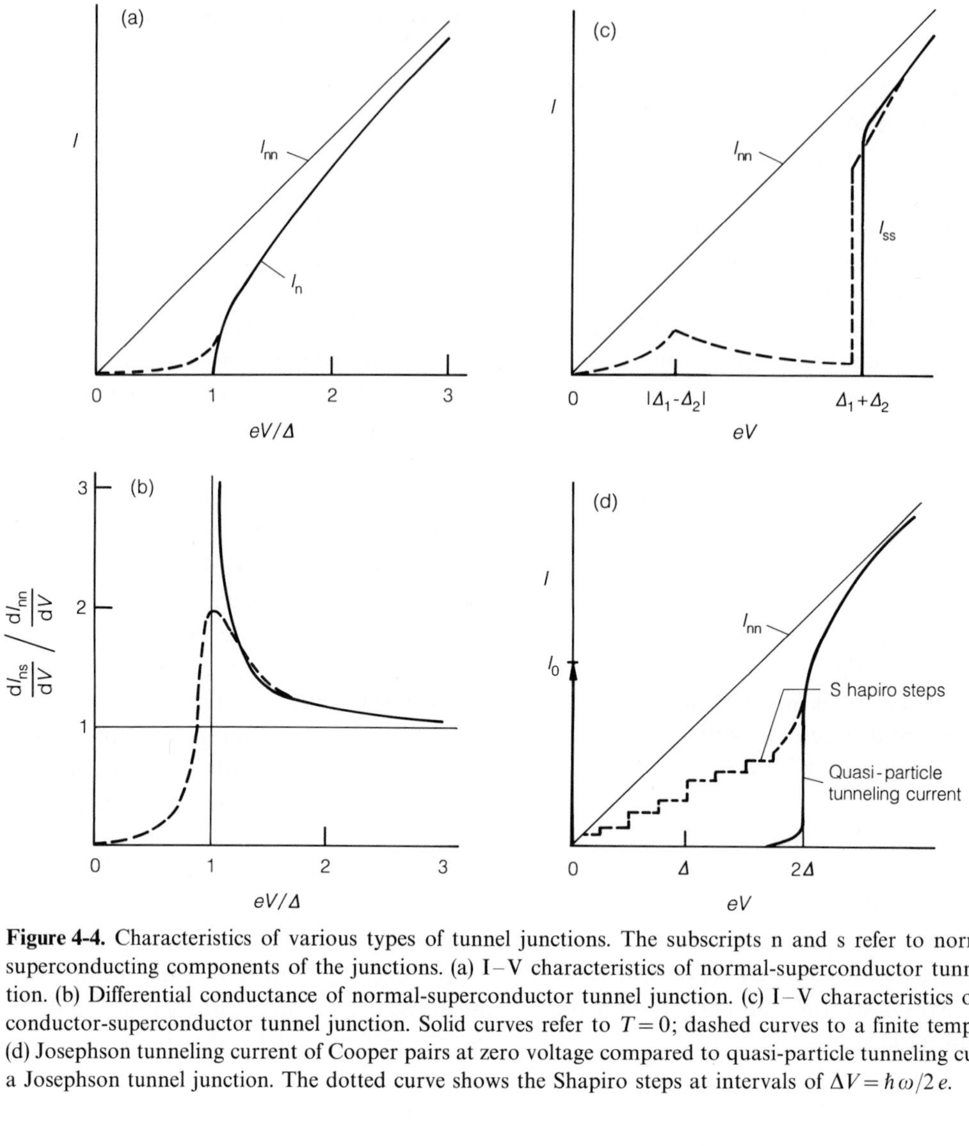

Figure 4-4. Characteristics of various types of tunnel junctions. The subscripts n and s refer to normal and superconducting components of the junctions. (a) I–V characteristics of normal-superconductor tunnel junction. (b) Differential conductance of normal-superconductor tunnel junction. (c) I–V characteristics of superconductor-superconductor tunnel junction. Solid curves refer to $T = 0$; dashed curves to a finite temperature. (d) Josephson tunneling current of Cooper pairs at zero voltage compared to quasi-particle tunneling current of a Josephson tunnel junction. The dotted curve shows the Shapiro steps at intervals of $\Delta V = \hbar\omega/2e$.

neling are proportional because of the coherence of the latter process, see Josephson (1965) and Anderson (1964). From first principles it was predicted that at zero-voltage over the junction a superconducting tunnel-current is generated by the phase difference $\Delta\phi = \phi_2 - \phi_1$ of the BCS wavefunctions in both superconductors S′ and S, see Fig. 4-4d. Ignoring the effect of a magnetic field the current density through the junction is

$$j = j_0 \sin \Delta\phi \tag{4-19}$$

with j_0 derived by Ambegoakar and Baratoff (1963):

$$j_0 = \frac{\pi \Delta(T)}{2 e R_n} \tanh \frac{\Delta(T)}{2 k_B T} \tag{4-20}$$

Here R_n is the normal-state resistance per unit area of junction. In presence of a magnetic field parallel to the junction the phase difference across the junction also depends on the vector potential giving rise to a Fraunhofer-like modulation of the maximum current with magnetic flux Φ contained in the junction. Zero's of j occur at multiple values of the flux quantum $\Phi_0 = 2.07 \times 10^{-15}$ Weber (see Sec. 4.3.4). This "Fraunhofer" pattern still forms one of the most convincing evidences of the dc Josephson effect.

The ac Josephson effect arises when a voltage is applied over the junction. Pair tunneling now involves an energy change $2eV$ which has to be compensated by the emission of a photon of energy $\hbar\omega = 2eV$ related to a fast oscillation of the phase difference $\Delta\phi$ according to

$$\frac{\partial \Delta\phi}{\partial t} = \frac{2eV}{\hbar} = \omega \tag{4-21}$$

Substitution in Eq. (4-19) shows that the application of a dc voltage gives rise to an ac current with a frequency of 483.593726 MHz per μV. The first experimental evidence for the ac Josephson effect was obtained by Shapiro (1963), who observed current steps in the dc tunneling characteristic of a Josephson junction subjected to microwave irradiation of frequency ω at voltages which were precisely integral multiples of $\hbar\omega/2e$, as seen in Fig. 4-4d. For more details, e.g., on field effects and practical applications of the Josephson effects see Solymar (1972) and Van Dutzer and Turner (1981). The exploitation of tunnel junctions for electron spectroscopy has been extensively reviewed by Wolf (1989).

4.2.7 Thermodynamics of Bulk Superconductors

In order to maintain the Meissner state with $B = 0$ a superconducting screening current is generated upon applying a magnetic field. At the thermodynamic critical field H_c the kinetic energy of the screening current equals the condensation energy. $H_c(T)$ therefore marks the phase boundary between superconducting and normal states. It can be obtained from a simple thermodynamic consideration. The Gibbs free energy *density* g of a bulk superconductor (dimensions $\gg \lambda$) in a field H (ignoring demagnetization effects) is given by

$$dg = -s\,dT - \mu_0 \boldsymbol{M} \cdot d\boldsymbol{H} \tag{4-22}$$

with s the entropy density. At the phase boundary we have $g_s = g_n$. After integration of Eq. (4-22) at constant T, using $M = -H$, and taking $M = 0$ in the normal state, it follows that

$$g_{sn}(T, H=0) = g_s(T,0) - g_n(T,0) = = -\tfrac{1}{2}\mu_0 H_c^2(T) \tag{4-23}$$

This is the same result as Eq. (4-6) because $f = g + \boldsymbol{B} \cdot \boldsymbol{H}$ and g_{sn} is deduced at $H = 0$. The entropy difference is determined using $s = -(\partial g/\partial T)_H$ which gives

$$s_s(T,0) - s_n(T,0) = \mu_0 H_c(T) \frac{dH_c}{dT} \tag{4-24}$$

Note $s_s = s_n$ both at $T = 0$ and $T = T_c$, and $s_s < s_n$ for $0 < T < T_c$. This means that the phase transition in zero field is of second order, i.e. no latent heat: $Q = T(s_n - s_s) = 0$ at T_c. The specific heat difference is obtained from $C = T(\partial s/\partial T)_H$:

$$C_s - C_n = \mu_0 T \left[\left(\frac{dH_c}{dT} \right)^2 + H_c(T) \frac{d^2 H_c}{dT^2} \right] \tag{4-25}$$

Evaluated at T_c Eq. (4-25) yields the specific heat jump at T_c (Sec. 4.2.3):

$$\frac{\Delta C}{\gamma T_c} = \frac{\mu_0}{\gamma} \left(\frac{dH_c}{dT} \right)^2_{T_c} \tag{4-26}$$

It follows from Eqs. (4-8) and (4-9) that $\Delta C/\gamma T_c$ should be a universal constant for a BCS superconductor, namely 1.43 (see Table 4-1).

4.2.8 Depairing Current

From the introduction of Sec. 4.2.7 it can be deducted that H_c/λ is related to the maximum current density a superconductor can sustain, i.e. the depairing current j_0. The coherence breaks down for Cooper pairs moving with a velocity $v_c = \Delta_0/(m\,v_F)$ corresponding to

$$j_0(0) \approx \left(\frac{2}{3\pi^2}\right)^{1/2} \frac{H_0}{\lambda_L(0)} \tag{4-27}$$

Only a fraction of the pairs are moving in the direction of the current. Nevertheless superconductivity is destroyed at a slightly (2 percent) larger current density than given by Eq. (4-27). The value of j_0 is of the order 10^{12} A/m^2. See Tinkham (1975) and Abrikosov (1988).

4.3 Ginzburg-Landau (GL) Theory

4.3.1 Intermediate State

The Meissner-Ochsenfeld effect provides a simple description for the magnetic behavior of a long cylindrical superconductor in an axial magnetic field: up to $H = H_c$ the magnetization grows linearly as $M = -H$, and it drops to zero at H_c. A more complicated situation occurs if the cylinder is placed in a transverse field. The demagnetization effect leads to an internal field

$$H_i = H - D\,M = H + D\,H_i \tag{4-28}$$

D is the demagnetizing factor (0.5 for a long cylinder in a transverse field). At $H = 0.5\,H_c$ the internal field has reached H_c. However, since the full condensation energy is not yet reached, this does not lead to the normal state. Instead an intermediate state is formed consisting of alternating normal-superconducting (ns) layers. At each ns interface the local field decays exponentially over λ, see Fig. 4-5, therefore gaining a field energy $-\lambda\,\mu_0\,H_c^2/2$ per unit area. It is unlikely that n_s at the interface would abruptly drop to zero. Instead, it is assumed that it gradually decays over a distance ξ at the cost of an amount $\xi\,\mu_0\,H_c^2/2$ of condensation energy. Making up the balance a net surface energy $(\xi - \lambda)\,\mu_0\,H_c^2/2$ is related to the interface. If this quantity is negative ($\lambda > \xi$) the intermediate state would be unstable. In fact, the superconductor would have assumed an entirely different state, the mixed state (Sec. 4.3.7 ff.), at a lower transition field H_{c1}. Situations in which n_s varies spatially can be treated in the framework of the phenomenological theory developed by Ginzburg and Landau (1950).

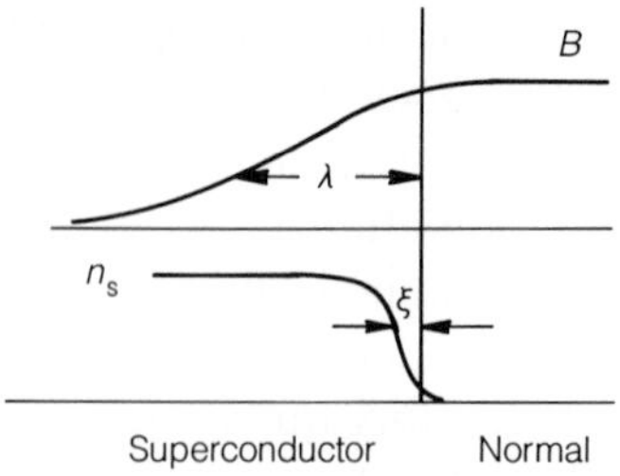

Figure 4-5. Variation of magnetic field and pair density at a normal-superconductor interface for the case $\lambda \gg \xi$.

4.3.2 Ginzburg-Landau Free Energy

At a second order phase transition the free energy can be written as a series expansion of an order parameter. For a superconductor this applies for $T \approx T_c$. A natural choice for the order parameter is the density of Cooper pairs $n_s/2$. By means of

$$n_s = 2\,|\psi|^2 \tag{4-29}$$

this is expressed in terms of a macroscopic wave function of the superconducting electrons

$$\psi(r) = \psi_0(r)\,e^{i\phi(r)} \tag{4-30}$$

In order to incorporate the stiffness of ψ against spatial fluctuations as well as the effect of local supercurrents and fields, Ginzburg-Landau (GL) proposed the following expression for the free energy density f_s

$$f_s - f_{n0} = \alpha|\psi|^2 + \frac{\beta}{2}|\psi|^4 + \frac{1}{4m}\left|\left(\frac{\hbar}{i}\nabla - 2eA\right)\psi\right|^2 + \frac{b^2}{2\mu_0} \tag{4-31}$$

Here, f_{n0} is the normal-state free energy density in zero field. The first two terms in Eq. (4-31) describe the situation of a uniform superconductor without fields. The physics of the problem requires $n_s \equiv 0$ for $T \geq T_c$, $n_s > 0$ for $T < T_c$, and $\partial f_s / \partial n_s > 0$ for $n_s \to \infty$, otherwise $n_s = \infty$ would always give the lowest energy. Therefore one should have

$$\alpha(T) = -\alpha_0(T_c - T), \quad \alpha_0 > 0,\ \beta > 0 \tag{4-32}$$

Minimization with respect to n_s and use of Eq. (4-23) leads to

$$\psi_\infty^2 \equiv -\frac{\alpha}{\beta} \tag{4-33}$$

$$\mu_0 H_c^2 = \alpha^2/\beta \tag{4-34}$$

It follows that β should only weakly depend on temperature, because $H_c \propto (T_c - T)$ for $T \approx T_c$.

The fourth and third term in Eq. (4-31) represent the energies related to fields, currents and gradients in ψ. At this point it should be noted that the form of Eq. (4-31) is such that gauge invariance is guaranteed. A gauge transformation $A' = A + \nabla\chi$ is compensated by a change of phase of ψ according to

$$\phi' = \phi + (2e/\hbar)\chi \tag{4-35}$$

Using Eq. (4-30) the third term can be rewritten as the sum of a gradient part and a kinetic-energy part:

$$\left|\left(\frac{\hbar}{i}\nabla - 2eA\right)\psi\right|^2 = \hbar^2(\nabla\psi_0)^2 + (\hbar\nabla\phi - 2eA)^2\psi_0^2 \tag{4-36}$$

By choosing a gauge in which ψ is real, the second term can be simply equated to the classical kinetic energy density leading to

$$(e^2/m)A^2\psi_0^2 = \tfrac{1}{2}m v_s^2 n_s = (m j_s^2 / 2e^2 n_s) \tag{4-37}$$

This results into Eq. (4-10) and a generalization of the definition of the penetration depth which now may depend on position and field (see next section for justification):

$$\lambda_{\mathrm{eff}}^2 = \frac{m}{2\mu_0 e^2|\psi|^2} = \frac{m}{2\mu_0 e^2 \psi_0^2} \tag{4-38}$$

Combining Eqs. (4-33), (4-34) and (4-38) and replacing ψ_∞ by ψ_0, expressions for $\alpha(T)$ and $\beta(T)$ can be derived

$$\alpha(T) = -\frac{2\mu_0^2 e^2}{m} H_c^2(T)\,\lambda_{\mathrm{eff}}^2(T) \tag{4-39}$$

$$\beta(T) = \frac{4\mu_0^3 e^4}{m^2} H_c^2(T)\,\lambda_{\mathrm{eff}}^4(T) \tag{4-40}$$

4.3.3 Ginzburg-Landau Equations

The GL expression depends on the material parameters α and β and on the "fields" $\psi(r)$ and $A(r)$. To find these "fields" the GL energy which is obtained by integrating Eq. (4-31) over the superconductor volume, is minimized with respect to ψ and A. Putting a surface term equal to zero (Sec. 4.3.13) GL derived two coupled differential equations in ψ and $\boldsymbol{A}$

$$\alpha\psi+\beta|\psi|^2\psi+\frac{1}{4m}\left(\frac{\hbar}{\mathrm{i}}\nabla-2e\boldsymbol{A}\right)^2\psi=0 \tag{4-41}$$

$$\boldsymbol{j}=\frac{1}{\mu_0}\nabla\times\boldsymbol{b}=\frac{e\hbar}{2\mathrm{i}m}(\psi^*\nabla\psi-\psi\nabla\psi^*)-\\-\frac{2e^2|\psi|^2}{m}\boldsymbol{A} \tag{4-42}$$

Clearly, there exists a large class of solutions depending on the physical circumstances and boundary conditions. Using Eq. (4-30) the second GL equation can be simplified to

$$\boldsymbol{j}=\frac{e}{m}(\hbar\nabla\phi-2e\boldsymbol{A})\psi_0^2 \tag{4-43}$$

For a real order parameter this leads to the definition of λ_{eff} in Eq. (4-38). It should be noted that the GL equations are local equations valid for $(T_c-T)\ll T_c$, although they are frequently used outside this regime without making a large mistake. So, they often provide a good starting point for theoretical investigations. However, the GL equations are inadequate in describing processes on a length scale $<\xi_0$, e.g., boundary effects. For this one has to rely on the microscopic form of these equations derived by Gor'kov in 1959 (Gor'kov, 1959, 1960) and reviewed by Werthamer (1969). The microscopic theory also includes the effects of impurity scattering.

4.3.4 Flux Quantization

Suppose one wants to know the flux enclosed by a hollow superconducting cylinder. If Eq. (4-43) is integrated along a contour in a region where the external field is totally screened off, i.e. $j=0$ and ψ_0 uniform, one obtains

$$\frac{\hbar}{2e}\oint\nabla\phi\cdot\mathrm{d}\boldsymbol{l}=\oint\boldsymbol{A}\cdot\mathrm{d}\boldsymbol{l}=\Phi \tag{4-44}$$

To guarantee ψ to be a single-valued function of r the phase should be determined modulo 2π, i.e. $\oint\nabla\phi\cdot\mathrm{d}\boldsymbol{l}=n\,2\pi$, which leads to $\Phi=n\,\Phi_0$ the flux quantum:

$$\Phi_0=\frac{2\pi\hbar}{2e} \tag{4-45}$$

Hence, the flux contained by the contour is an integer multiple of Φ_0. However, if the contour is taken close to the surface, $j\neq0$, giving rise to fluxoid quantization:

$$\Phi+\mu_0\lambda_{\text{eff}}^2\oint\boldsymbol{j}\cdot\mathrm{d}\boldsymbol{l}=n\,\Phi_0 \tag{4-46}$$

4.3.5 Characteristic Length Scales

We have seen that λ_{eff} is the length scale governing variations of field, vector potential and supercurrent. λ_{eff} can become very large in cases where ψ_0^2 is small (see Sec. 4.3.10). The length scale for variations of the order parameter follows from Eq. (4-42) ignoring field effects by putting $A=0$. By defining a reduced order parameter $f(r)\equiv\psi/\psi_\infty$ one obtains $(\hbar^2/4m|\alpha|)\cdot\nabla^2f+f-f^3=0$. The solution decays over a distance $\xi(T)$, the GL coherence length, which is defined by

$$\xi^2(T)=\frac{\hbar^2}{4m|\alpha|} \tag{4-47}$$

It should be noted that $\xi(T)$ differs from the Pippard or BCS coherence length ξ_0 which is related to the lifetime of a Cooper pair $\tau_c\approx\xi_0/v_F$. In impure superconductors with $\ell\ll\xi_0$ the phase coherence of the Cooper pairs is not destroyed by elastic scattering, but the distance over which superconducting order is maintained is now determined by a diffusion length $(D\,\tau_c)^{1/2}\approx(\ell\,\xi_0)^{1/2}$. Therefore, $\xi\approx(\ell\,\xi_0)^{1/2}$ in dirty superconductors.

A convenient expression relating the different parameters is obtained by combining Eqs. (4-39), (4-45) and (4-47):

$$\Phi_0 = 2^{3/2}\,\pi\,\mu_0\,H_c(T)\,\xi(T)\,\lambda_{\rm eff}(T), \quad T \approx T_c \tag{4-48}$$

A similar expression can be derived for $T=0$

$$\Phi_0 = (2/3)^{1/2}\,\pi\,\mu_0\,H_0\,\xi_0\,\lambda_L(0) \tag{4-49}$$

Now we can make use of Eqs. (4-8 b), (4-16) and (4-18) to get

$$\lambda_{\rm eff}(t) = \lambda_L(t) = \frac{\lambda_L(0)}{[2(1-t)]^{1/2}};$$

$$\xi(t) = \frac{0.74\,\xi_0}{(1-t)^{1/2}} \quad \text{(pure)} \tag{4-50}$$

$$\lambda_{\rm eff}(t) = \left[\frac{\lambda_L^2(0)\,\xi_0}{2.66\,\ell\,(1-t)}\right]^{1/2};$$

$$\xi_0 = \frac{0.855\,(\xi_0\,\ell)^{1/2}}{(1-t)^{1/2}} \quad \text{(dirty)} \tag{4-51}$$

With increasing impurity concentration λ increases and ξ decreases. Alloying therefore may give rise to a sign change in the energy of an ns interface with quite different physical properties as result (Sec. 4.3.6). This change in character can be more directly described by a dimensionless parameter which does not diverge at T_c, namely the GL parameter $\varkappa$ defined as

$$\varkappa \equiv \frac{\lambda_{\rm eff}}{\xi} =$$

$$= \begin{cases} 0.96\,\lambda_L(0)/\xi_0 \equiv \varkappa_0 & \text{(pure)} \quad (4\text{-}52) \\ 0.715\,\lambda_L(0)/\ell & \text{(dirty)} \quad (4\text{-}53) \end{cases}$$

The exact impurity dependence of $\varkappa$ has been derived by Gor'kov (1958 a, b; 1959 a, b; 1960) for weak coupling superconductors. He deduced

$$\varkappa = \varkappa_0\,\chi^{-1}(\varrho_d) \tag{4-54}$$

where $\varrho_d = 0.882\,\xi_0/\ell$ and $\chi(\varrho_d)$ is the Gor'kov function ($\chi = 1$ for $\varrho_d = 0$ and $1.33\,\ell/\xi_0$ for $\varrho_d \gg 1$). A useful empirical relation has been given by Goodman (1964): $\varkappa \approx \varkappa_0 + 2.4\times 10^6\,\varrho\,\gamma^{1/2}$ with ϱ in Ωm and γ in $\rm J\,m^{-3}\,K^{-2}$. It is accurate within 4 percent.

4.3.6 Nucleation Fields

Nucleation of superconductivity as a second-order phase transition in a field should follow from the linearized form of Eq. (4-41), i.e. neglecting the β term. The linearized equation is identical to the Schrödinger equation of a charged particle in a uniform field along the z axis with energy eigenvalues

$$-\alpha = \frac{1}{2}\,m\,v_z^2 + \left(n + \frac{1}{2}\right)\frac{\hbar\,e\,\mu_0\,H}{m}, \quad n \geq 0 \tag{4-55}$$

The lowest eigenvalue $-\alpha > 0$ at the largest value of H is obtained by the choice $v_z = 0$ and $n = 0$. Using Eq. (4-47) the nucleation field can be expressed as

$$\mu_0\,H_{c2} = \frac{\Phi_0}{2\,\pi\,\xi^2(T)} \tag{4-56}$$

and by substituting Eq. (4-48) as

$$H_{c2} = \varkappa\,\sqrt{2}\,H_c \tag{4-57}$$

In decreasing field superconductivity will spontaneously nucleate when $H = H_{c2}$. If $\varkappa > 2^{-1/2}$ then $H_{c2} > H_c$ and a state is favored in which $|\psi|^2$ gradually increases from zero. For $\varkappa < 2^{-1/2}$ a first-order phase transition characterized by a jump of the order parameter to ψ_∞^2 at H_c, preceeds the second-order transition. However, undercooling may delay the transition down to $H_{c2} < H_c$. At this point it becomes clear that two types of superconductors should be distinguished: type I with $\varkappa < 0.7$ and a magnetic phase diagram as discussed in Sec. 4.2.7, and type II with $\varkappa > 0.7$. The properties of type II superconductors is the subject of the following sections.

In concluding this section it should be remarked that in ideal situations nucleation at the surface can take place at an even higher field H_{c3}. It gives rise to surface superconductivity in a layer of thickness ξ. This subject is extensively described in Saint-James et al. (1969). The effect only occurs if the field is parallel to the surface which should also be of high quality. For a cylinder the nucleation field is found to be $H_{c3} = 1.70\,H_{c2}$.

4.3.7 The Abrikosov Vortex Lattice

The solution of the GL equations of type II superconductors in fields $H \lesssim H_{c2}$ has been given by Abrikosov in 1957, see also Abrikosov (1988). For $\varkappa > 0.7$ the energy of an ns interface is negative and for that reason a pattern of alternating ns regions should have the lowest energy. In the n regions $|\psi|^2 = 0$, in the s regions it is finite but much smaller than ψ_∞^2. For reasons of symmetry a structure of ns laminae is not very likely. In fact, it turns out to have a larger energy than the structure proposed by Abrikosov which consists of a periodic lattice of lines of $|\psi|^2 = 0$ parallel to the applied field. At the position of the lines the field penetrates and decays over a distance λ by the action of supercurrents circulating around the lines. Therefore the name vortex lattice (VL) has been introduced.

The geometry of the VL is yet to be determined. It can be shown that the current pattern coincides with the lines of equal $|\psi|^2$ and that the related field b_s is given by

$$b_s = -(\mu_0 H_{c2}/2\varkappa^2)\, f^2 \qquad (4\text{-}58)$$

with $f^2 = |\psi|^2/\psi_\infty^2$. After averaging over a VL cell Eq. (4-58) yields the induction $B = \mu_0 H + \bar{b}_s$ and the magnetization $M = \mu_0^{-1} B - H$. Because of the periodicity each line is enclosed by a contour of $j_s = 0$ and therefore contains a quantized amount of flux, in fact one flux quantum. Consequently, the names flux line and flux-line lattice are often used as well. By minimizing the free energy Abrikosov arrived at the relation

$$\overline{f^4}\left(1 - \frac{1}{2\varkappa^2}\right) - \overline{f^2}\left(1 - \frac{H}{H_{c2}}\right) = 0 \qquad (4\text{-}59)$$

Introducing the geometry parameter

$$\beta_A = \overline{f^4}/(\overline{f^2})^2 \qquad (4\text{-}60)$$

the magnetization and free energy density can now be easily obtained

$$-M = \frac{H_{c2} - H}{\beta_A(2\varkappa^2 - 1)} = \frac{B_{c2} - B}{\mu_0[1 + \beta_A(2\varkappa^2 - 1)]} \qquad (4\text{-}61)$$

$$g_s(H) = g_n(H_{c2}) + \frac{1}{2}\mu_0(H_{c2}^2 - H^2) - \\ - \frac{\mu_0(H_{c2} - H)^2}{2\beta_A(2\varkappa^2 - 1)} \qquad (4\text{-}62)$$

The $\varkappa$ in these expressions is generally indicated as $\varkappa_2$; that in Eq. (4-57) as $\varkappa_1$. At low temperatures they may differ slightly from each other as follows from the microscopic theory.

From Eq. (4-62) it follows that the VL geometry with the smallest β_A also has the lowest energy. Saint-James et al. (1969) show how β_A changes with the shape of the unit cell of the VL. For a square lattice with lattice parameter $a_\square = (\Phi_0/B)^{1/2}$ it is $\beta_{A\square} = 1.18$. It monotonuously decreases to $\beta_{A\triangle} = 1.16$ for a hexagonal lattice with lattice parameter

$$a_\triangle \equiv a_0 = (2/\sqrt{3})^{1/2}(\Phi_0/B)^{1/2} \qquad (4\text{-}63)$$

Note that the energy difference of both configurations is only 2 percent. It means that a VL shears easily. Decoration experiments (Brandt and Essmann, 1987) have indeed proven that the hexagonal lattice is preferred. Sometimes, at low fields in material with a cubic crystal-lattice, the square lattice can be detected. Usually, a lot of disorder is also observed. The latter is caused by defects in the superconductor crystal lattice. These defects locally change the free energy giving rise to pinning of the VL (Sec. 4.4). To estimate the force of a defect (pinning center) it is crucial to know the spatial variation of $|\psi|^2$. It turns out that in very good approximation the Abrikosov solution may be written as (Saint-James et al., 1969):

$$f^2 \approx \left(1-\frac{B}{B_{c2}}\right)\left\{1-\frac{1}{3}\left[\cos k_0\left(x-\frac{y}{\sqrt{3}}\right)+ +\cos k_0\frac{2y}{\sqrt{3}}+\cos k_0\left(x+\frac{y}{\sqrt{3}}\right)\right]\right\} \tag{4-64}$$

where the x axis is taken along the closed packed direction of the VL and $k_0 = 2\pi/a_0$.

4.3.8 Single-Vortex Properties and Lower Critical Field

At low fields the Meissner state prevails up to a value H_{c1}, the lower critical field. At H_{c1} the first vortex is created at the surface and enters the material. This field H_{c1} is determined by $G_M = G_{ms}$ (M, ms denote Meissner and mixed state, respectively). If the energy per unit length of a single vortex is U_v and L is the length of the vortex, it follows from $G = F - \int \boldsymbol{H}\cdot\boldsymbol{B}\,d^3r$ that $F_M/L = F_M/L + U_v - H_{c1}\Phi_0 L$. Hence,

$$H_{c1} = U_v/\Phi_0 \tag{4-65}$$

U_v can be obtained from Eq. (4-31). If the contribution of the core may be neglected, i.e. if $\xi \ll \lambda$ (London limit), only two terms remain: the field energy and the kinetic energy of the supercurrents. These terms should be integrated over the volume of the superconductor excluding the vortex core ($r = \xi$). The final expression contains a product of b and $db/dr \propto j_s$ determined at $r = \xi$. They follow from the extended London equation

$$\boldsymbol{b} + \lambda^2 \nabla\times\nabla\times\boldsymbol{b} = \boldsymbol{\Phi}_0\,\delta_2(\boldsymbol{r}) \tag{4-66}$$

The right-hand side expresses the fact that each vortex contains one flux quantum. The solution of Eq. (4-66) in cylindrical coordinates is

$$\boldsymbol{b}(r) = \frac{\Phi_0}{2\pi\lambda^2}K_0\left(\frac{r}{\lambda}\right), \quad r > \xi \tag{4-67}$$

where $K_0(z)$ is the zero-order Hankel function of imaginary argument. $K_0 \sim \ln z$ for $z \to 0$ and $K_0 \sim z^{-1/2}\,e^{-z}$ for $z \gg 1$. The solution is sketched in Fig. 4-6 together with the current distribution and the reduced order parameter. The final result is

$$U_v = \frac{\Phi_0^2}{4\pi\mu_0\lambda^2}(\ln\varkappa + 0.50) \tag{4-68}$$

where the second term in the right-hand side contains a small core contribution. Note that, because $U_v \propto \Phi_0^2$, one flux quant

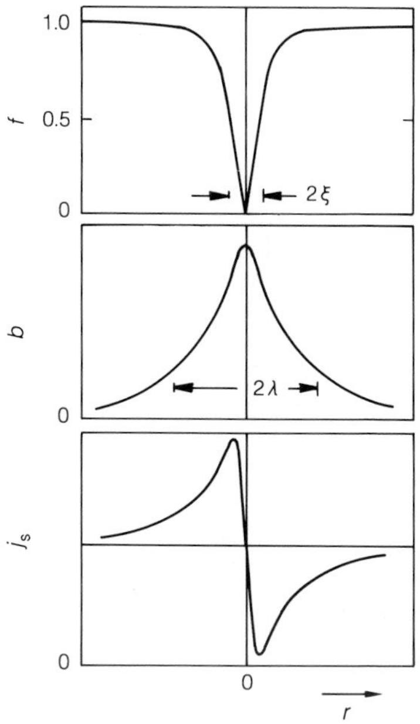

Figure 4-6. Variation of the reduced order parameter f (upper frame), the local field b (middle frame), and the current density j_s (lower frame) for an isolated vortex ($B < 0.2\,B_{c2}$).

per vortex gives the lowest energy. Finally, we get

$$H_{c1} = \frac{\Phi_0}{4\pi \mu_0 \lambda^2} (\ln \varkappa + 0.50) = $$

$$= \frac{H_c}{\varkappa \sqrt{2}} (\ln \varkappa + 0.50) \quad (4\text{-}69)$$

Combining this with Eq. (4-57) we find $H_{c1} \cdot H_{c2} \approx H_c^2 \ln \varkappa$. This $\varkappa$ is usually called $\varkappa_3$. From Eqs. (4-67) and (4-69) it follows that the field in the vortex core is about twice the applied field, namely $b(0) \approx 2\mu_0 H_{c1}$. Exact solutions, also for $\xi \approx \lambda$, are obtained by Kramer, Pesch, Watts-Tobin and Brandt, see Brandt and Essmann (1987) for further details.

4.3.9 Vortex Interaction and the Magnetization Curve

The interaction between two parallel vortices in the high-kappa limit can be obtained by adding the field profiles and integrating the extended London equation excluding the vortex cores. The result is

$$U_{12} = \frac{\Phi_0 b_1(r_2)}{\mu_0} = \frac{\Phi_0^2}{2\pi \mu_0 \lambda^2} K_0 \left(\frac{r_{12}}{\lambda} \right) \quad (4\text{-}70)$$

where $b_1(r_2)$ is the field of vortex 1 at the core position r_2 of the other vortex and r_{12} is the distance between them. The interaction is repulsive; the force is obtained by taking the gradient which leads to the very general expression:

$$\boldsymbol{f} = \boldsymbol{j}_s \times \boldsymbol{\Phi}_0 \quad (4\text{-}71)$$

This relation also describes the driving force on the vortices by a transport current through a superconductor in the mixed state.

The magnetization can be derived by adding the interaction terms to the Gibbs free energy at H_{c1} and by minimizing with respect to B. For fields just above H_{c1} only nearest neighbor interactions have to be taken into account which yields

$$B \approx \frac{2\Phi_0}{3^{1/2}\lambda^2} \left\{ \ln \left[\frac{3\Phi_0}{\mu_0 \lambda^2 (H - H_{c1})} \right] \right\}^{-2}, \quad H \approx H_{c1} \quad (4\text{-}72)$$

For slightly larger fields more terms are involved (because $\lambda > a_0$) and one arrives at the expression

$$M \approx H_{c1} \frac{\ln(\beta \mu_0 H_{c2}/B)}{\ln \varkappa}, \quad H \gtrsim H_{c1} \quad (4\text{-}73)$$

with $\beta = 0.381$ (De Gennes, 1966; p. 71). Equation (4-72) shows that $B \to 0$ when $H \to H_{c1}$. The phase transition for $\varkappa \gg 1$ is of second order. However, for $\varkappa \approx 1$ the vortex interaction becomes attractive at a certain distance d_0 which leads to a first order phase transition characterized by a jump from $B = 0$ to $B_0 = (3\Phi_0/4 d_0^2)^{1/2}$ as observed by Auer and Ullmaier (1973) in TaN alloys. In Fig. 4-7 $M(H)$ curves for several $\varkappa$-values as obtained by Brandt from numerical solutions of the GL equations, are given. For large $\varkappa$ the linear

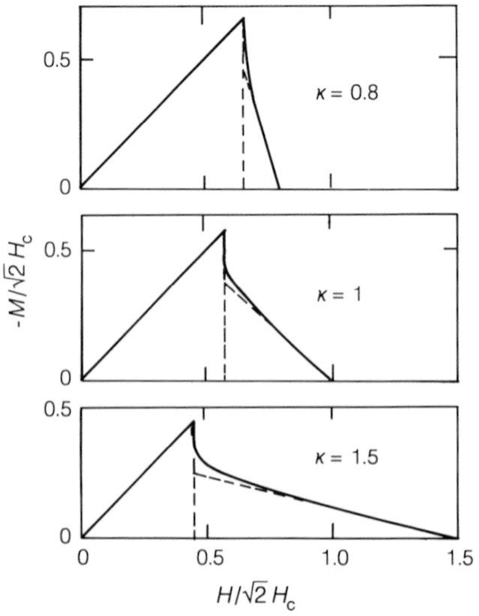

Figure 4-7. Magnetization curves of type II superconductors with different $\varkappa$ values.

Abrikosov solution Eq. (4-61) extends as far down as to about $2H_{c1}$. For a further reading on the theory and experiments on type II superconductors we refer to Fetter and Hohenberg (1969) and Serin (1969), respectively.

4.3.10 Elastic Constants of the Vortex Lattice

The interaction between vortices can be expressed in terms of the elastic tensor, see Campbell and Evetts (1972). Because of the symmetry there are three independent elastic moduli: c_{11} for compressional deformations, c_{44} for tilt and c_{66} for shear distortions. For uniform deformations of large flux bundles (size $> \lambda_h$) it is easy to show that

$$c_{11}(k=0) \simeq B^2 \frac{\mathrm{d}H}{\mathrm{d}B} \simeq BH \tag{4-74}$$

$$c_{44}(0) = BH \tag{4-75}$$

$$\lambda_h = \lambda_L/(1-b)^{1/2} \tag{4-76}$$

with $b \equiv B/B_{c2}$ and k^{-1} the wavelength of the deformation field. The shear modulus can be computed exactly for $B \approx 0$ and $B \approx B_{c2}$. A useful interpolation formula has been given by Brandt, see Brandt and Essmann (1987). For large $\varkappa$ it yields

$$c_{66} \simeq \frac{B_c^2}{4\mu_0} b(1-0.29b)(1-b)^2, \quad \varkappa \gg 1 \tag{4-77}$$

A comparison with the other moduli gives for $b > 0.01$

$$\frac{c_{66}}{c_{11}(0)} \simeq \frac{c_{66}}{c_{44}(0)} \simeq \frac{(1-b)^2}{9\varkappa^2 b} \tag{4-78}$$

For $\varkappa > 10$, it is clear that shear deformations are predominant, because they cost less energy.

The situation changes quite drastically if small-wavelengths deformations are considered. It is important to realise that the deformations are defined as the displacements of the vortex cores. At wavelengths smaller than λ_h the displacement of the cores is decoupled from the field- and current profiles of the vortices. Brandt showed that for $\varkappa \gg 1$:

$$c_{11}(k) \approx c_{11}(0)[(1+\lambda_h^2 k^2)(1+\lambda_\psi^2 k^2)]^{-1} \tag{4-79}$$

$$c_{44}(k) \approx c_{44}(0)(1+\lambda_h^2 k^2)^{-1} \tag{4-80}$$

$$\lambda_\psi^2 = \lambda_h^2 (2\varkappa^2)^{-1} \tag{4-81}$$

For very small wavelengths ($\approx a_0$) this leads to

$$c_{11}(k_{Br}) \approx c_{11}(0)(1-b)^2/(2\varkappa^2 b) \tag{4-82}$$

$$c_{44}(k_{Br}) \approx c_{44}(0)(1-b)/(2\varkappa^2 b) \tag{4-83}$$

$$k_{Br}^2 = (8\pi/3^{1/2})\, a_0^{-2} \tag{4-84}$$

where k_{Br} is the radius of the circular approximation for the first Brillouin zone. The moduli have strong dispersion and are considerably reduced as follows from a comparison with c_{66}

$$\frac{c_{66}}{c_{11}(k_{Br})} \approx \frac{2}{9} \tag{4-85}$$

$$\frac{c_{66}}{c_{44}(k_{Br})} \approx \frac{2}{9}(1-b) \tag{4-86}$$

Finally, the dispersion of c_{66} is small, because B does not change for any wavelength of shear deformations.

4.3.11 Critical Properties of Thin Films

The critical current and critical field of a thin film are quite different compared to the bulk critical-properties, especially if the film thickness d is smaller than both ξ and λ (see Tinkham (1975) and Abrikosov (1988) for discussions of intermediate regimes). The gradient term and, in parallel fields, also the field term in the free energy Eq. (4-31) may now be neglected; $|\psi|^2$ and j_s are uniform. Minimization with respect

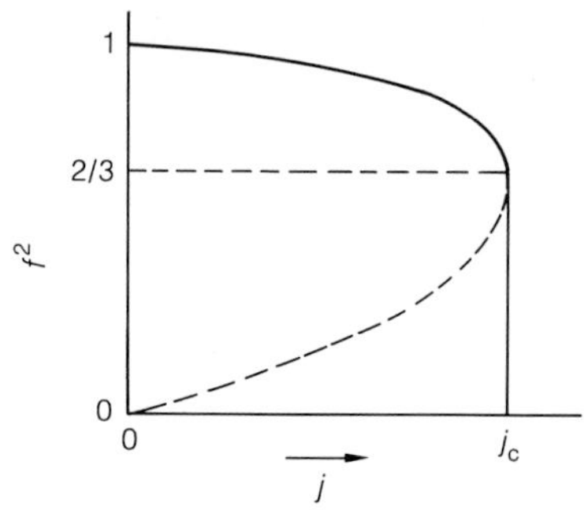

Figure 4-8. Reduced order parameter as a function of current density in a thin film. At the critical current f^2 falls abruptly to zero.

to $|\psi|^2$ at fixed v_s leads to

$$f^2 = |\psi|^2/\psi_\infty^2 = 1 - (m/|\alpha|)\, v_s^2 \tag{4-87}$$

$$j_s = 2e\,|\psi|^2 v_s = \\ = 2e\,\psi_\infty^2 (|\alpha|/m)^{1/2} f^2 (1-f^2)^{1/2} \tag{4-88}$$

The maximum current density j_c is reached for $m v_s^2 = |\alpha|/3$ and $f^2 = 2/3$. In Fig. 4-8 it is shown how f^2 gradually decreases if j is increased up to j_c where f^2 abruptly falls to zero. Using Eqs. (4-47) and (4-48) one obtains

$$j_0(T) = \left(\frac{2}{3}\right)^{3/2} \frac{H_c(T)}{\lambda_{\text{eff}}(T)}, \quad T \approx T_c \tag{4-89}$$

which shows that $j_0 \propto (1-t)^{3/2}$. Note that a different numerical constant in $j_0(0)$ is given in Eq. (4-27).

The nucleation field perpendicular to a thin film is $H_{c\perp} = H_{c2}$. Usually, H_{c2} is larger than the bulk value because the electron mean free path is reduced by surface scattering which increases $\varkappa$. For the parallel case the nucleation field is enhanced because the much smaller (by an amount $(d/\lambda)^2$) diamagnetic screening energy required. A similar approach as sketched above, but retaining the field energy, yields

$$f^2 = 1 - \frac{d^2 H^2}{24\lambda^2 H_c^2} \tag{4-90}$$

For $f^2 = 0$ the film becomes normal, i.e. at a field

$$H_{c\|} = 2\sqrt{6}\,\frac{\lambda H_c}{d} \tag{4-91}$$

Clearly, $H_{c\|} \gg H_c$ of the bulk, also for type I materials. The transition is second order as follows from Eq. (4-90). However, it becomes first order for $d > \sqrt{5}\,\lambda(T)$. The behavior of f as a function of $H/H_{c\|}$ and d/λ is given in Fig. 4-9 assuming $d \ll \xi(T)$. For an angle θ between field direction and film surface the nucleation field $H_c(\theta)$ is implicitely determined by

$$\left|\frac{H_c(\theta)\sin\theta}{H_{c\perp}}\right| + \left(\frac{H_c(\theta)\cos\theta}{H_{c\|}}\right)^2 = 1 \tag{4-92}$$

This expression gives rise to a cusp of $H_c(\theta)$ at $\theta = 0$ in good agreement with experiment.

4.3.12 Microscopic Background and Proximity Effect

Although the Ginzburg-Landau theory is very valuable, it is not able to describe phenomena on a length scale smaller than ξ_0, nor the effects of (magnetic) impurities and boundaries. Gor'kov was able to give a microscopic justification of the GL theory thereby providing the framework for

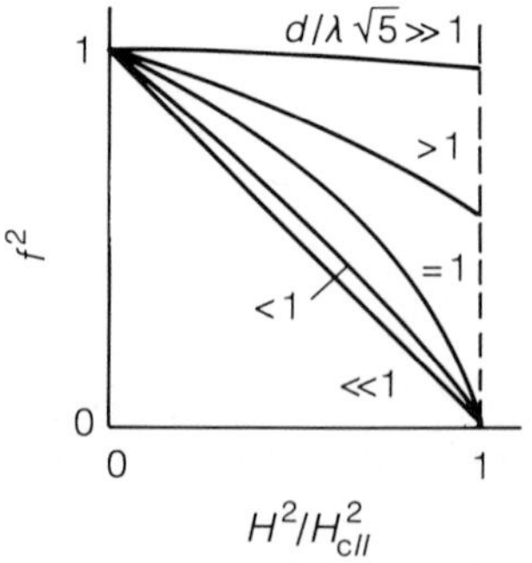

Figure 4-9. Reduced order parameter as a function of reduced magnetic field for thin films of various thicknesses.

the treatment of a rich variety of properties of type II superconductors. For instance, the limitation of H_{c2} by the Pauli paramagnetism of the electrons. In a strong field the electron spins like to line up with the field breaking up the paired states. The pair-breaking field is

$$\mu_0 H_p(0)/T_c = 1.84 \text{ (T/K)} \tag{4-93}$$

This limitation can be partially relaxed by the effect of spin-orbit scattering (Saint-James et al., 1969). Careful measurements of $H_{c2}(T)$ provides a way to estimate the relaxation times of such scattering effects. An other example is the slope of H_{c2} at T_c which in the dirty case ($\ell \ll \xi$) is related to the electron diffusion constant $D = v_F \ell/3$, the resistivity ϱ_0 at $T=0$, and $N(0)$ according to

$$-\mu_0 \left(\frac{dH_{c2}}{dT}\right)_{T_c} = \frac{4 k_B}{\pi e D} = \frac{8}{\pi} e k_B N(0) \varrho_0 \tag{4-94}$$

Another well-known effect is the rapid suppression of T_c by addition of a few percent of paramagnetic impurities.

The free energy difference between superconducting and normal states derived from the microscopic theory and expressed in terms of the pair potential $\Delta(\boldsymbol{r})$ and the vector potential $\boldsymbol{A}(\boldsymbol{r})$ is

$$F_{sn} = \int d^3 r \left[A|\Delta|^2 + \frac{B}{2}|\Delta|^4 + C|\partial\Delta|^2 + \frac{b^2}{2\mu_0} \right] \tag{4-95}$$

with $A = N(0)(1-t)$, $B = 0.098\, N(0) \cdot (k_B T_c)^{-2}$, $C = 0.55\, \xi_0^2 N(0) \chi(\varrho_d)$, and $\partial = -i\nabla - (2e/\hbar)\boldsymbol{A}$. The pair potential is proportional to the order parameter according to

$$C|\Delta|^2 \equiv \frac{\hbar^2}{4m} |\psi|^2 = \mu_0 H_c^2 \xi^2 |f|^2 \tag{4-96}$$

Equation (4-95) turns out to be very useful to compute the pin-energy of a vortex at a crystal defect (Sec. 4.4.2). The microscopic equivalent of Eq. (4-41) is a nonlocal self-consistent expression for $\Delta(\boldsymbol{r})$ which for small Δ takes the form (De Gennes, 1966; p. 210 ff.)

$$\Delta(\boldsymbol{r}) = \int d^3 s\, K(\boldsymbol{r}-\boldsymbol{s})\, \Delta(\boldsymbol{s}) \tag{4-97}$$

The kernel K has a range ξ_0 in the pure limit and $(\xi_0 \ell)^{1/2}$ in the dirty limit. It is now interesting to study the situation at a boundary.

In deriving the GL equations it has been assumed that at the surface $\boldsymbol{n} \cdot \partial\psi = 0$, where $\boldsymbol{n}$ is a unit vector normal to the surface. Substituting $\psi = \psi_0 e^{i\phi}$ one obtains

$$\boldsymbol{n} \cdot \nabla\psi_0 = 0 \tag{4-98}$$

$$\boldsymbol{n} \cdot (\hbar\nabla\phi - 2e\boldsymbol{A})\psi_0 = 0 \tag{4-99}$$

The first condition means that the gradient of the order parameter should be perpendicular to the surface, the second condition requires that there is no current flow through the surface. For a vacuum- (or insulator-) superconductor interface this is realistic, but for a normal-superconductor interface it is not acceptable, because supercurrents may flow through this interface. For an isolated superconductor cladded with a normal-metal layer De Gennes showed that the GL theory is applicable up to the surface with the boundary condition

$$\boldsymbol{n} \cdot (-i\hbar\nabla\phi - 2e\boldsymbol{A})\psi = i\hbar l^{-1} \psi \tag{4-100}$$

where l is a length related to the gradient of ψ_0 according to

$$\frac{d\psi_0}{dx}(x=0) = l^{-1} \psi_0(x=0) \tag{4-101}$$

The interface has been assumed at $x = 0$. In Fig. 4-10a the situation for $\psi_0(x)$ is sketched. In Fig. 4-10b the microscopic reality is displayed. Cooper pairs may penetrate in the normal layer up to a distance of order $\xi_n = \hbar v_{Fn}/k_B T$, whereas normal elec-

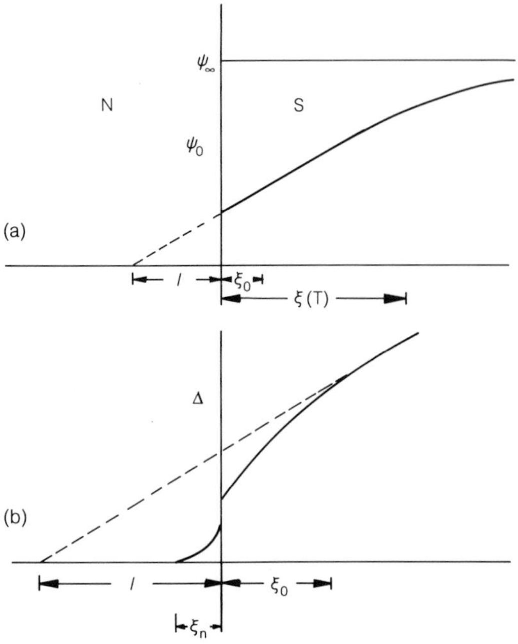

Figure 4-10. Variation of the order parameter at a normal metal-superconductor interface. (a) The phenomenological description for $\xi(T) \gg l$, (b) the microscopic situation.

trons penetrate the superconductor suppressing the pair potential over a distance of order $\xi_s = 0.18\,\hbar v_{Fs}/k\,T_c$. This effect is called the proximity effect and is described in more detail by Deutscher and De Gennes (1969). It is clear that only a nonlocal expression Eq. (4-97) for $\Delta(\boldsymbol{r})$ can describe this situation.

4.3.13 Anisotropic Superconductors

The crystal symmetry is generally reflected in the superconducting properties of single-crystalline samples, e.g. the energy gap may be anisotropic. A fast-growing interest in the last few decades concerns the research on single crystals of a special class of superconductors characterized by a layered structure of superconducting planes separated by normal or insulating layers. Among these are the high-temperature superconducting oxides and organic superconductors. Much interest is focussed also on Nb and Ta-dichalcogenides intercalated with organic molecules, and on artificially grown multilayers. Depending on the distances between the superconducting planes and the nature of the intermediate material, the coupling between the planes may become very weak, leading to a crossover of dimensionality at a temperature T^* determined by

$$\xi_\perp(T^*) = s/\sqrt{2} \tag{4-102}$$

Here $\xi_\perp$ is the GL-coherence length in the direction perpendicular to the planes. In the following the subscript $\|$ will denote the in-plane parameters; the possible anisotropy within the plane is neglected here. For a more complete analysis, see Clem (1989), and Song and Ketterson (this Volume, Chap. 6). The anisotropy can be expressed by the parameter

$$\Gamma = m_\perp / m_\| \tag{4-103}$$

where $m_\perp$ and $m_\|$ are effective masses for tunneling between the planes and motion in the planes. Since it follows from Eqs. (4-38) and (4-47) that $\lambda^2 \propto m$ and $\xi^2 \propto m^{-1}$, one gets

$$\xi_\perp / \xi_\| = \Gamma^{-1/2} \tag{4-104}$$

$$\lambda_\perp / \lambda_\| = \Gamma^{1/2} \tag{4-105}$$

We should note here that the subscripts in Eq. (4-105) refer to the directions of the screening currents. For large Γ values T^* can be very close to T_c. In the region $T^* < T < T_c$ the properties are described by the three-dimensional (3D) GL theory for anisotropic superconductors, e.g.

$$\mu_0 H_{c2\perp} = \frac{\Phi_0}{2\pi\,\xi_\|^2}; \quad \mu_0 H_{c2\|} = \frac{\Phi_0}{2\pi\,\xi_\|\,\xi_\perp} \tag{4-106}$$

$$\mu_0 H_{c1\perp} = \frac{\Phi_0}{4\pi\,\lambda_\|^2} \ln(\varkappa_\perp + 0.50);$$

$$\mu_0 H_{c1\parallel} = \frac{\Phi_0}{4\pi\lambda_\parallel\lambda_\perp}\ln(\varkappa_\parallel + 0.50) \quad (4\text{-}107)$$

$$\varkappa_\perp = \lambda_\parallel/\xi_\parallel, \quad \varkappa_\parallel = (\lambda_\parallel\lambda_\perp/\xi_\parallel\xi_\perp)^{1/2} \quad (4\text{-}108)$$

$$1 - \frac{T^*}{T_c} = \frac{2\xi_\parallel^2(0)}{\Gamma s^2} \quad (4\text{-}109)$$

Below T^* the planes are decoupled and superconductivity is maintained by virtue of Josephson tunneling. In the above expressions $\xi_\perp$ should be replaced by s. For $H_{c2}(t)$ the dimensional crossover reveals itself by a sharp upturn according to $H_{c2} \propto (1 - T/T^*)^{1/2}$, see Takahashi and Tachiki (1986). In addition, the angular dependence of H_{c2} in the 2D regime obeys Eq. (4-92) with a cusp at $H_{c2\parallel}$. (Note θ is the angle between field and planes.) In the 3D regime the $H_{c2}(\theta)$ curve is bell-shaped in accord with

$$\left(\frac{H_{c2}(\theta)\sin\theta}{H_{c2\perp}}\right)^2 + \left(\frac{H_{c2}(\theta)\cos\theta}{H_{c2\parallel}}\right)^2 = 1 \quad (4\text{-}110)$$

The behavior of $H_{c1}(\theta)$ is rather complicated and still the subject of numerous investigations.

The vortex lattice in an anisotropic superconductor has properties quite different from the Abrikosov lattice. The screening currents are predominantly located in the planes (except for $H_\parallel$) giving rise to a magnetic moment M perpendicular to the planes and a torque $\propto \boldsymbol{M} \times \boldsymbol{H}$. From the angular dependence of the torque Γ can be accurately determined. The lattice for $H_\parallel$ in the 2D regime consists of Josephson vortices which have no normal core, since the order parameter can only be a function of position within the superconducting planes. In the 3D regime and with $H_\parallel$ there is a normal core of ellipsoidal shape which is, like the VL unit cell, stretched in the direction of the planes and compressed in the perpendicular direction. For $H_\perp$ the Abrikosov vortices break up into 2D "pancake" vortices in the planes which are weakly coupled depending on Γ. The segmentation due to the weak coupling effects the tilt modulus $c_{44}(k)$ resulting in a reduction from Eq. (4-80) by a factor Γ. For $H_\parallel$ the shear modulus c_{66} is predicted to become very anisotropic which is related to the stretched vortex structure. All these features give rise to novel effects with regard to flux pinning and creep which is presently the subject of intensive studies (Kes and Van der Beek, 1991).

4.3.14 Fluctuations

At a second-order phase-transition thermal fluctuations temporally create superconducting regions above T_c and normal regions below T_c. The typical size of these droplets is $\xi(|T_c - T|)$ both below and above T_c. Above T_c fluctuations give rise to an enhancement both of the conductivity and the diamagnetism. Fluctuations will also smooth the specific-heat jump at T_c. Interesting effects in the temperature dependence are to be expected when one or more of the sample dimensions is smaller than $\xi(T)$. Each time a factor $\xi(T)$ is replaced by a sample dimension a factor $|1-t|^{-1/2}$ is erased in the final result, e.g. for the dc paraconductivity

$$\sigma'_{3D}(0) = \frac{G_0}{32\,\xi(0)}\left(\frac{T}{T-T_c}\right)^{1/2}, \quad 3D \quad (4\text{-}111)$$

$$\sigma'_{2D}(0) = \frac{G_0}{16\,d}\left(\frac{T}{T-T_c}\right), \quad 2D \quad (4\text{-}112)$$

$$\sigma'_{1D}(0) = \frac{\pi G_0\,\xi(0)}{16\,A}\left(\frac{T}{T-T_c}\right)^{3/2}, \quad 1D \quad (4\text{-}113)$$

Where $G_0 \equiv e^2/\hbar = 2.43 \times 10^{-4}\,\Omega^{-1}$, d is the film thickness and A the wire cross section. In 3D the conductivity enhancement is usually very small and hard to observe experimentally. However, the diamagnetism in 3D is presently very clearly observable with sensitive SQUID magnetometers. Within the framework of the GL theory the result for the dc susceptibility is

$$\chi_{3D} = -\frac{\mu_0 k_B T}{24 \Phi_0^2} \xi(T) \approx 10^{-7} (t-1)^{1/2} \tag{4-114}$$

$$\chi_{2D} \approx -\frac{\mu_0 k_B T \xi^2}{4 \Phi_0^2 d} \approx \frac{\xi}{d} \chi_{3D} \propto (t-1)^{-1} \tag{4-115}$$

However, as has been discussed before, the GL theory cannot adequately account for short-wavelength (shorter than ξ_0) fluctuations and overestimates their contribution severely. Fluctuation effects have been studied in great detail and are now well understood. For an extensive overview we refer to Skocpohl and Tinkham (1975).

4.3.15 Time-Dependent Ginzburg-Landau Theory and Flux Flow

Flux flow occurs when a current is applied to a perfectly uniform superconductor in the mixed state. According to Eq. (4-71) a driving force per unit length $\boldsymbol{f}_d = \boldsymbol{j} \times \boldsymbol{\Phi}_0$ is exerted on each flux line, where j now represents the density of the applied current. As a consequence the vortex lattice (VL) moves uniformly in the direction of the force (there is a small Hall effect related to the vortex motion which we shall ignore here). The vortex motion leads to a continuous change of both phase and amplitude of the order parameter at any point in the superconductor. A continuous acceleration and slowing down of supercurrents and conversion of Cooper pairs and normal excitations give rise to a viscous drag force and dissipation. In dynamic equilibrium the VL moves with velocity v_L given by

$$v_L = f_d / \eta \tag{4-116}$$

where η is the flux-flow viscosity. The generated electric field follows from

$$\boldsymbol{E} = -\boldsymbol{v}_L \times \boldsymbol{B} \tag{4-117}$$

For mutually perpendicular field, current, and force this leads to a flux-flow resistivity $\varrho_f = E/j$ given by

$$\varrho_f = \frac{\Phi_0 B}{\eta} \tag{4-118}$$

The theoretical derivation of ϱ_f (or rather $\sigma_f \equiv \varrho_f^{-1}$) has only been carried out in a few limiting cases. A recent review is given by Larkin and Ovchinnikov (1986). Solutions are mainly obtained in the dirty limit using the microscopic formulation of the time-dependent Ginzburg-Landau theory. Close to T_c the solution takes the form

$$\frac{\sigma_f}{\sigma_n} - 1 = \frac{1}{(1-t)^{1/2}} \frac{\mu_0 H_{c2}}{B} \tilde{f}\left(\frac{B}{\mu_0 H_{c2}}\right), \quad T \approx T_c \tag{4-119}$$

$$\tilde{f}(x) \equiv 4.04 - x^{1/4}(3.96 + 2.38\,x) \tag{4-120}$$

$\tilde{f}(x)$ follows from a numerical calculation approximating the VL cell by a circle. It should be noted that $\tilde{f}(x)$ already decreases for very small x. For fields close to H_{c2} the result is

$$\frac{\sigma_f}{\sigma_n} - 1 = \tilde{\alpha}\left(1 - \frac{B}{\mu_0 H_{c2}}\right), \quad H \approx H_{c2} \tag{4-121}$$

The function $\tilde{\alpha}(T)$ ranges between 2 and 4. At low temperatures and small fields Gor'kov and Kopnin (1975, 1976) derived

$$\frac{\sigma_f}{\sigma_n} = 0.9 \frac{\mu_0 H_{c2}(0)}{B}, \quad T \ll T_c, \; H \ll H_{c2} \tag{4-122}$$

In their respective regimes of validity these expressions have shown to agree well with experiment.

4.3.16 Concluding Remarks

The GL theory, its extensions and applications to many problems, has proven to be extremely useful, also in the regimes where it is not strictly valid. Many experiments could be very well explained and understood in terms of this theory. In cases of disagreement, e.g. $\varkappa_2(t)$ of Nb and V is roughly a factor 2 larger than the theory predicts, it can be often attributed to details related to the materials under investigation like band-structure effects. Even the properties of rather exotic superconductors like the heavy-fermion compounds, the high-temperature oxidic superconductors, organic superconductors, and multilayers, turn out to be at least qualitatively in agreement with the basic GL results. Most of the issues treated in the preceding sections are indispensable for the comprehension of vortex-lattice pinning in inhomogeneous superconductors which is the subject of the following sections. For further reading on the magnetic properties of type II superconductors we refer to Huebener (1979).

4.4 Flux Pinning

4.4.1 Irreversible Magnetic Properties, the Critical State and the Bean Model

The reversible magnetization curves of Fig. 4-7 are only observed when the vortices can freely move through the superconductor. However, inhomogeneities (defects) in the material cause deviations of the free energy density from the form valid for homogeneous superconductors, Eq. (4-31). The VL will adapt to the distribution of defects by assuming a configuration of minimum free energy determined by the sum of the deformation energy and the interaction energy with the defects. The net result is that the VL is pinned by the defects, they act as pinning centers (henceforth pins). The collective nature of pinning is demonstrated by the fact that pinning occurs for both attractive and repulsive pins. Pinning prevents the vortices from moving and dissipating energy (see Sec. 4.3.15). Therefore it is crucial for applications of superconductors in fields above H_{c1}.

The driving force (density) by a transport current with current density j is given by [compare with Eq. (4-71)]

$$\boldsymbol{F}_\mathrm{d} = \boldsymbol{j} \times \boldsymbol{B} \tag{4-123}$$

Up to the critical current density j_c the pinning is able to resist the driving force. For $j > j_\mathrm{c}$ flux flow sets in leading to a differential resistivity $\partial E/\partial j = \varrho_\mathrm{f}$ equal to the flux-flow resistivity discussed in Sec. 4.3.15. The relation

$$F_\mathrm{p} = j_\mathrm{c} B \tag{4-124}$$

defines the macroscopic bulk or volume pinning force (density) F_p. When $j = j_\mathrm{c}$ the superconductor is said to be in the critical state. This obviously is a metastable state, because thermal or other fluctuations may trigger vortex motion known as flux creep.

In response to a magnetic-field change vortices are driven into or out of the superconductor. Pinning then leads to flux density gradients and irreversible magnetization curves. The gradient can be related to a current density by

$$\boldsymbol{j} = \mu_\mathrm{A}^{-1} \nabla \times \boldsymbol{B} \tag{4-125}$$

where $\mu_\mathrm{A} = \mu_0 (1 + \chi_\mathrm{A})$ with $\chi_\mathrm{A} = \partial M_\mathrm{rev}/\partial H$, the slope of the reversible magnetization curve. Depending on the sign of the field increment the critical state is characterized by $j_\pm = \pm j_\mathrm{c}$. For a simple geometry, e.g. a solid cylinder with radius R in an axial field, the magnetization follows directly from

$$M_{\pm} = \frac{1}{R^2} \int_0^R \mathrm{d}r\, r^2 j_{\pm}(r) \quad (4\text{-}126)$$

If it is assumed that $j(r)$ is uniform and equal to j_c [the Bean model, see Bean, (1964)], the irreversibility of the magnetization curves $\Delta M = |M_+ - M_-|$ yields

$$j_c = \frac{3}{2} \frac{\Delta M}{R} \quad (4\text{-}127)$$

For a slab of thickness $2d$ one obtains

$$j_c = \frac{\Delta M}{d} \quad (4\text{-}128)$$

The Bean model is a convenient approximation which is reasonably accurate for large $\varkappa$ and fields $H \gg H_{c1}$. Under these conditions the effect of the reversible magnetization may be neglected, i.e., $\mu_A \approx \mu_0$. In addition, the radius or thickness of the sample should be small enough to guarantee that j is uniform, although j_c itself may depend on B. The Bean model is a rather poor approximation in case $\varkappa$ is small and especially if $H \lesssim 2 H_{c1}$. This is illustrated in Fig. 4-11 where the flux density inside a Nb slab ($\varkappa = 1$) is computed for a simple pinning model, namely $j_c(x) \propto B_{c2} - B(x)$, taking $\mu_A(B)$ into account as well. The Bean model can be improved by expanding $j[B(x)]$ as a Taylor series in $(R-r)$ or $(d-x)$. To first order one gets $M_{rev} = (M_+ + M_-)/2$ and Eqs. (4-127) or (4-128), respectively; see Kes et al. (1973).

The situation of a thin film in a field normal to the surface needs special attention. In this geometry the demagnetization effect is large and it follows from Eq. (4-125) that the currents are predominantly determined by a curvature of the flux lines instead of a flux density gradient. Since Eq. (4-126) is generally valid, the critical current density can still be estimated from

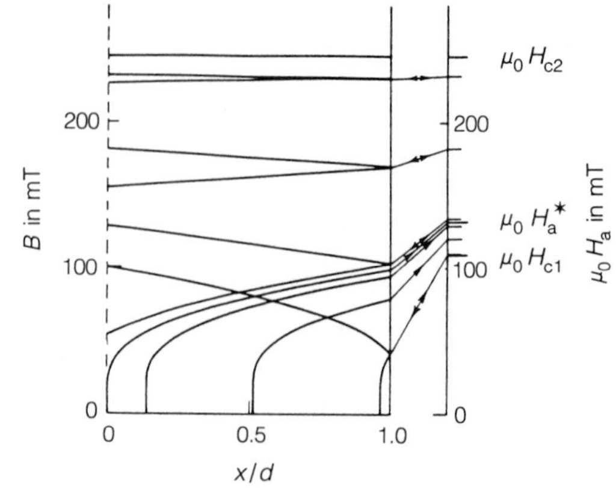

Figure 4-11. Flux density distribution in a slab as a function of distance x from the central plane in increasing and decreasing fields demonstrating deviations from the Bean model. The computations are carried out for a Nb foil of thickness $2d$ and $\varkappa = 1$ at $t = 0.56$ for a simple pinning model in which $j_c(x) \propto [B_{c2} - B(x)]$.

Eq. (4-127), usually with an effective radius, if the film is not circular in shape. Finally, it should be noted that the curvature required to sustain even large current densities is very small, e.g., for $j_c = 10^{10}\,\mathrm{Am}^{-2}$, $B = 1\,\mathrm{T}$ and a film thickness of 1 µm the maximum deviation from the normal is less than 0.4°.

4.4.2 Vortex-Defect Interaction

The changes of material parameters at defects, e.g. density, elastic constants, electron mean free path, give rise to local variations of T_c, ξ, and λ. As a consequence the parameters A, B, and C in the free energy expression Eq. (4-95) vary at the location of the defects resulting in variations of F_{sn}. The latter may be identified as the pin potential of the defects and is given by

$$\Omega_p(\boldsymbol{r}_d) = \int \mathrm{d}^3 r\, [\delta A(\boldsymbol{r} - \boldsymbol{r}_d) |\Delta|^2 + \delta C(\boldsymbol{r} - \boldsymbol{r}_d) |\partial \Delta|^2] \quad (4\text{-}129)$$

The variations δA and δC differ from zero only at the positions of the defects $\boldsymbol{r}_d$; the effect of δB is considered to be relatively small. The origin is assumed to coincide with the vortex core. It depends on the character of the pin whether the change in A or C is predominant. As a result one discriminates between δT_c and $\delta\varkappa$-pinning, respectively. The latter pin mechanism is caused by a change of the mean free path by scattering of the electrons at the defect and gives rise to a change in both $\varkappa$ and C. Moreover, it follows from Eq. (4-129) that the pin potential depends on several length scales. That is in the first place the size of the pin, but also the length scale over which the order parameter and the vector potential change. For isolated vortices ($B < 0.2\,B_{c2}$) this is ξ and λ, respectively. For overlapping vortices ($B > 0.2\,B_{c2}$) this is the vortex-lattice parameter a_0, defined in Eq. (4-63). For small pins (relevant size $< \xi$, λ or a_0) the pin potential can in principle be obtained from Eq. (4-129) retaining for Δ and $\boldsymbol{A}$ the solutions of the homogeneous superconductor for the simple reason that F_{sn} is supposed to be minimized with respect to Δ and $\boldsymbol{A}$. For large pins the boundary conditions at the interface become important.

Considering the above discussion it must be clear that the vortex-defect interaction cannot be evaluated by following a general recipe. For every pin mechanism one first has to decide which of the above considerations is most relevant. This will be illustrated in the following sections where some examples are discussed.

4.4.2.1 Surface Pinning

The sample surface in parallel field can be considered to be a pinning center, because it forms an energy barrier for flux entry and exit. For $B \approx 0$ this can be understood in terms of the boundary conditions for the screening currents around the vortex core when the core is at a distance $x < \lambda$ from the surface. The boundary conditions can be simulated by assuming an anti-vortex outside the superconductor at $-x$. The vortex and its image attract each other, while the London surface-currents repel the vortex. As described by de Gennes (1966), the net effect is a potential barrier, known as the Bean-Livingstone barrier, which leads under ideal circumstances to first flux entry at a field $H_s \approx H_c$ much larger than H_{c1}. The effect can be large in low-$\varkappa$ superconductors like Nb and V.

More or less related to surface pinning are pinning by thickness variations in a thin film with the field normal to the surface, and pinning at the interface of a large precipitate in the superconductor. A third example is interface pinning in an artificial multilayer with layer thicknesses larger than λ. A rough estimate for the pinning force per unit length on a single vortex is $f_p/L \approx \Phi_0\, M/\lambda$. The common feature of this pin mechanism is the coupling to the field variations in the mixed state.

4.4.2.2 Core Pinning

Most pins couple to the variation of the order parameter, a mechanism classified as core pinning. This kind of pinning is described by Eq. (4-129) and can be further devided into δT_c and $\delta\varkappa$-pinning as being related to the first and second term in Eq. (4-129), respectively. As argued by Kes and Van den Berg (1990) this distinguishment is reflected in a different temperature and field dependence of the pinning forces. The pin energy for δT_c-pinning is proportional to $(\delta A/A)\, A|\Delta|^2$ which can be written using Eq. (4-96) as $(\delta A/A)\,\mu_0\, H_c^2\, |f^2|$, where $f \equiv \psi/\psi_\infty$. On the other hand, the pin energy for $\delta\varkappa$-pinning contains the factor $(\delta C/C)\,\mu_0\, H_c^2\, \xi^2\, |\partial f|^2$. For small fields $|\partial f|^2 \sim \xi^{-2} |f|^2$, while for large fields, when the vortices overlap, $|\partial f|^2 \sim (2\pi/a_0)^2 |f|^2 \propto (b/\xi^2)|f|^2$, as follows from Eq. (4-64). Note that $|f|^2 \propto (1-b)$. The distinction between both pin mechanisms is given by the temperature dependence of $\delta A/A$, as will be specified below, whereas $\delta C/C$ is temperature independent. In addition, $\delta\varkappa$-pinning has a significantly different field dependence in the large field regime. The pin energies differ by a factor $b = B/B_{c2}$ and the pinning force by a factor $b^{3/2}$, because it follows from the gradient of the pin energy.

The distinction between temperature and field dependence helps to identify the predominant pin mechanism. However, the size, shape and orientation of the pins should be considered as well, since according to Eq. (4-129) one has to integrate over the volume of the defects. Therefore, point, line or planar defects cannot be treated on the same footing, but small dislocation loops (δT_c-pinning) may be compared with small voids ($\delta\varkappa$-pinning) and grain boundaries with interfaces.

4.4.2.3 $\delta\varkappa$-Pinning

Typical examples of pins operating by enhanced electron scattering are voids, vacancies, dielectric precipitates, argon or helium bubbles, grain boundaries, twin planes, etc. The general theory for this mechanism has been worked out by Thuneberg (1989). The concept is illustrated by deriving the pin potential from Eq. (4-129)

$$\Omega_p(\boldsymbol{r}_d) = \mu_0 H_c^2 \xi^2 g(\varrho_d) \xi_0 \int d^3 r' |\partial f(\boldsymbol{r}')|^2 \, \delta\left[\frac{1}{\ell_{tr}(\boldsymbol{r}_d - \boldsymbol{r}')}\right] \tag{4-130}$$

with $g(\varrho_d) = 0.882\,(d\ln\chi(\varrho_d)/d\varrho_d)$. χ is the Gor'kov function (Sec. 4.3.5) and $g(0) = 0.85$ in the clean limit and $g(\infty) = \ell/\xi_0$ in the dirty limit. The extra scattering at the defect causes a local decrease of ξ. Therefore, this pin mechanism always results in an attractive interaction, since less condensation energy is missing if the normal vortex core coincides with the defect.

For an extended defect the integral over the entire defect should be evaluated, as has been worked out by Kes and Van den Berg (1990) for a twin plane. The case of a small point pin with diameter $D < (\xi_0^{-1} + \ell^{-1})^{-1}$ is much simpler, because $\delta(\ell^{-1})$ can be replaced by a delta-function times the scattering cross-section $\pi D^2/4$. The final result is proportional to the condensation energy multiplied by the defect volume *enhanced* by a factor ξ_0/D in the clean limit and ℓ/D in the dirty limit. The maximum pinning force f_p is obtained from the gradient of Ω_p. A typical value for $D = 1$ nm, $\mu_0 H_c = 0.1$ T, and $b = 0.5$ is $f_p \approx 0.3 \times 10^{-14}$ N. The corresponding pin energy for $\xi_0 = 10$ nm is $\Omega_p \approx 0.2$ meV. Good agreement with experiment has been observed for pinning by voids in Nb and V samples by Van der Meij and Kes (1984).

4.4.2.4 δT_c-Pinning

A typical example of this kind of core interaction is pinning by dislocations, since the change in ℓ by these defects is usually small. The formalism used in the literature differs somewhat from the one described above, e.g., see Campbell and Evetts (1972). Two effects are distinguished, both related to the normal character of the vortex core which has a larger density (ΔV-effect) and larger elastic constants (ΔE-effect). These effects, though very small, give rise to a periodic stress and elasticity field, both connected to the variation of $|\psi|^2$. The pin energy arises via the coupling to the strain fields around the defects and is linear in the strain for the ΔV-effect and quadratic for the ΔE-effect. Accordingly, these effects are referred to as paraelastic or first-order and dielastic or second-order interactions. Typical values of $\Delta V/V$ and $\Delta E/E$ are 3×10^{-7} and 5×10^{-4}, respectively. The latter value refers to the variation of the crystal shear-modulus which is by far the largest. The ΔE effect, being quadratic in the strain is always a repulsive interaction because the vortex core is slightly stiffer. For the ΔV-effect, both repulsion and attraction can occur, depending on whether the density at the defect is smaller or larger. The maximum pinning force in the large-field or overlapping-vortex regime is given quite generally by

$$f_p = C_d \mu_0 H_c^2 \alpha \xi^{-1} b^{1/2} (1-b) V_D \qquad (4\text{-}131)$$

C_d is a constant depending on material parameters, the kind of dislocation, and the orientation of the dislocation with respect to the VL. V_D is the characteristic volume of the dislocations, e.g., for line dislocations parallel to the vortex $V_D = \boldsymbol{b}_B^2 L$, with $\boldsymbol{b}_B$ the Burgers vector and L the length of the dislocation; for dislocation loops of diameter D it is $\approx D^2 \boldsymbol{b}_B$. The parameter α represents the local variation of T_c at the defect and is for the first-order interaction expressed by $\partial \ln T_c/\partial\varepsilon_v$ and $T_c^{-1}\partial^2 T_c/\partial\varepsilon_v^2$ (the latter related to $\delta B/B$ in the formulation given in Sec. 4.4.2), where $\varepsilon_v = \Delta V/V$ is the volume dilatation. To get an impression of the size of this pin interaction in case of an isolated vortex taking $\mu_0 H_c = 0.1$ T and $\xi_0 = 10$ nm some estimates can be given:

(i) edge dislocation perpendicular to the vortex $f_p \approx 10^{-13}$ N,
(ii) edge and screw dislocations parallel to the vortex $f_p/L \approx 5\times10^{-10}$ and 10^{-5} N/m, respectively,
(iii) dislocation loop, $D = 10$ nm, perpendicular to the field direction $f_p \approx 3\times10^{-14}$ N.

4.4.2.5 Precipitates

Precipitates are often found to be a predominant pin mechanism. Both electron scattering and the strain field around a precipitate play a role, so that both types of core interaction may act simultaneously. In addition, the proximity effect should be taken into account when the precipitate is a (super-) conductor. Obviously, a precipitate with high T_c forms a repelling pinning center and vice versa. When the defects are very large, the pinning is predominantly coming from the edges and interfaces. All these complications make that an estimate of the pinning force is difficult to give.

4.4.3 Summation Problems

The pinning centers in real materials have a random site and size distribution. In order to compute the total force on the VL, it is usually assumed that all defects have the same size, so that only the positional distribution has to be considered. Consequently, there will be a distribution of pinning forces determined by the mutual positions of the pinning centers and the vortices. The positions of the vortices are defined by the zero's of the order parameter. In addition, it is assumed that the pins are small ($D < \xi$) in at least one direction perpendicular to the field.

For a perfectly uniform VL the effect of a random force distribution will average to zero. One may therefore conclude that the deviations from perfect periodicity are essential for a net effect. This has been pointed out by Labusch (1969) in a first attempt to solve the statistical summation problem taking into account the elastic deformations of the VL in presence of pinning centers. Labusch suggested that only pins with f_p larger than a certain threshold value f_{th} could be effective, because they are able to produce plastic defects in the

VL and disrupt the long range order, see also Campbell and Evetts (1972). However, this idea leads to a paradox, since large critical current densities were observed in materials with pinning centers for which $f_p \ll f_{th}$, as shown by Kramer (1978). It turns out that the threshold paradox disappears when one accounts for the collective effects which occur in dense pinning systems.

The elastic response of the VL to the force exerted by a specific pinning center is mainly determined by the shear and tilt moduli c_{66} and c_{44} (Sec. 4.3.10). The effect of the other pins is to keep the VL fixed. Such a mean field approach is only reasonable, if the range of the VL interaction is small compared to the distance between the pins. However, because of the electromagnetic origin, the interaction has a range λ which can be very large, especially in high-$\varkappa$ materials. Moreover, real pin systems are often very dense ($n_p^{1/3} \ll \lambda$, n_p the concentration of pinning centers), which suggests that a collective treatment of the pinning forces is more appropriate. Especially, in case of weak pins, for which the individual defect-flux-line interactions only lead to elastic deformations with strains $\ll 1$, the summation problem can be solved. This has been shown for the first time by Larkin and Ovchinnikov in their theory of collective pinning, see Larkin and Ovchinnikov (1979) and (1986).

4.4.3.1 Collective Pinning Theory

The local displacements due to weak pins will be very small ($\ll a_0$), so that around each flux line an hexagonal VL can be well-defined. Over longer distances, however, the displacements may accumulate as in a random-walk process causing a gradual breakdown of the order in the VL. This is expressed by the monotonous growth of the displacement correlation function

$$g(\boldsymbol{r}) \equiv \langle [\boldsymbol{u}(\boldsymbol{r}) - \boldsymbol{u}(0)]^2 \rangle \qquad (4\text{-}132)$$

with distance $|\boldsymbol{r}|$. Here $\boldsymbol{u}$ is the displacement in the $x\,y$-plane (perpendicular to the field) and the origin may be located at the core of any flux line. The average is taken over an ensemble of pin configurations. When $g(\boldsymbol{r})$ is of order a_0^2, the positional long range order is destroyed. However, for the summation of pinning forces, it is more relevant to know the criterion for which the pin interaction becomes unpredictable. This happens when $g(\boldsymbol{r}) = r_f^2$, where r_f is the distance in which $f(x, y)$ changes from $-f_p$ to f_p, i.e. $r_f \approx \xi$ for $b < 0.2$ and $r_f \approx a_0/2$ for $b > 0.2$.

The function $g(\boldsymbol{r})$ has been first derived by Larkin and Ovchinnikov (1979) (LO) for an infinitely large system in which the order is destroyed in the directions both parallel and perpendicular to the field. This is called three-dimensional (3D) disorder. The corresponding longitudinal and perpendicular correlation lengths L_c and R_c are defined by

$$g_3(0, L_c) = g_3(R_c, 0) = r_f^2 \qquad (4\text{-}133)$$

L_c and R_c are functions of c_{66} and c_{44} and of the pin correlation function (pin "strength") which for small point pins is given by

$$W = n_p \langle f^2 \rangle \approx 0.5\, n_p\, f_p^2 \qquad (4\text{-}134)$$

where f now denotes the actual force exerted by a pin. The average is taken for a uniform distribution of pinning centers over a unit cell of the VL. The relation between L_c and R_c depends in a quite natural way on the elastic moduli

$$L_c = \left(\frac{c_{44}}{c_{66}}\right)^{1/2} R_c \qquad (4\text{-}135)$$

It should be noted that all these quantities depend in a specific way on the temperature and field, as well as the defect morphology.

The physical meaning of the correlation lengths is illustrated by deriving qualitative expressions for L_c and R_c. Instead of a continuous displacement field, one may consider correlated regions with volume $V_c \approx R_c^2 L_c$ which are weakly coupled to each other so that they can move independently over a distance of order r_f. Inside V_c the VL distortions may be ignored. The size of the correlated volume is determined by minimizing the free energy consisting of two terms: the elastic energy related to the mutual displacements of the correlated regions and the work done by the pinning centers in V_c to affect this displacement. This model readily provides the pinning-force density as well, since in order to unpin the entire VL, each correlated volume has to be unpinned independently. A statistical argument tells us that the fluctuation (per unit volume) in the net effect of $n_p V_c$ pinning centers is

$$F_p = j_c B \approx \left(\frac{W}{V_c}\right)^{1/2} \tag{4-136}$$

The situation becomes very simple for a thin-film amorphous superconductor in a perpendicular field. Since $L_c \gg R_c$ and W is small, L_c can be easily made larger than the thickness d of the film. In this case the VL deformations in the field direction may be ignored, so that only two-dimensional disorder in the $x\,y$-plane remains. One then has $V_c \approx \mathrm{d}R_{c,\,2D}^2$ and

$$R_{c,\,2D} = \frac{(8\,\pi)^{1/2}\, r_f\, c_{66}}{(W/d)^{1/2} \ln(w/R_{c,\,2D})} \tag{4-137}$$

$$F_{p,\,2D} = \frac{W \ln(w/R_{c,\,2D})}{(8\,\pi)^{1/2}\, c_{66}\, r_f\, d} \tag{4-138}$$

where w is the width of the thin film. The only parameter to be determined is W, thereby offering a means to get information about f_p.

In the LO theory it is assumed that elastic deformations solely determine the correlation lengths. The effect of dislocations in the VL is not considered, although their existence is known from decoration experiments. In fact, almost equivalent expressions can be derived by modelling the disorder in 2D by a square array of edge dislocations. This illustrates that the concept of computable correlation lengths is less settled. The statistical argument that leads to Eq. (4-136), has a more general validity and may be used to obtain V_c from experiment, if W is known.

4.4.3.2 Collective Pinning Experiments

Clear evidence for the existence of collective pinning has been found for the first time in thin films of amorphous Nb_3Ge. A typical series of $F_p(b)$ curves is shown in Fig. 4-12[2]. By using Eqs. (4-137) and (4-138) both R_c/a_0 and W could be determined as functions of b. Especially, the behavior of R_c/a_0 is worth to be reminded here. The R_c/a_0 data shows a dome-shaped behavior with a maximum at $b = 0.3$, typically of the order of 40, but values as large as 200 have been determined for more homogeneous films. Both at small fields and close to H_{c2}, R_c decreases demonstrating the increase of disorder related to the softening of the VL, i.e. decrease of c_{66} to zero. It could be demonstrated experimentally

2 It has been shown very recently by Berghuis et al. (1990) that the field at which the pinning force goes to zero actually is not the upper critical field, but rather a melting field B_m. Above B_m there is still a vortex liquid which crosses over to the normal state at B_{c2} which can be significantly larger than B_m depending on the film thickness. It turns out, however, that this does not really affect the interpretation of the results.

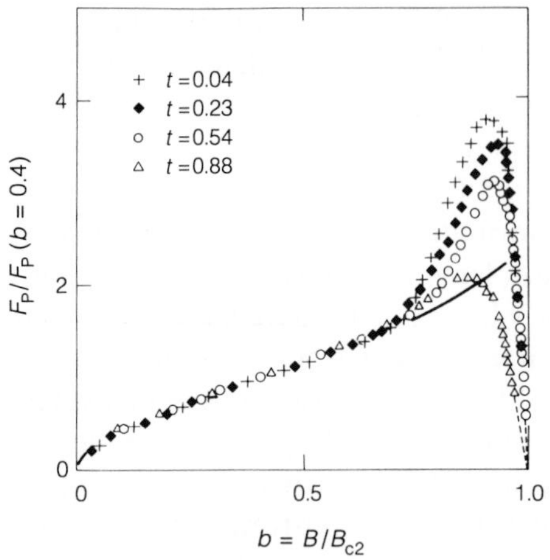

Figure 4-12. Normalized bulk pinning-force versus reduced field for a thin a-Nb_3Ge film (1.24 μm, T_c = 3.81 K) at various values of $t = T/T_c$. The 2DCP behavior is represented by the solid line (Wördenweber and Kes, 1989).

that on approaching H_{c2} at a value of R_c/a_0 between 20 and 15 the VL becomes unstable, see Wördenweber and Kes (1989). In the $F_p(b)$ curves this structural transition is marked by the onset of the peak effect. For these values of R_c/a_0 the shear strength may be locally surpassed giving rise to the creation of edge-dislocations in the VL and history effects around the onset field. The presence of edge-dislocations makes the VL softer, so that it can easier adapt to the pin distribution causing an enhancement of the pinning force. Eventually, very close to H_{c2} at the maximum of F_p, the disorder reaches saturation yielding an amorphous VL with $R_c/a_0 \approx 2$. For an amorphous VL with 2D disorder F_p is simply given by $F_p \approx (W/a_0^2 d)^{1/2}$.

From Eqs. (4-135) and (4-137) L_c can be calculated to be about 5 μm. For thick enough films a transition to 3D disorder is expected at $L_c/d = 0.5$. This crossover in the dimension of the disorder was indeed observed and is shown in Fig. 4-13. At the transition field b_{co} the pinning force increases by almost an order of magnitude. Further experiments reveal a different thickness dependence below and above the transition confirming the crossover in dimensionality. Below the transition the 2D collective pinning theory (2DCP) describes the results very well and even predicts the transition field correctly, i.e. when a specific choice is made for c_{44} in Eq. (4-135) (see discussion below and Wördenweber and Kes, 1989). Above the transition, however, the 3DCP expressions derived by LO for elastic deformations result in a much larger correlated volume than is determined from the data using Eq. (4-136). The far too small value of V_c suggests that the disorder is determined by dislocations. Especially, screw-dislocations created at the film surface triggering the 2D–3D transitions, may move freely along the flux lines, since they have no Peierls potential. This model is justified by the observed IV curves and is nicely picturized in Fig. 4-14.

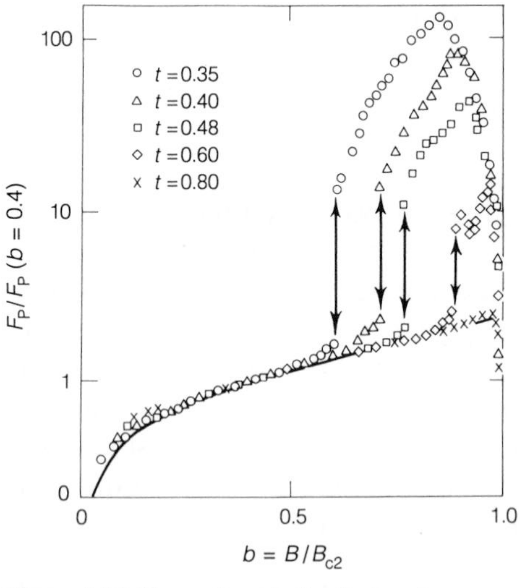

Figure 4-13. Normalized bulk pinning-force versus reduced field for a thick a-Nb_3Ge film (18 μm, T_c = 4.3 K) at various reduced temperatures t displaying the crossover from 2D disorder (low fields, solid line) to 3D disorder (high fields) at B_{co} indicated by the arrows, Wördenweber and Kes (1989).

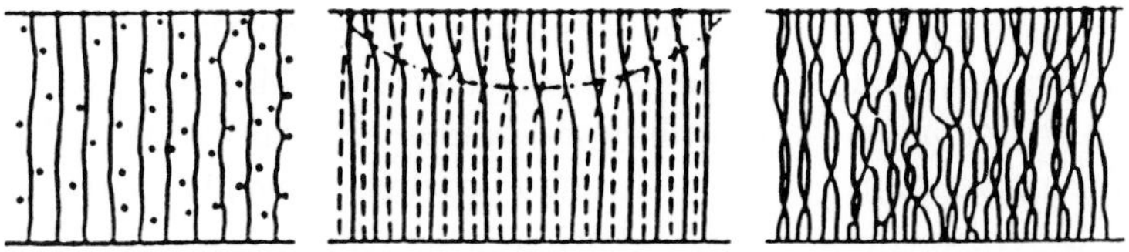

Figure 4-14. Schematic vortex-lattice disorder due to pinning in films in perpendicular field. (a) 2D-disorder ($B < B_{c0}$), the pins are indicated by dots [not shown in (b) and (c)], the curvy (dashed) lines display the vortices. (b) $B = B_{c0}$, a screw dislocation (dash-dotted line) nucleates at the surface. (c) 3D-disorder ($B > B_{c0}$), many screw dislocations have penetrated (Brandt and Essmann, 1987).

The fact that these features are demonstrated in materials with weak pinning centers, does suggest that the "pure" 3DCP situation is not stable and therefore not observable at all. Instead, a highly defective VL seems to be a better picture to bear in mind, certainly in the case of stronger pinning centers. One might speculate that for strong pinning the VL contains many edge and screw-dislocations, dislocation loops, dislocation dipoles, and other defects. Obviously, dislocations can be created or annihilated by flux flow giving rise to new flux-line arrangements and history effects, see Küpfer and Gey (1977). In the thin-film geometry with straight flux lines, it is easy to visualize how dislocation glide can make the VL softer or more disordered, thereby giving it the opportunity to adapt itself more effectively to the distribution of pinning centers evidenced by peak effects. For 3D disorder one might have to allow for flux-line cutting in order to model the VL rearrangements. On the other hand, it should be realized that screw dislocations and screw-dislocation loops (causing a twist of the VL around the field direction) can only occur in virtue of the presence of pinning centers and, in addition, that they easily glide along the flux lines. Therefore, it might be easier to create and annihilate dislocations in a vortex lattice than in a crystal lattice.

4.4.3.3 The Ultimate Pinning Force

For the ultimate limit of disorder, i.e. an amorphous VL, the CP theory offers an estimation for the correlation length referred to as the single flux-line approximation. In order to be more specific about L_c in this limit (denoted by L_{cA}) the behavior of $c_{44}(k_\perp, k_z)$ should be known. Here the wave vector of the deformation field is denoted by $k_\perp^2 = k_x^2 + k_y^2$ and k_z. As discussed by Kes and Van den Berg (1990) c_{44A} in the amorphous-limit for anisotropic superconductors is given by

$$c_{44A} = c_{44}(k_{Br}, L_{cA}^{-1}) \approx c_{44}(0) \frac{1}{\Gamma} \frac{k_h^2}{k_{Br}^2} \tag{4-139}$$

where $k_h = \lambda_h^{-1}$ and k_{Br} is the radius of the Brillouin zone, see Sec. 4.3.10, and Γ is the anisotropy parameter, Eq. (4-103).

For $b > 0.2$ the expression for L_{cA} is

$$L_{cA} \approx 1.52 \left[\frac{1}{\Gamma^2} \frac{c_{44}^2(0)\, r_f^2\, k_h^4\, a_0^2}{W k_{Br}^4} \right]^{1/3} \tag{4-140}$$

Because $W = C(T)\, b^p (1-b)^2$. $L_{cA} = f(T) \cdot b^{-p/3}$ with $p = 1$ or 3 for δT_c- and $\delta\varkappa$-pinning, respectively. The expression for F_p in the amorphous limit for $b > 0.2$ becomes

$$F_p \approx 2.4 \left[\frac{1}{\Gamma} \frac{\mu_0 \lambda^2 W^2}{a_0^7 B^2 (1-b)} \right]^{1/3} \approx \tag{4-141}$$

$$\approx 2.0 \left[\frac{1}{\Gamma} \frac{\mu_0 \lambda^2 B_{c2}^{3/2} C^2(T)}{\Phi_0^{7/2}} \right]^{1/3} \cdot$$

$$\cdot\, b^{(4p+3)/6} (1-b)$$

which has a maximum at $b = 5/7$ for $p = 3$ and at $b = 7/13$ for $p = 1$. Because this bulk pinning force corresponds to the smallest possible correlated volume for a given pinning strength W, Eq. (4-141) describes the ultimate pinning force. Such a dome-shape behavior is often observed, see Kramer (1973).

Under certain cirumstances the summation of the pinning forces is very simple. This depends on the morphology of the predominant pin mechanism. In the HTS $YBa_2Cu_3O_{7-\delta}$ we may assume, for instance, that the twin planes are strong pinning centers if H ∥ c-axis. Because the flux lines are attracted to the twin planes they line up along the planes and the other flux lines between the planes take rather random positions in such a way that the flux-density gradient is minimum (the VL has a large compression modulus). This condition holds the better if the distance between the twin planes $L \gg a_0$, i.e. $B \gg 1$ T for $L \approx 50$ nm. If, in addition, the twin planes are mutually perpendicular, a simple direct summation is applicable leading to $F_p \approx f_1 / L a_0$, where f_1 is the elementary interaction between the twin plane and one unit length of a flux line.

4.4.4 Saturation of the Pinning Force

Features observed in strong pinning materials are dome-shaped $F_p(B)$ curves which close to B_{c2} behave either like $(1-b)$ or $(1-b)^2$. Scaling is frequently seen, i.e. $F_p(B, T) \propto B_{c2}^n(T)\, b^p (1-b)^q$ with $n \approx 2\text{–}2.5$, $p \approx 1\text{–}2.5$, and $q = 1$ or 2, but also "saturation", e.g. in NbTi by Matsushita and Küpfer (1988). (It should be noted that these authors use a different definition for saturation.) It means that the $F_p(B)$ curves are changing in shape and increase up to a certain envelope curve when the pinning strength is enhanced. The envelope curve is dome-shaped and represents the optimum pinning force achievable for the defect morphology under consideration. The non-saturated curves typically show a much sharper peak effect closer to B_{c2}. Also history effects are often reported in these cases.

In a pioneering paper Kramer (1973) pointed out that pin breaking gives rise to a $(1-b)$ behavior, since it is characterized by a uniform motion of the VL and because all elementary pinning forces decrease according to $(1-b)$ when B_{c2} is approached. On the other hand, a very inhomogeneous pinning-force distribution would lead to a shear flow of less-strongly pinned regions of the VL along strongly pinned (bundles of) flux lines. Being proportional to c_{66}, the shear-flow mechanism yields a $(1-b)^2$ behavior. Although "Kramer-plots" have been very useful in describing $F_p(B)$ curves, a more profound comprehension of the mechanism for bulk materials is still needed. Recent experiments in *thin films* which simulated the shear-flow behavior, proved that the VL for all fields behaves like an incompressible fluid with a small, but finite shear modulus (see Pruijmboom et al., 1989). It was shown that the shear modulus is considerably reduced by the effect of edge dislocations in the VL. These results support the view that pin breaking occurs in case of extended defects perpendicular to the direction of flux flow, whereas flux-line shear would occur for smaller defects with a more inhomogeneous distribution, or along grain boundaries parallel to the Lorentz force, or due to the activation of Frank-Read sources of flux line dislocations (Dew-Hughes, 1987).

4.4.5 Pin Potential

As described above, the modelling of the VL disorder in terms of correlated volumes of greatly flexible size and shape depending on the pinning strength, can explain many experimental features. The correlated regions can be identified with the flux bundles which have to be evoked in order to describe flux-creep or flux-flow-noise measurements. The flux bundles are the entities considered to be located in a pin potential which is generated by the net effect of the pinning centers within the bundles themselves. By applying an increasing driving force, which can be realized depending on the experimental geometry by either a transport current or a change of the external field or by a temperature gradient, the flux bundles move up in their pin potentials, until they eventually break away entering the flux-flow regime. Campbell (1971) introduced this picture and showed how to derive force-displacement $[F(u)]$ curves from experiment. A more precise description has been recently given by Yamafuji et al. (1989).

As seen is from Fig. 4-15 $F(u)$ curves look very similar to the stress-strain curves of solids. The slope at $u=0$ is the Labusch parameter α_L which is the curvature of the average pin potential per unit volume. In the linear regime the displacement from the equilibrium position is given by $\boldsymbol{u}=\alpha_L^{-1}\,\boldsymbol{j}\times\boldsymbol{B}$. The resulting change of induction follows from a London-type equation Eq. (4-14) with

$$\lambda_C=(B^2/\mu_0\,\alpha_L)^{1/2} \tag{4-142}$$

instead of λ_L. Here λ_C is Campbell's penetration depth for compressional and tilt waves into an elastically pinned vortex lattice, Campbell (1971), Campbell and Evetts (1972) and Brandt (1990). The transition to the flux flow regime at $F=j_c\,B$ takes place gradually revealing the effect of flux-line dislocation. $j_c\,B/\alpha_L$ determines the range of the pin potential. In case of "pure" collective pinning this should be r_f. However, in presence of many flux line dislocations this distance can be a much smaller fraction of a_0, because of the reduced flow stress in a defective lattice.

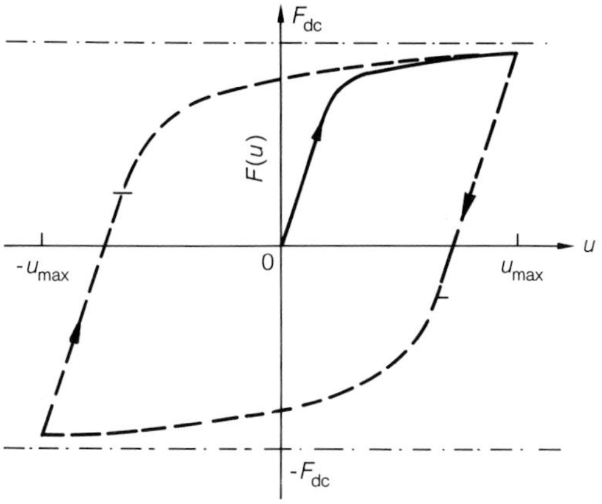

Figure 4-15. Force-displacement diagram of a vortex lattice uniformly displaced over a distance u by a driving force density $F(u)=j\,B$. Such a diagram can be deduced from ac susceptibility measurements in a small ac field superposed on a large dc field, Campbell (1971). $F_{dc}=j_c\,B$ denotes the macroscopic pinning force. From the linear behavior at small u one obtains $dF/du=\alpha_L$, the curvature of the average pin potential.

4.4.6 Technological Superconductors

A survey of the conventional superconductors which have potential possibilities for applications in view of their high T_c and $B_{c2}(0)$ values, is given in Fig. 4-16. Four of them (NbTi, Nb_3Sn, V_3Ga and Nb_3Al) have reached the state of commercialization with NbTi by far as the most common technological superconductor. A very useful overview of the relevant superconducting metallurgical, and mechanical properties together with information on fabrication technology is presented by Wilson (1983). Almost all today's superconducting magnets are made of NbTi for fields up to 10 T at 4.2 K and 12 T at 2.2 K. If higher fields are requested insert coils of Nb_3Sn

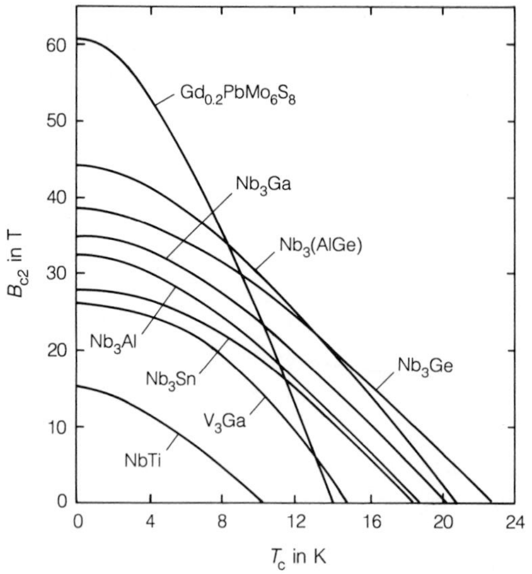

Figure 4-16. Upper critical field versus temperature of conventional high-field superconductors.

are most commonly used to upgrade the field to 20 T at the present technology. Regarding the intrinsic superconducting properties, Fig. 4-16 shows that other A15 compounds, like Nb_3Ge, and the Chevrel phases might be better candidates for applications. Such conductors do not yet exist, because their fabrication requires essential processing steps which are unfavorable for the formation of sufficient and strong enough pinning centers, so that their j_c's are far below the economical lower limit of about 2×10^2 A/mm^2. Creating enough strong pinning centers during the wire fabrication is therefore of major importance, e.g. in the 20 T Nb_3Sn wire a third element, Ti, is added to increase the pinning at high fields. In Fig. 4-17 the solid lines display typical critical current densities as functions of field at 4.2 K of four conventional superconductors.

Since NbTi is a ductile alloy, wires are drawn out of the superconductor as the starting material, whereas the brittle compound Nb_3Sn is not formed until the final heat treatment. Consequently, their pinning centers are quite different in character. In NbTi, depending on the composition, the pins are either elongated networks of dislocations or very long α-Ti precipitates. This pin morphology is favorable, because it is oriented perpendicular to the driving force, so that $F_p \propto (1-b)$ in accord with the pin breaking description (Sec. 4.4.4). In Nb_3Sn small grain boundaries are considered to be the predominant pins. Because their distribution and size is not so uniform, a path of weaker pinning may exist between regions of stronger pinning. This gives rise to a bulk pinning force

$$F_p \approx \frac{2\tau}{W}\frac{l_p}{w} \tag{4-143}$$

where τ is the shear strength of the VL, $\tau \propto c_{66}$, W is the width of the weak-pinning path, l_p its length, and w the width of the wire or film. Clearly, $l_p \geq w$ depending on how the path wanders from one side of the material to the other. In addition, $W \geq a_0$ because at least one row of vortices should be unpinned. For grains larger than a_0, it seems not unreasonable to assume, that W is given by the grain size D which is in agreement with the experimental observation $F_p \propto D^{-1}$ for $D > 30$ nm. The typical field dependence for this kind of defect morphology is $F_p \propto (1-b)^2$ and the IV curves are highly non linear because the unpinned fraction of the VL is growing with increasing driving force. The role of the added third element Ti is to form precipitates on the grain boundaries thereby suppressing the above shear mechanism, see Dew-Hughes (1987).

In addition to a large j_c, there are several other requirements to be met for a reliable conductor design. Mechanical stability against the stress of the Lorentz force is most important, especially for the brittle A15 compounds. Protection against flux

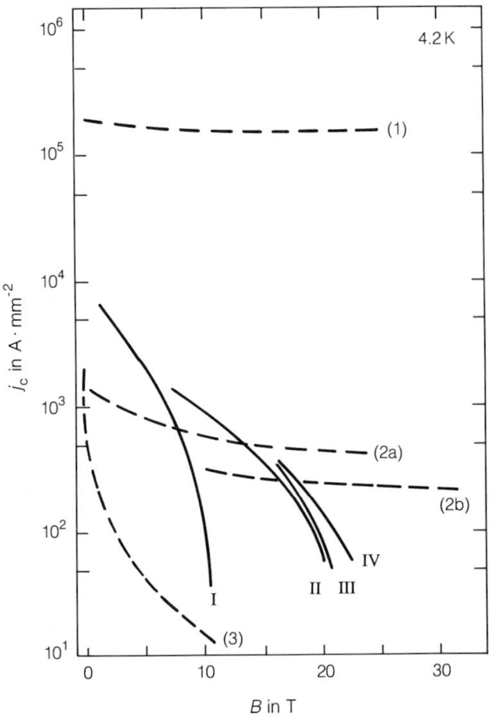

Figure 4-17 a. Typical critical current densities as functions of field for various conventional and high-temperature superconductors at 4.2 K: (1) Y: 123 film on MgO substrate, (2 a) Bi: 2212 Ag-sheathed tape, (2 b) Bi: 2212 Ag-sheathed wire, (3) Y: 123 film on Ag tape, (I) NbTi, (II) Nb_3Sn, (III) V_3Ga, and (IV) $(Nb,Ti)_3Sn$. Source: Nikkei Superconductors, Japan, February 1990.

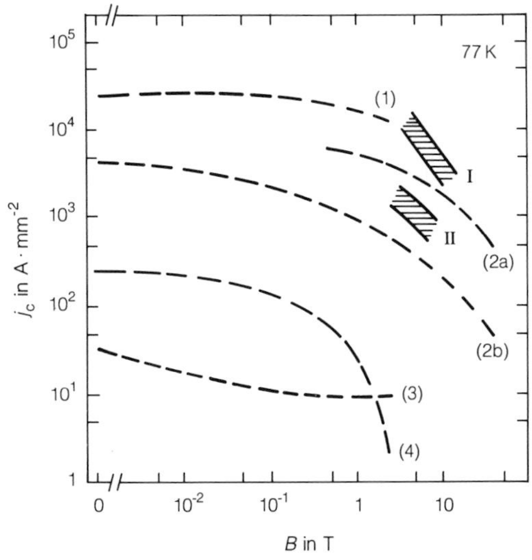

Figure 4-17 b. Comparison of current densities of various HTS at 77 K and conventional materials at 4.2 K. Note the change to logarithmic scale of B. (I) Nb_3Sn, (II) NbTi. (1) Sputtered Y: 123 film, (2 a) CVD deposited Y: 123 film, $H \parallel ab$-planes, (2 b) same as (2 a), $H \parallel c$-axis, (3) Y: 123 wire, (4) Bi: 2212 tape, $H \parallel ab$-planes. Source: Nikkei Superconductors, Japan, December 1989.

jumping is another problem to be solved. For an extensive discussion of these issues we refer to Wilson's book (1983).

4.4.7 High-Temperature Superconductors (HTS)

In all respects the HTS form a challenge for science. After three years of intensive research the microscopic puzzle of the superconducting mechanism is still not resolved. The phenomenology faces new questions regarding the vortex structure and its thermodynamic phase transitions. For the materials sciences and engineering there is still a long way to go before large scale applications are common practice. Within the scope of this section some aspects of the second issue will be briefly discussed in relation to flux pinning and the current-carrying capacity. We will concentrate here on the three most studied systems among the class of ceramic superconductors: $YBa_2Cu_3O_7$ (Y: 123) with $T_c = 93$ K, $Bi_2Sr_2Ca_nCu_{n+1}O_{6+2n}$ (Bi: 22n $n+1$), and $Tl_2Ba_2Ca_nCu_{n+1}O_{6+2n}$ (Tl: 22n $n+1$) with T_c's between 85 K and 125 K for $n = 1$ and 2.

4.4.7.1 Pinning in High-Temperature Superconductors

The common factor in these materials is the double, or triple layer of CuO_2 planes which are believed to be the superconducting planes. The coherence length along the planes is less than 2 nm. In case of Y: 123

this leads to the "weak-link problem", probably because T_c of this material decreases considerably with oxygen deficiency which especially occurs at grain boundaries. If the deficient layer at the boundaries is thicker than ξ the grains are only weakly coupled and have a correspondingly small j_c.

The distance between the superconducting layers is relatively large and is responsible for the large, to very large anisotropy of the superconducting parameters (Sec. 4.3.13). Consequently, the pinning in these materials is also very anisotropic, partially because $F_p \propto B_{c2}^n(T)\, b^p (1-b)^q$ and $B_{c2\,\parallel} \gg B_{c2\,\perp}$, but also because of the very anisotropic nature of the VL itself, see Kes and Van den Berg (1990). When the decoupling between the layers is large, as in the Bi and Tl-compounds, the supercurrents are confined to the layers and the vortices are segmented into vortex "pancakes" with weak electro-magnetic and Josephson coupling between them. This is demonstrated in Fig. 4-18 where two different representations are given for a flux line in an applied field making an angle θ with the normal to superconducting layers (c-axis). The segmentation is indicated by the current loops in the CuO_2 (ab-) planes. The magnetic moment related

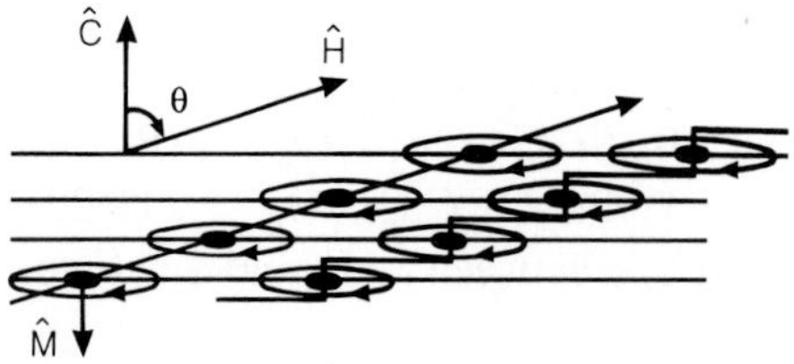

Figure 4-18. Schematic view (not to scale) of two representations of flux lines intersecting CuO_2 double layers (horizontal lines) when the field H makes an angle θ with the c-axis. The ellipses (lying in the $a\,b$-plane) depict superconducting screening currents. To the left is the continuous flux-line picture for very anisotropic superconductors; to the right, the segmented flux-line picture with pancake vortices valid for "Josephson" coupled superconducting layers. Note that the magnetization is directed along the c-axis.

with the screening currents is oriented along the normal which leads to a net torque on the sample when θ deviates from either $\theta = 0$ or $\pi/2$. This provides an accurate way to measure the anisotropy parameter Γ. The zero's of the order parameter and the vortex cores are depicted by the black dots. The representation at left shows the vortex in the anisotropic 3D picture, while the zigzag line at right represents the vortex in the quasi-2D case. The latter applies below the crossover temperature T^* defined in Eq. (4-102). The vortex consists of 2D pancake vortices in the layers and Josephson vortices along the layers. Flux-line cutting in such a system of vortices is relatively easy and leads to quasi-2D behavior above a crossover field B_{2D} defined by

$$B_{2D} = \frac{\Phi_0}{\Gamma s^2} \tag{4-144}$$

For Bi: 2212 this amounts to $B_{2D} \approx 0.3$ T, for Tl: 2212 to about 10 mT according to $s = 1.5$ nm and $\Gamma = 3 \times 10^3$ and 10^5, respectively, see Farrell et al. (1989) and (1990). In many respects the properties of the VL for $B > B_{2D}$ may be considered as if a thin superconducting film of thickness s in a perpendicular field is studied. We think that the different behavior of j_c in Y: 123 and Bi: 2212 is caused by the large difference in anisotropy of both materials, see also Palstra et al. (1990, 1991).

Fig. 4-17a shows a comparison of $j_c(B)$ data of various HTS and conventional superconductors at 4.2 K. Curve (3) for Y: 123 wire demonstrates the weak-link problem. This problem is absent in high-quality Y: 123 thin films deposited by either laser ablation or sputtering, curve (1). The critical current in Bi: 2212 tape (2a) and wire (2b) is also much below the Y: 123 film results, but a lot of progress is still to be made, although there might be a fundamental limitation due to the small pinning energy of a quasi-2D vortex lattice, see Sec. 4.4.7.2. Very promising is of course the independence of j_c on B, which makes high-field applications by means of HTS accessible. In this case a major problem will be the protection of the wire against the hoop stress caused by the magnetic pressure.

In Fig. 4-17 b the performance of HTS at 77 K is compared with that of NbTi and Nb_3Sn at 4.2 K. It is seen that Y: 123 films still compare favourably to the conventional materials. The role of the anisotropy is clearly displayed by the difference in curves (2 a) and (2 b) for c-axis oriented (c perpendicular to substrate) films of Y: 123 prepared by chemical vapour deposition (CVD). This effect is even more predominant for highly textured Bi: 2212 tape, curve (4). The reason is that only the pancake vortices can be pinned effectively, while the Josephson vortices cannot be pinned if the force is directed along the ab-planes. The pinning is supposed to be predominantly exerted by defects in the superconducting layers. Oxygen vacancies are obvious candidates for pinning centers. Using the formalism of Sec. 4.4.2.3 the elementary pinning force on an isolated vortex is found to be

$$f_p \approx \mu_0 H_c^2 \xi_0 D^2/\xi_{\|} \quad (4\text{-}145)$$

which amounts to about 7×10^{-14} N when $\mu_0 H_c(0) = 1$ T and $D = 0.3$ nm is assumed. Other effective pinning centers may be the twin planes in Y: 123. The same formalism as for vacancies can be used. Among the parameters determining f_p is the scattering probability of electrons at a twin plane. This parameter may depend on the amount of other defects which aggregate on twin planes explaining why different experimental evidence is found for the relevance of this pin mechanism. An interesting property of layered superconductors is that the zero of the order parameter for vortices parallel to the layers preferentially lies between these layers. This effect gives rise to *intrinsic* pinning for motion perpendicular to the layers, see Ivlev and Kopnin (1990). For arbitrary orientations of field with respect to the layers and with the currents being transported through the layers, the driving force has parallel and perpendicular components. The perpendicular component will be compensated by the very strong intrinsic pinning force, the parallel component by the pinning force on the pancakes vortices. Effectively, it is the field component in the c-direction which provides the important driving force. With the field along the ab-planes this component is small and therefore j_c is relatively large.

Of technological importance is the considerable increase of the critical current of sintered Y: 123 at 77 K by heat treatment of an off-stoichiometric composition, the melt-process, melt-quench (MPMQ) procedure, introduced by Murakami et al. (1989). This leads to the formation of Y_2BaCuO_5 precipitates which are supposed to be responsible for the improved pinning force. In general, the pinning force in thin films at 77 K is some orders of magnitude larger than in single crystals. One can think of several explanations for this effect, but convincing evidence is lacking at present. It should be noted that at low temperatures the difference is less dramatic. Moreover, the reported values of j_c of order 10^5 A/mm^2 can be shown to be not far below an optimal value determined by the density and strength of pins taking into account that a too large density of pins results into a degradation of T_c. This may be illustrated by the fact that the creation of pins by several means of irradiation damage only leads to an order of magnitude increase of j_c at elevated temperatures, see for a review Weber (1989).

4.4.7.2 Flux Creep in High-Temperature Superconductors

The critical state is in fact a metastable state. Thermal fluctuations may give rise to flux motion (flux creep) comparable to diffusion in solids. In low temperature superconductors this effect usually is small and has been greatly ignored. Because of the large temperature range, flux creep is very obvious in HTS, not only in polycrystalline materials where it is related to breaking of weak links, but also in high-quality single crystals and thin films. Apart from the large thermal energy available, it is the small pin potential which leads to the "giant" creep effects. It is manifest in magnetization measurement where M decays with time, and in the broad resistance transitions observed in a magnetic field. It gives rise to the "mobility line" which separates a high T, B regime, where pinning is negligible, from a low T, B regime where the critical current can be very large. The location of the mobility line can be detected reliably by several ac techniques, such as magnetic ac susceptibility, vibrating reed, or ultrasonic absorption. A characteristic observation is the shift of the mobility line to higher T when the frequency of the probing technique is increased. This frequency dependence distinguishes the mobility line from the melting line which characterizes the transition from a high-T vortex liquid to a low-T vortex solid. Because the long range order in the solid is destroyed by collective pinning, the name vortex glass has been introduced for this solid vortex phase. It is quite possible that the occurrence of a mobility line actually displays the plasto-elastic properties of the vortex liquid under the constraints of many pinning centers. However, at present, there is no consensus yet on this issue.

Undeniable, though, is the correlation between the width of the resistive transition, the position of the mobility line and the magnetic decay time on the one hand and the anisotropy of the HTS on the other. The larger the anisotropy ($\Gamma \approx 29$, 3000 and 10^5 for Y:123, Bi:2212 and Tl:2212, respectively), the larger are the thermal effects, especially in an applied field. Directly related to the amount of anisotropy is the tilt modulus of the VL. As mentioned in Sec. 4.3.13, c_{44} is reduced by a factor Γ which leads to flux cutting and segmentation of the vortices. Consequently, the correlated regions can be as small as one pancake vortex: i.e. $V_c \approx a_0^2 s$, where s is the separation between the CuO_2 layers. Since $F_p = j_c B = (W/V_c)^{1/2}$, the low temperature j_c can be very large. However, the pin potential $U_p = F_p V_c r_f = (W V_c)^{1/2} r_f$, can be very small, so that at elevated temperatures a fast decay prevents the observation of the potential j_c. Instead, a much lower j is observed at which the decay time is comparable to or larger than the practical, experimental time scales. By increasing the defect density, e.g. by means of irradiation, one can raise U_p only slightly. The observed current density, however, may go up by one or two orders of magnitude, because it scales with $\exp(-U_p/k_B T)$. Therefore, it remains to be seen if and which of the HTS will in the future be applied at liquid nitrogen temperatures. More details may be found in the recent review articles on flux creep by Malozemoff (1989) and Hagen and Griessen (1989), and the theoretical paper on vortex phases, collective pinning and collective creep by Fisher et al. (1991).

4.5 References

Abrikosov, A. A. (1988), *Fundamentals of the Theory of Metals*. Amsterdam: North Holland.

Ambegoakar, V., Baratoff, A. (1963), *Phys. Rev. Lett. 10,* 486; *11,* 104 (*E*).

Anderson, P. W. (1964), in: *Lectures on the Many-Body Problem:* Caiamiello E. R. (Ed.). New York: Academic Press.

Auer, J., Ullmaier, H. (1973), *Phys. Rev. B 7,* 136.

Bardeen, J., Cooper, L. N., Schrieffer, J. R. (1957), *Phys. Rev. 106,* 162; *108,* 1175.

Bean, C. P. (1964), *Rev. Mod. Phys. 36,* 36.

Berghuis, P., van der Slot, A. L. F., Kes, P. H. (1990), *Phys. Rev. Lett. 65,* 2583.

Brandt, E. H. (1990), *Z. Phys. B 80,* 167.

Brandt, E. H., Essmann, U. (1987), *Phys. Stat. Sol. (b) 144,* 13–38.

Campbell, A. M. (1971), J. Phys. C.: *Solid State Phys. 4,* 3186–3198.

Campbell, A. M., Evetts, J. E. (1972), *Advan. Phys. 21,* 199–428.

Clem, J. R. (1989), *Physica C162–164,* 1137–1142.

De Gennes, P. G. (1966), *Superconductivity of Metals and Alloys*. New York: Benjamin.

Deutscher, G., De Gennes, P. G. (1969), in: *Superconductivity:* Parks, R. D. (Ed.). New York: Marcel Dekker, Ch. 17.

Dew-Hughes, D. (1987), *Phil. Mag. B55,* 459.

Farrell, D. E., Bonham, S., Foster, J., Chang, Y. C., Jiang, P. Z., Vandervoort, K. G., Lam, D. J., Kogan, V. (1989), *Phys. Rev. Lett. 63,* 782.

Farrell, D. E., Beck, R. G., Booth, M. F., Allen, C. J., Bukowski, E. D., Ginsberg, D. M. (1990), *Phys. Rev. B 42,* 6758.

Fetter, A. L., Hohenberg, P. C. (1969), in: *Superconductivity:* Parks, R. D. (Ed.). New York: Marcel Dekker, Ch. 14.

Fisher, D. S., Fisher, M. P. A., Huse, D. A. (1991), *Phys. Rev. B 43,* 130.

Ginsberg, D. M., Hebel, L. C. (1969), in: *Superconductivity:* Parks, R. D. (Ed.) New York: Marcel Dekker, Ch. 4.

Ginzburg, V. L., Landau, L. D. (1950), *Zh. Eksperim. i Teor. Fiz. 20,* 1064.

Gladstone, G., Jensen, M. A., Schrieffer, J. R. (1969), in: *Superconductivity:* Parks, R. D. (Ed.) New York: Marcel Dekker, Ch. 13.

Goodman, B. B. (1964), *Rev. Mod. Phys. 36,* 67.

Gor'kov, L. P. (1958 a), *Zh. Eksperim. i Teor. Fiz. 34,* 735.

Gor'kov, L. P. (1958 b), *Soviet Phys. JETP 7,* 505.

Gor'kov, L. P. (1959 a), *Zh. Eksperim. i Teor. Fiz. 36,* 1918; *37,* 1407.

Gor'kov, L. P. (1959 b), *Soviet Phys. JETP 9,* 1364.

Gor'kov, L. P. (1960), *Soviet Phys. JETP 10,* 998.

Gor'kov, L. P., Kopnin, N. B. (1975), *Usp. Fiz. Nauk 116,* 413.

Gor'kov, L. P., Kopnin, N. B. (1976), *Sov. Phys. Usp. 18,* 496.

Gorter, C. J., Casimir, H. B. G. (1934 a), *Phys. Z. 35,* 963.

Gorter, C. J., Casimir, H. B. G. (1934 b), *Z. Tech. Phys. 15,* 539.

Hagen, C. W., Griessen, R. (1989), in: *Studies of High Temperature Superconductors,* Vol. 3: Narlikar, A. (Ed.). New York: Nova Science Publishers, p. 159.

Huebener, R. P. (1979), in: *Magnetic Flux Structures in Superconductors*. Berlin: Springer-Verlag.

Ivlev, B. J., Kopnin, N. B. (1990), *Phys. Rev. Lett. 64,* 1828.

Josephson, B. D. (1965), *Advan. Phys. 14,* 419–451.

Kes, P. H., Van den Berg, J. (1990), in: *Studies of High Temperature Superconductors,* Vol. 5: Narlikar, A. V. (Ed.). New York: Nova Science Publishers, 83–117.

Kes, P. H., Van der Beek, C. J. (1991), *Physica B 169,* 80.

Kes, P. H., van der Klein, C. A. M., de Klerk, D. (1973), *J. Low Temp. Phys. 10,* 759.

Kramer, E. J. (1973), *J. Appl. Phys. 44,* 1360–1370.

Kramer, E. J. (1978), *J. Nuclear Materials 72,* 5.

Kuper, C. G. (1968), *Introduction to the Theory of Superconductivity*. Oxford: Oxford University Press.

Küpfer, H., Gey, W. (1977), *Phil. Mag. 36,* 859.

Labusch, R. (1969), *Crystal Lattice Defects 1,* 1.

Larkin, A. I., Ovchinnikov, Yu. N. (1979), *J. Low Temp. Phys. 34,* 409.

Larkin, A. I., Ovchinnikov, Yu. N. (1986), in: *Nonequilibrium Superconductivity:* Langenberg, D. N., Larkin, A. I. (Eds.). Amsterdam: North Holland, Ch. 11.

London, F. (1950), *Superfluids,* Vol. II. New York: Wiley.

Malozemoff, A. P. (1989), in: *Physical Properties of High Temperature Superconductors:* Ginsberg, D. M. (Ed.). Singapore: World Scientific.

Matsushita, T., Küpfer, H. (1988), *J. Appl. Phys. 63,* 5048–5059.

McMillan, W. L., Rowell, J. M. (1969), in: *Superconductivity:* Parks, R. D. (Ed.) New York: Marcel Dekker, Ch. 11.

Meservey, R., Schwartz, B. B. (1969), in: *Superconductivity:* Parks, R. D. (Ed.) New York: Marcel Dekker, Ch. 3.

Murakami, M., Monta, M., Doi, K., Miyamoto, K. (1989), *Jap. J. Appl. Phys. 28,* 1189.

Nikkei Superconductors, February 1990, Japan.

Palstra, T. T. M., Batlogg, B., Van Dover, R. B., Schneemeyer, L. F., Wasczak, J. V. (1990), *Phys. Rev. B 41,* 6621.

Palstra, T. T. M., Batlogg, B., Schneemeyer, L. F., Wasczak, J. V. (1991), *Phys. Rev. B 43,* 3756.

Phillips, J. C. (1989), *Physics of High-T_c Superconductors*. San Diego: Academic Press.

Pruijmboom, A., Kes, P. H., Van der Drift, E., Radelaar, S. (1989), *Cryogenics 29,* 232–235.

Rickayzen, G. (1964), *Theory of Superconductivity*. New York: Wiley.

Roberts, B. W. (1978), *NBS Technical Note 983*. Washington D.C.: U.S. Government Printing Office.
Saint-James, D., Sarma, G., Thomas, E. J. (1969), *Type II Superconductivity*. Oxford: Pergamon.
Scalapino, D. J. (1969), in: *Superconductivity:* Parks, R. D. (Ed.) New York: Marcel Dekker, Ch. 10.
Schrieffer, J. R. (1964), *Theory of Superconductivity*. New York: Addison-Wesley.
Serin, B. (1969), in: *Superconductivity:* Parks, R. D. (Ed.) New York: Marcel Dekker, Ch. 15.
Shapiro, S. (1963), *Phys. Rev. Lett. 11,* 80.
Skocpol, W. J., Tinkham, M. (1975), *Rep. Progr. Phys. 38,* 1049–1097.
Solymar, L. (1972), *Superconductive Tunnelling and Applications*. London: Chapman and Hall.
Takahashi, S., Tachiki, M. (1986), *Phys. Rev. B 34,* 3162.
Thuneberg, E. (1989), *Cyrogenics 29,* 236–244.
Tinkham, M. (1975), *Introduction to Superconductivity*. New York: McGraw-Hill.
Van der Meij, G. P., Kes, P. H. (1984), *Phys. Rev. B 29,* 6233.
Van Dutzer, T., Turner, C. W. (1981), *Principles of Superconductive Devices and Circuits*. New York: Elsevier.
Weber, H. W. (1989), in: *Studies of High Temperature Superconductors,* Vol. 3: Narlikar, A. (Ed.). New York: Nova Science Publishers, p. 197.
Werthamer, N. R. (1969), in: *Superconductivity:* Parks, R. D. (Ed.) New York: Marcel Dekker, Ch. 6.
Wilson, M. N. (1983), *Superconducting Magnets*. Oxford: Oxford University Press.
Wolf, E. L. (1989), *Principles of Electron Tunelling Spectroscopy*. Oxford: Oxford University Press.
Wördenweber, R., Kes, P. H. (1989), *Cryogenics 29,* 321.
Yamafuji, K., Fujiyoshi, T., Toko, K., Matsushita, T. (1989), *Physica C159,* 743.

General Reading

Abrikosov, A. A. (1988), *Fundamentals of the Theory of Metals*. Amsterdam: North Holland.
Brandt, E. H., Essmann, U. (1987), *Phys. Stat. Sol. (b) 144,* 13–38.
Campbell, A. M., Evetts, J. E. (1972), *Advan. Phys. 21,* 199–428.
De Gennes, P. G. (1966), *Superconductivity of Metals and Alloys*. New York: Benjamin.
Huebener, R. P. (1979), in: *Magnetic Flux Structures in Superconductors*. Berlin: Springer-Verlag.
Rickayzen, G. (1964), *Theory of Superconductivity*. New York: Wiley.
Saint-James, D., Sarma, G., Thomas, E. J. (1969), *Type II Superconductivity*. Oxford: Pergamon.
Schrieffer, J. R. (1964), *Theory of Superconductivity*. New York: Addison-Wesley.
Solymar, L. (1972), *Superconductive Tunneling and Applications*. London: Chapman and Hill.
Tinkham, M. (1975), *Introduction to Superconductivity*. New York: McGraw-Hill.
Van Dutzer, T., Turner, C. W. (1981), *Principles of Superconductive Devices and Circuits*. New York: Elsevier.
Wilson, M. N. (1983), *Superconducting Magnets*. Oxford: Oxford University Press.

5 Magnetic Properties of Metallic Systems

Damien Gignoux

Laboratoire Louis Néel, Centre National de Recherche Scientifique, Grenoble, France

List of Symbols and Abbreviations

$\boldsymbol{B}$	magnetic flux density
$\mathscr{B}_J(x)$	Brillouin function
E_F	Fermi energy
e	electron charge
g_J	gyromagnetic ratio
$\boldsymbol{H}$	magnetic field strength
$\boldsymbol{H}_a$	applied field
$\boldsymbol{H}_c$	coercive field
$\boldsymbol{H}_d$	demagnetizing field
$\boldsymbol{H}_i$	internal field
H	Hamiltonian
H_C	crystal field Hamiltonian
H_{ij}	exchange Hamiltonian
H_{Zee}	Zeeman Hamiltonian
h	Planck constant
J	current
$\boldsymbol{J}$	total angular momentum
J_{ij}	exchange term
k	Boltzmann constant
$\boldsymbol{L}$	orbital angular momentum
$\mathscr{L}(x)$	Langevin function
$\boldsymbol{M}$	magnetization
m	magnetic moment
N_d	demagnetizing factor
n	total number of electrons per atom
$n(E)$	density of states
n_+, n_-	number of electrons with spin up and down
q	charge
$\boldsymbol{S}$	spin angular momentum
T	temperature
T_C	Curie temperature
T_f	freezing temperature
T_K	Kondo temperature
T_{sf}	spin-fluctuation temperature
U	interaction energy
U_{ij}	Coulomb terms
$V(\boldsymbol{r})$	electrostatic potential
Z	average atomic number
γ	domain wall energy per surface unit
ε_0	permittivity constant of vacuum
η	chemical potential
θ_P	paramagnetic Curie temperature

Λ	Russel-Saunders coupling constant
λ	spin-orbit coupling constant
$\boldsymbol{\mu}$	magnetic moment
μ	chemical potential
μ_B	Bohr magneton
μ_r	relative permeability
μ_0	permeability constant of vacuum
ϱ	resistivity
ϱ_m	magnetic resistivity
ϕ	electronic wave function
χ	susceptibility
χ	spin state
ψ	orbital state
CEF	crystalline electric field
INS	inelastic neutron scattering
NMR	nuclear magnetic resonance
RKKY	Ruderman-Kittel-Kasuya-Yosida exchange interaction
vbs	virtual bound state

5.1 Introduction

This chapter is the first of the two chapters of this Volume to be devoted to magnetism. That is why Secs. 5.2 and 5.3 describe the main concepts and basic properties which are necessary to understand magnetism in metallic systems as well as in insulators. Subsequently, we present the basic models of itinerant magnetism which characterize the metallic state (Sec. 5.4). In the two following sections we discuss metallic substances formed with the two main series of magnetic elements, namely the 3d (iron group) elements and the 4f (rare earth) elements. The former need an itinerant description whereas for the latter the localized model is quite appropriate. Section 5.7 is devoted to the intermediate situation between localized and itinerant magnetism, a field which has attracted much attention in recent years because of its fundamental interest for the understanding of magnetism. In such a chapter it is natural to mention disordered magnetic systems (Sec. 5.8), a topic which is also fashionable, mainly because of the fundamental concepts which come into play and which also concern other scientific areas. A short survey of magnetoelastic properties is presented in Sec. 5.9. The purpose of Sec. 5.10 is to describe alloys with both 3d and 4f elements. These compounds belong to one of the most important categories of magnetic materials for their fundamental interest as well as for their technical applications. Finally, Sec. 5.11 is devoted to a semi-microscopic description of ferromagnetic materials, especially their magnetization processes, which are the basis for most technical applications, also briefly presented at the end of this section.

5.2 Definitions and Free Ion Magnetism

5.2.1 Definitions and Units

A magnetic moment $\boldsymbol{\mu}$ is equivalent to a current loop $\boldsymbol{\mu} = I\,\boldsymbol{S}$, where I is the current intensity, and $\boldsymbol{S}$ is the vector oriented normal to the surface S of the loop ($|\boldsymbol{S}| = s$); it is therefore expressed in $\mathrm{A\,m^2}$, the SI unit used throughout.

Magnetization $\boldsymbol{M}$ is the magnetic moment per unit volume, expressed in $\mathrm{A\,m^{-1}}$. It is the sum of all the individual moments carried by the atoms. Magnetization, especially in e.m.u., is also expressed per unit mass: the specific magnetization.

Magnetic induction:

$$\boldsymbol{B} = \mu_0 (\boldsymbol{H} + \boldsymbol{M}) \tag{5-1}$$

with $\mu_0 = 4\pi\,10^{-7}$. $\boldsymbol{B}$ is expressed in Tesla. This formula is valid in any point of the space, inside the material as well as outside (in this latter case $\boldsymbol{M}$ is zero). The magnetic field $\boldsymbol{H}$ (in $\mathrm{A\,m^{-1}}$) is the sum of the field originating from the true electrical currents and of the dipolar field produced by the magnetic moments. This latter field created by a moment $\boldsymbol{\mu}$, can be written as: $\boldsymbol{h} = [3\,\boldsymbol{r}(\boldsymbol{\mu}\,\boldsymbol{r})/r^5 - \boldsymbol{\mu}/r^3]/4\pi$ where $\boldsymbol{r}$ is the vector joining the moment to the considered point. Inside a magnetic substance the average of this field over the interatomic spacing is opposite to the magnetization and is called the "demagnetizing field".

The response of a substance to a static field is expressed as: $\boldsymbol{M} = [\boldsymbol{\chi}]\,\boldsymbol{H}$; where $[\boldsymbol{\chi}]$ is the susceptibility tensor, one of the main quantities used to characterize the magnetic properties of a material.

Correspondence between the SI unit system and the e.m.u. one: 1 Tesla $= 10^4$ Gauss, $1\ \mathrm{A\,m^{-1}} = 4\pi\,10^{-3}$ Oe.

5.2.2 Dipolar Field – Demagnetizing Field

A magnetic field $\boldsymbol{H}$ is produced by electric currents and the magnetic moments of each atom. This latter contribution is dipolar and can be estimated by replacing the material by *magnetic poles* on its surface with a density proportional to the magnetization. The $\boldsymbol{H}$ lines produced by the poles begin on the north pole and end on the south pole. As an example, the $\boldsymbol{H}$ lines of a bar magnet in zero applied field are shown in Fig. 5-1.

Inside the magnet the $\boldsymbol{H}$ lines constitute a field $\boldsymbol{H}_d$ opposite to the magnetization: it is the *demagnetizing* field which tends to demagnetize the magnet. Generally $\boldsymbol{H}_d$ is nonuniform inside the magnet. However, for an ellipsoid the magnetization and the demagnetizing field are uniform inside the material. In particular, when the magnetization is along a symmetry axis of the ellipsoid, $\boldsymbol{H}_d$ is proportional to the magnetization: $\boldsymbol{H}_d = -N_d \boldsymbol{M}$, where the demagnetizing factor or coefficient, N_d, mainly depends on the shape of the material. In particular: (i) $N_d = 1/3$ for a sphere, (ii) $N_d \approx 0$ for an elongated ellipsoid when magnetization is parallel to the long dimension (Fig. 5-2a), (iii) $N_d = 1$ for a slab when its thickness is small compared to its area and when magnetization is perpendicular to its surface (Fig. 5-2b).

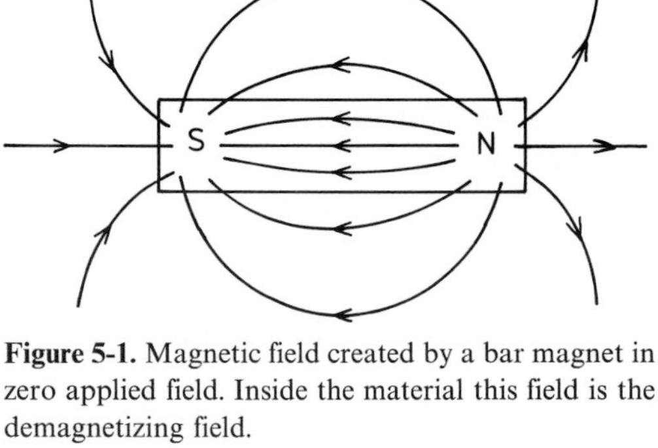

Figure 5-1. Magnetic field created by a bar magnet in zero applied field. Inside the material this field is the demagnetizing field.

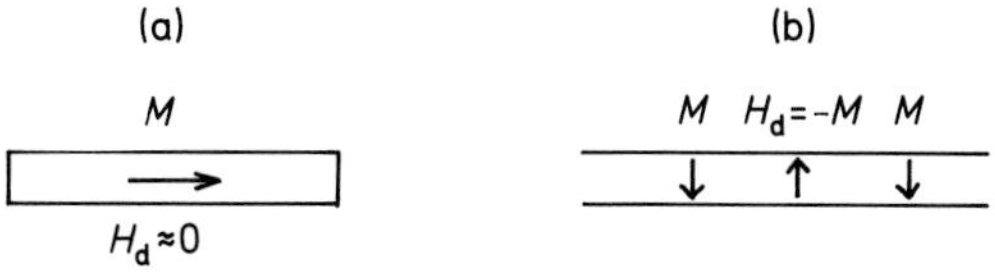

Figure 5-2. Demagnetizing field in an elongated bar (a) and in a slab (b).

We will see later that demagnetizing field effects are of great importance for domain structures in ferromagnets and for permanent magnets.

The local field $\boldsymbol{H}$, or internal field $\boldsymbol{H}_i$, acting on magnetic moments can be written as: $\boldsymbol{H} = \boldsymbol{H}_i = \boldsymbol{H}_a + \boldsymbol{H}_d$, where $\boldsymbol{H}_a$ is the applied field. Generally $\boldsymbol{H}_d$ is weak and $\boldsymbol{H}_a$ is often used whereas $\boldsymbol{H}_i$ should be used, in particular when the magnetization is large.

5.2.3 Free Ion Magnetism

5.2.3.1 Origin of Magnetism at the Atomic Scale

At the atomic scale, magnetism comes from electronic motions which are of two different types, orbital and spin. The orbital magnetic moment originates from the motion of the electrons around the nucleus and may be visualized as a current in a circular loop of wire having no resistance. In Bohr's original (1913) quantum description of the atom, the electron is assumed to move with velocity v in a circular orbit of radius r. The orbital magnetic moment is then:

$$\mu = S I = \pi r^2 e v / 2 \pi r = e v r / 2.$$

Moreover, in this description the angular momentum of the electron must be an integral multiple of $\hbar = h/2\pi$ (h = Planck's constant): $m v r = n \hbar$ (m = electron mass). Therefore: $\mu_{\text{orbital}} = e\hbar/2m = \mu_B$ for the magnetic moment of the electron on the first ($n = 1$) Bohr's orbit. *Bohr magneton* μ_B is the fundamental quantity of magnetic moment. It has been shown that the magnetic moment due to electron spin has exactly the same value $\mu_{spin} = 2s\mu_B$ where $s = 1/2$ (see Sec. 5.2.3.2).

The nucleus can also carry a small magnetic moment, but it is insignificantly small compared to that of electrons, and it does not affect the gross magnetic properties.

5.2.3.2 Quantum Description of Magnetic Moment of Atoms

The magnetic moment of a free atom (or ion) is the sum of the moments of all electrons. The wave function describing the state of each electron in an average potential (Hartree approximation) due to the nucleus and the other electrons can be written as:

$$\Psi_{n,l,m,s}(\boldsymbol{r}) = R_{n,l}(r)\, Y_l^m(\theta, \phi)\, \chi_s \qquad (5\text{-}2)$$

The $R_{n,l}(r)$ are the radial wave functions and the $Y_{l,m}(\theta, \phi)$ are the spherical harmonics. The quantum numbers l and m characterize the orbital angular momentum. χ_s, where $s = \pm 1/2$ characterizes the spin angular momentum state.

The energy of each electron $\varepsilon_{n,l}$ does not depend on the quantum numbers m and s. At this stage of approximation, the energy of one atom is only characterized by the quantum numbers n_i and l_i of each electron. The state of the system is called a *configuration*. For each atom the lowest energy "configuration" is obtained by considering that electrons are fermions (particles with half-integral spin). As a consequence the total wave function must be totally antisymmetric with respect to particle interchange. In particular, this means that two electrons cannot be in the same quantum state at the same time (Pauli principle). For each atom, the lowest energy configuration is that given in the periodic table of elements. For example iron has an electron configuration of: $1s^2\ 2s^2\ 2p^6\ 3s^2\ 3p^6\ 4s^2\ 3d^6$. The electrons are divided into shells with different energies. On each shell the total spin and angular momenta are:

$$\hbar \boldsymbol{L} = \sum_i \hbar \boldsymbol{l}_i \quad \text{and} \quad \hbar \boldsymbol{S} = \sum_i \hbar \boldsymbol{s}_i .$$

For a closed shell: $L = 0$ and $S = 0$. Hence only the incomplete shells may have an orbital and spin angular momentum and thus a magnetic moment given by:

$$\boldsymbol{\mu} = \mu_B(\boldsymbol{L} + 2\boldsymbol{S}).$$

At the next stage of approximation one has to take into account the existence of correlations and electrostatic interactions between electrons (in a configuration only the electrostatic energy between each electron and the nucleus is taken into account) which split the configuration energy levels and gives rise to *terms* characterized by the quantum numbers L and S. Separation between terms is about 10 eV [$\approx 10^5$ K (Kelvin) if $E = kT$ (k = Boltzmann's constant)]. The L and S values of the ground state term are given by Hund's rules: (i) S has the maximum value compatible with the Pauli principle, (ii) L has the maximum value compatible with the S value. The multiplicity of each term is $(2L+1)(2S+1)$.

It is worth nothing that maximum S is associated with a fully antisymmetric spatial wave function. Then each electron (or hole if the shell is more than half filled) belongs to a different orbital and the repulsive electrostatic energy is thus lowered. Thus the correlations between electrons give rise to "ferromagnetism" inside the atom. In fact, ferromagnetism in the solid has the same origin.

Note that L and S characterize only the incomplete shell (as an example, for elements of the iron group this concerns 3d electrons for which $n = 3$ and $l = 2$).

In the atom there is a coupling between the spin and orbital magnetic moments:

each electron in its own reference system sees the nucleus moving and is therefore submitted to a magnetic field acting on the spin. This perturbation can be written: $H_{s.o.} = \Lambda \boldsymbol{L} \cdot \boldsymbol{S}$ (spin-orbit or Russell-Saunders coupling) where $\Lambda(L, S)$ depends upon the L and S values of the term. Each term is thus split into *multiplets* characterized by the new quantum number J ($\boldsymbol{J} = \boldsymbol{L} + \boldsymbol{S}$, where $\boldsymbol{J}$ is the total angular momentum).

The ground state multiplet is such that $J = |L - S|$ if the shell is less than half full and $J = L + S$ in the other case. The spin orbit coupling leads to a distance between multiplets of some 10^2 meV (10^3 K). The heavier the atoms in the periodic table of the elements, the stronger is the LS coupling. As an example, the splitting of the $3d^3$ configuration of the Cr^{3+} ion is shown in Figure 5-3. The ground state multiplet characterized by $L = 3$, $S = 3/2$ and $J = 3/2$ is written as: ${}^4F_{3/2}$ ($4 = 2S + 1$, $3/2 = J$, F corresponds to $L = 3$).

The multiplicity of each multiplet is $2J + 1$ and the wave functions of the usual basis are $|J, M_J\rangle$ ($M_J = J, J - 1, \ldots -J + 1, -J$) such that $J_z |J, M_J\rangle = M_J |J, M_J\rangle$ where $\hbar J_z$ is the total angular momentum operator.

$\boldsymbol{\mu} = \mu_B(\boldsymbol{L} + 2\boldsymbol{S})$ and $\boldsymbol{J} = \boldsymbol{L} + \boldsymbol{S}$ are not collinear. However within a multiplet all the matrix elements of μ_i and J_i ($i = x, y, z$) are proportional such that one can generally write $\boldsymbol{\mu} = g_J \mu_B \boldsymbol{J}$ where g_J is the Landé factor which can be shown to be:

$$g_J = 1 + [J(J+1) + S(S+1) - L(L+1)] / 2J(2J+1).$$

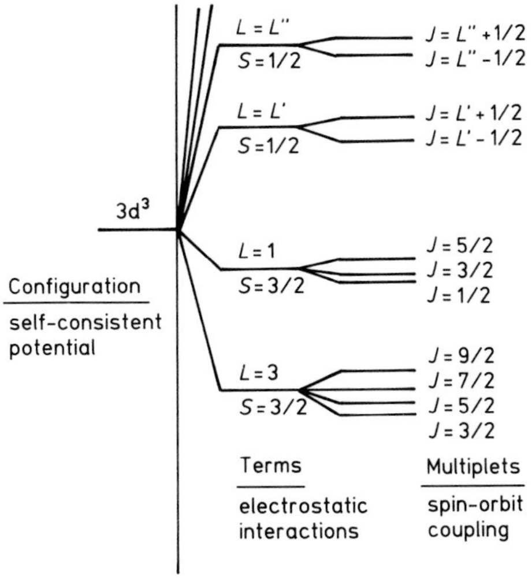

Figure 5-3. Splitting of the $3d^3$ configuration of the Cr^{3+} ion.

5.2.3.3 The Two Most Important Series of Magnetic Elements

Two series of elements play a fundamental role in magnetism, namely the iron group (3d) and the rare earths (4f). Electron configuration and ground state multiplet of the elements in these two series are reported in Tables 5-1 and 5-2. These two series are important because the unfilled shells are not the outer shells and in the solids the 3d (respectively 4f) shell can remain unfilled, leading to magnetism.

Another series, namely the actinides (5f), must also be mentioned although it is less studied because a majority of the elements are radioactive and because its interest is more fundamental than applied. At the moment the number of studies devoted to this series is increasing.

With regard to the two series of elements characterized by the filling of the 4d and 5d shells, these shells are rather delocalized. Therefore the electrons of these shells can participate in the conduction band and their contribution to magnetism is generally very weak.

In compounds of elements different from those quoted above, the outer shells can be filled by binding or by electrons of the conduction band, leading to closed shells which are not magnetic.

Table 5-1. Electron configurations and ground state multiplets of free atoms in the iron group.

	^{21}Sc	^{22}Ti	^{23}V	^{24}Cr	^{25}Mn	^{26}Fe	^{27}Co	^{28}Ni
Ar envelope +	$3d\,4s^2$	$3d^2\,4s^2$	$3d^3\,4s^2$	$3d^4\,4s^2$	$3d^5\,4s^2$	$3d^6\,4s^2$	$3d^7\,4s^2$	$3d^8\,4s^2$
	$^2D_{3/2}$	$^3F_2$	$^4F_{3/2}$	$^7S_3$	$^6S_{5/2}$	$^5D_4$	$^4F_{9/2}$	$^3F_4$

Table 5-2. Electron configurations and ground state multiplets of the free rare earth atoms.

	^{58}Ce	^{59}Pr	^{60}Nd	^{61}Pm	^{62}Sm	^{63}Eu	^{64}Gd
Xe envelope +	$4f^2\,6s^2$	$4f^3\,6s^2$	$4f^4\,6s^2$	$4f^5\,6s^2$	$4f^6\,6s^2$	$4f^7\,6s^2$	$4f^7\,5d\,6s^2$
	$^3H_4$	$^4I_{9/2}$	$^5I_4$	$^6H_{5/2}$	^7F	$^8S_{1/2}$	$^9D_2$
	^{65}Tb	^{66}Dy	^{67}Ho	^{68}Er	^{69}Tm	^{70}Yb	
Xe envelope +	$4f^8\,5d\,6s^2$	$4f^{10}\,6s^2$	$4f^{11}\,6s^2$	$4f^{12}\,6s^2$	$4f^{13}\,6s^2$	$4f^{14}\,6s^2$	
	$^8H_{17/2}$	$^5I_8$	$^4I_{15/2}$	$^3H_6$	$^2F_{7/2}$	$^1S_9$	

Let us come back to the 3d and 4f series. The spatial extent of the electrons in these two series is different. The radial densities $r^2|R_{n,l}(r)|^2$ can be calculated using the radial part of the electronic wave function; these are compared in Fig. 5-4. It can be seen that the 4f shell is more localized than the 3d one, which, as we will see later, has important consequences when atoms are embedded in the solid. The other important difference between these groups concerns the spin orbit coupling. Indeed it varies as Z^4 (Z = atomic number) and is larger in 4f compounds ($\approx 10^4$ K) than in 3d compounds ($\approx 10^2$ K).

5.2.3.4 Diamagnetism

A diamagnetic substance is characterized by a negative susceptibility. To under-

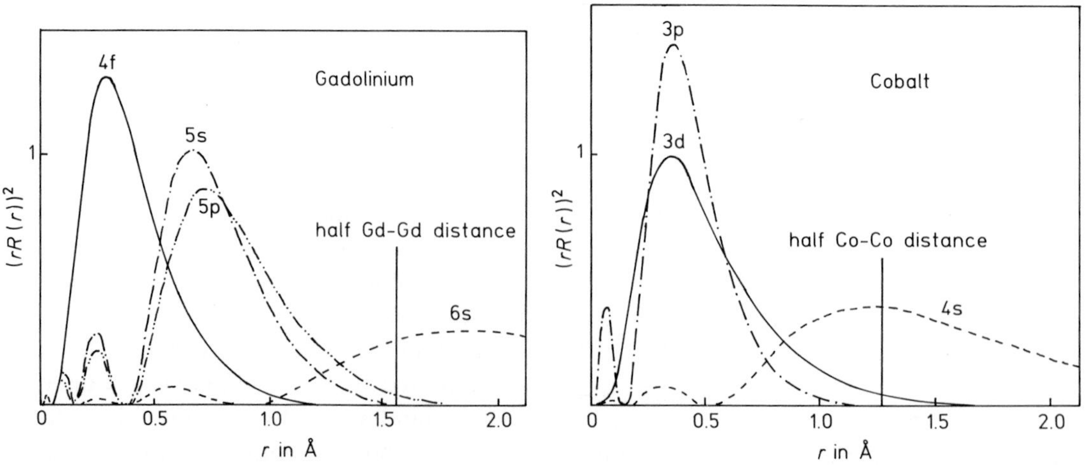

Figure 5-4. Radial densities of the outer electronic orbitals of two elements of the 4f (Gd) and 3d (Co) series.

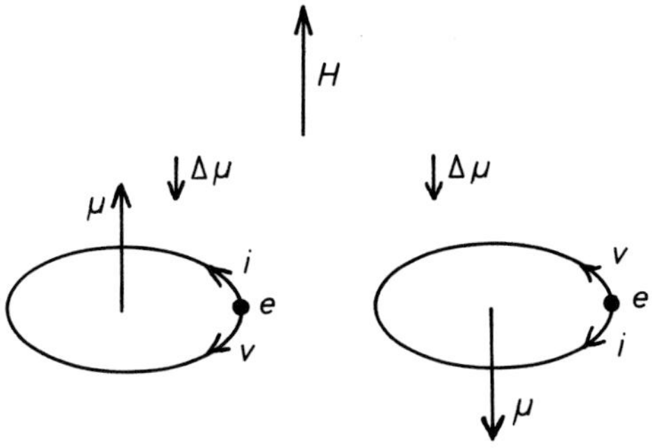

Figure 5-5. Simplified picture of diamagnetism.

stand what occurs, let us consider two electrons on the same orbit perpendicular to the applied field and moving in opposite directions (Fig. 5-5). We also assume that the orbit is unaffected by the field and is equivalent to a current loop without resistance. When the field is applied, the Faraday law leads to an electromotive force: $V = -\mathrm{d}\Phi/\mathrm{d}t$. The resulting electric field acting on each electron moving around the orbit of radius r is then: $E = V/2\pi r = -(\mu_0 \pi r^2/2\pi r)\,\mathrm{d}H/\mathrm{d}t$. The resultant acceleration is $\mathrm{d}v/\mathrm{d}t = (\mu_0 e r/2m)\,\mathrm{d}H/\mathrm{d}t$ and then $\Delta v = \mu_0 e r H/2m$ leading to a change of the electronic orbital moment $\Delta\mu = -\pi r^2 e \Delta v/2\pi r = -\mu_0 e^2 r^2 H/4m = \chi_d H$. Both electrons have the same contribution. Therefore there is an induced magnetization resulting from the variation $\Delta\mu$ of the magnetic moment opposite to the applied field for both electrons. To get a better evaluation of the susceptibility χ_d one has to take into account: (i) more realistic orbits, (ii) the contribution of all the electrons of each atom, and finally (iii) the number of atoms per unit volume.

The main characteristics of diamagnetic susceptibility are as follows:

(i) as observed experimentally χ_d is temperature independent,

(ii) χ_d is very small, on the order of 10^{-3} to 10^{-5},

(iii) χ_d is the only contribution to the susceptibility in substances where all the electrons are accommodated in closed shells (monoatomic rare gases, polyatomic gases such as H_2, N_2 ..., almost all organic compounds ...).

(iv) because of the absence of electrical resistance a superconductor is a perfect diamagnet; this is the origin of the Meissner effect.

5.2.3.5 Paramagnetism – Curie Law

Those substances where atoms carry intrinsic magnetic moments which do not interact with each other exhibit a paramagnetic behavior: the susceptibility is positive and is observed to vary inversely with temperature: $\chi = C/T$. This is the Curie law. Such a law results from the competition between magnetic energy $E = -\mu_0 \boldsymbol{M} \cdot \boldsymbol{H}$ which tends to align the moments parallel to the applied field and the thermal effects which favor disorder of magnetic moments. The result is only partial alignment in the field direction, and therefore a small positive susceptibility results. The effect of increasing temperature is to increase the randomizing effect of the thermal agitation and therefore to decrease the susceptibility. We will see in Sec. 5.3 that this C/T law is a special case of a more general law: $\chi = C/(T-\theta)$ called the Curie-Weiss law, where θ characterizes interatomic interactions.

Classical Model: The Langevin Function

In this classical model, atoms carry an intrinsic moment m_0 which can have any direction. Magnetic energy in an applied field $\boldsymbol{H}$ can be written as: $-\mu_0 m_0 H \cos\theta$ where θ is the angle between $\boldsymbol{m}_0$ and $\boldsymbol{H}$. Using the Boltzmann statistics, the proportion of magnetic moments whose direction belongs to the solid angle $\mathrm{d}\Omega = 2\pi \cdot$

$\sin\theta.\ d\theta$ is $dN = \exp(\mu_0 m_0 H u/kT)\,du/Z$ where $u = \cos\theta$ and Z is the partition function:

$$Z = \int_{-1}^{+1} \exp(\mu_0 m_0 H u/kT)\,du = \frac{2\sinh(x)}{x}$$

$$\text{with} \quad x = \frac{\mu_0 m_0 H}{kT} \tag{5-3}$$

The average value of the magnetic moment m along the field is $m = \langle m_0 \cos\theta \rangle = kT\,\partial \mathrm{Ln}\,Z/\partial H$. This leads to the magnetization: $M = M_0\,\mathscr{L}(x)$ where

$$\mathscr{L}(x) = \coth x - 1/x \tag{5-4}$$

is the Langevin function which is shown in Fig. 5-6. For small x values, $\mathscr{L}(x) \propto x/3$ and M varies linearly with H: $M = N\mu_0 m_0^2 H/3kT = \chi_0 H$, where N is the number of atoms per unit volume. Therefore $\chi_0 = C/T$. This is the Curie law where C is the Curie constant $C = N\mu_0 m_0^2/3k$.

At room temperature the paramagnetic susceptibility, on the order of 1, is much larger than the diamagnetic susceptibility.

Quantum Model: The Brillouin Function

In the quantum model the projection of the magnetic moment along the quantum axis (Oz) can have only discrete values. Under the application of a magnetic field each multiplet (see Sec. 5.2.3.2) is split into $2J+1$ levels. These energy levels E_i are the eigenstates of the perturbing Hamiltonian $\mathscr{H} = -\mu_0 g_J \mu_B J_z H_z$:

$$E_i = -\mu_0 g_J \mu_B H_z \langle J, M_J | J_z | J, M_J \rangle = -\mu_0 g_J \mu_B H_z M_J \tag{5-5}$$

At $T=0$, only the ground state is occupied and the magnetization per atom is $\mu = g_J \mu_B J$. At $T \neq 0$, we have to consider the thermal population of the excited states. Boltzmann statistics leads to a magnetization $M = M_0\,\mathscr{B}_J(x)$ where

$$\mathscr{B}_J(x) = (2J+1)/2J \coth[(2J+1)x/2J] - 1/2J \coth(x/2J) \tag{5-6}$$

is the Brillouin function. This function is drawn in Fig. 5-6 for different values of J. When x is small, i.e. when in H is small and T is high, we use the expansion: $\coth x = 1/x + x/3 + \ldots$ then: $M = CH/T$. The Curie constant is: $C = N\mu_0 g_J^2 J(J+1)\mu_B^2/3k$, where $g_J \mu_B \sqrt{J(J+1)}$ is the effective paramagnetic moment. Note that unlike the classical description leading to the Langevin law, the effective paramagnetic moment is slightly larger than the saturated moment $g_J \mu_B J$.

As is well known, classical effects can be considered as a limit of quantum effects when quantum number becomes infinite. It

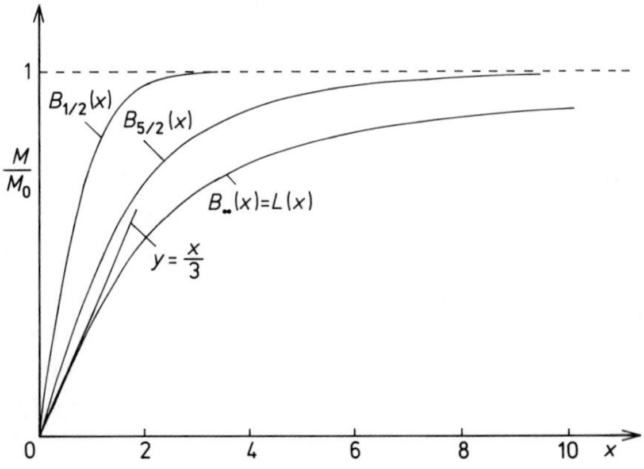

Figure 5-6. Representation of the Langevin $\mathscr{L}(x) = \mathscr{B}_\infty(x)$ and of two Brillouin $\mathscr{B}_{1/2}(x)$ and $\mathscr{B}_{5/2}(x)$ functions. The initial slope of the Langevin function is also reported.

is readily seen that: $\mathscr{L}(x) = \lim_{J \to \infty} \mathscr{B}(x)$. In both descriptions the thermal dependence of the reciprocal susceptibility follows a linear Curie law in the whole temperature range. So far we did not take into account two effects, namely the *exchange interactions* between localized moments and the *magnetocrystalline anisotropy* originating from the electrical potential due to the surrounding ions. Actually, in most substances these effects are important and we will see in Sec. 5.3 the modification of the Curie law arising from these two effects.

The Brillouin law is particularly well verified in gadolinium salts such as $Gd_2(SO_4)_3 \cdot 8H_2O$, where, the Gd^{3+} ion being in an S state ($L=0$), there is no magnetocrystalline anisotropy and where interatomic interactions are very small because the Gd^{3+} ions are rather diluted.

Literature on this section: textbook by Cullity (1972), p. 1–21 and 85–110.

5.3 Localized Moment in the Solid

In the previous section we considered an assembly of isolated ions. Actually in a solid, magnetic atoms interact with each other and with the electric potential of the surroundings. This section deals with these two effects assuming localized magnetic moments.

Magnetic interactions between localized moments favor magnetic ordering which competes with thermal disorder effects. Each material has a critical temperature (or ordering temperature) which is a measure of the strength of magnetic interactions. Above this temperature, thermal effects dominate and the material is paramagnetic; below this temperature, magnetic interactions dominate and the material is magnetically ordered. The interaction with the electric potential of the surroundings, by linking the orbit (and hence the spin through $\boldsymbol{L} \cdot \boldsymbol{S}$ coupling) to the lattice, leads to the main part of magnetic anisotropy, namely to magnetocrystalline anisotropy.

5.3.1 Magnetic Interactions and the Molecular Field Model

5.3.1.1 Origin of Magnetic Interactions: Exchange Coupling

The significant interaction between localized moments, named *exchange interaction*, has the same origin as the one which occurs between electrons inside the atom: the correlations between two electrons lead to different energies for the parallel and antiparallel spin states. This is a result of the Pauli principle which imposes the antisymmetry of the total wave function, i.e. a change of sign when permuting two electrons (exchange interactions deal with electrons belonging to two different atoms). An antisymmetric spin state is associated with a symmetric orbital state and vice versa. Correlation effects which are of electrostatic origin lead to the splitting of energies of the antisymmetric and symmetric orbital states and hence to the energy splitting of the symmetric (↑↑) and antisymmetric (↑↓) spin states. When the magnetic orbitals of two neighboring atoms are sufficiently extended to produce a direct overlap between them, the above process leads to an effective interaction between the spins of these atoms which is called *direct exchange*. This direct exchange, which occurs in 3d intermetallic compounds, is the largest interatomic interaction and it is, in particular, responsible for the high ordering temperatures found in the ferromagnets used for most technological applications. The variation of this interaction as a function of distance is illustrated by the Slater-Néel curve shown in Fig. 5-7.

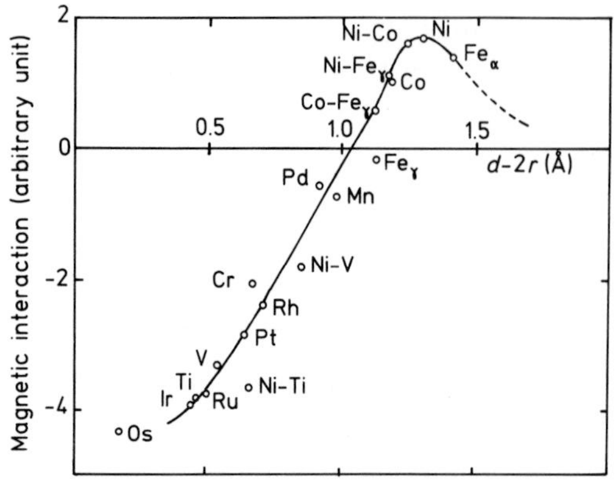

Figure 5-7. Slater-Néel curve: exchange interaction versus distance between magnetic shells (d = distance between two atoms, r = magnetic shell radius) in 3d metals and alloys.

When magnetic orbitals of two neighboring atoms are too localized to overlap, as is the case for the 4f series, the exchange process can occur through conduction electrons if the system is metallic. This leads to the *RKKY indirect exchange interaction* which will be emphasized in Sec. 5.5.

Finally, if there are no conduction electrons, as in ceramics where magnetic atoms are separated by non-magnetic atoms like oxygen, the external electrons of the latter participate in covalent binding and mediate the exchange interaction. This is the *superexchange interaction* presented in Chap. 6.

In all cases the effective exchange interaction energy or Hamiltonian between two ions can be written as:

$$H_{ij} = -J_{ij}\boldsymbol{S}_i \boldsymbol{S}_j \tag{5-7}$$

In 3d based materials: $J_{ij} \approx 10^2$ to 10^3 K. For indirect interactions between 4f atoms: $J_{ij} \approx 10$ to 10^2 K.

5.3.1.2 The Molecular Field Model

Because of spin orbit coupling, the spin-spin interaction between atoms is equivalent to an interaction between magnetic moments:

$$H_{ij} = -\mu_0\, n_{ij}\, \boldsymbol{m}_i \boldsymbol{m}_j \tag{5-8}$$

The interaction energy for the whole solid is then considered:

$$H = -\tfrac{1}{2}\mu_0 \sum_{i,j\neq i} n_{ij}\, \boldsymbol{m}_i \boldsymbol{m}_j = -\frac{\mu_0}{2}\sum_i \boldsymbol{m}_i \boldsymbol{H}_i \tag{5-9}$$

with

$$\boldsymbol{H}_i = \sum_{j\neq i} n_{ij}\, \boldsymbol{m}_j$$

$\boldsymbol{H}_i$ can be considered as a local magnetic field called *molecular field.* It fluctuates with time and depends on the instantaneous values of magnetic moments $\boldsymbol{m}_j$. To solve this complicated system, Weiss proposed in 1906 the very simple and ingenious *mean field model* which allowed important progresses in magnetism. This model is based on the powerful assumption which consists of neglecting the fluctuating character of $\boldsymbol{H}_i$. The average energy:

$$\langle H\rangle_T = \frac{\mu_0}{2}\sum_i \langle \boldsymbol{m}_i \boldsymbol{H}_i\rangle_T \tag{5-10}$$

then becomes:

$$\langle H\rangle_T = \frac{\mu_0}{2}\sum_i \langle \boldsymbol{m}_i\rangle_T \langle \boldsymbol{H}_i\rangle_T \tag{5-11}$$

where $\langle \boldsymbol{H}_i\rangle_T = \boldsymbol{H}_m = \sum_{j\neq j} n_{ij}\langle \boldsymbol{m}_j\rangle_T$ is the molecular field which depends on the magnetic moments of the neighboring ions. If all the magnetic moments are identical $\langle \boldsymbol{m}_j\rangle_T = \langle \boldsymbol{m}\rangle_T$ and

$$\boldsymbol{H}_m = \frac{1}{\mu_0} W \boldsymbol{M} \tag{5-12}$$

where $\boldsymbol{M} = N\langle \boldsymbol{m}\rangle_T$ is the magnetization per volume unit which contains N atoms, and $W = \sum_{j\neq j} n_{ij}/N$.

n_{ij} is a local molecular field coefficient reduced to one atomic magnetic moment. As the molecular field H_i is in A m^{-1} and

m_j in A m^2, n_{ij} corresponds to the inverse of a volume. W is dimensionless.

5.3.2 Molecular Field Model of Magnetically Ordered Materials

In this section we present the major features characterizing the behavior of ferro-, ferri-, antiferro-, heli- and modulated magnetic structures.

5.3.2.1 Ferromagnetism

Ferromagnetism occurs in compounds where exchange interactions favor a parallel alignment of atomic moments. At 0 K, the alignment is complete, leading to a spontaneous magnetization $M_S(T=0) = M_0$ which reaches the maximum possible value. As temperature is increased, $M_S(T)$ decreases and vanishes at the so-called *Curie temperature* T_C above which the compound is paramagnetic. We will see in Sec. 5.11 that, because of the existence of domains in which the magnetization is parallel to different directions, a bulk ferromagnet generally has not net magnetization. But a large magnetization is easily induced by an applied field (see, for instance, the field dependence of the magnetization schematized in Fig. 5-15a and those shown in Figs. 5-17 and 5-18 for single crystals of pure Fe and Co). Experimentally, a ferromagnet is characterized by: (i) at low temperature, a large susceptibility in low applied field (which corresponds to an infinite susceptibility in low internal field H_i, see Sec. 5.2.2), (ii) a drastic drop of this susceptibility at T_C. These features may be seen in Figs. 5-15a and 5-16a to be discussed later.

Let us come back to the microscopic aspect of ferromagnetism. The simplest model uses the Langevin law (see Sec. 5.2.3.5). As, in the molecular field theory, the exchange interactions are equivalent to an applied magnetic field, they are taken into account simply by replacing H in Eq. (5-3) by $H + WM$, which leads to:

$$x = \mu_0 m_0 (H + WM)/kT \tag{5-13}$$

and hence to:

$$y = \frac{M}{M_0} = \frac{NkT}{\mu_0 W M_0^2}\, x - \frac{H}{WM_0} \tag{5-14}$$

Moreover:

$$y = \mathscr{L}(x) \tag{5-15}$$

For each value of T and H, M is obtained from the set of Eqs. (5-14) and (5-15) which can be solved graphically (see Fig. 5-8). In zero applied field ($H = 0$), the straight line y vs. x of Eq. (5-14) passes through the origin 0. At low temperature the slope of this straight line is small and a non-trivial solution corresponds to point A giving rise to a finite value of y, i.e. of the spontaneous magnetization $M_S(T)$, different from zero. When temperature is increased the slope of the OA straight line increases and A moves leftward on the Langevin curve (points A′, A″...). $M_S(T)$ then decreases and vanishes at T_C when the slope of the straight line (D) is equal to the derivative of the Langevin

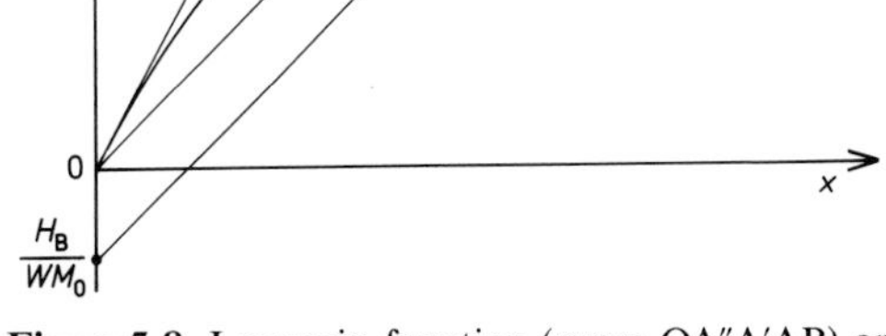

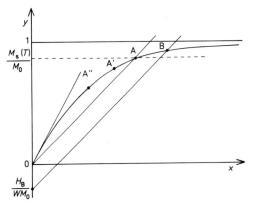

Figure 5-8. Langevin function (curve OA″A′AB) and graphical determination of magnetization for any applied field and temperature in a simple ferromagnet.

function $\mathscr{L}(x)$ near the origin. From the expansion of $\mathscr{L}(x)$ around $x=0$:

$$y = M/M_0 = x/3 - x^3/45 + \ldots \qquad (5\text{-}16)$$

one obtains $T_C = CW$.

At a given temperature (T_1 for instance) smaller than T_C, when the field is applied the straight line OA moves rightward with the same slope (line CB for $H = H_1$ for instance). Magnetization then increases with the field, this increase being inasmuch slow as the temperature is small and accordingly the spontaneous magnetization is large. Above T_C, i.e. in the paramagnetic range, the slope of CB is larger than that of (D) and in low applied field, magnetization is small. The low field susceptibility obtained from Eq. (5-16) is:

$$\chi = (\mathrm{d}M/\mathrm{d}H)_{H=0} = C/(T - T_C) \qquad (5\text{-}17)$$

This is the paramagnetic susceptibility which follows the so-called *Curie Weiss law*: As expected, χ diverges at T_C. The reciprocal susceptibility is then: $1/\chi = 1/\chi_0 - W$, showing that exchange interactions simply lead to a shift of the reciprocal susceptibility without interaction $1/\chi_0 = T/C$ (Sect. 5.2.3.5), (Fig. 5-9).

The results obtained with the Langevin function are weakly modified if we use the Brillouin function. The best illustrations of these models are observed in Gd-based ferromagnetic alloys where no extra contributions to the Hamiltonian have to be taken into account. In that case the $\mathscr{B}_{7/2}$ function has to be considered. This model gives a rather good account for Fe metal as well. At the first stage of the approximation, the Fe magnetization can be considered as localized and is rather well accounted for by the model. For other materials the Brillouin function gives only a qualitative description either because moments are not localized enough (3d metals) or because anisotropic effects are of the same order of magnitude as exchange effects (most of the rare earths).

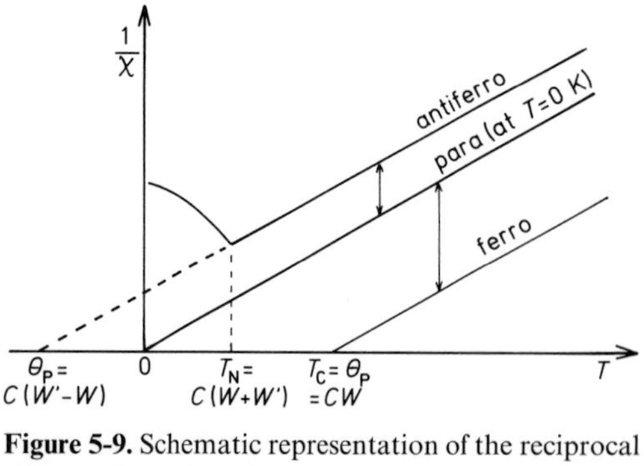

Figure 5-9. Schematic representation of the reciprocal susceptibilities for paramagnetic (down to 0 Kelvin), ferromagnetic and antiferromagnetic compounds.

5.3.2.2 Antiferromagnetism

In a simple antiferromagnet, magnetic atoms can be divided into two sublattices with their magnetization equal but antiparallel. The net magnetization is then zero at any temperature. The susceptibility is different if the field is applied parallel or perpendicular to the direction of magnetic moments in zero field, as is shown in Fig. 5-10.

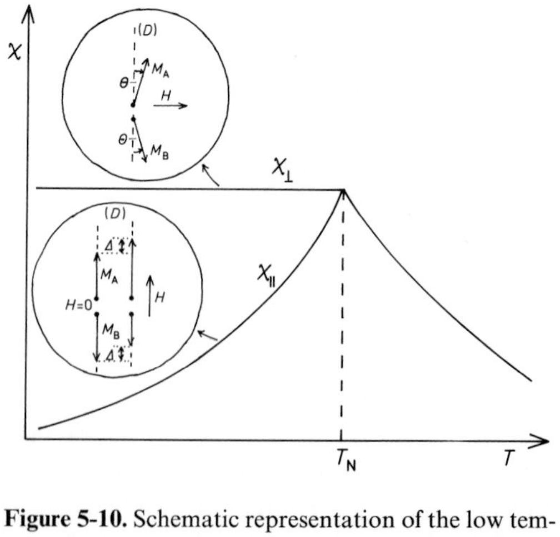

Figure 5-10. Schematic representation of the low temperature parallel and perpendicular susceptibilities of an antiferromagnet.

Néel Temperature

Let $W_{AB} = W_{BA} = -W\,(W>0)$ and $W_{AA} = W_{BB} = W'\,(W'>0)$ be the molecular field coefficients between the two sublattices and within each sublattice, respectively. The total fields acting on the A and B sublattices are:

$$H_{\substack{A\\B}} = \boldsymbol{H} - W\boldsymbol{M}_{\substack{B\\A}} + W'\boldsymbol{M}_{\substack{A\\B}} \qquad (5\text{-}18)$$

At each temperature, M_A and M_B are given by:

$$M_{\substack{A\\B}} = M^0_{\substack{A\\B}}\,\mathscr{B}_J\left(\frac{\mu_0 m_0 |H_{\substack{A\\B}}|}{kT}\right) \qquad (5\text{-}19)$$

When the argument of $\mathscr{B}_J$ is small, i.e. when field is small and temperature large enough, one has:

$$M_{\substack{A\\B}} = \frac{C}{T}(H - WM_{\substack{B\\A}} + W'M_{\substack{A\\B}}) \qquad (5\text{-}20)$$

with $C = N\mu_0 m_0^2/3k$, where N is the number of A (or B) atoms per volume unit. For $H=0$, both Eqs. (5-20) have a solution with $M_A \neq 0$ and $M_B \neq 0$ only at or below the ordering temperature T_N or *Néel temperature* so that:

$$\left(1 - \frac{CW'}{T_N}\right)^2 - \left(\frac{CW}{T_N}\right)^2 = 0 \qquad (5\text{-}21)$$

leading to $T_N = C(W+W')$. T_N is then a measure of the sum of the absolute values of the molecular field coefficients.

Paramagnetic Susceptibility

Above the Néel temperature for small values of H:

$$M = M_A + M_B = 2CH/(T-\theta) \qquad (5\text{-}22)$$

leading to

$$1/\chi = (T-\theta_P)/2C \qquad (5\text{-}23)$$

with $\theta_P = C(W'-W)$. One obtains a Curie Weiss behavior where, contrary to the ferromagnetic case θ_P is smaller than T_N; especially it can be negative. As well, at T_N, χ has a finite value. This behavior is illustrated in Fig. 5-9.

Perpendicular Susceptibility

When the field is applied perpendicular to the antiferromagnetic direction (***D***), magnetization of each sublattice is rotated by a small angle θ (Fig. 5-10) until the decrease of the Zeeman energy E_Z is counterbalanced by the increase of exchange E_{ex} and anisotropy E_A (see Sec. 5.2.3) energies. With $|M_A| = |M_B| = M$, these three energies can be written: $E_Z = -2\mu_0 MH\theta$, $E_{ex} = -\mu_0 WM^2(1-2\theta^2)$ and $E_A = K\theta^2$ where K is the anisotropy constant. Minimizing the total energy leads to: $\theta = \mu_0 MH/(2\mu_0 WM^2 + K)$. The perpendicular susceptibility is then:

$$\chi_\perp = \frac{2M\theta}{H} = \left[W\left(1 + \frac{K}{2\mu_0 WM^2}\right)\right]^{-1} \qquad (5\text{-}25)$$

As W is almost temperature independent and $K/2M^2$ weakly temperature dependent, this susceptibility is almost constant below T_N (Fig. 5-10).

Parallel Susceptibility

As shown in Fig. 5-10 the applied field and the moments are collinear. Under an applied field H, magnetizations of parallel and antiparallel sublattices are $M+\Delta$ and $M-\Delta$, respectively (if $T \neq 0$ for classical moments). For small H and Δ values one obtains:

$$\chi_\parallel = \frac{2Cb(T)}{T + C(W-W')b(T)} \qquad (5\text{-}26)$$

with

$$C = \frac{\mu_0 m_0 M_0}{kT}\mathscr{B}'_J(0) \quad \text{and}$$

$$b(T) = \frac{\mathscr{B}'_J\left[\dfrac{m_0}{kT}(W+W')M_T\right]}{\mathscr{B}'_J(0)}$$

$\chi_{\parallel}$ decreases with T and vanished at 0 K (Fig. 5-10). As Eq. (5-26) is valid at any temperature, it is readily seen that there is only a slope discontinuity at T_N.

Metamagnetism

In an antiferromagnet below T_N, $\chi_{\perp}$ is larger than $\chi_{\parallel}$ which means that the state with moments almost perpendicular to $\boldsymbol{H}$ is of lower energy than that with moments collinear with $\boldsymbol{H}$. Thus, in this latter case, there is a tendency for magnetic moments to take the former configuration. However this effect competes with magnetocrystalline anisotropy which favors the configuration stable in zero field. In the simple model of antiferromagnetism presented above, two cases have to be considered:

– If the anisotropy energy is small enough (Fig. 5-11), when the field is parallel to the antiferromagnetic direction (***D***) (curve b), as H increases beyond a critical value the moments *flop over* to the configuration shown in the Figure, causing a sudden increase of magnetization; in higher field the moments rotate slightly towards the direction of the field.

– If the uniaxial anisotropy is large enough (Fig. 5-12), when the field is parallel to (***D***) (curve b), magnetization remains in the same direction. As the field is increased beyond a critical value, the magnetic moments antiparallel to the field *flip over* into parallelism leading to a step in the magnetization curve.

In both cases, when the field is perpendicular to (***D***), magnetization increases linearly until saturation (curves (a) in Figs. 5-11 and 5-12). This variation corresponds to the progressive rotation of the moments towards the field direction.

Whatever the strength of the anisotropy is, a transition takes place in the magnetization curve of a single crystal when $\boldsymbol{H}$ is parallel to the antiferromagnetic direction and hence when measurement is performed on a polycrystal. Such a behavior is called a *metamagnetic transition*. It is observed in all the antiferromagnetic compounds as soon as the applied field is large enough to overcome the antiferromagnetic coupling. Depending on the compound, a large variety of metamagnetic transitions can be observed. Especially in strongly anisotropic uniaxial compounds the field-induced fer-

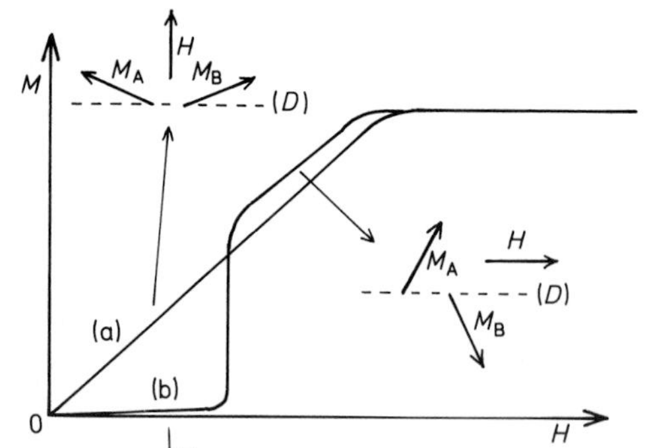

Figure 5-11. Schematic representation of the low temperature magnetization processes in an antiferromagnet with weak magnetocrystalline anisotropy. (a) $\boldsymbol{H} \perp (D)$, (b) $\boldsymbol{H} \parallel (D)$: metamagnetic transition.

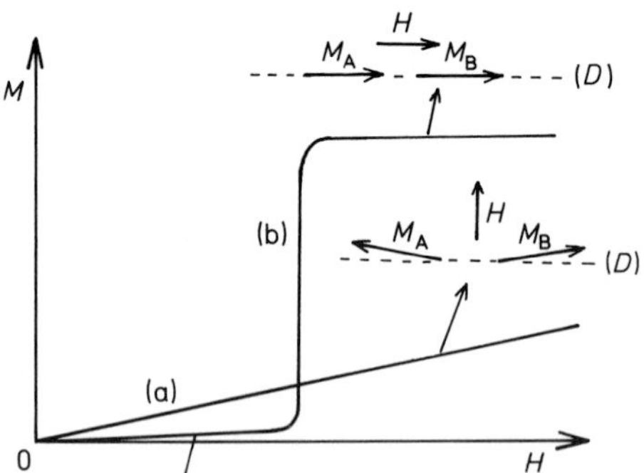

Figure 5-12. Schematic representation of low temperature magnetization processes in an antiferromagnet with strong magnetocrystalline anisotropy. (a) $\boldsymbol{H} \perp (D)$, (b) $\boldsymbol{H} \parallel (D)$: metamagnetic transition.

romagnetic state can be reached after several metamagnetic transitions (several threshod fields with magnetization jumps equal to a fraction of the saturated magnetization), each one corresponding to the flip of only one fraction of the moments antiparallel to the field.

5.3.2.3 Ferrimagnetism

A ferrimagnet is an antiferromagnet in which the different sublattice magnetizations are not equal. This requires non-equivalent magnetic sublattices and/or atoms. The first prediction and theoretical description was given by Néel. We consider two different types of magnetic sites, A and B, having different moments ($M_A \neq M_B$). The total field acting on atoms A (or B) is:

$$H_{\substack{A\\B}} = H + W_{\substack{AB\\BA}} M_{\substack{B\\A}} + W_{\substack{AA\\BB}} M_{\substack{A\\B}} =$$

$$= H - W(M_{\substack{B\\A}} - \underset{\beta}{\alpha} M_{\substack{A\\B}}) \tag{5-27}$$

where $W = -W_{AB} = -W_{BA} \, (W > 0)$ and $\underset{\beta}{\alpha} W = W_{\substack{AA\\BB}}$.

In order to simplify this complex problem, as a first step one assumes $\alpha = \beta = 0$. Using Eq. (5-27) one obtains the ordering temperature $T_C = W\sqrt{C_A C_B}$ where $C_{\substack{A\\B}} = \mu_0 N_{\substack{A\\B}} m_{\substack{A\\B}}^2 / 3K$. At T_C a spontaneous magnetization $|M_S| = |M_A - M_B|$ appears such that $M_A = \sqrt{C_A/C_B} \cdot M_B$. At lower temperature M_A and M_B are no longer proportional and the thermal dependence of M_S can have different shapes as schematized in Fig. 5-13. The most interesting situation is the case (N) in Fig. 5-13 where the bulk spontaneous magnetization vanishes not only at T_C but at a lower temperature, the *compensation temperature* T_{Co}. This occurs because, at this temperature the two sublattices exactly compensate each other. Conditions for such a behavior are:

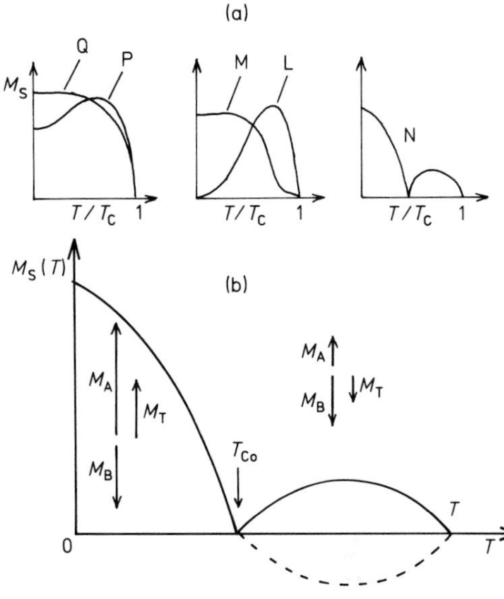

Figure 5-13. (a) Different types of thermal dependences of spontaneous magnetization in a ferrimagnet as they were proposed by Néel (1978). (b) Details of case *N*.

(i) $N_A m_A > N_B m_B$ ($M_A > M_B$ at 0 K) and
(ii) $N_A m_A^2 < N_B m_B^2$ ($M_A < M_B$ for $T_{Co} < T < T_C$).

Paramagnetic Susceptibility

Above T_C the reciprocal susceptibility can be written as:

$$\frac{1}{\chi} = \frac{T + \theta_P}{C} - \frac{\gamma}{T - \theta} \quad \text{where } C = C_A + C_B$$

is the mean value of the Curie constant of the assembly of m_A and m_B moments, and where θ, γ and θ_P are constants that depend on W, α, β, C_A and C_B. The thermal variation of the reciprocal susceptibility has two asymptotes: one for infinite T crossing the temperature axis at $T = -\theta_P$, called the *asymptotic Curie point*; the other for $T = \theta$.

Note that from bulk magnetic measurement a ferrimagnetic compound is not very different from a ferromagnetic one and at first glance it is not very easy to distinguish between both possibilities, inasmuch when the magnetization of one site is much smaller than that of the other site. For both situations one observes a spontaneous magnetization at low temperature with a field dependence of the magnetization characteristic of a ferromagnet. The reciprocal susceptibility is linear at high temperature (Curie Weiss behavior) and becomes zero at T_C. The main differences may arise from the thermal dependences of the spontaneous magnetization.

5.3.2.4 Helimagnetism and Sine Wave Modulated Structure

Helimagnetism

The classes of magnetic structures considered above are the simplest ones, involving collinear magnetic sublattices. Neutron diffraction experiments have revealed a much wider variety of structures. For particular interactions, incommensurate long period magnetic structures can become stabilized, among which the helimagnetic and sine wave modulated ones are the simplest examples. Let us consider a uniaxial crystal (hexagonal, tetragonal) in which the magnetic atoms are in identically parallel layers, and let us assume that the anisotropy is such that the moments lie in the layers. We also assume that in each layer all the moments are parallel giving rise to a magnetization M. Considering only the exchange interactions inside the layers (W_0) and between 1st (W_1) and 2nd (W_2) nearest neighbor layers, the energy of the configuration shown in Fig. 5-14a where $\varphi_n = n\,\varphi$ can be written as:

$$E_{ex} = -\frac{M^2}{2}(W_0 + 2\,W_1 \cos\varphi + 2\,W_2 \cos 2\varphi) \tag{5-28}$$

The minimization of E_{ex} leads to three magnetic configurations: (i) ferromagnetism ($\varphi = 0$), (ii) antiferromagnetism ($\varphi = \pi$), (iii) helimagnetism with θ given by $\cos\varphi = -W_1/4\,W_2$. The helimagnetic structure is the most stable when $W_2 < 0$ and $|W_2| > |W_1/4|$. As for all the compounds with a zero bulk magnetization, the bulk magnetic properties of a helimagnet are similar to those of a simple antiferromagnet: the susceptibility χ is always small and the Néel temperature T_N is characterized by a maximum of χ. In addition, below T_N a metamagnetic transition, although less pronounced than in a commensurate antiferromagnet, occurs.

Sine Wave Modulated Structure

In the above case, the anisotropy is such that the moments are within the layers ($K < 0$ if the anisotropy energy is written as $E_A = K \sin^2\alpha$ if α is the angle between the uniaxial direction and the moment direction, see Sec. 3.3.1). If we have the same competition between positive and negative interactions with, in addition, a strong uni-

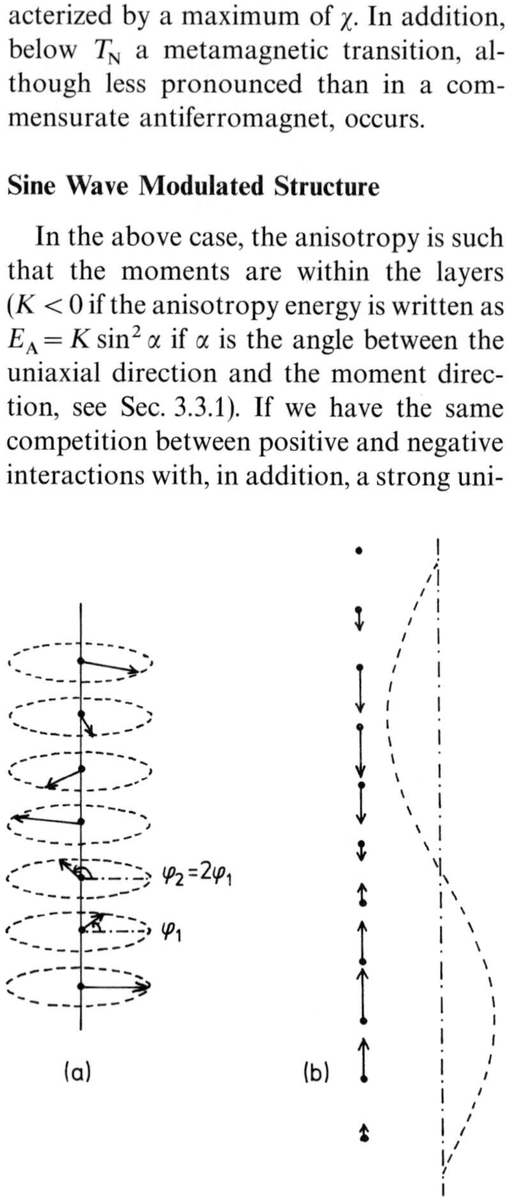

Figure 5-14. Helimagnetic (a) and sine wave modulated (b) structures.

axial anisotropy favoring the axis perpendicular to the layers ($K > 0$), moments are always along this axis but their amplitude varies sinusoidally with the distance with the same periodicity as in the corresponding helimagnetic structure having the same value of W_1/W_2 (Fig. 5-14 b).

Note that in such a structure, the moment of a large number of the atoms is reduced. Assuming that the moments can be described by a Langevin or a Brillouin function, the reduction of the magnetic moments can occur only if temperature is not too small. It is the reason why sine wave modulated structures are generally not stable at low temperature, especially at 0 K (Elliott, 1961; Kaplan, 1961). Indeed for $T \rightarrow 0$ K, in Eq. (5-14), $x \rightarrow \infty$ and whatever the value of the field, all moments must have their maximum value. Then at low temperature the magnetic structure transforms into a state with equal magnetic moments. However, when in the absence of any applied or exchange field, the ground state is a singlet, the magnetic moment at $T = 0$ K is induced by the field (this is a quite general and fundamental result in magnetism) and generally increases slowly with it. Then at 0 K, if the amplitude of the molecular field is modulated with the distance, it will induce a modulated magnetization. Therefore a sine wave modulated structure can be stable at 0 K (Gignoux et al., 1972 a). We will see later (Secs. 5.6.3.2 and 5.7.3) two examples of such structure where the singlet ground state arises from two different interactions.

5.3.2.5 Summary

When characterizing, from magnetic measurements, a polycrystalline material where exchange interactions are present, one can observe two types of behavior: (i) a mainly ferromagnetic behavior, i.e. with a spontaneous magnetization, (ii) a mainly antiferromagnetic behavior, i.e. without spontaneous magnetization.

Ferromagnetic and antiferromagnetic behavior can be characterized by the two graphs shown in Fig. 5-15 and Fig. 5-16 respectively. In any case, neutron diffraction experiments are necessary to determine the true magnetic structure: ferro collinear or canted, ferri, antiferro collinear, helimagnetic or modulated structure (Bacon, 1975; Rossat-Mignod, 1986).

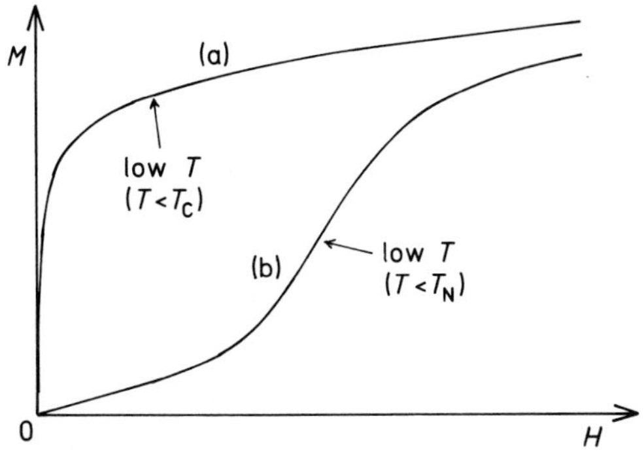

Figure 5-15. Low temperature magnetization curves in the cases of ferro- (a) and antiferromagnetic (b) type behaviors.

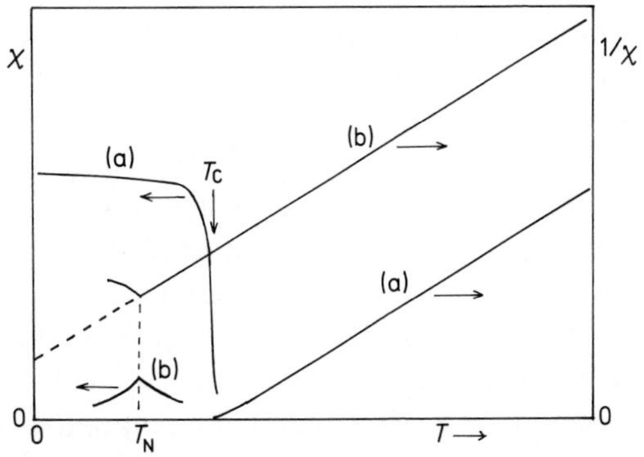

Figure 5-16. Low temperature susceptibilities and paramagnetic reciprocal susceptibilities in the cases of ferro (a) and antiferromagnetic (b) type behaviors.

5.3.3 Magnetocrystalline Anisotropy – the Crystalline Electric Field (CEF)

Most substances exhibit magnetic anisotropy which plays a predominant role in their magnetic properties. Among the different types of anisotropy, the magnetocrystalline anisotropy originating from the Crystalline Electric Field (CEF) is dominant and is the subject of this section. Actually the CEF gives rise not only to magnetocrystalline anisotropy but also to other very interesting effects. We will see later that CEF effects are of major importance with regard to the magnetic properties of most materials, especially those containing rare earth elements. Other types of magnetic anisotropy such as shape anisotropy and exchange anisotropy will be dealt with more briefly in other sections. As we will see in what follows, the CEF has quite a number of consequences for the magnetic properties previously described. Especially M vs. H and χ vs. T strongly depend on the orientation when measurements are performed on single crystals.

5.3.3.1 Classical Approach to Magnetocrystalline Anisotropy

Figures 5-17 and 5-18 show magnetization curves for single crystals of iron and cobalt. Iron has a cubic structure and magnetization is measured along the [001], [101] and [111] axes which are the main symmetry directions. Along [001], magnetization is the largest and reaches its maximum value in a very small applied field. This means that the spontaneous magnetization is along [001], which is then the *easy magnetization axis*. When a field is applied along [101] and [111], magnetization does not immediately turn toward H. At the beginning magnetic moments are still parallel to [001] and the magnetizations measured in zero internal field, $M_{[101]}$ and $M_{[111]}$ correspond to the projection of the spontaneous magnetization on these directions: this is the so-called *phase rule*. When the field is increased the magnetization turns progressively toward the applied field direction and, above $40 \times 10^3\ \mathrm{A\,m^{-1}}$, magnetization has become parallel to the field. The area between the magnetization curves measured along these so-called *hard magnetization axes* and that measured along the easy axis determines the anisotropy energy. In cubic

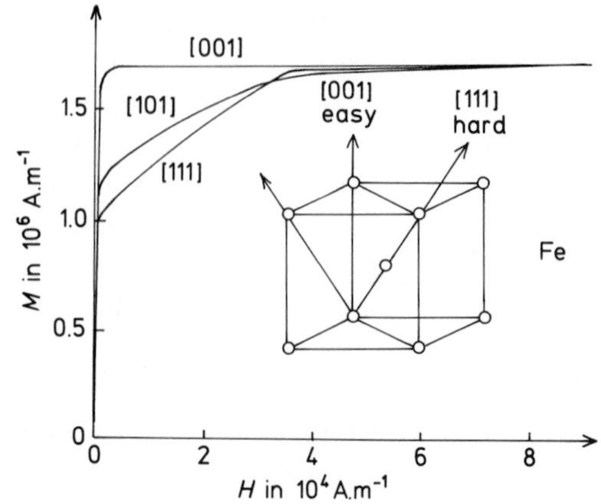

Figure 5-17. Magnetization curves of a single crystal of iron (cubic symmetry).

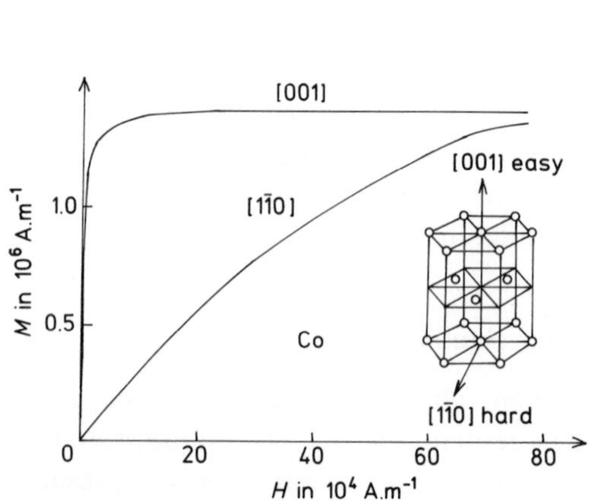

Figure 5-18. Magnetization curves of a single crystal of cobalt (hexagonal symmetry).

symmetry this energy can be written as:

$$E_A = K_1(\alpha_1^2\alpha_2^2 + \alpha_2^2\alpha_3^2 + \alpha_3^2\alpha_1^2) + + K_2\alpha_1^2\alpha_2^2\alpha_3^2 + \ldots \quad (5\text{-}29)$$

where α_1, α_2, α_3 are the cosines of the angles of magnetization with the [100], [010] and [001] directions, respectively. K_1, K_2, ... are anisotropy constants which generally decrease as temperature is increased.

Cobalt is hexagonal and the magnetization in Fig. 5-18 is measured parallel and perpendicular to the ***c*** ([001]) axis (within the basal plane perpendicular to ***c*** the anisotropy is very weak). One observes a very large anisotropy. Both directions of measurement being perpendicular, magnetization in zero field is zero perpendicular to ***c***. For such a symmetry the anisotropy energy can be written as:

$$E_A = K_0\alpha_3^2 + K_1\alpha_3^4 + \ldots \quad (5\text{-}30)$$

Actually such an approach of magnetocrystalline anisotropy is phenomenological. Especially it is independent of the value of the magnetization and it is not able to explain a reduction of the magnetization along the easy direction and of the anisotropy of the magnetization ΔM, as is often observed. Here ΔM is the difference of the magnetization between two axes in a given field large enough to align the moments in its direction. However, this classical description is a good approximation when the exchange interaction energy is much larger than the anisotropy energy, as is the case for both examples described above.

5.3.3.2 The Crystalline Electric Field (CEF)

General Remarks About the CEF

Because of electrostatic interactions, the orbitals are linked to the lattice. For instance, in Fig. 5-19 the state (a) is energetically favored in comparison with state (b). The orbital moment and hence the spin moment, through spin orbit coupling, are then also linked to the lattice. This is the origin of magnetocrystalline anisotropy. These electrostatic interactions, by destroying the spherical symmetry of the ion, remove the degeneracy of the multiplets of the free ion.

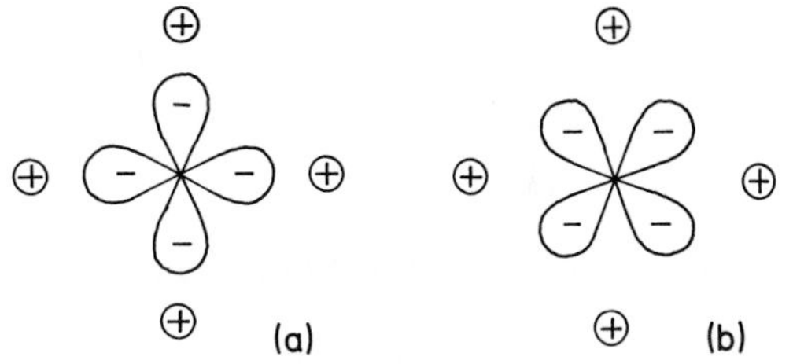

Figure 5-19. d orbitals of the xy type in the CEF created by point charges.

At this stage, two very different cases must be considered. As a matter of fact, in the transitional series 3d (Fe group) and 4f (rare earth atoms) the relative importances of spin orbit coupling and of the crystal field are reversed (Fig. 5-20). The $L-S$ coupling is much larger in the 4f series than in the 3d one, since the nuclei of the rare earths carry a larger electric charge and the localization of the magnetic 4f electrons is larger. On the other hand, because of this latter characteristic, the crystal field is weaker in the 4f series.

As illustrated in Fig. 5-19, the origin of the CEF results from the interaction between the non-spherical orbitals and the electrostatic field of the environment, which is also non-spherical. Below we will consider the case of a 4f ion, where CEF effects are one order of magnitude smaller than the spin orbit coupling.

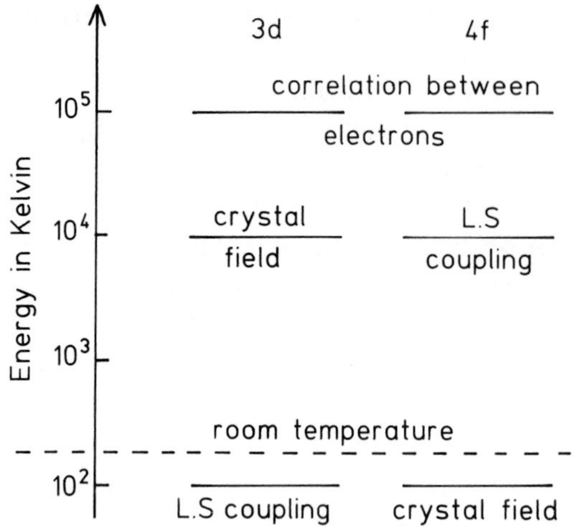

Figure 5-20. CEF and spin orbit coupling in 3d and 4f ions.

Splitting of the Free Ion Multiplet by CEF

Each multiplet of the free ion is characterized by the quantum number J and is associated with the D_J representation of the rotation group. This means that wave functions $|J, M_J\rangle$ transform according to the $D_J(R)$ matrix as one performs a rotation R. Let G be the point symmetry of the site where the studied ion is located. The perturbing electric potential is invariant with respect to the symmetry operations of $G \cdot D_J$ (see Appendix 1) is then reducible into the irreducible representations Γ_i of the G group. The degeneracy of the multiplet is then removed. With each representation Γ_i is associated an energy level, the degeneracy of which is equal to the order of Γ_i. As explained in Appendix 1, group theory allows one to predict the splitting of each multiplet by CEF. However, it does not allow the determination of the position of the energy levels E_i within the split multiplet.

The Perturbing Hamiltonian

Let $V(\boldsymbol{r}_i)$ be the electrostatic potential acting on the electron i the position of which is $\boldsymbol{r}_i$. The perturbing Hamiltonian due to CEF can be written as:

$$H_C = \sum_i q_i V(\boldsymbol{r}_i) = -e \sum_i V(\boldsymbol{r}_i) \tag{5-31}$$

The sum is restricted to electrons in unfilled shells (4f in case of rare earths), as CEF does not affect closed shells to the first order of perturbation. $V(\boldsymbol{r}_i)$ can be expanded in spherical harmonics Y_l^m:

$$V(\boldsymbol{r}_i) = \sum_l r_i^l \sum_{m=-l}^{+l} A_l^m Y_l^m(\theta_i, \phi_i) \tag{5-32}$$

In this expression the number of terms is limited by the site symmetry of the ion considered. The A_l^m's are the CEF parameters which characterize the compound. Terms with $l > 6$ are not considered because, due to selection rules which would need further development, they do not act on the state of the magnetic shell. As an example, for an ion occupying a hexagonal site:

$$\begin{aligned} V(\boldsymbol{r}_i) = {} & A_2^0 r_i^2 Y_2^0 + A_4^0 r_i^4 Y_4^0 + \\ & + A_6^0 r_i^6 Y_6^0 + A_6^6 r_i^6 (Y_6^6 + Y_6^{-6}) \end{aligned} \tag{5-33}$$

In this hexagonal case the number of A_l^m parameters is four, while it is only two in the cubic case. In general the number of parameters required to describe $V(\boldsymbol{r}_i)$ is inasmuch important as the symmetry is low.

The Point Charge Model. In this oversimplified model, one assumes that the electrostatic potential of the surroundings can be described by electric point charges located at the center of the atoms of the crystal. At a point $\boldsymbol{r}$ the electrostatic potential can then be written as:

$$V(r, \theta, \phi) = \sum_j \frac{1}{4\pi\varepsilon_0} \frac{q_j}{|\boldsymbol{R}_j - \boldsymbol{r}|} \tag{5-34}$$

where q_j is the charge of the surrounding ion j and $\boldsymbol{R}_j$ is its position. If q_j is known, this potential can be calculated exactly. Although this model sometimes gives the correct sign of the A_l^m coefficients, it is known to fail in most cases. Actually it neglects the finite extent of the charges on the ions, the overlap of the magnetic wave functions with those of neighboring ions, and the complex effects due to the screening of magnetic electrons by the outer electron shells of magnetic ions. Moreover, in metallic compounds, this model does not take into account the conduction electrons which produce a strong contribution to the CEF Hamiltonian.

Calculation of the Matrix Elements of the CEF Hamiltonian

In the $|J, M_J\rangle$ basis of a multiplet, the matrix elements are of the type:

$$\langle J, M_J | H_C | J, M_J' \rangle = \\ = \langle J, M_J | -e \sum_i V(\boldsymbol{r}_i) | J, M_J' \rangle \qquad (5\text{-}35)$$

The free ion $|J, M_J\rangle$ states are wave functions obtained from determinental product states involving single electron wave functions $\Phi_i = \Psi_i(\boldsymbol{r}_i)\,\chi_i(s_i)$ (Ψ_i = orbital state, χ_i = spin state), on which the corresponding terms $V(\boldsymbol{r}_i)$ of H_C act. The matrix elements expressed in spherical coordinates therefore reduce to sums of terms of the form:

$$\int \Psi_i^*(\boldsymbol{r}_i)\, r_i^l\, Y_l^m(\theta_i, \phi_i)\, \Psi_i(\boldsymbol{r}_i)\, \mathrm{d}^3 \boldsymbol{r}_i \qquad (5\text{-}36)$$

Because Ψ is the product of an orbital term and a spin term, this integral is also the product of two parts:

a radial part

$$\langle r_{l'} \rangle = \int [R_{nl'}(r_i)]^2\, r_i^l\, r_i^2\, \mathrm{d}r_i \qquad (5\text{-}37)$$

an orbital part

$$\int Y_{l'}^{m'}(\theta_i, \phi_i)\, Y_l^m(\theta_i, \phi_i)\, Y_{l'}^{m''}(\theta_i, \phi_i) \cdot \\ \cdot \sin\theta_i\, \mathrm{d}\theta_i\, \mathrm{d}\phi_i \qquad (5\text{-}38)$$

For rare earth ions the radial parts were calculated and are tabulated.

A quite convenient method, called the Stevens Operator Equivalent method (Stevens, 1952), is generally used for evaluating the orbital part of the matrix elements. It eliminates the need to go back to single electron wave functions by the use of an equivalent operator consisting of angular momentum operators which act on the angular part of the wave function. This is an application of the Wigner-Eckart theorem (see Appendix 1). Due to the fact that the angular momentum $\boldsymbol{J}$ and the position of an electron $\boldsymbol{r}$ are both vectors, the position variables x, y and z of H_C are replaced by J_x, J_y and J_z respectively. As an example

$$\sum_i 3 z_i^2 - r_i^2 \equiv \alpha_J \langle r^2 \rangle [3 J_z^2 - J(J+1)] = \\ = \alpha_J \langle r^2 \rangle O_2^0 \qquad (5\text{-}39)$$

Therefore:

$$\langle J, M_J | \sum_i (3 z_i^2 - r_i^2) | J, M_J' \rangle \equiv \\ \equiv \alpha_J \langle r^2 \rangle \langle J, M_J | [3 J_z^2 - J(J+1)] | J, M_J' \rangle \qquad (5\text{-}40)$$

The multiplicative factor α_J is a constant which depends on the rare earth (for 4th and 6th order terms, β_J and γ_J are used respectively). These factors characterize the orbitals of the 4f electrons for each multiplet. In Sec. 5.5 the link between 4f orbitals and the $|J, M_J\rangle$ will be emphasized.

Finally, the perturbing CEF Hamiltonian acting on the $|J, M_J\rangle$ states of a multiplet can be simply written as:

$$H_C = \alpha_J \sum_{m=-2}^{2} A_2^m O_2^m + \beta_J \sum_{m=-4}^{4} A_4^m O_4^m + \\ + \gamma_J \sum_{m=-6}^{6} A_6^m O_6^m \qquad (5\text{-}41)$$

Table 5-3 lists H_C for the symmetry of a lot of studied compounds when the quantization axis z is along the [001] direction. Note that the number of CEF parameters A_l^m increases when the symmetry is lowered. Note also, and this has important consequences for magnetic properties, that

Table 5-3. CEF Hamiltonian in cubic, hexagonal and tetragonal symmetries.

Point symmetry	H_C
Cubic	$\beta_J A_4^0 (O_4^0 + 5 O_4^4) + \gamma_J A_6^0 (O_6^0 + 21 O_6^4)$
Hexagonal	$\alpha_J A_2^0 O_2^0 + \beta_J A_4^0 O_4^0 + \gamma_J A_6^0 O_6^0 + \gamma_J A_6^6 O_6^6$
Tetragonal	$\alpha_J A_2^0 O_2^0 + \beta_J A_4^0 O_4^0 + \beta_J A_4^4 O_4^4 + \gamma_J A_6^0 O_6^0 + \gamma_J A_6^4 O_6^4$

there is no second order term for cubic symmetry.

Kramers, Non-Kramers ions. Depending on whether the number of electrons is even or odd, the degeneracy of the free ion multiplets is odd (non-Kramer's ion) or even (Kramer's ion), and it can be shown that the removal of degeneracy by CEF leads to levels which are at least doublets in the latter case, whereas they can be singlets in the former.

The CEF, by removing the degeneracy of the multiplets, has many quite different effects on magnetic properties. Especially it leads to modifications of the different types of behavior described in Sec. 5.3.2 where the field (applied plus molecular field) acting on the unsplit multiplet always gives rise to the maximum moment ($g_J \mu_B J$) at $T = 0$ K. The magnetic properties now become dependent also on the splitting due to CEF and especially on the wave functions associated with the ground state. In order to illustrate some types of behavior due to CEF we will examine a very simply case: the effect of a small magnetic field at $T = 0$ K for different types of CEF ground states and the effect of different directions of the field. Let us consider a uniaxial symmetry (such as hexagonal, tetragonal ...) where the second order term of the CEF is preponderant. Then the CEF perturbing Hamiltonian is simply:

$$H_C = B_2^0 O_2^0 = B_2^0 [3 J_z^2 - J(J+1)] \quad (5\text{-}42)$$

To the first order, the perturbing Zeeman Hamiltonian acting on the CEF ground state is:

$$H_{Zee} = -\mu_0 g_J \mu_B \boldsymbol{H} \cdot \boldsymbol{J} \quad (5\text{-}43)$$

Three different situations are worth studying depending on the nature of the CEF ground state.

(i) $B_2^0 < 0$ and J is an integer or half integer, $J = 4$ for instance. The resulting CEF splitting is schematized in Fig. 5-21.

A $T = 0$ K, one only has to look at the diagonalization of H_{Zee} in the space of the two states ($|4\rangle$and $|-4\rangle$) of the ground state doublet.

When $\boldsymbol{H}$ is along the quantization axis Oz, the matrix representing H_{Zee} on the above basis is reduced to:

	$\|4\rangle$	$\|-4\rangle$
$\langle 4\|$	$-4 \mu_0 g_J \mu_B H$	0
$\langle -4\|$	0	$4 \mu_0 g_J \mu_B H$

It is diagonal and leads to the splitting shown in Fig. 5-21. The magnetic moment is then along Oz and reaches its maximum value, i.e. $4 g_J \mu_B$.

When $\boldsymbol{H}$ is along Ox (or Oy), the matrix representing H_{Zee} has all elements equal to zero. Then the doublet is not split and

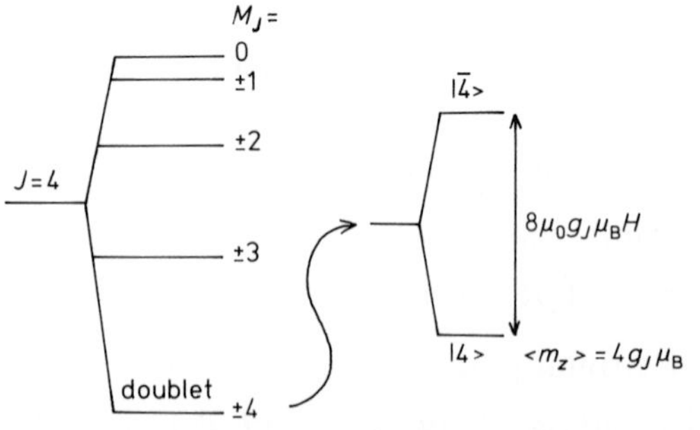

Figure 5-21. Splitting of a $J = 4$ multiplet by the second order term $B_2^0 O_2^0$ of the CEF Hamiltonian with $B_2^0 < 0$ (left-hand part) and Zeeman splitting of the associated $|\pm 4\rangle$ doublet ground state by a magnetic field along the quantization axis Oz (right-hand part).

there is no intrinsic moment along Ox (or Oy) (i.e., perpendicular to Oz) to the first order of perturbation.

Hence: At low temperature magnetization is along Oz and there is a huge anisotropy between Oz and the plane perpendicular to it. At higher temperature the conclusions are not drastically modified and one observes an anisotropy of the thermal dependence of the susceptibility in favor of the Oz axis.

(ii) $B_2^0 > 0$ and J is a half integer, $J = 5/2$ for instance. The resulting CEF splitting is shown in Fig. 5-22.

When $\boldsymbol{H}$ is along Oz, the matrix representing H_{Zee} in the $|\pm 1/2\rangle$ basis is:

	$\|1/2$	$\|-1/2$
$\langle 1/2\|$	$\dfrac{-\mu_0\, g_J\, \mu_B\, H}{2}$	0
$\langle -1/2\|$	0	$\dfrac{\mu_0\, g_J\, \mu_B\, H}{2}$

It is diagonal and leads to the splitting shown in Fig. 5-22a. The magnetic moment along Oz is then $g_J\, \mu_B/2$.

When $\boldsymbol{H}$ is perpendicular to Oz, along Ox for instance, the states that diagonalize H_{Zee} are:

$$|a\rangle = \frac{1}{\sqrt{2}}(|1/2\rangle + |-1/2\rangle) \quad \text{and}$$

$$|b\rangle = \frac{1}{\sqrt{2}}(|1/2\rangle - |-1/2\rangle)$$

The matrix is then:

	$\|a\rangle$	$\|b\rangle$
$\langle a\|$	$-\frac{3}{2}\mu_0\, g_J\, \mu_B\, H$	0
$\langle b\|$	0	$\frac{3}{2}\mu_0\, g_J\, \mu_B\, H$

and the Zeemann splitting is shown in Fig. 5-22b. Hence the magnetic moment along Ox is $3\, g_J\, \mu_B/2$.

Comparing this splitting to that of Fig. 5-22a, one can conclude that the anisotropy favors the plane perpendicular to Oz (lower energy and larger magnetic moment than along Oz).

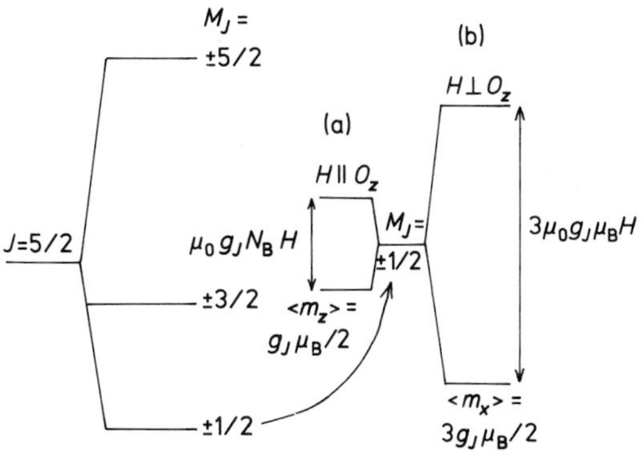

Figure 5-22. Splitting of a $J = 5/2$ muliplet by the second order term $B_2^0 O_2^0$ of the CEF Hamiltonian with $B_2^0 > 0$ (left-hand part). The right-hand part shows the Zeeman splitting of the associated $|\pm 1/2\rangle$ doublet ground state by a magnetic field parallel (a) and perpendicular (b) to the quantization axis Oz.

(iii) $B_2^0 > 0$ and J is an integer, $J = 4$ for instance. In that case, one easily works out that CEF splitting leads to a singlet ground state. There is no magnetic moment in any direction to the first order of perturbation. Actually applying a field, a small magnetic moment and accordingly a small susceptibility are induced because of higher order terms in perturbation involving matrix elements between the ground state and the excited states of the CEF splitting. In this case the anisotropy favors the plane perpendicular to Oz.

Review articles and books on CEF: Hutchings (1964), Stevens (1952), Heine (1964).

Books on this section: Barbara et al. (1988), Chikazumi (1978), Cullity (1972), Herpin (1968), Néel (1978).

Appendix 1: Symmetry and Quantum Theory

Let us consider a physical system which is invariant under some symmetry operations; these operations form a group of transformations such as translation, rotation, inversion center, time reversal, particle permutation, etc. Since the physical system under investigation is invariant with respect to symmetry operations, the Schrödinger equation is also invariant under the same operations. This means that, when applying a symmetry transformation, the wave functions corresponding to a given energy level transform into linear combinations of one another: they give an irreducible representation of the group. In other words, if we define a space spanned by the states of the quantum system, the substates associated with one given energy level E_n form an invariant subspace. Each energy level is associated with an irreducible representation and its degeneracy is equal to the order of the representation. As a consequence, if the symmetry group of the Hamiltonian is known, we can work out the number and degeneracy of irreducible representations, which in turn give us all the possible degeneracies and symmetries of the eigenstates E_n.

Furthermore, let us consider an unperturbed Hamiltonian H_0 which is invariant under a group G_0; we then apply a perturbation V invariant under G: since G is less symmetric than G_0, a given energy level $E_{n,0}$ may split into new levels whose separation will be given by perturbation theory. Group theory can tell us how it may split. The eigenfunctions associated with $E_{n,0}$ give an irreducible representation Γ_0^n of G. But Γ_0^n is no longer an irreducible representation of G_0: it is reducible and decomposes into irreducible representations of G. This decomposition can be easily worked out using character tables; they tell us the number and degeneracies of the energy levels that arise from $E_{0,n}$ when activating the perturbation V.

5.4 Itinerant Magnetism

Opposite to the localized model is the itinerant (or band) model which considers that each magnetic carrier (electron or hole) is itinerant through the solid. It moves in the potential of other electrons and ions, and the corresponding atomic levels form energy bands as shown in Chap. 1 of this Volume. Ordered magnetic states, stabilized by electron-electron interactions (see Sec. 5.5), are characterized by the difference of the number of electrons (or holes) with up and down spins. Whereas the localized model is mainly applied to insulators and also to metals in the case of rare earths, the band model is more relevant to metals containing d-elements where the unfilled d shell is rather extended. However, as we will show later, contrary to the localized situation, the itinerant model is a simplified description which is more phenomenological and generally cannot give account quantitatively of the observed properties. Indeed, in the localized picture, using the mean field approximation, the number of parameters is limited, leading to models which allow a quantitative analysis of the observed magnetic properties. Conversely, itinerant magnetism involves a many-body system which can be treated only thanks to extremely crude approximations.

The purpose of this section is to briefly present the simplest model of itinerant magnetism, namely the *Stoner model* (Stoner, 1938), which has mainly been used to account for the existence of ferromagnetism in itinerant systems.

5.4.1 The Stoner Model

5.4.1.1 The Stoner Criterion for the Onset of Ferromagnetism

The interaction Hamiltonian which leads to magnetism can be written as:

$$H = U\, n_+\, n_- \qquad (5\text{-}44)$$

where n_+ and n_- represent the number of electrons per atom in the considered band (Fig. 5-23) with their spin up and down, respectively. U is the interaction which tends to increase the number of electrons with ↑ spin. It originates from the Coulomb interaction between the electrons and a more detailed description of this interaction is presented in Sec. 5.5. In this model the magnetic moment has no orbital contribution and is only of spin origin. Moreover the existence of a magnetic moment is associated with the onset of magnetic order, whereas in the localized picture, the magnetic moment persists above the ordering temperature.

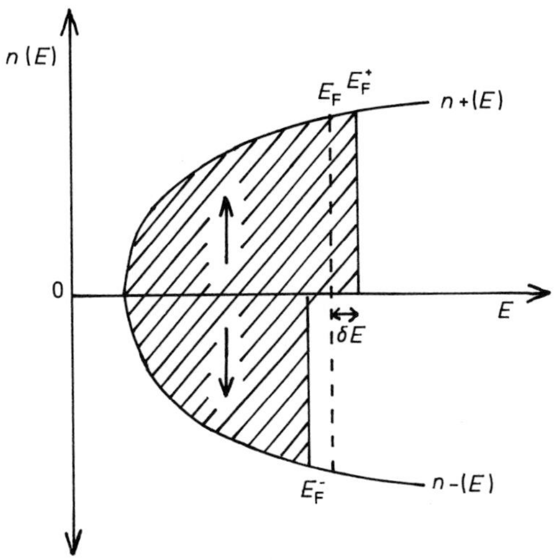

Figure 5-23. Stoner model: schematic representation of the up and down spin bands. E_F^+ and E_F^- are the associated Fermi levels of the kinetic energy. Their difference corresponds to the gain of the Coulomb interaction energy which occurs when the bands are split.

Conditions for the Stability of a Ferromagnetic State

Let us define $n(E)$ as the density of states per spin, and $(n_+ - n_-)\,\mu_B$ as the magnetization per atom. The variation of the magnetic interaction energy between the magnetic and non magnetic states is then:

$$\Delta E_M = U\, n_+\, n_- - U\,\tfrac{1}{4} n^2 = -U\,\tfrac{1}{4} n^2\, m^2 \qquad (5\text{-}45)$$

where $n = n_+ + n_-$ is the total number of electrons per atom in the band and $m = (n_+ - n_-)/n$. To this energy gain ΔE_M, is opposed the variation of energy ΔE_k (kinetic energy) due to the occupation of higher energy states in the band (see Fig. 5-23). If U is large enough, a shift of the up and down spins bands occurs and the Fermi levels are shifted by $\pm\delta E$. To the first order, for a small change δE, ΔE_k is written as:

$$\Delta E_k = (n_+ - n_-)\,\delta E/2 = n\, m\, \delta E/2 \qquad (5\text{-}46)$$

Therefore, the total variation in energy is:

$$\Delta E_T = \Delta E_k + \Delta E_M = -U\, n^2\, m^2/4 + n\, m\, \delta E/2 \qquad (5\text{-}47)$$

since:

$$n(E_F)\,\delta E = (n_+ - n_-)/2 \qquad (5\text{-}48)$$

hence:

$$\Delta E_T = n^2\, m^2\,(1 - U\, n(E_F))/4\, n(E_F) \qquad (5\text{-}49)$$

If $[1 - U\, n(E_F)] > 0$, the state of lowest energy is obtained for $m = 0$. The material has no spontaneous magnetization: it is paramagnetic.

If $[1 - U\, n(E_F)] < 0$, there will be a gain in energy when band splitting occurs leading to ferromagnetism. This is the classical *Stoner criterion.* The conditions favoring a magnetic ordering are then a large value of U, but also a large value for $n(E_F)$, the density of states at the Fermi level.

Paramagnetic Susceptibility

In the same order of approximation, let us calculate the magnetic susceptibility at zero Kelvin in a field H when the ferromagnetic state is not stable. The quantity (Zeeman energy) $-\mu_0 n m \mu_B H$ must be introduced in the expression of the total variation of the energy of Eq. (5-47). The equilibrium value is such that $d(\Delta E_T)/dm = 0$, which leads to the susceptibility per volume unit:

$$\chi = N n m \mu_B / H = R \chi_0 \tag{5-50}$$

where (i) $\chi_0 = 2 N \mu_0 \mu_B^2 n(E_F)$ is the susceptibility *without interaction*, (ii) N is the number of magnetic atoms per volume unit and (iii) $R = [1 - U n(E_F)]^{-1}$, the Stoner *enhancement factor*. We will see later that this susceptibility is generally almost temperature independent. The compounds which are paramagnetic down to 0 K are *Pauli paramagnets*. This susceptibility is very weak in systems where magnetic atoms are of s or p type: it is the largest in d compounds where the d wave functions are more localized, i.e. where $n(E_F)$ and U are large.

5.4.1.2 The Stoner Model at Finite Temperature

The above section presents the simplest model to describe a ferromagnetic state in an itinerant system. Let us now look at a more quantitative approach by taking into account thermal effects and assuming that magnetization is weak, i.e. the shift between the Fermi levels of the up and down spin bands is weak. Considering single particle excitations, thermal effects are taken into account through Fermi Dirac statistics which, for an ideal electron gas in thermal equilibrium, gives the probability for a state of energy E to be occupied:

$$f(E) = 1/\{\exp[(E-\eta)/kT] + 1\} \tag{5-51}$$

η being the chemical potential; it is such that the total number of electrons per atom is always equal to n. $f(E)$ and its derivative $f'(E)$ are plotted in Fig. 5-24 for $kT = 0$ and $kT = \eta/20$.

The electron number for both spin states is:

$$n_-^+ = \int_0^\infty f(E, \eta_-^+) \, n(E) \, dE \tag{5-52}$$

where

$$\eta_-^+ = \mu_-^+ k\theta' m_-^+ \mu_0 \mu_B H \tag{5-53}$$

$m = m(H,T) = M(H,T)/nN\mu_B$ is the relative magnetization, $M(H,T)$ (or M) is the magnetization associated with N atoms per unit volume, μ is the chemical potential in the absence of interaction and magnetic field, $n(E)$ is the density of states per atom per spin, and $k\theta' = nU/2$ is the interaction parameter (at $T = 0$, $\eta^+ = E_F^+$ and $\eta^- = E_F^-$). In fields H which are not excessively strong, it is reasonable to assume that $m(H,T) \ll 1$. In the integration by parts of Eq. (5-52), the derivative $f'(E)$ appears, which, as shown in Fig. 5-24, is different from zero only in the vicinity of η. It is

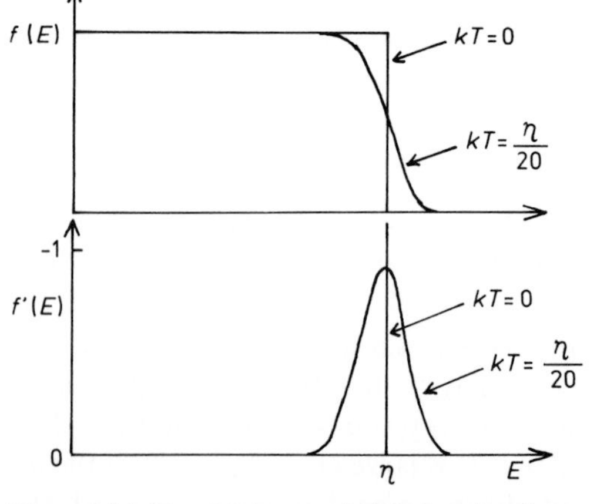

Figure 5-24. Fermi-Dirac statistical function (upper part) and its derivative (lower part) at 0 K and at finite temperature.

then appropriate to use an expansion of $n(E)$ near E_F over the temperature range $T/T_F \ll 1$ ($kT_F = E_F$). One then obtains:

$$\chi_0 H = M[1 - U n(E_F) + a T^2] + b M^3 \quad (5\text{-}54)$$

where

$$a = \frac{\pi^2 K^2}{6}(v_1^2 - v_2) \quad \text{and}$$

$$b = \frac{\mu_0^2 \mu_B^2}{2\chi_0^2}\left(v_1^2 - \frac{v_2}{3}\right)$$

with

$$v_1 = \frac{n'(E_F)}{n(E_F)} \quad \text{and} \quad v_2 = \frac{n''(E_F)}{n(E_F)}$$

Depending on the values of the coefficients of M and M^3, this equation can give a spontaneous magnetization, i.e. a non-zero value of M in zero applied field. When the system is paramagnetic, i.e. when M is zero in a zero applied field, this equation can be rewritten as:

$$M = \frac{\chi_0 R H}{1 + R a T^2} - \frac{R^4 \chi_0^3 b H^3}{(1 + R a T^2)^4} \quad (5\text{-}55)$$

From Eqs. (5-54) and (5-55) it is possible to look at different types of possible behavior by letting the product $U n(E_F)$ increase, starting from small values.

Pauli Paramagnetism

If $R > 1$, at $T = 0$ the Stoner criterion is not satisfied and the compound is paramagnetic. From Eq. (5-55), one can deduce the low field susceptibility:

$$\chi = \chi_0 R/[1 + \pi^2 K^2 R (v_1^2 - v_2) T^2/6] \quad (5\text{-}56)$$

Depending on whether $v_1^2 - v_2$ is positive or negative, χ decreases or increases, respectively, as temperature is increased. In the latter case χ passes through a maximum because for large enough temperature χ always decreases.

Collective Electron Metamagnetism

Such a type of behavior has been predicted (Wohlfarth et al., 1962) to occur in substances that are close to the onset of ferromagnetism and where: (i) R is large and positive, i.e. $U n(E_F) <$ and ≈ 1, (ii) "b" and accordingly "a" are negative. As a consequence, the thermal variation of the enhanced susceptibility passes through a maximum (Fig. 5-25) and at $T = 0$ the field dependence of magnetization [see Eq. (5-55)] presents a positive curvature anticipating a transition towards a state of large magnetization. It can be shown that for a critical field H_C a transition, called *metamagnetic transition*, toward a high magnetization state occurs (see Fig. 5-27). The negative value of "b", associated with this behavior, shows that it depends critically on the shape of $n(E)$ around E_F. Especially this condition is fulfilled when the density of state has a sufficiently positive curvature around E_F, the best situation occurring when E_F lies in a minimum of $n(E)$.

Very Weak Itinerant Ferromagnetism

When $U n(E_F)$ is barely larger than 1, the Stoner criterion is fulfilled but magnetization is weak and we have the so-called *very weak itinerant ferromagnetism*. From

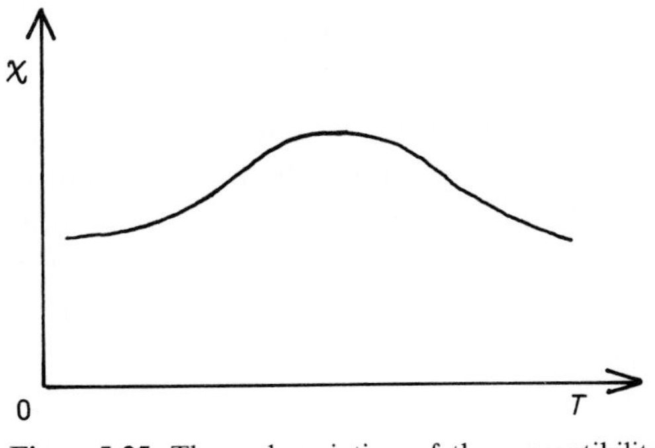

Figure 5-25. Thermal variation of the susceptibility associated with collective electron metamagnetism.

Eq. (5-54) it is possible to deduce the following results:

(i) at $T=0$, $M^2(0,0)=[U\,n(E_F)-1]/b$. Hence a positive value of "b" is a necessary but not sufficient condition for stable very weak itinerant ferromagnetism.

(ii) defining the zero field differential susceptibility at $T=0$ as: $\chi(0,0)=\mathrm{d}M/\mathrm{d}H$ for $H=0$ and $T=0$ (actually to get χ one calculates $\chi^{-1}=\mathrm{d}H/\mathrm{d}M$ for $M=M(0,0)$ and $T=0$), from Eq. (5-54) one then obtains: $\chi(0,0)=-R\chi_0/2$.

(iii) again from Eq. (5-54) one can deduce the Curie temperature for which $\chi(0,T)$ diverges ($\mathrm{d}M/\mathrm{d}H=0$):

$$T_C=\{[U\,n(E_F)-1]/a\}^{1/2}$$

(iv) from the above formulae, Eq. (5-54) can be rewritten as:

$$\left[\frac{M(H,T)}{M(0,0)}\right]^3-\frac{M(H,T)}{M(0,0)}\left[1-\frac{T^2}{T_C^2}\right]=$$
$$=\frac{2\chi(0,0)\,H}{M(0,0)} \qquad (5\text{-}57)$$

which leads to:

$$M(H,T)^2=$$
$$=M(0,0)^2\left[1-\frac{T^2}{T_C^2}+\frac{2\chi(0,0)\,H}{M(H,T)}\right] \qquad (5\text{-}58)$$

Hence the theory predicts that the isotherms of M^2 vs. H/M are parallel lines, in particular the isotherm at T_C passes through the origin. Plots of this type are called *Arrott plots* and are frequently used to determine the ordering temperature from magnetization measurements. In Fig. 5-26 such isotherms are drawn for the compound $ZrZn_2$ (Blythe, 1968), which is one of the best examples of very weak itinerant ferromagnet. Even at low temperature, this type of variation leads to a strong dependence of magnetization on the field.

(v) from Eq. (5-58), one obtains the thermal dependence of the zero field susceptibility $\chi(0,T)$: $\chi=\chi(0,0)[1-(T/T_C)^2]^{-1}$ for $T<T_C$ and $\chi=2\chi(0,0)[(T/T_C)^2-1]^{-1}$ for $T_F \gg T>T_C$.

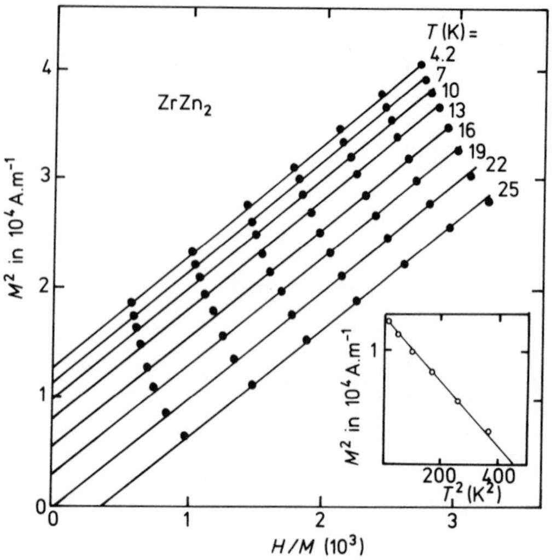

Figure 5-26. Arrott plots in $ZrZn_2$. The inset shows the square of the spontaneous magnetization as a function of T^2.

(vi) finally from Eq. (5-58) one can deduce the thermal variation of the spontaneous magnetization:

$$M^2(0,T)=M^2(0,0)[1-(T/T_C)^2].$$

Therefore, considering individual excitations, $M^2(0,T)$ vs. T^2 must be a straight line. Such variation is reported in the inset of Fig. 5-26 for $ZrZn_2$.

Ferromagnetism

When $U\,n(E_F)$ is well above 1, at $T=0$ ferromagnetism is well established, with a quite reasonable value of the spontaneous magnetization. Expansions near the Fermi level are not valid and the formulae worked out for the latter case do not apply. However, in the Stoner model, one can still use Eq. (5-55) and determine $n^\pm$ numerically by taking into account the true shape of the density of state $n(E)$.

The first order zero field susceptibility dM/dH, still given at first approximation by the expression written in the previous section, becomes infinite at $T = T_C$. Two situations have to be considered:

– If both the two subbands of opposite spin are not completely filled up, the compound is a *weak ferromagnet*: this is, for instance, the case for Fe metal (Fig. 5-37);

– If the subband with up spin is completely full, the compound is a *strong ferromagnet*: this is, for instance, the case for Ni and Co metals (Fig. 5-37).

These terms (weak and strong ferromagnets) are only definitions. In particular they do not reveal anything about the moment magnitude. Indeed, in Fe the magnetization is 2.4 μ_B/Fe whereas in Co and Ni it is 1.9 and 0.6 respectively. Moreover, weak ferromagnetism is quite different from the very weak itinerant ferromagnetism which is described in the previous section.

Antiferromagnetism

The Stoner model can be extended to describe band antiferromagnetism. In the simplest case of a compound formed by two identical sublattices obtained from each other by a translation vector, one considers that magnetizations of each sublattice are equal and opposite. It will be seen in Sec. 5.5 that antiferromagnetism tends to appear when the band is half filled whereas ferromagnetism is favored in almost empty or almost full bands.

Some Comments About the Stoner Model

Orbital contribution to magnetism is not taken into account. Actually as is discussed in Sec. 5.5, although it is rather weak in cubic transition metals, this contribution affects both the value of magnetization and the magnetocrystalline anisotropy.

The Stoner model is based on the mean field approximation and, as in the localized model, the decrease of magnetization when temperature is increased corresponds to individual excitations (reversal of a spin independently from the others corresponding to the hopping of an electron from one subband to the other). Actually, except in the case of very weak itinerant ferromagnetism, the Stoner model fails to account for thermal effects. These limitations of the Stoner model are discussed in Sec. 5.5.7.

5.4.2 Free Energy Approach to Band Magnetism

This is another way to work out the Stoner model and to understand the different aspects of band magnetism described above. Assuming that magnetization is not too large and using the molecular field approximation, the magnetic contribution F_M to the free energy of a solid can be expressed using the so-called Landau expansion of the magnetization M:

$$F_M = AM^2/2 + BM^4/4 + CM^6/6 + \ldots - \mu_0 MH \tag{5-59}$$

In this expansion A, B, C, ... can be related to the coefficients previously used in the Stoner model but, whatever the model, when M is small enough this expansion is always valid. Specifically, it is always valid near T_C. Minimizing F_M with respect to M yields the equilibrium value of M for a given field H:

$$M^2 = -A/B + \mu_0 H/BM \tag{5-60}$$

The identification of this formula with Eq. (5-57) leads, in the Stoner model, to:

$$A = \mu_0[(T/T_C)^2 - 1]/2M(0,0)\chi_0 \quad \text{and}$$
$$B = \mu_0/2M^2(0,0)\chi_0$$

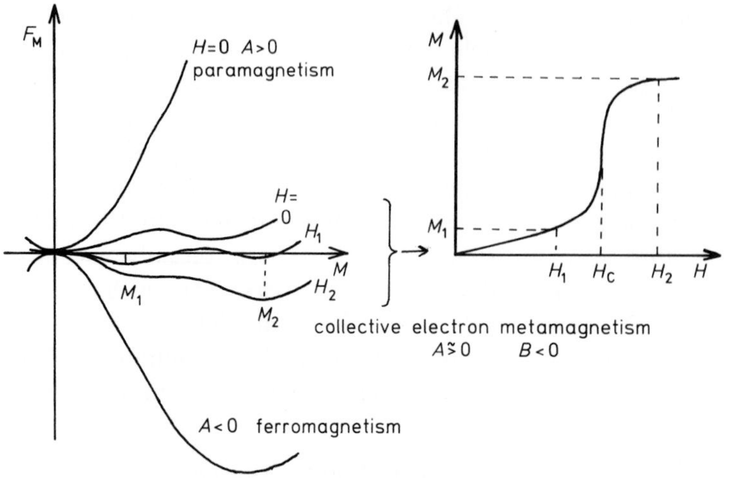

Figure 5-27. Schematic magnetization dependence of F_M in the cases of Pauli paramagnetism, ferromagnetism and collective electron metamagnetism. In this latter case the field dependence of the magnetization (at $T = 0$ K) is also presented.

The different behaviors described above are illustrated in Fig. 5-27 where the M dependences of F_M are schematized:

– Pauli paramagnetism: A is positive and inasmuch large as the susceptibility is weak.

– Collective Electron Metamagnetism (Wohlfarth et al., 1962; Shimizu, 1982): A is still positive but close to zero (Stoner criterion almost satisfied). B is negative leading to a minimum for $M \neq 0$ of larger energy than that at $M = 0$. When a field is applied, except for $M = 0$, all the curves are lowered and the second minimum becomes of lower energy for fields larger than a critical field leading to the transition.

– Ferromagnetism: A is negative; the Stoner criterion is satisfied and in zero field the compound is spontaneously magnetized.

Books and review articles on itinerant magnetism: Gautier (1982) pp. 4–18; Barbara et al. (1988), Chap. 10; Shimizu (1981).

5.5 3d Metallic Magnetism

Whereas the previous section described the simplest approach to itinerant magnetism, the present one deals with a more realistic description of 3d magnetism, especially its origin. The relatively detailed descriptions of the 3d band and the origin of its magnetism would allow a better understanding of the main characteristics of 3d magnetism. The main aspects of this section are based on articles by Friedel (1969), pp. 340–411, and Gautier (1982), pp. 18–244.

5.5.1 Localized Versus Band Magnetism

3d magnetism, as it is observed in Cr, Mn, Fe, Co and Ni, is intermediate between the localized (Sec. 5.3) and the completely itinerant (Sec. 5.4) descriptions. Indeed, the observation of Curie Weiss susceptibilities suggests that magnetic carriers (d electrons or holes) are localized. However: (i) contrary to the case of rare earths, it is impossible to assign, from the experimental value of the saturation magnetization, an integer number of 3d electrons per atom contributing to magnetism (Table 5-4); (ii) magnetic moments deduced from the Curie constants are larger than the saturation magnetization.

It must then be considered that magnetism in these transition elements mainly arises from d electrons, the energies of which form a narrow band which is unfilled. As sketched in Fig. 5-28 and discussed further below, the itinerant electrons fill the 3d and 4s bands like communicating vessels. The number of d electrons which contribute to magnetism may then become non-integer, the contribution of s electrons to magnetism being very weak.

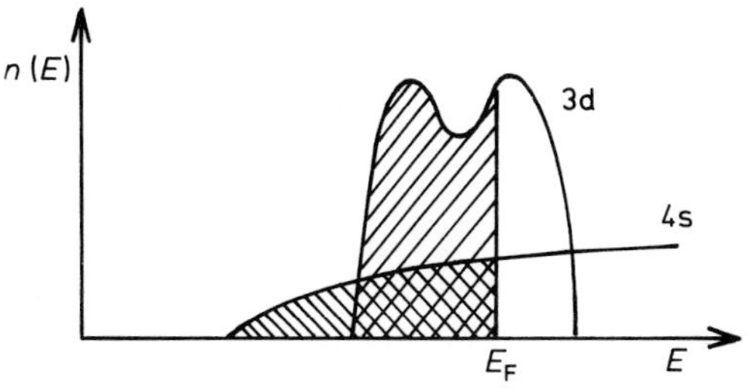

Figure 5-28. Schematic representation of the 3d and 4s density of states.

5.5.2 The d Band

The band calculation described here is performed using the Hartree approximation: wave functions and energies of an electron are studied in the electrostatic potential of the metal. The latter is taken as a sum of atomic potentials V_i centered on the various lattice sites i: $V \approx \sum_i V_i$. For each site there are five d atomic orbitals, written as $|im\rangle$ where m is the projection of the orbital moment which can take five different values: 2, 1, 0, -1 and -2. The wave functions $|\Psi\rangle$ are then taken as linear combinations of such atomic orbitals. This so-called LCAO (Linear Combinations of Atomic Orbitals) method is valid since the s and d bands are distinct, and also the interactions responsible for cohesion in the metal are weak compared with electrostatic potential on each site (tight binding approximation). The matrix elements of the Hamiltonian are then:

$$\langle\Psi|\left(T+\sum_i V_i\right)|\Psi\rangle \tag{5-61}$$

where T is the kinetic energy and $|\Psi\rangle = \sum_{i,m} a_{im}|im\rangle$. Furthermore $(T+V_i)|im\rangle = E_0|im\rangle$, $\langle im|jm'\rangle \approx \delta_{ij}\delta_{mm'}$ and $\sum_{i,m}|a_{im}|^2 = 1$.

Among the $\langle im|V_k|jm'\rangle$'s only the two-center integrals between first (or second) neighbors are retained. The potential energy $\langle\Psi|\sum_i V_i|\Psi\rangle$ is then the sum of atomic terms: $\langle im|V_i|im\rangle$ and of supplementary terms written as:

$$\alpha_{im} = \langle im|\sum_{j\neq i} V_j|im\rangle$$

$$\beta_{im}^{jm'} = \langle im|V_j|jm'\rangle$$

The atomic states $|im\rangle$ are not modified by the α terms, the effect of which is only to shift the energies of atomic levels. The *transfer integrals* β mix the atomic states into molecular states extending over the whole solid. Going from the atomic state to the metal, the energy change can be

Table 5-4. 3d electron contribution per atom to magnetization[a].

	q_s	q_c	q_c/q_s
Fe	2.2	2.3	1.05
Co	1.7	2.3	1.35
Ni	0.6	0.9	1.50
Sc_3In	0.06	0.22	3.86

[a] In d elements it is often assumed that the orbital contribution to magnetism is negligible. So that in a localized picture the saturation magnetization is $m_s = 2S\,\mu_B = q_s\,\mu_B$; assuming an integer number of electrons on the 3d shell, q_s is also expected to have an integer value, which is not observed. Furthermore, the effective paramagnetic moment is expected to reach $m_{eff} = \sqrt{2S(2S+2)} = \sqrt{q_c(q_c+2)}$; q_c then should have the same value as q_s. Experimentally they are different. The system is far from the localized picture when q_s is small and the q_c/q_s ratio large.

written as:

$$E = \sum_{i,m} |a_{im}|^2 \alpha_{im} + \sum_{\substack{i,m \\ j \neq i, m'}} a_{im}^* a_{jm'}^* \beta_{im}^{jm'} \quad (5\text{-}62)$$

The α and β integrals, corresponding to attractive potential for electrons, are mainly negative. The contribution of β integrals to E varies with the values of the coefficients a_{im}, from a minimum value where most (or all) of the β terms are negative to a maximum value where most (or all) of them are positive. These two energy states E_b and E_a correspond to the formation of *bonding* and *antibonding* states respectively (Fig. 5-29).

In the *bonding* states the electronic density is increased along the bonds, as compared with the atomic density; in the *antibonding* states the density is decreased. The $5N$ atomic d states $|im\rangle$ give rise to a band of $5N$ levels which are distributed quasi-continuously between these two extremes in energy: the β integrals give rise to the width w of the band and the α integrals give rise to the shift s of the atomic levels (Fig. 5-30).

The shape of the d band obtained in this way clearly depends on the values taken for the α and β integrals. These are difficult to compute accurately. The main difficulty arises from the definition of the atomic potentials V_i. In principle, the lattice potential $\sum_i V_i$ should be computed in a self-consistent way. In fact, the V_i are usually taken as those of the positive ions. Typical values computed are 5 to 10 eV for the width w of the band and 1 to 2 eV for the shift s. These values are in agreement with experimental results. As an example, the density of states of Ni metals is shown in Fig. 5-31.

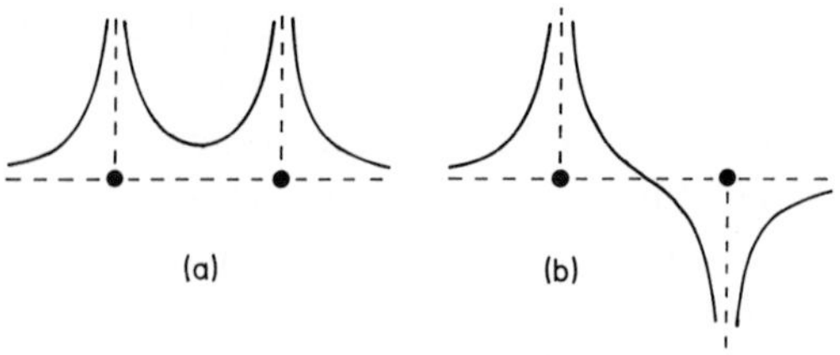

Figure 5-29. Bonding (a) and antibonding (b) states in the d band: schematic representation of the wave functions.

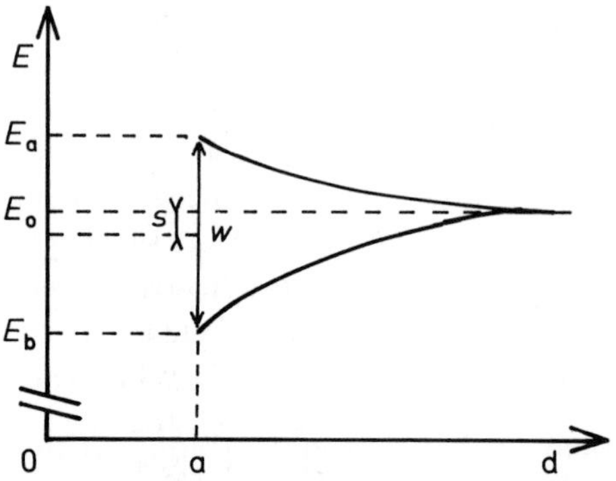

Figure 5-30. Energy shift (s) and width (w) of the d band as a function of atomic spacing d. a is the atomic spacing in the considered metal. Large values of d correspond to free atoms.

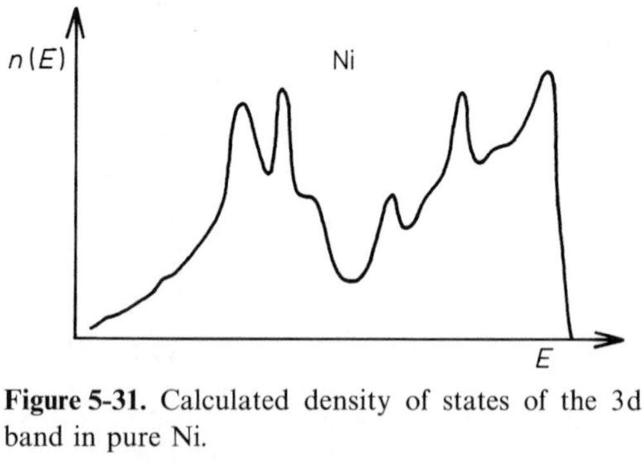

Figure 5-31. Calculated density of states of the 3d band in pure Ni.

Possible Splittings of the d Band

In the above description each state has a five-fold degeneracy corresponding to the 5d angular orbitals. Here we describe three contributions which can lift this degeneracy. We will see in Sec. 5.6.6 that these contributions are of utmost importance for the magnetocrystalline anisotropy.

The α Integrals: CEF Splitting. CEF is present in the α integrals. For instance, in a cubic environment, these integrals are different for the so-called e_g (x^2-y^2, $3z^2-r^2$) and t_{2g} (xy, yz, zx) sets of orbitals. Each state is then split; however due to the smallness of the α integrals compared to the β ones, i.e. the bandwidth, there is no CEF splitting of the band. Only the top and bottom of the band have definite (e_g or t_{2g}) character (Fig. 5-32).

The β integrals. This contribution is related to the itinerant character of electrons. It leads to a spread in energy of the collective states which is different for each d orbital. One example can illustrate this effect. Let us assume a hexagonal compound in which the distance between 3d atoms along $\boldsymbol{c}$ is much smaller than the distances in the plane perpendicular to $\boldsymbol{c}$ (quasi-one-dimensional system). Then the dispersion in energy of the states $|2,0\rangle$, i.e. the $3z^2-r^2$ orbitals, which are elongated along $\boldsymbol{c}$, is larger than for the other 3d states. Thus, as sketched in Fig. 5-33, the states at the ends of the band will have the $|2,0\rangle$ character.

Spin Orbit Coupling. In the tight-binding approximation used above, one considers spin orbit coupling such as arises from the interaction of the spin of each electron with its own orbital moment. The matrix elements between spin states σ and σ′ belonging to the same orbital are then: $\lambda\langle im,\sigma|l\cdot s|im,\sigma'\rangle$, where λ is the spin orbit coupling constant. In cubic symmetry, $\langle im,\sigma|l_z|im,\sigma\rangle=0$ and the trace of the spin orbit matrix is zero. This means that the mean value of the energy levels is not affected by the spin orbit interaction. This latter then contributes only to the width of the d band. If λ was much larger than w, the d band would be split into two subbands with nearly pure $j=l+s=5/2$ and $j=l-s=3/2$ character ($D_2\times D_{1/2}=D_{5/2}+D_{3/2}$).

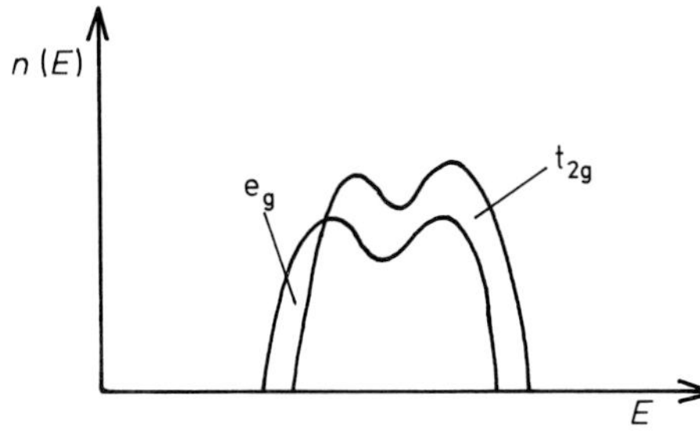

Figure 5-32. Sketch of CEF on the 3d band.

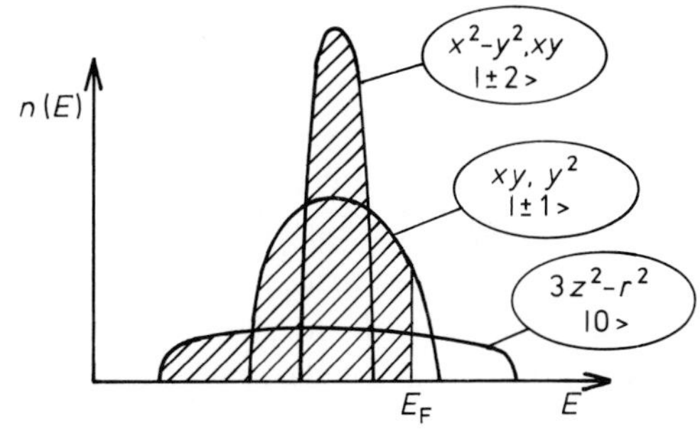

Figure 5-33. Sketch of the 3d density of states in the hexagonal compound $La_2Co_{1.7}$ where the Co–Co distance is much smaller along $\boldsymbol{c}$ than in the plane perpendicular to this axis.

As $\lambda\approx10^{-2}$ eV and $w\approx5$ eV, the situation is opposite and the spin orbit coupling does not alter the d band very much, except near the top and the bottom.

5.5.3 Origin of Magnetism: Coulomb Interaction

In the Hartree approximation the total wave function $\Psi(1,2,\dots i,\dots N)$ can be written as the product of single electron functions: $\Psi\approx\prod_{i=1}^{N}\Psi_i(i)$, each wave function corresponding either to a *molecular orbital* or to an *atomic orbital*. This wave function does not obey the Pauli principle, which implies the full antisymmetry of the total wave function (space and spin) when permuting 2 electrons: $\Psi(1,\dots i,j,\dots N)=-\Psi(1,\dots j,i,\dots N)$. On the contrary, the wave function built into the "Hartree-Fock" model obeys the Pauli principle. This wave function is the determinant obtained from the matrix:

$$\Psi = \begin{bmatrix} \varphi_1(1)\chi_1(1) & \varphi_1(2)\chi_1(2)\ldots & \varphi_1(N)\chi_1(N) \\ \varphi_2(1)\chi_2(1) & \ldots\ldots\ldots\ldots & \varphi_2(N)\chi_2(N) \\ \vdots & & \vdots \\ \varphi_N(1)\chi_N(1) & \ldots\ldots\ldots\ldots & \varphi_N(N)\chi_N(N) \end{bmatrix}$$

where $\varphi_i(j)\chi_i(j) = \Psi_i(j)$ are one electron wave functions $\varphi_i(j)$ = space functions, and $\chi_i(j)$ = spin functions χ_+ and χ_- which are eigenfunctions of s_z with eigenvalues 1/2 and −1/2, respectively.

For example, in the case of two electrons 1 and 2, an antisymmetric space wave function, associated with a symmetric spin function (spin state $S=1$), is written as:

$$\varphi_A = \frac{1}{\sqrt{2}}[\varphi_{im}(1)\varphi_{jm'}(2) - \varphi_{im}(2)\varphi_{jm'}(1)] \quad (5\text{-}63)$$

Similarly, an antisymmetric spin wave function (spin state $S=0$) is associated a symmetric space function:

$$\varphi_S = \frac{1}{\sqrt{2}}[\varphi_{im}(1)\varphi_{jm'}(2) + \varphi_{im}(2)\varphi_{jm'}(1)] \quad (5\text{-}64)$$

We are going to see how the electrostatic interaction energy between electrons splits the energy of these two states. The energy gap then corresponds to the energy difference between the antiparallel and parallel spin states.

Introducing the perturbing Hamiltonian, for the symmetrical ↑↑ spin wave function, the energy of the electrostatic interaction is (for sake of simplicity we omit in what follows the quantity $e^2/4\pi\varepsilon_0$):

$$E_{Aij}^{mm'} = \left\langle \varphi_A \left| \frac{1}{r_{12}} \right| \varphi_A \right\rangle =$$

$$= \frac{1}{2}\Bigg[\left\langle \varphi_{im}(1)\varphi_{jm'}(2) \left| \frac{1}{r_{12}} \right| \varphi_{im}(1)\varphi_{jm'}(2) \right\rangle + \left\langle \varphi_{im}(2)\varphi_{jm'}(1) \left| \frac{1}{r_{12}} \right| \varphi_{im}(2)\varphi_{jm'}(1) \right\rangle - \left\langle \varphi_{im}(1)\varphi_{jm'}(2) \left| \frac{1}{r_{12}} \right| \varphi_{im}(2)\varphi_{jm'}(1) \right\rangle - \left\langle \varphi_{im}(2)\varphi_{jm'}(1) \left| \frac{1}{r_{12}} \right| \varphi_{im}(1)\varphi_{jm'}(2) \right\rangle\Bigg]$$

$$= U_{ij}^{mm'} - J_{ij}^{mm'} \quad (5\text{-}65)$$

and similarly for the antisymmetric ↑↓ spin wave function:

$$E_{Sij}^{mm'} = \left\langle \varphi_S \left| \frac{1}{r_{12}} \right| \varphi_S \right\rangle = U_{ij}^{mm'} + J_{ij}^{mm'} \quad (5\text{-}66)$$

The $J_{ij}(>0)$ are called the *exchange* terms and the $U_{ij}(>0)$ are called the *Coulomb* terms. Note that for φ_S one can have both electrons on the same orbital ($i=j$, $m=m'$), whereas this cannot occur for φ_A (indeed if $i=j$ one must have $m \neq m'$). Due to the strongly localized character of the wave functions, it is obvious that the interatomic terms U_{ij} or $J_{ij}(i \neq j)$ are much weaker than the intraatomic terms U_{ii} and J_{ii}. Taking into account only the intraatomic terms, different situations have to be considered keeping in mind that d electrons can belong to five different orbitals:

(i) Due to the Pauli principle it is impossible for two electrons with parallel spins to lie in the same orbital.

(ii) For two electrons with antiparallel spins lying in the same orbital one obtains:

$$\left\langle \varphi_S \left| \frac{1}{r_{12}} \right| \varphi_S \right\rangle = 2J_{ii}^{mm} = 2U_{ii}^{mm} = 2U \quad (5\text{-}67)$$

This U term which increases the energy does not favor this situation.

(iii) For two electrons on different orbitals one has:

$$\left\langle \varphi_A \left| \frac{1}{r_{12}} \right| \varphi_A \right\rangle - \left\langle \varphi_S \left| \frac{1}{r_{12}} \right| \varphi_S \right\rangle = \quad (5\text{-}68)$$

$$= U_{ii}^{mm'} - J_{ii}^{mm'} - U_{ii}^{mm'} - J_{ii}^{mm'} = -2J_{ii}^{mm'}$$

J being positive, φ_A is of lower energy, and the situation in which the electrons have their spin parallel is favored. This result can be easily understood in the following way: φ_A (and hence $|\Psi|^2$) vanishes when $|\boldsymbol{r}_1 - \boldsymbol{r}_2| \to 0$; this means that both electrons tend to avoid each other because of the exchange and cannot be at the same place at the same time (Pauli principle). There is then a so-called *exchange hole.* The two electrons then avoid the situation where the electrostatic interaction is the less favorable ($1/r_{12} \to \infty$). The energy is then smaller than in the other situation, for which $|\varphi_S|^2$ has a slight maximum around $|\boldsymbol{r}_1 - \boldsymbol{r}_2| = 0$ leading to an increase of the electrostatic energy between both electrons.

Therefore the intraatomic terms U and J favor two electrons to be in different orbitals with parallel spins. This is nothing other than the first part of Hund's rules already seen in Sec. 5.2.2.2. One then defines an average exchange energy per pair of electrons per atom U which represents the interactions favoring the creation of magnetic moments. Computation leads to $U \approx 5$ eV. This value, on the order of the bandwidth, is strongly overestimated. Actually, electrons do not move independently; they are strongly correlated. These correlations tend to favor electrons to be far from each other and therefore they reduce the probability for two electrons to lie in the vicinity of the same point at the same time. As a result U is strongly reduced leading to reasonable values on the order of 0.5 to 1 eV.

Whereas U represents the interactions favoring the creation of magnetic moments, the transfer integrals β responsible for the bandwidth work in the opposite direction, since creating a magnetic moment implies the occupation of states of higher "kinetic" energies in the band. There is a competition between U and β which is accounted for in the Stoner model previously worked out. As shown in Sec. 5.4 the balance is in favor of the creation of magnetic moments only if U and $n(E_F)$ are large enough; the latter condition occurs when the β integrals are small, i.e. when the band is narrow. These conditions are fulfilled only for the elements at the end of the 3d series, the spatial extent of which is rather weak leading to a narrow band and to large enough values of U.

5.5.4 Sign of Magnetic Interactions: Antiferro- to Ferromagnetism

The sign of magnetic interactions in transition metals has been worked out from band theory (Friedel et al., 1961) and is discussed in a very simplified way below.

If the d band is nearly empty or nearly filled up (in this latter case holes instead of electrons have to be considered), the intraatomic Coulomb interaction discussed in Sec. 5.5.3 favors ferromagnetism. Indeed, as shown in the crude sketch of Fig. 5-34 for a nearly empty band, because of the itineracy, electrons hop from site to site and the energy is smaller (i) if, because of the U_{ii}^{mm} term [Eq. (5-67)], the hopping electron goes in an empty orbital of the host site, and (ii) if, because of the $J_{ii}^{mm'}$ term [Eq. (5-68)], the spin of this electron is parallel to those of the electrons of the host site. This is the situation for Fe, Co and Ni, which are among the best examples of ferromagnetism.

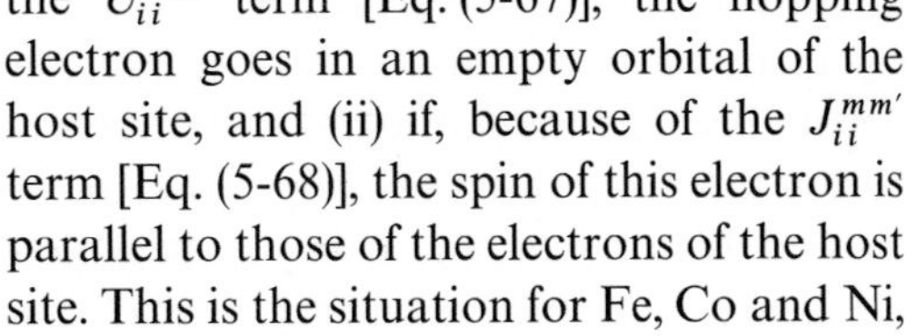

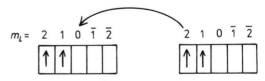

Figure 5-34. Simplified picture to explain why ferromagnetism is favored when the d band is almost empty or filled.

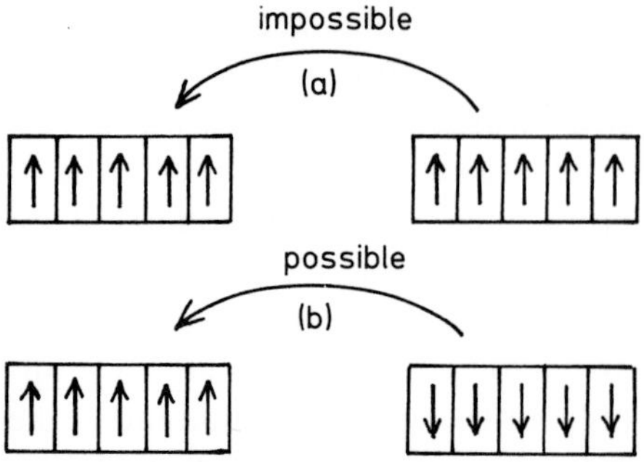

Figure 5-35. Simplified picture to explain why antiferromagnetism is favored when the d band is nearly half filled. For a better understanding, the case of an exactly half filled band is considered.

On the contrary, for a half-filled band antiferromagnetism is favored. This can be understood from the simplified sketch of Fig. 5-35. If the spins of two neighboring atoms are parallel (case a), the electrons cannot hop from site to site on account of the Pauli principle and the system is then an insulator. Such a situation is not energetically favored, because it is in contradiction with the itinerant character of electrons. Actually, the structure of lowest energy is antiferromagnetic (case b). Such a configuration satisfies the itinerant character of the electrons through the β integrals in spite of the U term, which favors the electrons to be localized (such a sitation occurs because the β integral magnitude is larger than the U magnitude). Indeed, in Cr and Mn metals, the d bands of which are nearly half-filled, negative interactions are present:

– Mn metal has a collinear antiferromagnetic structure;

– Cr metal has a sinusoidal antiferromagnetic structure.

5.5.5 Magnetization Amplitude – Weak and Strong Ferromagnetism

Using a more realistic approach than that of Stoner, we will discuss in this section the magnetization amplitude of 3d metals. In the ferromagnetic case, the magnetic energy $E(m)$ can be directly computed for large atomic moments from the knowledge of the density of state:

$$\Delta E(m) = \int_{E_F}^{E_F^+} n(E)\,\mathrm{d}E - \int_{E_F^-}^{E_F} n(E)\,\mathrm{d}E - \\ - \frac{1}{2}\left(n_+^2 - n_-^2 - \frac{n^2}{2}\right) U \qquad (5\text{-}69)$$

with

$$n_-^+ = \int^{E_F^\pm} n(E)\,\mathrm{d}E, \quad n = n_+ + n_- \quad \text{and} \quad m = (n_+ - n_-)\,\mu_B$$

Ferromagnetism is obtained when at point M (Fig. 5-36): $\mathrm{d}\Delta E(m)/\mathrm{d}m \leqq 0$ which leads to:

$$\bar{n}_\pm = \frac{n_+ - n_-}{E_F^+ - E_F^-} = \\ = \frac{1}{E_F^+ - E_F^-} \int_{E_F^-}^{E_F^+} n(E)\,\mathrm{d}E \geqq \frac{1}{U} \qquad (5\text{-}70)$$

This equation leads to the true Stoner criterion for ferromagnetism:

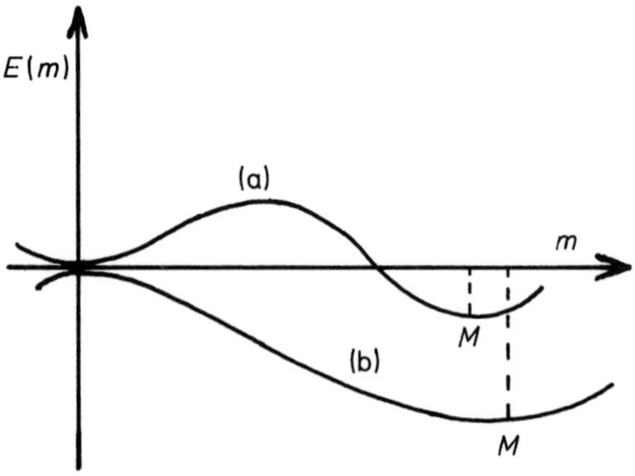

Figure 5-36. Example of two variations of the magnetic energy in the case of itinerant ferromagnetism. In (a) the true Stoner criterion [Eq. (5-70)] is satisfied whereas the simplified Stoner criterion [$U n(E_F) \geqq 1$] is not satisfied. Indeed $E(m)$ has a minimum at the origin but the minimum in M for a finite value of m is of lower energy. In (b) both criteria are satisfied.

$\bar{n}\,U \geqq 1$ where $\bar{n}$ is the mean value of the density of states between E_F^+ and E_F^-. Contrary to the approximation developed in Sec. 5.4 $[U\,n(E_F) \geqq 1]$, this criterion can predict ferromagnetism when the paramagnetic phase is metastable, as is schematized in case (a) of Fig. 5-36. For a given number n of d electrons per atom, m has a maximum value equal to $n\,\mu_B$ (Fig. 5-37 a) for less than half-filled bands, or to $(10-n)\,\mu_B$ (Fig. 5-37 b) for more than half-filled bands. As a result, two situations may arise:

(i) $\Delta E\,(m)$ has a minimum for a value of m smaller than this maximum value (Fig. 5-37 c). The condition of Eq. (5-70) is then fulfilled with the equality sign. Both halves of the d band with opposite spin directions are partially filled leading to the weak ferromagnetism presented in Sec. 5.4.

(ii) $\Delta E\,(m)$ has no minimum for m smaller than its maximum value. This maximum $m = n\,\mu_B$ or $(10-n)\,\mu_B$ is then the equilibrium value. The condition of Eq. (5-70) then applies with the inequality sign. This is the case of the strong ferromagnetism presented in Sec. 5.4.

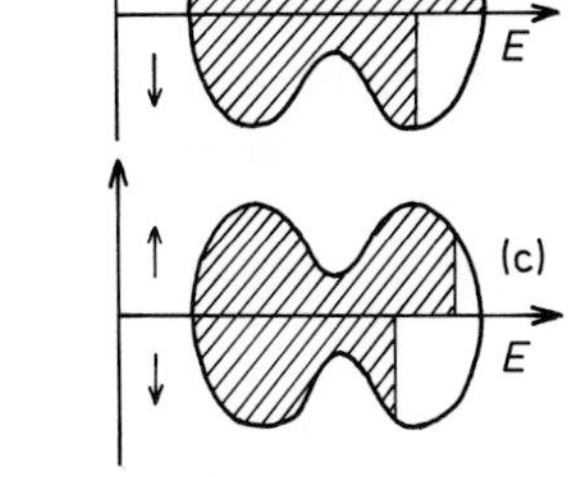

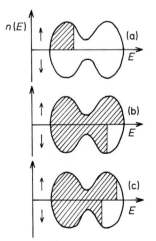

Figure 5-37. Ferromagnetism of the d band: (a) strong ferromagnetism, (b) strong ferromagnetism (Co and Ni), (c) weak ferromagnetism (Fe).

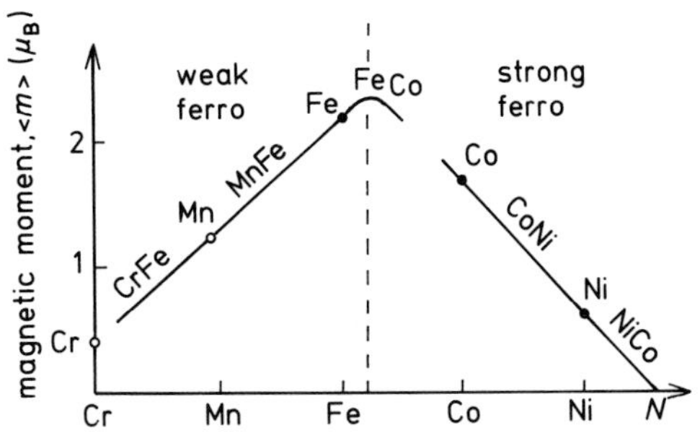

Figure 5-38. Pauling-Slater curve.

For a nearly filled or nearly empty band, where usually $\mathrm{d}n(E)/\mathrm{d}E \neq 0$ and $\mathrm{d}^2 n(E)/\mathrm{d}E^2 \leq 0$, an increase of U leads progressively from a paramagnetic state, through a weak, to a strong ferromagnetic one. In such an evolution, logically the magnetic moment is smaller in the case of weak ferromagnetism than in the case of strong ferromagnetism. Besides, for a given value of U, upon increasing the number of carriers (electrons or holes), there might come a moment when strong ferromagnetism is no longer possible and is replaced by weak ferromagnetism. As seen in Sec. 5.4, in such a situation the magnetic moment in the weak ferromagnetic state can be larger than in the strong ferromagnetic state. In Fig. 5-38 we show the so-called "Pauling-Slater" curve which gives the experimental average magnetic moment m of 3d metals and alloys versus their average atomic number Z, i.e. the number of d electrons. This curve gives an order of magnitude of the maximum possible moment of 3d elements in the metallic state. The peak, near the Fe atomic number, separates the weak from the strong ferromagnetism and reaches $2.5\,\mu_B$. Such a curve can be understood if one considers that in these metals the d half-bands are fairly well divided into two sub-bands as shown in Fig. 5-31 and sketched in Fig. 5-37. Strong ferromagnetism occurs when the number of holes increases from zero to 2.5, corresponding to the situation in which the upper sub-band of down spin is empty. Weak ferromagnetism starts for a larger number of holes and magnetic moment decreases steadily as this number increases. Indeed the two sub-bands are so well separated that, due to the condition of Eq. (5-70) in which one considers that U is small enough compared to the width w of the d band, the upper sub-band with spin up empties completely before the lower sub-band with spin down begins to empty. Such a limitation to the ratio of U/w is consistent with the values of w (≈ 7 eV) and U (≈ 1 eV) of the estimates discussed above. If U were much larger, the maximum moment could be $5\,\mu_B$ whatever the shape of the band.

One can notice that, although pure Mn and Cr are not ferromagnetic, their magnetic moments fall on the Pauling-Slater curve. Note also that the emptying of the lower d sub-band does not lead to any magnetism in the early 3d elements. This could be due to the size and overlap increase of the 3d shells, which leads to a broadening of the d band and then lowers $n(E)$. As well, the stronger mixing with the s band should reduce U. Both effects work against the Stoner criterion Eq. (5-70) for ferromagnetism.

5.5.6 Magnetocrystalline Anisotropy

As quoted previously (Sec. 5.3.3.2), magnetocrystalline anisotropy depends on the one hand on the orbital character of the states under consideration and on the other hand on the spin orbit coupling. Contrary to the case of the 4f electrons of rare earths, in 3d ions the CEF interaction (≈ 1 eV) is much larger than the spin orbit coupling ($\approx 10^{-2}$ eV). As shown above, only the states at the ends of the band can have an orbital character due to CEF effects through the α integral and/or due to the β integrals which causes the dispersion. Fe, Ni and Co belong to this category. The spin orbit coupling is then studied as a perturbation and leads to energy shifts:

$$\Delta E_{k,\sigma} = -\lambda \langle k\sigma | \boldsymbol{l}\cdot\boldsymbol{s} | k\sigma\rangle + \\ + \lambda^2 \sum_{k',\sigma'} \frac{|\langle k\sigma|\boldsymbol{l}\cdot\boldsymbol{s}|k'\sigma'\rangle|^2}{E_k - E_{k'}} \qquad (5\text{-}71)$$

where σ and σ' refer to the two spin directions and k is the wave vector of the Bloch function describing the orbital states. The $|k\rangle$ states are indeed the states $|\Psi\rangle$ of Eq. (5-71). The total correction due to the spin-orbit coupling is then:

$$\Delta E_{\text{total}} = \sum_{\substack{k \text{ occupied} \\ \text{or inoccupied}}} (\Delta E_{k\sigma} + \Delta E_{k\sigma'}) \qquad (5\text{-}72)$$

A phenomenological approach leads also [see Eq. (5-87)], in cubic symmetry, to:

$$\Delta E_{\text{total}} = K_1 (\alpha_1^2\alpha_2^2 + \alpha_2^2\alpha_3^2 + \alpha_3^2\alpha_1^2) + \\ + K_2\,\alpha_1^2\alpha_2^2\alpha_3^2 + \ldots \qquad (5\text{-}73)$$

Starting from Eqs. (5-73) and (5-71) and (5-72) it is possible to show that $K_n \propto \lambda(\lambda/w)^{2n+1}$ (the CEF effects on the orbital moment entering in the proportionality coefficient). In 3d metals $\lambda \approx 0.07$ eV and $w \approx 7$ eV. Hence the values of the K_n anisotropy constants decrease rapidly as n increases. In cubic symmetry the magnetic anisotropy energy is then mainly determined by the K_1 coefficient; as an order of magnitude one then expects $|K_1| \approx 10^3$ J/m^3, a value consistent with those measured in Fe and Ni metal have a cubic structure (see Table 5-5).

In non-cubic metals the second order term in the α_i's ($K_0\,\alpha_3^2$ in hexagonal symmetry for instance) contributes to the anisotropy. This term should be on the order of $\lambda^2/w \approx 10^7$ J/m^3, leading to an anisotropy larger than in cubic metals. This

Table 5-5. Main characteristics of magnetism in Fe, Co and Ni metals, as well as in YCo_5.

	Structure	T_C (K)	$M_S(\mu_B)$ at 4.2 K	$H_a(A \cdot m^{-2})$ at 300 K	$E_a(J/m^3)$ at 300 K
Fe	cubic	1043	2.22	3.1×10^4	0.13×10^5
Co	hexagonal	1393	1.72	7.9×10^5	7×10^5
Ni	cubic	631	0.60	1.6×10^4	0.02×10^5
YCo_5	hexagonal	980	1.65	11.1×10^6	60×10^5

value is consistent with those measured in Co metal and the compound YCo_5, both having hexagonal crystal symmetry (see Table 5-5).

Actually, the values of the K_n coefficients through the $\langle k\sigma|\boldsymbol{l}\cdot\boldsymbol{s}|k'\sigma'\rangle$ matrix elements strongly depend on the orbital character of the considered states. As shown in the appendix, this strongly enhances the anisotropic character of non-cubic symmetry compared to cubic.

Therefore the conditions which favor magnetic anisotropy are:

(i) A 3d band nearly empty or nearly filled up, favoring a maximum of states with definite orbital character.

(ii) These states must have an anisotropic orbital character as large as possible. Then in cubic symmetry the e_g states have no orbital character. On the contrary, in hexagonal symmetry these states are strongly anisotropic (see Appendix 2).

(iii) A uniaxial symmetry (especially hexagonal or tetragonal) rather than a cubic symmetry. Indeed the largest K_n anisotropy constant, entering in the expression of the anisotropy energy, is larger in the former case (K_0) than in the latter (K_1).

It is worth giving two examples of anisotropy which can be understood, at least qualitatively, thanks to the β integrals.

Let us first consider the example quoted above of a hexagonal intermetallic compound in which the distance between 3d atoms is much smaller along $\boldsymbol{c}$ than distances in the layers perpendicular to $\boldsymbol{c}$. The density of state is that schematized in Fig. 5-33 (Weinert et al., 1983 and Ballou and Yamada, 1990). If E_F is near the upper part of the band, carriers (holes in that case) have a strongly $|2,0\rangle$, $|2,1\rangle$ and $|2,-1\rangle$ orbital character, leading to an orbital moment perpendicular to $\boldsymbol{c}$ ($\langle l_x\rangle = \langle l_y\rangle$ larger than $\langle l_z\rangle$) and, through spin orbit coupling, to a spin moment also perpendicular to $\boldsymbol{c}$. Then anisotropy favors the basal plane. The compound $La_2Co_{1.7}$ is an excellent illustration of such a situation (Ballou and Yamada, 1990).

The second example concerns a hexagonal intermetallic compound in which the situation is opposite to the previous one: the distances between 3d atoms in the layers are shorter than along $\boldsymbol{c}$. The density of state is that schematized in Fig. 5-39. If E_F is near the top of the band, carriers have a strongly $|2,\pm 2\rangle$ orbital character which leads to an orbital moment and hence a spin moment along $\boldsymbol{c}$. The anisotropy then

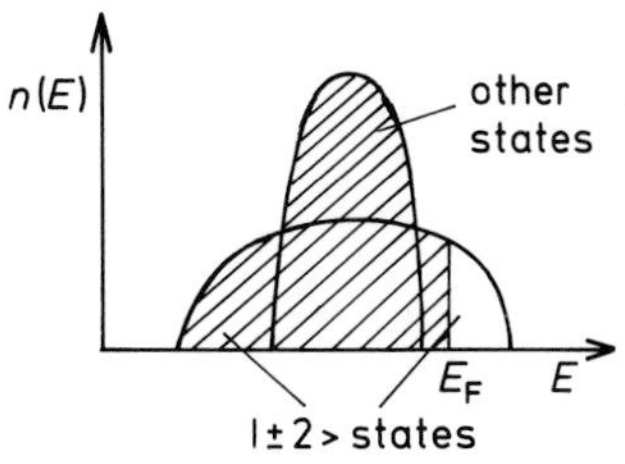

Figure 5-39. Sketch of the density of 3d states in YCo_5.

favors the *c* axis. This situation is observed in YCo_5 (Ballou and Lemaire, 1988) and leads to an anisotropy one order of magnitude larger than in pure cobalt, as is shown in Table 5-5. Note that in the other RCo_5 (R = rare earth) compounds in which R is magnetic, the Co anisotropy is close to that of YCo_5.

5.5.7 Thermal Effects – Ordering Temperature

The Stoner model gives a satisfactory description of the magnetic properties of 3d metallic magnetism at $T = 0$ K. As shown in Sec. 5.3, it also accounts for thermal effects in very weak itinerant ferromagnets. However this model fails to account for thermal effects in systems with rather large moments, especially Fe, Co and Ni metals. Indeed, in this model magnetic ordering is associated with the existence of a magnetic moment. T_C is then expected to be on the order of U, i.e. on the order of 10^4 K, a value which is one order of magnitude larger than the experimental ones. The Stoner model only considers individual excitations, i.e. longitudinal fluctuations of magnetic moments as schematized in Fig. 5-40b, which leads to a decrease of the magnetic moment amplitude. Actually, collective excitations, especially transverse fluctuations (i.e., fluctuations associated with a rotation of the atomic moments as schematized in Fig. 5-40c) play a preponderant role on T_C. There are different ways to take into account these collective excitations, or "spin waves", in the Stoner model and calculations then lead to much more realistic T_C values. Note that the Curie temperatures of Fe, Co and Ni metals, although smaller than those predicted in the crude Stoner model, are the highest known among ferromagnetic materials. This situation is quite different from that of rare earth-based intermetallics where the indirect exchange interactions lead to ordering temperatures below room temperature.

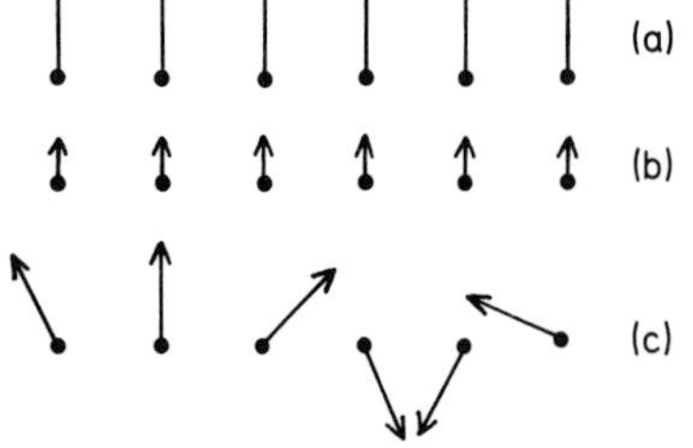

Figure 5-40. Crude picture of spin fluctuations. (a) spin arrangement at $T = 0$; (b) longitudinal fluctuations at $T \neq 0$; (c) transverse fluctuations at $T \neq 0$.

Articles on this subsection: Murata et al. (1972), Moriya and Kawabata (1973), Moriya and Takahashi (1978), Moriya (1979), Lonzarich (1984), Lonzarich and Taillefer (1985).

Book: Moriya (1987).

5.5.8 Summary

In conclusion, the above description of d metallic magnetism:

(i) explains why only the metals Cr, Mn, Fe, Co and Ni of the d series present a well established magnetism;

(ii) explains why Cr and Mn metals are antiferromagnetic, whereas Fe, Co and Ni metals are ferromagnetic;

(iii) gives an account of the values of the magnetic moments in these metals and their alloys;

(iv) explains why magnetocrystalline anisotropy is very small in cubic symmetry whereas it can reach rather large values in non-cubic symmetry such as in hexagonal symmetry;

(v) finally explains why the exchange interactions and accordingly the ordering temperatures are one order of magnitude larger than in rare earths.

Book on this section: Herring (1966).

Appendix 2: CEF Effects on One 3d Electron ($l=2$, $s=1/2$) in the Absence of Spin Orbit Coupling

Contrary to the case of 4f electrons in rare earths, the CEF interaction in 3d ions (≈ 1 eV) is much larger than the spin-orbit coupling (≈ 10 meV). It is then normal, in the perturbation acting on the electronic state that CEF is considered before applying spin-orbit coupling. Here we consider two rather characteristic situations, namely the hexagonal and cubic symmetries.

Hexagonal Symmetry. $H_{\mathrm{CEF}} = B_2^0 O_2^0 + B_4^0 O_4^0$, where the Stevens equivalent operators are expressed as a function of the orbital kinetic moment operator $\boldsymbol{l}$: $O_2^0 = 3\,\boldsymbol{l}_z^2 - l(l+1)$. Note that the sixth order terms do not appear in H_{CEF} with 3d electrons. The matrix elements of the CEF perturbing Hamiltonian are:

$$\begin{array}{cccc} & |\pm 2\rangle & |\pm 1\rangle & |0\rangle \\ \langle +2| & 6B_2^0+12B_4^0 & 0 & 0 \\ \langle \pm 1| & 0 & -3B_2^0-48B_4^0 & 0 \\ \langle 0| & 0 & 0 & -6B_2^0+72B_4^0 \end{array}$$

Using for example $B_2^0=-1$ and $B_4^0=-0.1$, one obtains the splitting of Fig. 5-41 a. Such a splitting leads to a strong orbital contribution along Oz ($\langle l_z\rangle = \pm 2$) and a strong uniaxial anisotropy ($\langle l_x\rangle = \langle l_y\rangle = 0$).

Cubic Symmetry. $H_{\mathrm{CEF}} = B_4(O_4^0 + 5\,O_4^4)$. In this case one obtains the splitting shown in Fig. 5-41 b or the reverse splitting depending on the sign of B_4. Each energy level is associated with the Γ_3 and Γ_5 representations of the cubic symmetry group. In the situation shown in the figure, where Γ_3 is the ground state, there is no orbital moment and hence no anisotropy to the first order. One says that the orbital moment is

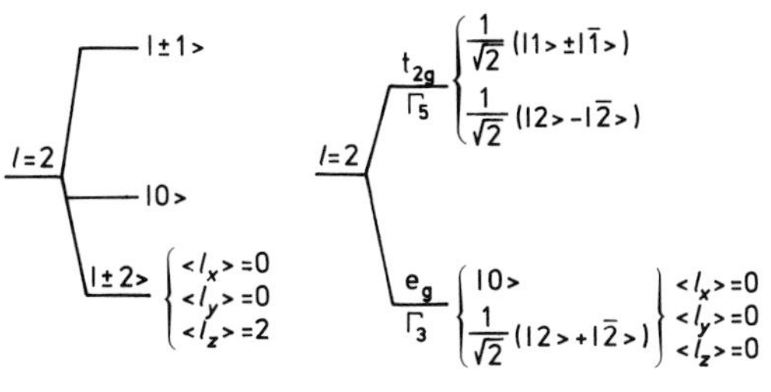

a) hexagonal b) cubic

Figure 5-41. Examples of CEF splitting of a d state ($l=2$) in hexagonal (a) and cubic (b) symmetries.

quenched. In the opposite situation, where Γ_5 is the ground state, $\langle \boldsymbol{l}\rangle$ is strongly reduced.

As a conclusion, in cubic symmetry CEF effects lead to a very weak anisotropy, whereas in hexagonal symmetry the anisotropy can be important.

5.6 Localized 4f Magnetism in Alloys

5.6.1 Introduction

As seen previously, the fourteen elements of the rare earth (R) group are characterized by the filling of the 4f shell. In most materials rare earths appear as tripositive ions. Two other elements are normally considered as rare earths because of their similar chemical properties: La and Y. The 4f electrons are tightly bound and reside inside the outer occupied shells of the atom. Therefore, their wave functions are only weakly modified when going from the free ion to the solid state. Two main interactions drive magnetic properties of rare earth based intermetallics, namely the crystalline electric field and the exchange interactions.

5.6.2 Indirect Exchange Interaction

In metallic alloys, due to the strongly localized character of the 4f shell, interactions occur mainly through conduction electrons: a spin $\boldsymbol{S}_i$, localized on atom i, interacts with conduction electrons of spin $\boldsymbol{s}(\boldsymbol{r})$ and leads to a spin polarization. Then, this polarization interacts with another spin $\boldsymbol{S}_j$, localized on atom j, and therefore creates an indirect interaction between the spins $\boldsymbol{S}_i$ and $\boldsymbol{S}_j$. It is the so-called RKKY interaction (Ruderman and Kittel, 1954; Kasuya, 1956; Yosida, 1957). The Hamiltonian, corresponding to the interaction between the spin of a conduction electron and that of a 4f localized one, is then:

$$H = -\sum_i \Gamma(\boldsymbol{r} - \boldsymbol{R}_i)\,\boldsymbol{s}(\boldsymbol{r})\,\boldsymbol{S}_i \tag{5-74}$$

It is generally assumed that this interaction is punctual so that $\Gamma(\boldsymbol{r}) = \Gamma\,\delta(\boldsymbol{r})$. Let us also assume that conduction electrons are free; this means that they can be described by a plane wave function $|l,\chi\rangle = \exp(\mathrm{i}\,k\,r)\,\chi_s$. The Fermi surface is then spherical and the density of state $n(E)$ is parabolic. For sake of simplicity one considers this Hamiltonian in the case of a collinear ($\parallel \mathrm{O}z$) configuration of 4f spins S_i^z. Taking into account this perturbation to the 1st and 2nd orders, calculations lead to a density of conduction electrons with up and down spins given by:

$$\varrho_\pm(\boldsymbol{r}) = n_\mp^- \frac{(2n)^2}{E_\mathrm{F}}\,\pi\,\Gamma \sum_i F(2\,k_\mathrm{F}|\boldsymbol{r} - \boldsymbol{R}_i|)\,S_i^z \tag{5-75}$$

where n is the average density of conduction electrons, $\boldsymbol{R}_i$ is the position of atom i, k_F is the wave vector at the Fermi level and $F(x) = (x\cos x - \sin x)/x^4$. This function

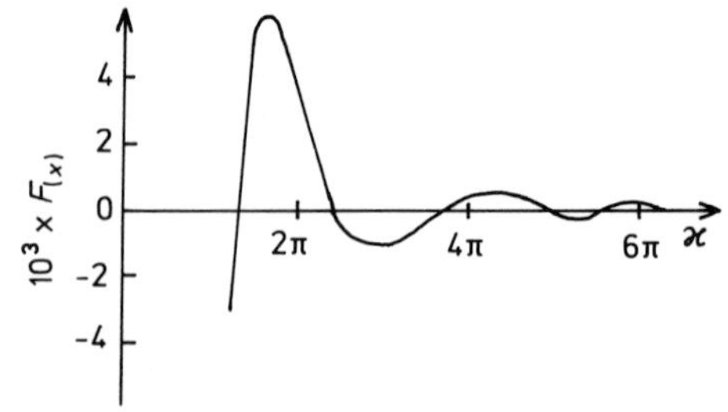

Figure 5-42. The oscillatory function $F(x)$ showing the variation of the RKKY interaction with distance in the free electron approximation.

(Fig. 5-42) is oscillatory and its amplitude decreases with distance. For 4f ions $\Gamma > 0$. Consequently the conduction electron polarization $(\varrho_+ - \varrho_-)$ in the vicinity of an atom is parallel to its spin S_i^z. The conduction electron polarization produced by the spin S_i^z interacts with the neighboring spins and accordingly creates an indirect interaction between rare earths. One can then write the interaction energy between spins as:

$$H_\mathrm{ex} = \frac{(3n)^2}{2E_\mathrm{F}}\,\pi\,\Gamma^2 \sum_{\substack{i,j\\ i\neq j}} \boldsymbol{S}_i\,\boldsymbol{S}_j\,F(2k_\mathrm{F}|\boldsymbol{R}_i - \boldsymbol{R}_j|) \tag{5-76}$$

Replacing $\boldsymbol{S}$ by its projection on $\boldsymbol{J}[\boldsymbol{S} = (g_J - 1)\,\boldsymbol{J}]$ one gets:

$$H_\mathrm{ex} = -\sum_{i,j\neq i} J_{ij}\,\boldsymbol{J}_i\,\boldsymbol{J}_j$$

where

$$J_{ij} = -\frac{(3\,n)^2}{2E_\mathrm{F}}\,\pi\,\Gamma^2 (g_J - 1)^2\,F(2k_\mathrm{F}|\boldsymbol{R}_i - \boldsymbol{R}_j|) \tag{5-77}$$

In rare earth based intermetallics, this interaction generally leads to magnetic ordering temperatures smaller than room temperature.

The oscillatory character of $F(x)$ can give rise to positive and negative interactions between the atoms according to their distance and to the filling of the conduction band (through k_F). One then expects a wide range of magnetic arrangements in rare earth intermetallics.

Textbook on this subsection: Coqblin (1977), pp. 12–22.

5.6.3 Crystal Field (CEF) Effects

5.6.3.1 CEF Effects on Paramagnetic Susceptibility

Systems Without Exchange Interaction

In this section we essentially deal with rare earth ions for which the orbital momentum $\boldsymbol{L}$ is different from zero. As seen previously, CEF effects on the free ion lead to a splitting of each multiplet. Within a multiplet (generally only the ground state multiplet is considered), the multiplicity of each CEF energy level E_n^0 is totally removed by the magnetic field (applied plus molecular field) and the new energy levels $E_{n,l}(H)$ can be expanded up to the second order of perturbation:

$$E_{n,l}(H) = E_n^0 + W_1^{n,l} H + W_2^{n,l} H^2 +$$

n is associated with the energy states when $H=0$, and l is used to distinguish the levels which have the same energy when $H=0$. Perturbation theory yields:

$$W_1^{n,l} = -\mu_0\, g_J\, \mu_B \langle n,l|J_z|n,l\rangle \quad \text{and}$$

$$W_2^{n,l} = -\mu_0^2\, g_J^2\, \mu_B^2 \sum_{l',n'\neq n} \frac{|\langle n',l'|J_z|n,l\rangle|^2}{E_{n'}^0 - E_n^0}$$

Within the Boltzmann statistics, the susceptibility of N atoms having an average value $\langle m\rangle_T$ of magnetic moment along the field at a given temperature T, can be written as: $\chi = \mu_0 N\, \mathrm{d}\langle m\rangle_T/\mathrm{d}H$ with:

$$\langle m\rangle_T = \sim \frac{\delta E}{\delta H} =$$

$$= -\frac{\sum_{n,l} (\delta E_{n,l}/\delta H) \exp(-E_{n,l}/kT)}{\mu_0 \sum_{n,l} \exp(-E_{n,l}/kT)},$$

which leads to:

$$\chi = -\frac{N}{\mu_0 Z}\left[\frac{1}{kT}\sum_{n,l} (W_1^{n,l})^2 \exp(-E_n^0/kT) - \right.$$
$$\left. -2\sum_{n,l} W_2^{n,l} \exp(-E_n^0/kT)\right] \qquad (5\text{-}78)$$

In this expression the first term is called the Curie term, whereas the second one is the Van Vleck term. Note that in the absence of CEF, only the Curie term is present and gives rise to the Curie law mentioned in Sec. 6.2. This formula leads to the following results:

– As expected from symmetry considerations, (i) χ is isotropic for cubic compounds ($\langle n,l|J_x|n',l'\rangle = \langle n,l|J_y|n',l'\rangle = \langle n,l|J_z|n',l'\rangle$), (ii) $\chi_{\parallel c} \neq \chi_{\perp c}$ in hexagonal and tetragonal systems and (iii) $\chi_x \neq \chi_y \neq \chi_z$ for orthorhombic systems.

– At high temperature the reciprocal susceptibility along each axis tends to be linear with the same slope giving rise to the same effective moment as the one deduced in Sec. 6.2.

– For a non-cubic compound, as shown in Fig. 5-43, the shifts between the high temperature susceptibilities is proportional to the second order CEF parameters.

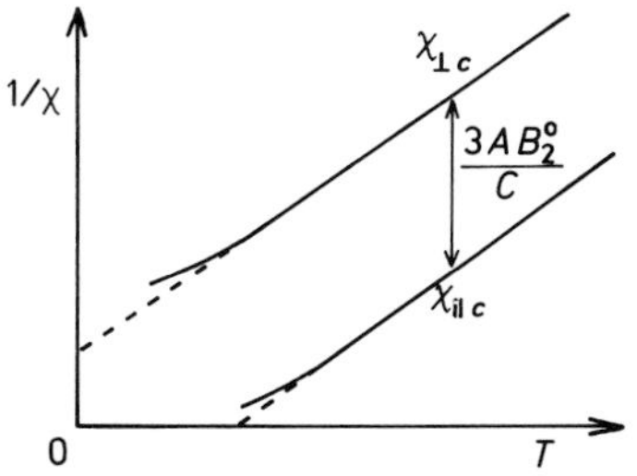

Figure 5-43. Schematic representation of the high temperature reciprocal susceptibility along and perpendicular to $\boldsymbol{c}$ of a hexagonal or a tetragonal rare earth based compound. In this example, $\boldsymbol{c}$ is the easy axis. If $\boldsymbol{c}$ were the hard axis, the variation of $\chi_{\perp c}$ would be the lowest. The shift between both variations is $3AB_2^0/C$ with $C=$ Curie constant and $A = (2J-1)(2J+3)/5k$.

– At low temperature, deviations from the Curie law take place; they are characteristic of the CEF ground state as sketched in Fig. 5-44 for two cubic compounds where exchange interactions are negligible:

(i) Pr^{3+} ion has two electrons in the 4f shell. Let us assume that the CEF yields a splitting such that the ground state, in this non-Kramers ion, is a singlet and hence it is non-magnetic. It can then easily be shown that when the temperature is decreased the Curie term vanishes, and through the Van Vleck term the susceptibility (and hence the reciprocal one) tends toward a finite value, as shown in Figure 5-44.

(ii) Nd^{3+} ion has three electrons in the 4f shell. It is a Kramers ion and the multiplicity of each CEF level is even. In a Nd based compound the ground state is always magnetic. When the temperature is decreased, the Curie term is predominant and, as for the free ion, the susceptibility becomes infinite (Fig. 5-44).

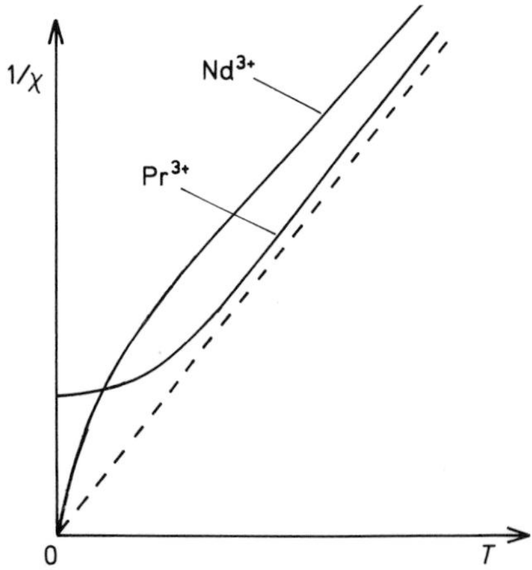

Figure 5-44. Schematic representation of CEF effects on the susceptibility for a non-magnetic singlet (Pr) and a magnetic (Nd) ground state. Dashed line would be the susceptibility without CEF. No units are given because the slope (Curie constant) is not the same for Pr and Nd.

Systems with Exchange Interactions

In the molecular field approximation, as for systems without CEF (see Sec. 5.3.2), the reciprocal susceptibility is simply: $\chi^{-1} = \chi_0^{-1} - n$, where n is the molecular field coefficient in the paramagnetic state. Especially $n = W$ and $n = W' - W$ in the case of a ferromagnet and of a simple antiferromagnet, respectively.

Then magnetic interactions yield a shift of the reciprocal susceptibility without interaction equal to the total molecular field coefficient.

5.6.3.2 CEF Effects and Low Temperature Magnetic Structure

CEF and Easy Magnetization Direction in Uniaxial Compounds

In this section we examine a situation which is often observed and which allows a more comprehensive approach of CEF.

Except for the cubic case, second order CEF terms (see Sec. 5.3.3) are present and preponderant. This is especially true for hexagonal and tetragonal symmetries. So let us consider the case where the CEF Hamiltonian is reduced to $H_C = \alpha_J A_2^0 O_2^0$. In this simple case diagonalization leads to pure $|J, M_J\rangle$ states as already discussed in Sec. 5.3. This term corresponds to the interaction between the quadrupole of the 4f shell with that of the surroundings. Let us assume that $A_2^0 < 0$. This case corresponds, for a given distance from the center of the considered ion, to a larger density of positive charges along Oz and hence favors a 4f orbital elongated along this axis as sketched in Fig. 5-45. If $\alpha_J > 0$, as is the case for Sm^{3+} ($J = 5/2$) for instance, the diagonalization of H_C shows that the energy is minimum for the $|\pm 5/2\rangle$ orbitals, which corresponds to a magnetic moment along Oz. On the contrary, if $\alpha_J < 0$, as it is the

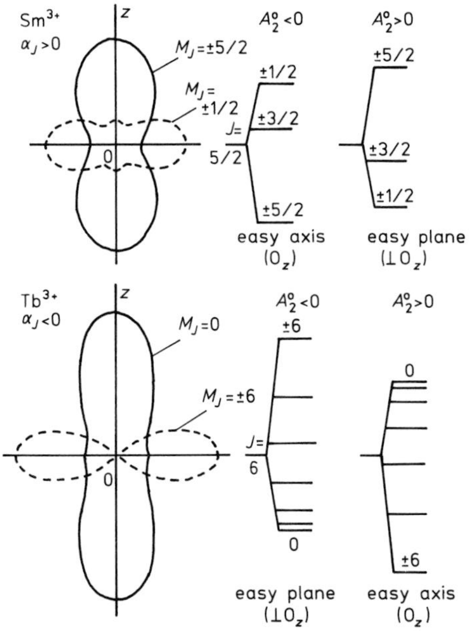

Figure 5-45. 4f orbitals and associated magnetic moment directions when only the main second order term of the CEF interaction in a uniaxial symmetric case is considered.

case for Tb^{3+} ($J=6$) for instance, it is easy to show that the energy is a minimum for the $|0\rangle$ orbital which favors the moment to lie in the plane perpendicular to Oz (actually, as discussed in Sec. 5.3.3.2, the CEF ground state in a non-magnetic singlet, but as soon as exchange interaction is present the induced moments is in this plane). Obviously, if $A_2^0 < 0$ the opposite situations occur.

These properties allow us to understand the change of easy magnetization direction of rare earths (R) through the series for compounds which often are prominent because of their technological applications (RCo_5 and $R_2Fe_{14}B$), as is shown in Table 5-6. Indeed, in the same series of compounds the potential of the surrounding, i.e. the A_l^m's, changes only weakly. The easy magnetization direction can then directly be associated with the sign of α_J (Buschow, 1988; Strnat, 1988).

Non-Collinear Magnetic Structures in Low Symmetry Compounds

In many compounds, which generally are of low symmetry, rare earths occupy low symmetry sites. As usual in such cases, second order CEF terms are preponderant and lead, for each atom, to an easy magnetization axis which is not along a symmetry axis. By applying the symmetry operations the easy axis of one part of the other atoms are different from the previous one. If the exchange interactions are smaller than the magnetocrystalline anisotropy, the magnetization direction is almost that imposed by this anisotropy, for a given site. The role of the exchange interactions is only to determine the relative orientation of moments along the easy axes. These effects lead to complex non-collinear magnetic structures as illustrated in Fig. 5-46 for the compounds DyNi and ErNi where Ni is non-magnetic. Note that due to the opposite signs of α_J for Dy and Er, moments are almost perpendicular for these two

Table 5-6. Easy magnetization directions of rare earth R moments in R metals and the RNi_5 and RCo_5 compounds. All these materials are hexagonal.

Ion	Pr	Nd	Sm	Tb	Dy	Ho	Er	Tm
Sign of α_J	−	−	+	−	−	−	+	+
R metal (hex.)				$\perp c$	$\perp c$	$\perp c$	$\parallel c$	$\parallel c$
RNi_5, RCo_5 (hex.)	$\perp c$	$\perp c$	$\parallel c$	$\perp c$	$\perp c$	$\perp c$	$\parallel c$	$\parallel c$

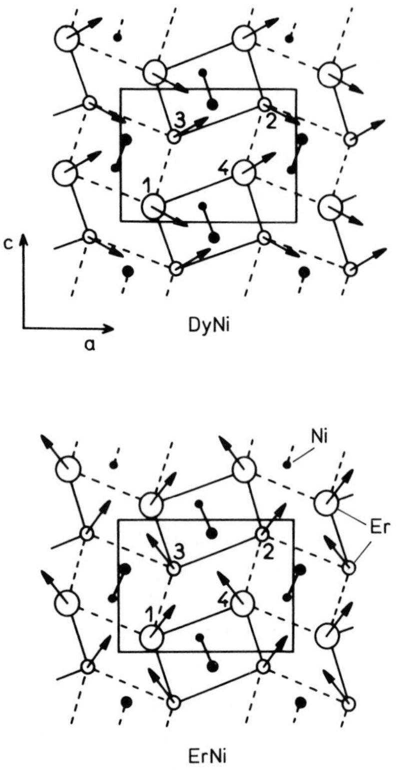

Figure 5-46. Non-collinear magnetic structures of the equiatomic compounds DyNi and ErNi where Ni is not magnetic.

ions (Barbara et al., 1973; Gignoux et al., 1972 b; Rossat-Mignod, 1986).

CEF and Sine Wave Modulated Structures at 0 K

We have seen (Sec. 5.2) that sine wave modulated magnetic structures can be stable down to 0 K when, in the absence of applied and/or exchange field, the ground state is a non-magnetic singlet. This conditions, which can be fulfilled in any compound with non-Kramers ions, is always satisfied for this type of ion in very low symmetry compounds, for which each CEF state is a non-magnetic singlet. We will see in Sec. 5.7 that in the Ce^{3+} Kramers ion a non-magnetic singlet ground state can arise through an additional mechanism.

CEF and Low Temperature Field Dependence of Magnetization

In this section we illustrate, for ferromagnetic materials of different symmetries, some CEF effects on *M* vs. *H* below T_C measured along the main symmetry axes of single crystals.

Ferromagnetic Cubic Compounds. In the compound $TbAl_2$ (Barbara et al., 1974), shown in Figure 5-47, [111] is the easy axis. Along the [100] and [110] axes, in zero internal field, the phase rule is observed. Along [110], the change of slope around 8×10^6 A m^{-1} corresponds to the anisotropy field along this axis, i.e. the field for which magnetization has become parallel to the applied field. In higher field there remains a difference between $M_{[111]}$ and $M_{[110]}$. This reflects the *anisotropy of magnetization*. The hatched area is the *anisotropy energy* for this axis. Along [100] the anisotropy field is larger than the maximum applied field. Note that the anisotropy fields are much larger than in 3d metals (see Figs. 5-17 and 5-18).

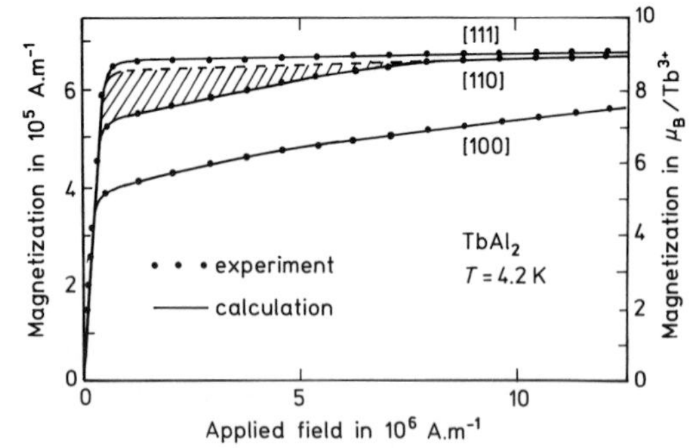

Figure 5-47. Magnetizations versus applied field at 4.2 K on a single crystal of the cubic compound $TbAl_2$ ($T_C = 105$ K). The hatched area is a measure of the anisotropy energy between the [110] and [111] axes.

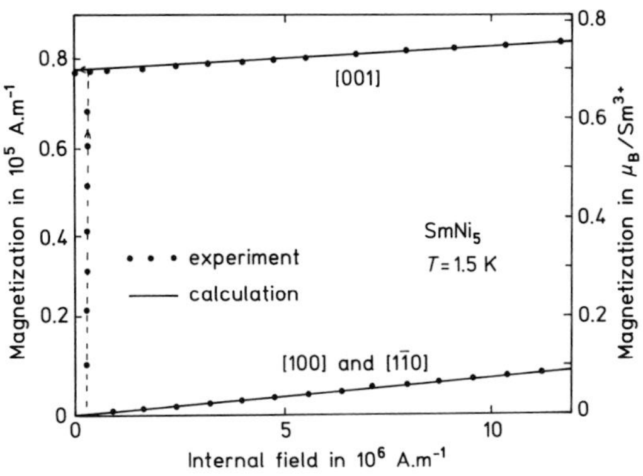

Figure 5-48. Magnetization versus internal field (applied minus demagnetizing field) at 1.5 K on a single crystal of the hexagonal $SmNi_5$ compound (T_C= 27.5 K).

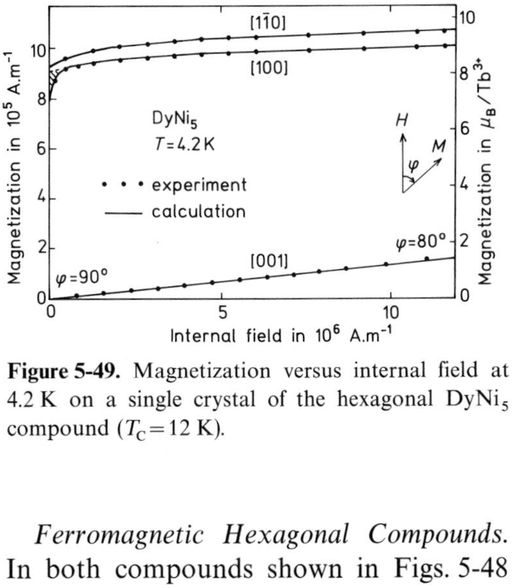

Figure 5-49. Magnetization versus internal field at 4.2 K on a single crystal of the hexagonal $DyNi_5$ compound (T_C = 12 K).

Ferromagnetic Hexagonal Compounds. In both compounds shown in Figs. 5-48 and 5-49, Ni is non-magnetic. In $SmNi_5$ (Ballou et al., 1988), ***c*** is the easy axis. The magnetization is much smaller in the basal plane than along ***c***, indicating a huge anisotropy. Note that $SmNi_5$ has the same structure as $SmCo_5$ which is one of the best permanent magnets. Due to the opposite sign of α_J in $DyNi_5$ (Aubert et al., 1981), ***c*** is the hard magnetization axis and a huge anisotropy is also observed. In the basal plane the anisotropy is weak, the [110] axis being easier than the [100] one. A field of 0.6×10^6 A m^{-1} is enough to align Dy moment along this latter axis. Note that also due to CEF, for higher fields a large anisotropy of magnetization occurs.

Quantitative Analysis: CEF and Exchange Parameter Determination

In rare earth based intermetallic compounds two experimental techniques are mainly used for the determination of CEF and exchange parameters, namely inelastic neutron scattering and magnetization measurements on single crystals. Other experiments, such as heat capacity, nuclear magnetic resonance (NMR), Mößbauer effect etc. are less direct and are commonly used to check the CEF parameters.

Inelastic Neutron Scattering (INS). This technique is used to directly observe transitions between CEF levels (Fulde and Loewenhaupt, 1985). These transitions originate from the interaction between the magnetic moment of 4f electrons and the spin of the neutrons. The energy lost or gained by neutrons corresponds to the energy of the transition. Moreover the intensity of each transition gives information about the wave functions of the states involved in the scattering process. Note that CEF states must not be widened by exchange interactions. So experiments, which must be performed in the paramagnetic domain, are made on samples with small ordering temperature. When this latter is large, experiments are performed on samples in which this temperature is lowered by replacing a large enough amount of magnetic rare earth by a non-magnetic one such as Y or La. As an example, INS spectra with the corresponding CEF splitting are reported in Fig. 5-50 for the cubic compound TmMg (Giraud et al., 1986). In cubic symmetry the CEF only depends on two parameters which are then generally

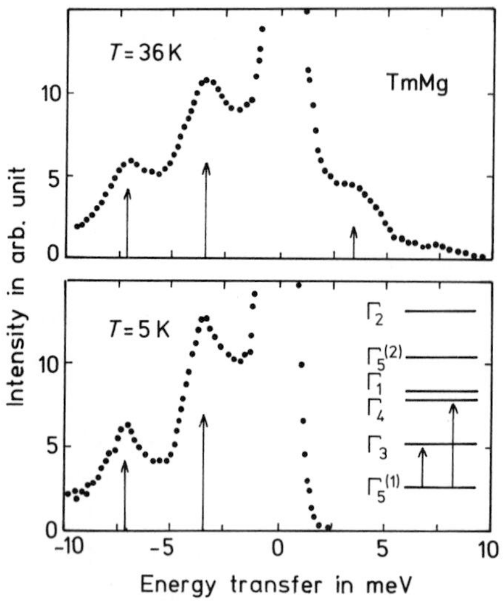

Figure 5-50. Inelastic neutron scattering spectra for the cubic TmMg compound. Arrows indicate the theoretical position and intensity of CEF transitions for the parameters deduced. The inset gives the corresponding level scheme.

determined unambiguously from INS. For lower symmetry the number of CEF parameters is larger: four in hexagonal and five in tetragonal symmetry, etc. Hence INS experiments alone are not sufficient for the determination of these parameters. Analyses of magnetization measurements have to be used as well.

Magnetization Measurements on Single Crystals. When M versus H is considered, CEF and exchange effects cannot be separated in the same way, as for the paramagnetic susceptibility. Magnetic behavior can be accounted for after diagonalization of the perturbing Hamiltonian which includes CEF and exchange interactions and after taking into account thermal effects. In the case of a collinear ferromagnetic compound the perturbing Hamiltonian is:

$$H = H_{\mathrm{CEF}} - \mu_0\, g_J\, \mu_{\mathrm{B}}\, \boldsymbol{J}(\boldsymbol{H} + n\, g_J\, \mu_{\mathrm{B}} \langle \boldsymbol{J} \rangle)$$

where the 1st term corresponds to CEF effects whereas the 2nd and 3rd terms are the energies due to applied and molecular fields, respectively. Calculation has to be performed in the three dimensional space and in a self consistent way until $\langle \boldsymbol{J} \rangle$ used in H is equal to its value resulting from the diagonalization. Finally Boltzmann statistics are used for the thermal effects.

CEF and exchange parameters, at least in ferromagnetic compounds, are determined by fitting INS data obtained on polycrystals and by fitting the temperature dependences of the paramagnetic susceptibility as well as the low temperature field dependences of the magnetization measured along the main symmetry axes of a single crystal.

Some examples of such fits are shown in Figs. 5-47, 5-48, and 5-49 for the cubic $TbAl_2$ and the hexagonal $SmNi_5$ and $DyNi_5$ compounds, respectively. The agreement is excellent, showing that the model used is well adapted to describe the magnetic properties of these compounds with good accuracy. Because of the small number of parameters, cubic compounds have mainly been studied. So far, fewer studies were devoted to hexagonal compounds and there are almost no quantitative studies on systems with lower symmetry.

General reviews on this section: Buschow (1977, 1979 and 1980), Kirchmayr and Poldy (1979).

5.7 Localized Magnetism Instabilities

This section presents very briefly several topics associated with metallic magnetism. These fields of investigation came up rather recently and are extensively studied in many laboratories. Their success arises from their fundamental interest but also from their technological applications, as is the case for amorphous intermetallics.

5.7.1 Virtual Bound State (vbs) and Magnetism of Impurities

The purpose of this section is to deal with the case of a d or f ion dissolved in a conduction electron sea. This situation is found for instance in non-magnetic host metals such as Cu or Al (Friedel, 1956; Anderson, 1961). Assuming that the d or f shell of the solute is partially filled and of lower energy than the Fermi level, in a crude picture one could expect a filling of the localized level (d or f) and accordingly a disappearance of the magnetic moment. Actually, the incompletely filled d or f shells can survive the alloying process to give the ion a net magnetic moment. To appreciate what occurs, let us assume a single atomic orbital (say a d orbital) with a one electron energy E_0 which lies within the conduction band of the host metal as shown in Fig. 5-51 a. Suppose the level is occupied by a single electron of spin up ($E_{d\uparrow} = E_0$). Only a spin down electron can be added to fill the orbital, but this will lead to supplementary energy U due to the Coulomb repulsion. The spin down state is, therefore, situated at an energy U above the spin up one ($E_{d\downarrow} = E_0 + U$) and above the Fermi level if the impurity is magnetic. This is the situation when there is no interaction between the impurity and the solute. Actually such an interaction exists and leads to an admixture or hybridization between the s states of the solute and d states of the impurity. As in all cases of interacting systems, the hybridization is important only between orbitals which overlap in space and energy. As a result there is an energy broadening of the two localized $E_{d\uparrow}$ and $E_{d\downarrow}$ states (Fig. 5-51 b). (This is characteristic of all resonant systems.) This means

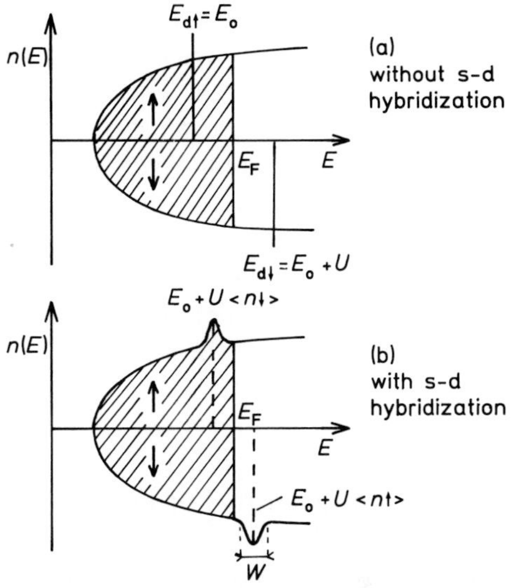

Figure 5-51. (a) Unperturbed energy levels in the absence of s-d hybridization. (b) Density of states in the magnetic case; humps at $E_0 + U\langle n_- \rangle$ and at $E_0 + U\langle n_+ \rangle$ are the virtual levels for up and down spin, respectively.

that the energies of the impurities are distributed about the values of the two humps of the up and down densities of states (Fig. 5-51 b). Note also that the gap between the centers of gravity of these two sets of energies is smaller than the gap (U) between the two unmixed localized states. This is due to the fact that, because of the broadening, a certain amount of spin down electrons at the Fermi level can be transferred to the d↑ initially occupied orbital, leading to a decrease of the associated magnetic moment ($m \propto \langle n\uparrow \rangle - \langle n\downarrow \rangle$); $\langle n\uparrow \rangle$ and $\langle n\downarrow \rangle$ being the relative amounts of electrons with up and down spin respectively, in the impurities ($\langle n\uparrow \rangle + \langle n\downarrow \rangle = 1$). Due to the Coulomb interaction this gives rise to an energy increase of the center of gravity of the lowest mixed localized states which becomes ($E_0 + \langle n\downarrow \rangle U$). For the same reason there is a decrease of the energy of the center of gravity of the upper mixed localized states which becomes ($E_0 + \langle n\uparrow \rangle U$). The decrease of the gap between both sets of energies is inasmuch large as they were initially closer to the Fermi energy. These new d states are called *virtually bound states* (vbs).

Let W be the width of the virtual bound states, which is a measure of the broadening of the localized d (or f) level by the s–d (or s–f) hybridization. Whether or not a virtual d or f state has a net magnetic moment mainly depends on the relative magnitude of U and W.

For 3d transition elements, the U and W values are such that the system is close to the onset of d magnetism. For instance 3d elements dissolved in Al give rise to non-magnetic virtual bound states, whereas magnetic states are observed for 3d elements dissolved in Cu. On the contrary, for most of the 4f elements U is large and the s–f mixing negligible, so that these elements have a well-defined net magnetic moment. However, there are three notable anomalies, namely Ce, Eu and Yb, where the 4f level E_{4f} often lies close to the Fermi level E_F in the alloys, leading to more complicated situations. Especially with Ce, depending on the band structure and on the distance $E_{4f}-E_F$, several situations can be encountered: magnetic Ce^{3+} state, Kondo effect, heavy Fermion behavior, intermediate valence state, non-magnetic state. All these cases are described below.

Book on a conceptual approach of the virtual bound state: Hurd (1975).

5.7.2 The Kondo Effect

Historically the Kondo effect was first discussed and observed in the case of diluted 3d impurities. But it is much more dramatic in Ce-based intermetallics which were studied more recently. That is the reason why we mainly discuss here the case of Ce. The situation of this element is rather simple in a sense that, in its 3^+ ionic state, it has only one 4f electron which, because of its spatial extent, can be transferred into the conduction band in metallic materials.

Before going further, a distinction in teminology can be useful: although *resonance, mixing, hybridization* and *scattering* are used for the same physical concept, one generally speaks of s–d (or s–f) mixing (or hybridization) when the vbs is well below the Fermi level, and of scattering when the hybridized electron state is involved in electronic transport.

The Kondo effect concerns the elastic scattering of itinerant electrons by an ion having a net magnetic moment and, as emphasized above, it occurs when the vbs is not far below the Fermi level, i.e. when magnetic moment is close to its instability. The scattering process can involve spin flip and non-spin flip of the itinerant electrons. For reasons which would need further explanation, the spin flip process becomes predominant when the temperature is low enough. This is the *Kondo effect* which can be described by a local negative exchange interaction between the spin $\boldsymbol{s}$ of the itinerant electron and the spin $\boldsymbol{S}$ of the ion:

$$H = -J_K \boldsymbol{s} \cdot \boldsymbol{S} \tag{5-79}$$

where the Kondo interaction J_K is negative. In a static picture, such a negative interaction gives rises to a screening of the local spin by a surrounding cloud of conduction electrons with opposite spin, which can give rise to a disappearance of magnetism at very low temperature. In other words, if $S=1/2$ (case of Ce^{3+}) this interaction removes the degeneracy of the system, giving rise to a non-magnetic singlet ground state and to an excited triplet state.

Schrieffer and Wolff (1966) worked out the following formula for the expression of J_K:

$$J_K = \frac{2|V_{kd}|^2}{E_{4f}-E_F} \tag{5-80}$$

where V_{kd} is the matrix element of the mixing between 4f (or 3d) state and conduction electrons at the Fermi level (it is the same matrix element which enters into the width of the vbs $W = \pi n(E_F) V_{kd}^2$).

The characteristic temperature below which this process becomes predominant is the *Kondo temperature*, T_K, given by:

$$T_K = \frac{1}{n(E_F)} e^{\frac{1}{J_K n(E_F)}} \quad (5\text{-}81)$$

Typically $J_K n(E_F) \approx 0.1$ and T_K is on the order of a few Kelvin. At high temperatures the influence of this process decreases and the moment recovers the characteristics of the free ion as it can be observed from the slope of the high temperature of the reciprocal susceptibility.

The first manifestation observed and also the clearest experimental signature of the Kondo effect concerns the electrical resistivity, which shows a logarithmic increase when the temperature decreases in the range near and below T_K:

$$\varrho(T) \propto J_K n(E_K) \ln(kT) \quad (5\text{-}82)$$

The other scattering processes give rise to an increase of the resistivity with temperature. Therefore the total resistivity of the substance under consideration shows a minimum (Fig. 5-52).

Some papers and general reviews on the Kondo effect: Kondo (1969); Heeger (1969); Nozières (1974); Brandt et al. (1984); Grüner et al. (1974); Hurd (1975).

5.7.3 The Kondo Lattice

As can easily be seen, the negative interaction J_K, which gives rise to the Kondo effect, is quite different from the positive exchange interaction ($-\Gamma \boldsymbol{s} \cdot \boldsymbol{S}$ with $\Gamma > 0$) discussed in Sec. 5.6.2 and which is associated with the Pauli principle which requires the antisymmetry of the total wave function through permutations of particles. Both interactions add to give rise to the indirect RKKY interaction between localized moments. Note that the sign of the RKKY interaction is independent of that of Γ or J_K because these quantities appear squared in the expression of this interaction, as shown in Eq. (5-77). In what follows we will use $J = \Gamma + J_K$. In normal rare earths J_K is very small and $J \approx \Gamma > 0$. On the contrary, with Ce, J_K can become preponderant at low temperature, so that $J \approx J_K < 0$. In Ce-based compounds there is then a competition between the RKKY interaction which favors magnetic ordering, and the Kondo-type interaction which acts in the opposite direction because it favors a non-magnetic singlet ground state. Because Ce ions are not diluted impurities but form a concentrated lattice, these compounds are called *Kondo lattices*.

$CeAl_2$: An Antiferromagnetic Kondo Lattice

The cubic compound $CeAl_2$ is the most prominent and perhaps the best example of a Kondo lattice. Indeed T_K is on the order of T_N and the resulting competition between the demagnetizing tendency of the Kondo effect and the moment-stabilizing tendency of the RKKY interaction is quite well evidenced through two aspects, namely electrical resistivity and low temperature magnetic structure.

(i) The total electrical resistivity ϱ exhibits a minimum around 16 K whereas the magnetic contribution ϱ_m to this resistivity present two logarithmic decreases when the temperature is increased (Fig. 5-52) (Buschow and Van Daal, 1969). The increase of ϱ_m with temperature between these two regions originates from CEF effects (Cornut and Coqblin, 1972).

(ii) Below $T_N = 3.8$ K, associated with a drop of the resistivity, $CeAl_2$ magnetically

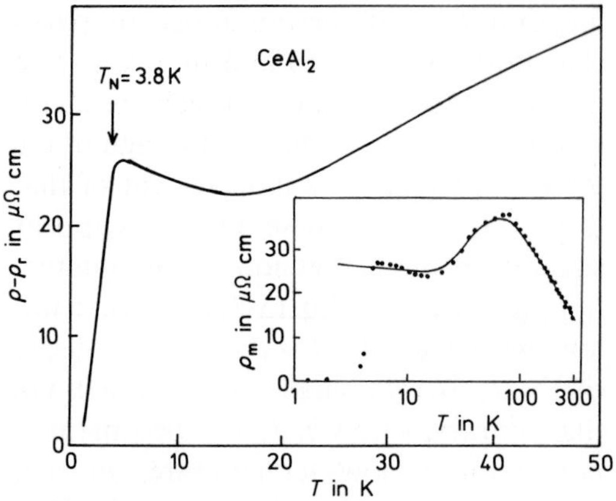

Figure 5-52. ϱ versus T in $CeAl_2$ (ϱ_r is the residual resistivity at 0 K). The inset shows ϱ_m (magnetic contribution obtained by subtracting the resistivity of the non-magnetic compound $LaAl_2$) versus $\ln(T)$.

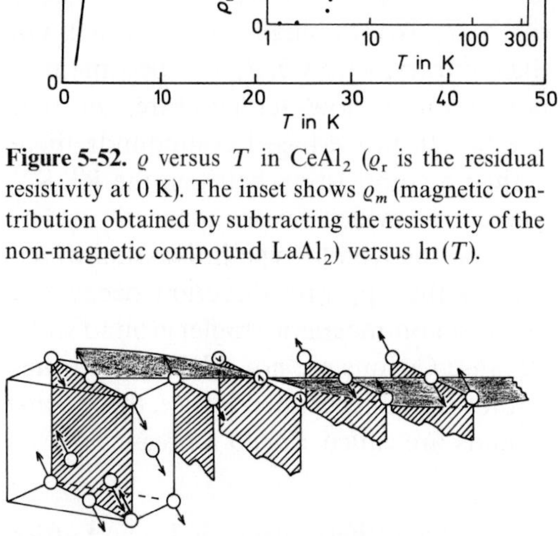

Figure 5-53. Low temperature magnetic structure of $CeAl_2$.

orders in a particularly exotic incommensurate sine wave modulated structure which is shown in Fig. 5-53 (Barbara et al., 1979). The key point is that the modulation persists down to very low temperatures (0.4 K) and hence is not thermally induced for in this latter case the structure should square up at $T = 0$ K. As previously discussed in Sec. 5.6.3.2, this means that the ground state is a non-magnetic singlet, in contradiction to a classical analysis in which CEF effects lead to a magnetic doublet ground state. This singlet ground state is then the signature of the Kondo effect.

Phase Diagram of the Kondo Lattice

Assuming that Γ is small compared to J_K, the Kondo lattice model (Doniach, 1977; Lacroix and Cyrot, 1979) predicts that the energy W_M of the magnetic state arising from the RKKY interaction and the Kondo energy W_K simultaneously increase when the product $|\varrho J| = |J\,n(E_F)|$ increases, as is shown in the upper part of Fig. 5-54 (note that the Symbol ϱ does not pertain to resistivity in the present context). Indeed $W_M \propto J^2 n(E_F)$ and $W_K = k T_K$ [see Eq. (5.7.3)]. When $|\varrho J|$ becomes larger than a critical value $|\varrho J|_C$, the system passes from a magnetic state with a magnetic moment reduced by the Kondo effect, to a non-magnetic pure Kondo state. This leads to the phase diagram of Fig. 5-54 (lower part) which shows the variations of T_K, the ordering temperature (T_N or T_C) and the magnetic Ce moment as a function of the product $|\varrho J|$. This predicted phase diagram has been observed in $Ce(Ni_xPt_{1-x})$ as seen in Fig. 5-55, but also in the systems $Ce(Si_xGe_{1-x})_2$ and $CeIn(Cu_xAg_{1-x})_2$ (Gignoux and Gomez-Sal, 1984; Lahiouel et al., 1987a and b), the

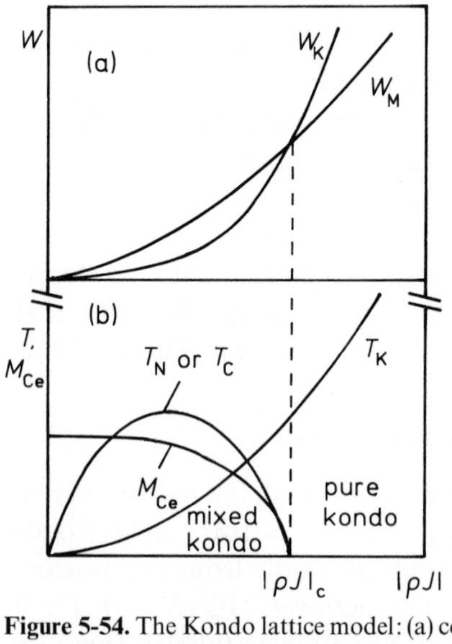

Figure 5-54. The Kondo lattice model: (a) comparison of the W_M and W_K energies. (b) Phase diagram; in the pure Kondo phase Ce is not magnetic, in the mixed Kondo phase Ce moment M_{Ce} is reduced.

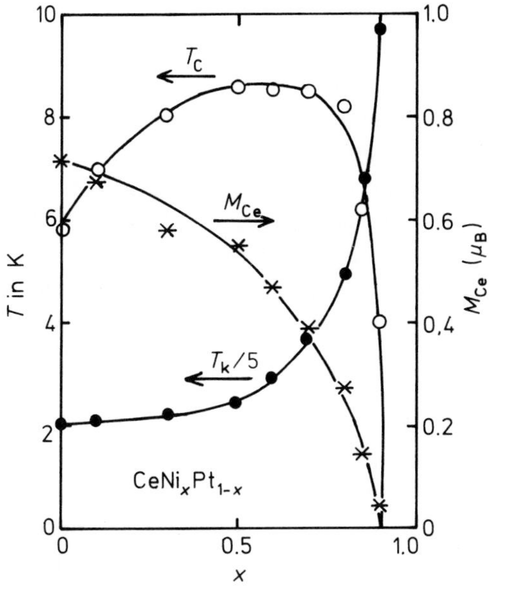

Figure 5-55. Phase diagram of pseudo-binary $CeNi_xPt_{1-x}$ compounds.

increase of the product $|\varrho J|$ being obtained either by changing x or by applying pressure. The main effect of these parameters is an increase of $|J|$ through the lattice volume decrease.

5.7.4 Other 4f Instabilities: Intermediate Valence State, Heavy Fermion Behavior

5.7.4.1 Intermediate Valence State

Valence fluctuation phenomena occur in rare earth compounds in which the proximity of the 4f level to the Fermi energy leads to instabilities of the charge configuration (valence) and hence of the magnetic moment. For cerium, E_{4f} and E_F have to be closer to each other than for the Kondo effect and $|E_{4f}-E_F|$ must be smaller than the width w of the vbs. The valence is then intermediate between 3+ and 4+. In Ce^{4+} the 4f shell is empty and accordingly non-magnetic. As for the two previous sections, these properties are illustrated here with a Ce-based compound, namely $CeSn_3$, which is one of the archetypes of intermediate valence systems. The main characteristics, which allow one to identify such a system as exhibiting an intermediate valence, are the lattice parameters (i) and the magnetic susceptibility (ii).

(i) As shown in the inset of Fig. 5-56, the cubic root of the unit cell volume in RSn_3 compounds as a function of the rare earth R^{3+} ionic radii exhibits an anomaly for Ce: the value is smaller than that corresponding to a 3+ valence state. The 4+ state being of smaller volume, this lattice characteristic is the first clue for an intermediate valence state.

(ii) The magnetic susceptibility of $CeSn_3$ (Buschow et al., 1979; Lawrence, 1979; and Béal-Monod and Lawrence, 1980) shown in Fig. 5-56 is characteristic of many valence fluctuating materials (Misawa, 1986 and 1988): it tends to follow a Curie-Weiss behavior at high temperature $[\chi = C/(T+T_{sf})]$, it exhibits a broad maximum around $T_{Max} = m_1\, T_{sf}$ and it is constant at low temperature $\chi_0 = C/m_2\, T_{sf}$. Here m_1 and m_2 are constants of order unity, and T_{sf} is the so-called *spin fluctuation temperature*. The temperature of the maximum corresponds to an energy $k\,T_{Max} \approx k\,T_{sf}$ characterizing the energy gap between the 3+ and 4+ states. At high temperature (i.e., when

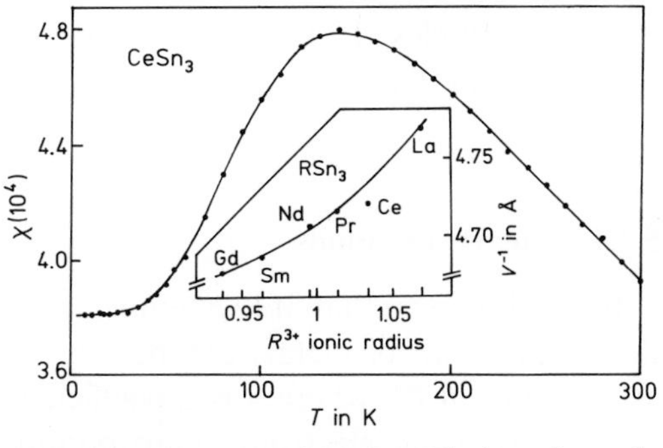

Figure 5-56. χ versus T in $CeSn_3$. The inset shows the cubic root of the volume cell versus the radius of the R^{3+} ion in the compounds RSn_3 (R = rare earth).

$T \gg T_{sf}$), due to the volume increase associated with the thermal expansion, Ce tends to become closer to the free Ce^{3+} ion state and the susceptibility becomes characteristic of this state (the Curie constant corresponds to the Ce^{3+} effective moment). On the contrary, at low temperature (i.e., when $T \ll T_{sf}$) Ce tends towards a non-magnetic state (closer to the Ce^{4+} state) leading to a constant susceptibility. Note that because of the strong charge and hence spin fluctuation character in the vicinity of T_{sf} large anomalies are also observed in the thermal dependence of many physical properties such as electrical resistivity, heat capacity, thermal expansion, elastic constants, etc. Note also that due to the strong correlation between the Ce volume and its ionic state (electron-phonon coupling), huge pressure effects are also observed (see Sec. 5.8). All these thermodynamic properties are well analyzed in the Fermi liquid theory in which charge and hence spin fluctuations are taken into account.

When the 4f energy is high enough, the 4f electron is so delocalized that Ce is no longer magnetic. Depending on the experimental technique used, the maximum valence state which can possibly be observed is either 4+ (lattice parameter analysis) or near 3.3+ (spectroscopy technique). This state is observed, for instance, in $CeRh_3$ and $CeNi_5$.

Review articles: Lawrence et al. (1981), Flouquet et al. (1982), Wohlleben et al. (1985).

5.7.4.2 Heavy Fermions

It is well known that at low temperature the specific heat of metals can be written as: $C = \gamma T + \beta T^3$, where γ is a coefficient characterizing the electronic contribution. This coefficient is proportional to the density of state at the Fermi level [$\gamma = \pi^2 k^2 n(E_F)/3$] and in the case of free electrons it reduces to $\gamma = Am$, where A is a constant and m the electron mass.

In a large number of Ce-based (and also U-based, see Sec. 5.7.5) intermetallics, γ strongly depends on temperature at low temperatures and reaches values which are two or three orders of magnitude larger than those expected for free electrons commonly observed in normal metals such as Cu or Zn (≈ 0.6 mJ mol^{-1} K^{-2}). Especially in $CeAl_3$ (Andres et al., 1975) and $CeCu_6$ (Fujita et al., 1985), γ values as large as 1620 and 1300 mJ mol^{-1} K^{-2} are observed below 1 Kelvin. Furthermore, the magnetic susceptibility of these materials continues to vary with temperature below 20 K and is some two or more orders of magnitude larger than the temperature-independent Pauli susceptibility observed in this region in an ordinary metal. This then means a large value of $n(E_F)$, in agreement with band structures in which the 4f and the Fermi energies are close together and where this large value arises from the 4f contribution to the density of states at the Fermi level. In other words, because of their huge scattering by Ce atoms, the conduction electrons are far from being free. Using the same picture as for free electrons leads one to consider an effective mass m^* of the electron much larger than the true mass m. As electrons are fermions they are called *heavy fermions*. Heavy fermions with the largest effective mass are generally found in systems at the boundary between Kondo lattice and intermediate valence state. This new class of materials represents a challenge for a better knowledge of the general many-body problem, inasmuch as some of these materials exhibit superconductivity at low temperature which seems unconventional.

Some papers and review articles on Kondo lattices and heavy fermions recommended for further reading are: Fisk et al. (1988), Fulde (1988), Steglich (1985), Stewart (1984), Varma (1985).

5.7.5 5f Magnetism

In this section devoted to localized moment instabilities it is necessary to mention the actinides, which are characterized by the filling of the 5f shell. Indeed, because 5f orbitals have a spatial extent larger than 4f orbitals, the actinide based intermetallics present magnetic properties at the boundary between localized and itinerant character. For this reason these materials exhibit a wide scope of unusual and still not well understood magnetic properties. So the number of studies devoted to this topic has increased greatly during the last few years. Among these materials, the uranium-based systems are especially interesting. In particular, some of them exhibit behaviors similar to those observed with Ce and they are among the best examples of heavy fermion systems. One of the most characteristic features of the 5f based intermetallic compounds is the following. Those compounds where the distance between nearest neighbor U atoms is relatively large exhibit magnetic properties rather similar to those of normal rare earths, in particular a large orbital contribution is observed. Oppositely, the compounds in which this distance is smaller than a critical distance, evaluated at 3.5 Å, display a Pauli paramagnetic behavior, the 5f electrons being then quite delocalized. Finally, in substances where the U–U distances are close to this critical distance, the instability of the 5f shell gives rise to the type of unusual magnetic properties which have been described above in the case of Ce compounds.

Review article on actinides based intermetallic compounds: Sechovsky et al. (1988).

Books on Section 5.6: Magnetism Vol. V, Suhl (1973); Hurd (1975); Theory of Heavy Fermions and Valence Fluctuations, Steglich (1985).

5.8 Disordered Systems

5.8.1 Spin Glasses

Spin glasses are magnetic systems in which the interactions between the magnetic moments are "in conflict" with each other due to some frozen-in structural disorder. This conflict arises when negative interactions are present and leads to *frustration*, i.e. to magnetic moment configurations where these interactions are not satisfied (Fig. 5-57). Thus, in these structurally-disordered systems, no classical long range magnetic order (of the ferro or antiferromagnetic type) can be stabilized. However, when the temperature is decreased, these systems exhibit a *freezing (or spin glass) transition* towards a state with a new kind of order (spin glass state) in which the spins point along random directions. In spite of a very large number of experimental as well as theoretical works, the essential questions about the nature of the spin glass transition and the spin glass state are still controversial. Spin glasses are still a challenge in solid state physics especially because their properties are fairly universal. Indeed, they are observed in a wide variety of different systems with competing inter-

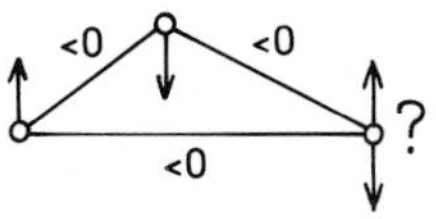

Figure 5-57. Schematic picture of the frustration of magnetic interactions. Assuming that the magnetic interactions between the three atoms are negative, if $\boldsymbol{m}_1$ and $\boldsymbol{m}_2$ are antiparallel, no direction of $\boldsymbol{m}_3$ can simultaneously satisfy the J_{13} and J_{23} interactions.

actions between spins: in crystalline metals (e.g., Fe_xAu_{1-x}) and in insulators (e.g., $Eu_xSr_{1-x}S$) as well as in amorphous (e.g., Gd_xAl_{1-x}). In metallic systems the interaction is the indirect RKKY one which has been discussed in Sec. 5.6.2. The oscillatory character of this interaction as a function of interatomic distance and the structural disorder give rise to the frustration.

There is no unique experiment but rather several characteristics used to definitely identify a substance as a spin glass. The main characteristics are as follow:

(i) The thermal dependence of the magnetization in a small applied field is that sketched in Fig. 5-58 and shows a large difference below the *freezing temperature* T_f between zero-field and non-zero-field cooling behaviors.

(ii) Below T_f the field dependence of the magnetization depends strongly on its magnetic history.

(iii) The alternate current (ac) susceptibility is characterized by a peak in low magnetic fields.

(iv) Neutron diffraction patterns do not present any magnetic Bragg peak, which means that the spin freezing is accompanied with no periodic long range order for $T \leqq T_f$.

This topic, which is still very new, would need much more space in order for all its aspects to be emphasized and to show its

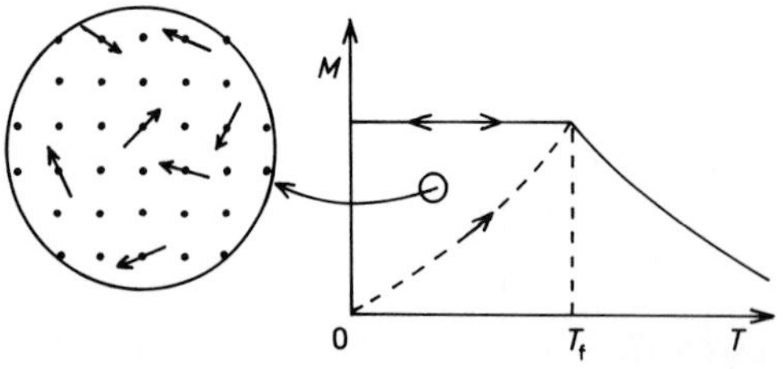

Figure 5-58. Temperature dependence of the magnetization of a spin glass measured in a small applied field. The dashed line indicates this dependence after zero field cooling.

importance in basic research. Indeed it has a large impact in statistical mechanics and can be applied to the understanding of disorder phenomena in quite different scientific areas.

Review article: Binder and Young (1986).

Text book: Maletta and Zinn (1989); Rammal and Souletie (1982).

5.8.2 Amorphous Metallic Alloys

Amorphous alloys can be divided into two main classes: (i) the metglas alloys between a 3d metal (Fe, Co, Ni) and metalloids (B, C, N, P, Si ...) these latter allowing the amorphous state to be obtained, (ii) rare earth-transition metal alloys. Note that pure metals do not seem to be stable in the amorphous state, at least at room temperature. The interest of these materials is twofold. First, they have important technological applications such as magnetic devices in the fields of electronics and energy (metglasses) and as constituents for recording devices (rare earth-transition metals). These materials are also very interesting in basic research because of the loss of periodicity and well-defined symmetry. As for the crystalline state, one can distinguish itinerant and localized magnetism.

5.8.2.1 Itinerant Electron Magnetism in Amorphous Alloys

We consider here the case of ferromagnetic materials. Structural disorder hardly affects the magnitude of the average magnetic moment. Moreover, the band concept is essentially conserved partly because of the existence of short range order. However, slight modifications, compared to the crystalline state of the same stoichiometry, arise because of the smaller density of the amorphous state which is then closer to the free ion limit. In metglasses, the result is a decrease of the Stoner factor $U\,n(E_F)$ and hence a slight reduction of the magnetic moment compared to the crystalline state.

Amorphous alloys are also of special interest for studying the onset of 3d magnetism, in particular in rare earth-transition metal alloys for which comparisons with the crystalline state can be achieved. Let us look for instance at the Y–Ni system (Gignoux et al., 1982). In the crystalline state, Y and Ni form a limited number of compounds as described in the beginning of Sec. 5.10. On the contrary, the amorphous state allows the study of alloys where the composition can be varied continuously. As discussed in Sec. 5.10 a decrease of the Ni moment with decreasing Ni concentration is expected. In amorphous Ni-rich alloys this decrease is faster than in the crystalline alloys, whereas on the contrary, in amorphous alloys, Ni moment vanishes at larger amounts of Y (see Fig. 5-65). This has been ascribed to environmental effects which are dominant in the decrease of Ni moment in the crystalline state. In amorphous alloys with a Ni content larger than 78%, although the Ni moment is not always small, the model of very weak itinerant ferromagnetism is well-adapted for the description of magnetic properties. When the Ni concentration is slightly less than this critical concentration for ferromagnetism, magnetic clusters in a strongly Pauli paramagnetic matrix are observed at low temperature. Such a property arises from the diversity of environments, the condition for the onset of ferromagnetism being locally satisfied.

5.8.2.2 Localized Magnetism in Amorphous Alloys

As for crystalline rare earth alloys and metallic spin glasses, magnetic interactions in these types of amorphous alloys are of RKKY type, i.e. long range and oscillatory with distance. However they are damped because of the highly incoherent interference between incident and scattered electronic waves of the conduction electrons. As in spin glasses these interactions lead to frustration effects. Furthermore, due to the very low symmetry of the environment of the magnetic atoms, the one ion anisotropy is uniaxial and distributed at random throughout the material. Such anisotropy effects due to the CEF are preponderent in rare earth based amorphous alloys.

Magnetic properties and magnetic structures mainly depend on the balance between exchange and anisotropy. Although much less information can be obtained experimentally for amorphous alloys than for the crystalline state, a classification of the different possible magnetic arrangements has been proposed (Coey and Readman, 1973; Coey et al., 1976) and is sketched in Fig. 5-59:

(a) Ferromagnetism is the simplest conceptual order which could exist in a non-periodic solid.

(b) Antiferromagnetism is more difficult to conceive since there is no obvious way to define two sublattices with antiparallel magnetization, except when two kinds of magnetic atoms are present. A random antiferromagnetic arrangement is called *speromagnetism:* magnetic moment directions are isotropically distributed. This structure is different from paramagnetism in the sense that moments are not fluctuating in time.

(c) In the *asperomagnetic* structure, magnetic moment directions are anisotropically distributed leading to a spontaneous magnetization. The origin of asperomagnetism is twofold: (i) It can occur in systems with small anisotropy where positive and negative interactions compete but where positive interaction are predominant. This situation, rather similar to that of spin

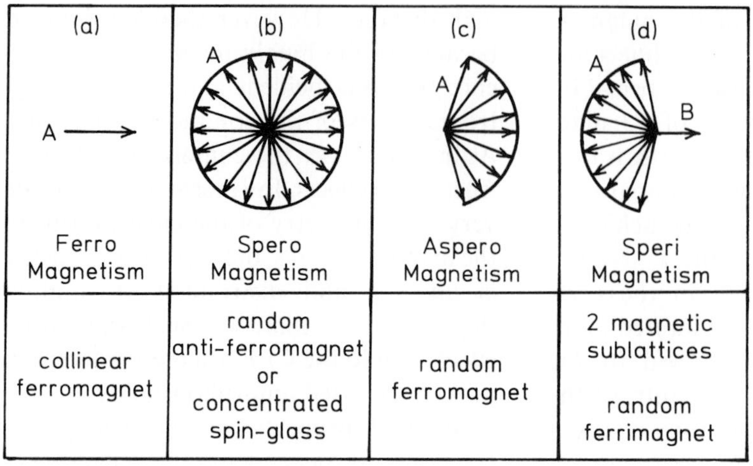

Figure 5-59. Schematic representation of the possible spin structures for an amorphous alloy: (a) ferromagnetism, (b) speromagnetism, (c) asperomagnetism, (d) sperimagnetism when A and B atoms are subjected to a strong and to a weak magnetocrystalline anisotropy, respectively.

glasses, is that of amorphous GdAg and $Y_{1-x}Fe_x$ alloys. (ii) The frustration is due to the competition between positive exchange interactions and a random anisotropy field; in that case the term *random ferromagnet* is also used. This situation is that which occurs in $Dy_{21}Ni_{79}$ where Ni is not magnetic.

(d) A great number of amorphous alloys, as rare earth-3d transition metal alloys, contain two types of magnetic atoms. Consequently one can have *sperimagnetism* which corresponds to a random ferrimagnet. Such a situation is typically encountered in amorphous alloys containing a rare earth and a 3d transition metal both carrying a magnetic moment. One of the best examples is amorphous $Dy_{23}Co_{77}$ where, as in all the rare earth-3d crystalline or amorphous alloys, the spins of rare earth and 3d atoms are antiferromagnetically coupled. The weak Co anisotropy and the dominant positive Co–Co interaction leads to the colinear arrangement of their moments. Because of the strong anisotropy, Dy moments are scattered within a hemisphere such that the resulting Dy magnetization is antiparallel to the Co moments. It is worth noting that, like in a ferrimagnet, a compensation temperature is observed at 230 K.

Text book on amorphous metallic alloys: Chappert (1982).

5.8.2.3 Random Anisotropy Systems

This topic concerns the asperomagnetic systems mentioned above. Until 1973 the theory based on the mean field approximation predicted for such magnetic structure a behavior very close to that of a crystalline ferromagnet (divergence of the initial susceptibility at T_C and appearance of a spontaneous magnetization below this temperature) (Harris et al., 1973). Apart from some minor deviations observed in low field, such behavior was confirmed by magnetic measurements. More recently, theoretical (Imry et al., 1975; Jayaprakash and Kirkpatrick, 1980) and experimental (von Molnar et al., 1982) works have actually shown that in these systems there is no long range magnetic ordering, i.e. no spontaneous magnetization. Indeed, at low temperature and in low field, neighboring magnetic moments tend to be parallel to comply with the exchange interactions but each moment is forced towards its anisotropy direction. The compromise is such that locally moments point in the same hemisphere. The spatial extent of such regions depends on the ratio D/J of the constants characterizing the random anisotropy and the exchange energy. Macroscopically one can expect no spontaneous magnetization in the absence of applied field since in some regions, the moments point toward a given hemisphere, but in other regions they point toward different directions. This field of research is still especially active and questions still remain: What is the symmetry of the low temperature phase? Does this symmetry depend on the spatial extent? How does the random anisotropy modify the nature of the phase transition with respect to a para-ferromagnetic transition? Note that this topic also concerns crystalline ferromagnetic alloys in which non-magnetic ions are diluted; the chemical disorder, if large enough, then induces a local anisotropy of random character.

Review articles on Sec. 5.8.2.3: Barbara et al. (1984, 1990); Chudnovsky (1988).

5.9 Magnetoelastic Properties

5.9.1 Introduction

All magnetic systems exhibit magnetoelastic properties. This means that the application of stress to a magnetic substance perturbs the magnetic properties and, conversely, the application of a magnetic field perturbs the elastic properties. The most obvious and also most studied manifestation of the presence of the coupling between magnetic and elastic properties is the magnetostriction, namely the lattice deformation induced by the formation of a magnetic state or by any change of this state. Forced and spontaneous magnetostrictions are used to distinguish the field induced deformation of the lattice from that observed in zero applied field, respectively. Obviously both effects vary with temperature. Another way to study magnetoelastic effects is to measure the modifications of the magnetic properties (magnetization, ordering temperature, etc.) induced by a stress (pressure or force). Magnetoelastic interactions are also often studied through dynamic elastic properties, such as the field and temperature dependences of the elastic constants.

The magnetoelastic energy depends on the magnetization (direction and amplitude), on the components of the strain and on magnetoelastic coefficients which characterize the material. The competition of this energy with the elastic energy leads to the equilibrium position of the lattice. The parameters which are used to describe the magnetostriction then depend on the elastic and magnetoelastic coefficients. As for the magnetocrystalline anisotropy, the number of these magnetostriction constants is inasmuch large as the symmetry is low. For the simplest case of a cubic symmetry, the relative change of length $\delta l/l$ in a direction $\boldsymbol{D}'$ associated with a direction $\boldsymbol{D}$ of magnetization of a ferromagnet is written in a classical description as:

$$\frac{\delta l}{l} = \frac{3}{2}\lambda_{100}(\alpha_1^2\alpha_1'^2+\alpha_2^2\alpha_2'^2+\alpha_3^2\alpha_3'^2-1/3)+ \\ +3\lambda_{111}(\alpha_1\alpha_1'\alpha_2\alpha_2'+\alpha_2\alpha_2'\alpha_3\alpha_3'+ \\ +\alpha_3\alpha_3'\alpha_1\alpha_1') \qquad (5\text{-}83)$$

where the α_i and α_i' are the cosines of the angles between $\boldsymbol{D}$ and $\boldsymbol{D}'$ and the directions [100], [010] and [001], respectively; λ_{100} and λ_{111} are the magnetostriction constants along the vertices and the main diagonals of the cube, respectively.

All these effects are present at different degrees in all types of magnetism described in the previous sections. They have been extensively studied, especially in systems with normal rare earths where models can be applied quantitatively and in the case of magnetic instabilities where the effects are often dramatically large. In this section we briefly discussed, along with some examples, the different physical origins of the magnetostriction. Moreover we restrict ourselves to the microscopic aspects. The macroscopic aspects associated with the existence of magnetic domains will be dealt with in the last section.

5.9.2 Dependence of the Exchange Energy on Interatomic Spacing

The sign and the amplitude of the exchange interactions depend on the distance between the magnetic atoms. This corresponds to the Slater-Néel curve (Fig. 5-7) for 3d elements and to the RKKY function for rare earth based intermetallics. Strains associated with these two-ion effects mainly depend on the magnetic moment amplitude rather than on its direction. In ferromagnets, when the magnetostriction constants are large enough magnetoelastic effects can give rise to ordering temperatures which are of first order (discontinuity and hysteresis of the temperature dependence of the magnetization and of the lattice parameters, etc.). These effects are especially large in compounds where the existence of negative interactions can lead to frustrated magnetic arrangements, i.e. where these interactions are not all satisfied. As one can expect from the Slater-Néel curve, this is the case in compounds with Fe and especially with Mn where negative interactions are present. Indeed, among the best examples of magnetoelastic effects found in this category it is worth to mention the compounds R_2Fe_{17} and RMn_2 (R = rare earth).

5.9.3 Crystal Field Effects

Looking at the CEF Hamiltonian as it is written in Eq. (5-41), strain effects can be introduced through the strain dependence of the A_l^m parameters that characterize the surroundings. Because these effects concern the interaction of each magnetic ion with its surroundings, one speaks of single-ion effects. The first order magnetoelastic effects depends on the derivative of these parameters with respect to the strain, leading to supplementary terms in the Hamiltonian, which couple strains with second order Stevens operators. This coupling gives rise to isotropic and anisotropic distortions. The latter has magnetic symmetry and is preponderant. As for magnetic properties, the model based on localized 4f ions and the mean field approximation is quite appropriate to give quantitative interpretations of the experimental results.

Physically, magnetoelastic effects arising from CEF are associated with the deformation of the spatial distribution of 4f electrons; the main effects being due to the quadrupolar term which is characterized by the second order Stevens operators. The largest effects are generally observed at the ordering temperature because of the onset or the modification of quadrupolar terms associated with appearance of magnetic moment. A very simple example can illustrate this (Fig. 5-60). Let us consider a ferromagnetic crystal of cubic symmetry where the effects are especially dramatic because there is no quadrupolar term in the paramagnetic state (except in very few substances where two-ion quadrupolar interactions are larger than the magnetic ones). Above T_C the magnetic ions appear to be cubic because in this symmetry the 4f orbitals associated with each CEF level correspond to cubic distributions of the 4f electrons. If the fourfold axis is the easy

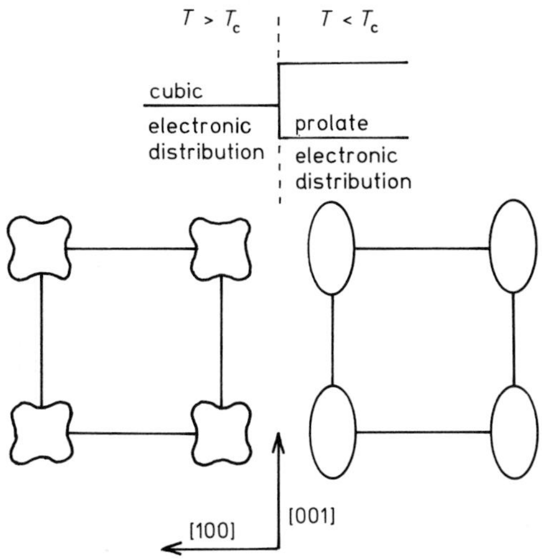

Figure 5-60. Schematic representation of the spontaneous distortion which occurs when a cubic ferromagnet is cooled below its Curie temperature.

magnetization direction, the magnetic symmetry is tetragonal below T_C. The second order CEF term introduced by this symmetry gives rise to a ground state associated with a 4f electron distribution which is no longer cubic. Assuming, for instance, a prolate shape, one observes a lattice expansion along [001] and a contraction along [100] and [010]. The contrary may also be observed, depending on the sign of the λ_{100} magnetoelastic constant.

Review article: Morin and Schmitt (1989).

5.9.4 Itinerant Electron Striction

In the itinerant electron model of magnetism, spin polarization causes a splitting of the bands and an increase in kinetic energy. (For small magnetic moments, in the Stoner model this kinetic energy increase is proportional to the inverse of the density of state at the Fermi level.) The system orders magnetically when the kinetic energy dissipated is more than balanced by the gain in exchange energy. A repulsive force then arises because the system can reduce the cost in kinetic energy due to magnetic ordering by undergoing a lattice expansion (Janak and Williams, 1976). Indeed this lattice expansion leads to a decrease of the bandwidth and hence to an increase of $n(E_F)$. In a more general picture, the character of the d states changes continuously from low kinetic energy (bonding) at the bottom of the band to high kinetic energy (antibonding) at the top of the band (see Sec. 5.2.2). When the energy of the majority spin band is lowered and electrons are transferred into it from the minority spin band, the result is always a transfer of electrons to higher kinetic energy (less bonding) orbitals and thus a net repulsion in order to decrease this energy (this effect is similar to what happens in a gas in which a decrease of pressure and hence of kinetic energy results from a volume increase).

Although this argument is based on ferromagnetic ordering, it also applies to cases where the ordering varies spatially, as for simple antiferromagnets or sine wave modulated structures. As far as it is possible to think in terms of a local density of states, the magnetic ordering will still lead to a repulsion, because the effect is independent of which band has its energy lowered. Therefore, in the itinerant model, magnetic ordering is accompanied by a repulsive force for all magnetic arrangements.

These effects are generally more pronounced in systems close to the onset of 3d magnetism such as those which exhibit itinerant electron metamagnetism. In particular, in the RCo_2 compounds (R = rare earth), the existence of such a behavior is responsible for the first order transition at T_C for R = Dy, Ho and Er. With the discontinuity of the magnetization at this temperature is associated a dramatic thermal expansion anomaly corresponding to a discontinuous jump of the volume cell ($dv/v \approx 4 \times 10^{-3}$) when temperature is lowered (Lee et al., 1976; Minakata et al., 1976; Inoue and Shimizu, 1982).

Review article on itinerant electron striction: Wasserman (1989).

5.9.5 4f Instabilities

As seen previously, in some rare earths the delocalization of 4f electrons leads to electronic and magnetic instabilities. As rare earth ions exhibit large changes in their ionic radii which accompany the changes in the ionization state, the effect of valence fluctuations in these compounds induces severe strain effects. Large magnetostrictions as well as softening of the elastic constants are observed in the temperature range where the charge and the concomitant spin fluctuations are a maximum. Furthermore, in some compounds pressure or temperature-induced first-order transitions are observed between two valence states, the higher valence corresponding to the smaller volume and being less magnetic (Figs. 5-61 and 5-62). This

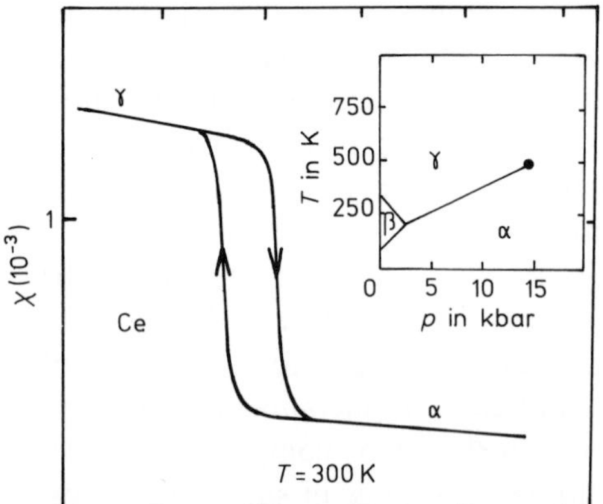

Figure 5-61. Pressure dependence of the magnetic susceptibility of pure Ce at 300 K (MacPherson et al., 1971). The inset shows the corresponding pressure-temperature phase diagram.

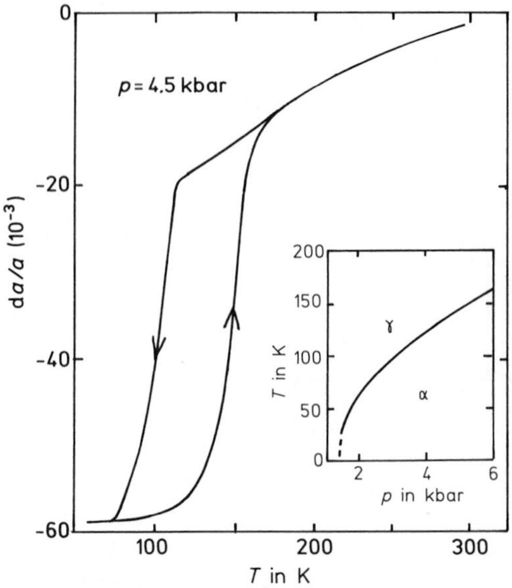

Figure 5-62. Thermal dependence of the relative thermal variation of the lattice parameter of the orthorhombic CeNi compound at 4.5 kbar. The inset shows the corresponding pressure-temperature phase diagram (Gignoux et al., 1987).

gives rise to pressure-temperature phase diagrams like that shown in the insets of Figs. 5-61 and 5-62 for pure Ce (for a review see Koskenmaki and Gschneider, 1978) and CeNi (Gignoux et al., 1987), respectively. In pure Ce the largest volume cubic γ phase corresponds to a valence state close to 3^+; the lowest volume α phase is also cubic with valence close to 4^+, i.e. without 4f magnetism, the huge relative volume change being around 15%. In addition to these static aspects, dynamic effects are also dramatically large in the vicinity of the spin (and charge) fluctuation temperature T_{sf}. Through electron-lattice (electron-phonon) coupling these electronic fluctuations can give rise to a large decrease (softening) of the elastic constants in this temperature range (Gignoux et al., 1985; Bömken et al., 1987).

Review articles on Sec. 5.9.5: Lüthi (1985); Lüthi and Yoshizawa (1987).

Review articles on Sec. 5.9: Lines (1979).

5.10 Magnetism of Alloys of Rare Earths with d-Transition Elements or s, p Elements

5.10.1 Introduction

As rare earth-transition metal alloys, one generally considers alloys which contain R and M elements. R is a rare earth (elements framed with a bold line in Table 5-7). M is one of the elements which are framed with a double line in Table 5-7. Most of these latter elements belong to the 3d, 4d or 5d series, but other elements of the right part of the periodic table such as Al, Ga, etc. are also included in this category. During the last twenty years investigations were mainly concentrated on binary systems. But in the meantime the number of studies devoted to ternary and even quaternary alloys has increased; the extra elements being frequently B, C, Si, and Ge.

These alloys form outstanding tools for the study of a wide scope of topics in magnetism such as 4f localized magnetism, 4f magnetic instabilities and 3d magnetism. Moreover, as we will see, some of these materials are of utmost importance for technical applications.

As shown in Table 5-7, R and M elements are characterized by a large electronegativity difference and by a large atomic radius difference. Therefore, as illustrated in Fig. 5-63 showing the Dy–Ni phase diagram, there exists a large number of well-defined compounds with different crystallographic structures (no solid solution). This leads to large possibilities to study the two main families of magnetic elements, namely the 3d and 4f elements. Indeed, for given R and M elements, changing the composition leads to modifications of:

(i) The band structure and hence the exchange interactions between magnetic ions

Table 5-7. Elements of the R and M based intermetallics with electronegativities and atomic radii as they are reported in the periodic table of the elements.

											Al	
											1.61	
											1.82	
3d				Cr	Mn	Fe	Co	Ni	Cu	Zn	Ga	
Electronegativity				1.66	1.55	1.83	1.88	1.91	1.90	1.65	1.81	
Atomic Radius				1.85	1.79	1.72	1.67	1.62	1.57	1.53	1.81	
4d	Y					Ru	Rh	Pd	Ag	Cd	In	Sn
	1.22					2.2	2.28	2.20	1.93	1.69	1.78	1.96
	2.28					1.89	1.83	1.79	1.75	1.71	2.00	1.72
5d	La						Ir	Pt	Au	Hg	Tl	Pb
	1.10						2.20	2.28	2.54	2.0	2.04	2.33
	2.74						1.87	1.83	1.79	1.76	2.08	2.81

4f	Ce	Pr	Nd	Pm	Sm	Eu	Gd	Tb	Dy	Ho	Er	Tm	Yb	Lu
	1.12	1.13	1.14	1.13	1.17	1.2	1.20	1.2	1.22	1.23	1.24	1.25	1.1	1.27
	2.70	2.67	2.64	2.62	2.59	2.56	2.54	2.51	2.48	2.47	2.45	2.42	2.40	2.25

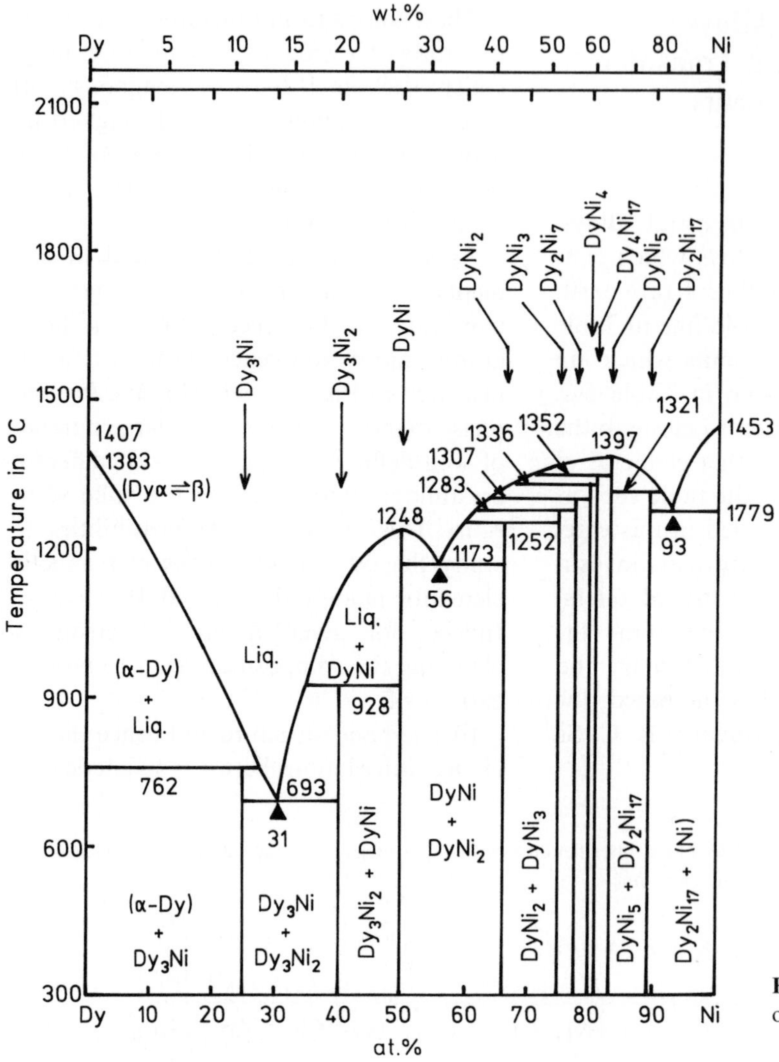

Figure 5-63. Phase diagram of the binary Dy–Ni system.

and the magnitude of the 3d magnetic moment;

(ii) the crystal structure and the corresponding anisotropy.

Furthermore, for a given composition and for one or several M elements, the same crystal structure generally exists throughout the whole series of R elements. For instance, this is the case for the hexagonal RM_5 with M = Co, Ni, Cu, Zn. Hence it is possible to study the effect of:

(i) the anisotropy (Gd with $L=0$ is isotropic whereas the other rare earths with $L \neq 0$ are anisotropic);

(ii) the change of the easy magnetization direction;

(iii) the change of the magnetic R moment.

In this section we illustrate the interest in R-M based intermetallics by means of rare earth-3d compounds.

Some comments: There also exist many intermetallics between the actinides and the same M metals as discussed above. Of the 5f series, Th, U, Np and Pu are studied almost exclusively because of the nuclear instability of the actinides. Moreover, because of their radioactivity and toxicity only few research centers are equipped to prepare and study compounds of these elements. Consequently, although the number of studies devoted to actinide based intermetallics is increasing, these materials have been much less studied than those of rare earths. Finally, they are very interesting for the fundamental understanding of magnetism, especially for its instabilities as described in Sec. 5.7.5, but so far they have not given rise to technical applications. For all these reasons we will not discuss the magnetic properties of these actinides based intermetallics in this chapter.

Review articles on actinide based intermetallics: Brodsky (1978); Sechovsky and Havela (1988).

5.10.2 Band Structure of the Rare Earth-3d Alloys

When these alloys are formed one is faced with the combination of a 3d band and a 5d (lanthanides) or 4d (Y) band, the latter two being of higher energy. A schematic representation of the densities of states before and after alloying the R and M metals is sketched in Fig. 5-64. The electronegativity difference between the constituents gives rise to a transfer of 4d (or 5d) electrons to the unfilled 3d band. Since the screenings of the nuclear potentials by these electrons are modified (decrease for 3d, increase for 4d or 5d), the two d bands draw nearer, leading to 3d–4d (or 3d–5d) hybridized states at the top of the 3d band and at the bottom of the 4d (or 5d) band (Cyrot and Lavagna, 1979; Shimizu et al., 1984). The Fermi level of the compounds often lies in this region where the density of states $n(E)$ varies strongly (Fig. 5-64). Starting from a pure 3d metal, the progressive introduction of rare earths leads to a filling of the 3d band and accordingly to an increase of the Fermi level as well as a decrease of the density of states at the Fermi level $n(E_F)$. As this process gives rise to a decrease in 3d magnetism, for a given range of concentrations the alloys are close to the conditions for the Stoner criterion $[U\,n(E_F) > 1]$ to be satisfied and magnetic instabilities can be observed, each behavior strongly depending on the fine structure of $n(E)$ near E_F. This is particularly true in the cases of Co and Ni as illustrated in Fig. 5-65 which shows the variations of the 3d moment as a function of the R amount in the Y–Co, Gd–Co and Y–Ni compounds. The critical concentration for the onset of 3d magnetism corresponds to the RCo_2 and RNi_5 compounds for cobalt and

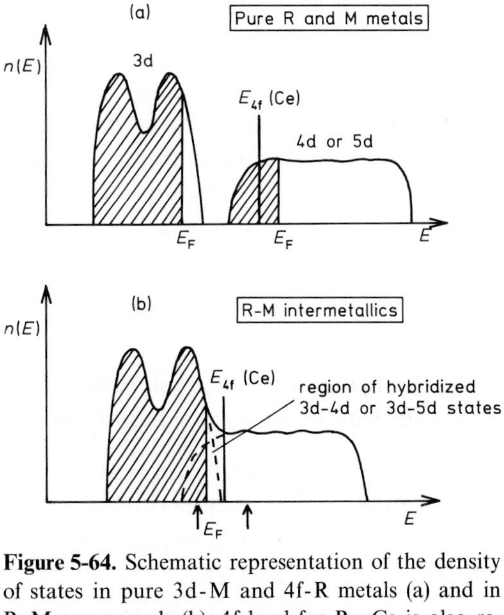

Figure 5-64. Schematic representation of the density of states in pure 3d-M and 4f-R metals (a) and in R-M compounds (b). 4f level for R = Ce is also reported. Arrows indicate the range of E_F according to the stoichiometry.

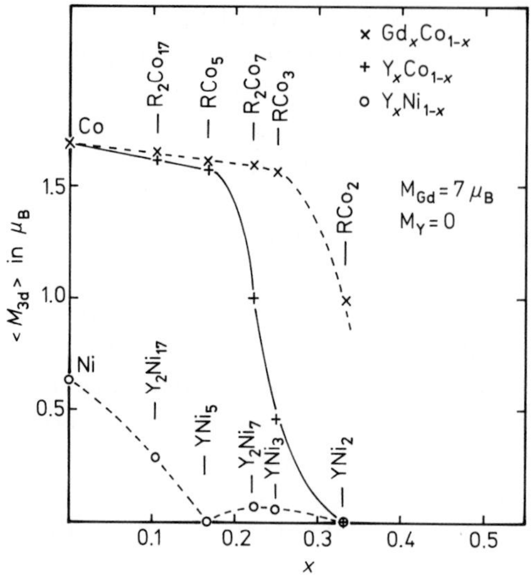

Figure 5-65. Mean values of the 3d moment in some R-M binary compounds as a function of the R content.

nickel, respectively (the properties of the RCo_2 compounds are described in Sec. 5.10.4). Note that because of the change of the band structure with the relative amount of R and M elements, in some cases the Fermi level can be very close to the 4f one giving rise to 4f magnetic instabilities. In the following sections we examine three quite different situations: (i) compounds with small R amount where 3d magnetism is well established, (ii) compounds near the onset of 3d magnetism and (iii) compounds showing 4f instability.

5.10.3 Compounds with Comparable 4f and 3d Contributions to Magnetism

One of the best examples of such a situation is found in the hexagonal RCo_5 compounds. The main characteristics of these compounds are as follows:

(i) The Co moment is close to its maximum value, i.e. that of pure Co metal (Fig. 5-65);

(ii) Thanks to the strength of the Co–Co exchange interaction, the Curie temperatures are high ($T_C > 950$ K);

(iii) As shown for YCo_5 in Table 5-5, the anisotropy is particularly large for a 3d element (one order of magnitude larger than in pure Co) and favors $\boldsymbol{c}$ as the easy magnetization axis. The origin of this anisotropy has been previously discussed in Sec. 5.5.6;

(iv) As shown in Table 5-6, CEF effects lead to the strong anisotropy of the rare earths and favor the $\boldsymbol{c}$ axis for R^{3+} ions with $\alpha_J > 0$ (R = Sm, Er, Tm), whereas $\boldsymbol{c}$ is the hard axis for R^{3+} ions with $\alpha_J < 0$ (R = Pr, Nd, Tb, Dy, Ho).

(v) As is the case in all rare earth-3d intermetallics, the R–Co exchange interaction (one order of magnitude smaller than the Co–Co interaction) favors an antiparallel coupling of the Co and R spins. Therefore this leads to a ferromagnetic coupling of R and Co moments for light rare earths where $\boldsymbol{m}_R \propto \boldsymbol{J} = \boldsymbol{L} - \boldsymbol{S}$ (R = Pr, Nd, Sm). Compounds with these latter rare earths are then ferromagnetic. On the contrary, with heavy rare earths where $\boldsymbol{m}_R \propto \boldsymbol{J} = \boldsymbol{L} + \boldsymbol{S}$, the coupling is antiferromagnetic between R and Co moments leading then to ferrimagnetism. Note that the R–R exchange interactions are still one order of magnitude smaller than the R–Co interactions and are generally not considered in the analysis of the magnetic properties of these RCo_5 compounds.

Keeping in mind these features, two compounds are worth presenting in more detail, namely $SmCo_5$ and $TbCo_5$.

5.10.3.1 $SmCo_5$: A Material with Intrinsic Outstanding Permanent Magnet Properties

$SmCo_5$ is the only compound in the series in which simultaneously: (i) $\boldsymbol{c}$ is the easy magnetization direction for Sm as well as for Co, (ii) the Sm–Co exchange interaction leads to a ferromagnetic coupling between $\boldsymbol{m}_{Sm}$ and $\boldsymbol{m}_{Co}$. Therefore, as we will see in Sec. 5.11, it satisfies the conditions for a ferromagnetic material to give rise to excellent permanent magnetic properties, namely: (i) a large Curie temperature thanks to the Co–Co interaction, (ii) a large magnetization and (iii) a large uniaxial anisotropy thanks to the hexagonal structure and to the large anisotropy of Sm. Indeed one of the best permanent magnet materials is obtained from this compound. To illustrate the magnetic properties of this compund, the low temperature field dependences of the magnetization of a $SmCo_5$ single crystal are shown in Fig. 5-66. For comparison, the same variations are also reported for a YCo_5 single crystal (Alameda et al., 1981).

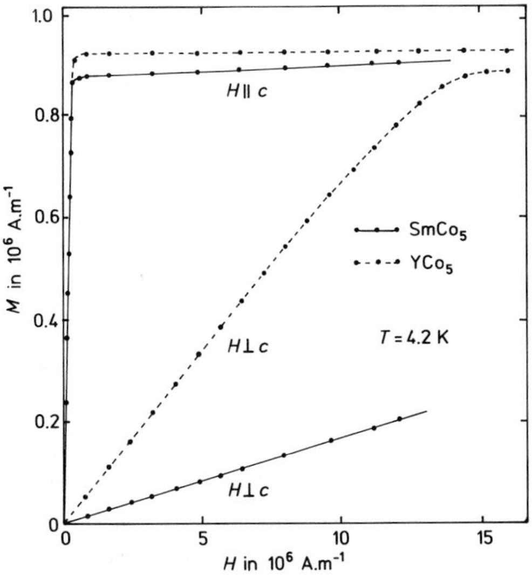

Figure 5-66. Field dependences of the magnetization measured at 4.2 K on single crystals of the hexagonal $SmCo_5$ and YCo_5 compounds. Although Sm has a small moment (much smaller than that of Co) whereas Y is not magnetic, the magnetization of YCo_5 is a little larger than that of $SmCo_5$. This originates from the fact that the density of YCo_5 is about 3% larger than that of $SmCo_5$.

Perpendicular to the sixfold axis, in a $12 \times 10^6\ \mathrm{A\,m^{-1}}$ applied field, the magnetization of $SmCo_5$ is far from being parallel to this field. In this case, the extrapolation of the field dependence of the magnetization leads to an anisotropy field of about $51 \times 10^6\ \mathrm{A\,m^{-1}}$ at 4.2 K, i.e. almost five times larger than in YCo_5.

It is worth noting that the same characteristics exist in $Nd_2Fe_{14}B$ based permanent magnets which currently are the best. The large Curie temperature of this ferromagnetic compound arises fom the Fe–Fe interactions. The large magnetocrystalline anisotropy comes from CEF effects on Nd and the uniaxial character arises fom the small amount of B which allows stabilization of a tetragonal structure. This last feature is important because no R–Fe binary compound has all the properties simultaneously.

Recommended for further reading: Strnat (1988).

5.10.3.2 $TbCo_5$: A Compound with Complex Magnetization Processes

In this compound, the Tb moments are coupled antiparallel to the Co moments and at low temperature the magnetization of the Tb sublattice is a little larger than that of the Co sublattice. Because the Co–Co exchange interactions are much larger than the R–Co interactions, the thermal variation of the Tb sublattice magnetization is faster than that of the Co sublattice. This leads to a compensation temperature T_{Co} around 110 K (see Sec. 5.3.2.3) as shown in Fig. 5-67. Moreover, the strong Tb anisotropy favors the Tb moment to lie in the basal plane perpendicular to the sixfold $\boldsymbol{c}$ axis, whereas the Co anisotropy favors the Co moment to be along this axis. However the value of the Tb–Co interaction is large enough to retain the colinear-

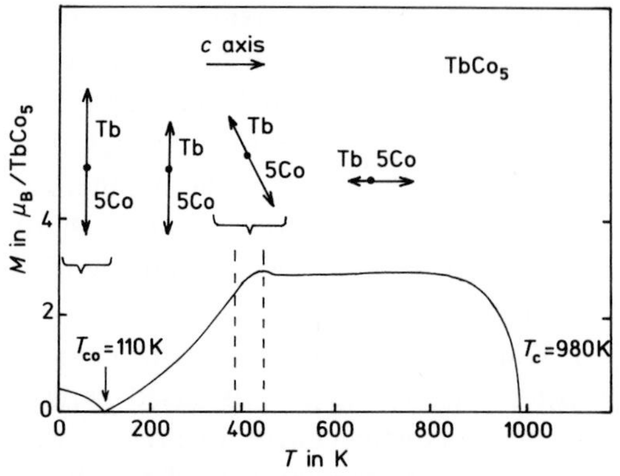

Figure 5-67. Schematic representation of the thermal variation of the spontaneous magnetization of $TbCo_5$. Moment configuration in the three characteristic ranges of temperatures are also sketched.

ity between the magnetizations of both sublattices, at least in the absence of an applied field. At low temperature the Tb anisotropy overcomes the Co anisotropy and all the moments are perpendicular to *c*. At high temperature the Co anisotropy becomes the largest due to the faster decrease of the Tb anisotropy (associated with the faster decrease of the Tb magnetization). So, between 390 and 440 K, magnetization progressively rotates toward the sixfold axis and then remains along this axis up to the ordering temperature (T_C = 980 K) as shown in Fig. 5-67 (Lemaire, 1966a). In the whole range of temperatures below T_C the magnetization processes are complex. They result from the interplay of the magnetocrystalline energy and the exchange interaction that links the Tb and Co moments together. The analysis of these processes measured along the main symmetry axes of a single crystal provides useful information for the determination of the exchange interactions and anisotropy parameters.

At this stage it is worth mentioning the R_2Fe_{17} and R_2Co_{17} compounds the crystal structure of which is derived from that of the previous compounds but is slightly more complex. The same type of anisotropy and exchange interactions lead also to complex magnetization processes (especially when moments are in the plane perpendicular to the sixfold axis) which were remarkably well accounted for (Franse et al., 1985) as is illustrated in Fig. 5-68 for the compound Dy_2Co_{17} (Franse et al., 1988).

5.10.4 Onset of 3d Magnetism

From Fig. 5-65 we have seen that the RCo_2 compounds are at the boundary of the onset of magnetism and they illustrate particularly well the collective electron metamagnetism previously described (Sec. 5.4.1.2 and 5.4.2). Indeed, when R is nonmagnetic (R = Y, Lu) the compound is a Pauli paramagnet at any temperature and when R is magnetic the compound is ferri or ferromagnetic and Co carries a moment close to 1 μ_B (Lemaire, 1966b). However in YCo_2 and $LuCo_2$ the metamagnetic transition is not observed directly because the critical field ($\approx$100 T) is much larger than the maximum field available (Fig. 5-69). A hint that such a transition may be present is given by the positive curvature of M vs. H at low temperatures around 35 T (Schinkel, 1978). With magnetic rare earths the transition is achieved thanks to the molecular field arising from the R–Co interaction. The substitution of a small amount of Al for Co modifies the band structure and leads to a decrease of the critical field such that the transition can be observed directly as shown in Fig. 5-69 (Sakakibara et al., 1987).

Collective electron metamagnetism has been observed in other R–M intermetallics where the 3d metal is close to the onset of magnetism: $Ce(Co_xNi_{1-x})_5$, Y_2Ni_{17}

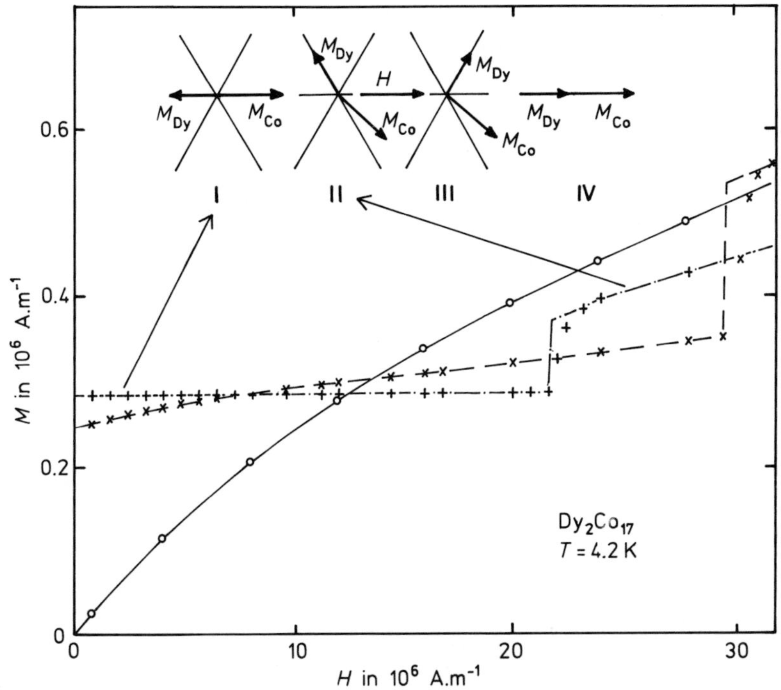

Figure 5-68. Field dependences of the magnetization of a Dy_2Co_{17} single crystal at 4.2 K; (o) [001] axis; (+) [100] axis; (×) [110] axis. The various lines represent the calculated variations. The upper part of the figure shows the different configurations predicted for the Dy and Co magnetizations when the field is applied along the easy direction ([100] axis) in the hexagonal plane. The magnetization jump along this direction corresponds to the transition from configuration I to configuration II. Configurations III and IV are not induced because they would occur in fields larger than those available.

(Gignoux et al., 1980) and $ThCo_5$ (Givord et al., 1979).

Near the onset of magnetism, if the Stoner criterion for ferromagnetism is satisfied, the 3d moment can be very small due to the weakness of the splitting between the spin up and spin down bands. The model of very weak itinerant ferromagnetism applies very well in the case of the compound YNi_3.

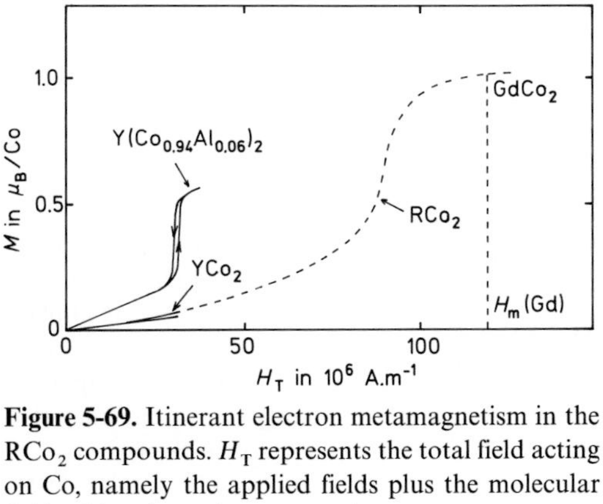

Figure 5-69. Itinerant electron metamagnetism in the RCo_2 compounds. H_T represents the total field acting on Co, namely the applied fields plus the molecular field arising from the rare earth.

5.10.5 4f Instability

In the Ce–M alloys, by changing the stoichiometry it is possible to change the relative position of the Fermi and the 4f levels and accordingly the large diversity of Ce behaviors discussed in Sec. 5.7 can be observed, including the Kondo effect, heavy fermion behavior, intermediate valence and behavior associated with a nonmagnetic Ce state (4+?). The Ce–Ni system provides a typical illustration of this diversity (Gignoux and Gomez-Sal, 1985). The sketched density of states of these alloys is shown in Fig. 5-64. Starting from pure Ni, the Fermi level progressively increases with the Ce content because of the filling of the 3d band. Hence Ce is expected to be in the lowest magnetic state in the compound richest in Ni, whereas it has the largest magnetic state for the opposite Ce and Ni concentrations. Indeed three quite different behaviors have been observed through the series namely (i) in $CeNi_5$, the compound richest in Ni, (ii) in the equiatomic CeNi compound and (iii) in Ce_7Ni_3, the compound richest in Ce.

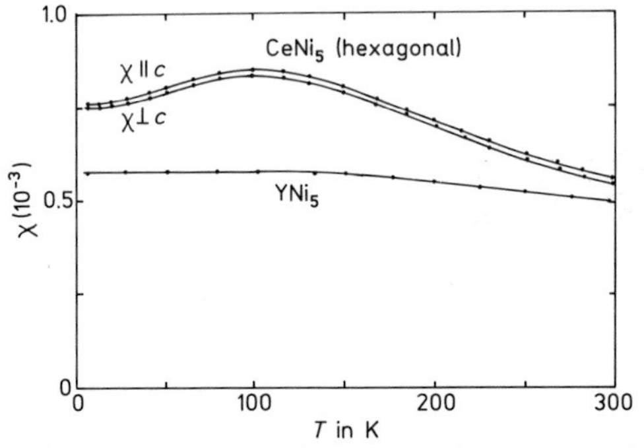

Figure 5-70. Thermal variations of the susceptibility of a single crystal of the hexagonal compound $CeNi_5$. For comparison, the susceptibility of YNi_5 is also reported. The slightly larger values and the broad maximum for $CeNi_5$ arise from the fact that Ni is closer to the onset of magnetism in this compound than in YNi_5.

5.10.5.1 $CeNi_5$

The hexagonal compound $CeNi_5$ is a Pauli paramagnet at all temperatures and its susceptibility, which is almost isotropic, shows a broad maximum around 100 K (Fig. 5-70). A microscopic analysis of this susceptibility led to the conclusion that Ce is not magnetic and that the field-induced magnetization comes from Ni, which is close to the onset of magnetism as suggested in Fig. 5-65. Indeed, a polarized neutron diffraction experiment, performed at 100 K in a 3.7×10^6 A m^{-1} applied field, led to the projection of the magnetization density shown in Fig. 5-71. At any temperature the magnetization is localized only on the Ni atoms whereas almost no contribution on the Ce atoms is observed. This result is consistent with the fact that the susceptibility is isotropic (if Ce were mag-

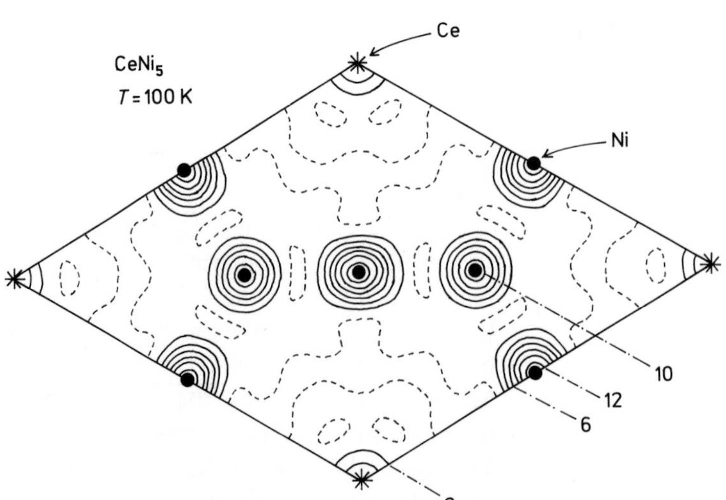

Figure 5-71. $CeNi_5$: projection of the magnetic density on the hexagonal plane induced at 100 K by a field of 3.7×10^6 A m^{-1} applied along ***c***. Contour lines of equal density are in $10^{-3}\,\mu_B\,\text{Å}^{-2}$. Dashed lines are those of zero density.

netic, a large anisotropy due to CEF would be observed in this uniaxial compound) and that lattice parameters correspond to the state having the smallest possible ionic radius (as quoted above a Ce^{4+} state is deduced from this analysis but a valence state close to 3.3 is derived from spectroscopy experiments). The susceptibility maximum arises from spin fluctuations which are characteristic of a system close to the onset of magnetism; in this compound this magnetism arises from Ni.

5.10.5.2 CeNi

Like $CeNi_5$, the orthorhombic CeNi compound is a Pauli paramagnet at all temperatures with a magnetic susceptibility which passes through a broad maximum around 140 K. However, unlike $CeNi_5$, this susceptibility is strongly anisotropic, especially the maximum is much more pronounced when the field is applied along the ***c*** direction of the orthorhombic cell (Fig. 5-72). The microscopic origin of the magnetization was also studied by polarized neutron diffraction (Fig. 5-73). Contrary to $CeNi_5$, the field-induced magnetization is localized on Ce atoms whereas no magnetic density is observed on Ni atoms. This result was confirmed by the quite satisfactory analysis of the high temperature susceptibility in terms of CEF effects on Ce. Then in agreement with the analysis of the lattice parameters and with thermal expansion experiments, CeNi, like $CeSn_3$ (see Sec. 5.7.4.1), is an intermediate valence compound in which the valence varies with temperature. The susceptibility maximum arises here from spin fluctuations on Ce which is close to the magnetic state, i.e. close to carrying a localized moment. Therefore the similarity in magnetic behavior of CeNi and $CeNi_5$ does not have the same origin.

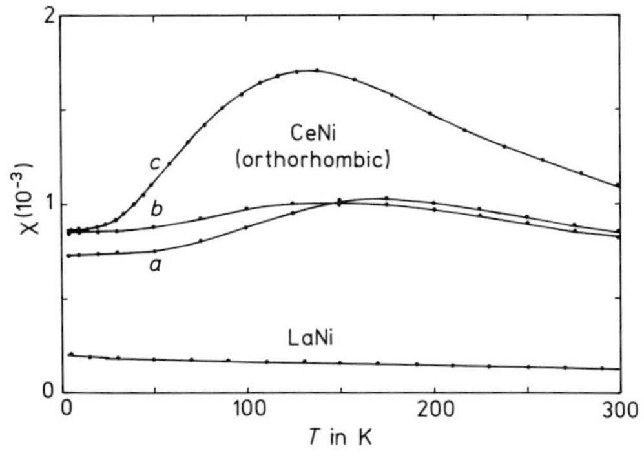

Figure 5-72. Thermal variations of the susceptibility of a single crystal of the orthorhombic compound CeNi. Note the large anisotropy compared to the very small one in $CeNi_5$. This is the clue of the predominant contribution of Ce to the susceptibility of CeNi.

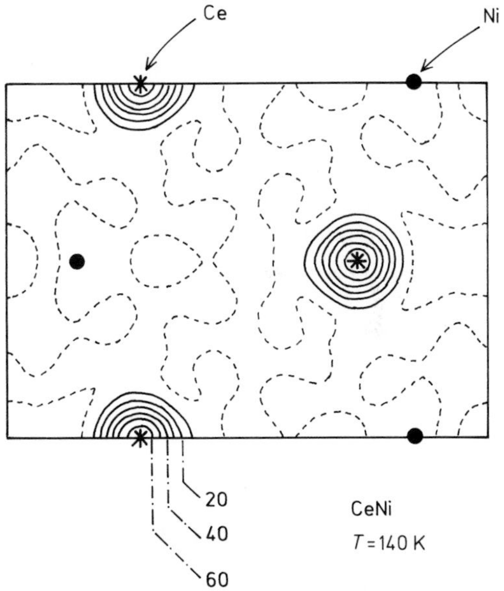

Figure 5-73. CeNi: projection of the magnetic density on the (***a***, ***b***) plane induced at 140 K by a field of $3.7\times10^6\,\mathrm{A\,m^{-1}}$ applied along ***c***. Contour lines of equal density are in $10^{-3}\,\mu_B\,\text{Å}^{-2}$. Dashed lines are those of zero density.

The comparison of these two compounds may give one an idea of the power of the polarized neutron diffraction technique in obtaining a microscopic insight into magnetism.

5.10.5.3 Ce_7Ni_3

This compound crystallizes in a complex hexagonal structure. From the analysis of the lattice parameters and the high temperature susceptibility (Curie-Weiss law with a Curie constant corresponding to the effective moment of Ce^{3+}), it is concluded that cerium is in the trivalent state. Ce_7Ni_3 orders antiferromagnetically at 1.8 K and, along the *c* axis the field dependence of magnetization at 1.5 K exhibits a metamagnetic transition in low fields. Moreover, a minimum of the resistivity is observed around 4 K showing that this compound is an antiferromagnetic Kondo lattice, like $CeAl_2$ (Sec. 5.7.3).

In conclusion, due to the large difference between the Fermi levels of pure Ce and Ni, the Ce–Ni system is especially well adapted for observing a wide range of magnetic behaviors of Ce.

Some review articles on rare earth based intermetallic compounds: Buschow (1977, 1979, and 1980); Kirchmayr and Poldy (1979).

5.11 Magnetization Processes in Ferromagnets – Applications

In this section we are concerned with a semi-microscopic approach to ferromagnetic materials, namely the response of a bulk ferromagnet to an applied magnetic field. This aspect of a ferromagnet is of utmost importance because it is the basis of most technical applications. It is also the reason why, after describing the mechanisms responsible for these magnetization processes, we present several main applications, in particular those using metallic materials.

5.11.1 Magnetization Processes in Ferromagnets

In ferro- (or ferri-) magnetic substances the magnetization curves, especially in low magnetic field, differ widely from sample to sample and are a function of the magnetic history of the sample, i.e. of the previous fields which have been successively applied.

5.11.1.1 Hysteresis

When a ferromagnetic material in the virgin state (i.e., when it has never been magnetized) is submitted to an increasing magnetic field, its magnetization begins to increase following the *first magnetization curve* (curve (A) in Fig. 5-74). The linear part corresponds to a *reversible* variation of magnetization, whereas the positive curvature is associated with an *irreversible* variation. When the field is large, M tends to the saturation M_s.

Subsequently, when the field is decreased down to a large negative value, the magnetization follows by the curve (B) down to $-M_s$. Then, increasing the field again, the magnetization follows the curve

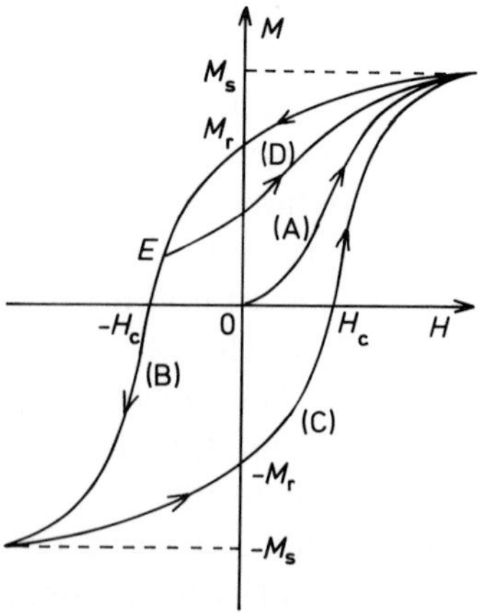

Figure 5-74. First magnetization curve and hysteresis loop in a ferromagnet.

(C), which can be transformed into curve (B) by using the origin O as a center of symmetry. These two curves, (B) and (C), constitute the *hysteresis loop*. The area of the loop is a measure of the energy which is spent when describing this loop and which is transformed into heat in the sample (magnetic losses).

The hysteresis loop is characterized by the values of (i) the *residual magnetization* M_r that the substance retains in zero field, and (ii) the *coercive field* H_c which is required for the magnetization to vanish. Depending on the material, the range of coercive fields is very large: H_c is on the order of $10^{-1}\ \mathrm{A\,m^{-1}}$ in very pure Fe but on the order of $10^5\ \mathrm{A\,m^{-1}}$ in permanent magnets such as $SmCo_5$ or $Nd_2Fe_{14}B$. If H_c is large, as in this latter case, the material is *magnetically hard.*

In some magnetization processes one is also interested in the maximum susceptibility χ_m, equal to the maximum value of M/H, or in the differential susceptibility $\chi_d = \mathrm{d}M/\mathrm{d}H$. In other applications one is interested in the relative permeability $\mu_r = 1 + \chi = B/\mu_0 H$, which can reach values larger than 10^4 in Fe–Ni alloys. When μ_r is very large in low fields and H_c is small (the smallest values are some tenths of $\mathrm{A\,m^{-1}}$), the material is said to be *magnetically soft.*

The hysteresis loop described above is the *major loop*, both extremities corresponding to very high fields. But one can also describe *minor loops*. For instance, if one stops in *E* (Fig. 5-74) and if the field is increased, the magnetization subsequently follows the curve (D). In particular, if the sample is subjected to an alternating field which is slowly decreased down to zero from an initial value large compared to H_c, the magnetization describes a series of minor loops the size of which decreases until the magnetization reaches zero in zero field

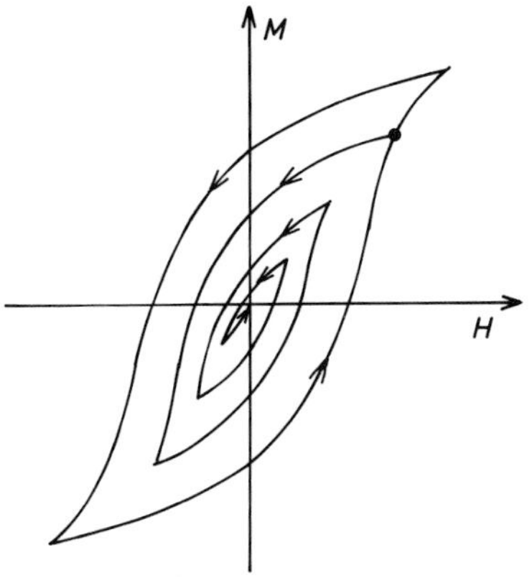

Figure 5-75. Demagnetization by cycling with an applied field of decreasing amplitude.

(Fig. 5-75). The sample has then been demagnetized: its residual magnetization is zero. The only other way to demagnetize a ferromagnetic material is to heat it above its Curie temperature. If it is then cooled in the absence of any applied field, it will be in the demagnetized state. It has been shown that the thermally demagnetized state is different from that obtained by cycling the material in a decreasing field.

Magnetization processes described above for a ferromagnet depend strongly on the existence of magnetic domains, which will be described in Sec. 5.11.1.3.

5.11.1.2 Magnetostatic Energy and Shape Anisotropy

In a sample with uniform magnetization, the dipolar interactions produce a demagnetizing field $H_d = -NM$ inside the sample (Sec. 5.2). Associated with this demagnetizing field is a *magnetostatic energy* given by:

$$E_{ms} = -\mu_0 \int_0^{H_d} M\,\mathrm{d}H_d = \frac{1}{2}\mu_0 N M^2 \qquad (5\text{-}84)$$

We know that for a spherical sample $N = 1/3$ and H_d does not depend on the magnetization direction. For a prolate ellipsoid H_d is larger when the magnetization is forced to be perpendicular to the long axis because the magnetic *poles* of opposite sign are nearer to each other. Two demagnetizing factors must be introduced in this case: $N_{\parallel}$ and $N_{\perp}$ when M is parallel or perpendicular to the long axis, respectively. The energy difference of the two configurations

$$K_f = \mu_0 (N_{\perp} - N_{\parallel}) M^2/2 \qquad (5\text{-}85)$$

is called *shape anisotropy*, which favors the magnetization to be parallel to the long axis. As Eq. (5-85) shows, the strength of this anisotropy increases (i) with increasing difference between the long and short dimensions of the sample and (ii) with increasing value of magnetization.

5.11.1.3 Magnetic Domains

In order to minimize the magnetostatic energy E_{ms} associated with the demagnetizing field, a single crystal of a ferromagnet is divided into *elementary domains*, small in size but each one still consisting of a large number of atoms. These elementary domains are called *Weiss domains*. Inside a domain, the magnetic moments are aligned, but from domain to domain the magnetic moments point toward different directions so that the net magnetization is zero. For instance, in a single Fe crystal of cubic structure there are six types of domains, each corresponding to one of the equivalent easy magnetization directions, namely the [100] axes (Fig. 5-76).

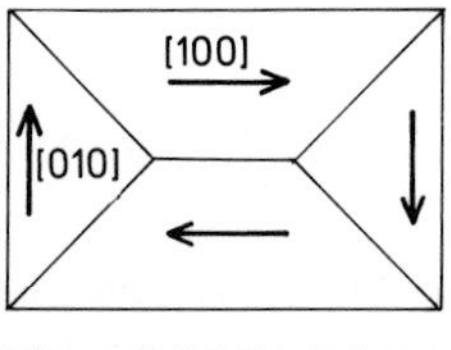

Figure 5-76. Domain structure in a cubic crystal having ⟨100⟩ easy axes.

As mentioned above, the existence of domains is necessary because a substance in a saturated magnetic state, i.e. when consisting of a single domain with all moments aligned in the same direction, would have a high magnetostatic energy. For instance, in a saturated sphere of Fe the energy density associated with the demagnetizing field is ten times larger than the magnetocrystalline anisotropy energy. The existence of domains makes it possible to have alternating zones of opposite polarity at the surface of the sample, as is sketched in Fig. 5-77 in the case of a uniaxial substance.

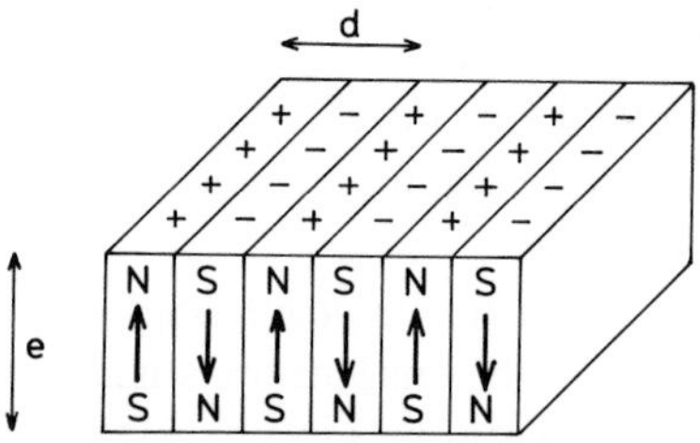

Figure 5-77. Schematic representation of the domain structure in a slab with uniaxial anisotropy.

Hence the demagnetizing field in the main part of the sample is almost completely cancelled. Such a process, if it were complete, would lead to monoatomic layer domains. This does not occur, however, because there is a prohibitive increase in the exchange energy.

5.11.1.4 Domain Walls or Bloch Walls

Domain walls are interfaces between regions in which the spontaneous magnetization has different directions. At or within the wall the magnetization changes its direction.

In 1932, F. Bloch showed that the transition between two domains takes place through a progressive rotation of moments. The simplest situation is that of a

180° wall between two antiparallel domains in an uniaxial compound (Fig. 5-78).

There is no simple way to determine the exact moment configuration in a domain wall, but we can make an approximate calculation in order to have an order of magnitude estimate of the domain wall width and energy. Let us consider that magnetization rotation is regular with an angle ϕ between two adjacent moments (Fig. 5-78). The exchange energy for a pair of atoms is: $E_{ex} = -2JS^2 \cos\phi = -4W \cos\phi$. If ϕ is small this can be written as $E_{ex} = 2W\phi^2 - 4W$. Disregarding the second term, which is independent of ϕ, the extra exchange energy due to the presence of the wall is $2W\phi^2$. Assuming for the sake of simplicity a simple cubic cell of edge a, if the wall is N atoms thick, the extra exchange energy per unit area of wall is: $\gamma_{ex} = NW\phi^2/a^2$.

With $\phi = \pi/N$ for a 180° wall, we have:

$$\gamma_{ex} = \frac{W\pi^2}{Na^2} \tag{5-86}$$

This energy, which decreases when N increases, favors a wall as thick as possible.

The anisotropy energy (per volume unit) for moments which make an angle θ with the easy magnetization axis (Fig. 5-78) is $K \sin^2\theta$ $(K > 0)$. To obtain an order of magnitude estimate of the total anisotropy energy we use a very simple calculation by replacing this energy by K (as if all moments of the wall were perpendicular to the easy axis). Therefore, per unit area of wall:

$$\gamma_{an} = KNa \tag{5-87}$$

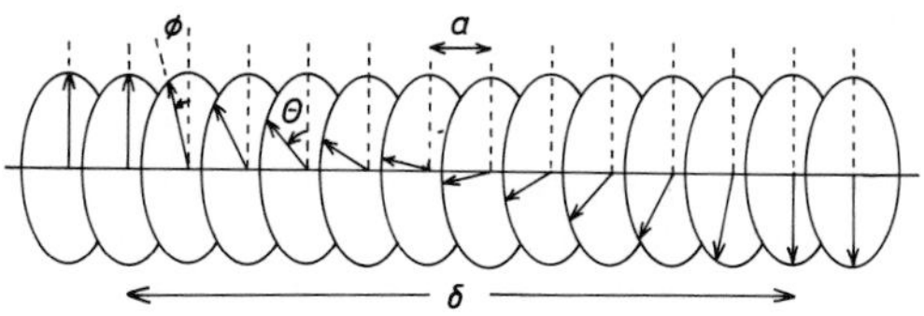

Figure 5-78. Structure of a 180° wall.

This expression shows that the anisotropy favors a wall as thin as possible and therefore works opposite to the exchange energy. The balance between the actions of exchange and anisotropy leads to an equilibrium state. Minimizing the total energy

$$\gamma = \gamma_{ex} + \gamma_{an} \tag{5-88}$$

with respect to N leads to the wall thickness:

$$\delta = Na = \pi \sqrt{\frac{W}{Ka}} \tag{5-89}$$

The order of magnitude of δ is some 10^2 Å in 3d materials and can become as small as a few (and at the limit only one) interatomic distances in rare earth materials where the anisotropy is very large. In this latter case the existence of *narrow walls* gives rise to an intrinsic magnetization process which is described at the end of this section.

Substituting Eq. (5-89) into Eq. (5-88) we find that the exchange and anisotropy energy densities are equal and that the wall energy (per unit area of wall), due to the existence of the wall, is:

$$\gamma = 2\pi \sqrt{\frac{WK}{a}} \tag{5-90}$$

This superficial wall energy is on the order of 10^{-3} to 10^{-2} J/m^2.

In cubic materials the shape of a 180° wall is more complex. Moreover, 90° walls (case of Fe with easy [100] axes), or 72° and 108° walls (case of Ni with easy [111] axes) are present in these materials.

Narrow Domain Walls and the Magnetization Process

It is possible to show that *narrow domain walls* have a shape like that sketched in Fig. 5-79 a. Especially to minimize the anisotropy energy the largest angle between

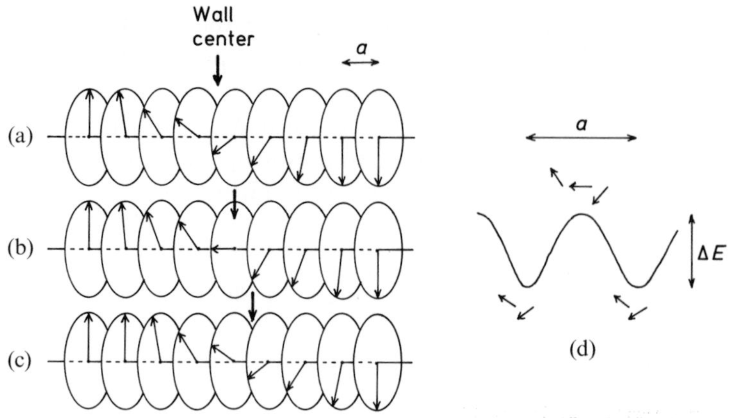

Figure 5-79. Narrow wall structure and propagation mechanism.

two consecutive moments occurs at the wall center. As shown in the Figure, in the absence of an applied field the wall is symmetrical and its energy γ is at a minimum. Applying a field, moments tend to rotate in the field direction and the wall becomes asymmetrical. Its energy increases and becomes a maximum when it is again symmetrical (Fig. 5-79 b). In fact, the potential energy of the wall varies sinusoidally with a period equal to the lattice periodicity in the direction of the wall propagation. The wall will move irreversibly when the force on the wall due to the applied field will overcome the energy barrier ΔE of the potential energy (Fig. 5-79 d). ΔE and accordingly the propagation field H_p is an increasing function of the ratio K/W. When this ratio is large enough one observes a first magnetization curve similar to a metamagnetic transition in an antiferromagnet because the walls can move only when $H \geqq H_p$. When the field is reversed, a coercive field of the same of magnitude as H_p is predicted. This particular magnetization process (Fig. 5-80, curve b), which is an intrinsic property of the pure material, is characteristic of narrow domain walls (Barbara et al., 1971; Van den Broek and Zijlstra, 1971). Unfortunately, no permanent magnet can be obtained with such materials, because this property occurs only at temperatures much lower than room temperature due to the low values of the exchange interaction ($T_C < 100$ K) and thermal activation effects.

5.11.1.5 Size of Elementary Domains

Let us consider the very simple situation of a monocrystalline slab of a uniaxial material (magnetization M_s, thickness e) with its surface perpendicular to the easy axis.

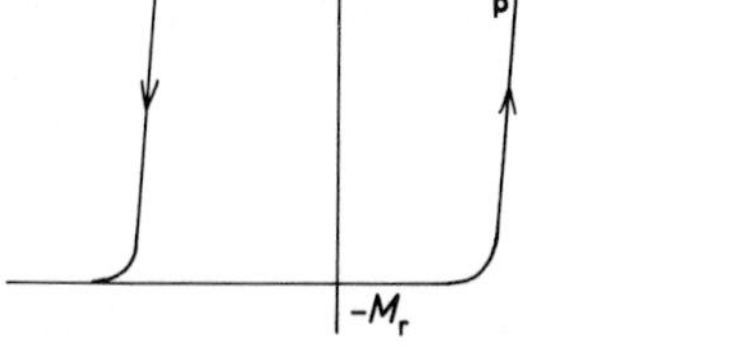

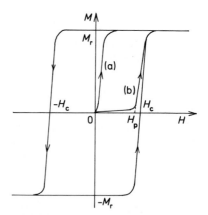

Figure 5-80. Two extreme first magnetization curves giving rise to the same hysteresis loops. These two curves make it possible to recognize which type of coercivity mechanism is present. Curve (b) is also characteristic of the intrinsic property referred to as narrow domain walls.

Let us also assume that the domains are parallel strips with periodicity d (Fig. 5-77). Per unit area of the slab, the wall energy is $2\gamma e/d$. It is possible to show that the demagnetizing energy due to the alternating distribution of magnetic poles of either sign is $0.85\ \mu_0 M_s^2 d$. Minimizing the total energy leads to the equilibrium periodicity d_0:

$$d_0 \approx \frac{1}{M_s}\sqrt{\frac{2\gamma e}{\mu_0}} \tag{5-91}$$

As an order of magnitude one finds $d_0 \approx 1\ \mu m$. Although schematic, this model shows that the domain structure is a result of the competition between pole (or magnetostatic) and wall energies which tend to increase and to decrease the number of domains, respectively.

Domain structures like that discussed above are currently observed in uniaxial materials with a relatively high anisotropy. But when the anisotropy is smaller, surface domains appear, called *closure domains*, which yield a decrease of the magnetostatic energy at the expense of the anisotropy energy. Closure domains always exist when there are several easy axis, in particular in cubic crystals where 90° walls are present (Fig. 5-76).

Among the different techniques used to observe domains and walls, the oldest and also the simplest one is called the *Bitter method*. It involves the application of an aqueous suspension of extremely fine (colloidal) particles of magnetite Fe_3O_4 to a polished surface of the specimen. The particles are attracted toward the regions of nonuniform field, namely toward those regions where the walls cross the surface. The particles then form lines which can be seen through a light microscope. Other techniques such as electron microscopy and optical methods involving the *Kerr* or *Faraday* effects are also commonly used. These latter techniques are based on the rotation of the polarization of a light beam by a magnetized specimen, during reflection or transmission, respectively.

5.11.1.6 Macroscopic Magnetostriction

Let us consider a ferromagnetic cubic compound in which the easy axis is the [100] direction (Fig. 5-81). In the magnetically ordered state, i.e. below T_C, the lattice of each domain is no longer cubic but tetragonal due to the slight distortion arising from the intrinsic spontaneous magnetostriction (see Sec. 5.9). For instance, in Fe metal the length is found to increase along the fourfold axis which is parallel to the magnetization. However, a bulk single crystal (and consequently also a polycrystal) does not undergo any distortion when it is cooled from $T > T_C$ to a temperature $T < T_C$. Indeed, the distortions of all domains cancel owing to the distribution of domains with magnetization parallel to the [100], [010] and [001] axes as it sketched in Fig. 5-81 a. If the material is subjected to an applied field (for instance parallel to the [100] axis), it undergoes a macroscopic tetragonal distortion because the sample becomes single domain with all moments

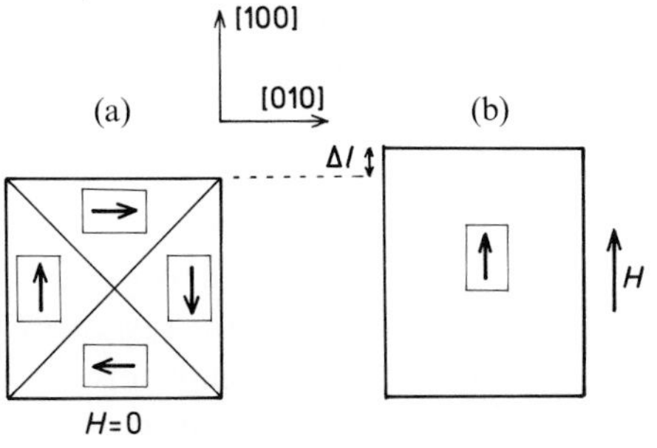

Figure 5-81. Intrinsic (spontaneous) (a) and extrinsic (field induced) (b) magnetostriction of a single cubic crystal where [100] axes are of easy magnetization. Arrows indicating the moment directions are contained in rectangles which show the tetragonal distortion of each domain enormously exagerated.

parallel to the field. We then have an extrinsic (or macroscopic) magnetostriction as is sketched in Fig. 5-81 b.

Note that in uniaxial compounds where there are only 180° walls, wall motion does not produce any magnetostrictive change of length.

Cubic compounds are then especially appropriate to obtain magnetostriction associated with domain wall motion. Among these compounds it is worth mentioning the cubic ferro (or ferri) magnetic RFe_2 compounds (R = rare earth) because of their technological applications (Clark, 1980). Indeed (i) thanks to the large Fe–Fe exchange interactions the Curie temperatures of these compounds are much larger than room temperature, and (ii) large magnetostriction effects mainly arise from crystal field effects on rare earths (Sec. 5-85). For instance, at room temperature $\lambda_{111} = 2.4 \times 10^{-3}$ in $TbFe_2$ whereas λ_{100} reaches only 2.1×10^{-5} in pure Fe [for the meaning of λ_{100} and λ_{100} see Eq. (5-83)]. These compounds are used as transducers to generate acoustic waves at low frequencies around 1 kHz (Sonar) or inversely, using the modifications of magnetic properties under applied stresses, as sensors for force or torque. For high frequency devices, metallic materials are not suitable because of the induced eddy currents. Non-metallic materials are then more appropriate.

5.11.1.7 Coercivity

In this section we present a survey of the different mechanisms which are responsible for coercivity. These mechanisms depend on many factors associated with the intrinsic and extrinsic properties of the material. Intrinsic properties are those which depend on crystal structure and chemical composition, for example the Curie temperature T_C, the spontaneous magnetization M_S, and the anisotropy energy K or the anisotropy field H_a. Extrinsic properties are those associated with the microstructure of the bulk material such as defects, grain size or grain shape. They are controlled by metallurgical treatments which can be complex and which are not always well understood.

Soft materials require a large permeability and a small coercive field. They usually work in low applied fields and are not saturated. Wall motion must then be as easy as possible. In compounds with very small intrinsic anisotropy and without defects, walls can be moved very easily. Consequently, in order to produce soft magnetic materials one tries to reduce the number of defects through annealing processes.

In hard materials, the coercivity which has to be considered is that which develops after the magnetization has been saturated. The simplest approach to a subsequent magnetization reversal process assumes coherent rotation of the magnetization in which the moments remain collinear. In a uniaxial single crystal particle this coherent rotation needs to overcome the energy barrier associated with both the intrinsic uniaxial magnetocrystalline anisotropy and the shape anisotropy. Actually, this process almost never occurs and the coercive field measured experimentally is always significantly smaller than that predicted with this model. However, the two types of anisotropy mentioned above are of utmost importance to obtain hard material. The experimental coercivity processes in hard materials can be classified into two broad categories:

(i) Collective processes in which the magnetization reversal simultaneously involved all magnetic moments. This takes place when the magnetocrystalline anisotropy is low and when the particles are small enough to be single domains. In elongated particles *curling* is expected to occur. Curling means that magnetization is not uniform and allows the magnetostatic energy associated with single domain and shape anisotropy to be decreased at the expense of the exchange energy. These processes seem to occur in fine particles of γ-Fe_2O_3 and Fe metal used for magnetic recording and in AlNiCo permanent magnets.

(ii) Non-collective processes in which the magnetization is reversed by domain wall nucleation and propagation. These processes are important in materials with large uniaxial magnetocrystalline anisotropy. They allow the magnetization to be reversed without all the moments having to pay the very large anisotropy energy at the same time.

The *nucleation* phenomenon refers to the formation of a small domain in a field H_n whose magnetization is opposed to that of the saturation magnetization. Once the nucleus is formed, it easily propagates through the whole volume at its disposal giving rise to total magnetization reversal if nothing restrains the wall motion. This is the case in homogeneous materials where the domain wall energy is independent of position and where therefore $H_c = H_n$. In normal crystallized materials, defects lead to very small values of H_n. However, experimentally H_n strongly increases as the grain size of the material decrease and can reach about $H_a/10$ (H_a = anisotropy field). This property is used in the fabrication of sintered $SmCo_5$ and $Nd_2Fe_{14}B$ magnets, the intrinsic properties of which have been discussed earlier (Sec. 5.10.3). In these materials the appropriate microstructure is obtained via the so-called powder metallurgy/sintering route. The powder obtained is pressed under an applied field which aligns the easy magnetization axes of the particles and is subsequently sintered at high temperatures to compact the magnet to almost full density. The obtained microstructure consists of individual grains which are magnetically isolated from each other. The performances of these magnets are shown in Fig. 5-82.

Domain wall *pinning* prevails in heterogeneous systems where the domain wall energy depends on position and where walls thus remain trapped in regions where their energy is reduced. In this case, domain wall nucleation occurs in small fields H_n, and is not important in the magnetization reversal process which proceeds via domain wall unpinning in a much larger field $H_p (= H_c)$, the propagation field. In order for the pinning to be effective, extended defects are in general considered to be necessary. H_p is the largest when these defects are of a well-defined thickness on the order of the domain wall width δ, and when interfaces with the main magnetic phase are sharp and discontinuous. This mechanism is used to obtain coercivity in

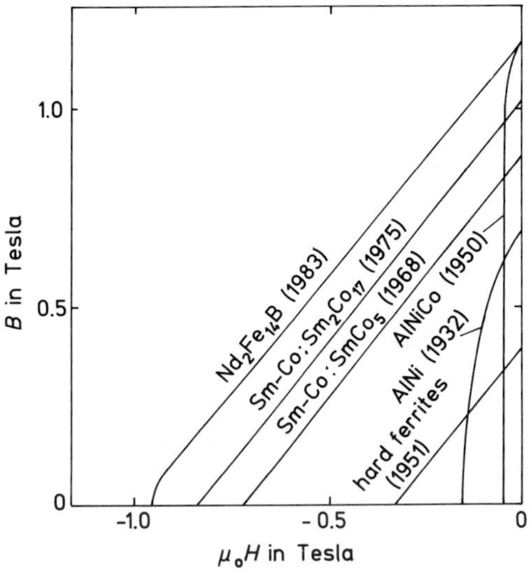

Figure 5-82. Demagnetizing curves of different permanent magnets with the year of their discovery (H is the internal field, i.e. the applied plus the demagnetizing field).

both $Sm(CoCu)_5$ and $Sm(CoCuFeZr)_{7.7}$ magnets which are derived from the compound $SmCo_5$. The substitution of other elements for Co and changes of the stoichiometry (for the second material) give rise to inhomogeneous regions of appropriate size. In both materials coercive fields up to $\mu_0 H_c = 3\,T$ and $(BH)_{max}$ values (see Sec. 5.11.2.2) larger than 270 $kJ\,m^{-3}$ have been obtained at room temperature, which represents an improvement with respect to single phase sintered $SmCo_5$.

As sketched in Fig. 5-80, the first magnetization curve allows the determination of the mechanism mainly responsible for the coercivity. Curve (a) means either that the crystallite size is below the critical size for single domain particles or that the domain walls are pinned. Microstructural studies may allow selection between these possibilities. Curve (b) means that free domain wall displacement occurs and that coercivity is mainly driven by the nucleation mechanism.

Books on coercivity: Zijlstra (1982); McCaig (1977).

5.11.2 Technical Applications

Most technical applications involve ferromagnetic materials and are directly associated with the characteristics of their hysteresis loop.

5.11.2.1 Soft Magnetic Materials

Soft magnetic materials are required to have a large permeability and a small coercive field in order for the wall motion to be as easy as possible. The conditions are: (i) Intrinsically small anisotropy and magnetoelastic energies, (ii) extrinsically a defect density as small as possible. Conditions (i) can be obtained in 3d based cubic or amorphous materials. Condition (ii) is obtained after extremely sophisticated metallurgical processes.

Most soft metallic materials are used as magnetic cores in electrical machines such as transformers, motors and generators where it is quite appropriate for these materials to have large magnetic induction and where low frequencies are used (≈ 50 Hz, sometimes 400 Hz). The most commonly used materials are Fe–Si steels (with a low percentage of Si) in which, during the last century, the hysteresis energy losses were improved by more than one order of magnitude. Fe–Ni alloys (with $\approx 80\%$ Fe), called *permalloys*, are used for specific applications such as in electrical motors for aeronautics, and as sensitive elements of differential circuit-breakers. Recently, amorphous alloys such as Co–Fe–B–Si metallic glasses with very small anisotropy and magnetostriction have been developed (Hansen, 1991).

5.11.2.2 Permanent Magnets

Permanent magnets require large coercive fields and their performances are closely related to their hysteresis loop, in particular to the demagnetization curve (Fig. 5-83). This figure illustrates the advantage of having the availability of a large H_c value. In case (b) it is possible to construct magnets having the shape of a thin slab with the magnetization perpendicular to its plane. Note that in case (a) the magnet must be elongated and have its magnetization parallel to the long dimension in order for the field (applied + demagnetizing) to be smaller than H_c. From the demagnetization curve one may also derive the maximum value of the product BH, which corresponds to the shape in which the magnet is energetically the most favorable to be used. Indeed, the energy stored in the magnet increases with the value of this product.

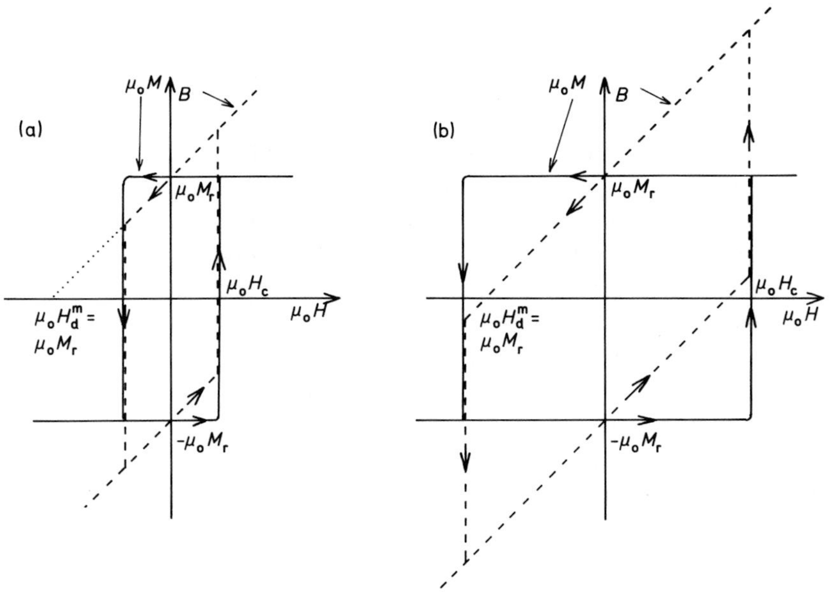

Figure 5-83. Schematic representation of hysteresis loops as a function of the internal field (applied plus demagnetizing field) for a AlNiCo type material (a) and for a modern material of $SmCo_5$ or $Nd_2Fe_{14}B$ type (b). Permanent magnet performances are large when the magnet can resist a large opposing field while retaining a large magnetization. $H_d^m = -M$ is the maximum demagnetizing field corresponding to a thin slab with magnetization perpendicular to its surface. In case (a) the magnet has to be elongated with magnetization parallel to the long dimension otherwise it is unstable. This limits the application range of such a magnet. In case (b) the magnet can have any shape for the magnetization is very hard to reverse.

The performances of the main types of permanent magnets used currently are illustrated in Fig. 5-82. The years of their discovery give an idea of the continuous and large improvements of these magnets in the last 50 years. It is worth noting the remarkably large performances of the modern permanent magnet of the type $SmCo_5$, Sm_2Co_{17} and $Nd_2Fe_{14}B$ which were discussed earlier in this chapter.

Applications of permanent magnets are extremely large and extended. About 35% of sales concerns motors and generators, 25% are used in telecommunications, information transmission devices, instrumentation, controls (switches, progressive wave tubes, sensors, measurement appliances), and 20% in audio systems such as loud speakers, listening devices, headphones. The remaining 20% concerns mechanical applications (lock, joining or suspension devices) as well as specific applications such as particle guiding (cyclotron, synchrotron).

Books on permanent magnets: Buschow (1988); McCaig (1977); Strnat (1988); Zijlstra (1982).

5.11.2.3 Magnetic Recording

Although magnetic recording uses a large amount of non-metallic magnetic materials, we will mention in this chapter this type of application at least as far as metallic substances are concerned. Indeed, this field has grown strongly owing to applications in audio-visual and computer devices. Magnetic recording is especially well adapted for permanent storage of large amounts of information due to its huge storage density (10^6 to 10^7 bits per cm^2 in 1988), low access time and low cost. Moreover magnetic recording allows for writing as well as for reading. In addition to magnetic recording on tapes and disks for audio-visual and computer devices, it is worth noting the strong growth of applications such as magnetic cards and magnetic tickets. In 1988, the world turnover of magnetic recording devices was larger than that of permanent magnets and transformer steels.

Magnetic recording media (tape, disk or card) consist of a support which is coated on one side with a thin layer of small ferromagnetic particles embedded in plastic. Depending on the specific application, the coercivity of the particle ranges between 20 and 120 $kA\,m^{-1}$.

In 1988, new techniques under development included perpendicular recording, magnetoresistive recording heads and magnetooptic recording. Whereas in traditional longitudinal recording the magnetization is parallel to the support, in perpendicular recording the magnetization is perpendicular to it. Magnetoresistive recording heads, which use the resistivity variation of a material as a function of its magnetization, have the advantage of be-

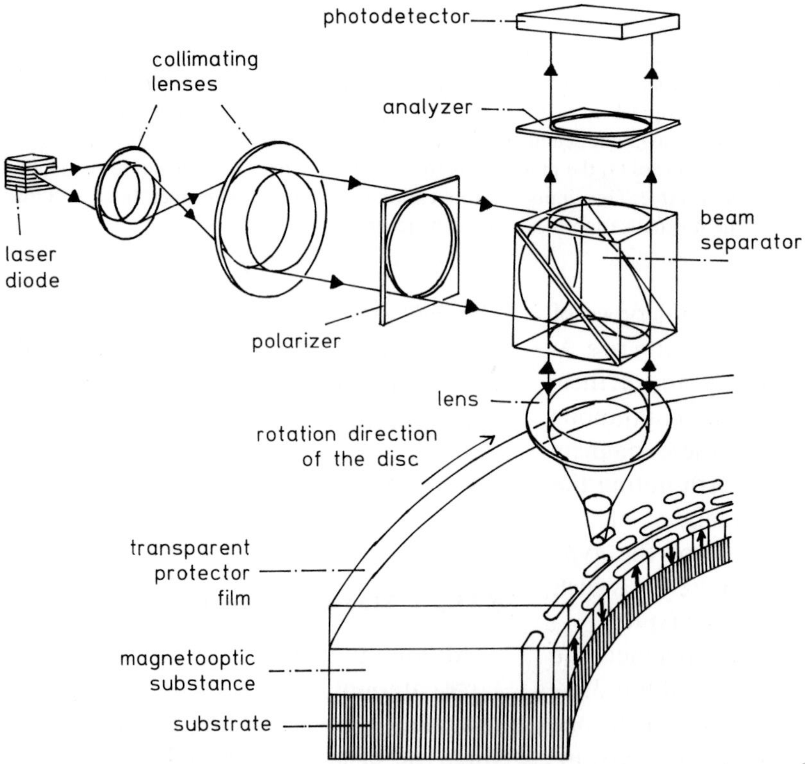

Figure 5-84. In magnetooptic recording, writing and reading is done by laser. Principles of writing, which woud need further development to be understood, are based on the fact that magnetic material has a compensation point near room temperature (Reim and Schoenes, 1989). The laser heats the magnetooptic material locally and changes its magnetic state. On reading, a polarized laser beam of smaller intensity is used; the sign of the polarization rotation by the magnetooptic material depends on the direction of magnetization. Subsequently an analyzer gives rise to a light intensity which depends on the polarization state and which is measured with a photodetector.

ing especially sensitive. In fact, these techniques (perpendicular recording and magnetoresistive heads) will have to prove themselves before replacing traditional recording which is used worldwide and which is likely to be improved.

The most dramatic event is the appearance of the magnetooptic disk which offers the two advantages of read-only laser compact disks (huge recording density – 6×10^7 bits per cm^2 in 1989 – and absence of mechanical contact during reading) and the advantages of magnetic disks, i.e. the possibility of writing. Recording is digital and perpendicular (Fig. 5-84). The recording medium is a ferrimagnetic material (amorphous Tb–Fe or Gd–Fe alloys or rare earth based oxides) such that the compensation temperature is near room temperature. This property allows one to write information by a simple local and very short heating of the recording media by means of a laser. Reading is based on the magnetooptic rotation of the polarization of a laser beam. This technique is schematized in Fig. 5-84.

Books on this section: Hansen (1991); Schoenes (1990).

5.11.2.4 Invar Alloys

In a ferromagnet, magnetoelastic effects give rise to spontaneous changes of lattice parameters with temperature which are at a maximum in the temperature range where the magnetization exhibits a large variation, i.e. below the Curie temperature. In some cases the volume change can exactly compensate the thermal expansion of non-magnetic origin: the dimensions of the substance are then temperature independent. This is the so-called *Invar effect*. This name comes from invar, a Fe–Ni alloys with 36% Ni where this effect occurs near room temperature. This is the reason why this material is used in time precision instruments. This material was used in a very different application in 1989, when television sets were equipped with tubes having invar masks. This innovation led to an important improvement in the contrast, the color purity and the resolution of the image.

Review articles on this section: Cullity (1972) (pp. 248–599); Givord et al. (1990a, b); Wassermann (1989).

5.12 References

Alameda, J. M., Givord, D., Lemaire, R., Lu, Q. (1981), *J. Appl. Phys. 52,* 2079.

Anderson, P. W. (1961), *Phys. Rev. 124,* 41–53.

Andres, K., Graebner, J. E., Ott, H. R. (1975), *Phys. Rev. Lett. 35,* 1779.

Aubert, G., Gignoux, D., Hennion, B., Michelutti, B., Nait-Saada, A. (1981), *Sol. Stat. Commun. 37,* 741.

Bacon, G. E. (1975), *Neutron Diffraction.* Oxford: Clarendon Press.

Ballou, R., Lemaire, R. (1988), *J. Physique 12,* C8, 523.

Ballou, R., Barthem, V. M. T. S., Gignoux, D. (1988), *Physica B 149,* 340.

Ballou, R., Yamada, H. (1990), to be published.

Barbara, B., Diény, B. (1984), *Physica B 130,* 245.

Barbara, B., Bècle, C., Lemaire, R., Paccard, D. (1971), *J. de Physique 32-C1,* 299.

Barbara, B., Gignoux, D., Givord, D., Givord, F., Lemaire, R. (1973), *Int. J. Magnetism 4,* 77.

Barbara, B., Rossignol, M. F., Purwins, H. G., Walker, E. (1974), *Phys. Stat. Solidi a 22,* 553–558.

Barbara, B., Rossignol, M. F., Boucherle, J. X., Schweizer, J., Buevoz, J. L. (1979). *J. Appl. Phys. 50,* 2300–2307.

Barbara, B., Gignoux, D., Vettier, C. (1988), *Lectures on Modern Magnetism, Science Press.* Berlin: Springer-Verlag.

Barbara, B., Diény, B., Filippi, J. (1990), *J. Appl. Phys.,* to appear. (Proceedings of the MMM Conference, Boston, 1989).

Béal-Monod, M. T., Lawrence, J. M. (1980), *Phys. Rev. B 21,* 5400.

Binder, K., Young, A. P. (1986), *Reviews of Modern Physics 58,* 801–976.

Blythe, H. J. (1968), *J. Phys. C, 1,* 1604.

Bömken, K., Weber, D., Yoshizawa, M., Assmus, W., Lüthi, B., Walker, E. (1987), *J. Magn. Magn. Mater. 63 & 64,* 315.

Brandt, N. B., Moshchalkov, V. V. (1984), *Advances in Physics 33,* 373–468.
Brodsky, M. B. (1978), *Rep. Prog. Phys. 41,* 1547.
Buschow, K. H. J. (1977), *Rep. Prog. Phys. 40,* 1179–1256.
Buschow, K. H. J. (1979), *Rep. Prog. Phys. 42,* 1373–1477.
Buschow, K. H. J. (1980), in: *Ferromagnetic Materials, Vol. 1:* Wohlfarth, E. P. (Ed.). Amsterdam: North-Holland, pp. 297–414.
Buschow, K. H. J. (1988), in: *Ferromagnetic Materials, Vol. 4:* Wohlfarth, E. P., Buschow, K. H. J. (Eds.). Amsterdam: North-Holland, pp. 1–130.
Buschow, K. H. J., Van Daal, H. J. (1969), *Phys. Rev. Lett. 23,* 408.
Buschow, K. H. J., Goebel, U., Dormann, E. (1979), *Phys. Stat. Solidi b,* 93, 607–615.
Chappert, J. (1982), in: *Magnetism of Metals and Alloys.* Cyrot, M. (Ed.). Amsterdam: North-Holland, pp. 487–533.
Chikazumi, S. (1978), *Physics of Magnetism.* New York: R. E. Krieger Publishing Company.
Chudnovsky, E. M. (1988), *J. Appl. Phys. 64,* 5770.
Clark, A. E. (1980), in: *Ferromagnetic Materials,* Vol. 1: Wohlfarth, E. P. (Ed.). Amsterdam: North-Holland, pp. 531–589.
Coey, J. M. D., Readman, P. W. (1973), *Nature 246,* 476.
Coey, J. M. D., Chappert, J., Rebouillat, J. P., Wang, T. S. (1976), *Phys. Rev. Lett. 36,* 1061.
Coqblin, B. (1977), *The Electronic Structure of Rare Earth Metals and Alloys: The Magnetic Heavy Rare Earths.* New York: Academic Press.
Cornut, B., Coqblin, B. (1972), *Phys. Rev. 5,* 4541.
Cullity, B. D. (1972), *Introduction to Magnetic Materials.* Reading (MA): Addison-Wesley Publishing Company.
Cyrot, M., Lavagna, M. (1979), *J. Physique 40,* 763.
Doniach, S. (1977), *Physica B 91,* 231.
Elliott, R. J. (1961), *Phys. Rev. 124,* 348.
Fisk, Z., Hess, D. W., Pethick, C. J., Pines, D., Smith, J. L., Thompson, J. D., Willis, J. O. (1988), *Science 239,* 33–42.
Flouquet, J., Haen, P., Vettier, C. (1982), *J. Magn. Magn. Mater. 29,* 159.
Franse, J. J. M., de Boer, F. R., Frings, P. H., Gersdorf, R., Menovsky, A., Muller, F. A., Radwanski, R. J., Sinnema, S. (1985), *Phys. Rev. B 31,* 4347.
Franse, J. J. M., Radwanski, R. J., Sinnema, S. (1988), *J. Physique 49,* C8, 505.
Friedel, J. (1956), *Canad. J. Phys. 34,* 1190.
Friedel, J., Leman, G., Olszewski, S. (1961), *J. Appl. Phys. 32,* 3255.
Friedel, J. (1969), in: *The Physics of Metals, 2. Electrons:* Zeeman, J. M. (Ed.). Cambridge: Cambridge University Press.
Fulde, P. (1988), *J. Phys. F 18,* 601.
Fulde, P., Loewenhaupt, M. (1985), *Advances in Physics 34,* 589.
Fujita, T., Satoh, K., Onuki, Y., Komatsubara, J. (1985), *J. Magn. Magn. Mater. 47–48,* 66.
Gautier, F. (1982), in: *Magnetism of Metals and Alloys:* Cyrot, M. (Ed.). Amsterdam: North-Holland, pp. 1–244.
Gignoux, D., Gomez-Sal, J. C. (1984), *Phys. Rev. B 30,* 3967.
Gignoux, D., Gomez-Sal, J. C. (1985), *J. Appl. Phys. 57,* 3125–3130.
Gignoux, D., Lemaire, R., Paccard, D. (1972a), *Phys. Letters 41 A,* 187.
Gignoux, D., Rossat-Mignod, J., Tcheou, F. (1972b), *Phys. Stat. Solidi 14a,* 483.
Gignoux, D., Lemaire, R., Molho, P. (1980), *J. Magn. Magn. Mater. 21,* 119.
Gignoux, D., Givord, D., Lienard, A. (1982), *J. Appl. Phys. 53,* 2321.
Gignoux, D., Givord, F., Hennion, B., Ishikawa, Y., Lemaire, R. (1985), *J. Magn. Magn. Mater. 52,* 421.
Gignoux, D., Vettier, C., Voiron, J. (1987), *J. Magn. Magn. Mater. 70,* 388.
Giraud, M., Morin, P., Rouchy, J., Schmitt, D. (1986), *J. Magn. Magn. Mater. 59,* 255–265.
Givord, D., Laforest, J., Lemaire, R. (1979), *J. Appl. Phys. 50,* 7489.
Givord, D., Lu, Q., Rossignol, M., Tenaud, P., Viadieu, T. (1990a), *J. Magn. Magn. Mater. 83,* 183.
Givord, D., Lu, Q., Tenaud, P., Viadieu, T. (1990b), in: *R–Fe–B Magnets:* Coey, J. M. D. (Ed.). Oxford: University Press.
Grüner, G., Zawadowski, A. (1974), *Rep. Prog. Phys. 37,* 1497.
Hansen, P. (1991), in: *Ferromagnetic Materials, Vol. 6:* Wohlfarth, E. P., Buschow, K. H. J. (Eds.). Amsterdam: North-Holland.
Harris, R., Plischke, M., Zuckermann, M. J. (1973), *Phys. Rev. Lett. 31,* 160.
Heeger, A. J. (1969), *Sol. Stat. Phys. 23,* 283.
Heine, V. (1964), *Group Theory in Quantum Theory.* Oxford: Pergamon Press.
Herring, C. (1966), *Magnetism, Vol. IV: Exchange Interactions among Itinerant Electrons:* Rado, G. T., Suhl, H. (Eds.). New York, London: Academic Press.
Herpin, A. (1968), *Théorie du Magnetisme:* Institut National des Sciences et Techniques Nucleaires (Ed.). Paris: P.U.F.
Hurd, C. M. (1975), *Electrons in Metals.* New York: John Wiley and Sons.
Hutchings, M. T. (1964), *Sol. Stat. Phys. 16,* 227.
Imry, Y., Ma, S. K. (1975), *Phys. Rev. Lett. 35,* 1399.
Inoue, J., Shimizu, M. (1982), *J. Phys. F: Metal. Phys. 12,* 1811.
Janak, J. F., Williams, A. R. (1976), *Phys. Rev. B 14,* 4199.
Jayaprakash, C., Kirkpatrick, S. (1980), *Phys. Rev. B 21,* 4072.
Kaplan, T. A. (1961), *Phys. Rev. 124,* 329.

Kasuya, T. (1956), *Prog. Theor. Phys. 16,* 58.

Kirchmayr, H. R., Poldy, C. A. (1979), *Handbook on the Physics and Chemistry of Rare Earth, Vol. 2:* Gschneidner, K. A. Jr., Eyring, L. (Eds.). Amsterdam: North-Holland, pp. 55–230.

Kondo, J. (1969), *Sol. Stat. Phys. 23,* 183.

Koskenmaki, D. C., Gschneidner, K. A. Jr. (1978), *Handbook on the Physics and Chemistry of Rare Earth, Vol. 1:* Gschneidner, K. A. Jr., Eyring, L. (Eds.). Amsterdam: North-Holland, pp. 337–377.

Lacroix, C., Cyrot, M. (1979), *Phys. Rev. B 20,* 1969.

Lahiouel, R., Galera, R. M., Pierre, J., Siaud, E., Murani, A. (1987a), *J. Magn. Magn. Mater. 63&64,* 98.

Lahiouel, R., Pierre, J., Siaud, E., Galera, R. M., Besnus, M. J., Kappler, J. P., Murani, A. P. (1987b), *Z. Phys. B – Condensed Matter 67,* 185–191.

Lawrence, J. M. (1979), *Phys. Rev. B 20,* 3770–3782.

Lawrence, J. M., Riseborough, P. S., Parks, R. D. (1981), *Rep. Prog. Phys. 44,* 1–84.

Lee, E. W., Pourarian, F. (1976), *Phys. Stat. Sol. (a) 33,* 483.

Lemaire, R. (1966a), *Cobalt 32,* 132–140.

Lemaire, R. (1966b), *Cobalt 32,* 201–211.

Lines, M. E. (1979), *Physics Reports 55,* 133.

Lonzarich, G. G. (1984), *J. Magn. Magn. Mater. 45,* 43.

Lonzarich, G. G., Taillefer, L. (1985), *J. Phys. C 18,* 4339.

Lüthi, B. (1985), *J. Magn. Magn. Mater. 52,* 70.

Lüthi, B., Yoshizawa, M. (1987), *J. Magn. Magn. Mater. 63&64,* 274.

MacPherson, M. R., Everett, G. E., Wohlleben, D., Maple, M. B. (1971), *Phys. Rev. Letters 26,* 20.

Maletta, H., Zinn, W. (1989), *Handbook on the Physics and Chemistry of Rare Earths, Vol. 12:* Gschneidner, H. A. Jr., Eyring, L. (Eds.). Amsterdam: North-Holland, pp. 213–356.

Mattis, D. C. (1981), *The Theory of Magnetism I, Solid State Sciences 17:* Fulde, P. (Ed.). Berlin: Springer-Verlag.

McCaig, M. (1977), *Permanent Magnets.* Pentech Press, printed in Great Britain by Billing and Sons Ltd., Guildford and London.

Minakata, R., Shiga, M., Nakamura, Y. (1976), *J. Phys. Soc. Japan 41,* 1435.

Misawa, S. (1986), *Solid State Commun. 58,* 63.

Misawa, S. (1988), *Physica B 149,* 162.

Morin, P., Schmitt, D. (1989), in: *Ferromagnetic Materials, Vol. 5:* Wohlfarth, E. P., Buschow, K. H. J. (Eds.). Amsterdam: North-Holland, p. 30.

Moriya, T. (1979), *J. Magn. Magn. Mater. 14,* 1.

Moriya, T. (1987), in: *Metallic Magnetism, Vol. 42, in the series: Topics in Current Physics:* Capellmann, H. (Ed.). Berlin: Springer-Verlag, pp. 15–54.

Moriya, T., Kawabata, A. (1973), *J. Phys. Soc. Japan 34,* 639 and *35,* 669.

Moriya, T., Takahashi, Y. (1978), *J. Phys. Soc. Japan 45,* 397.

Murata, K. K., Doniach, S. (1972), *Phys. Rev. Lett. 29,* 285.

Néel, L. (1978), *Oeuvres Scientifiques:* C.N.R.S. (Ed.). Paris: C.N.R.S.

Nozières, P. (1974), *J. Low. Temp. Phys. 17,* 31.

Rammal, R., Souletie, J. (1982), in: *Magnetism of Metals and Alloys:* Cyrot, M. (Ed.). Amsterdam: North-Holland, pp. 379–485.

Reim, W., Schoenes, J. (1989), in: *Ferromagnetic Materials, Vol. 5:* Wohlfarth, E. P., Buschow, K. H. J. (Eds.). Amsterdam: North-Holland.

Rossat-Mignod, J. (1986), in: *Neutron Scattering in Condensed Matter Research:* Sköld, K., Price, D. L. (Eds.). London, New York: Academic Press, Chap. 20.

Ruderman, M. A., Kittel, C. (1954), *Phys. Rev. 96,* 99.

Sakakibara, T., Goto, T., Yoshimura, K., Shiga, M., Nakamura, Y., Fukamichi, K. (1987), *J. Magn. Magn. Mater. 70,* 126–128.

Schinkel, C. J. (1978), *J. Phys. F: Met. Phys. 8,* L87.

Schoenes, J. (1990), in: *Materials Science and Technology, Vol. 3: Electronic and Magnetic Properties of Metals and Ceramics:* Buschow, K. H. J. (Ed.). Weinheim: VCH-Verlag, Ch. 7.

Schrieffer, J. R., Wolff, P. A. (1966), *Phys. Rev. 149,* 491.

Sechovsky, V., Havela, L. (1988), in: *Ferromagnetic Materials, Vol. 4:* Wohlfarth, E. P., Buschow, K. H. J. (Eds.). Amsterdam: North-Holland, pp. 309–491.

Shimizu, M. (1981), *Rep. Prog. Phys. 44,* 331–409.

Shimizu, M. (1982), *J. Physique 43,* 155.

Shimizu, M., Inoue, J., Nagasawa, S. (1984), *J. Phys. F: Met. Phys. 14,* 2673–2687.

Steglich, F. (1985), *Theory of Heavy Fermions and Valence Fluctuations.* Solid State Sciences 62. Kasuya, T., Saso, T. (Eds.). Berlin: Springer-Verlag, pp. 23–44.

Stevens, K. W. H. (1952), *Proc. Phys. Soc. (London) A 65,* 209.

Stewart, G. R. (1984), *Rev. Mod. Phys. 56,* 755–787.

Stoner, E. C. (1938), *Proc. Roy. Soc. A 165,* 372.

Strnat, K. J. (1988), in: *Ferromagnetic Materials, Vol. 4:* Wohlfarth, E. P., Buschow, K. H. J. (Eds.). Amsterdam: North-Holland, pp. 131–210.

Suhl, H. (Ed.) (1973), *Magnetism, Vol. V: Magnetic Properties of Metallic Alloys.* New York, London: Academic Press.

Van den Broek, J. J., Zijlstra, H. (1971), *I.E.E.E. Trans. Magn. 7,* 226.

Varma, C. M. (1985), *Comments Solid State Phys. 11,* 221–243.

Von Molnar, S., Barbara, B., McGuire, T. R., Gambino, R. J. (1982), *J. Appl. Phys. 53,* 1350.

Wassermann, E. F. (1989), in: *Ferromagnetic Materials, Vol. 5:* Wohlfarth, E. P., Buschow, K. H. J. (Eds.). Amsterdam: North-Holland.

Weinert, M., Freeman, A. J. (1983), *J. Magn. Magn. Mater. 38,* 23–33.

White, R. M. (1970), *Quantum Theory of Magnetism, Advanced Physics Monograph Series*. New York: McGraw-Hill.

Wohlfarth, E. P., Rhodes, P. (1962), *Philos. Mag. 7,* 1817.

Wohlleben, D., Wittershagen, B. (1985), *Advances in Physics 34,* 403.

Yoshida, K. (1957), *Phys. Rev. 106,* 893.

Zijlstra, H. (1982), in: *Ferromagnetic Materials, Vol. 3:* Wohlfarth, E. P. (Ed.). Amsterdam: North-Holland, pp. 37–105.

General Reading

Barbara, B., Gignoux, D., Vettier, C. (1988), *Lectures on Modern Magnetism,* Science Press. Berlin: Springer-Verlag.

Bacon, G. E. (1975), *Neutron Diffraction*. Oxford: Clarendon Press.

Chikazumi, S. (1964), *Physics of Magnetism*. New York, London, Sydney: John Wiley and Sons.

Coqblin, B. (1977), *The Electronic Structure of Rare Earth Metals and Alloys: The Magnetic Heavy Rare Earths*. New York: Academic Press.

Cullity, B. D. (1972), *Introduction to Magnetic Materials*. Reading, Mass.: Addison-Wesley Publishing Company.

Ferromagnetic Materials, Vol. 1 to 6: Wohlfarth, E. P., Buschow, K. H. J. (Eds.). Amsterdam: North-Holland.

Friedel, J. (1969), *The Physics of Metals, Vol. 1: Electrons:* Zeeman, J. M. (Ed.). Cambridge: Cambridge University Press, Ch. 8.

Kasuya, T., Saso, T. (Eds.) (1985), *Theory of Heavy Fermions and Valence Fluctuations,* in: *Solid State Sciences 63*. Berlin: Springer-Verlag.

Magnetism of Metals and Alloys: Cyrot, M. (Ed.). Amsterdam: North-Holland.

Mattis, D. C. (1981), *The Theory of Magnetism I,* in: *Solid State Sciences 17:* Fulde, P. (Ed.). Berlin: Springer-Verlag.

6 Ultrathin Films and Superlattices

Shengnian N. Song and John B. Ketterson

Department of Physics and Astronomy and Materials Research Center, Northwestern University, Evanston, IL, U.S.A.

List of Symbols and Abbreviations

a	thickness of material A and B
a_{nm}	expansion coefficients
A	cross section of the wire
A	absolute energy difference between sites
$\boldsymbol{A}(\boldsymbol{r})$	vector potential
A_{12}	exchange constant
A_{e}	extremal-cross-sectional areas of the Fermi surface
b, b_{r}	nuclear scattering length (of atom r)
b_{n}	modulation function
B	Bloch constant
B	magnetic field
c	lattice constant along the c-axis
$\bar{c}$	velocity of electromagnetic waves propagating in the junction
$c(x)$	composition profile
C	constant
$C_{\mathrm{A}}(n)$	concentration modulation for the A-layer
$\mathscr{C}$	charge conjunction operator
d, d_1, d_2	film thickness
d_0	monolayer thickness
d_0	thickness of non-magnetic layer
d_{cr}	Cr layer thickness
D	effective dimensionality
D	exchange stiffness constant
$D, D_{\mathrm{A}}, D_{\mathrm{B}}$	electron diffusion coefficient
D_z	effective diffusion coefficient along z
e	electronic charge
E	total energy density
E_{c}	critical energy
E_{g}	Coulomb gap
E_{F}	Fermi energy
$E_{m m'}$	absolute energy difference between sites
f	angle-dependent form factor
F	electric field
F	thermodynamic free energy
F_j	structure factor
F_n	thermodynamic free energy in the normal state
$F_{\mathrm{M}}(+Q)$	structure factor associated with magnetic scattering
$F_{\mathrm{N}}(\boldsymbol{Q})$	structure factor associated with nuclear scattering
$F(z)$	pair amplitude
$\tilde{F}_{\sigma}$	correction factor due to Hartree terms
g	Landé splitting factor
g_1	coupling constant
$g_n(\boldsymbol{r})$	normalized eigenfunctions

g_ξ	function
$\mathcal{g}_\xi(r, r', \Omega)$	unnormalized correlation function
$g(\boldsymbol{r}, \boldsymbol{r}', t)$	Fourier transform of g_ξ
$\mathcal{G}_\omega$	one particle Green's function
h	rms fluctuations in film thickness
H	Hall coefficient
H	external field, along the z-direction
H_c	coercivity
H_i	internal field
H_s	saturation field
$\mathcal{H}'$	perturbing Hamiltonian
I_s	scattering intensity
j_m	parameter of the Josephson equation
j_{sf}	s-f exchange integral
J	interlayer exchange coupling
J	quantum number corresponding to the total angular moment of the atom
$J(\boldsymbol{q})$	exchange coupling function
k_F	Fermi wave vector
k_z	wave vector (normal of the film plane)
$\boldsymbol{k}_i$	wave vector of incident light
K_s	surface anisotropy
K_u	uniaxial anisotropy constant
K_v	volume anisotropy
$K(\boldsymbol{r}, \boldsymbol{r}')$	kernel
$\mathcal{K}$	Wigner time reversal operator
$\tilde{\mathcal{K}}$	transpose of $\mathcal{K}$
l, l_1, l_2	mean free path
L	size of the system
L_i	inelastic-scattering length
L_H	cyclotron orbit diameter
L_{th}	thermal length
m	index for the bilayer where the atom resides
$\mathbf{m}$	effective mass tensor
$\left(\frac{1}{\mathbf{m}}\right)$	reciprocal effective mass tensor
M_s	saturation magnetization
$\hat{\boldsymbol{M}}$	unit vector along the direction of the atomic moment
n	electron density
n	metal volume fraction
n_c	threshold value of n
N	number of bilayers
N_F	number of occupied subbands
$N(E_F)$	density of localized states
N_x, N_y, N_z	demagnetization factors
N_Λ	number of atomic planes per bilayer

$N(\xi, \boldsymbol{r})$	local density of states
$N(\xi)$	density of states
p, q	specularity parameter
p	temperature exponent in weak localization
p_n	modulation function
p, p_r	magnetic scattering length (of atom r)
P	oscillation period of the SdH oscillations
$P(m, n)$	magnetic amplitudes
$\boldsymbol{q}$	scattering vector
Q_j	peak position
$\boldsymbol{Q}$	scattering vector
$\hat{\boldsymbol{Q}}$	unit vector along the direction of the scattering vector
$\boldsymbol{r}$	vector denoting the position of the atom in the crystal
$r(m, n)$	atom position along the c-axis
R	hopping distance
R	flipping ratio
$R_\square$	sheet resistance
R_g	grain boundary refraction coefficient
$R_m, R_{m'}$	localized sites
$\boldsymbol{R}$	position operator
s	layer thickness
t	conductivity onset exponents
T	temperature
T_0	constant
T_c	critical temperature
T_c, T_{c0}	transition temperatures
$T_{c\infty}$	bulk transition temperature
$T(\boldsymbol{H})$	field dependent transition temperature
V, V_A, V_B	cut-off in the BCS interaction
$V(\boldsymbol{A})$	pairing potential
w	mean grain width
w	energy separation between two states
$w_n(\boldsymbol{r})$	real eigenfunctions of the normal state mean field Hamiltonian
W_{mm}	hopping probability
$z(n)$	relative position of the nth atom
$1/\alpha$	localization length
α	incident angle
β	critical exponent, diffusion length
γ	gyromagnetic ratio
$\Delta(z)$	gap function
ε	dielectric constant
θ	electron angle of incidences to the plane
θ	angle between magnetization and surface normal
θ, ϕ	polar coordinates of the magnetization

θ_D	Debye temperature
λ	de Broglie wavelength
λ	shift exponent
λ	wavelength
$\lambda(n_z)$	dimensionless eigenvalue
Λ	superlattice modulation wavelength
μ_B	Bohr magneton
μ_l	local moment
$\bar{\mu}$	conduction electron polarization
ν	Matsubara integers
ν_{ph}	phonon frequency
ξ	roughness-coherence length
ξ	Ginzburg-Landau coherence length
ξ_0	zero temperature coherence length
ξ_n	energies of the normal state
ϱ_f	resistivity of the film
ϱ_g	contribution to the resistivity from grain boundary scattering
ϱ_∞	bulk resistivity
$\sigma, \sigma_1, \sigma_2$	bulk conductivity
σ_2, σ_z	Pauli matrix
τ	electron relaxation time
τ_i, τ_s	inelastic and spin scattering time
τ_{so}	spin-orbit scattering time
τ_ϕ	de-phasing time
ϕ_0	flux quantum
$\phi_n(\boldsymbol{r})$	one particle self-consistent eigenfunctions
$\phi_{N\alpha}$	eigenfunctions of the normal state mean field Hamiltonian
Φ	total magnetic phase shift
Φ	phase difference
$\Delta\Phi$	amplitude phase variation
$\chi(\boldsymbol{q})$	wave vector dependent susceptibility
$\chi^{(P)}$	band susceptibility
ψ	complex order parameter
ω	resonance frequency
ω	Matsubara frequencies
ω_D	Debye cut-off energy
ω_o	Josephson plasma frequency
ω_p	plasma frequency
Ω_n	eigenvalues
Ω_L	spin Lamour frequency
AF	antiferromagnetic
AP	atomic plane
BCS	Bardeen-Cooper Schrieffer
DE	Damon-Eshback

FMR	ferromagnetic resonance
FWHM	peakwidth of half maxima
GL	Glinzburg-Landau
JK	Jin and Ketterson (1986)
LS	light scattering
MBE	molecular beam epitaxy
MSA	magnetic surface anisotropy
QSE	quantum size effects
RE	rare earth
RH	Rado and Hicken
RKKY	Ruderman-Kittel-Kasuya-Yosida
rms	root mean square
SdH	Shubnikov-de Haas
SDF	spin-density-functional
SI	international system of units
S-I	superconductor-insulator
SM	surface mode
SQUID	superconducting quantum interference device

6.1 Introduction

This chapter reviews the transport, magnetic and superconducting properties of thin films and artificial multilayers, at least one constituent of which is a metal. One of the major scientific motivations to study thin films is to observe property changes when some fundamental electronic length scale becomes comparable to the film thickness. For transport properties some relevant length scales include the de Broglie wavelength, the mean free path, various length scales characterizing electron localization and (in a magnetic field) the cyclotron radius. In a magnetic system the relevant length scale is the range of the exchange interaction which is of order an interatomic spacing. In a superconductor there are two relevant length scales: the London penetration depth and the coherence length (the range of the pairing interaction itself is of order an interatomic spacing). Much of the technological interest in the electronic properties of thin films centers around their present and future uses in the electronics and recording industries: conductors connect semiconductor devices, magnetic films function as memory elements and superconductors promise to be useful for all three (devices, interconnects, memories). (Although the high T_c superconductors are most relevant to the latter they will be treated elsewhere in the series.)

Artificial multilayers are structures prepared by sequentially depositing two (or more) materials from independent sources on a suitable substrate. To be specific, suppose that we have two substances, called A and B, contained in two deposition sources which, in turn, produce vapor fluxes (or beams) of the materials. These beams are directed towards the substrate. We further suppose that above the sources we have some sort of shutter or beam interrupt mechanism. The opening and closing of this shutter is timed so that an average thickness a of material A deposits followed by an average thickness b of material B; the whole process is then repeated N times. We designate the structure by writing A/B. The thicknesses a and b are referred to as the sublayer thicknesses and the layer thickness $a + b = \Lambda$ as the superlattice (or composition) wavelength. The resulting structure will be referred to as an artificial multilayer. Other names have been given to such structures including artificial superlattices, composition modulated alloys (or foils), coherent-modulated structures, and layered ultrathin coherent structures. The thickness a and b may range from a few atomic (or molecular) planes to thicknesses of hundreds or thousands of atomic planes. Materials with very large sublayer thicknesses belong more naturally to the field of composite materials and will not be of interest to us here.

An ideal structure might be envisaged as one in which the thicknesses a and b are strictly constant and where no interdiffusion takes place between layers. If each sublayer consisted of an integral number of atomic planes and the interatomic spacings and lattice symmetry at the layer interfaces of the two constituents were sufficiently compatible, the artificial superlattice would be a single crystal with an appropriately enlarged unit cell. This idealized limit is intuitively very appealing but is seldom realized in practice. The well known GaAs/AlAs semiconductor superlattice approximates this limit; however, most of the artificial superlattice systems which have been prepared depart rather strongly form this ideal. In fact, materials having a large difference in the in-plane interatomic spacings and even different in-plane symmetries will often form rather uniformly layered superlattices. As an example (Schuller, 1980), we cite the case of a superlattice consisting of (111) planes of Cu on Nb (110) (the lattice constants of Cu and Nb are $a = 0.361$ nm and $a = 0.329$ nm respectively). In contrast high quality artificial superlattices of Nb and Ta form (Durbin et al., 1981 and 1982) and the close matching of the lattice constant (0.329 nm as opposed to 0.330 nm) suggest that this metallic system approximates a single crystal with a large unit cell.

There is no guarantee that two constituents will form a high quality multilayer structure. Interlayer diffusion (or even a chemical reaction) may obliterate the interface. One would like the two constituents to wet each other sufficiently well, at least metastabally, so that they will mutually spread on one another, although we must keep in mind that, at the macroscopic level, if material A wets material B, B will not wet A. Island growth is a common mode, particularly for metals on insulators. Hence it is generally easier to make a metal/metal multilayer than a metal/insulator. Multilayers offer the possibility of preparing materials in metastable structures. If an unstable structure has only a slightly higher free energy than its stable form it may be possible to stabilize the former through interface coherency with a lattice compatible (symmetry and spacing) substrate (called interface coherency stabilization or pseudo-morphic growth).

In an artificial multilayer we may *tune* the sublayer thicknesses to match the various thin film electronic length scales. For magnetic properties, the availability of a larger number of interfaces (relative to a thin film) produces larger signals.

The larger number of parameters and greater total quantity of material give multilayers certain advantages over thin films, particularly for recording or superconducting applications.

Our primary emphasis in this chapter will be in describing the material electronic properties (transport, magnetic and superconducting) and the underlying physics of thin films and multilayers. We will also mention a few of the possible applications.

6.2 Transport Properties

6.2.1 Size Effects

The finite dimensions of a thin film can, under appropriate conditions, have rather drastic effects, called size effects, on the electronic structure and transport properties. Size effects in thin films have been extensively discussed by Tellier and Tosser (1982). The length scales of interest are the effective carrier de Broglie wavelength, λ, the mean free path, l, and the film thickness, d (or for a superlattice, the modulation wavelength Λ). Depending on the relative magnitudes of these quantities, transport phenomena in thin films or superlattices can be divided into several regimes: (i) the classical regime where $\lambda < l \ll d$, and the electron is considered to be a classical particle with the transport properties calculated using the Boltzmann transport equation; (ii) the classical size-effect regime where $\lambda \ll d < l$; and (iii) the quantum size-effect regime: $d \leq \lambda$.

6.2.1.1 Classical Size Effects

In the classical size-effect regime, the surface of the film contributes significantly to the scattering of electrons. The problem was first discussed by Thompson (1901). Later, Fuchs (1938) and Sondheimer (1952) improved Thomson's formulation. The Fuchs theory gives the following expression for the ratio of the resistivity of a film to the bulk resistivity, ϱ_∞:

$$\frac{\varrho_\infty}{\varrho_f} = 1 - \frac{3}{2\varkappa} \cdot \int_0^1 du\,(u-u^3)\,\frac{(1-p)\,[1-\exp(-\varkappa/u)]}{1-p\exp(-\varkappa/u)} \qquad (6\text{-}1)$$

where $\varkappa = d/l$, l is the bulk mean free path, and p is the specularity parameter. The thin film and thick film limits are given by

$$\varrho_f = \varrho_\infty\left[1 + \frac{3}{8}(1-p)\frac{l}{d}\right] \quad (d \gg l) \qquad (6\text{-}2a)$$

which shows a simple $1/d$ dependence of the film resistivity and

$$\varrho_f = \varrho_\infty\,\frac{4}{3}\,\frac{1-p}{1+p}\,\frac{l}{d}\,\frac{1}{\ln(l/d)} \quad (d \ll l) \qquad (6\text{-}2b)$$

The Fuchs theory oversimplifies the problem by specifying the scattering in terms of a fixed specularity parameter. In fact, specularity depends upon the electron angle of incidences, θ, to the plane. Soffer (1967) introduced an angularly dependent specularity parameter

$$p(\cos\theta) = \exp\left[-\left(\frac{4\pi h}{\lambda}\cos\theta\right)^2\right] \qquad (6\text{-}3)$$

where h is the rms roughness of the boundary. Equation (6-3) is then substituted into Eq. (6-1) with $u = \cos\theta$.

It is often the case that a substrate-metal interface will have an entirely different character from that of a metal-vacuum interface. Two different specularities, p and q, have been introduced to account for this situation (Lucas, 1965). The asymptotic expressions are

$$\varrho_f = \varrho_\infty\left[1 + \frac{3}{8}\left(1 - \frac{p+q}{2}\right)\frac{l}{d}\right] \quad d \gg l \qquad (6\text{-}4a)$$

and

$$\varrho_f = \varrho_\infty\,\frac{4}{3}\,\frac{1-pq}{(1+p)(1+q)}\,\frac{l}{d}\,\frac{1}{\ln(l/d)} \quad d \ll l \qquad (6\text{-}4b)$$

Comparing these equations with Eq. (6-2) shows that the parameter p in the Fuchs-Sondheimer expressions is an *effective* specularity.

Except for epitaxial or single crystal films, thin films usually contain a large number of grain boundaries. Mayada and Shatzkes (1970) have discussed the effects of grain boundary scattering in thin films by assuming free-electron-like behavior and modeling the grain boundaries as randomly placed partially reflecting surfaces perpendicular to the film. The parameters involved are the grain boundary reflection coefficient, R_g, and the mean grain width, w. In the limit of large l and small w, the contribution to the resistivity from grain boundary scattering, ϱ_g, is given by

$$\frac{\varrho_\infty}{\varrho_g} = 1 - \frac{3}{2}\,\frac{R_g}{1-R_g}\,\frac{l}{w} \qquad (6\text{-}5)$$

It is often found experimentally that the grain size rises roughly linearly with the film thickness and consequently the grain boundary scattering gives a resistivity which varies inversely with the film thickness, i.e., with the same thickness dependence as that predicted by the Fuchs theory. In view of this fact, caution must be exercised when using the Fuchs theory to deduce the important material parameter $\varrho_\infty l$. Measuring the behavior over a wide temperature range will help to differentiate between grain boundary and surface scattering. All surface scattering theories lead to a markedly temperature-dependent contribution from the surface scattering (the so-called deviation from Matthiessen's rule) whereas grain boundary scattering theories give no such temperature dependence (Sambles, 1983). Since a discontinuity or islandization has remarkable effects on the transport properties of thin films. Extensive structural characterization of a thin film must be carried out first in order to meaningfully apply the above theories.

For a superlattice with its modulation wavelength, Λ, in the classical size-effect regime, a theory analogous to Fuchs' theory for thin films can be formulated by introducing the analogous probability for coherent passage across an interface and ignoring the refraction of the electron wave. Carcia and Suna (1983) derived the following expression for the conductivity σ of an infinite layered structure composed of metals 1 and 2, having layer thicknesses d_1 and d_2, bulk conductivities σ_1, σ_2, and mean free paths l_1, l_2:

$$\sigma \Lambda = d_1 \sigma_1 + d_2 \sigma_2 + + [(l_2 \sigma_1 + l_1 \sigma_2) p - l_1 \sigma_1 - l_2 \sigma_2] \cdot \cdot I - (1-p^2) [\sigma_1 l_1 J_1 + \sigma_2 l_2 J_2] \quad (6\text{-}6)$$

where the integrals I, J_i are defined as

$$I = (3/2) \int_0^1 \mathrm{d}u \, u (1-u^2)(1-e_1)(1-e_2)/ (1-p^2 e_1 e_2) \quad (6\text{-}7\,\text{a})$$

$$J_1 = (3/2) \int_0^1 \mathrm{d}u \, u (1-u^2)(1-e_1) e_2/ (1-p^2 e_1 e_2) \quad (6\text{-}7\,\text{b})$$

$$J_2 = (3/2) \int_0^1 \mathrm{d}u \, u (1-u^2)(1-e_2) e_1/ (1-p^2 e_1 e_2) \quad (6\text{-}7\,\text{c})$$

with

$$e_i = \exp(-d_i/l_i u) \quad (6\text{-}7\,\text{d})$$

The resistivity of Pd/Au multilayer films at 77 K has been fitted to Eq. (8-6) (Fig. 6-1).

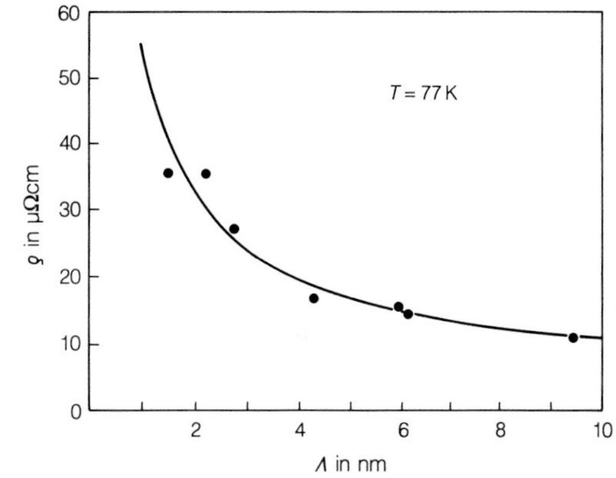

Figure 6-1. Electrical resistivity of Pd/Au films at 77 K. The solid curve is calculated according to Eq. (6-6) (after Carcia and Suna, 1983).

6.2.1.2 Quantum Size Effects

When the thickness of a film is comparable to the de Broglie wavelength, λ, the conduction and valence bands will break into subbands and the wave vector k_z (normal to the film plane) will be quantized approximately as $N\pi/d$. This leads to the so-called quantum size effects (QSE) which have been predicted theoretically (Lifshitz and Kosevich, 1953) and have been reported experimentally for many phenomena involving electrons in thin films. As the film thickness increases, the adjacent subband levels will successively cross the Fermi surface. Accordingly, the QSE causes oscillatory variations with film thickness of the Fermi energy E_F, electron density at E_F, the work function, the electrical resistivity, and various thermodynamic quantities. The QSE in the Fermi energy and density of states of thin films has been discussed by Rogers et al. (1987) based on infinite, as well as finite, square well models. The typical thickness dependence of the Fermi energy is shown in Fig. 6-2. Numerical results suggest that (for both the models) the oscillation period is given by $\Delta d = \lambda/2 = \pi/k_F$, which appears to be independent of film thickness. More sophisticated calculations have been done by Schulte (1976), who applied the density-functional formalism with the jellium approximation to calculate self-consistently the dependence of the work function on

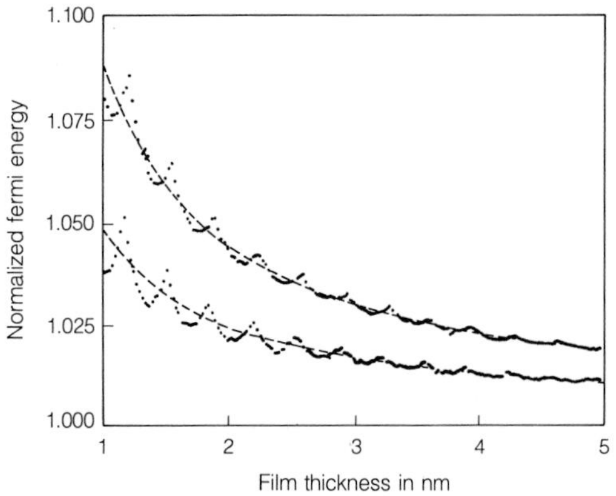

Figure 6-2. The normalized Fermi energy as a function of film thickness. The square well depth is 10 eV (upper curve) and 5 eV (lower curve) respectively (after Rogers et al., 1987).

film thickness in thin metallic films. The discrete lattice has been included in a fully self-consistent way in calculations by Feibelman and Hamann (1984), Ciraci and Batra (1986), and Rogers et al. (1989).

In a superlattice which consists of alternating conducting and insulating layers, the QSE should also exist when the insulators provide barriers which are sufficiently high and wide to confine the carriers in the conducting layers and the reciprocal lifetimes are small relative to the level splittings. Ivanov and Pollmann (1979) used a method based on a Green function formulation of scattering theory to study the QSE in a superlattice system. They calculated the band structure of a model GaAs/$Ga_{1-x}Al_xAs$ superlattice with different barrier widths. The results are shown in Fig. 6-3, where we see that the zone-folded superlattice bands evolve into discrete energy states for barrier thickness larger than 8.5 nm, which clearly demonstrates the presence of the QSE in a superlattice. Experimentally, the presence of the QSE in GaAs/GaAlAs was first observed by Tsu and Esaki (1973) in a tunneling experiment.

The QSE is expected to be more generally associated with semiconductors and semimetals (for a review see Ando et al., 1982) rather than with metals. For the latter, the reason for the elusiveness of the QSE is the difficulty of preparing sufficiently thin continuous metals films which are bounded by parallel surfaces with a large specularity parameter p. The success of a QSE experiment depends, therefore,

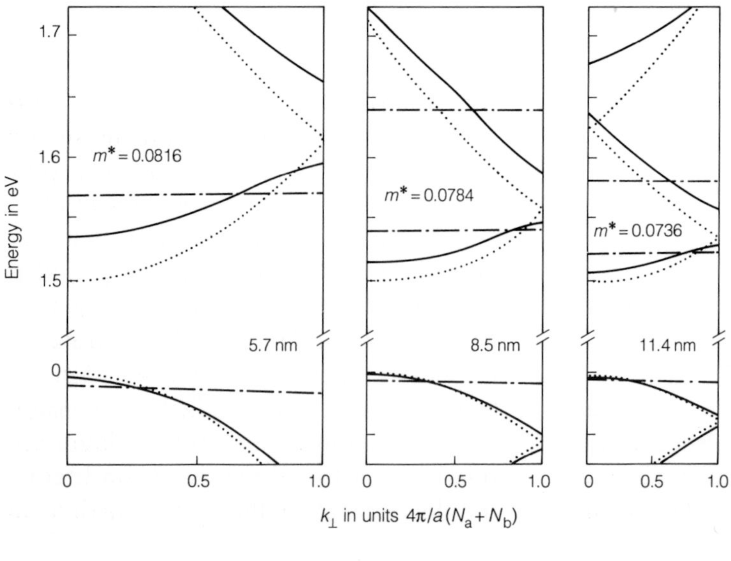

Figure 6-3. Superlattice bands in the direction perpendicular to the (100) layers for different GaAs layer thicknesses in comparison with the corresponding (folded) bulk bands (dotted lines). Solid lines represent the superlattice bands for a 1.1 nm-thick barrier (8 bilayers) and the dashed-dot lines show the superlattice bands for an 8.5 nm-thick barrier (60 bilayers) (after Ivanov and Pollmann, 1979).

on the preparation techniques for the film. Usually, epitaxial films are grown in two modes. Metals with a low surface energy can sometimes be grown on high-surface-energy metals in a monolayer-by-monolayer mode (Frank-van der Merwe mode) (Beuer and van der Merwe, 1986). On insulating or semiconducting substrates, metals as a rule grow via the formation of isolated nuclei (Volmer-Weber mode). The film becomes continuous only after a critical thickness has been reached which depends upon the metal, substrate, and deposition conditions. In any case, the film grows in steps of (at least) one monolayer, rather than increasing continuously. The oscillatory resistivity may consist of two components: that arising from surface-induced oscillations, due to the oscillatory variation of the effective specularities p and q with thickness, and that associated with the true QSE. As d can change only in multiples of the monolayer thickness d_0, and as d_0 and λ are in general incommensurate, the QSE condition is fulfilled only over limited thickness ranges $m_1 d_0 \leq d \leq m_2 d_0$ for which $m d_0 = n \lambda/2$. The frequently used picture of regular, saw-tooth-like oscillations in ϱ (which is based on a continuously changing film thickness) is not realistic. Figure 6-4 shows the oscillatory resistivity for a Pb film grown on the Si(111) surface (Jalochowski and Bauer, 1988). No oscillations are seen during the growth of the first 4 monolayers. After a structural transition at this thickness, weak but well-pronounced oscillations with a period of 1 monolayer occur. In the following we summarize studies on the electrical transport in ultra-thin metallic films.

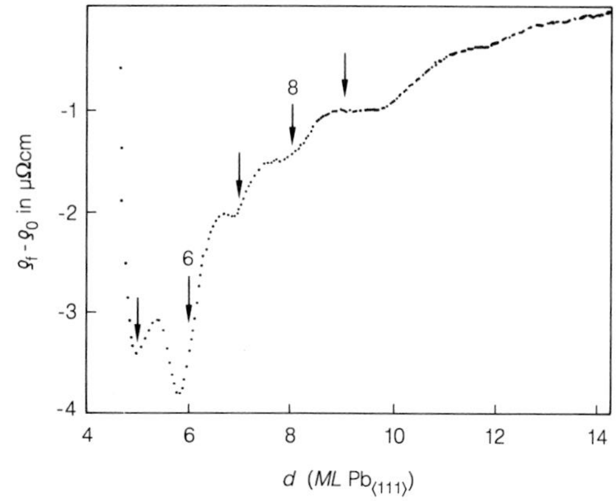

Figure 6-4. Specific resistivity difference ($\varrho_f - \varrho_0$) as a function of thickness of a Pb film grown on Si(111) (7 × 7) at 95 K (after Jatochowski and Bauer, 1988).

Tešanović et al. (1986) discussed the effect of surface scattering on quantum transport in thin films. In their model, the variations in the surface profile, characterized by h (the rms fluctuations in film thickness) and ξ (the roughness-coherence length), are transformed into a set of pseudo potentials acting on the quantum states of a system with the same average thickness but with a smooth surface. The resistivity of the film is given by

$$\frac{\varrho_\infty}{\varrho} = \frac{1}{N_F} \sum_{n=1}^{N_F} \left[1 + \frac{l}{l_{max}} n^2 \right]^{-1} \tag{6-8}$$

where N_F is the number of occupied subbands ($N_F \cong k_F d/\pi$) and $l_{max} = 6\pi (N_F^2/k_F^2 h^2) d$. This expression describes the crossover from the impurity-scattering-dominated region ($d/l \gg 1$) to the surface-scattering-dominated region ($d/l \ll 1$). Figure 6-5 shows the comparison between the theory [Eq. (6-8)] and the experimental results of Hensel et al. (1985), where the resistivity of epitaxially grown, single-crystal $CoSi_2$ films with $l \sim 100$ nm has been measured as a function of thickness in a wide region, $6 \leq d \leq 110$ nm. The model of Tešanović et al. has been extended, by including spin-dependent interfacial roughness scattering, to determine the origin of the giant magnetoresistance observed in Fe/Cr magnetic superlattices (Levy et al., 1990a).

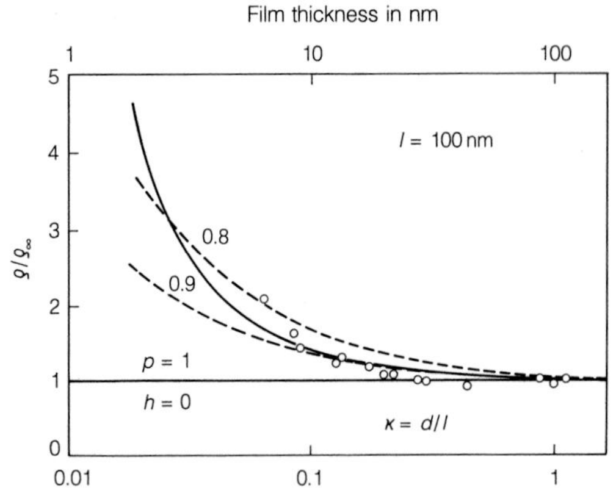

Figure 6-5. Resistivity vs. thickness for $CoSi_2$ films. The experimental values are shown as open circles. The solid line is calculated from Fuchs theory with different values of the specularity, p (after Tešanović et al., 1986).

At present our knowledge of quantum transport in the intermediate region (where several size-quantized subbands are relevant) is still limited. Experiments based on resistivity measurements are inevitably complicated by the detailed scattering mechanisms. The Shubnikov-de Haas (SdH) effect is a powerful way to explore dimensional crossover phenomena in quantum transport since the oscillation period of the SdH oscillations depends only on the topological structure of the Fermi surface and is given by $P = \Delta(1/B) = 2\pi e/\hbar A_e$, where A_e are the extremal-cross-sectional areas of the Fermi surface perpendicular to the field B. Song et al. (1990) have carried out systematic studies of the SdH oscillations in MBE grown gray tin films in the size effect quantized regime as a function of the film thickness, temperature and magnetic field (both magnitude and direction). The film thickness ranged from 80 nm to 160 nm, corresponding to 4–10 size-effect-quantized-subbands. One of the key tests of 2D character is the behavior of the oscillations as the field is titled away from the film plane normal. A strictly 2D system will have oscillatory properties which only involve the *normal component* of the field (Fang and Stiles, 1968). In Fig. 6-6 the values of $1/B$ at which maxima occur in the oscillations are plotted versus the quantum number n for the field at 0° and 45° with respect to the plane normal respectively. For the 160, 100, and 80 nm-thick samples, the ratios $P(45°)/P(0°)$ are 1.0, 0.82 and 0.74, respectively, which approach the value $(P(45°)/P(0°) = \cos\theta = 0.707)$ expected for a strictly 2D band. Therefore the data indicate that a gradual 3D-2D transition occurs as the film thickness is reduced.

6.2.2 Localization/Interaction Effects

6.2.2.1 Introduction

Various degrees of disorder exist in solids, ranging from weak disorder to the strongly disordered limit of an amorphous phase. As a result, the electronic structure and the transport properties of solids can be drastically modified. In these disordered materials, the mean free path of the elec-

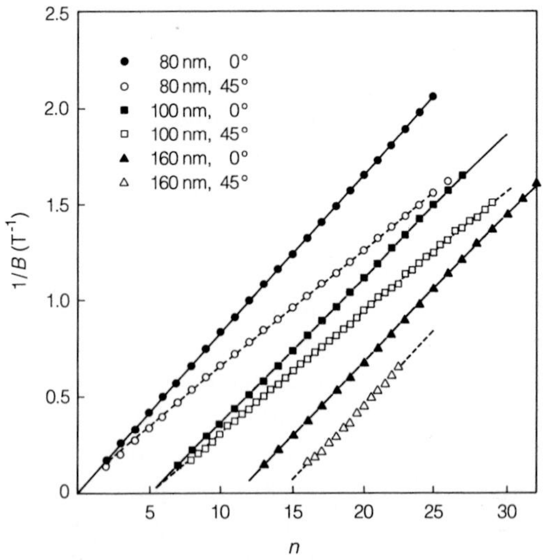

Figure 6-6. Nodal position of the SdH oscillations vs. quantum number n for three gray tin films having different thicknesses (after Song et al., 1990).

trons may approach an interatomic spacing; the Boltzmann equation is no longer valid in this limit.

One important consequence of disorder is that the electrons may become *localized,* i.e., spatially confined to some region (that can be as small as a single atomic site), as first pointed out by Anderson (1958). Mott (1967) later developed the concept of a mobility edge based on the Ioffe and Regel (1960) limit of minimum metallic conductivity ($k_F\, l \sim 1$). He suggested that the localized states are separated from the extended states by a critical energy E_c, called the mobility edge, which marks the transition between a conductor and an insulator (Mott and Davis, 1979). It was argued that there is a minimum metallic conductivity and that in 3D it is given by

$$\sigma_{\min}^{3D} \approx \left(\frac{e^2}{3\hbar a}\right) \Gamma^2 \tag{6-9}$$

where $\Gamma \sim 1/3$ (but depends on the coordination number) and in the Ioffe-Regel limit $a \sim 1/k_F$. In 2D,

$$\sigma_{\min}^{2D} \approx 0.1\, e^2/\hbar \tag{6-10}$$

In 1D systems, all states are localized even by weak disorder ($k_F\, l \gg 1$) owing to the quantum interference effect (weak localization), which is a *different* quantum-mechanical effect from that of strong localization (Mott and Twose, 1961).

For many years Mott's claim was not seriously challenged except by Webman and Cohen (1975) who viewed the metal-insulator transition as classical percolation of metallic regions imbedded in an insulating medium, rather than an intrinsically quantum phenomenon. Percolation theory implied an onset of $\sigma(0) \propto (n - n_c)^t$ with $t \approx 1.8$, where n is the metal volume fraction and n_c the corresponding threshold value. The scaling theory of the localization transition (Abrahams et al., 1979) has also cast some doubt on the concept of a minimum metallic conductivity. For 3D, $\sigma(0)$ has a continuous, albeit critical, onset: $\sigma(0) = \sigma_0\,(n/n_c - 1)^t$ with t being estimated to be 1. Various proposed models for the conductivity onset are shown in Fig. 6-7.

An important result of the scaling theory is that there is no true metallic behavior; in a 2D system, no matter how small the disorder is, the conductivity decreases either logarithmically (weak localization) or exponentially (strong localization), when the size L of the system is scaled down. The crossover from weak localization to strong localization in 2D is not as sharp as the metal-insulator transition in 3D systems, however, it may resemble a *mobility edge* for experimental purposes. After the scaling theory of Abrahams et al., numerous theoretical and experimental studies were devoted to the topic and comprehensive reviews are available (Lee and Ramakrishnan, 1985; Bergmann, 1984; Kromer et al., 1985).

The criterion for the existence of localized states is a finite localization length (decay length of the wave function) and a

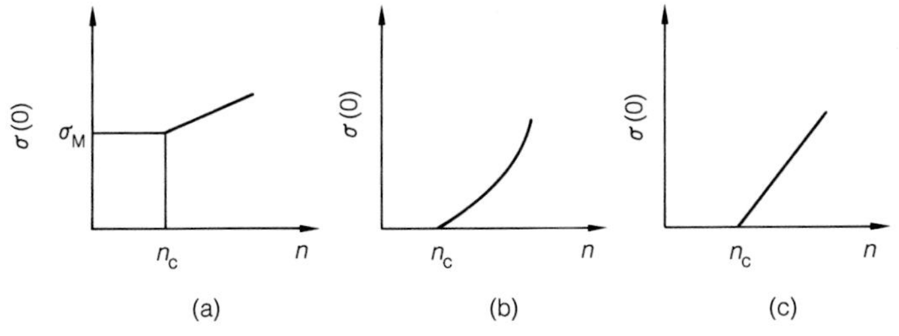

Figure 6-7. Various proposed models for the conductivity onset: (a) Mott, (b) classical percolation, and (c) scaling theory.

vanishing ensemble average of the DC conductivity as $T \to 0$. Depending on the degree of disorder, there are essentially two limiting cases: the strong localization regime and the weak localization regime.

6.2.2.2 Strong Localization Regime

In the strong localization regime, the conduction electrons experience confinement due to the disorder. Phonons or other electrons can supply the necessary matrix element and energy reservoir for electrons to hop between localized states having an energy difference of the order of $k_B T$. We can distinguish the following three transport processes (Nagels, 1985; Imry, 1984).

Firstly, an electron can be activated to the mobility edge which will yield

$$\sigma \propto \exp\left[-(E-E_c)/k_B T\right] \tag{6-11}$$

Secondly, the activation may occur between neighboring localized states where the *nearest neighbor* activated conductivity has the form

$$\sigma \propto \exp(-w/k_B T) \tag{6-12}$$

Here w is a typical energy separation between such states and, in D-dimensions, $w \sim \alpha^D/N(E_F)$; $N(E_F)$ is the density of localized states and $1/\alpha$ the localization length.

Finally, we have variable range hopping which is explained as follows. As the distance from a localized state increases, there is a higher probability of finding another state with nearly the same energy. Hopping between these states is energetically favorable. The electronic transport is governed essentially by the hopping probability $W_{mm'}$, that an electron hops from a localized site at R_m to another localized site of $R_{m'}$. This is given by the product of the tunneling and phonon terms, i.e.,

$$W_{mm'} \sim \exp\left(-2\alpha R_{mm'} - \frac{E_{mm'}}{k_B T}\right) \tag{6-13}$$

where $R_{mm'} = |\boldsymbol{R}_m - \boldsymbol{R}_{m'}|$ and $E_{mm'}$ is the absolute energy difference between sites. For the 3D case, the temperature dependence of the conductivity is given by

$$\sigma(T) = \sigma_0(T)\exp(-A/T^{1/4}) \tag{6-14}$$

with

$$A = 2.1\,[\alpha^3/k_B N(E_F)]^{1/4} \tag{6-15}$$

and

$$\sigma_0(T) = \frac{e^2}{(32\pi)^{1/2}}\,\nu_{ph}\left[\frac{N(E_F)}{\alpha k_B T}\right]^{1/2} \tag{6-16}$$

where ν_{ph} is a typical phonon frequency. For the 2D case one has

$$\sigma(T) = \sigma_0 \exp\left[-(T_0/T)^{1/3}\right] \tag{6-17}$$

T_0 is a constant given by

$$T_0 = C^3\alpha^2/k_B N(E_F) \tag{6-18}$$

where C is a constant of order unity.

Variable range hopping has been observed in materials such as a-Si or a-Ge (Knotek, 1975). Using $1/\alpha \sim 1$ nm, values for $N(E_F) \sim 10^{18} - 10^{19}/\text{cm}^3 \cdot \text{eV}$ are commonly obtained for a-Si. The hopping distance, $R = [9/8\pi\alpha N(E_F) k_B T]^{1/4}$, is estimated to be $R \sim 8$ nm at 100 K.

Variable range hopping gives rise (in the absence of the Coulomb interaction) to a Hall coefficient which is proportional to temperature in 2D (Boettger and Brykskin, 1985).

In the strong localization regime, the effect of the Coulomb interaction is significant. It introduces a so-called *Coulomb gap* in the density of states (Pollak, 1980). The one-electron hopping probability is reduced by the Coulomb gap. In Eq. (6-13) for the hopping probability one has to replace $E_{mm'}$ by the Coulomb gap E_g. Nevertheless, according to Pollak, a multiple hop can have a nonvanishing transition rate in the presence of the electron-electron interaction and the conductivity is given by an expression similar to Eq. (6-17) (Boettger and Brykskin, 1985).

6.2.2.3 Weak Localization Regime

Localization Effects

One of the chief characteristics of weak localization is a decreasing conductivity with decreasing temperature. Weak localization arises when two wave packets traversing the same closed path within the medium, but in opposite directions, interfere constructively to enhance the probability amplitude for returning to the origin (compared with the Drude value). Weak localization makes a contribution to the conductance which depends on dimensionality, magnetic field, and various electron scattering times. In zero-field the dimension-dependent conductivity is predicted as follows (Lee and Ramakrishnan, 1985):

$$\sigma_{3D}(T) = \sigma_0 + \frac{e^2}{\hbar\pi^3}\frac{1}{a}T^{p/2} \qquad (6\text{-}19\,a)$$

$$\sigma_{2D}(T) = \sigma_0 + (\alpha p)\frac{e^2}{2\hbar\pi^2}\ln\left(\frac{T}{T_0}\right) \qquad (6\text{-}19\,b)$$

$$\sigma_{1D}(T) = \sigma_0 - \frac{a\,e^2}{\hbar\pi}T^{-p/2} \qquad (6\text{-}19\,c)$$

where p is an exponent used to parameterize the temperature dependence of the inelastic-scattering length L_i through $L_i \propto T^{-p/2}$. The value of p is usually between 1 (attributed to electron-electron scattering) and 2 (attributed to electron-phonon scattering) (Uren et al., 1981). In the presence of strong spin-orbit scattering the quantum diffusion yields a reduced echo probability, an effect called weak anti-localization (Bergmann, 1984). As a result $\alpha = -\frac{1}{2}$ in Eq. (6-19b) while $\alpha = 1$ if spin-orbit scattering is negligible.

For thin 2D films exhibiting weak localization, the general case of the perpendicular magneto-conductance was derived by Hikami et al. (1980), and further studied by Maekawa and Fukuyama (1981). The magnetoresistance for the parallel field case was treated by Fukuyama (1981).

Interaction Effects

Soon after Abrahams et al. (1979) proposed their scaling theory of localization, Altshuler and Aronov (1979) predicted another disorder-related quantum correction to the conductivity based on a completely different theory involving electron-electron correlations. The resultant corrections are summarized as follows (Lee and Ramakrishnan, 1985).

$$\delta\sigma_{1D}^{I} = -\frac{1}{A}\frac{e^2}{2\pi\hbar}\left(4-\frac{3}{2}\tilde{F}_\sigma\right)\left(\frac{D}{2T}\right)^{1/2} \qquad (6\text{-}20\,a)$$

$$\delta\sigma_{2D}^{I} = \frac{e^2}{4\pi^2\hbar}\left(2-\frac{3}{2}\tilde{F}_\sigma\right)\ln(T\tau_e) \qquad (6\text{-}20\,b)$$

and

$$\delta\sigma_{3D}^{I} = -\frac{e^2}{4\pi^2\hbar}\frac{1\cdot 3}{\sqrt{2}}\left(\frac{4}{3}-\frac{3}{2}\tilde{F}_\sigma\right)\left(\frac{T}{D}\right)^{1/2} \qquad (6\text{-}20\,c)$$

where A is the cross-section of the wire, and $\tilde{F}_\sigma$ is a correction factor due to *Hartree* terms, which measures the degree of screening (being equal to unity when the screening is complete and zero when it is absent). Therefore, the Coulomb interaction in 2D also produces a $\ln T$ dependence of the conductance. The origin of this anomaly is the fact that the electron-electron interaction in disordered metals is retarded; a sudden change of the charge distribution in a disordered metal cannot be rapidly screened. In a clean metal the *free* electrons can respond (screen) in a time of order ω_p^{-1} (the plasma frequency); in a very dirty metal the response is retarded by the slow diffusion of the electrons. Electron-electron interaction in disordered metals results in nontrival corrections which are small in the λ/l parameter and depend on the temperature, external fields, and sample size. Such corrections have to be applied not only to transport, but also to thermodynamic quantities and to the electron density of states at the Fermi level. This is quite different from the pure metal case where the interaction at low temperatures manifests itself only as a renormalization of the electron spectral parameters.

In a real system, both localization and interaction effects are present. In lowest order, the two effects are believed to be simply additive (Lee and Ramakrishnan, 1985). It is difficult, therefore, to distinguish between these two effects simply from the temperature dependence of the resistance. However, the two theories predict a different behavior of the magnetoresistance and Hall effect. For instance, the localization effect (in the absence of spin-orbit coupling) gives a negative magnetoresistance, contrary to the prediction of the interaction effect. A very high magnetic field also serves to separate these two effects by suppressing the weak localization effect and making the cyclotron orbit diameter $[L_H = (\hbar/eH)^{1/2}]$ the relevant length scale instead of the inelastic scattering length, L_i. In 2D the Hall coefficient remains unchanged in the presence of weak localization but a correction $\delta R_H/R_H = 2\,\delta\sigma_I/\sigma$ does arise from the Coulomb anomaly (Altshuler et al., 1980).

In the weak localization theory, dimensionality plays an important role. The essential requirement for quantum diffusion is that the electron wave functions have some kind of phase coherence. Hence the condition for 2D behavior is that the film thickness, d, be shorter than the relevant length scale. For the localization effect, the condition is $d < L_\phi = (D\,\tau_\phi)^{1/2}$ where L_ϕ is the phase breaking length, and the de-phasing time (τ_ϕ) is given by $1/\tau_\phi = 1/\tau_i + 2/\tau_s$. Here τ_i and τ_s are inelastic and spin scattering times, respectively. For the interaction effect, the condition is $d < L_{th} = (\hbar D/k_B T)^{1/2}$; L_{th} is the thermal length. Therefore, with increasing film thickness, temperature, or magnetic field, the above conditions may be violated, resulting in a 2D–3D crossover.

By combining Eqs. (6-19) and (6-20), the conductivity correction due to the interaction and localization effects in 2D can be written as

$$\sigma(T) = \sigma_0 + C\,\frac{e^2}{2\pi^2\hbar}\ln\left(\frac{T}{T_0}\right) \qquad (6\text{-}21)$$

where the coefficient $C = \alpha p + 1 - \frac{3}{4}\tilde{F}_\sigma$ characterizes the strength of localization and interaction. Experimentally, a $\ln T$ dependence is observed in various materials and over a wide range of $R_\square$ values, ranging from ten's of $\Omega/\square$ to several $K\Omega/\square$, as shown in Fig. 6-8 (White and Bergmann, 1989).

The magnetoresistance can serve as a useful diagnostic tool to distinguish the localization and interaction effects and to determine different scattering times. An excellent example is shown in Fig. 6-9, which shows the measured magneto-conductance in Mg films covered with a sub-monolayer of Au (to modify the spin-orbit interaction). The inelastic time and its temperature dependence can be deduced from the magnetoresistance measured at different temperatures. A typical result is shown in Fig. 6-10 for Bi films (Komori et al., 1983). The general trend is consistent with the expectation that at low temperatures $p \sim 1$ due to electron-electron interaction and $p \sim 2$ at higher temperatures where electron-phonon scattering dominates.

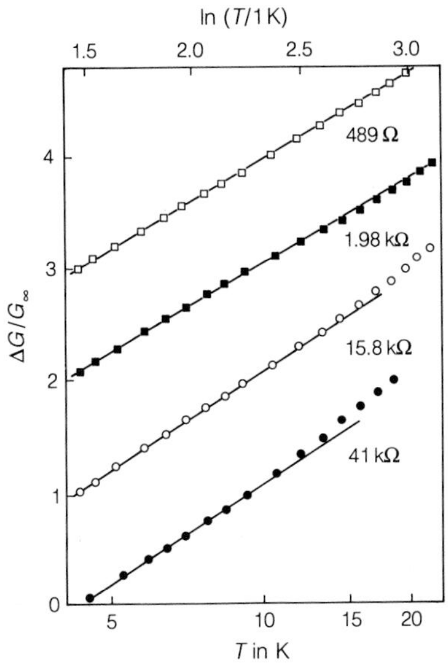

Figure 6-8. Conductance vs. $\ln(T)$ plot for four Cu films. Note that the conductance scale has been shifted (after White and Bergmann, 1989).

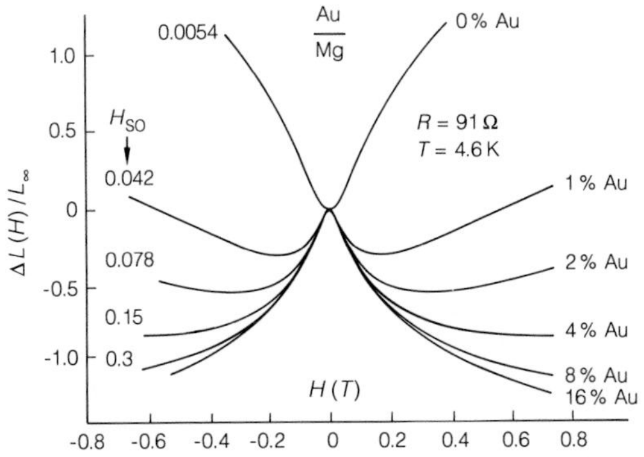

Figure 6-9. Magnetoresistance of Mg films covered with a sub-monoatomic layer of Au. The numerical designations denote the % of a monolayer (after Bergmann, 1984).

Localization and interaction effects have also been observed in various superlattice systems such as PbTe/Bi (Shin et al., 1984), CdTe/Bi (Divenere et al., 1988), NbTi/Ge (Jin and Ketterson, 1986), and HgTe/CdTe (Moyle et al., 1987). Because of the lack of a complete theory, most data were analyzed based on a simple normalized-layer

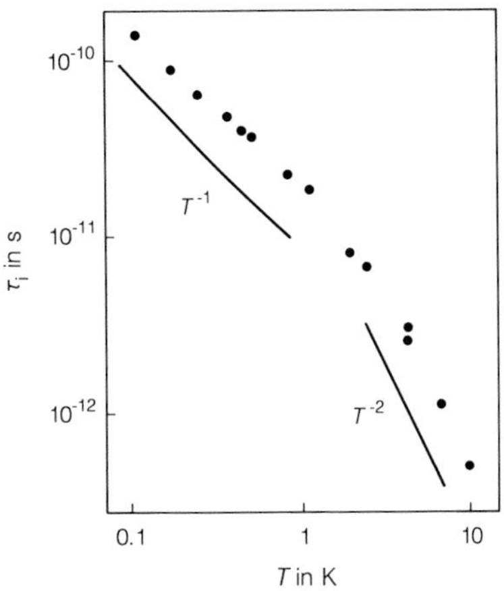

Figure 6-10. Energy relaxation time in Bi films (after Komori et al., 1983).

assumption. In some cases, in order to compare the experimental results with the theories, an active layer number (less than the actual superlattice period number) had to be assumed (Szott et al., 1989). One of the advantages offered by a superlattice structure is that 2D–3D crossover can be probed by tuning the interlayer coupling; with increasing interlayer coupling, the effective dimensionality increases, resulting in a dimensional crossover in localization/interaction phenomena.

Anisotropic Systems

A general formalism, involving an anisotropic diffusion tensor, $D_{\mu\nu}$, for transport in disordered anisotropic metallic systems has been described by Bhatt et al. (1985). For the scale-dependent conductivity, due to the localization effect, and to lowest order in the electron-electron interaction they found that the correction terms exhibit the same anisotropy as the Boltzmann conductivity. Thus, the anisotropy can be factored out and absorbed into a *renormalized* diffusion constant. For example, in 2D the $\ln T$ term in the conductivity arising from the localization effect is

$$\Delta\sigma_{\mu\mu}(T) = \frac{p}{2}\frac{e^2}{\pi^2\hbar}\frac{\sigma_{\mu\mu}}{\sigma}\ln T \qquad (6\text{-}22)$$

where $\mu = x, y$ and $\bar{\sigma} = (\sigma_x \sigma_y)^{1/2}$. Similarly, in a perpendicular magnetic field, the differential magnetoconductance is the same for the x and y directions; i.e.,

$$\frac{\delta\sigma_{xx}(H)}{\sigma_{xx}(0)} = \frac{\delta\sigma_{yy}(H)}{\sigma_{yy}(0)} \qquad (6\text{-}23)$$

The equality of the anisotropy in the Boltzmann and correction terms is consistent with the experimental results of Bishop et al. (1984) on anisotropic Si metal-oxide-semiconductor field-effect transistors.

The weak localization correction in the parallel and vertical conductivity of superlattices was discussed by Szott et al. (1989). The *miniband* of the superlattice is described by a tight-binding model with a bandwidth of w. It is found that the correction to the conductivity has a form analogous to the results predicted by the theory based on an effective diffusion tensor (Bhatt et al., 1985). However, the effective diffusion coefficient along the z (superlattice axis) direction, D_z, has a different character:

$$D_z = \frac{\tau}{8}\left(\frac{w\Lambda}{\hbar}\right)^2 \qquad (6\text{-}24)$$

which is independent of the carrier density, provided that $E_F > w$. For this case, we have the following useful scaling relation for the localization correction:

$$\frac{\delta\sigma_{\parallel}}{\sigma_{\parallel}} = \frac{\delta\sigma_z}{\sigma_z} \qquad (6\text{-}25)$$

The localization transition in predominantly layered systems with anisotropic diagonal disorder was treated by Xue et al. (1989). Li et al. (1989) have studied an anisotropic tight-binding model using the finite-size scaling method.

The experimental verification of the above theories remains a challenge. It depends on carefully choosing the model system.

6.2.3 Vertical Transport

Since the original proposal of superlattices by Esaki and Tsu (1970), the vertical transport in these structures, especially semiconductor superlattices, has been the subject of intense investigation (Chang and Giessen, 1985).

In an ideal superlattice, consisting of a periodic repetition of quantum wells separated by narrow potential barriers, the electronic states are completely delocalized and their energies are distributed in minibands that result from the strong coupling among the wells. Interesting transport properties in such superlattice minibands, such as Bloch oscillations, are based on the assumption that the Bloch states are *well defined*. In a real superlattice, however, deviations from perfect periodicity (caused by well-width or barrier height fluctuations) reduce the coupling and induce a degree of localization, thus decreasing the coherence length of the superlattice wave function. Under strong localization, coherence will be reduced to a few periods and, in the limit, to a single quantum well (Lang and Nishi, 1985; Littleton and Camley, 1986).

Applying an electric field, F, in the superlattice direction can also induce localization; some different transport regimes are commonly identified. At low fields, the current increases linearly with field (mobility regime). According to Esaki and Tsu (1970), the current is expected to decrease with increasing field when the electron distribution probes the *negative-mass* region of the miniband. The condition for negative differential conductance to occur is

$$F > \hbar/e\Lambda\tau \qquad (6\text{-}26)$$

where τ is the scattering time and Λ is the modulation wavelength. Using the classical Boltzmann approach and a tight-binding band model, such as $E_z = E_0 - w/2 \cos(k_z \Lambda)$, the current density along the superlattice direction is given by (Lebwohl and Tsu, 1970)

$$j_z = n \left[\frac{e^2 F}{m(0)}\right]\left[1 + \left(\frac{2\pi e F \tau}{\hbar k}\right)^2\right]^{-1} h \quad (6\text{-}27)$$

where n is the electron density, $m(0) = 2(\hbar k/2\pi)^2/w$, and

$$h = \frac{1}{2}\frac{x - \sin x \cos x}{\sin x - x \cos x} \quad (6\text{-}28\,a)$$

for $E_0 - w/2 \le E_F \le E_0 + w/2$

and

$$h = \frac{1}{4}\frac{w}{E_F - E_0} \quad (6\text{-}28\,b)$$

for $E_F > E_0 + w/2$

with $x = \cos^{-1}[2(E_0 - E_F)/w]$.

In an electric field the electronic wave functions extend over a number of periods (of the order of $w/e\Lambda F$) and are separated in energy by eF. This is the so-called Stark ladder (Fig. 6-11 a) (Wannier, 1960; Kazarinov and Suris, 1972; Leo and Movaghar, 1988). Thus, as the field is increased, the wave functions become increasingly localized in space, up to an extreme where they shrink to a single well when

$$F > w/e\Lambda \quad (6\text{-}29)$$

In this limit, the superlattice consists of a *ladder* of identical isolated quantum wells (Fig. 6-11 b). A decrease in the current is expected throughout this regime, since the spatial overlap between the Stark-ladder states decreases with increasing field and with it the hopping matrix element (Tsu and Döhler, 1975). Complete localization, leading to Stark-ladder quantization, was observed in optical experiments (Mendez et al., 1988; Voisin et al., 1988). The I–V characteristics, which show a negative differential conductance, are shown in Fig. 6-12 for a heterojunction superlattice structure (Baltram et al., 1990).

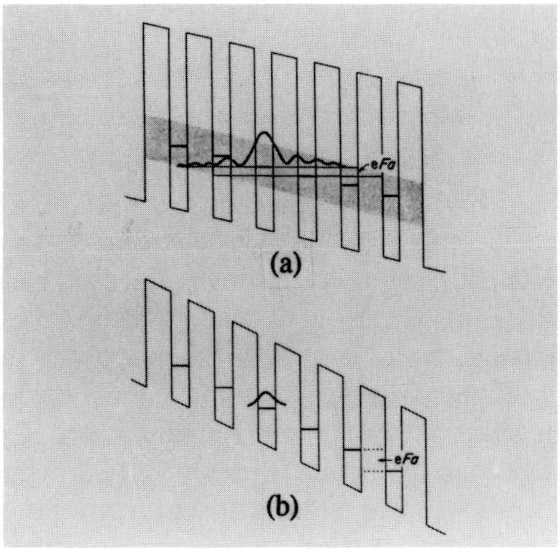

Figure 6-11. Schematic conduction-band diagram of a heterojunction superlattice in an applied field: (a) electronic states extend over several periods; (b) at very high biases where the electronic states are confined to single wells (after Baltram et al., 1990).

In contrast to semiconductor superlattices, vertical transport in metallic superlattices has received little attention, due mainly to the technical difficulties associ-

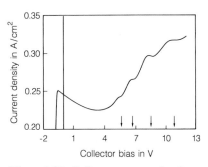

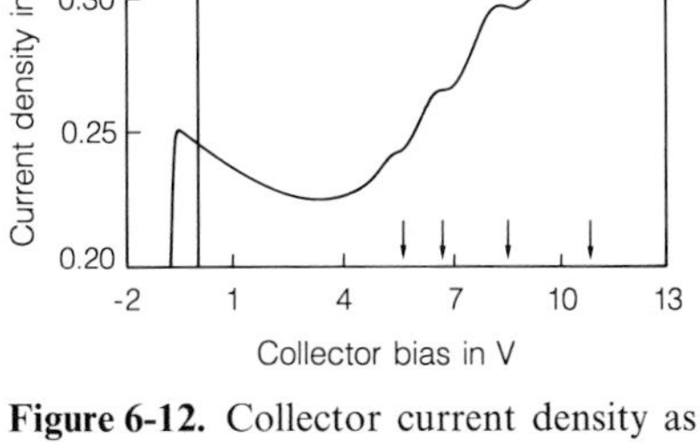

Figure 6-12. Collector current density as a function of the collector bias at a constant-emitter current measured at 15 K. The arrows indicate the calculated bias positions of the resonances (after Baltram et al., 1990).

ated with these structures (structural irregularity, small junction resistance, etc.), even though the importance of the subject is obvious. Song and Ketterson (1990) studied the vertical electrical conductance in a series of Si/Nb multilayers. It is found that the vertical conductance changes its behavior systematically with silicon layer thickness and with temperature, i.e., changes from phonon-assisted hopping to weak localization as the thickness of the Si layers is reduced, and is dominated by quantum mechanical tunneling at low temperatures. The measured anisotropy ratio between the vertical and in-plane resistivities can be over 10^5 (Fig. 6-13). An anisotropic percolation behavior of the vertical and in-plane resistive transition has also been observed in Si/Nb multilayers with and without initiating and terminating thick Nb electrodes (Song and Ketterson, 1990).

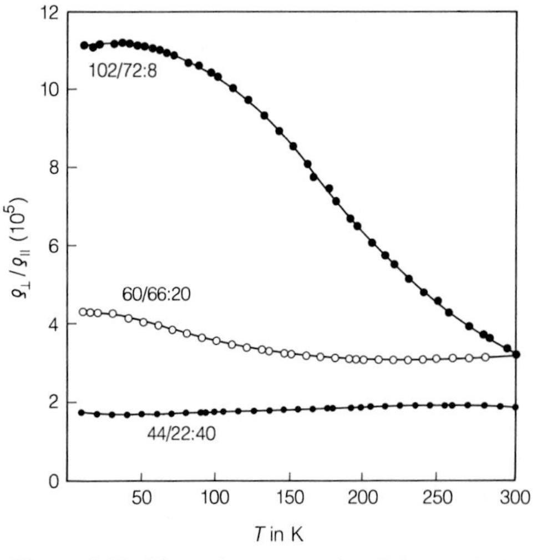

Figure 6-13. The anisotropy ratio of the vertical and in-plane resistivities between 10 K and 300 K for three Si/Nb multilayer films (after Song and Ketterson, 1990).

6.3 Magnetic Properties

6.3.1 Introduction

The magnetic properties of thin films and superlattices are of interest in understanding the physics of magnetism, and also for various technological applications. Possible applications include thin film recording heads and a variety of recording media (longitudinal, perpendicular and magneto-optic, etc.), all of which impose special requirements on the magnetic properties. Magneto-optic properties of thin films are discussed in Chapter 7 of this volume.

Magnetic superlattices (involving at least one magnetic component) are, at present, the most extensively studied type of superlattice system. The first magnetic measurements, on compositionally modulated Cu-Ni materials, were reported by Hirsch et al. (1964). Later, stimulated by work on Cu/Ni multilayers by Thaler et al. (1978), a large literature emerged, involving both theoretical and experimental studies. This rapid rise of interest in magnetic superlattices is due to the following:

(1) The large number of interfaces favors the observation of surface magnetism and surface anisotropy.

(2) One can study ultra-thin magnetic layers, and the additional length scale (the intervening layer thickness) introduced facilitates the study of possible 2D magnetism and various interlayer coupling mechanisms.

(3) There is the possibility of studying magnetic collective modes of the superlattice structure as a whole.

(4) The many possible combinations of materials (magnetic, non-magnetic, metallic, insulating and superconducting), and the ability to prepare superlattices with nearly atomically smooth interfaces, offers

many opportunities to create new magnetic materials of practical importance, potentially with tailor-made properties.

In a magnetic superlattice, the magnetic component can be either a 3d transition metal (Cr, Fe, Co, Ni) or a rare-earth. The heavy rare-earth metals have very large magnetic moments. Five of the six magnetic rare earth metals – Tm, Er, Ho, Tb and Dy – become ferromagnetic at low temperatures only after intermediate antiferromagnetic phases; Gd goes directly into a ferromagnetic phase at a Curie temperature $T_c \approx 16\,°C$. Since the 4f electrons of the rare earths, which give rise to the orbital and spin moments, are well shielded by the outer-shell conduction electrons, the magnetism is of a local nature. Studies of magnetic structure and interlayer coupling form a large fraction of the current research.

Magnetic properties can be divided into two categories: static properties (magnetization, hysteresis, etc.) and dynamic properties (various resonance phenomena, elementary excitations, etc.). The static properties may be characterized by various magnetometers while low frequency dynamic properties may be studied with ferromagnetic resonance (FMR) and light scattering (LS). Neutron diffraction is a powerful tool for determining the magnetic structure of superlattices. These techniques will be discussed in the present section.

For magnetic properties, the dominant characteristic length scale is the exchange coupling length, which is only a few atomic spacings. Therefore interface perfection is very crucial in obtaining reproducible results. This was the main limiting factor in early experimental observations. In this context, the rapid progress of computational physics is very encouraging, allowing the simulation of the actual process of epitaxial growth and the calculation of the detailed electronic structure of superlattices. Such theoretical predictions can serve as a guide to experiments.

6.3.2 Static Properties

6.3.2.1 Magnetization

The rotation of the magnetization between domains of a ferromagnet takes place within a transition zone called the domain wall. The properties of domain walls and their motion are key to many phenomena in magnetic films. As devices (such as thin-film recording heads) are made smaller, control of the domain wall motion and configurational stability become central to optimizing the signal to noise ratio for reading and writing. The complex domain structure is believed to be responsible for the observed *giant* magnetoresistance in Fe/Cr multilayers (Baibich et al., 1988).

For thin films with the domains magnetized in-plane, three basic types of walls may occur due to the competition between the exchange and anisotropy energies: the Néel wall, the cross-tie wall, and the Bloch wall (Cullity, 1972). The boundaries are shown in Fig. 6-14 as a function of the film thicknesses for NiFe/SiO_2/NiFe sandwich films (Middelhoek, 1966). The vast majority of thin-film domain studies have been

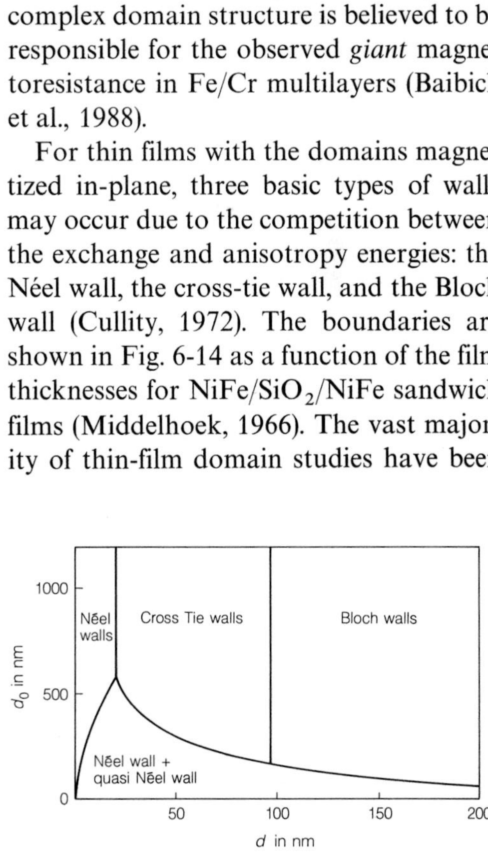

Figure 6-14. Diagram indicating the type of wall as a function of d and d_0 (after Middelhoek, 1966).

performed on Fe-Ni alloys (permalloy) (Herman et al., 1987); it is observed that a metallic spacer can result in single domain structure (no walls), vastly improving the read/write performance. Studies of domains in Tb/Cu multilayers were performed with Lorentz microscopy (Draaisma and de Jonge, 1987). A microscopic magnetic antiphase domain structure has been observed in a single-crystal Gd/Y superlattice by neutron diffraction (Majkrzak et al., 1986).

A low coercivity is crucial for recording applications. Néel noted that when the thickness of a film is reduced to ~ 100 nm or less, variations in the thickness of a film have an important effect on the coercive force. Moog and Bader (1985) have used the surface magneto-optic Kerr effect to study the magnetism of Fe (100) films in the monolayer range which were epitaxially grown on Au (100) in a UHV chamber. For films in the 1–10 atomic plane (AP) range they observed a broad peak in the coercivity (H_c) around 5 atomic planes which they attributed to roughness effects. Variation of the deposition temperature was found to change H_c appreciably. H_c can also be modified by layering. A drastic reduction of H_c (by a factor of ~ 10) is achieved for simple bi-layers of NiFe and Si (or SiO, SiO_2) (Herd and Ahn, 1979). Domain energy calculations show that walls in bi-layers are expected to be both lower in energy and wider than those in single films; both characteristics contributing to a lower value of H_c. This was also confirmed by studies of the perpendicular coercivity in Co/Cr multilayers (Nakagawa et al., 1988). The multilayer technique is thus a new method to control the coercivity of the Co/Cr thin films (in addition to the substrate temperature).

The saturation magnetization, M_s, as a long-range order parameter, plays a central role in phase transition phenomena. The temperature and thickness dependence of M_s, as well as a possible moment enhancement at interfaces, are all of interest.

According to spin-wave theory, the temperature dependence of M_s for a bulk ferromagnet should follow the Bloch's $T^{3/2}$ law:

$$M_s(T) = M_s(0)\,(1 - B\,T^{3/2} + C\,T^{5/2} \ldots) \tag{6-30}$$

The Bloch constant, B, depends on the exchange integral and the lattice symmetry. For the 2D case, numerical calculations (Klein and Smith, 1951) show that $M_s(T)$ exhibits a linear temperature dependence below a critical thickness $d_c \sim 25$ atomic layers. This linear temperature dependence was frequently used as the only criterion in the early efforts to verify the existence of 2D magnetism; an $M_s(T)$ vs. T plot for Fe (110) films grown epitaxially on Ag (111) in UHV is shown in Fig. 6-15. The temperature dependence of the saturation magnetization changes from a $T^{3/2}$ law to linear at a critical thickness $d_c \sim 0.5$ nm. The values of d_c are strongly dependent on the

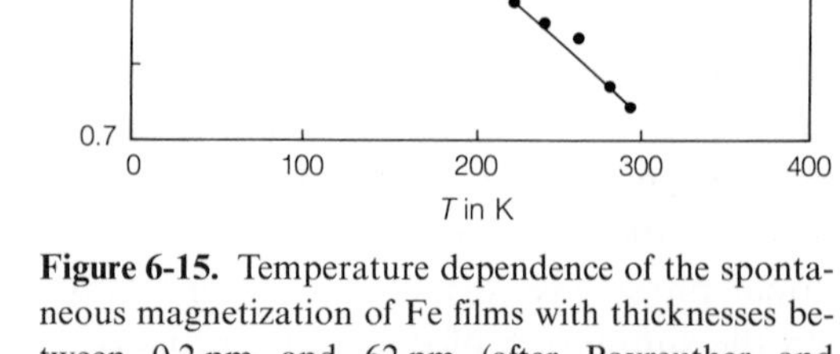

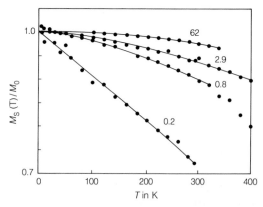

Figure 6-15. Temperature dependence of the spontaneous magnetization of Fe films with thicknesses between 0.2 nm and 62 nm (after Bayreuther and Lugert, 1983).

film-substrate system and the growth conditions. One must be cautious in attributing an observed linear temperature dependence to 2D magnetism. As a matter of fact, for Fe films thinner than 0.5 nm, Mössbauer spectra indicate a pronounced island structure which results in superparamagnetism at room temperature. The lack of magnetic surface anisotropy and a reduced absolute value of M_s support this picture. In this case the observed linear M_s vs. T behavior is connected in some way with the island structure.

A number of superlattice systems also exhibit a linear temperature dependence of $M_s(T)$ for magnetic layer thicknesses below some critical value d_c. For Fe/V, $d_c \sim 0.39$ nm (Wong and Ketterson, 1986), for Fe/Y, $d_c \sim 1.2$ nm (Morishta et al., 1986).

Measurements of the critical exponents is a powerful means to determine the dimensionality of a system. Near the transition temperature, T_c, the magnetization can be characterized by a simple power law $M(T) \propto (1 - T/T_c)^{\beta}$. Here β is the critical exponent, which is a universal quantity and depends only on the dimensionality of the system and the symmetry of the order parameter. Theoretical calculations yield $\beta \sim 0.38$ in 3D (Suter and Hohenemser, 1979), and $\beta \sim 0.1 - 0.15$ in 2D (Wu, 1982). The value of β for the 2D Ising model is $\frac{1}{8}$ (Yang, 1952).

A truly 2D magnetic phase transition was reported recently by Dürr et al. (1989). They measured the temperature dependence of the long range order parameter, M, in thin (1–3 atomic plane) films of b.c.c. Fe on Au(100) using two complementary techniques: spin-polarized low energy electron diffraction and spin-polarized secondary electron emission spectroscopy. They found that (i) all films undergo a second-order magnetic phase transition at a

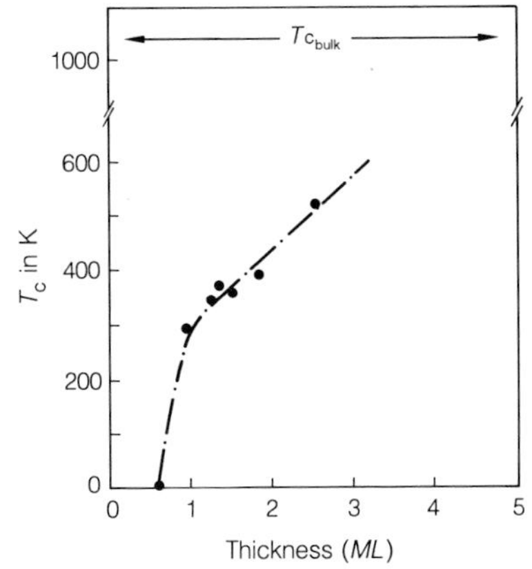

Figure 6-16. Thickness dependence of the Curie temperature T_c as determined by spin-polarized, secondary-electron emission; the dashed line is a guide to the eye (after Dürr et al., 1989).

well-defined Curie temperature, T_c, which depends on the film thickness (Fig. 6-16); and (ii) the critical exponent β is independent of the film thickness (Fig. 6-17). The onset of long range magnetic order occurred at a thickness of about 0.6 atomic plane. The thickness independence indicates that the phase transition in this system is a truly two-dimensional one. The critical exponent β is found to be 0.22 ± 0.05, deviating from the expected value of $\frac{1}{8}$ for a 2D Ising-type system.

In ultrathin films, the transition temperature becomes a function of thickness-dependent corrections. With the number of monolayers, n, as the scaling factor, it has been shown that the approach of $T_c(n)$ to the bulk value $T_{c\infty}$ can be characterized by a shift exponent λ defined by

$$\frac{1}{T_c(n)} - \frac{1}{T_{c\infty}} \approx A\, n^{-\lambda} \tag{6-31}$$

Depending on whether free-surface or periodic boundary conditions are assumed, λ is predicted to lie between 1.0 and 2.0 (Allan, 1970; Ritchie and Fisher, 1973). Analysis of the thickness and temperature dependence

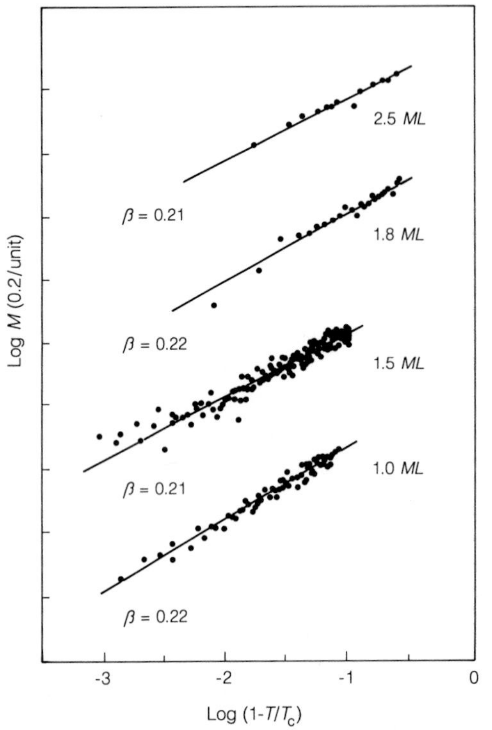

Figure 6-17. Temperature dependence of the magnetization for Fe films having different thicknesses, plotted in a $\log M$ vs. $\log(1-T/T_c)$ representation (after Dürr et al., 1989).

of the spontaneous magnetization for Ni films of different thicknesses on Cu(111) (Ballentine et al., 1990) yields a shift exponent $\lambda \approx 0.24$, which is in good agreement with the result of Dürr et al.

The classical (mean-field) ground state as a function of magnetic field for superlattices consisting of alternating layers of ferromagnetic and antiferromagnetic material has been studied theoretically by Hinchey and Mills (1986). The ground state depends critically on the number of atomic planes within each antiferromagnetic sublayer. An odd number of atomic planes is most likely to produce a ground state where all spins line up either parallel or antiparallel to an applied field; an even number of layers often produces a twisted state where the individual spins are at some angle relative to the magnetic field. The difference in ground states will be manifested in the spin-wave excitation spectrum. The phase diagram and magnetization for superlattices formed from two different ferromagnetic (Camley and Tilley, 1988; Schwenk et al., 1988) and from two antiferromagnetic (Diep, 1989) constituents has also been investigated.

6.3.2.2 Magnetic Anisotropy at Surfaces and Interfaces

One of the remarkable properties of surface or interface magnetism is the anisotropy caused by the reduced local symmetry. The corresponding contribution to the surface energy is called magnetic surface anisotropy (MSA). Néel (1953) discussed the leading term, which is a function of the angle, θ, between the magnetization, $\boldsymbol{M}_s$, and surface normal, $\boldsymbol{n}$, according to $E_{an} = K_s \sin^2\theta$.

To explore the effects of MSA, it is convenient to introduce a total (uniaxial) anisotropy constant, K_u, which includes contributions from the surface anisotropy K_s, volume anisotropy K_v, and shape anisotropy $M_s^2/2\mu_0$. If the in-plane anisotropy can be neglected, we can write the anisotropy energy per unit volume of a superlattice as

$$E = K_u \sin^2\theta \tag{6-32}$$

where

$$K_u = -(2K_s + K_v d + M_s^2 d/2\mu_0)/\Lambda \tag{6-33}$$

here d is the thickness of the magnetic layers and Λ is the bilayer periodicity. For a single layer magnetic film, Λ is replaced by d in Eq. (6-33). Note that a positive K_u means that the easy axis of magnetization is along the film plane normal in the present convention. If K_s has the opposite sign of $(K_v + M_s^2/2\mu_0)$, Eq. (6-33) predicts that there is a critical film thickness, $d_c = 2K_s/(K_v + M_s^2/2\mu_0)$, at which the magnetization direction switches from the film plane to the surface normal. The physical origin of this switching behavior is the competition between the dipolar-shape anisotropy (which favors M_s lying within the plane) and the magnetic surface anisotropy energy; the latter, arising mainly from the spin-orbit interaction, prevails at small thicknesses due to its $1/d$ dependence.

To date, a comprehensive microscopic theory of MSA is still missing. Néel's model is based on a localized moment approximation and therefore is not well suited for itinerant ferromagnets. It is thus desirable to study MSA within the framework of band theory. Takayama et al. (1976) studied the magnetic surface anisotropy caused by the spin-orbit interaction in 3d transition metals; the spin-orbit interaction was treated by perturbation theory. It was found that the effect of a surface or an interface on the local electronic structure extends for only one or two atomic layers. Therefore, the surface-magnetic anisotropy is also confined within the same region. Gay and Richter (1986) calculated the MSA for monolayer films of V, Fe and Ni based on spin-density-functional band calculations. The calculated values of K_s are -0.61, 6.4 and 0.096 (in 10^{19} J/atom) for Fe, Ni and V, respectively. In units of μ_B, the magnetic moments are 3.20 (Fe), 1.04 (Ni), and 3.00 (V). As the shape anisotropy is about 0.48×10^{-19} J/atom for Fe, it is possible that the magnetization is aligned perpendicular to the film surface.

Magnetic thin films and superlattices with perpendicular anisotropy are currently of great interest as possible vertical recording media. Vertical recording promises increased storage density by reducing

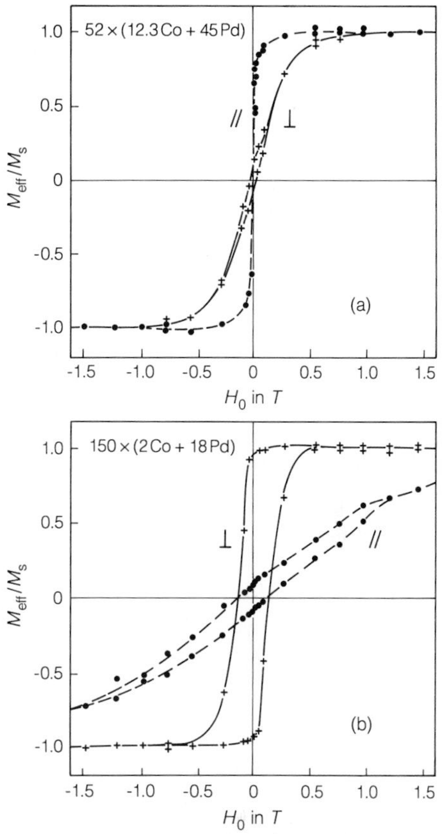

Figure 6-18. Magnetization curves in parallel and perpendicular fields for: (a) multilayer 52 × (1.23 nm Co + 45 nm Pa); (b) multilayer 150 × (0.2 nm Co + 1.8 nm Pd) (after Draaisma et al., 1987).

the self-demagnetization that limits high information density in conventional longitudinal recording. MSA can be deduced by FMR or by polar Kerr-effect measurements (Liu et al., 1988). The total anisotropy, K_u, is often estimated from the ratio of the areas enclosed by the perpendicular and parallel magnetization curves. Figure 6-18 shows the magnetization curves of two Co/Pd multilayer films measured with the magnetic field perpendicular and parallel to the film plane respectively. Clearly, the multilayer with 0.2 nm-thick Co layers is more easily magnetized with the field perpendicular to the film. The critical thickness, d_c, as well as the interface anisotropy, K_s, may be determined by plotting $K_u \Lambda$ vs. d [see Eq. (6-33)], as shown in Fig. 6-19 for Co/Pd multilayers. The easy axis of magnetization switches from in-plane to out-of-plane as the Co layer thickness decreases to less than 0.8 nm. The $d=0$ intercept yields $K_s = -1.6 \times 10^{-4}$ J/m^2. The volume anisotropy, K_v, can be estimated from the slope, provided that one knows the saturation magnetization.

Perpendicular anisotropy has also been observed in systems such as Co/Au sandwiches (Chappert and Bruno, 1988), Co/Cr superlattices (Sato, 1987), sputtered CoEr films (Nawate et al., 1988), Fe/Tb multilayers (Shan and Sellmyer, 1990), and in Dy/Co multilayers (Shan et al., 1989).

Generally, the samples available for experiments are far from perfect and it is important to study the influence of defects. For example, the interface roughness is shown to give rise to a *dipolar surface anisotropy* and to reduce the magnetocrystalline surface anisotropy (Chappert and Bruno, 1988). Intermixing can be studied by annealing experiments; intermixing tends to suppress the perpendicular anisotropy. However, den Broeder et al. (1988) have demonstrated that the diffused interfaces of ion-beam sputtered Co/Au multilayers can be made sharper while maintaining the periodic structure by annealing at 250 °C to 300 °C. As a result, the magnetic interface anisotropy is enhanced and the easy magnetization direction becomes perpendicular to the film plane for Co layers thinner than about 1.4 nm.

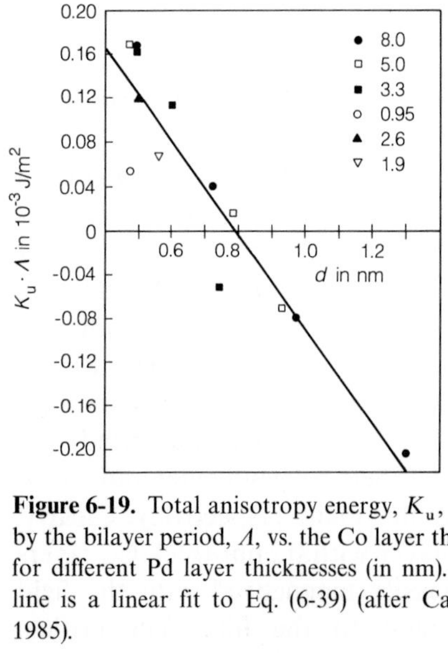

Figure 6-19. Total anisotropy energy, K_u, multiplied by the bilayer period, Λ, vs. the Co layer thickness, d, for different Pd layer thicknesses (in nm). The solid line is a linear fit to Eq. (6-39) (after Carcia et al., 1985).

6.3.2.3 Magnetism at Surfaces and Interfaces

Spin-density-functional (SDF) band calculations for metallic superlattices show that the electronic and magnetic structures of a given atomic layer are generally affected by only the first and second nearest-neighbor layers, characterized by a length scale, ξ. Interfaces are relevant if the length scale characterizing the composition modulation along the z-direction is comparable to ξ.

Stimulated by the first experimental study by Thaler et al. (1978), spin-density-functional band calculations have been performed to study the behavior of Ni near the interface of Cu/Ni multilayers (Jahlborg and Freeman, 1980). The results show that the magnetic moment of a Ni atom at the interface is 0.37 μ_B, reduced by about 30% from the bulk value (0.54 μ_B); however the reduction is confined to only the first layer of the interface. In the calculation, a square-wave composition modulation was assumed. In the experiments, the atomic configuration of films with monolayer thicknesses is not ideally flat. Therefore one should be cautious about making a direct comparison between theory and experiments.

The most striking result of the calculations for ferromagnetic, antiferromagnetic and some normally nonmagnetic surfaces and interfaces is the prediction of enhanced magnetic moments. This effect mainly arises from the band narrowing (which results in an increased density of states at the Fermi energy) caused by the reduced coordination number at the surface and the reduced symmetry of these atoms. Surface states may also play a role. The collective nature of the magnetic behavior is demonstrated by the effect of the dimensionality (D) on the predicted moments (in μ_B/atom) (Freeman and Fu, 1986) (as listed in Table 6-1).

Table 6-1. The magnetic moments of Ni and Fe in different dimensions.

D	Zero (free atom)	One (chain)	Two [(100) surface]	Three (bulk)
Ni	2.0	1.1	0.68	0.56
Fe	4.0	3.3	2.96	2.27

The enhancement of the surface moment is predicted to be 30% for Fe(100) and 20% for Ni(100). The moment increase is mostly confined to the surface layer. The moments for each layer in a 7 layer slab are listed in Table 6-2 (Freeman and Fu, 1987).

Table 6-2. The calculated layer dependence of the moments for Fe and Ni.

	Fe (100)	Ni (100)	Fe (110)
S	2.96	0.68	2.64
S-1	2.35	0.60	2.37
S-2	2.39	0.59	2.28
C	2.27	0.56	2.25

The Fe(110) surface shows a smaller enhancement, probably due to the greater number of nearest neighbors (6) and higher surface packing density. The enhanced magnetism was observed for Fe (Gradmann et al., 1983).

Another example is chromium. Bulk Cr is well-known for its spin-density-wave ground state with a period of about 22 unit cells and a maximum magnetic moment of 0.59 μ_B (Shirane and Takei, 1962). The Neél temperature is 312 K. Applying the functional integral method to the finite temperature problem (within the static approximation and the local-saddle point approximation), the ferromagnetic ordering at a Cr film surface was shown to persist for temperatures up to about 900 K. Figure 6-20 shows the temperature dependence of magnetization for the first four layers from the surface together with that for the bulk (Hasegawa, 1986).

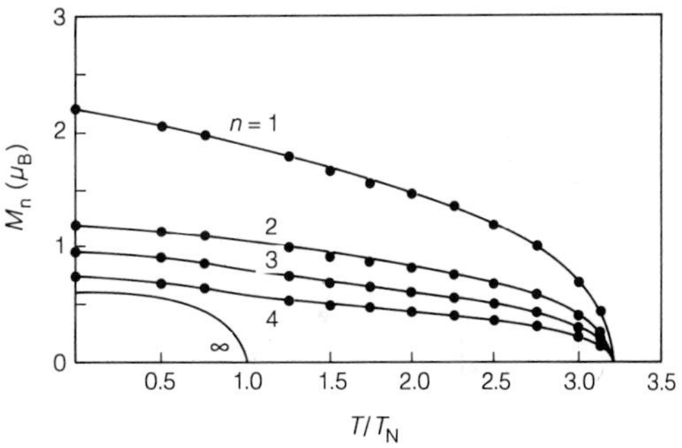

Figure 6-20. Temperature dependence of the average magnetization of the first four layers and in the bulk. T_N is the bulk Néel temperature (Hasegawa, 1986).

In magnetic superlattices the magnetic moment distribution may be deduced from Mössbauer spectrum measurements. Mössbauer spectra for Fe/V superlattice samples have been reported by Jaggi et al. (1985) and Hosoito et al. (1984). The two groups assumed different models to interpret the spectra and hence arrived at different conclusions about the magnetic moment distribution in the Fe layers.

The hyperfine fields in Fe/V superlattices, as demonstrated by Jaggi et al., show the following important features.

(1) They are primarily determined by nearest- and next-nearest neighbor atomic environments.

(2) The dependence on the local environment is identical with that known for Fe-V alloys.

Therefore it is reasonable to assume an alloy model for the interfacial composition profile: $c(x) = [1 + \tanh(\beta x)]/2$, which is predicted from the theory of the diffuse interface (Cahn and Hilliard, 1985).

The composition profile was then used to compute the distribution of Fe (p, q) configuration for each Fe layer; here p and q are the number of nearest and second-nearest V atoms of the Fe atom. By using the computed distribution Fe (p, q) and an alloy hyperfine field model (Jaggi et al., 1985, and references therein), they were able to fit the experimental Mössbauer spectra quite satisfactorily. The diffusion length β^{-1} that best fitted the Mössbauer spectra was in the range of 0.82–1.2 interatomic spacings for the samples studied. The short diffusion length implies a nearly rectangular-like composition profile. The same model provided a good fit to the spectra of surface-selectively enriched samples that were measured by Hosoito et al. (1984).

It is well known that the dominant contribution to the hyperfine fields in this system comes from core polarization. The sign and the magnitude of the conduction electron polarization is less well understood. Jaggi et al. estimated the Fe moment with the following equation, which treats conduction electron polarization in an average fashion:

$$H F = a \mu_1 + b \bar{\mu} \tag{6-34}$$

here μ_1 is the local moment and $\bar{\mu}$ is the conduction electron polarization (Freeman and Watson, 1965). They find a large reduction (30–55%) in the interface Fe moments which is in reasonable agreement with the band structure calculation by Hamada et al. (1983).

While the overwhelming importance of local environment in determining spin-density and hyperfine field perturbations seems to be well established, the exact nature of this perturbation is still an open question.

Elzain and Ellis (1987) performed theoretical calculations of the electronic and magnetic properties of Fe/V superstructures. Contrary to the previous experimental interpretations and the band structure calculations, these researchers found that, even if the magnetic moment on a central Fe atom increases, the hyperfine field decreases when the number of nearest-neighbor V atoms increases. Elzain and Ellis suggest that the sign of the polarization of the 4s conduction electrons could lead to an opposing contribution, and a lowering of the hyperfine fields.

The indirect nature of Mössbauer experiments combined with the complications due to the possibility of negatively induced V moments render the absolute determination of local moments very difficult. These problems may hopefully be resolved by the application of careful neutron diffraction and force balance measurements.

6.3.2.4 Magnetic Coupling in Multilayer Films

One of the basic questions in magnetic multilayers is the origin and nature of the magnetic coupling between adjacent layers (across an intervening layer). As the thickness of the intervening layer is increased, the coupling may change from direct exchange coupling to an indirect Ruderman-Kittel-Kasuya-Yosida (RKKY) like coupling (via polarization of the conduction electrons) or a dipolar coupling.

Recent neutron diffraction experiments (see Sec. 6.3.3) show that in rare-earth multilayers the coupling of ferromagnetic rare-earth layers across interlayers can become antiferromagnetic. This happens for Gd across Y (Majkrzak et al., 1986), and with a slight canting, for Dy across Y (Salamon et al., 1986). In these rare-earth metals, the localized moments arise from the unfilled 4f-shell. The 4f electrons are exchange-coupled to the 5d and 6s conduction electrons, leading to an (enhanced) RKKY exchange interaction between the localized spins. The coupling of Gd or Dy across Y is very likely of the RKKY type. In this model the magnetic atoms (Gd, Dy) polarize the conduction electrons in the Y layer (which has a similar band structure and wave-vector dependent susceptibility) and therefore couples the magnetic layers.

Instead of going to a first principles band-structure calculation for the superlattice, Yafet et al. (1988) have presented a simplified model to calculate the magnitude and spatial dependence of the RKKY interaction in Gd/Y superlattices; it is valid when the thickness of the RE layers is much smaller than that of Y layers. They assume the exchange coupling function, $J(\boldsymbol{q})_{\text{Gd-Y}}$, of Gd ions in a Y matrix has the form

$$J(\boldsymbol{q})_{\text{Gd-Y}} = |j_{\text{sf}}(\boldsymbol{q})_{\text{Gd}}|\, \chi(\boldsymbol{q})_{\text{Y}} \tag{6-35a}$$

where j_{sf} is the s-f exchange integral and χ the wave vector dependent susceptibility. In real space the interlayer exchange interaction, $J(N)$, between layers separated by N atomic layers is then given by

$$J(N) = \tag{6-35b}$$
$$= \frac{c}{2\pi} \int_0^{2\pi/c} J(\boldsymbol{q})_{\text{Gd-Y}} \cos[(N+1)\, c\, q/2]\, \mathrm{d}q$$

where c is the lattice constant along the c-axis of a superlattice with an h.c.p. structure. The calculated result is shown in Fig. 6-21. The coupling is fairly long range and is consistent in sign with the ordering observed in Gd/Y superlattices (solid circles). A similar analysis shows that the RKKY coupling can lead to a long-range coherence, a helical ordering, in Dy/Y multilayers.

A strong antiferromagnetic coupling between Fe layers separated by Cr has been found in Fe/Cr/Fe sandwiches (Grünberg, 1985) and Fe/Cr multilayers (Sellers et al., 1986 and 1990; Baibich et al., 1988). Grünberg measured the light scattered spectra from sandwiches with two Fe (100) layers separated by a Cr, V, Cu, Ag, or Au layer. The exchange coupling can be deduced from the frequency shift in the light scattering spectra. The results are shown in Fig. 6-22 as a function of the intervening layer thickness, d_0. The coupling length is short ($\sim$1 nm) for a Cr or V interlayer, and relatively long ($\sim$3 nm) for Cu, Ag, and Au. The difference was interpreted as due to the different crystal structure and charge

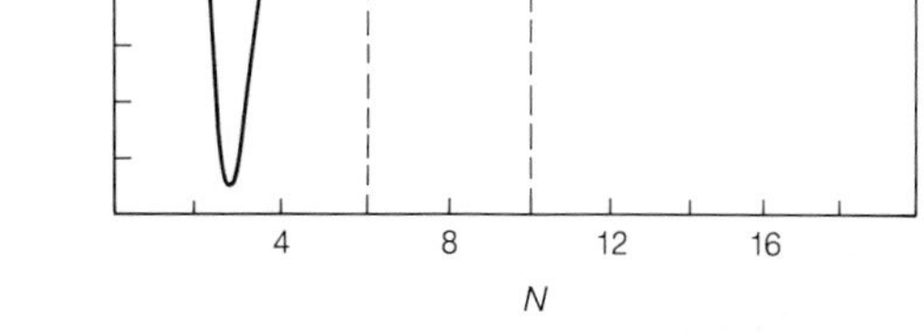

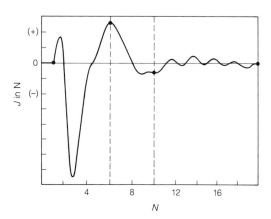

Figure 6-21. Range function of the RKKY interaction along the c-axis for Gd ions embedded in Y vs. the number of Y atomic layers separating the Gd. One vertical division corresponds to 0.025 meV (after Yafet et al., 1988).

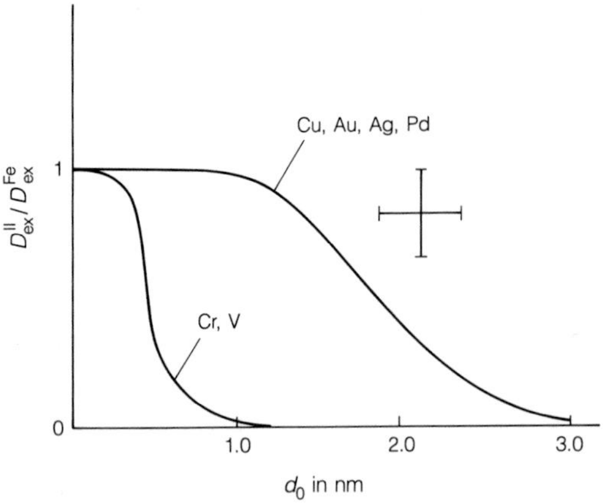

Figure 6-22. Frequency shift due to the coupling of modes for various interlayer materials as a function of d_0 (after Grünberg, 1985).

transfer effects. Recently, Parkin et al. (1990) have studied the exchange coupling in three types of superlattice structures: Fe/Cr, Co/Cr, and Co/Ru. The samples were grown on Si(111) wafers by d*c* magnetron sputtering. High resolution micrographs show that the Co/Ru superlattice has a high structural perfection with flat interfaces. Similar micrographs on Fe/Cr superlattices show a much less perfect structure with rumpled, disconnected layers. A striking result is that in these systems both the magnitude of the interlayer magnetic exchange coupling and the saturation magnetoresistance oscillate with the Cr or Ru spacer-layer thickness with a period ranging from 1.2 nm in Cu/Ru to ~1.8–2.1 nm in the Fe/Cr and Co/Cr systems. The results for Fe/Cr are shown in Fig. 6-23. The saturation field H_s is defined as the field at which the magnetization curve first departs from the high-field slope. The strength of the antiferromagnetic (AF) interlayer exchange coupling, J, is given by $J = -H_s M_s d_{cr}/4$. Since the oscillation period is relatively long, it appears difficult to reconcile this result with a RKKY coupling mechanism, which corresponds to a much smaller oscillation period (see Fig. 6-21).

At the present stage, the origin of the strong AF coupling between Fe layers separated by Cr is not clear. Since the magnetic state of the chromium atom is sensitive to the local environment, it might be perturbed by the adjacent Fe layers. Band structure calculations of Fe/Cr superlattices as a function of d_{cr} are desirable in order to derive the interlayer coupling. Calculations based on the augmented-spherical wave method have been per-

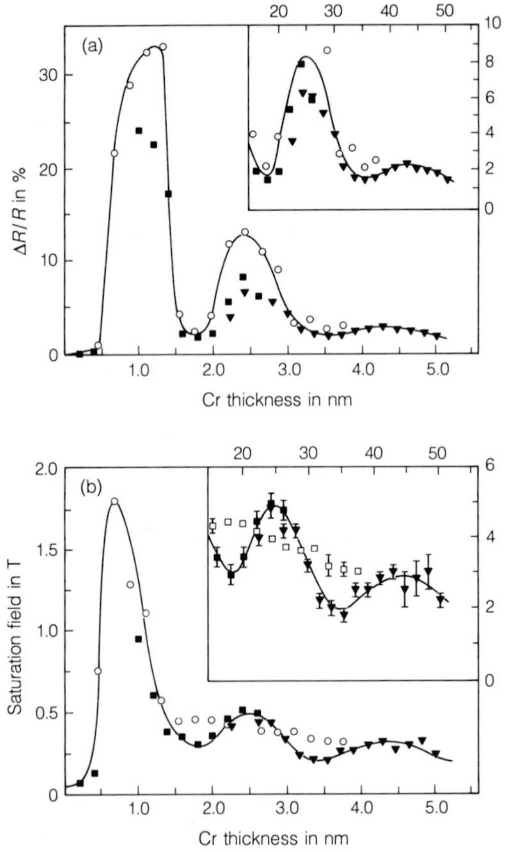

Figure 6-23. (a) Transverse-saturation magnetoresistance (4.5 K) and (b) saturation field (4.5 K) vs. Cr layer thickness for structures of the form Si(111)/10 nm Cr/[(2 nm) Fe/d_{Cr}] N/(5 nm) Cr, deposited at temperatures of ▼, ■, 40 °C (N = 30); ○, □, 125 °C (N = 20) (after Parkin et al., 1990).

formed on three Fe(m)/Cr(n) structures (Levy et al., 1990b). The results show that the AF couplings are too strong to be accounted for by dipolar interactions and have to be ascribed to exchange interactions through the Cr layers, which are different from the RKKY interaction.

6.3.3 Neutron Scattering

The scattering of neutrons by a system of atoms with magnetic moments furnishes an important means of obtaining information about their magnetic order. A general discussion of neutron diffraction can be found in Bacon (1975).

Earlier neutron diffraction work on multilayer system has been summarized by Endoh (1982) and Endoh et al. (1983), and by Majkrzak (1986). The recent developments have been described by Erwin et al. (1987), by Rhyne et al. (1987), and by Majkrzak et al. (1988).

6.3.3.1 Kinematic Theory

In a solid, neutron scattering occurs from the nuclei as well as from electronic magnetic moments; the neutron-nucleus interaction is a relatively strong, short-range, isotropic interaction. The nuclear scattering length, b, for neutrons is, in general, different for different isotopes.

The magnitude of the magnetic scattering length, p, is given by

$$p = (0.27 \times 10^{-14}\,\mathrm{m})\, g\, J\, f \tag{6-36}$$

where g is the Landé splitting factor, J is the quantum number corresponding to the total angular momentum of the atom, and f is an angle-dependent form factor.

Values of b and p for various nuclei and ions are tabulated in the book by Bacon (1975).

The interaction between the magnetic moment of a neutron and various atoms depends on the magnitudes of the moments, their relative orientation, and the scattering vector $\boldsymbol{Q}$. The latter dependence is characterized by a vector $\boldsymbol{q}$ defined by

$$\boldsymbol{q} = (\hat{\boldsymbol{Q}} \cdot \hat{\boldsymbol{M}})\, \hat{\boldsymbol{Q}} - \hat{\boldsymbol{M}} \tag{6-37}$$

where $\hat{\boldsymbol{Q}}$ and $\hat{\boldsymbol{M}}$ are unit vectors along the directions of the scattering vector and the atomic moment, respectively.

In a superlattice, the nuclear and magnetic scattering lengths depend on the compositional modulation. Neutron diffraction from superlattices is thus similar to X-ray diffraction. Both of the techniques can be used to determine the compositional profile of a superlattice. In this sense, the kinematic theory of X-ray diffraction from multilayer films can be easily applied to the neutron case with an extension to include the scattering from the magnetic moments. General discussions on X-ray diffraction from multilayers are given by de Fontaine (1966), McWhan (1985), and Segmüller and Blakeslee (1973). For a review, see Jin and Ketterson (1989).

According to the kinematic theory, the structure factor of a solid is given by

$$F(\boldsymbol{Q}) = \int_{V} \varrho(\boldsymbol{r}) \exp(\mathrm{i}\, \boldsymbol{Q} \cdot \boldsymbol{r})\, \mathrm{d}^3 r \tag{6-38}$$

where the scattering vector $\boldsymbol{Q}$ is defined as $\boldsymbol{Q} = \boldsymbol{k}_\mathrm{f} - \boldsymbol{k}_\mathrm{i}$ and $k_\mathrm{f} = k_\mathrm{i} = 2\pi/\lambda$. Here λ, $\boldsymbol{k}_\mathrm{i}$, and $\boldsymbol{k}_\mathrm{f}$ are the wavelength of the incident beam, and the incident and reflected wavevectors respectively; $\varrho(\boldsymbol{r})$ is the appropriate scattering density.

Following Erwin et al. (1987), for neutron diffraction involving an unpolarized beam the scattering intensity is given by

$$I_\mathrm{s}(\boldsymbol{Q}) \propto |F_\mathrm{N}(\boldsymbol{Q})| + |F_\mathrm{M}(\boldsymbol{Q})|^2 = \tag{6-39}$$
$$= \left|\sum_r b_r \exp(\mathrm{i}\, \boldsymbol{Q} \cdot \boldsymbol{r})\right|^2 + \left|\sum_r p_r \exp(\mathrm{i}\, \boldsymbol{Q} \cdot \boldsymbol{r})\right|^2$$

The vector $\boldsymbol{r}$ denote the positions of the atoms in the crystal, while b_r and p_r are the nuclear-scattering and magnetic-scattering lengths of that atom, respectively. The atom positions along the c-axis can be expressed as $r(m,n) = m\Lambda + z(n)$, where m indexes the bilayer where the atom resides and $z(n)$ is the relative position of the nth atom within the bilayer. The structure factor associated with the nuclear scattering is written as

$$|F_N(\boldsymbol{Q})|^2 = \left|\sum_m e^{\mathrm{i}\, m Q \Lambda}\right|^2 \left|\sum_n b_n\, e^{\mathrm{i}\, Q z(n)}\right|^2 \quad (6\text{-}40)$$

Since the magnetic modulation is not necessarily the same as the chemical modulation, nor even commensurate with it, the magnetic amplitudes can be expressed as

$$P(m,n) = \quad (6\text{-}41)$$
$$= p_n\{\hat{x}\cos[m\Phi + \phi(n)] \pm \hat{y}\sin[m\Phi + \phi(n)]\}$$

[the phase shift corresponding to $r(m,n)$ is $m\Phi + \phi(n)$]. Therefore, the structure factor associated with the magnetic scattering is

$$|F_M(+\boldsymbol{Q})|^2 = \frac{1}{2}\left|\sum_m e^{\mathrm{i}\, m(Q\Lambda + \Phi)}\right|^2 \cdot \left|\sum_n e^{\mathrm{i}[Q z(n) + \phi(n)]}\, p_n\right|^2 \quad (6\text{-}42)$$

From Eqs. (6-40) and (6-42) we see that the structure factors are determined by the modulation functions b_n and p_n. Since we cannot measure the large number of Fourier components necessary to extract directly the real-space modulation functions, we have to resort to a model for the interface. The most simple one is the step model, that involving a square-wave (or in general a rectangular-wave) modulation of the concentrations of the constituents, the lattice spacings, and the magnetic phase shifts. For this simple model the structure factors can be written as (Erwin et al., 1987)

$$|F_N(\boldsymbol{Q})|^2 = \frac{\sin^2(N Q \Lambda/2)}{\sin^2(Q\Lambda/2)}\, \frac{\sin^2(N_A Q\, d_A/2)}{\sin^2(Q\, d_A/2)}\, b_A^2 \quad (6\text{-}43)$$

and

$$|F_M(\pm\boldsymbol{Q})|^2 = \frac{\sin^2[N(\pm Q_A + \Phi)/2]}{\sin^2[(\pm Q_A + \Phi)/2]} \cdot \frac{\sin^2[N_A(\pm Q + k_A)\, d_A/2]}{\sin^2[(\pm Q + k_A)\, d_A/2]}\, \frac{1}{2}\, p_A^2 \quad (6\text{-}44)$$

Here N is the number of bilayers of thickness Λ, $\Phi = N_A k_A d_A + N_B k_B d_B$ is the total magnetic phase shift across the bilayer, N_A is the number of atomic planes in a magnetic A layer, and k_A and k_B are the magnetic modulation wave vector in the A-layer and B-layers, respectively. In writing Eqs. (6-43) and (6-44) it is assumed that the non-magnetic B-layer scattering amplitudes (b_B and p_B) are zero.

In Eq. (6-43), the first factor corresponds to the usual Bragg diffraction function, producing nuclear Bragg peaks at $Q = \tau_N = l\left(\frac{2\pi}{\Lambda}\right)$ for integer l with peak widths at half maxima (FWHM's) of approximately $2\pi/(N\Lambda)$. The second factor is the unit-cell structure factor, which has peaks at $Q = l(2\pi/d_A)$.

Equation (6-44) shows that the addition of incommensurate magnetic order produces additional principal scattering peaks of a solely magnetic origin. The magnetic Bragg peaks are located at $Q = \tau_N \mp \Phi/\Lambda$.

The peak positions and their widths yield the following information: (1) the bilayer thickness (Λ); (2) the total magnetic phase shift across the bilayer (Φ); (3) the number of atomic planes per bilayer (N_Λ); and (4) the nuclear and magnetic coherence lengths. The specific modulation profiles can only be determined by the relative peak intensities.

A strictly rectangular modulation is rarely achieved in a real superlattice due to (1) a compositional modulation (interdiffusion and island formation), (2) a strain modulation (resulting in a non-uniform d-spacing), and (3) a magnetic phase-angle modulation. These imperfections may be described by proper modulation profile models. If the concentration modulation for the A-layer is $C_A(n)$, the effective scattering lengths can be written as

$$b_n = C_A(n)\, b_A + [1 - C_A(n)]\, b_B \tag{6-45 a}$$

and

$$p_n = C_A(n)\, p_A + [1 - C_A(n)]\, p_B \tag{6-45 b}$$

The d-spacing modulation and magnetic-phase shift per atomic plane can be expressed in a similar manner.

In order to extract the magnetic structure, it is necessary to distinguish the nuclear and magnetic contributions. In principle, the detailed chemical modulation can be determined from a combination of X-ray and neutron diffraction data taken about a number of (0001) reciprocal lattice points.

Another approach is to use polarized neutron scattering. Experimentally, the magnetic structure factor may be determined more directly by measuring the flipping ratio, R, defined as

$$R = \frac{I_{s\uparrow}}{I_{s\downarrow}} = \left| \frac{F_N + F_M}{F_N - F_M} \right|^2 \tag{6-46}$$

where the arrow $\uparrow(\downarrow)$ means that the neutron spin is parallel (antiparallel) to the magnetization. According to the model calculation by Endoh et al. (1983), if the flipping ratio is independent of the order of the superlattice reflection, then the magnetic modulation should be the same as the concentration modulation, i.e., no interface anomaly exists.

When translating the observed Bragg-peak intensities into values for the structure factor, corrections for extinction, and the scattering geometry (Lorentz factor) must be considered (Bacon, 1975). For low-angle diffraction, or large enough M, the extinction effect may be significant (Majkrzak et al., 1985), and a dynamical theory of diffraction is preferred.

6.3.3.2 Experimental Results

Neutron diffraction studies have been performed on multilayer systems such as Ni/Cu (Felcher et al., 1980), Fe/Sb, Fe/Pd, Fe/V (Endoh, 1982), Fe/Ge (Majkrzak et al., 1985), Dy/Y (Salamon et al., 1986), Gd/Y (Majkrzak et al., 1986), and Co/Cu (Cebollada et al., 1989).

Fe/Ge

Low-angle polarized neutron diffraction measurements have been made on Fe/Ge multilayer films (Majkrzak et al., 1985). The low angle $\theta - 2\theta$ scan with $\boldsymbol{Q}$ perpendicular to the film plane is shown in Fig. 6-24. For small angle diffraction, the scattering density profile of a bilayer can be treated as a continuum. If the scattering density can be expanded as a Fourier cosine series, then the structure factor, F_j, evaluated at a low angle superlattice peak position given by $Q_j = j\,2\pi/\Lambda$, is proportional to the jth coefficient of the Fourier series expansion. The neutron and X-ray diffraction data were fitted to various models for the structure factor. The analysis reveals that there is a significant reduction in the Fe moment, which was attributed to an FeGe alloy-interface region between adjacent Fe and Ge layers. In the presence of interdiffusion, it becomes difficult to separate the effects of alloying and reduced dimensionality on the magnetization at the interface.

Dy/Y

With the advances in modern thin film synthesis techniques, one can now produce single-crystal, rare-earth magnetic superlattices of high quality. Gd/Y and Dy/Y

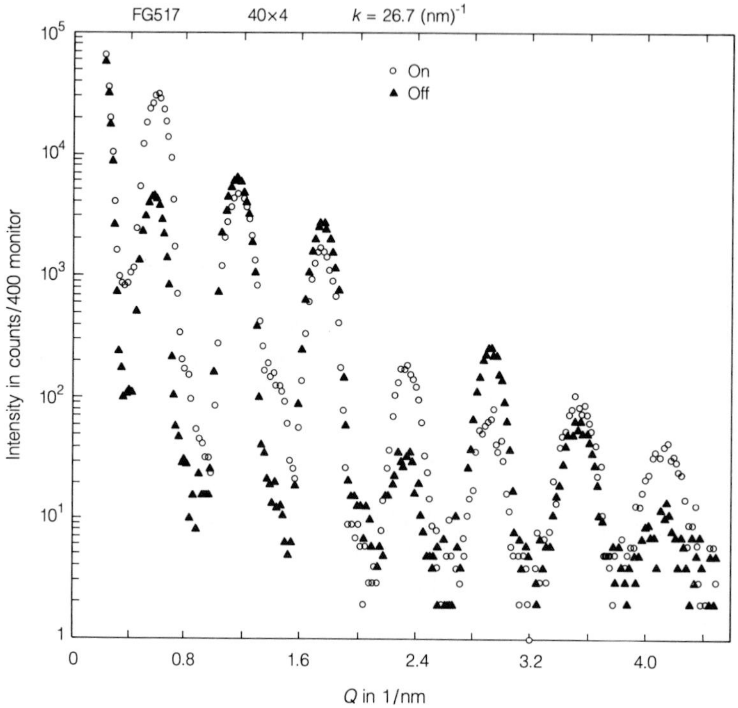

Figure 6-24. Polarized neutron diffraction data at low Q for an Fe/Ge multilayer. Seven diffraction peaks at values of $Q = m\,2\pi/\Lambda$ are shown. The "ON" ("OFF") data points correspond to incident neutrons in the "+" ("−") spin eigenstate (after Majkrzak et al., 1985).

superlattices are well-known examples. In these superlattices, the interfaces between the two constituents are nearly perfect on the atomic scale. These sharp chemical boundaries are ideal for studies of dimensionality and interface effects, and the control of the layer thickness makes it possible to study the details of interactions propagating through the interlayers.

The first evidence for the presence of long-range incommensurate magnetic order with Fourier components at the multilayer periodicity was reported in Dy/Y multilayers by Salamon et al. (1986).

Figure 6-25 shows the neutron diffraction scans around the (0002) principal Bragg peak for the $[Dy_{16}Y_{20}]_{89}$ multilayer for several temperatures below $T_c = 167$ K. The (0002) nuclear Bragg peak intensity is found to be independent of temperature. Its FWHM yields a crystalline width of approximately 42 nm. The Q^- and Q^+ magnetic satellites, and the accompanying

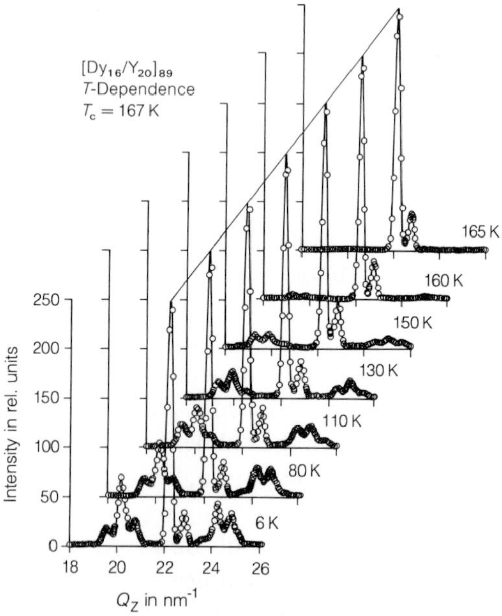

Figure 6-25. Neutron diffraction scans along the (0001) direction in $[Dy_{16}/Y_{20}]_{89}$ for several temperatures. The (0002) peak at $Q_z = 22.15\ \mathrm{nm}^{-1}$ is temperature independent. The magnetic satellites are temperature dependent (after Rhyne et al., 1987).

harmonics, are observed to decrease in amplitude as T increases. This behavior was attributed to the spreading and decrease of the order parameter at higher temperatures. The sharp magnetic satellites indicate that the magnetic coherence length is over 25 nm. The nature of the ordering, including the phase and chirality coherence, suggests that the order is propagated via a spin-density wave in the Y conduction bands stabilized by the Dy 4f spins. Applying a magnetic field in the basal plane results in the destruction of the helix and its conversion to ferromagnetic order (Fig. 6-26). The approach of this phase transition is revealed by a continuous drop in the magnetic satellite intensities and a growing intensity of the (0002) Bragg peak as the magnetic field increases.

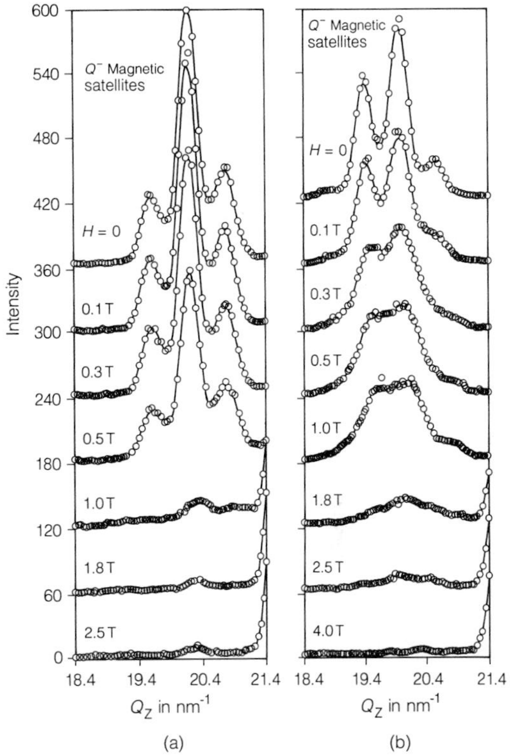

Figure 6-26. Field dependence of the Q^- magnetic satellite in $[Dy_{16}/Y_{20}]_{89}$ at $T = 10$ K (left-hand side). Satellite intensity of $T = 130$ K (0.79 T_c) showing the broadening of the peaks with field corresponding to a loss in long-range coherence (right-hand side) (after Rhyne et al., 1987).

Gd/Y

A microscopic magnetic antiphase domain structure has been observed in a single-crystal Gd/Y superlattice by neutron diffraction (Majkrzak et al., 1986). Below the Curie temperature and in low fields, the ferromagnetic Gd layers tend to align antiferromagnetically relative to one another; for $N_{Gd} = N_Y = 10$ there is a microscopic antiphase domain structure that is coherent over more than seventeen superlattice periods. For $N_Y = 6$ or 20, however, a single long-range ferromagnetic order is observed. The oscillatory behavior is consistent with a recent theoretical speculation (Yafet et al., 1988) that the Gd moments are coupled through the intervening Y via the RKKY interaction.

Gd/Dy

Neutron diffraction studies of a number of different Gd/Dy superlattice systems, with various combinations of Gd and Dy layer thicknesses, have yielded interesting results (Majkrzak et al., 1988). At 300 K, only the first-order chemical modulation satellite is observed. For temperatures between 200 and 130 K, a relatively weak incommensurate peak with a temperature-dependent wave vector coexists with commensurate magnetic reflections at multiples of one-half the chemical-modulation wave vector, indicating a doubling of the magnetic unit cell. Below about 130 K, the magnetic structure appears to be entirely commensurate. The moments of the Gd and Dy were found to be 6 μ_B and 9 μ_B. A schematic representation of the basal-plane component directions in three super-

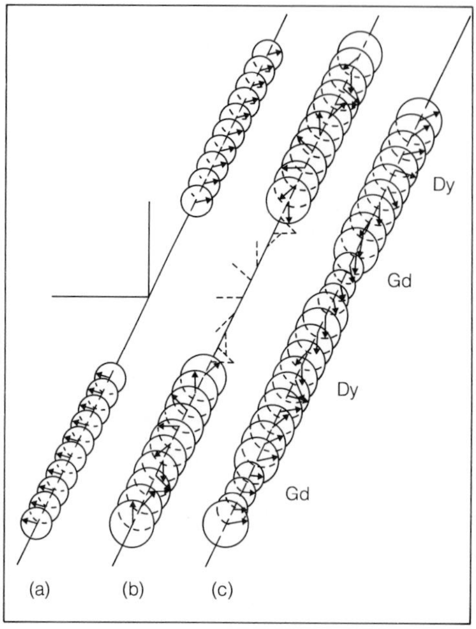

Figure 6-27. Schematic representation of the basal-plane component directions in three RE superlattice systems: (a) Gd/Y canted-antiphase domain structure; (b) Dy/Y coherent-incommensurate spiral; and (c) Gd/Dy *asymmetric* state (after Majkrzak et al., 1988).

lattice systems (Gd/Y, Dy/Y, Gd/Dy) is shown in Fig. 6-27.

Co/Cu

It is also of interest to grow single crystal superlattices involving 3D magnetic elements, since in 3D magnetic transition metals the magnetic moments are smaller and the electrons are itinerant (in contrast with the localized rare-earth systems). Single-crystalline Co/Cu superlattices have been prepared and characterized by Cebollada et al. (1989). The relative intensity and peak position for the different satellites in the neutron diffraction can be well accounted for by a simple step model. The appearance of extra satellites, with half-integer index $m = \frac{1}{2}, \frac{3}{2}, \ldots$, implies that at zero-field the adjacent Co layers are coupled antiferromagnetically via the intervening Cu layer, resulting in a doubling of the modulation wavelength. Measurements of polarized neutron diffraction show that the intensity of the half-integer satellites decreases with magnetic field and falls to zero at the saturation field. The origin of the observed antiferromagnetic ground state remains to be explored.

6.3.4 Ferromagnetic Resonance

The ferromagnetic resonance (FMR) technique is a powerful tool for exploring the average magnetization of thin films, and the interlayer coupling in multilayers.

For a homogeneous sample, the FMR frequency is given by the expression (Kittel, 1948)

$$\left(\frac{\omega}{\gamma}\right)^2 = \mu_0^2 [H + (N_y - N_z) M] \cdot \cdot [H + (N_x - N_z) M] \qquad (6\text{-}47)$$

where γ is the gyromagnetic ratio, ω is the resonance frequency, and N_x, N_y and N_z are demagnetization factors. The external field, H, is along the z-direction. For a thin film, the demagnetization factors perpendicular and parallel to the film surface are 1 and 0 respectively, and Eq. (6-47) reduces for H parallel to the film to

$$\omega/\gamma = \mu_0 [H_\parallel (H_\parallel + M)]^{1/2} \qquad (6\text{-}48)$$

and for H perpendicular to the film to

$$\omega/\gamma = \mu_0 (H_\perp - 4\pi M) \qquad (6\text{-}49)$$

The effect of an inhomogeneous distribution of the magnetic moments, which is typical of a multilayer sample, was considered by including an interlayer exchange term in the equation of motion (White and Herring, 1980). The resulting expressions are

$$\omega/\gamma = \mu_0 [H_\parallel (H_\parallel + \langle M \rangle \eta)]^{1/2} \qquad (6\text{-}50)$$

for H parallel to the film and

$$\omega/\gamma = \mu_0 (H_\perp - \langle M \rangle \eta) \tag{6-51}$$

for H perpendicular to the film. Here $\eta \equiv \langle M^2 \rangle / \langle M \rangle^2$ and the brackets represent averages over the film thickness. According to Eqs. (6-50) and (6-51), the quantity actually measured is $\langle M \rangle \eta$, which for a complex moment distributions (e.g., one involving regions with antiparallel moments) can appear to be larger than the maximum magnetic moment of the sample.

If anisotropy factors are included, the resonance field can be calculated by using the general equation derived by Smit and Beljers (1955):

$$\left(\frac{\omega}{\gamma}\right)^2 = \frac{1}{M^2 \sin^2\theta} \left[\frac{\partial^2 E}{\partial \theta^2} \frac{\partial^2 E}{\partial \phi^2} - \left(\frac{\partial^2 E}{\partial \theta \, \partial \phi}\right)^2\right] \tag{6-52}$$

where the polar coordinates of the magnetization, M, are θ and ϕ and the field is in the z-direction. The total energy density, E, contains the Zeeman energy, the magnetostatic energy and the anisotropy energy (for more details see Chappert et al., 1986; Flevaris et al., 1982).

For a superlattice structure, the FMR field and the complete line shape can be obtained by solving Maxwell's equations and the Landau-Lifshitz equation (1935) of motion for the magnetization with appropriate boundary conditions. This has been discussed by Spronken et al. (1977) for an arbitrary field direction and interlayer exchange coupling. Vittoria (1985) introduced the transfer function matrix method (which relates the microwave fields at the two opposite surfaces of magnetic and dielectric layers); with a perpendicular configuration they obtained a simplified numerical formalism to calculate the FMR absorption power spectra. The real part, $\mathrm{Re}\, Z_s$, of the surface impedance plotted as

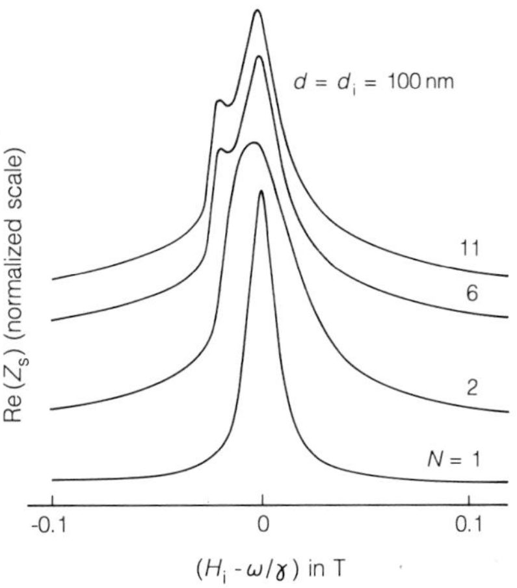

Figure 6-28. Real part of the surface impedance $\mathrm{Re}(Z_s)$ is plotted as a function of $H_i - \omega/\gamma$. The vertical scale has been shifted for comparison purposes. $H_i = \omega/\gamma$ is the FMR field for uniform procession of the magnetic moments. N is the number of bilayers (after Vittoria, 1985).

a function of $\mu_0 H_i - \omega/\gamma$ is shown in Fig. 6-28, where $H_i = H - M$ is the internal field. From this figure we see that, in the limit of a few layer pairs, only one FMR excitation is predicted to occur. As the number N of bilayers increases, a subsidiary FMR line appears at fields where $\mu_0 H_i < \omega/\gamma$, which is attributed to a spin wave mode. The FMR line shape for magnetic multilayer structures of alternating iron and iron-alloy layers was calculated by McKnight and Vittoria (1987). The model assumes abrupt interfaces and that the spins on adjacent layers are ferromagnetically coupled with an exchange constant A_{12}. The surface anisotropy fields are ignored in the calculations. As A_{12} is increased, satellite lines in the surface impedance split off from the main ferromagnetic resonance lines of the component materials. The number of satellites is one more than the number of the bilayers. The peak positions, widths and oscillator strengths are also found to depend sensitively on the relative layer thicknesses. This fact makes it possible to measure the magnetic interlayer coupling.

The coupled resonance modes of a set of MBE grown single-crystal Fe/Cr/Fe(001) sandwiches have been observed by means of multiple-frequency (2–14 GHz) FMR (Krebs et al., 1989 and 1990). When combined with the magnetization and magnetoresistance data, it is found that samples with $1.2\,\text{nm} < d_c < 2.5\,\text{nm}$ show anti-aligned Fe layers in zero magnetic field (Fig. 6-29). The AF-coupling may be modeled by adding a term $J\hat{\boldsymbol{M}}_1 \cdot \hat{\boldsymbol{M}}_2$ to the total energy density E such that

$$E = \tfrac{1}{2}(E_1 + E_2) + J\hat{\boldsymbol{M}}_1 \cdot \hat{\boldsymbol{M}}_2 \tag{6-53}$$

where $E_1(E_2)$ is the energy density of the Fe film 1(2). The FMR data reveal two resonance modes which correspond to in-phase and out-of-phase precessions of $\boldsymbol{M}_1$ and $\boldsymbol{M}_2$. The coupling, J/M, is peaked at $d_{cr} = 1.6\,\text{nm}$, in contrast to the monotonic variation in J found from the analysis of Cr/Fe(001) multilayers with $0.9\,\text{nm} < d_{cr} < 3\,\text{nm}$ (van Dau et al., 1988). This is reasonable because when d_c vanishes, the direct exchange between the Fe layers will cause them to be ferromagnetically coupled. The Cr thickness dependence of J, and the physical origin of J, are not understood.

The magnetic coupling of 20 nm-thick Ni and $Ni_{78}Fe_{22}$ films through an intervening Ag film with thicknesses ranging between 0 and 20 nm has been investigated by FMR at 33 GHz (Layadi et al., 1990). From the fits of the FMR-mode-positions, intensities, and line widths, the interlayer coupling is found to obey the empirical relation $J \propto \exp(-d_0/2.86)$, where d_0 is the Ag layer thickness (in nm).

FMR studies have also been performed for Co/Ni (Krishnan, 1985), Co/Cr (Ramakrishna et al., 1987), and Ni/C (Krishnan et al., 1981) multilayers.

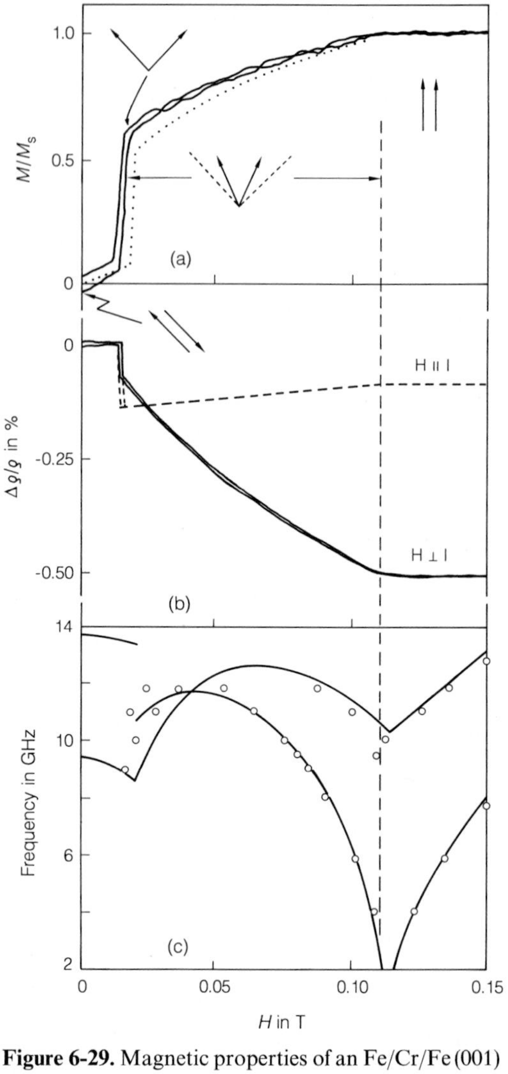

Figure 6-29. Magnetic properties of an Fe/Cr/Fe(001) film width $d_{Cr} = 1.3$ nm. (a) Magnetization field dependence for $\boldsymbol{H} \parallel [110]$, the relative orientations of layer moments are also shown; (b) longitudinal and transverse magnetoresistance; (c) magnetic resonance data for $\boldsymbol{H} \parallel [110]$ (after Krebs et al., 1989).

Before leaving the discussion of FMR we note that a surface-wave-vector specific geometry has been proposed by Bogacz and Ketterson (1985). It utilizes a meander-line-like *antenna*, the spacing of which defines the in-plane wave vector. It was shown that this technique will couple to both the bulk spin waves and the Damon Eshbach mode (see next section). To date this technique has not been applied.

6.3.5 Light Scattering

6.3.5.1 Introduction

The linear response of a material to an external probe is controlled by the nature of the spectrum of its elementary excitations. In magnetic superlattices, elementary excitations such as spin waves are collective excitations of the structure as a whole and accordingly can have properties distinctly different from the modes associated with any one constituent. While static magnetization, neutron diffraction, and ferromagnetic resonance studies performed to date can be explained in terms of the effects of a single film or a large number of interfaces, the collective spin wave excitations are the first and the only dynamic collective features of superlattices reported so far.

Collective spin-wave modes in magnetic superlattices have been studied extensively in the past decade. Camley et al. (1983) have investigated spin waves and the light scattering spectrum of a system consisting of alternating ferromagnetic and non-magnetic layers, taking into account the Zeeman and dipolar interactions (but neglecting exchange). Similar systems were also studied by Grünberg and Mika (1983). Hillebrands (1988) later generalized the treatment to take into account the exchange and surface anisotropy contributions. Magnetic polarizations in systems of magnetic layers separated by dielectric layers have also been studied (Thibandeam and Caillé, 1985; Zhou and Gong, 1989). The first detailed experimental studies of the collective spin-wave modes of superlattice structures were the Brillouin scattering studies on Ni/Mo metallic superlattices reported by Grimsditch et al. (1983). Since this early work, numerous other experimental light scattering studies have been reported in magnetic superlattice structures of various sorts. Comprehensive reviews on the subject are available (Cardona and Güntherodt, 1982, 1989).

6.3.5.2 Spin Wave Excitations in Magnetic Superlattices

The dispersion relation of the spin waves in magnetic media is determined by long range fields (the magnetic dipole fields of the spins) and short range exchange couplings between spins on the lattice. The contribution of the dipole fields to the frequency of the wave is on the order of $\gamma \mu_0 M_s$, where M_s is the saturation magnetization and γ the gyromagnetic ratio. This is a frequency which lies in the microwave regime for nearly all materials. An applied Zeeman field, $\boldsymbol{H}_0$, contributes an amount $\gamma \mu_0 H_0$ to the frequency of the wave. The short range interactions between the spins are the exchange interactions (of quantum mechanical origin) and are characterized by $\gamma D q^2$, where D is the *exchange stiffness* constant. Including all these factors, the long wavelength dispersion relation for a spin wave in an isotropic ferromagnet of infinite spatial extent is given by the well-known Holstein-Primakoff (1940) expression

$$\omega_B(\boldsymbol{q}) = \gamma \mu_0 [(H_0 + D q^2) \cdot (H_0 + M_s \sin^2\theta + D q^2)]^{1/2} \quad (6\text{-}54)$$

where θ is the angle between $\boldsymbol{q}$ and $\boldsymbol{M}_s$. As $\boldsymbol{q} \to 0$, ω_B has the limiting values

$$\omega_B = \gamma \mu_0 H_0, \qquad \theta = 0 \quad (6\text{-}55\text{a})$$

$$\omega_B = \gamma (B_0 B)^{1/2}, \qquad \theta = \frac{\pi}{2} \quad (6\text{-}55\text{b})$$

where $B_0 = \mu_0 H_0$ and $B = \mu_0 (H_0 + M_s)$; these are in agreement with Eq. (6-54) for a film geometry.

Apart from bulk modes, at the surface of a ferromagnet an additional excitations is present which is known as the Damon-Eshback (DE) mode (1961) or surface magnon. In the absence of exchange, the frequency, ω_s, of a surface magnon is given by

$$\omega_s = \frac{\gamma}{2}(B_0/\sin\theta + B\sin\theta) \quad (6\text{-}56)$$

There is a critical angle $\theta_c = \sin^{-1}(B_0/B)^{1/2}$ which corresponds to $\omega_s = \gamma(B_0 B)^{1/2}$. Beyond this angle, the DE mode decays spontaneously into bulk spin waves. The most striking characteristic of the DE spin waves is the *nonreciprocal* propagation. Propagation is only possible for a certain range of directions. In the presence of exchange, for $\theta = 90°$, one has

$$\omega_s'(q) = \gamma\mu_0[H_0 + \tfrac{1}{2}M_s + D'q^2] \quad (6\text{-}57)$$

An isolated ferromagnetic film of thickness d has surface spin waves which propagate along the film and are localized on the two surfaces as shown in Fig. 6-30. The two waves propagate in opposite directions. When $\boldsymbol{M}_s$ and $\boldsymbol{H}_0$ are both parallel to the surface and $\boldsymbol{q}$ is perpendicular to $\boldsymbol{M}_s$, the frequency of these waves is given by

$$\omega_s'(q_\parallel) = \gamma\mu_0[(H_0 + \tfrac{1}{2}M_s)^2 - \\ - \tfrac{1}{4}M_s^2\exp(-2q_\parallel d)]^{1/2} \quad (6\text{-}58)$$

The bulk-like modes in the film are still given by Eq. (6-54), but have a standing-wave character with the allowed values of q restricted to $q = n\pi/d$, where n is an integer.

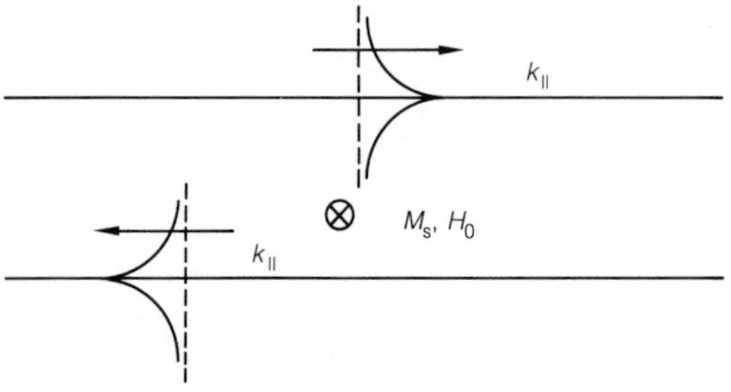

Figure 6-30. Schematic illustration of surface spin waves in thin ferromagnetic films.

Now we discuss the case of an infinitely extended superlattice consisting alternatively of ferromagnetic layers of thickness d and non-magnetic layers of thickness d_0. Collective excitations of the structure as a whole may be excited. In this type of superlattice, each interface supports a surface mode. These surface modes, coupled by their dipolar fields, split off into a band as d_0 is reduced. As a result, we have Bloch states which may transport energy normal to the interface.

Theoretical calculations have been performed by Camley et al. (1983). The results show that the standing spin waves do not couple and remain at the frequency given by Eq. (6-54). The surface-like modes (ω_s') couple and form a band of modes given by

$$\omega(q_\parallel, q_\perp) = \\ = \gamma\mu_0[H_0(H_0 + M_s) + \tfrac{1}{4}M_s^2 w]^{1/2} \quad (6\text{-}59)$$

where

$$w = \frac{2\sinh(q_\parallel d)\sinh(q_\parallel d_0)}{\cosh[(d+d_0)q_\parallel] - \cosh[(d+d_0)q_\perp]} \quad (6\text{-}60)$$

For a semi-infinite superlattice, the theory reveals the existence of a surface-like magnon of the superlattice as a whole, provided that $d > d_0$. Its frequency is given by $\omega_s = \gamma\mu_0(H_0 + \tfrac{1}{2}M_s)$, equal to that of a surface wave in a bulk material of magnetization M_s [as opposed to the average magnetization of the superlattce, $M_s d/(d+d_0)$].

The theory mentioned above is valid only in the dipolar limit, where the dipolar modes are well separated in frequency from the exchange modes. Contributions from exchange (as well as anisotropy) have been ignored. Although the theory is able to produce the salient features of the collective spin wave modes in superlattices, recent light scattering studies on ultra thin

Fe films show the significance of interface anisotropies for small layer thicknesses (Hillebrands et al., 1987). For thickness below 60 nm the frequency shift of the Brillouin line is found to increase with decreasing film thickness. Hillebrands et al. attributed this observation to the effects of magnetic surface anisotropies and have made calculations based on the DE magnetostatic surface mode (by replacing the magnetic surface anisotropy by an effective volume anisotropy). Their results are shown in Fig. 6-31. The problem has been re-examined by Rado and Hicken (RH) (1988). The latter calculations are based on a solution of the equations of motion of the magnetization, $\boldsymbol{M}_s$, in the magnetostatic limit. The surface anisotropies have been introduced by means of the general exchange boundary conditions of Rado and Weertman (1959), although this boundary condition is not universally accepted. Their calculations yield practically the same results as that given by Hillebrands et al. The calculations by Rado and Hicken show that for film thicknesses $d \leq 1.5$ nm, the film thickness is small compared to the spin wavelength, and therefore the film is, essentially, uniformly magnetized. Exchange effects may therefore be neglected and the use of an effective volume anisotropy is reasonable.

Another interesting result of the RH theory is that the calculated Brillouin shift contains additional branches of spin-wave modes which would not exist in the absence of exchange.

The RH theory has been extended to treat two types of multilayer structures by Hillebrands (1988): (i) conventional structures, consisting of alternating ferromagnetic and nonmagnetic layers, and (ii) all-ferromagnetic multilayered structures, consisting of magnetic layers with different magnetic properties (Fe/Ni). For

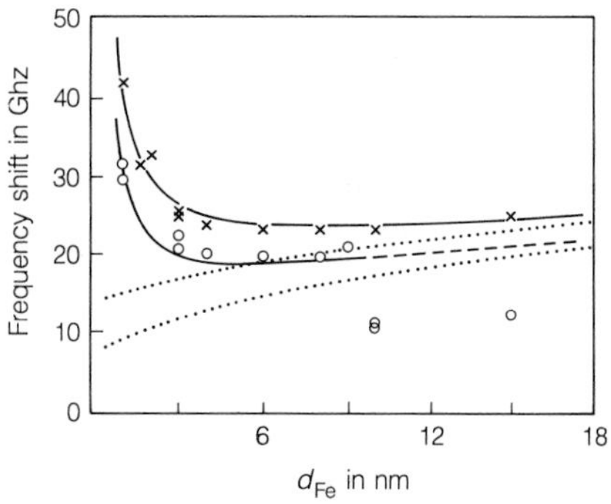

Figure 6-31. Spin-wave frequencies as a function of the Fe film thickness in applied fields of 0.1 T (crosses) and 3×10^{-3} T (open circles), respectively. The solid and dotted lines are the theoretical results with and without surface anisotropy respectively (after Hillebrands et al., 1987).

$d < 30$ nm, the dipolar modes exhibit a characteristic increase in frequency, since interface anisotropies become dominant in this regime. In the regime where the exchange modes [which are characterized by their typical $1/d^2$ behavior, see Eq. (6-54)] cross with the dipolar modes, a gap occurs which is determined primarily by the amount of interface anisotropy. In the case of all-magnetic multilayer structures, new types of collective spin waves (i.e., coupled exchange modes) are predicted. The mode splitting depends strongly on the interface exchange constant, A_{12}; therefore a measurement of the spin wave frequencies in the region of interest might allow for the determination of A_{12}.

6.3.5.3 Light Scattering Spectra

Incident light of wave vector $\boldsymbol{k}_i$ and frequency ω_i is scattered into a state $\boldsymbol{k}_s$, ω_s by a spin wave of vector $\boldsymbol{q}$ and frequency ω such that

$$\boldsymbol{k}_s - \boldsymbol{k}_i = \pm \boldsymbol{q} \tag{6-61}$$

and

$$\omega_s - \omega_i = \pm \omega \tag{6-62}$$

where the + sign refers to absorption (the anti-Stokes process) and the − sign to emission (the Stokes process) of the magnon. It has been shown that the dominant interaction responsible for light scattering from spin waves is via fluctuations produced in the dielectric constant ε by spin-orbit coupling (Fleury and Loudon, 1968). The direct interaction between the spin fluctuations and the magnetic vector of the light is relatively weak.

Brillouin scattering is particularly suited to investigate spin waves in thin films and superlattices. For the back-scattering geometry, the wave vector of the detected spin wave is determined by the wavelength, λ, and the incident angle, α, of the Laser beam according to

$$q = \frac{4\pi}{\lambda} \sin\alpha \qquad (6\text{-}63)$$

For $\lambda \sim 500$ nm, the maximum q is thus on the order of $2 \times 10^7 \, m^{-1}$, which is small enough for the magnetostatic approximation to be valid, but large enough to give a significant contribution of the q-dependent terms to the spin-wave frequency. A typical Brillouin spectrum for a 106 nm-thick amorphous $Fe_{80}B_{20}$ film is shown in Fig. 6-32. The surface mode (SM) is only observed on the Stokes side (lower scale). This situation reverses upon reversal of the external field. The peaks labeled $A_1 - A_4$ and $S_1 - S_3$ are the aS (upper scale) and S bulk-standing modes respectively. The observed S-aS asymmetry arises from the lack of reflection symmetry of the magnon at the surface and is the same asymmetry responsible for the nonreciprocity in the DE mode propagation (Camley et al., 1981). The experimental observations are in good agreement with the theory.

Currently the LS technique is the only method to detect the standing modes in very thin films. From Eq. (6-54) we see that

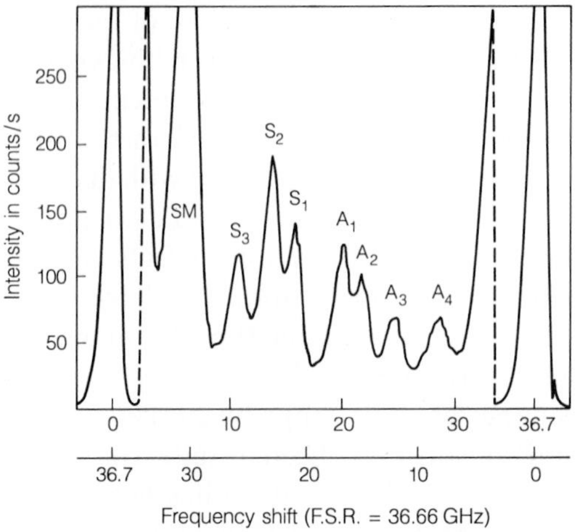

Figure 6-32. Brillouin spectrum of magnons in an amorphous sample of $Fe_{80}B_{20}$ of thickness of 106 nm, with a magnetic field of 0.26 T in the plane of the sample. The peak labeled SM is the Stokes surface magnon while $A_1 - A_4$ and $S_1 - S_3$ label the anti-Stokes and Stokes substructure of the bulk magnons, respectively (after Grimsditch et al., 1983).

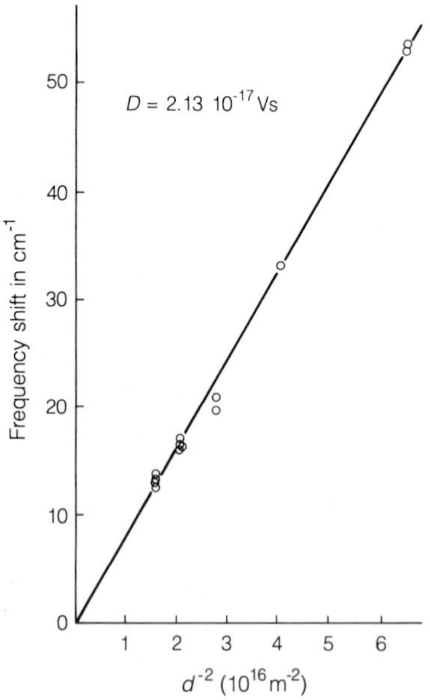

Figure 6-33. Standing spin wave mode frequencies obtained by Raman spectroscopy vs. d ($1 \, cm^{-1} \simeq 30$ GHz) (after Blamenröder et al., 1985).

for small d the exchange term dominates, leading to $\omega \propto 1/d^2$. The mode frequency vs. d^{-2} plot is shown in Fig. 6-33 from an 8 nm-thick Fe film. The quantity $\gamma D(n\pi)^2$ is then determined by the slope. The observed mode is the one with $n=2$. The $n=1$ mode may be hidden in the Rayleigh tail.

In multilayers a band of surface-like character is formed. However, if the number of bilayers is finite, the band of modes actually splits into separate peaks, as originally pointed out by Grünberg and Mika (1983). This is confirmed by the measurements on Co/Nb multilayers for two samples (Fig. 6-34) (Rupp et al., 1985). The peaks labeled P and M correspond to phonons and magnons, respectively. The positions of these peaks as a function of applied field are well accounted for by the theory.

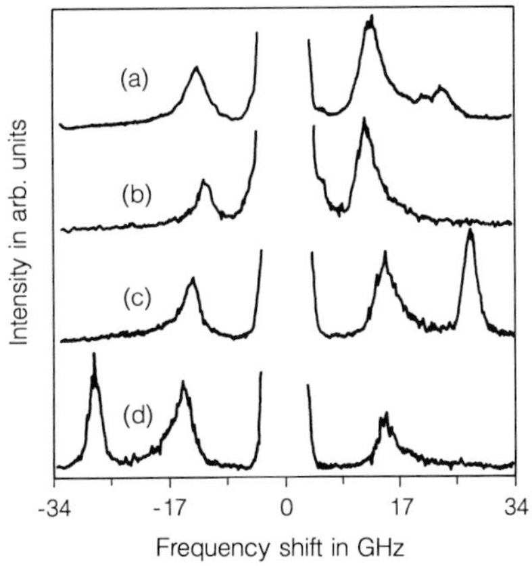

Figure 6-35. Room-temperature Brillouin spectra of Fe/Pd superlattices in an applied field at 0.1 T: (a) $d=2.19$ nm and $d_0=2.43$ nm, (b) $d=4.17$ nm and $d_0=13.87$ nm, and (c), (d) $d=4.1$ nm and $d_0=0.91$ nm. In (d) the direction of the applied field has been reversed compared to (c) (after Hillebrands, 1988).

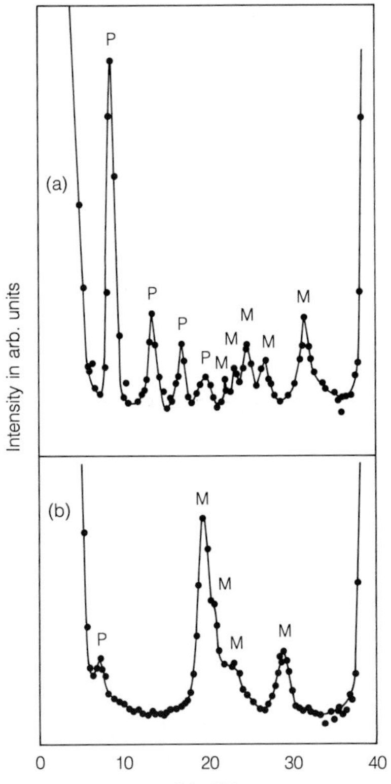

Figure 6-34. Brillouin scattering spectra (Stokes) for two Co/Nb multilayers: (a) $d_{Co}/d_{Nb}=21$ nm/14 nm; (b) $d_{Co}/d_{Nb}=11$ nm/10 nm. Magnon (M) as well as phonons (P) are observed (after Rupp et al., 1985).

Figure 6-35 shows typical Brillouin spectra of Fe/Pd superlattices (Hillebrands, 1988). For $d \approx d_0$ (trace a), a broad surface-like band of modes can be identified by their strong S-aS asymmetry. For $d \ll d_0$ (trace b) the spin-wave band becomes narrower due to the reduced dipolar coupling across the Pd layers. For $d_0 \ll d$ (trace c), a very intense, discrete mode is observed in the aS spectrum, which is identified as a superlattice surface spin wave mode. (It should merge with the collective spin wave band for $d=d_0$.) An inverted field changes the S and aS parts of the spectra (trace d).

The only observation of standing spin waves in a superlattice was reported by Kueng et al. (1984). In a Mo/Ni superlattice having $d=d_0 \approx 55$ nm, apart from the peak due to the band, an additional peak at a higher frequency was also observed.

By fitting the magnetic field dependence of this extra peak to Eq. (6-54), these authors suggested that this peak corresponds to the lowest standing-spin-wave mode in each layer and that coupling between the layers does not drastically modify its position relative to that in a single isolated film.

LS spectra in Fe/Cr multilayers with antiferromagnetic interlayer exchange have been reported by Saurenbach et al. (1988). Magnons in Cu/Ni superlattices have been studied by Mattson et al. (1990).

From spectra such as those shown above, together with the formalism outlined in Sec. 6.3.5.2, magnetic parameters can be determined. The parameters obtained so far from LS experiments have been tabulated by Grünberg (1989).

6.4 Superconducting Properties

The superconducting properties of artificial superlattices are especially interesting because the modulation wavelength may be made comparable to various length scales which characterize the superconducting state. Many of the observed properties may be described by the phenomenological Ginzburg-Landau (GL) theory (St. James et al., 1969). However we must keep in mind that, in a uniform superconductor, the range of validity of the GL theory is usually limited to temperatures in the immediate vicinity of the bulk (zero field) transition temperature. In a metallic superlattice involving components with widely differing bulk transition temperatures, and interfaces with uncertain boundary conditions, the theory may be of semiquantitative or qualitative value only. More quantitatively reliable results require various extensions of the BCS theory to inhomogeneous systems. Although the most powerful theoretical treatment involves the use of thermal Green's functions (Abrikosov et al., 1963; Lifshitz and Pitaevskii, 1980), de Gennes (1966) has shown that most of the pertinent results can be extracted using the self-consistent field approach of Bogoliubov (1959). We begin our discussion with a review of those behaviors which may be understood on the basis of the GL theory. A number of good discussions of this theory are available (de Gennes, 1966; Tinkham, 1980), with which we will assume some familiarity; we will use the notation of Lifshitz and Pitaevskii (1980).

6.4.1 Ginzburg-Landau Theory

The GL theory postulates (and the microscopic theory confirms) the existence of a complex *order parameter*, ψ, which close to the transition temperature, T_c, governs the change in the thermodynamic free energy, F, relative to its value in the normal state, F_n. A variational procedure yields the following two Ginzburg-Landau equations (see Chapter 4 by Kes of this Volume).

$$-\frac{\hbar^2}{4m}\left(\nabla - \frac{2\mathrm{i}e}{\hbar}\boldsymbol{A}\right)^2 \psi + a\psi + b|\psi|^2 = 0 \qquad (6\text{-}64)$$

$$\boldsymbol{j} = -\frac{\mathrm{i}e\hbar}{2m}(\psi^* \nabla\psi - \psi\nabla\psi^*) - \frac{2e^2}{m}|\psi|^2 \boldsymbol{A} \qquad (6\text{-}65)$$

where $\boldsymbol{A}$ is the vector potential.

We will generally be interested in the behavior very close to the transition where $|\psi^2|$ is small and we can take b as a constant and write $a = \alpha(T - T_c)$. In these cases the last term in Eq. (6-64) may be neglected. We then have a linearized GL equation, which is formally identical to the time independent Schrödinger equation,

$$-\frac{\hbar^2}{4m}\left(\nabla - \frac{2\mathrm{i}e}{\hbar}\boldsymbol{A}\right)^2 \psi + a\psi = 0 \qquad (6\text{-}66)$$

For $A=0$ and $a<0$, solutions of the form e^{ikx} exist with $k^2=-\frac{4ma}{\hbar^2}=\frac{4m\alpha(T_c-T)}{\hbar^2}$. We are thus led to introduce a fundamental length, the Ginzburg-Landau coherence length, $\xi=\left[\frac{\hbar^2}{4m\alpha(T_c-T)}\right]^{1/2}$. For cases where the spatial variations of ψ can be neglected Eq. (6-65) reduces to London's equation

$$j=-\frac{2e^2}{m}|\psi|^2 A \tag{6-67 a}$$

which, on substitution into the fourth Maxwell equation (and using $\nabla\cdot B=0$, $\boldsymbol{B}=\mu_0\boldsymbol{H}$), yields

$$\nabla^2\boldsymbol{H}=\frac{1}{\lambda_L^2}\boldsymbol{H} \tag{6-67 b}$$

where $\lambda_L^2\equiv\frac{m}{2\mu_0 e^2|\psi|^2}=\frac{mb}{2\mu_0 e^2\alpha(T_c-T)}$. Equation (6-67 b) has a solution with the dependence e^{-x/λ_L}; thus a second length, the London penetration depth λ_L, enters the phenomenological GL theory. Note that both λ and ξ diverge as $(T_c-T)^{-1/2}$ as we approach the transition (from below); thus their ratio, defined as $\varkappa\equiv\frac{\lambda_L}{\xi}$, is constant. The parameter $\varkappa$ is of paramount importance in determining the magnetic properties of bulk superconductors.

6.4.1.1 The Anisotropic Superconductor

We now begin a discussion of the GL theory of artificial superlattices. In some situations we may regard an artificial superlattice as an anisotropic superconductor. Anisotropy may be incorporated into the phenomenological GL theory simply by introducing an *effective mass tensor* into the *kinetic energy* term of Eq. (6-64) in the form

$$\frac{1}{2}\left[-i\hbar\left(\nabla-\frac{2ie}{\hbar}A\right)\right]\cdot\left(\frac{\mathbf{1}}{\mathbf{2m}}\right)\cdot \tag{6-68}$$
$$\cdot\left[-i\hbar\left(\nabla-\frac{2ie}{\hbar}A\right)\right]\psi+a\psi+b|\psi|^2\psi=0$$

For a superlattice which may be regarded as isotropic in the plane of the film (due to either its textural properties or crystallographic symmetry), the reciprocal effective mass tensor may be written

$$\left(\frac{\mathbf{1}}{\mathbf{m}}\right)=\begin{bmatrix}\frac{1}{m_x} & 0 & 0\\ 0 & \frac{1}{m_x} & 0\\ 0 & 0 & \frac{1}{m_z}\end{bmatrix} \tag{6-69}$$

where the x and y axes lie in the plane of the film and the z axis is the direction of the composition modulation. The inverse of Eq. (6-60) yields the (effective) mass tensor, $\mathbf{m}$.

We now give an expression for the upper critical field $H_{c2}(T)$. Near the field-dependent-transition temperature the order parameter is vanishingly small and we may neglect the nonlinear term. If we define two coherence lengths

$$\xi_x\equiv\left(\frac{\hbar^2}{4m_x\alpha(T_c-T)}\right)^{1/2}$$

and

$$\xi_z=\left(\frac{\hbar^2}{4m_z\alpha(T_c-T)}\right)^{1/2}$$

and introduce the *flux quantum* $\phi_0=\frac{h}{2e}$, then the upper critical field is given by[1]

$$H_{c2}(\theta)=\frac{\phi_0}{2\pi[\sin^2\theta\,\xi_x^2\xi_z^2+\cos^2\theta\,\xi_x^4]^{1/2}} \tag{6-70}$$

[1] For convenience we have omitted μ_0 in the following expressions for the upper critical field. However, it is understood that H is in Tesla rather than in ampere per meter.

here θ measures the angle of the magnetic field from the z-axis. In particular, for the field parallel and perpendicular to the plane of the film we have, respectively,

$$H_{c2\parallel} = \frac{\phi_0}{2\pi\,\xi_x\,\xi_z} \tag{6-71 a}$$

and

$$H_{c2\perp} = \frac{\phi_0}{2\pi\,\xi_x^2} \tag{6-71 b}$$

The angular dependence of the upper critical field is thus characterized by the two parameters ξ_x and ξ_z (Lawrence and Doniach, 1971). Note that Eqs. (6-71 a) and (6-71 b) both lead to a linear behavior of the upper critical field in $T_c - T$.

Equation (6-67 a) can be generalized to the case of an anisotropic superconductor by replacing m^{-1} by $\left(\frac{\mathbf{1}}{\mathbf{m}}\right)$ from Eq. (6-69); an anisotropic London penetration depth follows from making the appropriate substitutions.

When the magnetic field is parallel to the plane of the film (and for some range of angles in the vicinity of this condition) the phenomena of surface superconductivity can be encountered as first discussed by Saint-James and de Gennes (1963). The upper critical field associated with this surface superconductivity, H_{c3}, is related to H_{c2} as

$$H_{c3} = 1.7\,H_{c2} \tag{6-72}$$

This effect can qualitatively alter the angular dependence of H_{c2} that is predicted by Eq. (6-70).

6.4.1.2 The Thin Superconducting Slab: Tinkham's Formula

A superconductor/insulator superlattice, in the limit where the (Josephson) coupling between superconducting layers is negligible, may be regarded as an isolated superconducting slab which we will take to have thickness d. As d is reduced, the effects of boundary scattering and reduced dimensionality usually cause T_c to fall (and ultimately change the character of the transition); however we will here regard T_c and d as independent parameters. If we consider the case of a parallel magnetic field, in the gauge $\boldsymbol{A} = -H\,z\,\boldsymbol{y}$, and seek a solution of Eq. (6-66) of the form $\psi = u(z)\,e^{i k_x x + i k_y y}$, we obtain the differential equation

$$-\frac{\hbar^2}{4m}u'' + m\,\omega_c^2\,(z - z_0)^2\,u + \left(a + \frac{\hbar^2 k_x^2}{4m}\right)u = 0 \tag{6-73}$$

where $\omega_c = \dfrac{e\,H}{m}$ and $z_0 \equiv \dfrac{\hbar\,k_y}{2\,e\,H}$.

We adopt the boundary condition

$$\left(\boldsymbol{\nabla}\Psi - \frac{2\,i\,e}{\hbar}\boldsymbol{A}\,\psi\right)\cdot\hat{\boldsymbol{n}} = 0 \tag{6-74}$$

where $\hat{\boldsymbol{n}}$ is a unit vector normal to the surface. This is not a general boundary condition; it turns out to be correct for a smooth, non magnetic superconductor/insulator or superconductor-vacuum interface. For $d \ll \xi$ we may regard the second term in Eq. (6-73) as a perturbation. The zeroth order solution satisfying the boundary condition is then $u = \text{const}$. With this wavefunctions and our perturbation we have

$$-a = \frac{m\,\omega_c^2\,d^2}{12} \tag{6-75}$$

where we have set $z_0 = 0$ (which minimizes a). This leads immediately to

$$H_{c2\parallel} = \frac{\sqrt{12}\,\phi_0}{2\pi\,\xi\,d} \tag{6-76}$$

The behavior of Eq. (6-76) is radically different from that of Eq. (6-71). The reduced dimensionality has the effect of replacing ξ_z

by $\frac{d}{\sqrt{12}}$. Since only one factor of ξ enters Eq. (6-76), the temperature dependence of $H_{c2\|}$ is altered from $H_{c2} \propto (T_c - T)$ to $H_{c2\|} \propto (T_c - T)^{1/2}$. We will refer to these two behaviors as three dimensional and two dimensional, respectively.

Let us now consider the case where the field is at an arbitrary angle with respect to the thin slab. We take the vector potential in the form

$$A = \hat{y}\,(x \cos\theta - z \sin\theta)\, H \tag{6-77}$$

Defining[2]

$$H_{c2\|} = \frac{\sqrt{12}\,\phi_0}{2\pi\,\xi\, d} \quad \text{and} \quad H_{c2\perp} = \frac{\phi_0}{2\pi\,\xi^2}$$

we obtain

$$1 = \frac{H_{c2}(\theta)}{H_{c2\perp}}\,|\cos\theta| + \frac{H_{c2}^2(\theta)}{H_{c2\|}^2}\,\sin^2\theta \tag{6-78}$$

note that the angular dependence of this expression, first suggested by Tinkham, is radically different from that of Eq. (6-70b); in particular it has a cusp for $\theta = \pi/2$.

6.4.1.3 Thin Slab with Ferromagnetic Boundaries

We now treat the problem where the boundary condition is altered such that $u(z)|_{z=\pm d/2} = 0$. In practice this condition is approximated by a superconducting film sandwiched between two films containing a large concentration of paramagnetic impurities (e.g., ferromagnetic films) where pair breaking drives the order parameter to zero within a small distance. [Weaker boundary conditions are possible as discussed by Tokuyasu et al. (1988); the problem has also been discussed by Millis et al. (1988).] For $H = 0$ ($\omega_c = 0$) the eigenfunction satisfying the boundary condition and the linearized GL equation is

$$\mu = \mu_0 \cos k t \tag{6-79}$$

with $k d = \pi$ for the ground state. This leads to a suppression of the transition temperature governed by

$$T_c = T_{c0} - \frac{\hbar^2 \pi^2}{4 m \alpha d^2} \tag{6-80}$$

here T_{c0} is the bulk transition temperature and T_c is the film transition temperature subject to the new boundary condition. Application of a parallel magnetic field induces an additional reduction of T_c which can be calculated from perturbation theory (Wong and Ketterson, 1986). The result is

$$H_{c2\|} = \frac{5.53\,\phi_0}{2\pi\,\xi\, d} \tag{6-81}$$

which is to be compared with Eq. (6-76). The angular dependence follows from inserting (6-81) into (6-78).

6.4.1.4 The Superconductor-Insulator Superlattice

The superconductor/insulator superlattice is an especially interesting system because of the Josephson effect. The simplest model is to assume infinitesimally thin superconducting layers which are coupled via *order parameter tunneling* (Josephson coupling) through insulating layers of thickness s. Following Lawrence and Doniach (1971), we can introduce the following modified GL equations.

$$-\frac{\hbar^2}{4m}\left(\nabla - \frac{2ie}{\hbar} A_\perp\right)^2 \psi_n - \\ -\frac{\hbar^2}{4m_z s^2}\left(\psi_{n+1}\, e^{-\frac{2ie}{\hbar}\bar{A}_z s} - 2\psi_n + \psi_{n-1} + \right. \\ \left. + e^{+\frac{2ie}{\hbar c}\bar{A}_z s}\right) + \\ + a\psi_n + b|\psi_n|^2\psi_n = 0 \tag{6-82}$$

[2] The coherence length involved here is actually the in-plane coherence length introduced in the previous section.

$$j_n = -\frac{i e \hbar}{2m}(\psi_n^* \nabla \psi_n - \psi_n \nabla \psi_n^*) - \\ -\frac{2e^2}{m} A_\perp |\psi_n|^2 \qquad (6\text{-}83)$$

and

$$j_{z\,n+1,n} = -\frac{i e \hbar}{2 m_z s}\left(\psi_{n+1}\, e^{-\frac{2ie}{\hbar}\bar{A}_z s}\, \psi_n^* - \\ - \psi_{n+1}^*\, e^{+\frac{2ie}{\hbar}\bar{A}_z s}\, \psi_n\right) \qquad (6\text{-}84)$$

here we have defined $\bar{A}_z = \frac{1}{s}\int_{ns}^{(n+1)s} A_z \,\mathrm{d}z$. Our order parameter $\psi_n(\boldsymbol{r})$ has a discrete dependence on the index n and a continuous dependence on $\boldsymbol{r} = x\hat{\boldsymbol{x}} + y\hat{\boldsymbol{y}}$; the total vector potential $\boldsymbol{A}_\perp + A_z \hat{\boldsymbol{z}}$ is, however, defined at all points. Equation (6-84) for the tunneling current flowing between planes n and $n+1$ is equivalent to the first Josephson's equation (written in a gauge invariant form).

Solutions of Eq. (6-82) were first examined in detail by Klemm et al. (1975).

For the case of the field perpendicular to the layers we again have a harmonic oscillator problem. For H parallel to the planes the solution is more complicated. Near the zero field transition temperature the upper critical field is given by $H_{c2\parallel} = \frac{\phi_0}{2\pi \xi \xi_z}$, where $\xi \equiv \left(\frac{\hbar^2}{4m|a|}\right)^{1/2}$ and $\xi_z \equiv \left(\frac{\hbar^2}{4m_z|a|}\right)^{1/2}$. For temperatures further from T_c, where the eigenvalue a and the associated parallel critical field are both larger, the order parameter becomes less concentrated at the potential minimum and our harmonic approximation breaks down. Performing a high field expansion of the resulting Mathieu function (Abramowitz and Stegun, 1970) yields

$$H_{c2\parallel} = \frac{\phi_0}{2\pi s^2 \left(1 - \frac{s^2}{2\xi_z^2}\right)^{1/2}} \left(\frac{m}{m_z}\right)^{1/2} \qquad (6\text{-}85)$$

Equation (6-85) is valid only for temperatures such that $\xi_z \cong \frac{s}{\sqrt{2}}$; however $H_{c2\parallel}$ is real only for $\xi_z > \frac{s}{\sqrt{2}}$. No (real) solution exists for $\xi_z < \frac{s}{\sqrt{2}}$ and hence the critical field becomes formally infinite (in this model at a temperature T^* such that $\xi_z(T^*) = \frac{s}{\sqrt{2}}$. In a real system this divergence would be removed by the effect of paramagnetic limiting (to be discussed later).

Deutcher and Entin-Wohlman (1978) generalized the above model to the case of a superlattice of thin slabs of thickness d separated by Josephson coupled insulating layers of thickness s.

We will only quote the results for the limiting behavior of this model. For $H \to 0\ (T \to T_c)$ we have

$$H_{c2\parallel} = \frac{\phi_0}{2\pi \xi \xi_z \left(\frac{\Lambda}{s}\right)} \qquad (6\text{-}86)$$

this we referred to as a three dimensional (3D) behavior with $H_{c2\parallel} \propto (T_c - T)$.

The high field behavior is governed by

$$1 - \frac{H^2}{H_\parallel^2} - 2\frac{\xi^2}{s^2}\frac{m}{m_z} = -\frac{\phi_0^2}{2\pi^2 H^2 s^4}\frac{\xi^2}{\Lambda^2}\left(\frac{m}{m_z}\right)^3 \qquad (6\text{-}87)$$

Here we see that as $H \to \infty$ we obtain $H_{c2\parallel} = H_\parallel$; i.e., we obtain the upper critical field of an isolated slab, where $H_{c2\parallel} \propto (T_c - T)^{1/2}$, which we referred to as two dimensional (2D) behavior. Thus as the temperature is lowered a 3D–2D *crossover* occurs. Note that the divergence encountered with the Lawrence-Doniach superlattice model involving infinitesimally thin superconducting layers is avoided in the Deutcher Entin-Wohlman model where $d \neq 0$.

6.4.1.5 The Superconductor-Normal Metal Superlattice

Although one can construct a Ginzburg-Landau theory for the transition temperature of metal/superconductor or superconductor/superconductor superlattices, the resulting reduction of T_c (caused by the component with a lower bulk T_c) can be large, which often places the system outside the range of validity of a quartic-order-parameter expansion. For this reason we usually use the microscopic theory to describe such systems so we will not list the GL expressions (for a discussion see Jin and Ketterson, 1986).

6.4.1.6 Wave Propagation in the Josephson Superlattice

Up to this point we have restricted the discussion to the static properties of superlattices. The dynamic (time dependent) properties are also of interest. We will limit our remarks to wave propagation in the Josephson coupled superconductor/insulator superlattice in the London limit, first discussed by Auvil and Ketterson (1987) and extended by Song et al. (1987); we will quote only the final results of the calculations, but with some physical arguments to clarify their meaning.

Wave propagation in an SIS sandwich was first discussed by Swihart (1961). The structure acts like a wave guide, but differs from the standard problem of propagation between parallel plates in that one must allow for the fact that the magnetic field penetrates a distance of order a London depth, λ_L [see Eq. (6-68)], into the superconductor. The electric field on the other hand is restricted to the insulating region, which we take to have thickness a (note $a \ll \lambda_L$). The velocity of electromagnetic waves propagating in the junction is given by

$$\bar{c} = \frac{1}{\left[\varepsilon \mu \left(1 + \frac{2\lambda_L}{a}\right)\right]^{1/2}} \tag{6-88}$$

where ε is the dielectric constant of the insulator and μ is the permeability (taken, for simplicity, to be the same in both media). If we take $\lambda_L \cong 50$ nm and $a \cong 1$ nm we see that the wave moves about ten times slower than in free space (assuming $\varepsilon_r = \mu_r = 1$). Qualitatively this behavior arises from the fact that the media is *softened* by the fact that the magnetic field of the wave can *escape* the narrow junction region, penetrating a distance λ_L on each side. If one repeats Swihart's calculation for an S/I superlattices, where the thickness of the superconducting layer is b, the resulting velocity (in the limit $b \ll \lambda_L$) is

$$\bar{c} = \frac{c}{\left[\varepsilon \mu \left(1 + \frac{b}{a}\right)\right]^{1/2}} \tag{6-89}$$

for a symmetric superlattice ($b = a$) the velocity is reduced by only a factor $\sqrt{2}$. The increased velocity of propagation implies a smaller impedance mismatch at the boundary between the superlattice (junction) and free space and hence a greater output of Josephson radiation; the presence of multijunctions, each radiating in phase, also contributes to the output.

The solution of London's equation in the superconductor and Maxwell's equation in the insulator, subject to the Maxwell boundary conditions at the interface and Josephson coupling between the superconductors, yields the following partial differential equation for the phase difference Φ

$$\left(\nabla_\perp^2 - \frac{1}{\bar{c}^2}\frac{\partial^2}{\partial t^2}\right)\Phi(\boldsymbol{r}, t) = \frac{1}{\delta_J^2}\sin[\Phi(\boldsymbol{r}, t)] \tag{6-90}$$

where $\boldsymbol{r}$ is the coordinate in the plane of the junction $\nabla_\perp^2$ is the 2D in-plane Laplacian, $\bar{c}$ was defined previously, and λ_J is the Josephson penetration depth given by

$$\lambda_J^{-2} = \frac{2\pi\mu_0 j_m}{\phi_0}(a + 2\lambda_L) \quad \text{(SIS sandwich)} \tag{6-91}$$

or

$$\lambda_J^{-2} = \frac{2\pi\mu_0 j_m}{\phi_0}(a + b) \quad \text{(SI superlattice, } b \ll \lambda_L\text{)} \tag{6-92}$$

with j_m the parameter of the Josephson equation ($j = j_m \sin\Phi$). Equation (6-90) with Eq. (6-91) was first derived by Kulik (1965); the static version was obtained earlier by Ferrell and Prange (1963). Equation (6-90) with Eq. (6-92) was obtained by Auvil and Ketterson (1987). Equation (6-90) is the Sine-Gordon equation which is completely integrable and possesses a host interesting soliton solutions, which we will not discuss here (see Lamb, 1980). The experimental study of the dynamic non-linear properties of Josephson superlattices is essentially virgin territory. The wave propagation and generation problem follows from solving Eq. (6-90) subject to Maxwell's equations (see Auvil and Ketterson, 1987); we note the following properties: for small amplitude phase variations, $\Delta\Phi$, Eq. (6-90) reduces to the Klein-Gordon form

$$\left(\nabla_\perp^2 - \frac{1}{\bar{c}^2}\frac{\partial^2}{\partial t^2}\right)\Delta\Phi = \frac{1}{\lambda_J^2}\Delta\Phi \tag{6-93}$$

which has a plane wave solution of the form

$$\Delta\Phi \sim e^{ikx - i\omega t} \tag{6-94}$$

where

$$\omega^2 = \omega_0^2 + \bar{c}^2 k^2 \tag{6-95}$$

Here $\omega_0 = \bar{c}/\lambda_J$ and is called the Josephson plasma frequency; it is a cut-off frequency below which waves will not propagate in the superlattice (or junction). The static version of Eq. (6-90) is

$$\nabla_\perp^2 \Phi(\boldsymbol{r}) = \frac{1}{\lambda_J^2}\sin\Phi(r) \tag{6-96}$$

In the limit where the phase shift across the junction is very small we may approximate $\sin\Phi \cong \Phi$ and the solutions take the form (for $H_{\parallel}\, y$)

$$\Phi(x) = \Phi(0)\, e^{-x/\lambda_J} \tag{6-97}$$

From Maxwell's equation, $\dfrac{\partial H_y}{\partial x} = j_z$, and Josephson's equation, $J = j_m \sin\Phi$, we see that the magnetic field also has the form $H_y(x) \sim e^{-x/\lambda_J}$; i.e. the field only penetrates into the junction (from an edge) a distance of order λ_J (and hence the name Josephson penetration depth). The superlattice (or junction) is then in a Meissner state. The length λ_J is generally much larger than λ_L. In the opposite limit where $\sin\Phi$ passes through many cycles in traversing the junction, we may write $\Phi \cong \dfrac{2\pi y}{l}$ where $l = \phi_0/(a+b)\,\mu_0 H$ when $b \gg \lambda_L$ (in the two different limits a flux quantum occupies an area $[l(a+b)$ or $l(a+2\delta)]$. In this limit the superlattice is in an intermediate state with an array of equally spaced flux lines centered on the insulating layers. The transition between these two regimes corresponds to H_{c1} for the superlattice.

The case of a four sublayer SI superlattice, where successive insulating layers have thickness a_1 and a_2 while successive conducting layers have thicknesses b_1 and b_2, was treated by Song et al. (1987).

6.4.2 Microscopic Theory

We will limit our discussion of the microscopic theory of the superconducting properties of artificial superlattices to the zero field transition temperature and the temperature dependence of the upper critical field for dirty superconductors. Provided the transition is second order, the transition temperature may be determined from the linearized self-consistency condition of Gorkov. This condition may be derived (without using Green's function techniques) from Bogoliubov's self-consistent potential theory (1959). The procedures involved have been reviewed, in detail, by Jin and Ketterson (1986) which we henceforth denote as JK; here we will only summarize the major components of the theory. The approach used is that of de Gennes (1964a, 1964b, and 1966), and Takahashi and Tachiki (1986a and 1986b). The problems associated with extending the theory to the clean limit have been discussed by Silvert (1975). Extension of the theory deep into the superconducting phase would require the application of various extensions of de Gennes' quasi-classical approach; the most powerful method is due to Eilenberger (1968) which has recently been reviewed by Serene and Rainer (1983) and by Alexander et al. (1985). Other theoretical approaches have been used by Biagi et al. (1985, 1986) and Menon and Arnold (1985).

$$g_\xi(\boldsymbol{r}, \boldsymbol{r}', \hbar\Omega) \equiv \frac{\sum\limits_{m,n} \langle n|\delta(\boldsymbol{R}-\boldsymbol{r})\mathscr{C}|m\rangle \langle m|\mathscr{C}^\dagger \delta(\boldsymbol{R}-\boldsymbol{r}')|n\rangle\, \delta(\xi-\xi_n)\,\delta(\xi_m-\xi_n-\hbar\Omega)}{\sum\limits_n \delta(\xi-\xi_n)} \tag{6-101}$$

6.4.2.1 The Linearized Self-Consistency Condition

Our treatment will begin by recalling the linear integral equation which determines the superconducting transition temperature, T_c, and the position dependence of the gap function, $\Delta(\boldsymbol{r})$, at the transition temperature; following de Gennes (1966) [rederived as Eqs. (A 35) and (A 39) of JK]

$$\Delta(\boldsymbol{r}) = \int K(\boldsymbol{r}, \boldsymbol{r}')\,\Delta(\boldsymbol{r}')\,\mathrm{d}^3 r' \tag{6-98}$$

where

$$K(\boldsymbol{r}, \boldsymbol{r}') = V(\boldsymbol{r})\, k_B T \sum_\omega \sum_{m,n} \frac{\phi_m^*(\boldsymbol{r}')\,\phi_n^*(\boldsymbol{r}')\,\phi_m(\boldsymbol{r})\,\phi_n(\boldsymbol{r})}{(\xi_m + \mathrm{i}\hbar\omega)(\xi_n - \mathrm{i}\hbar\omega)} \tag{6-99}$$

Here $\phi_n(\boldsymbol{r})$ and ξ_n are the one particle self-consistent eigenfunctions and energies of the normal state (in the absence of the pairing potential), $V(\boldsymbol{r})$ is the (generally position dependent) pairing potential and ω are the *Matsubara* frequencies, $\hbar\omega_\nu = 2\pi k_B T(\nu + \frac{1}{2})$, where ν is the set of all positive and negative integers[3].

The transition temperature is that temperature for which Eq. (6-98) first has a non-trivial solution [i.e., $\Delta(\boldsymbol{r}) \neq 0$] and $\Delta(\boldsymbol{r})$ is the position dependent gap function [since Eq. (6-98) is a linear equation, $\Delta(\boldsymbol{r})$ is determined only within a multiplicative factor]. Through some additional manipulations the kernel $K(\boldsymbol{r}, \boldsymbol{r}')$ may be rewritten in the form

$$K(\boldsymbol{r}, \boldsymbol{r}') = L^3 k_B T \sum_\omega \int \frac{N(\xi)\, V\, \mathrm{d}\xi\, \mathrm{d}\xi'}{(\xi - \mathrm{i}\hbar\omega)(\xi' + \mathrm{i}\hbar\omega)} \cdot g_\xi(\boldsymbol{r}, \boldsymbol{r}', \hbar\Omega) \tag{6-100}$$

where

with

$$\hbar\Omega \equiv \xi' - \xi$$

$\mathscr{C}$ is the charge conjugation operator and $\boldsymbol{R}$ is the position operator. We will assume

[3] For those familiar with Gorkov's (Abrikosov et al., 1963) theory of superconductivity, K may be written as $K(\boldsymbol{r}, \boldsymbol{r}') = V k_B T \sum_\omega \mathscr{G}_\omega(\boldsymbol{r}, \boldsymbol{r}')\,\mathscr{G}_{-\omega}(\boldsymbol{r}, \boldsymbol{r}')$ where $\mathscr{G}_\omega$ is the one particle Green's function.

that for the range of energies relevant to the pairing interaction that the function g_ξ is independent of ξ. It is instructive to examine the Fourier transform of Eq. (6-101)

$$g(\boldsymbol{r},\boldsymbol{r}',t) = \int e^{-i\Omega t}\, g(\boldsymbol{r},\boldsymbol{r}',\Omega)\, d\Omega \qquad (6\text{-}102)$$

We can obtain an alternate form of $K(\boldsymbol{r},\boldsymbol{r}')$ by inserting Eq. (6-102) into Eq. (6-100) and carrying out the two integrations over $d\xi$ and $d\xi'$

$$K(\boldsymbol{r},\boldsymbol{r}') = \frac{2\pi}{\hbar} L^3 k_B T N(0) V \cdot \cdot \sum_\omega \int_0^\infty dt\, e^{-2|\omega| t}\, g(\boldsymbol{r},\boldsymbol{r}',t) \qquad (6\text{-}103)$$

with further manipulations we may rewrite $g(\boldsymbol{r},\boldsymbol{r}',t)$ as

$$g(\boldsymbol{r},\boldsymbol{r}',t) = \frac{\sum_n \langle n|\delta[\boldsymbol{R}(t)-\boldsymbol{r}]\mathscr{C}(t)\mathscr{C}^\dagger(0)\delta[\boldsymbol{R}(0)-\boldsymbol{r}']|n\rangle\, \delta(\xi-\xi_n)}{\sum_n \delta(\xi-\xi_n)} \qquad (6\text{-}104)$$

The time dependence of $\mathscr{C}$ is given by the usual Heisenberg equation of motion

$$\frac{d\mathscr{C}}{dt} = \frac{i}{\hbar}[\mathscr{H}',\mathscr{C}] = = -\frac{e}{m}[\nabla\cdot\boldsymbol{A} + \boldsymbol{A}\cdot\nabla]\mathscr{C} \cong \cong \frac{-2e}{\hbar}\boldsymbol{A}\cdot\boldsymbol{V} \qquad (6\text{-}105)$$

where $\mathscr{H}'$ is the perturbing Hamiltonian resulting from a vector potential $\boldsymbol{A}(\boldsymbol{r})$ and $\boldsymbol{V}$ is the velocity operator. Semiclassically $\frac{2e}{\hbar}\boldsymbol{A}\cdot\boldsymbol{V}$ is the rate of change of phase of the Cooper pair; i.e., $\frac{d\mathscr{C}}{dt} = -\frac{d\phi}{dt}\mathscr{C}$ or $\mathscr{C}(t) = e^{-i\phi(t)}\mathscr{C}(0)$.

When $A=0$, $\mathscr{C}(t)$ is time independent, $\mathscr{C}\mathscr{C}^\dagger = 1$, and $g(\boldsymbol{r},\boldsymbol{r}',t)$ has a simple interpretation as a *correlation function:* it measures the amplitude that a particle of energy ξ located at $\boldsymbol{r}'$ at $t=0$ will be at the point $\boldsymbol{r}$ at time t. In calculating this amplitude we sum over all states n with energy ξ; the denominator in Eq. (6-104) normalizes the amplitude by the total density of such states.

In a homogeneous system, Eq. (6-98) may be written

$$\Delta(\boldsymbol{r}) = \int K(\boldsymbol{r}-\boldsymbol{r}')\,\Delta(\boldsymbol{r}')\, d^3r' \qquad (6\text{-}106)$$

A perfect crystal is invariant only under discrete translations and an alloy has no microscopic translational symmetry at all; i.e., neither is homogeneous. However for most situations the distance scales of interest in superconductivity are of the order of a coherence length or larger, i.e., much larger than an interparticle separation. We will assume that we may take an *average* of Eq. (6-106) in the form

$$\overline{\Delta(\boldsymbol{r})} = \int \overline{K(\boldsymbol{r},\boldsymbol{r}')}\;\overline{\Delta(\boldsymbol{r}')}\, d^3r' \qquad (6\text{-}107)$$

and that $\overline{K(\boldsymbol{r},\boldsymbol{r}')} \to K(\boldsymbol{r}-\boldsymbol{r}')$ and $\overline{\Delta(\boldsymbol{r})} \to \Delta(\boldsymbol{r})$ to yield Eq. (6-106); the details of this averaging procedure are complex, and we refer the reader to Abrikosov et al. (1963) for details. (When dealing with superlattices we will still have to be concerned about macroscopic inhomogeneities resulting from layering.) The Fourier transform of Eq. (6-107) is given by

$$\Delta(\boldsymbol{q}) = K(\boldsymbol{q})\,\Delta(\boldsymbol{q}) \qquad (6\text{-}108)$$

which has the solutions $K(\boldsymbol{q})=1$ or $\Delta(\boldsymbol{q})=0$. A uniform system (not a superlattice) corresponds to $\boldsymbol{q}=0$. The temperature for which $K(\boldsymbol{q})=1$ corresponds to the transition temperature. If boundary conditions require $\boldsymbol{q}$ to have some non-zero value then the condition $K(\boldsymbol{q}\neq 0)=1$ will occur at a suppressed transition temperature $T_c(\boldsymbol{q}) < T_c(\boldsymbol{q}=0)$.

6.4.2.2 The Self-Consistency Condition in the Diffusion Dominated Limit

Our discussion will be restricted to so-called dirty superconductors where the mean free path is less than a coherence length; i.e., $l < \xi$. At the same time, if we are to avoid localization–interaction effects we must require $k_F l \gg 1$ (where k_F is the Fermi wavevector). For times short compared with an inelastic relaxation time but long compared to an elastic scattering time the electron moves diffusively but at constant energy; by Fourier transformation the diffusion equation which then governs $g(\boldsymbol{r}, t)$ it is easy to show that

$$g(\boldsymbol{q}, t) = \mathrm{e}^{-D q^2 |t|} \tag{6-109}$$

where D is the electron diffusion coefficient $D = \frac{1}{3} V_F^2 \tau$ with τ being the (elastic) electron relaxation time. $K(\boldsymbol{q})$ is then given by

$$K(\boldsymbol{q}) = \frac{2\pi}{\hbar} k_B T N(0) V \sum_{\omega} \frac{1}{2|\omega| + D q^2} \tag{6-110}$$

As it stands the sum in Eq. (6-110) is divergent; this behavior arises because we have neglected the cut-off in the BCS interaction, V. We may avoid this divergence using a trick; we rewrite $K(\boldsymbol{q})$ in the form

$$K(\boldsymbol{q}) = K(\mathbf{0}) + [K(\boldsymbol{q}) - K(\mathbf{0})] \tag{6-111}$$

Note

$$K(\mathbf{0}) = \int \frac{\mathrm{d}\xi\, N(\xi)\, V(\xi)}{2\xi} \tanh \frac{\beta \xi}{2} \tag{6-112}$$

The right side is the usual integral occurring in the BCS theory of the transition temperature and is given by

$$K(\mathbf{0}) = N(0) V \ln \left(\frac{1.14 \hbar \omega_D}{k_B T} \right) \tag{6-113}$$

where ω_D is the Debye cut-off energy. The second term can be written as

$N(0) V \chi(\xi^2 q^2)$ where

$$\xi^2 \equiv \frac{\hbar D}{2\pi k_B T} = \frac{\hbar V_F l}{6\pi k_B T} \quad \text{and}$$

$$\chi(z) \equiv \psi\left(\frac{z}{2} + \frac{1}{2}\right) - \psi\left(\frac{1}{2}\right) \tag{6-114}$$

$\psi(z)$ is the digamma function (Abramowitz and Stegun, 1970, p. 259). Combining the above we have

$$K(\boldsymbol{q}) = N(0) V \left[\ln \left(\frac{1.14\, \theta_D}{T} \right) - \chi(\xi^2 q^2) \right] \tag{6-115}$$

where $\theta_D \equiv \dfrac{\hbar \omega_D}{k_B}$.

6.4.2.3 The Diffusion Dominated Limit at Finite Magnetic Field

As mentioned earlier $g(\boldsymbol{r}, t)$ satisfies the diffusion equation which we write as

$$\left(D \nabla^2 - \frac{\partial}{\partial |t|} \right) g(\boldsymbol{r}, \boldsymbol{r}', t) = 0 \tag{6-116}$$

where the Laplacian operator may be taken with respect either $\boldsymbol{r}$ or $\boldsymbol{r}'$; in addition g must satisfy any boundary conditions at the sample surface and the initial condition $\lim_{t \to 0} g(\boldsymbol{r}, \boldsymbol{r}', t) = \delta(\boldsymbol{r} - \boldsymbol{r}')$. To satisfy gauge invariance in the presence of a vector potential we must write $\boldsymbol{\nabla} \to \boldsymbol{\nabla} - \frac{2ie}{\hbar} \boldsymbol{A}(\boldsymbol{r})$. This initially surprising procedure [given that Eq. (6-146) is the diffusion equation and not the Schroedinger equation] can be justified microscopically (see de Gennes, 1965 or JK). Thus, in the presence of a magnetic field, Eq. (6-116) becomes [4]

$$\left\{ D \left[\boldsymbol{\nabla}' - \frac{2ie}{\hbar} \boldsymbol{A}(\boldsymbol{r}') \right]^2 - \frac{\partial}{\partial |t|} \right\} g(\boldsymbol{r}, \boldsymbol{r}', t) = 0 \tag{6-117}$$

[4] When $\boldsymbol{A}(\boldsymbol{r}) \neq 0$, g may be a complex function and, as such, does not have the simple interpretation as a correlation function, which must be positive definite. It may still be thought of as an *information function* with its complex nature arising from the fact that the superconducting order parameter has both an amplitude and phase in the GL sense.

when operating on $\boldsymbol{r}$ we would use the complex conjugate of the operator in brackets in Eq. (6-117).

The solution of the integral Eq. (6-100) for Δ is most easily accomplished by finding the normalized eigenfunctions, $g_n(\boldsymbol{r})$, and eigenvalues, Ω_n, of the operator $D\left[\boldsymbol{\nabla}-\frac{2ie}{\hbar}\boldsymbol{A}(\boldsymbol{r})\right]^2$ subject to the prescribed boundary conditions; i.e.,

$$D\left[\boldsymbol{\nabla}-\frac{2ie}{\hbar}\boldsymbol{A}(\boldsymbol{r})\right]^2 g_n(\boldsymbol{r})=-\Omega_n g_n(\boldsymbol{r}) \quad (6\text{-}118)$$

We observe that the function

$$g(\boldsymbol{r},\boldsymbol{r}',t)=L^{-3}\sum_n g_n^*(\boldsymbol{r}')\,g_n(\boldsymbol{r})\,\mathrm{e}^{-\Omega_n|t|} \quad (6\text{-}119)$$

satisfies the diffusion equation (with respect to $\boldsymbol{r}$ or $\boldsymbol{r}'$) and, from the completeness relation, reduces to $\delta(\boldsymbol{r}-\boldsymbol{r}')$ at $t=0$. Fourier transforming with respect to the time variable yields

$$g(\boldsymbol{r},\boldsymbol{r}',\Omega)=\frac{1}{2\pi L^3}\sum_n g_n^*(\boldsymbol{r}')\,g_n(\boldsymbol{r})\,\frac{2\Omega}{\Omega^2+\Omega_n^2} \quad (6\text{-}120)$$

Inserting Eq. (6-120) into Eq. (6-118) gives

$$\Delta(\boldsymbol{r})=\frac{2\pi}{\hbar}k_{\mathrm{B}}T\,N(0)\,V\cdot\sum_\omega\sum_n\frac{g_n(\boldsymbol{r})}{2|\omega|+\Omega_n}\int \mathrm{d}^3r'\,g_n^*(\boldsymbol{r}')\,\Delta(\boldsymbol{r}') \quad (6\text{-}121)$$

From Eq. (6-121) we see immediately that the eigenfunctions of the integral Eq. (6-98) are the g_n's themselves; since the eigenvalue of Eq. (6-98) is unity we must have

$$1=\frac{2\pi}{\hbar}k_{\mathrm{B}}T\,N(0)\,V\sum_\omega\frac{1}{2|\omega|+\Omega_n} \quad (6\text{-}122)$$

Note the smallest eigenvalue, Ω_0, yields the highest transition temperature. From the discussion surrounding Eq. (6-115), we have immediately

$$\ln\frac{T(0)}{T(H)}=\chi\left(\frac{\hbar\,\Omega_0}{2\pi\,k_{\mathrm{B}}T}\right) \quad (6\text{-}123)$$

where $T(H)$ is the field dependent transition temperature. For a bulk superconductor Eq. (6-118) is equivalent to the *linearized Ginzburg-Landau equation* provided we identify D with $\hbar/4m$. The lowest eigenvalue is

$$\Omega_0=\frac{1}{2}\frac{eH}{m}=\frac{2eDH}{\hbar}$$

or

$$\ln\frac{T(0)}{T(H)}=\chi\left(\frac{eDH}{\pi k_{\mathrm{B}}T}\right)=\chi\left(\frac{\hbar DH}{\phi_0 k_{\mathrm{B}}T}\right) \quad (6\text{-}124)$$

which is the well known Maki-de Gennes-Werthamer result for the upper critical field of a bulk, type II superconductor.

6.4.2.4 Spin-Dependent Potentials

A more general form of the kernel is required if we are to include the effects of spin paramagnetism, spin-orbit coupling, and magnetic impurities (the first and last of these act to depair condensate electrons and are referred to as *pair breaking* effects). The generalized kernel is given by[5]

$$K(\boldsymbol{r},\boldsymbol{r}')=\frac{1}{2}V k_{\mathrm{B}}T\sum_{\alpha\beta\gamma\delta}\sum_{M,N}\sum_\omega\frac{\phi_{N\alpha}^*(\boldsymbol{r})\,\mathscr{K}_{\alpha\beta}^\dagger\,\phi_{M\beta}(\boldsymbol{r})\,\phi_{M\gamma}^*(\boldsymbol{r}')\,\mathscr{K}_{\gamma\delta}\,\phi_{N\delta}(\boldsymbol{r}')}{(\xi_M+\mathrm{i}\hbar\omega)(\xi_N-\mathrm{i}\hbar\omega)} \quad (6\text{-}125)$$

[5] One must be careful to distinguish spin coordinates (or components) of a wavefunction and spin quantum numbers. In particular we write $u_N=\begin{pmatrix}u_{N1}\\u_{N2}\end{pmatrix}$. The upper and lower components correspond to $\alpha=1$, 2 (or $\uparrow$, $\downarrow$). With $N=(n,\nu)$, there will be two two-component wavefunctions, corresponding to $\nu=1$, 2, for each value of n. In what follows we will use the letters ν, μ for spin quantum numbers and α, β, γ, δ for spin wavefunction components.

The substantial modification is that an operator $\mathscr{K}$ (the Wigner time reversal operator) replaces $\mathscr{C}$;

$$\mathscr{K} \equiv \mathrm{i}\,\sigma_2\,\mathscr{C} \tag{6-126}$$

where σ_2 is the Pauli matrix. The $\phi_{N\alpha}$ are the eigenfunctions of the normal state mean field Hamiltonian. We introduce a correlation function analogous to Eq. (6-101)

$$g_\xi^{(\nu)}(\boldsymbol{r},\boldsymbol{r}',t) = -\frac{1}{2}\sum_n \delta(\xi-\xi_{n,\nu})\,\frac{\langle n,\nu\,|\,\delta[\boldsymbol{r}-\boldsymbol{R}(t)]\,\mathscr{K}^\dagger(t)\,\mathscr{K}(0)\,\delta[\boldsymbol{r}-\boldsymbol{R}(0)]\,|\,n,\nu\rangle}{\sum_n \delta(\xi-\xi_{n,\nu})} \tag{6-127}$$

where we will again suppress the ξ subscript on g. In terms of this correlation function the Kernel entering the gap equation is

$$K(\boldsymbol{r},\boldsymbol{r}') = L^3\,k_\mathrm{B}\,T\sum_\omega\sum_\nu\int\frac{N\,V\,\mathrm{d}\xi\,\mathrm{d}\xi'}{(\xi-\mathrm{i}\,\hbar\,\omega)\,(\xi'+\mathrm{i}\,\hbar\,\omega)}\,g^{(\nu)}[\boldsymbol{r},\boldsymbol{r}',(\xi'-\xi)/\hbar] \tag{6-128}$$

or, in a form analogous to Eq. (6-103).

$$K(\boldsymbol{r},\boldsymbol{r}') = \frac{2\pi}{\hbar}\,L^3\,k_\mathrm{B}\,T\,N\,V\sum_\omega\sum_\nu\int_0^\infty \mathrm{d}t\,\mathrm{e}^{-2|\omega|t}\,g^{(\nu)}(\boldsymbol{r},\boldsymbol{r}',t) \tag{6-129}$$

We will treat the effect of various pair breaking terms entering the mean potential by considering their effect on the time dependent operator $\mathscr{K}(t)$ which is defined as

$$\mathscr{K}(t) = \mathrm{e}^{\mathrm{i}\mathscr{H}'t/\hbar}\,\mathscr{K}(0)\,\mathrm{e}^{-\mathrm{i}\mathscr{H}'t/\hbar} \tag{6-130}$$

Formally, the spin-orbit term in the Hamiltonian commutes with $\mathscr{K}$. It nonetheless has an indirect effect which we will neglect for the moment. The time derivative of $\mathscr{K}(t)$ follows from the Heisenberg equation of motion and is given by

$$\dot{\mathscr{K}}(t) = \left[-\mathrm{i}\frac{\mathrm{d}\phi}{\mathrm{d}t} + \frac{2\mathrm{i}}{\hbar}\,I\,\sigma_z - \frac{1}{\tau_\mathrm{s}}\right]\cdot\mathscr{K}(t) \tag{6-131}$$

The first term, arising from the vector potential, was obtained earlier following Eq. (6-105) and the second term, which accounts for spin paramagnetism follows directly on evaluating the commutator of $\mathscr{K}$ with the operator $I\,\sigma$; here $I \equiv \mu_\mathrm{B}\,H\,\chi/\chi^{(\mathrm{P})}$ where χ is the true susceptibility and $\chi^{(\mathrm{P})}$ is the band susceptibility (i.e., $\chi/\chi^{(\mathrm{P})}$ is the exchange enhancement factor associated with the spin paramagnetism). The last term accounts for uncorrelated (or ergotic) spin-flip scattering caused by magnetic impurities and has been incorporated in a phenomenological manner; we refer the reader to the original work of Abrikosov et al. (1963) for a formal justification. The term in τ_s leads to an exponential decay of $\mathscr{K}(t)$, a behavior which is referred to as ergodic. The term in I on the other hand leads to a periodic behavior of $\mathscr{K}(t)$; i.e., $\mathscr{K}(t\to\infty) \neq 0$ and this behavior is termed nonergodic.

We have already discussed the effect of the vector potential on the upper critical magnetic field, leading to Eq. (6-124). We now discuss, separately, the effect of electron paramagnetism and impurity spins. Solving Eq. (6-131) for $\mathscr{K}(t)$ the quantity $\mathscr{K}^\dagger(t)\,\mathscr{K}(0)$ entering Eq. (6-127) is given by the unitary operator

$$\mathscr{K}^\dagger(t)\,\mathscr{K}(0) = \begin{pmatrix}\mathrm{e}^{-\mathrm{i}\,\Omega_\mathrm{L}t} & 0\\ 0 & \mathrm{e}^{\mathrm{i}\,\Omega_\mathrm{L}t}\end{pmatrix} \tag{6-132}$$

where Ω_L is the spin-Larmour frequency, $2I/\hbar$ [the $\mathscr{K}(t)$ operator remains antiunitary for all times, as it should].

We thus have the following implicit equation for the transition temperature

$$1 = \frac{2\pi}{\hbar} N V k_B T \sum_{\omega} \int_0^{\infty} dt\, e^{-2|\omega| t} \cdot \frac{1}{2}[e^{-i\Omega_L t} + e^{+i\Omega_L t}]$$

or

$$1 = \frac{\pi}{\hbar} N(0) V k_B T \cdot \sum_{\omega} \left[\frac{1}{2|\omega| + i\Omega_L} + \frac{1}{2|\omega| - i\Omega_L}\right] \tag{6-133}$$

Using Eq. (6-114) we have

$$1 = N(0) V \left\{\ln \frac{1 \cdot 14 \hbar \omega_D}{k_B T} - \frac{1}{2}\left[\chi\left(\frac{i\hbar\Omega_L}{2\pi k_B T}\right) + \chi\left(\frac{-i\hbar\Omega_L}{2\pi k_B T}\right)\right]\right\} \tag{6-134}$$

or, using the property $\operatorname{Im}\psi(\frac{1}{2}+iy) = \frac{\pi}{2}\tanh(\pi y)$, we have

$$\ln\left(\frac{T_c}{T}\right) = \operatorname{Re}\chi\left(\frac{i\hbar\Omega_L}{2\pi k_B T}\right) \tag{6-135}$$

The reduction of the transition temperature due to the effect of Pauli paramagnetism was first examined by Chandrasekav (1962) and Clogston (1962).

Next we consider the isolated effect of dilute magnetic impurities [6]. The time dependence of $\mathscr{K}(t)$ arising from the last term on the right of Eq. (6-131) is

$$\mathscr{K}(t) = e^{-t/\tau_s} \mathscr{K}(0) \tag{6-136}$$

or $\mathscr{K}^\dagger(t)\mathscr{K}(0) = e^{-t/\tau_s}$. Following arguments similar to those leading to Eq. (6-134) yields

$$1 = \frac{2\pi}{\hbar} N V k_B T \sum_{\omega} \frac{1}{2|\omega| + \frac{1}{\tau_s}} \tag{6-137}$$

or

$$\ln\left(\frac{T_c}{T}\right) = \chi\left(\frac{\hbar}{2\pi k_B T \tau_s}\right) \tag{6-138}$$

We now return to the question of spin-orbit coupling [7]. Our approach here will be phenomenological. By introducing an ad-hoc spin-orbit relaxation term in Eq. (6-131), we can reproduce the Fulde-Maki result (1966). The term we add must have the property that, in the absence of a time reversal breaking perturbation, it results in no evolution of the time reversal operator, $\mathscr{K}(t)$, from its form at $t=0$. A relaxation contribution proportional to the sum of $\mathscr{K}$ and its transpose $\tilde{\mathscr{K}}$ satisfies this requirement. In what immediately follows we limit ourselves to only the Larmor and spin-orbit terms and Eq. (6-131) becomes

$$\dot{\mathscr{K}} = i\Omega_L \sigma_z \cdot \mathscr{K}(t) - \frac{1}{\tau'}(\mathscr{K} + \tilde{\mathscr{K}}) \tag{6-139}$$

where τ' is a phenomenological spin-orbit scattering time. We must solve Eq. (6-177) subject to the initial condition $\mathscr{K}(0) = \begin{pmatrix} 0 & 1 \\ -1 & 0 \end{pmatrix} \mathscr{C}$. The calculations are straight forward, though tedious, and we only give the final result for $\operatorname{Tr} \overset{+}{\mathscr{K}}(t)\mathscr{K}(0)$ which determines the transition temperature through Eq. (6-129)

[6] As impurities become more concentrated they begin to process due to their mutual interaction. Alternatively, in the presence of an external magnetic field, the classical spin vectors undergo Larmour precession. Such dynamic effects greatly complicate the theory.

[7] The effect of spin orbit coupling was introduced by Ferrell (1959) to account for the results of Knight shift experiments and was further developed by Anderson (1958) and Abrikosov et al. (1963). Its effect on reducing the pair breaking arising from Pauli paramagnetism has been studied by Fulde and Maki (1966), Werthamer (1963) and Maki (1966) using the Green's function approach.

$$\mathrm{Tr}\,\overset{+}{\mathscr{K}}(t)\,\mathscr{K}(0) = = \left(1 + \frac{1}{(1-\Omega_L^2\tau'^2)^{1/2}}\right)\mathrm{e}^{-a_+ t} + + \left(1 - \frac{1}{(1-\Omega_L^2\tau'^2)^{1/2}}\right)\mathrm{e}^{-a_- t} \quad (6\text{-}140)$$

where

$$a_\pm = -\frac{1}{\tau'}[\pm(1-\Omega_L^2\tau'^2)^{1/2} - 1] \quad (6\text{-}141)$$

Carrying out the time integration and evaluating the ω sum in (6-129) as we have in previous calculations we obtain the Fulde-Maki (1966) expression for the transition temperature

$$\ln\left(\frac{T_c}{T}\right) = \frac{1}{2}\left[1 + \frac{1}{(1-\Omega_L^2\tau'^2)^{1/2}}\right]\chi\left(\frac{\hbar a_+}{\pi k_B T}\right) + \frac{1}{2}\left[1 - \frac{1}{(1-\Omega_L^2\tau'^2)^{1/2}}\right]\chi\left(\frac{\hbar a_-}{\pi k_B T}\right) \quad (6\text{-}142)$$

In order to make contact with the more microscopic approach we must define $\tau' = 3\tau_{so}$ where τ_{so} is the spin-orbit scattering time introduced by Abrikosov and Gorkov. Equation (6-142) reduces to Eq. (6-135) in the limit $\frac{1}{\tau'} \to 0$, as it must. In the limit $\tau' \to 0$ the transition temperature is not depressed.

We conclude by giving the results of including the effects of the vector potential. It turns out the behavior is still given by Eq. (6-142), however with

$$a_\pm = \frac{1}{\tau'}[1 \mp \sqrt{1-\Omega_L^2\tau'^2}] + \frac{\Omega_0}{2} \quad (6\text{-}143)$$

where Ω_0 is given by the equation preceding Eq. (6-124).

6.4.2.5 Macroscopically Inhomogeneous Superconductors

Up to this point the only macroscopic inhomogeneity that we have treated involved an external magnetic field described by a position dependent vector potential $\boldsymbol{A}(\boldsymbol{r})$. However one of the reasons for studying superconductivity in artificial metallic superlattices is to observe phenomena resulting from the inhomogeneities generated by layering constituents with different superconducting properties. To treat such situations we must reformulate the theory discussed in the previous three sections. We will limit our discussion to dirty superconductors only, since clean superconductors require a much more complete specification of the electronic structure than is generally available. We may capture the basic physics of layered dirty superconductors with three parameters for each constituent: the diffusion coefficient, D, the density of states, N and the BCS pairing interaction, V (one might also want to assign different Debye temperatures, θ_D).

We begin by rewriting Eq. (6-99) in the form

$$K(\boldsymbol{r},\boldsymbol{r}') = k_B T V(\boldsymbol{r}) \sum_\omega Q_\omega(\boldsymbol{r},\boldsymbol{r}') \quad (6\text{-}144)$$

where

$$Q_\omega(\boldsymbol{r},\boldsymbol{r}') \equiv \sum_{m,n} \frac{\phi_m^*(\boldsymbol{r}')\,\phi_n^*(\boldsymbol{r}')\,\phi_m(\boldsymbol{r})\,\phi_n(\boldsymbol{r})}{(\xi_m + \mathrm{i}\hbar\omega)(\xi_n - \mathrm{i}\hbar\omega)} \quad (6\text{-}145)$$

We will need the integral of $Q_\omega(\boldsymbol{r},\boldsymbol{r}')$ over either $\boldsymbol{r}$ or $\boldsymbol{r}'$, and we now limit the discussion to zero field ($H=0$) where we may choose the eigenfunctions $\phi_n(\boldsymbol{r})$ to be real and write them as $w_n(\boldsymbol{r})$. From the orthonormality of the w_n

$$\int Q_\omega(\boldsymbol{r},\boldsymbol{r}')\,\mathrm{d}^3r' = \sum_n \frac{[w_n(\boldsymbol{r})]^2}{\xi_n^2 + (\hbar\omega)^2} \quad (6\text{-}146)$$

We define a *local* density of states by

$$N(\xi,\boldsymbol{r}) = \sum_n [w_n(\boldsymbol{r})]^2\,\delta(\xi - \xi_n) \quad (6\text{-}147)$$

We then have, from the approximation $N(\xi,\boldsymbol{r}) \approx N(\boldsymbol{r})$, that

$$\int Q_\omega(\boldsymbol{r},\boldsymbol{r}')\,\mathrm{d}^3r' = = \int \frac{\mathrm{d}\xi}{\xi^2+(\hbar\omega)^2}\sum_n [w_n(\boldsymbol{r})]^2\,\delta(\xi-\xi_n) \approx N(\boldsymbol{r})\int \frac{\mathrm{d}\xi}{\xi^2+(\hbar\omega)^2} = \frac{\pi}{\hbar|\omega|}N(\boldsymbol{r}) \quad (6\text{-}148)$$

As discussed earlier, the behavior of a dirty (or clean) superconductor is governed by a correlation function. However the correlation function defined by Eq. (6-101) is normalized by the total density of states which, for a macroscopically inhomogeneous system (like an artificial superlattice), is not appropriate. We define an **unnormalized** correlation function by

$$\mathscr{g}_\xi(\boldsymbol{r},\boldsymbol{r}',\Omega) \equiv \sum_{n,m} w_n(\boldsymbol{r})\,w_m(\boldsymbol{r})\,w_n(\boldsymbol{r}')\,w_m(\boldsymbol{r}')\cdot \cdot\,\delta(\xi-\xi_n)\,\delta[(\xi_n-\xi_m)/\hbar+\Omega] \quad (6\text{-}149)$$

The subscript ξ signifies that $\mathscr{g}$ is a function of the energy. However, the range of energies relevant to superconductivity is restricted to the immediate vicinity of the Fermi energy and we will therefore assume $\mathscr{g}$ is independent of ξ and delete the ξ subscript in what follows. The Fourier transform of Eq. (6-149) can be written as

$$\mathscr{g}(\boldsymbol{r},\boldsymbol{r}',t) = \quad (6\text{-}150)$$
$$= \sum_n \langle n|\,\delta[\boldsymbol{R}(t)-\boldsymbol{r}]\,\delta[\boldsymbol{R}(0)-\boldsymbol{r}']\,|n\rangle\,\delta(\xi-\xi_n)$$

Using procedures similar to those discussed earlier we can obtain

$$Q_\omega(\boldsymbol{r},\boldsymbol{r}') = \frac{2\pi}{\hbar}\int_0^\infty \mathrm{d}t\;\mathrm{e}^{-2|\omega|t}\,\mathscr{g}(\boldsymbol{r},\boldsymbol{r}',t) \quad (6\text{-}151)$$

In the diffusion approximation the behavior of $\mathscr{g}$ is governed by [8]

$$\frac{\partial \mathscr{g}(\boldsymbol{r},\boldsymbol{r}',t)}{\partial t} - D\nabla^2\,\mathscr{g}(\boldsymbol{r},\boldsymbol{r}',t) = 0\,; \quad t>0 \quad (6\text{-}152)$$

[8] In an inhomogeneous anisotropic media we would replace $D\nabla^2$ by $\boldsymbol{\nabla}\cdot\mathbf{D}(\boldsymbol{r})\cdot\boldsymbol{\nabla}$.

The function $\mathscr{g}(\boldsymbol{r},\boldsymbol{r}',t)$ is singular at $t=0$. The form and strength of the singularity follow from taking the Fourier time transform of Eq. (6-192) setting $t=0$, and using the closure property

$$\lim_{\boldsymbol{r}\to\boldsymbol{r}'} \mathscr{g}(\boldsymbol{r},\boldsymbol{r}',t=0) = = \delta(\boldsymbol{r}-\boldsymbol{r}')\sum_n |w_n(\boldsymbol{r})|^2\,\delta(\xi-\xi_n) = = N(\boldsymbol{r})\,\delta(\boldsymbol{r}-\boldsymbol{r}') \quad (6\text{-}153)$$

From Eq. (6-192) and using Eqs. (6-151) and (6-153) we have

$$2|\omega|\,Q_\omega(\boldsymbol{r},\boldsymbol{r}') - D\nabla^2 Q_\omega(\boldsymbol{r},\boldsymbol{r}') = = \frac{2\pi}{\hbar}N(\boldsymbol{r})\,\delta(\boldsymbol{r}-\boldsymbol{r}') \quad (6\text{-}154)$$

Equation (6-154) is the basic equation describing the behavior of a macroscopically in-homogeneous superconductor.

Our present interest is in artificial superlattices which, in the dirty limit, will be regarded as homogeneous in the two directions which are perpendicular to the superlattice plane normals and we choose the associated coordinates as x and y; we will use z for the coordinate along the plane normal. We may formally perform an in-plane average over x and y by Fourier transforming the variables $x-x'$ and $y-y'$ into q_x and q_y and then setting q_x and $q_y=0$. This transformed version of Eq. (6-154) is

$$2|\omega|\,Q_\omega(z,z') - D\,\frac{\mathrm{d}^2}{\mathrm{d}z^2}\,Q_\omega(z,z') = = \frac{2\pi}{\hbar}N(z)\,\delta(z-z') \quad (6\text{-}155)$$

where $N(z) = \frac{1}{L^2}\int N(\boldsymbol{r})\,\mathrm{d}x\,\mathrm{d}y$ with L^2 the area of the interface. $Q_\omega(z,z')$ is a shorthand for $Q_\omega(z,z',q_x=0,q_y=0)$.

By definition, the function Q_ω satisfying Eq. (6-155) must be symmetric with respect to the variables z and z', and the differential operator may be applied to either variable.

To apply Eq. (6-155) to artificial superlattices we require boundary conditions for the interfaces separating the successive layers. We will adopt the de Gennes boundary conditions

$$\frac{\Delta_A}{N_A V_A} = \frac{\Delta_B}{N_B V_B} \tag{6-156}$$

and

$$\frac{D_A}{V_A}\frac{d\Delta_A}{dz} = \frac{D_B}{V_B}\frac{d\Delta_B}{dz} \tag{6-157}$$

where the subscripts "A" and "B" denote values in medium A and medium B, respectively; for a derivation we refer the reader to JK or the original work of de Gennes (1964a and 1964b).

6.4.2.6 The Zero Field Transition

In this section we use the differential equation for Q_ω, Eq. (6-155), and the de Gennes boundary conditions to solve the gap equation for the zero field transition temperature of an artificial metallic superlattice; we will also assume there is no spin flip scattering. Our approach is to use a variant of the eigenfunction technique discussed earlier. The superlattice is assumed to consist of alternate layers of two metals, A and B, which, respectively, are characterized by the diffusion constants and pairing potentials D_A, V_A and D_B, V_B. The particular approach used here is due to Takahashi and Tachiki (1986), which is the most general treatment to date. Earlier treatments, which were more restrictive, were reviewed by Ruggiero and Beasley (1985) and include work by Dabrosavljevic (1973), Dabrosavljevic and Kulik (1978), Ami and Maki (1975), Galaiko and Bezuglyi (1973), Nabutoskii and Shapiro (1981) and Xing and Gong (1981).

We seek a solution to Eq. (6-155) for $Q_\omega(z,z')$ in the *symmetrical* form

$$Q_\omega(z,z') = \tag{6-158}$$
$$= [N(z)\,N(z')]^{1/2} \sum_{n,m} a_{nm}\,\psi_n^*(z')\,\psi_m(z)$$

where the $\psi_n(z)$ are eigenfunctions of the equation

$$D(z) = \frac{d^2}{dz^2}\psi_n(z) + \Omega_n \psi_n(z) = 0 \tag{6-159}$$

here $D(z) = D_A$ or D_B and $N(z) = N_A$ or N_B in media A or B respectively, and the a_{nm} are expansion coefficients yet to be determined.

The $\psi_n(z)$ will be normalized according to the condition

$$\int_0^\Lambda \psi_n^*(z)\,\psi_m(z)\,dz = \delta_{nm} \tag{6-160}$$

where Λ is the repeat distance.

Note the ψ_n could be chosen to be real in which case Q_ω is completely symmetrical in z and z' (as assumed in the previous section). Our writing ψ_n^* in Eqs. (6-158) and (6-160) is in anticipation of the field dependent case where the Q_ω (and hence Δ) are no longer necessarily real. Inserting Eq. (6-158) into Eq. (6-155) we can show that

$$Q_\omega(z,z') = \frac{2\pi}{\hbar}[N(z)\,N(z')]^{1/2} \sum_n \frac{\psi_n^*(z')\,\psi_n(z)}{2|\omega| + \Omega_n} \tag{6-161}$$

Inserting Eq. (6-121) into the gap equation, we have

$$\Delta(z) = \frac{2\pi}{\hbar} k_B T V(z)\,[N(z)]^{1/2} \cdot \tag{6-162}$$
$$\cdot \sum_\omega \sum_n \int \frac{[N(z')]^{1/2}\,\psi_n^*(z')\,\psi_n(z)}{2|\omega| + \Omega_n}\,\Delta(z')\,dz'$$

We define the quantity $F(z) = \frac{\Delta(z)}{V(z)}$, which is referred to as the pair amplitude. We expand $F(z)$ in terms of the $\psi_n(z)$ according to

$$F(z) = [N(z)]^{1/2} \sum_{n'} c_{n'} \psi_{n'}(z) \tag{6-163}$$

Substituting Eq. (6-163) into Eq. (6-162) we can show that the eigenvalues are determined by

$$\left| \delta_{nm} - \frac{2\pi}{\hbar} k_B T \sum_{\omega} \frac{\langle m|NV|n\rangle}{2|\omega| + \Omega_m} \right| = 0 \tag{6-164a}$$

where we have defined the matrix element

$$\langle m|NV|n\rangle \equiv$$
$$\equiv \int \psi_n^*(z) N(z) V(z) \psi_n(z) \, dz \tag{6-164b}$$

Equation (6-164) is the formal solution to the problem we have posed and the highest temperature, T_c, for which it has a solution corresponds to the physical transition temperature. The uniform superconductor is a special case of Eq. (6-164) corresponding to $\langle m|NV|n\rangle = NV\delta_{mn}$; i.e., the determinant is diagonal and every eigenvalue Ω_n leads directly to a zero of the determinant. Since the smallest eigenvalue $\Omega_n = 0$ leads to the highest transition temperature, we have

$$1 = \frac{2\pi}{\hbar} k_B T N V \sum_{\omega} \frac{1}{2|\omega|} \tag{6-165}$$

which is equivalent to the BCS expression for the transition temperature [9].

The eigenfunctions of Eq. (6-159) are constructed by joining together the solutions in the two separate media, A and B, subject the boundary conditions, Eqs. (6-156) and (6-157). From Eq. (6-158) it is clear that $Q_\omega(z,z')$ will satisfy these boundary conditions if $\psi(z)/[N(z)]^{1/2}$ and $[N(z)]^{1/2} D(z) \frac{d\psi(z)}{dz}$ are both continuous at the interfaces [10].

We assume media A and B occupy $0<z<a$, and $b<z<0$, respectively, after which the system periodically reproduces itself; we seek solutions of the form $e^{\pm i k_A z}$ and $e^{\pm i k_B z}$ in these two regions respectively where $k_A = (\Omega/D_A)^{1/2}$ and $k_B = (\Omega/D_B)^{1/2}$. Applying the above boundary conditions and Bloch theorem yields the following equation:

$$\cos(k\Lambda) = \cos(k_A a)\cos(k_B b) -$$
$$- \gamma \sin(k_A a)\sin(k_B b) \tag{6-166}$$

where

$$\gamma \equiv \frac{D_A^2 N_A^2 k_A^2 + D_B^2 N_B^2 k_B^2}{2 D_A D_B N_A N_B k_A k_B} \tag{6-167}$$

This restricts $k = \dfrac{2\pi n}{\Lambda}$ and Eq. (6-166) for the eigenfrequencies becomes Ω

$$\cos(k_A a)\cos(k_B b) -$$
$$- \gamma \sin(k_A a)\sin(k_B b) = 1 \tag{6-168}$$

(where $k_A^2 = \Omega/D_A$ and $k_B^2 = \Omega/D_B$).

With some trigonometric manipulations Eq. (6-168) may be rewritten in the form

$$N_A D_A^{1/2} \tan\left(\frac{k_A a}{2}\right) =$$
$$= - N_B D_B^{1/2} \tan\left(\frac{k_B b}{2}\right) \tag{6-169}$$

It is worth noting that the Cooper-de Gennes limit (de Gennes, 1964) is obtained from our general formalism in the limit $\Lambda \to 0$. In this limit $\Omega_n \to \infty$ for $n>0$ (Ω_0 is always zero); all the terms in the determinant vanish except the 00 term. One can show that the matrix element

$$\langle 0|NV|0\rangle = \frac{N_A^2 V_A a + N_B^2 V_B b}{N_A a + N_B b} \equiv \varrho \tag{6-170}$$

[9] A superconductor which has an inhomogeneous diffusion constant, $D(z)$, but with a constant $N(z)\,V(z)$ also has an unaltered transition temperature (at zero field only).

[10] We may equivalently require the logarithmic derivative $N(z)\,D(z) \frac{d}{dz} \ln \psi(z)$ to be continuous.

which yields

$$k_B T_c = 1.14 \hbar \omega_D e^{-1/\varrho} \qquad (6\text{-}171)$$

6.4.2.7 Transition Temperature in a Magnetic Field

In this section we discuss the behavior of an artificial metallic superlattice in a magnetic field along the lines developed by Takahashi and Tachiki. We must generalize our expression for $Q_\omega(\boldsymbol{r}, \boldsymbol{r}')$ for the case of a finite magnetic field; for later-required generality we also include the effects of Pauli paramagnetism, spin orbit coupling and spin-flip scattering. From Eq. (6-125) we have

$$Q_\omega(\boldsymbol{r},\boldsymbol{r}') = \frac{1}{2} V k_B T \sum_{\alpha\beta\gamma\delta} \sum_{MN} \frac{\phi^*_{N\alpha}(\boldsymbol{r}) \mathscr{K}^\dagger_{\alpha\beta} \phi_{M\beta}(\boldsymbol{r}) \phi^*_{M\gamma}(\boldsymbol{r}') \mathscr{K}_{\gamma\delta} \phi_{N\delta}(\boldsymbol{r}')}{(\xi_M + i\hbar\omega)(\xi_N - i\hbar\omega)} \qquad (6\text{-}172)$$

In terms of a correlation function, we have from Eq. (6-129)

$$Q_\omega(\boldsymbol{r},\boldsymbol{r}') = \frac{2\pi}{\hbar} \int_0^\infty dt \cdot e^{-2|\omega|t} [g^{(1)}(\boldsymbol{r},\boldsymbol{r}',t) + g^{(2)}(\boldsymbol{r},\boldsymbol{r}',t)] \qquad (6\text{-}173)$$

where

$$g^{(\nu)}(\boldsymbol{r},\boldsymbol{r}',t) = -\tfrac{1}{2} \sum_n \delta(\xi - \xi_{n,\nu}) \langle n,\nu |\, \delta[\boldsymbol{r} - \boldsymbol{R}(t)] \cdot \mathscr{K}^\dagger(t) \mathscr{K}(0)\, \delta[\boldsymbol{r} - \boldsymbol{R}(0)] \,| n,\nu\rangle \qquad (6\text{-}174)$$

The time evolution of $\mathscr{K}(t)$ is governed by Eq. (6-131) or Eq. (6-139)

$$\dot{\mathscr{K}} = \left[-i \frac{d\phi}{dt} + \frac{2i}{\hbar} I \sigma_z - \frac{1}{\tau_s} \right] \cdot \mathscr{K}(t) - \frac{1}{\tau'} (\mathscr{K} - \tilde{\mathscr{K}}) \qquad (6\text{-}175)$$

and, referring to the discussion of Sec. 8.4.2.4, we may write

$$g^{(\nu)}(\boldsymbol{r},\boldsymbol{r}',t) = \frac{1}{2} g(\boldsymbol{r},\boldsymbol{r}',t) \cdot \exp\left(-\frac{2ie}{\hbar c} \int_{\boldsymbol{r}(0)}^{\boldsymbol{r}(t)} \boldsymbol{A}(\boldsymbol{s}) \cdot d\boldsymbol{s} - \frac{t}{\tau_s} \right) \cdot [e^{-i\Omega_L t} \delta_{\nu 1} + e^{+i\Omega_L t} \delta_{\nu 2}] \qquad (6\text{-}176)$$

where $g(\boldsymbol{r},\boldsymbol{r}',t)$ is given by Eq. (6-150). According to the discussion in Sec. 6.4.2.3, we must modify the differential operator $\boldsymbol{\nabla}$ to the (gauge invariant) form $\left[\boldsymbol{\nabla} - \frac{2ie}{\hbar} \boldsymbol{A}\right]$. Including the other contributions to the time evolution of $\mathscr{K}$, we must then solve the following pair of equations of motion for $g^{(\nu)}(\boldsymbol{r},\boldsymbol{r}',t)$

$$\left[\frac{\partial}{\partial t} + i\Omega_L(\boldsymbol{r}) + \frac{1}{\tau_s(\boldsymbol{r})} + \frac{1}{\tau'(\boldsymbol{r})}\right] g^{(1)} - \frac{1}{\tau'(\boldsymbol{r})} g^{(2)} - D(\boldsymbol{r}) \left[\boldsymbol{\nabla} - \frac{2ie}{\hbar} \boldsymbol{A}(\boldsymbol{r})\right]^2 g^{(1)} = 0 \qquad (6\text{-}177a)$$

and

$$\left[\frac{\partial}{\partial t} - i\Omega_L(\boldsymbol{r}) + \frac{1}{\tau_s(\boldsymbol{r})} + \frac{1}{\tau'(\boldsymbol{r})}\right] g^{(2)} - \frac{1}{\tau'(\boldsymbol{r})} g^{(1)} - D(\boldsymbol{r}) \left[\boldsymbol{\nabla} - \frac{2ie}{\hbar} \boldsymbol{A}(\boldsymbol{r})\right]^2 g^{(2)} = 0 \qquad (6\text{-}177b)$$

where we take the complex conjugate of the operator in the square brackets containing $\boldsymbol{\nabla}$ when acting on $\boldsymbol{r}'$. Equations (6-177a) and (6-177b) are solved subject to the initial condition

$$g^{(\nu)}(\boldsymbol{r},\boldsymbol{r}',0) = \tfrac{1}{2} N(\boldsymbol{r})\, \delta(\boldsymbol{r} - \boldsymbol{r}') \qquad (6\text{-}178)$$

Introducing an auxiliary function

$$R_\omega(\boldsymbol{r},\boldsymbol{r}') = \frac{2\pi}{\hbar} \int_0^\infty dt \cdot e^{-2|\omega|t} [g^{(1)}(\boldsymbol{r},\boldsymbol{r}',t) - g^{(2)}(\boldsymbol{r},\boldsymbol{r}',t)] \qquad (6\text{-}179)$$

we obtain the following coupled set of differential equations for Q_ω and R_ω

$$\left[2|\omega| + \frac{1}{\tau_s(\boldsymbol{r})} + \mathscr{L}\right] Q_\omega(\boldsymbol{r},\boldsymbol{r}') + i\Omega_L(\boldsymbol{r}) R_\omega(\boldsymbol{r},\boldsymbol{r}') = \frac{2\pi}{\hbar} N(\boldsymbol{r})\, \delta(\boldsymbol{r} - \boldsymbol{r}') \qquad (6\text{-}180a)$$

and

$$\left[2|\omega| + \frac{1}{\tau_s(\boldsymbol{r})} + \frac{2}{\tau'(\boldsymbol{r})} + \mathscr{L}\right] R_\omega(\boldsymbol{r},\boldsymbol{r}') + \\ + \mathrm{i}\,\Omega_L(\boldsymbol{r})\, Q_\omega(\boldsymbol{r},\boldsymbol{r}') = 0 \tag{6-180b}$$

where we have defined the operator

$$\mathscr{L} = -D(\boldsymbol{r})\left[\boldsymbol{\nabla} - \frac{2\mathrm{i}e}{\hbar}\boldsymbol{A}(\boldsymbol{r})\right]^2 \tag{6-181}$$

Analogous to the treatment in Sec. 6.4.2.6 we expand Q_ω and R_ω in terms of the eigenfunctions of the (gauge invariant) differential equation

$$\left[\mathscr{L} + \frac{1}{\tau_s(\boldsymbol{r})}\right]\psi_n(\boldsymbol{r}) = \Omega_n\,\psi_n(\boldsymbol{r}) \tag{6-182}$$

here the Ω_n are the eigenvalues and we observe that $\frac{1}{\tau_s(\boldsymbol{r})}$ behaves as a potential in the Schrödinger equation analogy. Q_ω and R_ω take the form

$$Q_\omega(\boldsymbol{r},\boldsymbol{r}') = [N(\boldsymbol{r})\,N(\boldsymbol{r}')]^{1/2}\sum_{mn} a_{mn}\psi_n^*(\boldsymbol{r}')\,\psi_m(\boldsymbol{r}) \tag{6-183a}$$

and

$$R_\omega(\boldsymbol{r},\boldsymbol{r}') = [N(\boldsymbol{r})\,N(\boldsymbol{r}')]^{1/2}\sum_{mn} b_{mn}\,\psi_n^*(\boldsymbol{r}')\,\psi_m(\boldsymbol{r}) \tag{6-183b}$$

If we define a matrix Γ having the elements

$$(\Gamma)_{ml} = (2|\omega| + \Omega_m)\,\delta_{ml} + \\ + \sum_k \frac{\langle m|\,\Omega_L\,|k\rangle\,\langle k|\,\Omega_L\,|l\rangle}{2|\omega| + \Omega_k} \tag{6-184}$$

and

$$\langle m|\Omega_L|k\rangle = \int \mathrm{d}^3 r\,\psi_m^*(\boldsymbol{r})\,\Omega_L(\boldsymbol{r})\,\psi_k(\boldsymbol{r}) \tag{6-185}$$

we then have the secular equation

$$\left|\delta_{mn} - \frac{2\pi}{\hbar}k_B T\sum_\omega\sum_l (\Gamma)^{-1}_{ml}\,\langle l|\,VN\,|n\rangle\right| = 0 \tag{6-186}$$

which is a generalization of Eq. (6-164a) to the case of finite magnetic fields, spin paramagnetism and spin flip scattering.

In the presence of a field the wavefunctions are not necessarily real. In what follows we will *adopt* a gauge invariant form of the boundary conditions Eqs. (6-156) and (6-157); specifically we will assume (with Takahashi and Tachiki, 1986a and 1986b) the continuity of the quantities [11]

$$\frac{\Delta(\boldsymbol{r})}{N(\boldsymbol{r})\,V(\boldsymbol{r})} = \frac{F(\boldsymbol{r})}{N(\boldsymbol{r})} \tag{6-187}$$

and

$$\frac{D(\boldsymbol{r})}{V(\boldsymbol{r})}\left[\boldsymbol{\nabla} - \frac{2\mathrm{i}e}{\hbar}\boldsymbol{A}(\boldsymbol{r})\right]\Delta(\boldsymbol{r}) = \\ = D(\boldsymbol{r})\left(\boldsymbol{\nabla} - \frac{2\mathrm{i}e}{\hbar}\boldsymbol{A}(\boldsymbol{r})\right)F(\boldsymbol{r}) \tag{6-188}$$

these conditions will be satisfied provided $[N(\boldsymbol{r})]^{1/2}\,\psi(\boldsymbol{r})$ satisfies the same boundary conditions as $F(\boldsymbol{r})$.

We require the eigenfunctions, $\psi_{\boldsymbol{n}}(\boldsymbol{r})$ of Eq. (6-182); in what follows $\boldsymbol{n}$ stands for n_x, n_y, n_z. We will treat only the case $\boldsymbol{H}\parallel\hat{\boldsymbol{z}}$ and $\boldsymbol{H}\parallel\hat{\boldsymbol{x}}$ and we will use the gauges $\boldsymbol{A} = H\,x\,\hat{\boldsymbol{y}}$ and $\boldsymbol{A} = -H\,z\,\hat{\boldsymbol{y}}$ respectively. The equations are then separable and we write the eigenfunctions in the form

$$\psi_{\boldsymbol{n}}(\boldsymbol{r}) = \mathrm{e}^{\mathrm{i}k_y y + \mathrm{i}k_z z}\,u_{n_x}(x)\,; \quad \boldsymbol{H}\parallel\hat{\boldsymbol{z}} \tag{6-189}$$

$$\psi_{\boldsymbol{n}}(\boldsymbol{r}) = \mathrm{e}^{\mathrm{i}k_x x + \mathrm{i}k_y y}\,w_{n_z}(z)\,; \quad \boldsymbol{H}\parallel\hat{\boldsymbol{x}} \tag{6-190}$$

which lead to the differential equations

$$\frac{\mathrm{d}^2 u_{n_x}(x)}{\mathrm{d}x^2} - \left[\frac{4e^2H^2}{\hbar^2}(x-x_0)^2 + \\ + k_z^2 - \frac{\Omega_{\boldsymbol{n}} - \tau_s^{-1}}{D}\right]u_{n_x}(x) = 0\,; \quad \boldsymbol{H}\parallel\hat{\boldsymbol{z}} \tag{6-191}$$

and

$$\frac{\mathrm{d}^2 w_{n_z}(z)}{\mathrm{d}z^2} - \left[\frac{4e^2H^2}{\hbar^2}(z-z_0)^2 + \\ + k_x^2 - \frac{\Omega_{\boldsymbol{n}} - \tau_s^{-1}}{D}\right]w_{n_z}(z) = 0\,; \quad \boldsymbol{H}\parallel\hat{\boldsymbol{x}} \tag{6-192}$$

---

[11] It is plausible that arguments similar to those in Sec. 8.4.2.5 could establish these boundary conditions rigorously.

where $x_0, z_0 = (\hbar k_y/2eH)$. For the case of $\boldsymbol{H} \parallel \hat{z}$ we may put $x_0=0$ (corresponding to k_y, or equivalently, $n_y=0$). From symmetry considerations in Eq. (6-191), the only relevant x dependent eigenfunction will be the Gaussian ground state harmonic oscillator wavefunction ($n_x=0$) which has the eigenvalue $\frac{2DeH}{\hbar}$ or D/ξ^2 where we define $\xi^2 \equiv \phi_0/2\pi H$. The values of k_z in media A and B take the values $k_{n_z\mathrm{A}}$ and $k_{n_z\mathrm{B}}$, with frequencies $D_\mathrm{A} k_{n_z\mathrm{A}}$ and $D_\mathrm{B} k_{n_z\mathrm{B}}$, which are determined from Eq. (6-169). Since only the n_z eigenfunctions are relevant, we may drop the suffix z in what follows. The total eigenfrequency is given by

$$\Omega_n = \tau_{\mathrm{sA}}^{-1} + \frac{D_\mathrm{A}}{\xi^2} + D_\mathrm{A} k_{n\mathrm{A}}^2 =$$
$$= \tau_{\mathrm{sB}}^{-1} + \frac{D_\mathrm{B}}{\xi^2} + D_\mathrm{B} k_{n\mathrm{B}}^2 \qquad (6\text{-}193)$$

Eq. (6-192) for the $\boldsymbol{H} \parallel \hat{\boldsymbol{x}}$ case is much more problematic in that the translational invariance is now lost in the z direction and we must use the boundary conditions to match the eigenfunctions w_{n_z} at each interface. We may set $k_x=0$ with no loss of generality. By moving the origin to z_0, defining a variable $x=\sqrt{2}z/\xi$ and defining a dimensionless eigenvalue $\lambda(n_z)$ as $\lambda+\frac{1}{2} \equiv \frac{\Omega-\tau_\mathrm{s}^{-1}}{(2D/\xi^2)}$, we may rewrite Eq. (6-181) in canonical form as

$$\left(-\frac{\mathrm{d}^2}{\mathrm{d}x^2}+\frac{1}{4}x^2\right) w_\lambda(x) =$$
$$= \left(\lambda+\frac{1}{2}\right) w_\lambda(x) \qquad (6\text{-}194)$$

Although this equation is identical to the (dimensionless) equation for the harmonic oscillator, the solutions differ because of the boundary conditions (harmonic oscillator functions vanish as $x \to \infty$). The eigenvalues λ, though discrete (for given boundary conditions) are no longer (necessarily) integers and furthermore (unlike the harmonic oscillator) there are two linearly independent solution, $w_\lambda(+x)$ and $w_\lambda(-x)$, for each λ. The total eigenfrequency is given by (where we write n for n_z)

$$\Omega_n = \tau_{\mathrm{sA}}^{-1} + \left(\lambda_\mathrm{A}+\frac{1}{2}\right)\frac{2D_\mathrm{A}}{\xi^2} =$$
$$= \tau_{\mathrm{sB}}^{-1} + \left(\lambda_\mathrm{B}+\frac{1}{2}\right)\frac{2D_\mathrm{B}}{\xi^2} \qquad (6\text{-}195)$$

As can be seen from the above discussion the application of the Takahashi-Tachiki formalism in the case of finite magnetic fields is relatively complex requiring extensive numerical analysis. To date it has not been used to specifically fit the data of any of the existing experiments. However, under the simplifying assumption that only one of the three quantities, $N(z)$, $D(z)$ or $V(z)$ varies (in a stepwise fasion), Takahashi and Tachiki have carried out the numerical calculations (in some cases using a variational approach); their analysis is complete enough to bring out most of the expected qualitative features.

We first discuss the case where the density of states is discontinuous at the interface. Figure 6-36 shows the parallel, $H_{\mathrm{c2}\parallel}$ (solid), and the perpendicular, $H_{\mathrm{c2}\perp}$ (dashed), upper critical field [normalized to the zero temperature upper critical field of the bulk superconductor, $H_{\mathrm{c2}}^{\mathrm{S}}(0)$] as a function of T/T_c (where T_c is the zero field transition temperature of the superlattice); here we have $N_\mathrm{N}/N_\mathrm{S} = 0.15$, $V_\mathrm{S} = V_\mathrm{N}$, $D_\mathrm{S} = D_\mathrm{N}$, and $a = b$. The numbers beside the different sets of curves denote the value of $a/\xi_\mathrm{S}(0)$ where $\xi_\mathrm{S}(0) \equiv [\phi_0/2\pi H_{\mathrm{c2}}^{\mathrm{S}}(0)]^{1/2}$. We note the following features. For T close to T_c the perpendicular critical field always behaves linearly in $T_\mathrm{c}-T$. However the behavior of the parallel critical field depends on the magnitude of $a/\xi_\mathrm{S}(0)$: when this

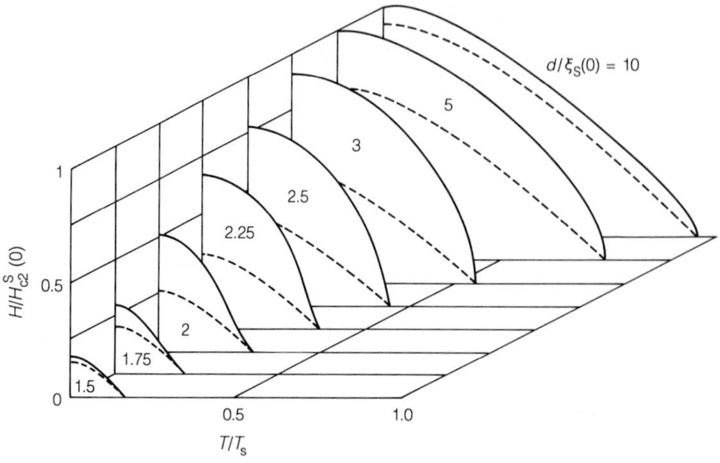

Figure 6-36. The parallel upper critical field $H_{c2\parallel}$ (———) and the perpendicular upper critical field $H_{c2\perp}$(– – –) in reduced units (see text); here all the quantities characterizing the sublayers are equal with the exception of the density of states for which $N_N/N_s = 0.15$ (after Takahashi and Tachiki, 1986a and 1986b).

quantity is small we obtain linear, or 3D like, behavior of the upper critical field, while for large values we have square root like, or 2D, behavior (except for a small region near T_c); for values near unity we have a 2D – 3D cross over analogous to the obtained in our analysis of the superconductor/insulator superlattice in the LD model.

In a later study Takahashi and Tachiki (1986a and 1986b) reexamined the behavior of the parallel upper critical field for the case of widely differing diffusion constants. They noted that by symmetry there are two possible solutions, one where nucleation occurs in the N layers and one where it occurs in the S layers. It turns out that these two nucleation temperatures (i.e., the temperatures satisfying our linearized integral equation for Δ) have a different field dependence, as shown in Fig. 6-37 for the case of $D_N/D_S = 12.5$, $a = b = \xi_N(0)$ and $V_N = V_S$. The dashed and dash-dot curves refer to solutions when $\Delta(r)$ is centered in the N (clean) and S (dirty) layers respectively. For T close to T_c nucleation first occurs in the superconducting layer (since nucleation will always occur at the highest field satisfying the linearized gap equation). However on passing the point H^*, T^*, the solution centered in the N (clean) layers has a smaller field. In the absence of pinning a second order phase transition will occur at this point. Figure 6-38 shows the parallel upper critical field [for $a = b = \xi_N(0)$ and $V_N = V_S$] as a function of D_N/D_S. Note the abrupt change in slope for the three curves for which the phase transition occurs. The possibility of a first order transition in superconducting superlattices was discussed by Buzdin and Simonov (1990).

Lastly, we examine the effect of a change in the BCS pairing potential. Figure 6-39

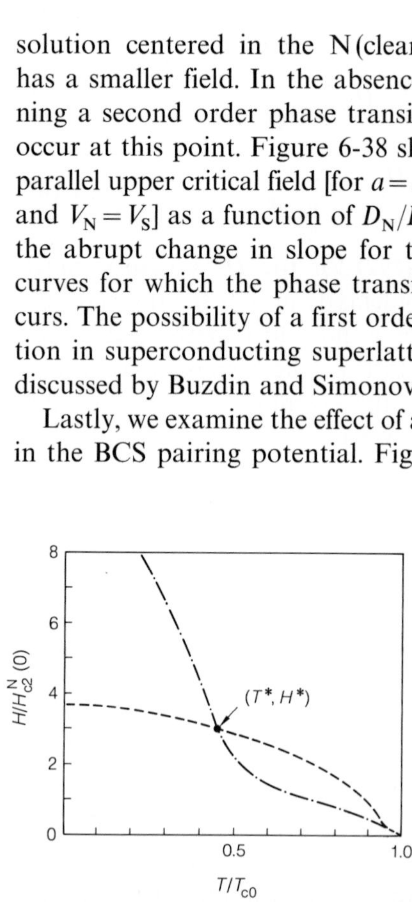

Figure 6-37. Plot of the temperature dependence for nucleation in the N (– – –) and S (— · —) layers. Note that of a temperature-field point T^*, H^*, the solutions are degenerate, implying a phase transition (after Takahashi and Tachiki, 1986a).

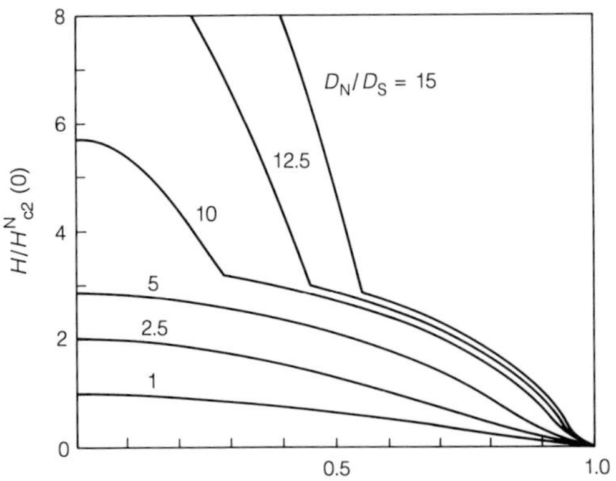

Figure 6-38. Temperature dependence of the upper critical field as a function of D_N/D_S (see text) (after Takahashi and Tachiki, 1986a).

shows the parallel and perpendicular upper critical fields for the case $a=b=d$, $V_S N_S = 0.3$ and $V_N N_N = 0$, $N_N = N_S$ and $D_N = D_S$. Crossover behavior in the parallel upper critical field is again observed. Takahashi and Tachiki also present graphs displaying the effect of spin paramagnetism on the upper critical field that will not be reproduced here.

Although, we have restricted all our discussion to the reversible behavior, we note in passing the work of Takahashi and Tachiki (1986a) on vortex pinning in superconductor/ferromagnetic superlattices.

Before concluding the theoretical discussion of the behavior of artificial superlattices in the presence of a field, we note that the occurrence of surface superconductivity has not been addressed. This interesting and complex problem could presumably be addressed with an extension of the aforementioned techniques.

6.4.2.8 The Werthamer Approximation

In an infinite homogeneous system at zero field, $K_0(\boldsymbol{r}, \boldsymbol{r}')$ may be written as a function of $|\boldsymbol{r}-\boldsymbol{r}'|$ only (see discussion leading up to Eq. (6-106)), and its Fourier transform, $K_0(q)$, is given by Eq. (6-115).

Fourier transforming Eq. (6-115) back to $\boldsymbol{r}$ space we have

$$K_0(\boldsymbol{r}-\boldsymbol{r}') = \tag{6-196}$$
$$= N V \left[\ln \frac{1.14 \hbar \omega_D}{k_B T_c} \delta(\boldsymbol{r}-\boldsymbol{r}') - X(\boldsymbol{r}-\boldsymbol{r}') \right]$$

where

$$X(\boldsymbol{r}-\boldsymbol{r}') \equiv \frac{1}{(2\pi)^3} \int d^3 q \, e^{i q \cdot (\boldsymbol{r}-\boldsymbol{r}')} \chi(\xi^2 q^2) \tag{6-197}$$

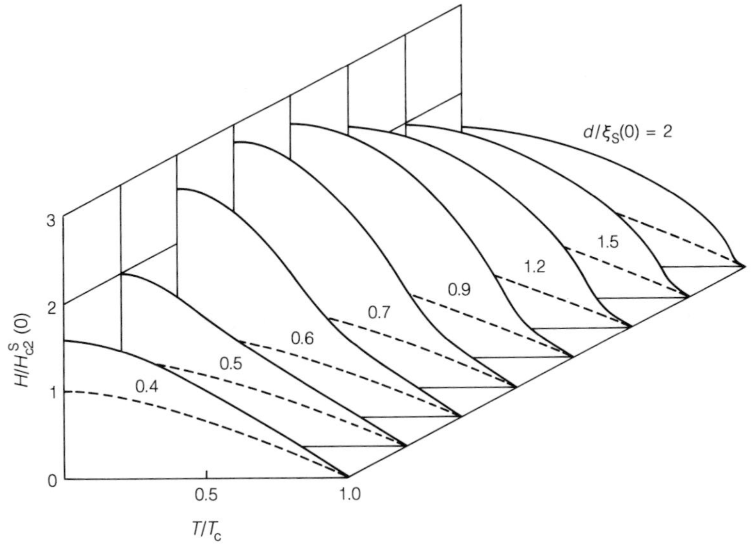

Figure 6-39. Temperature dependence of the parallel upper critical field $H_{c2\parallel}$ (———) and perpendicular upper critical field $H_{c2\perp}$ (– – –) for varying values of $d/\xi_s(0)$. Here $V_S N_S = 0.3$, $V_N N_N = 0$, $N_S = N_N$ and $D_S = D_N$ (see text) (after Takahashi and Tachiki, 1986a and 1986b).

Werthamer (1963) makes the *assumption* that in an inhomogeneous system (such as a sandwich or artificial superlattice) $K(\boldsymbol{r},\boldsymbol{r}')$ takes the form [12]

$$K(\boldsymbol{r},\boldsymbol{r}') = N(\boldsymbol{r})\,V(\boldsymbol{r})\left[\ln\frac{1.14\,\hbar\,\omega_{\mathrm{D}}}{k_{\mathrm{B}}\,T}\,\delta(\boldsymbol{r}-\boldsymbol{r}') - X(\boldsymbol{r}-\boldsymbol{r}')\right] \tag{6-198}$$

Defining $T_{\mathrm{c}}(\boldsymbol{r})$ through $\ln\left[\frac{1.14\,\hbar\,\omega_{\mathrm{D}}}{k_{\mathrm{B}}\,T_{\mathrm{c}}(\boldsymbol{r})}\right] = N(\boldsymbol{r})\,V(\boldsymbol{r})$, Werthamer has shown that

$$\chi(-\xi^2\nabla^2)\,\Delta(\boldsymbol{r}) = \ln\left[\frac{T_{\mathrm{c}}(\boldsymbol{r})}{T}\right]\Delta(\boldsymbol{r}) \tag{6-199}$$

Rather than an integral equation for $\Delta(\boldsymbol{r})$, one obtains an infinite order differential equation for $\Delta(\boldsymbol{r})$. Two of the (many) solutions to Eq. (6-199) in media A and B are $\Delta^{(\mathrm{A})}(z) = \mathrm{e}^{\pm\mathrm{i}\,q_{\mathrm{A}}\,z}$ and $\Delta^{(\mathrm{B})}(z) = \mathrm{e}^{\pm\mathrm{i}\,q_{\mathrm{B}}\,z}$ where $q_{\mathrm{A,\,B}}$ satisfies

$$\chi(\xi^2\,q_{\mathrm{A}}^2) = \ln\left(\frac{T_{\mathrm{cA}}}{T}\right) \quad (z \text{ in layer A}) \tag{6-200a}$$

and

$$\chi(\xi^2\,q_{\mathrm{B}}^2) = \ln\left(\frac{T_{\mathrm{cB}}}{T}\right) \quad (z \text{ in layer B}) \tag{6-200b}$$

In addition Werthamer imposes de Gennes boundary conditions which amount to

$$N_{\mathrm{A}}\,D_{\mathrm{A}}\tan\left(\frac{q_{\mathrm{A}}\,a}{2}\right) = = N_{\mathrm{B}}\,D_{\mathrm{B}}\tanh\left(\frac{q_{\mathrm{B}}\,b}{2}\right) \tag{6-201}$$

Equations (6-200a, b) and (6-201) must be solved simultaneously to obtain the three unknowns q_{A}, q_{B} and T (the sandwich or superlattice transition temperature).

We must emphasize that the above procedure is only an approximation since, in general, it is not possible to write the kernel $K(\boldsymbol{r},\boldsymbol{r}')$ in the form Eq. (6-162) [see Deutcher and de Gennes (1969) for further discussion]. It has, however, been widely applied with little regard to its limitations.

[12] Werthamer originally assumed $N(\boldsymbol{r})$ was constant.

6.4.2.9 Comparison of Werthamer's Procedure and a Numerical Solution

It is instructive to compare the transition temperature obtained for Werthamer's procedure and that obtained by applying the Takahashi-Tachiki formalism described in Sec. 6.4.2.6, as first done by Auvil and Ketterson (1987). Since our only goal here is to test the range of validity of the Werthamer procedure, we will pick the simplest possible case, that of a superconductor-metal superlattice with equal layer thicknesses, $a = b = \Lambda/2$, diffusion constants, D, and densities of states, N; in the metal layer we choose $V=0$.

Table 6-3 shows the results of the calculation [note we use reduced units $\frac{\Lambda}{4\,\xi}$ and T/T_{c} where $\xi = \left(\frac{\hbar\,D}{2\,\pi\,k_{\mathrm{B}}\,T_{\mathrm{c}}}\right)^{1/2}$ and $k_{\mathrm{B}}\,T_{\mathrm{c}} = 1.14\,\hbar\,\omega_{\mathrm{D}}\,\mathrm{e}^{-1/NV}$ where T_{c} and V are the bulk transition temperature and the pairing potential of the superconductor].

Table 6-3. Transition temperature vs. superlattice wavelength.

$\Lambda/4\,\xi$	0	0.1	0.5	1	2	5	10
T/T_{c} (Werthamer)	0.25	0.25	0.26	0.29	0.43	0.83	0.95
T/T_{c} (exact)	0.03	0.03	0.03	0.11	0.36	0.83	0.95

The above results also apply to a sandwich with equal layer thickness, Λ, if we replace $\frac{\Lambda}{4\,\xi}$ by $\frac{\Lambda}{\xi}$.

Note that for our parameters the de Gennes-Cooper $\Lambda\to 0$ limit, which follows from Eqs. (6-170) and (6-171), is given $T/T_{\mathrm{c}} = 0.03$, in agreement with our numerical calculations. Note also that the Werthamer procedure does give reasonably accurate results for $4\Lambda/\xi > 3$.

6.4.2.10 Microscopic Theory of the Josephson Superlattice

In Sec. 6.4.1.5 we discussed the superconductor-insulator superlattice in the Landau-Ginzburg model. We may carry over most of this discussion to the microscopic case, provided we restrict ourselves to the diffusion dominated (dirty) limit. For a uniform system (or a stepwise discontinuous media) the correlation function, $g(\boldsymbol{r},\boldsymbol{r}',t)$ is governed by a diffusion equation. As a model for the superconductor-insulator (Josephson) superlattice, rather than the 3D diffusion equation, we can use the corresponding form of the Lawrence-Doniach equation, Eq. (6-105). We introduce the eigenvalue equation

$$-D\left(\boldsymbol{\nabla}-\frac{2\mathrm{i}e}{\hbar c}\boldsymbol{A}_{\perp}\right)^2 g_m^{(n)}(\boldsymbol{r}) - \frac{J}{2}\left[g_m^{(n+1)}(\boldsymbol{r})\,\mathrm{e}^{-\frac{2\mathrm{i}e}{\hbar c}\bar{A}_z} - 2g_m^{(n)}(\boldsymbol{r}) + g_m^{(n-1)}(\boldsymbol{r})\,\mathrm{e}^{\frac{2\mathrm{i}e}{\hbar c}\bar{A}_z}\right] = \Omega_m^{(n)}\,g_m^{(n)}(\boldsymbol{r}) \tag{6-202}$$

where the eigenfunctions $g_m^{(n)}(\boldsymbol{r})$ satisfy the ortho-normality condition

$$\sum_n \int \mathrm{d}^2 r\; g_m^{(n)}(\boldsymbol{r})\, g_{m'}^{(n)}(\boldsymbol{r}) = \delta_{mm'} \tag{6-203}$$

We introduce a correlation function, $g^{n,n'}(\boldsymbol{r},\boldsymbol{r}',t)$, which measures the gauge invariant *probability* that a particle initially in layer n' with position $\boldsymbol{r}'$ will later be found in layer n with position $\boldsymbol{r}$. We expand $g^{n,n'}(\boldsymbol{r},\boldsymbol{r}',t)$ in the form

$$g^{n,n'}(\boldsymbol{r},\boldsymbol{r}',t) = L^{-2}\sum_m g_m^{(n')*}(\boldsymbol{r}')\, g_m^{(n)}(\boldsymbol{r})\,\mathrm{e}^{-\Omega_m|t|} \tag{6-204}$$

In place of Eq. (6-98), the gap equation in this model takes the form

$$\Delta^{(n)}(\boldsymbol{r}) = \sum_{n'}\int \mathrm{d}^2 \boldsymbol{r}'\, K^{n,n'}(\boldsymbol{r},\boldsymbol{r}')\,\Delta^{(n')}(\boldsymbol{r}') \tag{6-205}$$

The analog of Eq. (6-103) is

$$K^{n,n'}(\boldsymbol{r},\boldsymbol{r}') = \frac{2\pi}{\hbar} L^2 k_\mathrm{B} T N(0) V \sum_\omega \mathrm{d}t\, \mathrm{e}^{-2|\omega|t}\, g^{n,n'}(\boldsymbol{r},\boldsymbol{r}',t) \tag{6-206}$$

Expanding the gap function in the form $\Delta^{(n)}(\boldsymbol{r}) = \sum_m a_m\, g_m^{(n)}(\boldsymbol{r})$, we obtain, on using Eq. (6-203),

$$\frac{2\pi}{\hbar} k_\mathrm{B} T N(0) V \sum_\omega \frac{1}{2|\omega| + \Omega_m} = 1 \tag{6-207}$$

Thus the problem of finding the transition temperature for a superconductor-insulator superlattice reduces to the same form as for a uniform superconductor; Ω_n is just the lowest eigenvalue of the linearized form of Eq. (6-82). For a more complete discussion of the various solutions see Klemm (1974) and Klemm et al. (1974). Pair breaking effects associated with paramagnetic impurities and Larmor precession (as modified by spin orbit coupling) may be simply included.

6.4.3 Experiments on Superconducting Thin Films and Superlattices

The experiments on the superconducting properties of artificial metallic superlattices may logically be divided into three categories in which one of the constituents is a superconductor and the other is either

(1) an insulator (or semiconductor);
(2) a normal metal (or a metal with a lower T_c);
(3) a magnetic metal (or one containing paramagnetic impurities).

A major source of interest in the metal/semiconductor superlattice has centered around the possibility of achieving higher transition temperatures via the excitonic mechanism, first discussed by Little (1964) and Ginzburg (1970). We have intentionally not discussed this topic in our theoretical review because at the present time there does not appear to be any experimental evidence for its existence, although the picture may change with the discovery of high-temperature superconductors [in, e.g., $La_{2-x}Ba_xCuO_4$ by Bednorz and Müller (1986) and in $YBa_2Cu_3O_7$ by Wu et al. (1987)]. We may loosely summarize the excitonic mechanism as follows. Suppose that we have a thin metallic layer located adjacent to a narrow band-gap semiconductor with the Fermi-level of the metal lying within the energy gap of the semiconductor. The wavefunction of an electron near the surface of the metal will penetrate a short distance into the semiconductor where it can polarize the chemical bonds of the semiconductor; this polarization can be regarded as a superposition of virtual electron-hole pairs (and hence the name exciton). A second electron penetrating into the same region a short time later will sense this polarization and this results in an attractive interaction. A model calculation of this effect was performed by Allender et al. (1973); the excitonic mechanism is also discussed at length in the book by Ginzburg and Kirzhnits (1982). An extensive series of experiments was carried out to test these ideas by Miller et al. (1976), however they were unable to clearly establish the mechanism. Another mechanism involving the exchange of phonons across a thin dielectric boundary was proposed by Cohen and Douglass (1967).

An alternate reason to study the superconductor-insulator superlattice is the hope of achieving an enhanced parallel upper critical field using the thin film effect discussed in Sec. 6.4.1 [Eq. (6-76)]. Alternatively one may hope that the insulating layers (or the interfaces) might enhance the critical current by providing pinning centers for the vortex lines.

The motivations to study metal/metal superconducting superlattices are somewhat less dramatic. Work on such superlattices was proceeded by the study of thin film metallic bilayers. Here a major reason was to use the proximity effect to probe for superconductivity in metals which had not yet displayed superconductivity in bulk. We will not attempt to review this aspect in any detail. (Some of the work involved films that either were not prepared with modern high-vacuum techniques or were improperly characterized.) However, historically, the careful work of Hilsch (1962) stimulated the theoretical analysis of de Gennes and Guyon (1963). Werthamer's (1963) procedure was applied in the work of Hauser et al. (1964) in the study of superconductor/normal-metal bilayers involving Pb/Cu and Pb/Pt in order to estimate the pairing potential in the normal metals. The effect of magnetic impurities was modeled within the Werthamer procedure in the work of Hauser et al. (1966).

The majority of our experimental discussion will center on the wavelength and composition dependence of the zero field transition temperature and the temperature and angular dependence of the upper critical magnetic field. Other aspects which will be briefly touched upon are the critical current and the penetration depth. We will not discuss tunneling. This is not because we don't think it is interesting; our feeling is that the theoretical understanding is not complete enough at this time to properly interpret the experimental results. Experimental tunneling spectra have been reported on Nb/Cu by Vaglio et al. (1987), on Nb/Al by Geerk et al. (1972), and on Nb/Ta by Hertel et al. (1982), and on Pb/Ag by Jatochowski (1984).

6.4.3.1 Superconductor/Insulator Superlattices

The Al/Ge system was studied by Haywood and Ast (1978). Of most interest here was their study of the dependence of the parallel and perpendicular upper critical fields on the reduced temperature, $t \equiv T/T_c$. Since Al has a coherence length of order 10^3 nm, we are clearly in a thin film regime. The parallel critical field displays the $(1-t)^{1/2}$ behavior characteristic of an isolated superconducting slab [Eq. (6-76)]. The perpendicular critical fields do not display the expected linear behavior in t. The unexpected up turn appears to be associated with the number of layers, as Haywood and Ast showed by a study of a film with a sublayer thickness of 18 nm and various numbers of layers ranging from $1-18$.

The Nb/Ge system was studied by Ruggiero et al. (1980). Figure 6-40 shows the behavior of the parallel and perpendicular upper critical fields on reduced temperature, $t = T/T_c$, for a number of ratios of the thicknesses of the sublayers, d_{Nb}/d_{Ge}.

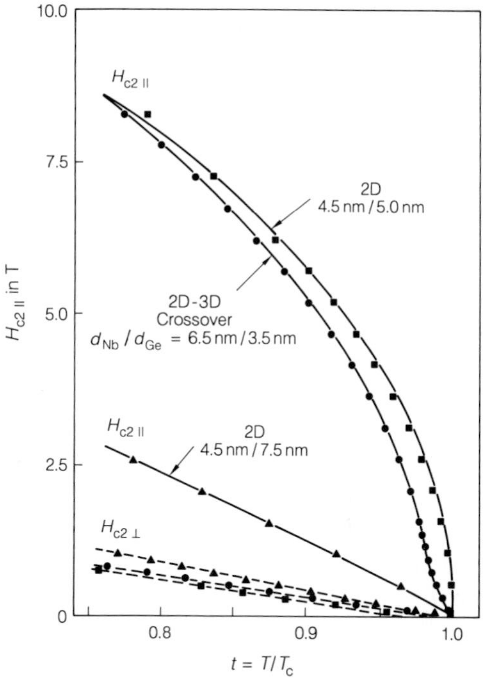

Figure 6-40. The parallel and perpendicular upper-critical magnetic fields for Nb/Ge multilayers with varying ratios d_{Nb}/d_{Ge} (after Ruggiero et al., 1980).

The sample with 45 nm/5 nm clearly shows the square root like behavior, $(1-t)^{1/2}$, of a 2D thin slab [Eq. (6-76)]. On the other hand the 4.5 nm/0.7 nm sample clearly has strong interlayer coupling because it has the linear $1-t$ behavior of a 3D system. The most interesting behavior of the field dependence is that of the 65 nm/3.5 nm sample which has three regimes: (1) a square root like behavior for $t \ll 1$, (2) a linear behavior for t near 1, and a $2D-3D$ crossover behavior separating these two regimes. Note the perpendicular critical field of all specimens studied shows a very nearly linear behavior in $1-t$. The Nb/Ge system should be a good model for the Josephson coupled superlattice discussed in Sec. 6.4.1. Ruggiero et al. (1980) also discussed the temperature dependence of the zero field resistance in the vicinity of the superconducting transition and compared it to theoretical predictions involving the Aslamazov-Larkin theory of fluctuations of the order parameter (including Maki-Thompson corrections) using the procedures of Skocpol and Tinkham (1975); this interesting aspect will not be discussed here.

Li et al. (1986) have studied the Nb/Si superlattice. The observed transition temperatures measured by Li et al. in Nb/Si are about $1-2$ K higher than those of Ruggiero et al. for the Nb/Ge system, which Li et al. attribute to the better quality of their e-gun deposited samples. An extensive structural characterization of Nb/Si superlattices was carried out by Song et al. (1989).

The superconducting properties of the Josephson coupled Pb/Ge system were studied by Locquet et al. (1987) (along with other systems). As a function of the Ge thickness, the behavior of the upper critical magnetic field was similar to that seen in the Nb/Ge system: For large Ge thicknesses 2D behavior was seen while for small thicknesses 3D behavior was observed; at intermediate thickness the data display a cross over behavior (which is more distinct than that of Nb/Ge).

An extensive set of measurements on the $Nb_{.53}Ti_{.47}$/Ge artificial superlattice system was carried out by Jin et al. (1985, 1986 and 1987). The original motivation for studying this system was simply to fabricate a very high performance material for high field applications by exploiting the thin film effect. The parallel upper critical field was shown to not be enhanced (due to paramagnetic limiting), however, in the course of the study, the system took on an intrinsic interest of its own as a model for the disordered (superconducting) alloy/insulator superlattice (see Sec. 6.4.3). The angular dependence of the upper critical field was studied by Jin et al. (1987). Although an attempt was made to model this behavior, the expressions used involved some ad hoc assumptions, due to uncertainties in the correct way to include the localization and electron-electron interactions effects. We therefore show the data fitted by the simple Tinkham expression in Fig. 6-41. The data for $d_{Ge} = 22.1$ nm show an isotropic 3D behavior. The remaining data show cusp-like behavior with varying levels of agreement with the Tinkham expression, depending on the thickness and temperature.

The behavior of $H_{c2}(\theta)$ for a Josephson superlattice has been examined by Glazmana (1989) using the method of adiabatic separation of variables in the linearized

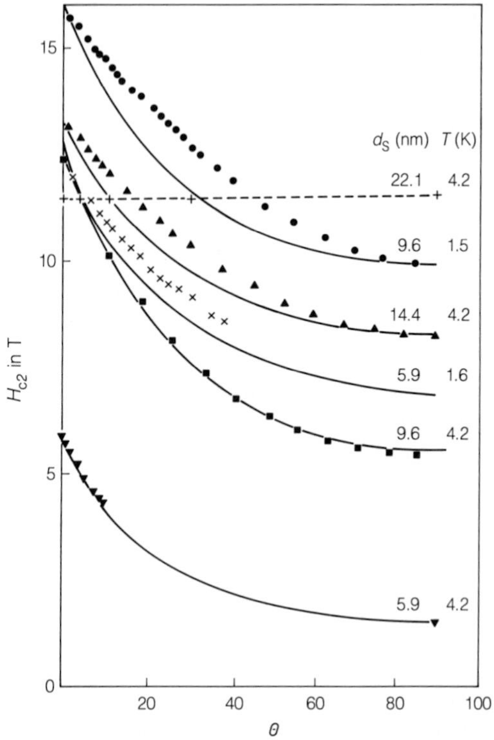

Figure 6-41. The angular dependence of H_{c2} for $Nb_{0.52}Ti_{0.48}$/Ge fitted using Tinkham's expression: ▾, 5.9 nm/3.8 nm (4.2 K); ×, 5.9/3.8 nm (1.6 K); ■, 9.6 nm/4.5 nm (4.5 K); ●, 9.6 nm/4.5 nm (1.5 K); ▴, 14.4 nm/3.2 nm (4.2 K); +, 22.1 nm/4.5 nm (4.2 K) (after Jin et al., 1987).

GL equation. At small angles, $H_{c2}(\theta)$ is shown to be determined by a modified Tinkham formula which describes a more rapid decrease in the critical field with angle than that described by Eq. (6-78). This is in agreement with the $H_{c2}(\theta)$ data taken on V/Si superlattices (Tovazhanyanskii et al., 1987).

Superconducting properties of the multilayers are affected by the crystallinity of the constituents, interfaces coherence, formation of alloy phases due to interdiffusion, and the periodicity of the sublayers. To distinguish these effects, T_c as a function of the modulation wavelength of Mo/Si multilayers was measured by Nakajima et al. (1989). The result is shown in Fig. 6-42. Since crystalline bulk Mo is a low T_c superconductor (with T_c as low as 0.9 K), the observed T_c enhancement in the small Λ region is attributed to a MoSi amorphous phase formed at the interfaces. When Λ is larger than 3.0 nm, the crystalline Mo phase appears, resulting in T_c depression through the proximity effect. These authors proposed a model similar to the Cooper limit discussed in Sec. 6.4.2.6 to explain the T_c enhancement in the small Λ region.

High critical current densities are a key factor for high field applications of superconductors. High J_c can be achieved by introducting crystalline imperfections or second phase dispersion to provide the necessary flux pinning centers. J_c can also be enhanced with thin film techniques; NbN films show an enhanced J_c down to 10 nm (Gavaler et al., 1971). However, the total current carrying capacity is limited by the film cross section. A natural solution is layering, as demonstrated in NbN/AlN multilayers.

NbN/AlN multilayers were first prepared and characterized by Murduck et al.

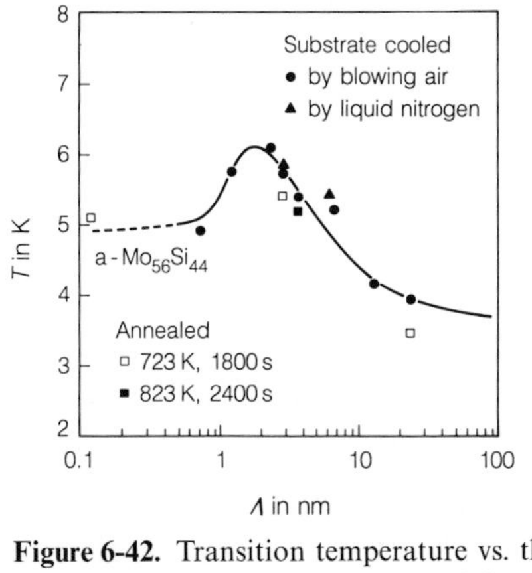

Figure 6-42. Transition temperature vs. the modulation wavelength of Mo/Si multilayers having $d_{Mo} = d_{Si}$ (after Nakajima et al., 1989).

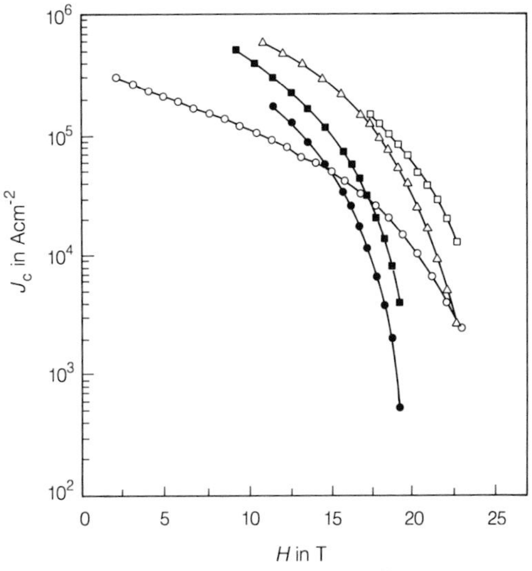

Figure 6-43. The magnetic field dependence of the critical current for NbN/AlN for various NbN thicknesses: □, 6.2 nm; △, 131.0 nm; ■, 26.3 nm; •, 35 nm; ○, an optimized thick film. The AlN layer thickness is 2 nm in all cases (after Murduck et al., 1988).

(1987). The rock-salt NbN and the wurtzite (hexagonal) AlN structures are well lattice matched for [111]–[0001] growth. Measurements of the high field, critical-current density (see Fig. 6-43) show that J_c and the flux-pinning density are significantly increased in NbN/AlN multilayers (Murduck et al., 1988); for a multilayer with $d_{NbN} = 6.2$ nm, J_c at 20 T is about 6 times higher than that of the optimized thick NbN film of Kampwirth et al. (1985). Copper cladded multilayers were shown to have higher J_c at high fields. By comparing the J_c values for both parallel and perpendicular fields, the main mechanism responsible for the J_c enhancement has been identified as periodic pinning by the insulating AlN layers.

There has also been an effort (Nagaoka et al., 1984) to grow epitaxial NbN/MgO multilayers; the system is a very promising candidate for achieving the multi-junction structure discussed in Sec. 6.4.1.6.

6.4.3.2 Superconductor/Normal-Metal Superlattices

Perhaps the first multilayer metallic systems were studied by Strongin et al. (1968); however the goal of this work was not to make regularly spaced superlattices but rather to probe the effect of the successive deposition of overlayers of varying thickness on the transition temperature. Since this early work a number of carefully controlled studies have been performed on a number of systems. We will not attempt to review all of this work in detail, but will rather discuss a number of model systems.

Nb/Cu

The superconducting properties of the Nb/Cu system have been studied by Schuller and Falco (1979), Banerjee et al. (1982 and 1983), Chun et al. (1984), and Locquet et al. (1987). Structural studies were performed by Schuller (1980) and Lowe et al. (1981).

Banerjee et al. (1982) performed a detailed study of the dependence of the zero-field transition temperature on the layer thickness for symmetrical superlattices ($d_{Nb} = d_{Cu}$). Samples were prepared by d.c. magnetron sputtering on sapphire. Since Nb and Cu are immiscible, the interfaces are quite sharp; the above cited structural work shows that the layers are also rather uniform in thickness and well textured (for thickness $\gtrsim 1$ nm).

Banerjee et al. (1982) interpreted their data using the Werthamer procedure, discussed in Sec. 6.4.2.8. Unlike the Takahasi and Tachiki theory this approach breaks down in the limit of thin layers, and ultimately does not approach the de Gennes-Cooper limit (see Sec. 6.4.2); the Werthamer transition temperature is always greater than the more precise predictions based on the full numerical solution of the Takahasi and Tachiki formalism. Auvil and Ketterson (1988a and 1988b) have fitted the data of Banerjee et al. (1982) to the Takahasi and Tachiki theory. In order to fit the data it is necessary to assume that the transition temperature of the Nb layers is suppressed as they become thinner. This behavior is rationalized by noting that the Fermi level of Nb is situated near a sharp peak in the energy dependence of the density of states of the 4d b.c.c. transition metals; interface scattering might then be expected to *smear out* this structure and hence lower the transition temperature. Figure 6-44 shows the data of Banerjee et al. (1982) and a fit of the theory. The dashed line shows the thickness dependence of T_c that was used to achieve the fit which was derived from independent considerations. Banerjee et al. also used a thickness dependent niobium transition temperature, however the Werthamer procedure requires a more dramatic suppression.

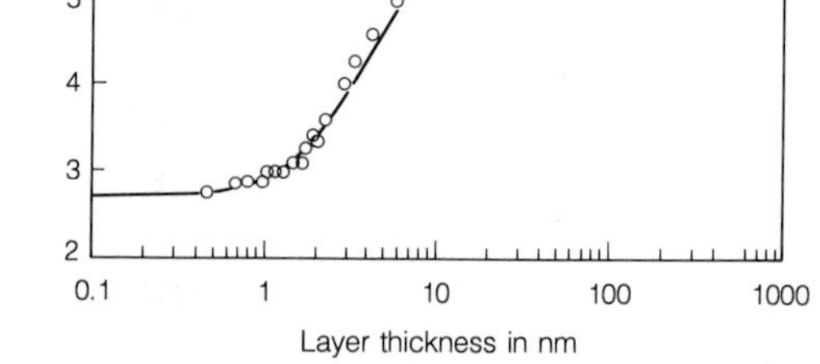

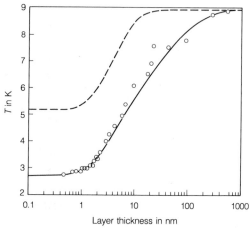

Figure 6-44. Dependence of the transition temperature of symmetric Nb/Cu superlattices on layer thickness (o) (after Bannerjee et al., 1982). The fit to the data of the theory of Takahashi and Tachiki (———) (after Auvil and Ketterson, 1988a), and the suppression of the transition temperature of *pure* Nb as a function of thickness used to achieve the fit, which was derived from independent considerations (– – –).

The magnetic field dependence of the transition temperature of the Nb/Cu system was also studied by the Argonne group. The temperature dependence of the parallel and perpendicular upper critical field for a Nb/Cu sample with $d_{Nb}/d_{Cu} = 17.2$ nm/33.3 nm clearly shows a dimensional cross over; other samples show 2D and 3D behavior, depending on the layer thicknesses. The angular dependence of the upper critical field was also examined in this work; again, depending on the layer thicknesses, it shows a singular cusp-like or a continuous behavior near the parallel field direction.

Nb/Ti

The Nb/Ti superlattice system was studied by Qian et al. (1982). Samples were prepared with d.c. magnetron sputtering on sapphire substrates. Since Nb and Ti are miscible, there may be some tendency to form an alloy at the interface. Such mixing, if present, must occur during the deposition process itself since no time dependent changes in the resistivity were subsequently observed. Figure 6-45a shows the dependence of the zero field transition temperature on the composition wavelength. The following features are noted. At long wavelengths the transition temperature approaches that of pure Nb, as expected. At short wavelengths, 3 – 20 nm, Qian et al. (1982) obtain a de Gennes-Cooper limiting behavior with a thickness independent T_c. Application of Eq. (6-171) yields a higher temperature (5.9 K) than the experimental value (3.7 K). However two effects operate to lower this value. Firstly, there is the aformentioned T_c reduction for thin Nb layers, and secondly, the X-ray data show the presence of some hexagonal Ti (the transition temperatures for Ti are 4.0 K and 0.39 K for the b.c.c. and h.c.p. phases

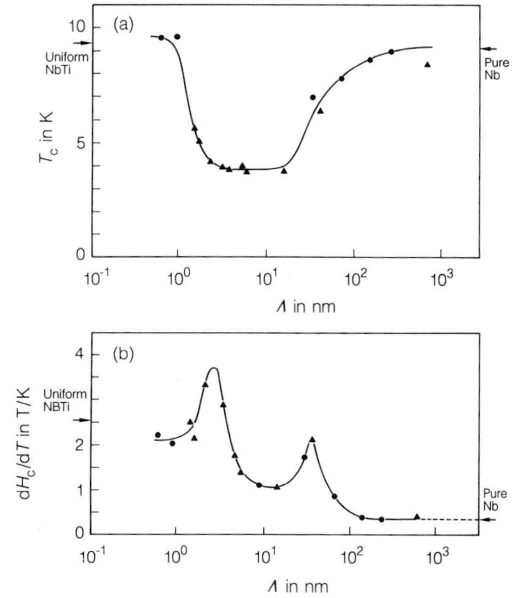

Figure 6-45. The wavelength dependence of the transition temperature of Nb/Ti superlattices: ▲, group A; ●, group B (after Qian et al., 1982). (b) The wavelength dependence of the slope of the temperature dependence of the upper-critical field for Nb/Ti superlattices; ▲, group A; ●, group B (after Qian et al., 1982).

respectively). For very short wavelengths T_c rises again due to intermixing, thus anticipating the uniform NbTi alloy (which has a T_c slightly higher than that of pure Nb).

Figure 6-45b shows the upper critical field slopes; the curve displays peaks near 2 nm and 40 nm. The rise with decreasing wavelength leading to the peak near 20 Å comes from the decreasing mean free path, as shown by the resistivity measurements of Zheng et al. (1981a and 1981b) on the same samples. The peak near 40 nm is not understood although Qian et al. (1982), have speculated on its origin.

Recently Karkut et al. (1988) have reported the observation of the phase transition predicted by Takahashi and Tachiki (1986b) and discussed in connection with Figs. 6-37 and 6-38. The system chosen was

$Nb/Nb_{0.6}Ti_{0.4}$ in which the diffusion constants differ greatly while T_c is minimally affected. The experimental results are in good agreement with the theory.

Nb/Ta

All of the metal-metal superlattice discussed previously have involved constituents that are not lattice matched. Nb and Ta on the other hand are exceedingly well matched and Durbin et al. (1981 and 1982) have grown epitaxial Nb/Ta superlattices using e-beam under UHV conditions. Since the constituents are mutually soluble there is potential for some interdiffusion; however X-ray structural studies (Durbin et al., 1981) have shown that the interdiffusion region is limited to 0.15 – 0.3 nm. Multilayers of nearly the same perfection have been grown by Hertel et al. (1982) using the sputtering technique. Different orientations have been achieved: [100] on [110] MgO; [110] on [11$\bar{2}$0] sapphire; [111] on [0001] sapphire and [211] on [10$\bar{1}$0] sapphire. An indication of the crystal quality was the observation of a residual resistance ratio ~11 for a specimen with a layer thickness of 2.5 nm; mean free paths well in excess of the layer thickness are also reported. The critical currents in sputtered Nb/Ta multilayers were measured by Broussard and Geballe (1988). J_c was found to be enhanced for the parallel fields, which is due to pinning by the multilayering similar to that observed in NbN/AlN system. The pinning force is mainly due to dislocations for larger bilayer periods, but decreases and changes over to a collective mechanism as the bilayer period decreases.

V/Ag

V/Ag multilayers were prepared by Kanoda et al. (1986) using dual e-gun deposition under UHV conditions onto Mylar substrates; samples began and ended with an Ag layer to suppress surface superconductivity. Figure 6-46 shows the temperature dependence of the parallel and perpendicular upper critical field for samples having d_V/d_{Ag} in the ratio 1/2. Kanoda et al. interpret both the A and B feature as arising from the a flux lattice which is commensurate with the superlattice. From the slope of the perpendicular critical field and the initial slope of the parallel critical field (as a function of temperature) one can deduce the coherence lengths, $\xi_{\parallel}$ and $\xi_{\perp}$, parallel and perpendicular to the layers us-

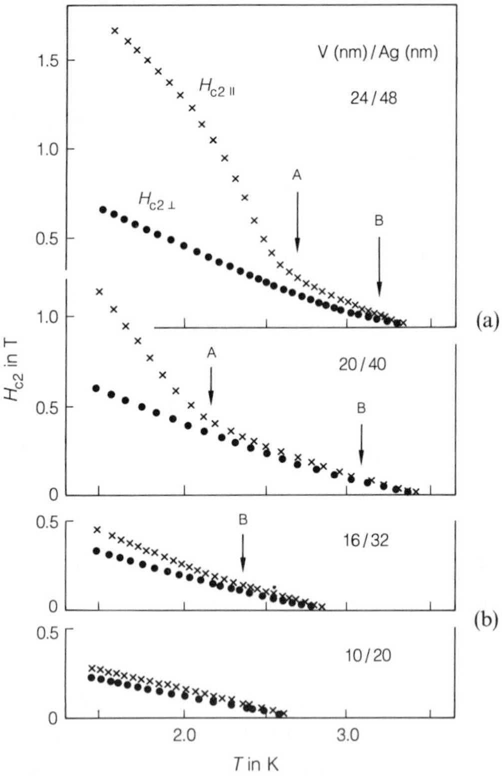

Figure 6-46. (a) Parallel and perpendicular upper-critical fields for the V/Ag superlattices. The letters A and B denote points where an upturn is observed. (b) Same as (a) but expanded to better illustrate the higher-temperature upturn (after Kanoda et al., 1986).

ing the Lawrence-Doniach theory. From these lengths they then calculate the temperature dependent height of the (isosceles) triangular flux lattice (with one long side parallel to the superlattice planes). They note that the A feature occurs when this height matches the superlattice period while the weaker B feature happens when the height is twice the superlattice period. Whether this behavior will arise from a detailed numerical solution of the Takahashi-Tachiki theory is not presently known.

In later work Kanoda et al. (1987) measured the London penetration depth of V/Ag multilayers. The temperature dependence of the penetration depth, is different from the prediction of either the BCS model in the dirty limit or the two-fluid model. A complete theory of the penetration depth in an artificial superlattice is not available at present, although there has been work along this line by Ye et al. (1990) and Buzdin et al. (1989).

Other metal-metal systems studied so far are Nb/Zr (Lowe and Geballe, 1984; Broussard et al., 1984), Nb/Al (Guimpel et al., 1986), V/Mo (Karkut et al., 1986), Pb/Bi (Raffy et al., 1972 and 1974; Raffy and Guyon, 1981; etc.). For a review see Jin and Ketterson (1986).

We note that multilayering is also a promising technique to stabilize metastable structures. For example, theoretical calculations (Pickett et al., 1981; Papaconstantopoulos et al., 1985) predict that the transition temperature of B1-MoN might be about 30 K. However, B1-MoN is a nonequilibrium phase and is difficult to synthesize by conventional methods. Kawaguchi and Shin (1990) and Lee and Ketterson (1991 a) have tried to stabilize the B1-MoN phase by multilayering with B1-TiN; the latter is an equilibrium phase and its lattice constant is approximately equal to the value predicted for B1-MoN (0.425 nm). Samples were epitaxially grown on MgO substrates by an alternate reactive deposition method. The epitaxial growth was revealed by X-ray diffraction and the satellite peak intensity profile analysis. The transition temperature of films showing B1-MoN structure is lower than 5 K, which may be related to the structural quality or the nitrogen composition.

Prior to the discovery of high T_c oxides, there was also interest at Stanford University and Northwestern in stabilizing metastable A15 structures such as Nb_3Ge and Nb_3Si, via coherency strain, by sandwiching it between a second A15 structure such as Nb_3Ir or Ti_3Au.

6.4.3.3 Superconductor/Magnetic-Metal Superlattices

Stimulated by the discovery of a ferromagnetic ($ErRh_4B_4$) and various antiferromagnetic superconductors, there has been some interest in studying superconductor/magnetic-metal superlattices. The *magnetic* metal may be 3d transition (or 4f rare earth) metals, or a normal metal containing paramagnetic impurities. If a magnetic metal is employed one may choose an element (or compound) which has a ferromagnetic, antiferromagnetic or a more complex (eg., spiral) magnetic structure. Much controversy still surrounds the correct description of a uniform magnetic superconductor, to say nothing about the problems that would be encountered in superconductor/magnetic-metal superlattice. If one takes the over simplified point of view that the magnetic effects can be parameterized via the Abrikosov-Gorkov paramagnetic spin-scattering time introduced in Eq. (6-131), then the picture is considerably simplified and one may attempt to fit experimental data using the modified Takahashi-Tachiki formalism discussed in Sec. 6.4.2.4 [the resulting time dependence of $\mathscr{H}(t)$ is ergodic]. One might alternatively model the behavior by introducing a (nonergodic) spin polarization through the parameter I. Both these procedures can cause strong pair breaking in the magnetic component. Alternatively, either pair breaking effect can be modeled using Werthamer's procedure by adding its contribution into the argument of the digamma function, as done by Hauser et al. (1966). Deutcher and de Gennes have remarked that perfect antiferromagnetic ordering should be far less destructive of superconductivity than ferromagnetic ordering (this is consistent with the fact that, in the uniform case, there are more antiferromagnetic than ferromagnetic superconductors); it was suggested that defects in the antiferromagnetic structure would be strong pair breakers.

One of the first superconductor/ferromagnet systems to be studied was V/Fe by Wong et al. (1984, 1985 a, 1985 b and 1986). Two interesting features have been observed in V_m/Fe_n superlattices regarding the competition between ferromagnetism and superconductivity; the problem has been treated theoretically by Radovic et al. (1988), and by Buzdin et al. (1991). The first feature is the anomalous superconducting T_c suppression due to ferromagnetic Fe layers. As shown in Fig. 6-47, for samples with thin V layer thickness ($m = 84$), T_c was quenched rapidly, even for a monolayer of Fe. However, for samples with V layer thicknesses larger than the coherence length ($\xi = 44$ nm for V), the quenching is smaller and an anomalous upturn occurs as the number of Fe monolayers, n, increases. This upturn was observed for different sets of samples and is unexpected within the picture of the proximity effect (Granquist and Claeson, 1979). However

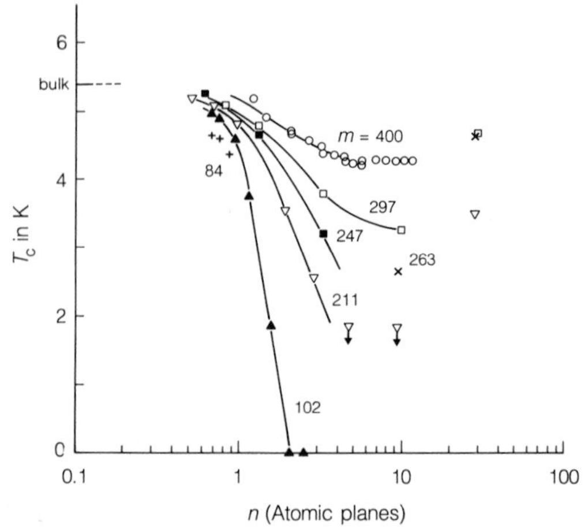

Figure 6-47. Transition temperature of V_m/Fe_n superlattices at $H = 0$. The data points with arrows represent T_c values lower than the cryostat limit. The data points with $T_c \approx 0$ K were obtained with a dilution refrigerator cryostat. The curves are not theoretical fits (after Wong and Ketterson, 1986).

Buzdin et al. (1991) have recently predicted that an oscillatory behavior of T_c is possible. The second feature is the coexistence of ferromagnetism and superconductivity in V/Fe superlattices. Figure 6-48 shows the temperature dependence of $H_{c2\parallel}$ which displays a $(1 - T/T_c)$ (3 D) behavior near the transition temperature and a $(1 - T/T_c)^{1/2}$ (2 D) behavior at lower temperatures. Since the Fe layers are ferromagnetic, a 3D superconducting behavior implies the coexistence of superconductivity and ferromagnetism in the Fe layers. This is perhaps the first direct evidence of the coexistence of these two phenomena.

The prototype, lattice-matched, b.c.c., 3D superconductor/antiferromagnetic system is V/Cr, which has been studied by Davis et al. (1988). Similar to the V/Fe system, there is a rise in the transition temperature for the relatively thick Cr layers. The origin of this effect is unclear; it might involve a coherent-incoherent transition or a reduction of spin fluctuations or spin dis-

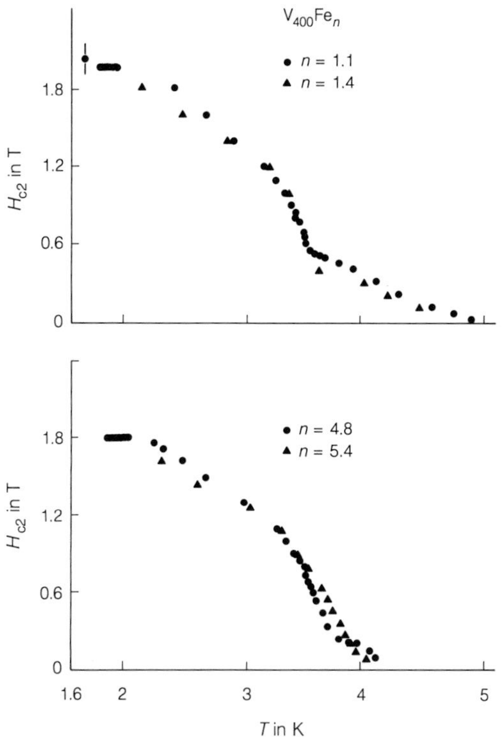

Figure 6-48. The temperature dependence of the parallel upper critical field for $V_{400}Fe_n$ superlattices (after Wong and Ketterson, 1986).

order with increasing Cr layer thickness. It may also be related to the phenomena of Buzdin et al. (1991).

The V/Ni superlattice is the structural analogue of Nb/Cu, involving mutually insoluble components and a [110] b.c.c./[111] f.c.c. stacking sequence; it has been extensively studied by Homma et al. (1986). Two symmetrical Mo/Ni superlattices with wavelengths of 1.66 and 1.38 nm were studied by Uher et al. (1986).

Greene et al. (1985a and 1985b) have prepared Nb/RE superlattices where the rare earth (RE) component was Er, Tm or Lu. Nb was selected as the superconductor because it does not form compounds with the RE elements, which should lead to sharp interfaces. The bulk ferromagnetic ordering temperatures of Er and Tm are 18 and 32 K respectively. These low Curie temperatures suggest the possibility of a competition between superconducting and magnetic ordering, the possibility of ultimately engineering superlattices exhibiting reentrant behavior and other unusual behavior seems a reasonable goal (as several groups have recognized).

Superlattice samples were prepared by magnetron sputtering onto sapphire. Characterization studies were performed using Rutherford back scattering and X-ray diffraction and we refer the reader to the original work for details.

Figure 6-49 shows the wavelength dependence of the supeconducting transition temperature of Nb/Er (magnetic) and Nb/Lu (nonmagnetic) superlattices. As expected, the transition temperature of the magnetic superlattice falls rapidly with decreasing wavelength. The authors state that a T_c reduction of the Nb layers is required to account for the data.

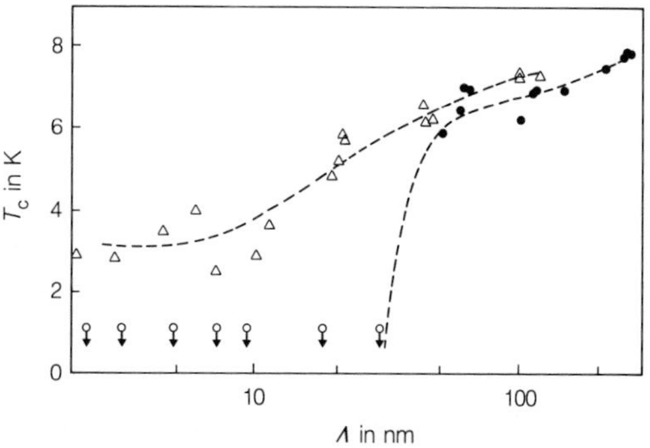

Figure 6-49. The transition temperature of Nb/Er (magnetic) (●) and Nb/Lu (non-magnetic) (△) superlattices as a function of wavelength Λ for samples grown at ambient temperature on sapphire (note the dramatic decrease in T_c of the magnetic system with decreasing wavelength): ♀, lowest temperature studied (1.8 K) and not a transition temperature; ---, a guide to the eye (after Greene et al., 1985a and 1985b).

6.4.3.4 Superconducting Transition in Ultrathin Films

A wide variety of novel and fundamental physical phenomena occur in two-dimensional superconducting films. A vortex-antivortex unbinding transition exists below the usual BCS transition (Beasley et al., 1979). In a perpendicular external field, a vortex melting transition occurs further below T_c (Huberman and Doniach, 1979). Of particular interest is the interplay between localization/interaction effects and superconductivity; the transition temperature of disordered thin films has been observed to decrease systematically as the film thickness is reduced. More specifically, Maekawa and Fukuyama (1981) studied the effects of localization/interaction on superconductivity in the 2D weak localization regime. Their perturbation theory result is

$$\ln\left(\frac{T_c}{T_{c0}}\right) = -\frac{1}{2}\beta\left[\ln\left(5.5\,\frac{\xi_0}{l}\,\frac{T_{c0}}{T_c}\right)\right]^2 - -\frac{1}{3}\beta\left[\ln\left(5.5\,\frac{\xi_0}{l}\,\frac{T_{c0}}{T_c}\right)\right]^3 \quad (6\text{-}208)$$

where

$$\beta = \frac{g_1\,N(0)\,e^2\,R_\square}{2\pi^2\,\hbar} \quad (6\text{-}209)$$

Here T_c and T_{c0} are the transition temperatures of the films with and without impurity scattering; ξ_0 is the zero temperature coherence length corresponding to T_{c0}, l is the mean free path, and g_1 is a (positive) coupling constant associated with the Coulomb interaction. The first term on the right hand side of Eq. (6-208) arises from a correction to the density of states and the second to an enhancement of the Coulomb repulsion between electrons due to impurities.

The T_c reduction with increasing sheet resistance, $R_\square$, observed in W-Re films (Raffy et al., 1983) and amorphous Mo-Ge films (Graybeal and Beasley, 1984), is consistent with Eq. (6-208) when $R_\square \leq 600\,\Omega/\square$. Beyond this $R_\square$ value, Eq. (6-208) always underestimates the transition temperature and hence it breaks down.

Superconductor/semiconductor multilayers are an ideal model system to identify mechanisms responsible for the T_c reduction in ultrathin films. The interlayer coupling can be systematically tuned by varying the semiconductor-layer thickness. The most extensively studied systems are NbTi/Ge (Jin and Ketterson, 1986) and V/Si (Kanoda et al., 1989) multilayers. A systematic enhancement over the single-film T_c was observed in $Mo_{79}Ge_{21}$/a-$Mo_{1-x}Ge_x$ multilayers (Missert and Beasley, 1989). By changing both the composition, x, and the normal-metal-layer thickness, the effect was attributed to increased diffusion between the superconducting layers, which suppresses the Coulomb interaction via a 2D$-$3D crossover.

The superconductor-insulator (S-I) transition in ultrathin films in the $T=0$ limit forms another focus of recent research. Experiments on homogeneous amorphous Bi films (Haviland et al., 1989) suggest that there is a universal threshold sheet resistance, $R_c = h/4e^2 = 6.45\ \mathrm{K}\Omega/\square$, which governs the onset of superconductivity in ultrathin films. It has been argued that for metallic ultrathin film at $T=0$, there are only two stable fixed points, $R(T=0)=0$ and $R(T=0)\rightarrow\infty$, with R_c separating the superconducting and insulating states (Pang, 1989). However, the presence of a universal sheet resistance has been questioned by experiments on other systems such as MoC (Lee and Ketterson, 1990). For this kind of disorder-tuned superconductor-insulator transitions, a realistic comparison of the experimental behavior and theoretical predictions may depend critically on the homogeneity of the films; the length scale characterizing the uniformity of disorder must likely be longer than that governing the superconducting behavior.

In addition to the disorder-tuned S-I transition, Fisher (1990) has predicted that tuning the magnetic field should induce a S-I transition of a fundamentally different nature. This has been confirmed by experiments on α-InO_x films (Hebard and Paalanen, 1990). As seen from Fig. 6-50, with increasing applied field, the slopes of the isomagnetic curves at low temperature change rapidly, which is consistent with the magnetic field-tuned S-I transition picture.

However recent experiments on thin MoC films by Lee and Ketterson (1991 b) display the conventional behavior.

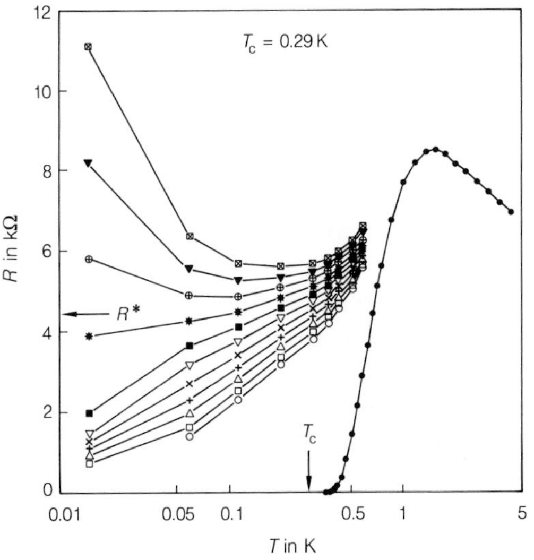

Figure 6-50. The resistive transition of α-InO_x films in zero field (●) and nonzero fields. The isomagnetic lines range from $B = 0.4\,T$(○) to $B = 0.6\,T$ (□) in 0.02-T steps. The arrows identify the critical resistance R^* and T_c, respectively (after Hebard and Paalanen, 1990).

6.5 Outlook

In this section we attempt some projections on the future evolution of thin film and multilayer science. However, having completed a relatively cautious scientific summary of various physical properties, we will be much less guarded in our outlook.

The study of thin metallic films should continue to be an area of intense scientific interest, primarily because of the curiosity about, and the lessons to be learned from, the physics of two dimensions. Of course the one dimensional limit, now addressable with high resolution lithography, is also of interest, but lies outside the scope of this review. We have addressed the transport, magnetic and superconducting properties of thin films emphasizing the two dimensional aspects. But there are many unanswered questions. As an example we have noted the uncertainty surrounding the ultimate state (or states) of a thin film material, which is superconducting in bulk, as its thickness is reduced. The electrical transport involves the strong and weak localization regimes and more work is needed on the transition between these limiting cases.

The electronic property that has the most technological relevance continues to be the magnetic behavior (although we must not ignore the obvious utility of thin film thermal sensors). Magnetoresistive (or Hall effect) field sensors are particularly important, e.g., magnetoresistive read heads, in magneticrecording applications. Surely new materials with still larger magneto resistances exist. Tailoring the magnetic anisotropy to achieve a vertical orientation of the magnetic domains is already a reality and the greater information storage density available in this configuration will spur further research. Optical recording where domain reversal is achieved by local (laser) heating of the film above the Curie temperature (typically using rare earth materials) in the presence of a locally-applied field is sure to evolve. Optical reading via the Kerr (or Faraday) effect could benefit from materials (likely involving strong spin-orbit coupling) with larger magneto-optic coefficients.

The present applications of superconducting thin films involve SQUID and other Josephson electronics and electronic interconnects. This is a universally targeted area of application for high T_c superconductors. Thick film superconducting tapes, of either conventional or high T_c materials, are relevant for applications involving magnets (conventional materials still displaying the best high field performance). Superconducting thin film memories appear to be a neglected area of study. While the competition between the exchange, field and anisotropy energies controls the size of a magnetic domain, a bit, the Abrikosov vortex is the ultimate bit of a superconducting memory. One can conceive of memories involving the presence or absence of a vortex, or oppositely paired (up, down) vortices. If the variation in the magnetic field is sensed, the bit density can approach $1/\lambda_L^2$ where the London depth, λ_L, ranges from ~ 50 nm to 1 μ, the lower limit corresponding to 4×10^{14} bits/m^2. If the normal core diameter were sensed (e.g., via tunneling microscopy) the bit density approaches $1/\xi^2$; coherence lengths range from $\sim 1-1000$ nm, the lower limit corresponding to a density of 10^{18} bits/m^2 (a rather impressive number). Possible advantages of coupled magnetic and superconducting films should not be overlooked.

Applications of metallic films (or multilayers) as coatings to enhance chemical resistance, hardness (e.g., transition metal carbides and nitrides), tribological properties, etc., are not the subject of this review but are clearly very important.

Important scientific aspects of the study of the electronic properties of artificial metallic superlattices or multilayers will continue to involve 2D − 3D dimensional cross-over phenomena (transport, magnetic, superconducting), the stabilization of metastable compounds (through heteroepitaxy), interlayer coupling effects (exchange or dipole magnetic coupling, Josephson superconducting tunneling, activated classical or quantum-mechanical normal electron tunneling, etc.). The ability to *turn on* the third dimension by adjusting the interlayer coupling in insulator/metal superlattices is only beginning to be appreciated.

The most important practical application of multilayers continues to be in X-ray optical systems (as mirrors, flat or concave). However the ability to tailor the magnetic anisotropy of a multilayer while still maintining an overall (total) film thickness sufficient to generate the needed signal is a significant advantage over a thin film magnetic recording head. Incorporating insulating layers which act as flux pinning centers in superconductor superlattices has already been demonstrated; what remains is to make macroscopic lengths of conductor and prototype high-field magnets (~ 25 T). Three and four component layering (e.g., containing superconductor/insulator/Cu stabilization/high strength alloy) could lead to the ideal conductor. Combining insulating, magnetic and superconducting constituents into three component, four sublayer superlattices should allow marrying the best characteristics of all three without deleterious interaction (e.g., Nb/Ge/Fe/Ge).

Magnetic multilayers where the magnetic properties of the individual sublayers are varied could lead to *multibit* storage at a given site: the number of layers flipped could be adjusted by the strength of the external magnetizing field.

At the time of this writing there is somewhat of a lull on activity in multilayer materials science. The reason is obvious: a sizable fraction of the thin film community has switched their activity to high T_c thin films. A measure of balance needs to be restored.

6.6 Acknowledgements

This work was supported by the NSF Materials Research Center (under DMR-85-20280) and the NSF Science and Technology Center for Superconductivity (under DMR-88-09854). We would like to thank Mrs. A. Jackson for typing the manuscript and F. L. Du for assistance.

6.7 References

Abrahams, E., Anderson, P. W., Licciardello, D. C., Ramakrishnan, T. V. (1979), *Phys. Rev. Lett. 42,* 673.

Abramowitz, M., Stegun, I. A. (1970), *Handbook of Mathematical Functions,* New York: Dover Publications.

Abrikosov, A. A., Gorkov, L. P., Dzyakoshinski, I. E. (1963), in: *Methods of Quantum Field Theory in Statistical Physics.* Englewood Cliffs, New Jersey: Prentice-Hall, Chap. 7.

Alexander, J. A. X., Orlando, T. P., Rainer, D., Tedrow, P. M. (1985), *Phys. Rev. B31,* 5811.

Allan, G. A. (1970), *Phys. Rev. B1,* 352.

Allender, D., Bray, J., Bardeen, J. (1973), *Phys. Rev. 137,* 1020.

Altshuler, B. L., Aronov, A. G. (1979), *JETP Lett. 30,* 514.

Altshuler, B. L., Khmelnitskii, D., Larkin, A. I., Lee, P. A. (1980), *Phys. Rev. B22,* 5142.

Ami, S., Maki, K. (1975), *Prog. Theor. Phys., Osaka: 53,* 1.

Anderson, P. W. (1958), *Phys. Rev. 109,* 1492.

Ando, T., Fowler, A. B., Stern, F. (1982), *Rev. Mod. Phys. 54,* 437.

Auvil, P. R., Ketterson, J. B. (1987), *J. Appl. Phys. 61,* 1957.

Auvil, P. R., Ketterson, J. B. (1988a), *Solid St. Commun. 67,* 1003.

Auvil, P. R., Ketterson, J. B. (1988b), *Superlattices and Microstructures 4,* 431.

Auvil, P. R., Ketterson, J. B., Song, S. N. (1989), *J. Low Temp. Phys. 74,* 103.

Bacon, G. E. (1975), *Neutron Diffraction, 3rd ed.,* Oxford: Clarendon Press.

Baibich, M. N., Broto, J. M., Fert, A., Nguyen van Dau, F., Petroff, F., Eitenne, P., Creuzet, G., Friederich, A., Chazelas, J. (1988), *Phys. Rev. Lett. 61,* 2472.

Ballentine, C. A., Fink, R. L., Araya-Pochet, J., Frskine, J. L. (1990), *Phys. Rev. B41,* 2631.

Baltram, F., Capasso, F., Sivco, D. L., Hutchinson, A. L., Chu, S. G., Cho, A. Y. (1990), *Phys. Rev. Lett. 64,* 3167.

Banerjee, I., Yang, Q. S., Falco, C. M., Schuller, I. K. (1982), *Solid St. Commun. 41,* 805.

Banerjee, I., Yang, Q. S., Falco, C. M., Schuller, I. K. (1983), *Phys. Rev. B28,* 5037.

Bayreuther, G., Lugert, G. (1983), *J. Mag. Mag. Mater. 35,* 50.

Beasley, M. R., Mooij, J. E., Orlando, T. P. (1979), *Phys. Rev. Lett. 42,* 1165.

Bednorz, J. G., Müller, K. A. (1986), *Z. Phys. B64,* 189.

Bergmann, G. (1984), *Phys. Reports 107,* 1.

Beuer, E., van der Merwe, J. H. (1986), *Phys. Rev. B33,* 3657.

Bhatt, R. N., Wolfle, P., Ramakrishnan, T. V. (1985), *Phys. Rev. B32,* 569.

Biagi, K. R., Kogan, V. G., Clem, J. R. (1985), *Phys. Rev. B32,* 7165.

Biagi, K. R., Clem, J. R., Kogan, V. G. (1986), *Phys. Rev. B33,* 3100.

Bishop, D. J., Dynes, R. C., Lin, B. S. J., Tsui, D. C. (1984), *Phys. Rev. B30,* 3539.

Blamenröder, S., Zirngiebl, E., Grünberg, P., Güntherodt, G. (1985), *J. Appl. Phys. 57,* 3684.

Boettger, H., Brykskin, V. V. (1985), *Hopping Conduction in Solids.* Florida: VCH, Deerfield Beach.

Bogacz, S. A., Ketterson, J. B. (1985), *J. Appl. Phys. 58,* 1935.

Bogoliubov, N. N., Tolmacher, V. V., Shirkov, D. V. (1959), *A New Method in the Theory of Superconductivity.* New York: Consultants Bureau.

den Broeder, F. J. A., Kuiper, D., van de Mosselaer, A. P., Hoving, W. (1988), *Phys. Rev. Lett. 60,* 2769.

Broussard, P. R., Geballe, T. H. (1988), *Phys. Rev. B37,* 68.

Broussard, P. R., Mael, D., Geballe, T. H. (1984), *Phys. Rev. B30,* 4055.

Buzdin, A. I., Simonov, A. J. (1990), *Physica C167,* 388.

Buzdin, A. I., Bujicic, B. U., Kuptsov, D. A. (1989), *Sov. Phys. JETP 69,* 621.

Buzdin, A. I., Kupriyanov, M. Y., Vujicic, B. (1991), *Physica C,* to be published.

Cahn, J. W., Hilliard, J. E. (1985), *J. Chem. Phys. 28,* 258.

Camley, R. E., Grimsditch, M. (1980), *Phys. Rev. B22,* 5420.

Camley, R. E., Tilley, D. R. (1988), *Phys. Rev. B37,* 3413.

Camley, R. E., Rahman, T. S., Mills, D. L. (1981), *Phys. Rev. B23,* 1226.

Camley, R. E., Rahman, T. S., Mills, D. L. (1983), *Phys. Rev. B27,* 261.

Carcia, P. F., Suna, A. (1983), *J. Appl. Phys. 54,* 2000.

Carcia, P. F., Meinhaldt, A. D., Suna, A. (1985), *Appl. Phys. Lett. 47,* 178.

Cardona, M., Güntherodt, G. (Eds.) (1982), *Light Scattering in Solids III;* (1989), *Light Scattering in Solids V.* New York: Springer-Verlag.

Cebollada, A., Martinez, J. L., Gallego, J. M., de Miguel, J. J., Mirando, R., Ferrer, S., Batallan, F., Fillion, G., Rebouillat, J. P. (1989), *Phys. Rev. B39,* 9726.

Chandrasekav, B. S. (1962), *Appl. Phys. Lett. 1,* 7.

Chang, L. L., Giessen, B. C. (1985), *Synthetic Modulated Structures.* New York: Academic Press.

Chappert, C., Bruno, V. (1988), *J. Appl. Phys. 64,* 5736.

Chappert, C., Le Dang, K., Beauvilain, P., Hurdeguint, H., Renard, D. (1986), *Phys. Rev. B31,* 92.

Chun, C. S. L., Zheng, G., Vincent, J. L., Schuller, I. K. (1984), *Phys. Rev. B29,* 4915.

Ciraci, S., Batra, P. (1986), *Phys. Rev. B33,* 4294.

Clogston, A. M. (1962), *Phys. Rev. Lett. 9,* 266.

Cohen, M. H., Douglass, D. H. Jr. (1967), *Phys. Rev. Lett. 19,* 118.

Cullity, B. D. (1972), *Introduction to Magnetic Materials.* London: Addison-Wesley.

Dabrosavljevic, L. (1973), *Phys. Stat. Sol. (b) 55,* 773.

Dabrosavljevic, L., Kulick, M. (1978), *J. Low Temp. Phys. 32,* 505.

Damon, R., Eshback, J. (1961), *J. Phys. Chem. Solids 19,* 308.

van Dau, F. N., Fert, A., Etienne, P., Baibich, M. N., Broto, J. M., Chazelaz, J., Friederich, A., Hadjoud, S., Hurdeguint, H., Radoules, J. P., Massies, J. (1988), *J. Phys. (Paris), Collog. 49,* C8-1633.

Davis, B., Zheng, J. Q., Auvil, P. R., Ketterson, J. B., Hilliard, J. E. (1988), *Superlattices and Microstructures 4,* 465.

Deutcher, G., Entrin-Wohlman, O. (1978), *Phys. Rev. B17,* 1249.

Deutcher, G., de Gennes, P. G. (1969), in: *Superconductivity:* Parks, R. D. (Ed.). New York: Marcel Dekker, p. 1005.

Diep, H. T. (1989), *Phys. Rev. B40,* 4818.

Divenere, A., Wong, H. K., Wong, G. K., Ketterson, J. B. (1985), *Superlattices and Microstructures 1,* 21.

Draaisma, H. J. G., de Jonge, W. J. M. (1987), *J. Mag. Mag. Mat. 66,* 351.

Durbin, S. M., Cunningham, J. E., Mochel, M. E., Flynn, C. P. (1981), *J. Phys. F11,* L223.

Durbin, S. M., Cunningham, J. E., Flynn, C. P. (1982), *J. Phys. F12,* L75.

Durbin, S. M., Cunningham, J. E., Flynn, C. P. (1987), *J. Phys. F17,* L59.

Dürr, W., Taborelli, M., Paul, O., Germar, R., Gudat, W., Pescia, D., Landolt, M. (1989), *Phys. Rev. Lett. 62,* 206.

Eilenberger, G. (1968), *Z. Phys. 214,* 195.

Elzain, M. E., Ellis, D. E. (1987), *J. Magn. Magn. Mater. 65,* 128.

Endoh, Y., Hosoito, N., Shinjo, T. (1983), *J. Mag. Mag. Mat. 35,* 93.

Endoh, Y. (1982), *J. de Phys. C17,* 159.

Erwin, R. W., Rhyne, J. J., Salamon, M. B., Borchers, J., Sinha, S., Du, R., Cunningham, J. E., Flynn, C. P. (1987), *Phys. Rev. B25,* 6808.

Esaki, L., Tsu, R. (1970), *IBM, J. Res. Dev. 4,* 61.

Fang, F. F., Stiles, P. J. (1968), *Phys. Rev. 174,* 823.

Feibelman, P. J., Hamann, D. R. (1984), *Phys. Rev. B29,* 6463.

Felcher, G. P., Cable, J. W., Zheng, J. Q., Ketterson, J. B., Hilliard, J. E. (1980), *J. Mag. Mag. Mat. 21,* L198.

Ferrell, R. A. (1959), *Phys. Rev. Lett. 3,* 262.

Ferrell, R. A., Prange, R. (1963), *Phys. Rev. Lett. 10,* 479.

Fisher, M. P. A. (1990), *Phys. Rev. Lett. 65,* 923.

Fleury, P. A., Loudon, R. (1968), *Phys. Rev. 166,* 514.

Fleveris, N. K., Ketterson, J. B., Hilliard, J. E. (1982), *J. Appl. Phys. 53,* 2439.

de Fontaine, D. (1966), in: *Local Atomic Arrangements Studied by X-Ray Diffraction:* Cohen, J. B., Hilliard, J. E. (Eds.). New York: Gordon and Breach, p. 51.

Freeman, A. J., Fu, C. L. (1986), in: *Magnetism at Surfaces and Interfaces:* Falicov, L. M., Moran-Loper, J. L. (Eds.). Berlin: Springer Verlag, pp. 16–23.

Freeman, A. J., Fu, C. L. (1987), *J. Appl. Phys. 61,* 3356.

Freeman, A. J., Watson, R. E. (1965), in: *Magnetism Vol. IIA:* Rado, C. T., Suhl, H. (Eds.). New York: Academic Press, p. 167.

Fuchs, K. (1938), *Proc. Cambridge Philos, Soc. 34,* 100.

Fukuyama, H. (1981), *J. Phys. Soc. Jpn. 50,* 3407.

Fulde, P., Maki, K. (1966), *Phys. Rev. 141,* 275.

Galaiko, V. P., Bezuglyi, E. V. (1973), *Sov. Phys. JETP 36,* 377.

Gay, J. G., Richter, R. (1986), *Phys. Rev. Lett. 56,* 2728.

Gavaler, J. R., Janocko, M. A., Patterson, A., Jones, C. K. (1971), *J. Appl. Phys. 42,* 54.

Geerk, J., Gurvitch, M., McWhan, D. B., Rowell, J. M. (1972), *Physica, B109–110,* 1775.

de Gennes, P. G. (1964a), *Rev. Mod. Phys. 36,* 226.

de Gennes, P. G. (1964b), *Phys. Kondens. Mat. 3,* 79.

de Gennes, P. G. (1966), *Superconductivity in Metals and Alloys.* New York: Benjamin.

de Gennes, P. G., Guyon, E. (1963), *Phys. Lett. 3,* 168.
Ginzburg, V. L. (1970), *Sov. Phys. Usp. 13,* 335.
Ginzburg, V. L., Kirzhnits, D. A. (1982), *High Temperature Superconductivity.* New York: Consultants Bureau.
Glazmana, L. I. (1989), *Sov. Phys. JETP 66,* 780.
Gradmann, U., Waller, G., Fedder, R., Tamura, E. (1983), *J. Mag. Mat. 31–34,* 883.
Granquist, C. G., Claeson, T. (1979), *Solid St. Commun. 32,* 531.
Graybeal, J. M., Beasley, M. R. (1984), *Phys. Rev. B29,* 4167.
Greene, L. H., Feldman, W. L., Rowell, J. M., Battlogg, B., Byorgy, E. M., Lowe, W. P., McWhan, D. B. (1985a), *Superlattices Microstructures 1,* 407.
Greene, L. H., Feldman, W. L., Rowell, J. M., Battlogg, B., Hull, R., McWhan, D. B. (1985b), in: *Layered Structures Epitaxy and Interfaces. Mat. Res. Soc. Symp. 37, Pittsburgh,* p. 523.
Grimsditch, M., Khan, M. R., Kueny, A., Schuller, I. K. (1983), *Phys. Rev. Lett. 51,* 498.
Grünberg, P., Mika, K. (1983), *Phys. Rev. B27,* 2955.
Grünberg, P. (1985), *J. Appl. Phys. 57,* 3673.
Grünberg, P. (1989), in: *Light Scattering in Solids V:* Cardona, M., Grüntherodt, G. (Eds.). New York: Springer Verlag, p. 303.
Guimpel, J., de la Cruz, M. E., de la Cruz, F., Fink, H. J., Laborde, O., Villegier, J. C. (1986), *J. Low Temp. Phys. 63,* 151.
Hamada, N., Terakura, K., Yanase, A. (1983), *J. Magn. Magn. Mater. 35,* 7.
Haseguwa, H. (1986), *J. Phys. F-16,* 1555.
Hauser, J. J., Theuerer, H. C., Werthamer, N. (1964), *Phys. Rev. 136,* 637.
Hauser, J. J., Theuerer, H. C., Werthamer, N. (1966), *Phys. Rev. 141,* 118.
Haviland, D. B., Liu, Y., Goldman, A. M. (1989), *Phys. Rev. Lett. 62,* 2180.
Haywood, T. W., Ast, D. G. (1978), *Phys. Rev. B18,* 225.
Hebard, A. F., Paalanen, M. A. (1990), *Phys. Rev. Lett. 65,* 927.
Hensel, J. C., Tung, R. T., Poste, J. M., Unterwald, F. C. (1985), *Phys. Rev. Lett. 54,* 1840.
Herd, S. R., Ahn, K. Y. (1979), *J. Appl. Phys. 50,* 2384.
Herman, D. A. Jr., Argyle, B. E., Petek, B. (1987), *J. Appl. Phys. 61,* 4200.
Hertel, G., McWhan, D. B., Rowell, J. M. (1982), in: *Superconductivity in d and f Band Metals.* Karlsruhe: Kernforschungszentrum, p. 497.
Hikami, S., Lakin, A. I., Nagaoka, Y. (1980), *Prog. Theor. Phys. 63,* 707.
Hillebrands, B. (1988), *Phys. Rev. B37,* 9885.
Hillebrands, B., Baumgart, P., Grüntherodt, G. (1987), *Phys. Rev. B36,* 2450.
Hilsch, P. (1962), *Phys. 167,* 511.
Hinchey, L. L., Mills, D. L. (1986), *Phys. Rev. B34,* 1689.
Hirsch, A. A., Friedman, N., Eliezer, Z. (1964), *Physica 30,* 2314.
Holstein, T., Primakoff, H. (1990), *Phys. Rev. 58,* 1098.
Homma, H., Chun, C. S. L., Zheng, G. G., Schuller, I. K. (1986), *Phys. Rev. B33,* 3562.
Hosoito, N., Kawaguchi, K., Shinjo, T., Takada, T., Endoh, Y. (1984), *J. Phys. Soc. Jpn. 53,* 2659.
Huberman, B. A., Doniach, S. (1979), *Phys. Rev. Lett. 43,* 950.
Imry, Y. (1984), in: *Percolation, Localization, and Superconductivity:* Goldman, A. M., Wolf, S. A. (Eds.). New York: Plenum, p. 189.
Ioffe, A. F., Regel, A. R. (1960), *Prog. Semicond. 4,* 237.
Ivanov, I., Pollmann, J. (1979), *Solid State Commun. 32,* 869.
Jaggi, N. K., Schwartz, L. H., Wong, H. K., Ketterson, J. B. (1985), *J. Magn. Mater 49,* 1.
Jahlborg, T., Freeman, A. J. (1980), *Phys. Rev. Lett. 45,* 653.
Jatochowski, M. (1984), *Z. Phys. B56,* 21.
Jatochowski, M., Bauer, E. (1988), *Phys. Rev. B38,* 5272.
Jin, B. Y., Ketterson, J. B. (1986), *Phys. Rev. B33,* 8797.
Jin, B. Y., Ketterson, J. B. (1989), *Adv. Phys. 38,* 189.
Jin, B. Y., Shen, Y. H., Yang, H. Q., Wong, H. K., Hilliard, J. E., Ketterson, J. B., Schuller, I. K. (1985), *J. Appl. Phys. 57,* 2543.
Jin, B. Y., Shen, Y. H., Hilliard, J. E., Ketterson, J. B. (1986), *Solid St. Commun. 58,* 189.
Jin, B. Y., Ketterson, J. B., Hilliard, J. E., McNiff, E. J. Jr., Foner, S. (1987), in: *Interfaces, Superlattices, and Thin Films. Mat. Res. Soc. Symp. 77, Pittsburg,* p. 145.
Kampwirth, R. T., Coponell, D. W., Gray, K. E., Vicens, A. (1985), *IEEE Trans. MAG-21,* 459.
Kanoda, K., Mazaki, H., Yamada, T., Hosoito, N., Shinjo, T. (1986), *Phys. Rev. B33,* 2052.
Kanoda, K., Mazaki, H., Hosoito, N., Shinjo, T. (1987), *Phys. Rev. B35,* 8413.
Kanoda, K., Mazaki, H., Mizutani, T., Hosoito, N., Shinjo, T. (1989), *Phys. Rev. B40,* 4321.
Karkut, M. G., Triscone, J. M., Ariosa, D., Fischer, O. (1986), *Phys. Rev. B34,* 4390.
Karkut, M. G., Matijasevic, V., Antognazza, L., Tirscone, J. M., Missert, N., Beasley, M. R., Fischer, O. (1988), *Phys. Rev. Lett. 60,* 1751.
Kawaguchi, K., Shin, S. (1990), *J. Appl. Phys. 67,* 921.
Kazarinov, R. F., Suris, R. A. (1972), *Sov. Phys. Semicond. 6,* 120.
Kittel, C. (1948), *Phys. Rev. 73,* 155.
Klein, M. J., Smith, R. S. (1951), *Phys. Rev. 81,* 378.
Klemm, R. A., Luther, A., Beasley, M. R. (1974), *J. Low Temp. Phys. 16,* 607.
Klemm, R. A., Luther, A., Beasley, M. R. (1975), *Phys. Rev. B12,* 877.
Knotek, M. L. (1975), *Solid State Commun. 17,* 1437.

Komori, F., Kobayashi, S., Sasaki, W. (1983), *J. Phys. Soc. Jpn. 52,* 4306.

Krishnan, R. (1985), *J. Mag. Mag. Mat. 50,* 189.

Krishnan, R., Youn, K. B., Sella, C. (1981), *J. Appl. Phys. 61,* 4073.

Krebs, J. J., Lubitz, P., Chaiken, A., Prinz, G. A. (1989), *Phys. Rev. Lett. 63,* 1645.

Krebs, J. J., Lubitz, P., Chaiken, A., Prinz, G. A. (1990), *J. Appl. Phys. 67,* 5920.

Kromer, B., Bergman, G., Bruynseraede, Y. (Eds.) (1985), *Localization, Interaction, and Transport Phenomena.* Berlin: Springer Verlag.

Kueng, A., Khan, M. K., Schuller, I. K., Grimsditch, M. (1984), *Phys. Rev. B 29,* 2879.

Kulik, I. O. (1965), *JETP Lett. 2,* 84.

Lamb, G. L. Jr. (1980), *Elements of Soliton Theory,* John Wiley & Sons.

Landau, L. D., Lifshitz, E. M. (1935), *Phys. Z. Sov. 8,* 153.

Lang, R., Nishi, K. (1985), *Appl. Phys. Lett. 45,* 98.

Layadi, A., Artman, J. O., Hoffman, R. A., Jensen, C. L., Saunders, D. A., Hall, B. O. (1990), *J. Appl. Phys. 67,* 4451.

Lawrence, W. E., Doniach, S. (1971), in: *Proceedings of the Sixteenth International Conference on Low Temperature Physics:* Kanda, E. (Ed.). Kyoto: Academic Press of Japan, p. 361.

Lebwohl, P. A., Tsu, R. (1970), *J. Appl. Phys. 41,* 2664.

Lee, S. J., Ketterson, J. B. (1990), *Phys. Rev. Lett. 64,* 3078.

Lee, S. J., Ketterson, J. B. (1991 a), to be published.

Lee, S. J., Ketterson, J. B. (1991 b), to be published.

Lee, P. A., Ramakrishnan, T. V. (1985), *Rev. Mod. Phys. 57,* 287.

Leo, J., Movaghar, B. (1988), *Phys. Rev. B38,* 8061.

Levy, P. M., Ounadjela, K., Zhang, S., Wang, Y., Sommers, C. B., Fert, A. (1990 a), *J. Appl. Phys. 67,* 5914.

Levy, P. M., Zhang, S., Fert, A. (1990 b), *Phys. Rev. Lett. 65,* 1643.

Li, C., Cai, X., Ye, Z., Zheng, H., Zheng, W., Xiong, G., Wu, K., Wang, S., Yin, D. (1986), in: *Layered Structures and Epitaxy. Materials Research Society Symposium Proceedings, 56:* Gibson, J. M. (Ed.). Pittsburg, PA: Materials Research Society, p. 177.

Li, Q., Soukoulis, C. M., Economou, E. M., Great, G. S. (1989), *Phys. Rev. B40,* 2825.

Lifshitz, E. M., Kosevich, A. K. (1953), *Dokl Akad. Mauk. USSR 91,* 795.

Lifshitz, E. M., Pitaevskii, L. P. (1980), *Statistical Physics, Part II,* Oxford: Pergamon.

Little, W. A. (1964), *Phys. Rev. A134,* 1416.

Littleton, R. K., Camley, R. E. (1986), *J. Appl. Phys. 59,* 2817.

Liu, C., Moog, E. R., Bader, S. D. (1988), *Phys. Rev. Lett. 60,* 2422.

Locquet, J. P., Sevenhaus, W., Bryunserade, Y., Homma, H., Schuller, I. K. (1987), *IEEE Trans. Magn. 23,* 1393.

Lowe, W. P., Geballe, T. H. (1984), *Phys. Rev. B29,* 4961.

Lowe, W. P., Barbee, T. W., Geballe, T. H., McWhan, D. B. (1981), *Phys. Rev. B24,* 6193.

Lucas, M. S. P. (1965), *J. Appl. Phys. 36,* 1632.

Maekawa, S., Fukuyama, H. (1981), *J. Phys. Soc. Jpn. 50,* 2516.

Majkrzak, C. F. (1986), *Physica 136 B,* 69.

Majkrzak, C. F., Axe, J. D., Böni, P. (1985), *J. Appl. Phys. 57,* 3657.

Majkrzak, C. F., Cable, J. W., Kwo, J., Hong, M., McWhan, D. B., Yafet, Y., Waszczak, J. W., Vettier, L. (1986), *Phys. Rev. Lett. 56,* 2700.

Majkrzak, C. F., Gibbs, D., Böni, P., Goldman, A. I., Kwo, J., Hong, M., Hsieh, T. C., Fleming, R. M., McWhan, D. B., Yafet, Y., Cable, J. W., Bohr, J., Grimm, H., Chien, C. L. (1988), *J. Appl. Phys. 63,* 3447.

Maki, K. (1968), *Prog. Theor. Phys., Osaka, 39,* 897.

Mattson, J., Robertson, W., Welp, V., Ketterson, J., Grimsditch, M. (1990), *Superlattices and Microstructures 7,* 47.

Mayadas, A. F., Shatzkes, M. (1970), *Phys. Rev. B1,* 1382.

McKnight, S. W., Vittoria, C. (1987), *Phys. Rev. B36,* 8574.

McWhan, D. B. (1985), in: *Synthetic Modulated Structures:* Chang, L. L., Giessen, B. C. (Eds.). New York: Academic Press, p. 43.

Mendez, E. E., Agullo-Rueda, F., Hong, J. M. (1988), *Phys. Rev. Lett. 60,* 2426.

Menon, M., Arnold, G. B. (1983), *Phys. Rev. B27,* 5508.

Menon, M., Arnold, G. B. (1985), *Superlattices and Microstruct. 1,* 451.

Middelhoek, S. (1966), *J. Appl. Phys. 37,* 1276.

Miller, D. L., Strongin, M., Kammerer, O. F. (1976), *Phys. Rev. B13,* 4834.

Millis, A., Rainer, D., Sauls, J. (1988), *Phys. Rev. B38,* 4504.

Minenko, E. V. (1983), *Sov. J. Low Temp. Phys. 9,* 535.

Missert, N., Beasley, M. R. (1989), *Phys. Rev. Lett. 63,* 672.

Moog, E. R., Bader, S. D. (1985), *Superlattices and Microstructures 1,* 543.

Morishta, T., Togami, Y., Tsuhima, K. (1986), *J. Mag. Mag. Mat. 54–57,* 789.

Mott, N. F. (1967), *Adv. Phys. 16,* 49.

Mott, N. F., Davis, E. A. (1979), *Electronic Processes in Non-Crystalline Materials,* 2nd ed. Oxford.

Mott, N. F., Twose, W. D. (1961), *Adv. Phys. 10,* 107.

Moyle, J. K., Cheung, J. T., Ong, N. P. (1987), *Phys. Rev. B35,* 5639.

Murduck, J. M., Vicent, J., Schuller, I. K., Ketterson, J. B. (1987), *J. Appl. Phys. 62,* 4216.

Murduck, J. M., Copone II, D. W., Schuller, I. K., Foner, S., Ketterson, J. B. (1988), *Appl. Phys. Lett. 52,* 504.

Nabatoskii, V. M., Shapiro, B. Y. (1981), *Solid St. Commun. 40,* 303.

Nakajima, H., Ikebe, M., Muto, Y., Fujimori, H. (1989), *J. Appl. Phys. 65,* 1637.

Nagaoka, S., Hamasaki, K., Yamashita, T., Komata, T. (1984), *Jpn. J. Appl. Phys. (1) 28,* 1367.

Nagels, P. (1985), in: *Amorphous Semiconductors:* Brodsky, M. H. (Ed.). New York: Springer Verlag, p. 113.

Nakagawa, S., Sumide, M., Kitamoto, Y., Niimura, Y., Nooe, M. (1988), *J. Appl. Phys. 63,* 2911.

Nawate, M., Tsunashima, S., Uchiyama, S. (1988), *J. Appl. Phys. 64,* 5437.

Néel, L. (1953), *Compt. Rend. 237,* 1468.

Pang, T. (1989), *Phys. Rev. Lett. 62,* 2176.

Papaconstantopoulos, D. A., Pickett, W. E., Klein, B. M., Boyer, L. L. (1985), *Phys. Rev. B31,* 752.

Parkin, S. S. P., More, N., Roche, K. P. (1990), *Phys. Rev. Lett. 64,* 2304.

Patton, C. E. (1984), *Phys. Reports 103,* 251

Pickett, W. E., Klein, B. M., Papacostantopoulos, D. A. (1981), *Physica 107B,* 667

Pollak, M. (1980), *Philos. Mag. B42,* 781.

Puszkarski, H. (1979), *Surf. Sci. 9,* 191.

Qian, Y. J., Zheng, J. Q., Sarma, B. K., Yang, H. Q., Ketterson, J. B., Hilliard, J. E. (1982), *Low Temp. Phys. 49,* 279.

Rado, G. T., Hicken, R. J. (1988), *J. Appl. Phys. 63,* 3885.

Rado, G. T., Weertman, J. R. (1959), *J. Phys. Chem. Solids 11,* 315.

Radovic, Z., Dobrosavljevic-Grujic, L., Buzdin, A. I., Clemi, J. R. (1988), B38, 2388.

Raffy, H., Guyon, E. (1981), *Physica B108,* 947.

Raffy, H., Renard, J. C., Guyon, E. (1972), *Solid St. Commun. 11,* 1679.

Raffy, H., Guyon, E., Renard, J. C. (1974), *Solid St. Commun. 14,* 427.

Raffy, H., Laibowitz, R. B., Chaudhari, P., Maekawa, S. (1983), *Phys. Rev. B28,* 6607.

Ramakrishna, B. L., Lee, C. H., Cheng, Y., Stearns, M. B. (1987), *J. Appl. Phys. 61,* 4290.

Rhyne, J. J., Erwin, R. W., Bochers, J., Sinha, S., Salamon, M. B., Du, R., Flynn, C. P. (1987), *J. Appl. Phys. 61,* 4043.

Ritchie, D. S., Fisher, M. E. (1973), *Phys. Reb. B7,* 480.

Rogers III, J. P., Cutler, P. H., Feuchtwang, T. E. (1987), *Surf. Sci. 181,* 436.

Rogers III, J. P., Nelson, J. S., Cutler, P. H., Feuchtwang, T. E. (1989), *Phys. Rev. B40,* 3638.

Ruggiero, S. T., Barbee, T. W. Jr., Beasley, M. R. (1980), *Phys. Rev. Lett. 45,* 1299.

Ruggiero, S. T., Beasley, M. R. (1985), in: *Synthetic Modulated Structures:* Chang, L. L., Giessen, B. C. (Eds.). New York: Academic Press, p. 365.

Rupp, G., Wettling, W., Jantz, W., Krishnan, R. (1985), *Appl. Phys. A37,* 73.

Saint-James, D. (1965), *Phys. Lett. 16,* 218.

Saint-James, D., de Gennes, P. G. (1963), *Phys. Lett. 7,* 306.

Saint-James, D., Sarma, G., Thomas, E. J. (1969), in: *Type II Superconductivity*. Oxford: Pergamon, Sec. 4.2–4.3.

Salamon, M. B., Sinha, S., Rhyne, J. J., Cunningham, J. E., Erwin, R. W., Borchers, J., Flynn, C. P. (1986), *Phys. Rev. Lett. 56,* 259.

Sambles, J. R. (1983), *Thin Solid Films 106,* 321.

Sato, N. (1987), *J. Appl. Phys. 61,* 1979.

Saurenbach, F. (1988), *J. Appl. Phys. 63,* 3473–3475.

Schuller, I. K. (1980), *Phys. Rev. Lett. 44,* 1597.

Schuller, I. K., Falco, C. M. (1979), in: *Inhomogeneous Superconductors, AIP Conference Proceedings No. 58:* Gubser, D. V., Francavilla, T. L., Wolf, S. A., Leibowitz, J. R. (Eds.). New York: AIP, p. 197.

Schulte, F. K. (1976), *Surf. Sci. 55,* 427.

Schwenk, D., Fishman, F., Schwabl, F. (1988), *Phys. Rev. B38,* 11618.

Segmüller, A., Blakeslee, A. E. (1973), *J. Appl. Crystallogr. 6,* 19, 413.

Sellers, C., Shiroishi, Y., Jaggi, N. K., Ketterson, J. B., Hilliard, J. E. (1986), *J. Magn. Mat. 54,* 787.

Sellers, C. H., Hilliard, J. E., Ketterson, J. B. (1990), *J. Appl. Phys. 68,* 5778.

Serene, J. W., Rainer, D. (1983), *Phys. Rep. 101,* 221.

Shan, Z. S., Sellmyer, D. J. (1990), *J. Appl. Phys. 67,* 5713.

Shan, Z. S., Sellmyer, D. J., Jaswal, S. S., Wang, Y. T., Shen, J. X. (1989), *Phys. Rev. Lett. 63,* 449.

Shirane, G., Takei, W. J. (1962), *J. Phys. Soc. Jpn. 17, Suppl. 13-111,* 35.

Shin, S. C., Hilliard, J. E., Ketterson, J. B. (1984), *J. Vac. Sci. Technol. A2,* 296.

Silvert, W. (1975), *J. Low Temp. Phys. 20,* 439.

Skocpol, W. J., Tinkham, M. (1975), *Rep. Prog. Phys. 38,* 1049.

Smit, J., Beljers, H. C. (1955), *Philips Res. Rep. 10,* 113.

Sondheimer, E. H. (1952), *Adv. Phys. 1,* 1.

Song, S. N., Ketterson, J. B. (1990), *Solid State Commun. 75,* 651.

Song, S. N., Ketterson, J. B. (1990), *Solid State Commun. 77,* 281.

Song, S. N., Auvil, P. R., Ketterson, J. B. (1987), *IEEE Trans. Mag. 23,* 1154.

Song, S. N., Li, D. X., Ketterson, J. B., Hues, S. (1989), *J. Appl. Phys. 66,* 5360.

Song, S. N., Yi, X. J., Zheng, J. Q., Zhao, Z., Tu, L. W., Wong, G. K., Ketterson, J. B. (1990), *Phys. Rev. Lett. 65,* 227.

Soffer, S. B. (1967), *J. Appl. Phys. 38,* 1710.

Spronken, G., Friedmann, A., Yelon, A. (1977), *Phys. Rev. B15,* 5141, 5151.

Strongin, M., Kammerer, O. F., Chow, J. E., Parks, R. D., Douglass, D. H. Jr., Jensen, M. A. (1968), *Phys. Rev. Lett. 21,* 1320.

Suter, R. M., Hohenemser, C. (1979), *J. Appl. Phys. 50,* 1814.

Swihart, J. C. (1961), *Phys. Rev. 32,* 461.

Szott, W., Jedrzejek, C., Kirk, W. P. (1989), *Phys. Rev. B,* 1790.

Takahashi, S., Tachiki, M. (1986a), *Phys. Rev. B33,* 4620.

Takahashi, S., Tachiki, M. (1986b), *Phys. Rev. B34,* 3162.

Takayama, H., Bohuen, K. P., Fulde, P. (1976), *Phys. Rev. B14,* 2287.

Tellier, C. R., Tosser, A. J. (1982), *Size Effects on Thin Films.* Elsevier: Amsterdam.

Tešanović, Z., Jaić, M., Maekawa, S. (1986), *Phys. Rev. Lett. 57,* 2760.

Thaler, B., Ketterson, J. B., Hilliard, J. E. (1978), *Phys. Rev. Lett. 41,* 336.

Thibaudeam, C., Caille, A. (1985), *Phys. Rev. B32,* 5907.

Thomson, J. J. (1901), *Proc. Cambridge Philos. Soc. 11,* 1120.

Tinkham, M. (1964), *Phys. Lett. 9,* 213.

Tinkham, M. (1980), *Introduction to Superconductivity.* New York: Krieger.

Tokuyasu, T., Sauls, J. A., Rainer, D. (1988), *Phys. Rev. B38,* 8823.

Tovazhnyanskii, V. L., Chekasova, V. G., Fogel, N. Y. (1987), *Sov. Phys. JETP66,* 787.

Tsu, R., Döhler, G. (1975), *Phys. Rev. B12,* 680.

Tsu, R., Esaki, L. (1973), *Appl. Phys. Lett. 22,* 562.

Uher, C., Cohn, J. L., Schuller, I. K. (1986), *Phys. Rev. B34,* 4906.

Uren, M. J., Davis, R. A., Kavah, M., Pepper, M. (1981), *J. Phys. C14,* 1395.

Vaglia, R., Cucolo, A., Falco, C. M. (1987), *Energy Res. Abstr. 12(2),* Abstr. 4271.

Vittoria, C. (1985), *Phys. Rev. B32,* 1679.

Voisin, P., Bleuse, J., Bouche, C., Gaillard, S., Alibert, C., Ragreny, A. (1988), *Phys. Rev. Lett. 61,* 1639.

Wannier, G. H. (1960). *Phys. Rev. 117,* 432.

Webman, I. J., Cohen, M. H. (1975), *Phys. Rev. B8,* 2885.

Werthamer, N. R. (1963), *Phys. Rev. 132,* 2440.

White, H., Bergmann, G. (1989), *Phys. Rev. B40,* 11594.

White, R. M., Herring, C. (1980), *Phys. Rev. B22,* 1465.

Wong, H. K., Ketterson, J. B. (1986), *J. Low Temp. Phys. 63,* 139.

Wong, H. K. (1984), Ph.D. Thesis, Northwestern University, Ann Arbor, Mich., U.S.A.

Wong, H. K., Yang, H. Q., Hilliard, J. E., Ketterson, J. B. (1985a), *J. Appl. Phys. 57,* 3660.

Wong, H. K., Jin, B. Y., Yang, H. Q., Hilliard, J. E., Ketterson, J. B. (1985b), *Superlattices and Microstructures 1,* 259.

Wong, H. K., Jin, B. Y., Yang, H. Q., Hilliard, J. E., Ketterson, J. B. (1986), *J. Low Temp. Phys. 63,* 307.

Wu, F. Y. (1982), *Rev. Mod. Phys. 54,* 235.

Wu, M. K., Ashburn, J. R., Torng, C. J., Hor, P. H., Meng, R. L., Gao, L., Huang, Z. J., Wang, T. Q., Chu, C. W. (1987), *Phys. Rev. Lett. 58,* 908.

Xue, W., Sheng, P., Chu, Q., Zhong, Z. (1989), *Phys. Rev. Lett. 63,* 2837.

Xing, D., Gong, C. (1981), *Physica B108,* 987.

Yafet, Y., Kwo, J., Hong, M., Majkrzak, C. F., O'Brien, T. (1988), *J. Appl. Phys. 63,* 3453.

Yamafuji, K., Kusayanagi, E., Iric, F. (1966), *Phys. Lett. 21,* 11.

Yang, C. N. (1952), *Phys. Rev. 85,* 808.

Yang, H. Q., Wong, H. K., Zheng, J. Q., Ketterson, J. B., Hilliard, J. E. (1984), *J. Vac. Sci. Technol. A.2,* 1.

Ye, Z., Umezawa, H., Teshima, R. (1990), *Solid State Commun. 74,* 1327.

Zheng, J. Q., Falco, C. M., Ketterson, J. B., Schuller, I. K. (1981a), *Appl. Phys. Lett. 38,* 424.

Zheng, J. Q., Ketterson, J. B., Falco, C. M., Schuller, I. K. (1981b), *Physica B–C 107,* 945; *J. Appl. Phys. 53,* 3150.

Zhou, C., Gong, C. D. (1989), *Phys. Rev. B39,* 2603.

General Reading

Atwood, D. T., Hamke, B. L. (Eds.) (1981), *Low Energy X-Ray Diagnostics, ATP Conf. Proc. No. 75,* p. 124.

Chang, L. L., Giessen, B. G. (Eds.) (1985), *Synthetic Modulated Structures.* New York: Academic Press.

de Gennes, P. G. (1966), *Superconductivity in Metals and Alloys.* New York: Benjamin.

Jin, B. Y., Ketterson, J. B. (1989), *Artificial Metallic Superlattices, Adv. Phys. 38,* 189.

Lifshitz, E. M., Pitaevskii, L. P. (1980), *Statistical Physics, Part II.* Oxford: Pergamon.

7 Fermi Surfaces in Strongly Correlated Electron Systems

Yoshichika Ōnuki

Institute of Materials Science, University of Tsukuba, Tsukuba, Ibaraki, Japan

Takenari Goto

Research Institute for Scientific Measurements, Tohoku University, Sendai, Japan

Tadao Kasuya

Department of Physics, Tohoku University, Sendai, Japan

List of Symbols and Abbreviations

a	lattice constant
$A(B)$	generalized extremal area encircled by the cyclotron motion
A_i	de Haas-van Alphen amplitude factor due to the i-th orbit
$A'(B_i)$	expansion function of $A(B)$
B	magnetic induction
c	phase velocity of light
C	specific heat
C_B	bulk modulus
C_{ij}	elastic constant
C_{Γ}^{osc}	oscillatory part of the elastic constant
e	electron charge
E_F	Fermi energy
E_{ex}	exchange splitting energy
E_{nl}, $E_{i,k}$	energy eigenstates
$E_{n,\boldsymbol{k}}$	energy of an electron at the state of $\psi_{n,\boldsymbol{k}}(\boldsymbol{r})$
F	de Haas-van Alphen frequency
F_i	F due to S_i
g_i	g-factor of the spin due to the i-th orbit
$\boldsymbol{G}_i(\boldsymbol{k})$	i-th reciprocal lattice vector
$\boldsymbol{H}$	magnetic field
H, K, L, M, N, R	points of high symmetry in the Brillouin zones
H_c	critical transition field
$\hbar$	Planck constant divided by 2π
H_{eff}	effective H including the exchange field
$\boldsymbol{J}$	current
J, J_z	total angular momentum, z-component
$\boldsymbol{k}$, k	wave vector, magnitude
k_B	Boltzmann constant
k_H	magnitude of wave vector along the field direction
$k_{1,2}$	two principal radii of an ellipsoid in the $\boldsymbol{k}$-space
k_{Fi}	Fermi vector along the three principal axes ($i=1, 2, 3$)
m_0	free electron mass
m_b, m_b	band mass, tensor
m^*	effective mass
m_c^*	cyclotron effective mass
M_{osc}	oscillatory component of magnetization
n	carrier number per unit volume of the conduction electron
$n(\boldsymbol{k})$	momentum density in the $\boldsymbol{k}$-space
n	band index
N	number of bodies in a many-body system
$N(E_F)$	density of states at E_F
$N(p_y, p_z)$	2D-angular correlation spectrum of the annihilation radiation
$N_b(E_F)$	$N(E_F)$ calculated from the band model

n_e, n_h	carrier number of electrons, holes
n_i	number of Landau levels within S_i
N_0	Avogadro number
$\boldsymbol{p}$	momentum of an electron-positron pair
p_x	momentum along the axis defined by the collinear 2γ decays
P_{nl}	wave function with the state nl in the atom
$p_{\gamma 1, \gamma 2}$	moments of an electron-positron annihilation generated phonon pair
$\boldsymbol{q}$, q	phonon wave vector, magnitude
$\boldsymbol{r}$, r, r'	radial coordinate vector, magnitude
r	number of higher harmonics at dHvA oscillation
$R(r)$	radial function
$R_{D,S,T}$	substitutional functions in the expression of A_i
S	extremal cross-sectional area of the Fermi surface
S_i	i-th extremal cross-sectional area of the Fermi surface
S_F	total area of the Fermi surface
S, T, W, X, Z	points of high symmetry in the Brillouin zones
T	absolute temperature
T_c	Curie temperature
T_D	Dingle temperature
T_K	Kondo temperature
T_N	Néel temperature
T_Q	quadrupolar ordering temperature
U_f	effective f correlation energy
$\|u_{ni}\|^2$	probability of the $\boldsymbol{k}+\boldsymbol{G}_i$ state
v	velocity of an electron
$V(r)$	external potential
V	crystal volume
$V_{xc}[\varrho(r)]$	exchange-correlation potential
$V_{eff}(r)$	effective potential
$v_{\parallel}$	v in direction perpendicular to the equi-energy surface in the $\boldsymbol{k}$-space
$v_{\perp}$	velocity component perpendicular to the Fermi surface or cyclotron orbit
v_F	Fermi velocity
$(v_F)^{-1}$	$=\langle v^{-1}\rangle_F$, average over inverse of v_F
Z	direction of the crystal axis
α	angle between the current direction and the open orbit direction in $\boldsymbol{k}$-space
β, γ	angles in the monoclinic, orthorhombic crystal structure
γ	electronic specific heat coefficient
Γ	point of high symmetry in the Brillouin zones
γ_b	electronic specific heat coefficient calculated from the band model
Δ, Λ, Σ	directions of the crystal axis
ε_{ij}	elastic strain ($i, j =$ x, y, z)
ε_B	elastic strain of bulk modulus

ε_Γ	elastic strain of special symmetry
θ	angle between field direction and symmetry or cylinder axis
θ, ϕ	angle of small deviation of an electron-positron annihilation-induced phonon pair from collinearity
$\theta(E_F - E_{n,\boldsymbol{k}})$	energy step function
λ	mass enhancement factor
λ_m	mass enhancement factor due to the electron-magnon interaction
λ_p	mass enhancement factor due to the electron-phonon interaction
μ_B	Bohr magneton
ϱ	electrical resistivity
ϱ	material density
$\varrho(\boldsymbol{p})$	momentum density in $\boldsymbol{p}$-space
$\varrho(r)$	local charge density at r
ϱ_0	residual resistivity
ϱ_{RT}	electrical resistivity at room temperature
$\Delta\varrho/\varrho$	magnetoresistance, $\Delta\varrho/\varrho = [\varrho(H) - \varrho(0)]/\varrho(0)$
$\sigma(r)$	local spin density at r
τ_i	scattering lifetime of the i-th orbit
$\phi(\boldsymbol{r})$	electron wave function
χ	magnetic susceptibility
$\chi_c(q, \omega)$	dynamical magnetic susceptibility of the conduction electron
χ_0	magnetic susceptibility extrapolated to $T=0$
$\psi_i(r)$, $\psi_{n,\boldsymbol{k}}(\boldsymbol{r})$	wave function of an electron
$\psi_+(\boldsymbol{r})$	wave function of a thermalized positron ($\boldsymbol{k}=0$)
ω_c	cyclotron frequency
$\Omega_{e,h}$	volume of ellipsoids in the $\boldsymbol{k}$-space (e: electron, h: hole)
Ω_g	electronic Grüneisen constant
$\uparrow$ (l), $\downarrow$ (c)	localized, conduction electron spin
ACAR	angular correlation of annihilation radiation
$AF_{1,2}$	antiferromagnetic states
APW	augmented plane wave
ARBIS	angle resolved BIS
ARPES	angle resolved photoemission spectroscopy
BIS	bremsstrahlung isochromat spectroscopy
dHvA	de Haas-van Alphen
f.c.c.	face centered cubic
FFT	fast Fourier transformation
HF	Hartree-Fock
IPES	inverse PES
KKR	Korringa-Kohn-Rostoker
LAPW	linearized augmented plane wave
LDF	local density functional
LDA	local density approximation
LMTO	linearized muffin-tin orbital

PES	photoemission spectroscopy
PVDF	polyvininilidene fluoride
RF	radio frequency
RKKY	Ruderman, Kittel, Kasuya and Yosida
SCR	self-consistent renormalization

7.1 Introduction

The measurement of de Haas-van Alphen effect (dHvA) was done at first for the s and p electron systems in which the detected cyclotron effective mass was small and almost the same as the band mass. The topology of the Fermi surface was explained well by the one-electron band picture. The dHvA experimental study contributed to the progress of the band calculations and played an important role in establishing the applicability of the band theory to these systems (Cracknell, 1971; Cracknell and Wong, 1973).

Next, our interest shifted to an interacting electron system based on the transition metals, to study whether or not the band theory was also applicable. Initially ferromagnetism in transition metals such as Fe and Ni was considered on the basis of the localized atomic 3d-electron model. Itinerant d electron ferromagnetism started with the Bloch theory of electron gas based on the Hartree-Fock (HF) approximation with the familiar picture of exchange-splitting between spin up and spin down bands.

Stoner (1936, 1938) studied the finite temperature properties of the itinerant electron model within the HF approximation. This type of theory for Fe, Co and Ni explained well the non-integral Bohr magneton per atom of the observed spontaneous magnetizations, their large cohesive energies and large specific heat coefficients at low temperature, etc., which were not compatible with the Heisenberg picture for the localized electron model. However, at finite temperatures the Stoner theory encountered a number of serious difficulties. It gave too high a Curie temperature T_c and too small a specific heat anomaly around T_c, and in particular it could not explain the spin wave excitations. These difficulties in the Stoner theory were, however, better understood by the Heisenberg picture.

Theoretical efforts have been concentrated on finding a way of reconciling the above mentioned mutually opposite pictures into a unified one, taking account of the effect of electron-electron correlations in the itinerant electron model. The spin wave model was proposed by Herring (1952) as a collective mode. This model explained well the experimentally observed magnon dispersion. However, the dHvA measurement in the transition metal was the most important and crucial study, showing that the band model is also applicable to this d electron system as the fundamental starting picture (Gold, 1974; Springford, 1980). The self-consistent renormalization (SCR) theory developed by Moriya (1983, 1985) to treat the spin fluctuation in the band electrons is now one of the leading theories for d electron magnetism, especially weak ferromagnetism in which the above mentioned difficulties in the Stoner model are removed.

It is, however, well known that in some compounds with strong ionicity, typically the halides and oxides, the system is usually an insulator and the 3d electrons are best treated as the atomic localized state. The superexchange interaction is the main exchange mechanism between the localized spins. The transition from the well localized 3d state to the itinerant 3d band state occurs with decreasing ionicity, the heavy chalcogenides and pnictides being situated at the boundary. In this sense it is interesting to study the dHvA effect for the compounds near the boundary. Some examples are reviewed in one of the following sections.

High T_c materials with the CuO_2 layered type structure also belong near this boundary. The dHvA effect has been detected in these materials, and angle resolved photoemission spectroscopy (ARPES) has also been done, giving some information on the Fermi surface. The topics on ARPES are reviewed briefly in the final section.

The study of rare earth compounds started practically in the 1950s with the development of a new technology to separate various rare earth elements. At first the pure rare earth metals were the main subject to be studied. Because of their strong atomic character, the 4f electrons were treated as typical examples of atomic-like localized states. The mutual magnetic interaction between the 4f electrons occupying different sites cannot be a direct one, such as in 3d metal magnetism, but should be of an indirect type, such as the so-called RKKY interaction (Ruderman and Kittel, 1954; Kasuya, 1956; Yosida, 1957) in which the c–f exchange interaction (c means the usual conduction band electron) takes a prominent position. The indirect f–f exchange interaction was determined essentially by the dynamical magnetic susceptibility for the conduction electron $\chi_c(q, \omega)$, and thus there are strong correlations between the transport and magnetic properties. The nesting character in the conduction electron band structure is typical for controlling both properties simultaneously. This was shown in the study of the heavy rare earth metals (Kasuya, 1966). Measurement of the Fermi surface was therefore very important. As expected from the atomic 4f-character, the Fermi surface of trivalent rare earth metals was similar to those of Y and La with a pair of fairly flat Fermi surfaces perpendicular to the hexagonal *c*-axis (Freeman et al., 1966).

Initially the intra-atomic s–f and/or d–f exchange interaction was considered as the main mechanism for the c–f exchange interaction (Kasuya, 1956, 1959). Actually, this is true theoretically and experimentally in the heavy rare earth metals. Later Anderson (1959) proposed a mixing model based on the intersite c–f interaction. The second order perturbation gave the same form as the intra-atomic c–f exchange model but with a negative sign indicating antiferromagnetic coupling. With this negative sign Kondo (1964) showed that the third order scattering due to the c–f exchange interaction diverges logarithmically with decreasing temperature, and clarified the origin of the long standing resistivity minimum problem which was found in Cu with a small amount of Fe impurity. This became the start of the Kondo problem, and it took ten years for theorists to solve this divergence problem at the Fermi energy (Wilson, 1975).

The many-body Kondo bound state is now understood as follows. For the simplest case of no orbital degeneracy, the localized spin $\uparrow$(l) is compensated by the conduction electron spin polarization $\downarrow$(c). Consequently the singlet state $\{\uparrow(\mathrm{l})\downarrow(\mathrm{c}) + \downarrow(\mathrm{l})\uparrow(\mathrm{c})\}$ is formed with the binding energy $k_B T_K$ relative to the magnetic state. Here T_K is called the Kondo temperature and is the single energy scale in the simple Kondo problem. The unitarity limit scattering of the phase shift at the lowest temperature comes from this virtually bound state of the conduction electrons. The real situation is of course much more complicated, but we believe that a similar type of singlet state is formed in this bound state.

In the 1960s the most typical examples for the Kondo effect were 3d impurities such as Fe and Mn, for which we were forced to study ppm ranges of dilute alloys to avoid the mutual interaction effect between the impurities (Rado and Suhl, 1973). This means that for 3d impurities the mutual interaction is much stronger than the one-site Kondo binding energy, and thus the latter is easily destroyed by the former.

In the 1970s, however, we suddenly discovered various anomalous rare earth compounds, typically Ce-compounds, in which the Kondo-like behavior was observed even in dense systems (Buschow et al., 1971; Parks, 1977; Falicov et al., 1981). In these materials, different from the 3d case, the Kondo effect seems to overcome the mutual interaction. In a purely periodic system, the ground state cannot be a scattering state but becomes a coherent Kondo lattice state.

It is well established that in the impurity Kondo state the low-energy excitation at low temperature is described by the local Fermi-liquid; a linear T dependence of the specific heat $C=\gamma T$, a T^2 dependence of the electrical resistivity $\varrho=\varrho_0-AT^2$ and the magnetic susceptibility $\chi=\chi_0-BT^2$, in which the ratios of A/γ^2 and χ_0/γ are determined from the Fermi-liquid relation. In particular χ_0/γ is called the Wilson ratio. The normal state of the coherent Kondo lattice system also shows the same Fermi-liquid behavior in the low-energy excitation at lower temperatures than T_K. Here T_K depends on the materials, ranging from 1 K to 300 K. The study of the Fermi surface in the coherent Kondo Fermi-liquid should be very interesting and crucial to understand this new kind of Fermi-liquid state. Note that because of the periodicity $\varrho_0=0$ the resistivity is given by $\varrho=AT^2$ in the pure Kondo lattice.

In the impurity Kondo system T_K is the only energy scale because of the local character. On the other hand, the situation seems to be more complicated in the Kondo lattice system because the wave vector $\boldsymbol{q}$ is added as a new parameter, and thus the Fermi surface exists as the most crucial physical observable. Here we have a famous Luttinger theorem (Luttinger, 1960). It can be said that when we start from the one-particle band model and introduce the mutual interaction, the volume of the Fermi surface does not change as far as the perturbation does not diverge and the Fermi-liquid character persists. This is the most crucial problem for the heavy fermion system. As an example, here is the case of LaSb and CeSb, which is treated in detail later.

LaSb is a compensated semi-metal with an equal carrier number of electrons and holes which is small in number, a few percent per mole. If we treat the 4f electrons as the itinerant band electrons, CeSb becomes a monovalent metal with one electron per mole, because the Ce atom has one 4f electron. Therefore CeSb should have a large Fermi surface as a monovalent metal. However, as mentioned before for the rare earth metals, if we treat the 4f electrons as the localized electrons, the Fermi surface should be the same as that of LaSb. This is a controversial issue to be discussed in the coherent Kondo Fermi-liquid.

We can scale the specific heat coefficient γ by T_K. Then γ should be proportional to T_K^{-1}. This means that $C=\gamma T_K(T/T_K)$ and γT_K is constant, for example, the order of 10^4 mJ/K · mol. When T_K is the order of the conduction band width, namely 10^4 K, γ adopts a usual value of 1 mJ/K^2 · mol. However, when T_K becomes 10 K, the γ-value increases by three orders of magnitude. This means that the mass of the coherent Kondo Fermi-liquid can be three orders of magnitude heavier than the free-electron mass. Actually, in some materials the γ-value becomes 2000 mJ/K^2 · mol. Therefore we call them heavy fermion (heavy electron) systems. One may be tempted to think that the heavy fermion with a small value of T_K belongs to the atomic 4f model, while the heavy fermion with a large value of T_K belongs to the 4f band model. This seems to be true, as treated later in detail, but remains a big puzzle.

The heavy fermion shows various kinds of phase transitions, entering into new symmetry-broken states, for example, very weak but stable magnetic ordering, gap states, and superconducting states. In particular, the anomalous superconducting state in $CeCu_2Si_2$ is attracting a lot of interest. Furthermore, the heavy fermion with a very low carrier concentration is also attracting attention because it seems to contradict the simple single Kondo problem. In the understanding of these interesting problems, the Fermi surface study is important to give the fundamental ground state.

The actinide compounds with 5f electrons now form a rapidly growing field. These compounds are particularly interesting in the sense that they bridge the gap between the 3d and the 4f compounds. Some of them show a behavior similar to the localized 4f electrons but also have features of itinerant 3d electrons. The properties of most U compounds have been understood on the basis of the 5f band model, similar to the 3d transition metals and their intermetallic compounds. The dHvA effect was measured in some of these U compounds (Arko et al., 1985). The 5f band model can explain the observed dHvA results very well. The discovery of heavy fermions in rare earth compounds encouraged the search and study of the heavy fermion states in the actinide compounds and we actually found some heavy fermion systems in the U compounds too. In particular UPt_3, UBe_{13} and URu_2Si_2 show anomalous superconductivity with unusual magnetic ordering (Kasuya and Saso, 1985; Willis et al., 1990). This coexistence of unusual superconductivity and magnetism accelerated furthermore the study of heavy fermion U compounds. In many respects, however, the heavy fermions in the U compounds are different from those in the rare earth compounds, and thus Fermi surface studies are interesting and very important. Such studies have actually been done in some of these materials and they offer interesting information. However, because of the complicated crystal structures and the incomplete observations of the heavy Fermi surfaces, the question of whether the Fermi surfaces in the U heavy fermions are explained by the usual 5f band model is not yet conclusively answered, as will be shown in more detail below.

Interest in the 5f electron systems also accelerated the study of transuranium compounds, in particular the Np, Pu and Am compounds. These are important substances in the sense that they form a connection between the U and rare earth compounds. So far, however, no detailed Fermi surface studies have been made on these materials.

In the following sections we review experimental and theoretical studies of the Fermi surfaces in transition metal, rare earth and uranium compounds, which should shed light on the basic understanding of the strongly correlated 3d, 4f and 5f electron systems.

7.2 Energy Band Structure and Mass Enhancement

7.2.1 Energy Band Structure

We will first describe briefly the electronic states in the Ni, Ce and U atoms and will then go on to present the procedure of band calculation. Usually the one-electron wave function in the atom is obtained by the Hartree-Fock approximation, which is thought to give the best result. To extend the same method to the lattice system, the local density functional (LDF) method is

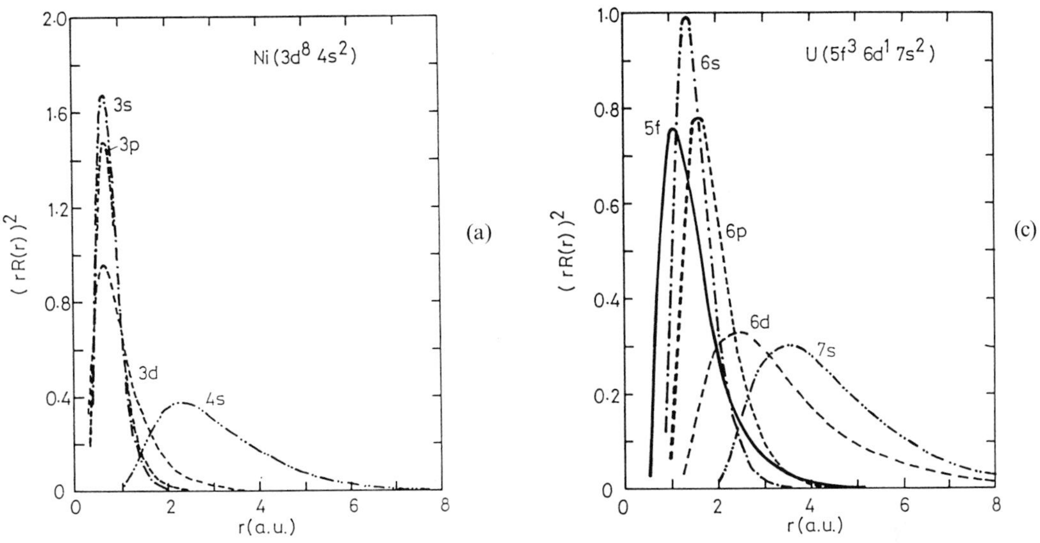

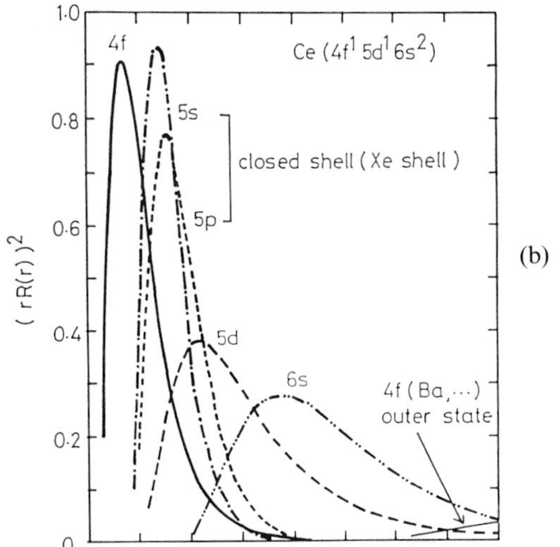

Figure 7-1. Effective radial charge densities of (a) Ni, (b) Ce and (c) U atoms (Kasuya, 1988).

much more convenient because it is now considered to be the most common approximation for the band calculation. For more details regarding band structure calculations the reader is referred to the corresponding chapter, Chap. 1 by Kübler and Eyert in this volume.

The radial function of the one-electron wave function in the atom satisfies the following differential equation:

$$\left[-\frac{d^2}{dr^2} + V_{\text{eff}}(r) + \frac{l(l+1)}{r^2}\right] \cdot P_{nl}(r) = E_{nl} P_{nl}(r) \quad (7\text{-}1)$$

where the suffix nl has its usual meaning, $n=4$ and $l=3$ in 4f systems, for example. Figures 7-1 a, b and c show the effective radial charge densities of Ni, Ce and U atoms, respectively (Kasuya, 1987). Here the radial function $R(r)$ is given by $P_{nl}(r)/r$. In the Ni atom the 3s and 3p electrons form the outermost closed shells and are relatively tightly bounded in the atom, while the 3d and 4s electrons are loosely bounded. Although the position of the 3d electrons is the same as those of the 3s and 3p shells, the 3d electron energies are higher than the 3s and 3p electron energies due to the centrifugal potential $l(l+1)/r^2$, where $l=2$ holds for 3d electrons. Therefore the 3d electrons spread widely to the outside of the 3s and 3p shells, favoring the itinerant character in the crystal.

On the other hand, the 4f electron in the Ce atom is pushed deep into the interior of the closed 5s and 5p shells because of the much stronger centrifugal potential for the 4f electron. This is the reason why the 4f electron possesses an atomic-like character in the crystal. However, the tail of its wave function spreads to the outside of the closed 5s and 5p shells, which is highly influenced by the potential energy and the distances of the surrounding atoms in the crystal. The valences of the 4f electrons in the rare earth compounds, especially in the Ce compounds, may change in different crystals. Note that the effective potential of the 4f electron has a peak around the 5d peak in energy. Therefore, the 4f wave function has two stable states, namely inside and outside of this peak. When the outside potential is lowered, the 4f wave function changes suddenly to the extremely extended state, as shown in Fig. 7-1 b. In the first stage of the valence fluctuation problem in the Ce-compounds, this sudden change was considered as the possible mechanism for the valence change.

The 5f electrons in the U atom have a character between the 3d and 4f electrons, located slightly inside the closed 6s and 6p shells. Therefore, they may possess both band-like and atomic-like characters, even in the crystal.

Next we will describe the standard band calculation. The elegant description of an interacting many electron system under the potential of the fixed nuclei is based on the density functional formalism originating from the work of Hohenberg, Kohn and Sham (Hohenberg and Kohn, 1964; Kohn and Sham, 1965). The density functional theory is built on the fundamental ansatz that the ground state properties of a system are given by the functional of the local charge density $\varrho(r)$ and the spin density $\sigma(r)$ if the ground state is magnetic. Hereafter we consider the case of a non-magnetic ground state for simplicity.

The observable static properties of the ground state are its total energy and the local charge density. To obtain these values, the self-consistent one-electron wave equation is solved,

$$\left\{-\frac{\mathrm{d}^2}{\mathrm{d}r^2} + V(r) + \int \frac{\varrho(r')}{|r-r'|}\,\mathrm{d}r' + V_{\mathrm{xc}}[\varrho(r)]\right\} \cdot \psi_i(r) = E_i\,\psi_i(r) \qquad (7\text{-}2)$$

where the first term is the kinetic energy, the second the external potential, and the third the classical Coulomb potential which includes the self-interaction. The last term represents the exchange-correlation potential which should be a complicated functional of $\varrho(r)$ and its exact form is unknown. The most essential point is how to find a reasonably good and simple form of the functional for the exchange-correlation potential. The local density functional (LDF) is a drastic approximation by which the band calculation becomes practically possible but has been proved to give sufficiently good results, at least for the usual s, p and d electrons except in some special cases. The most well-known shortcoming of LDF is to give a too narrow band gap. The reason for this is also well known. The exchange interaction is a typical example of the non-local character, acting differently for the occupied and unoccupied states, while the LDF approximation cannot treat the exchange interaction correctly. As the shortcoming of LDF is clear in origin, there are various methods to correct it. Some cases are shown in detail in Sec. 7-4. Even with the drastic simplification of LDF it is still not easy to solve the self-consistent equations for a complicated lattice and band structure. Usually the muffin-tin sphere approximation is employed. This is also a drastic simplification and there are various methods to improve this simplified model.

In order to calculate the energy band structure of rare earth and uranium compounds, relativistic effects should be taken into account because for high atomic numbers the extremely strong nuclear potential extends into the core region of the atom. In this case we use the Dirac one-electron wave equation instead of Eq. (7-2). The standard techniques to calculate the band structure self-consistently within a required accuracy are the Green's function or Korringa-Kohn-Rostoker (KKR) method, the linearized muffin-tin orbital (LMTO) method, and the augmented plane wave (APW) method or linearized APW (LAPW) method, etc. Readers can find details and the standard method for the practical calculation in papers published by Bennett and Waber (1967) and Mattheiss et al. (1968).

Here we wish to mention a recent improvement of the band calculation for the f electrons. Loucks (1967) derived a relativistic APW method for the f electrons, by which major relativistic effects such as the relativistic energy shifts, the relativistic screening and the spin-orbit interaction can be fully taken into account in the energy band structure calculation. Loucks' method is a natural extension of Slater's non-relativistic APW method to the relativistic theory. In his method, however, the symmetrization of the basic functions is not taken into account nor is the method a self-consistent one. Yamagami and Hasegawa (1990) recently improved these shortcomings by incorporating into their calculation the symmetrization of the relativistic APW bases with double space groups and using a very precise self-consistent treatment. Similar improvement is also achieved by Yanase and Harima (1990).

The spin-orbit splittings of the 4f state in the Ce atom and the 5f state in the U atom are 30 mRy and 60 mRy in magnitude, respectively. Since the magnitudes of these spin-orbit splittings are of the same order of magnitude as the hybridization energy of the 4f and 5f electrons with other electrons, the relativistic effect must be taken into account, including the spin-orbit interaction, in a self-consistent way.

It should be noted that in the many-body system the physically measurable quantity for one-electron spectroscopy is the one-particle Green's function. It described the $N \pm 1$ body spectra when one electron is added to or subtracted from the N body system, which is experimentally observed in photoemission and bremsstrahlung isochromat spectra (PES-BIS). Inverse photoemission spectrum (IPES) is also used instead of BIS. When the Fermi liquid model works, the lifetime of the quasi-particle at the Fermi energy becomes infinite with $T \rightarrow 0$, and thus well-defined quasi-particles exist at the Fermi energy, as seen by the dHvA measurement. However, there is no proof that the one-particle state calculated by the density functional method corresponds to this quasi-particle. As we have no reliable methods to calculate the one-particle Green's function in the many-body system, we have to compare the experimentally observed results with the results of the band calculation. Then the meaning of the difference between them and the improvement of the band calculation are considered.

7.2.2 Mass Enhancement

The effective mass m_c^* determined by experiments is usually different from the band mass m_b, in particular in the strongly correlated electron system. In some heavy fermions the ratio m_c^*/m_b is more than 1000.

Note that the band mass m_b is a tensor defined by $(m_b^{-1})_{ij} = \hbar^{-2}\,\partial^2 E_k/\partial k_i\,\partial k_j = \partial v_i/\partial \hbar k_j$, in which i and j correspond to x, y and z components and v is the velocity. The absolute value of the velocity v is given by $v_\parallel$ in which $\parallel$ means the direction perpendicular to the equi-energy surface in $\boldsymbol{k}$-space. The inverse mass is a measure for acceleration. The density of states is given by $N(E_F) = S_F/8\pi^3\,\hbar\,v_F$, where $(v_F)^{-1}$ is the average inverse velocity over the Fermi surface, $(v_F)^{-1} = \langle v^{-1}\rangle_F$ and S_F is its total area. There is no direct relation between the band mass and either $N(E_F)$ or the γ value in the linear specific heat term without specifying the model. For example, in the free electron model with mass m_0 one has

$$\gamma = \frac{2\pi^2}{3}\,k_B\,N(E_F) = \\ = \pi^2\,k_B^2\,m_0\,N_0\,\frac{(3\pi^2 n)^{2/3}}{\hbar^2} \qquad (7\text{-}3)$$

where N_0 is Avogadro's number and n is the carrier number per unit volume of the conduction electron. Usually, however, the ratio between the observed value γ or $N(E_F)$ and the corresponding value calculated from the band model γ_b or $N_b(E_F)$ is called the mass enhancement, $1+\lambda$. This is true only when the band is shrink uniformly. As seen from the form of $N(E_F)$, it might better be called the inverse v_F ratio, in which the area with the small velocity has an enhanced contribution to the density of states. This means that small velocity is translated into large mass.

The effective mass defined from the dHvA measurement, the so-called effective cyclotron mass m_c^*, is given as follows;

$$m_c^* = \frac{\hbar}{2\pi}\oint \frac{1}{v_\perp}\,dk = \\ = \frac{\hbar^2}{2\pi}\,\frac{\partial S}{\partial E} \qquad (7\text{-}4)$$

where S is the extremal area of the plane perpendicular to the applied magnetic field, $v_\perp$ is the velocity component in the plane which is perpendicular to the Fermi surface or the cyclotron orbit and ∂E means the change in Fermi energy on this plane. Therefore, the inverse average of the velocity component $v_\perp$ along the cyclotron orbit is observed as m_c^*. For the spherical dispersion of $\hbar^2 k^2/2m^*$, m_c^* gives m^*. Again the ratio of the experimentally determined mass m_c^* to the corresponding average band mass m_b is called mass enhancement in the same sense as with $N(E_F)$ or γ. When the band is not known, the ratio of m_c^* to the free electron mass m_0 is simply called the mass enhancement. This difference should be checked carefully.

The mass enhancement mechanism of the usual band electron is fairly well known. The most common mechanism is due to the electron-phonon interaction and this part of λ is written as λ_p (Goy and Castaing, 1973). The theoretical treatment is rather simple because of the applicability of Migdal's theorem (Migdal, 1958) which guarantees applicability of the lowest order perturbation. Note that λ_p is also a measure of the usual s-wave superconductivity and is usually less than one because a large λ_p value induces lattice instability. There are many other enhancement and reduction mechanisms, such as charge fluctuations, magnetic fluctuations and exchange interactions. Of these the magnetic fluctuations, which include spin wave, paramagnon and Kondo type fluctuation, are thought to be the most important ones. The Migdal theorem is not proved to be applicable. The mass enhancement factor due to these magnetic fluctuations, namely, λ_m can be much larger than one because there is no instability problem. Usually λ_m acts to destroy s-wave superconductivity and is believed to stabilize some kinds of non s-wave superconductivity.

For the magnon enhancement mechanism there are at least two different types. One of these is typically observed in the common rare earth compounds with well-localized 4f states. In this case, the conduction band mass is enhanced by the 4f spin fluctuation through the c–f interaction, similar to the electron-phonon case. For example, the mass enhancement of the conduction electrons in Gd due to the 4f ferromagnetic spin waves was considered many years ago (Kasuya, 1966; Nakajima, 1967). A large mass enhancement of about 4 in terms of γ/γ_b was observed in Pr (Forgan, 1981) and was explained by the magnetic excitation from the crystal field singlet ground state to the excited levels (White and Fulde, 1982; Fulde and Jensen, 1983). Note that due to the RKKY type exchange interaction between different Pr sites the crystal field levels are strongly modulated and the singlet ground state is close to the instability of magnetic order. Therefore, a strong magnetic field dependence is observed for the γ value, being reduced by as much as 25% in an applied field of 40 kOe. The cyclotron mass determined by the dHvA effect (Lonzarich, 1988) confirmed the field reduction mechanism.

Apart from the electron-phonon case, λ_m depends in some cases strongly on the magnetic field as shown for Pr. For the ferromagnetic paramagnon or spin waves, the effect of the magnetic field is a direct one and reduces λ_m strongly when the Zeeman energy exceeds the characteristic energy of the spin fluctuations. For the antiferromagnetic case, the effect is more complicated and in some cases, for example in CeB_6, as seen later, λ_m increases with increasing field because of an induced magnon softening. However, when the Zeeman energy is much stronger than the characteristic energy of the magnetic fluctuations, λ_m decreases rather rapidly with increasing field.

Another case of the magnon enhancement mechanism is found typically in the transition metals and their intermetallic compounds, where the freedom of charge transfer of the 3d electrons appears in the form of the 3d band model but the freedom of spin fluctuation of the same 3d electrons appears as spin fluctuation spectra. These two freedoms are nearly independent and so the situation is not so much different from the one considered before.

The magnetic field dependence was studied in some detail for the γ values as well as for the magnetic susceptibilities of the transition metals (Gschneidner and Ikeda, 1983). Because of the fairly large characteristic energy of the spin fluctuation, the magnetic field dependence is not clearly observed. For example, the effective cyclotron mass m_c^* in Pd is independent of magnetic fields up to 150 kOe (Joss et al., 1984), although the specific heat is reported to be reduced somewhat in magnetic fields. Probably much higher fields are necessary to reduce the cyclotron mass. Effects of spin fluctuations on the specific heat and on the magnetic susceptibility are discussed quantitatively by using the self-consistent renormalization (SCR) theory (Konno and Moriya, 1987).

The situation for the Kondo lattice or for the heavy fermion in the f electron system is more complicated and seems to be situated between the above two cases. However, even though the mass enhancement factor is very large, the former mechanism seems to be applicable for the case where the Fermi surface is described by the localized f electron model, while the latter mechanism seems to be applicable to the Fermi surface in the itinerant f electron model, as will be discussed in detail below.

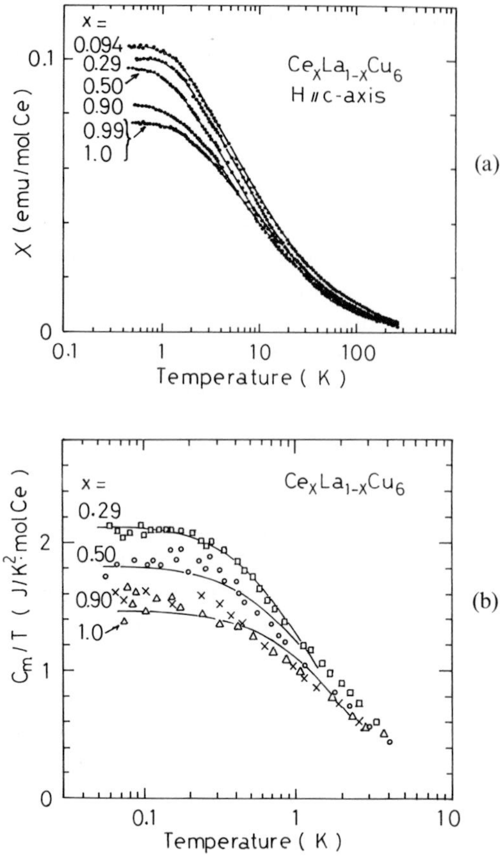

Figure 7-2. Temperature dependence of (a) the magnetic susceptibility and (b) the magnetic specific heat coefficient in $Ce_xLa_{1-x}Cu_6$ (Ōnuki and Komatsubara, 1987; Satoh et al., 1989).

A typical example of a non-magnetic Kondo lattice material is $CeCu_6$ (Ōnuki and Komatsubara, 1987; Satoh et al., 1989), as shown in Fig. 7-2. Its specific heat coefficient and magnetic susceptibility are almost as large as those in the dilute systems $Ce_xLa_{1-x}Cu_6$. These experimental results seem to suggest that in $CeCu_6$ there is a simple superposition of the Ce Kondo impurities. The Kondo lattice is, however, very different from the dilute Kondo impurity system in some other aspects. As shown in Fig. 7-3, the electrical resistivity in $CeCu_6$ increases with decreasing temperature, forms a maximum around 15 K and decreases rapidly at lower temperatures (Sumiyama et al., 1986). This behavior is in contrast to the dilute system characterized by a resistivity-minimum and a so-called unitarity limit value at the lowest temperature. The γ value of 1.6 J/K^2 · mol in $CeCu_6$ may be compared to the value 8 mJ/K^2 · mol in $LaCu_6$. Namely, the effective mass in $CeCu_6$ is two hundred times larger than the one in $LaCu_6$. A large magnetic field dependence of the γ value as well as of the cylcotron mass are also observed. This is due to the small value of the characteristic energy $T_K = 4$ K, which corresponds

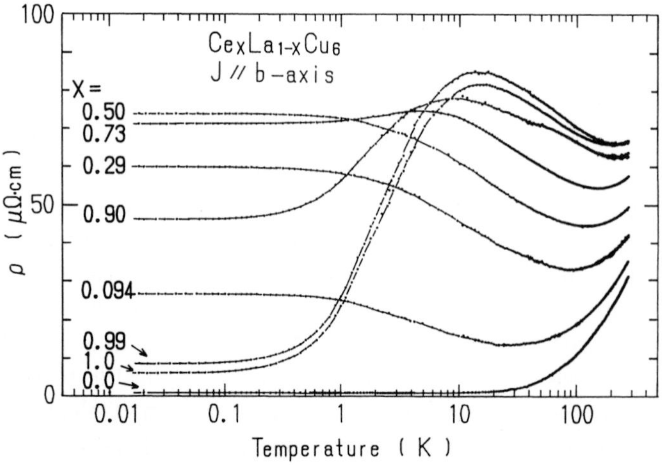

Figure 7-3. Temperature dependence of electrical resistivity in $Ce_xLa_{1-x}Cu_6$ (Sumiyama et al., 1986).

to the magnetic excitation energy of the Kondo singlet states to the magnetic states. Note that most of the Ce Kondo lattice compounds become antiferromagnetic below 10 K due to RKKY interaction, but the masses are still extremely large compared to those of the reference La-compounds.

7.3 Probes to Determine Fermi Surfaces: Transverse Magnetoresistance and de Haas-van Alphen Effect

7.3.1 Transverse Magnetoresistance

The high field transverse magnetoresistance $\Delta\varrho/\varrho = [\varrho(H) - \varrho(0)]/\varrho(0)$, in which the directions of the magnetic field and the current are perpendicular to each other, provides important information on the overall topology of the Fermi surface (Fawcett, 1964) even though the experimental technique is simple. Under the high-field condition of $\omega_c \tau \gg 1$, it is possible to know whether the sample under investigation is a compensated metal with an equal carrier number of electrons and holes, $n_e = n_h$, or an uncompensated metal, $n_e \neq n_h$, and whether the open orbit exists or not. Here, $\omega_c = eH/m_c^* c$ is the cyclotron frequency, τ the scattering lifetime, m_c^* the effective cyclotron mass defined before and $\omega_c \tau / 2\pi$ is the number of cyclotron cycles performed by the carrier without being scattered. The characteristic features of the high field magnetoresistance are summarized as follows for $\omega_c \tau \gg 1$:

(1) For a given field direction, when all of the cyclotron orbits are closed orbits, (a) for the uncompensated metal the magnetoresistance saturates ($\Delta\varrho/\varrho \sim H^0$), and (b) for the compensated metal the magnetoresistance increases quadratically ($\Delta\varrho/\varrho \sim H^2$).

(2) For a given magnetic field direction, when some of the cyclotron orbits are not closed but form open orbits, the magnetoresistance increases quadratically and depends on the current direction as $\Delta\varrho/\varrho \sim H^2 \cos^2\alpha$, where α is the angle between the current direction and the open orbit direction in $\boldsymbol{k}$-space. This is true regardless of the state of compensation.

In Fig. 7-4 we show this transverse magnetoresistance behavior for a metal with a partially cylindrical Fermi surface whose cylinder axis is in the k_z-plane and deviates by an angle α from the k_x-axis. Here, the current $\boldsymbol{J}$ is directed along the k_x-axis and the magnetic field $\boldsymbol{H}$ rotates in the k_x-plane.

If we count the number of valence electrons of various rare earth and uranium compounds in the unit cell, most of them are even in number, meaning that they are compensated metals. In this case the transverse magnetoresistance increases as H^n $(1 < n \leqq 2)$ for a general direction of the field. When the magnetoresistance is saturated for a particular field direction, often a symmetrical direction, some open orbits exist whose directions are parallel to $\boldsymbol{J} \times \boldsymbol{H}$, namely $\alpha = \pi/2$ in $\boldsymbol{k}$-space. As the magnetoresistance in the general direction is roughly equal to $(\omega_c \tau)^2$, we can estimate the $\omega_c \tau$ value.

Experimentally, the current direction is fixed to a crystal symmetry axis of the sample and the sample is slowly rotated in a constant magnetic field which is perpendicular to the current direction. The presence of open orbits is revealed by (a) spikes against a low background for the uncompensated metal and (b) dips against a large background for the compensated metal, as shown in Fig. 7-4.

Magnetic breakdown, sometimes observed in the magnetoresistance, is able to change closed orbits into open orbits, and

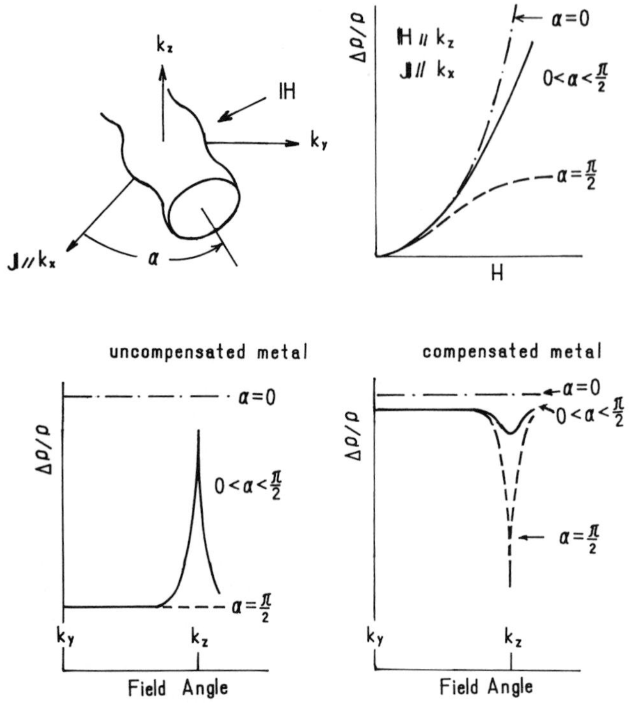

Figure 7-4. Schematic picture of the transverse magnetoresistances in uncompensated and compensated metals when a partially cylindrical Fermi surface exists. The magnetic field H rotates in the k_x-plane.

vice versa, and this gives rise to a change in the field dependence of the magnetoresistance between low and high fields.

We should be careful when applying the above general rule to the heavy fermion system. For example, a given heavy fermion material such as CeSb with a strongly anisotropic mass enhancement might show a different behavior, as will be shown later.

7.3.2 de Haas-van Alphen Effect

Under a strong magnetic field the orbital motion of the conduction electron is quantized and forms Landau levels. Therefore various physical quantities show a periodic variation with H^{-1} since increasing the field strength H causes a sharp change in the free energy of the electron system when a Landau level crosses the Fermi energy. In a three-dimensional system this sharp structure is observed at the extremal areas in the $\boldsymbol{k}$-plane, perpendicular to the field direction and enclosed by the Fermi energy because the density of states also becomes extremal. From the field and temperature dependences of various physical quantities, we can obtain the extremal area S, the cyclotron mass m_c^* and the scattering lifetime τ for this cyclotron orbit (Shoenberg, 1984). The magnetization or the magnetic susceptibility is the most common one of these physical quantities, and its periodic character is called the de Haas-van Alphen effect (1930, 1932). It provides one of the best tools for the investigation of Fermi surfaces of metals. Other physical quantities are sometimes also measured; for example, torque measurements, static strain measurements, ultrasonic velocity measurements, and magnetoresistance measurements, etc. The last type of measurement is called the Shubnikov-de Haas effect. All these techniques are also useful for determining Fermi surfaces. Ultrasonic velocity measurement is treated in the next section.

The theoretical expression for the oscillatory component of magnetization M_{osc} due to the conduction electrons was given by Lifshitz and Kosevich (1955, 1956) as follows:

$$M_{\text{osc}} = \sum_r \sum_i \frac{(-1)^r}{r^{3/2}} A_i \sin\left(\frac{2\pi r F_i}{H} + \beta_i\right) \tag{7-5}$$

$$A_i \propto H^{1/2} \left|\frac{\partial^2 S_i}{\partial k_{\text{H}}^2}\right|^{-1/2} R_{\text{T}} R_{\text{D}} R_{\text{S}} \tag{7-6}$$

$$R_{\text{T}} = \frac{\lambda r m_{ci}^* T/H}{\sinh(\lambda r m_{ci}^* T/H)} \tag{7-7}$$

$$R_{\text{D}} = \exp(-\lambda r m_{ci}^* T_{\text{D}}/H) \tag{7-8}$$

$$R_{\text{S}} = \cos(\pi g_i r m_{ci}^*/2 m_0) \tag{7-9}$$

$$\lambda = 2\pi^2 c k_{\text{B}}/e\hbar \tag{7-10}$$

Here the magnetization is periodic on $1/H$ and has a dHvA frequency F_i:

$$F_i = \frac{\hbar}{2\pi e} S_i \tag{7-11}$$

which is directly proportional to the i-th extremal (maximum or minimum) cross-sectional area S_i at zero field. The magnetization is a sum of the contributions of many extremal areas S_i ($i = 1, \ldots, n$). The amplitude factor A_i is related to the thermal damping at finite temperatures T and the Landau level broadening T_{D}. Here T_{D} is due to both the lifetime broadening and inhomogeneous broadening caused by impurities, crystalline imperfections or strains. T_{D} is called the Dingle temperature and is given by:

$$T_{\text{D}} = \frac{\hbar}{2\pi k_{\text{B}}} \frac{1}{\tau_i} \tag{7-12}$$

We can determine the effective cyclotron mass m_{ci}^* from the slope of a plot of $\ln\{A_i[1-\exp(-2\lambda m_{ci}^* T/H_0)]/T\}$ versus T at constant field H_0 by using a method of successive approximation and can obtain the Dingle temperature T_{D} or the scattering lifetime τ_i from the field dependence of the amplitude A_i.

The amplitudes of the higher harmonics ($r \geqq 2$) are small, and the fundamental one ($r = 1$) becomes dominant in the usual dHvA measurements. However, when the effective cyclotron mass is not large and the temperature becomes lower than 1 K, the higher harmonics become detectable. To distinguish the higher harmonics from the fundamental one it is necessary to check carefully the magnitude, intensity and angular dependence of the dHvA frequencies and their cyclotron masses.

When the cyclotron mass becomes large it is necessary to keep the temperature low in order to detect the dHvA signal. Even for high purity samples of rare earth compounds, a temperature of 0.4 K is required (which is attained in a He^3-cryostat; Windmillar and Ketterson, 1968) to observe the dHvA oscillation with a cyclotron mass of $10\, m_0$. For carriers with masses larger than $10\, m_0$ the temperature must be one order of magnitude lower (40 mK). The recent top-loading 20 mK/15 T cryomagnetic system (Reinders et al., 1987) is a powerful tool to detect a large mass of about $100\, m_0$ in $CeCu_6$ and UPt_3.

The quantity $|\partial^2 S/\partial k_{\text{H}}^2|^{-1/2}$ is called the curvature factor. The rapid change of the cross-sectional area around the extremal area along the field direction diminishes the dHvA amplitude for this extremal area. The neck-type orbit in LaB_6 or the nearly spherical Fermi surface with bumps in $CeSn_3$ is discussed from this point of view in the next chapter.

The term $\cos(\pi g_i r m_{ci}^*/2 m_0)$ is called the spin factor. When $g_i = 2$ (free electron value) and $m_c^* = 0.5\, m_0$, this term becomes zero for $r = 1$, and the fundamental oscillation and the dHvA oscillation vanish for all values of the field. This is called the zero spin-splitting situation in which the up and down spin contributions to the oscillation cancel out, and this can be useful for determining the value of g_i. Note that in this situation the second harmonics for $r=2$ should have a full amplitude.

Many rare earth or uranium compounds show a magnetically ordered state at low temperatures. Most of the Ce Kondo lattice compounds become antiferromagnetic around 5 K. Conduction electrons in this system have different Zeeman and exchange energies depending on the up and down spin states. However, when the extremal area changes linearly with increasing external field, their dHvA frequencies have the same value, so giving the extremal area for zero field. The antiferromagnetic AF_1 state of these compounds often changes into a different antiferromagnetic AF_2 state or into the field-induced ferromagnetic (paramagnetic) state. In these cases, we usually get different Fermi surface areas for the up and down spin states when the field is increased to above the transition field H_c, as shown in Fig. 7-5. The spin factor becomes

$$R_S = \cos\left[\frac{\pi r m_{ci}^*}{2m_0}\left(g_i + \frac{H_{eff}}{H}\right)\right] \quad (7\text{-}13)$$

where H_{eff} is defined by the exchange splitting energy $E_{ex} = \mu_B H_{eff}$. In ferromagnetic systems it is possible to obtain information of the different Fermi surface areas associated with the up and down spin states in zero field.

For antiferromagnetic ordering, in general the size of the Brillouin zone is reduced to a smaller Brillouin zone based on the magnetic unit cell, or gaps appear on the Fermi surfaces in the corresponding paramagnetic Brillouin zone. Usually these new gaps, which are induced by the magnetic order, are small compared to the already existing gaps and thus magnetic breakdown occurs rather easily in weak magnetic fields. In this case the paramagnetic Brillouin zone is representative even for the antiferromagnetically ordered state. This should be carefully checked, however, by means of the magnetic field dependence of the dHvA oscillation.

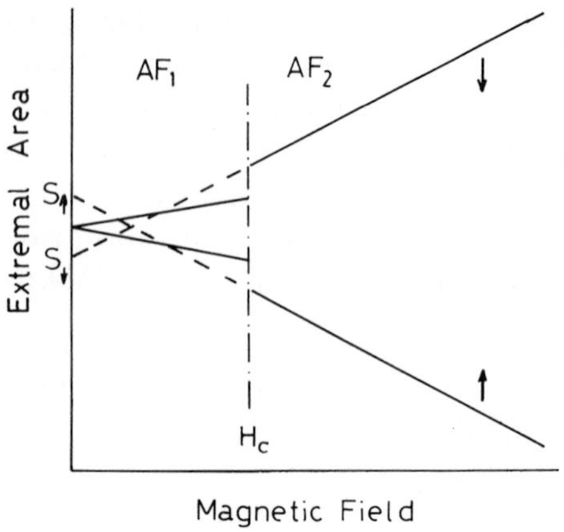

Figure 7-5. Schematic picture of the change in extremal area depending on the up and down spin states. AF_1 and AF_2 mean different antiferromagnetic states, and H_c is the spin-flop field. $S\uparrow$ and $S\downarrow$ are the extremal areas for the up and down spin electrons, respectively, obtained from the dHvA measurement in the AF_2 region. (Harima, 1988).

7.3.3 Acoustic de Haas-van Alphen Effect

The acoustic dHvA effect, namely the quantum oscillation of the sound velocity of the ultrasonic wave, is also an excellent probe for the investigation of the Fermi surface. Sound velocity measurements with the pulse echo method are particularly recommended because of the high resolution of the sound velocity change in a field. As this method is a new technique to detect the dHvA oscillation we will introduce it here in some detail (Heil et al., 1984; Goto et al., 1989).

In Fig. 7-6 a block diagram is shown of an ultrasonic apparatus for sound velocity measurements based on the phase comparison method. The sine wave of the synthesized signal generator is divided into two different paths, namely the ultrasonic driving signal and the reference signal for the phase detector. The phase difference between the reference signal and the n-th pulse echo signal, $\phi_n = 2\pi(2n-1)lf/v$, is converted into a dc signal by the phase detector. Here $(2n-1)l$ is the effective sample length for the n-th echo, f is the frequency of the signal generator for the sound wave in the frequency range of 10 MHz to 1 GHz and v is the velocity of sound. In the apparatus shown in Fig. 7-6, the dc signal of the phase detector is given as a feedback into the signal generator to keep a constant value of the phase difference. The relative change of the sound velocity as a function of field strength corresponds to the change of the frequency because the relation $\Delta f/f = \Delta v/v$ is satisfied. As the wave vector of the ultrasound is kept constant, this measurement condition implies a constant k method. The relative change of the sound velocity in a field is detected with a resolution better than 10^{-7} even if field modulation is not applied. Figure 7-7 shows a typical example of an acoustic dHvA oscillation for the longitudinal C_{11} mode in YCu_2 (Settai et al., 1991).

In practice the piezoelectric transducer of $LiNbO_3$ with a large electromechanical coupling coefficient provides an improvement in S/N ratio compared to the conventional quartz transducers. For the high frequency ultrasonic experiments up to

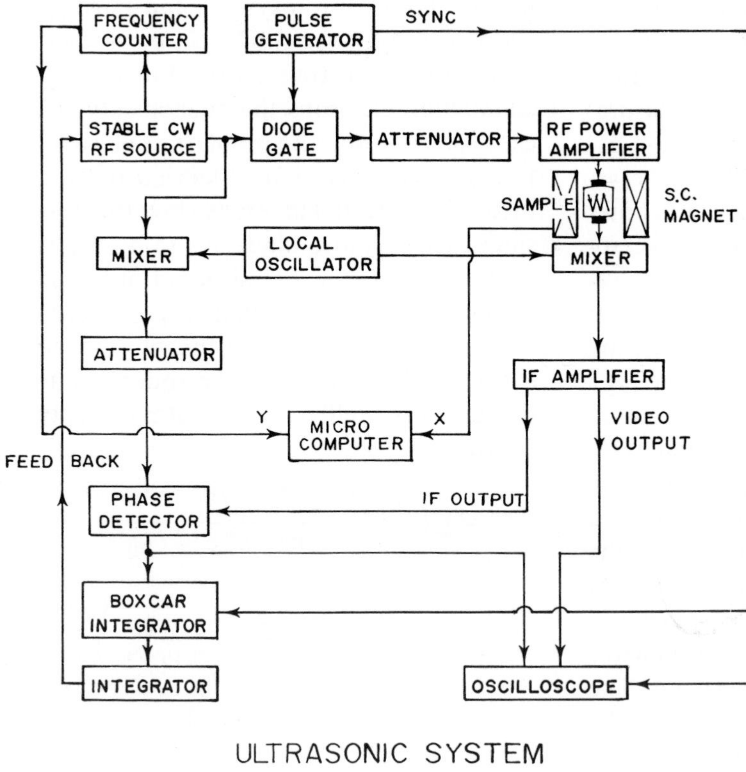

Figure 7-6. Block diagram to measure the sound velocity based on the phase comparison method (Goto et al., 1989). The high resolution of the sound velocity change in magnetic fields leads to the observation of the acoustic dHvA effect.

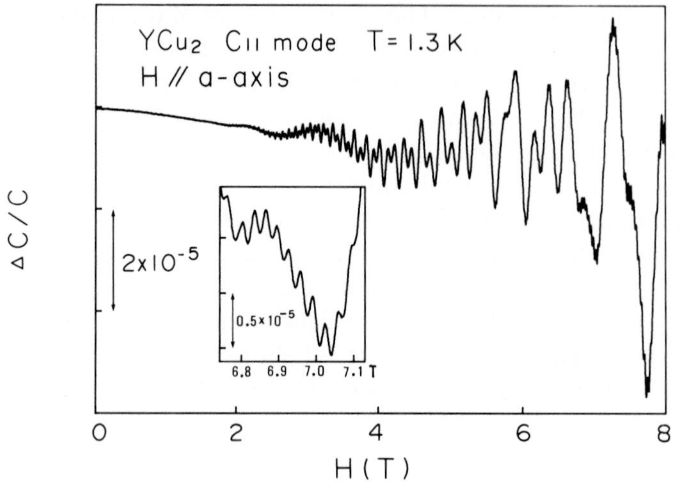

Figure 7-7. Acoustic dHvA oscillations associated with the longitudinal C_{11} mode in YCu_2 (Settai et al., 1991). An enlarged picture of part of the oscillations is given in the insert.

1 GHz, a piezoelectric semiconductor film of ZnO or of the polymer PVDF (polyvininilidene fluoride) is recommended for the ultrasonic transducers. The introduction of transducers with high efficiency is important for the acoustic dHvA effect of the heavy fermion compounds in dilution refrigerators because generation of thermal energy should be kept to a minimum. Actually a small input power of less than 0.1 μW was realized in the real measurement.

The elastic stiffness constant of the crystal is defined as $C = \varrho v^2$. Here ϱ denotes the density of the material and v is the appropriate sound velocity. In a cubic crystal the elastic constant C_{11} corresponds to the longitudinal sound wave propagating along the [100] direction. The $(C_{11} - C_{12})/2$ mode is obtained by the transverse sound wave of the [110] propagation and $[1\bar{1}0]$ polarization. C_{44} is also related to the transverse mode of [001] propagation and [010] polarization. It is convenient that the transverse $(C_{11} - C_{12})/2$ and C_{44} modes are responsible for the pure symmetric elastic strain $\varepsilon_{xx} - \varepsilon_{yy}$ with Γ_3 symmetry and $(\varepsilon_{yz}, \varepsilon_{zx}, \varepsilon_{xy})$ with Γ_5, respectively. The C_{11} mode is the combination of the $(C_{11} - C_{12})/2$ mode associated with the strain $(2\varepsilon_{zz} - \varepsilon_{xx} - \varepsilon_{yy})/\sqrt{3}$ and the bulk modulus $C_B = (C_{11} + 2C_{12})/3$ with the strain $\varepsilon_B = \varepsilon_{xx} + \varepsilon_{yy} + \varepsilon_{zz}$ of Γ_1 symmetry.

When the f electrons are well described by localized atomic states, the interaction between the acoustic wave and the f electron system is given by the quadrupolar field of the f electron system, and thus the effect is given by the quadrupolar susceptibility of the f electron system. For normal conduction electrons the phenomenological Lifshitz-Kosevich formula is applicable. For the heavy fermion system no reliable theoretical treatment exists, but the Lifshitz-Kosevich formula seems to be applicable as far as the Fermi liquid model is applicable. Assuming that the extremal cross-sectional area S_i is a function of the elastic strain induced by the sound wave, one gets the oscillatory part of the elastic constant:

$$C_\Gamma^{\mathrm{osc}} = \sum_i \left(\frac{e}{c\hbar}\right)^{3/2} \frac{2^{1/2} e\hbar H^{1/2}}{\pi^{5/2} m_{ci}^* c} \left|\frac{\partial^2 S_i}{\partial k_H^2}\right|^{-1/2} \cdot R_T R_D R_S \cdot \left[\left(\frac{\partial F_i}{\partial \varepsilon_\Gamma}\right)^2 \cos\left(\frac{2\pi F_i}{H} + \beta_i\right) + + \frac{H}{2\pi}\left(\frac{\partial^2 F_i}{\partial \varepsilon_\Gamma^2}\right) \sin\left(\frac{2\pi F_i}{H} + \beta_i\right)\right] \tag{7-14}$$

where the higher harmonics are neglected and R_T, R_D and R_S are defined in the previous section. In Eq. (7-14) there are two oscillatory terms which resemble the Curie and the van Vleck terms in the strain susceptibility for a localized 4f electron system. The first term, which is proportional to the square of the first derivative of the area F_i with respect to ε_Γ, is usually the dominant oscillatory term. Note that $F_i = \hbar S_i / 2\pi e$ is the area which is converted to the field strength and $F_i/H = n_i$ is the number of Landau levels within the extremal area. Therefore, roughly speaking, the ratio of the first and the second terms in Eq. (7-14) is of the order of n_i. The ratio becomes smaller for a small value of S_i and with increasing field strength. Furthermore, the first derivative $\partial F_i / \partial \varepsilon_\Gamma$ frequently vanishes as a result of symmetry for a special ε_Γ. Then the second derivative should be the leading term. A detailed treatment is given in the section dealing with LaB_6 in which the acoustic dHvA measurement plays an important role.

The energy of the band electron circulating in the appropriate orbit is shifted as a function of the elastic strain induced by the sound wave. Therefore these processes are essentially related to the electron-phonon coupling constant. As the acoustic dHvA oscillation of the elastic constant is measured as a function of field without a field modulation technique it is easy to determine the absolute value of the oscillatory amplitude. In practice, the fast Fourier transformation (FFT) is used to obtain the dHvA frequencies and their amplitudes from the oscillation. The temperature and field dependencies of the oscillation provide the effective mass m_c^* and the Dingle temperature T_D of the sample. When one has knowledge of the reduction factor R_T from the effective mass m_c^* and also R_D from T_D, one can derive the degree of deformation potential of the appropriate orbit.

Many of the heavy fermion compounds are characterized by an extremely large electronic Grüneisen constant $\Omega_g = 100$ (Lüthi and Yoshizawa, 1987), which is contrasted by a normal metal such as Cu with $\Omega_g = 1$. Such a large Grüneisen parameter originates from the strain modulation effect on the narrow band of the heavy fermion state. This coupling mechanism is related to various anomalous behaviors in the thermal expansion, elastic constant and ultrasonic attenuation. Observation of an oscillatory amplitude in the acoustic dHvA effect may clarify the details of the large Grüneisen coupling parameter.

7.3.4 Sample Preparation

The techniques most commonly used to obtain large usable single crystals (congruent and peritectic compounds) are (1) the RF floating zone technique, (2) the RF horizontal zone melting method, (3) the Czochralski pulling method and (4) the Bridgman technique. The review written by Abell (1989) gives very useful details for growing single crystals.

The first method was applied to grow the high purity RB_6 single crystals (Tanaka et al., 1975). There is no crucible available to hold molten RB_6 because of the high melting point (about 2400 °C) and the high reactivity. So single crystals of RB_6 were grown by the RF floating zone technique under a pressurized argon gas atmosphere of about 15–30 kg/cm^2, which prevents vaporization and dissociation of RB_6. Impurities have been qualitatively analyzed by emission spectrography. Of the impurities Co, Si, Mn, Cr, Fe, Mo, Ti, Zr and Ni in the initial raw material, only Si, Mg and Fe were detected in the crystals that had passed the floating zone once. For the triply passed crystals no impurities were detected by emission spectrography. The corresponding residual resistivity ratio $\varrho_{RT}/\varrho_{4.2K}$ was about 20 for the singly passed crystals and 200–450 for the triply passed crystals. A typical size of the single crystal ingot was 7 mm in diameter and 60 mm in length.

The horizontal zone melting technique was applied to grow UPt_3 and $CeRu_2Si_2$ single crystals, which were prepared under an ultra high vacuum of 10^{-9}–10^{-10} torr by zone-melting a rod of the compound, contained in a water-cooled copper crucible, with RF heating (Lonzarich, 1988). The heating temperature attained in this method is below 2000 °C. The $\varrho_{RT}/\varrho_{1.2K}$ ratio of UPt_3 was higher than 400. Single crystals 0.5 mm thick and having diameters of 3.0 mm were cut from the purest parts of the ingots. This technique was also applied to obtain single crystals of transition metal intermetallic compounds Ni_3Ga and MnSi.

The third method uses a W-crucible or a water-cooled copper crucible. The former was used to grow single crystals of most rare earth compounds such as RIn_3, RAl_2, RCu_2, RGa_2, RCu_6 and RSn_3 under pressurized argon or in a helium gas atmosphere (Ōnuki et al., 1984). The residual resistivities were about 0.5 μΩcm or less for these compounds, and 0.1 μΩcm for $CeSn_3$ (Umehara et al., 1990). Typical samples usually had a diameter of 7–10 mm and a length of 100 mm. Furthermore, the CeNi and LaNi single crystal rods grown by this method were purified by the solid-state electro-transport method at pressures of 10^{-9} to 10^{-10} torr (Maezawa et al., 1989). The residual resistivities of CeNi and LaNi were reduced to 0.2 and 0.1 μΩcm, respectively. In the latter crucible, the so called "tri-arc" Czochralski pulling method is most commonly used (Menovsky and Franse, 1983). There is no temperature limit in principle, although tungsten, used as the torch, is inserted into the melt at temperatures higher than 2000 °C. This procedure is suitable for highly reactive 3d, 4f and 5f intermetallic compounds such as UPt_3, URu_2Si_2, UB_{12} and Ho_2Co_{17} that have rather low vapor pressures at the melting temperatures. The quality of the "tri-arc" grown crystal is comparatively low due to the large temperature gradient in the melt and the non-uniform manner of heating.

The last method, the Bridgman technique, was applied to rare earth compounds with high vapor pressure (Spirlet and Vogt, 1982). The starting materials, which are sealed in the W or Mo container, are passed through a large temperature gradient. This method was used to grow single crystals of CeSb and CeBi having a melting point of 2400 °C (Kitazawa et al., 1983).

There are several other methods, such as the electron-beam melting technique, which are used for obtaining UIr_3 and URh_3 (Arko et al., 1975). Flux growth in a Bi solution was used for UGe_3 (Arko and Koelling, 1978) and an iodine vapor transport was used for preparing U_3As_4 and U_3P_4 (Henkie and Markowski, 1977).

7.4 Experimental Results and Comparisons with Band Calculations

7.4.1 Transition Intermetallic Compounds – 3d Systems

The dHvA measurements for the monoatomic metals, such as the transition metal Fe and the rare earth metal Gd, were done in the 1960s because it is not very difficult to grow good single crystals from pure metals. It is, however, very difficult to grow good single crystals of intermetallic compounds. Here the precise control of stoichiometry is a delicate and difficult problem. The first dHvA measurement on a magnetic compound was done by the Hirahara group (Ohbayashi et al., 1976) on MnP after many years' study of this material.

At the lowest temperature and for the magnetic field applied parallel to the b-axis, MnP orders as a screw structure up to 6 kOe and then changes to a fan structure up to 40 kOe, above which simple ferromagnetic order exists. Figure 7-8 shows a typical dHvA oscillation in the fan phase. The induced magnetic moment along the b-axis does not show a simple linear dependence on the field strength. Consequently the extremal area in the Fermi surface varies in a complicated way with the magnetic induction B or the external magnetic field H. In this situation, the usual plots of B^{-1} vs. positive integer n of the dHvA oscillation may be meaningful only in small local regions of the field, and the effective extremal areas obtained by such analysis depend on B, as illustrated in Fig. 7-9 (Kasuya, 1976). Here the generalized extremal area $A(B)$ encircled by the cyclotron motion of the conduction electrons is quantized via the relation

$$\frac{A(B)}{B} = n + n_0 \tag{7-15}$$

where n_0 is a constant. $A(B)$ may be expanded as

$$A(B) = A(B_i) + A'(B_i)(B - B_i) \tag{7-16}$$

and Eq. (7-15) can be rewritten by using Eq. (7-16) as

$$\frac{A(B_i) - A'(B_i)B_i}{B} = n + n_0 - A'(B_i) \tag{7-17}$$

For $A(B_i) = A'(B_i)B_i$, the expansion of the extremal area is just equal to the movement of the Landau level, and thus no dHvA oscillation is observed. For $A(B_i) > A'(B_i)B_i$, the usual dHvA oscillation is realized in which the Landau levels cross the Fermi energy from the lower energy side with increasing field strength. On the other hand, for $A(B_i) < A'(B_i)B_i$, the expansion of the extremal area with field strength is stronger than the expansion of the Landau levels, and thus the number of Landau levels in the extremal area increases with in-

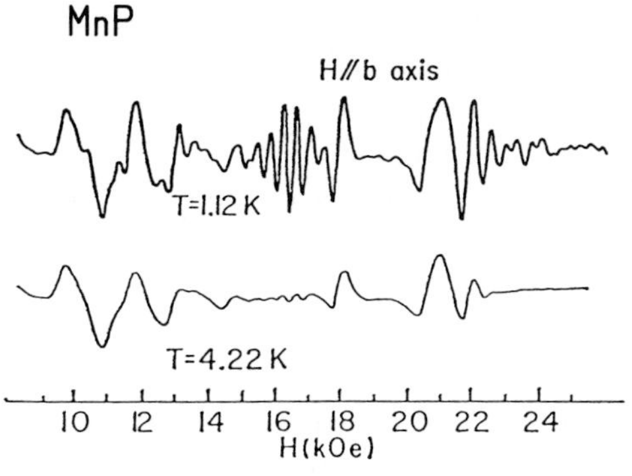

Figure 7-8. dHvA oscillations in the fan phase of MnP (Ohbayashi et al., 1976).

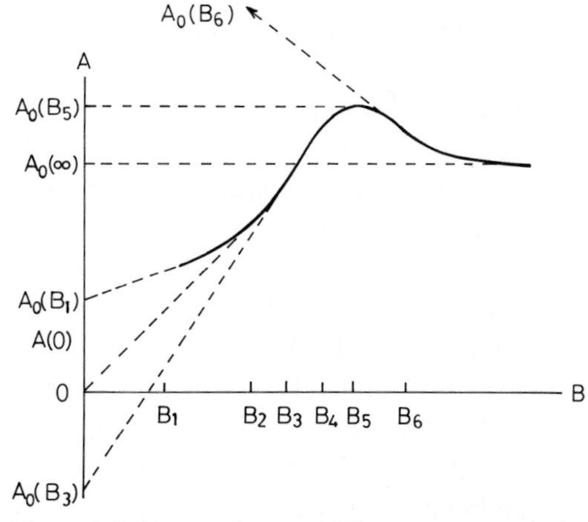

Figure 7-9. Extremal area $A(B)$ versus magnetic induction B (solid curve). The broken lines indicate the tangential lines at B_n, $n = 1$ to 6. $A_0(B_n)$ is the observed effective area at B_n determined by the local dHvA measurement (Kasuya, 1976).

creasing field strength. In this case the effective extremal area determined by the B^{-1} vs. n relation is given by $|A(B_i) - A'(B_i)\,B_i|$. The antiferromagnetic case discussed in the previous chapter is a simple example of the general case discussed above.

The complicated dHvA behavior observed in MnP indicates that the above described situation can be realized. As mentioned before, MnP is an interesting material in the sense that it seems to form the boundary between the itinerant and localized 3d electron models. More precise experimental studies are needed to determine its Fermi surface.

In the following sections we show the Fermi surface properties of three materials that can be described by the 3d band model.

7.4.1.1 Ni_3Ga and Ni_3Al

Ni_3Ga and Ni_3Al crystallize in the simple-cubic Cu_3Au type structure. The former substance is a strongly exchange-enhanced paramagnet, while the latter is a weak ferromagnet with a Curie temperature of 41 K. The electronic specific heat coefficients of Ni_3Ga and Ni_3Al are 24 mJ/K^2 · mol and 26 mJ/K^2 · mol, respectively (de Dood and de Chatel, 1973).

Figure 7-10 shows the angular dependence of the dHvA frequency in Ni_3Ga (Lonzarich, 1984; Hayden and Lonzarich, 1986). The solid lines in the figure are theoretical results obtained from band calculations. Note that the lines connecting the data are only guides. The electronic energy band structure has been calculated by means of the LMTO technique for the local exchange correlation potential. The self-consistent calculations are based on a lattice constant of $a = 3.575$ Å for the Ni_3Ga unit cell, and include s, p and d partial waves and corrections to the atomic sphere approximation. Four bands which cross the Fermi level are of predominant d character, and the corresponding four hole sheets are centered at the Γ point, as shown in Fig. 7-11.

The two lowest branches α and γ have masses in the range 0.4 m_0 to 0.7 m_0 and are nearly isotropic, which is consistent with two nearly spherical pockets of the Fermi surface in bands 14 and 15, respectively. The ζ branch, which has a mass of 5.5 m_0 at ⟨100⟩, is observed within an angular range of 30° around ⟨100⟩ and its dHvA frequency varies roughly as $1/\cos\theta$, as expected for a cylindrical surface, which corresponds to the X-centered necks of band 17. Finally, the η branch, which has a mass of 3.6 m_0 at ⟨100⟩, is observed over a range of approximately 20° around ⟨100⟩ and varies more rapidly than $1/\cos\theta$, as expected for a fluted or hypoboloidal surface,

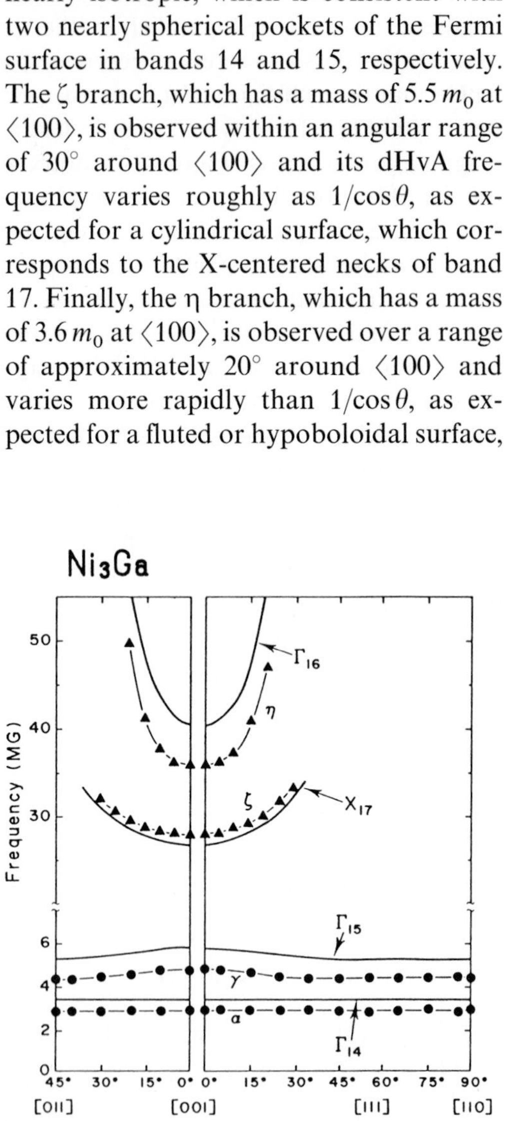

Figure 7-10. Angular dependence of the dHvA frequency of Ni_3Ga (Hayden and Lonzarich, 1986). Solid lines show the theoretical results of the LMTO band calculation. The thin lines connecting the data are guides to the eyes.

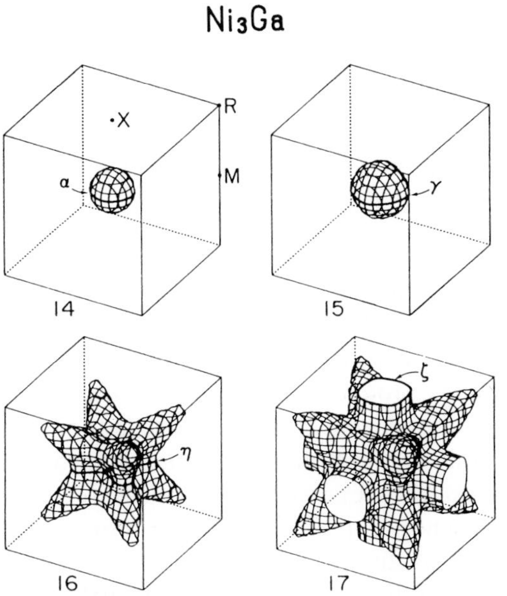

Figure 7-11. Fermi surfaces of Ni_3Ga (Hayden and Lonzarich, 1986). Spherical Fermi surfaces in bands 14 and 15 are expanded for clarity.

Table 7-1. Mass enhancement factor m_c^*/m_b for each orbit and γ/γ_b in Ni_3Ga and MnSi.

	m_c^*/m_b				γ/γ_b
Ni_3Ga	α	γ	ζ	η	
	1.9	1.7	2.6	1.8	1.6
MnSi	α	η	μ		
	4.8	3.3	4.8		5.4

which leads us to identify it with the Γ-centered waist of band 16.

As shown in Table 7-1, the measured masses are larger than the band masses. The mass enhancement factor is in the range of 1.7 to 2.6.

The Fermi surface of Ni_3Al (Sigfusson et al., 1984) is very similar both in topology and frequency to that of Ni_3Ga. The mass enhancement in Ni_3Al is somewhat higher than in Ni_3Ga but exhibits a similar anisotropy. In Ni_3Al each α branch is spontaneously split into two close branches due to ferromagnetic exchange. This exchange splitting is also found in Ni_3Ga at rather high fields of 30–65 kOe, meaning non-linearity in the field-induced exchange splitting.

7.4.1.2 MnSi

MnSi orders magnetically below 30 K into a helical structure in zero magnetic field and changes to a ferromagnetic state with an unsaturated magnetic moment of 0.4 μ_B/Mn in a field of 6.2 kOe. The electronic specific heat coefficient is large, being 36 mJ/K^2 · mol (Fawcett et al., 1970).

The band calculation was done using the standard LMTO method within the framework of the local density approximation (Taillefer et al., 1986). Self-consistent spin polarized calculations predict a stable magnetic moment of 0.25 μ_B/Mn, in fair agreement with the measured value. The bands near the Fermi energy consist mainly of the ten very flat 3d-bands of manganese in the ferromagnetic state.

Figure 7-12 shows the typical Fourier spectrum of the dHvA oscillations in the ferromagnetic region. Data were taken only in the ⟨100⟩ field direction. Fifteen

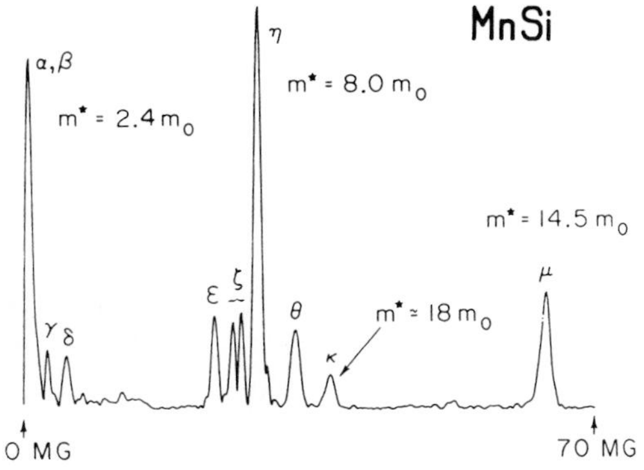

Figure 7-12. FFT spectrum of the dHvA oscillation at 0.35 K in MnSi. The field direction is ⟨100⟩ and the field range is from 102.8 kOe to 124.8 kOe (Taillefer et al., 1986).

branches are detected (Taillefer et al., 1986). Three strong branches α, η and μ are assigned, in comparison with the band calculations shown in Fig. 7-13, to a small electron pocket centered on Γ (Γ_{22}), to a hole neck centered on X (X_{21}) and to a large electron loop centered on M (M_{21}), respectively. The detected cyclotron masses are large, extending from $2.4\,m_0$ to $18\,m_0$. The average mass enhancement factor is about 5, and is comparable to the enhancement of the linear coefficient of the specific heat which is 5.4.

de Haas-van Alphen measurements (Newcombe and Lonzarich, 1988) and band calculations (Mattheiss and Hamann, 1988) were reported also for another silicide, $CoSi_2$. Three bands of s and p characters cross the Fermi level to give rise to three hole Fermi surfaces centered on Γ point of the f.c.c. brillouin zone. The cyclotron masses are thus small, ranging from $0.78\,m_0$ to $2.47\,m_0$.

We summarize in Table 7-1 the mass enhancement m_c^*/m_b for each orbit and γ/γ_b in Ni_3Ga and MnSi. Even though the γ-values are fairly large in the 3d-compounds, the usual 3d band model seems to be applicable to all materials reported in this section for describing the topology of the Fermi surface.

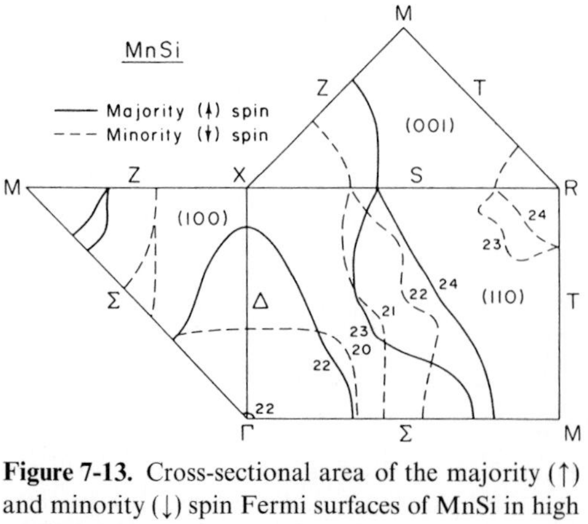

Figure 7-13. Cross-sectional area of the majority (↑) and minority (↓) spin Fermi surfaces of MnSi in high symmetry planes predicted by the band calculation (Taillefer et al., 1986).

7.4.2 Rare Earth Intermetallic Compounds – 4f Systems

For the rare earth compounds it is also difficult to grow good quality single crystals with well controlled stoichiometry. dHvA experiments were first done for $CeSn_3$ (Johanson et al., 1981) and CeSb (Kitazawa et al., 1983). Since then the technique for sample preparation has made great progress, and dHvA measurements have been carried out for many more materials, as discussed below.

A study of the series of rare earth compounds from La to Yb is interesting because systematics of the f electron behavior can be obtained. The presence of the f electrons alters the Fermi surface of conduction electrons through the following mechanisms: hybridization with the conduction electrons, the f electron contribution to the effective crystal potential, a change in the Brillouin zone to a new magnetic Brillouin zone or the formation of magnetic energy gaps formed in the paramagnetic Brillouin zone when the f moments order. The first two mechanisms may affect both the Fermi surface topology and the cyclotron mass, especially in the Ce compounds. At present the well known models used to treat the 4f state are either the localized atomic 4f model or the 4f band model. The third effect occurs, for example, in the antiferromagnetically ordered Kondo lattice substances such as CeSb, CeBi, $CeIn_3$, $CeAl_2$, CeB_6 and $CeCu_2$. Note that non-magnetic Kondo lattice substances exist such as $CeCu_6$, $CeRu_2Si_2$, CeNi and $CeSn_3$.

These antiferromagnetic Kondo lattice substances and the non-magnetic substances $CeCu_6$ and $CeRu_2Si_2$ show meta-

Table 7-2. Typical properties of various Ce compounds of different crystal structure, Néel temperature T_N, Curie temperature T_C, quadrupolar ordering temperature T_Q, superconducting transition temperature T_0, electronic specific heat coefficient γ and magnetic field at the metamagnetic transition H_c.

Substances	Crystal structure	T_N (K)	T_C (K)	γ (mJ/K^2 · mol)	H_c (kOe)
CeSb	cubic	16.2		20	38
$CeIn_3$	cubic	10.2		130	150
$CeAl_2$	cubic	3.8		135	53
CeB_6	cubic	2.3 (T_Q=3.2 K)		220	15
$CeCu_2$	orthorhombic	3.4		82	18
$CeGa_2$	hexagonal	11.4	8.2		
$CeCu_6$	orthorhombic			1600	20
$CeCu_2Si_2$	tetragonal	0.7 (T_0=0.7 K)		1000	70
$CeRu_2Si_2$	tetragonal			350	80
CeNi	orthorhombic			65–85	
$CeSn_3$	cubic			53	

magnetic transitions at critical fields H_c above which the exchange interaction dominates the Kondo effect. The systems are then transformed into the ferromagnetic state. In this ferromagnetic phase the Brillouin zone is the same as the old paramagnetic one. Usually, the magnetic energy gaps associated with the magnetic ordering in the 4f electron system are small enough and thus the electrons undergoing cyclotron motions can tunnel through these gaps and follow the orbits on the paramagnetic Fermi surface, even if the magnetic field is not very strong. This should, however, be checked carefully.

In Table 7-2 the fundamental properties of Ce compounds, which are treated in this section, are shown from a viewpoint of the Fermi surface properties.

7.4.2.1 RX (Monopnictides)

Of the rare earth monopnictides RX with the NaCl type crystal structure, especially the CeX compounds in which X means N, P, As, Sb and Bi, have attracted a particular interest because of various anomalous magnetic and transport properties. LaX, the reference compound with no occupied 4f electrons, is expected to be a semi-metal with a small and equal number of electrons and holes. A similar semimetallic character is observed also in CeX, except for CeN. However, these compounds are Kondo lattice substances having anomalous magnetic properties at low temperatures. Even though the carrier number is small, the Kondo effect is strong and the γ-value is large, about 20 mJ/K^2 · mol in both CeBi and CeSb.

Figure 7-14 shows the angular dependence of the external cross-sectional area of LaSb (Kitazawa et al., 1983; Kasuya et al., 1987). In the figure the solid lines show the result of relativistic APW band calculations made with the local density approximation (Hasegawa, 1985). Later, Sakai et al. (Sakai et al., 1985; Kasuya et al., 1987) made a more detailed LMTO band calculation and obtained a very good fit with the experimental results when the 5d band was shifted upwards relative to the valence Sb-p band, in order to reach a proper overlap between the valence Sb-p band and the conduction La-5d band. The 4f band was also shifted upwards in order

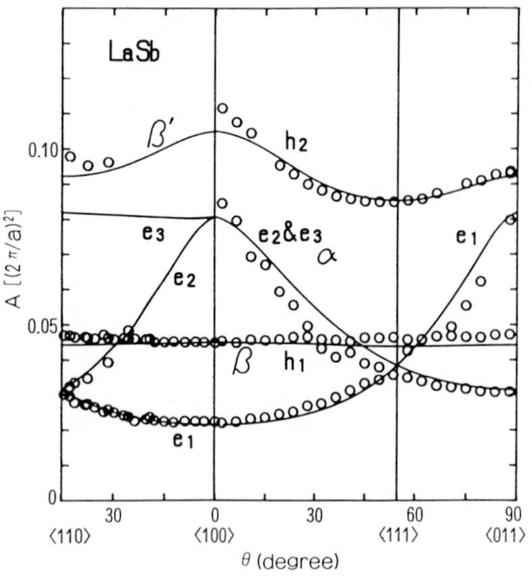

Figure 7-14. Extremal cross-sectional area of the Fermi surface in LaSb (Kitazawa et al., 1983). The solid lines indicate the calculated results (Hasegawa, 1985).

to be consistent with the BIS experiment. Note that the local density functional potential cannot fit well for the band gap and the 4f level position. Therefore they have to be adjusted to fit the experimental results.

The Fermi surface consists of three kinds of sheets, as shown in Fig. 7-15. Band 2 forms a small and nearly spherical hole Fermi surface centered at the Γ point, denoted by β or h_1. Band 3 also forms a small hole Fermi surface that is larger than the band 2 hole Fermi surface centered at the Γ point, denoted by β′ or h_2. It is slightly stretched in the ⟨100⟩ direction. On the other hand, band 4 consists of three equivalent, nearly spheroidal electron Fermi surfaces centered at the X points, denoted by α or e_i ($i = 1, 2, 3$).

LaSb is thus a semi-metal. Namely, the top of the valence band with the dominant Sb-p character is located at the Γ point where the $J = 3/2$ spin-orbit coupled quartet of Γ_8 symmetry exists. The bottom of the conduction band of mostly La-5d character is located at the X points. The bands slightly overlap each other. Narrow f bands lie a few eV above the Fermi level, and thus mixing of the f states into the valence band states is not large. The cyclotron masses are thus supposed to be small, but they have not been experimentally determined yet.

LaBi is also a semi-metal. The Fermi surface is quite similar to that of LaSb. The cyclotron masses are in the range of $0.18\,m_0$ to $0.36\,m_0$. The experimental electronic specific heat coefficients of LaSb and

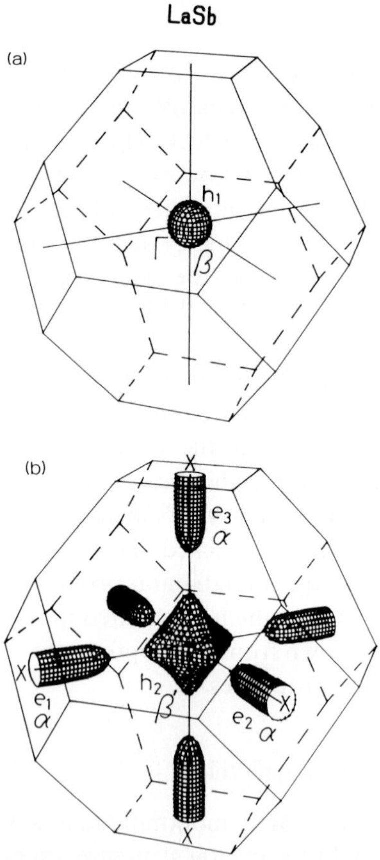

Figure 7-15. Fermi surface of LaSb (Hasegawa, 1985): (a) the band 2 hole Fermi surface centered at Γ, (b) the band 3 hole Fermi surface centered at Γ and the band 4 three electron Fermi surfaces centered at X.

LaBi are 0.80 and 0.95 mJ/K^2 · mol, respectively. This is in agreement with the calculated values, 0.50 and 0.85 mJ/K^2 · mol, respectively (Hasegawa, 1985, Sakai et al., 1987).

CeSb and CeBi, which show anomalously large magnetic anisotropy and a very complex antiferromagnetic phase diagram called devil's stair (Rossat-Mignod et al., 1985), belong to Kondo lattice systems with a small carrier concentration. Even though the cubic Γ_7 state is the ground state in the paramagnetic phase, anisotropy in the ordered phase is very strong. The ordered phase shows an Ising-like character with ferromagnetic order within the (001) or z-plane while the nearly full moment of $J_z = 5/2$ is oriented along the [001] or z-axis with Γ_8 character. In CeSb the non-magnetic planes, which are thought to arise predominantly as a consequence of the Γ_7 Kondo state, are intercalated between the magnetic planes. These interesting properties are explained by the p-f mixing model (Takahashi and Kasuya, 1985). Because of the strong p-f mixing between the 4f (Γ_8) and the Sb-p (Γ_8) states at the Γ point, the 4f (Γ_8) state becomes more stable than the 4f (Γ_7) state at low temperatures in the ferromagnetic ordered state. This feature is strongly enhanced by the non-linear effects due to the small Fermi energy. Here the 4f (Γ_7) state mixes rather weakly with the electrons in the Ce 5d conduction band.

The strong p-f mixing effect appears very clearly and strongly in the photoemission spectra of CeX as the two peak structure of the 4f states (Hillebrecht et al., 1985). This was theoretically explained by Takeshige et al. (Takeshige et al., 1985a and b) in terms of many-body bonding and antibonding peaks due to the p-f mixing, without any adjustable parameters except information from the band calculations.

Figure 7-16 shows the angular dependence of the extremal cross-sectional area in the ferromagnetic (F) phase of CeSb (Kitazawa et al., 1983; Kasuya et al., 1988;

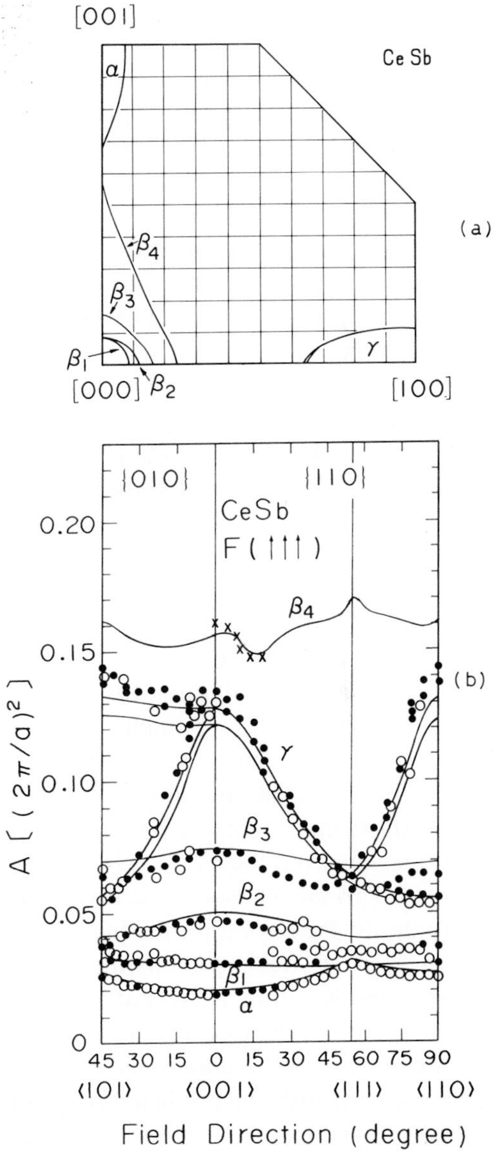

Figure 7-16. (a) Calculated cross-sectional areas of the Fermi surface in the ferromagnetic phase of CeSb and (b) the angular dependence of the extremal cross-sectional area in the ferromagnetic phase of CeSb (Kitazawa et al., 1983; Kasuya et al., 1988; Aoki et al., 1985). The solid lines show the calculated results (Kasuya et al., 1987).

Aoki et al., 1985). The solid lines show the results of band calculation. The band structure of CeSb in the ferromagnetic state was calculated by Sakai et al. (Sakai et al., 1985; Kasuya et al., 1987) by using the same parameters as in LaSb and putting the $J_z = 5/2$ 4f state 1 eV below the Fermi energy so as to be consistent with the photoemission spectra mentioned above, without any adjustable parameters. Agreement between the experimental data and the calculated results is surprisingly good. Due to the ferromagnetic order, the quartet hole bands split into four different hole bands, β_1, β_2, β_3 and β_4. On the other hand, the electron bands at the X points are split by the 5d-4f intra-atomic exchange interaction. Because of the strong anisotropic exchange interaction, the exchange splitting in the α branch is so strong that only the down spin state is occupied. On the other hand, the exchange splitting in the γ branch is not that large so that both up and down spin bands are occupied. Note the good agreement between experiments and calculations for the exchange splitting of the γ branch.

It is clear from the above description that the β_4 branch is the most important band with strong p-f mixing. In earlier experiments, however, the β_4 branch was not observed. This fact caused some investigators to doubt the strong p-f mixing model and to apply a weak p-f mixing model (Norman and Koelling, 1986). Later, the β_4 branch was observed by a careful acoustic dHvA measurement and analysis, in excellent agreement with the predicted result of the above mentioned band calculation in both magnitude and dispersion.

Generally speaking, it may be difficult to detect the β_4 branch by the usual susceptibility-dHvA experiment because the small amplitude of the FFT spectrum for the β_4 branch with its large mass is masked by the huge amplitudes of higher harmonics for the other branches with small masses. This difficulty in observing the β_4 branch is thus ascribed to its large mass. As mentioned before, the mass enhancement in the γ value of CeSb is very large, more than twenty. However, the cyclotron masses in CeSb, except the β_4 branch, are not that large. They are in the range of $0.3\,m_0$ to $0.9\,m_0$, of which the mass enhancement factor is two or three (Kitazawa et al., 1988; Aoki et al., 1985). Therefore the mass of the β_4 branch should be a very large one. Unfortunately, it has not yet been clearly determined by the experiment mentioned. The acoustic dHvA measurement of the β_4 branch was performed only in a small angle range, as shown in Fig. 7-16. More detailed experiments are needed. Note that the transverse magnetoresistance of CeSb is very large in magnitude, more than 70 in a field of 10 T and even at 1.5 K, but the β_4 branch can only be observed below 0.1 K in the dHvA experiment. This discrepancy is due to the fact that the Dingle temperature T_D is important in the scattering mechanism of the magnetoresistance, while both the measuring temperature T and the value of T_D are important in the dHvA experiment.

The important physical results obtained in LaSb and CeSb can be summarized as follows. For the band calculation it is necessary to shift the 5d and 4f levels to fit the gap or to give overlap between the valence and the conduction bands, as well as to fit the results of the PES and BIS experiments. The thus obtained results of the band calculations give a very good fit with the experimental results on the Fermi surface, even for the heavy fermion state with a large mass enhancement. The large mass enhancement in CeSb is also explained by the usual electron-magnon interaction, at least qualitatively. For a quantitative explanation one needs more information on the magnons, which are not detected fully by neutron scattering. CeSb is actually a very well compensated semi-metal. This was checked by the dHvA measurement as well as by the transverse magnetoresistance. This means that no 4f bands cross the Fermi level.

Next is a brief mention of dHvA measurements in other RX compounds. The dHvA measurement of CeBi was made by Kasuya et al. (1987). The result is not perfect, but the observed α, β_1 and β_2 branches agree very well with the band calculation. The dHvA measurement was also done for SmSb. The Fermi surface obtained is similar to that of LaSb. Schubnikov-de Haas oscillations were detected in CeAs. For a good stoichiometric sample of CeAs, the carrier number is one order of magnitude smaller than that of CeSb. Compared to CeSb, it is much more difficult to grow a good quality single crystal of CeAs. The dHvA measurements are also made for YbP and YbAs. These are particularly interesting materials in the sense that they are Kondo lattice systems which compete strongly with magnetic fluctuations. At low temperatures of about 0.6 K they order antiferromagnetically. The dHvA signals were observed both above and below the Néel temperature without a change of the dHvA frequency but with a sudden decrease of dHvA amplitude below the Néel temperature. This is thought to be due to the magnetic breakdown effect (Suzuki, 1990).

Many theorists believe that the Fermi surfaces of heavy fermion systems in Ce compounds should be constructed on the basis of the 4f band model, following the Luttinger theorem, because the Ce compound possesses one more electron than the La compound. This is not true for the RX Kondo lattice compounds, as shown here. The dHvA experiments indicate that the Fermi surfaces of the Ce compounds are almost the same as those of the corresponding La compounds and that there are no differences between the Fermi surfaces of their paramagnetic and antiferromagnetic states.

7.4.2.2 RB_6

The rare earth hexaborides RB_6 crystallize in the cubic (CaB_6 type) structure which possesses a CsCl type arrangement of R atoms and B_6 octahedra. LaB_6 is a reference non-f electron compound. CeB_6 is a typical Kondo lattice compound which undergoes two magnetic ordering transitions at the quadrupolar ordering temperature $T_Q = 3.2$ K and the Néel temperature $T_N = 2.3$ K (Effantin et al., 1985; Komatsubara et al., 1983). PrB_6 ($T_N = 7.0$ K) and NdB_6 ($T_N = 7.89$ K) are typical localized 4f systems with magnetic ordering.

(a) LaB_6

The measurements of the dHvA effect in LaB_6 (Arko et al., 1976, Ishizawa et al., 1977, 1980), shown in Fig. 7-17, revealed that the Fermi surface consists of a set of three equivalent, nearly-spherical ellipsoids centered at the X points and connected by necks. This topological property of the Fermi surface is well explained by the calculated band structure (Hasegawa and Yanase, 1977), as shown in Fig. 7-18a. It is characterized by the occupied wide B-sp bonding band and the unoccupied B-sp antibonding band. The La-d (e_g) band, which is mixed strongly with this antibonding band, exists around the energy gap between the bonding and antibonding bands, and one electron per mole occupies this conduction band, resulting in the La-d character. In this sense this material is a monovalent metal, and the volume of each

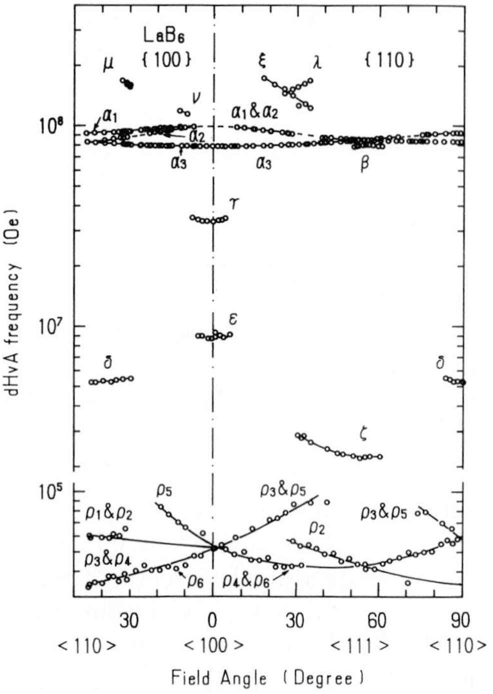

Figure 7-17. Angular dependence of the dHvA frequency in LaB_6 (Ishizawa et al., 1977, 1980). The solid and broken lines connecting the data are guidelines.

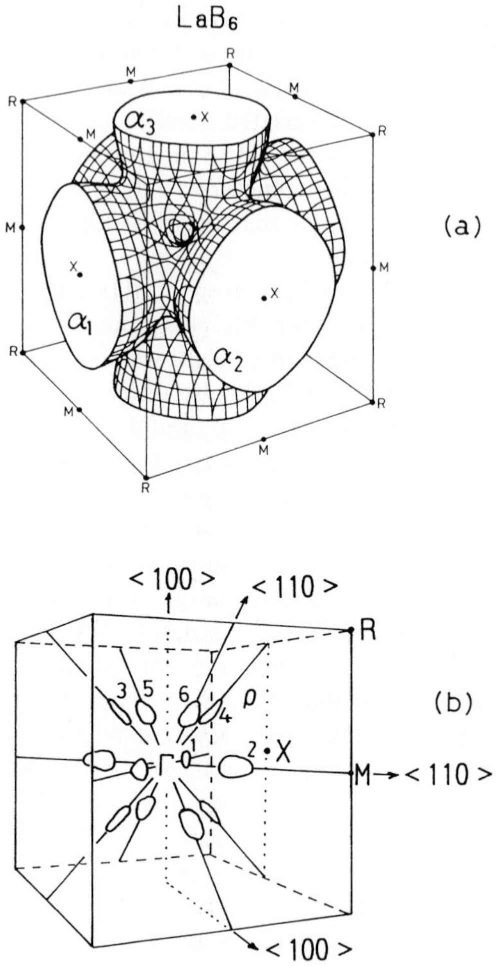

Figure 7-18. (a) The main three multiply connected ellipsoidal Fermi surfaces (Hasegawa and Yanase, 1977) and (b) twelve pocket Fermi surfaces in LaB_6 (Harima et al., 1988). The pocket Fermi surface is enlarged for visual convenience.

ellipsoidal Fermi surface is rigorously determined.

The neck orbit was, however, not detected in the above mentioned dHvA measurements. Later the ϱ_i branches shown in Fig. 7-17 were detected by the torque method (Ishizawa et al., 1980) and were attributed to the necks because the angular dependence of the ϱ_i branches is consistent with the topologies of the necks. This Fermi surface of the neck is, however, thin and rather cylindrical, inconsistent with the short and thick neck constructed from the unobserved region of the ellipsoidal α_i branches. This puzzle was solved later by a combination of the improved ultrasonic dHvA measurement by Suzuki et al. (1988) and the careful band calculation by Harima et al. (1988). It was shown that the ϱ_i branches are not due to the necks but due to the small and flat electron pockets. New band calculations were made by shifting the unoccupied 4f levels upwards by an amount of 0.10 Ry. This shift is based on the same reason as mentioned before for LaSb or CeSb, and leads a new band to cross the Fermi energy very slightly. In Fig. 7-18 b twelve pocket Fermi surfaces are shown. The neck orbit is not detected experimentally because of the rapid varia-

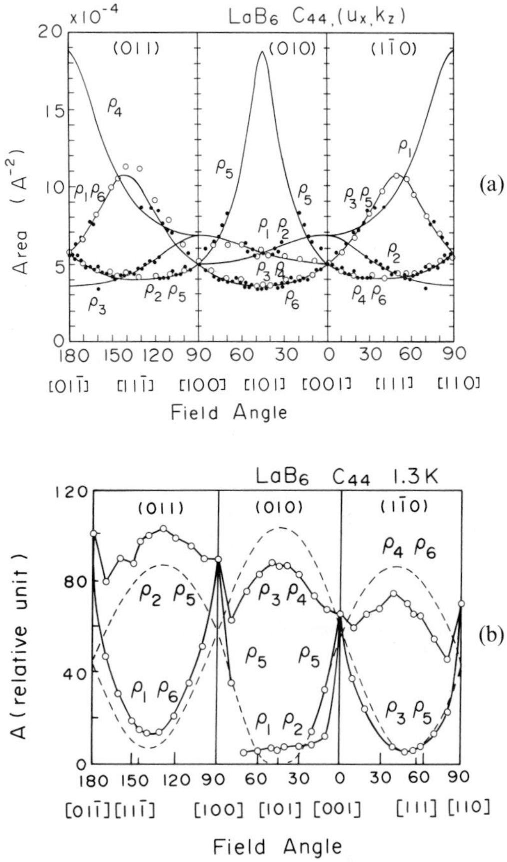

Figure 7-19. (a) Angular dependence of the cross-sectional areas for the ϱ_i branches determined by the transverse C_{44} mode (open circles) and the torque method (solid circles) (Suzuki et al., 1988; Ishizawa et al., 1980). The following relation holds: $F\,(\mathrm{Oe}) = 1.047 \times 10^8\, S\,(\mathrm{A}^{-2})$. The solid lines are the results of band calculation (Harima et al., 1988). (b) Angular dependences of the acoustic dHvA amplitudes of ϱ_i branches. The dotted lines are the results of band calculation.

tion of the cross-sectional area around the extremal neck orbit, implying a large curvature factor. This problem can be studied much better by acoustic dHvA measurement because it does not use a modulation field and thus the absolute intensity of the oscillation can be measured much more accurately.

Figure 7-19 shows the angular dependences of the cross-sectional areas and the acoustic dHvA amplitudes of the ϱ_i branches (Suzuki et al., 1988; Ishizawa et al., 1980). Here the experimental results of torque measurements and acoustic measurements for the C_{44} mode are shown by the solid and open circles, respectively. The solid lines in Fig. 7-19a are the results of band calculation for the electron pocket Fermi surface (Harima et al., 1988), in good agreement with the experimental results. The complete observation of the ϱ_3 and ϱ_5 branches in the $(1\bar{1}0)$ plane by the acoustic C_{44} mode is clear evidence for the closed pocket Fermi surface and is inconsistent with the previous slender neck model. For the amplitude shown in Fig. 7-19b the general agreement is also good. Note that the experimentally observed amplitude on the symmetry axis is the sum of two branches. It is also clear that the higher order effect mentioned before is important.

The acoustic intensity of the C_{44} mode for this pocket Fermi surface depends on $S^{-1/2}$. On the other hand, the torque method probes S in a different manner, namely as $S^{-5/2}\,(\partial S/\partial \theta)$ via the amplitude, where θ specifies the direction of the magnetic field with respect to the symmetry axis. Therefore the latter amplitude becomes quite weak for a small Fermi surface compared to the acoustic method. The acoustic dHvA technique is quite powerful for small portions of the Fermi surface because dHvA frequencies of less than 10^5 Oe are not detected by the usual susceptibility-dHvA method.

Anyway, the study of LaB_6 has led to the following important instructions. First, the angular dependence of the intensity should be checked carefully to determine which part of the Fermi surface is not observed. Otherwise one would obtain a completely different topology for the Fermi surface.

Secondly the position of the 4f level should be chosen carefully because the c-f mixing is not that weak and can cause an important effect on the Fermi surface topology.

(b) CeB_6 and PrB_6

Similar Fermi surface topologies were obtained in CeB_6 (Ōnuki et al., 1989 b; Joss et al., 1987 and 1989; Goto et al., 1988 a; Suzuki et al., 1987 a; van Deursen et al., 1985) and PrB_6 (Ōnuki et al., 1985 b, 1989 d; van Deursen et al., 1985), as shown in Figs. 7-20 and 7-21, respectively. Judging from the values of the α_i branches in the $\langle 100 \rangle$ direction, the main Fermi surface is more spherical in CeB_6 than in LaB_6 and PrB_6. The ratio of the maximum to minium areas of the ellipsoidal Fermi surface is about 1.16 in CeB_6, 1.24 in PrB_6 and 1.27 in LaB_6.

We also show in Fig. 7-22 the cross-sectional area of the Fermi surface for the electron pocket deduced from the ϱ_i branches. This is flat in character. It is approximately an ellipsoidal Fermi surface. The Fermi wave vectors k_{Fi} ($i = 1, 2$ and 3) along the three principal axes are $k_{F1} = 0.023$ $(2\pi/a)$, $k_{F2} = 0.0044$ $(2\pi/a)$ and $k_{F3} = 0.012$ $(2\pi/a)$ in LaB_6.

In PrB_6 two kinds of pockets ϱ_i and ϱ_i' as well as the ellipsoids α_3 and α_3' are found,

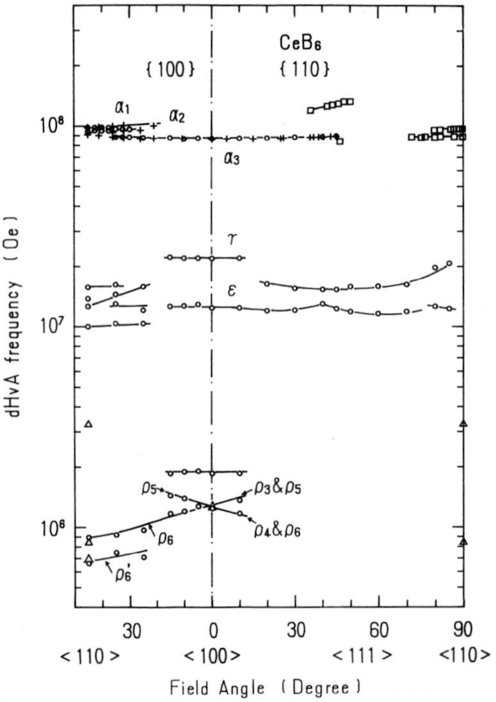

Figure 7-20. Angular dependence of the dHvA frequency in CeB_6. Data shown by circles, crosses, squares and triangles are cited from Ōnuki et al. (1989 b), van Deursen et al. (1985), Joss et al. (1987, 1989), and Suzuki et al. (1987 a) and Goto et al. (1988 a), respectively. The solid lines connecting the data are guidelines.

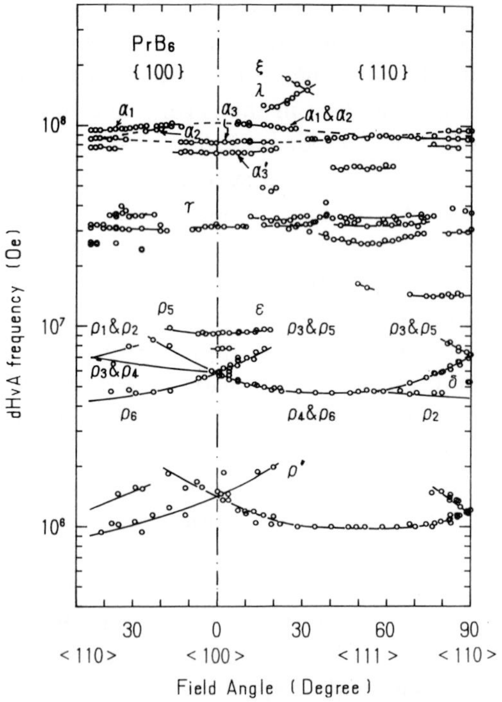

Figure 7-21. Angular dependence of the dHvA frequency in PrB_6 (Ōnuki et al., 1985 b, 1989 d). The solid and broken lines connecting the data are guidelines.

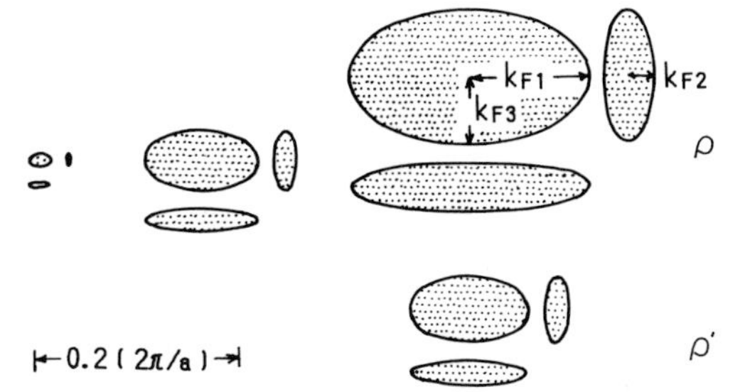

Figure 7-22. Cross-sections of the Fermi surface pockets in (a) LaB_6, (b) CeB_6 and (c) PrB_6. These electron pockets are flat in character (Ōnuki et al., 1989b).

as shown in Fig. 7-21. The cross-sectional areas of ϱ_i and ϱ_i' are 118 and 28 times larger than in LaB_6, respectively. The pocket of CeB_6 denoted by ϱ_i has the same size as the lower pocket in PrB_6. The existence of two kinds of Fermi surfaces in PrB_6 is explained by an exchange splitting of the up and down spin states of the conduction electrons, as shown schematically in Fig. 7-5. This is due to a change of the antiferromagnetic spin structure at 20 kOe in PrB_6. The up and down spin states have different effective Fermi surface areas and cyclotron masses. For example, the α_3 and α_3' branches in PrB_6 have the values of 8.19×10^7 Oe ($1.95\,m_0$) and 7.25×10^7 Oe ($2.52\,m_0$) at $\langle 100 \rangle$, respectively. A similar spin splitting of the effective Fermi surfaces is expected in CeB_6 because the antiferromagnetic state of the so-called phase III changes into that of the phase II (quadrupolar ordering) at about 15 kOe. Goto (1990) has confirmed that the ϱ_6' branch shown in Fig. 7-20 is due to Fermi surface pockets of different spin states.

The cyclotron masses in LaB_6, CeB_6 and PrB_6 are summarized in Fig. 7-23. All masses in CeB_6 are heavily renormalized compared to those of LaB_6 and PrB_6. The cyclotron masses in PrB_6 are also three times larger than those in LaB_6, which should be attributed to the usual electron-magnon interaction.

The cyclotron mass of the α_3 orbit in CeB_6 shows a striking variation as a function of magnetic field, as shown in Fig. 7-24 using experimental data represented by circles, triangles and a square. Quite a different field dependence is observed in the low temperature specific heat coefficient (Müller et al., 1988), as shown also in Fig. 7-24. Here the masses shown by the solid line through the crosses in Fig. 7-24 were estimated from the specific heat coefficient of CeB_6 by using the relation

$$m_c^*(CeB_6)/m_c^*(LaB_6) = \gamma(CeB_6)/\gamma(LaB_6),$$

where $m_c^*(LaB_6) = 0.61\,m_0$ for the α_3 orbit and $\gamma(LaB_6) = 2.6\ \mathrm{mJ/K^2 \cdot mol}$. A clear discrepancy exists between the results of the two types of experiment. The reason for this is thought to be as follows. When one compares the Fermi surface of CeB_6 to

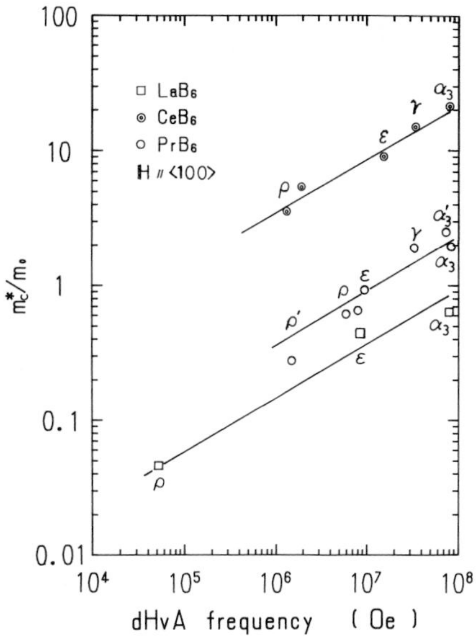

Figure 7-23. dHvA frequencies and the corresponding cyclotron masses in LaB_6 (Ishizawa et al., 1977, 1980), CeB_6 (Ōnuki et al., 1989b) and PrB_6 (Ōnuki et al., 1985b, 1989d). The solid straight lines are guidelines.

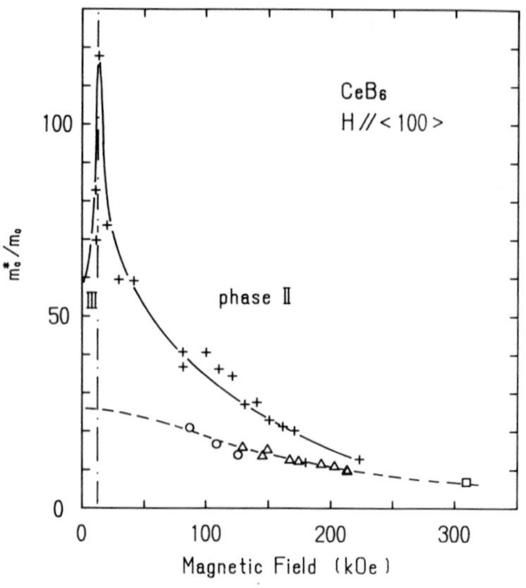

Figure 7-24. Field dependence of the cyclotron mass of the α_3 orbit. The data are shown by circles (Ōnuki et al., 1989b), triangles (Joss et al., 1987, 1989) and a square (van Deursen et al., 1985). The cyclotron mass estimated from the linear low temperature specific heat coefficients at various fields are shown by crosses (Müller et al., 1988). The solid and dotted lines are guidelines.

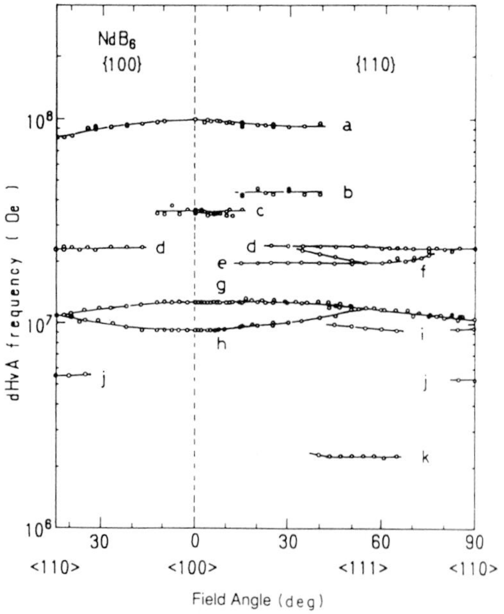

Figure 7-25. Angular dependence of the dHvA frequency in NdB_6 (Ōnuki et al., 1989d). The solid lines are guidelines.

that of PrB_6, one is led to believe that the observed α_3 Fermi surface in CeB_6 corresponds to the α_3 branch in PrB_6. In PrB_6 the mass of the α_3' branch is larger than that of the α_3 branch. Therefore it is natural to assume that the α_3' branch in CeB_6 also has a larger mass and hence could not be observed experimentally. Therefore, the missing α_3' branch is the main cause of the present discrepancy. From the discrepancy shown in Fig. 7-24 the mass of α_3' is estimated to be about twice as large as that of α_3. It should be noted that the effective extremal area used when constructing Fig. 7-21 does not correspond to the real area in the magnetic field applied and it is possible that the real area of α_i' is larger than that of α_i. This is the explanation for the opposite behavior of the ϱ_i branch, in which the ϱ_i branch with the larger effective area has a larger mass. It should be noted, however, that the missing angular region of the α_1 and α_2 branches around the $\langle 100 \rangle$ direction indicates accurately the size of neck. This is also true for the γ and ε branches in which the observed dHvA angular area indicates the size and topology of the real Fermi surface accurately. Of course the change in the dHvA intensity should be checked carefully, as already shown in the case of LaB_6.

The dHvA frequencies observed in NdB_6 (Ōnuki et al., 1989d) are substantially different from those of LaB_6 and PrB_6, as shown in Fig. 7-25. The a branch seems to correspond to the α_1 and α_2 branches in shape and magnitude, but it is observed around $\langle 100 \rangle$. There are no branches for electron pockets of the Fermi surface. One main reason for the discrepancy between the dHvA branches in NdB_6 and LaB_6 seems to be the large magnetic gap in NdB_6 due to a larger number of occupied 4f electron states, and thus antiferromagnetic Fermi surfaces in a new magnetic

Brillouin zone have to be considered in NdB_6. Kubo (1990) has calculated the Fermi surface from the above standpoint and has explained well the experimental result.

To conclude this sub-section we summarize the physics obtained from the above experiments. First it was shown that the Fermi surface in CeB_6 is very similar to those of LaB_6 and PrB_6 corresponding to one conduction electron per mole. The shape of the Fermi surface in CeB_6 shows no indication of changes in magnetic fields between 4 T and 30 T, even though the effective mass changes by more than a factor of ten. Furthermore, the field dependence of the γ-value is nearly inversely proportional to the field strength, which is consistent with the usual electron-magnon enhancement mechanism. The peak of the γ-value at the phase boundary is due to magnon softening, again consistent with this enhancement mechanism. These results are consistent with the view that the heavy fermion behavior in CeB_6 is due to the usual electron-magnon enhancement of the conduction electrons while the 4f electrons behave as localized, even though the enhancement is very large. Note that the field dependence of the mass of α_3 should be checked more carefully in the weak field region.

7.4.2.3 RX_3 (X = In and Sn)

There are many compounds of RX_3 with the $AuCu_3$ type cubic structure. Of them $CeSn_3$, $CeIn_3$ and their solid solution $Ce(Sn_{1-x}In_x)_3$ have been studied intensively owing to their interesting magnetic properties. $CeIn_3$ is a well known Kondo lattice compound showing antiferromagnetic ordering at 10 K. The ordered moment 0.65 μ_B per cerium atom is comparable to the value 0.71 μ_B expected from the Γ_7 ground state (Lawrence and Shapiro, 1980). Nevertheless $CeIn_3$ possesses the large electronic specific heat coefficient of 130 $mJ/K^2 \cdot mol$ at low temperatures, indicating a heavy fermion system (Nasu et al., 1971). Therefore $CeIn_3$ is considered to be in a similar situation to CeB_6, possessing a low Kondo temperature and displaying magnetic order at low temperatures. The ground crystal field state is, however, the Γ_7 doublet and thus no quadrupolar ordering is expected.

On the other hand, $CeSn_3$ is thought at low temperatures to be in the so-called valence fluctuation regime with a Kondo temperature of about 200 K. Therefore it is interesting to study the alloy system $Ce(In_{1-x}Sn_x)_3$, clarifying how the Kondo regime changes to the valence fluctuating regime. Actually experimental work has been reported (Benoit et al., 1985). There are some mysteries hidden in $CeSn_3$ itself (Gschneidner and Ikeda, 1983). The magnetic susceptibility shows an anomalous sharp increase at low temperatures, where neutron scattering reveals an anomalous component of 4f magnetic moment that is much more extended in space than at high temperatures. Some anomalies are attributed to non-stoichiometry effects and/or to the inclusion of other phases such as Ce_2Sn_5, but most of the anomalies are not yet well understood. In this sense it is interesting to study the Fermi surface of these compounds.

In Fig. 7-26 the angular dependences of the dHvA frequencies in $LaIn_3$ are shown (Umehara et al., 1991 a). The solid lines represent the results of the APW band calculations (Kletowski et al., 1987; Kitazawa et al., 1985).

Branch a originates from the band 7 electron Fermi surface centered at the R point, while the others originate from the band 6 hole Fermi surface, as shown in

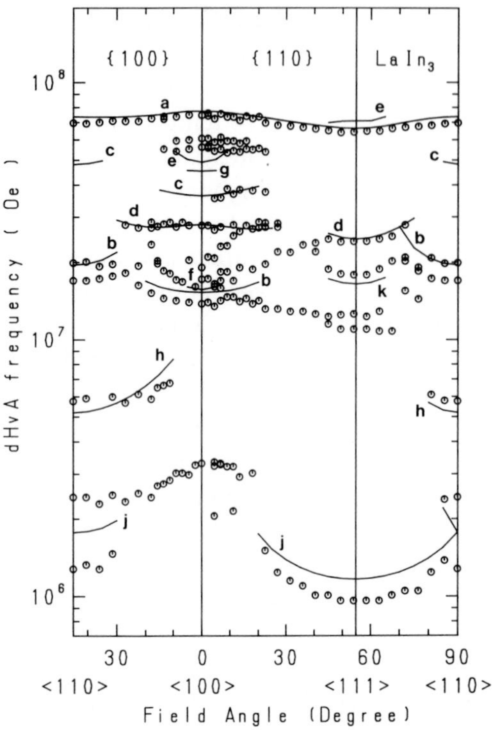

Figure 7-26. Angular dependence of dHvA frequency in $LaIn_3$ (Umehara et al., 1991 a). The solid lines represent the results of band calculations (Kletowski et al., 1987; Kitazawa et al., 1985).

Fig. 7-27. The latter Fermi surface consists of three kinds of major parts, which are centered at the Γ, R and X points. The Fermi surface, denoted as the d orbit in band 6, centered at Γ, is essentially spherical, bulges slightly along the ⟨100⟩ axis and connects with another part of the Fermi surface centered at R and comprising slender arms along the ⟨111⟩ direction.

The dHvA measurement in $CeIn_3$ was done by Kurosawa et al. (Kurosawa et al., 1990). The result is shown in Fig. 7-28. Only one branch was observed. This branch is very similar to the d branch in $LaIn_3$. Here we note that $CeIn_3$ is antiferromagnetic even in the highest applied field of 150 kOe. The cyclotron masses 2.02 m_0 at ⟨100⟩ and 2.88 m_0 at ⟨111⟩ in $CeIn_3$ are five to eight times larger than the corresponding values 0.40 m_0 and 0.37 m_0 in $LaIn_3$, respectively. This may be compared to the electronic specific heat coefficients of 130 $mJ/K^2 \cdot mol$ in $CeIn_3$ and 6.3 $mJ/K^2 \cdot mol$ in $LaIn_3$. Carriers with much larger cyclotron masses should be present in $CeIn_3$ but were not observed for this very reason. Fairly large magnetic gaps also seem to reduce the dHvA signal. Here we note that the band masses of the d orbit in $LaIn_3$ are 0.53 m_0 at ⟨100⟩ and

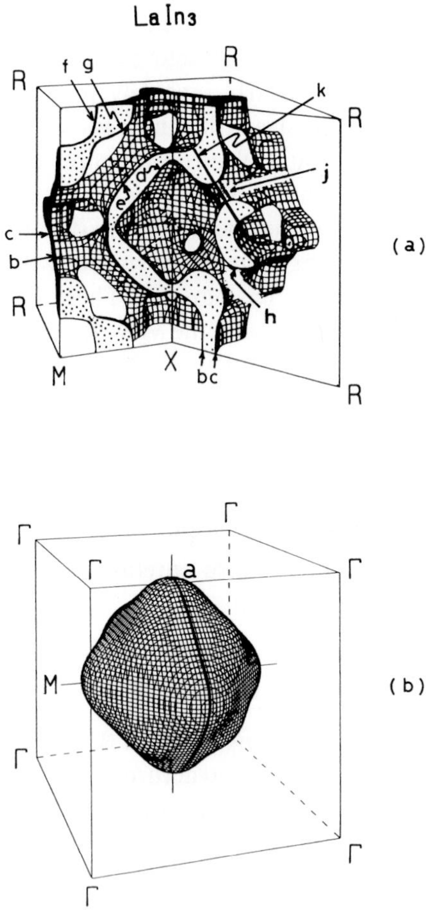

Figure 7-27. Fermi surfaces of $LaIn_3$ in (a) the band 6 hole and (b) band 7 electron (Kletowski et al., 1987; Kitazawa et al., 1985). In the band 6 hole Fermi surfaces, the dotted regions indicate the cross-sections of the hole Fermi surfaces in the {100} and {110} planes.

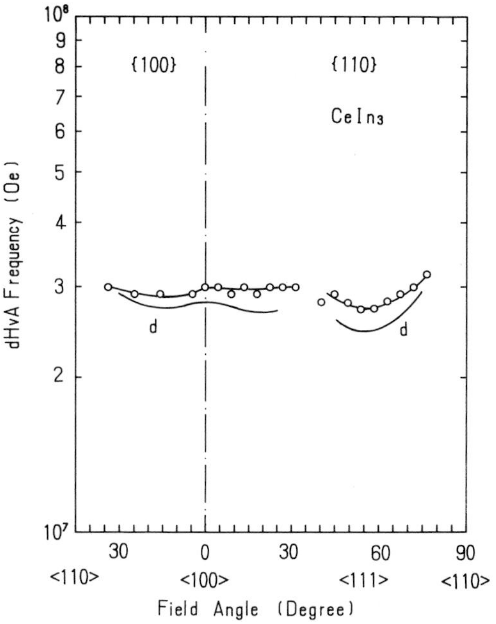

Figure 7-28. Angular dependence of dHvA frequency in $CeIn_3$ (Kurosawa et al., 1990). The solid lines connecting the data are guidelines. The other solid lines represent results of band calculations for the d-orbits in $LaIn_3$ (Kitazawa et al., 1985).

0.39 m_0 at ⟨111⟩, being in good agreement with the experimental values.

The magnetoresistances of $LaIn_3$ and $CeIn_3$ increase in a wide field region (Umehara et al., 1991 a; Kurosawa et al., 1990). This result suggests that these compounds are compensated metals with an equal number of electrons and holes. This is expected from the total number of valence electrons, 12 per unit cell, if the La and Ce ions are trivalent and the 4f electron is of localized character. A different behavior of the magnetoresistance is observed in $GdIn_3$, where the magnetoresistance saturates in all field directions in the {100} plane (Kletowski et al., 1985). For $GdIn_3$ it is necessary to consider the Fermi surface in the antiferromagnetic state.

The dHvA branches a, d, b and h in $LaIn_3$ are also found in $PrIn_3$ (Kitazawa et al., 1985). The Fermi surface of $PrIn_3$, which has a singlet ground state, is similar to that of $LaIn_3$.

In powdered $YbIn_3$ one dHvA branch of 4.50×10^6 Oe is detected in high fields up to 30 T, suggesting that a small spherical portion of the Fermi surface exists (Meyer et al., 1973).

To summarize, even though the information is not very complete, the Fermi surface in $CeIn_3$ is thought to be very similar to that of $LaIn_3$, in agreement with the cases of LaB_6 and CeB_6. To draw more precise conclusions, more detailed measurement and analysis including the magnetic breakdown effect are needed.

Figure 7-29 shows the angular dependence of dHvA frequency in $LaSn_3$ (Umehara et al., 1991 b; Johanson et al., 1983; Boulet et al., 1982). Main branches consist of two or three closed Fermi surfaces denoted α and β, and three branches of γ_1,

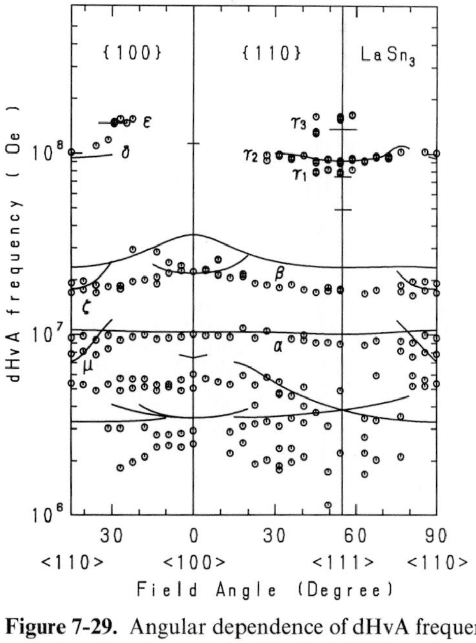

Figure 7-29. Angular dependence of dHvA frequency in $LaSn_3$ (Umehara et al., 1991 b). Solid lines show the results of band calculations (Hasegawa and Yamagami, 1990 b).

γ_2, and γ_3 centered at ⟨111⟩. Figure 7-30 shows the angular dependence of the magnetoresistance in $LaSn_3$ (Umehara et al., 1991 b). The sharp dips and spikes around the ⟨100⟩ direction indicate that the ⟨100⟩ direction is a singular field direction, and the spikes mean open orbits resulting from the uncompensated nature of $LaSn_3$.

Following the band calculations done by Hasegawa (1981) and Koelling (1982), the α and β branches correspond to bands 7 and 8 hole Fermi surfaces centered at Γ, respectively. Band 8 also contains a large portion of the Fermi surface centered at the R point, which is a complicated network of arms connected in the ⟨100⟩ direction through M. A hollow is seen through

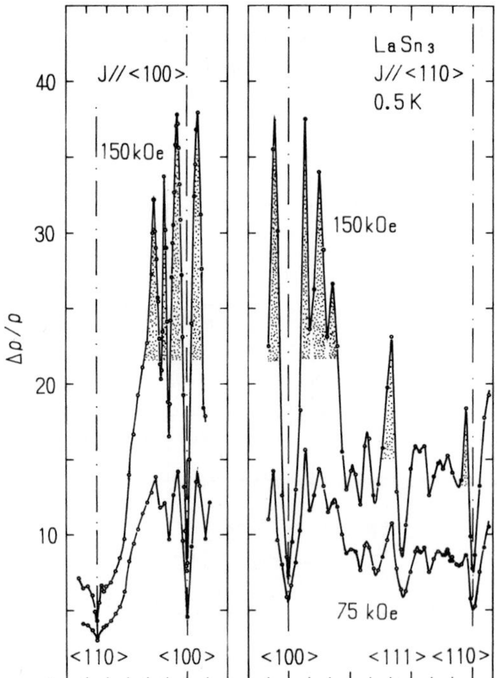

Figure 7-30. Angular dependence of magnetoresistance in $LaSn_3$ (Umehara et al., 1991 b). Open orbits occur in the dotted region. The solid lines connecting the data are guidelines.

the arms in the Γ-M-R plane. This is not a closed piece but a tunnel. The γ_1, γ_2 and γ_3 branches are large, nearly circular orbits in this Fermi surface, centered at the R point.

The results of the band calculations mentioned above can explain the experimental result qualitative, but not quantitatively. The reason for this is mainly the neglect of the spin-orbit interaction. Hasegawa and Yamagami (1991) have made energy band calculations, taking into account the spin-orbit interaction. The spherical closed hole Fermi surfaces in bands 7 and 8 are almost the same as found in previous calculations, but the complicated hole Fermi surface in band 8 is different. To clarify the topology of the band 8 Fermi surface, Hasegawa and Yamagami have divided it into two parts, namely a large distorted sphere with small and short necks and a network constructed from many arms, as shown in Fig. 7-31 a and b, respectively. The solid lines in Fig. 7-29 are the results of these band calculations. The detected dHvA branches are well explained by the present band calculation and, moreover, these multiply connected Fermi surfaces favor the experimentally observed open orbits.

Johanson et al. (1983) observed the dHvA effect for $CeSn_3$, showing that some Fermi surface parts seem to be similar but other parts are considerably different from those of $LaSn_3$, and the detected cyclotron masses are roughly five times larger in $CeSn_3$ than in $LaSn_3$, which is in agreement with the γ values.

From the observed γ value, $CeSn_3$ seems to belong to the valence flutuation regime where the 4f band model is applicable as a starting point. However, various unusual properties in $CeSn_3$ suggested that it is not a typical valence fluctuating material but may be located at the boundary between the valence fluctuation and the Kondo

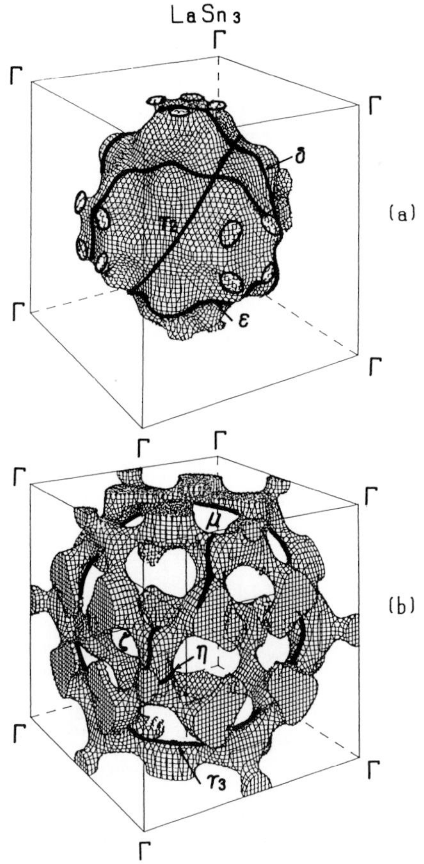

Figure 7-31. Band 8 hole Fermi surfaces in $LaSn_3$ are shown in two separated parts; (a) the large distorted sphere and (b) the network (Hasegawa and Yamagami, 1991).

regimes. In this respect the study of the Fermi surface in $CeSn_3$ is very interesting.

There are no well established ways to calculate the band or, more rigorously, the one-particle 4f Green function in the boundary region, but many physicists agree with the following point. In the Ce compounds the 4f quasi-particle bands are given by the renormalized 4f bands, whose center is situated at about $k\,T_K$ above the Fermi energy and whose width is given by a reduced c-f mixing interaction. Therefore, its band width is very narrow. The damping of the quasi-particle due to many-body interaction is of the same order as the band width. However, due to the Fermi liquid character, the damping around the Fermi energy is reduced by a factor $(E - E_F)^2$ and thus well defined quasi-particles exist at the Fermi energy. The band calculations based on this model were done by several physicists (Strange and Newns, 1986; d'Ambrumenil and Fulde, 1985; Zwicknagl et al., 1990). In particular, Strange and Newn's calculation for $CeSn_3$ showed better agreement with experiment than the usual 4f band model, both as regards the shape of the Fermi surface and the effective mass.

On the other hand, a similarity is present, in the main Fermi surfaces, between $LaSn_3$ and $CeSn_3$. Namely, the large dHvA frequency of 9×10^7 Oe in $CeSn_3$ is almost the same as for the γ_2 branch of $LaSn_3$. Harima and Kasuya (1985) emphasized this similarity because the bands around the R point are very similar between $CeSn_3$ and $LaSn_3$. In view of this situation they suggested that for the narrow 4f model mentioned above the system of conduction electrons in $CeSn_3$, which are equal in number of that of $LaSn_3$, may fit the Fermi surface better.

Umehara et al. (1990) have succeeded in clarifying the Fermi surface in $CeSn_3$ by using a much better quality single crystal and by measuring the high field magnetoresistance and the de Haas-van Alphen effect for magnetic field directions corresponding to both the $\{100\}$ and $\{110\}$ planes, as shown in Figs. 7-32 and 7-33 respectively.

The magnetoresistance of $CeSn_3$ increases in all field directions, indicating compensated carriers and the absence of open orbits in $CeSn_3$, which may be compared to the uncompensated state of $LaSn_3$.

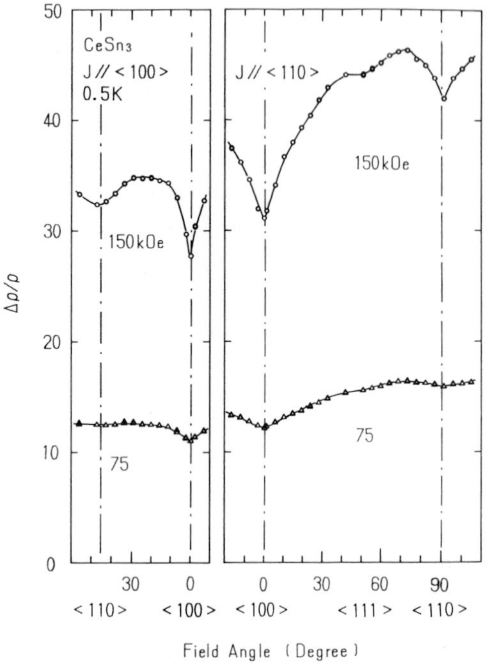

Figure 7-32. Angular dependence of magnetoresistance in $CeSn_3$ (Umehara et al., 1990). The solid lines connecting the data are guidelines.

Hasegawa et al. (1990) have calculated the energy band structure for $CeSn_3$ self-consistently by the relativistic APW method, including spin-orbit effects. Figure 7-34 shows the energy band structure in the vicinity of E_F. The 4f bands split into two groups which correspond to the total angular momentum $J = 5/2$ (lower bands) and 7/2 (upper bands). The magnitude of the splitting is nearly equal to the spin-orbit splitting of the atomic 4f state. It is clear that the overall feature of this dispersion in the vicinity of E_F is very similar to that considered by Harima and Kasuya. However, the spin-orbit interaction causes a gap between the crossing bands, which in turn leads to an important change in the topology of the Fermi surface, in particular around the Γ point. It should also be noted that the 8^- and 7^- states at the Γ point have strong p-f mixing which produces the

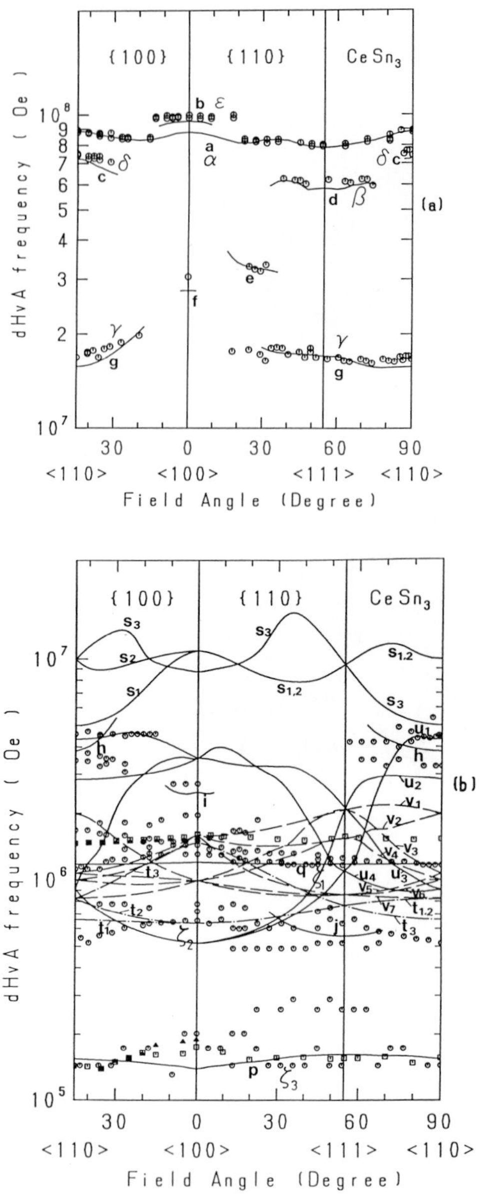

Figure 7-33. Angular dependence of dHvA frequency in $CeSn_3$; (a) a part for the higher dHvA frequency and (b) for the lower one (Umehara et al., 1990). The solid lines show the results of band calculation (Hasegawa et al., 1990).

unoccupied region around the Γ point. It is a delicate problem as to whether the 8^- and 7^- states are above the Fermi energy or not. This may also cause an important

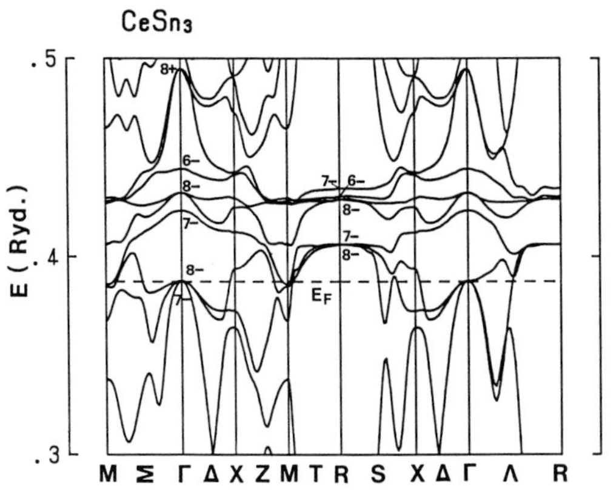

Figure 7-34. Energy band structure in $CeSn_3$ (Hasegawa et al., 1990).

change in the topology of the Fermi surface. As shown in Fig. 7-35 a, the calculated Fermi surfaces mainly consist of a large band 8 hole Fermi surface centered at R and a large band 9 electron surface centered at Γ. The origin of the similarity in Fermi surfaces between $CeSn_3$ and $LaSn_3$ is now clear. Namely, the large distorted spherical hole Fermi surface of band 8 in $LaSn_3$ is similar to the band 8 hole Fermi surface in $CeSn_3$. Here we note again that the large electron sheet in $CeSn_3$ has no occupied states along the ⟨111⟩ directions, as shown in Fig. 7-35 a. This result is compared to the previous calculation done by Koelling, (1982) shown in Fig. 7-35 b, where this electron Fermi surface looks like a sphere centered at Γ, having deep concavities in the ⟨111⟩ direction.

The solid lines in Fig. 7-33 indicate the results of the band calculation done by Hasegawa et al. (1990). In both the magnitude of the cross-sectional area and the observed range of angle the theoretical branches agree reasonably well with those of the experimental ones, supporting the validity of both the large hole and the large electron Fermi surfaces predicted by the calculation. Figure 7-36 a and b show the large hole Fermi surface (band 8) and the large electron Fermi surface (band 9), respectively.

Branch a in Fig. 7-36, which is due to the large hole Fermi surface, corresponds undoubtedly to branch α although it disappears at angles in the vicinity of the ⟨100⟩ direction. The reason for the disappearance of branch α is due to the combined effect of the curvature factor and the cyclotron mass. The hole Fermi surface is not a perfect sphere but bulges appreciably towards the ⟨100⟩ directions. In the ⟨111⟩ direction the orbit does not pass on any bulges. Therefore, as shown in Fig. 7-37,

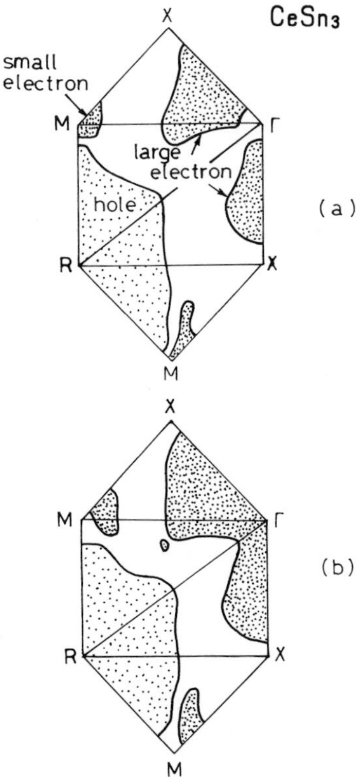

Figure 7-35. Cross sectional areas of the Fermi surface in $CeSn_3$ calculated by (a) Hasegawa et al. (1990) and (b) Koelling (1982). The R centered large hole Fermi surface comes from band 8, and the Γ centered large electron from band 9, and the M centered small electron Fermi surface from band 9.

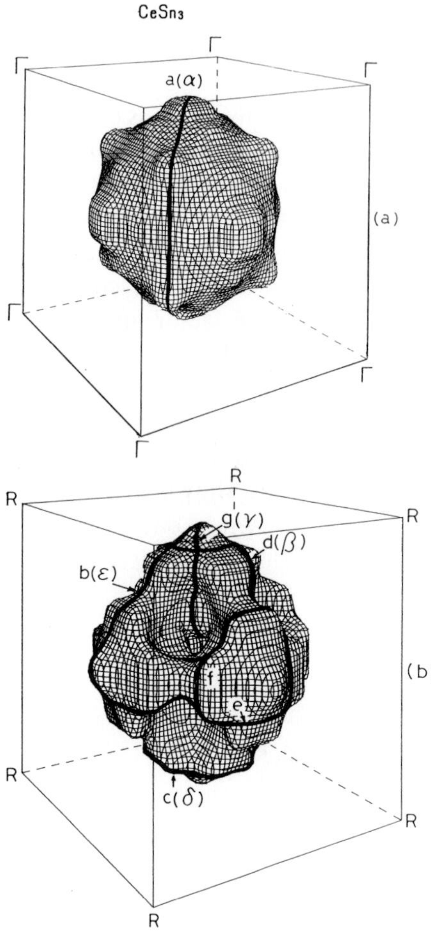

Figure 7-36. (a) The large hole Fermi surface in band 8 and (b) the large electron Fermi surface in band 9 in $CeSn_3$ (Hasegawa et al., 1990).

$|\partial^2 S/\partial k_H^2|^{-1/2}$ has a sharp peak along $\langle 111\rangle$ and becomes small along $\langle 100\rangle$ and $\langle 110\rangle$. The minimum value is about 1/20 of the maximum. Another possible origin is an increase of the cyclotron effective mass m_c^* for branch a (α) because it becomes maximum along $\langle 100\rangle$ and minimum along $\langle 111\rangle$. Here we note that the probability amplitude of the 4f state averaged over the orbits perpendicular to the $\langle 100\rangle$, $\langle 110\rangle$ and $\langle 111\rangle$ directions is found to be 70%, 65% and 60% respectively. The intensity of the oscillation of the de Haas-van Alphen effect is thus dampened by the factor $[\sinh(\lambda m_c^* T/H)]^{-1}$. Figure 7-37 shows the angular dependence of the curvature factor multiplied by this factor, which is calculated for branch a (α) by assuming that $T = 0.5$ K and $H = 150$ kOe. We see that this combined damping factor explains the experimental result for the intensity of branch a (α) qualitatively.

The large hole sheet in band 8 contains 0.44 holes/cell and the large electron sheet in band 9 contains the compensating number of electrons. The number of holes and electrons contained in other small sheets is fairly small, in total about 0.1 per cell.

The b (ε), c (δ), d (β), e, f and g (γ) branches originate from the large electron Fermi surface and exist in the limited range of angles, as shown in Fig. 7-36 b. The theoretical dHvA branches which originate from other small Fermi surfaces are compared to the experimental results. Of them, branch p, which corresponds to the small hole Fermi surface in band 7, is the candi-

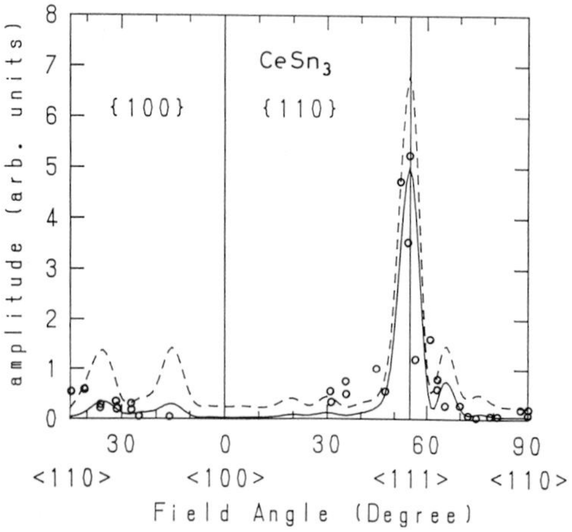

Figure 7-37. Angular dependence of dHvA amplitude for branch a (α) in $CeSn_3$ (Hasegawa et al., 1990). The dashed curve shows the curvature factor $|\partial^2 S/\partial k_H^2|^{-1/2}$, and the solid curve includes another factor $[\sinh(\lambda m_c^* T/H)]^{-1} \cdot |\partial^2 S/\partial k_H^2|^{-1/2}$. All data refer to $T = 0.5$ K and $H = 150$ kOe. The circles are the experimental results.

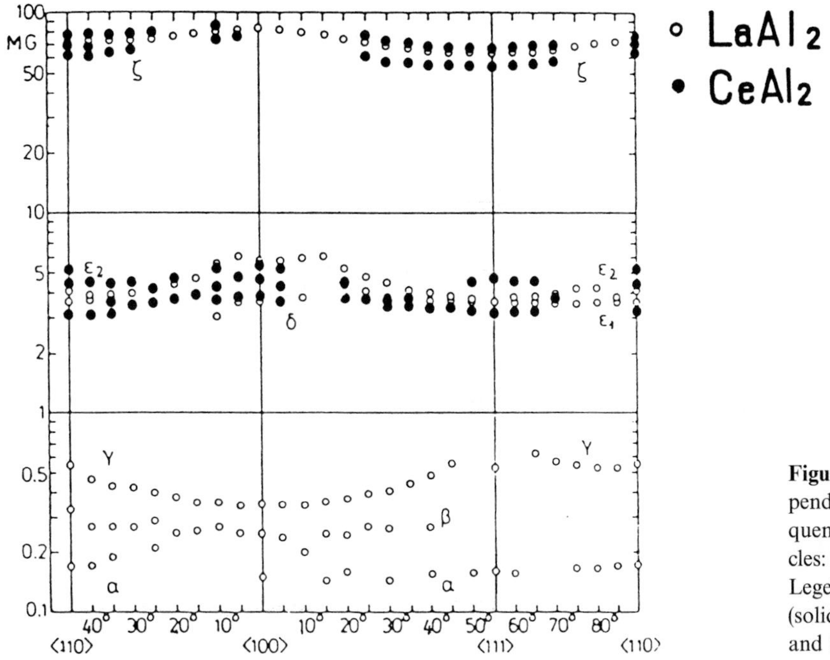

Figure 7-38. Angular dependence of dHvA frequency in $LaAl_2$ (open circles: from Seitz and Legeler, 1979) and $CeAl_2$ (solid circles: from Reinders and Springford, 1989).

date for the smallest experimental branch ζ_3. The branch q, which corresponds to the small hole sheet in band 8, is the candidate for ζ_1. The acoustic dHvA effect also supports the validity of this interpretation of the ζ_1 and ζ_3 branches (Suzuki et al., 1987b).

The cyclotron masses are in the range of 0.4 m_0 to 6.3 m_0. All cyclotron masses are larger than the band masses. The mass enhancement factor ranges from 2 to 4.

To conclude the dHvA effect in $CeSn_3$, it is remarkable that the 4f band model with the spin-orbit interaction included fits the experimental results very well. This is consistent with a relatively small mass enhancement, indicating that $CeSn_3$ is a typical material belonging to the valence fluctuation regime.

Finally, in powdered $YbSn_3$, one dHvA branch of 1.68×10^6 Oe has been detected in high fields up to 30 T, suggesting that a small spherical portion of the Fermi surface exists in $YbSn_3$ (Klaasse et al., 1980).

7.4.2.4 RAl_2

Rare earth compounds of the type RAl_2 possess the cubic Laves structure. $CeAl_2$ provides a good example of a Kondo lattice system with Γ_7 ground state and an incommensurate, sinusoidally modulated, antiferromagnetic structure below 3.8 K (Barbara et al., 1979).

First we show in Fig. 7-38 the angular dependence of dHvA frequencies in $LaAl_2$ by open circles (Seitz and Legeler, 1979, Reichelt and Winzer, 1978). These dHvA branches are well explained by the results of APW band calculation by Hasegawa and Yanase (1980), as shown in Fig. 7-39.

The main spherical branch is due to the band 10 electron Fermi surface which is a sphere centered at the Γ point but has bumps in the ⟨100⟩ direction. The other branches are ascribed to the multiply connected hole Fermi surface associated with band 9. The latter "junglegym" Fermi surface favors the ⟨100⟩ and ⟨110⟩ open or-

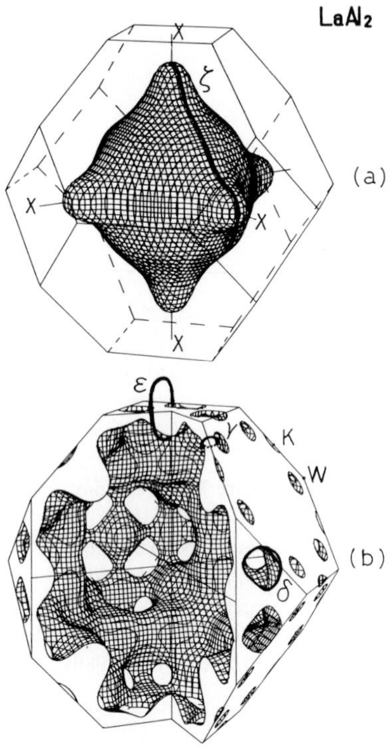

Figure 7-39. Fermi surfaces of (a) the band 10 electron and (b) the band 9 hole in $LaAl_2$ (Hasegawa and Yanase, 1980).

bits, consistent with the experimental magnetoresistance results (Reichelt and Winzer, 1978). At the Fermi energy the conduction electron states consist mainly of the La-5d and Al-3p states while the La-4f components have only a small contribution, less than 3%.

Figure 7-40 shows the typical dHvA oscillation with the magnetic field oriented 45° from ⟨100⟩ in the {110} plane of $CeAl_2$ (Reinders and Springford, 1989). The metamagnetic transition from the antiferromagnetic to the field-induced ferromagnetic (or paramagnetic) state occurs at 5.27 T at this orientation. One dHvA frequency of 3.10×10^6 Oe is seen in the antiferromagnetic phase. The same frequency is also seen above the transition, changing its amplitude to a huge value.

The angular dependence of dHvA frequency in $CeAl_2$ is shown in Fig. 7-38 by solid circles. The dHvA branches in $CeAl_2$ are similar to those in $LaAl_2$. Both the electron branch ζ and the hole branches, ε_1, ε_2 and δ are split into up and down spin branches due to the field induced, ferromagnetic exchange. The exchange splitting energy defined by Eq. (7-13) is estimated to be 10 meV.

The measured effective cyclotron masses in $CeAl_2$ are close to $1.3\,m_0$ for the ε orbit and $16\,m_0$ for the ζ orbit. Enhancement over the $LaAl_2$ value is seen to vary and falls between 5 and 10 for the orbits studied. Here the specific heat coefficient of

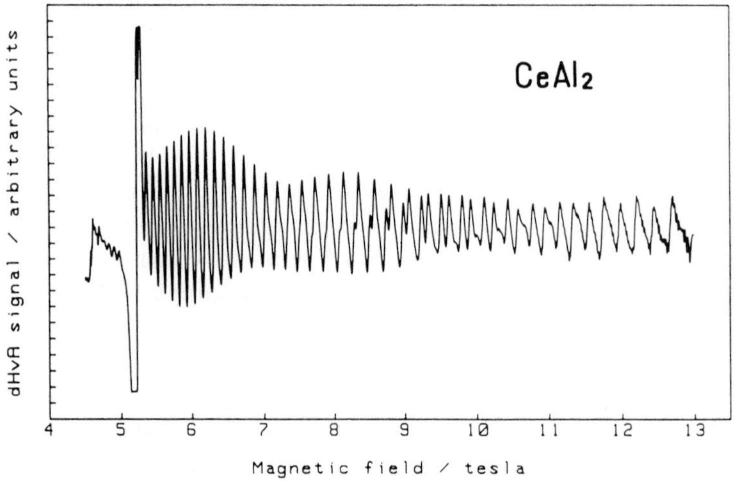

Figure 7-40. dHvA oscillation taken with the field oriented 45° from ⟨100⟩ in the {110} plane of $CeAl_2$ (Reinders and Springford, 1989). The dHvA oscillations are seen above and slightly below the magnetic phase transition, which occurs at 5.27 T at this field orientation.

$CeAl_2$ at low temperatures (Bredl et al., 1978) is 135 mJ/K^2 · mol, while that of $LaAl_2$ is 11 mJ/K^2 · mol (Hungsberg and Gschneidner, 1972). Therefore the enhancements of m_c^* and γ in $CeAl_2$ over their values in $LaAl_2$ are of comparable magnitudes.

$CeAl_2$ has been shown to be another example for which the Fermi surface topology of the Kondo regime material is similar to that of the corresponding La-compound. Again no visible difference was observed. It is also interesting to note that the Fermi surface does not change at the metamagnetic transition.

7.4.2.5 RCu_2

The crystal structure of RCu_2 is orthorhombic (Larson and Cromer, 1961). This structure can be thought of as a distorted hexagonal AlB_2 structure because the orthorhombic *b*-axis approximately corresponds to the hexagonal *c*-axis, and the relation $c \sim \sqrt{3}\,a$ holds. In fact, only $LaCu_2$ possesses the hexagonal AlB_2 structure. Therefore, instead of $LaCu_2$, YCu_2 becomes a non-f reference material for RCu_2. $CeCu_2$ can be classified as an antiferromagnetic Kondo lattice substance with the Néel temperature $T_N = 3.4$ K (Ōnuki et al., 1985a, 1990a; Gratz et al., 1985). Below T_N the magnetization of $CeCu_2$ shows a metamagnetic behavior around 17 kOe when the magnetic field is applied along the *a*-axis. The magnetic susceptibility and magnetization show a large anisotropy at low temperatures, reflecting the orthorhombic character of the structure.

Figure 7-41 shows the angular dependence of the dHvA frequency in YCu_2 (Ōnuki et al., 1989c; Settai et al., 1991). About twenty branches are observed in YCu_2. Many cylindrical arms are detected

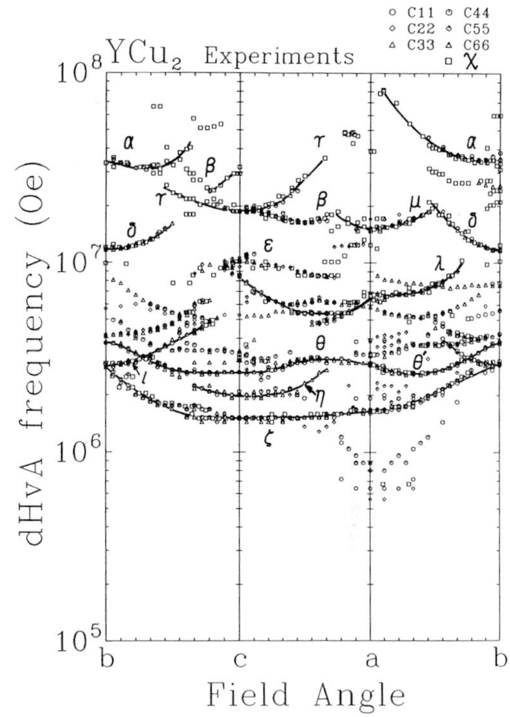

Figure 7-41. Angular dependence of the susceptibility-dHvA frequencies denoted by squares (Ōnuki et al., 1989c) and acoustic dHvA frequencies denoted by circles, diamonds and triangles (Settai et al., 1991) in YCu_2. The solid lines connecting the data are guidelines.

nearly showing $1/\cos\theta$ behavior for the angular dependence. Here θ is the angle between the direction of the cylinder axis and the field direction. We note that the cylinder axes of the α and δ branches do not coincide with the symmetry axes. They deviate by 15° and 4° from *b*-axis and *a*-axis, respectively. The β and γ branches also deviate by 40° in the *b*-plane and by 10° in the *a*-plane from the *c*-axis, respectively. The ζ branch is a closed orbit corresponding to a nearly ellipsoidal Fermi surface along the *b*-axis.

The LAPW band calculation was done by Harima et al. (Harima et al., 1990a). The origins of the dHvA branches are shown in Fig. 7-42. YCu_2 is a compensated

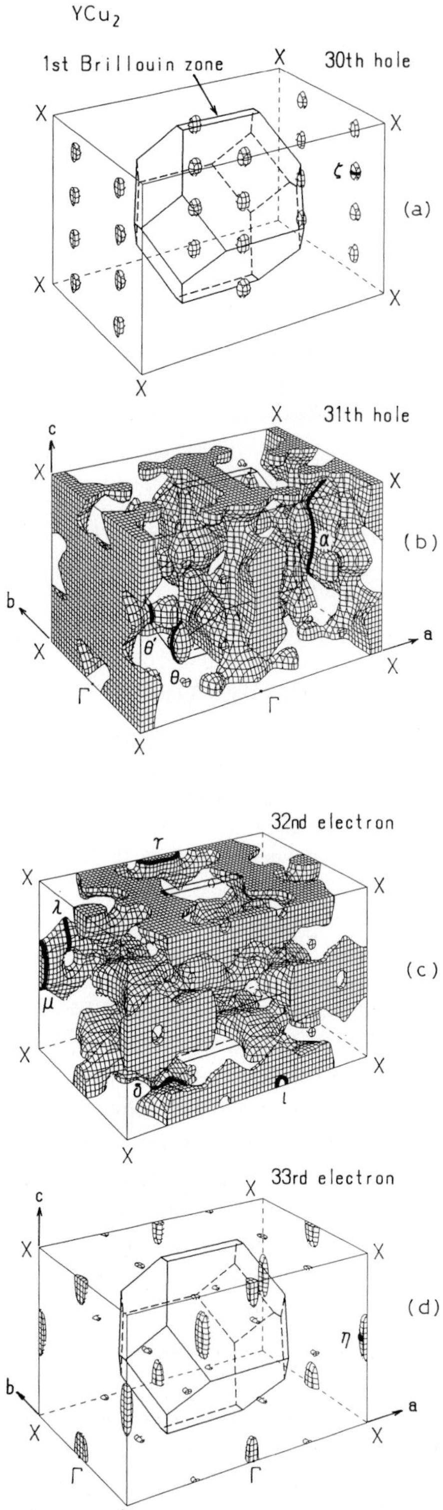

metal. Band 30 consists of small hole ellipsoids. Band 31 consists of ellipsoidal hole Fermi surfaces centered at the W point and a multiply connected Fermi surface stretching in the *a*-plane. The latter Fermi surface favors the presence of open orbits along the *b*- and *c*-axes. Band 32 is also a multiply connected electron Fermi surface which favors open orbits along the *b*-axis. Finally band 33 consists of small electron-ellipsoids centered at the Γ point. These results are consistent with the results of magnetoresistance measurements (Ōnuki et al., 1989c). The reason why the minimum of the branch is off-symmetrical is due to the non-cubic crystal structure.

On the other hand, as shown in Fig. 7-43, about ten kinds of dHvA branches are observed around the *a*-axis in $CeCu_2$, whereas only one branch is detected around the *c*-axis (Satoh et al., 1990). As the dHvA experiment was done above 60 kOe this substance is in the paramagnetic (or field-induced ferromagnetic) state for the field along the *a*-axis. On the other hand, it is antiferromagnetic in the *a*-plane. In general, the detectable number of dHvA branches is small in the antiferromagnetic state due to the magnetic breakdown effect, as seen also in $CeAl_2$ and $CeIn_3$.

When we compare the Fermi surface of $CeCu_2$ to that of the reference material YCu_2, it is difficult to say whether they are similar or not because too many branches are thought to not be observed in $CeCu_2$. More detailed experiments will be needed. It is also necessary to check the 4f band calculation for $CeCu_2$.

Figure 7-42. (a) Band 30 hole, (b) band 31 hole, (c) band 32 electron and (d) band 33 electron Fermi surfaces in YCu_2. The solid lines show the first Brillouin zone (Harima and Yanase, 1990a). The thick solid lines correspond to the observed orbits.

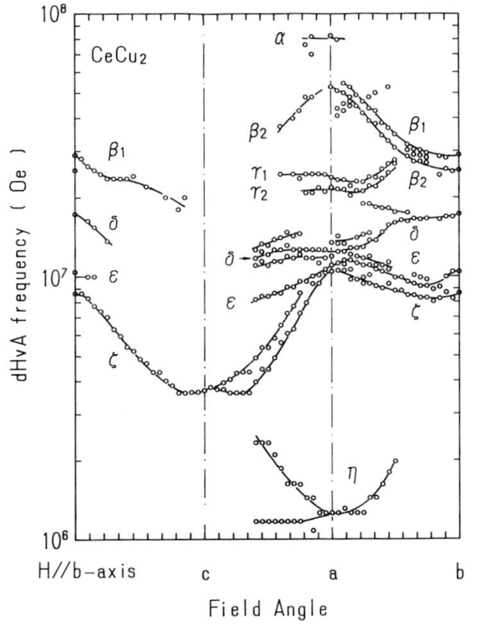

Figure 7-43. Angular dependence of the dHvA frequency in $CeCu_2$ (Satoh et al., 1990). The solid lines connecting the data are guidelines.

The effective cyclotron masses in $CeCu_2$ range from $0.5\,m_0$ to $5.3\,m_0$, and are larger than those in YCu_2, $0.1-0.7\,m_0$. This mass enhancement is roughly consistent with that of the low-temperature specific heat coefficient ratio, namely $50\ \mathrm{mJ/K^2 \cdot mol}$ measured at 80 kOe in $CeCu_2$ (Bredl, 1987) and $6.7\ \mathrm{mJ/K^2 \cdot mol}$ measured in YCu_2 (Luong et al., 1985).

Acoustic dHvA investigations were also made for studying the Fermi surfaces of YCu_2 and $CeCu_2$ (Settai et al., 1991). The results for YCu_2 are included in Fig. 7-41 and show that the acoustic dHvA measurement can detect the small Fermi surfaces better than the susceptibility method. Here the transverse C_{44} and C_{66} modes in YCu_2 exhibit pronounced softenings in their temperature dependences, as shown in Fig. 7-44. These behaviors may be related to the band Jahn-Teller effect due to the coupling between the elastic strain ε_{yz}, ε_{xy} and the conduction electrons of the appropriate orbits. To examine the origin of the softenings the angular dependences of the oscillatory amplitudes are important. As shown in Fig. 7-45, the α and δ branches reveal remarkable oscillatory intensities only in the soft modes C_{44} and C_{66}. This behavior may suggest that the electrons in the α and δ branches play a role in the band Jahn-Teller effect of the C_{44} and C_{66} modes.

The band Jahn-Teller effect is also found in LaAg (Niksch et al., 1987). The $(C_{11}-C_{12})/2$ mode shows a remarkable softening. The acoustic dHvA oscillation of $(C_{11}-C_{12})/2$ in LaAg indicates coupling to the ellipsoidal electron pockets at the X point. The lifting of the degeneracy of these pockets under the elastic strain $(\varepsilon_{xx}-\varepsilon_{yy})$ is thought to be the origin of the band Jahn-Teller instability.

Lastly we show in Fig. 7-46 the field dependence of the longitudinal C_{22} mode in

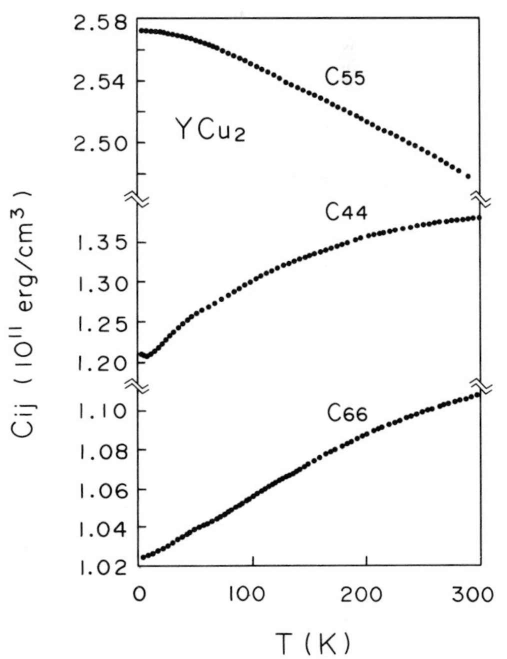

Figure 7-44. Temperature dependence of transverse C_{44}, C_{55} and C_{66} modes in YCu_2 (Settai et al., 1991).

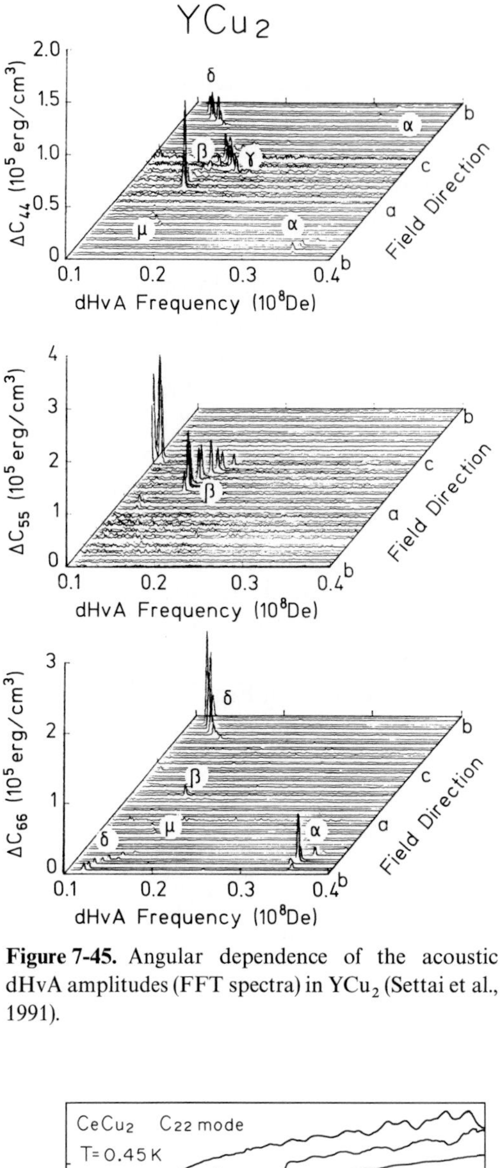

Figure 7-45. Angular dependence of the acoustic dHvA amplitudes (FFT spectra) in YCu_2 (Settai et al., 1991).

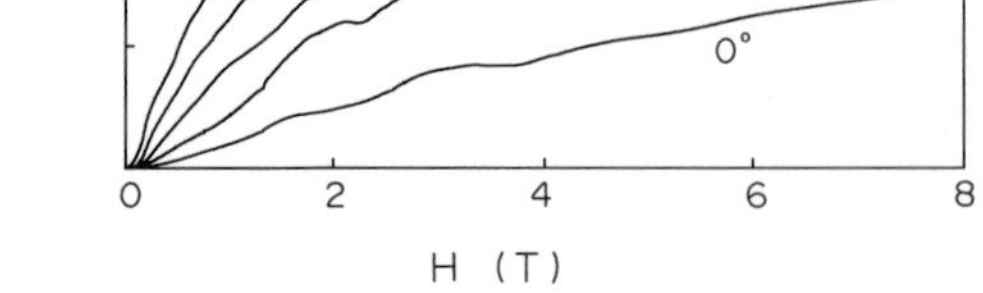

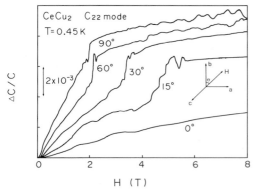

Figure 7-46. Field dependence of the longitudinal C_{22} mode of $CeCu_2$ (Settai et al., 1991).

$CeCu_2$ (Settai et al., 1991). The oscillations above the critical field for the metamagnetic transition reveal the usual dHvA oscillation proportional to $1/H$. The angular dependence of the cross-sectional areas agrees well with the dHvA data shown in Fig. 7-43. The oscillations below the critical field, however, have the unconventional period without a $1/H$ proportionality. Similar behavior is also found in UPt_3 and $CeRu_2Si_2$ (Kouroudis et al., 1987). The origin is not clear at present.

7.4.2.6 RGa_2

RGa_2 compounds crystallize in the simple hexagonal AlB_2-type structure. $CeGa_2$ is a highly anisotropic ferromagnet with the easy axis in the basal plane. The magnetic phase diagram is not simple, indicating the ferromagnetic state below 8.4 K and two or three complicated antiferromagnetic states in the temperature region between 8.4 K and 11.4 K (Jerjini et al., 1988, Takahashi et al., 1988 a). From the resistivity data and other magnetic properties it is concluded that $CeGa_2$ is not a Kondo lattice compound but a usual f localized compound.

We show in Figs. 7-47 and 7-48 the angular dependence of dHvA frequencies in $LaGa_2$ (Sakamoto et al., 1990) and $CeGa_2$ (Umehara et al., 1991 c). The dHvA branches of $CeGa_2$ are similar to those of $LaGa_2$ although all branches in $CeGa_2$ are split into up and down spin states due to the ferromagnetic exchange interaction. The f eletron is thus localized in $CeGa_2$. The ferromagnetic exchange interaction is estimated to be about 17 meV from the spin factor.

The α and β branches correspond to ellipsoidal Fermi surfaces of revolution along the c-axis. It is not easy to clarify the topologies of the γ, ε and δ branches from the present data.

The cyclotron mass in $CeGa_2$ is in the range $0.32\,m_0$ to $1.44\,m_0$. The masses of the α and β branches in $CeGa_2$ are twice those of $LaGa_2$. On the other hand, the masses of the other branches in $CeGa_2$ are almost the same as in $LaGa_2$. The mass enhancement is anisotropic, depending on the band.

The dHvA experiment of antiferromagnetic $SmGa_2$ ($T_N = 20$ K) shows that its Fermi surface is also almost the same as that of $LaGa_2$ (Sakamoto et al., 1990). The masses of $SmGa_2$ are the same as those of $CeGa_2$. The mass enhancements of $CeGa_2$ and $SmGa_2$ are due to the usual electron-magnon interaction.

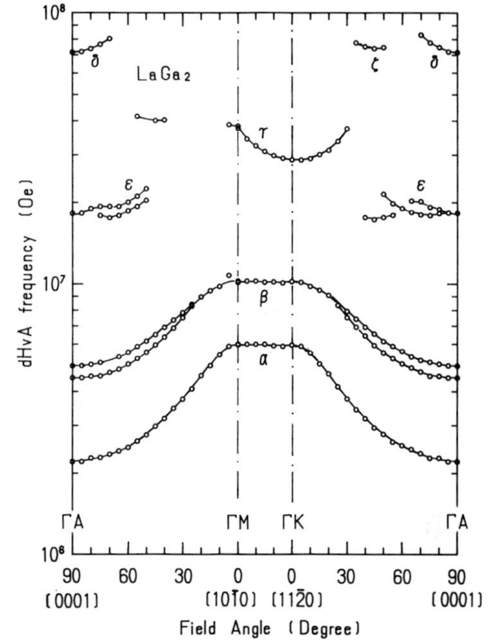

Figure 7-47. Angular dependence of the dHvA frequency in $LaGa_2$ (Sakamoto et al., 1990). The solid lines connecting the data are guidelines.

7.4.2.7 RCu_6

The f electrons in RCu_6 (R: Ce, Pr, Nd and Sm) show a variety of magnetic behaviors. $CeCu_6$ is a typical non-magnetic Kondo lattice compound with a Kondo temperature of abut 4 K (Ōnuki and Komatsubara, 1987). $PrCu_6$ is a nuclear cooling material with a singlet ground state (Takayanagi et al., 1988). $NdCu_6$ ($T_N = 6.1$ K) is a metamagnetic substance with four discontinuous steps in the magnetization curve (Ōnuki et al., 1986; Takayanagi et al., 1990) and $SmCu_6$ ($T_N = 9.6$ K) shows a Van Vleck susceptibility due to the Sm^{3+} ion (Ōnuki et al., 1990b).

The RCu_6 compounds possess a monoclinic structure at low temperatures which is a slight modification of the orthorhombic $CeCu_6$ structure. For example, the orthorhombic structure of $CeCu_6$ with $a = 8.105$ Å, $b = 5.105$ Å and $c = 10.159$ Å at room temperature changes into the monoclinic one with $a = 5.080$ Å, $b = 10.121$ Å, $c = 8.067$ Å and $\beta = 91.36°$ at 65 K (Asano et al., 1986). The structural transition temperature is roughly 200 K. This transition is associated with a com-

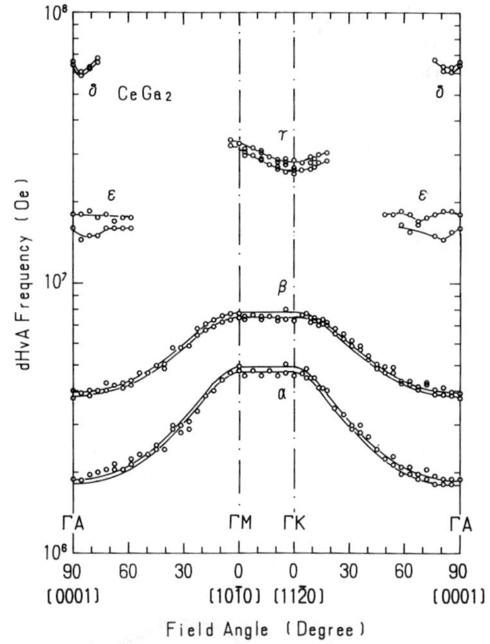

Figure 7-48. Angular dependence of the dHvA frequency in $CeGa_2$ (Umehara et al., 1991c). The solid lines connecting the data are guidelines.

plete softening of the transverse elastic constant C_{66} (Suzuki et al., 1985). Here, a, b, c and γ (= 90°) in the orthorhombic notation are converted into c, a, b and β in the monoclinic one. As the crystal distortion is small we use the orthorhombic notation.

Figures 7-49 to 7-53 show the angular dependence of the dHvA frequency in RCu_6, where R represents La, Ce, Pr, Nd and Sm (Ōnuki et al., 1987; Springford and Reinders, 1988; Chapman et al., 1990b; Endoh et al., 1987; Ōnuki et al., 1991a). The band calculation for $LaCu_6$ made by Harima et al. (Harima et al., 1991) shows a lot of extremal cross sections. The observed signals in $LaCu_6$ are really quite numerous but they represent only a part of the signals expected from the band calculation. However, the main observed signals correspond to the results of the band calculation. In the same sense the signals in $PrCu_6$ and $NdCu_6$ seem to correspond to the band calculation. The common dHvA branches are denoted by Greek letters. The detected dHvA branches are many in number, concentrating at the symmetry axes. The largest branch is α which consists of two or three branches with dHvA frequencies of about 2×10^7 Oe.

The Fermi surface of $SmCu_6$ is much different from that of $LaCu_6$. In $CeCu_6$ we find similar but also different branches along the c-axis when compared to $NdCu_6$ or $LaCu_6$.

As mentioned above, Harima et al. (Harima et al., 1991) have calculated the band structure of $LaCu_6$ with the orthorhombic structure by an LAPW method. The Cu-3d band and La-4f band are well localized and are separated from the Fermi level. The conduction bands are mainly due to the 4s components of Cu. Figure 7-54 shows their calculated Fermi surfaces. As the non-cubic unit cell contains four molecules of $LaCu_6$, about 1200 basis functions are needed for each point especially for 120 d-bands due to the 24 Cu atoms. $LaCu_6$ is a compensated metal with hole Fermi surfaces in bands 149 and 150 and electron Fermi surfaces in bands 151 and 152. The theoretical Fermi surfaces are many in number and are characterized as corrugated Fermi surfaces with concave and convex curvature, elongated along the c-axis. From the magnitudes and angular dependences of the extremal orbits most of the experimental dHvA branches can be identified, as shown in Fig. 7-54. The largest orbit, which is a belly orbit on a "catcher mitt" due to the band 151-electron, corresponds to the α branch. The theoretical dHvA frequency of 2.06×10^7 Oe in the b-axis is in good agreement with the experimental value of about 2×10^7 Oe.

The calculated electronic specific heat coefficient of 6.2 mJ/K^2 · mol is almost the same as the experimental value of 8 mJ/K^2 · mol for $LaCu_6$. The detected masses are thus small and fall into the range $0.076\,m_0$ to $2.50\,m_0$, reflecting the main 4s components of Cu.

On the other hand, the cyclotron masses of $PrCu_6$ and $CeCu_6$ are twice and forty times larger than that of $LaCu_6$, respectively. No mass enhancement is found in $NdCu_6$ and $SmCu_6$, as shown in Fig. 7-55 (Ōnuki et al., 1988). Here the electronic specific heat coefficients of $PrCu_6$ and $CeCu_6$ at zero external field are 16 and 1600 mJ/K^2 · mol, respectively. These values may be compared to 8 mJ/K^2 · mol for $LaCu_6$ (Takayanagi et al., 1988; Satoh et al., 1989).

The large mass due to the Kondo lattice property of $CeCu_6$ seems to be strongly reduced by a magnetic field. The specific heat coefficient γ at low temperature strongly depends on the magnitude and direction of the field (Amato et al., 1987; Satoh et al., 1989). Experimental results

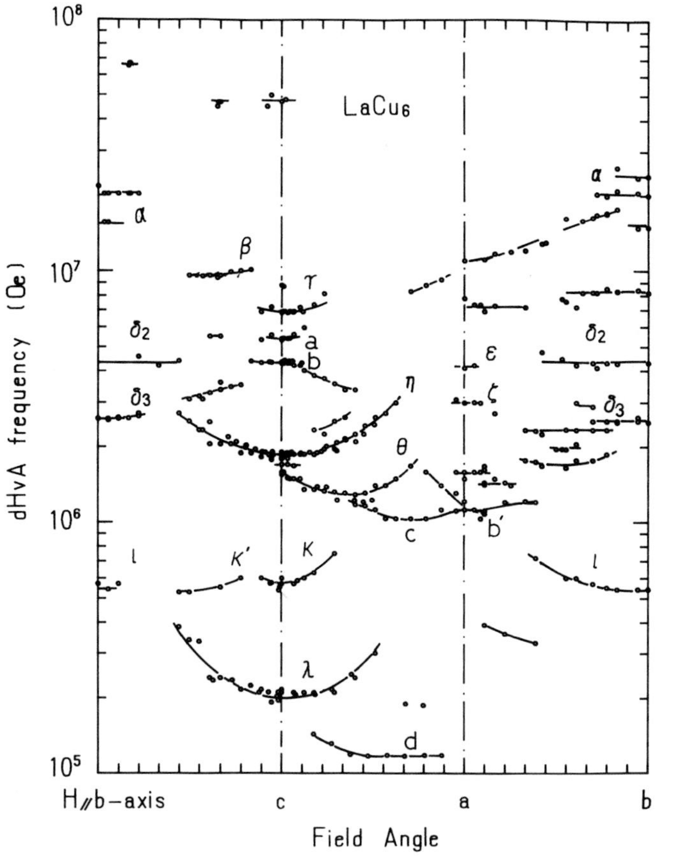

Figure 7-49. Angular dependence of the dHvA frequency in $LaCu_6$ (Ōnuki et al., 1987). The solid lines connecting the data are guidelines.

measured by Amato et al. for the field along the *c*-axis, which is the most sensitive direction, are shown in Fig. 7-56. Note that the field effect along the hard axes of the *a*- and *b*-axes is weak. For the highest applied field of 7.5 T, which almost corresponds to the initial field for the dHvA measurement, the γ value is reduced to 500 $mJ/K^2 \cdot mol$. This value is roughly consistent with the mass enhancement ratio of 40 mentioned above.

The field dependence of the cyclotron mass was studied by Chapman et al. (1990a) for two dHvA branches with the field along the *c*-axis. The mass of 11 m_0 measured at 4.1 T is reduced to 6.0 m_0 at 10.6 T for the dHvA branch of 1.2×10^6 Oe. For another branch of 1.03×10^7 Oe, 34.6 m_0 at 10.3 T is reduced to 29.6 m_0 at 12.2 T. These results are shown in Fig. 7-56. Here a zero field cyclotron mass was assumed for which the ratio $m_c^*(H)/m_c^*(0)$ fits the field dependence of $\gamma(H)$. Because of the limited, narrow range of the field it is difficult to establish whether both field dependences are the same or not. Note that the field effect on the cyclotron mass for the hard axes has to be weak in order to be consistent for $\gamma(H)$.

Chapman et al. have concluded that the Fermi surface of $CeCu_6$ is not well described by LMTO band calculations performed within the local density approximation, in which the f-electrons are included as either band or core states. Other kinds of band calculations and more exper-

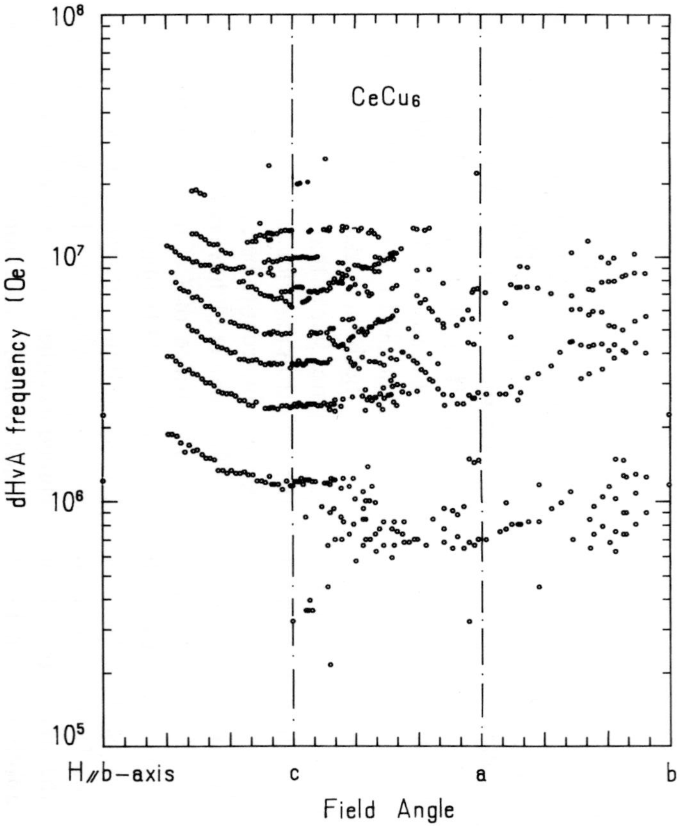

Figure 7-50. Angular dependence of the dHvA frequency in $CeCu_6$ (Chapman et al., 1990b).

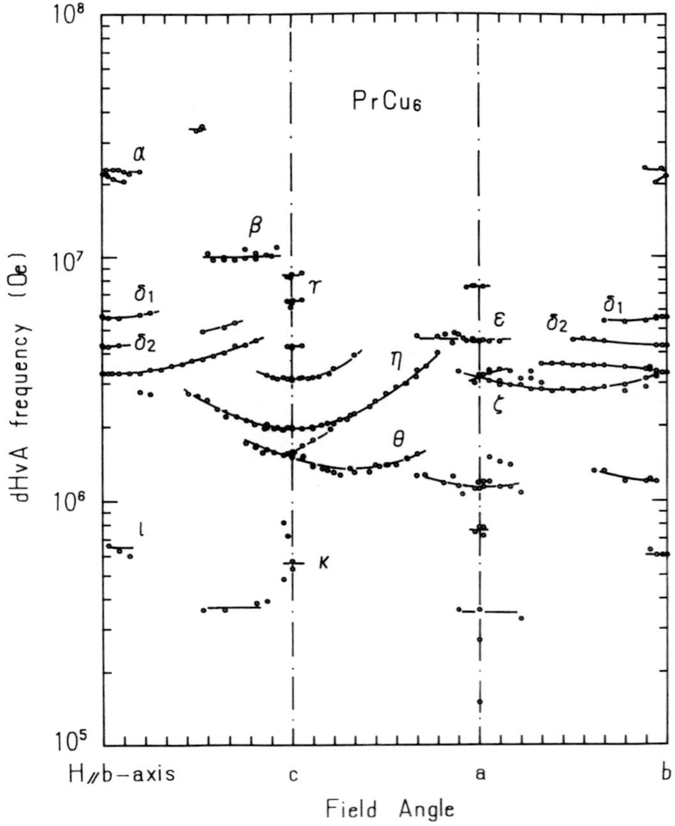

Figure 7-51. Angular dependence of the dHvA frequency in $PrCu_6$ (Ōnuki et al., 1987). The solid lines connecting the data are guidelines.

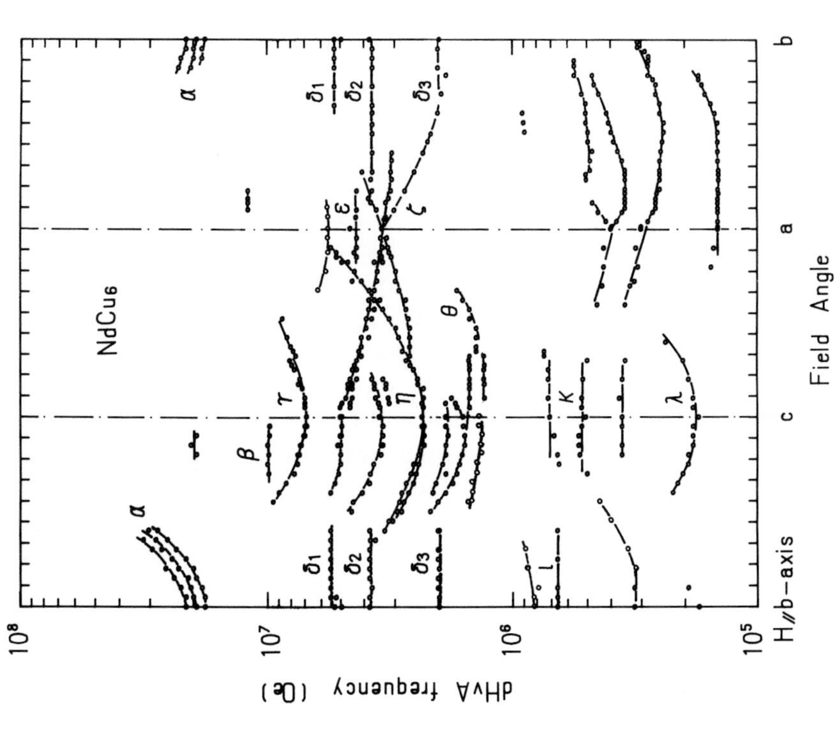

Figure 7-52. Angular dependence of the dHvA frequency in $NdCu_6$ (Ōnuki et al., 1991a). The solid lines connecting the data are guidelines.

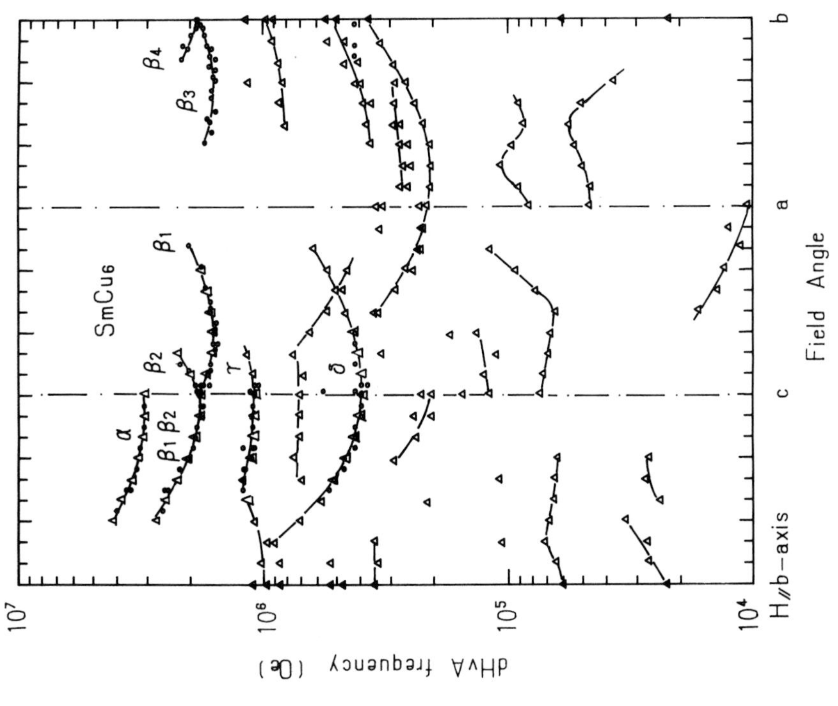

Figure 7-53. Angular dependence of the dHvA frequency in $SmCu_6$ (Endoh et al., 1987; Ōnuki et al., 1990b). The solid lines connecting the data are guidelines.

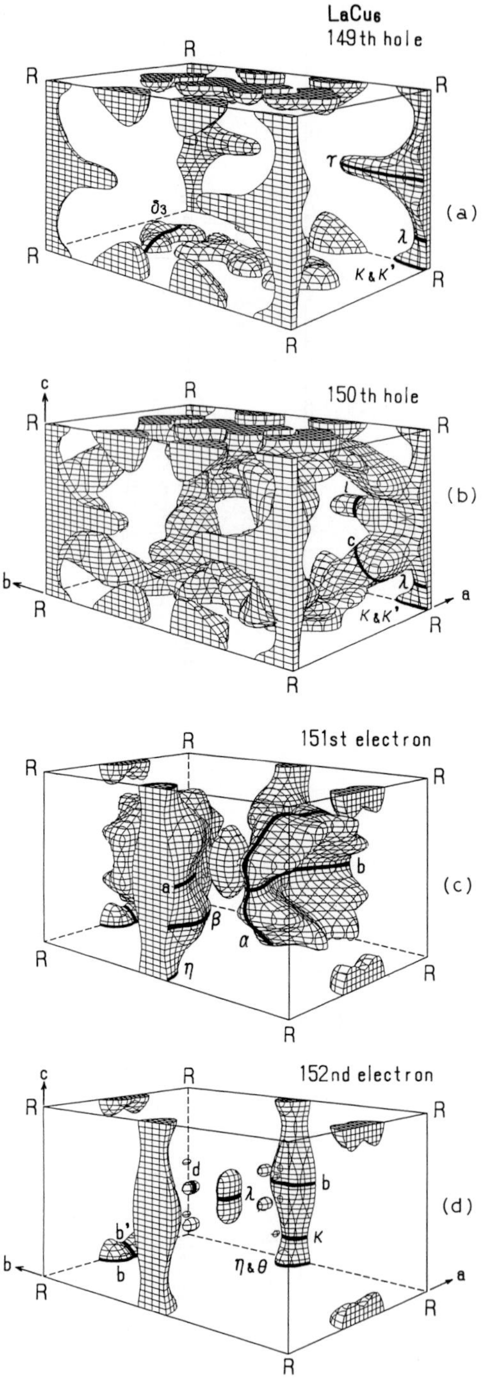

Figure 7-54. Fermi surfaces associated with the (a) band 149 hole, (b) band 150 hole, (c) band 151 electron and (d) band 152 electron in $LaCu_6$ (Harima et al., 1991). The thick solid lines correspond to the observed orbits.

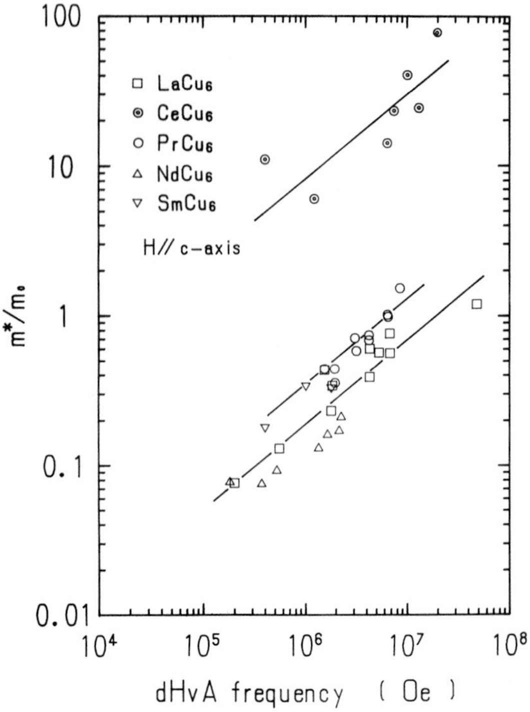

Figure 7-55. Cyclotron masses vs. dHvA frequencies in RCu_6 (Ōnuki et al., 1988). The solid straight lines are guidelines.

imental dHvA data, in particular around the *b*-axis, are desirable.

The temperature dependence of the elastic constant above the Kondo temperature of about 5 K in $CeCu_6$ exhibits a softening due to the quadrupole-strain interaction associated with the crystalline electric field states. Below 5 K the longitudinal C_{11} and C_{33} modes again show softening, reflecting the development of the heavy fermion state. As shown in Fig. 7-57, the field dependence of C_{33} mode exhibits a remarkable oscillatory behavior at low temperatures below 500 mK (Goto et al, 1988 b). The anomaly at 2 T relates to the metamagnetic transition. The period of the oscillation above 4 T up to 8.5 T is not proportional to $1/H$. The origin is not clear, as mentioned when discussing the $CeCu_2$ data.

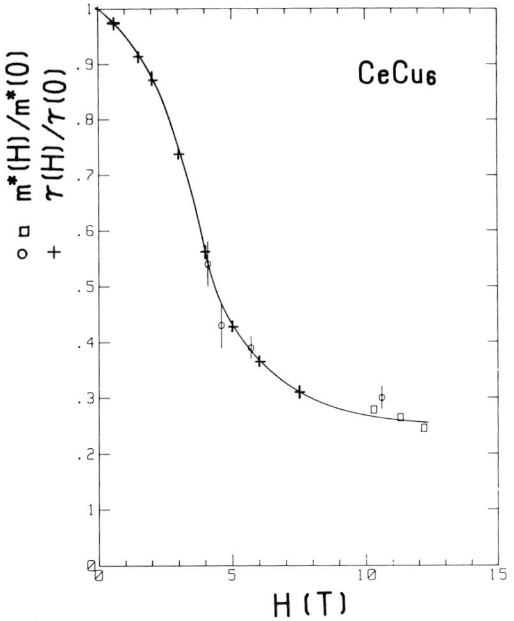

Figure 7-56. Field dependence of the normalized cyclotron masses $m^*(H)/m^*(0)$ (Chapman et al., 1990a) and the specific heat coefficients $\gamma(H)/\gamma(0)$ (Amato et al., 1987). The cyclotron masses at zero field are estimated as $m^*(0) = 20$ for a dHvA frequency $F = 1.20 \times 10^6$ Oe (open circles) and $m^*(0) = 120$ for $F = 1.034 \times 10^7$ Oe (squares). The results of specific heat coefficient are denoted by crosses, and the solid line is a guideline.

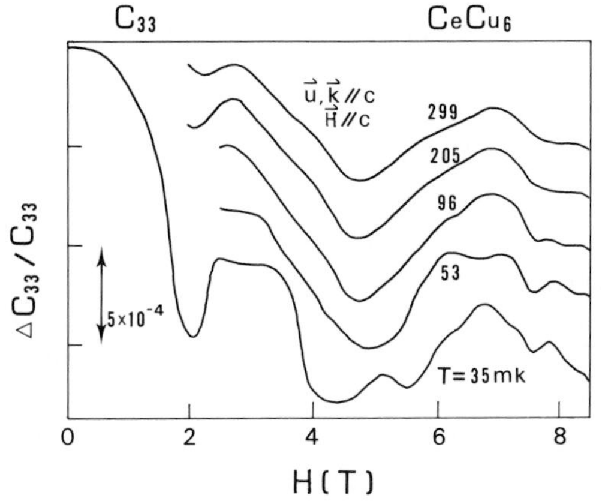

Figure 7-57. Field dependence of the elastic constant C_{33} mode in $CeCu_6$ (Goto et al., 1988b).

The Fermi surface of $SmCu_6$ is different from that of $LaCu_6$. One possibility is that the Fermi surface of $SmCu_6$ consists of small surfaces reflecting the small magnetic Brillouin zone. Alternatively, the main Fermi surface could be a complicated one without simple extremal cross-sections. In this case, the present results would show only a part of the Fermi surface.

7.4.2.8 $CeRu_2Si_2$

$CeRu_2Si_2$ possesses the tetragonal $ThCr_2Si_2$-type crystal structure with one molecule per primitive cell. This material is thought to be a non-magnetic Kondo lattice compound. Reflecting a rather low value of Kondo temperature, about 20 K, the electronic specific heat coefficient is large, being 350 mJ/K^2 · mol (Besnus et al., 1985).

Figure 7-58 shows the angular dependence of dHvA frequency (Lonzarich, 1988). The β and γ branches are observed over the whole angle region, indicating nearly spherical Fermi surfaces. The corresponding cyclotron masses are small, being of the order of the free mass m_0. On the other hand, the cyclotron mass of the ε branch, which is observed around the a-axis, possesses a large value, equal to $20\,m_0$. The metamagnetic transition occurs at about 80 kOe when the field is directed along the c-axis. However, this brings about no appreciable change in Fermi surface topology for the β and γ branches. As no dHvA data of $LaRu_2Si_2$ are available, no discussion can be given of the relation of the topologies between the Fermi surface in $CeRu_2Si_2$ and the one in $LaRu_2Si_2$. It is, however, clear that the Fermi surfaces with large masses, corresponding to the observed large γ value, are not yet observed.

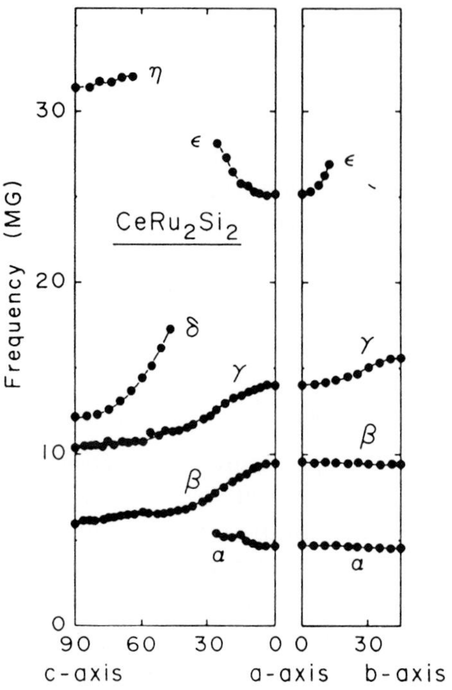

Figure 7-58. Angular dependence of dHvA frequency in $CeRu_2Si_2$ (Lozarich, 1988). The solid lines connecting the data are guidelines.

The band calculation made by Zwicknagl et al. (1990) suggests four closed hole Fermi surfaces centered around Z and one complicated, multiply connected surface. The former three small closed Fermi surfaces were assigned to the α, β and γ branches, while the δ, ε and η branches are neck orbits associated with the multiply connected surface. Here the Ce 4f electrons are strongly hybridized with the Ru d electrons.

Note that dHvA oscillations have been observed in the typical Kondo lattice compound $CeCu_2Si_2$ (Hunt et al., 1990), having the same crystal structure as $CeRu_2Si_2$. Superconductivity and very weak antiferromagnetic order co-exist in this material below 0.7 K. Although only a part of the Fermi surfaces is observed, the detected carriers possess small masses of $5\,m_0$. Judging from the large value of the electronic specific heat coefficient, $1000\ \mathrm{mJ/K^2 \cdot mol}$, carriers with much larger masses should exist in this material. Band calculations were made for $CeCu_2Si_2$ by Sticht et al. (1986) using a Kondo lattice ansatz for the cerium 4f state and LDA potential parameters. Also, Harima and Yanase (1991) have calculated the energy band structure for $LaCu_2Si_2$ and $CeCu_2Si_2$. As regards the observed part of the Fermi surface, the calculated Fermi surface of $LaCu_2Si_2$ seems to fit the experimental results better than the itinerant f electron-Fermi surface for $CeCu_2Si_2$. Precise experiments are important to clarify the real situation.

7.4.2.9 RNi

RNi crystallizes in an orthorhombic structure. CeNi can be characterized as an interesting valence fluctuating system similar to $CeSn_3$ or $CePd_3$. Its Kondo temperature is about 150 K (Gignoux et al., 1983).

Figures 7-59 and 7-60 show the typical angular dependences of the transverse magnetoresistances in LaNi and CeNi, respectively (Maezawa et al., 1989; Ōnuki et al., 1989a). The magnetoresistance of CeNi is found to be almost the same as in LaNi. The magnetoresistance of LaNi at 75 kOe is similar to that in CeNi at 150 kOe regarding its shape as well as its magnitude. In other planes the behavior between LaNi and CeNi is also similar. The magnetoresistance increases with increasing field over a wide angle region, except for several particular configurations of field and current. These behaviors suggest that LaNi and CeNi are compensated metals with similar Fermi surfaces. The open orbits exist along the *b*- and *c*-axes.

Figures 7-61 and 7-62 show the angular dependences of dHvA frequencies in LaNi and CeNi respectively (Maezawa

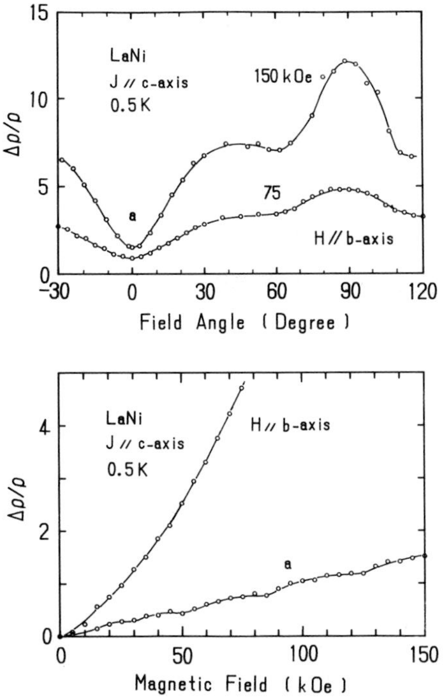

Figure 7-59. Angular and field dependences of the magnetoresistance in LaNi (Ōnuki et al., 1989 a). The solid lines connecting the data are guidelines.

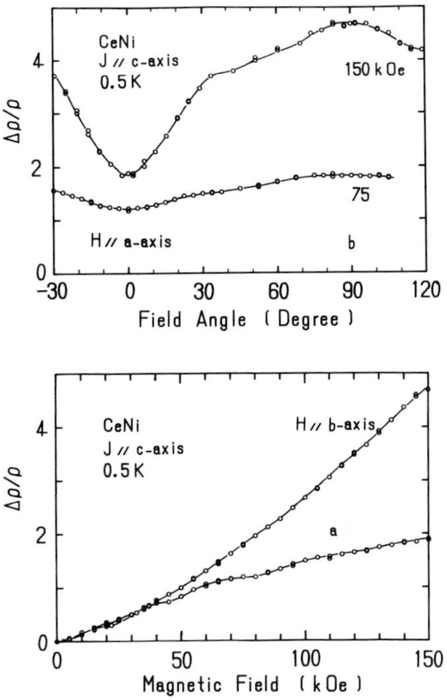

Figure 7-60. Angular and field dependence of the magnetoresistance in CeNi (Ōnuki et al., 1989 a). The solid lines connecting the data are guidelines.

et al., 1989; Ōnuki et al., 1989 a). The two branches, α at 4.10×10^7 Oe and β at 1.40×10^7 Oe for which the field is along the *b*-axis in the LaNi may correspond to those in CeNi, although the angular region for the α and β branches are different in both compounds. Moreover, two other branches indicated as γ and δ are observed along the *b*-axis in CeNi. From these experimental results, Ōnuki et al. have concluded that the Fermi surface of CeNi is similar to that of LaNi. Consequently the f electron is almost localized in CeNi. The cyclotron masses of the α and β branches in CeNi are 10.3 m_0 and 8.91 m_0, respectively. The corresponding masses in LaNi are 1.73 m_0 and 0.93 m_0, respectively. The cyclotron masses of CeNi are about ten times larger than those of LaNi. This is roughly consistent with the electronic specific heat coefficient. It is 65 or 85 $mJ/K^2 \cdot mol$ in CeNi, while it is only 5 $mJ/K^2 \cdot mol$ in LaNi (Gignoux et al., 1983; Ishikawa et al., 1987).

Yamagami and Hasegawa (1991 a) have calculated the energy band structure of LaNi and CeNi by the relativistic APW method within the framework of the local density approximation, taking into account the spin-orbit interaction from the start through the self-consistent iteration process. The solid curves in Figs. 7-61 and 7-62 are theoretical results. The experimentally observed branches are almost in agreement with the calculated ones. Here the f electron in CeNi is treated as an itinerant electron. The origins of each branch for LaNi and CeNi are shown in Figs. 7-63 and 7-64, respectively. The Fermi surfaces in bands 19 and 20 in LaNi refer to hole

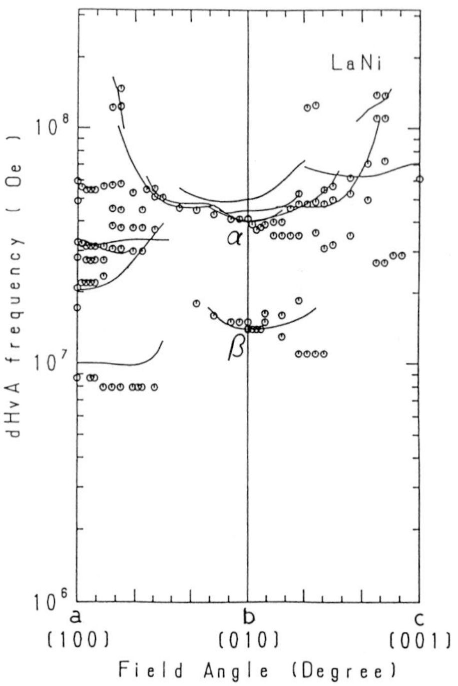

Figure 7-61. Angular dependence of the dHvA frequency in LaNi (Maezawa et al., 1989). The solid lines represent the results of band calculations (Yamagami and Hasegawa, 1991 a).

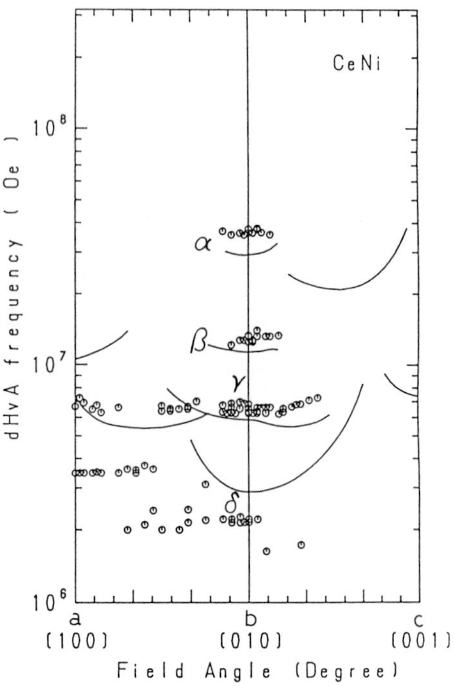

Figure 7-62. Angular dependence of the dHvA frequency in CeNi (Ōnuki et al., 1989 a). The solid lines represent the results of band calculations (Yamagami and Hasegawa, 1991 a).

and electron, respectively. Bands 20 and 21 in CeNi refer to hole and electron, respectively. Both substances are compensated metals with an equal carrier number of electrons and holes and possess the multiply connected Fermi surfaces which favor the experimentally observed open orbits along the *b*- and *c*-axes. The fact that LaNi and CeNi have almost the same dHvA frequencies for the α and β branches is thus accidental because the topologies of the Fermi surfaces are different for LaNi and CeNi. As the main dHvA branches are thus explained by the present band calculation, the f electron in CeNi is considered to form an itinerant 4f band in the same way as in $CeSn_3$.

The dHvA experiment of the anisotropic ferromagnetic PrNi ($T_c = 20$ K) has shown that the Fermi surface is roughly similar to that of LaNi (Maezawa, 1990).

The dHvA study in LaNi and CeNi represents an important lesson. When the number of Ce atoms per unit cell is even, the band structures of the La and Ce com-

Table 7-3. Mass enhancement factor m_c^*/m_b for each orbit and γ/γ_b in $CeSn_3$ and CeNi.

	m_c^*/m_b					γ/γ_b
$CeSn_3$	γ	β	δ	α	ε	
	1.9	2.7	2.7	2.4–4.2	1.6	3.4
CeNi	δ	γ	β	α		
	2.6	3.4	2.4	3.9		4.6–6.0

pounds can be similar even in the itinerant 4f band model. Also the transverse magnetoresistance is very similar in character. Careful band calculations and precise experimental studies are essential.

We summarize in Table 7-3 the mass enhancement factor m_c^*/m_b for each orbit and γ/γ_b in the valence fluctuating materials $CeSn_3$ and CeNi. Even though the band masses and the degree of the 4f character are different in $CeSn_3$ and CeNi, the mass enhancement factor is similar and not so large, less than 5. In this sense the valence fluctuating Ce compounds are different from the Kondo lattice compounds where the 4f bands are not observed.

7.4.3 Uranium Intermetallic Compounds – 5f Systems

Initial dHvA experiments for some uranium compounds had been performed by the group at the Argonne National Laboratory. The Fermi surfaces of URh_3, UIr_3, UGe_3 and U were clarified by the band calculations. Some experimental data were also obtained for U_3As_4, U_3P_4, UPd_3 and UAl_2. For details see the excellent review written by Arko et al. (Arko et al., 1985).

Later more extended dHvA experiments in uranium compounds, including the heavy fermion materials, were made in England and Japan. In this section the

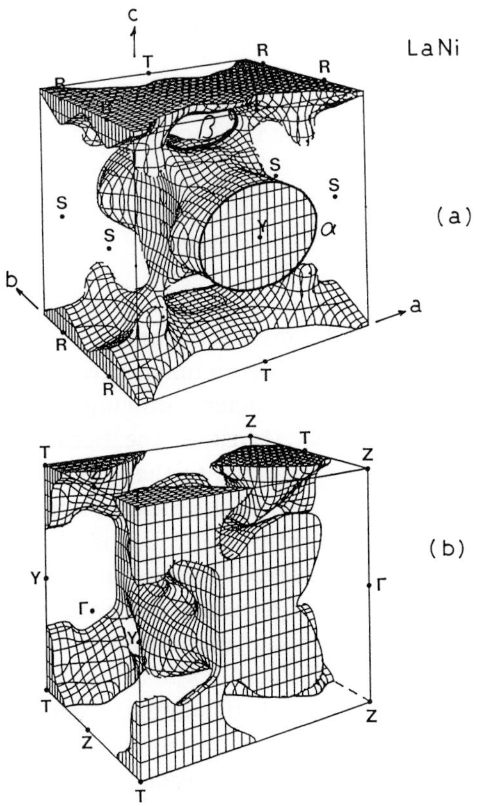

Figure 7-63. Hole and electron Fermi surfaces in (a) band 19 and (b) band 20 in LaNi, respectively (Yamagami and Hasegawa, 1991 a).

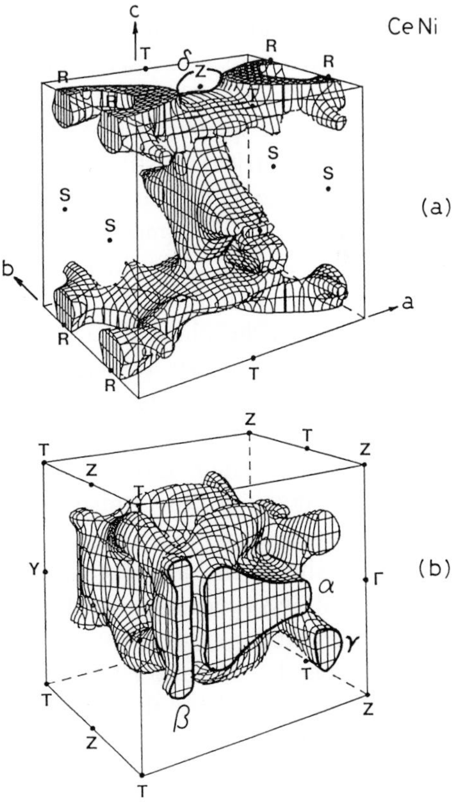

Figure 7-64. Hole and electron Fermi surfaces in (a) band 20 and (b) band 21 in CeNi, respectively (Yamagami and Hasegawa, 1991 a).

Fermi surface properties of UPd_3, UPt_3, UB_{12} and UC are described. Note that careful band calculations were done for UGe_3 (Hasegawa, 1984) and U (Yamagami and Hasegawa, 1990). By dHvA experiments on U_3P_4 a new branch was detected (Takeda, 1990).

7.4.3.1 UX_3 (X = Pd and Pt)

UPd_3 possesses the hexagonal (Ni_3Ti-type) structure with 16 atoms per unit cell. This substance is one of the U compounds whose crystalline field excitations have been observed, indicating the possibility of localized 5f electrons. UPd_3 shows a magnetic phase transition at 6.5 K which is thought to be due to a quadrupolar ordering (Buyers et al., 1980).

UPt_3 also crystallizes in the hexagonal ($MgCd_3$-type) structure. It is a well known spin fluctuator similar to UAl_2, possessing a $T^3 \ln T$ term in the specific heat (Stewart et al., 1984). The electrical resistivity is normal compared to the Kondo-like behavior in $CeCu_6$ or in UBe_{13}. The most interesting property of this substance is focused on possible triplet superconductivity, indicating multi-phases in the phase diagram of the superconducting upper critical field (Taillefer, 1990). To clarify the antiferromagnetic spin fluctuation and superconductivity, the alloy system $U(Pt_xPd_{1-x})_3$ was studied in detail by the Amsterdam group (de Visser et al., 1988).

Figure 7-65 shows the angular dependence of the dHvA frequency in UPd_3 (Ubachs et al., 1986).

A LMTO band calculation has been performed on UPd_3 where the $5f^2$ electrons have been treated as core states (Norman et al., 1987). The solid curves in Fig. 7-65 are the results of band calculation. The large branch α corresponds to the Γ-centered hole in band 2. One of the small

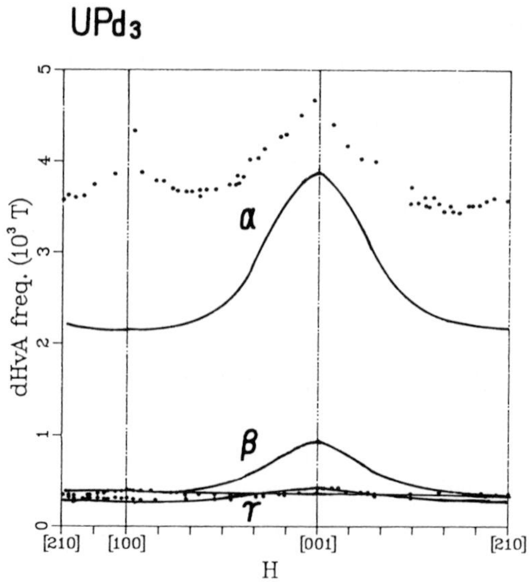

Figure 7-65. Angular dependence of the dHvA frequeny in UPd_3 (Ubachs et al., 1986). The results of band calculation are shown by the solid curves (Norman et al., 1987).

branches β must be the Γ-centered hole in band 1. The only other closed branch γ is the electron around H for band 5. In all three branches it is necessary to move the Fermi energy down to reach agreement. The shifts are 10 mRy for band 1, 4.5 mRy for band 2 and 5 mRy for band 5.

The detected cyclotron masses are of the usual d-band character, ranging from $0.50\,m_0$ to $2.0\,m_0$. Most of the occupied density of states are due to the Pd d electrons. The occupied U-d states are spread over the entire width of the Pd bands. In principle the observed band is similar to that of $ThPd_3$. To get a better quantitative agreement it seems to be necessary to include c-f mixing effects.

Figure 7-66 shows the angular dependence of the dHvA frequency in UPt_3 (Taillefer et al., 1987; Lonzarich, 1988). The dHvA experiment has been performed in the low temperature range between 150 mK and 20 mK and in high field rang-

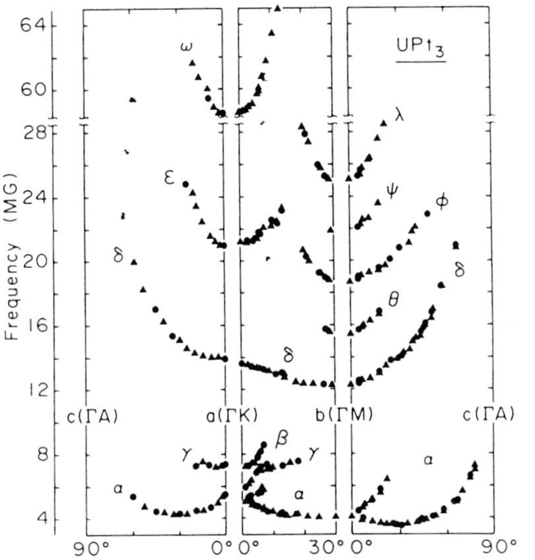

Figure 7-66. Angular dependence of the dHvA frequency in UPt_3 (Lonzarich, 1988).

ing from 40 kOe to 115 kOe in order to detect carriers with large masses. The dHvA branches are detected with the field in the basal plane, but they are not detected with the field along the *c*-axis probably due to the larger masses and/or curvature factor problems.

Figure 7-67 shows the cross-sectional areas of the Fermi surface for the LAPW calculation in the high-symmetry planes (Oguchi et al., 1987). This material is a compensated metal and the calculated Fermi surface consists of three closed electron surfaces centered at Γ (orbits 1, 2 and 3), one closed electron surface around K (orbit 3) and two hole tubes around the middle point of the upper plane AHL (orbits 4 and 5). The electron surfaces around Γ are dominantly of U f character, while there is a strong hybridization with Pt d states on the hole tube surfaces near H and L. The hybridization on the electron surface around K is intermediate between them.

The detected dHvA branches α, γ, δ, ε and ω are considered to correspond to Γ(e1), L(e4), K(e3), Γ(e2) and Γ(e3), respectively. The main features of the Fermi surfaces seem to be consistent with the results of band calculations, although some of the predicted Fermi surfaces are not observed experimentally. The main reason is due to the large mass because the specific heat coefficient is 420 $mJ/K^2 \cdot mol$. Later various more sophisticated techniques of band calculations were employed to improve the band calculation for UPt_3 (Wang et al., 1987; Christensen et al., 1988). It is, however, still necessary to shift the Fermi

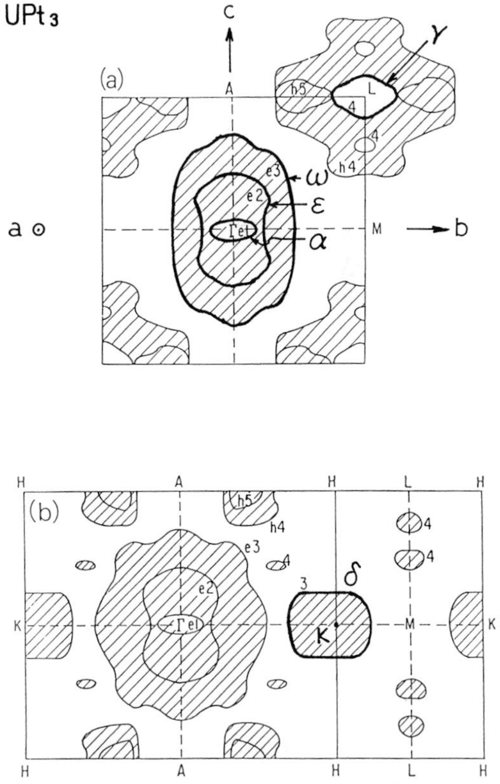

Figure 7-67. Cross-sections of the Fermi surface in UPt_3 showing three electron surfaces centered at Γ, one electron surface around K and two hole surfaces around the outer regions of the Brillouin zone (Oguchi et al., 1987).

energy up to 2 mRy to reach qualitative agreement between theory and experiment. As we have seen in CeSb and $CeSn_3$, more theoretical and experimental efforts are needed to draw a conclusion on whether the usual 5f band calculation can explain the observed Fermi surfaces or not.

In UPt_3 the cyclotron masses are in the range of 25 m_0 to 90 m_0. Since the field dependence of the linear coefficient of the specific heat is small, up to a field strength of 100 kOe, no appreciable field dependence of the masses is expected. The mass enhancement factor is about 20, homogeneous and consistent with γ/γ_b as shown in Table 7-4.

Table 7-4. Mass enhancement factor m_c^*/m_b for each orbit and γ/γ_b in UPt_3, UB_{12} and UC.

	m_c^*/m_b					γ/γ_b
UPt_3	α	γ	δ	ε	ω	
	19	40	23	24	21	21
UB_{12}	γ	η	ζ	β	α	
	0.9	1.6	1.1	1.2	1.0	1.1
UC	e_1	h_1				
	1.4	1.7				4.2

The metamagnetic transition of UPt_3 occurs around 20 T (Kouroudis et al., 1987). The longitudinal sound wave of C_{11} shows a remarkable softening around this transition. The oscillation of the transverse C_{66} has an unconventional period without a $1/H$ dependence around this transition. The reason for this is not clear.

7.4.3.2 UB_{12}

UB_{12} is one of the cubic RB_{12} type compounds, where R is a heavy rare earth element, U, Pu, Np, Y or Zr. This material possesses a high melting point of 2235 °C and is a congruently melting material. The previous magnetic susceptibility data show a Pauli-paramagnetic behavior, although the distance between the U atoms of 5.28 Å exceeds the Hill limit value of 3.4 Å (Troc et al., 1971).

As the magnetoresistance increases in a wide field region, this material was thought to be a compensated metal (Ōnuki et al., 1990 d). The magnetoresistance along the ⟨110⟩ direction, however, saturates at high fields, following a $H^{0.7}$-dependence. This result implies that an open orbit exists and that it is directed along the ⟨100⟩ direction.

Figure 7-68 shows the angular dependence of the dHvA frequency in the {100} and {110} planes (Ōnuki et al., 1991 b; Matsui et al., 1991). About fifteen branches are observed in a wide frequency range between 10^6 Oe and 10^8 Oe.

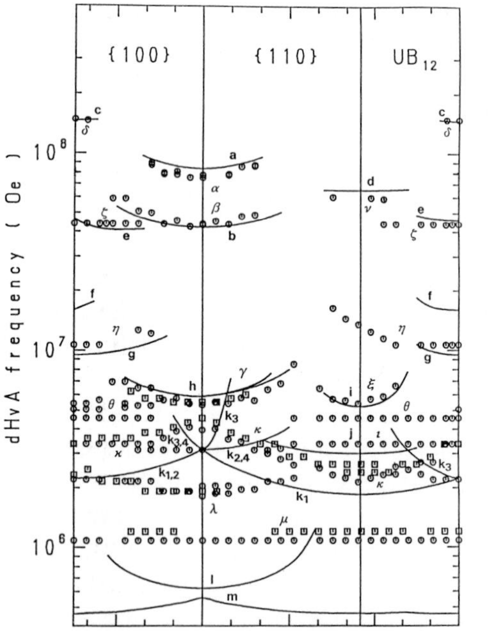

Figure 7-68. Angular dependence of the dHvA frequency in UB_{12} (Ōnuki et al., 1991 b). The solid lines represent the results of band calculations (Yamagami and Hasegawa, 1991 b).

The experimental results are compared with the energy band structure. Two band calculations have been reported (Harima et al., 1990 b; Yamagami and Hasegawa, 1991 b), but the latter band calculation shows better agreement with the experimental result. Therefore we use the result of the latter band calculation. The U-5f bands have a large band width of about 0.2 Ry because of their strong hybridization with the B-2p electrons. The Fermi energy is crossed by two wide bands due to the itinerant 5f electrons, which have mainly $J = 5/2$ components. These two bands make UB_{12} a compensated metal, consistent with the result for the magnetoresistance. As shown in Fig. 7-69 a, one hole Fermi surface in band 24, which is a cube with tunnels along the ⟨100⟩ direction, exists at the Γ point in the fcc Brillouin zone and possesses slender arms elongated along the ⟨111⟩ direction thus forming a multiply connected Fermi surface. On the other hand, the electron Fermi surface in band 25 is also a multiply connected one, as shown in Fig. 7-69 b. The cylindrical Fermi surface parts with an empty tunnel centered at X are mutually connected by flat arms. Here the axes of the arms are directed along ⟨110⟩, which favor an open orbit along the ⟨100⟩ direction. Moreover, a small spherical Fermi surface, centered at Γ, is an electron Fermi surface in band 25. The origins of the dHvA branches are shown in these figures. The solid lines in Fig. 7-68 are the results of band calculations by Yamagami and Hasegawa. These are in very good agreement with the experimental data.

The cyclotron masses are in the range $0.6\,m_0$ to $2.8\,m_0$. For example, the α and β branches are $1.88\,m_0$ and $1.93\,m_0$, respectively. The corresponding theoretical masses are $1.9\,m_0$ and $1.6\,m_0$, respectively. The experimental cyclotron masses

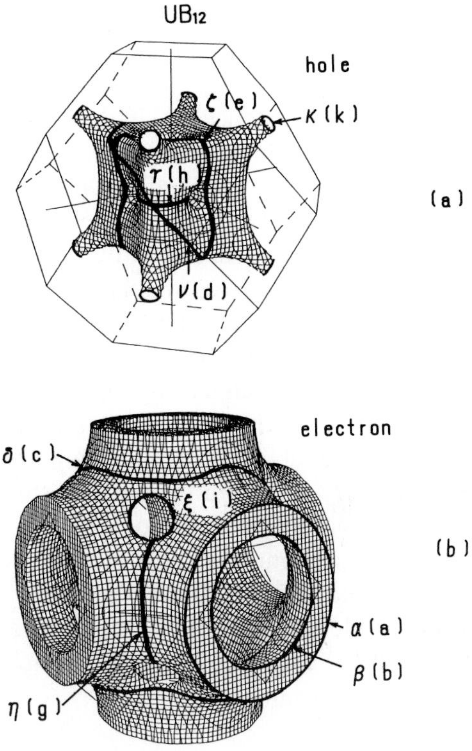

Figure 7-69. Calculated (a) hole and (b) electron Fermi surfaces in UB_{12} (Yamagami and Hasegawa, 1991).

are nearly equal to the theoretical ones. The electronic specific heat coefficient is $20\ mJ/K^2 \cdot mol$ (Kasaya et al., 1990), while the calculated density of states corresponds to $18\ mJ/K^2 \cdot mol$. The mass enhancement factor is very small in UB_{12}. The 5f electrons in UB_{12} behave like the 3d itinerant electrons in 3d intermetallic compounds.

7.4.3.3 UC

The NaCl-type uranium carbide UC, which is well known as nuclear fuel, has attracted interest because of its unusual properties, such as a high melting point of 2507 °C, high hardness, large brittleness and metallic conductivity (Holleck and Kleykamp, 1987). Owing to its simple crystal structure, this compound is appropriate for a fundamental study of the nature of the 5f electrons.

The magnetoresistance of UC increases with increasing field in the whole angle range, showing a $H^{1.7}$ dependence (Ōnuki et al., 1990c). It follows from this behavior that UC is a compensated metal.

Figure 7-70 shows the angular dependences of the dHvA frequency. The data shown by circles and triangles represent different branches defined by h_i and e_i ($i = 1$, 2 and 3). As the h_i branches are degenerate in the ⟨111⟩ direction, they are thought to originate from three Fermi surfaces having the shape of ellipsoids of revolution which are centered at three X points of the fcc Brillouin zone, the axes of revolution being along the Δ-axis. Two principal radii of the ellipsoid, k_1 and k_2, which are parallel and perpendicular to the Δ-axis, are determined in order to reproduce the dHvA branches: $k_1 = 0.141\,(2\pi/a)$ and $k_2 = 0.275\ (2\pi/a)$. The volume of an ellipsoid, Ω_h, is $0.0447\,(2\pi/a)^3$. Therefore the total carrier concentration contained in three ellipsoid is 0.067/cell.

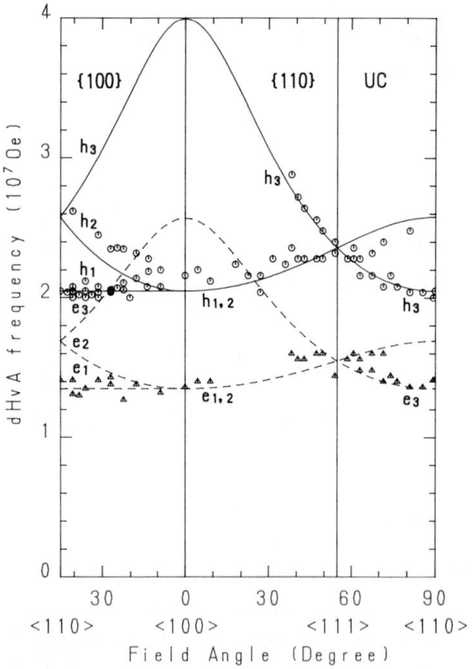

Figure 7-70. Angular dependence of the dHvA frequency in UC. Solid and broken lines show the calculated results based on the ellipsoidal Fermi surfaces (Ōnuki et al., 1990c).

The e_i branches have similar symmetry. Ōnuki et al. (1990e) tentatively assumed that other ellipsoids of revolution exist at each X point having the radii: $k_1 = 0.116$ $(2\pi/a)$ and $k_2 = 0.221\,(2\pi/a)$. The volume of a single ellipsoid, Ω_e, is $0.0236\,(2\pi/a)^3$. Note that the ratio Ω_e/Ω_h is 0.528, very close to 1/2. It is not clear whether this small deviation from 1/2 is a simple error due to crudeness of the ellipsoid model, or if it implies the existence of other Fermi surfaces which have not been observed. If it is assumed that Ω_e/Ω_h equals 1/2, to maintain compensation, the number of e_i ellipsoids should be six.

The energy band structure and the Fermi surface for UC was calculated by Hasegawa and Yamagami (1990) by a symmetrized, self-consistent relativistic APW method with the exchange and correlation potential described in the local density approximation. UC is a semi-metal which has 0.068 holes/cell and the compensating number of electrons, which agrees well with the experimental value. This result is different from previous band calculations (Freeman and Koelling, 1977; Weinberger et al., 1979). The reason for this is not clear.

The Fermi surface consists of three hole Fermi surfaces in the C-2p valence band which are centered at the X points and six electron Fermi surfaces in the U-5f conduction band which are centered at the W points, as shown in Fig. 7-71. The hole Fermi surface looks almost like an ellipsoid of revolution with the Δ-axis as the axis of revolution. The electron Fermi surface also looks like an ellipsoid of revolu-

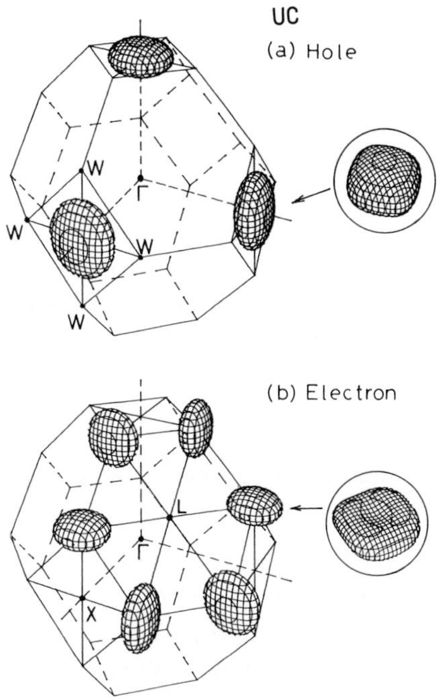

Figure 7-71. Ellipsoidal (a) hole and (b) electron Fermi surfaces in UC (Ōnuki et al., 1990c). Encircled Fermi surfaces are based on the band calculation of Hasegawa and Yamagami (1990).

tion which has its axis of revolution along the Z-axis, but has a dip in the middle of the surface, as shown in the circles of Fig. 7-71.

The two detected cyclotron masses, $0.84\,m_0$ (2.05×10^7 Oe) and $2.67\,m_0$ (1.35×10^7 Oe), which were observed at $\langle 110\rangle$, originate obviously from the hole and the electron Fermi surfaces, respectively. The corresponding band masses at $\langle 110\rangle$ are $0.50\,m_0$ (1.98×10^7 Oe) and $1.9\,m_0$ (1.26×10^7 Oe), respectively. Although the experimental cyclotron masses are larger than the theoretical ones, the observed trend is correctly predicted by the band theory. The calculated total density of states is 28 states/Ry cell which corresponds to the electronic specific heat coefficient of 4.8 mJ/$K^2\cdot$mol. This value is about 25% of the experimental value of 20 mJ/$K^2\cdot$mol (Westrum et al., 1965), implying that the average enhancement factor for the effective mass is about four. This means that the unobserved parts of the dHvA signals for electrons around $\langle 100\rangle$ have large masses, more than $10\,m_0$. This is the main reason why they were not observed in the present experiment.

Table 7-4 summarizes the mass enhancement factor m_c^*/m_b for each orbit and γ/γ_b in UPt_3, UB_{12} and UC.

7.5 Other Methods to Probe Fermi Surfaces

7.5.1 Positron Annihilation

The positron annihilation technique is also useful for obtaining information on the band structures, electron wave functions and topologies of the Fermi surfaces in metals. When a positron is injected into a metal, it looses its kinetic energy very quickly and is thermalized with an energy of $k_B T$. The thermalization time is of the order of 10^{-12} s at room temperature, which is two orders of magnitude shorter than the position lifetime of 10^{-10} s. The merit of the positron annihilation technique is that it is applicable to not so pure samples. However, the positron is easily trapped at crystal defects such as vacancies or negatively charged impurities. Actually the positron annihilation technique is used most commonly to check crystalline defects. It is thus necessary to use high-quality samples for this purpose.

When the electron density is low, it is energetically favorable for the positron to form a positronium, a hydrogen-like bound state with an electron. This state is easily confirmed because it shows a sharp spectrum due to the positronium annihilation.

As shown in Fig. 7-72, the injected positron annihilates an electron, and then a pair of photons (2γ) are emitted in nearly opposite directions, $p_{\gamma 1}$ and $p_{\gamma 2}$, to satisfy the momentum conservation law. When the involved electron and positron have finite moments, the total momentum of the two emitted photons is also finite. As the

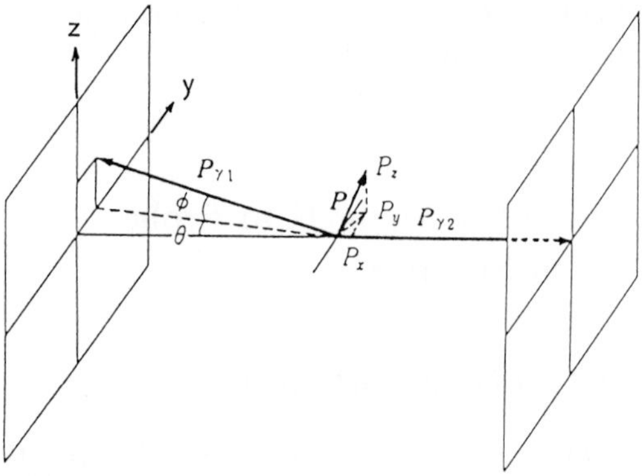

Figure 7-72. General outline of the geometry of the 2D-ACAR apparatus.

energy for the positron annihilation $2m_0c^2$ is much larger than the Fermi energy, an electron-positron pair with a momentum $\boldsymbol{p}$ can be determined by detecting a small deviation from collinearity between the directions of two photons by means of two detectors placed symmetrically on either side of the central sample in Fig. 7-72 and connected to read-out electronics equipped with a coincidence trigger. This makes it possible to obtain integrated information for the momentum in the photons direction. Namely, the two-dimensional angular correlation spectrum of the annihilation radiation (2D-ACAR), $N(p_y, p_z)$, is a projection of the momentum density distribution $\varrho(\boldsymbol{p})$ of the total electron-positron pair momentum onto the $p_y - p_z$ plane. This 2D-angular correlation can be expressed by the relation

$$N(p_y, p_z) = \int \varrho(\boldsymbol{p})\, \mathrm{d}p_x \tag{7-18}$$

where p_x is the momentum along the axis defined by the collinear 2γ decays, and $\phi = p_z/mc$, $\theta = p_y/mc$ are the angles measuring the small deviation of the two photons from collinearity.

From such 2D-ACAR spectra for several different projections we can reconstruct the three-dimensional momentum density distribution $\varrho(\boldsymbol{p})$ by using the direct Fourier reconstruction technique. Here the momentum density can be described as

$$\varrho(\boldsymbol{p}) \propto \sum_{n,\boldsymbol{k}} \left| \int_V \mathrm{d}\boldsymbol{r}\, \psi_+(\boldsymbol{r})\, \psi_{n,\boldsymbol{k}}(\boldsymbol{r}) \exp(-\mathrm{i}\boldsymbol{p}\cdot\boldsymbol{r}) \right|^2 \tag{7-19}$$

where $\psi_{n,\boldsymbol{k}}(\boldsymbol{r})$ is the wave function of an electron, n is the band index, $\boldsymbol{k}$ is the wave vector, $\psi_+(\boldsymbol{r})$ is the wave function of a thermalized positron ($\boldsymbol{k} \sim 0$), V is a crystal volume and the summation is taken over all occupied electron states.

To get information on the Fermi surface topology, we reduce the $\boldsymbol{p}$-space distribution $\varrho(\boldsymbol{p})$ to the $\boldsymbol{k}$-space distribution $n(\boldsymbol{k})$. Namely, the $\boldsymbol{p}$-space distribution should be packed back to the Brillouin zone by adding the reciprocal wave vectors. This is known as a Lock-Crisp-West folding procedure (Lock et al., 1973), and is a periodical superposition of $\varrho(\boldsymbol{p})$ on every reciprocal lattice point, as given by

$$n(k) = \sum_{\boldsymbol{G}_i} \varrho(\boldsymbol{k} + \boldsymbol{G}_i) \tag{7-20}$$

where $\boldsymbol{G}_i$ is the i-th reciprocal lattice vector and $\boldsymbol{k}$ is the wave vector defined within the first Brillouin zone. Using the Bloch theorem, $n(\boldsymbol{k})$ can be described as

$$n(\boldsymbol{k}) \propto \sum_{n,\boldsymbol{k}} \theta(E_\mathrm{F} - E_{n,\boldsymbol{k}}) \cdot \int_{\mathrm{Cell}} \mathrm{d}\boldsymbol{r}\, |\psi_+(\boldsymbol{r})|^2\, |\psi_{n,\boldsymbol{k}}(\boldsymbol{r})|^2 \tag{7-21}$$

where $E_{n,\boldsymbol{k}}$ is the energy of the electron in the state $\psi_{n,\boldsymbol{k}}(\boldsymbol{r})$ and $\theta(E_\mathrm{F} - E_{n,\boldsymbol{k}})$ is a step function as follows:

$$\theta(E_F - E_{n,k}) = \begin{cases} 1 & E_F \geqq E_{n,k} \\ 0 & E_F < E_{n,k} \end{cases} \quad (7\text{-}22)$$

Therefore, $n(\boldsymbol{k})$ should have breaks, corresponding to the integral in Eq. (7-21), between the inside and the outside of the Fermi surface. If the density of positrons can be assumed to be uniform in space, the integral in Eq. (7-21) equals unity for occupied electron states.

Figure 7-73 a shows a typical spectrum for Nb (Kubota et al., 1990). Here the 45° angle is the deviation of the projection from [100]. Figure 7-73 b is an obtained contour plot of the reduced zone momentum density $n(\boldsymbol{k})$ for Nb in the first Brillouin zone. The open circles are the Fermi surface edges, and the dark patterns have lower density when compared with the values of $n(\boldsymbol{k})$ on the open circles. Here the wave function of the positron $\Psi_+(\boldsymbol{r})$ is assumed to be uniform in a whole region. Three hole Fermi surfaces, namely the ellipsoids centered at N and the multiply connected arms which overlap with an octahedron centered at Γ, are clearly identified, and are compared with the results of the band calculation in Fig. 7-73 c (Mattheiss, 1970). The resolution in 2D-ACAR is roughly equal to ten percent of the distance of the Brillouin zone.

Fermi surface studies for the monoatomic metals and their alloys were almost all carried out by the 2D-ACAR method. As for the intermetallic compounds, successful analyzes are few in number. For example, Fig. 7-74 shows a hole Fermi surface for V_3Si obtained by using the procedure described above (Manuel, 1982). This result is consistent with the result of a LMTO band calculation.

Usually the thermalized positron is treated as a simple, free particle, plane wave, as mentioned above. For delicate

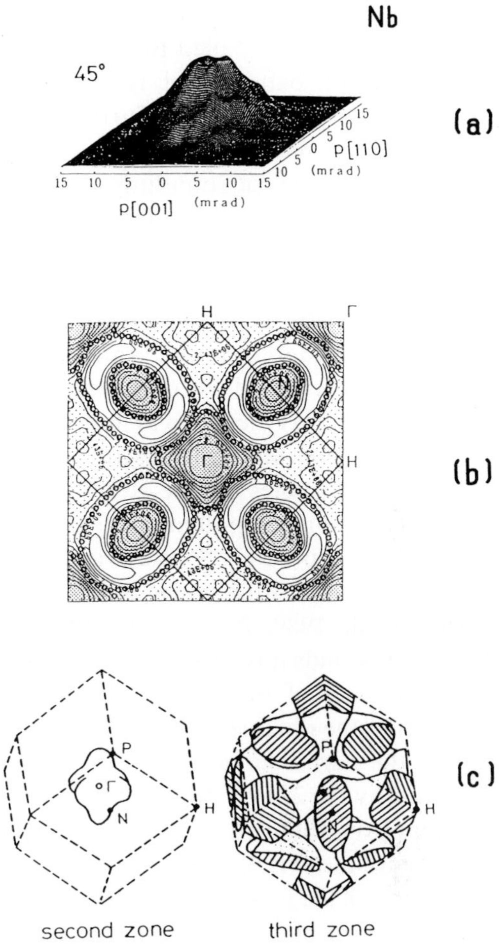

Figure 7-73. (a) Typical 2D-ACAR spectrum for Nb. The angle denotes the deviation of the projection from [100]. (b) Contour plots of the reduced zone momentum density $n(\boldsymbol{k})$ for Nb (Kubota et al., 1990). (c) Calculated Fermi surface of Nb (Matheiss, 1970).

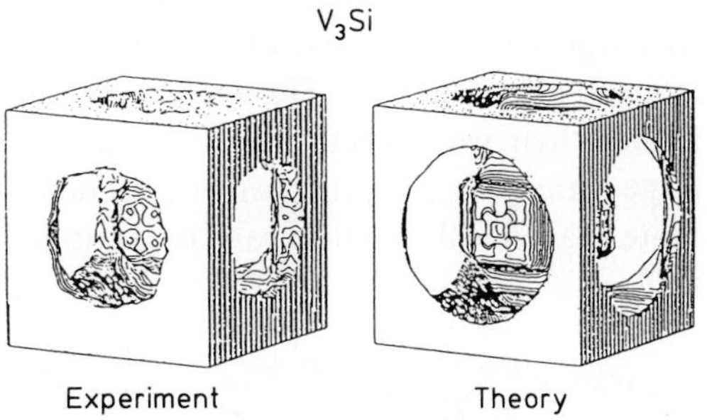

Figure 7-74. A hole sheet of the V_3Si Fermi surface determined by 2D-ACAR experiments and band structure calculations (Manuel, 1982).

cases such as when d and f electrons are involved, the more detailed Bloch state, on the basis of the band model, is needed for describing the positron state. Furthermore, many-body correlation effects between the positron and the surrounding electrons should be checked carefully. This is important for systems with a low electron density. A limiting case corresponds to positronium formation, as mentioned above. Therefore, in order to study Fermi surfaces for strongly correlated electron systems, in particular for heavy fermion systems, careful measurements and analyzes are needed.

2D-ACAR studies for the strongly correlated f electron compounds such as CeB_6 and URu_2Si_2 or high T_c materials were also carried out (Tanigawa et al., 1989; Rozing et al., 1989; Peter et al., 1989). In these compounds it is difficult to determine the topologies of the Fermi surfaces directly from the obtained $n(\boldsymbol{k})$ data. For example, the Fermi surface model for CeB_6 obtained by the 2D-ACAR method is very different from that of LaB_6. This is inconsistent with the dHvA result. As mentioned above, one of the reasons for this is that the wave function of the positron is not uniform. The structures in $\varrho(k)$ are very strongly distorted according to how the positron wave function overlaps with the electron wave functions.

In this sense it might be difficult to observe accurately the relatively well-localized states, such as those of 3d and f electrons, by this method. The reason for this is that their wave vectors spread into the large $\boldsymbol{k}$-range in the extended zone scheme. Note that the Bloch state can be expanded as

$$\psi_{n,\boldsymbol{k}}(\boldsymbol{r}) = \mathrm{e}^{\mathrm{i}\boldsymbol{k}\cdot\boldsymbol{r}} \sum_{\boldsymbol{G}_i} u_{ni}\, \mathrm{e}^{-\mathrm{i}\boldsymbol{G}_i\cdot\boldsymbol{r}} \quad (7\text{-}23)$$

where $\boldsymbol{G}_i$ is the reciprocal wave vector and the $\boldsymbol{k}+\boldsymbol{G}_i$ state appears in the probability $|u_{ni}|^2$. Furthermore, the positron annihilation intensity of Eq. (7-19) is weak because the positron wave function has smaller amplitudes around the positive ions where 3d and f electrons are located.

However, the techniques for obtaining stronger positron sources are improving very rapidly. Actually, instead of electrons, positrons are now circulating in the storage ring of synchrotron radiation centers, and the positron diffraction technique opens a new era. In this sense positron annihilation measurement is also expected to be a more useful technique in the near future.

7.5.2 Photoemission and Bremsstrahlung Isochromat Spectroscopies

Other important methods are angle resolved photoemission spectroscopy (ARPES) and Bremsstrahlung isochromat spectroscopy (ARBIS). These methods can obtain the one-particle Green's function directly and thus the energy dispersion in $\boldsymbol{k}$-space is also determined directly. Unfortunately, the resolution is not very good, 20 meV in energy at best, and thus the dispersion in the heavy fermion bands in the f electron systems has not yet been observed by PES and BIS. However, these methods provide important data for the f electron systems, as will be discussed next. The dispersion in a high T_c material was, however, measured successfully, although there are controversies in the experimental results, as will also be discussed below.

In f electron systems, most of the data are obtained through angle integrated PES and BIS. Here PES is one of the powerful methods for investigating the density of states of occupied states, which can be determined by analyzing the kinetic energy of the emitted electron through the photoelectric effect when the energy of the incident photon is transferred to the electron in the sample. On the other hand, BIS can be considered as a time reversed photoemission process with an incident electron and an outgoing photon. BIS is complementary to PES, probing the density of states of unoccupied states. When the angle between the incident photon or electron and the single crystal plane is changed, the energy dispersion of the electronic states can be obtained. This method corresponds to the angle resolved PES of BIS, respectively.

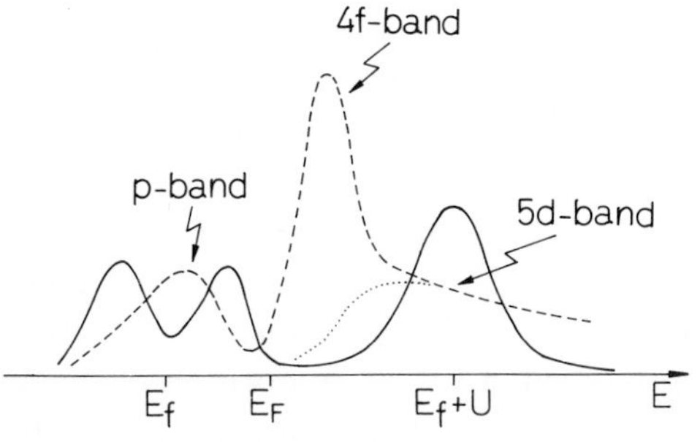

Figure 7-75. Schematic PES and BIS of 4f electrons in CeSb, shown by the solid curve. PES is magnified in area. The occupation number of 4f electrons is one below E_F and thirteen above E_F. The density of states of p, 4f and 5d bands in the usual band calculation are shown by the broken and dotted lines, respectively.

Schematic 4f PES and BIS for CeSb are shown in Fig. 7-75 by the solid curve as an example. Note that it is possible to get the 4f character selectivity by the resonance PES technique. The density of states obtained for LaSb by the usual band calculation is shown in Fig. 7-76 for comparison. Usually PES and BIS for the s, p and d electrons are fairly well described by the usual band model, and thus also for LaSb. On the other hand, PES and BIS for the 4f electron are substantially different from the density of states calculated by the band model because of the strong correlation effects.

When the correlation is introduced as a many-body effect, a certain amount of the 4f intensity spills into the satellite peaks, as shown in Fig. 7-77a. On increasing the correlation effects, the satellite intensity gradually increases in magnitude and finally becomes a so-called atomic peak, as shown in Fig. 77-7c, in which the PES

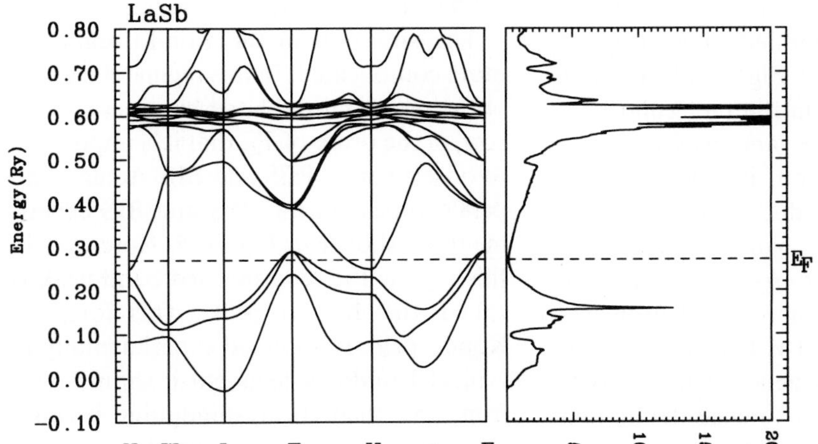

Figure 7-76. (a) Energy band structure and (b) density of states in LaSb. The numerous narrow bands around 0.6 Ry are 4f bands. The occupied (electron-) bands at X are 5p (Sb) bands and the unoccupied (hole-) bands at Γ are mainly 5d (La) bands.

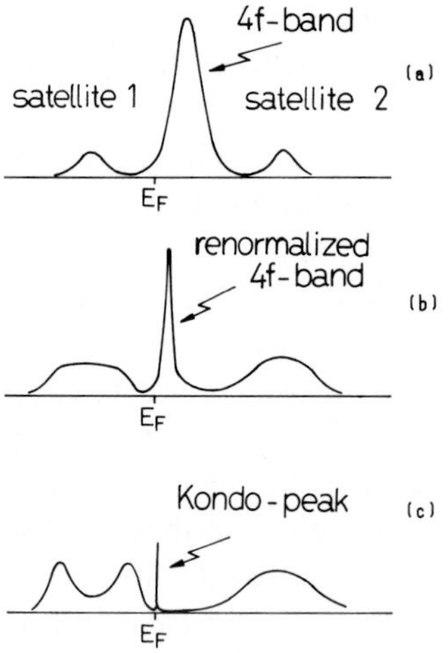

Figure 7-77. Schematic density of states of 4f-electrons in the sense of the one-particle Green's function: (a) The 4f band-like or near valence fluctuation regime, (b) the renormalized 4f band-like or valence fluctuation regime and (c) the Kondo regime with a fairly large T_K. PES is enlarged in area.

peak corresponds to the transition of $4f^1$ to $4f^0$, the BIS peak corresponds to the one of $4f^1$ to $4f^2$ and the energy difference between these peaks corresponds to the effective 4f correlation energy U_{4f}.

Let us again consider the situation where the 4f band intensity is decreased. The 4f band is then termed a renormalized 4f band, as shown in Fig. 7-77b. Usually this band is treated in the molecular field approximation in the same way as in the usual band calculation, in which the c-f mixing intensity is reduced due to the reduced weight of the 4f band. Therefore, the width and thus the inverse of the effective mass of the renormalized band decreases when the peak height is kept nearly constant. This model is believed to be fairly good, at least for the valence fluctuation regime because it seems to be consistent with the observed PES and BIS. Note, however, that no systematic experimental studies have been done so far to relate the mass enhancement obtained from the dHvA measurement to the narrowing effect in the renormalized 4f band, as seen in BIS.

As for the band calculation of the Fermi surface, the situation is not simple. The renormalized band calculation in $CeSn_3$ was done by Strange and Newns (1986), in principle, on the basis of the band calculation by Koelling (1982). They showed that the original band structure is modified substantially, in such a way that there is better agreement with experiment. However, the results of the band calculation done by Hasegawa et al. (1990) agree perfectly with the dHvA experiment without any such renormalized calculation. Therefore, the renormalized band calculation based on the results of Hasegawa et al. is expected to disagree with the experiment. A further step is needed for the renormalized band calculation to show that it can explain the experimental result for $CeSn_3$. Otherwise the method to calculate the Fermi surface by the usual band model, and to calculate the mass enhancement by taking into account the low frequency spin and charge fluctuations, may be a better way to describe valence fluctuation systems.

The situation in the Kondo regime is more complicated. In this regime the width of the renormalized band becomes so narrow, of the order of T_K, that it is impossible to observe it by PES and BIS. Actually no peaks appear in the PES and BIS experiment, as shown in Fig. 7-75. Note that in this regime the renormalized band peak is called the Kondo peak. Therefore, no Kondo peak is observed experimentally in typical Kondo systems but it should exist from the theoretical standpoint because the Luttinger theorem and the impurity Kondo picture are, in principle, applicable to dense Kondo regimes.

In some Yb compounds, such as $YbAl_3$, the Kondo peak is observed in PES because of the electron-hole symmetry (Oh et al., 1988). Analyzes based on the impurity Kondo model seem to be applicable, although there are some controversies. Whether $YbAl_3$ belongs to the Kondo regime or the valence fluctuation regime should be checked by dHvA measurement because T_K is fairly high, more than 300 K.

It is expected, however, that the energy region where the quasiparticle picture with a well-defined dispersion is applicable is limited to a region very near the Fermi energy. In the most regions the quasiparticle picture is no longer valid because of the short lifetime due to various types of interactions. Therefore the simple renormalized 4f band picture without the lifetime effect should be improved. Note also the double peak structure of the 4f Green's function in PES as seen in Fig. 7-75. This double peak structure was explained by Sakai et al. (Sakai et al., 1984; Takeshige et al., 1985) as follows. When a 4f hole or the f^0 state in a given Ce-site is created, it is screened immediately by 5d or by other 4f electrons. In this process the p-d or p-f mixing is important (for example in monopnictides RX) because it causes charge transfer to the Ce-site. The importance of the 5d screening effect was theoretically studied in detail by Takeshige (1990), and the experimental evidence is now accumulating.

Many theorists believe that the Luttinger theory should be applicable even for the Kondo regime because the Kondo peak is recognized as a continuous change in the renormalized 4f band picture. Experimental results show clearly, however, that there is a big difference between the Kondo regime and the valence fluctuation regime and that no 4f bands are observed in the Kondo regime. It is suspected that the 4f band disappears very quickly, even for a very weak magnetic field. Experimental data show, however, that the γ value changes rather slowly in a weak field. Furthermore, in the high field region where the γ value changes rapidly, the topology of the Fermi surface does not depend on the field strength. On the other hand, the field dependence of the γ value is explained fairly well by the field dependence of the mass enhancement of the conduction electrons. Therefore, even if the 4f band is assumed to exist in some parts of the Fermi surface, but is not observed for some reason, the mass should not be larger than that of the conduction band since the experimental γ value is already accounted for by the large mass of the conduction band. If the mass is not too large, the 4f band Fermi surface should be observed experimentally, again contradicting the experiment. Anyway, there is no evidence of a renormalized 4f band in the Kondo regime, even in zero magnetic field. A possible explanation may be found to be a very strong impurity effect on the renormalized 4f band. More detailed experimental studies are needed in the Kondo regime as well as the boundary between the Kondo and valence fluctuation regimes.

The situation regarding the U 5f electrons is also somewhat confusing. Most U compounds are rather well described by the 5f band model. In some materials, such as in UB_{12}, the mass enhancement factor is very small, of the order of 10%. Because of the weaker effective f-f correlation energy (U_{5f} is 1–2 eV) the 5f PES and BIS shapes for such materials are also described fairly well by the 5f band model. These shapes are characterized by an additional satellite structure which is not separated from the 5f band peak but appears as a shoulder. The same type of PES and BIS structure is observed even in typical heavy fermion compounds, such as UBe_{13}, in which the main 5f band-like peak is broadened, in contrast to the usual narrowing in the 4f systems. However, when the resolution is increased, a sharp peak structure should appear just at the Fermi energy. The situation is thus clearly different from the 4f case, even though the low frequency and low temperature behaviors are similar to each other. It is also not clear whether dilute U systems have to be described as the Kondo states or not. Anyway, the study of 5f systems is not so extensive as that of 4f systems and more efforts are needed, in particular for the heavy fermion U compounds.

Angle resolved PES and BIS studies on the 3d compounds have been made in some detail because the low energy photon has a sufficiently large cross-section for 3d electrons, and the energy resolution increases for the lower energy photons. The same situation exists in BIS too, and angle resolved BIS for the 10 eV emitted photon has been also performed. Even if the materials are less pure than those used in the usual dHvA measurement, angle resolved PES and BIS experiments can be done to study the Fermi surface as well as the overall dispersion. Typical examples are the high T_c superconducting Cu-oxide materials. Figures 7-78 and 7-79 show the results for $Bi_2Sr_2CaCu_2O_8$, which were obtained by Takahashi et al. (1988b, 1989) for the first time. However, the resolution is not as good in this experiment. Later measurements with better resolutions were performed by other groups (Olson et al., 1989; Flipse et al., 1989), giving different results. Takahashi et al. emphasize that the observed dispersion is not explained by band calculations and their sample is of better quality than those used in the other experiments. The latter groups claim that one of the dispersion relations observed by Takahashi et al. does not exist and that the obtained Fermi surface is consistent with the band calculations. It is well known that the

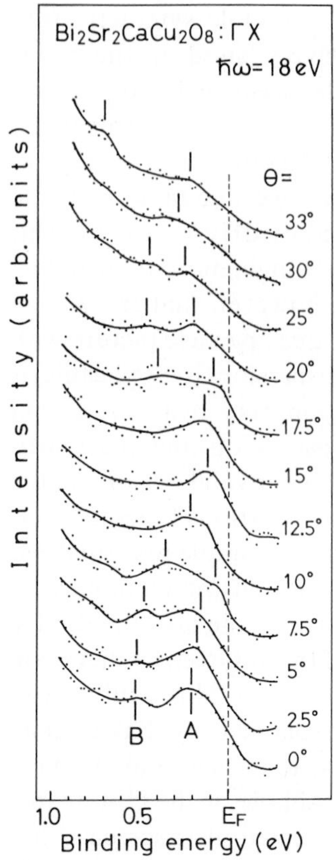

Figure 7-78. Angle resolved photoemission spectra of $Bi_2Sr_2CaCu_2O_8$ in the vicinity of E_F (Takahashi et al., 1989).

surface of the high T_c oxides is very unstable and thus very careful treatment is necessary. Furthermore it is very difficult to grow a good quality-single crystal of $Bi_2Sr_2CaCu_2O_8$. In this sense more study is necessary to determine the Fermi surface of high T_c material.

There are also the dHvA measurements made on $YBa_2Cu_3O_7$ under a pulsed magnetic field of 100 T (Smith et al., 1990). The high field is necessary to destroy superconductivity. Several small Fermi surfaces were found and one of these was also observed by Kido (1990) in the mixed superconducting state up to 20 T.

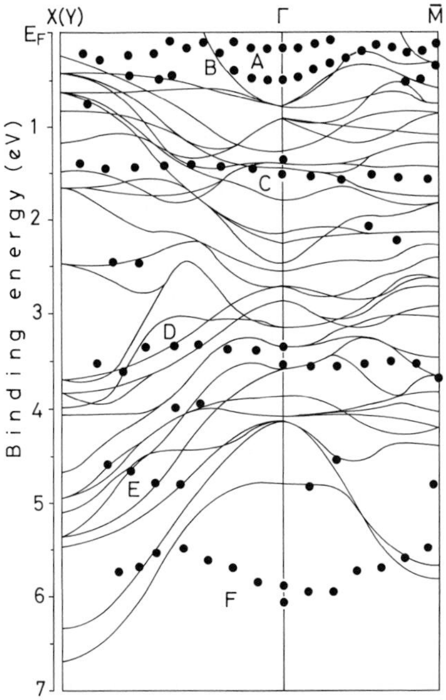

Figure 7-79. Band structures of $Bi_2Sr_2CaCu_2O_8$ determined by the angle-resolved photoemission study (Takahashi et al., 1989). The solid lines represent the results of band calculations (Massidda et al., 1988).

7.6 Concluding Remarks

The experimental Fermi surface studies of strongly correlated 3d, 4f and 5f electron systems and the concomitant band calculations have led to the following situation:

(1) For the transition metal compounds, the dHvA measurement has only been done for a limited number of materials which have not too large masses. The Fermi surfaces observed are consistent with the results of band calculation.

(2) For the rare earth compounds, for the Ce-compounds, in particular, many dHvA measurements have been made, including heavy fermion systems. The results can be summarized as follows. For the materials belonging to the valence fluctuation regime in which the renormalized 4f peak is observed by BIS near the Fermi energy, the Fermi surface can be fit very well by advanced band calculations. The narrowing effect on the renormalized 4f band, expected by the molecular field theory, is not clearly observed. The mass enhancement is fairly homogeneous and not very large. For the materials belonging to the Kondo regime the renormalized 4f peak or the Kondo peak is not observed by BIS because of the limited resolution. The Fermi surfaces can be fit very well by a generalized Hartree-Fock type band, that is, by putting one 4f state below the Fermi energy in the case of Ce-compounds, which is consistent with the PES peak. The other unoccupied 4f bands above the Fermi energy, which are separated by U_f from the occupied 4f state, are also consistent with the BIS 4f peak. Therefore the Fermi surface is essentially similar to that of the La-compound, as well as to those of well localized 4f systems, such as the Pr compounds. Because of the gradual change of the γ value under an applied magnetic field, the above situation is expected to be valid even in the absence of a magnetic field and thus in the absence of an induced moment. This means that the Luttinger theory starting from the 4f band picture is not applicable in the Kondo regime. The mass enhancement factor for the conduction band in the Kondo regime is very large, more than 100 and is anisotropic in some cases. It becomes larger for bands of stronger c-f mixing. On the other hand, the mass enhancement factor for the 4f band in the valence fluctuation regime is not very large, less than 5, because the 4f band mass is already fairly large.

(3) For most uranium compounds, the Fermi surfaces can be explained very well by the usual 5f band calculations. For the heavy fermion type U compounds, the

dHvA measurements are not very complete. A typical example is UPt_3. The observed Fermi surfaces are claimed to be explained by the 5f band model with some artificial modifications. As for the overall features of PES and BIS, no essential differences are observed between the normal and heavy U compounds. This feature is essentially different from those of rare earth compounds, reflecting the weaker 5f-5f correlation energy. Therefore, it is not clear whether the Kondo state seen in the Ce and Yb compounds exists in the heavy U compounds, even in the dilute limit.

7.7 Acknowledgements

We are very grateful to A. Hasegawa, A. Yanase, H. Harima, O. Sakai, S. Tanigawa and K. Ueda for helpful discussions, and to Y. Fujimaki for typing the present manuscripts.

7.8 References

Abell, T. S. (1989), in: *Handbook on the Physics and Chemistry of Rare Earths,* Vol. 12: Gschneidner Jr., K. A., Eyring, E. (Eds.). Amsterdam: Elsevier Science Publishers B. V., pp. 1–51.

Amato, A., Jaccard, D., Flouquet, J., Lapierre, F., Tholence, J. L., Fisher, R. A., Lacy, S. E., Olsen, J. A., Philips, N. E. (1987), *J. Low. Temp. Phys. 68,* 371.

Anderson, P. W. (1959), *Phys. Rev. 115,* 2.

Aoki, H., Crabtree, G. W., Joss, W., Hulliger, F. (1985), *J. Mag. Mag. Mat. 52,* 389.

Arko, A. J., Koelling, D. D. (1978), *Phys. Rev. B17,* 3104.

Arko, A. J., Brodsky, M. B., Crabtree, G. W., Karim, D., Koelling, D. D., Windmiller, L. R. (1975), *Phys. Rev. B12,* 4102.

Arko, A. J., Crabtree, G. W., Karim, D., Mueller, F. M., Windmiller, L. R., Ketterson, J. B., Fisk, Z. (1976), *Phys. Rev. B13,* 5240.

Arko, A. J., Koelling, D. D., Schirber, J. E. (1985), in: *Handbook on the Physics and Chemistry of the Actinides,* Vol. 2: Freeman, A. J., Lander, G. H. (Eds.). Amsterdam: Elsevier Science Publishers B. V., pp. 175–238.

Asano, H., Umino, M., Ōnuki, Y., Komatsubara, T., Izumi, F., Watanabe, N. (1986), *J. Phys. Soc. Jpn. 55,* 454.

Barbara, B., Rossignol, M. F., Boucherle, J. X., Schweizer, J., Buerzo, J. L. (1979), *J. Appl. Phys. 50,* 2300.

Bennett, L. H., Waber, J. T. (Eds.) (1967), *Energy Bands in Metals and Alloys.* New York: Gorden and Breach, Science Publishers.

Benoit, A., Boucherle, J. X., Flouquet, J., Sakurai, J., Schweizer, J. (1985), *J. Mag. Mag. Mat. 47&48,* 149.

Besnus, M. J., Kappler, J. P., Lehmann, P., Meyer, A. (1985), *Solid State Commun. 55,* 779.

Boulet, R. M., Jan, J-P., Skriver, H. L. (1982), *J. Phys. F: Met. Phys. 12,* 293.

Bredl, C. D. (1987), *J. Mag. Mag. Mat. 63&64,* 355.

Bredl, C. D., Steglich, F., Schotte, K. D. (1978), *Z. Phys. B29,* 327.

Buschow, K. H. J., van Daal, H. J., Maranzana, F. E., van Aken, P. B. (1971), *Phys. Rev. 3,* 1662.

Buyers, W. J. L., Murray, A. F., Holden, T. M., Svensson, E. C., Duplessis, de V., Langer, G. H., Vogt, O. (1980), *Physica 102 B,* 291.

Chapman, S. B., Hunt, M., Meeson, P., Springford, M. (1990 a), *Physica B163,* 361.

Chapman, S. B., Hunt, M., Meeson, P., Reinders, P. H. P., Springford, M., Norman, M. (1990 b), submitted to *J. Phys. Cond. Mat.*

Christensen, N. E., Andersen, O. K., Gunnarsson, O., Jepsen, O. (1988), *J. Mag. Mag. Mat. 76&77,* 23.

Cracknell, A. P. (1971), *The Fermi Surfaces of Metals.* London: Taylor & Francis LTD.

Cracknell, A. P., Wong, K. C. (1973), *The Fermi Surface.* Oxford: Clarendon Press.

d'Ambrumenil, N., Fulde, P. (1985), *J. Mag. Mag. Mat. 47&48,* 1.

de Haas, W. J., van Alphen, P. M. (1930), *Leiden Comm. 208 d,* 212 a.

de Haas, W. J., van Alphen, P. M. (1932), *Leiden Comm. 220 d.*

de Dood, W., de Chatel, P. F. (1973), *J. Phys. F: Metal Phys. 3,* 1039.

de Visser, A., Haen, P., Vorenkamp, T., van Sprang, M., Menovsky, A. A., Franse, J. J. M. (1988), *J. Mag. Mag. Mat. 76&77,* 112.

Effantin, M., Rossat-Mignod, J., Burlet, P., Bartholin, H., Kunii, S., Kasuya, T. (1985), *J. Mag. Mag. Mat. 47&48,* 145.

Endoh, D., Goto, T., Suzuki, T., Fujimura, T., Ōnuki, Y., Komatsubara, T. (1987), *J. Phys. Soc. Jpn. 56,* 4489.

Falicov, L. M., Hanke, W., Maple, M. B. (Eds.) (1981), *Valence Fluctuation in Solids.* Amsterdam: North-Holland.

Fawcett, E. (1964), *Adv. Phys. 13,* 139.

Fawcett, E., Maita, J. P., Wernick, J. H. (1970), *Intern. J. Magn. 1,* 29.

Flipse, C. F. J., Edvardsson, S., Kadowaki, K. (1989), *Physica C 162–164,* 1389.
Forgan, E. M. (1981), *Physica 107 B,* 65.
Freeman, A. J., Koelling, P. D. (1977), *Physica B 86–88,* 16.
Freeman, A. J., Dimmock, J. O., Watson, R. E. (1966), *Phys. Rev. Lett. 16,* 94.
Fulde, P., Jensen, J. (1983), *Phys. Rev. B 27,* 4085.
Gignoux, D., Givord, F., Lemaire, R., Tasset, F. (1983), *J. Less-Common Metals 94,* 165.
Gold, A. V. (1974), *J. Low Temp. Phys. 16,* 3.
Goto, T. (1990), private commun.
Goto, T., Suzuki, T., Ohe, Y., Sakatsume, S., Kunii, S., Fujimura, T., Kasuya, T. (1988 a), *J. Phys. Soc. Jpn. 57,* 2885.
Goto, T., Suzuki, T., Ohe, Y., Fujimura, T., Sakatsume, S., Ōnuki, Y., Komatsubara, T. (1988 b), *J. Phys. Soc. Jpn. 57,* 2612.
Goto, T., Suzuki, T., Tamaki, A., Ohe, Y., Nakamura, S., Fujimura, T. (1989), *Bulletin of the Research Institute for Scientific Measurements.* Sendai: Tohoku University, p. 37, 65 (in Japanese).
Goy, P., Castaing, B. (1973), *Phys. Rev. B 7,* 4409.
Gratz, E., Bauer, E., Barbara, B., Zemirli, S., Steglich, F., Bredl, C. D., Lieke, W. (1985), *J. Phys. F: Met. Phys. 15,* 1975.
Gschneidner, Jr., K. A., Ikeda, K. (1983), *J. Mag. Mag. Mat. 31–34,* 265.
Harima, H. (1988), thesis at Tohoku University.
Harima, H., Kasuya, T. (1985), *J. Mag. Mag. Mat. 52,* 179.
Harima, H., Yanase, A. (1991), *J. Phys. Soc. Jpn. 60,* 21.
Harima, H., Sakai, O., Kasuya, T., Yanase, A. (1988), *Solid State Commun. 66,* 603.
Harima, H., Miyahara, S., Yanase, A. (1990 a), *Physica B 163,* 205.
Harima, H., Yanase, A., Ōnuki, Y., Umehara, I., Kurasawa, Y., Satoh, K., Kasaya, M., Iga, F. (1990 b), *Physica B 165 & 166,* 343.
Harima, H., Yanase, A., Hasegawa, A. (1991), *J. Phys. Soc. Jpn. 59,* 4054.
Hasegawa, A. (1985), *J. Phys. Soc. Jpn. 54,* 677.
Hasegawa, A. (1984), *J. Phys. Soc. Jpn. 53,* 3929.
Hasegawa, A. (1981), *J. Phys. Soc. Jpn. 50,* 3313.
Hasegawa, A., Yamagami, H. (1990), *J. Phys. Soc. Jpn. 59,* 218.
Hasegawa, A., Yamagami, H. (1991), *J. Phys. Soc. Jpn. 60,* 1654.
Hasegawa, A., Yamagami, H., Johbettoh, H. (1990), *Phys. Soc. Jpn. 59,* 2457.
Hasegawa, A., Yanase, A. (1977), *J. Phys. F 2,* 1245.
Hasegawa, A., Yanase, A. (1980), *J. Phys. F 10,* 847.
Hayden, S. M., Lonzarich, G. G. (1986), *Phys. Rev. 33,* 4977.
Heil, J., Kouroudis, I., Lüthi, B., Thalmeier, P. (1984), *J. Phys. C: Solid State Phys. 17,* 2433.
Henkie, Z., Markowski, P. J. (1977), *J. Cryst. Growth 41,* 303.
Herring, C. (1952), *Phys. Rev. 87,* 60.
Hillebrecht, F. U., Gudat, W., Martensson, N., Sarma, D. D., Campagna, M. (1985), *J. Mag. Mag. Mat. 47 & 48,* 221.
Hohenberg, P., Kohn, W. (1964), *Phys. Rev. B 136,* 864.
Holleck, H., Kleykamp, H. (1987), *Gmelin Handbook of Inorganic Chemistry, C 12.* Berlin: Springer-Verlag.
Hungsberg, R. E., Gschneidner, K. A. (1972), *J. Phys. Chem. Solids 33,* 401.
Hunt, M., Meeson, P., Probst, P-A., Reinders, P., Springford, M., Assmus, W., Sun, W. (1990), *Physica B 165 & 166,* 323.
Ishikawa, Y., Mori, K., Mizushima, T., Fujii, A., Takeda, H., Sato, K. (1987), *J. Mag. Mag. Mat. 70,* 385.
Ishizawa, Y., Tanaka, T., Bannai, E., Kawai, S. (1977), *J. Phys. Soc. Jpn. 42,* 112.
Ishizawa, Y., Nozaki, H., Tanaka, T., Nakajima, T. (1980), *J. Phys. Soc. Jpn. 48,* 1439.
Jerjini, M., Bonnet, M., Burlet, P., Lapertot, G., Rossat-Mignod, J., Henry, J. Y., Gignoux, D. (1988), *J. Mag. Mag. Mat. 76 & 77,* 405.
Johanson, W. R., Crabtree, G. W., Edelstein, A. S., McMasters, D. D. (1981), *Phys. Rev. Lett. 46,* 504.
Johanson, W. R., Crabtree, G. W., Edelstein, A. S., McMasters, O. D. (1983), *J. Mag. Mag. Mat. 31–34,* 377.
Joss, W., Hall, L. N., Crabtree, G. W. (1984), *Phys. Rev. B 30,* 5637.
Joss, W., van Ruitenbeek, J. M., Crabtree, G. W., Tholence, J. L., van Deursen, A. J. P., Fisk, Z. (1987), *Phys. Rev. Lett. 59,* 1609.
Joss, W., van Ruitenbeek, J. M., Crabtree, G. W., Tholence, J. L., van Deursen, A. J. P., Fisk, Z. (1989), *J. de Physique C 8,* 747.
Kasaya, M., Iga, F., Katoh, K., Kasuya, T. (1990), *J. Mag. Mag. Mat. 90 & 91,* 521.
Kasuya, T. (1956), *Prog. Theor. Phys. 16,* 45.
Kasuya, T. (1959), *Prog. Theor. Phys. 22,* 227.
Kasuya, T. (1966), in: *Magnetism,* Vol. II B: Rado, G. T., Suhl, H. (Eds.). New York: Academic, pp. 215–294.
Kasuya, T. (1976), *J. Phys. Soc. Jpn. 40,* 1086.
Kasuya, T. (1987), *Butsuri 42,* 722 (in Japanese).
Kasuya, T., Saso, T. (Eds.) (1985), *Theory of Heavy Fermions and Valence Fluctuations.* Berlin: Springer-Verlag.
Kasuya, T., Sakai, O., Tanaka, J., Kitazawa, H., Suzuki, T. (1987), *J. Mag. Mag. Mat. 63 & 64,* 9.
Kasuya, T., Sakai, O., Harima, H., Ikeda, M. (1988), *J. Mag. Mag. Mat. 76 & 77,* 46.
Kido, G. (1990), to be published.
Kitazawa, H., Suzuki, T., Sera, M., Oguro, I., Yanase, A., Hasegawa, A., Kasuya, T. (1983), *J. Mag. Mag. Mat. 31–34,* 421.
Kitazawa, H., Gao, Q. Z., Shida, H., Suzuki, T., Hasegawa, A., Kasuya, T. (1985), *J. Mag. Mag. Mat. 52,* 286.

Kitazawa, H., Kwon, Y. S., Oyamada, A., Takeda, N., Suzuki, H., Sakatsume, S., Satoh, T., Suzuki, T., Kasuya, T. (1988), *J. Mag. Mag. Mat. 76&77,* 40.

Klaasse, J. C. P., Meyer, R. T. W., de Boer, F. R. (1980), *Solid State Commun. 33,* 1001.

Kletowski, Z., Iliew, N., Stalinski, B. (1985), *Physica 130 B,* 84.

Kletowski, Z., Glinski, M., Hasegawa, A. (1987), *J. Phys. F: Met. Phys. 17,* 993.

Koelling, D. D. (1982), *Solid State Commun. 43,* 247.

Komatsubara, T., Sato, N., Kunii, S., Oguro, I., Furukawa, Y., Ōnuki, Y., Kasuya, T. (1983), *J. Mag. Mag. Mat. 31–34,* 368.

Kondo, J. (1964), *Prog. Theor. Phys. 32,* 37.

Konno, R., Moriya, T. (1987), *J. Phys. Soc. Jpn. 56,* 3270.

Kohn, W., Sham, L. J. (1965), *Phys. Rev. A 140,* 1133.

Kouroudis, I., Weber, D., Yoshizawa, M., Lüthi, B., Puech, L., Haen, P., Flouquet, J., Bruls, G., Welp, U., Franse, J. J. M., Menovsky, A., Bucher, E., Hufnagl, J. (1987), *Phys. Rev. Lett. 58,* 820.

Kubo, Y. (1990), private commun.

Kubota, T., Kondo, H., Watanabe, K., Murakami, Y., Cho, Y.-K., Tanigawa, S., Kawano, T., Bahng, G.-W. (1990), *J. Phys. Soc. Jpn. 59,* 4494.

Kurosawa, Y., Umehara, I., Kikuchi, M., Nagai, N., Satoh, K., Ōnuki, Y. (1990), *J. Phys. Soc. Jpn. 59,* 1545.

Larson, A. C., Cromer, D. T. (1961), *Acta Crystallogr. 14,* 73.

Lawrence, J. M., Shapiro, S. M. (1980), *Phys. Rev. B22,* 4379.

Lifshitz, I. M., Kosevich, R. M. (1955), *J. exp. theor. Phys. 29,* 730.

Lifshitz, I. M., Kosevich, R. M. (1956), *Soviet Phys. JETP 2,* 636.

Lock, D. G., Crisp, V. H. C., West, R. N. (1973), *J. Phys. F3,* 561.

Lonzarich, G. G. (1984), *J. Mag. Mag. Mat. 45,* 43.

Lonzarich, G. G. (1988), *J. Mag. Mag. Mat. 76&77,* 1.

Loucks, T. L. (1967), *Augmented Plane Wave Method.* New York: Benjamin.

Luong, N. H., Franse, J. J. M., Hien, T. D. (1985), *J. Mag. Mag. Mat. 50,* 153.

Lüthi, B., Yoshizawa, Y. (1987), *J. Mag. Mag. Mat. 63&64,* 274.

Luttinger, J. M. (1960), *Phys. Rev. 119,* 1153.

Maezawa, K. (1990), private commun.

Maezawa, K., Kato, T., Ishikawa, Y., Sato, K., Umehara, I., Kurosawa, Y., Ōnuki, Y. (1989), *J. Phys. Soc. Jpn. 58,* 4098.

Manuel, A. A. (1982), *Phys. Rev. Lett. 49,* 1525.

Massidda, S., Yu, J., Freeman, A. J. (1988), *Physica C52,* 251.

Matsui, H., Goto, T., Kasaya, M., Iga, F. (1991), to be published.

Mattheiss, L. F. (1970), *Phys. Rev. B1,* 373.

Mattheiss, L. F., Hamann, D. R. (1988), *Phys. Rev. B37,* 10623.

Mattheiss, L. F., Wood, J. H., Switendick, A. C. (1968), in: *Methods in Computational Physics,* Vol. 8: Fernbach, S., Rotenberg, M. (Eds.). New York: Academic Press, p. 63.

Menovsky, A., Franse, J. J. M. (1983), *J. Crystal Growth 65,* 286.

Meyer, R. T. W., Roeland, L. W., de Boer, F. R., Klaasse, J. C. P. (1973), *Solid State Commun. 12,* 923.

Migdal, A. B. (1958), *Sov. Phys. JETP 7,* 996.

Moriya, T. (1983), *J. Mag. Mag. Mat. 31–34,* 10.

Moriya, T. (1985), *Spin Fluctuations in Itinerant Electron Magnetism.* Berlin: Springer-Verlag.

Müller, T., Joss, W., van Ruitenbeek, J. M., Welp, U., Wyder, P. (1988), *J. Mag. Mag. Mat. 76&77,* 35.

Nakajima, S. (1967), *Prog. Theor. Phys. 38,* 23.

Nasu, S., van Diepen, A. M., Neumann, H. H., Craig, R. S. (1971), *J. Phys. Chem. Solids 32,* 2773.

Newcombe, G. C. F., Lonzarich, G. G. (1988), *Phys. Rev. B37,* 10619.

Niksch, M., Lüthi, B., Kübler, J. (1987), *J. Phys. B-Cond. Mat. 68,* 291.

Norman, M. R., Koelling, D. D. (1986), *Phys. Rev. 33,* 6730.

Norman, M. R., Oguchi, T., Freeman, A. J. (1987), *J. Mag. Mag. Mat. 69,* 27.

Oguchi, T., Freeman, A. J., Crabtree, G. W. (1987), *J. Mag. Mag. Mat. 63&64,* 645.

Oh, S.-J., Suga, S., Kakizaki, A., Taniguchi, M., Ishii, T., Kang, J. S., Allen, J. W., Gunnarsson, O., Christensen, N. E., Fujimori, A., Suzuki, T., Kasuya, T., Miyahara, T., Kato, H., Sconhammer, K., Torikachvili, M. S., Maple, M. B. (1988), *Phys. Rev. B37,* 2861.

Ohbayashi, M., Komatsubara, T., Hirahara, E. (1976), *J. Phys. Soc. Jpn. 40,* 1088.

Olson, C. G., Liu, R., Yang, A.-B., Lynch, D. W., Arko, A. J., List, R. S., Viel, B. W., Chang, Y. C., Jiang, P. Z., Paulilkas, A. P. (1989), *Science 245,* 731.

Ōnuki, Y., Komatsubara, T. (1987), *J. Mag. Mag. Mat. 63&64,* 281.

Ōnuki, Y., Furukawa, Y., Komatsubara, T. (1984), *J. Phys. Soc. Jpn. 53,* 2734.

Ōnuki, Y., Machii, Y., Shimizu, Y., Komatsubara, T., Fujita, T. (1985 a), *J. Phys. Soc. Jpn. 54,* 3562.

Ōnuki, Y., Nishihara, M., Sato, M., Komatsubara, T. (1985 b), *J. Mag. Mag. Mat. 52,* 317.

Ōnuki, Y., Ina, K., Nishihara, M., Komatsubara, T., Takayanagi, S., Kameda, K., Wada, N. (1986), *J. Phys. Soc. Jpn. 55,* 1818.

Ōnuki, Y., Umezawa, A., Kwok, W. K., Crabtree, G. W., Nishihara, M., Komatsubara, T., Maezawa, K., Wakabayashi, S. (1987), *Jap. J. Appl. Phys. 26,* 509.

Ōnuki, Y., Kurosawa, Y., Omi, T., Komatsubara, T., Yoshizaki, R., Ikeda, H., Maezawa, K., Waka-

bayashi, S., Umezawa, A., Kwok, W. K., Crabtree, G. W. (1988), *J. Mag. Mag. Mat. 76&77,* 37.
Ōnuki, Y., Kurosawa, Y., Maezawa, K., Umehara, I., Ishikawa, Y., Sato, K. (1989a), *J. Phys. Soc. Jpn. 58,* 3705.
Ōnuki, Y., Komatsubara, T., Reinders, P. H., Springford, M. (1989b), *J. Phys. Soc. Jpn. 58,* 3698.
Ōnuki, Y., Omi, T., Kurosawa, Y., Satoh, K., Komatsubara, T. (1989c), *J. Phys. Soc. Jpn. 58,* 4552.
Ōnuki, Y., Umezawa, A., Kwok, W. K., Crabtree, G. W., Nishihara, M., Yamazaki, T., Omi, T., Komatsubara, T. (1989d), *Phys. Rev. B40,* 11195.
Ōnuki, Y., Fukada, A., Ukon, I., Umehara, I., Satoh, K., Kurosawa, Y. (1990a), *Physica B163,* 600.
Ōnuki, Y., Umezawa, A., Kwok, W. K., Crabtree, G. W., Nishihara, M., Ina, K., Yamazaki, T., Omi, T., Komatsubara, T., Maezawa, K., Wakabayashi, S., Takayanagi, S., Wada, N. (1990b), *Phys. Rev. B41,* 568.
Ōnuki, Y., Umehara, I., Kurosawa, Y., Satoh, K., Matsui, H. (1990c), *J. Phys. Soc. Jpn. 59,* 229.
Ōnuki, Y., Umehara, I., Nagai, N., Kurosawa, Y., Satoh, K. (1991a), *J. Phys. Soc. Jpn. 60,* 1022.
Ōnuki, Y., Umehara, I., Kurasawa, Y., Nagai, N., Satoh, K., Kasaya, M., Iga, F. (1991b), *J. Phys. Soc. Jpn. 59,* 2320.
Parks, R. D. (Ed.) (1977), *Valence Instabilities and Related Narrow Band Phenomena.* New York: Plenum Press.
Peter, M., Hoffmann, L., Manuel, A. A. (1989), in: *Positron Annihilation:* Dorikens-Vanpraet, L., Dorikens, M., Segers, D. (Eds.). Singapore: World Scientific, pp. 197–203.
Rado, G. T., Suhl, H. (Eds.) (1973), *Magnetism,* Vol. V. New York: Academic Press.
Reichelt, J., Winzer, K. (1978), *Phys. Stat. Sol. (b) 89,* 489.
Reinders, P. H. P., Springford, M., Hilton, P., Kerley, N., Killoran, N. (1987), *Cryogenics 27,* 689.
Reinders, P. H. P., Springford, M. (1989), *J. Mag. Mag. Mat. 79,* 295.
Rossat-Mignod, J., Effantin, J. M., Burlet, P., Chattopadhyay, T., Regnault, L. P., Bartholin, H., Vettier, C., Vogt, O., Ravot, O., Achart, J. C. (1985), *J. Mag. Mag. Mat. 52,* 111.
Rozing, G. J., Rabou, L. P. L. M., Mijnarends, P. E. (1989), in: *Positron Annihilation:* Dorikens-Vanpraet, L., Dorikens, M., Segers, D. (Eds.). Singapore: World Scientific, pp. 233–235.
Ruderman, M., Kittel, C. (1954), *Phys. Rev. 96,* 99.
Sakai, O., Takahashi, H., Takeshige, M., Kasuya, T. (1984), *Solid State Commun. 52,* 997.
Sakai, O., Takeshige, M., Harima, H., Otaki, K., Kasuya, T. (1985), *J. Mag. Mag. Mat. 52,* 18.
Sakai, O., Kaneta, Y., Kasuya, T. (1987), *Jap. J. Appl. Phys. 26,* 477.
Sakamoto, I., Miura, T., Miyoshi, E., Sato, H. (1990), *Physica B165&166,* 339.
Satoh, K., Fujita, T., Maeno, Y., Ōnuki, Y., Komatsubara, T. (1989), *J. Phys. Soc. Jpn. 58,* 1012.
Satoh, K., Umehara, I., Kurosawa, Y., Ōnuki, Y. (1990), *Physica B165&166,* 329.
Seitz, E., Legeler, B. (1979), *J. de Phys. Coll. C5, suppl. 5, 40,* 76.
Settai, R., Goto, T., Ōnuki, Y. (1991), to be published in *J. Phys. Soc. Jpn.*
Shoenberg, D. (1984), *Magnetic Oscillations in Metals.* Cambridge: Cambridge University Press.
Sigfusson, T. I., Bernhoeft, N. R., Lonzarich, G. G. (1984), *J. Phys. 14,* 2141.
Smith, J. L., Foeler, C. M., Freeman, B. L., Hults, W. L., King, J. C., Mueller, F. M. (1990), *Proc. 3rd Int. Symp. on Superconductivity,* Sendai.
Spirlet, J. C., Vogt, O. (1982), *J. Mag. Mag. Mat. 29,* 31.
Springford, M., Reinders, P. H. P. (1988), *J. Mag. Mag. Mat. 76&77,* 11.
Springford, M. (Ed.) (1980), *Electrons at the Fermi Surface.* Cambridge: Cambridge University Press.
Stewart, G. R., Fisk, Z., Wills, J. O., Smith, J. L. (1984), *Phys. Rev. Lett. 52,* 679.
Sticht, J., d'Ambrumenil, N., Kübler, J. (1986), *Z. Phys. B. Cond. Matt. 65,* 149.
Stoner, E. C. (1936), *Proc. Roy. Soc. A154,* 656.
Stoner, E. C. (1938), *Proc. Roy. Soc. A165,* 372.
Strange, D., Newns, D. M. (1986), *J. Phys. F: Metal Phys. 16,* 335.
Sumiyama, A., Oda, Y., Nagano, H., Ōnuki, Y., Shibutani, K., Komatsubara, T. (1986), *J. Phys. Soc. Jpn. 55,* 1294.
Suzuki, T., Goto, T., Tamaki, A., Fujimura, T., Ōnuki, Y., Komatsubara, T. (1985), *J. Phys. Soc. Jpn. 54,* 2367.
Suzuki, T., Goto, T., Sakatsume, S., Tamaki, A., Kunii, S., Kasuya, T. (1987a), *Jap. J. Appl. Phys. 26,* 511.
Suzuki, T., Goto, T., Tamaki, A., Fujimura, T., Kitazawa, H., Suzuki, T., Kasuya, T. (1987b), *J. Mag. Mag. Mat. 63&64,* 563.
Suzuki, T., Goto, T., Ohe, Y., Fujimura, T., Kunii, S. (1988), *J. de Physique C8,* 799.
Suzuki, T. (1990), to be published.
Taillefer, L. (1990), *Physica B163,* 278.
Taillefer, L., Lonzarich, G. G., Strange, P. (1986), *J. Mag. Mag. Mat. 54&57,* 957.
Taillefer, L., Newbury, R., Lonzarich, G. G., Fisk, Z., Smith, J. L. (1987), *J. Mag. Mag. Mat. 63&64,* 372.
Takahashi, H., Kasuya, T. (1985), *J. Phys. C18,* 2697, 2709, 2721, 2731, 2745, 2755.
Takahashi, M., Tanaka, H., Satoh, T., Kohgi, M., Ishikawa, Y., Miura, T., Takei, H. (1988a), *J. Phys. Soc. Jpn. 57,* 1377.
Takahashi, T., Matsuyama, H., Katayama-Yoshida, H., Okabe, Y., Hosoya, S., Seki, K., Fujimoto, H., Sato, M., Inokuchi, H. (1988b), *Nature 334,* 691.

Takahashi, T., Matsuyama, H., Katayama-Yoshida, H., Okabe, Y., Hosoya, S., Seki, K., Fujimoto, H., Sato, M., Inokuchi, H. (1989), *Phys. Rev. B39,* 6636.

Takayanagi, S., Wada, N., Watanabe, T., Ōnuki, Y., Komatsubara, T. (1988), *J. Phys. Soc. Jpn. 57,* 3552.

Takayanagi, S., Furukawa, E., Wada, N., Ōnuki, Y., Komatsubara, T. (1990), *Physica B163,* 574.

Takeda, N. (1990), private commun.

Takeshige, M. (1990), thesis at Tohoku University.

Takeshige, M., Sakai, O., Kasuya, T. (1985 a), *J. Mag. Mag. Mat. 52,* 363.

Takeshige, M., Sakai, O., Kasuya, T. (1985 b), in: *Theory of Heavy Fermions and Valence Fluctuations:* Kasuya, T., Saso, T. (Eds.). Berlin: Springer-Verlag, pp. 120–131.

Tanaka, T., Bannai, E., Kawai, S., Yamane, T. (1975), *J. Crystal Growth 30,* 193.

Tanigawa, S., Kurihara, T., Osawa, M., Komatsubara, T., Ōnuki, Y. (1989), in: *Positron Annihilation:* Dorikens-Vanpraet, L., Dorikens, M., Segers, D. (Eds.). Singapore: World Scientific, pp. 233–235.

Troc, R., Trzebiatowski, W., Pippek, K. (1971), *Bull. Acad. Pol. Sci. Ser. Sci. Chim. 19,* 427.

Ubachs, W., van Deursen, A. P. J., de Vroomen, A. R., Arko, A. J. (1986), *Solid State Commun. 60,* 7.

Umehara, I., Kurosawa, Y., Nagai, N., Kikuchi, M., Satoh, K., Ōnuki, Y. (1990), *J. Phys. Soc. Jpn. 59,* 2848.

Umehara, I., Nagai, N., Ōnuki, Y. (1991 a), *J. Phys. Soc. Jpn. 60,* 591.

Umehara, I., Nagai, N., Ōnuki, Y. (1991 b), *J. Phys. Soc. Jpn. 60,* 1294.

Umehara, I., Nagai, N., Ōnuki, Y. (1991 c), *J. Phys. Soc. Jpn. 60,* 1464.

van Deursen, A. J. P., Pols, R. E., de Vroomen, A. R., Fisk, Z. (1985), *J. Less-Common Metals 111,* 331.

Wang, C. S., Norman, M. R., Albers, R. C., Boring, A. M., Pickett, W. E., Krakauer, H., Christensen, N. E. (1987), *Phys. Rev. B35,* 7260.

Weinberger, P., Podloucky, R., Mallet, C. P., Neckel, A. (1979), *J. Phys. C12,* 801.

Westrum, E. F., Takahashi, Y., Stout, N. D. (1965), *J. Phys. Chem. 69,* 1520.

White, R. M., Fulde, P. (1982), *J. Appl. Phys. 53,* 1994.

Willis, J. O., Thompson, J. D., Guertin, R. P., Crow, J. E. (Eds.) (1990), *Proc. of Int. Conf. on Physics of Highly Correlated Electron Systems, Physica B163.*

Windmillar, L. R., Ketterson, J. B. (1968), *Rev. Sci. Instrum. 39,* 1672.

Wilson, K. G. (1975), *Rev. Mod. Phys. 47,* 773.

Yamagami, H., Hasegawa, A. (1990), *J. Phys. Soc. Jpn. 59,* 2426.

Yamagami, H., Hasegawa, A. (1991 a), *J. Phys. Soc. Jpn. 60,* 1011.

Yamagami, H., Hasegawa, A. (1991 b), *J. Phys. Soc. Jpn. 60,* 987.

Yanase, A., Harima, H. (1990), private commun.

Yosida, K. (1957), *Phys. Rev. 106,* 893.

Zwicknagl, G., Runge, E., Christensen, N. E. (1990), *Physica B163,* 97.

Index

Distribution:

VCH, P.O. Box 101161, D-6940 Weinheim (Federal Republic of Germany)

Switzerland: VCH, P.O. Box, CH-4020 Basel (Switzerland)

United Kingdom and Ireland: VCH (UK) Ltd., 8 Wellington Court, Cambridge CB1 1HZ (England)

USA and Canada: VCH, Suite 909, 220 East 23rd Street, New York, NY 10010-4606 (USA)

ISBN 3-527-26816-2 (VCH, Weinheim)
Set ISBN 3-527-26813-8 (VCH, Weinheim)

ISBN 0-89573-691-8 (VCH, New York)
Set ISBN 1-56081-190-0 (VCH, New York)